Sugar Beet Cultivation, Management and Processing

Varucha Misra • Santeshwari Srivastava •
Ashutosh Kumar Mall
Editors

Sugar Beet Cultivation, Management and Processing

Volume 1

Springer

Editors
Varucha Misra
ICAR-Indian Institute of Sugarcane Research
Lucknow, India

Santeshwari Srivastava
ICAR-Indian Institute of Sugarcane Research
Lucknow, India

Ashutosh Kumar Mall
ICAR-Indian Institute of Sugarcane Research
Lucknow, India

ISBN 978-981-19-2729-4 ISBN 978-981-19-2730-0 (eBook)
https://doi.org/10.1007/978-981-19-2730-0

This Springer imprint is published by the registered company Springer Nature Singapore Pte Ltd.
The registered company address is: 152 Beach Road, #21-01/04 Gateway East, Singapore 189721, Singapore

To
Late (Mr.) N.K. Misra
(F/O Dr. Varucha Misra)

Foreword

The book entitled *Sugar beet Cultivation, Management and Processing* is a very useful compilation of excellent work on sugar beet that may prove to be the torchbearer in research field of this subject. The editors of this book have cast a broad net to assure how each topic is addressed by some of the world's foremost experts of respective disciplines. Sugar beet is farmed and researched all over the world, as seen by the countries that made a significant contribution in this area. During the last two centuries, sugar seems to have become a staple diet with a yearly intake of 20 kg per person around the globe, and consumption is expected to rise in the near future. Domestic demand for ethanol is also increasing, with projections of 14.5 billion gallons in 2021. Because of its short life span and tremendous potential, sugar beet is increasingly attracting agriculturalists, millers, and researchers for the manufacturing of ethanol. The editors have taken a keen interest in compiling information in the form of chapters on all essential aspects for high and quality production. In a straightforward, crisp, and succinct style, the book summarizes recent achievements and research in this crop. The book's information is simple to comprehend for the reader and will be of great use to those interested in this crop. Furthermore, the book's contributors have provided up-to-date and high-quality information, enhancing the value of this edited volume.

Birsa Agricultural University, Kanke
Ranchi, Jharkhand, India

Onkar Nath Singh

Foreword

Sugar beet (*Beta vulgaris* L.) is cultivated worldwide in many temperate countries like Europe, North America, Asia, South America, and North Africa. It is an annual vegetative crop that has become a key source of sugar production in temperate countries around the world. Sugar beet varieties can adapt to a variety of habitats and growing circumstances. Now its cultivation is being emerged in India after a long halt. It is utilized in diverse industries as raw materials, such as in bioethanol and molasses production and pectin extraction.

Considering the diverse usage and application in sugar and other agro-industries, this crop has been researched meticulously for the benefit of farmers and sugar millers. Several research papers on different aspects are being published, but it is very surprising that only limited information compiled in the form of a book is available in the public domain. I am delighted to hear about this compilation entitled *Sugar beet Cultivation, Management and Processing* by Varucha Misra, Santeshwari Srivastava, and A.K. Mall (eds) devoted specifically to sugar beet which will be published by Springer, Singapore. This book examines sugar beet production, processing, and management from a variety of perspectives, with a particular focus on sugar beet studies in the Indian subcontinent and Southeast Asian countries. This will aid in boosting the long-term development of sugar beets around the world. An array of topics on this crop like crop production and management, mechanization in sugar beet, seed production, weed management, foliar and root diseases, diverse uses of by-products, insect pests, abiotic stress management, and application of artificial intelligence have been covered in this book which is of interest for the readers. One of the highlighting sections in each chapter is of future prospects that give a new route to sugar beet research.

I appreciate the work of the editors who have cast a broad net to guarantee that each topic is addressed by sugar beet researchers worldwide. This edited book is a compilation of 48 chapters by experienced sugar beet experts from continents of Asia, Europe, Africa, North and South America in different fields. This book contains in-depth information on a variety of contemporary sugar beet development and establishment practices and technologies giving a glimpse of recent advancements. Researchers and practitioners interested in understanding the essential dynamics of sugar beet cultivation, processing, and diverse usage for its sustainable development will find the book particularly useful and interesting. This book will serve as a reference book for them. The editor and authors should be commended for compiling so much current material in one spot. Sugar beet will be an important part of the industry's knowledge base and a springboard for further advancement for the next one or two decades, serving practitioners and scholars alike.

I wish this edited book a great success.

Institute of Field and Vegetable Crops
National Institute of the Republic of Serbia
Novi Sad, Serbia

Jegor Miladinović

Preface

Sugar beet is known as an alternative crop for sugar production and is a renowned crop of temperate regions. It runs behind sugarcane in production of sugar contributing to 20% of sugar in the world. It is a biennial and short-duration crop with less water requirement capability compared to sugarcane. Historically, this crop is considered to be a European crop and is known to cultivated in many other countries worldwide.

India is also among such fortunate countries where this crop has been introduced many years ago. Exhaustive research work on this crop has been done in India during 1970s. ICAR-Indian Institute of Sugarcane Research, Lucknow, has always been in the leading role of sugar beet research in India along with other government institutes and sugar industries. The search for the right location for seed production of indigenous and exogenous varieties is a great effort by Indian sugar beet researchers. Successful development of indigenous varieties like LS 6, IISR Comp 1 is another achievement to the determination of the researchers. Agronomic practices for crop production and protection of Indian subtropical and tropical conditions have also been developed and standardized through several attempts. Identification and management of various diseases and insect pests on sugar beet in Indian conditions has also been thoroughly explored. In the current time, germplasm maintenance, seed production, post-harvest quality analysis, drought and saline tolerance, and ethanol production are some of the ongoing research areas. The chapters on seed and crop production, crop adaptability to Indian subtropical climatic conditions, identification of foliar and root diseases, insect pests, bioethanol production, and management measures for diseases and insects have discussed the efforts of Indian research in brief.

Improvement in crop production technologies with the involvement of advanced machines and artificial intelligence has further strengthened the base of this crop in the world. The increased usage of agricultural applications of drones for monitoring and surveillance of diseases, insects, and application of insecticidal sprays has also benefitted this crop. Image processing techniques and deep learning techniques, for disease detection, have great potential benefits for sugar beet protection technologies. The chapter on crop production and management of various countries like Serbia, Turkey, mechanization for this crop, and new technologies for disease monitoring has thrown light on it. As improvement in crop production has always been done for the farmer's benefit, intercropping of sugar beet with other crops has been a good option for doubling farmers' income. The cultivation package and practices have been clearly discussed in the chapter on the intercropping of sugar beet with different agricultural crops.

Climate change is another key ongoing issue that is giving birth to several abiotic stresses. Soil salinity and recurrent droughts are to name a few. The halophytic natural endowment helps sugar beet to not only grow but reclaim saline soil for cultivation, where other crops lag behind. The increasing frequency of water shortage in the world has strongly impacted sugar beet cultivation all over the world. Research have been performed and conclusive results have been obtained. The chapters on Sugar beet production under changing climate: Opportunities and Challenges Improving Sugar beet Production under Salinity Conditions, and Drought stress Management in this crop have covered the worldwide research work.

Furthermore, sugar beet is identified as a promising feedstock for ethanol production. The ethanol blending with petrol in many countries has been the reason behind catching their eyes towards this crop. One ton of sugar beet has been known to attain 100–120 L of ethanol, and this may exceed even through the recent advanced technologies being developed all over the world. Furthermore, this crop has shown its versatile nature by its diverse usage in different fields. The molasses obtained after juice extraction is rich in betaine and is being used as a feed supplement. The non-sucrose dry matter and fleshy leaves are also used as fodder for lactating animals like cattle. The vinasse obtained is being used in the production of petrochemicals. Additionally, sugar beet pectin is now emerging as a natural ingredient for food and allied industries like pharmaceutical, cosmetic, and polymer industries. The chapters based on production and utilization of sugar beet molasses, prospects and scope of this crop as cattle feed, diversified usage of sugar beet pulp and pectin throw complete insight on the expanded application of by-products and recent progress in this field.

This book *Sugar Beet Cultivation, Management and Processing* is a complete package of sugar beet cultivation covering every essential aspect for high production and good yield of sugar beet cultivation, development, and crop management, such as origin, breeding, seed production, physiology, pathology, entomology, biotechnology, and post-harvest technology, in a concise way. This book will give readers an overview of sugar beet crop, beginning with their origins. The book endows with the recent advancements and up-to-date knowledge and research on this crop. We hope this book will serve as a significant reference for students, researchers, scientists, industrialists, and entrepreneurs associated with this field and will emerge out with newer research ideas and work on this issue.

Lucknow, Uttar Pradesh, India

Varucha Misra
Santeshwari Srivastava
Ashutosh Kumar Mall

Acknowledgment

Opportunity is missed by most people
because it is dressed in overalls and looks like work.
—Thomas Edison

A famous quote from Thomas Edison ran into my mind while penning down an acknowledgment note. On behalf of all the editors of this book, I express my gratitude first to the contributors of the chapters who are experts in the sugar beet field of different aspects worldwide. Their experience and research have added value to this book. I would like to thank all of them who in the era of COVID 19 could make up in writing the chapters for this book. Their efforts and cooperation play a vital part in the successful publication of this book.

I would also like to especially thank the people who were always supportive and encouraging behind the curtain. To begin with, I would like to thank my father, Late Mr. N. K. Misra for being there with me when I was down during the compilation of this book. His words of inspiration and motivation have always boosted me up and when I have lost him recently his words made me strong to reinitiate and finish this book. This would not be possible without the strong support from my mother and brother, Er Varun Misra, during this tough time in my life.

I would also like to express my gratitude to Late Dr. T.P. Mall for his blessings during the tenure. Furthermore, I would like to thank the respective family members of editors for their continuous support, motivation, encouragement, and cooperation during the compilation and editing of this book.

Last but not least, I would like to thank God for letting us sail out safely through the hurdles in the path of completion of this book.

I like to conclude the thanksgiving by the renowned line of footballer, Mr. Pele,

Success is no accident.
It is hard work, perseverance, learning, studying, sacrifice and most of all, love of what you are doing or learning to do.

Varucha Misra

Contents

Contents for Volume 2

Editors and Contributors

About the Editors

Varucha Misra is working in ICAR-Indian Institute of Sugarcane Research, Lucknow, as a Research Fellow. She received a Ph.D. degree in Biotechnology from Amity University, Uttar Pradesh, India. Recently, her thesis work was awarded "SAAR Best Thesis Award 2018–2019" for her outstanding work in the field of frontier sciences. Her thesis work was also appreciated and published in the renowned newspaper of *Dainik Jagran*, Uttar Pradesh, India, in 2020. She is a lifetime member of many professional societies such as Biologix Research and Innovation Centre Pvt. Ltd. (BRICPL), India; the Society for Scientific and Social Development (SSSD), Meerut, UP, India; and the Society for Community Mobilization for Sustainable development, India. She has published 34 research papers and 18 review papers in national and international journals of repute. She has also published 15 book chapters with reputed publishers like Springer. She has published three books, two edited books, and three e-bulletins focusing on sugarcane and sugar beet. She has also published several popular articles in *Hindi* language for benefit of farmers in several magazines and has also recently published two articles in *Prabhat Khabar* newspaper, Muzaffarpur, Bihar (India), for the welfare of farmers. She was also appreciated for her reviewing work from Biosciences Biotechnology Research Asia. She is also reviewer of many reputed Journals like Scientific Reports.

In 2021, she received the **AWSAR Award** for Popular Writing from the Department of Science and Technology. In 2020, she got the Plantica Society's **Young Scientist Award**, and in 2018, she received the **Young**

Biochemist Award from Biologix Research and Innovation Centre Pvt. Ltd.. In 2018, she received a **Doctoral Research Fellow Award** from the Society for Scientific and Social Development (SSSD) in Meerut, Uttar Pradesh, India. In 2017, she was awarded **Young Scientist Award** from Samgra Vikas Welfare Society. She was also awarded **Subject Expert** in Microbiology, Agriculture, and Nutrition from the International Consortium for Clinical Research Excellence, Ethics and Education and appreciated the study of "Effect of Sugar Intake towards Human Health" as a noble cause of advancement of the art and science of clinical research in 2017. Further, she received **Prof HS Shrivastava Young Scientist Award (Gold medal)** in National Conference on CCSD-2015 for her oral presentation; **Silver medal** from poster presentation in the International Conference on Impact of Environment on Women's Health organized by Amity University and the National Institute of Environmental Health Sciences, North California, USA, in 2017; and **Best Paper** in the National Conference on Women in Sugarcane Agriculture and Industry in 2013. In addition, she received the **Best Worker Award** in the Research Fellow Category from ICAR-Indian Institute of Sugarcane Research, Lucknow, in 2020 and 2021.

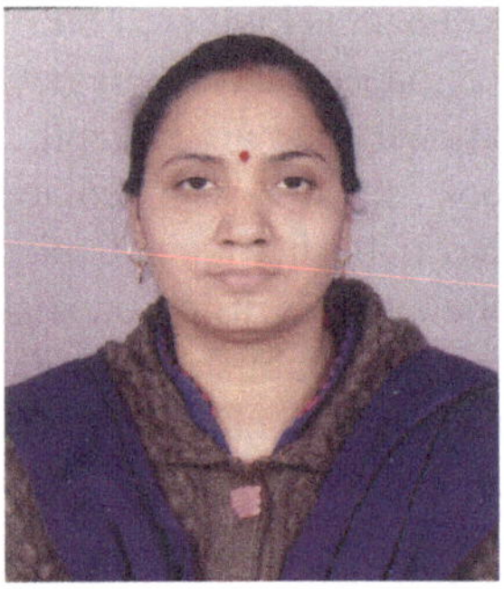

Santeshwari Srivastava is working as Research Fellow at ICAR-Indian Institute of Sugarcane Research, Lucknow and has a M.Sc. and Ph.D. in Zoology. She has an experience of 10 productive years in the research field of entomology as a Research Fellow at the Department of Entomology and Agricultural Zoology, Institute of Agricultural Sciences, Banaras Hindu University, Varanasi. Currently, her research is mainly focused on the economic implications of insect pests on sugar beet and their management. She is actively involved in research in various fields and her research interests include insect taxonomy, insecticidal applications, soil arthropods identification, etc. She has published her 22 research outputs in reputed peer-reviewed journals having citations. Considering her expertise in entomology, she has published 10 research papers in reputed journals. In addition to it, she has 10 popular articles in the field of agriculture. As an editor of books, she has four books to her credit and published more than

12 book chapters with recognized national publishers. She has compiled and edited souvenirs of international conferences during her research tenure. She has presented her studies in various national and international conferences. She has been awarded **Scientist Associate Award** from **Astha Foundation**, Meerut (UP), in 2016 and **Young Achiever Award 2020** from InSc for her excellent contribution in the field of research. For her exemplary work in international conference, she received **Appreciation Award** from the Department of Extension Education, Banaras Hindu University, Varanasi, in 2016. Recently, she was honored **Appreciation Award** from the Society for Agriculture and Allied Research (SAAR) for her marvellous association and inputs in organizing international conference.

Ashutosh Kumar Mall is working as Principal Scientist (Genetics and Plant Breeding), ICAR-Indian Institute of Sugarcane Research, Lucknow, completed his B.Sc. (Ag.) from VBSP University, Jaunpur, UP, in 2001 and both M.Sc. (Ag.) and Ph.D. (Ag.) in Genetics and Plant Breeding from NDUA&T, Ayodhya (UP), during 2003 and 2006, respectively. He has done his post-doctorate from the International Rice Research Institute, Philippines. After serving for 3 years as Scientist with IRRI, Philippines, he served as Senior Scientist at ICAR-Indian Grassland and Fodder Research Institute, Jhansi (UP), and then joined ICAR-IISR, Lucknow. Dr. Mall has research and managerial responsibilities at IISR, Lucknow, as In-charge, IISR Regional Center, Motipur and IISR Sugar beet Outpost, Mukteswar (Nainital). He has developed 15 rice varieties for different ecologies and 13 varieties in different forage crops, namely Guinea grass, Oat, Cowpea, Pearl millet, Tall fescue, Lucerne, and *Sewan* grass, and one variety of sugarcane in collaboration with All India Coordinated Research Project. Through several cross combinations, he has identified 11 germplasm for sugarcane and five germplasm for sugar beet. He has lifetime membership of 27 professional societies.

Dr. Mall significantly contributed to the sugarcane improvement program and is working for seed replacement through Breeder Seed Production under private-public partnership mode in Bihar, which is recognized by task force for improving sugarcane productivity and sugar recovery by the Government of Bihar. He is also working on varietal development of sugar beet for Indian sub-tropical conditions and maintaining nearly 200 germplasm. He has received many prestigious awards. To site a few "Best Scientist Award" by the Society for Scientific and Social Development; "Young Professional Award-2017" by the Society for Community Mobilization for Sustainable Development; and "Dr. Basant Ram Eminent Young Scientist Award-2017" by Uttar Pradesh Academy of Agricultural Sciences, Lucknow. He has published more than 90 research articles in international and national journals of repute, 60 popular articles and manuals, 69 abstracts published in national/international seminar/symposium, published 9 technical bulletins, 3 leaf folders, 3 e-technical bulletins, 3 e-folders, 1 inventory, 3 e-books, 24 e-annual reports and e-proceedings, 10 annuals reports and 7 workshop proceedings compiled and edited to his credit. He has written three books, two edited souvenirs cum abstract books, and written 25 book chapters including publishers like Springer, etc. and handled five externally funded and eight institutional research projects with ongoing one external project and six institutional projects.

Contributors

Mohamed S. Abbas Natural Resources Department, Faculty of African Postgraduate Studies, Cairo University, Giza, Egypt

Mohamed Abdel-Raheem Pests & Plant Protection Department, Agricultural and Biological Research Institute, National Research Centre, Cairo, Egypt

Anuja Rameshchand Adbhai Vasantdada Sugar Institute, Pune, Maharashtra, India

Arkadiusz Artyszak Institute of Agriculture, Warsaw University of Life Sciences—SGGW, Warsaw, Poland

Arun Baitha Division of Crop Protection, ICAR-Indian Institute of Sugarcane Research, Lucknow, Uttar Pradesh, India

Makhan Singh Bhullar Farm Science Centre, Guru Angad Dev Veterinary and Animal Sciences University, SAS Nagar, Punjab, India

Radmila Bojović Faculty of Agriculture, University of Belgrade, Zemun-Belgrade, Serbia

Navjot Singh Brar Farm Science Centre, Guru Angad Dev Veterinary and Animal Sciences University, SAS Nagar, Punjab, India

Atun Roy Choudhury Fecal Sludge Management Division, Banka BioLoo Limited, Hyderabad, Telangana, India

Intiaz Amin Chowdhury Department of Plant Pathology, North Dakota State University, Fargo, ND, USA

Mihajlo Ćirić Institute of Field and Vegetable Crops, Novi Sad, Serbia

Siddhartha Das Department of Plant Pathology, M.S. Swaminathan School of Agriculture, Centurion University of Technology and Management, Paralakhemundi, Odisha, India

Sujaya Dewanjee RRSS, Bidhan Chandra Krishi Viswavidyalaya, Purulia, West Bengal, India

Marijana Djordjević Department of Carbohydrate Food Engineering, Faculty of Technology Novi Sad, University of Novi Sad, Novi Sad, Serbia

Miljana Djordjević Department of Carbohydrate Food Engineering, Faculty of Technology Novi Sad, University of Novi Sad, Novi Sad, Serbia

Ayman Esh Agricultural Research Center, Sugar Crops Research Institute, Giza, Egypt

Parviz Fasahat Sugar Beet Seed Institute (SBSI), Agricultural Research, Education and Extension Organization (AREEO), Karaj, Iran

M. B. Flores-Romero Faculty of Economic and Administrative Sciences, Michoacan University of San Nicolas of Hidalgo, Morelia, Mexico

A. E. Ghaly Department of Agricultural Engineering, Faculty of Agriculture, Cairo University, Giza, Egypt

Meenakshi Goyal Department of Plant Breeding and Genetics, Punjab Agricultural University, Ludhiana, Punjab, India

Md. Alamgir Hossain Department of Crop Botany, Bangladesh Agricultural University, Mymensingh, Bangladesh

Mostafa Hosseinpour Sugar Beet Seed Institute (SBSI), Agricultural Research, Education and Extension Organization (AREEO), Karaj, Iran

M. M. Ibrahim Department of Agricultural Engineering, Faculty of Agriculture, Cairo University, Giza, Egypt

Md. Jahirul Islam Department of Bio-Health Convergence, College of Biomedical Science, Kangwon National University, Chuncheon, Republic of Korea
Physiology and Sugar Chemistry Division, Bangladesh Sugarcrop Research Institute, Pabna, Bangladesh

Navdeep S. Jamwal PAU, Regional Research Station, Kaurthala, Punjab, India

J. D. Janse Institute of Plant Protection NAAS, Kyiv, Ukraine

L. A. Janse Institute of Plant Protection NAAS, Kyiv, Ukraine

D. Jiménez-Islas Renewable Energy Division, Higher Technological Institute of Huichapan, Huichapan, Mexico

Zoran Jovović Biotechnical Faculty, University of Montenegro, Podgorica, Montenegro

Mozhdeh Kakueinezhad Sugar Beet Seed Institute (SBSI), Agricultural Research, Education and Extension Organization (AREEO), Karaj, Iran

K. A. Kalatur Institute of Bioenergy Crops and Sugar Beet NAAS, Kyiv, Ukraine

Nilima Karmakar Department of Soil Science and Agricultural Chemistry, NMCA, Navsari Agricultural University, Navsari, Gujarat, India

Aanchaldeep Kaur Department of Plant Breeding and Genetics, Punjab Agricultural University, Ludhiana, Punjab, India

Simerjeet Kaur Department of Agronomy, Punjab Agricultural University, Ludhiana, Punjab, India

Rıza Kaya Department of Plant Protection, Sugar Institute, Etimesgut, Ankara, Turkey

Sepideh Kazemi Department of Horticultural Science, Science and Research University of Tehran, Tehran, Iran

Mohammed Khan Department of Plant Pathology, North Dakota State University, Fargo, ND, USA

Samar Khayamim Seed and Plant Improvement Institute, Agricultural Research, Education and Extension Organization, Karaj, Iran

Amit Kumar Plant Breeding Section, Division of Crop Science, ICAR-RC-NEH Region, Umiam, Meghalaya, India

Dharmendra Kumar ICAR-CPRI, Shimla, Himachal Pradesh, India

Rajeev Kumar ICAR-IISR, Lucknow, Uttar Pradesh, India

Santosh Kumar ICAR-IARI, Hazaribag, Jharkhand, India

Ram Ji Lal ICAR-Indian Institute of Sugarcane Research, Lucknow, Uttar Pradesh, India

Young-Seok Lim Department of Bio-Health Convergence, College of Biomedical Science, Kangwon National University, Chuncheon, Republic of Korea

Nataša Ljubičić Biosense Institute, University of Novi Sad, Novi Sad, Serbia

A. K. Mall Division of Crop Improvement, ICAR-Indian Institute of Sugarcane Research, Lucknow, Uttar Pradesh, India

Ashutosh Kumar Mall ICAR-Indian Institute of Sugarcane Research, Lucknow, Uttar Pradesh, India

Nikola Maravić Department of Carbohydrate Food Engineering, Faculty of Technology Novi Sad, University of Novi Sad, Novi Sad, Serbia

Mohammad Mehdizadeh Department of Agronomy and Plant Breeding, Faculty of Agriculture and Natural Resources, University of Mohaghegh Ardabili, Ardabil, Iran
Ilam Science and Technology Park, Ilam, Iran

Mostafa Mehdizadeh Faculty of Agriculture, University of Guilan, Rasht, Iran

Zoleikha Mehdizadeh Department of Biology, Payame Noor University, Tehran, Iran

Manish Chandra Mehta Department of Entomology and Agricultural Zoology, Institute of Agricultural Sciences, Banaras Hindu University, Varanasi, India

Anja Milosavljević Research and Development Center Sunoko, Indjija, Serbia

Varucha Misra Division of Crop Improvement, ICAR-Indian Institute of Sugarcane Research, Lucknow, Uttar Pradesh, India

Hemat S. Mohamed Animal Nutrition Department, Animal Production Research Institute, Agricultural Research Center, Cairo, Egypt

Masuma Akter Mou Department of Agronomy, Bangladesh Agricultural University, Mymensingh, Bangladesh

Barbara Magdalena Muir Tereos Corporate Technology Centre, Aalst, Belgium

José M. Mulet Biochemistry and Molecular Biology, IBMCP, Universitat Politècnica de València-CSIC, Valencia, Spain

Mona Nagargade ICAR-IISR, Lucknow, Uttar Pradesh, India

Violeta Oro Institute for Plant Protection and Environment, Belgrade, Serbia

Mehmet Metin Ozguven Department of Biosystems Engineering, Faculty of Agriculture, Tokat Gaziosmanpaşa University, Tokat, Turkey

Kamlesh G. Patel Department of Soil Science and Agricultural Chemistry, NMCA, Navsari Agricultural University, Navsari, Gujarat, India

A. D. Pathak ICAR-Indian Institute of Sugarcane Research, Lucknow, Uttar Pradesh, India

Sanjay Patil Department of Alcohol Technology and Biofuels, Vasantdada Sugar Institute, Pune, Maharashtra, India

Sudeepta Pattanayak Division of Plant Pathology, ICAR—Indian Agricultural Research Institute, New Delhi, India

Swapan Kumar Paul Department of Agronomy, Bangladesh Agricultural University, Mymensingh, Bangladesh

M. E. Pérez-Romero Faculty of Economic and Administrative Sciences, Michoacan University of San Nicolas of Hidalgo, Morelia, Mexico
Business Management Division, Higher Technological Institute of Huichapan, Huichapan, Mexico

W. S. Philanim Plant Breeding Section, Division of Crop Science, ICAR-RC-NEH Region, Umiam, Meghalaya, India

Mohammed Arif Sadik Polash Department of Crop Botany, Khulna Agricultural University, Khulna, Bangladesh

Dragana Popović Faculty of Economics in Subotica, University of Novi Sad, Novi Sad, Serbia

Vera Popović Institute of Field and Vegetable Crops, Novi Sad, Serbia

Aiming Qi Centre for Agriculture, Food and Environmental Management Research, School of Life and Medical Sciences, University of Hertfordshire, Hatfield, Hertfordshire, UK

Vuk Radojević Institute of Field and Vegetable Crops, Novi Sad, Serbia
Faculty of Agriculture, University of Novi Sad, Novi Sad, Serbia

M. Raghuraman Department of Entomology and Agricultural Zoology, Institute of Agricultural Sciences, Banaras Hindu University, Varanasi, India

Abazar Rajabi Agricultural Research, Education and Extension Organization, Sugar Beet Seed Institute (SBSI), Karaj, Iran

Md. Abdur Razzak Faculty of Life and Environmental Science, Shimane University, Matsue, Japan

Javad Rezaei Sugar Beet Research Department, Khorasan Razavi Agricultural and Natural Resources Research and Education Center, Agricultural Research, Education and Extension Organization (AREEO), Mashhad, Iran

Alessandro Romano Council for Agricultural Research and Economics, Plant Protection and Certification Centre, Lonigo, Italy

Md. Arif Sakil Department of Biochemistry and Molecular Biology, Bangladesh Agricultural University, Mymensingh, Bangladesh
Department of Regulatory Biology, Saitama University, Saitama, Japan

Gulzar S. Sanghera PAU, Regional Research Station, Kaurthala, Punjab, India

Prabhjot S. Sanghera Punjab Agricultural University, Ludhiana, Punjab, India

Shahida Sazia Department of Horticulture, Bangladesh Agricultural University, Mymensingh, Bangladesh

Mile Sečanski Maize Research Institute "Zemun Polje", Belgrade, Serbia

Zita Šereš Department of Carbohydrate Food Engineering, Faculty of Technology Novi Sad, University of Novi Sad, Novi Sad, Serbia

Mohamed A. Shahba Natural Resources Department, Faculty of African Postgraduate Studies, Cairo University, Giza, Egypt

Nivedita Shettigar Plant Breeding Section, Division of Crop Science, ICAR-RC-NEH Region, Umiam, Meghalaya, India

A. K. Shrivastava ICAR-Indian Institute of Sugarcane Research, Lucknow, Uttar Pradesh, India

Shahida Anusha Siddiqui DIL e.V.–German Institute of Food Technologies, Quakenbrück, Germany
Technical University of Munich Campus Straubing for Biotechnology and Sustainability, Straubing, Germany

Preeti Singh ICAR-IARI, Hazaribag, Jharkhand, India

Pankaj Kumar Sinha ICAR-IARI, Hazaribag, Jharkhand, India

Archana Siraree Division of Seed Production, Genda Singh Sugarcane Breeding and Research Institute, Kushinagar, Uttar Pradesh, India

Dragana Šoronja-Simović Department of Carbohydrate Food Engineering, Faculty of Technology Novi Sad, University of Novi Sad, Novi Sad, Serbia

Velibor Spalević Biotechnical Faculty, University of Montenegro, Podgorica, Montenegro

Sangeeta Srivastava ICAR-Indian Institute of Sugarcane Research, Lucknow, Uttar Pradesh, India

Santeshwari Srivastava Division of Crop Protection, ICAR-Indian Institute of Sugarcane Research, Lucknow, Uttar Pradesh, India

Marijenka Tabaković Maize Research Institute "Zemun Polje", Belgrade, Serbia

Shadia Taghian Agricultural Research Center, Sugar Crops Research Institute, Giza, Egypt

Dariush Taleghani Agricultural Research, Education and Extension Organization, Sugar Beet Seed Institute (SBSI), Karaj, Iran

Koko Tampubolon Program Study of Agrotechnology, Faculty of Agriculture and Animal Husbandry, Universitas Tjut Nyak Dhien, Medan, Indonesia

Paul Townson Lion Seeds Ltd., Woodham Mortimer, Maldon, Essex, UK

Nenad Trkulja Institute for Plant Protection and Environment, Belgrade, Serbia

Koç Mehmet Tuğrul Department of Biosystems Engineering, Faculty of Agriculture, Eskişehir Osmangazi University, Eskişehir, Turkey

I. Ventura-Cruz Mechatronic Division, Higher Technological Institute of Huichapan, Huichapan, Mexico

Vivek Chandra Verma DIHAR-DRDO, Chandigarh, India

Yusuf Yanar Department of Plant Protection, Faculty of Agriculture, Tokat Gaziosmanpaşa University, Tokat, Turkey

Guiping Yan Department of Plant Pathology, North Dakota State University, Fargo, ND, USA

List of Figures

List of Tables

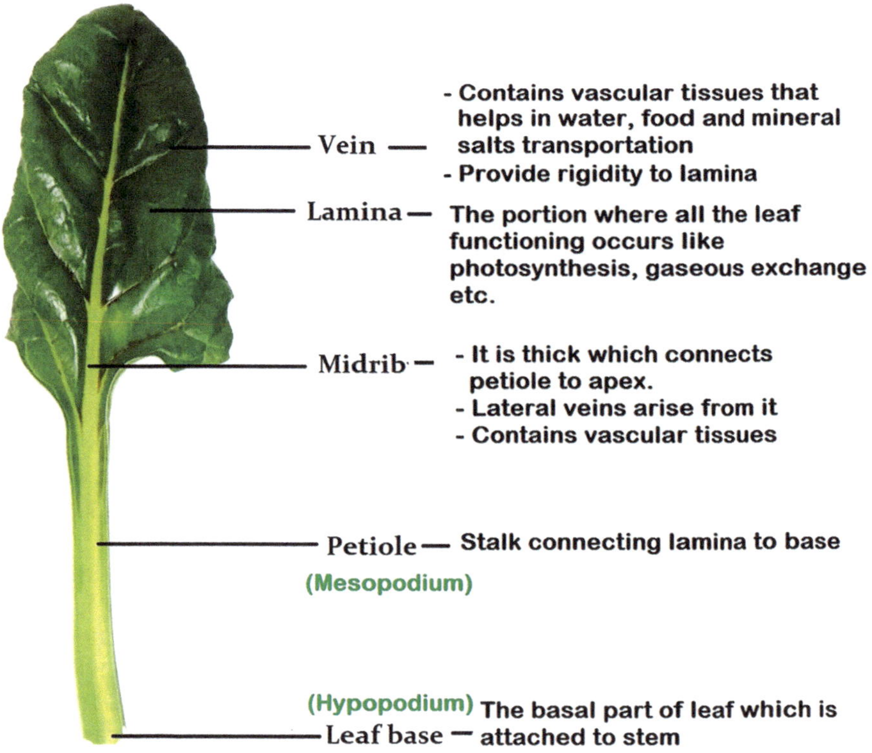

Fig. 2.2 Parts of a sugar beet leaf

2.3.2.1 Stomata

Stomata help in switching off-gases from the atmosphere into the cells *via* these structures. Stomata are present on both, abaxial and adaxial sides of the leaf. The epidermis of the leaf contains a number of stomata's excluding the region of collenchyma. This structure also regulates the uptake of CO_2 from the atmosphere into the cell which helps in the process of photosynthesis. The closing and opening of this aperture also govern the liberation of water vapor depending on physiological and environmental situations (Wyse 1982). On an average per square millimeter, stomatal number is 91–114 located on the upper while on the lower side it is 144-162 (De Vries 1879; Dryosen 1877). The size of the stomata also varies on the upper and lower surfaces of the leaves. The size of stomata in adaxial surface is around 23 × 32 micron and on the abaxial side, these are comparatively smaller in size (De Vries 1879; Dryosen 1877). The old leaves, however, have large stomata approaching a maximum of 45 microns (Artschwager 1926).

2.3.3 Root

The root of a sugar beet plant is a white tap root in conical shape (thicker on the upper end and tapering at the lower end). They grow deep in the soils (up to a depth of 1–2 m) with a crown portion of the root exposed out. The primary root is diarch. The roots have two vertical grooves in which lateral roots emerge out. Sucrose gets accumulated in roots majorly in vascular rings of phloem (Fig. 2.3). On an average, roots of sugar beet contain 20% sugar, 5% pulp, 2.6% non-sugars, and the rest 75% water (Artschwager 1926). There are three parts of roots: crown, neck, and true root. *Crown* is virtually a compressed stem comprising leaf buds and acts as a support to the leaves (Wyse 1982). Milford and Houghton (1999) have described it as a part of the root situated above the last leaf scar. Although its size depends on variety, it is also predisposed by the nitrogen usage (Zielke 1970). The *neck* is the narrow zone found adjacent to the crown region and is the broadest portion of the root which comprises the thickened hypocotyl (Artschwager 1926). Dividing the rest part into two halves, the upper portion is resultant from the seedling hypocotyl while the lower one (true root) is formed by a number of cambial rings (Artschwager 1926). The organization of concentric rings is present at the transition zone between hypocotyl and crown portion. An increase in the size of the root is mainly due to cell division and cell enlargement occurring in the vascular rings (Wyse 1982). In general, the vascular rings in the root, range from 8 to 13 in number at the time of harvest, although some rings were formed when the root is just 1 cm, in its size (Artschwager 1926; Milford 1973). Sucrose concentration is found to be most in the true root portion (Table 2.1). The dry matter in roots comprises 17% marc—the insoluble cell wall material (cellulose, pectins, and hemicellulose), 73% sucrose and 10% other soluble compounds. The pectin present in the cell wall material works as a fixing agent like cement that helps in binding the two cells with each other

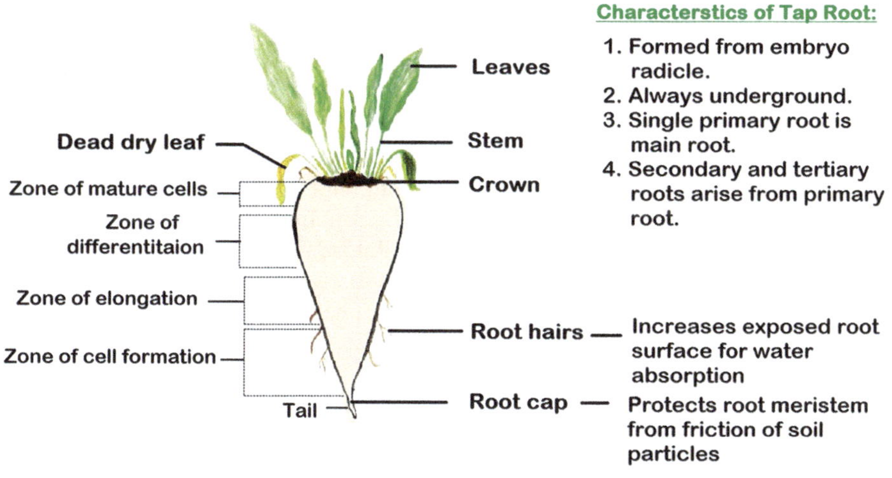

Fig. 2.3 Sugar beet root system

Table 2.1 Sucrose concentration in different parts of roots

Part of root	Sucrose concentration (%)
True root	16–20
Hypocotyl	15–16
Upper portion of crown	7–9
Lower portion of crown	13–14

(McCready 1966). The hemicellulose and cellulose works, as a supporting system (McCready 1966). Decrease in sucrose content in sugar beet is associated with an increasing amount of potassium, sodium, amino-nitrogen compounds and reducing sugar (Harvey and Dutton 1993), which interfere with its processing in sugar mills, particularly during crystallization. Variations in chemical composition in different parts of the root have been observed. The crown portion has 1–3% lower sucrose content than the true root and a larger amount of non-sucrose contents. The sodium, potassium, and amino acids are relatively higher in the crown region than in the true root (Fort and Stout 1948; Zielke 1970) with sodium and amino acid contents being 125% and potassium being 60% higher (Carruthers et al. 1960). Application of nitrogen (and also the variety used) also influences it (Zielke 1970). The sugar yield in sugar beet has also been reported to enhance when leaf dry matter decreased which further reduces the marc content (Hoffmann et al. 2005; Kenter and Hoffmann 2009). This is considered as one of the criteria for developing new varieties of sugar beet having high yield (Hoffmann and Kenter 2018).

2.3.3.1 Composition of Sugar Beet Roots and Its Molasses

The sugar beet root on harvest contains sugars (15–20%) and non-sugars (2.6%) along with water (75–76%) and leftover pulp (4–6%). When one ton of sugar beet (freshly harvested) is processed in mills, it yields sugar (121 kg) besides molasses (38 kg) and pulp (50 kg). The molasses comprised 18.2 kg sugar with impurities of 12.1 kg and 7.8 kg water (Shrivastava et al. 2013). Out of sugar present in fresh sugar beet roots, nearly 83% is recovered as white sugar, 12.5% is lost in molasses, and 4.4% is lost in some other ways. Beet roots also contain 0.3% betaine, an important osmoregulant. Molasses obtained after processing sugar beets (as % of dry weight) contains 66.5% sugars (sucrose 63.5, raffinose 1.5 and other sugars 1.5%), 23% other organic compounds (Glutamic acid + Pyrrolidine carboxylic acid 4.0%, other amino acids 3.0%, Betaine 5.5%, Pectins 5.0%), and 10.5% inorganic compounds with 6% K_2O (Clarke and Godshall 1988). Beet pulp consists of water-insoluble fibrous materials which are: pectins (2.4%), cellulose (1.2%), hemicellulose (1.1%), proteins (0.1%), saponins (0.1%), and minerals (0.1%) (Clarke and Godshall 1988; Mosen 2007). It has been observed that high-sugar beet roots have a relatively higher amount of pulp (van der Poel et al. 1998).

2.3.3.2 Fangy Roots

Occasionally, morphological deviation in sugar beet roots has been observed which is known as fangy roots wherein the roots get forked which also laid its name as forked/sprangled roots. It is also termed as overdeveloped secondary roots which

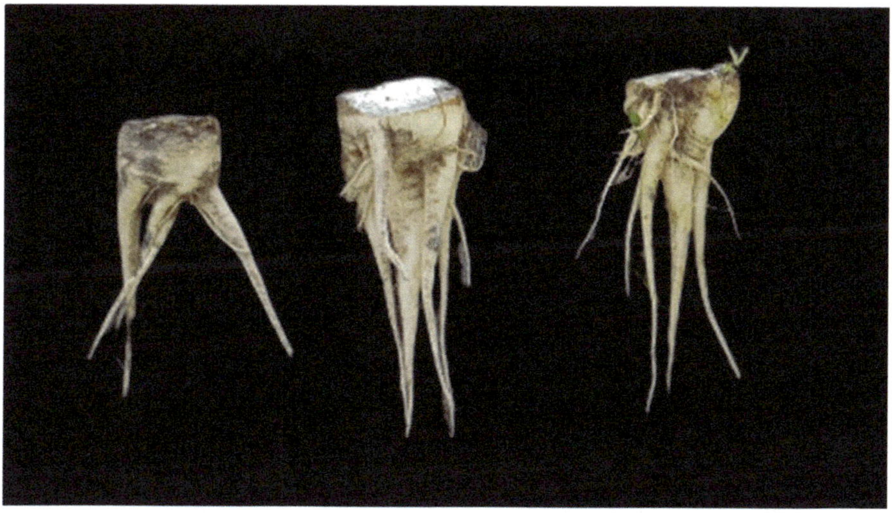

Fig. 2.4 Fangy roots

arise along the side of tap root (Fig. 2.4). These roots are developed due to the occurrence of diseases, soil conditions, or weather alterations. The different soil conditions like poor soil quality, shallow soils, soil plough, or acidity lead to fangy roots. Nematodes such as stubby root nematodes (*Trichodorus* spp. and *Paratrichodorus* spp.), etc. is another reason for their formation as during the early stages of growth the true root is damaged. The lateral roots take up the purpose of the main taproot which causes fangy or furcated storage roots to form. Besides, it is known that any damage to the main tap root system such as *Rhizoctonia* infection, mechanical damage, chemical damage, and water logging condition may cause roots to become forked or fangy.

2.3.3.3 Problems of Fangy Roots

Several difficulties occur due to the formation of fangy roots, which makes it also differ from normal beet roots. These are:

1. Difficulty in lifting and have high dirt tare.
2. Increase in harvester losses at the time of lifting.
3. Lower sugar content.
4. Higher impurity than the body of the beet.
5. Prior to processing, difficulty in the cleaning of roots while during processing in steps like diffusion.
6. Higher breakage of the beets prior to diffusion. This results in sugar loss from broken portions/surfaces.
7. Causes losses due to a reduction in the fine pulp.

2.3.4 Flowers

For flowering in sugar beet, a prerequisite condition is vernalization followed by long day length which mainly corresponds to overwintering and an increase in day length during spring. The requirement for vernalization in this plant is obligate where the shoots of those plants persist for the production of newer leaves devoid of flowers, even for many years. The favorable temperature for it is 5–10 °C. There is a need of 40 days cold temperature exposure of the plant for it to enter into its reproductive stage (Sparkes 2003). Flowers are small in size and are produced in a dense spike-like manner (Fig. 2.5). The inflorescence is basally interrupted. In one to three flowered glomeruli, very small flowers occur. The glomeruli are present on the axils of short bracts or at times in the upper portion of the inflorescence which lack bracts. There are male, female, and hermaphrodite flowers. The hermaphrodites are urn-shaped, green in color and consist of five tepals (basally connate perianth segments). The size of the flower varies from 3–5 × 2–3 mm. The flowers have five stamens and semi-inferior ovary with 2–3 stigmas. The perianths of neighboring flowers are often fused. The flowers are mainly wind-pollinated but at times can be insect-pollinated too. During the flowering phase, plant energy is no longer being used for developing the root; it is rather diverted towards flower production followed by seed production. The color of flowers varies from green to white. The flowers are clustered at one place (having two to eight flowers in one cluster) or solitary in form (Smith 1980). Flowers are bisexual in nature implying male and female reproductive organs in one single flower and are pentamerous. The flowering is of the asynchronous type where the flower does not bloom at the same time. The flowering period of the beet is about 28 days or even longer. In monogerm and multigerm varieties, the flowers are borne singly or in clusters on a stalk (Artschwager 1927; Smith 1980).

2.3.5 Fruits

The fruit of sugar beet is a nut. The glomeruli of the flowers become hard in texture and connate in shape to form the fruit. The perianth surrounds the fruit where the fruit is situated at its swollen base. The perianth is leathery and incurves in shape (Shultz 2003). Normally, a fruit contains two to five seeds (Ahlawat 2008).

2.3.6 Seeds

Sugar beet seed is circular, small in size (1–2 mm diameter), lighter in weight ranges from 1.5 to 6 g (for a 1000 seeds), and dark brown in color (Fig. 2.6). Seed comprises seed ball which consists of either two seeds or more seeds (OECD 1993). Smith (1980) defined the seed ball as a dry body formed by floral receptacles having an irregular shape. Sugar beet seeds are of two types, *viz.*, monogerm and multigerm

Fig. 2.5 (**a**) Sugar beet inflorescence bearing buds. (**b**) Enlarged view of inflorescence with flowers

seeds (Mall et al. 2021). Monogerm seeds possess a solitary embryo while multigerm seeds possess more than one embryo. When we plant monogerm seeds, manual thinning is not required which reduces the cost of sugar beet cultivation (Biancardi et al. 2010).

Fig. 2.6 Sugar beet seeds: (**a**) Monogerm seed (**b**) Multigerm/polygerm seed

2.3.7 Pollination

Pollination in sugar beet occurs through cross pollination and can intercross freely. The small pollens are dispersed through air and wind (Down and Lavis 1930). It may also be transported by vectors like thrips and insects (Shaw 1916). Archimowitsch (1949) revealed that the pollens dispersed through wind can reach a minimum distance of 4500 m with a height of 5000 m. Pollen drift has been recorded to travel up to 5 km from the field (Smith 1980) while pollen from an airborne source can travel up to 8 km (Harding and Harris 1994).

2.4 Photosynthesis and Sucrose Synthesis

Sugar beet is a C_3 plant implying the first product is a three-carbon compound—Phosphoglyceraldehyde (PGAl) produced as primary carboxylation product utilizing RuDP carboxylase enzyme. Arulanantham et al. (1990) revealed that this process is regulated by ribulose 5-phosphate kinase. A study conducted at the Indian Institute of Sugarcane Research, Lucknow, using multigerm (*Romanskaya*, LS 6 and IISR Comp. 1) and monogerm varieties (*Shubhra*, IN 11 and IN 12) showed that the rates of photosynthesis in sugar beet ranges from 38 to 52 $\mu mol/m^2/second$ while carotenoids and chlorophyll content ranged from 3.6 to 7.76 mg/g and 1.20–1.75 mg/g fresh weight of the leaf, respectively. When the conditions of light and temperature are favorable and ideal for sugar beet plant, the efficiency of photosynthesis is about 10% implying that this much amount of light is transformed into chemical energy by the leaves during the process of photosynthesis (Went 1954). Sucrose is synthesized in the leaves by the activity of transglycosylases (UDP-glucose, D-fructose 6 phosphate, 2 D-glucosyl-transferase (sucrose phosphate synthetase, SPS), and UDP-glucose, D-Fructose 2 D-glucosyl transferase (sucrose synthetase, SS)). Translocation of sucrose in the leaves is by symplastic pathway (i.e., movement from cell to cell takes place through plasma membranes and plasmodesmata). It is not inhibited by parachloro-mercuribenzene sulfonic acid (PCMB) (Schmalstig and Geiger 1985). In young and mature storage roots, sucrose translocates *via* apoplastic pathway (i.e., movement from cell to cell through spaces in the cell walls). It is inhibited by PCMB (Lemoine et al. 1988). There are evidence for circadian regulation of starch and sucrose synthesis in the sugar beet leaves (Li et al. 1992). However, the movement of sucrose from conducting tissue into the cytoplasm of storage tissue and what prevents apoplastic sucrose from re-entering root xylem cells and being transported back is yet to be fully understood.

2.5 Development of Cambial Rings in Roots and Their Activity

The activity of cambial tissues is mainly responsible for sucrose storage in roots. Elliott and Weston (1993) revealed that after 14 days of shoot emergence, the development of primary cambium occurs whereas when the plant is of 42 days the secondary cambium ring develops. This ring develops in phloem cells of parenchyma tissue found in the endodermis. The consecutive ring is differentiated in the outer layer of parenchyma. Zamski and Azenkot (1981) found that the first and foremost cambium ring develops from pro-cambium in parenchyma cells. The innermost cambium ring is formed at the transition of primary xylem and phloem cells. The cambial rings are known to keep growing in a centrifugal manner, but the mystery remains as to which tissue these rings originate. When the plant age is of 42 days, it possesses 12–13 leaves with six cambial rings and two more cambial rings are added when the plant is 56 days old, and when the plant is fully grown (around 182 days or so), there could be 12–15 cambial rings (Artschwager 1926). The ring-like structure seen in the transverse section of a mature root is the resultant of alternate zones of vascular and parenchymatous tissues (Fig. 2.7). The cambial rings are also known as *annular zones*. These zones are relatively at an equidistant to each other excluding the zone near to periphery (the place where the rings are relatively much closer to each other). The annular rings are decreased at the tapering

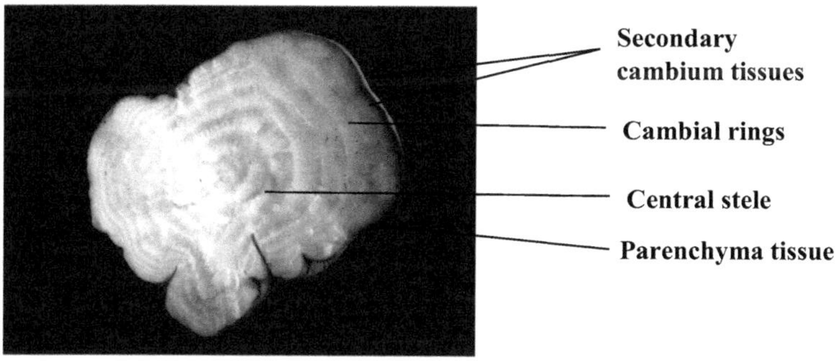

Fig. 2.7 Transverse section of sugar beet root depicting the cambial rings formation

zone of the root as these rings steadily combine into each other. The annular zone comprises the ring of vascular tissue (consisting of collateral bundles) and parenchymatous cells. Medullary ray tissue separates the collateral bundles from one to other and these tissues differ from each other in width (Artschwager 1926). In mature rings, vascular bundles are widest apart from each other in the cambium region. The bundles taper steadily at the phloem and xylem pole which is the reason of double wedge appearance. Fibers are present in the oldest rings of the hypocotyls (Zamski and Azenkot 1981).

2.6 Process of Sucrose Storage

The sucrose content produced in the leaves reaches the root through phloem cells and the vacuolar region of parenchyma cells is the storage house of the root. Sucrose concentration is maximum in the center portion of the root where root diameter is maximum. The sucrose content is enhanced in proportion with cell volume up to a cell size of 10–15 $\times$ 10^{-8} cm^3. Sucrose storage is an active uptake process that is inhibited by a number of chemicals like gramicidin D, carbonyl cyanide m-chlorophenyl hydrazone at a certain concentration. This process is followed by another transportation mechanism occurring via ions where at 95 mM concentration of the alkali cations, uptake sucrose content increased to 2.1–4 folds (Saftner and Wyse 1980). High sucrose concentration and dry weight of root can be obtained in sugar beet due to large storage organs. During the vegetative growth of sugar beet, high temperatures are favored while high sugar yields when night temperatures range between 15 and 20 °C whereas day temperatures between 20 and 25 °C. However, if the temperature, at this time, increases more than 30 °C then sugar yields are reduced.

The parenchymatous cells in the roots are the sites of sucrose storage. These cells are thin-walled and highly vacuolated. The cytoplasm comprises organelles and enzymes (responsible for cellular metabolism) and is thin-walled structure. The vacuole which makes the rest of the volume of the cell is the site where exactly sucrose content gets stored. Doll et al. (1979) revealed that cells nearer vascular bundles had higher sucrose content. This also forms the base that roots having a higher number of cambial rings are more likely to have higher sucrose content (Milford 2008). But this concept was not justified and was even not further used for the selection of breeding lines.

2.7 Future Prospects

Sugar beet is known for its high sucrose content and short duration of the life cycle. It is used for the production of sugar in many European countries. Sugar beet is gaining importance in biofuel industries, and it has a good amount of ethanol content. There is a need to identify and develop indigenous varieties focusing on high ethanol production, particularly for Indian condition. Identification of its multifarious uses is also important for the future. The use of biotechnological and molecular tools will help in unraveling the pest and disease resistance mechanism and genes in some tolerant varieties. Also, the incorporation of certain tolerant genes from other plants to sugar beet will also add up in developing new varieties. With the advancement of science, the complete utilization of sugar beet potential will help in attaining maximum yield and quality.

2.8 Conclusion

This chapter, *per se*, introduces the sugar beet crop and its physiology. Like most other crops, plant of a sugar beet is divided into three parts, *viz*., roots, stem, and leaves. Root is the economical part while leaves also serve as fodder for the cattle. The stem attains a height of about 1.5–2 m. The tap root system often shows abnormal behavior in its growth resulting in fangy roots. Root contains 20% sugar, 5% pulp, 2.6% non-sugars, and the rest 75% water. For flowering in sugar beet, a prerequisite condition is cold vernalization followed by long day length. The root parenchymatous cells are the location where sucrose storage happens. Sucrose concentration is maximum in the center portion of the root where root diameter is maximum. In mature rings, vascular bundles are widest apart from each other in cambium region. The cambial ring develops in phloem cells of parenchyma tissue found in the endodermis.

References

Ahlawat IPS (2008) Sugar beet. http://nsdl.niscair.res.in/bitstream/123456789/514/1/SUGAR%20BEET%20-%20Formatted.pdf

Archimowitsch A (1949) Control of pollination in sugar beet. Bot Rev 15:613–628

Artschwager E (1926) Anatomy of the vegetative organs of the sugar beet. J Agric 33(2):143–176

Artschwager E (1927) Development of flowers and seed in the sugar beet. J Agric Res 34:1–25

Arulanantham AR, Rao IM, Terry N (1990) Limiting factors in photosynthesis. VI. Regeneration of ribulose 1,5-bisphosphate limits photosynthesis at low photochemical capacity. Plant Physiol 93:1466–1475

Biancardi E, McGrath JM, Panella LW, Lewellen RT, Stevanato P (2010) Sugar beet. In: Bradshaw JE (ed) Root and tuber crops (Handbook of plant breeding). Springer, New York; Dordrecht; Heidelberg; London, pp 173–219

Carruthers A, Oldfield JFT, Teague HF (1960) The removal of interfering ions in the determination of betaine in sugar-beet juices and plant material. Analyst 85:272–274

Clarke MA, Godshall MA (1988) Chemistry and processing of sugar beet and sugarcane. In: Proceedings of the Symposium on the chemistry and processing of sugar beet, Denver, Colorado, April 6, 1987 and the Symposium on the Chemistry and Processing of Sugarcane, New Orleans, Louisiana, September 3–4, 1987, vol Volume 9, p 406

Doll S, Rodier F, Willenbrink I (1979) Accumulation of sucrose in vacuoles isolated from red beet tissue. Planta 144:407–411

Down EE, Lavis CA (1930) Studies on methods for control of pollination in sugar beets. J Am Soc Agron 22(1):1–9

Dryosen KJ (1877) Beitrage zue anatomie und entwickulungsgeschichte der zuckerrube, p 37 Halle a. S (Inaug. Diss.)
Elliott MC, Weston GD (1993) Biology and physiology of the sugar beet plant. In: Cooke DA, Scott RK (eds) The sugar beet crop. Chapman and Hall Publishers, pp 37–66
Elliott MC, Weston GD (1995) Biology and physiology of sugar beet plant. In: Cooke DA, Scott RK (eds) The sugarbeet crop Science to practice. Chapman and Hall Publications, pp 37–66
Fort CA and Stout M (1948) American Society of Sugarbeet Technology. Proceedings of 5th General Meeting, p 651–659
Harding K, Harris PS (1994) Risk assessment of the release of genetically modified plants: a review. Ministry of Agriculture, Fisheries and Food, London (ed), pp 54
Harvey CW, Dutton JV (1993) Root quality and processing. In: Cooke DA, Scott RK (eds) Sugar beet crop. Chapman and Hall, London, pp 571–617
Hoffmann CM, Kenter C (2018) Yield potential of sugar beet - have we hit the ceiling? Front Plant Sci 9:289. https://doi.org/10.3389/fpls.2018.00289
Hoffmann CM, Kenter C, Bloch D (2005) Marc concentration of sugar beet (*Beta vulgaris* L.) in relation to sucrose storage. J Sci Food Agric 85:459–465. https://doi.org/10.1002/jsfa.2002
Joanna B, Binczarski M, Dziugan P, Wilkowska A (2018) Sugar beet pulp as a source of valuable biotechnological products. In: Holban AM, Grumezescu AM (eds) Advances in biotechnology for food industry handbook of food bioengineering. Academic Press, pp 359–392
Kenter C, Hoffmann CM (2009) Ursachen der verringerung des markgehaltes von zuckerrüben und auswirkungen auf menge und qualität der schnitzelerzeugung. Sugar Ind 134:246–254
Lemoine R, Daie J, Wyse R (1988) Evidence for the presence of a sucrose carrier in immature sugar beet tap roots. Plant Physiol 86(2):575–580
Li B, Geiger DR, Shieh WJ (1992) Evidence for circadian regulation of starch and sucrose synthesis in sugar beet leaves. Plant Physiol 99:1393–1399
Mall AK, Misra V, Santeshwari, Singh BD, Pathak AD (2021) Quality seed production of sugar beet in India. In: Tiwari AK (ed) Advances in seed production and management, pp 139–159
McCready RM (1966) Polysaccharides of sugar beet pulp, a review of their chemistry. J Am Soc Sugarbeet Technol 14:260–270
Milford GFJ (1973) The growth and development of the storage root of sugar beet. J Ann Appl Biol 75:427–438
Milford GFJ (2006) Plant structure and crop physiology. In: Draycott AP (ed) Sugar beet, pp 33–49
Milford GFJ, Houghton BJ (1999) An analysis of the variation in crown size in sugar beet (*Beta vulgaris*) grown in England. Ann Appl Biol 134(5):225–232
Milford GFJ, Jarvis PJ, Jones J, Barraclough PB (2008) An agronomic and physiological re-evaluation of the potassium and sodium requirements and fertilizer recommendations for sugar beet. J Agric Sci 146:429–443. https://doi.org/10.1017/S0021859607007630
Mosen A (2007) Beet–sugar handbook. John Wiley & Sons, p 884
OECD (1993) Traditional crop breeding practices: an historical review to serve as a baseline for assessing the role of modern biotechnology. OECD, Paris. isbn:92-64-14047-6
Pathak AD, Kapur R (2013) Current status of sugarbeet research in India. In: Kumar S, Singh PK, Swapna M, Pathak AD (eds) Souvenir of IISR Industry interface on research and development initiatives for sugar beet in India, pp 9–14
Saftner RA, Wyse RE (1980) Alkali cation/sucrose co transport in the root sink of sugar beet. Plant Physiol 66:884–889
Schmalstig JG, Geiger DR (1985) Phloem unloading in developing leaves of sugar beet: evidence for pathway through the symplast. Plant Physiol 79(1):237–241
Shaw HB (1916) Self, close and cross fertilization of beets. Mem New York Bot Gard 6:149–152
Shrivastava AK, Sawnani A, Shukla SK, Solomon S (2013) Unique features of sugarbeet and its comparison with sugarcane. In: Kumar S, Singh PK, Swapna M, Pathak AD (eds) Souvenir of IISR Industry interface on research and development initiatives for sugar beet in India, pp 36–39
Shultz LM (2003) *Beta vulgaris*. In: Flora of North America Editorial Committee (eds) Flora of North America North of Mexico, volume 4: "Magnoliophyta: Caryophyllidae", part 1. Oxford University Press, New York, pp 266–267

Smith GA (1980) Sugar beet. In: Hybridization of crop plants. American Society for Agronomy – Crop Science Society of America, Madison, WI, pp 601–616

Sparkes D (2003) Growth and development, field crops. Encyclopedia of Applied Plant Sciences, pp 595–600

van der Poel, Schiweck H, Schwartz T (1998) Sugar technology. Beet and cane sugar manufacture, p 1120

Vries HDe (1879) Beitrage Zue Speciellen Physiologie Landwirthschaft Licher Culturpflanzen VII. Wachsthumsgeschichte der Zuckerrube. Landw. Jahrb. 8:417-498, Illus

Went FW (1954) The physiology of the growth of sugar beets. American Society of Sugar beet technologists pp 319-324

Wyse RE (1982) The sugar beet and sucrose formation. In: McGinnis (ed) Beet sugar technology. Third edition RA, pp 17–24

Zamski E, Azenkot A (1981) Sugarbeet vasculature I Cambial development and the tree dimensional structure of the vascular system. Bot Gaz 142(3):334–343

Zielke R (1970) Yield and selected chemical constituents of sugarbeet root and crown. PhD thesis Michigan State University, East Lansing, Michigan

Physiological and Molecular Aspects of Sucrose Accumulation in Sugar Beet

3

Parviz Fasahat, Mostafa Hosseinpour, Mozhdeh Kakueinezhad, and Paul Townson

Abstract

Like many higher plant species, sucrose plays an important role in the growth of the sugar beet (*Beta vulgaris* L.). Sucrose is stored as the most important form of transport and storage of sugar in many plant species, especially plants with high economic value such as sugarcane, carrot, melon, tomato, and beet root. Sugar beet cultivation as a sugar crop in the eighteenth and nineteenth centuries through selection among fodder beets with considering morphological and physiological traits is a successful effort in plant breeding. At the beginning, the overall goal was to increase the sugar concentration to a suitable level for efficient processing and extraction while maintaining the yield level. Advances in this field intensified after the introduction of syrup concentration measurements and polarimetry machines. Following repeated selections, the sucrose content in sugar beet has risen from 6% to over 18% (fresh weight) in today's hybrids. Improving the sugar yield per unit area was the foremost imperative goal of sugar beet breeding after promoting its cultivation. Recent advances in increasing the extraction coefficient of sugar confirm a promising future for this crop. In this chapter, the factors affecting the increase in sugar content in sugar beet, from the physiological and molecular aspects, are discussed.

Keywords

Sucrose accumulation · Sugar beet · Storage root

P. Fasahat (✉) · M. Hosseinpour · M. Kakueinezhad
Sugar Beet Seed Institute (SBSI), Agricultural Research, Education and Extension Organization (AREEO), Karaj, Iran

P. Townson
Lion Seeds Ltd., Woodham Mortimer, Maldon, Essex, UK

V. Misra et al. (eds.), *Sugar Beet Cultivation, Management and Processing*,
https://doi.org/10.1007/978-981-19-2730-0_3

Abbreviations

ATPase	Adenine Transport Phosphatase
CAPS	Cleaved Amplified Polymorphic Sequences
EST markers	Expressed Sequence Tag Markers
HA	Heteroduplex Analysis
QTL	Quantitative Trait Loci
SNP	Single Nucleotide Polymorphism
SSCP	Single-Strand Conformation Polymorphism

3.1 Introduction

Sugar yield is of importance agronomic trait of sugar beet calculated by the percentage of sucrose in the weight of harvested roots minus any sugar losses during post-harvest storage (McGrath and Trebbi 2007). The sucrose content in the root is generally measured as a root fresh weight percentage owing to the ease of measuring specific gravity and the application of refractometric and then polarimetry (McGrath and Fugate 2012). In freshly harvested roots, the amount of water has a significant influence on sugar content as well as root yield. A negative correlation is reported between the sugar percentage in freshly harvested roots and root yield in breeding lines, so that improvement in one trait will often lead to a reduction of the other (Pritchard 1916; Powers 1957; Bergen 1967; Doney et al. 1981; Carter 1987). Simmonds (1994) called this relationship a false correlation. In 1987, Carter reviewed field data from 11 years of research on root weight, sugar content (based on fresh and dry weight), and a combination of sugar beet and fodder beet yields and reported that the water content in the root has a major influence on the inverse relationship between sugar yield and root yield and considered it as a component that affects sugar content and root yield. By measuring the root water content, McGrath and Townsend (2015) showed that this trait has genetic diversity and that the commercial hybrids tested had lower water content than the USDA-ARS germplasm. Intriguingly, most of the possible quantitative trait loci (QTL) for both sugar content and sugar yield traits segregate with the QTLs for root water content (Trebbi 2005). According to this definition, the amount of sucrose in the dry matter is also inheritable, and modification for the dry matter content of sugar beet will be a breeding priority. Fortunately, there is a high correlation (0.7–0.8) between sugar content and total dry matter in sugar beet (Theurer 1979; Hoffmann et al. 2005). When the sucrose content is measured by dry weight, there is often little difference between species due to water loss in the roots (Bergen 1967). Bergen (1967) compared high-yield cultivars (E-type) with sugar cultivars (Z-type) and found that although there was significant difference between cultivars based on fresh weight, the difference based on dry weight was only significant for the last harvest time (Goodman 1966; Follett et al. 1970). According to Theurer (1979), sugar beet breeders usually select high-sugar cultivars based on fresh weight, but selection based on dry weight will be more accurate. There is a negative relationship between sucrose content and root yield even in new genotypes, but the degree of correlation is less than before (Campbell and Kern 1983; Harms and Schulz 2015; Fasahat et al. 2018, 2021).

According to Shimamoto and Hosokawa (1967, 1969), in young plants, root diameter correlates better with root yield than root length. One of the reasons for this correlation is the conical shape of sugar beet. An increase in the diameter of the cone has a greater effect on the total volume of the cone than a similar increase in its length. Doney (1979) showed this relationship in 3-week-old young plants. In a pot experiment, it was found that after reaching the maximum root diameter, the amount of sugar begins to decrease (Schnepel and Hoffmann 2015). One possible explanation for this observation could be that the roots must apply pressure to the soil to expand further into the pot, with limited space. In this manner, the photosynthetic materials need increases and the produced sugar is consumed. This is consistent with the results of Gemtos et al. (2000) which showed that high pressure on the soil reduces dry matter yield, diameter, and root length.

3.2 Physiology of Sugar Accumulation During Growth Stages

3.2.1 Allocation of Photosynthetic Materials

The sugar accumulation depends on the material quantity produced by photosynthesis, their efficient transport and storage, and also processes that compete for the use of carbohydrates stored in the sink (root). Photosynthetic substances are constantly transferred to the roots of sugar beet during the growing season. The proposed priority for photosynthetic material allocation includes respiration, upper root growth, capillary root growth, and root storage growth which involves sugar accumulation (Fick et al. 1973). However, the amount of available photosynthetic material allocated to each part of the plant during the season varies constantly subjected to the ratio of "sink capacity" of the plant. This type of long-term division of material during the season is called "balanced" as opposed to the "stage" division that occurs in potatoes, corn, wheat, and some other crops (Loomis and Rapoport 1976). Photosynthetic material allocated to the root is divided between its growth and sugar storage. The yield formation is on the base of root development as well as sugar production by the leaves, considered as source organ (Schnepel and Hoffmann 2015). The sugar product is transported down the plant through the phloem to respond to the consuming organ (sink) demand (Zamski and Azenhot 1981). In stored roots, sugar is transported from the phloem to the parenchyma cell vacuoles through a diffusion mechanism (Milford 1973). In 1979, Wyse proposed the sucrose-slope hypothesis (Wyse 1979a). According to this hypothesis, sucrose enters the root through the phloem (Giaquinta 1979), is transported into the apoplast (Richter and Ewald 1983), and is stored in the vacuum of undifferentiated parenchyma cells (without breakdown of glucose and fructose prior transport across cell membranes). However, this trend is opposite to sugarcane, in which sucrose is hydrolyzed and regenerated before being transferred to the cell vacuole (Giaquinta 1977). Parenchyma cells are located between cambium rings that form during secondary thickening of the root. Thus, the cambium ring formation is indispensable for the development of sugar beet yield (Artschwager 1926; Milford 1973). Each ring contains an additional set of xylem and phloem tissues. The presence of phloem tissue in each ring makes it possible to distribute sugar and store it throughout the roots. Milford (1973) and Rapoport and Loomis (1986) illustrated that the cambium ring formation begins early in growth, in such a way that the primary cambium is completed within 2 weeks of root emergence. The next two secondary cambiums form in the third week. The rings formation is continued rapidly and 6 weeks after seedling emergence, at the time that 12–13 leaves have been produced, six rings will be visible. At this stage, the root diameter is between 1.0 and 1.5 cm. By the end of the eighth week, two more rings will form, but after that the speed of ring formation will decrease. At the time of harvest, the maximum number of rings is 12–15. Although about 12–15 cambium rings are formed, only about half of the cambium rings play a significant role in root expansion. Most root growth occurs in rings 1 and 2, while rings 3–8 usually show less activity. Rings 1–6 contain about 75% of the root. Therefore, the rings that have the greatest effect on the final root yield from the 22nd week onwards are the rings formed in the first 6 weeks. It is noteworthy to mention that growth occurs simultaneously in all of these rings (Elliott et al. 1984). Cambium ring formation is stimulated by leaf formation (Zamski and Azenhot 1981). Although extending the growth duration leads to continuous leaf formation, it is anticipated that the number of cambium rings also increase. In addition, the number of cambium rings and the distance between adjacent rings are related to the sugar content in sugar beet (Artschwager 1930; Milford 1973; Wyse 1979a, b). Therefore, it can be assumed that increasing the number of cambium rings increases the sugar content. In contrast to this hypothesis, several researchers have noted that the number of cambium rings in autumn sugar beet does not extend despite a long growing period (Hoffmann and Kluge-Severin 2011; Reinsdorf et al. 2014). Dry matter production also decreases in autumn due to reduced radiation intake (Milford et al. 1988). Although the distance between cambium rings increases during growth, mainly the development of five internal rings is closely related to yield formation (Milford 1973; Hoffmann 2010a). Even at high plant densities in the field or greenhouse (such as two plants in a pot), the number of cambium rings remain constant (Milford 1976). Moreover, the fact that the number of cambium rings does not extend over a long growing period suggests strong genetic influence (Milford 1973; Hoffmann and Kluge-Severin 2011). The highest sucrose accumulation occurs in the five inner rings of the central cortex (out of a total of 12–15 rings) at the site of the maximum root diameter (Elliott and Weston 1993). This trait begins within 3 weeks after germination and at 10 weeks of growth can cause significant differentiation between genotypes. Sucrose accumulation increases during the growing season (Kenter and Hoffmann 2006; Trebbi and McGrath 2009). Most researchers have found a weak positive correlation between ring number and ring density (number of rings divided by root radius) with sucrose content (Artschwager 1926;

Artschwager 1930; Pack 1930; Doney and Theurer 1976; Heinisch and Bohme 1959). Since the cambium formation in fodder beet, leaf beet and sugar beet (three of so different subspecies of *Beta vulgaris* L.) is similar, no effect of this factor on dry matter and sucrose accumulation has been reported (Lohaus et al. 1994).

The sucrose concentration decreases from the root to the stem because of a sharp decline in the number and volume of parenchyma cells and consequently the storage tissue in the upper part of the root. In addition, the sucrose concentration in sugar beet depends entirely on the size of the root parenchyma cells (Milford and Watson 1971; Milford 1973). Large parenchyma cells have lower sucrose concentration than smaller ones (Vukov 1972; Milford 1973). However, the correlation between sugar content and cell size is not entirely linear. Some exceptions that suggest other factors such as osmotic pressure may also affect sucrose accumulation. High turgor pressure in the cell inhibits plasma membrane adenine transport phosphatase (ATPase) and ultimately reduces sucrose uptake (Wyse et al. 1986). Milford and Watson (1971) reported that the number of rings in heavy sugar beet roots fed with nitrogen fertilizer was akin to the number of rings in the roots receiving less nitrogen, and that the expansion in root diameter was because of an increase in the width of each ring. Cell number was not influenced by nitrogen fertilizer but increased the average cell volume by 40%. According to Doney (1983), progress in increasing sucrose production relies on the breeders' ability to identify lines or hybrids that produce more cells (quicker cell division) instead of larger cells.

Various studies have shown that the sucrose percentage is closely tied with the total area of the cell surface and not the storage capacity. However, the propagation distance of the material can also explain this phenomenon. For instance, in large cells, the diffusion distance increases as the distance between the vascular rings increases. Vivien (1920) suggested that having the highest number of rings and the narrowest parenchyma areas are to be prominent features of sugar beet roots with high sugar content. As sucrose moves in the free spaces by diffusion, its concentration would be higher near the unloading site of the vascular bundles. Also, since the rate of sucrose uptake by parenchyma cells is strongly associated with the sucrose concentration in the adjacent free spaces, these parenchyma cells contain the highest sucrose concentration. Cells farther away from the vascular area are exposed to low sucrose concentrations in the free space, and therefore less sucrose accumulates in their vacuoles. Roots with cells mostly concentrated near the vascular system are those with high sucrose concentrations (Z-type). Within the intercellular region, cells adjacent to the phloem are younger and smaller, and cells farther away are larger and older. Since the adsorption capacity of sucrose in fodder beet and sugar beet is similar, the controlling agent for sucrose uptake is not the cell size, but the distance of the cell from the vascular system. Cell size or number of cells determines this distance. Therefore, it is necessary to expand the field operations that cause the formation of narrow rings in the roots and consequently high sucrose concentrations. For example, excessive application of nitrogen fertilizer increases root size by enlarging the cell. Narrowing row space or high density reduces root size by limiting cell expansion (Milford and Watson 1971). In a study by Pack (1930), the associations between 50 traits were examined and it was shown that sucrose concentration was positively correlated with high plant density, high dry matter content, tissue strength, root length, and skin hardness. In sugar beet experiments, it was found that applying nitrogen fertilizer at the end of the growing season contrasted with sucrose accumulation. A number of researchers have shown that fodder beet generally has a higher ionic content (mainly potassium and nitrate) and a much lower sugar content than sugar beet (Burba et al. 1984; Barbier-Brygoo et al. 1987). The ionic content, in addition to sucrose, also affects the osmotic potential of sugar beet cells. Negative relationship between non-mineral ions and sugar content can affect sucrose accumulation in the vacuole; leaf vacuoles accelerate sucrose-facilitated uptake (Kaiser and Heber 1984) while inorganic ions are actively adsorbed (for more sources, refer to Martinoia 1992). Therefore, the effect of genotype, environment, and nitrogen fertilizer on sucrose concentration can be described by limited or controlled release of photosynthetic material in sugar beet root (Fasahat et al. 2019, 2020). The older leaves of the sugar beet plant are peripheral and the youngest are in the center. Photosynthetic material from old leaves is mainly sent to the main (central) area of the root, while younger leaves first feed the outer parts of sugar beet root with photosynthetic material and nitrogen (Haeder and Beringer 1987). The parenchyma cells of the phloem and the meristem cells of the cambium are fed by their proximity to the phloem first, and then the stored parenchyma cells between the cambium and finally the central part (middle) of the sugar beet root. However, because the central part is fed for a longer period of time, its sugar content is higher (Fieuw and Willenbrink 1987).

3.3 Sugar Storage Stage

After seed sowing, the germination takes between 3 and 5 days subjected to the temperature. About 3 days after sowing, the germinating seeds produce radicle and cotyledon appears by the fifth day. Growth is very slow in the next 5–7 days until the true leaves are formed. The first true leaves begin to appear about 10–12 days after planting, and during the growing season, about two to four leaves appear each week (Doney 1979). After 30 days of plant growth, the number of true leaves reaches to 6–10. The roots do not begin to thicken until the first true leaves have formed. Once cambium layers formed, cell divisions occur which results in new cells differentiated into xylem, phloem, and parenchyma cells. After 40 days of plant growth, cell growth and proliferation occur simultaneously in all rings. The genetic identity of a sugar beet plant has been obtained at this time. This means that it is possible to measure growth parameters at the seedling growth stage, rather than waiting until harvest. The leaves grow rapidly at first and until the root diameter reaches about 1 cm (when all the rings are formed). Then root growth increases and because of the formation of more meristematic tissue in the roots, more photosynthetic material is needed for cell division and growth (Doney 1979). Studies by Milford et al. (1985a, b, c) showed that up to about 12 leaves, mature leaves become increasingly larger, but subsequent leaves have a smaller final size. The first leaves die in the order of their emergence, and when the largest leaves reach their full size, the leaf area index reaches a maximum. The leaves emerge and expand in a linear relationship with thermal time.

Leaves and petioles are the first priority for metabolic products during the growing season until plant growth conditions become favorable. During the first few weeks of growth, leaves and petioles form the major structure of the plant as well as the dry matter (Terry 1968; Follett et al. 1970; Loomis et al. 1971; Storer et al. 1973). Around the sixth week, the rate of dry matter accumulation in the root increases compared with leaves and petioles. Since then, the root shows accelerated linear accumulation throughout the season, while the dry matter content of the petiole and blade increases at a constant rate. This trend is shown by data from the evaluation of 24 hybrids and inbred lines cultivated in Logan in 1974 (Theurer 1979). The earlier the canopy grows, the better the chance of producing sucrose because the roots receive a large amount of photosynthetic material for a long time instead of the foliar. Root growth occurs by cell division and increase in cell size, through which cultivars may differ in the ratio of each of these two processes (Theurer 1979). Greenhouse experiments showed that sugar beet plants can achieve very high root yield with low leaf dry matter (Hoffmann 2010b; Schnepel and Hoffmann 2016). Because plants usually have a higher leaf area index than is needed for material formation, reducing leaf dry matter after complete canopy formation can help improve root yield. However, this alteration in the allocation of photosynthetic materials requires an increase in the capacity of the storage root reservoir.

In 1979, the results of a trial showed that increasing CO_2 levels up to 700 ppm increased root yield by 21% but reduced sugar content from 15.4 to 15.1. Additional photosynthetic material was therefore used to store sucrose. This conclusion was also confirmed in the research of Watson et al. (1972) in which shading was used to reduce the supply of photosynthetic materials. Shading reduced root dry weight yield but did not change the ratio of sucrose to dry weight. Therefore, photosynthetic material transferred to the root was balanced between root growth and sucrose storage which was independent of the supply of photosynthetic material. The concept of balanced division is crucial to understanding the relationship between sucrose and performance. Theurer (1979) research showed that a significant difference in sugar content (based on fresh weight) is evident on 7 weeks after emergence, and the genes responsible for this increase are apparently expressed 5 weeks after emergence (Trebbi and McGrath 2003). Milford (1973) reported that 6 weeks after seedling emergence, relatively high sucrose concentrations reached 9% by weight. In a study by Bellin (2005), in the ninth week after planting, the amount of sucrose was measured at 10.17% of fresh weight in 2001, while in its initial measurement in 2002, the sucrose concentration in fresh weight was 7.54%. Sucrose concentration in leaves was also reported to be less than 2% by fresh weight. From these observations it can be concluded that sucrose accumulation is a process that begins early in growth and from the very beginning, root cells must adapt to high levels of osmotic stress. From the ninth week onwards, the amount of sucrose in the roots increases linearly with time (Milford 1973; Bellin 2005). However, considering the sucrose concentration, it is found that in the first part of root development, the sucrose concentration increases linearly, but in the second part of development, saturation is obtained. This associates with the negative relationship between root yield and sucrose concentration (Bellin 2005). Studies performed under controlled conditions (Ulrich 1952, 1955, 1956, 1961; Loomis et al. 1971) illustrated that at late growing season, temperature is responsible for the rapid increase in sugar content of beet roots (Akeson 1981). Low temperatures during the day are associated with high sugar content (Ulrich 1952) although night temperatures are more important for high sugar content than daily temperatures (Ulrich 1955, 1961). By increasing the night temperature from 4 to 30 °C, sugar content decreased linearly (Ulrich 1955).

3.4 The Relationship Between Source and Sink

Wardlaw (1990) proved that source and sink activities are highly correlated. In addition, sink activity has been shown to affect source photosynthetic activity (Paul and Foyer 2001). In 1974, Kursanov emphasized the sink capacity as a determinant of sucrose transport through activating enzymes (Pavlinova and Prasolova 1972) or changes in the membrane transmission system (Turkina and Sokolova 1972). According to some studies, the restriction in the growth of sugar beet root storage is not influenced by the source but is controlled by sink capacity since yield was not increased with an increase in CO_2 level (Burkart et al. 2009; Manderscheid et al. 2009). On the other hand, most previous studies show that sugar storage is limited by the sink organ. Given these results, it can be expected that sugar content of the sugar beet will not increase with an increase in growth period but will probably remain stable. It is widely acknowledged that the long-distance transport process does not in itself limit sugar storage or plant growth (Hawker 1985).

Farrar and Minchin (1991) stated that the material unloading by the phloem and the plant sink activity is limited after the sugar concentration in the sink reaches a certain value, which sends a negative signal to the source organs (leaves). This stimulates leaf aging and reduces photosynthesis and leaf function over time. It is shown by Clark and Loomis (1978) that depending on the age of the plant, the size of fresh sugar beet leaves decreases. The decrease in source capacity is clearly owing to the lack of demand for new photosynthetic materials from the main sink. As previously noted, one of the ways to increase sugar yield is to produce cultivars that reach the maximum leaf area at the beginning of the growing season and then do not exceed the leaf area index for optimal growth. Transferring the photosynthetic material to the root and early formation of a sugar storage sink in the root is essential to achieve high sugar yield.

3.5 Impurities (Non-Sugar Components) in the Root

Although many factors in the root affect the calculation of sugar yield, quality assessment formulas around the world consider only certain qualitative components such as potassium, sodium, and amino-nitrogen, and in a number of instances glucose (Mahn et al. 2002). These components are considered as a representative of beet composition and for daily analysis of raw syrup in the factory (Kenter and Hoffmann 2006). According to Burba (1980), in sugar beet leaves, the sodium and potassium concentration is five and ten times higher than the root, respectively. In a study by Mahn et al. (2002) and Winner and Feyerabend (1971), the sodium and potassium concentration increased from root to stem at harvest time in October. Sodium and potassium concentrations can be even higher in delayed harvest because these components move from dead leaves at the end of the growing season (Burcky et al. 1978). The total concentration of dissolved nitrogen and amino-nitrogen in the upper part of the stem was three times higher than the root (Mahn et al. 2002). In comparison, the betaine and nitrate concentrations showed less change. Nitrogen metabolism occurs mainly in plant leaves, where nitrate absorbed through the capillary root system is metabolized by enzymatic reactions to produce amino acids, proteins, nucleotides, and other nitrogenous compounds. Betaine acts as a cytosolic osmotic pressure regulator and increases significantly under drought conditions to keep the cell turgor pressure constant (Lüttge et al. 1988). All nitrogen compounds are transported from the leaves to the roots through the phloem. Therefore, the highest concentration of total soluble nitrogen and amino-nitrogen is observed in the upper part of the beet.

The negative effect of non-sugar compounds during processing can be related to several factors (Hoffmann 2010a). The most important thing is that non-sugar compounds increase the sugar in molasses because they increase the solubility of sucrose and thus reduce crystallization. These compounds are called molasses because of the above-mentioned characteristics (Schneider et al. 1961). Each kilogram of impurities prevents 1.5–1.8 kg of sucrose from crystallizing, which is lost through molasses (Alexander 1971; Dutton and Huijbregts 2006). Doxtator and Galton (1950) reported that potassium and sodium content were to a certain degree associated with high performance and a negative correlation with sucrose. The sodium percentage has a positive correlation with potassium. Dahlberg (1950) stated that selection for low sodium content may be effective in decreasing sugar content changes and eliminating single roots with low-sugar content. Finkner and Bauserman (1956) found that selection for sodium alone had little effect on sucrose concentration and concluded that selection for low sodium was of little value in most cases. In contrast, Wood et al. (1958) presented evidence that sugar content can be effectively increased by selecting the high sucrose and low sodium compared with selecting only the sucrose concentration. Dudley and Powers (1960) concluded that it was possible to produce lines with low concentrations of potassium and sodium. Two stages of selection for low amino-nitrogen from a heterogeneous population resulted in a 36% decrease in amino-nitrogen concentration (Campbell and Fugate 2013). However, quite the reverse selection (i.e., for high concentrations) augmented the amino-nitrogen concentration by 93% (Smith and Martin 1989). Continuous selection from this population yielded a slight change in amino-nitrogen concentration (Campbell and Fugate 2012). Coe (1987) observed that the roots selected for the lowest concentrations of non-sugar solutions were generally the smallest roots and suggested that size should be taken into account to minimize reduced root yield. The sucrose concentration has changed a lot over time. This change was mostly because of increased root yield followed by a decrease in impurities (Hoffmann and Kluge-Severin 2011). Sodium and amino-nitrogen concentrations have decreased over the past 36 years, but the most significant change was a 30–50% decrease in potassium concentrations (30–40% of impurity reduction is allocated to improvement in breeding programs, Campbell and Fugate 2015). Schnepel and Hoffmann (2014) estimated that 11% of changes in sugar losses and 12% of invert sugar accumulation were due to genotype.

3.6 Post-Harvest Root Physiology

Sugar beet varieties differ in terms of sugar loss and increase in non-sugar content during storage (Kenter and Hoffmann 2009) which seems to be an interesting option for breeders (Hoffmann 2010a). During storage, sugar levels are reduced to maintain vital physiological processes in beets, such as respiration and root injuries (Burba 1976). In carbohydrate metabolism, sucrose is broken down by enzymatic activity into fructose and glucose (Berghall et al. 1997). Most of these hexoses are oxidized by respiration, but a significant portion accumulates in cells as invert sugar (Klotz et al. 2006). Both of these processes increase with storage period as well as temperature resulting in the excess of invert sugar content (Berghall et al. 1997; Klotz et al. 2006). Respiration is in charge of 70–80% of sucrose losses throughout storage (Wyse 1970). Sucrose is one of the foremost sources of carbohydrates for carbon dioxide formation which is used in respiration (Barbour and Wang 1961). The respiration process begins with the breakdown of sucrose into hexose sugars, possibly controlled by the enzyme sucrose synthase (Echeverria 1998; Etxeberria and Gonzalez 2003). However, hydrolysis of sucrose by alkaline invertase and acid invertase is also possible (Wyse 1974; Berghall et al. 1997). Enzyme activity is temperature dependent and most enzyme activity occurs at temperatures around 40 °C (Klotz and Finger 2001). Respiration requires sufficient oxygen to convert sucrose into carbon dioxide and water. Some of the energy generated during this heat reaction is stored as an energy-rich molecule, adenosine triphosphate (ATP), and the rest is converted to heat. Amino acids contribute to respiration when broken down into organic acids. If oxygen is restricted, ethanol is formed by fermentation. This causes more damage and loss of sugar because anaerobic respiration requires 15–16 times more sucrose to produce the equivalent amount of ATP than aerobic respiration (Zhang and Greenway 1994). The ratio of glucose to fructose in freshly harvested sugar beets is 2.5 on average. At harvest, the fructose ratio is lower than glucose but increases during storage. Thus, the ratio of glucose to fructose storage decreases during storage (Campbell and Klotz 2006). The quality of beets can also reduce *via* the formation of raffinose during storage (Kenter and Hoffmann 2009). Storage of roots with a surface contamination of >25% must be carried out with caution. In storage piles where healthy and rotten roots are mixed, the temperature rises because of the high rate of respiration in the diseased roots (Campbell and Klotz 2006), which can increase respiration among nearby healthy roots. The rate of the harmful effects of diseased roots on adjacent healthy roots is subjected to the extent of heat generated from the pile (Campbell et al. 2014).

During storage, insoluble nitrogen compounds such as proteins are hydrolyzed to amino acids and the amino-nitrogen concentration in beet roots increases (Vukov and Hangyal 1985; Jaggard et al. 1997; Martin et al. 2001). As with other enzymatic processes, this conversion is presumably to augment with raise in temperature. Plants usually respond to stress factors such as wounds, drought, or infections caused by pathogens through increasing the accumulation of invert sugar (Roitsch 1999). A negative correlation has been reported between sugar concentration and disease severity (Bugbee 1973; Hemayati et al. 2017). However, resistance to phoma has been reported in some low-sugar genotypes. In this regard, sugar content showed a negative correlation with disease severity in plants infected with the *Fusarium* root rot (Lukezic et al. 1969). In 1974, Vidhyasekaran studied the role of carbohydrates in resistance to *Helminthosporium nodulosum* in millet leaves and reported only glucose linked with resistance and that low glucose content was associated with susceptibility. Fructose content increased in tomato seedlings resistant to *Pythium aphanidermatum* (Muthusamy et al. 1974). In numerous other cases, however, no association was found between carbohydrates and disease resistance (Gibbs and Wilcoxson 1972; Sindhamathar and Vidhyasekaran 1978). When the weight loss in beets reaches more than 25–30%, the roots can no longer resist microbial growth (Bugbee 1993). One of the main usages of biotechnology in sugar beet breeding is to reduce post-harvest sugar loss and impurities that affect sugar extraction (Bosemark 1993). Many attempts have been made to better understand the molecular contexts of sucrose degradation in sugar beet in order to identify candidate genes to reduce post-harvest sucrose loss. A suitable target to decrease the invertase enzyme activity is the gene that encodes the invertase inhibitor (Harms and Schulz 2015). Invertase inhibitors were first identified in potato (Schwimmer et al. 1961) and then in sugar beet, tomato, and tobacco (Pressey 1968, 1994; Weil et al. 1994; Rausch and Greiner 2004). Using these polypeptides, the sucrose concentration in sugar beet can be increased. Thus, genes that encode polypeptides for reduced enzymatic activity can be easily obtained or identified through breeding techniques as well as genetic engineering.

3.7 Increasing the Sugar Content *via* Breeding Techniques

3.7.1 Inheritance

Heterosis for root yield is widely reported in sugar beet literature and justifies the use of hybrids in commercial production. Heterosis reported for sucrose content is lower than sugar yield. Often, the sucrose content of a hybrid is between its parents. Heterosis usually increases with wider genetic diversity of parents; therefore, long-term hybrid development programs require the use of unique genetic breeding lines with relatively high sucrose content (Campbell 1990; Fasahat et al. 2018).

It has been reported in a number of studies about the reciprocal cross of Z-type with E-type beet. The results showed that the offspring produced by the Z-type parent had at least 0.4% higher sugar content than the offspring produced by the E-type parent, which clearly indicates the maternal effect (Savitsky 1940c). Similarly, Schlosser (1949) reported the effect of maternal cytoplasm on sugar content. Sliwinska et al. (1998) reported that 75–90% of mother roots with above-average sugar content produced offspring with above-average sugar content. Some researchers have reported the presence of dominant genes and the occurrence of heterosis for sugar content (Powers et al. 1959; Takebe and Izumiyama 1977; MacLachlan 1972c). According to Powers et al. (1959), sugar content in hybrids showing heterosis was in most cases closer to the maternal parent; however, this pattern was not observed when the maternal parent had a much lower sugar content than the pollinating parent. The results obtained by Rush and Oldemeyer (in Powers et al. 1959) showed a greater effect of maternal parent, in such a way that the mean difference in sugar content between offspring (hybrid), the maternal, and pollinator parent was 0.4% and 0.6%, respectively. However, the results of the above observations on the heterosis occurrence for sugar beet were cross-examined by other researchers in terms of sample size and parent's homogeneity. In different studies (Savitsky 1940c; Schlosser 1949; Jassem et al. 2000), it was shown that the maternal parent is more effective than the pollinator on the sugar content of offspring. Sugar content is the main qualitative trait in determining the value of sugar beet, but it should be noted that the maternal effect on the inheritance of potassium concentration has also been observed. According to the above-mentioned results, the selection for high sugar content and low potassium in the development of cytoplasmic sterile parent should be considered, while the improvement of hybrid root yield mainly counts on the combinability of specific parental lines (Fasahat et al. 2016).

The majority of sugar beet breeding programs have at least two themes: the breeding of monogerm O-type male parents with genetically equivalent cytoplasmic male sterile females, and the breeding of multigerm pollinating parents (Jassem et al. 2000). Hybrids grow faster than their parents and produce more cells. It is the number of cells (more cells) rather than the cell size that is responsible for causing heterosis. When parents with high heterosis potential are crossed for root yield, the offspring have an average cell size equal to the average parent, but with more cells. Therefore, the offspring have higher root yield without reducing the sugar content. For this reason, sugar beet breeders have moved towards hybrid production because total sugar yield can be increased without adversely affecting sugar content (Doney 1983). In 1908, the study of sugar content inheritance in sugar beet was started by Andrilk et al. and continued by many other researchers (Vilmorin 1923; Stehlik 1933; Savitsky 1940c; Culbertson 1942). Although the results of all studies are not conclusive, the prevailing belief is that sugar content is a quantitative and multigene trait that is mainly inherited under high genetic heritability (Savitsky 1940b; Culbertson 1942; Powers 1957; Powers et al. 1963; Zhao et al. 1997; McGrath and Trebbi 2007). Estimation of sugar content heritability varies widely conditional to the studied materials and calculation method (Jassem et al. 2000). MacLachlan (1972a, b, c) estimated the sugar content heritability to 0.19–0.6. The coefficient of variation calculated by Savitsky (1940c) was between 4.4 and 7.2 and was often ten times lower than the coefficient of variability of root weight. Smith et al. (1973) reported a negative correlation (−0.68) between root weight (yield) and sugar content, which is inconsistent with the expression of heterosis observed in some hybrids (Powers et al. 1959; MacLachlan 1972a).

Root yield is controlled by additive and non-additive gene effects (Smith et al. 1973; Helmerick et al. 1963). Approximately half of the genetic diversity of root yield is because of the additive effect of genes and half of it is due to the non-additive (epistatic) effect of genes. By making specific crosses that exploit epistasis, sugar beet breeders are being able to produce hybrids with high root yield without reducing sucrose content. However, these specific genetic combinations are very difficult to identify and require extensive crosses and field experiments.

3.7.2 Genome Mapping

Identification of the genetic factors involved in sucrose accumulation is not only of scientific importance, but also of economic importance for improving the sugar content of sugar beet. In this respect, different breeding methods (e.g., mass selection) has been generally used, but to study each genetic factor at the molecular level requires accurate fine mapping, which is undoubtedly a very time-consuming process. Important agricultural traits are mainly controlled by several QTLs. Genomic mapping and identification of markers attached to these QTLs facilitate crop breeding. The primary goal of all mapping methods is to identify the QTLs involved, to estimate their chromosomal position, their effects, and the variance ratio of the genotype expressed by them. In such case, the method of characterized genes can be used, which involves selecting genes based on their predicted function from genomic sequence studies, but other criteria such as gene expression levels are also used to identify unknown genes. In 1940, Savitsky estimated three to four gene loci responsible for sugar content expression (Savitsky 1940a). The first applied map of the sugar beet genome was on the basis of Single Nucleotide Polymorphism (SNP) markers that used different methods such as single-strand conformation polymorphism (SSCP), cleaved amplified polymorphic sequences (CAPS), and heteroduplex analysis (HA) (Schneider et al. 1999). On this map, 42 genes involved in carbohydrate and nitrogen construction were examined in five segregating populations and assigned to nine linkage groups. Weber et al. (2000) identified QTLs involved in quantitatively similar traits for sugar, potassium, and nitrogen on chromosomes 2, 7, and 8, but these loci were not associated with different populations and environments. Schneider et al. (2002) studied an F_3 population for qualitative traits in different environments. Twenty-one QTLs involved in quantitative traits for amino-nitrogen (chromosomes 3 and 4) and ion balance control (chromosomes 5 and 9) were mapped by expressed sequence tag (EST) markers. In addition, five chromosomal regions associated with QTLs for sugar content were identified. Among those involved in quantitatively known traits, only four QTLs were stable on chromosomes 2 (potassium content), 4 (modified root and sugar yield), and

9 (sugar content) and were reported in different environments (Biancardi et al. 2010). In other studies, association mapping was used to map QTLs involved in quantitative traits related to sugar content (Stich et al. 2008a, b; Wurschum et al. 2011a, b). It seems that association mapping has promising prospect for a more efficient detection of QTLs involved in agronomic traits and disease.

3.7.3 Sucrose Transporter Genes

The massive accumulation of sucrose in the phloem vessels of sugar beet indicates the presence of an active, highly efficient, and energy-dependent transporter in the plasma membranes of some phloem cells. Using antisense probe, the mRNA of sucrose transporter called BvSUT1 was found in the phloem vessels of sugar beet (Vaughn et al. 2002). Sucrose has been identified as the preferred material of BvSUT1. BvSUT1 does not transport only glucose and fructose monosaccharides but also lactose disaccharide or raffinose trisaccharide. The high dependence of BvSUT1 on sucrose is clearly linked to other sucrose transporters involved in phloem loading, such as SoSUT1 of spinach, StSUT1 of potato, and AtSUC2 or PmSUC2 of plantago major (Riesmeier et al. 1992, 1993; Gahrtz et al. 1994; Sauer and Stolz 1994). Under experimental conditions, BvSUT1 also accepts maltose as a substitute for sucrose, even with almost twice lower the transfer capacity. However, maltose is unlikely to play a role as a substrate for BvSUT1 under physiological conditions because this disaccharide is not found in the phloem juice of sugar beet (Lohaus et al. 1994). In addition, BvSUT1 shows little transfer capacity for arbutin. Because BvSUT1 is also capable of transmitting esculin, it can be classified as a type 1 sucrose transporter (Gora et al. 2012; Reinders et al. 2012). BvSUT1 loads sucrose specifically into the phloem vessels of cell in leaves in line with its expression pattern. Although the transport of sugar across vacuole membrane has been studied for four decades, the molecular nature of carriers has been identified over the past 10 years. Thus, our images of sugar carriers across the tonoplast canal have recently taken on a real form (Hedrich et al. 2015). The *bv*TST2.1 vacuole transporter for sucrose loading works as an inverse transporter of proton-coupled sucrose (Jung et al. 2015). From a thermodynamic standpoint, assuming complete sucrose pairing with H^+, *bv*TST2.1 can accumulate up to 100 times more sugar. In fact, sugar beet root vacuoles contain up to one millimolar of sugar. Therefore, there is a lot of attention for plant production with an increase in the activity of TST2.1 and vacuole P3-ATPases to further conduct of sucrose and storage in cells. The BvSUT1 gene family in the plasma membrane of sugar beet is different from the BvTST2.1 sucrose transporter gene family in tonoplasts, which evolved from the monosaccharide transporter superfamily (Jung et al. 2015). Compared with BvSUT1, members of the TST family act as reverse transmitters coupled to protein to store sugar in the vacuole (Schulz et al. 2011; Jung et al. 2015). Unlike BvSUT1, BvTST2.1 is mainly expressed at the root. In this storage tissue, BvTST2 represents the major sucrose receptor in the vacuole (Jung et al. 2015; Fasahat et al. 2018). Thus, the BvSUT1 activity in the source leaves provides the basis for sucrose to be transported to the root via the phloem for storage under BvTST2.1 regulation.

3.8 Future Prospects

Root selection accordant with this trait in preliminary breeding programs may be done with simple selection methods, but breeding beyond the usual physical limitation requires more complex and expensive methods. The molecular markers application and also linkage maps have opened the door for the swift breeding of sugar beet.

3.9 Conclusion

As early in the growing season, the sucrose accumulation proceeds and augment directly with time in the first half of growth and saturation occurs in the second half. Optimal hybrids are expected to have small cell root tissue and rapid cell division and also the characteristics of high-sugar cultivars. Qualitative traits in sugar beet are potential goals of breeding programs but are strongly influenced by environmental factors and to some extent genotype-by-environment interaction. Actual yield is always less than yield potential (optimal environment without any restrictions on water, food or pest, disease, weed or other stresses) because climatic conditions are usually not favorable and management of crop operations, regions, and many other factors limit sugar beet yield on commercial field (Trimpler et al. 2017). The gap between the achievable yield estimated under value for cultivation and use trials and the actual yield on fields in some countries is more than 30% (Jaggard et al. 2012). In various studies, it was shown that sugar content, as the most important quality parameter, is evidently affected by the parent with incremental inheritance.

References

Akeson WR (1981) Relationship of climate and sucrose content of sugar beet roots. J Sugar Beet Res 21(1):27–41

Alexander JT (1971) Factors affecting quality. In: Johnson RT, Alexander JT, Rush GE, Hawkes GR (eds) Advances in sugarbeet production: principles and practices. Iowa State Univ. Press, Ames, Iowa, pp 371–381

Artschwager E (1926) Anatomy of the vegetative organs of the sugar beet. J Agric Res 33:143–176

Artschwager E (1930) A study of the structure of sugar beets in relation to sugar content and type. J Agric Res 40:867–915

Barbier-Brygoo H, Mailer P, Mathien Y, Guern J (1987) Ionic contents of sugar and fodder beets in relation to sucrose accumulation in the vacuole. International Institute for Sugar Beet Research, 50th Winter Congress, Bruxelles, p 149–62

Barbour RD, Wang CH (1961) Carbohydrate metabolism of sugar beets I. Respiratory catabolism of mono and disaccharides. J Sugar Beet Res 11(5):436–442

Bellin D (2005) Macroarray analysis of gene transcription during sucrose accumulation in sugar beet (*Beta vulgaris* L.) root: identification of developmental and metabolism related candidate genes. Doctoral dissertation, Universität zu Köln

Bergen P (1967) Seasonal patterns of sucrose accumulation and weight increase in sugar beets. J Amor Sec Sugar Beet Technol 14:538–545

Berghall S, Briggs S, Elsegood SE, Eronen L, Kuusisto JO, Phili EJ, Theobald TC, Walliander P (1997) The role of sugar beet invertase and related enzymes during growth, storage and processing. Zuckerindustrie 122:520–530

Biancardi E, McGrath JM, Panella LW, Lewellen RT, Stevanato P (2010) Sugar beet. In: Bradshaw JE (ed) Root and tuber crops, Handbook of plant breeding, vol 7. Springer Science Business Media, New York, pp 173–219

Bosemark NO (1993) Genetics and breeding. In: Cooke DA, Scott RK (eds) The sugar beet crop: science into practice. Chapman & Hall, London, pp 67–119

Bugbee WM (1973) Sucrose and cell walls as factors affecting *Phoma* storage rot of sugar beet. Phytopathology 63:480–484

Bugbee WM (1993) Storage. In: Cooke DA, Scott RK (eds) The sugar beet crop: science into practice. Chapman and Hall, London, pp 551–570

Burba M (1976) Atmung und Saccharosestoffwechsel lagernder Zuckerrüben. Zuckerindustrie 26(10):647–658

Burba M (1980) Kalium und Natrium im Stoffwechsel der Zuckerrübe. Zuckerrübe 29:16–19

Burba M, Nitzschke U, Rltterbusch R (1984) Die N-Assimilation der Pflanze unter besonderer Berucksichtigung der ZuckerrObe. Zuckerindustrie 109(7):613–628

Burcky K, Beiß U, Winner C, Drath L, Schiweck H (1978) Versuche zur Bedeutung des Nahrstoffangebotes fur die Qualitat der Zuckerrube. 2. Stickstoff und Kalium. Zuckerindustrie 103:190–200

Burkart S, Manderscheid R, Weigel HJ (2009) Canopy CO_2 exchange of sugar beet under different CO_2 concentrations and nitrogen supply: results from a free-air CO_2 enrichment study. Plant Biol 11:109–123

Campbell LG (1990) Sugarbeet germplasm selected from the USDA collection. North Dakota Farm Res 47(6):32–34

Campbell LG, Fugate KK (2012) Registration of F1025, F1026, and F1027 sugarbeet genetic stocks with low concentrations of sodium, potassium, or amino-nitrogen. J. Plant Registrations 7:250–256

Campbell LG, Fugate KK (2013) Divergent selection for amino-nitrogen concentration in sugarbeet roots. J Sugar Beet Res 50(3&4):1–13

Campbell LG, Fugate KK (2015) Relationships among impurity components, sucrose, and sugarbeet processing quality. J Sugar Beet Res 52(1&2):2–21

Campbell LG, Kern JJ (1983) Relationships among components of yield and quality of sugarbeets. J Am Soc Sugar Beet Technol 22:135–145

Campbell L, Klotz K (2006) Storage. In: Draycott AP (ed) Sugar beet. Blackwell Publishing, London, pp 387–408

Campbell LG, Windels CE, Fugate KK, Brantner JR (2014) Postharvest losses associated with severity of Rhizoctonia crown and root rot of sugarbeet at harvest. J Sugar Beet Res 51(1&2): 31–51

Carter JN (1987) Sucrose production as affected by root yield and sucrose concentration of sugarbeets. J Am Soc Sugar Beet Technol 24:14–31

Clark AE, Loomis RS (1978) Dynamics of leaf growth and development in sugar beets. J Amer Soc Sugar Beet Technol 20:97–112

Coe GE (1987) Selecting sugarbeets for low content of nonsucrose solubles. J Am Soc Sugar Beet Technol 24:41–48

Culbertson JO (1942) Inheritance of factors influencing sucrose percentage in *Beta vulgaris*. J Ag Res 64:153–172

Dahlberg HW (1950) Chemical methods for breeding sugar beets. Proc Am Soc Sugar Beet Tech 6: 137–138
Doney DL (1979) Seedling physiology and sugarbeet yield. J Am Soc Sugar Beet Technol 20(4): 399–418
Doney D (1983) Sugarbeet root yield sucrose concentration: physiology and genetics. Sugarbeet Res Extension Rep 14:216–219
Doney DL, Theurer JC (1976) Hypocotyl diameter as a predictive selection criterion in sugarbeet. Crop Sci 16:513–515
Doney DL, Wyse RE, Theurer JC (1981) The relationship between cell size, yield, and sucrose concentration of the sugarbeet root. Can J Plant Sci 61:447–453
Doxtator CW, Galton FR (1950) Sodium and potassium content of sugar beet varieties in some western beet growing areas. Proc Am Soc Sugar Beet Tech 6:144–151
Dudley JW, Powers L (1960) Population genetic studies on sodium and potassium in sugar beets (*Beta vulgaris* L.). J Am Soc Sugar Beet Technol 9(2):97–127
Dutton J, Huijbregts T (2006) Root quality and processing. In: Draycott AP (ed) Sugar beet. Blackwell Publishing Ltd., Oxford, pp 409–442
Echeverria E (1998) Acid invertase (sucrose hydrolysis) is not required for sucrose mobilization from the vacuole. Physiol Plant 104:17–21
Elliott MC, Weston GD (1993) Biology and physiology of the sugar-beet plant. In: Cooke DA, Scott RK (eds) The sugar beet crop: science into practice. Chapman and Hall, London, pp 37–66
Elliott ME, Hosford DJ, Lenton JR, Milford GFJ, Pocock TO, Smith JE, Lawrence DK, Firby DJ (1984) Hormonal control of storage root growth. In: Jackson MB, Stead T (eds) Growth regulators in root development. British Plant Regulator Group, Wantage, Monograph no. 10, pp 25–35
Etxeberria E, Gonzalez P (2003) Evidence for a tonoplast-associated from of sucrose synthase and its potential involvement in sucrose mobilization from the vacuole. J Exp Bot 54(386): 1407–1414
Farrar RF, Minchin PEH (1991) Carbon partitioning in split root systems of barley: relation to metabolism. J Exp Bot 42:1261–1268
Fasahat P, Rajabi A, Rad JM, Derera J (2016) Principles and utilization of combining ability in plant breeding. Biometrics Biostatistics Int J 4(1):1–24
Fasahat P, Aghaeezadeh M, Hosseinpour M, Hemayati SS (2021) Correlation between root weight and sugar content; do we have to continue traditional hypothesis? Plant Physiol Rep 26(1): 188–191
Fasahat P, Aghaeezadeh M, Jabbari L, Hemayati SS, Townson P (2018) Sucrose accumulation in sugar beet: from fodder beet selection to genomic selection. Sugar Tech 20(6):635–644
Fasahat P, Aghaeezadeh M, Kakueinezhad M, Jabbari L (2020) A meta-analysis of genotype × environment interactionon sugar beet performance. Biometrical Lett 57(2):221–236
Fasahat P, Khayamim S, Soltani IJ, Darabi S, Pedram A, Hasanai M, Jalilian A, Babaei B (2019) Stability analysis of genotype× environment interaction effect on sugar yield in sugar beet hybrids. J Crop Breed 11(32):33–40
Fick GW, Williams WA, Loomis RS (1973) Computer simulation of dry matter distribution during sugar beet growth. Crop Sci 13:413–417
Fieuw S, Willenbrink J (1987) Sucrose synthase and sucrose phosphate synthase in sugar beet plants (*Beta vulgaris* L. spp. altissima). J Plant Physiol 131:153–161
Finkner RE, Bauserman HM (1956) Breeding sugar beets with reference to sodium, sucrose, and raffinose content. J Am Soc Sugar Beet Technol 9:170–177
Follett RF, Schmehl WR, Viets FG Jr (1970) Seasonal leaf area, dry weight, and sucrose accumulation by sugarbeets. J Am Soc Sugar Beet Technol 16:235–253
Gahrtz M, Stolz J, Sauer N (1994) A phloem specific sucrose-H+ symporter from *Plantago major* L. supports the model of apoplastic phloem loading. Plant J 6:697–706

Gemtos TA, Papathanassiou J, Kylindris T, Chr K (2000) Instrumentation of a tractor to measure the power absorbed by implements. In: Proc 2nd National Conference of the Hellenic Society of Agricultural Engineers, Volos, Greece

Giaquinta R (1977) Sucrose hydrolysis in relation to phloem translocation in *Beta vulgaris*. Plant Physiol 60(3):339–343

Giaquinta RT (1979) Sucrose translocation and storage in the sugar beet. Plant Physiol 63:828–832

Gibbs AF, Wilcoxson RD (1972) Effect of sugar content of Poa pratensis on Helminthosporium leaf disease. Physiol Plant Pathol 2:279–287

Goodman PJ (1966) Effect of varying plant populations on growth and yield of sugar beet. Agric Prog 41:89–107

Gora PJ, Reinders A, Ward JM (2012) A novel fluorescent assay for sucrose transporters. Plant Methods 8:13

Haeder HE, Beringer H (1987) Distribution of 14C assimilates in the storage root of sugar beet plants after import from leaves of different age. Physiol Plant 70(4):590–596

Harms K, Schulz B (2015) Method for increasing sucrose yield in agricultural production of sugar beet and sugar cane. U.S. Patent 9,029,635, issued May 12, 2015

Hawker JS (1985) Sucrose. In: Day PM, Dixon RA (eds) Biochemistry of storage carbohydrates in green plants. Academic Press, New York, pp 1–51

Hedrich R, Sauer N, Neuhaus HE (2015) Sugar transport across the plant vacuolar membrane: nature and regulation of carrier proteins. Curr Opin Plant Biol 25:63–70

Heinisch O, Bohme L (1959) The question of the correlation between the number of rings of vascular bundles and the percentage dry matter content of beets. In: Contributions on beet research No. 2, Wiss Abh Dtsch Akad Landwiss Berlin 38:7–37

Helmerick RH, Finkner RE, Doxtator CW (1963) Variety crosses in sugar beet (*Beta vulgaris L.*) I. Expression of heterosis and combining ability. J Am Soc Sugar Beet Technol 12:573–584

Hemayati SS, Akbar MRJE, Ghaemi AR, Fasahat P (2017) Efficiency of white mustard and oilseed radish trap plants against sugar beet cyst nematode. Appl Soil Ecol 119:192–196

Hoffmann CM (2010a) Root quality of sugarbeet. Sugar Tech 12:276–287

Hoffmann CM (2010b) Sucrose accumulation in sugar beet under drought stress. J Agron Crop Sci 196:243–252

Hoffmann CM, Kluge-Severin S (2011) Growth analysis of autumn and spring sown sugar beet. Europ J Agron 34:1–9

Hoffmann CM, Kenter C, Bloch D (2005) Marc concentration of sugar beet (*Beta vulgaris* L.) in relation to sucrose storage. J Sci Food Agric 85:459–465

Jaggard KW, Koch HJ, Sanz JA, Cattanach A, Duval R, Eigner H, Legrand G, Olsson R, Qi A, Thomsen JN, Van Swaaij N (2012) The yield gap in some sugar beet producing countries. Int Sugar J 114(1363):496

Jaggard KW, Clark CJA, May MJ, McCullagh S, Draycott AP (1997) Changes in the weight and quality of sugarbeet (*Beta vulgaris*) roots in storage clamps on farms. J Agric Sci 129:287–301

Jassem M, Sliwinska E, Pilarczyk W (2000) Maternal inheritance of sugar concentration. J Sugar Beet Res 37:41–53

Jung B, Ludewig F, Schulz A, Meißner G, Wöstefeld N, Flügge U, Pommerrenig B, Wirsching P, Sauer N, Koch W, Sommer F, Mühlhaus T, Schroda M, Cuin T, Graus D, Marten I, Hedrich R, Neuhaus H (2015) Identification of the transporter responsible for sucrose accumulation in sugar beet taproots. Nat Plant 1:14001

Kaiser G, Heber U (1984) Sucrose transport into vacuoles isolated from barley mesophyll protoplasts. Planta 161:562–568

Kenter C, Hoffmann CM (2006) Seasonal patterns of sucrose concentration in relation to other quality parameters of sugar beet (*Beta vulgaris* L.). J Sci Food Agric 86:62–70

Kenter C, Hoffmann C (2009) Changes in the processing quality of sugar beet (*Beta vulgaris* L.) during long-term storage under controlled conditions. Int J Food Sci Technol 44(5):910–917

Klotz K, Finger F (2001) Sucrose metabolism in postharvest sugarbeet roots: activities and properties of the major sucrolytic enzymes. Sugarbeet Res Ext Rep:147–149

Klotz KL, Finger FL, Anderson MD (2006) Wounding increases glycolytic but not sucrolytic activities in stored sugarbeet root. Postharvest Biol Technol 41:48–55
Kursanov AL (1974) Assimilattransport und Zuckerspeicherung in der Zuckerrube. Z Zuekerindust 24:478–487
Lohaus G, Burba M, Heldt HW (1994) Comparison of the contents of sucrose and amino acids in the leaves, phloem sap and taproots of high and low sugar-producing hybrids of sugar beet (*Beta vulgaris* L.). J Exp Bot 45(277):1097–1101
Loomis RS, Rapoport H (1976) Productivity of root crops. In: Cock J, Maclntyre R, Graham M (eds) Proceedings of the fourth symposium of the international society for tropical root crops. The International Society for Tropical Root Crops, pp 70–84
Loomis RS, Ulrich A, Terry N (1971) Environmental factors. In: Johnson RT, Alexander JT, Rush GE, Hawkes GR (eds) Advances in sugarbeet production. Iowa State University Press, Ames, pp 19–48
Lukezic FC, Bloom JR, Carroll RB (1969) Development of *Fusarium* root rot and its effect on root and crown carbohydrate levels of clipped alfalfa plants grown in a gnotobiotic environment. Phytopathology 59:1575
Lüttge U, Kluge M, Bauer G (1988) Botanik: Ein grundlegendes Lehrbuch. VCH, Weinheim
MacLachlan JB (1972a) Estimation of genetic parameters in a population of mono germ sugar beet *(Beta vulgaris)*. 1. Sib-analysis of mother line progenies. Ir 1. Agric Res 11:237–224
MacLachlan JB (1972b) Estimation of genetic parameters in a population of monogerm sugar beet *(Beta vulgaris)*. 2. Offspring/parent regression analysis of mother line progenies. Ir 1. Agric Res 11:319–325
MacLachlan JB (1972c) Estimation of genetic parameters in a population of monogerm sugar beet *(Beta vulgaris)*. 3. Analysis of a diallel set of crosses among heterozygous populations. Ir J Agric Res 11:327–338
Mahn K, Hoffmann C, Marlander B (2002) Distribution of quality components in different morphological sections of sugar beet (*Beta vulgaris* L.). Eur J Agron 17:29–39
Manderscheid R, Pacholski A, Weigel HJ (2009) Effect of free air carbon dioxide enrichment combined with two nitrogen levels on growth, yield and yield quality of sugar beet: evidence for a sink limitation of beet growth under elevated CO_2. Eur J Agron 32:228–239
Martin SS, Narum JA, Chambers KH (2001) Sugarbeet biochemical changes during pile storage. Part I sugars. J Sugar Beet Res 38:35–53
Martinoia E (1992) Transport processes in vacuoles of higher plants. Botanica Ada 105:232–245
McGrath JM, Fugate KK (2012) Analysis of sucrose from sugar beet. In: Preedy VR (ed) Dietary sugars chemistry, analysis, function and effects, food and nutritional components in focus, vol 3. Royal Society of Chemistry Publishing, Cambridge, pp 526–545
McGrath JM, Townsend BJ (2015) Sugar beet, energy beet, and industrial beet. In: Industrial crops: breeding for bioenergy and bioproducts. Springer, New York, pp 81–99
McGrath JM, Trebbi D (2007) Genetics of water content in sugarbeet roots. Proceedings of American Society of Sugar Beet Technologists 2007 Biennial Meeting 44(135)
Milford GFJ (1973) The growth and development of the storage root of sugar beet. Ann Appl Biol 75:427–438
Milford GFJ (1976) Sugar concentration in sugar beet: varietal differences and the effect of soil type and planting density on the size of the root cells. Ann Appl Biol 83:251–257
Milford GFJ, Pocock TO, Riley J, Messem AB (1985a) An analysis of leaf growth in sugar beet. III Leaf expansion in field crops. Ann Appl Biol 106:187–203
Milford GFJ, Pocock TO, Jaggard KW, Biscoe PV, Armstrong MJ, Last PJ, Goodman PJ (1985b) An analysis of leaf growth in sugar beet. IV The expansion of the leaf canopy in relation to temperature and nitrogen. Ann Appl Biol 107:335–347
Milford GFJ, Pocock TO, Riley J (1985c) An analysis of leaf growth in sugar beet. II: Leaf appearance in field crops. Ann Appl Biol 106:173–185
Milford GFJ, Travis KZ, Pocock TO, Jaggard KW, Day W (1988) Growth and dry-matter partitioning in sugar beet. J Agric Sci Camb 110:301–308

Milford GF, Watson DJ (1971) The effect of nitrogen on the growth and sugar content of sugar beet. Ann Bob 35:287–300
Muthusamy P, Vidhyasekaran P, Soumini Rajagopalan CK (1974) Disease resistance in tomato against damping-off. Indian Phytopathol 27:182
Pack DA (1930) Selection characters as correlated with percent of sucrose, weight, and sucrose content of sugar beet. J Agric Res 40:523–546
Paul MJ, Foyer CH (2001) Sink regulation of photosynthesis. J Exp Bot 52:1383–1400
Pavlinova OA, Prasolova MF (1972) Physiological role of sucrose synthetase in sugar beet root. Fiziol Rast 19:920–925
Powers L (1957) Identification of genetically-superior individuals and the prediction of genetic gains in sugar beet breeding programs. J Am Soc Sugar Beet Technol 9:408–432
Powers LR, Finkner RE, Rush GE, Wood RR, Peterson DF (1959) Genetic improvement of processing quality in sugar beets. J Am Soc Sugar Beet Technol 10:578–593
Powers L, Schmehl WR, Federer WT, Payne MG (1963) Chemical genetic and soil studies involving thirteen characters in sugar beets. J Am Soc Sugar Beet Technol 12:393–448
Pressey R (1968) Invertase inhibitors from red beet, sugar beet, and sweet potato roots. Plant Physiol 43(9):1430–1434
Pressey R (1994) Invertase inhibitor in tomato fruit. Phytochemistry 36(3):543–546
Pritchard FJ (1916) Correlation between morphological characters and the saccharine content of sugar beet. Am J Bot 3:361–376
Rapoport HF, Loomis RS (1986) Structural aspects of root thickening in *Beta vulgaris* L.: comparative thickening in sugarbeet and chard. Bot Gaz 147:270–277
Rausch T, Greiner S (2004) Plant protein inhibitors of invertases. Biochim Biophys Acta (BBA)-Proteins Proteomics 1696(2):253–261
Reinders A, Sivitz AB, Ward JM (2012) Evolution of plant sucrose uptake transporters. Front Plant Sci 3:22
Reinsdorf E, Koch HJ, Loel J, Hoffmann CM (2014) Yield of bolting winter beet (Beta vulgaris L.) as affected by plant density, genotype and environment. Eur J Agron 54:1–8
Richter E, Ewald R (1983) Apoplastic mobility of sucrose in storage parenchyma of sugar beet. Physiol Plant 58:263–268
Riesmeier J, Hrner B, Frommer WB (1993) Potato sucrose transporter expression in minor veins indicates a role in phloem loading. Plant Cell 5:1591–1598
Riesmeier JW, Willmitzer L, Frommer WB (1992) Isolation and characterization of a sucrose carrier cDNA from spinach by functional expression in yeast. EMBO J 11:4705–4713
Roitsch T (1999) Source-sink regulation by sugar and stress. Curr Opin Plant Biol 2:198–206
Sauer N, Stolz J (1994) SUCI and SUC2: Two sucrose transporters from *Arabidopsis thaliana*; expression and characterization in baker's yeast and identification of the histidine tagged protein. Plant J 6:67–77
Savitsky H (1940a) Cytology and embryology of sugar beets. In: Monograph for biology genetics and breeding of sugar beet, vol 1, pp 454–548
Savitsky YE (1940b) Genetics of sugar beets. USDA-National Agricultural Library Translation 5825. Translated from Russian, original source unavailable, p 135
Savitsky VF (1940c) Sugarbeet genetics (in Russian). In Svieklovodstvo, Kijew, p 549–684
Schlosser LA (1949) Ober plasmatische Vererbung auf polyploiden Stufen. Planta 37:533–564
Schneider K, Borchardt DC, Schäfer-Pregl R, Nagl N, Glass C, Jeppsson A, Gebhardt C, Salamini F (1999) PCR-based cloning and segregation analysis of functional gene homologues in *Betavulgaris*. Mol Gen Genet 262:515–524
Schneider F, Emmerich A, Reinefeld E, Walter E, Kelm W (1961) Auswirkung der Nichtzuckerstoffe der Rübe, insbesondere auf die Melassebildung. Teil 1. Zuckerindustrie 14: 208–212
Schneider K, Schafer-Pregl IL, Borchardt DC, Salamini F (2002) Mapping QTLs for sucrose content, yield and quality in a sugar beet population fingerprinted by EST-related markers. Theor Appl Genet 104:1107–1113

Schnepel K, Hoffmann CM (2014) Genotypic variability in storage losses of sugar beet. Sugar Ind 139(5):302–310
Schnepel K, Hoffmann CM (2015) Effect of extending the growing period on yield formation of sugar beet. J Agro Crop Sci:1–12
Schnepel K, Hoffmann CM (2016) Genotypic differences in storage losses of sugar beet – causes and indirect criteria for selection. Plant Breed 135:130–137
Schulz A, Beyhl D, Marten I, Wormit A, Neuhaus E, Poschet G, Buttner M, Schneider S, Sauer N, Hedrich R (2011) Proton-driven sucrose symport and antiport are provided by the vacuolar transporters SUC4 and TMT1/2. Plant J 68:129–136
Schwimmer S, Makower RU, Rorem ES (1961) Invertase & invertase inhibitor in potato. Plant Physiol 36(3):313
Shimamoto Y, Hosokawa S (1967) Statistical genetical analysis of root growth in sugarbeet. I. Changes in genetical parameters with growth. Bull Sugar Beet Res Suppl Mo 9: 118–122
Shimamoto Y, Hosokawa S (1969) Genetic variability of certain characters at various growth stages in sugar beet. Jap J Breeding 19:100–105
Simmonds NW (1994) Yield and sugar content in sugar beet. Int Sugar J 96:413–416
Sindhamathar A, Vidhyasekaran P (1978) Physiology of resistance to rust in sunflower. Indian Phytopathol 31:289
Sliwinska E, Zurawska K, Jassem M (1998) Efficiency of individual selection in sugar-beet breeding. Plant Breed Seed Sci 42(2):57–64
Smith GA, Hecker RJ, Maag GW, Rasmuson DM (1973) Combining ability and gene action estimates in an eight-parent diallel cross of sugarbeet. Crop Sci 13:312–316
Smith GA, Martin SS (1989) Effect of selection for sugarbeet purity components on quality and extractions. Crop Sci 29:294–298
Stehlik V (1933) L'heredite de la richesse en sucre et du poids chez les hybrides de la betterave sucriere et de la betterave fouragere. IIRB Compte rendu 3:74–109
Stich B, Melchinger A, Heckenberger M, Möhring J, Schechert A, Piepho HP (2008a) Association mapping in multiple segregating populations of sugar beet (*Beta vulgaris* L.). Theor Appl Genet 117:1167–1179
Stich B, Piepho HP, Schulz B, Melchinger A (2008b) Multi-trait association mapping in sugar beet (*Beta vulgaris* L.). Theor Appl Genet 117:947–954
Storer KR, Schmehl WR, Hecker RJ (1973) Growth analysis studies of sugarbeet. Colorado State University Experiment Station. Techn Bull 118:1–69
Takebe M, Izumiyama Y (1977) The relationship between the sugar content and root weight of the sugar beet. Proc Sugar Beet Res Assoc 19:251–258
Terry N (1968) Developmental physiology of the sugar beet. I The influence of light and temperature on growth. J Exp Bot 19:795–811
Theurer JC (1979) Growth patterns in sugarbeet production. J Am Soc Sugar Beet Technol 20(4): 343–367
Trebbi D (2005) Genetic analysis of sucrose accumulation in sugar beet (Beta vulgaris L.). Unpublished Ph.D. Dissertation. Michigan State University, East Lansing, MI
Trebbi D, McGrath JM (2003) Sucrose accumulation during early sugar beet development. In: Proceedings of the American Society of Beet Sugar Technologists. Joint Meeting of the International Institute for Beet Research and the American Society of Beet Sugar Technolgists, San Antonio, TX, vol 26, pp 267–271
Trebbi D, McGrath JM (2009) Functional differentiation of the sugar beet root system as indicator of developmental phase change. Physiol Plant 135(1):84–97
Trimpler K, Stockfisch N, Marlander B (2017) Efficiency in sugar beet cultivation related to field history. Eur J Agron 91:1–9
Turkina MV, Sokolova SV (1972) A study on membrane transfer of sucrose in the plant tissue. Fiziol Rast 19:912–919

Ulrich A (1952) The influence of temperature and light factors on the growth and development of sugar beets in controlled climatic, environments. Agron J 44:66–73
Ulrich A (1955) Influence of night temperatures and nitrogen nutrition on the growth, sucrose accumulation, and leaf minerals of sugar beet plants. Plant Physiol 30:250–257
Ulrich A (1956) The influence on antecedent climates upon the subsequent growth and development of the sugar beet plant. J Am Soc Sugar Beet Technol 9:97–109
Ulrich A (1961) Variety climate interactions of sugar beet varieties in simulated climates. Proc Am Soc Sugar Beet Technol 11:376–387
Vaughn MW, Harrington GN, Bush DR (2002) Sucrose-mediated transcriptional regulation of sucrose symporter activity in the phloem. Proc Natl Acad Sci USA 99:10876–10880
Vidhyasekaran P (1974) Finger millet helminthosporiose, a low sugar disease. Zeit f Pflanzenkrankheiten u Pflanzenschutz 81:28
Vilmorin JE (1923) L' Heredite chez la Betterave Cultivee. Gaughiers-Villars, Paris
Vivien A (1920) Saccharogenic content and selection of sugar beets. Bull Assoc Chem Sucr Dist 38:143–163
Vukov K (1972) Auswirkungen der Nichtzuckerkomponenten auf Weisszucker-Ausbeute und Melasse in Phisik und Chemie Der Zuckermbe, Budapest, p 367
Vukov K, Hangyal K (1985) Sugar beet storage. Sugar Technol Rev 12:143–265
Wardlaw IF (1990) Tansley Review No. 27 the control of carbon partitioning in plants. New Phytol 116:341–381
Watson DJ, Motomatsu T, Loach K, Milford GFJ (1972) Effects of shading and seasonal differences in weather on the growth, sugar content and sugar yield of sugarbeet crops. Ann Appl Biol 71:159–185
Weber WE, Borchardt DC, Koch G (2000) Marker analysis for quantitative traits in sugar beet. Plant Breed 119:97–106
Weil M, Krausgrill S, Schuster A, Rausch T (1994) A 17-kDa Nicotiana tabacum cell-wall peptide acts as an in-vitro inhibitor of the cell-wall isoform of acid invertase. Planta 193(3):438–445
Winner C, Feyerabend I (1971) Ein Beitrag zur Morphologie und technischen Qualitat des Rubenkopfes. 2. Verteilung qualitatbestimmender in der Rube. Zuckerindustrie 24:36–43
Wood RR, Bush HL, Oldemeyer RK (1958) The sucrose-sodium relationship in selecting sugar beets. J Am Soc Sugar Beet Technol 10:133–137
Wurschum T, Maurer H, Kraft T, Janssen G, Nilsson C, Reif J (2011a) Genome-wide association mapping of agronomic traits in sugar beet. Theor Appl Genet 123:1121–1131
Wurschum T, Maurer HP, Schulz B, Mohring J, Reif JC (2011b) Genome-wide association mapping reveals epistasis and genetic interaction networks in sugar beet. Theor Appl Genet 123:109–118. https://doi.org/10.1007/s00122-011-1570-3
Wyse R (1974) Enzymes involved in the postharvest degradation of sucrose in *Beta vulgaris* L. root tissue. Plant Physiol 53(3):507–508
Wyse RE (1970) Factors influencing the respiration rate of sugar beet roots. IIRB Winter Congress Session II (2.9), p 1–9
Wyse RE (1979a) Parameters controlling sugar content and yield of sugar beet. J Am Soc Sugar Beet Technol 20:268–385
Wyse RE (1979b) Sucrose uptake by sugar beet tap root tissue. Plant Physiol 64:837–841
Wyse RE, Zamski E, Tomos AD (1986) Turgor regulation of sucrose transport in sugar beet taproot tissue. Plant Physiol 81:478–481
Zamski E, Azenhot A (1981) Sugarbeet vasculature: I. Cambial development and the three dimensional structure of the vascular system. Bot Gaz 142:334–343
Zhang Q, Greenway H (1994) Anoxia tolerance and anaerobic catabolism of gaed beetroot storage tissues. J Exp Bot 45(274):567–575
Zhao B, Mackay IJ, Caligari PDS, Mead R (1997) The efficiency of between and within full-sib family selection in a recurrent selection programme in sugar beet *(Beta vulgaris* L). Euphytica 95:355–359

Shaping the Sugar Beet of Tomorrow: Current Advances in Sugar Beet Biotechnology and New Breeding Techniques

4

José M. Mulet

Abstract

Sugar beet genome was first published in 2014. Before this, there were GMO cultivars of sugar beet commercially available. The possibility of genetic transformation, together with the complete characterization of the genome facilitates the use of molecular techniques as well as new breeding techniques for sugar beet improvement. Another important aspect is the fact that sugar beet is a recently domesticated crop, and there is a huge wild genetic diversity, which constitute a great genetic pool which can be used as a source of useful genes for introgressions or for designing new GMO crops. In this chapter, we will review all the knowledge available in GMO sugar beet, the recent advances and applications of biotechnology and novel breeding techniques in sugar beet improvement, and the use of sugar beet genes in other crops as well as the future prospects.

Keywords

Abiotic stress · Beta vulgaris · CRISPR/Cas9 · Genetic transformation · GMO · Pest management · Sugar yield

Abbreviations

QTL	Quantitative Trait Locus
ALS	Acetolactate Synthase
AsA	Ascorbic Acid

J. M. Mulet (✉)
Biochemistry and Molecular Biology, IBMCP, Universitat Politècnica de València-CSIC, Valencia, Spain
e-mail: jmmulet@ibmcp.upv.es

V. Misra et al. (eds.), *Sugar Beet Cultivation, Management and Processing*, https://doi.org/10.1007/978-981-19-2730-0_4

BCTV	Beet Curly Top Virus
BNYVV	Beet Necrotic Yellow Vein Virus
BWYV	Beet Western Yellows Virus
BYV	Beet Yellows Virus
D-DNA	Defective DNAs
DHA	Monodehydroascorbate
ds-DNA	Double-Stranded DNA
GA	Gibberellic Acid Biosynthetic
GMO	Genetically Modified Organism
ICPs	Insecticidal Crystal Proteins
MDHAR	Monodehydroascorbate Reductase
mtlD	Mannitol-1-Phosphate Dehydrogenase
PAT	Phosphinothricin N-Acetyltransferase
PEG	Polyethylene Glycol
PvPGIP2	Polygalacturonase-Inhibiting Protein 2
RNAi	Interference RNA
scFv	Single-Chain Variable Fragment
SPS	Sucrose-Phosphate Synthase
StGCS-GS	γ-Glutamylcysteine Synthetase-Glutathione Synthetase
TILLING	Targeting Induced Local Lesions In Genomes

4.1 Introduction

The world population is nowadays about 7.5 billion and it is increasing. Some models expect numbers of about nine billion in 2050. To feed this growing population, the agricultural yield must increase concomitantly. Sugar beet (*Beta vulgaris* L.) is one of the world's main suppliers of calories in the diet, together with sugarcane. Worldwide 80% of sugar comes from sugarcane and 20% comes from sugar beet. In the 2019/20 period, the production was about 166.18 million tons and is expected to be about 182 million tons in 2020/21, and this produces about 35 million tons of sugar per year. Over 120 countries produce sugar. While sugarcane cultivation is concentrated in the tropics, sugar beet is cultivated in temperate climates mainly in the northern hemisphere in regions of Europe, North America, and some countries in Asia. The Russian Federation, France, the United States, Germany, and Turkey represent the main producers in 2018 (FAOSTAT 2018). Besides its pivotal role as a source of sugar, sugar beet cultivation has other important outcomes. *Beta vulgaris* is used as an energetic crop (Zabed et al. 2014), for animal feed (Evans and Messerschmidt 2017) and some varieties (i.e., Cicla) are cultivated as green-leaf vegetables, commonly known as Swiss chard.

Environmental stress and several pests greatly threat affect yield. Sugar beet is grown in temperate areas of the northern hemisphere where the crop is usually sown in early spring. This allows improving the root production and avoids summer drought and excessive heat at the plant maturity. In warmer climates or subtropical climates like southeast Asia, it is also possible to sow sugar beet in autumn (autumn sowing) or in other seasons, in order to anticipate harvest and escape drought and other environmental stresses. Sugar beet is also prone to pest of several kinds. Its cultivation is affected by weeds like the Redroot Pigweed (*Amaranthus retroflexus*), virus as beet necrotic yellow vein virus (BNYVV), bacteria such as *Pseudomonas aptata*, fungal disease as *Cercospora*, nematodes as *Ditylenchus dipsaci,* or insects like *Aphis fabae*; therefore, there is a great need to breed for more resistant *Beta vulgaris* cultivars.

There are pivotal achievements that have been obtained by classical breeding. Sugar beet is a recently domesticated plant created for the production of sugar (sucrose) from the beet used for animal feed in the 1800s. Classical breeding raised its sugar content from about 4-6% to the current 18% in 200 years (Mcgrath et al. 2018). Classical breeding has also made great achievements in traits of agronomical interest like the discovery and fixation of the monogerm gene-seed character (Savitsky 1950) which avoids multiple seeds germinating from a single fruit (utricle). Another major breeding achievement was the cytoplasmic male sterility for commercial varieties, which enabled the hybrid seed production, a trait that depends on a mitochondrial locus (Kitazaki et al. 2015). Hybrids present heterosis in root yield, among other traits (Ćurčić et al. 2017; Schwegler et al. 2014). Another achievement was the semitropical beet germplasm by increasing heat and bolting resistance which enables the cultivation in huge areas of India and southeast Asia (Srivastava 1995).

Abiotic stress is also a main limiting factor for yield. Soil salinity is a great constraint for productivity in many areas of cultivation (i.e., the Mediterranean basin, Iran, India, Southeast Asia, or the western United States). On the other hand, the advantage is that sugar beet is phylogenetically related to halophytic wood plants, being the most important the sea beet (*Beta vulgaris* ssp. *Maritima)* (Munns and Tester 2008). Sugar beet can stand concentrations up to 250 mM of sodium chloride in soil, which induces a 50% reduction in yield. In other major crops like beans, a 60 mM concentration of NaCl produces a similar reduction (Taïbi et al. 2021). Also, the defense system against salt stress is quite effective, for instance, in response to 300 mM sodium chloride solutions, beets are reported to suppress the generation of reactive oxygen species by transcriptional regulation (Hossain et al. 2017). The problem faced by farmers is that sugar beet is sensitive to salt stress at the seedling stage, and therefore, getting early vigor for salt stress tolerance is a great objective (Khayamim et al. 2014). The genotype “EL56” (PI 663211) is able to germinate in soils in up to 150 mM sodium chloride (McGrath 2011). Drought is also a problem faced in many cultivation areas. There are at least three loci described by quantitative trait locus (QTL) analysis suggested to be important for drought tolerance, but so far, the advances in sugar beet tolerance to drought stress are very limited (Rajabi and Borchardt 2015).

Also, classical breeding has focused on introgression disease and stress tolerance into novel germplasm and genotypes to offer solutions to farmers (Doney 1995; Panella et al. 2015). Thanks to that there are cultivars available with resistance or tolerance to predominant seedling diseases such as *Pythium*, *Aphanomyces*, and

Rhizoctonia. There are also some cultivars resistant to the main root disease Rhizomania "crazy root," due to the viral pathogen BNYVV (beet necrotic yellow vein virus). The main source of disease resistance genes for classical breeding, again, is the sea beet although some resistance has been found in wild beets, in fact, four different resistant genes (Rz1-5) isolated from different sources have been introgressed into commercial cultivars (Pavli et al. 2011b) although there are some strains that have broken the resistance (Biancardi and Tamada 2016). There have been some advances in resistance to nematodes and insects by classical genetic means (Zhang et al. 2008b). Therefore yield, pest resistance, and abiotic stress tolerance are the main objectives for the biotechnological improvement of sugar beet.

4.2 Sugar Beet Transformation

As we have seen classical plant breeding has improved the effectiveness of sugar beet for yield, pest, and to a very minor extent abiotic stress tolerance. It becomes clear that there is a need for biotechnological improvement. The first genetically engineered plant is considered to be tobacco and was reported in 1983 (Bevan et al. 1983). Attempts to transform sugar beet started almost immediately. The first was not successful as plants prove to be difficult to regenerate from transformed calluses (Harpster et al. 1988; Krens et al. 1988; Jacq et al. 1993) or cells (Wozniak and Owens 1989; Kallerhoff et al. 1990). Although transgene was incorporated in the plant genome, in most of these early reports stable transformation and reproducibility were not confirmed (Gurel et al. 2008).

Lindsey and Gallois (1990) described the first successful regeneration of a transformed sugar beet. Authors co-cultivated shoot-base tissues with *Rhizobium radiobacter* (formerly *Agrobacterium tumefaciens*) strain LBA4404 transformed with the gene *nptII* which confers resistance to the antibiotic kanamycin and either the chloramphenicol acetyltransferase (*cat*) gene (antibiotic resistance) or the β-glucuronidase (*gusA*) gene (enables staining). Shortly after the first genetically modified organism (GMO) sugar beet expressing an agronomically important trait (Herbicide tolerance) was described by D'Halluin (D'Halluin et al. 1992). In this case, the authors used callus derived from several organs transformed with *R. radiobacter*. As a selectable marker, they used (*bar*) (herbicide and bialaphos resistance) or a mutated acetolactate synthase (*ALS*) gene (herbicide resistance). These protocols were lengthy (2 years to obtain shoots) and showed a great genotype dependency. In fact, several years later it was described that easiness to transformation is a heritable trait (Kagami et al. 2016). The first genotype-independent protocol was described by (Fry et al. 1991) using cotyledonary explants inoculated with *R. radiobacter* containing different genes for herbicide or antibiotic tolerance. In this report, they confirmed the presence of transgenes, their mendelian segregation, and their functionality and the phenotype conferred (tolerance to herbicide). But the specific protocol was not published. Since then, other protocols based on callus transformation have been described (Kagami et al. 2015). Transformation with other techniques has been described, specifically, particle bombardment (Saunders et al. 1992; Hashimoto and Shimamoto 1999); polyethylene glycol (PEG)-mediated transformation of protoplasts (Hall et al. 1996), petiole explants of haploid and diploid sugar beet genotypes using *Rhizobium rhizogenes* and sonication (Klimek-Chodacka and Baranski 2014). Most of these techniques took advantage of the advances in somatic embryogenesis (Zhang et al. 2008a) and regeneration (Ninković et al. 2010). Transformation techniques are so advanced that there are also several molecular tools available to the community, like root-specific promoters (Padmanaban et al. 2016), a promoter from Swiss chard that directs petioles and roots preferential expression (Yu et al. 2015) and a promoter that drives expression specific to nematode-feeding sites (Thurau et al. 2003). There are other tools to modulate gene expression such as the lox gene (Wang et al. 2003), and there is also a description of a set of transgenic plants and cell lines of sugar beet transformed with the maize transposable elements Spm/dSpm which allows gene-tagging (Kishchenko et al. 2010). Moreover, heterologous expression of the *Arabidopsis thaliana* GRF5 (AtGRF5) in sugar beet callus cells greatly increases transformation efficiency (Kong et al. 2020).

4.3 Plastid Transformation

Another point of interest has been transforming plastids into sugar beet. This technique has many advantages, among them, the fact that pollen does not have chloroplast, and therefore, gene flow is prevented. The first description dates back to 2009 although a previous communication can be found in the literature (De Marchis et al. 2007) and was achieved from petioles of the line Z025 by biolistic bombardment. Chloroplast was transformed with a vector that directs genes to the rrn16/rps12 intergenic region, employing the *aadA* and the *GFP* genes as markers (De Marchis et al. 2009). The technique has been improved but is not a routine technique yet (De Marchis and Bellucci 2021), and all the available literature comes from the same laboratory.

4.4 Transient Expression

Although the transient expression is not a routine technique in agronomy, it is very useful for investigation purposes, mostly related to basic science. There is a recent description of a transient expression in sugar beet using *Rhizobium radiobacter* (Moazami et al. 2018). One of the most common strategies for transient expression is agroinfiltration in leaves with bacterial cultures of *Rhizobium* transformed with the transgene. This is the election technique, for instance, for subcellular localization of a given protein using confocal microscopy (Locascio et al. 2019b). In our laboratory, we have transiently transformed sugar beet leaves using the published protocol for lettuce and tomato (Wroblewski et al. 2005) with positive results (our unpublished observations) (Fig. 4.1).

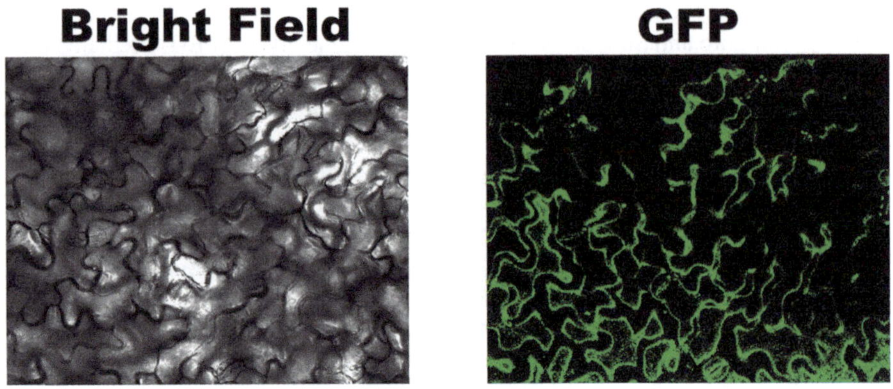

Fig. 4.1 Transient expression of GFP protein in sugar beet leaves

4.5 Heterologous Genes Transformed in Sugar Beet

We have just reviewed the different methods for sugar beet transformation. It becomes clear that inserting exogenous genes in the sugar beet genome poses no technical problem. Transgenic sugar beets have existed for almost three decades with a wide variety of traits with different objectives.

4.5.1 GMO Sugar Beet for Improvement of Biotic Stress

4.5.1.1 Herbicide Tolerance

Beta vulgaris culture is strongly affected by weeds (May and Wilson 2006). Standard control techniques require herbicide spraying at different periods, so controlling weeds in sugar beet cultivation is difficult, expensive and increases the carbon and the water footprint of the cultivation, this reducing sustainability (Klenk et al. 2012; Gerbens-Leenes and Hoekstra 2012).

As mentioned, the first stable transformed GMO sugar beet was resistant to herbicide (D'Halluin et al. 1992). There is only three commercial sugar beet approved and all contain genes conferring tolerance to herbicides. Liberty Link™ sugar beet (name: T120-7; code: ACS-BVØØ1-3) by Bayer crop science contains the gene *pat* encoding the phosphinothricin N-acetyltransferase (PAT) enzyme from *Streptomyces viridochromogenes* (resistance to glufosinate/phosphinothricin) and the *nptII* from *Escherichia coli* Tn5 transposon which confers tolerance to neomycin and kanamycin antibiotics. This variety was authorized in the United States in 1998, in Canada in 2000, and in Japan in 2001 (James 2011). The Roundup Ready™ sugar beet by Monsanto (name: H7-1; code: KM-ØØØH71-4) contains a single transgene, the cp4 epsps (aroA:CP4) from *Rhizobium radiobacter* strain CP4 conferring tolerance to glyphosate. This variety was approved in many countries worldwide for food, specifically Australia, Canada, China, Colombia, European Union, Japan, Mexico, New Zealand, Philippines, Russia, Singapore, South Korea, Taiwan, and the United States, but for cultivation only Canada and the United States (since 2005) and Japan (since 2007). Finally, the InVigor™ sugar beet (name: GTSB77 (T9100152); code: SY-GTSB77-8), by Novartis and Monsanto which contains three different genes, the mentioned cp4 epsps (aroA:CP4), the glyphosate oxidase goxv247 from *Ochrobactrum anthropi* strain LBAA that confers tolerance to glyphosate and the marker gene uidA (GUS) gene from *Escherichia coli*. This was authorized in the United States in 1998, New Zealand, and Australia in 2002 and Japan in 2003. From an economic and environmental perspective, the use of these varieties (mostly the Roundup Ready) has saved millions of tons of herbicide. Usually, a glyphosate-resistant GMO sugar beet requires about 2 kg ha^{-1}, while non-GMO cultivars require about 6 kg ha^{-1} or higher, depending on the amount of weeds (Märländer 2005).

In 2012, there was a partial ban on the use of GMO sugar beet in the United States motivated by the case of Center for Food Safety v. Vilsack (CFS v. Vilsack), set in the U.S. District Court for the Northern District of California (CA District Court). The point was the presence of species like *Beta macrocarpa* and *Beta vulgaris ssp. maritima* in areas of California where RR sugar beet was planted, raising concerns on the possibility of transfer of the genes conferring tolerance to herbicides (McGinnis et al. 2010). Finally, following the different scientific reports, the USDA deregulated RR Sugar beet (Khan 2014). GMO sugar beets account for about 95% of the sugar beet cultivated in the United States (United States Department of Agriculture 2019). Given that the US imports 30% of sugar and produces 70%, and that sugar beet production is mainly for the sugar industry, which means that 66.5% of the sugar consumed in the United States comes from glyphosate-resistant GMO sugar beet. Similar numbers apply in Canada (Dillen et al. 2013). There are no reports of any adverse effects on human or animal health related to the use of GMO sugar beet. Studies show that after 2 weeks of the herbicide application only trace amounts of glyphosate are detected in roots or shoots, and in the crystalline sugar glyphosate is below the levels of detection (Barker and Dayan 2019).

Although not commercially available, there are also reports of transgenic sugar beet resistance to phosphinothricin with the *bar* gene (D'Halluin et al. 1992; Mishutkina et al. 2010), to imidazolinone by transformation with the *als* gene from *Arabidopsis thaliana* (a mutated acetolactate synthase). This mutation confers tolerance to some commercial herbicides such as imazethapyr (Pursuit®, BASF) (Kishchenko et al. 2011). There are also reports of crop trials of novel sugar beet varieties with tolerance to a specific acetolactate synthase (ALS) inhibitor herbicide bred by KWS using standard breeding techniques and branded Conviso Smart, but these varieties have not been marketed yet, and there are no published reports of the results of the assays (https://www.fwi.co.uk/arable/herbicide-tolerant-sugar-beet-trialled-uk). The herbicide-resistant sugar beet has been a great advance for sugar beet farmers but is the only one available in the market. Even though, there is a lot of investigation being performed in other traits.

4.5.1.2 Insect Resistance

Transgenic insect-resistant plants are in the market for a long time with great success in crops such as maize, cotton, soy (Koch et al. 2015), or brassicas (Poveda et al. 2020). These varieties have been developed mainly by transformation with the *Bacillus thuringiensis* insecticidal crystal proteins (ICPs) encoded by the cry gene family. Sugar beet plants transformed with cry1Ab (Shimamoto and Domae 2000), proved to be very effective protection against lepidopterans (Jafari et al. 2008) and the Egyptian leafworm (*Spodoptera littoralis*) (Sedighi et al. 2011). Sugar beet has also been transformed with cry1C (Kimoto and Shimamoto 2001), or with cry1C and cry2A (Lytvyn et al. 2014). A joint transformation with Cry1Ab and Cry1C protected against cabbage armyworm (*Mamestra brassicae*) (Kimoto and Shimamoto 2001).

4.5.1.3 Nematode Resistance

Sugar beet is very sensitive to nematode attack, among them *Heterodera schachtii*, the sugar beet cyst nematode (SBCN) is the one causing more damage to farmers (Villarías Moradillo 1999). The standard treatment depends on the use of chemicals with nematicide activity, but most of them are being banned or have undergone a strict regulation in many countries due to their environmental toxicity. Other strategies rely on resistant cultivars obtain by breeding, crop rotations (Joersbo 2007), or trap crops (Lathouwers et al. 2005). Another problem is that nematode pests are recalcitrant, as eggs can stay latent for years, even in adverse environments or in the absence of the host crop.

There was some success using classical breed to introduce the $Hs1^{pro-1}$ gene from *Beta procumbens* in commercial sugar beet cultivars (Jung 1998), but this causes a severe yield penalty, perhaps due to linked genes that correlated with decreased yield (Heller et al. 1996). Using genetic engineering is a way to override the problem with the linked gene (Ali et al. 2017). The $Hs1^{pro-1}$ gene from *Beta procumbens* was transformed in sugar beet, and the observed resistance was higher than in resistant cultivars developed by conventional breeding. GMO sugar beet expressing the SpTI-1 gene (a Kunitz-type trypsin inhibitor) exhibited similar phenotypes (Cai et al. 2003).

4.5.1.4 Fungal Resistance

The most important fungal diseases in sugar beet are *Cercospora* leaf spot (*Cercospora beticola*), root and crown rot (*Rhizoctonia solani*) and powdery mildew (*Erysiphe betae*, syn. E. polygoni) (Villarías Moradillo 1999). So far traditional control is based on integrated approaches, using both culture techniques, chemicals, or resistant varieties. There is not any GMO variety in the market with resistance to any fungal disease, indicating that from the biotechnological point of view there is still a need for improvement. There are several strategies which target the increase of the systemic acquired resistance by increasing the salicylic acid production or the hypersensitive response (with elicitors like the ones encodes by the *avr* genes). There are also attempts of transforming sugar beet with different defense proteins. The expression of a chitinase gene from pumpkin was able to increase chitinase activity and in some cases even suppressed some symptoms of fungal infection, as this protein is able to degrade the fungus cell wall (Hashimoto and Shimamoto 2001). Following a similar strategy, the polygalacturonase-inhibiting protein 2 *(PvPGIP2)* of common bean (*Phaseolus vulgaris*) protects GMO sugar beet against *Rhizoctonia solani* and *Alternaria alternata* (Goudarzi et al. 2015). This protein inhibits fungal polygalacturonase and therefore slows downs the ability of the fungus to degrade the plant cell wall (Mohammadzadeh et al. 2012). A recent report describes that overexpressing the native enzyme of *Beta vulgaris* (*BvPGIP2*) in hairy roots confers resistance to *Fusarium oxysporum* (Li and Smigocki 2019). The transformation with the chloroplastic and the cytosolic superoxide dismutase from tomato also increased resistance to *C. beticola* (Tertivanidis et al. 2004). Even a gene from a different species of this fungus, the cercosporin toxin export (*CFP*) gene from *Cercospora kikuchii*, has been transformed into sugar beet, but to date, there are no reports of whether transformed plants exhibited any phenotype (Kuykendall and Upchurch 2004). Another strategy consisted of the transformation of sugar beet with the Mannitol-1-Phosphate Dehydrogenase (mtlD) from *E. coli*. This increased the resistance to *Alternaria alternata*, *Botrytis cinerea* and *Cercospora beticola* although it depended on the transgenic line and the conditions of the assay (Goudarzi et al. 2016).

4.5.1.5 Viral Resistance

There are GMO commercial crops in the market whose trait is viral resistance (Fuchs and Gonsalves 2007), but this is not the case for sugar beet yet. Sugar beet is prone to viral diseases that cause great yield losses like rhizomania, caused by beet necrotic yellow vein virus (BNYVV), or can be infected by viruses such as beet western yellows virus (BWYV), beet yellows virus (BYV), or beet curly top virus (BCTV). There are descriptions of viral-resistant varieties using classical means (Scholten and Lange 2000). Genetic engineering has been used either expressing transgenes or using double-stranded DNA (ds-DNA) or interference RNA (RNAi) to interfere with the viral infection and dissemination, and therefore gain resistance (Lennefors et al. 2006). The expression of viral coat protein proved to be effective against BNYVV as early as in 1990 (Kallerhoff et al. 1990) and in hairy roots cultures (Ehlers et al. 1991), in protoplasts from transformed suspension cells (Mannerlöf et al. 1996) or in protoplast derived from guard cells (Lathouwers et al. 1997). Expression of the hrpZ gene of *Pseudomonas syringae pv. Phaseolicola* also conferred rhizomania resistance (Pavli et al. 2011a). In some cases, the virus resistance does not correlate with the level of expression of the transgene (Kerr 2005).

Another method to induce resistance is to disrupt the viral cell-to-cell movement. The usual strategy is to disrupt the P42, P13, and P15, triple gene cluster from BNYVV (Lauber et al. 1998; Erhardt et al. 2000). This has been done by designing an RNAi with a three-intron hairpin construct carrying parts of the BNYVV replicase gene (Pavli 2010) or a dsRNA derived from the replicase gene. This construct was able to confer protection to the roots of transformed plants (Pavli et al. 2010). Constructions using a similar strategy, but against other sites of the virus induced stronger resistance (Zare et al. 2015). Gene silencing against rhizomania has proved to be effective even in field tests (Safar et al. 2021). There is also a report describing the use of a single-chain variable fragment (scFv) specific to a major coat protein of virus, p21, with a construct that targeted the antibody to different organelles. After mechanical infection with BNYVV, the cytoplasmic construct showed the greatest efficiency in preventing the infection (Jafarzade et al. 2019). A different strategy has been found against BCTV. This infection provokes the generation of defective DNAs (D-DNA) which is suspected to be a plant response to attenuate symptoms. Transgenic plants expressing synthetic D-DNA presented attenuated symptoms (Horn et al. 2011).

4.6 Abiotic Stress

4.6.1 Drought Tolerance

In the current context of anthropogenic global warming, and the concomitant change in the precipitation regime the global aridity is increasing and affecting many sugar beet producing areas. Sugar beet is more tolerant to abiotic stress than most major crops, but under suboptimal water concentrations yield decreases drastically (Rajabi et al. 2007). There is no drought-resistant phenotypes of sugar beet identified so the use of classical breeding is very limited (Ebmeyer et al. 2021). A strategy to generate drought tolerance in crops is the transformation with genes able to increase the cellular content of osmolytes or osmoprotectans (small hydrophilic molecules that plants and other organism accumulates under drought stress conditions in order to avoid turgor loss) although in some cases with a severe yield penalty that compromises its market value and the utility for farmers (Van Camp 2005). There are only two cultivars in the market whose trait conferred by the transgene is drought tolerance: the DroughtGard® maize (Wang et al. 2015) and very recently the soybean expressing the HB4 transcription factor from sunflower (Ribichich et al. 2020). GMO plants transformed with the *Bacillus subtilis* gene, *SacB* accumulated fructans to low levels (0.5% of dry weight) but plants performed better under drought stress (dry weight +25–35% than control plants) and with no yield penalty under control conditions (Pilon-Smits et al. 1999).

4.6.2 Salt Tolerance

There are very few reports on GMO sugar beet resistant to salt stress. The first was authored by Yang et al. (2005). They introduced they introduced the *AtNHX1* gene (a sodium proton vacuolar exchanger from *Arabidopsis thaliana*) and reported salt tolerance. Sugar beet plants transformed with a paralogue of the same gene (*AtNHX3*) accumulate more soluble sugar and less salt in storage roots (Liu et al. 2008). There is also a report in which authors describe the transformation of sugar beet with *AVP1*, the vacuolar H^+-pyrophosphatase of *Arabidopsis thaliana* (Wang et al. 2012). And also, the co-expression of the sodium proton vacuolar antiporter

(*ZxNHX*) and the vacuolar proton pyrophosphatase (*ZxVP1-1*) from the xerophyte plant *Zygophyllum xanthoxylum* increased the osmoregulatory mechanism and diminished sodium toxicity, and also accumulated a higher content of sugars (Wu et al. 2015). And although not directly related to salt stress, overexpression of the γ-glutamylcysteine synthetase-glutathione synthetase (StGCS-GS) from *Streptococcus thermophilus*, and enzyme of the glutathione biosynthetic pathway increased the tolerance sugar beet to grow under high cadmium, copper, and zinc (alone or in combination) and also the ability to accumulate these heavy metals, providing a useful tool for bioremediation (Liu et al. 2015).

4.7 Development and Metabolism

4.7.1 Bolting Resistance

Bolting resistance is a prime objective for breeders as in some areas such as California and the Mediterranean sugar beet is a winter crop, but as they do not suffer freezing temperatures, bolting is a serious problem. This is also a problem in subtropical or warm areas of India and southeast Asia. Bolting depends on the expression of flowering genes, most of them regulated or part of the gibberellic acid biosynthetic (GA) pathway. The use of GA inhibitors like paclobutrazol is a standard strategy to avoid bolting (Sadeghi-Shoae et al. 2014). Therefore, there is a huge interest in modulating GA response and production in sugar beet to avoid bolting, as this will expand the growing season and permit plants to accumulate higher yields. There have been some studies coping with the modulation of bolting-related genes, and some important QTL have been identified (Pfeiffer et al. 2014). The first attempts to develop bolting resist sugar beet by biotechnology consisted of the transformation with the repressor of gibberellin biosynthesis from *Arabidopsis thaliana gai* or by or deactivation of gibberellic acid by heterologous expression of the scarlet runner bean (*Phaseolus coccineus*) gene *GA2ox1* (Mutasa-Gottgens et al. 2009).

B. vulgaris genome holds a vernalization-responsive FLC homolog (*BvFL1*), but opposite to what happens in other plants, its role is not important in bolting regulation, as plants transformed with iRNA to knock down *BvFL1* did not eliminate the requirement for vernalization of biennial beets and did not have a major effect on bolting time after vernalization. Overexpression of *BvFL1* only had a minor effect delaying the bolting after vernalization for about 1 week. Genes *BvFT1* and *BvFT2* are targets of regulation of BvFL1, and their protein product forms a module of regulation. These genes have antagonistic roles in the control of flowering (Pin et al. 2010). When these modules are downregulated, the phenotype is a several weeks delay in bolting, indicating some kind of redundancy in the BvFL1 function (Vogt et al. 2014). But so far, genetic engineering has not succeeded in creating a bolting-resistant genotype.

4.7.2 High Sucrose Yield

As sugar is the main product from sugar beet, increasing the yield is an obvious objective. This has been sought for long. First attempts consisted of knocking down the homolog of the SNF1 gene, a negative regulator of sucrose biosynthesis (Monger et al. 1995) or a maize sucrose-phosphate synthase (SPS) gene (Hashimoto and Shimamoto 1999), but the desired sugar yield increase was not obtained. Elliott et al. (1996) hypothesized that increasing the levels of cytokinin, a hormone related to development and mass accumulation, may increase sugar yield. They expressed a bacterial cytokinin biosynthesis gene, the isopentenyl transferase (ipt) (Snyder et al. 1999; Ivic et al. 2001). There was an effective increase in cytokinin accumulation, but also an inhibition of the taproot development and a minor sucrose accumulation.

4.7.3 Fructan Production

Sugar beet can be a source of molecules of industrial interest. For instance, fructans a fructose polysaccharide with many industrial applications (Turk and Smeekens 1999; Sévenier et al. 2002). There is a great interest in finding a cheap source of fructans, which includes the possibility to produce them from GMO sugar beet. First attempts consisted of using a gene from the traditional source of fructans, the Jerusalem artichoke (*Helianthus tuberosus*). The 1-sst gene of this crop, encoding a l-sucrose:sucrose fructosyl transferase was able to convert native sugar beet fructose to low molecular weight fructans in tap root cells and to a minor extent in leaves (Sévenier et al. 1998). Similar results were obtained when sugar beet was transformed with the 1-sst orthologue of onion (*Allium cepa*) and an additional gene of the fructan biosynthetic pathway (6g-fft, a fructan:fructan 6G-fructosyl transferase) (Weyens et al. 2004). Pilon-Smits et al. (1999) also tried to produce fructans in sugar beet by expressing the sacB bacterial gene, but as mentioned in a previous section, with very low yield and the drought tolerance phenotype was more interesting. There is also a report from transgenic beet transformed with PpFT1 and PpFT2, (two homologous sucrose:fructan 6-fructosyltransferases from timothy (*Phleum pratense*)) to produce levan, a molecule of the fructan family. Levans were successfully produced in GMO sugar beet, but the polymerization was much shorter than the polymerization obtained when levan was obtained from microorganisms (Matsuhira et al. 2014).

Finally, red beet has been used as a biofactory to develop test vaccines against type-I diabetes (Santoni et al. 2019), and there are also some reports of GMO crops designed for basic science, like the gene silencing of CYP76AD1, that blocks the red pigmentation of beets and induces a yellow coloration due to the accumulation of a betaxanthin pigment (Hatlestad et al. 2012) also the maize Ac transposase was expressed in sugar beet to confirm its alternative splicing in an heterologous system (Lisson et al. 2010).

4.8 Sugar Beet as a Source of Genes for Biotechnological Applications

Sugar beet genome was completed in 2014 (Dohm et al. 2014). This facilitates the identification of sugar beet genes useful for biotechnological applications, which has been a field of research interest in the last decades with some remarkable results.

4.8.1 Biotic Stress

There are some sugar beet genotypes that are resistant to pests. It means that could be a source of genes to transfer the resistance to plants in which interbred is not possible. This strategy has been used with the genes for nematode resistance *BvcZR3* and *BvHs1*$^{pro\text{-}1}$ that have been transformed into canola (*Brassica napus*) (Zhong et al. 2019) and tobacco obtaining resistance to nematodes (Sönmez et al. 2014). There are some sugar beet genes that have been transferred to other plants in order to get insect resistance. The serine proteinase inhibitor gene (*BvSTI*) has been expressed in *Nicotiana benthamiana* (Smigocki et al. 2008). In another report, Nicotiana transgenic plants overexpressing this gene were bioassayed against five lepidopteran pests with disparate effectiveness (Smigocki et al. 2013). There are also woody plants transformed with sugar beet genes. The silver birch (*Betula pendula*) has been transformed with the chitinase IV gene. The observed phenotype has been an increase in the fungal (Pappinen et al. 2002; Pasonen et al. 2004; Vihervuori et al. 2013) although in field trials transgenic trees suffered higher attacks from aphids than control plants (Vihervuori et al. 2008).

The gene codifying a germin-like protein (*BvGLP-1*), involved in nematode resistance, was transformed in *Arabidopsis thaliana* and conferred resistance to several pathogenic fungi (*Verticillium longisporum* and *Rhizoctonia solani*), without affecting beneficial fungus such as *Piriformospora indica* (Knecht et al. 2010). Rhizoctonia is a major problem of fungal origin for sugar beet cultivation. The overexpression of BvMLP1 and 3 major latex proteins in *Arabidopsis thaliana* resulted in less infectivity (Holmquist et al. 2021). Another strategy has been the overexpression of *BvPGIP1* and *BvPGIP2* Polygalacturonase-inhibiting proteins (PGIPs) that contain 11 leucine-rich repeat domains, contrary to most of the plants that only have 10 leucine-rich repeats. This construction conferred resistance in *Nicotiana benthamiana* (Li and Smigocki 2019).

4.8.2 Abiotic Stress

Sugar beet is a semi-domesticated crop that belongs to the Amaranthaceae family. Its closest cultivated relatives are spinach (*Spinacia oleracea*) and quinoa (*Chenopodium quinoa*). All three crops are suitable for poor soils or are able to resist environmental stress. It means that may be a source of useful genes to develop novel crops resistant to abiotic stress. One classical strategy is to test genes that have demonstrated the ability to confer salt stress in other organisms or which have some previous evidence that relates this gene to abiotic stress. For instance, S-adenosylmethionine decarboxylase is an enzyme involved in the biosynthesis of polyamines, and its overexpression increases the amounts of spermine and spermidine. This enzyme was identified in a proteomic study for M14, a salt-tolerant monosomic addition line obtained from the intercross between *Beta vulgaris L.* and *Beta corolliflora Zoss* proteins with a differential expression upon a salt treatment (Yang et al. 2013). The overexpression of this enzyme from conferred salt stress tolerance to *Arabidopsis thaliana* (Ji et al. 2019). Similarly, the overexpression of another enzyme of the same pathway, S-Adenosyl-l-Methionine Synthetase 2 increased abiotic stress tolerance (salt and oxidation) (Ma et al. 2017). Some other genes cloned from this germplasm have been assayed for their biotechnological potential against abiotic stress. The cysteine protease inhibitor, cystatin, increased salt tolerance when overexpressed in *Arabidopsis* (Wang et al. 2012) and also a monodehydroascorbate reductase (MDHAR), an enzyme that reduces monodehydroascorbate (DHA) to ascorbic acid (AsA), when overexpressed in *Arabidopsis thaliana* confers salt stress, longer roots, higher chlorophyll content, and higher AsA/DHA ratios (Li et al. 2020).

The overexpression of the glyoxalase I gene from the same cultivar increased tolerance to pleiotropic abiotic stresses (salt, drought, oxidation) in *E. coli* and tobacco (Wu et al. 2013). A proteinase inhibitor BvSTI has been expressed in the forage legume *Lotus corniculatus L.* increasing the resistance to salt stress salt and altering plant architecture (Savić et al. 2019).

There is a strategy that has been very useful to identify sugar beet genes, the screening in a heterologous system such as yeast (Locascio et al. 2019a). There are several reports of such screenings which have been performed using *Beta vulgaris* cDNA libraries (Serrano et al. 2003). This is a fast and straightforward methodology that can unveil novel genes to identify limiting factors for abiotic stress tolerance. This strategy allowed the cloning of *BvCK2*, the catalytic subunit of the casein kinase which conferred salt tolerance upon overexpression in yeast (Kanhonou et al. 2001), the translation initiation factor BveIF1A which conferred salt tolerance by overexpression in yeast and Arabidopsis thaliana (Rausell et al. 2003) and *BvSATO1* (RNA-binding protein with RGG and RE/D motifs), *BvSATO2* (paralogous to *BvSATO1*), *BvSATO4* (RNA-binding protein), *BvSATO5* (RNA-binding protein), and *BvU2AF* (U2snRNP AF protein) (Téllez et al. 2020), all of them conferred tolerance to salt stress when overexpressed in yeast.

The same library was overexpressed in an osmosensitive yeast strain and screened for osmotic stress tolerance. The results were a serine acetyl-transferase, an enzyme involved in the biosynthesis of cysteine form serine (Mulet et al. 2004) and class 2 non-symbiotic plant hemoglobin (BvHb2) (Salort et al. 2010). *BvHb2* conferred tolerance to drought stress in yeast and *Arabidopsis thaliana*. When transformed in a horticultural crop, tomato (*Solanum lycopersicum* L.) also conferred resistance to drought induced withering. Another interesting aspect is that it also altered iron content by overexpression, increasing it in leaves and decreasing in fruit (Gisbert et al. 2020). *BvHb2* has also been expressed in wild field cress (*Lepidium campestre*) and increased the seed oil content without altering its composition (Ivarson et al. 2017).

A different screening of the mentioned *Beta vulgaris* cDNA library for genes able to improve growth under cold conditions (10 °C) identified (*BvCOLD1*), a novel aquaporin gene not conserved in *Arabidopsis thaliana* and other model plants, which could only be found in evolutionarily related crops such as spinach or chinoa. Overexpression of *BvCOLD1* conferred pleiotropic abiotic stress tolerance and increased growth under poor boron medium (Porcel et al. 2018). In the same screening, several genes related to endosomal vesicle transport (Salort and Salom 2009), a ring finger protein (Sanz Molinero et al. 2009), a protein from the *PATELLIN* family (Molinero et al. 2014), and a growth-regulating factor (Reuzeau et al. 2017) were also identified.

In some cases, transformation has been performed to investigate the role of a given gene. This has allowed knowing that *BvCMO* the gene encoding a choline monooxygenase is required for salt stress tolerance, as the transformation of sugar beet with an antisense construction to block the expression of this gene conferred a phenotype of salt sensitivity (Yamada et al. 2015). Similarly, *Arabidopsis* transformed with the tonoplast glucose exporter *BvIMP* exhibit decreased freezing tolerance and germination (Klemens et al. 2014).

4.8.3 Development and Metabolism

Most of the studies found in the literature are not interested in the biotechnological improvement of sugar beet but in gaining knowledge on the molecular mechanisms in sugar beet. For instance, the role of *BvCOL1* as a photoperiod regulator was determined by its ability to complement *Arabidopsis* AtCOL1 and AtCOL2 mutants (Chia et al. 2008). Genetic engineering has been used as a tool to investigate sugar accumulation in sugar beets. *BvSUT1*, the gene that encodes the protein responsible for sucrose loading to the phloem, was transformed into yeast and expressed in oocytes of *Xenopus laevis* to investigate its mechanism (Nieberl et al. 2017). To investigate the subcellular localization of the betalain biosynthetic enzymes, those were expressed in tobacco (Chen et al. 2017). The sugar beet enzyme CYP716A was expressed in yeast to evaluate its ability to oxidize triterpenoids, molecules with several pharmacological and industrial applications (Suzuki et al. 2018).

The line M14 has also been a source of genes for studying the molecular biology of *Beta vulgaris*. For instance, the BvM14-MADSBOX was overexpressed in tobacco and led to severe phenotypic changes (Ma et al. 2011). The male sterility of the phenotype of some *Beta vulgaris* cultivars was mimicked in tobacco by expressing the mitochondrial ORF129 in tobacco with a mitochondrial targeting pre-sequence and a promoter for expression in flowers (Yamamoto et al. 2008). Transgenic sugar beet overexpressing bvORF20, a nuclear factor known as a restorer of fertility, was able to partially restore pollen fertility when overexpressed in cytoplasmic male sterile plants (Matsuhira et al. 2012) and has a complex, post-translational regulation which includes the interaction with the preSATP6 protein (Kitazaki et al. 2015).

Another strategy is that during harvesting the plant suffers wounds and this induces the expression of invertase genes which degrades sucrose, and therefore reducing yield. Overexpression of sugar beet invertase inhibitor BvC/VIF to block this postharvest decrease proved to be ineffective (Jansen 2009).

4.9 New Breeding Techniques in Sugar Beet

New breeding techniques are the hyperonym for a set of techniques that go beyond classical breeding (hybridization, mutagenesis, grafting) but cannot be considered GMO as do not involve the transformation of a gene into another species. Among these techniques, the most popular is CRISPR/Cas9 that consists of using a bacterial defense system to edit the genome of the host species in a specific site of the genome, defined by the guide RNA. Although there should be intense research in this field, there are not many descriptions published yet the use of CRISPR in sugar beet, nor there is information on a field trial with CRISPR sugar beet. There are descriptions of the design of a CRISPR system based in the BNYVV that allows transient expression of four different proteins in different tissues of the plant. This system has been used to transform *Nicotiana benthamiana*, *Beta macrocarpa*, and *Beta vulgaris* (Jiang et al. 2019). Also, considered a new breeding technique the TILLING (Targeting Induced Local Lesions in Genomes), the platform is a reverse genetics technology that is being used in *Beta vulgaris* to increase its agro-diversity (Kornienko and Butorina 2013).

4.10 Future Prospects

We have summarized in this book chapter the main advances in genetic engineering in sugar beet. It is clear that new breeding techniques are still starting and there is not much development yet. CRISPR, TILLING, and other strategies may be implemented. In a similar manner, we have described a lot of advances on developing GMO sugar beet for avoiding bolting, sugar yield increase, abiotic stress tolerance, or pest management, but farmers currently only have access to herbicide-resistant cultivars. There is a long way ahead, not only for scientists, breeders, and agro-bio companies but also for politicians to enable the use of these varieties that are already in hand. Let's hope that in the close future some of the advances described in this chapter or some unexpected could help produce more and better food.

References

Ali MA, Azeem F, Abbas A, Joyia FA, Li H, Dababat AA (2017) Transgenic strategies for enhancement of nematode resistance in plants. Front Plant Sci 8:750. https://doi.org/10.3389/FPLS.2017.00750

Barker AL, Dayan FE (2019) Fate of glyphosate during production and processing of glyphosate-resistant sugar beet (*Beta vulgaris*). J Agric Food Chem 67:2061–2065. https://doi.org/10.1021/ACS.JAFC.8B05672

Bevan MW, Flavell RB, Chilton MD (1983) A chimaeric antibiotic resistance gene as a selectable marker for plant cell transformation. Nature 304:184–187. https://doi.org/10.1038/304184A0

Biancardi E, Tamada T (2016) Rhizomania. Springer International Publishing Switzerland. https://doi.org/10.1007/978-3-319-30678-0. isbn:978-3-319-30676-6

Cai D, Thurau T, Tian Y, Lange T, Yeh KW, Jung C (2003) Sporamin-mediated resistance to beet cyst nematodes (*Heterodera schachtii Schm.*) is dependent on trypsin inhibitory activity in sugar beet (*Beta vulgaris* L.) hairy roots. Plant Mol Biol 516(51):839–849. https://doi.org/10.1023/A:1023089017906

Chen N, Yu ZH, Xiao XG (2017) Cytosolic and nuclear co-localization of betalain biosynthetic enzymes in tobacco suggests that betalains are synthesized in the cytoplasm and/or nucleus of betalainic plant cells. Front Plant Sci 8(831):1–10. https://doi.org/10.3389/fpls.2017.00831

Chia TYP, Müller A, Jung C, Mutasa-Göttgens ES (2008) Sugar beet contains a large CONSTANS-LIKE gene family including a CO homologue that is independent of the early-bolting (B) gene locus. J Exp Bot 59:2735–2748. https://doi.org/10.1093/JXB/ERN129

Ćurčić Ž, Taški-Ajduković K, Nagl N (2017) Relationship between hybrid performance and genetic variation in self-fertile and self-sterile sugar beet pollinators as estimated by SSR markers. Euphytica 213:108. https://doi.org/10.1007/s10681-017-1897-1

D'Halluin K, Bossut M, Bonne E, Mazur B, Leemans J, Botterman J (1992) Transformation of sugarbeet (*Beta vulgaris* l.) and evaluation of herbicide resistance in transgenic plants. Biotechnology 10:309–314. https://doi.org/10.1038/nbt0392-309

De Marchis F, Bellucci M (2021) Plastid transformation in sugar beet: an important industrial crop. In: Methods in Molecular Biology, pp 283–290

De Marchis F, Wang Y, Bellucci M and Arcioni S (2007) Developing a method for sugar beet chloroplast transformation. Proceedings of the 51st Italian Society of Agricultural Genetics Annual Congress. Riva del Garda, Italy – 23/26 September, 2007

De Marchis F, Wang Y, Stevanato P, Arcioni S, Bellucci M (2009) Genetic transformation of the sugar beet plastome. Transgenic Res 18:17–30. https://doi.org/10.1007/s11248-008-9193-4

Dillen K, Demont M, Tillie P, Rodriguez Cerezo E (2013) Bred for Europe but grown in America: the case of GM sugar beet. New Biotechnol 30:131–135. https://doi.org/10.1016/J.NBT.2012.11.004

Dohm JC, Minoche AE, Holtgräwe D, Capella-Gutierrez S, Zakrzewski F, Tafer H, Rupp O, Sorensen TR, Stracke R, Reinhardt R, Goesmann A, Kraft T, Schulz B, Stadler PF, Schmidt T, Gabaldon T, Lehrach H, Weisshaar B, Himmelbauer H (2014) The genome of the recently domesticated crop plant sugar beet (*Beta vulgaris*). Nature 505:546–549. https://doi.org/10.1038/nature12817

Doney DL (1995) USDA-ARS sugarbeet releases. J Sugar Beet Res 32:229–257. https://bsdf-assbt.org/wp-content/uploads/2017/04/JSBRVol32No4P229to257USDAARSSugarbeetReleases.pdf

Ebmeyer H, Fiedler-Wiechers K, Hoffmann CM (2021) Drought tolerance of sugar beet – Evaluation of genotypic differences in yield potential and yield stability under varying environmental conditions. Eur J Agron 125:126262. https://doi.org/10.1016/J.EJA.2021.126262

Ehlers U, Commandeur U, Frank R, Landsmann J, Koenig R, Burgermeister W (1991) Cloning of the coat protein gene from beet necrotic yellow vein virus and its expression in sugar beet hairy roots. Theor Appl Genet 81:777–782. https://doi.org/10.1007/BF00224989

Elliott MC, Chen DF, Fowler MR, Kirby MJ, Kubalakova M, Scott NW, Slater A (1996) Transgenesis-a scheme for improving sugar beet productivity. Russ J Plant Physiol 43:544–551

Erhardt M, Morant M, Ritzenthaler C, Stussi-Garaud C, Guilley H, Richards K, Jonard G, Bouzoubaa S, Gilmer D (2000) P42 movement protein of beet necrotic yellow vein virus is targeted by the movement proteins P13 and P15 to punctate bodies associated with plasmodesmata. Mol Plant-Microbe Interact MPMI 13:520–528. https://doi.org/10.1094/MPMI.2000.13.5.520

Evans E, Messerschmidt U (2017) Review: Sugar beets as a substitute for grain for lactating dairy cattle. J Anim Sci Biotechnol 8:25. https://doi.org/10.1186/s40104-017-0154-8

FAOSTAT (2018) http://www.fao.org/faostat/en/?#data/QC/visualize. Accessed 22 May 2020

Fry JE, Barnason AR, Hinchee M (1991) Genotype-independent transformation of sugarbeet using Agrobacterium tumefaciens. In: Molecular biology of plant growth and development, Third International Congress of Plant Molecular Biology, p 384

Fuchs M, Gonsalves D (2007) Safety of virus-resistant transgenic plants two decades after their introduction: lessons from realistic field risk assessment studies. Annu Rev Phytopathol 45:173–202. https://doi.org/10.1146/ANNUREV.PHYTO.45.062806.094434

Gerbens-Leenes W, Hoekstra AY (2012) The water footprint of sweeteners and bio-ethanol. Environ Int 40:202–211. https://doi.org/10.1016/J.ENVINT.2011.06.006

Gisbert C, Timoneda A, Porcel R, Ros R, Mulet JM (2020) Overexpression of *BvHb2*, a Class 2 non-symbiotic hemoglobin from sugar beet, confers drought-induced withering resistance and alters iron content in tomato. Agronomy 10:1754. https://doi.org/10.3390/agronomy10111754

Goudarzi A, Safaei N, Jafari M, Mahmoudi SB, Far MT (2015) Transformation of sugar beet with a bean chitinase gene and enhanced resistance to Alternaria alternata. Agric Biotechnol J 7:175–200. https://doi.org/10.22103/JAB.2015.1358

Goudarzi A, Jafari M, Safaie N, Mohammad Jafari S (2016) Transgenic sugar beet expressing a bacterial mannitol-1-phosphate dehydrogenase (mtlD) gene shows enhanced resistance to fungal pathogens. Sugar Tech 18:192–203. https://doi.org/10.1007/s12355-015-0379-9

Gurel E, Gurel S, Lemaux PG (2008) Biotechnology applications for sugar beet. CRC Crit Rev Plant Sci 27:108–140. https://doi.org/10.1080/07352680802202000

Hall RD, Riksen-Bruinsma T, Weyens GJ et al (1996) A high efficiency technique for the generation of transgenic sugar beets from stomatal guard cells. Nat Biotechnol 14:1133–1138. https://doi.org/10.1038/nbt0996-1133

Harpster MH, Townsend JA, Jones JDG, Bedbrook J, Dunsmuir P (1988) Relative strengths of the 35S cauliflower mosaic virus, 1′, 2′, and nopaline synthase promoters in transformed tobacco sugarbeet and oilseed rape callus tissue. Mol Gen Genet 2121(212):182–190. https://doi.org/10.1007/BF00322463

Hashimoto R, Shimamoto Y (1999) Growth of transgenic sugar beet plants with sucrose-phosphate synthase (SPS). In: Proc. Jpn. Soc. Sugar Beet Technol, pp 85–89

Hashimoto R, Shimamoto Y (2001) Transgenic sugar beet plants harboring a pumpkin chitinase gene demonstrating improved resistance to *Rhizoctonia solani.* In: Proc Jpn Soc Sugar Beet Technol, pp 24–28

Hatlestad GJ, Sunnadeniya RM, Akhavan NA, Gonzalez A, Goldman IL, McGrath M, Lloyd AM (2012) The beet R locus encodes a new cytochrome P450 required for red betalain production. Nat Genet 44:816–820. https://doi.org/10.1038/ng.2297

Heller R, Schondelmaier J, Steinrücken G, Jung C (1996) Genetic localization of four genes for nematode (*Heterodera schachtii Schm.*) resistance in sugar beet (*Beta vulgaris* L.). Theor Appl Genet 92:991–997. https://doi.org/10.1007/BF00224039

Holmquist L, Dölfors F, Fogelqvist J, Cohn J, Kraft T, Dixelius C (2021) Major latex protein-like encoding genes contribute to *Rhizoctonia solani* defense responses in sugar beet. Mol Gen Genomics 296:155–164. https://doi.org/10.1007/s00438-020-01735-0

Horn J, Lauster S, Krenz B, Krausb J, Frischmutha T, Jeskea H (2011) Ambivalent effects of defective DNA in beet curly top virus-infected transgenic sugarbeet plants. Virus Res 158:169–178. https://doi.org/10.1016/j.virusres.2011.03.029

Hossain MS, Persicke M, ElSayed AI, Kalinowski J, Dietz KJ (2017) Metabolite profiling at the cellular and subcellular level reveals metabolites associated with salinity tolerance in sugar beet. J Exp Bot 68:5961–5976. https://doi.org/10.1093/JXB/ERX388

Ivarson E, Leiva-Eriksson N, Ahlman A, Kanagarajan S, Bülow L, Zhu LH (2017) Effects of overexpression of WRI1 and hemoglobin genes on the seed oil content of Lepidium campestre. Front Plant Sci 7:2032. https://doi.org/10.3389/fpls.2016.02032

Ivic S, Sicher R, Smigocki A (2001) Growth habit and sugar accumulation in sugarbeet (*Beta vulgaris* L.) transformed with a cytokinin biosynthesis gene. Plant Cell Rep 208(20):770–773. https://doi.org/10.1007/S002990100389

Jacq B, Lesobre O, Sangwan RS (1993) Sangwan-Norreel BS (1993) Factors influencing T-DNA transfer in Agrobacterium-mediated transformation of sugarbeet. Plant Cell Rep 1211(12): 621–624. https://doi.org/10.1007/BF00232811

Jafari M, Norouzi P, Malboobi MA, Ghareyazie B, Valizadeh M, Mohammadi SA, Mousavi M (2008) Enhanced resistance to a lepidopteran pest in transgenic sugar beet plants expressing synthetic cry1Ab gene. Euphytica 165:333–344. https://doi.org/10.1007/S10681-008-9792-4

Jafarzade M, Ramezani M, Hedayati F, Mokhtarzade Z, Zare B, Sabet MS, Norouzi P, Malboobi MA (2019) Antibody-mediated resistance to rhizomania disease in sugar beet hairy roots. Plant Pathol J 35:692–697. https://doi.org/10.5423/PPJ.OA.04.2018.0073

James C (2011) Brief 42: global Status of Commercialized Biotech/GM Crops: 2010. ISAAA Br

Jansen A (2009) Modifying post-harvest sucrose loss in sugar beet: Assessment of transgenic approaches—heiDOK. Ph.D. Dissertation. https://doi.org/10.11588/heidok.00010077

Ji M, Wang K, Wang L, Chen S, Li H, Ma C, Wang Y (2019) Overexpression of a S-Adenosylmethionine decarboxylase from sugar beet M14 increased Arabidopsis salt tolerance. Int J Mol Sci 20:1990. https://doi.org/10.3390/ijms20081990

Jiang N, Zhang C, Liu JY, Guo ZH, Zhang ZY, Han CG, Wang Y (2019) Development of beet necrotic yellow vein virus-based vectors for multiple-gene expression and guide RNA delivery in plant genome editing. Plant Biotechnol J 17:1302–1315. https://doi.org/10.1111/pbi.13055

Joersbo M (2007) Sugar Beet. In: Pua E-C, Davey MR (eds) Transgenic crops IV. Springer, Berlin Heidelberg, pp 355–379

Jung C (1998) Cloning and breeding utility of the gene Hs1 for nematode resistance from Beta procumbens. In: Comptes-Rendus des Congres de l'Institut International de Recherches Betteravieres (Belgium)

Kagami H, Kurata M, Matsuhira H, Taguchi K, Mikami T, Tamagake H, Kubo T (2015) Sugar beet (*Beta vulgaris* L.). Methods Mol Biol 1223:335–347. https://doi.org/10.1007/978-1-4939-1695-5_27

Kagami H, Taguchi K, Arakawa T, Kuroda Y, Tamagake H, Kubo T (2016) Efficient callus formation and plant regeneration are heritable characters in sugar beet (*Beta vulgaris* L.). Hereditas 153:12. https://doi.org/10.1186/s41065-016-0015-z

Kallerhoff J, Perez P, Bouzoubaa S, Ben Tahar S, Perret J (1990) Beet necrotic yellow vein virus coat protein-mediated protection in sugarbeet (*Beta vulgaris* L.) protoplasts. Plant Cell Rep 94(9):224–228. https://doi.org/10.1007/BF00232185

Kanhonou R, Serrano R, Palau RR (2001) A catalytic subunit of the sugar beet protein kinase CK2 is induced by salt stress and increases NaCl tolerance in *Saccharomyces cerevisiae*. Plant Mol Biol 47:571–579

Kerr SP (2005) Performance of rhizomania resistant sugar beet varieties in UK trials. Asp Appl Biol 76:21

Khan MFR (2014) The beet goes on-roundup ready sugar beet deregulated again in USA. Int Sugar J 116:131–133. http://www.agra-net.com/portal2/isj/

Khayamim S, Tavkol Afshari R, Sadeghian SY, Poustini K, Roozbeh F, Abbasi Z (2014) Seed germination, plant establishment, and yield of sugar beet genotypes under salinity stress. J Agric Sci Technol 16:779–790

Kimoto Y, Shimamoto Y (2001) Difference in toxicity to larvae of cabbage armyworm between transgenic sugar beet lines with Cry1Ab and Cry1C. Proc Jan Soc Sugar Beet Technol 43:20–23

Kishchenko EM, Komarnitskii IK, Kuchuk NV (2010) Transposition of the maize transposable element dSpm in transgenic sugar beets. Cytol Genet 444(44):200–205. https://doi.org/10.3103/S009545271004002X

Kishchenko EM, Komarnitskii IK, Kuchuk NV (2011) Transgenic sugar beet tolerant to imidazolinone obtained by Agrobacterium-mediated transformation. Cytol Genet 453(45): 148–152. https://doi.org/10.3103/S0095452711030030

Kitazaki K, Arakawa T, Matsunaga M, Yui-Kurino R, Matsuhira H, Mikami T, Kubo T (2015) Post-translational mechanisms are associated with fertility restoration of cytoplasmic male sterility in sugar beet (*Beta vulgaris*). Plant J 83:290–299. https://doi.org/10.1111/tpj.12888

Klemens PAW, Patzke K, Trentmann O, Poschet G, Büttner M, Schulz A, Marten I, Hedrich R, Ekkehard Neuhaus H (2014) Overexpression of a proton-coupled vacuolar glucose exporter impairs freezing tolerance and seed germination. New Phytol 202:188–197. https://doi.org/10.1111/nph.12642

Klenk I, Landquist B, De Imaña OR (2012) The product carbon footprint of EU beet sugar (part I). Zuckerindustrie 137:169–177. https://doi.org/10.36961/SI12784

Klimek-Chodacka M, Baranski R (2014) A protocol for sonication-assisted Agrobacterium rhizogenes-mediated transformation of haploid and diploid sugar beet (*Beta vulgaris* L.) explants. Acta Biochim Pol 61:13–17. https://doi.org/10.18388/ABP.2014_1916

Knecht K, Seyffarth M, Desel C, Thurau T, Sherameti I, Lou B, Oelmüller R, Cai D (2010) Expression of *BvGLP-1* encoding a germin-like protein from sugar beet in Arabidopsis thaliana leads to resistance against phytopathogenic fungi. Mol Plant-Microbe Interact 23:446–457. https://doi.org/10.1094/MPMI-23-4-0446

Koch MS, Ward JM, Levine SL, Baum JA, Vicini JL, Hammond BG (2015) The food and environmental safety of Bt crops. Front Plant Sci 6:283. https://doi.org/10.3389/FPLS.2015.00283

Kong J, Martin-Ortigosa S, Finer J, Orchard N, Gunadi A, Batts LA, Thakare D, Rush B, Schmitz O, Stuiver M, Olhoft P, Pacheco-Villalobos D (2020) Overexpression of the transcription factor growth-regulating factor5 improves transformation of dicot and monocot species. Front Plant Sci 11:1389. https://doi.org/10.3389/FPLS.2020.572319

Kornienko AV, Butorina AK (2013) Induced mutagenesis in sugar beet (*Beta vulgaris* L.): obtained results and prospects for use in development of TILLING project. Biol Bull Rev 32(3):152–160. https://doi.org/10.1134/S2079086413020059

Krens FA, Zijlstra C, Wvd M, Jamar D, Huizing HJ (1988) Transformation and regeneration in sugar beet (*Beta vulgaris* L.) induced by 'shooter' mutants of Agrobacterium tumefaciens. Euphytica 393(39):185–194. https://doi.org/10.1007/BF00043382

Kuykendall LD, Upchurch RG (2004) Expression in sugar beet of the introduced cercosporin toxin export (CFP) gene from *Cercospora kikuchii*, the causative organism of purple seed stain in soybean. Biotechnol Lett 269(26):723–727. https://doi.org/10.1023/B:BILE.0000024096.22105.C3

Lathouwers J, Vandercappellen L, Lommel M, Rosquin I, Denys P, Bruyne E, Lefebvre M, Weyens G, Bleykasten C, Bouzoubaa S, Guilley H, Richards K, Jonard G (1997) Sugar beet transformation for rhizomania resistance: introduction and expression of different viral sequences. In: Comptes-Rendus des Congres de l'Institut International de Recherches Betteravieres (Belgium), vol 60, pp 491–495

Lathouwers J, Weyens G, Lefebvre M (2005) Transgenic research in sugarbeet. Genet Modif Sugar Beet:5–24

Lauber E, Bleykasten-Grosshans C, Erhardt M, Bouzoubaa S, Jonard G, Richards KE, Guilley H (1998) Cell-to-cell movement of beet necrotic yellow vein virus: I. Heterologous complementation experiments provide evidence for specific interactions among the triple gene block proteins. Mol Plant-Microbe Interact 11:618–625. https://doi.org/10.1094/MPMI.1998.11.7.618

Lennefors BL, Savenkov EI, Bensefelt J, Wremerth-Weich E, van Roggen P, Tuvesson S, Valkonen JPT, Gielen J (2006) dsRNA-mediated resistance to beet necrotic yellow vein virus infections in sugar beet (*Beta vulgaris* L. ssp. vulgaris). Mol Breed 184(18):313–325. https://doi.org/10.1007/S11032-006-9030-5

Li H, Smigocki AC (2019) Suppression of Fusarium oxysporum with recombinant polygalacturonase inhibiting proteins (BvPGIPs) extracted from sugar beet roots. Plant Cell Tissue Organ Cult (PCTOC) 136:197–203. https://doi.org/10.1007/s11240-018-1496-4

Li J, Li H, Yang N, Ma C, Li H (2020) Overexpression of a monodehydroascorbate reductase gene from sugar beet M14 increased salt stress tolerance. Sugar Tech 23:45–56. https://doi.org/10.1007/S12355-020-00877-0

Lindsey K, Gallois P (1990) Transformation of Sugarbeet (Beta vulgaris) by Agrobacterium tumefaciens. J Exp Bot 41:529–536. https://doi.org/10.1093/JXB/41.5.529

Lisson R, Hellert J, Ringleb M, Machens F, Kraus J, Hehl R (2010) Alternative splicing of the maize Ac transposase transcript in transgenic sugar beet (*Beta vulgaris* L.). Plant Mol Biol 74: 19–32. https://doi.org/10.1007/s11103-010-9651-2

Liu D, An Z, Mao Z, Ma L, Lu Z (2015) Enhanced heavy metal tolerance and accumulation by transgenic sugar beets expressing Streptococcus thermophilus StGCS-GS in the presence of Cd, Zn and Cu alone or in combination. PLoS One 10(6):e0128824. https://doi.org/10.1371/journal.pone.0128824

Liu H, Wang Q, Yu M, Zhang Y, Wu Y, Zhang H (2008) Transgenic salt-tolerant sugar beet (*Beta vulgaris* L.) constitutively expressing an Arabidopsis thaliana vacuolar Na^+/H^+ antiporter gene, *AtNHX3*, accumulates more soluble sugar but less salt in storage roots. Plant Cell Environ 31: 1325–1334. https://doi.org/10.1111/j.1365-3040.2008.01838.x

Locascio A, Andrés-Colás N, Mulet JMJM, Yenush L (2019a) Saccharomyces cerevisiae as a tool to investigate plant potassium and sodium transporters. Int J Mol Sci 20:2133. https://doi.org/10.3390/ijms20092133

Locascio A, Marqués MCMC, García-Martínez G, Corratgé-Faillie C, Andrés-Colás N, Rubio L, Fernández JA, Véry AA, Mulet JM, Yenush L (2019b) BCL2-ASSOCIATED ATHANOGENE4 Regulates the KAT1 potassium channel and controls stomatal movement. Plant Physiol 181:1277–1294. https://doi.org/10.1104/pp.19.00224

Lytvyn DI, Syvura VV, Kurylo VV, Olenieva VD, Yemets AI, Blume YB (2014) Creation of transgenic sugar beet lines expressing insect pest resistance genes cry1C and cry2A. Cytol Genet 48:69–75. https://doi.org/10.3103/S0095452714020078

Ma C, Wang Y, Gu D, Nan J, Chen S, Li H (2017) Overexpression of S-adenosyl-L-methionine synthetase 2 from sugar beet M14 increased Arabidopsis tolerance to salt and oxidative stress. Int J Mol Sci 18:847. https://doi.org/10.3390/ijms18040847

Ma C, Wang Y, Wang Y, Wang L, Chen S, Li H (2011) Identification of a sugar beet BvM14-MADS box gene through differential gene expression analysis of monosomic addition line M14. J Plant Physiol 168:1980–1986. https://doi.org/10.1016/j.jplph.2011.05.027

Mannerlöf M, Lennerfors B-L, Tenning P (1996) Reduced titer of BNYVV in transgenic sugar beets expressing the BNYVV coat protein. Euphytica 903(90):293–299. https://doi.org/10.1007/BF00027479

Märländer B (2005) Weed control in sugar beet using genetically modified herbicide-tolerant varieties – a review of the economics for cultivation in Europe. J Agron Crop Sci 191:64–74. https://doi.org/10.1111/J.1439-037X.2004.00135.X

Matsuhira H, Kagami H, Kurata M, Kitazaki K, Matsunaga M, Hamaguchi Y, Hagihara E, Ueda M, Harada M, Muratasu A, Yui-Kurino R, Taguchi K, Tamagake H, Mikami T, Kubo T (2012) Unusual and typical features of a novel Restorer-of-fertility gene of sugar beet (*Beta vulgaris* L.). Genetics 192:1347–1358. https://doi.org/10.1534/genetics.112.145409

Matsuhira H, Kichi T, Tamagake H, Sato Y, Anzai H, Yoshida M (2014) High production of plant type levan in sugar beet transformed with timothy (*Phleum pratense*) 6-SFT genes. J Biotechnol 192:215–222. https://doi.org/10.1016/j.jbiotec.2014.09.025

May MJ, Wilson RG (2006) Weeds and weed control. Sugar beet:359–386

McGinnis EE, Meyer MH, Smith AG (2010) Sweet and sour: a scientific and legal look at herbicide-tolerant sugar beet. Plant Cell 22:1653. https://doi.org/10.1105/TPC.110.077198

McGrath JM (2011) Notice of release of EL56 sugar beet germplasm with high levels of tolerance to salinity during germination. USDA-ARS Germplasm Release

Mcgrath JM, Panella L, Mitchell J (2018) Plant breeding reviews. Plant Breed Rev. https://doi.org/10.1002/9781119521358

Mishutkina YV, Kamionskaya AM, Skryabin KG (2010) The creation of sugar beet transgenic plants expressing bar gene. Appl Biochem Microbiol 46:80–86. https://doi.org/10.1134/S000368381001014X

Moazami K, Mortazavi SE, Heidari B, Nouroozi P (2018) Agrobacterium-mediated transient assay of the gus gene expression in sugar beet. Annu Res Rev Biol 30:1–7. https://doi.org/10.9734/ARRB/2018/46049

Mohammadzadeh R, Zamani M, Motallebi M, Norouzi P, Jourabchi E, Benedetti M, De Lorenzo G (2012) *Agrobacterium tumefaciens*-mediated introduction of polygalacturonase inhibiting protein 2 gene (*PvPGIP2*) from *Phaseolus vulgaris* into sugar beet (*Beta vulgaris* L.). Aust J Crop Sci 6:1290–1297

Molinero AIS, Reuzeau C, Frankard V, Salom RS and Salort JMM (2014, August 21) Plants having enhanced yield-related traits and a method for making the same US patent 14175402, 21 August 2014

Monger W, Barrington PJ, Acaster AM, Halford NG, Ainsworth C, Thomas H (1995) A transgenic approach towards investigating carbon metabolism in sugar beet. Proc IIRB Congress 58:193–196

Mulet JM, Alemany B, Ros R, Calvete JJ, Serrano R (2004) Expression of a plant serine O-acetyltransferase in Saccharomyces cerevisiae confers osmotic tolerance and creates an alternative pathway for cysteine biosynthesis. Yeast 21:303–312. https://doi.org/10.1002/yea.1076

Munns R, Tester M (2008) Mechanisms of salinity tolerance. Annu Rev Plant Biol 59:651–681. https://doi.org/10.1146/annurev.arplant.59.032607.092911

Mutasa-Gottgens E, Qi A, Mathews A, Thomas S, Phillips A, Hedden P (2009) Modification of gibberellin signalling (metabolism & signal transduction) in sugar beet: analysis of potential targets for crop improvement. Transgenic Res 18:301–308. https://doi.org/10.1007/s11248-008-9211-6

Nieberl P, Ehrl C, Pommerrenig B, Graus D, Marten I, Jung B, Ludewig F, Koch W, Harms K, Flügge UI, Neuhaus HE, Hedrich R, Sauer N (2017) Functional characterisation and cell specificity of BvSUT1, the transporter that loads sucrose into the phloem of sugar beet (*Beta vulgaris* L.) source leaves. Plant Biol 19:315–326. https://doi.org/10.1111/PLB.12546

Ninković S, Djordjević T, Vinterhalter B, Uzelac B, Cingel A, Savić J, Radović S (2010) Embryogenic responses of *Beta vulgaris* L. callus induced from transgenic hairy roots. Plant Cell Tissue Organ Cult 1031(103):81–91. https://doi.org/10.1007/S11240-010-9757-X

Padmanaban S, Li H, Puthoff D, Smigocki A (2016) *Beta vulgaris* promoters for directed tissue-specific root transcription. Am. Soc. Sugar Beet Tech. Proc. 36. http://assbtproceedings.org/ASSBT2011Proceedings/Posters/Physilogy%20and%20Biotechnology/Padmanaban%20_Smigocki_%20_2_.pdf. Accessed 3 Mar 2016

Panella L, Campbell LG, Eujayl IA, Lewellen RT, McGrath JM (2015) USDA-ARS Sugarbeet releases and breeding over the past 20 years. J Sugar Beet Res 52. https://doi.org/10.5274/jsbr.52.3.40

Pappinen A, Degefu Y, Syrjälä L, Keinonen K, von Weissenberg K (2002) Transgenic silver birch (*Betula pendula*) expressing sugarbeet chitinase 4 shows enhanced resistance to *Pyrenopeziza betulicola*. Plant Cell Rep 20:1046–1051. https://doi.org/10.1007/S00299-002-0449-9

Pasonen H-L, Seppänen S-K, Degefu Y, Rytkönen A, von Weissenberg K, Pappinen A (2004) Field performance of chitinase transgenic silver birches (*Betula pendula*): resistance to fungal diseases. Theor Appl Genet 1093(109):562–570. https://doi.org/10.1007/S00122-004-1650-8

Pavli OI (2010) Molecular characterization of beet necrotic yellow vein virus in Greece and transgenic approaches towards enhancing rhizomania disease resistance. Wageningen University and Research. https://library.wur.nl/WebQuery/wurpubs/386941

Pavli OI, Kelaidi GI, Tampakaki AP, Skaracis GN (2011a) The HrpZ gene of Pseudomonas syringae pv. phaseolicola enhances resistance to rhizomania disease in transgenic Nicotiana benthamiana and sugar beet. PLoS One:6. https://doi.org/10.1371/journal.pone.0017306

Pavli OI, Panopoulos NJ, Goldbach R, Skaracis GN (2010) BNYVV-derived dsRNA confers resistance to rhizomania disease of sugar beet as evidenced by a novel transgenic hairy root approach. Transgenic Res 19:915–922. https://doi.org/10.1007/s11248-010-9364-y

Pavli OI, Stevanato P, Biancardi E, Skaracis GN (2011b) Achievements and prospects in breeding for rhizomania resistance in sugar beet. F Crop Res 122:165–172

Pfeiffer N, Tränkner C, Lemnian I, Grosse I, Müller AE, Jung C, Kopisch-Obuch FJ (2014) Genetic analysis of bolting after winter in sugar beet (*Beta vulgaris* L.). Theor Appl Genet 127:2479–2489. https://doi.org/10.1007/s00122-014-2392-x

Pilon-Smits EAH, Terry N, Sears T, Van Dun K (1999) Enhanced drought resistance in fructan-producing sugar beet. Plant Physiol Biochem 37:313–317. https://doi.org/10.1016/S0981-9428(99)80030-8

Pin PA, Benlloch R, Bonnet D, Wremerth-Weich E, Kraft T, Gielen JJL, Nilsson O (2010) An antagonistic pair of FT homologs mediates the control of flowering time in sugar beet. Science 80(330):1397–1400. https://doi.org/10.1126/science.1197004

Porcel R, Bustamante A, Ros R, Serrano R, Mulet-Salort JM (2018) BvCOLD1: a novel aquaporin from sugar beet (*Beta vulgaris* L.) involved in boron homeostasis and abiotic stress. Plant Cell Environ 41:2844–2857. https://doi.org/10.1111/pce.13416

Poveda J, Francisco M, Cartea ME, Velasco P (2020) Development of transgenic brassica crops against biotic stresses caused by pathogens and arthropod pests. Plan Theory 9:1664. https://doi.org/10.3390/PLANTS9121664

Rajabi A, Borchardt D (2015) QTL mapping for root yield and leaf traits in sugar beet (*Beta vulgaris* L.) under drought stress condition. Iran J Crop Sci 17(1):46–62

Rajabi A, Griffiths H, Ober ES, Kromdijk W, Pidgeon JD (2007) Genetic characteristics of water-use related traits in sugar beet. Euphytica 1602(160):175–187. https://doi.org/10.1007/S10681-007-9520-5

Rausell A, Kanhonou R, Yenush L, Serrano R, Ros R (2003) The translation initiation factor eIF1A is an important determinant in the tolerance to NaCl stress in yeast and plants. Plant J 34:257–267. https://doi.org/10.1046/j.1365-313x.2003.01719.x

Reuzeau C, Molinero AIS, Frankard V, Salom RS, Salort JMM (2017) Plants having increased yield-related traits by expressing a growth-regulating factor (GRF) polypeptide and method for making the same US Patent 9617557, 11 April 2017

Ribichich KF, Chiozza M, Ávalos-Britez S, Cabello JV, Arce AL, Watson G, Arias C, Portapila M, Trucco F, Otegui ME, Chan RL (2020) Successful field performance in warm and dry environments of soybean expressing the sunflower transcription factor HB4. J Exp Bot 71: 3142–3156. https://doi.org/10.1093/jxb/eraa064

Rosa Téllez S, Kanhonou R, Castellote Bellés C, Serrano R, Alepuz P, Ros R (2020) RNA-binding proteins as targets to improve salt stress tolerance in crops. Agronomy 10:250. https://doi.org/10.3390/agronomy10020250

Sadeghi-Shoae M, Habibi D, Taleghani DF, Paknejad F, Kashani A (2014) Evaluation the effect of paclobutrazol on bolting, qualitative and quantitative performance in autumn sown-sugar beet genotypes in Moghan region. Int J Biosci 5:345–354. https://doi.org/10.12692/ijb/5.1.345-354

Safar S, Bazrafshan M, Khoshnami M, Behrooz AA, Hedayati F, Maleki M, Mahmoudi SB, Malboobi MA (2021) Field evaluation for rhizomania resistance of transgenic sugar beet events based on gene silencing. Can J Plant Pathol 43:179–188. https://doi.org/10.1080/07060661.2020.1783575

Salort JMM, Molinero AIS, Salom RS (2010) Method for producing transgenic plants with increased yield, comprising expressing of haemoglobin from Arabidopsis. U.S. Patent 7674953B2, 9 March 2010

Salort JMM, Salom RS (2009.) Protein for use in modifying abiotic stress tolerance in yeast US Patent 7612177, 3 November 2009

Santoni M, Bertini E, Zampieri R, Cuccurullo A, Commisso M, Gecchele E, Avesani L (2019) Transient expression in red beet of a biopharmaceutical candidate vaccine for type-1 diabetes. J Vis Exp. https://doi.org/10.3791/59298

Sanz Molinero AI, Reuzeau C, Frankard V, Mulet Salort JM and Salom RS (2009) Plants Having Enhanced Yield-Related Traits and a Method for Making the Same WO Patent WO/2009/068564. 4 June 2009

Saunders JW, Acquaah G, Renner KA, Doley WP (1992) Monogenic dominant sulfonylurea resistance in sugarbeet from somatic cell selection. Crop Sci 32:1357–1360. https://doi.org/10.2135/CROPSCI1992.0011183X003200060010X

Savić J, Nikolić R, Banjac N, Zdravković-Korać S, Stupar S, Cingel A, Ćosić T, Raspor M, Smigocki A, Ninković S (2019) Beneficial implications of sugar beet proteinase inhibitor BvSTI on plant architecture and salt stress tolerance in *Lotus corniculatus* L. J Plant Physiol 243:153055. https://doi.org/10.1016/j.jplph.2019.153055

Savitsky VF (1950) Monogerm sugar beets in the United States. Proc Am Soc Sugar Beet Technol 6:156–159

Scholten OE, Lange W (2000) Breeding for resistance to rhizomania in sugar beet: a review. Euphytica 1123(112):219–231. https://doi.org/10.1023/A:1003988003165

Schwegler DD, Gowda M, Schulz B, Miedaner T, Liu W, Reif JC (2014) Genotypic correlations and QTL correspondence between line per se and testcross performance in sugar beet (*Beta vulgaris* L.) for the three agronomic traits beet yield, potassium content, and sodium content. Mol Breed 341(34):205–215. https://doi.org/10.1007/S11032-014-0030-6

Sedighi L, Rezapanah M, Aghdam HR (2011) Efficacy of Bt transgenic sugar beet lines expressing cry1Ab Gene Against *Spodoptera littoralis* Boisd. (Lepidoptera: Noctuidae). J Entomol Res Soc 13:61–61

Serrano R, Montesinos C, Gaxiola R, Ríos G, Forment J, Leube M, Mulet JM, Naranjo MA, Roldán M, Vicente O, Kanhonou RA, Rausell A, Ros R (2003) Functional genomics of salt tolerance: the yeast overexpression approach. In: Proceedings of the International Symposium on managing Greenhouse Crops in Saline Environement. International Society Horticultural Science, pp 31–38. https://doi.org/10.17660/ActaHortic.2003.609.2

Sévenier R, der Meer IM, Bino R, Koops AJ (2002) Increased production of nutriments by genetically engineered crops. J Am Coll Nutr 21:199S–204S. https://doi.org/10.1080/07315724.2002.10719266

Sévenier R, Hall RD, van der Meer IM, Hakkert HJC, van Tunen AJ, Koops AJ (1998) High level fructan accumulation in a transgenic sugar beet. Nat Biotechnol 169(16):843–846. https://doi.org/10.1038/nbt0998-843

Shimamoto Y, Domae T (2000) Resistance to larvae of cabbage armyworm (Mamestra brassicae L.) in ICP gene transductant of sugarbeet [*Beta vulgaris*]. Proc Sugar Beet Res Association (Japan) 41:90–98

Smigocki AC, Ivic-Haymes S, Li H, Savić J (2013) Pest protection conferred by a *Beta vulgaris* serine proteinase inhibitor gene. PLoS One:8. https://doi.org/10.1371/journal.pone.0057303

Smigocki AC, Ivic-Haymes SD, Puthoff DP, Zuzga S (2008) Recent advances in functional genomics for sugar beet (*Beta vulgaris* L.) improvement: progress in determining the role of BvSTI in pest resistance in roots. Sugar Tech 10:91–98. https://doi.org/10.1007/S12355-008-0016-Y

Snyder GW, Ingersoll JC, Smigocki AC, Owens LD (1999) Introduction of pathogen defense genes and a cytokinin biosynthesis gene into sugarbeet (*Beta vulgaris* L.) by Agrobacterium or particle bombardment. Plant Cell Rep 18:829–834. https://doi.org/10.1007/S002990050669

Sönmez Ç, Elekciğolu IH, Yücel AM, Öktem HA (2014) Transgenic *Nicotiana tabacum* cultivar Samsun plants carrying the wild sugar beet Hs1pro1 gene have resistance to root-knot nematodes. Turk J Biol 38:200–207. https://doi.org/10.3906/biy-1306-35

Srivastava HM (1995) Sugarbeet pre-breeding in India. J Sugarbeet Res 32:99–111. https://doi.org/10.5274/JSBR.32.2.99

Suzuki H, Fukushima EO, Umemoto N, Ohyama K, Seki H, Muranaka T (2018) Comparative analysis of CYP716A subfamily enzymes for the heterologous production of C-28 oxidized triterpenoids in transgenic yeast. Plant Biotechnol 35:131. https://doi.org/10.5511/PLANTBIOTECHNOLOGY.18.0416A

Taïbi K, Ait Abderrahim L, Boussaid M, Bissoli G, Taïbi F, Achir M, Souana K, Mulet JM (2021) Salt-tolerance of *Phaseolus vulgaris* L. is a function of the potentiation extent of antioxidant enzymes and the expression profiles of polyamine encoding genes. South Afr J Bot 140:114–122. https://doi.org/10.1016/J.SAJB.2021.03.045

Tertivanidis K, Goudoula C, Vasilikiotis C, Hassiotou E, Perl-Treves R, Tsaftaris A (2004) Superoxide dismutase transgenes in sugarbeets confer resistance to oxidative agents and the fungus *C. beticola*. Transgenic Res 13:225–233. https://doi.org/10.1023/B:TRAG.0000034610.35724.04

Thurau T, Kifle S, Jung C, Cai D (2003) The promoter of the nematode resistance gene Hs1 pro−1 activates a nematode-responsive and feeding site-specific gene expression in sugar beet (*Beta vulgaris* L.) and *Arabidopsis thaliana*. Plant Mol Biol 52:643–660. https://doi.org/10.1023/A:1024887516581

Turk S, Smeekens SCM (1999) Genetic modification of plant carbohydrate metabolism. Appl Plant Biotechnol 71:100

United States Department of Agriculture (2019) 2017 Census of Agriculture United States summary and state data volume 1 - geographic area series - Part 51

Van Camp W (2005) Yield enhancement genes: seeds for growth. Curr Opin Biotechnol 16:147–153

Vihervuori L, Lyytikäinen-Saarenmaa P, Lu J, Pasonen HL (2013) Effects on lepidopteran herbivores of feeding on leaves of transgenic birch (*Betula pendula*) expressing the sugar beet chitinase IV gene. Eur J Entomol 110:253

Vihervuori L, Pasonen HL, Lyytikäinen-Saarenmaa P (2008) Density and composition of an insect population in a field trial of chitinase transgenic and wild-type silver birch (*Betula pendula*) clones. Environ Entomol 37:1582–1591. https://doi.org/10.1603/0046-225X-37.6.1582

Villarías Moradillo JL (1999) Compendio práctico del cultivo de la remolacha azucarera. Ediciones Agrotécnicas, S.L., Madrid

Vogt SH, Weyens G, Lefèbvre M, Bork B, Schechert A, Müller AE (2014) The FLC-like gene *BvFL1* is not a major regulator of vernalization response in biennial beets. Front Plant Sci 5:146. https://doi.org/10.3389/fpls.2014.00146

Wang C, Burzio LA, Koch MS, Silvanovich A, Bell E (2015) Purification, characterization and safety assessment of the introduced cold shock protein B in DroughtGard^{TM} maize. Regul Toxicol Pharmacol 71:164–173. https://doi.org/10.1016/j.yrtph.2014.12.014

Wang H, Kraus J, Dettendorfer J, Chua NH, Nehls R (2003) Marker gene elimination from transgenic sugarbeet by a chemically regulated Cre-lox system. Plant Biotechnol 2002 Beyond:229–231. https://doi.org/10.1007/978-94-017-2679-5_44

Wang X, Sun Y, Li G (2012) Agrobacterium-mediated transformation of sugar beet with vacuolar H^+-*ppase avp1* gene. Mol Plant Breed 9:370–375. https://agris.fao.org/agris-search/search.do?recordID=CN2012002147

Weyens G, Ritsema T, Van Dun K, Meyer D, Lommel M, Lathouwers J, Rosquin I, Denys P, Tossens A, Nijs M, Turk S, Gerrits N, Bink S, Walraven B, Lefèbvre M, Smeekens S (2004) Production of tailor-made fructans in sugar beet by expression of onion fructosyltransferase genes. Plant Biotechnol J 2:321–327. https://doi.org/10.1111/J.1467-7652.2004.00074.X

Wozniak CA, Owens LD (1989) Transformation of sugarbeet cell suspension cultures mediated by Agrobacterium tumefaciens. J Cell Biochem Suppl D 13:272

Wroblewski T, Tomczak A, Michelmore R (2005) Optimization of Agrobacterium-mediated transient assays of gene expression in lettuce, tomato and Arabidopsis. Plant Biotechnol J 3:259–273. https://doi.org/10.1111/J.1467-7652.2005.00123.X

Wu C, Ma C, Pan Y, Gong S, Zhao C, Chen S, Li H (2013) Sugar beet M14 glyoxalase I gene can enhance plant tolerance to abiotic stresses. J Plant Res 126:415–425. https://doi.org/10.1007/s10265-012-0532-4

Wu GQ, Feng RJ, Wang SM, Wang CM, Bao AK, Wei L, Yuan HJ (2015) Co-expression of xerophyte Zygophyllum xanthoxylum *ZxNHX* and *ZxVP1-1* confers enhanced salinity tolerance in chimeric sugar beet (*Beta vulgaris* L.). Front Plant Sci 6:581. https://doi.org/10.3389/FPLS.2015.00581

Yamada N, Takahashi H, Kitou K, Sahashi K, Tamagake H, Tanaka Y, Takabe T (2015) Suppressed expression of choline monooxygenase in sugar beet on the accumulation of glycine betaine. Plant Physiol Biochem 96:217–221. https://doi.org/10.1016/J.PLAPHY.2015.06.014

Yamamoto MP, Shinada H, Onodera Y, Komaki C, Mikami T, Kubo T (2008) A male sterility-associated mitochondrial protein in wild beets causes pollen disruption in transgenic plants. Plant J 54:1027–1036. https://doi.org/10.1111/j.1365-313X.2008.03473.x

Yang AF, Duan XG, Gu XF, Gao F, Zhang JR (2005) Efficient transformation of beet (*Beta vulgaris*) and production of plants with improved salt-tolerance. Plant Cell Tissue Organ Cult 83:259–270. https://doi.org/10.1007/S11240-005-6670-9

Yang L, Zhang Y, Zhu N, Koh J, Ma C, Pan Y, Yu B, Chen S, Li H (2013) Proteomic analysis of salt tolerance in sugar beet monosomic addition line M14. J Proteome Res 12:4931–4950. https://doi.org/10.1021/PR400177M

Yu ZH, Han YN, Xiao XG (2015) A PPO promoter from betalain-producing red Swiss chard, directs petiole- and root-preferential expression of foreign gene in anthocyanins-producing plants. Int J Mol Sci 16:27032–27043. https://doi.org/10.3390/IJMS161126011

Zabed H, Faruq G, Sahu JN, Azirun MS, Hashim R,1 and Boyce AN (2014) Bioethanol production from fermentable sugar juice. Sci World J 2014:957102 https://doi.org/10.1155/2014/957102

Zare B, Niazi A, Sattari R, Aghelpasand H, Zamani K, Sabet MS, Moshiri F, Darabie S, Daneshvar MH, Norouzi P, Kazemi-Tabar SK, Khoshnami M, Malboobi MA (2015) Resistance against rhizomania disease via RNA silencing in sugar beet. Plant Pathol 64:35–42. https://doi.org/10.1111/ppa.12239

Zhang CL, Chen DF, Kubalakova M, Zhang J, Scott NW, Elliott MC, Slater A (2008a) Efficient somatic embryogenesis in sugar beet (*Beta vulgaris* L.) breeding lines. Plant Cell Tissue Organ Cult 93:209–221. https://doi.org/10.1007/S11240-008-9364-2

Zhang CL, Xu DC, Jiang XC, Zhou Y, Cui J, Zhang CX, Chen DF, Fowler MR, Elliott MC, Scott NW, Dewar AM, Slater A (2008b) Genetic approaches to sustainable pest management in sugar beet (*Beta vulgaris*). Ann Appl Biol 152:143–156. https://doi.org/10.1111/J.1744-7348.2008.00228.X

Zhong X, Zhou Q, Cui N, Cai D, Tang G (2019) *BvcZR3* and $BvHs1^{pro\text{-}1}$ genes pyramiding enhanced beet cyst nematode (*Heterodera schachtii* schm.) resistance in oilseed rape (*Brassica napus* L.). Int J Mol Sci 20:1740. https://doi.org/10.3390/ijms20071740

Biotechnological Approaches in Sugar Beet Development

5

W. S. Philanim, Amit Kumar, and Nivedita Shettigar

Abstract

Sugar beet provides looming potential for sugar production globally supplementing sugarcane in the current scenario. The crop with the efficacy for bioethanol production from its pulp and molasses, minimal water requirement for its growth and shorter life cycle as compared to sugarcane is gaining importance. Its performance is influenced by various environmental and agronomic factors that ultimately decide the sugar yield. Genetic erosion of sugar beet is evident from the vast and prolonged use of varieties derived from similar parents. This hinders the selection process and renders it non-rewarding. The genetic diversity of the crop can be increased by the introgression of new alleles from its wild ancestors and wild relatives. Biotechnological tools like transgenics can help transfer the foreign gene even between two non-cross incompatible species. Effective genetic and genomic tools to screen and identify molecular tags conferring for important traits will help in the development of useful breeding material of sugar beet. Efforts to develop tolerance to biotic and non-biotic stress especially drought and cold is palpable. Genome sequencing through NGS and SMRT approaches helps in annotation of individual genes and deciphering phylogenic relationships among individuals. Incorporation of genetic transformation and *in vitro* technologies have been pertinent in producing salt-tolerant, herbicide-tolerant, disease-resistant, and pest-resistant cultivars.

Keywords

Genetic manipulations · OMICS · Regenerants · Sugar beet · Transgenics · Transcriptomics

W. S. Philanim (✉) · A. Kumar · N. Shettigar
Plant Breeding Section, Division of Crop Science, ICAR-RC-NEH Region, Umiam, Meghalaya, India

V. Misra et al. (eds.), *Sugar Beet Cultivation, Management and Processing*,
https://doi.org/10.1007/978-981-19-2730-0_5

Abbreviations

AFLP	Amplified fragment length polymorphism
BNYVV	Beet necrotic yellow vein virus
CLS	Cercospora leaf spot
GOX	Glyphosate oxidoreductase
GSH	Glutathione
LD	Linkage disequilibrium
NGS	Next-generation sequencing
PIC	Polymorphism information content
RAPD	Random amplified polymorphic DNA
RFLP	Restriction fragment length polymorphism
SBP	Sugar beet pulp
SSR	Microsatellite
StGCS-GS	*Streptococcus thermophilus* γ-glutamyl cysteine synthetase-glutathione synthetase

5.1 Introduction

Sugar beet (*Beta vulgaris* L.) is a crop of global importance that stands second in prominence after sugarcane (Brar et al. 2015) and contributes 20% to the world sugar production (FAO 2009). Cultivated beets belonging to family Chenopodiaceae is thought to have originated from its wild progenitor "sea beet" scientifically called *B. vulgaris subsp. maritima* (Biancardi et al. 2012). Formally, sugar beet was likely domesticated as a pot herb and consumed for its leaves as the first harvest from its wild progenitor, sea beet [*B. vulgaris* L. subsp. *maritima* (L.) *Arcang*] for food (Biancardi et al. 2012; Ford-Lloyd et al. 1975; Lange et al. 1999). Later, the roots were used both as medicinal herbs and vegetables (Biancardi et al. 2012; Goldman and Navazio 2008). Root type sugar beet and its enlarged root was earlier documented in the Near East (Turkey, Iran, and Iraq) and eventually spread to the west (Europe) (Zossimovich 1940). Sugar beet is becoming an essential biofuel alternative to fossil fuel energy (Zhang et al. 2008). Sugar is widely used as livestock feed supplement that is largely produced by the sugar industry along with sizeable amounts of molasses and sugar beet annually as by-products (Olmos and Hansen Zúñiga 2012; Kracher et al. 2014). Sugar beet pulp (SBP) and molasses hold great potential for the production of energy-efficient bioethanol due to its high content of readily fermentable sugars (Rodriguez et al. 2010; Maung and Gustafson 2011). The crop further provides useful feedstock for alcohol, yeast, and pharmaceutical companies. Sugar beet is considered to be originated from indigenous Mediterranean *B. maritima*, a relatively young crop possessing a narrow genetic base (van Geyt et al. 1990) and has undergone significant genetic improvements since its cultivation about 200 years ago (Draycott 2006). Wild beets have 4–6% of sucrose content whereas fodder beets have 12% sucrose content from which sugar beet was selected. The presently developed and cultivated cultivars have a much higher sugar content of 20% attributed to further improvements in the crop through conventional breeding. In India, it offers good potential to bridge the gap between projected and actual sugar production because of the high sugar content and production of useful by-products (Pathak et al. 2014). On a more recent development, advanced biotechnological methods alongside classical breeding approaches have been used to develop herbicide-tolerant, disease and pest-resistant cultivars. Sugar beet contest with sugarcane in sugar production at the global market. To compete with sugarcane and meet the high sugar demands of the global consumers, effective novel breeding technology and biotechnological interventions apart from the redundant breeding strategy are a necessity. Sugar beet diversity needs to be broadened by integrating wild alleles for useful traits from the wild species through skillful biotechnological methods as there exist crossability barriers between the cultivated and wild sugar beet species for effective selection and high-throughput molecular work establishment (Frese et al. 2001).

5.2 Molecular Studies and Advances in Sugar Beet

5.2.1 Genetic Diversity in Sugar Beet

Exhaustive selection over time and widespread adoption of a genetically uniform crop varieties resulted in genetic stagnation and loss of genetic diversity in cultivated crops that hamper further crop improvement programs. The wild ancestors and wild relatives carry important traits including pest and disease resistance, drought tolerance, cold tolerance, salt tolerance, and nutraceutical properties that are essentially needed by the crops for its survival and good performance (Zhang et al. 2016). It is therefore imperative to replenish the lost alleles from the breeding pools through introgression of useful genes from its wild species counterpart (Ordon et al. 2005). Understanding the genetic diversity of a crop helps in framing appropriate selection strategy and breeding schemes for the overall refinement of the crop. Total genetic diversity of sugar beet along with other *Beta* species including other cultivated beet crops and its wild relatives is fairly high (Fievet et al. 2007). The genetic diversity of sugar beet is established hitherto through morphological traits, isozymes, and molecular marker study. Study of the sugar beet diversity with DNA marker systems such as RFLP (Fragment Length Polymorphism), RAPD (Random Amplified Polymorphic DNA), and AFLP (Amplified Fragment Length Polymorphism) have been done in the early and mid-1990s (Jung et al. 1993; Barzen et al. 1995; Schondelmaier et al. 1996). Earlier attempts were made to understand the genetic relationship in *Beta Vulgaris* including table beet, sugar beet, and Swiss chard crop types using RAPD markers that revealed that table beet inbred lines clustered in an intermediate position between standard table beet germplasm and breeding lines of sugar beet, probably due to their origin from an introgression program designed to incorporate sugar beet genes (Wang and Goldman 1999). Linkage drag from introgressed genes from sugar beet to table beet during the 1950s and 1960s might have caused a larger genetic distance between inbred lines derived from sugar beet and standard table beet (Goldman 1996).Genetic diversity study of 14 individual sugar beet plants within each parent analyzed using 18 microsatellites (SSR) markers revealed 75.5% of total phenotypic variation explained by the first two principal components (43 and 32.6% PV) for agro-morphological traits that could distinguish salinity-tolerant and drought-tolerant parents. Molecular analysis through SSR revealed 104 total alleles with 5.7 average number of alleles per primer pair and an average polymorphism information content (PIC) of 0.64 with the highest PIC belonging to ESTSSR *FDSB502* (Abbasi et al. 2014). A total of 243 amplicons were obtained which were further grouped into 88 alleles with an average of 17.36 amplicons/primer with distinct molecular weight ranging from 124 to 1222 bp and 4–10 alleles/SSR locus with moderate to high PIC ranging from 0.625 to 0.851 (Srivastava et al. 2017). Efforts were made to understand the genetic diversity of sugar beet pollinators. The total of alleles obtained were 129 alleles with an average of 3.2 alleles per SSR marker. The observed heterozygosity ranged from 0.00 to 0.87 (mean = 0.30). Expected heterozygosity and Shannon's information index and expected heterozygosity were highest for markers SB15s and FDSB502s and lowest for marker BQ590934.The same markers with PIC values of 0.70 and 0.69, respectively, were found most informative and were able to distinguish between genotypes. Maximum private alleles were identified in pollinator EL0204; two private alleles in C51pollinator; and one allele in NS1pollinators, C93035, and FC221. Intrapopulation variability (variation within the population) govern 77.34% of the total genetic variation resulting from molecular variance analysis (Taški-Ajduković et al. 2017). Extensively shared, non-unique genetic variation among different species of beets was attributed to the distribution of genetic variation in sugar beet. The phenomenon of apomorphy deciphered shared lineages within each species while differentiation within strong crop types was supported by principal components analysis. Sharing common ancestor and gene flow among the crop types through time indicated sharing of genome variation likely for important phenotypic characters that concealed a good demarcation of different species of beets. Table beet revealed greater genetic differentiation within the crop types. Table beet groups were well differentiated in comparison to the sugar beet species (Galewski and McGrath 2020).

5.2.2 OMICS Approaches in Sugar Beet

OMICS techniques encompass genomics, transcriptomics, proteomics, and metabolomics that functions to realize the molecular and biochemical structure and pathways of a plant genotype and effectively improve the crop for its overall usability (Fig. 5.1). In recent times, genomics evidence based on Next-Generation Sequencing (NGS), gene silencing, gene-editing systems, and over-expression methods have given a huge repository of genetic output to aid in deciphering both biotic and abiotic tolerance mechanisms in plants (Saad et al. 2013; Shan et al. 2013; Yin et al. 2014). An OMICS-driven unearthing of novel genes, proteins, and

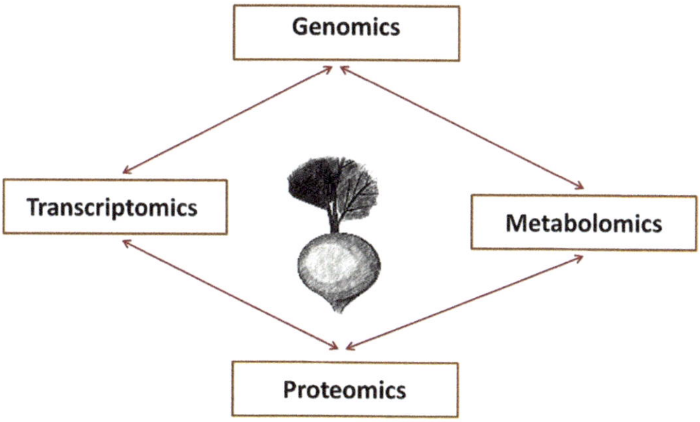

Fig. 5.1 Outline of OMICS studies in sugar beet

metabolites in sugar beet has aid in understanding the complex mechanisms underlying phenomena such as apomixis and tolerance to biotic and abiotic stresses. The knowledge harnessed is valuable for improving the tolerance of *B. vulgaris* to biotic and abiotic stresses and yield improvement of sugar beet for energy and food production (Zhang et al. 2016).

5.2.2.1 Genome Mapping for Useful Traits in *B. vulgaris*

Beta vulgaris is a diploid plant of $2n = 18$ chromosome number with an estimated genome size of 714–758 megabases. Efforts to genome map the chromosomes of sugar beet have been carried out (Laurent et al. 2007). The first reported linkage map in *B. vulgaris* was on the inheritance of the morphological markers for hypocotyl color (genes R and Y) and bolting behavior (B, annual vs. biennial), widely known as R–Y–B linkage association (Keller 1936; Owen and Ryser 1942), which is now mapped on Chromosome 2 of the Butterfass chromosome series. The crop shares an ancient genome triplication with other eudicot plants. The phylogenetic study revealed losses of gene family according to their lineages and further expansions and differentiation of Caryophyllales prior to the split of asterids and rosids (Dohm et al. 2014). The first linkage map with wide crosses in *B. vulgaris* between sugar beet and table beet mapped 23 new SSR makers (McGrath et al. 2007).

Leaf spot is known as one of the most widespread and devastating foliar diseases of sugar beet. It destroys the plant foliar structure and function and causes necrotic lesions (Holtschulte 2000). Further sugar recovery and yield of the sugar beet are greatly decreased by the disease. Four QTLs *viz*., qcr1, qcr4 qcr2, and qcr3 on chromosomes 3, 9, 4, and 6 underlying resistance to *Cercospora leaf spot* (CLS) was revealed through Composite Interval Mapping of RILs developed from a cross between a resistant line ("NK-310 mm-O") and a susceptible line ("NK-184 mm-O") (Taguchi et al. 2011). Another serious disease in sugar beet is Rhizomania, caused by Beet necrotic yellow vein virus (BNYVV) that lessens the sugar content and yield of beet. *Rz4,* a major QTL conferring resistance to BNYVV that explained

78% of the observed phenotypic variation was deciphered. RAPD marker Rz1 was mapped close to Rz4 in chromosome 3 which is also the previously identified mapped location for BNYVV resistance genes *Rz1*, *Rz2*, and *Rz3* (Lewellen et al. 1987; Paul et al. 1993; Scholten et al. 1996; Grimmer et al. 2007).

Association mapping is budding as a novel molecular tool in plant genomics (Myles et al. 2009) and is currently used in the molecular analysis of populations from applied breeding programs (Reif et al. 2010; Würschum et al. 2011). The technique helps in identifying major and minor QTLs that confers the traits of interest. It will be pivotal to acknowledge the existence of inherent population structure in the plant populations that may pose a potential problem while running the analysis. Presence of any non-functional correlations between the population structure and the trait will be projected as QTL (Zhao et al. 2007). Association mapping is based on the concept of linkage disequilibrium (LD), a non-random association of alleles of different loci between the QTL, and examined molecular markers associated with the trait. Linkage disequilibrium is an accurate indicator of the population genetic forces that structure a genome. Association mapping for traits is anticipated to have higher mapping resolution in contrast to classical linkage mapping as it excavates all the historical recombination events in the mapping population. The strength and extent of LD is dependent on the structure of the population, therefore, is population-specific and influenced by many genetic factors (Flint-Garcia et al. 2003). Moreover, the LD strength is highly variable across the genome. The extent of association between the QTL and marker determines the power and precision in detecting QTL conferring for the trait. The association is measured by r^2 value which establishes the marker and QTL correlation. Lower r^2 values will only allow the discernment of QTL with large effects (major QTLs) whereas high r^2 values are requisite to detect medium and small size QTL. LD is expected to be higher in the plant breeding population in contrast to the natural populations on account of the shorter history of the germplasms and selection of favorable genotypes over time. Trait associated markers with explained genotypic variance and QTL in *B. vulgaris* for important characters viz., nitrogen content, sodium content, potassium content, the proportion of impurities, sugar content, white sugar content, beet yield, root yield, sugar yield, and white sugar yield were studied (Weber et al. 1999; Schneider et al. 2002; Reif et al. 2010; Stich et al. 2008a, b, Würschum et al. 2011).

5.2.2.2 Next-Generation Sequencing and Other Sequencing Applications in Sugar Beet

The NGS technology has provided a platform for locating molecular tags of trait phenotype accurately. It has effectively aided forward genetics in the discerning causative variation of a phenotype easy and precise. NGS technologies have made molecular study easier offering high-throughput sequencing data as compared to Sanger sequencing with a 99% read accuracy. NGS also reduces the cost incurred in sequencing in comparison to sangers making the genomic study more affordable. The whole-genome sequencing of sugar beet was completed and reported by Dohm et al. (2014). Based on transcription data and sequence homology annotation of the genome, a total of 27,421 protein-coding genes were envisaged (Dohm et al. 2014). Reports on the complete sequence of mitochondrial genome ((Kubo et al. 2000) and chloroplast genome (Li et al. 2014; Stadermann et al. 2015) of sugar beet (*Beta vulgaris* L.) are available. The genome size of Mt is about 368,799 bp encompassing 29 proteins, 25 Trna, and 5 Rrna, also found in *Arabidopsis thaliana.* A novel $tRNA^{cys}$gene (trnc2-GCA) was deciphered that actually transcribes into mature Trna unlike the native $tRNA^{cys}$ gene (trnc1-GCA) that functions as a pseudogene (Kubo et al. 2000). SMRT sequencing of the sugar beet chloroplast genome revealed 79 genes encoding for an mRNA (i.e., proteins), 7 encode rRNA, and 28 encoding tRNAs in a total of 114 individual genes. Nine genes were located within the inverted repeat (IR) regions that conferred 5 mRNAs, 3 tRNAs, and 1 rRNA (Stadermann et al. 2015).

5.2.2.3 Transcriptomics and Proteomics Study in Sugar Beet

Transcriptomics and proteomics study revealed differentially expressed proteins involved in several processes and various biological pathways (Li et al. 2009; Zhu et al. 2009). A study on salt stress through proteomics revealed the involvement of cystatin (Wang et al. 2012), glyoxalase I (Wu et al. 2013), CCoAOMT, and thioredoxin peroxidase (Zhang et al. 2016) in salt resistance mechanism of M14, a high salt tolerance monosomic addition line of sugar beet. Proteins regulating drought stress through oxidative stress, signal transduction, and redox regulation were identified (Hajheidari et al. 2005). Genetic and non-genetic SSR has been deciphered in sugar beet through transcriptomics that has a good amount of polymorphism and demarcates clearly between genotypes. Forty of such primer-pairs were revealed with high polymorphic distinguished diversity present among eight diverse *B. vulgaris* genotypes. The transcriptomic data and identified SSR markers will make useful public domain genomic resources for understanding functional elements of the genome of sugar beet. It will further facilitate RNA-sequencing-based expression research, enable the discovery of novel genes, and propel selective breeding and genetic research in sugar beet (Fugate et al. 2014).

5.2.2.4 Genetic Manipulation Through Transgenics in *Beta vulgaris*

Non-crossability among different species has driven the wheel of transgenics where a foreign gene of interest is transported through a medium like bacterial pathogen *Agrobacterium tumefaciens* to the genome targeted for incorporation and expression of the trait in the host plant. Stable integration and safe transformation of the transferred DNA are essential in the plant nucleus for the successful expression of the trait. Alternatively, transient transformation may occur wherein the foreign DNA does not integrate but transiently remain in the nucleus and is transcribed to produce desirable gene products. *Agrobacterium tumefaciens* is an essential core tool of plant biotechnology and numerous interactions with plants studied and elucidated (Hwang et al. 2017). In sugar beet, transformation is achieved for some traits and illustrated by different studies for *A. tumefaciens* transformation (D'halluin et al. 1992; Elliott et al. 1996; Krens et al. 1996) and peg-mediated guard cell protoplast transformation (Hall et al. 1995). Progress through transformation techniques using *A. tumefaciens*-

mediated transformation has found success in sugar beet (Fry et al. 1991; Konwar 1994). Stable transformation is shown to be dependent on different factors including genotype (von Wordragen and Dons 1992) acetosyringone or phenolic compounds present in the plant tissue (Jacq et al. 1992). Expression of the introduced gene is determined by the transgene copy number that further enables their positive or negative association (Hobbs et al. 1993; Linn et al. 1990; Matzke and Matzke 1993).

Sugar beet is moderately salt tolerant. Lack of efficient gene transformation has limited the breeding of varieties in saline conditions for salt tolerance. Positive transformation of *GUS* gene in sugar beet is reported and has shown effective expression through *Agrobacterium*-mediated transformation (Lindsey and Gallois 1990; Krens et al. 1996; Hisano et al. 2004). Further, improved salt tolerance was observed in transgenic sugar beets expressing *AtNHX1* gene (Yang et al. 2005). The constitutive expression of *AtNHX3 gene in* sugar beet provided salt tolerance and improved sugar synthesis in transgenic plants.

Efforts have been put forth to develop glyphosate resistance sugar beet through genetic transformation. The chemical name of glyphosate is N-(phosphonomethyl)glycine, an active ingredient for the herbicide Roundup. Two transformants (HIAB1: 1 and HIAB2: 2) introduced with *CP4 EPSPS* gene showed high tolerance to Roundup that did not manifest any phytotoxic or morphological effects after treatment with the maximum dose of glyphosate (Mannerlöf et al. 1997). Reports on the transformation of *glyphosate oxidoreductase* (GOX) for tolerance to herbicide were also given (Steen and Pedersen 1993; Steen and Pedersen 1995a, b; Brants et al. 1995; Tenning et al. 1995; Mannerlöf et al. 1997).

Heavy metal pollution poses a serious environmental threat globally. The phytoremediation process is viewed as an ideal curbing mechanism to ameliorate heavy metal pollution given its high efficiency and absence of secondary environmental pollution. Phytoremediation should have higher proliferation rates in vivo, high biomass, and faster growth. Three transgenic sugar beet (*Beta vulgaris* L.) lines (s2, s4, and s5) introduced with novel *Streptococcus thermophilus* γ-glutamyl cysteine synthetase-glutathione synthetase (StGCS-GS) that synthesizes *glutathione* (GSH) gives enhanced tolerance to different concentrations of zinc, cadmium, and copper. These transformed lines have increased root length, biomass, and relative growth in comparison to wild-type plants (Liu et al. 2015).

5.2.3 Plant Tissue Culture Techniques in Sugar Beet

Plant tissue culture is an indispensable component of plant biotechnology. Tissue culture is becoming an alternative in vitro means to vegetative propagation of plants. As in vitro plants are propagated in sterile conditions, it is essentially free from bacterial and fungal diseases and can be reproduced at a faster rate in cultures. The individual plants produced through tissue culture are highly uniform within a clone population that allows commercial production of clonal cultivars (Krishna and Singh 2013). The presence of genetic variation however is seen in isolated protoplasts, undifferentiated cells, calli, tissues, and morphological characters of in vitro-raised plants (Bairu et al. 2011; Currais et al. 2013). Apart from being a useful biotechnological tool, plant tissue culture approaches have gained industrial importance in recent years for plant propagation, plant improvement, production of secondary metabolites, and disease elimination (Hussain et al. 2012). Further, in vitro cultures can help understand the physiological mechanism of injury caused by environmental stress (Dix et al. 1983; Van Swaaij et al. 1986).

5.2.3.1 Sugar Beet Micropropagation

Micropropagation can be obtained within a short period of time in a confined space (Krishna et al. 2008). In sugar beet, limited in vitro culture techniques are available despite the importance of the crop which is unfortunate. Shoot cultures maintained in vitro (Hussey and Hepher 1978), but regenerated from callus (Saunders and Daub 1984; Tetu et al. 1987; Freytag et al. 1988; Ritchie et al. 1989) tends to be inconsistent, occurring at low frequency and strongly cultivar dependent that limits its usability either for in vitro selection or clonal propagation. Success however has been reported in some cultivars where it was possible to obtain regenerated lines from hormone-treated autonomous cell cultures (Van Geyt and Jacobs 1985). Most of the undifferentiated culture regeneration is seen from adventitious shoot initiation and seldom from somatic embryos (Freytag et al. 1988). Protoplast culture and plant regeneration have also been seen rarely as the process is highly genotype dependent. The first successful culture has been reported in diploid beet (Krens et al. 1990). Direct organogenesis has been reported as the most effective way to produce true-type regenerants in sugar beet (Bekheet et al. 2007). Micropropagation of sugar beet has been carried successfully with a good percentage of regenerants (Mikami et al. 1985; Goska and Szota 1992; Sullivan et al. 1993; Grieve et al. 1997; Bekheet et al. 2007; Morsi et al. 2019).

5.2.3.2 Somaclonal Variation in Sugar Beet

Somaclonal variation, a term coined by Larkin and Scowkraft in 1981, denotes plant variants derived from any form of cell or tissue culture. Genetic variability is obtained quicker through tissue culture without any sophisticated technology. An added advantage is that the screening for desirable traits can be obtained in lesser time and space. Somaclones have ample applications in genetic improvements and recovery of novel variants with enhanced characteristics. Suitable in vitro selection might further aid the recovery of novel variants (Jain 2001; Lestari 2006). Somaclonal variants in sugar beet are most commonly seen through indirect regeneration from callus derived from petiole, leaf lamina, or hypocotyl explants (Saunders and Doley 1986; Brears et al. 1989; Jacq et al. 1992). There are reports also on protoplasts regeneration (Steen et al. 1986; Lenzner et al. 1995; Jazdzewska et al. 2000) and direct regenerants from explants (Harms et al. 1983; Dikalova et al. 1993; Zhong et al. 1993). Somaclonal variation in sugar beet for root rot resistance *F. oxysporum var. orthoceras* was reported with a frequency of shoot depending on the genotype of 1.0–12.5% and multiple shoot formations on the explants (Urazaliev et al. 2013). Rearrangements of mitochondrial DNA induced by cell suspension, culture, and regeneration were also reported.

5.3 Future Prospects

The OMICS information can further be applied to improve sugar beet stress tolerance and enhance yield and energy output (bioethanol) with an accumulation of useful metabolites, for example, betalains and glycine betaines.

5.4 Conclusion

Biotechnological intervention and the genomic study provide in-depth information on the whole genome of sugar beet and the structure and functions of genes underlying useful agronomical traits. OMICS study helps understand the molecular workings and biosynthetic pathways involved in response to tolerance to biotic and abiotic stress in sugar beet. Genomic information helps facilitate and engineer important metabolites. Apomixis and stress tolerance mechanism has been studied to great extend in unique sugar beet germplasm M14 through proteomics and transcriptomics to identify the genes and proteins underlying this traits. Transformation study has been successful in constitutive expression of *AtNHX3* gene for salt tolerance, CP4 EPSPS gene for tolerance to Roundup, and novel StGCS-GS that synthesizes GSH for phytoremediation. However, poor transformation success, expression of the gene due to low regeneration, genotype dependency, and practical applications of in vitro culture technologies in sugar beet being still at nascent stage limits sugar beet research and improvement. This stipulates ample scope for the application of sugar beet, a high economic value crop in food, bioenergy, and pharmaceutical industries through progressive genetic study and effective biotechnological protocol.

References

Abbasi Z, Arzani A, Majidi MM (2014) Evaluation of genetic diversity of sugar beet (*Beta vulgaris* L.) crossing parents using agro-morphological traits and molecular markers. J Agric Sci Technol 16(6):1397–1411

Bairu MW, Aremu AO, Staden JV (2011) Somaclonal variation in plants: causes and detection methods. Plant Growth Regul 63:147–173

Barzen E, Mechelke W, Ritter E, Schulte-Kappert E, Salamini F (1995) An extended map of the sugar beet genome containing RFLP and RAPD loci. Theor Appl Genet 90:89–193

Bekheet SA, Taha HS, Matter MA (2007) In vitro regeneration of sugar beet propagules and molecular analysis of the regenerants. Arab J Biotechnol 10(2):321–332

Biancardi E, Panella LW, Lewellen RT (2012) Beta maritima, the origin of beets. Springer Science +Business Media, New York, ISBNs 978-1-4614-0841-3, 978-1-4614-0842-0, 978-1-4899-9961-0. https://doi.org/10.1007/978-1-4614-0842-0

Brants I, Steen P, Bisgaard S, Pedersen HC (1995) Roundup ready sugar beet. Proc IIRB Congress 58:557–559

Brar NS, Dhillon BS, Saini KS, Sharma PK (2015) Agronomy of sugar beet cultivation-a review. Agric Rev 36(3):184–197

Brears T, Curtis GJ, Lonsdale DM (1989) A specific rearrangement of mitochondrial DNA induced by tissue culture. Theor Appl Genet 77:620–624

Currais L, Loureiro J, Santos C, Canhoto JM (2013) Ploidy stability in embryogenic cultures and regenerated plantlets of tamarillo. Plant Cell Tissue Organ Cult 114:149–159

D'Halluin K, Bossut M, Bonne E, Mazur B, Leemans J, Botterman J (1992) Transformation of sugar-beet (*Beta vulgaris* L.) and evaluation of herbicide resistance in transgenic plants. Biotechnology 10:309–314

Dikalova AE, Dudareva NA, Kubalakova M, Salganik RI (1993) Rearrangements in sugar beet mitochondrial DNA induced by cell suspension, callus cultures and regeneration. Theor Appl Genet 86:699–704

Dix PJ, Plunkett A, Toth I (1983) The use of tissue cultures for investigating the physiology of salt stress. In: Cassels AC, Kavanagh JA (eds) Plant tissue culture in relation to biotechnology. Royal Irish Academy, Dublin, pp 95–104

Dohm JC, Minoche AE, Holtgräwe D, Capella-Gutiérrez S, Zakrzewski F, Tafer H, Rupp O, Sörensen TR, Stracke R, Reinhardt R, Goesmann A (2014) The genome of the recently domesticated crop plant sugar beet (*Beta vulgaris*). Nature 505(7484):546–549

Draycott AP (2006) Introduction. In: Draycott AP (ed) Sugar beet. Blackwell Publishing Ltd., pp 1–8

Elliott MC, Chen DF, Fowler MR, Kirby MJ, Kubalakova M, Scott NW, Slater A (1996) Transgenesis—a scheme for improving sugar beetproductivity. Russ J Plant Physiol 43:544–551

FAO (2009) Sugar beet white sugar. agribusiness hanbook. European Bank and FAO, Rome

Fievet V, Touzet P, Arnaud JF, Cuguen J (2007) Spatial analysis of nuclear and cytoplasmic DNA diversity in wild sea beet (*Beta vulgaris* ssp. *maritima*) populations: do marine currents shape the genetic structure? Mo l Ecol 16:1847–1864

Flint-Garcia SA, Thornsberry JM, Buckler ES IV (2003) Structure of linkage disequilibrium in plants. Annu Rev Plant Biol 54(1):357–374

Ford-Lloyd BV, Williams ALS, Williams JT (1975) A revision of *Beta* section *vulgares* (*Chenopodiaceae*), with new light on the origin of cultivated beets. Bot J Linn Soc 71:89–102. https://doi.org/10.1111/j.1095-8339.1975.tb02448.x

Frese L, Desprez B, Ziegler D (2001) Potential of genetic resources and breeding strategies for base-broadening in Beta (Chapter 17). In: Cooper HD, Spillane C, Hodgkin T (eds) Broadening the genetic base of crop production. IPGRI/FAO, London, pp 295–309

Freytag AH, Anand SC, Rao-Areili AP, Owens LD (1988) An improved medium for adventitious shoot formation and callus induction in *Beta vulgaris* L. in vitro. Plant Cell Rep 7:30–34

Fry JE, Barnason AR, Hinchee M (1991) Genotype independent transformation of sugar beet using *Agrobacterium tumefaciens*. Third International Congress of plant mol. biol., Tuscon, Arizona, USA

Fugate KK, Fajardo D, Schlautman B, Ferrareze JP, Bolton MD, Campbell LG, Wiesman E, Zalapa J (2014) Generation and characterization of a sugar beet transcriptome and transcript based SSR markers. The Plant Genome 7(2):2013–2011

Galewski PJ, McGrath JM (2020) Genetic diversity among cultivated beets (*Beta vulgaris*) assessed via population-based whole genome sequences. BMC Genomic 21:189

Goldman IL (1996) Inbred line and open-pollinated population releases from the University of Wisconsin beet breeding program. Hort Sci 31:880–881

Goldman IL, Navazio JP (2008) Table beet. In: Prohens J, Nuez F (eds) Vegetables I. Asteraceae, Brassicaceae, Chenopodicaceae, and Cucurbitaceae. Handbook of plant breeding. Springer, New York, pp 219–238

Goska M, Szota M (1992) Micropropagation of sugar beet [*Beta vulgaris L.*] trisomics in vitro culture. Genetica Polonica 33(2):115–118

Grieve TM, Gartland KMA, Elliott MC (1997) Micropropagation of commercially important sugar beet cultivars. Plant Growth Regul 21(1):15–18

Grimmer MK, Trybush S, Hanley S, Francis SA, Karp A, Asher MJC (2007) An anchored linkage map for sugar beet based on AFLP, SNP and RAPD markers and QTL mapping of a new source of resistance to beet necrotic yellow vein virus. Theor Appl Genet 114(7):1151–1160

Hajheidari M, Abdollahian-Noghabi M, Askari H, Heidari M, Sadeghian SY, Ober ES (2005) Proteome analysis of sugar beet leaves under drought stress. Proteomics 5:950–960. https://doi.org/10.1002/pmic.200401101

Hall RD, Verhoeven HA, Krens FA (1995) Computer-assisted identification of protoplasts responsible for rare division events reveals guard-cell totipotency. Plant Physiol 107:1379–1386

Harms CT, Baktir I, Oertli JJ (1983) Clonal propagation in vitro of red beet (*Beta vulgaris* ssp.) by multiple adventitious shoot formation. Plant Cell Tissue Organ Cult 2:93–102

Hisano H, Kimoto Y, Hayakawa H, Takeichi J, Domae T, Hashimoto R, Abe J, Asano S, Kanazawa A, Shimamoto Y (2004) High frequency *Agrobacterium*-mediated transformation and plant regeneration via direct shoot formation from leaf explants in *Beta vulgaris* and *Beta maritima*. Plant Cell Rep 22:910–918

Hobbs SLA, Warkentin TD, De Long CMO (1993) Transgenic copy number can be positively or negatively associated with transgene expression. Plant Mol Biol 21:17–26

Holtschulte B (2000) *Cercospora beticola*–worldwide distribution and incidence. Cercospora beticola 2:5–16

Hussain A, Qarshi IA, Nazir H, Ullah I (2012) Plant tissue culture: current status and opportunities. In: Recent advances in plant in vitro culture, pp 1–28

Hussey G, Hepher A (1978) Clonal propagation of sugar beet plants and the formation of polyploids by tissue culture. Ann Bot 42:477–479

Hwang HH, Yu M, Lai EM (2017) Agrobacterium-mediated plant transformation: biology and applications. Arabidopsis Book 2017(15):e0186. The American Society of Plant Biologists. https://doi.org/10.1199/tab.0186

Jacq B, Tetu T, Sangwan RS, De Laat A, Sangwan-Norreel BS (1992) Plant regeneration from sugar beet (*Beta vulgaris* L.) hypocotyls cultured in vitro and flow cytometric nuclear DNA analysis of regenerants. Plant Cell Rep 11:329–333

Jain SM (2001) Tissue culture-derived variation in crop improvement. Euphytica 118:153–166

Jazdzewska E, Sadoch Z, Niklas A, Majewska-Sawka A (2000) Plant regeneration from sugar beet leaf protoplasts: analysis of shoots by DNA fingerprinting and restriction fragment length polymorphism. Can J Bot 78(1):10–18

Jung C, Pillen K, Frese L, Fahrand S, Melchinger AE (1993) Phylogenetic relationships between cultivated and wild species of the genus *Beta revealed* by DNA "fingerprinting". Theor Appl Genet 86:449–457

Keller W (1936) Inheritance of some major colour types in beets. J Agric Res (Washington, DC) 52: 27–38

Konwar BK (1994) *Agrobacterium tumefaciens*-mediated genetic transformation of sugar beet (*Beta vulgaris* L.). J Plant Biochem and Biotech 3:37–41

Kracher D, Oros D, Yao W, Preims M, Rezic I, Haltrich D et al (2014) Fungal secretomes enhance sugar beet pulp hydrolysis. Biotechnol J 9:483–492. https://doi.org/10.1002/biot.201300214

Krens FA, Jamar D, Rouwendal GJA, Hall RD (1990) Transfer of cytoplasm from new Beta CMS sources to sugar beet by asymmetric fusion. I. Shoot regeneration from mesophyll protoplasts and characterisation of regenerated plants. Theor Appl Gen 79:390–396

Krens FA, Trifonova A, Keizer LCP, Hall RD (1996) The effect of exogenously applied phytohormones on gene transfer efficiency in sugar beet (*Beta vulgaris* L.). Plant Sci 116:97–106

Krishna H, Singh D (2013) Micropropagation of lasora (CordiamyxaRoxb.). Indian J Hortic 70: 323–327

Krishna H, Sairam RK, Singh SK, Patel VB, Sharma RR, Grover M, Nain L, Sachdeva A (2008) Mango explants browning: effect of ontogenic age: mycorrhization and pre-treatments. Sci Hortic 118:132–138

Kubo T, Nishizawa S, Sugawara A, Itchoda N, Estiati A, Mikami T (2000) The complete nucleotide sequence of the mitochondrial genome of sugar beet *(Beta vulgaris L.)* reveals a novel gene for tRNACys (GCA). Nucleic Acids Res 28(13):2571–2576

Lange W, Brandenburg WA, de Bock TSM (1999) Taxonomy and cultonomy of beet (*Beta vulgaris* L.). Bot J Linn Soc 130:81–96. https://doi.org/10.1111/j.1095-8339.1999.tb00785.x

Larkin P, Scowcroft W (1981) Somaclonal variation—a novel source of variability from cell cultures for plant improvement. Theor Appl Genet 60:197–214

Laurent V, Devaux P, Thiel T, Viard F, Mielordt S, Touzet P, Quillet MC (2007) Comparative effectiveness of sugar beet microsatellite markers isolated from genomic libraries and GenBank ESTs to map the sugar beet genome. Theor Appl Gen 115(6):793–805

Lenzner S, Zoglauer K, Schieder O (1995) Plant regeneration from protoplasts of sugar beet (Beta vulgaris). Physiol Plant 94:342–350

Lestari EG (2006) In vitro selection and somaclonal variation for biotic and abiotic stress tolerance. Biodiversitas 7:297–301

Lewellen RT, Skoyen IO, Erichsen AW (1987) Breeding sugar beet for resistance to rhizomania: evaluation of host-plant reactions and selection for and inheritance of resistance. In: Proceedings of the IIRB 50th Congress, pp 139–156

Li H, Cao H, Wang Y, Pang Q, Ma C, Chen S (2009) Proteomic analysis of sugar beet apomictic monosomic addition line M14. J Proteome 73:297–308. https://doi.org/10.1016/j.jprot.2009.09.012

Li H, Cao H, Cai YF, Wang JH, Qu SP, Huang XQ (2014) The complete chloroplast genome sequence of sugar beet *(Beta vulgaris ssp. vulgaris)*. Mitochondrial DNA 25:205–211

Lindsey K, Gallois P (1990) Transformation of sugar beet (*Beta vulgaris* L.) by *Agrobacterium tumefaciens*. J Exp Bot 41:529–536

Linn F, Heidmann I, Saedler H, Meyer P (1990) Epigenetic changes in the expression of the maize A1 gene in petunia hybrida: role of numbers of integrated gene copies and state of methylation. Mol Gen Genet 222:329–336

Liu D, An Z, Mao Z, Ma L, Lu Z (2015) Enhanced heavy metal tolerance and accumulation by transgenic sugar beets expressing *Streptococcus thermophilus* StGCS-GS in the presence of Cd, Zn and Cu alone or in combination. PLoS One 10(6):e0128824

Mannerlöf M, Tuvesson S, Steen P, Tenning P (1997) Transgenic sugar beet tolerant to glyphosate. Euphytica 94(1):83–91

Matzke MA, Matzke AJM (1993) Genomic imprinting in plants: parental effects and trans-inactivation phenomena. Annu Rev Plant Physiol Plant Mol Biol 44:53–76

Maung TA, Gustafson CR (2011) The economic feasibility of sugar beet biofuel production in central North Dakota. Biomass Bioenergy 35:3737–3747. https://doi.org/10.1016/j.biombioe.2011.05.022

McGrath JM, Trebbi D, Fenwick A, Panella L, Schulz B, Laurent V, Barnes S, Murray SC (2007) An open source first generation molecular genetic map from a sugar beet× table beet cross and its extension to physical mapping. Crop Sci 47:S-27

Mikami T, Kinoshita T, Saito H (1985) Clonal propagation of sugar beet plants by apical meristem culture. J Fac Agric Hokkaido Univ 62(3):325–331

Morsi NA, El-Gabry YA, Abu-Ellail FF (2019) Indirect regeneration tissue culture and molecular characterization for some sugar beet *(Beta vulgaris L.)* genotypes. Middle East J 8(1):187–199

Myles S, Peiffer J, Brown P, Ersoz E, Zhang Z, Costich D, Buckler E (2009) Association mapping: critical considerations shift from genotyping to experimental design. Plant Cell 21(8): 2194–2202

Olmos JC, Hansen Zúñiga ME (2012) Enzymatic depolymerization of sugar beet pulp, production and characterization of pectin and pectic-oligosaccharides as a potential source for functional carbohydrates. Chem Eng J 192:29–36. https://doi.org/10.1016/j.cej.2012.03.085

Ordon F, Ahlemeyer J, Werner K, Köhler W, Friedt W (2005) Molecular assessment of genetic diversity in winter barley and its use in breeding. Euphytica 146:21–28

Owen FV, Ryser GK (1942) Some Mendelian characters in *Beta vulgaris* L. and linkages observed in the Y-R-B group. J Agric Res (Washington, DC) 65:153–171

Pathak AD, Kapur R, Solomon S, Kumar R, Srivastava S, Singh PR (2014) Sugar beet: a historical perspective in Indian context. Sugar Tech 16:125–132

Paul HB, Henken B, Scholten OE, Lange W (1993) Use of zoospores of *Polymyxa betae* in screening beet seedlings for resistance to beet necrotic yellow vein virus. Neth J Plant Pathol 99(3):151–160

Reif JC, Liu W, Gowda M, Maurer HP, Möhring J, Fischer S, Schechert A, Wurschum T (2010) Genetic basis of agronomically important traits in sugar beet (Beta vulgaris L.) investigated with joint linkage association mapping. Theor Appl Genet 8:1489–1495

Ritchie GA, Short KC, Davey MR (1989) In vitro shoot regeneration from callus, leaf axils and petioles of sugar beet *(Beta vulgaris* L.). J Exp Bot 40:277–284

Rodriguez LA, Toro ME, Vazquez F, Correa-Daneri ML, Gouiric SC, Vallejo MD (2010) Bioethanol production from grape and sugar beet pomaces by solid-state fermentation. Int J Hydrog Energy 35:5914–5917. https://doi.org/10.1016/j.ijhydene.2009.12.112

Saad ASI, Li X, Li HP, Huang T, Gao CS, Guo MW, Cheng W, Zhao GY, Liao YC (2013) A rice stress-responsive NAC gene enhances tolerance of transgenic wheat to drought and salt stresses. Plant Sci 203:33–40

Saunders JW, Daub ME (1984) Shoot regeneration from hormone-autonomous callus from shoot culture of several sugar beet *(Beta vulgaris* L.) genotypes. Plant Sci Lett 34:219–223

Saunders JW, Doley WP (1986) One-step shoot regeneration from callus of whole plant leaf explants of sugar beet lines and a somaclonal variant for in vitro behaviour. J Plant Physiol 124:473–479

Schneider K, Schäfer-Pregl R, Borchardt C, Salamini F (2002) Mapping QTLs for sucrose content, yield and quality in a sugar beet population fingerprinted by EST-related markers. Theor Appl Genet 104:1107–1113

Scholten OE, Jansen RC, Keizer LCP, De Bock TSM, Lange W (1996) Major genes for resistance to beet necrotic yellow vein virus (BNYVV) in *Beta vulgaris*. Euphytica 91:331–339

Schondelmaier J, Steinrücken G, Jung C (1996) Integration of AFLP markers into a linkage map of sugar beet (*Beta vulgaris* L.). Plant Breed 115:231–237

Shan Q, Wang Y, Li J, Zhang Y, Chen K, Liang Z, Zhang K, Liu J, Xi JJ, Qiu JL, Gao C (2013) Targeted genome modification of crop plants using a CRISPR-Cas system. Nat Biotechnol 31(8):686–688

Srivastava S, Pathak AD, Kumar R, Joshi BB (2017) Genetic diversity of sugar beet genotypes evaluated by microsatellite DNA markers. J Environ Biol 38(5):777

Stadermann KB, Weisshaar B, Holtgräwe D (2015) SMRT sequencing only *de novo* assembly of the sugar beet (*Beta vulgaris*) chloroplast genome. BMC Bioinformatics 16:295. https://doi.org/10.1186/s12859-015-0726-6

Steen P, Pedersen HC (1993) Gene transfer for herbicide resistance. J Sugar Beet Res 30:267–274

Steen P, Pedersen HC (1995a) Yield and quality characters in transgenic herbicide tolerant sugar beet *(Beta vulgaris* L.). Proc IIRB Congress 58:185–188

Steen P, Pedersen HC (1995b) Strategies in creating transgenic herbicide tolerant sugar beet *(Beta vulgaris* L.) varieties. Proc IIRB Congress 58:189–192

Steen P, Keimer B, D'Halluin K, Pedersen HC (1986) Variability in plants of sugar beet (*Beta vulgaris* L.) regenerated from callus, cell-suspension and protoplasts. In: Genetic manipulation in plant breeding, pp 633–635

Stich B, Melchinger AE, Heckenberger M, Möhring J, Schechert A, Piepho HP (2008a) Association mapping in multiple segregating populations of sugar beet (*Beta vulgaris* L.). Theor Appl Genet 117(7):1167–1179

Stich B, Piepho HP, Schulz B, Melchinger AE (2008b) Multi-trait association mapping in sugar beet (*Beta vulgaris* L.). Theor Appl Genet 117(6):947–954

Sullivan CF, Finch I, Dix PJ, Burke JI (1993) Studies of in vitro propagation systems for sugar beet. Irish J Agric Food Res 32:27–35

Taguchi K, Kubo T, Takahashi H, Abe H (2011) Identification and precise mapping of resistant QTLs of Cercospora leaf spot resistance in sugar beet (*Beta vulgaris* L). G3 (Bethesda) 1:283–291. https://doi.org/10.1534/g3.111.000513

Taški-Ajduković K, Nagl N, ĆurčićŽ ZM (2017) Estimation of genetic diversity and relationship in sugar beet pollinators based on SSR markers. Electron J Biotechnol 27:1–7

Tenning P, Bensefelt J, Fouillard P, Mannerlof M, Tuvesson S (1995) Glyphosate tolerance in transgenic sugar beet. Proc IIRB Congress 58:183

Tetu T, Sangwan RS, Sangwan-Norreel BS (1987) Hormonal control of organogenesis and embryogenesis in *Beta vulgaris* callus. J Exp Bot 39:506–517

Urazaliev K, Abekova A, Bazylova T, Bersimbaeva G, Daniyarova A, Massonichich-Shotunova R (2013) Somaclonal variation of sugar beet resistant to pathogenic root rot *Fusarium oxysporum var. orthoceras*. Genetika 45(3):629–640

Van Geyt JPC, Jacobs M (1985) Suspension culture of sugar beet (*Beta vulgaris L.*). Induction and habituation of dedifferentiated and self-regenerating lines. Plant Cell Rep 4:66–69

Van Geyt JPC, Lange W, Oleo M, de Bock TSM (1990) Natural variation within the genus beta and its possible use for breeding sugar beet: a Review. Euphytica 49:57–76

Van Swaaij AC, Jacobsen E, Kiel JA, Feenstra WJ (1986) Selection, characterisation and regeneration of hydroxyproline-resistant cell lines of Solanum tuberosum: tolerance to NaCl and freezing stress. Physiol Plant 68:359–366

von Wordragen MF, Dons HJM (1992) *Agrobacterium tumefaciens*-transformation of recalcitrant crops. Plant Mol Biol Rep 10(1):12–36

Wang M, Goldman IL (1999) Genetic distance and diversity in table beet and sugar beet accessions measured by randomly amplified polymorphic DNA. J Am Soc Hortic Sci 124(6):630–635

Wang Y, Zhan Y, Wu C, Gong S, Zhu N, Chen S (2012) Cloning of a cystatin gene from sugar beet M14 that can enhance plant salt tolerance. Plant Sci 191–192:93–99. https://doi.org/10.1016/j.plantsci.2012.05.001

Weber WE, Borchardt DC, Koch G (1999) Combined linkage maps and QTLs in sugar beet (*Beta vulgaris* L.) from different populations. Plant Breed 118:193–204

Wu C, Ma C, Pan Y, Gong S, Zhao C, Chen S, Li H (2013) Sugar beet M14 glyoxalase I gene can enhance plant tolerance to abiotic stresses. J Plant Res 26:415–425. https://doi.org/10.1007/s10265-012-0532-4

Würschum T, Maurer HP, Kraft T, Janssen G, NilssonC RJC (2011) Genome-wide association mapping of agronomic traits in sugar beet. Theor Applied Genet 123(7):1121

Yang AF, Duan XG, Gu XF, Gao F, Zhang JR (2005) Efficient transformation of beet (*Beta vulgaris*) and production of plants with improved salt-tolerance. Plant Cell Tissue Organ Cult 83:259–270

Yin F, Gao J, Liu M, Qin C, Zhang W, Yang A (2014) Genome-wide analysis of water-stress-responsive microRNA expression profile in tobacco roots. Funct Integr Genomics 14:319–332. https://doi.org/10.1007/s10142-014-0365-4

Zhang CL, Xu DC, Jiang XC, Zhou Y, Cui J, Zhang CX, Chen DF, Fowler MR, Elliott MC, Scott NW, Dewar AM, Slater A (2008) Genetic approaches to sustainable pest management in sugar beet (*Beta vulgaris*). Ann Appl Biol 152:143–156

Zhang Y, Nan J, Yu B (2016) OMICS technologies and applications in sugar beet. Front Plant Sci 7:900

Zhao K, Aranzana MJ, Kim S, Lister C, Shindo C, Tang C, Toomajian C, Zheng H, Dean C, Marjoram P, Nordborg M (2007) An *Arabidopsis* example of association mapping in structured samples. PLoS Genet 3(1):4

Zhong Z, Smith HG, Thomas TH (1993) In vitro culture of petioles and intact leaves of sugar beet (*Beta vulgaris*). Plant Growth Regul 12:59–66

Zhu H, Bi YD, Yu LJ, Guo DD, Wang BC (2009) Comparative proteomic analysis of apomictic monosomic addition line of *Beta corolliflora* and *Beta vulgaris* L. in sugar beet. Mol Biol Rep 36:2093–2098. https://doi.org/10.1007/s11033-008-9421-2

Zossimovich VP (1940) Wild species and origin of cultivated beets. Sveklovodstvo, Kiev, pp 17–44

Seed Production and Certification in Sugar Beet

6

Alessandro Romano

Abstract

The production of high-quality seeds in agriculture is one of the most vital fields as the seed is the base of a healthy plant. The quality level is dependent on cultural practices adopted during the process of seed production as well as the health of the parent crop. The health of the crop besides using proper agronomic practices is further dependent on several other factors. Sugar beet seed production undergoes two different phases, *viz.* a vegetative stage from basic seed to stecklings and seed production on vernalized plants. To meet high standards, seed production is exclusively done by farmers under the constant supervision of seed companies. In this chapter, sugar beet seed production through direct and indirect methods, seed processing, and certification has been discussed.

Keywords

Certification · Monogerm · Seed · Steckling · Sugar beet

Abbreviations

CMS Cytoplasmic Male Sterility
USDA United States Department of Agriculture

A. Romano (✉)
Council for Agricultural Research and Economics, Plant Protection and Certification Centre, Lonigo, Italy
e-mail: alessandro.romano@crea.gov.it

V. Misra et al. (eds.), *Sugar Beet Cultivation, Management and Processing*,
https://doi.org/10.1007/978-981-19-2730-0_6

6.1 Introduction

The use of high-quality seeds in agriculture is crucial for the success of any crop, since its performances depend not only on the agronomic practices adopted but also on the source material, represented by performing varieties adapted to the areas where they are grown, with desirable characteristics that will ensure high yield, and on the availability of high-quality seeds, with satisfying germination and a reduced damping-off in presence of adverse conditions. Seed quality is a general term that encompasses aspects relative to crops such as genetic purity and uniformity, and features relative to seeds like viability, vigour, and health. The quality of seeds is also determined by other characteristics such as chemical composition or tolerance to certain diseases or pests. In the case of sugar beet (*Beta vulgaris* L.), the goal of growers is to obtain a high sugar yield, which is also linked to the amount of root produced per hectare. Consequently, the highest germination ability and vigour are the key factors to ensure an early and high level of field emergence (Kockelman et al. 2010).

Seed certification schemes, through the monitoring of seed production stages, will basically warrant sugar beet growers with two accomplishments: (1) the certainty that the varieties they cultivate are conformed for identity and purity requirements to those registered in official lists; (2) high-quality standards of seeds, mostly in terms of germination capability, and purity, requisite that will assure the absence or a limited presence of other species, especially weeds. Seed certification is also an effective marketing tool to the benefit of sellers since it represents an added value that increases the potential sale of crop. National Authorities are designed to lead the process and release final certifications and the proper labels to seed producers.

6.2 Production Regions

Sugar beet is a biennial species, with two distinct growth stages: a vegetative phase taking place in the first year, and the reproductive one in the following year provided a thermal induction, vernalization, is satisfied. For this reason, seed production requires peculiar climatic conditions that could be different from those requested for the ordinary sugar crop whose cycle is limited to the vegetative stage. Good growing conditions in suitable production areas are essential to assure high seed quality and stable seed yields. Particularly, in choosing the regions where plants will be grown it should be paid special attention to the climatic conditions during flowering, maturation, and seed harvest (Salimi and Boelt 2019). The aforementioned conditions can be usually found around the 40–45° latitude in the northern and the southern hemisphere. The traditional areas with favourable conditions where the production of certified seed, mainly monogerm, is concentrated are the south of France and northern Italy for the European market. By far smaller areas can be found in Austria, Denmark, and Germany. Other areas can also be found in Europe but with a very neglectable hectarage (ESCAA 2020). Seed production for multigerm varieties can be found in England and Denmark (Kockelmann and Meyer 2006).

For the North American demand, seed is mainly produced in Oregon and Washington states (USDA 2015). Worldwide, significant areas can also be found in the northern regions of Turkey mainly for their national sugar production (Aksoy et al. 2017), and Chile both for national market and for export to Europe and North America (Kockelmann and Meyer 2006). Finally, India is proving to be a potential sugar beet seed producer, which may attract multinational companies because of the low production costs (Mall et al. 2020, 2021). The mentioned areas do not usually have marked continental features and are characterized by optimal climatic conditions for all steps of the seed production, that is overwintering, vegetative growth, flowering, and maturing of plants. The environmental conditions required for sugar beet seed production consist mainly in several weeks of low temperatures during the vegetative phase to induce complete and homogeneous bolting of seed bearers in the second year, but at the same time not so low to bring about possible losses owing to frost damages. The flowering period should be preferably concentrated into a relatively short time to ensure a uniform maturation of the seed. Ideally, there should be little probability of rain during harvest. Furthermore, climatic conditions with a probability of temperatures higher than 15 °C immediately after vernalization should be avoided, because of possible devernalization of plants that could cause undesired bolting in the subsequent sugar beet crop (Longden 1986). Air humidity plays an important role as well, indeed it should not be too high especially during the flowering period, with optimal values around 60–70% matched with temperature not exceeding 35 °C (Wood et al. 1980).

6.3 Plant Material

Early breeding techniques for sugar beet were developed by the United States Department of Agriculture (USDA) and were aimed to obtain cytoplasmic male sterility (CMS), monogerm seeds, and hybrid vigour (Panella 1996). Currently, most cultivars are represented by monogerm hybrids. The use of monogerm sugar beet seed has brought several advantages such as a drastic reduction of the need to thin clusters of seedlings, a requirement in case of multigerm seed sowing (Smith 1987). In sugar beet, cytoplasmic male sterility, as a consequence of mutations of mitochondrial DNA, leading to pollen sterility in combination with particular nuclear genes (Mikami et al. 2011), allows the breeder to obtain male sterile or female parental lines. These lines are a key factor in the breeding of hybrid cultivars (Mann et al. 1989; Bosemark 1993; Ran and Michaelis 1995; Forster et al. 1997; Ducos et al. 2001). The maintenance of maternal lines is genetically realized by maintainer lines, also called O-types, which have the same nuclear genotypes as maternal lines but are male fertile due to normal mitochondria (Arakawa et al. 2018). They will be hybridized with the equivalent monogerm male sterile of another line to produce a monogerm male sterile F1. Then, the parental lines combined for commercial seed production may also be hybrids, therefore, hybrid varieties are three-way or multiple way hybrids (Märländer et al. 2011). As well as for hybrid seed production, special attention must be paid to the spatial isolation from cultivated beet stands, because of the higher relevance of contamination of basic seed. Because of the genetic constitution of the hybrids (male sterile parental line), bolters originating directly from the seed in sugar beet crops are also sterile or semi-sterile and the seed, especially in triploid varieties, can have a reduced capacity to germinate. Sugar beet varieties exhibit different ploidy levels. Originally, sugar beet was diploid with 18 chromosomes (2x); anisoploid varieties became available in Europe since 1940s. Such varieties, which included diploid, triploid, and tetraploid individuals, were obtained crossing diploid and tetraploid parents (Forster et al. 1997). A topic role for the production of triploid varieties was played by colchicine (Rasmusson and Levan 1939). Colchicine is a mutagen that works by preventing the formation of the microtubules. As microtubules play a role in chromosome segregation, colchicine induces polyploidy by preventing the segregation of chromosomes during meiosis that results into half of the gametes (sex cells) containing double the chromosome number than usual. The second half of the gametes do not contain any chromosomes and produce embryos with doubled chromosome numbers (Manzoor et al. 2019). With this technique, tetraploid breeding material could be developed and used as a pollinator for diploid sterile female parents. The use of CMS along with polyploidy allowed the production of triploid varieties. Both diploid and triploid varieties are cultivated. Triploid varieties mostly show more vigorous growth and can have higher beet yields (Brykczynski 1960). However, for the selection for monogenic inherited traits, such as resistance, diploid breeding material is preferred.

6.4 Seed Production

To meet high standards, seed production is exclusively done by farmers under a constant supervision of seed companies. Seed of monogerm hybrid varieties is produced by crossing different parental lines consisting of male sterile monogerm plants, which will bear the seeds, and the corresponding diploid or tetraploid multigerm pollinators. Both hybrid components need to be managed with distinct care by seed producers during the seed production process to ensure optimal flowering synchronization, maximum hybridization during flowering and to guarantee that the seeds produced will be carrying the desired traits from both parent lines (Kockelmann et al. 2010). To ensure cross-pollination, a hybrid scheme is required for seed production. The parental lines are usually set up in blocks consisting in six to eight rows of male sterile and two rows of pollinator. After a full pollination and before harvesting the seed-bearing crop, male components will be removed from the field, whereas male sterile are grown until the full seed maturation and then harvested. The basic seed for monogerm varieties in western European and American production regions are usually sown in July and August, whereas seed harvest takes place in the summer of the following year. As a whole, in the seed production cycle two different phases can be distinguished: a vegetative stage from basic seed to stecklings, that is young plants that satisfied vernalization requirements and the subsequent seed production on vernalized plants. Both phases can take place in the same field (direct production method or overwintering) or in two different specialized fields as two different crops (transplanting method). In the direct method, sowing takes place in late summer, and plants are kept in the same field over winter and for the whole production cycle until the harvest. In the transplanting method, young plants are grown in the first season and transplanted in separate fields to produce seeds in the second season. In any case, despite the biennial cycle in sugar beet, the seed production is often completed within 12 months. Because of the lower risk and the higher flexibility, the indirect method is currently the standard method (Bornscheuer et al. 1993), especially for monogerm seed production.

6.5 Indirect Method

6.5.1 Generality

The most widespread production system in western Europe and Turkey is the transplanting method. In this method, basic seed is sown in late summer of the first year to produce the stecklings of hybrid parents. The vegetative phase happens in specialized nurseries supervised by breeding companies (Fig. 6.1).

The stecklings of both hybrid components, male sterile and pollinator, are grown in separate plots, in order to provide them with appropriate treatments, which can be different based on their genetic features. After a period of vernalization, stecklings

Fig. 6.1 Stecklings production in specialized nurseries (Italy) (photo courtesy of Piergiorgio Stevanato)

are harvested and processed for transplanting into the seed production fields. In Italy and France, it happens from February to March. In case of adverse weather conditions, the young plants are stored properly and transplanted lately. Despite some drawbacks, such as the high costs for steckling, harvesting, and transplanting and the large number of working hours needed, there are several advantages that make this method preferred to the direct method:

- A lower number of basic seeds sown compared to the direct method, with a better efficiency and reduction of costs. Basic seed requirements for direct-sown crops can be 1.5–5 times higher compared to the stecklings production method although with a variability depending on the growing region (Kockelmann and Meyer 2006).
- Suitability for steckling fields to be protected properly against severe frost, or alternatively for stecklings to be stored in small storage piles (clamps) after harvest.
- A better control of volunteer beets: in transplanted fields, the distances between and within rows are well-defined, permitting the operators to check for the presence of beet plants not belonging to the cultivated ones and weed them out easily. This allows a better flexibility about the necessary lag time between two seed production fields, which is usually fixed in 5 years.
- More flexibility for breeders in variating their seed production area based on performance of their sugar beet varieties and trial results and thus regulate seed production to meet the latest demands of the market.

6.5.2 Stecklings Production

6.5.2.1 Field Requirements

The conditions of soils are of the outmost importance to assure a better efficiency of the stecklings harvest especially when the process is mechanized. In northern Italy, stecklings production systems have been established on sandy soils along the Adriatic coast, also considering the tolerance of this species to salinity (Wang et al. 2017). Other eligible areas can be found in south-west France. Stecklings production requires a lag time of at least 4 years between two sugar beet nurseries. All fields, where sugar beet, both for seeds and for sugar, or any other *Beta* species have been grown in the past are excluded from the rotation for stecklings production (Kockelmann and Meyer 2006). To avoid undesired effects, which would affect the varietal purity in the subsequent rotation, it is necessary to look for and remove volunteer plants that could produce and shed seed in the rotation between two stecklings crops.

In order to reduce the risk of infection by virus transmitted from aphids such as BMYV and BYV, it would be crucial to maintain a safety distance between steckling plots and possible other *Beta* species. An important practice, common in several intensive production regions, such as in Italy, is also checking soils before sowing the basic seeds in order to detect the presence of resting spores of *Polymyxa betae* carrying rhizomania virus and beet cyst nematode (*Heterodera schachtii*). In case of positive results of testing, such soils should not be used for the production to avoid infection of young plants and transmission into seed production fields. The features of the preceding crop should be considered as well, with respect to hosts for beet pathogens or the possibility of herbicide residues. Usually, the preferred crops to put in rotation with beet are cereal and vegetable crops, while crops belonging to *Brassica* genus should be avoided because they can host cyst nematode (Gratwick 1992), apart from *Raphanus* and *Sinapis*, resistant to nematodes (Raaijmakers 2014). To this respect, it is noticeable the research about the use of winter oilseed rape, radish and mustard as trap crops for nematodes (Daub 2020; Smith et al. 2004).

6.5.2.2 Sowing of Basic Seed

The preliminary phases consist in preparing carefully the seedbed in order to enable a proper sowing and make seed germinate rapidly by the increase of seed-soil contact and promote a uniform field emergence (Blunk et al. 2018). To this respect, after deep ploughing, soils are levelled to reduce the risk of water logging and then worked in order to obtain a well-structured seedbed that will promote a deep root development (Håkansson et al. 2011; Durrant et al. 1988).

The opportunity of soil rolling should be carefully evaluated on the basis of the moisture level of the soil; it should be limited in wet conditions since it destroys soil structure and increases soil bulk density, which would result in soil crust formation, whereas it could be applied in dry condition when it increases the capillary moisture movement necessary for seed germination (Romaneckas et al. 2009). The most common scheme for producing monogerm hybrid seed by stecklings method consists in planting a reduced number of male component rows between a larger number of female ones (Bornscheuer et al. 1993; Kockelmann and Meyer 2006). Sowing is done with precision drilling machines in order to obtain a homogenous field establishment of stecklings. Basic seeds are calibrated and treated with chemicals to minimize damping-off due to diseases and early infestation from aphids. The goal is managing to produce stecklings with an optimal top diameter of 2–4 cm and a weight of 40–80 g without disease symptoms. Good nursery production should produce 300,000–400,000 plantable stecklings per hectare. The most common distances are 3–4 cm within the row, and 20–25 cm between the rows (Kockelmann and Meyer 2006). The depth to which seed is sown should not be more than 3–4 cm. The optimal temperature for germination is over 10–12 °C, requirement satisfied in summer as long as matched with a proper irrigation (Giordani 2013). However, the number and size of stecklings can be influenced by the date of sowing. On the one hand, a delay in sowing could cause a reduction in weight of stecklings with difficulties in transplanting, forcing growers to use a bigger number of seeds to achieve a proper number of uniform stecklings, on the other hand, a too early sowing could more easily expose the young plants to foliar fungal diseases; to this respect, sowing in July-August, in Italy, should allow plants a high recovery of stecklings at the correct stage of development when they are harvested at the end of winter (Kockelmann and Meyer 2006). In case of late sowing, the number of seed used should be lowered by about 10–20% in order to reduce the competition between plants and promote an optimal development of single plants. Drills should be cleaned carefully to avoid any possible contamination of hybrid components. Male and female components are sown in separate plots and their positions reported on proper field maps.

6.5.2.3 Irrigation

The sowing of basic seeds usually done in the warmest period of the year needs to be matched with a proper constant irrigation. The irrigation method adopted will have to prevent the seedbed from capping and crusting, very likely to happen during the summer months. The early development stages are the most sensitive to the environmental constraints, thus, in those phases will result crucial keeping the soil surface to reduce wind erosion and seedling damage quite likely on sandy soils (Rinaldi and Horemans 2012). Depending on climatic conditions, two to five irrigations with small quantities of water (15–25 mm) may be necessary to ensure homogenous crop development (Kockelmann et al. 2010).

6.5.2.4 Mineral Nutrition

Another important factor to take into consideration with the aim of obtaining well-calibrated stecklings is mineral nutrition. Any effective fertilization plan should be predated by the analysis of nutritional status of the soil to adapt the rates of fertilizer accordingly (Draycott and Christenson 2003). Some elements, such as potassium, phosphorus, and about 50% of nitrogen, should be applied before sowing, whereas the remainder of nitrogen when the young plants have formed four to eight true leaves. Among microelements, boron plays an important role for the development of sugar beet, promoting cell wall formation, carbohydrate metabolism, and being associated with sugar translocation. Dordas et al. (2007) found an improvement in mean seed weight, number of seeds/plant, and seed yield per plant following the application of foliar boron. Moreover, they also found a lower number of abnormal seedlings. Usually, this element is applied in the 4–6 or 6–8 leaf stage (Kockelmann et al. 2010).

6.5.2.5 Weed and Pest Control

Sugar beet is quite susceptible to the presence of weeds. At emergence, seedlings have a similar or slightly greater leaf area and a larger root system than the most small-seeded weeds. Thus, the sugar beet absolute growth rate is initially greater, and it usually remains in such condition for at least several weeks. Because small-seeded weed species tend to have higher rate of root elongation, the weeds tend to rapidly occupy the soil volume limiting the nutrient availability of sugar beet plants (Nørremark and Griepentrog 2004). The best practices should be adopted to avoid both nutritional competition with stecklings and the risk of spreading diseases owing to the capability of some weeds to carry pests, such as nematodes and aphids, both in the upper and lower parts of the plants. The weed control programme is essentially the same as for the sugar beet root crop (Bornscheuer et al. 1993). Usually, chemical treatments are done before sowing and before emergence in the nurseries. To avoid damage to young plants, it is necessary to previously carry out trials on new sensitive hybrid lines for testing possible specific interaction with herbicides (OECD 2003). It is also crucial to maintain a high health and nutrition status to assure a high recovery of stecklings. A special care is necessary to protect young plants from pests during their development. In addition to treatments directly applied to basic seeds before sowing, to ensure a protection in the early stages of development, it is important to control insects and fungal diseases such as leaf spot disease caused by *Cercospora beticola* by spray applications with insecticides and fungicides (Weiland and Koch 2004).

6.5.2.6 Overwintering and Steckling Harvest

Cultivated sugar beet commonly requires a period of 15–20 weeks with temperatures of 4–10 °C about (Wood et al. 1980; Longden 1986). In the main growing areas, such as France and Italy, stecklings stay over winter in the nurseries, where temperatures are low enough to accomplish vernalization requirements of plants, but not so low to damage the crop. Healthy and vigorous sugar beet stecklings can tolerate frost, with differences due to genotype, development stage, leaf mass, and duration of the frost exposition (Reinsdorf et al. 2013). However, if temperatures drop under 6–8 °C about, irreversible damage to plants can happen with losses estimate from 20% of plants to the whole field (Reinsdorf and Koch 2013) owing to the necrosis of leaves and crown tissues specially for genotypes with a reduced cold tolerance (Kirchhoff et al. 2012). In that case, an additional protection from critical temperatures is required. A common method consists in covering the steckling plots in advance with a fleece to keep temperatures below the fleece 2–5 °C higher than the environment ones above the covering. The fleece applied is permeable to air and water, letting at the same time the sunbeams transfer, which creates a greenhouse effect that promotes stecklings growing (Giordani 2013). Before taking out the plants from soil, the foliage is chopped mechanically in order to obtain a height variable between 2 and 8 cm above the crown depending on the cultivation regions and available techniques, taking care of the apical meristem, that must not be damaged. The ideal length of stecklings should be 8–12 cm to make roots having a good soil contact. After harvest stecklings are calibrated to a top diameter of 2–4 cm using proper devices. In the most traditional areas of production such as France and Italy, harvest is often completely mechanized, with adapted sugar beet or vegetable harvesters. Damaged plants will be discarded as well as those not clearly belonging to the cultivated variety. Stecklings should be planted just after the harvest in order to reduce water losses and respiration which could cause a non-uniformity in seed producing plants. However, in case of weather condition that could not allow a prompt planting, stecklings should be stored in cool conditions at temperature not exceeding 6–8 °C (Nardi 1998). In production areas with extremely cold winters or continental climate, the stecklings harvested in autumn are stored in field clamps or storage buildings for overwintering at a constant temperature of 2–3 °C and a maximum relative air humidity above 90% that will prevent desiccation of stecklings and a consequent not uniform growth of the seed crop. Furthermore, these environmental conditions should minimize the development of clamp rots caused by pathogenic fungi and bacteria.

6.6 Seed Production Through Steckling Method

6.6.1 Field and Rotation Requirements

Field requirements for seed production differ from those for stecklings production. The most suitable soils for seed production are deep and fertile with a high capability in retaining water necessary to promote plant growth during dry periods. The best soils are loamy, with 40–70% of silt and 20–50% of clay, with neutral to slightly alkaline pH reaction (Kockelmann and Meyer 2006). A proper clay content will also allow adult plants to be strictly anchored to soil. In some production regions, as seen for stecklings production, fields are systematically tested for nematode presence (*Heterodera schachtii*) since an infection will cause water imbalances in young plants with a consequent slow growth and death (Hafez and Seyedbagheri 1997). In case of positive result, soil will not be suitable for seed production. The presence of rhizomania virus will have to be ruled out as well. In order to accomplish the requirement of varietal purity, stecklings can only be transplanted in fields where no seed sugar beet or other *Beta* species have been grown for at least 4 years. Besides the rotation rules, it is not advisable planting *Medicago* or annual crops before stecklings transplanting because they could make the soil preparation difficult. Another crucial requirement is given by the distances between seed crops and other possible *Beta*; to this regard, in order to prevent cross-pollination, there are minimum distances prescribed by law that could be even stricter depending on rules applied by single breeding companies (see seed certification). A proper distance from other beet root crops cultivated nearby seed production fields is also important to reduce possible virus transmission by insects.

6.6.2 Stecklings Transplanting

Soil preparation plays a key role in determining a high seed yield and seed quality. In fact, a proper soil structure enables plants to develop an efficient root system and preserve soil moisture. To this respect, it is important to avoid any operation causing surface crust formation. Stecklings for production of monogerm hybrid varieties are transplanted respecting specific ratios of female and male plants, which can depend on various factors like flowering time, ploidy, and fertility of male components. The most common ratio varies from 2:1 to 4:1. The female component is usually planted in blocks of four, six, or eight rows, whereas the pollinator in two rows (Fig. 6.2).

Planting density is one of the crucial factors in determining the amount of seed produced and depends on several elements such as the productivity of the variety, the vigour, and the branching ability of female components. To this respect, seed companies before releasing new cultivars test the parentals to evaluate if they meet specific local agro-technical requirements. The usual ratio adopted between stecklings area and seed production area is 1:10, that is one hectare of stecklings every ten of seed producing field.

Fig. 6.2 Strips of male sterile components alternating with strips of pollinator (Italy) (photo courtesy of Piergiorgio Stevanato)

Total plant population varies between 30,000 and 40,000 plants per ha, both for female and male components with spacing of 70–75 cm between rows, and 35–45 cm within rows. Distances between female rows and pollinator rows could be increased to avoid damage to seed plants during male components removal. The most appropriate distances depend on machinery used and harvesting technique. Spring transplanting should be in February–March to promote a long vegetative period. Early planting ensures both a better growth of roots and an earlier start for leaves development, bolting, flowering, and maturity of seed crops and thus higher seed yields (Kockelmann and Meyer 2006). In case of susceptible components, a treatment with fungicides before transplanting could be necessary in order to avoid rot arising because of *Phoma betae*, *Rhizoctonia* and *Sclerotium* infections. This could be particularly necessary in case of storage of steckling before transplanting. Planting is mostly done with semi-automatic planting machines, which are often produced locally tailored to peculiar soil and growing conditions (Kazmeinkhah 2007). Irrigation immediately after transplanting is usually applied to promote a rapid establishment and root development.

6.6.3 Crop Management

A proper fertilization plan should be set after a careful soil analysis and taking into account the preceding crop and regional experiences in production. Usually, experience locally gained will provide guidelines that will support companies and farmers in choosing the type of fertilizer, time, and number of applications. Phosphorus and potassium are normally provided to the soil before transplanting, whereas nitrogen is subdivided during the different development stages of the crop, with 35–50% applied at transplanting in spring, 30% at the vegetative development end of March/mid-April, and the remainder at bolting at the end of April/mid-May, considering that an excess of nitrogen especially if applied late could delay maturity with a reduction of seed quality. In addition to the main nutrients, sugar beet plants are also provided with boron, at the stages of topping and at the beginning of flowering. Lehnhardt and Bonk (1991) reported the important role of boron on pollination and fruit setting in sugar beet seed production. With respect to water demand, Noli et al. (2007) highlighted the increase in seed yield and quality following irrigation during and at the end of flowering. Drip irrigation for sugar beet seed production is used in Italy because of the high flexibility and the possibility of giving small water amounts over the production cycle also in windy conditions (Tognetti et al. 2003) (Fig. 6.3).

Fig. 6.3 Drip irrigation in sugar beet seed production field (Italy) (photo courtesy of Piergiorgio Stevanato)

6.7 Direct Method

In the direct production method, plants stay on the same field both for vegetative and for reproductive stages. For seed production of monogerm varieties, this method is mainly used in Oregon (USA) and Chile. A great importance has to be given to sowing density, since it affects plant size range, which, in its turn, determines the capability of plant to tolerate low temperatures during overwintering. A density that warrants an average top diameter of 1–2 cm is preferred, since larger plants, because of a lower sowing density and wide spacing between them, will be so prone to being killed by frost. Furthermore, according to Pospisil and Mustapic (1999) and Hemayati et al. (2008), an inverse relationship between plant density and cluster size would exist. Usually, to fulfill these necessities, a number of 600,000 seeds per ha is considered optimal. The ideal sowing period should be within August, in order to give plants the time to develop 10–12 leaves and facing the winter sufficiently grown. Plants with fewer leaves are more susceptible to be infected by downy mildew (*Peronospora farinosa*), which will cause no bolting in the second year. Older leaves of beet in the second year may carry infections like *Alternaria* or *Phoma* from overwintering. In that case, a fungicide treatment may be necessary at the beginning of May to reduce disease development (Srivastava 2004).

6.7.1 Field and Rotation Requirements

The field requirements for direct-sown crops are basically the same as for the second year's growth in case of indirect method. The lapsed time between two beet crops should be at least 10 years, since detection and elimination of weed beet are more difficult especially within the rows because of high plant density. Moreover, their botanical vicinity to the crop makes the control by herbicides almost impossible in sugar beet fields (Landová et al. 2010). A long rotation reduces the occurrence of emerging beet plants and thus the risk of cross-pollination with undesired pollen (Bond and Turner 2004). The preferred preceding crops are wheat or other early-harvested crops that allow sufficient time for field preparation. Also, weed beet plants occurring in a closed stand of cereals are not able to compete with these crops and contribute to the soil seed bank (Landová et al. 2010). In France, the USA, and Chile, for production of monogerm seed, sowing takes place in August with a female: male ratios equal to 3:1 or 4:1. Common row spacings are 50–80 cm between rows and 6–14 cm within the row. Water is provided soon after sowing to improve seedling emergence and plant development.

6.8 Flowering

After a period of vernalization with temperature ranging from 1 to 15 °C, and with an increase of light hours, plants switch from the vegetative phase to the reproductive one (Wood et al. 1980). Bolting usually starts in the middle of April, and flowering at

6.10 Seed Harvest

There are two ways to harvest sugar beet seed: (1) Putting cut branches in swath followed by threshing. (2) Direct threshing after desiccation of seeds.

6.10.1 Cutting on Swath

After harvesting, the cut seed branches are laid in swath in order to let them achieve a uniform ripening. Seeds that fall off on the soil while being cut will germinate through tillage. During seed production, it is possible to distinguish plants emerging from stecklings and plants that emerge from the seed bank and therefore mechanically eliminated. Immediately after harvesting, the seed is delivered to seed stations located in the production areas and belonging to the seed companies where it will undergo complex processing before pelleting and storage. If necessary, seeds are dried for conservation (Meyerholz 1999). Cutting of seed bearers onto a swath is usually done mechanically. The cutting technique promotes an efficient ventilation and drying of the seed plants and seeds making them suitable for subsequent threshing. Choosing the optimal maturity stage of seed is crucial, because cutting too early could result in a poor germination capability owing to an incomplete maturation of seeds, whereas cutting too late will cause a loss of over ripened seeds for shattering (TeKrony 1969). The best time for cutting is early in the morning or the evening in order to reduce seed shattering (Giordani 2013). Seed should be threshed about 4–5 days after swathing; however, times could be different based on weather conditions and vegetative mass on the ground. Unlike cutting, threshing should be avoided early in the morning when dew could have increased seed humidity over 11-12%. Indeed, the optimal moisture content of seeds when threshed should be below 12%. Threshing capacities can take up to 4–6 ha per day.

6.10.2 Direct Threshing After Desiccation

Direct threshing following desiccation appears the most suitable method in case of hybrid components with a lower tendency for branching and higher tendency for seed shattering. Desiccation method is also preferred with a risk of rain at harvesting. Desiccation of seeds is attained by the use of a chemical desiccant once the moisture content of seed is not more than 30–40%. Thus, a first application of desiccant happens few days after cutting onto the swath. A second application could be necessary 2–3 days after the first one. Seeds are threshed 5–7 days after the last treatment. Since moisture content of seed at threshing could have a value of 25% with differences depending on the production region (Sliwinska 2003), an immediate drying of seed is of foremost importance to preserve seed quality. Direct threshing after desiccation takes longer times compared to swathing (Thibaud 2002). In Italy, the seed production is settled, on average, at about 25 q/ha of seeds with a size of 3.25–6.00 mm; however, higher values can be reached. Values lower than 18 q/ha could be caused by genetic source material or nutritional, health, and climatic issues (Giordani 2013).

6.11 Seed Processing

The physiological quality of seeds is determined by the growing conditions, especially during flowering and maturation stages of the seed. In general, the basic physiological quality of single seeds cannot be improved by processing. Therefore, the main aims of seed processing are making a selection to obtain fractions with the best seed quality within a seed lot and make seeds even-sized for pelleting, to promote water uptake and to improve germination by the removal of growth inhibitors located in the pericarp.

6.12 Processing of Cleaned Seed

The preparation of sugar beet seed prior to pelleting is a complex process, consisting of three stages: calibration, polishing, and gravity separation (Fig. 6.5).

The combination and intensity of each stage can be different depending on specific physical characteristics of every single seed lot. The whole processing scheme can be described as follows:

- Calibration of cleaned seed lots of 3.25–6.00 mm into different size fractions.
- Polishing of each fraction separately to remove the pericarp. This operation leads to a reduction in the pericarp thickness through parenchyma removal, and to a seed size to an optimal grade suitable for pelleting. Removing the pericarp also exposes the operculum and the basal pore which are proposed as the main water and oxygen entry points during germination and remove germination inhibitors, there located, promoting an increase in the speed of germination (Orzeszko-Rywka and Podlaski 2003; Ignatz et al. 2019) Modifying the pericarp may also reduce possible infestation with pathogens there localized (Fukui 1994).
- Several calibrations with multilevel sieve with round holes to remove small particles of pericarp, and those seeds too small or still too big; in the second case, seeds will be polished again (Tuğrul and Kaya 2020). The seed units are passed through a sieve with oval holes, and those with multiple embryos are separated (Bornscheuer et al. 1993). The remaining seeds are separated by weight and assessed by an x-ray test; those that are 100% full are then sent to the pelleting. Pellet sizes of 3.50-Ø4.75 and 3.75-Ø4.75 mm are widely preferred in Europe; only Finland and Sweden use pelleted seeds in the range of 4.00–5.00 mm units (Tuğrul and Kaya 2020).

Fig. 6.5 Sugar beet seed cleaning facility (photo courtesy of Piergiorgio Stevanato)

6.12.1 Pelleting

After calibration, seeds are pelleted. The aim of this operation is to give the seed a uniform size and a round shape suitable for sowing machines. Seed pelleting consists basically of subsequent steps: moistening or washing, pelleting material addition, usually a mixture of clay, wood flour, and adhesive (Tuğrul and Kaya 2020), drying, and calibration of pelleted seeds. Seed companies adopt different pelleting techniques to improve seed quality and germination capability (Afzal et al. 2020).

6.12.2 Coating

The last operation before placing seeds on the market is the treatments with different fungicides and insecticides, which can differ based on the specific needs of different countries. In this process, a specific suspension, containing fungicides as well as insecticides, is applied to the pelleted seed, along with a colour, which is usually peculiar to a breeding company. To this respect, it is noticeable to specify that the outdoor use of three neonicotinoid substances, clothianidin, thiamethoxam, and imidacloprid, also employed to coat sugar beet seeds to protect them from pests, was banned since 2018 mainly for their effects on bees (EU regulations No 783/2018a, 784-2018b, 785/2018c). Although derogations have been released for some countries, it appears important to find valid alternatives for sugar beet seeds. A few studies focused on assessing the actual presence of residues on sugar beet plants (ViricGasparic et al. 2020) and testing the efficacy of alternative products (ViricGasparic et al. 2021; Hauer-Jákli et al. 2016). The same chemical treatment is also applied for non-pelleted seed to produce finished encrusted seed. The overall process is often continuously controlled by computer technology used to monitor seed flow and seed treatment application (Danielson and Gaul 2011).

6.13 Seed Certification

Seed certification is a program aimed to maintain and make available to the public high-quality seeds and propagating materials of genetically distinct crop varieties (Copeland and McDonald 2001). As a general rule, excluding specific cases, sugar beet seed should be marketed only if the variety which it belongs to is included in the National List or in the Common Catalogue of Varieties and if it has been officially examined and certified as basic seed or certified seed in accordance with the rules for certification. Seed certification provided by the official authorities warrants every actor involved in seed production chain and final seed buyer that the seeds produced satisfy fixed requirements relating to varietal identity and purity and own quality characteristics examined by seed testing that will ensure a high emergence and give a good final establishment in the field.

The seed certification process is articulated in different steps starting with planting eligible seed stocks, field inspection of the growing crop, proper seed conditioning or cleaning, representative sampling from homogeneous lot, laboratory analysis, and final labelling. Thus, certification involves both the inspection of the crop in the field as well as analyses on samples of the harvested seed. If seed of a variety is produced in separate fields, each field, whose seed correspond to a lot, can be certified or rejected irrespective of the seeds belonging to other lots of the same variety. Both field and seed inspections from each unit or lot are done by inspectors and technicians of National Designated Authority or by authorized technicians under official supervision.

Certification procedures are regulated by laws, which can be different depending on the country. In the Europe Union, the Council Directive 2002/54/EC of 13 June 2002 on the marketing of beet seed prescribes rules and conditions about certification for the member States, including seed sampling, packaging, sealing, labelling, and minimum conditions for analytical purity, germination, and moisture content. In addition to this directive, there are national rules for seed marketing, adopted by each member state of the European Union.

In North America, there are no official minimum standards for seed quality and no certification system for seed lots is provided to growers. Therefore, seed companies themselves are responsible to make seed meet market-specific demands for varietal identity and seed quality. However, independent certifying agencies, such as those that are members of the Association of Official Seed Certifying Agencies (AOSCA), can provide a certification on a voluntary basis. AOSCA has certifying agencies located in North and South America, Australia, New Zealand and South Africa (OECD 2012).

Furthermore, OECD provides a set of rules and regulations, named "OECD Seed Schemes", for the varietal certification or the control of seed moving in international trade for a broad range of cultivated species, including sugar beet. OECD Seed Scheme, which is overall comparable with the European Union system, is open to all Members of the Organization, as well as to any member of the United Nations, its Specialized Agencies or the World Trade Organization desiring to take part in. Schemes can also be implemented by the Authorities designated on purpose by, and responsible to, the Governments of the States adhering to the Scheme (OECD 2021). The general conditions briefly described here refer to the EU regulations (Council Directive 2002/54/EC). Certification authorities of single states may adopt technical peculiarities.

6.14 Certification Scheme

The certification process mainly consists of field inspections and seed analyses. Controls to the processing plants are also done. Sugar beet seed producers will submit all the multiplications of seed of a given variety for examination by the certification authority aimed to determine whether they meet the requirements set out in the official directives. Basically, there are two kinds of inspectors: official inspectors, who work for the official authority, and licensed inspectors, allowed to conduct inspection following a special training with a final exam provided by official authority. A crop from which basic seed is to be produced must be examined by an official crop inspector, since an inspection by a non-official licensed inspector is not allowed in that case. Varieties that are required to be enlisted and are undergoing the first 2 years of trials must also be inspected by officials. A crop from which certified seed of an already listed variety has to be produced may also be examined by a licensed inspector, provided the seed sown to produce the crop is subject to satisfactory post control. Crop inspections must be carried out when the conditions of the field and the phenological phase of the crop let identity and varietal purity be easily checked. For sugar beet, the inspections are performed both on the nurseries where stecklings are grown and on seed producing plants fields. At least one field inspection for each of seed production stages must be carried out.

6.15 Crop Inspections

6.15.1 Nurseries Inspections

The inspector will make sure that the previous cropping of the field will not have been incompatible with the production of seeds of the variety of the crop, and the field will be sufficiently free from plants that are volunteers from the previous cropping.

Most importantly, seed production fields will be accepted only if there is assurance that there are no volunteer plants of the genus *Beta*. To this respect, sugar beet nurseries must take place only in those fields where no kind of beet cultivation (both to produce seeds or sugar) has been previously cultivated for at least 4 years.

The crop must sufficiently match the identity and purity of the variety. The inspector will recommend the refusal of any fields for the production of certified seed that can be shown not to be entirely planted with the basic seed supplied or where the plants show a different appearance from that expected of the variety.

The varietal identity is checked by comparing the morpho-physiological features of stecklings in fields with those listed in descriptive form available for that variety.

The possible off-types, e.g. plants belonging to other varieties, will have to be removed from the field before the inspection. The off-types for different shape or colour of the root will be removed at the harvest of the whole crop.

The inspection will also ascertain the health status of crops with particular respect to the presence of Beet Yellow Virus (BYV), Beet Mild Yellowing Virus (BMYV), and Beet Necrotic Yellow Vein Virus (BNYVV). Symptoms of these viruses will be tolerated only if present in traces. In case of doubt, specific exams to reveal a possible presence of the nematode *Heterodera schachtii* should be made.

6.15.2 Seed Bearing Crops Inspections

In the case of seed-bearing crop fields, just one variety is admitted for every producer apart from the case of properly separate fields and different tools to handle seeds belonging to different varieties. Before the inspection, the producer will have to remove individuals with morpho-physiological characteristics different from those typical of the tested variety (off-types), plants with a possible presence of diseases transmissible by seeds, plurigerm plants in case of production of genetic monogerm, and spontaneous plants. A field with the presence of weeds that prevent the inspector from checking properly will not be partially or totally admitted to the certification. If bad weather conditions or diseases affected the normal development of plants, so that they cannot be carefully examined, the relative part of the field will not be admitted to certification. As well as the steckling fields, no beet cultivation must have been cultivated the previous 4 years.

In order to avoid unwanted pollination by other plants belonging to the genus *Beta*, proper distances are required. These distances can be not respected in case of sufficient protection from any extraneous pollinator and in the case of seed crops with the same pollinator.

The ploidy of both components, male sterile and pollinator, is established by reference to the Common Catalogue or a National List. If such information is not included in these documents, the ploidy of the components is regarded as unknown for the purposes of isolation distances. In that case, the minimum required distance must be 600 m for certified seeds and 1000 m for the basic ones. The same requirement must be fulfilled in the case of neighbouring sugar beet crops grown

Fig. 6.6 Wild beet plants with annual habits (photo courtesy of Piergiorgio Stevanato)

for sucrose extraction but that can flower at the same time as the seed-bearing cultivar.

Plants not belonging to the variety and weeds whose seeds cannot be easily removed in the following mechanical selection can be tolerated only in low quantities, whereas other wild plants belonging to *Beta* genus must be weed out before bolting (Fig. 6.6). Plants affected by viruses must be readily removed and taken off the field, whereas no presence of *Heterodera schachtii* is tolerated.

6.15.3 Seed Analyses

In order to obtain an official certification, it is also necessary to carry out specific analyses on seeds to make sure they meet the minimum standards for basic and certified seed. The analysis process on representative seed samples is performed by an official seed testing station or by officially authorized laboratories.

6.15.4 Sampling

The starting point for seed testing is a representative sample of a seed lot with a fixed minimum weight. The lot itself cannot exceed a maximum weight. Seed samples are drawn from the seed lot by an official or by a licensed seed sampler; however, in a modern seed processing plant, beginning with cleaning, up to packaging, automatic seed samplers are common. Small samples are taken continuously during processing at short intervals out of the complete seed stream to get a representative sample of the whole seed lot (Kockelmann and Meyer 2006). This method is especially preferred for non-pelleted sugar beet seeds since they differ in size and weight and therefore tend to segregate during handling and/or transport. Indeed, larger seeds tend to move towards the outer and upper areas of the container, while smaller seeds concentrate in the central and lower position. Thus, an automatic sampler would assure a better homogeneity of the sample (Kockelmann and Meyer 2006).

6.15.5 Laboratory Analyses

Methods of seed testing should represent the field value of individual seed lots, be reproducible, and give comparable results between testing stations. The International Seed Testing Association (ISTA) (ISTA 2021) provides the most used laboratory seed testing methods. For the American market, other methods described by the Association of Official Seed Analysts (AOSA) are used (AOSA 2019). The analyses that must be done are the determination of analytical purity, germination percentage, and moisture content. Other analyses, such as ploidy level, are not compulsory but performed on demand.

6.15.6 Analytical Purity

The minimum analytical purity of both basic and certified seed must be 97% by weight without considering, when present, granulated pesticides, pelleting substances, or other solid additives. The weight of seeds not belonging to the species of the sample examined must not exceed 0.3%. In the case of basic seed, the weight of inert matter must not exceed 1.0% whereas in the case of certified seed, the value must not exceed 0.5%. In case of the pelleted seed of both categories, to verify the fulfillment of these conditions, an official sample of seed drawn from processed seed that has undergone partial decortications but has not yet been pelleted will have to be examined without prejudice to the official examination of the minimum analytical purity of the pelleted seed.

6.15.7 Germination Test

A germination test is necessary to establish whether seeds meet the minimum germination percentage required by law, equal to 80%. In the case of monogerm seed, at least 90% of the germinated clusters must give single seedlings. Clusters giving three or more seedlings are accepted as long as they do not exceed 5% of the germinated clusters. In the case of precision seed, at least 70% of germinated clusters must give single seedlings, and clusters giving three or more seedlings must not exceed 5% of the germinated clusters (Fig. 6.7).

6.15.8 Moisture Content

The maximum moisture content of seed belonging to both categories must be 15% by weight excluding, where appropriate, granulated pesticides, pelleting substances, or other solid additives. The test for moisture content must be carried out by an Official Seed Testing Station.

6.15.9 Other Conditions

The seed will have sufficient identity and purity of variety. Diseases that reduce the usefulness of the seed will be at the lowest possible level.

6.15.10 Labelling

Once the certification process has been accomplished with a positive result, seed lots will be provided with an official label reporting a set of information about the characteristics of the seed lot that will ensure the buyer that the seeds meet a specific standard level of high genetic purity and identity, high germinating capability, and minimum amounts of other seeds and inert matter.

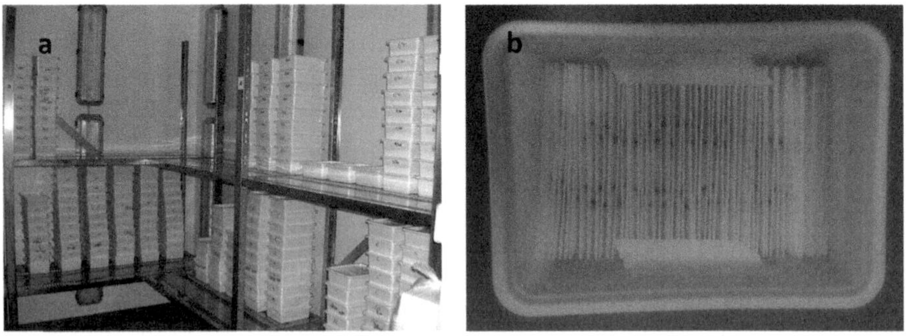

Fig. 6.7 Germination test in growth chamber (**a**) and detail of a tray with seeds in blotting folded paper (**b**) (photo courtesy of Piergiorgio Stevanato)

6.16 Future Prospects

Seed production, as well as the certification tool, appears to be susceptible to improvements in order, respectively, to put in the market more performing varieties and make the certification system more efficient in terms of seed lot identification and traceability. In general, this sector will face two different challenges in the future: improvement of seed production *sensu stricto* and improvement of certification schemes. In the first case, besides the breeding research, the efforts will have to be made to find new suitable areas worldwide where sugar beet seeds can be produced in addition to the traditional countries, especially for those varieties with the potential of being locally cultivated. To this respects, interesting results have been already obtained; an example is represented by India, where varieties adapted to tropical climates have been developing (FAO 2009) for years or in the Northern Caucasus region where experimentation is made in order to exploit the local varieties for seed production (Shevchenko et al. 2021).

In the case of certification schemes, for almost a decade, the OECD has been considering innovative ways to increase the effectiveness and integrity of seed certification, with the aim of making labelling and certification more reliable for supporting a high volume of internationally traded seeds and preventing the risk of fraud (OECD 2020). Moreover, another goal is to improve data collection and provide governments and other subjects involved in the scheme with more accurate and updated information on the production, certification, and international trade of seed. The future prospects involving OECD Seed Schemes explore the use of digital data management technologies, which have the potential to enhance the transparency of transactions and traceability. The prompt availability and exchange of data with different stakeholders would also allow rapid identification of problems within the supply chain, offering National Designated Authorities the means to validate the whole seed certification process (OECD 2020).

A few countries have already adopted a digitalization for their certification systems opening to opportunities of developing an international network, whereas other technologies, such as blockchain could further support more effective and resilient data tracking in international supply chains and provide greater transparency and traceability (OECD 2020).

To sum up, it appears clear that the seed production and certification sector in sugar beet, as well as in other species, give a broad range of interesting opportunities to be developed in the near future.

6.17 Conclusions

Having available high-quality seed in sugar beet is pivotal for farmers to get a successful production both quantitatively and qualitatively. To this purpose, the efforts of seed companies have been following two directions, on the one hand, they develop and use innovative plant breeding methods focusing more and more on obtaining varieties more suitable to trade requirements in terms of productivity, diseases tolerance, and adaptation to different environments; on the other hand, for seed reproduction in the field, the efforts are directed at the adoption of the most appropriate agronomic techniques, at the selection of the most suitable production sites and at the use of innovative technologies aimed to seed processing in order to preserve it from diseases and pests that would compromise the germination capability and the subsequent growth of crop.

The seed certification program proves to be of topic importance to assure both the correspondence of the morpho-physiological features of varieties to those codified by the breeder upon the registration of official lists and the stability that is the preservation of these characteristics over time. An official certification will also warrant that the traded seeds fulfill the requirements by law in terms of purity, germinability, and the absence of other seeds. To this respect, regulations lie on inspections over the whole production process that include field, seed selection, and processing plants, and eventually on laboratory analyses, performed by official and licensed inspectors and seed analysts. Certified seed ensures farmers several advantages, a higher yield, and a reduced number of treatments mainly those aimed to prevent weeds spread with a consequent reduction of costs.

References

Afzal I, Javed T, Amirkhani M, Taylor A (2020) Modern seed technology: seed coating delivery systems for enhancing seed and crop performance. Agriculture:15. https://doi.org/10.3390/agriculture10110526

Aksoy A, Demir N, Kaymak HC, Sari M (2017) Seed sector of Turkey in terms of sustainable agriculture. Atatürk Univ J Agric Faculty 48(2):133–138

Alcaraz G, Genter T, Laillet G, Rageot D (1998) Sugar beet pollen biology. Proceedings of the 61st IIRB Congress. Brussels, p 393–399

AOSA (2019) AOSA rules for seed testing—principles and procedures. https://www.analyzeseeds.com/product/aosa-rules-volume-1-principles-procedures-2019/

Arakawa T, Uchiyama D, Ohgami T, Ohgami R, Murata T, Honma Y, Hamada H, Kuroda Y, Taguchi K, Kitazaki K, Kubo T (2018) A fertility-restoring genotype of beet (*Beta vulgaris* L.) is composed of a weak *restorer-of-fertility* gene and a modifier gene tightly linked to the *Rf1* locus. PLoS One 13(6):e0198409. https://doi.org/10.1371/journal.pone.0198409

Blunk S, Heer M, Sturrock C, Mooney S (2018) Soil seedbed engineering and its impact on germination and establishment in sugar beet (*Beta vulgaris* L.) as affected by seed–soil contact. Seed Sci Res 28:1–9. https://doi.org/10.1017/S0960258518000168

Bond W, Turner R (2004) The biology and non-chemical control of Weed Beet (*Beta vulgaris* L.). HDRA the organic organization, Ryton Organic Gardens, CV8, 3LG, UK, Version: December 2004. http://www.organic weeds.org.uk

Bornscheuer E, Meierholz K, Wunderlich KH (1993) Seed production and quality. In: Cooke DA, Scott RK (eds) The sugar beet crop. Chapman & Hall, London, pp 121–153

Bosemark NO (1993) Genetics and breeding. In: Cooke DA, Scott RK (eds) The sugar beet crop: science into practice. Chapman and Hall, London, pp 67–119

Brykczynski J (1960) Efficient production of triploids in sugar beet. Euphytica 9:196–198. https://doi.org/10.1007/BF00022222

Copeland LO, McDonald MB (2001) Seed certification. In: Principles of seed science and technology. Springer, Boston, MA. https://doi.org/10.1007/978-1-4615-1619-4_14

Crane E, Walker P (1984) Pollination directory for world crops. International Bee Research Association, London, p 183. ISBN 9780860981435

Culley TM, Weller SG, Sakai AK (2002) The evolution of wind pollination in angiosperms. Trends Ecol Evol 17:361–369

Danielson B, Gaul A (2011) Category 4, seed treatment—Iowa Commercial Pesticide Applicator Manual, CS16; Iowa State University Extension and Outreach: Ames, IA, USA; p 40

Daub M (2020) Effect of winter oil seed rape cropping on the development of the sugar beet cyst nematode, *Heterodera schachtii*, and control of volunteer plants as a trap crop method. Agronomy 10(3):355

Dordas C, Apostolides G, Goundra O (2007) Boron application affects seed yield and seed quality of sugar beets. J Agric Sci 145:377–384. https://doi.org/10.1017/S0021859607006879

Draycott AP, Christenson DR (2003) Nutrients for sugar beet production: soil–plant relationships. CABI Publishing, Wallingford (UK). xvi + 242 pp

Ducos E, Touzet P, Saumitou-Laprade P, Vernet P, Cuguen J (2001) Nuclear effect on mitochondrial protein expression of the CMS Owen cytoplasm in sugar beet. Theor Appl Genet 102: 1299–1304. https://doi.org/10.1007/s001220100578

Durrant MJ, Loads AH (1990) Some changes in sugar-beet seeds during maturation and after density grading. Seed Sci Technol 18:11–21

Durrant MJ, Dunning RA, Jaggard K, Bugg RB, Scott RK (1988) A census of seedling establishment in sugar-beet crops. Ann Appl Biol 113:327–345

European Seed Certification Agencies Association ESCAA (2020) Field production area for seeds (escaa.org)

Food and Agriculture Organization of the United Nations (FAO) (2009) Responding to the challenges of a changing world: the role of new plant varieties and high quality seed in agriculture. In: Proceedings of the second World Seed Conference. FAO Headquarters, Rome, September 8–10, 2009. https://www.fao.org/3/am490e/am490e.pdf

Forster V, Marrese D, Staska K, Trinks, K, Stander JR (1997) Petition for determination of nonregulated status: Glufosinate tolerant sugar beet, transformation event T1207. Agr Evo USA Company, Wilmington, Delaware. 67 pp

Fukui R (1994) Growth patterns and metabolic activity of Pseudomonads in sugar beet spermospheres: relationship to pericarp colonization by Pythium ultimum. Phytopathology 84:1331

Giordani G (2013) Guida alla moltiplicazione del seme di barbabietola.CC-BY-NC-ND 3.0

Gratwick M (1992) Beet cyst nematode. In: Gratwick M (ed) Crop pests in the UK. Springer, Dordrecht. https://doi.org/10.1007/978-94-011-1490-5_78

Grimwalde JA, Grierson D, Whittington WJ (1987) The effect of differences in time to maturity on the quality of seed produced by different varieties of sugar beet. Seed Sci Technol 15:135–145

Hafez SL, Seyedbagheri MM (1997) Sugar beet cyst nematode: impact on sugarbeet production in Idaho and Eastern Oregon. University of Idaho, College of Agriculture, Cooperative Extension System, Agricultural Experiment Station

Håkansson I, Arvidsson J, Rydberg T (2011) Effects of seedbed properties on crop emergence: 2. Effects of aggregate size, sowing depth and initial water content under dry weather conditions. Acta Agric Scand Sect B Soil Plant Sci 61:469–479. https://doi.org/10.1080/09064710.2010.506447

Hauer-Jákli M, Hansen AL, Manderyck B, Olsson Å, Raaijmakers E, Hanse B, Stockfisch N, Märländer B (2016) Neonicotinoids in sugar beet cultivation in Central and Northern Europe: efficacy and environmental impact of neonicotinoid seed treatments and alternative measures. Crop Prot:93. https://doi.org/10.1016/j.cropro.2016.11.034

Hemayati SS, Taleghani DF, Shahmoradi S (2008) Effects of steckling weight and planting density on sugar beet (*Beta vulgaris* L.) monogerm seed yield and qualitative traits. Pak J Biol Sci 11(2): 226–231

Hermann K, Meinhard J, Dobrev P, Linkies A, Pešek B, Hess B, Machácková I, Fischer U, Leubner G (2007) 1-Aminocyclopropane-1-carboxylic acid and abscisic acid during the germination of sugar beet (Beta vulgaris L.): a comparative study of fruits and seeds. J Exp Bot 58:3047–3060. https://doi.org/10.1093/jxb/erm162

Ignatz M, Hourston JE, Turečková V, Strnad M, Meinhard J, Fischer U, Steinbrecher T, Leubner-Metzger G (2019) The biochemistry underpinning industrial seed technology and mechanical processing of sugar beet. Planta 250(5):1717–1729. https://doi.org/10.1007/s00425-019-03257-5

ISTA (2021) International Rules for Seed Testing. Bassersdorf, Switzerland. https://doi.org/10.15258/istarules.2021

Kazmeinkhah K (2007) Determination of energetic and ergonomic parameters of a semi-automatic sugar-beet steckling transplanter. J Agric Sci Technol 9:191–198

Kirchhoff M, Svirshchevskaya A, Hoffmann CM, Schechert A, Jung C, Kopisch-Obuch FJ (2012) High degree of genetic variation of winter hardiness in a panel of *Beta vulgaris* L. Crop Sci 52: 179–188. https://doi.org/10.2135/cropsci2011.04.0185

Kockelmann A, Meyer U (2006) Seed production and quality. In: Draycott AP (ed) Sugar beet. Blackwell Publishing Ltd, pp 89–113. https://doi.org/10.1002/9780470751114.ch5

Kockelmann A, Tilcher R, Fischer U (2010) Seed production and processing. Sugar Tech 12:267–275. https://doi.org/10.1007/s12355-010-0039-z

Landová M, Hamouzová K, Soukup J, Jursík M, Holec J, Squire GR (2010) Population density and soil seed bank of weed beet as influenced by crop sequence and soil tillage. Plant Soil Environ 56:541–549

Lehnhardt L, Bonk M (1991) Untersuchungenzum Einflusseiner Bordungung bluhender Zuckerrubenp flanzenauf Entwicklung und Eigenschaftenihrer Samen. Brandenburgische Landeshochschule Potsdam 35:65–71

Longden PC (1986) Influence of the seed crop environment on the quality of sugar beet seed. Proceedings of the 49th IIRB Congress. Brussels, p 1–16

Longden PC, Clarke NA, Thomas TH (1995) Control of bolting and flowering in sugar beet (*Beta vulgaris*). Proc 58th Congr Int Inst Sugar Beet Res, June 1995, Beaune, p 71–81

Mall A, Misra V, Santeshwari, Pathak AD (2020) Quality seed production of sugar beet in India. In: Tiwari AK (ed) Advances in seed production and management, pp 139–159. https://doi.org/10.1007/978-981-15-4198-8_7

Mall AK, Misra V, Santeshwari PAD, Srivastava S (2021) Sugar beet cultivation in India: prospects for bioethanol production and value added co-products. Sugar Tech 23:1218–1234

Mann V, McIntosh L, Theurer C, Hirschberg J (1989) A new cytoplasmic male sterile genotype in the sugar beet *Beta vulgaris* L.: a molecular analysis. Theor Appl Genet 78(2):293–297. https://doi.org/10.1007/BF00288814

Manzoor A, Ahmad T, Bashir MA, Hafiz IA, Silvestri C (2019) Studies on colchicine induced chromosome doubling for enhancement of quality traits in ornamental plants. Plan Theory 8(7): 194. https://doi.org/10.3390/plants8070194

Märländer B, Lange T, Wulkow A (2011) J Kult pflanzen 63(11):349–373

Meyerholz KH (1999) Erzeugung von Zuckerrübensaatgut. In: Keller ER, Hanus H, Heyland KU (eds) Handbuch des Pflanzenbaus Band 3: Knollen- und Wurzelfrüchte, Körner- und Futterleguminosen. Eugen Ulmer, Stuttgart

Mikami T, Yamamoto M, Matsuhira H, Kitazaki K, Kubo T (2011) Molecular basis of cytoplasmic male sterility in beets: an overview. Plant Genet Res 9:284–287. https://doi.org/10.1017/S1479262111000177

Nardi L (1998) Repiquage des plants: bien implanter, por bien multiplier. Bull Semences 142:33–34

Noli E, Montanari M, Rossi Pisa P (2007) Sugarbeet seed's yield and quality as affected by irrigation. Ital J Agron 2(s2):207–216. https://doi.org/10.4081/ija.2007.s2.207

Nørremark M, Griepentrog H (2004) Analysis and definition of the close-to-crop area in relation to robotic weeding. 6th Workshop of the EWRS Working Group "Physical and Cultural Weed Control", 8 March 2004, Lillehammer, Norway

OECD (2003) Guideline for the testing of chemicalsproposal for a new guideline 227. Organisation for Economic Co-operation and Development

OECD (2012) A synthesis of international regulatory aspects that affect seed trade. https://www.oecd.org/agriculture/seeds/documents/international-regulatory-aspects-seedtrade.pdf
OECD (2020) Feasibility study on the digitalization of OECD Seed Schemes labelling and certification processes. Digital technologies for seed trade, traceability and certification (oecd.org)
OECD (2021) Schemes for the varietal certification or the control of seed moving in international trade. http://www.oecd.org/agriculture/code/seeds.htm
Official Journal of the European Union (2002) Council Directive 2002/54/EC of 13 June 2002 on the marketing of beet seed
Official Journal of the European Union (2018a) Commission Implementing Regulation (EU) 2018/785. https://eur-lex.europa.eu/legal-content/EN/TXT/?uri=CELEX%3A32018R0785
Official Journal of the European Union (2018b) Commission Implementing Regulation (EU) 2018/784. https://eur-lex.europa.eu/legal-content/EN/TXT/?uri=CELEX%3A32018R0784
Official Journal of the European Union (2018c) Commission Implementing Regulation (EU) 2018/783. https://eur-lex.europa.eu/legal-content/EN/TXT/?uri=CELEX%3A32018R0783
Orzeszko-Rywka A, Podlaski S (2003) The effect of sugar beet seed treatments on their vigour. Plant Soil Environ 49:249–254
Panella L (ed) (1996) Report on the status of Beta Germplasm in the United States. Prepared by the Sugar beet Crop Germplasm Committee, p 41
Pospisil M, Mustapic Z (1999) Effect of stand density and nitrogen fertilization on the yield and quality of sugar beet seed. Rostlinna Vyroba 45(7):305–309
Raaijmakers E (2014) Use of green manure crops and sugar beet varieties to control *Heteroderabetae*. Commun Agric Appl Biol Sci 79:309–320
Ran Z, Michaelis G (1995) Mapping of a chloroplast RFLP marker associated with the CMS cytoplasm of sugar beet (*Beta vulgaris*). Theor Appl Genetics 91:836–840. https://doi.org/10.1007/BF00223889
Rasmusson J, Levan A (1939) Tetraploid sugar beets from colchicine treatments. Hereditas 25:97–102
Reinsdorf E, Koch HJ (2013) Modeling crown temperature of winter sugar beet and its application in risk assessment for frost killing in Central Europe. Agric For Meteorol 18:21–30. https://doi.org/10.1016/j.agrformet.2013.08.001
Reinsdorf E, Koch HJ, Märländer B (2013) Phenotype related differences in frost tolerance of winter sugar beet (*Beta vulgaris* L.). Field Crop Res 151:27–34
Rinaldi M, Horemans S (2012) Sugar beet. In: FAO Irrigation and drainage paper 66. Crop yield response to water. Food and Agriculture Organization of the United Nations, Rome (Italy)
Romaneckas K, Pilipavicius V, Šarauskis E, Sakalauskas A (2009) Effect of sowing depth on emergence and crop establishment of sugar beet (*Beta vulgaris* L.). J Food Agric Environ 7(2): 571–575 ref.15
Roquigny C, Lejosne M (1988) Betteravesucrièreportegraine: contribution à la détermination du stade optimal de récolte.Deleplanque et Cie, Maisons-Lafitte, France
Salimi Z, Boelt B (2019) Optimization of germination inhibitors elimination from sugar beet (*Beta vulgaris* L.) seeds of different maturity classes. Agronomy 9(11):763. https://doi.org/10.3390/agronomy9110763
Scott RK (1970) The effect of weather on the concentration of pollen within sugar-beet seed crops. Ann Appl Biol 66:119–127
Shevchenko AG, Logvinov AV, Mishchenko VN, Koshkin SS (2021) State, problems and prospects for the development of sugar beet seed production in the Krasnodar Territory. E3S Web Conf 285:02005. https://doi.org/10.1051/e3sconf/202128502005
Sliwinska E (2003) Cell cycle and germination of fresh, dried and deteriorated sugarbeet seeds as indicators of optimal harvest time. Seed Sci Res 13:131–138. https://doi.org/10.1079/SSR2003131
Smit AL (1983) Influence of external factors on growth and development of sugar-beet (*Beta vulgaris* L.). Agricultural Research Report Wageningen 914

Smith GA (1987) Sugar beet. In: Fehr WR (ed) Principles of cultivar development, Macmillan, vol 2. New York, NY, pp 577–625
Smith HJ, Gray FA, Koch DW (2004) Reproduction of *Heterodera schachtii* Schmidt on resistant mustard, radish, and sugar beet cultivars. J Nematol 36(2):123–130
Snyder FW (1971) Relation of sugar beet germination to maturity and fruit moisture at harvest. J Assoc Am Sugar Beet Technol 16:541–551
Srivastava SN (2004) Management of sugarbeet diseases. In: Mukerji KG (ed) Fruit and vegetable diseases. Disease management of fruits and vegetables, vol vol 1. Springer, Dordrecht. https://doi.org/10.1007/0-306-48575-3_9
TeKrony DM (1969) Seed development and germination of monogerm sugar beets (*Beta vulgaris* L.) as affected by maturity. Ph.D. thesis, Oregon State University
Thibaud N (2002) Itineraires de recolte: la tendence est a la dessicationchimique. Bull Semences 166:31–32
Tognetti R, Palladino M, Minnocci A, Delfine S, Alvino A (2003) The response of sugar beet to drip and low-pressure sprinkler irrigation in southern Italy. Agric Water Manag 60:135–155. https://doi.org/10.1016/S0378-3774(02)00167-1
Tuğrul KM, Kaya R (2020) The effect of seed coating thickness on sugar beet (Beta vulgaris L.) yield and quality under different irrigation conditions. Appl Ecol Environ Sci 18:6969–6979. https://doi.org/10.15666/aeer/1805_69696979
USDA (2015) USDA coexistence fact sheets. Sugar Beets. Office of Communications. United States Department of Agriculture
ViricGasparic H, Grubelic M, Verica DU, Bažok R, Cacija M, DrmićZ LD (2020) Neonicotinoid residues in sugar beet plants and soil under different agro-climatic conditions. Agriculture 10: 484. https://doi.org/10.3390/agriculture10100484
ViricGasparic H, Lemic D, Drmić Z, CacijaM BR (2021) The efficacy of seed treatments on major sugar beet pests: possible consequences of the recent neonicotinoid ban. Agronomy 11:1277. https://doi.org/10.3390/agronomy11071277
Wang Y, Stevanato P, Yu L, Zhao H, Sun X, Sun F, Li J, Geng G (2017) The physiological and metabolic changes in sugar beet seedlings under different levels of salt stress. J Plant Res 130(6): 1079–1093. https://doi.org/10.1007/s10265-017-0964-y
Weiland J, Koch G (2004) Sugarbeet leaf spot disease (*Cercosporabeticola* Sacc.). Mol Plant Pathol 5:157–166
Wood DW, Scott RK, Longden PC (1980) The effects of mother plant temperature on seed quality in *Beta vulgaris* L. (sugar beet). In: Hebblethwaite PD (ed) Seed production. Butterworths, London, pp 257–270

India's Sugar Beet Seed Technology and Production

7

Ashutosh Kumar Mall, Varucha Misra, Santeshwari Srivastava, and A. D. Pathak

Abstract

Sugar beets are a temperate crop with a lot of potential for growing in tropical and subtropical climates. The Indian Institute of Sugarcane Research has been exploring this crop for over seven decades, and the researchers have presented evidence for sugar beet growth and development in India's tropical and subtropical climates. With the release of numerous commercial cultivars for Indian agro-climatic conditions, along with the development of agro-technologies pertaining to crop production, protection, and machinery, seed is the main ingredient that plays a significant role in sugar beet establishment. The findings of many research projects conducted at IISR resulted in the selection of seed production sites and fine-tuning of technologies in India. Seed to seed and steckling methods have been developed by IISR to achieve optimal production in this aspect. Sugar beet seed production is the most pressing challenge in India's commercialization of this crop. In the current prevailing bioethanol production scenario, for making this crop a choice of farmers for bioethanol production, the utmost constrain is the seed availability. Designing a contract farming business model in a mission mode with mutual consent of the relevant government policy, industrial entrepreneurship, and a dedicated agriculture department can be an effective step toward spreading this crop in the Indian context.

Keywords

Indigenous · Multigerm · Seed · Sugar Beet · Variety

A. K. Mall (✉) · V. Misra · S. Srivastava · A. D. Pathak
ICAR-Indian Institute of Sugarcane Research, Lucknow, India
e-mail: Ashutosh.Mall@icar.gov.in

V. Misra et al. (eds.), *Sugar Beet Cultivation, Management and Processing*,
https://doi.org/10.1007/978-981-19-2730-0_7

7.1 Introduction

Sugar beet cultivation is being done worldwide for the production of sugar and other by-products. India is among the sugar beet producing countries worldwide; however, the production currently is not on large scale. Different agro-climatic conditions occurring in India are the key to the establishment of this crop in tropical and subtropical regions. Deol and Kanwar (1975) had shown that the end of September to mid of October is the favourable time for sugar beet cultivation in Indian climatic conditions. Soils having neutral to slightly alkaline in nature with a favourable pH of 7.0–8.5 favour the flourishing growth of sugar beet. The black cotton soils found in the Deccan tracts of Maharashtra, Karnataka, Andhra Pradesh, and Tamil Nadu are also good for sugar beet cultivation (Mall et al. 2021). A fairly cool climate with an adequate amount of rainfall, bright sunshine during different phases of growth helps in flourishing healthy sugar beet crops (Pathak et al. 2014).

For the cultivation of a successful crop, the seed is a master key per unit area. It is a basic unit of embryonic plant in agriculture and an important phenomenon of biological existence on which future plant develops. It is a living organism embedded in the supporting or the food storage tissue. Seed is the cheapest input used by the farmers but its efficiency increases the factor of crop production. Seed quality is an important aspect of good yield and production for any crop (Misra et al. 2020). Kanwar and Pawar (2017) had projected that 20–25% of productivity is affected by seed quality. Quality seeds are defined as having high germination percentage with varietal purity, disease, and pest-resistant with proper weight and moisture content. Production of quality seeds in sugar beet crop too plays the same impact as does the quality seed in other crops (Mall et al. 2020). This chapter discusses the seed production process in India, emphasizing multigerm seed production and the development of indigenous sugar beet varieties for Indian agro-climates.

7.2 Seed Production in Indian Hills

A prerequisite condition for flowering and seed production in sugar beets is heat induction (Kapur et al. 1986). Sugar beet grown in the plains lacks this need, preventing it from flowering in such conditions. As a result, seed production has been standardized at higher elevations (>5000 feet) where good climatic conditions exist. Srinagar (Jammu and Kashmir), Mukteswar & Ranichauri (Kumaon Hills), Darjeeling (West Bengal), Shimla & Kalpa (Himachal Pradesh), Auli (Garhwal Hills) are the chosen sites where seed production technologies have been established (Pathak et al. 2011). The Srinagar valley in Jammu and Kashmir provides the finest location for sugar beet seed production compared to the other chosen sites due to no rainfall during the ripening stage of seeds coinciding in months of July–August. Ramonskaya-06 (R-06), a Russian variety, was the first variety whose seeds were successfully produced in India. Pathak et al. (2014) had shown that with the advance in sugar beet research in India, development and production of indigenous varieties were also initiated with achievements.

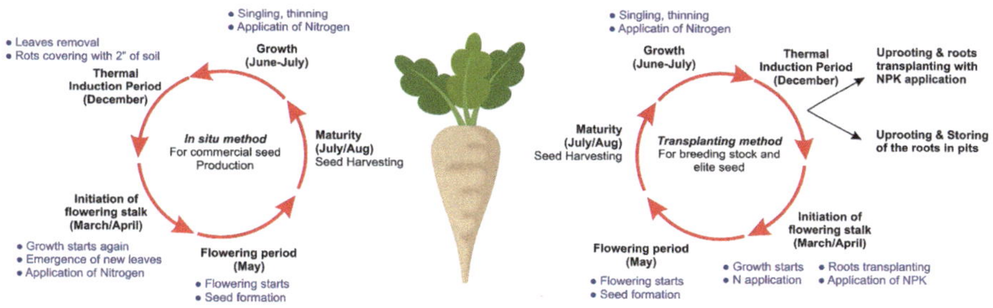

Fig. 7.1 Overview of sugar beet seed production methods. Different stages and duration with cultural practices in transplanting and in situ method in Indian conditions

Sugar beet seeds are grown either in situ or through transplantation. The in situ approach necessitates overwintering conditions for the beet to blossom and generate seeds. As overwintering is essential, this method owes its name to it. In the transplanting process, in the first season, selected steckling (or young roots) is uprooted and placed in trenches for overwintering conditions before being thermally inducted. The steckling is replanted in the next season during overwintering conditions to generate flowers and seeds. Sugar beets cultivated in India's plains are commonly subjected to this procedure (Kapur et al. 1986). Cutting one-third of the top and root area prior to planting is required for this process. Maintaining only 10 cm of leaves is a crucial habit. According to Balan et al. (1991), the transplanting procedure requires a root weight of 7–16 g with a pair of five to ten leaves and a plant height of 22–26 cm. Seed development takes place inside the sugar beet fruit (OCED 2006). The pericarp and operculum of the seed are joined by an ovary cap (Mall et al. 2021). The breeder and foundation seed can be produced *via* the transplanting method, which involves moving 2-month-old plants (stecklings) cultivated in the lowlands to the hills before winter. This method provides for the selection of roots as well as the annualization of seed production, whereas the in situ method allows for the production of certified seed. Seed production technology has been standardized. A flow diagram of the two seed production process undergoing in India is depicted in Fig. 7.1.

In India, multigerm seeds had been produced at various centres. Regular production of foundation seeds of multigerm type was reported in Kalpa centre, Himachal Pradesh. National Seeds Corporation Ltd., Govt of India, was producing and multiplying these seeds on a mass basis in 1970–80 (Pathak et al. 2014). ICAR-IISR, Lucknow (UP) is currently producing the unpelleted multigerm seeds of indigenous commercial varieties (Mall et al. 2020, 2021).

7.3 Multigerm Seeds and Its Development

The production of multigerm seed balls through fused petals in flowers grown in clusters gives birth to multigerm seeds (Anonymous 2010). In a multigerm plant, the number of flowers in a cluster may vary, irrespective whether the flower cluster is from the same inbred line or the same clone. Besides, the fruit size has also been reported to distinguish from each other depending on the growing condition. Both these attributes, *viz.* change in flower size in seed ball and number in cluster complicate its genetic study (Savitsky 1983). The germs in these seeds also vary in their weights. The large-sized germs cannot be removed from small sized with the help of segregation and seed cleaning methods.

The understanding of the development of multigerm seed in sugar beet plants is important from a breeder's point of view (Fig. 7.2). Meristematic tissues make the apex of the floral axis, and cells of these are small with dense protoplasm, having the capability of quick growth and cell division. The meristematic tissues are the main character of the story of multigerm seed. In proportion to the seed stalk, these tissues make new primordia of bracts on side of the growing stalk. Primordia of flowers (in small size) are formed on the edges of the apex of these tissues while seed balls are in the axils of the bracts. Rapid growth in the protuberances that separate receptacle and peduncle was seen. Flower production occurs on each receptacle. The primordium of the second flower in the seed is produced either prior to the sepal's protuberance arising on the first receptacle or when they are about to arise. The edges of the meristem of the first receptacle move towards the lower side of the peduncle so that it could cover the peduncle. An inflammation, small in size, is

Fig. 7.2 The formation of multigerm seeds in sugar beet. The bottoms of all flower's ovaries are buried in the same peduncle tissue. The upper side of the buds grows over the peduncle which looks as if each bud is progressing distinctly, however, it is not so

formed with cells of meristematic tissues on side of peduncle. This develops into new receptacle *via* protuberance formation. The 2-4 protuberances develop from these receptacles on a single peduncle, corresponding to the number of flowers in the seed ball which develop at the same time. The bottoms of all flowers' ovaries are buried in the same peduncle tissue. The tissue of the peduncle is the connecting link between the flowers which produces seed ball; however, the upper side of the buds grows over the peduncle (initiating from sepals) which looks as if each bud is progressing distinctly. It is assumed that though flowers are borne at the same time, yet they don't progress together (Savitsky 1950).

The complex and tedious part of the multigerm seed of sugar beet is its processing. Abrasion of the seed is required to destroy the multigerms found in these types of seeds. This will increase the monogermity of the seed. This must be accomplished without causing too much impact on the germination percentage. Some germination damage is expected, and this must be remedied using a gravity separation procedure for parting light weight from non-germinating seeds (Anonymous 1975).

7.4 Multigerm Seed Grading

Multigerm seed grading begins with the removal of inert material present in these seeds with the help of air screen cleaner. Subsequently, usage of a draper belt aids in the eradication of further left over any stem or inert material. Later on, the left over multigerm seed is sorted out through the use of a thickness grader. This grader discards off the large-sized multigerm seeds and only allows the monogerm sized seeds. Then, to improve seed quality and shelf life, fungicidal and insecticidal treatments are applied.

7.5 Sugar Beet Indigenous Varieties for Commercial Production in India

Approximately 300 genotypes were tested for economic qualities and adaptability in the early years. These included diploid open-pollinated variants, anisoploid varieties, hybrid triploids, and diploid hybrids, among other things. Ramonskaya-06 (R-06), a diploid open-pollinated multigerm Russian cultivar, was deemed viable for commercial production after thorough root crop studies. Maribo Magnapoly, Maribo Resistapoly, Tribel, Hh Raspoly, and triploids Martus, Solid, and Kristal are anisoploid while Martus, Solid, and Kristal in triploids varieties were developed for growth in Indian conditions. In addition, the varieties PAC 60006, PAC 60008, Felicita, Calixta, Arriba have recently been identified as promising for the Indian agro-climates. Srivastava (1995) had shown that diploid multigerm germplasm, *viz.* LS 6, has been developed and released for commercial purposes in India. IISR 2 and LS 7 were the two further promising diploid elite lines produced at GBPUAT, Pantnagar, Uttarakhand for cultivation in India. Both elite lines were chosen using their business. In doing so the seed availability becomes assured. A contract farming business model must be designed, in which diverse stakeholders commit to their specialized roles in the overall operation. Such time of contracts has been made in past but due to non-support of government policies and other issues, this crop was not able to sustain in India as it can be. The sudden rise in demand for sugar beet will now aid in developing the components of this model. The majority of the components have already been established, and a mission mode under mutual consent of the right government policy, industrial entrepreneurship, and a dedicated agriculture department will do wonders.

7.9 Conclusion

Crop improvement advances can be transferred to the farmer's field by sowing genetically pure seed of improved varieties. Sugar beet seed production performs well in this regard. Sugar beet is a biannual crop that can only be seeded on high-altitude mountains. The two viable seed generation strategies are in situ and trans-plantation. Sugar beet clones can be established from crown buds or seed stalk cuttings, and vegetative propagation provides a distinct advantage. Direct sowing affects days to bolting, stalk formation, first flowering, sub-branches, stalk length, test weight, germination, and yield. Planting on a ridge with correct plant to plant distance is found to be the best method for obtaining quality seed.

The problem will not be solved by relying solely on one region. The development of a technique to sustain seed production in areas where the cold temperature no longer exists is an economically and environmentally desirable answer. Depending on the eco-geographical region, the procedure of seed production and the selection of a suitable location differ. Temperature, rainfall, and microclimatic conditions all have a role in determining the suitability of a specific method.

References

Anonymous (1975) Sugar beet seed production and marketing in the United Kingdom. Outlook Agric 8(5):284–289. https://doi.org/10.1177/003072707500800507

Anonymous (1988) All India Co-Ordinated Research Project on Sugar Beet (ICAR). General recommendations

Anonymous (2010) Environment directorate joint meeting of the chemicals committee and the working party on chemicals, pesticides and biotechnology, p 17

Anonymous (2016–17) Annual Report ICAR-IISR, Lucknow

Anonymous (2017–18) Annual Report ICAR-IISR, Lucknow

Anonymous (2018–19) Annual Report ICAR-IISR, Lucknow

Anonymous (2019–20) Annual Report ICAR-IISR, Lucknow

Balan VM, Kirichenko VN, Zhovtonochuk LY (1991) Stand density for seed production sugar beet grown without transplanting. Vestinik Selskogozyaistvennoi Nauki Moskva 7:97–100

Deol DS, Kanwar RS (1975) Effect of the dates of sowing and harvesting on the yield and quality of sugarbeet. Crop Improv 2:105–111

Kanwar HS, Pawar R (2017) Current status of vegetable seed industry in India and future prospects. In: Advances in quality seed production of vegetable crops. Centre of Advanced Faculty Training in Horticulture (Vegetables), Solan, pp 1–6

Kapur R, Srivastava BL, Srivastava HM (1986) A new technique for vegetative propagation of sugarbeet seed crop. Indian Sugar Crops J 4:16–17

Mall AK, Misra V, Santeshwari, Singh BD, Pathak AD (2020) Quality seed production of sugar beet in India. In: Tiwari AK (ed) Advances in seed production and management. Springer Nature Singapore Pte Ltd., pp 139–159

Mall AK, Misra V, Santeshwari PAD, Srivastava S (2021) Sugar beet cultivation in India: prospects for bioethanol production and value added coproducts. Sugar Tech 23:1218–1234

Misra V, Mall AK, Pathak AD (2020) Sugarbeet: a sustainable crop for salt stress conditions. In: Mirza H (ed) Agronomic Crops. Springer, Singapore, pp 40–62

Mukhopadhyay AN (1971) Sclerotium root rot of sugarbeet in India. Mycopathol Applicanta 44: 265–370

OECD (2006) Section 8-sugar beet (*Beta vulgaris* L.). In: Safety Assessment of Transgenic Organisms. OECD, Paris, p 1. ISBN 9789264095380

Pathak AD, Kapur R, Kumar R, Vishwakarma MK (2011) Impact of different vernalization treatments on flowering and seed production in sugarbeet (*Beta vulgaris* L.). Indian J Sugarcane Technol 26(1):24–27

Pathak AD, Kapur R, Solomon S, Kumar R, Srivastava S, Singh PR (2014) Sugar beet: a historical perspective in Indian context. Sugar Tech 16(2):125–132. https://doi.org/10.1007/s12355-014-0304-7

Savitsky H (1950) Embryology of mono and multigerm fruits in the genus Beta L. Proc 6th General Meet ASSBT 6:160–164

Savitsky VF (1983) Inheritance of the number of flowers in flower clusters of *Beta vulgaris* L. Proc 8th General Meet Am Soc Sugar Beet Technol 8:4–15

Solomon S, Sharma AK, Shukla SK (2014) Technorama IISR. ICAR-Indian Institute of Sugarcane Research, Lucknow, pp 57–58

Srivastava HM (1990) Development of diploid and polyploid hybrids of sugar beet for sub-tropical climate and their seed production potentials. In: International Conference on seed Science and technology. International conference on seed science and technology, New Delhi Feb 21–25, 1991, p 8

Srivastava HM (1991) Breeding and Improvement of sugar beet for Sub Tropical Climate. In 2nd International Beta Genetic Resources Workshop. FAL, Braunschweig, Germany, June 24–28, 1991. International Crop Network Series, No. 7, IBPGR 70-71

Srivastava HM (1995) Sugar beet pre-breeding in India. J Sugar Beet Res 32(2 & 3):99–110

Artificial Seed Technology

8

Archana Siraree

Abstract

Artificial seeds are produced by encapsulating somatic embryos, shoot tips, or any other micropropagule which have the ability to convert into a plant in vitro or ex vitro. The need of artificial seed production was felt due to failed seed propagation in some crop species due to very small seed size, seed heterozygosity, reduced endosperm, no germination in the absence of seed–mycorrhizal association as in case of orchids and also time-consuming vegetative means of propagation in some seedless varieties of crops such as *Citrullus lanatus* and *vitis vinifera,* etc. Effective seed coating of micropropagules is done using different gelling agents such as alginate, agar, carrageenan, gellan gum, sodium pectate and carboxy methyl cellulose. However, sodium alginate has been documented as most frequently used gelling agent. The absence of seed coat and endosperm in somatic embryos necessitates the encapsulation matrix to be supplemented with nutrients and growth regulators such as 0.5 mg/L indoleacetic acid (IAA), 0.5 mg/L naphthalene acetic acid (NAA), 2 mg/L 6-benzyl aminopurine (BA), 2 mg/L Fe-EDTA and 30 g/L sucrose. In many plant species such as *Allium sativum, Ananas comosus, Dioscorea bulbifera, Cineraria maritima, Cucumis sativus,* etc. genetic stability of the plants derived from artificial seeds has also been examined with the help of biochemical and molecular markers and found them genetically consistent.

Keywords

Artificial seed · Encapsulation · Sodium alginate · Somatic embryos

A. Siraree (✉)
Division of Seed Production, Genda Singh Sugarcane Breeding and Research Institute, Kushinagar, Uttar Pradesh, India

V. Misra et al. (eds.), *Sugar Beet Cultivation, Management and Processing*, https://doi.org/10.1007/978-981-19-2730-0_8

Abbreviations

BA Benzyl Aminopurine
IAA Indoleacetic Acid
NAA Naphthalene Acetic Acid

8.1 Introduction

Encapsulated somatic embryos (raised from tissue culture), shoot tips, embryonic calluses, axillary buds, or any other micropropagules which can be planted as seed and have the ability to develop in to a whole plant in vitro and ex vitro are called as artificial or synthetic seed (Capuano et al. 1998; Ara et al. 2000; Rihan et al. 2011). The concept of coating somatic embryos and using them with the same ease as a normal seed was first given by Murashige in the year 1977. He defined artificial seed as "an encapsulated single somatic embryo". Later on Gray et al. (1991) defined artificial seed as "a somatic embryo that is engineered for the practical use in commercial plant production". Initially artificial seed production was only confined to somatic embryos and therefore has been utilized in only those plant species in which successful production of somatic embryos could be well exhibited (Rihan et al. 2017). Later, with the report of shoot tip coating by Bapat et al. (1987) in *Morus indica*, the definition of artificial seed has been modified as "an encapsulated somatic embryo or in vitro raised other plant propagules which are capable to develop in to a plant when grown in vitro or ex vitro with the characteristic of prolonged storage" (Daud et al. 2008; Saiprasad 2001; Ara et al. 2000). The objective of producing artificial seed was to promote cost effective and large-scale multiplication of superior plant genotypes or commercially valuable plant species (Reddy et al. 2012; Saiprasad 2001). The need of technological interventions in vitro was felt so that the problems such as failed seed propagation in some crop species due to very small seed size, heterozygosity of seed, reduced endosperm, no germination in the absence of seed–mycorrhizal association as in case of orchids could be addressed. Also, the time-consuming vegetative means of propagation in some seedless varieties of crops such as *Citrullus lanatus* and *vitis vinifera,* etc. (Saiprasad 2001) could be supplemented to certain extent. The prevailing traditional breeding system in coniferous forest species is cumbersome due to their prolonged life cycle. Attainment of better progeny is not always possible because of the heterogeneous nature of coniferous forest species. Artificial seeds can play a very important role in cloning of these trees at reduced cost and time (Desai et al. 1997). Hybrid seed production by hand pollination in some vegetable crops like tomato and seedless watermelon is very labour intensive and therefore responsible for the increased seed cost. Similarly, vegetative means of propagation is also very time consuming. Presence of cleistogamous flowers in cotton and soyabean increases the production cost of hybrid seeds at commercial level since pollination is done by hand. Significant reduction in the cost may be expected by developing synthetic seeds in such species by economizing labour and also, time and space constraints may also be dealt with (Chee and Cantliffe 1992; Tian and Brown 2000). Plants raised from artificial seeds have also been tested for their genetic stability in various plant species such as *Allium sativum*, *Ananas comosus*, *Dioscorea bulbifera*, *Cineraria maritima*, *Cucumis sativus,* etc., using biochemical and molecular tools and found them genetically consistent (Srivastava et al. 2009; Gangopadhyay et al. 2005; Narula et al. 2007; Tabassum et al. 2010; Bekheet 2006).

8.2 Advantages of Artificial Seed

Germplasm conservation through cryopreserving artificial seeds particularly in desiccation sensitive species like mango, cocoa, coconut, etc., utilization in hybrid seed production, i.e. use of artificial seed in propagating those plants which exhibit male or female sterility, multiplication of polyploid species, freedom from pathogens, easy handling during storage, transport feasibility, long term storage with no viability loss, and maintenance of the clonal nature of plants by using genetically identical somatic embryos, suitable medium to deliver the novel plant lines obtained through biotechnological means straight to the greenhouse or field, cost effective large scale propagation of superior plant varieties, etc. are some of the advantages of artificial or synthetic seed (Saiprasad 2001). Artificial seeds have been produced in various plant species including vegetable crops, fruit crops, medicinal plants, cereals, orchids, sugar crops, and forest trees (Siong et al. 2012; Masri et al. 2019; Shallal et al. 2020; Ismail et al. 2016; Rslan 2018; Bekheet 2006; Tsai and Saunders 1999; Bapat et al. 1987; Jain et al. 2018; Roy and Mandal 2008; Nieves et al. 2003).

8.3 Production of Artificial Seeds: The Prerequisites

8.3.1 Explants

The most commonly used explants are somatic embryos because they contain apical and basal meristem which gives rise to shoot and root (Ara et al. 2000), reproduction level is more; plants raised from somatic embryos are proficient and retain their regenerative capacity for a longer duration resulting in to uniform plant population (Leroy et al. 2000). Artificial seed production with the help of somatic embryos has been reported in *Gentiana kurroo* (Kotvi et al. 2016), *Daucus carota* (Kitto and Janick 1982), *Medicago sativa* (Gupta and Durzan 1987), *Vitis vinifera*, *Mangifera indica* (Ara et al. 1999), *Citrus reticulate* (Antonietta et al. 1999), *Saccharum spp.* hybrid (Nieves et al. 2003), *Oryza sativa* (Kumar et al. 2005), *Plumbago zeylanica* L. (Jain et al. 2018), etc.

Somatic embryos are the bipolar structures developed from somatic cells, instead of zygotes by means of somatic embryogenesis and thus used for clonal propagation

(Saiprasad 2001). There are two routes, i.e. direct and indirect, by which somatic embryogenesis can be induced in vitro. Formation of somatic embryos takes place at the side of an explant in direct embryogenesis, whereas in case of indirect way it takes place through the growth of an unorganized mass of cells called callus (Quiroz-Figueroa et al. 2006).

8.3.2 Shoot Tips, Axillary Buds, Internode Cuttings, Microshoots

Production of artificial seed has also been a success in other micropropagules used as explants. Some of the examples are coating of shoot tips (Aida et al. 2012; Masri et al. 2019; Ismail et al. 2016) and internode cutting of *Beta vulgaris* (Ismail et al. 2016), microshoots in *Saintpaulia ionantha* wendl. (Daud et al. 2008) and *Brassica oleracea* var. *botrytis* (Siong et al. 2012), axillary and apical buds of *Manihot esculenta* Crantz (Hegde et al. 2016), nodal segments and shoot tips of *Mimosa pudica* L. (Banu et al. 2014), nodal segments of *C. angustifolia* (Bukhari et al. 2014) and bulblets of *Allium sativum* L. (Bekheet 2006).

8.3.3 Encapsulation of Explants

Different gelling agents are used for the effective seed coating of micropropagules such as alginate, agar, carrageenan, gellan gum, sodium pectate and carboxy methyl cellulose. Based on the properties of being soluble at room temperature and to form hydrogel with calcium chloride, the sodium alginate has been recognized as most frequently used gelling agent (Bapat et al. 1987; Kikowska and Thiem 2011). Encapsulation of embryos, pro-embryos and embryo-like structures of androgenic origin in rice (Roy and Mandal 2008), cauliflower (Siong et al. 2012) and sugar beet was done using sodium alginate as gelling material. Tragacanth gum, carrageenan, polyox, agar, carboxy methylcellulose, guar gum, gelrite, sodium pectate ethyl cellulose and nitrocellulose, agarose, polyacrylamide, polyco 2133, alginate were examined for their suitable use in the synthetic seed production (Ara et al. 2000; Saiprasad 2001; Lambardi et al. 2006). Polyox and resins soluble in water have been found most appropriate for somatic embryos coating (Kitto and Janick 1982). Sodium alginate was found most suitable gelling agent in celery, cauliflower, alfalfa, and carrot (Redenbaugh et al. 1984; Redenbaugh et al. 1986). Encapsulation of explants is done using two types of solutions. One is polymeric solution, and the other is a solution which contains divalent metal ions. Polymeric solution when comes in contact with the solution containing divalent metal forms hydrogel due to the cross-linking reaction. Explants are dipped in the solution of sodium alginate followed by their dropwise placement into the calcium chloride solution for at least 30 min. When drops of sodium alginate touch the calcium chloride, ion exchange occurs between Na^{+} with Ca^{2+} resulting into bead formation. Each bead represents the one explant. Beads so produced are then taken out from the solution of calcium chloride and washed two to three times using sterilized distilled water (Hegde et al.

2016; Banu et al. 2014; Kotvi et al. 2016). These artificial seeds are then transferred to the petri plates containing germination medium enriched with macro and micronutrients from MS medium with additional 30 g/L of sucrose and 7 g/L of agar agar. These plates are kept at 25 °C in complete dark in the culture room (Pond and Cameron 2003). Different combinations of sodium alginate and calcium chloride concentrations have been tested for making artificial or synthetic seeds to achieve best encapsulation efficiency in different plant species. The firmness, size, texture, and shape of the beads are the deciding factors in selecting most suitable combination of sodium alginate and calcium chloride concentrations for encapsulation. In general, the procedure in which mixing of explants is done with 3% concentration of sodium alginate followed by their exposure to 100 mM of calcium chloride has been used most widely. The firm and round beads were produced when encapsulation of nodes of in vitro derived cassava variety and microshoots of African violet (*Saintpaulia ionantha* Wendl.) was done using 3% (w/v) sodium alginate and 100 mM calcium chloride ($CaCl_2$. $2H_2O$) solutions (Hegde et al. 2016; Daud et al. 2008). However, germination frequency under in vitro condition was on the higher side when the encapsulations were made in combinations of 2% and 3% of sodium alginate and 75 mM and 100 mM of $CaCl_2.2H_2O$ in case of cassava (Hegde et al. 2016). In sugar beet, instead of calcium chloride 100 mM calcium nitrate was used as combining solution along with 4% sodium alginate for the encapsulation of microshoots (Masri et al. 2019). Successful encapsulation of somatic embryos in an important medicinal plant *Gentiana kurroo* has also been reported using the combination of 3% sodium alginate and 100 mM calcium chloride (Kotvi et al. 2016). The viability of artificial seed is very much dependent on the material of gel matrix used for encapsulating the plant material. Longer viability of artificial seed must be ensured if the coating material provides protection to explants, exhibits proficiency in inclusion of nutrients, facilitates the storage, handling, and germination of the artificial seed, is non-toxic and compatible with biological and chemical system (Khor and Loh 2005), and sodium alginate was found to be the most suitable seed coating material containing all these characteristics (Saiprasad 2001).

8.3.4 Artificial Endosperm

Unlike zygotic embryos, somatic embryos are devoid of protective seed coat and endosperm which necessitates the coating material to be supplemented with nutrients and growth regulators such as 0.5 mg/L indoleacetic acid (IAA), 0.5 mg/L naphthalene acetic acid (NAA), 2 mg/L 6-benzyl aminopurine (BA), 2 mg/L Fe-EDTA and 30 g/L sucrose, which acts like artificial endosperm (Murashige and Skoog 1962). These added nutrients and growth regulators in the encapsulation material contribute by enhancing germination efficiency and viability of the somatic embryos. It has been suggested that storage of artificial seeds at 4 °C may help to retain their viability for a longer duration, i.e. up to 6 months. Addition of fungicides, pesticides, antibiotics, and microorganism such as rhizobia into the coating material has also been recommended to protect somatic embryos from desiccation and mechanical damage (Saiprasad 2001). Artificial seed production is considered successful only when the produced seeds are vigorous and their conversion efficiency is high. It has been reported that activated charcoal when added into the coating material, the vigour and conversion efficiency of coated somatic embryos were enhanced. When sodium alginate breaks up in the presence of charcoal, an increase in the respiration of somatic embryos occurs. Besides, activated charcoal withholds the nutrients within the gel matrix and also responsible for their moderate release to the growing somatic embryo (Saiprasad 2001).

8.4 Steps of Producing Artificial Seeds

Although the production of artificial seeds may vary among different species as per their type, need and economic viability, but in general, a process of making artificial seed has been outlined by Redenbaugh et al. (1987). The process includes (1) technological and commercial potentiality based crop selection, (2) establishment of species specific procedure for the development of somatic embryo, (3) protocol standardization for the clonal production system in order to obtain viable, mature embryos with the ability to convert in to normal plants, (4) self-regulated embryo production, (5) post-treatment, i.e. induction of quiescence in mature embryos, (6) embryo coating, (7) standardization of artificial endosperm, (8) extensive production of seeds, (9) streamlining the procedures required for plant growth in green house and field conditions, (10) pests and disease control, if any.

8.5 Types of Artificial Seeds

Artificial seeds can be categorized as (1) desiccated or hydrated (Ara et al. 2000; Bapat and Mhatre 2005) and (2) uncoated quiescent or uncoated non-quiescent (Grey 2003).

8.5.1 Desiccated or Hydrated Artificial Seeds

Such seeds are somatic embryos non-coated or coated with polyethylene glycol which are desiccated afterwards. Somatic embryos become quiescent on desiccation resulting into the hardening of protective cover. Handling of such seeds is easy if stored under unsophisticated conditions for a longer duration. On rehydration, protective hard cover softens, and somatic embryos resume growth. Drying is possible by two methods, i.e. rapid and slow. Rapid drying takes place by keeping seeds overnight in open petriplates, whereas slow drying of seeds is attained by reducing the relative humidity over a prolonged period under controlled condition. Such types of artificial seeds can be produced only when somatic embryos are desiccation tolerant (Sharma et al. 2013). Induced desiccation tolerance in somatic embryos was achieved by using maturation medium of high osmotic potential (Sundararaj et al. 2010). The osmotic potential of the medium can be increased by incorporating different osmotic agents such as sucrose, mannitol, etc. or by increasing the strength of the gel. However, attainment of desiccation tolerant somatic embryos has also been reported when some stresses like low temperature or nutrient distress were applied (Pond and Cameron 2003). Somatic embryos coated in hydrogel are called as hydrated seeds, which can be produced in recalcitrant plant species (Ara et al. 2000). It has been reported that encapsulation is important for transferring the micropropagules to the field, provided the material to be used for encapsulation helps in promoting germination (Latif et al. 2007). Also, encapsulation can be seen as a best way for protecting, as well as converting tissue culture derived micropropagules into artificial or synthetic seeds (Redenbaugh 1993).

8.5.2 Uncoated Non-Quiescent or Uncoated Quiescent Artificial Seed

Non-quiescent somatic embryos can be used in crops being raised through micropropagation, whereas quiescent seeds can be stored as germplasm.

8.6 Production of Artificial Seeds in Sugar Beet

Sugar beet (*Beta vulgaris* L.) improvement is primarily done by conventional means but now modern techniques particularly genetic transformation have also been introduced in sugar beet breeding (Ivic et al. 2001). With reference to genetic transformation and tissue culture, sugar beet is considered as recalcitrant species (Elliott et al. 1996; Krens et al. 1996). Also, some superior genotypes of sugar beet have reportedly been propagated through shoots regenerated in vitro (Grieve et al. 1997; Zhong et al. 1993) but it was observed that explants selected from different genotypes showed regenerative variations (Saunders and Tsai 1999). Protocol to produce synthetic seed in sugar beet has been developed for prolonged storability, minimizing cost of production and to facilitate seed handling (Ghosh and Sen 1994).

In sugar beet, shoot tip coating in 4% sodium alginate followed by its multiplication in the medium containing 2 mg/L BA gave best results in terms of leaf and shoot count. However, maximum shoot length was obtained in a medium which contained kinetin @2 mg/L. There was a significant improvement in germination when MS medium along with 30 g/L sucrose was added to the encapsulated seeds. Effective root formation was achieved by using 2 mg/L NAA. Media was also supplemented with osmotic agents, mannitol or sorbitol to examine artificial seed storage. Increased plant survival was reported when media was supplemented with these osmotic agents at 0.5 M concentration. The revelation that sugar beet genotype Francesca was found better than the Toro genotype in reference to its suitability for encapsulation, also suggests that not all the genotypes of a particular species are suitable for developing artificial seeds. Molecular analysis showed that not only the application of mannitol or sorbitol but also their interaction with the genotypes plays a crucial role in the storage of artificial seeds.

Masri et al. (2019) have reported their procedure as cost effective, time saving and suitable for long term conservation of sugar beet germplasm based on the recovery of plants resulted from the artificial seeds stored at 4 °C for a period of 2 months. 4% sodium alginate along with 100 mM $Ca(NO_3).24H_2O$ was used as gel matrix. Vitality of artificial seeds based on the germination was found better in the solution in which 1.3 BAP was added in addition to MS, 3% sucrose, 4% sodium alginate, 2% sorbitol and 2% mannitol.

Ismail et al. (2016) have suggested the solution containing sodium alginate @4%, 1.2% agar, 1.5 mg BA/L and 3% sucrose as best encapsulating solution for artificial seed production. According to them, encapsulation matrix should be enriched with MS, sucrose and BA in order to ensure the germination of artificial seeds. Based on the results of no germination of those artificial seeds (1) coated with only sodium alginate solution or/and (2) coated with sodium alginate and sucrose dissolved in water, they suggested that the coating of artificial seed with sodium alginate and sucrose dissolved in MS is required by the seed to germinate. However, artificial seed exhibited no change in germination frequency when coating was done with 4% sodium alginate dissolved in MS medium containing either 1.3 mg/L BA or 40 g/L sucrose. It has been reported that encapsulated shoot tips germinated earlier than the internode cutting in the sugar beet cultivar Frida, which emphasized that the nature of explants has an important role to play both in production and storability of the synthetic seeds. Shoot tip derived synthetic seeds remained viable for a longer duration as compared to synthetic seeds produced from internode and lost their viability after 2 months. Also, in *M. arvensis* when germination behaviour was compared between encapsulated shoot tip and nodal segment, the highest shoot formation was obtained from the artificial seeds developed from shoot tips (Islam and Bari 2012).

8.7 Future Prospects

Artificial seed production specifically with somatic embryos in comparison to other micropropagules as explants has been recommended in various studies as they possess both radical and plumule and can be coated in dried as well as in hydrated form (Kitto and Janick 1982). However, apical shoot buds/apical shoot tips have also been used successfully in producing synthetic seeds in many plant species such as *Actinidia deliciosa* (Kiwi fruit), *Arachis hypogaea* (Groundnut), *Brassica campestris* (Mustard), *Daucus carota* (Carrot), *Malus Pumila (Mill)* (Apple root stock M. 26), *Mangifera indica* L. (Mango cv. Amrapali), *Solanum melongena* (Egg plant), *Vitis Vinifera* (Grape), *Zingiber Officinale Pose* (Ginger), and Cucumber *(Cucumis sativus)* (Ara et al. 2000; Latif et al. 2007; Tabassum et al. 2010). The significant success in artificial seed production at commercial level can be achieved if rate of conversion of artificial seeds into vigorous plantlets will be high. Not only high conversion rate but also uniform transformation is also essential for making their use for clonal plant propagation (Magray et al. 2017). The self-incompatible behaviour exhibited by most of the fruit species has limited their mode of propagation mainly to vegetative means. Germplasm conservation of these species in the form of artificial seeds in a small space through cryopreservation would be the best way to minimize the cost of maintaining field gene banks and also the risk of adverse environmental conditions on germplasm can be avoided (Towill 1988). Extensive hybrid seed production in cross pollinated species particularly in maize by traditional breeding method is time absorbing and resource exhausting process due to the maintenance of parental lines. Use of artificial seeds in hybrid seed production may help in the commercialization of new hybrids and superior genotypes can be propagated in less time and cost as the step of maintaining parental line would be eliminated. Also, artificial seed production may be an alternative of conserving those forest species in which vegetative propagation is not possible and which are in the verge of extinction due to increasing desertification (Desai et al. 1997).

8.7.1 Limitations Associated with Artificial Seed Production

Although use of somatic embryos in artificial seed production has been reported in various plant species (Sharma et al. 2013), some limitations have been encountered in terms of asynchrony in somatic embryos development, non-uniform maturity, reduced conversion efficiency, unsuitability for long term storage (Reddy et al. 2012), reduced viability and attainment of low plant recovery if stored at low temperature (Makowczynska and Andrzejewska-Golec 2006) which need to be addressed in order to make artificial seed production system more efficient. Other than these factors some of the problems enlisted in the applicability of artificial seed technology are (1) high cost production of somatic embryos at commercial level, (2) extra care is required to prevent the somatic embryos from mechanical injury, microbial attack, desiccation, etc., (3) insufficient oxygen and nutrients, if not supplied properly, may adversely affect the germination of seeds, (4) somaclonal variations, (5) artificial seeds cannot be implanted directly in the substrates like vermiculite and compost, etc. (Singh et al. 2020).

8.8 Conclusion

Successful production of artificial seed by way of encapsulating micropropagules has been reported in vegetables, medicinal plants, fruit crops, cereals, orchids, sugar crops and forest trees. Protocols were standardized to obtain best combination of gel matrix to produce artificial seeds with enhanced vigour and high conversion efficiency. Studies have also shown positive effect of growth regulators and nutrients on germination behaviour and viability of somatic embryos when applied in the encapsulation material. This technology has made possible the germplasm conservation in desiccation sensitive species. Multiplication of polyploid species, easy handling during storage, transport feasibility, cost effective propagation of superior plant varieties are the advantages of artificial or synthetic seed. However, certain limitations such as asynchrony in somatic embryos development, non-uniform maturity, reduced conversion efficiency, etc. need to be resolved to enhance the efficiency and applicability of artificial seed technology in different plant species.

References

Aida AR, Badr-Elden AA, Ottai AM, El-Sayed M, Nasr MI, Esmail MNM (2012) Development of artificial seed technology and preservation in sugar beet. Sugar Tech 14(3):312–320

Antonietta GM, Emanuele P, Alvaro S (1999) Effects of encapsulation on *Citrus reticulata* Blanco somatic embryo conversion. Plant Cell Tissue Organ Cult 55:235–237

Ara H, Jaiswal U, Jaiswal VS (1999) Germination and plantlet regeneration from encapsulated somatic embryos of mango (*Mangifera indica* L.). Plant Cell Rep 19:166–170

Ara H, Jaiswal U, Jaiswal V (2000) Synthetic seed: prospects and limitation. Curr Sci 78:1438–1444

Banu LA, M-Or-Rashid H, Bari Miah MA (2014) Development of artificial seed and preservation in *Mimosa pudica* L. an important medicinal plant in Bangladesh. J Biosci 22:89–99

Bapat VA, Mhatre M (2005) Bioencapsulation of somatic embryos in woody plants. In: Protocol for somatic embryogenesis in woody plants. Springer, Dordrecht, pp 539–552

Bapat VA, Mhatre M, Rao PS (1987) Propagation of *Morus indica* L. (Mulberry) by encapsulated shoot buds. Plant Cell Rep 6:393–395

Bekheet SA (2006) A synthetic seed method through encapsulation of in vitro proliferated bulblets of garlic (*Allium sativum* L.). Arab J Biotech 9:415–426

Bukhari N, Siddique I, Perveen K, Siddiqui I, Alwahibi M (2014) Synthetic seed production and physio-biochemical studies in *Cassia angustifolia* Vahl-a medicinal plant. Acta Biol Hung 65(3):355–367

Capuano G, Piccioni E, Standardi A (1998) Effect of different treatments on the conversion of M.26 apple rootstock synthetic seeds obtained from encapsulated apical and axillary micropropagated buds. J Hortic Sci Biotechnol 73:299–305

Chee RP, Cantliffe DJ (1992) Improved Procedures for production of sweet potato somatic Embryos for a synthetic seed system. Hort Sci 27:1314–1316

Daud M, Taha MZ, Hasbullah AZ (2008) Artificial seed production from encapsulated micro shoots of *Sainpaulia ionantha* Wendl. (African Violet). J Appl 8:4662–4667

Desai BB, Kotecha PM, Salukhe DK (1997) Seeds handbook - biology. In: Production, processing and storage, pp 91–113

Elliott MC, Chen DF, Fowler MR, Kirkby MJ, Kubalakova M, Scott NW, Zhang CL, Slater A (1996) Towards the perfect sugar beet via gene manipulation. Sugar Crops China 1:23–30

Gangopadhyay G, Bandyopadhyay T, Poddar R, Gangopadhyay SB, Mukherjee KK (2005) Encapsulation of pineapple micro shoots in alginate beads for temporary storage. Curr Sci JSTOR 88:972–977

Ghosh B, Sen S (1994) Plant-regeneration from alginate encapsulated somatic embryos of Asparaguscooperi-baker. Plant Cell Rep 13:381–385

Gray DJ, Purohit A, Triglano RN (1991) Somatic embryogenesis and development of synthetic seed technology. Crit Rev Plant Sci 10:33–61

Grey DJ (2003) Artificial seed. In: Seed development/artificial seed. Elsevier Ltd.

Grieve TM, Gartland KMA, Elliott MC (1997) Micropropagation of commercially important sugar beet cultivars. Plant Growth Regul 21:5–18

Gupta PK, Durzan DJ (1987) Somatic embryos from protoplasts of Loblolly-pine proembryonal cells. Nat Biotechnol 5:710–712

Hegde V, Makeshkumar T, Sheela MN, Visalakshi CC, Koundinya AVV, Anil SR, Muthuraj R, Darshan S (2016) Production of synthetic seed in cassava (*Manihot esculenta* Crantz). J Root Crops 42(2):5–9

Islam MS, Bari MA (2012) *In vitro* regeneration protocol for artificial seed production in an important medicinal plant *Menthaarvensis* L. J Bio-Sci 20:99–108

Ismail RM, Raslan WM, Hussein GMH (2016) *In Vitro* propagation of sugar beet cultivar Frida, through encapsulated different explants. Am J Agric Sci 3(3):27–34

Ivic SD, Sicher RC, Smigocki AC (2001) Growth habit and sugar accumulation in sugarbeet (*Beta vulgaris* L.) transformed with a cytokinin biosynthesis gene. Plant Cell Rep 20:770–773

Jain P, Danwra K, Sharma HP, Mahato D (2018) In vitro tissue culture studies and synthetic seed production from *Plumbago zeylanica* L. Indian J Exp Biol 56:769–773

Khor E, Loh CS (2005) Artificial seeds. In: Nedovic V, Willaert R (eds) Applications of cell immobilization biotechnology. Springer, pp 527–537

Kikowska M, Thiem B (2011) Alginate-encapsulated shoot tips and nodal segments in micropropagation of medicinal plants. A review. Herba Polonic 57(4):45–57

Kitto SK, Janick J (1982) Polyox as an artificial seed coat for asexual embryos. Hort Sci 17:488–490

Kotvi P, Vashist E, Sharma S, Sood H (2016) Optimization of culture conditions for the production and germination of artificial seed in an important medicinal plant, *Gentiana kurroo*. Innovare J Agric Sci 4(4):13–16

Krens FA, Trifonova A, Keizer LCP, Hall RD (1996) The effect of exogenously-applied phytohormones on gene transfer efficiency in sugar beet (*Beta vulgaris* L.). Plant Sci 116:97–106

Kumar MBA, Vakeswaran V, Krishnasamy V (2005) Enhancement of synthetic seed conversion to seedlings in hybrid rice. Plant Cell Tissue Organ Cult 81:97–100

Lambardi M, Benelli C, Ozudogru EA (2006) Synthetic seed technology in ornamental plants. In: Teixeira da silva JA (ed) Floriculture, ornamental and plant biotechnology, vol 2. Global Science Books, pp 347–354

Latif Z, Nasir IA, Riazuddin S (2007) Indigenous production of synthetic seeds in Daucus carota. Pak J Bot 39:849–855

Leroy XJ, Leon K, Charles G, Branchard M (2000) Cauliflower somatic embryogenesis and analysis of regenerant stability by ISSRs. Plant Cell Rep 19:1102–1107

Magray MM, Wani KP, Chatto MA, Ummyiah HM (2017) Synthetic Seed Technology. Int J Curr Microbiol App Sci 6(11):662–674

Makowczynska J, Andrzejewska-Golec E (2006) Somatic seeds of *Plantago asiatica* L. Acta Soc Bot Pol 75:17–21

Masri MI, Ismail RM, Rslan WM, Abdallah (2019) *In vitro* synthetic seed and herbicide resistant sugar beet varieties (*Beta vulgaris* L.). Biosci Res 16(2):1913–1924

Murashige T (1977) Plant cell and organ cultures as horticultural practices. In: Proceedings of the symposium on tissue culture for horticultural purposes, Ghent, Belgium, 6–9 September

Murashige T, Skoog F (1962) A revised medium for rapid growth and bioassays with tobacco tissue cultures. Physiol Plant 15:473–497

Narula A, Kumar S, Srivastava PS (2007) Genetic fidelity of in vitro regenerants, encapsulation of shoot tips and high diosgenin content in *Dioscorea bulbifera* L., a potential alternative source of diosgenin. Biotechnol Lett Springer 29(4):623–629

Nieves N, Zambrano Y, Tapia R, Cid M, Pina D, Castillo R (2003) Field performance of artificial seed derived sugarcane plants. Plant Cell Tissue Organ Cult 75:279–282

Pond S, Cameron S (2003) Artificial seeds. In: Tissue Culture, Elsevier Ltd, pp 1379–1388

Quiroz-Figueroa FR, Monforte-González M, Galaz-Ávalos RM, Loyola-Vargas VM (2006) Direct somatic embryogenesis in *Coffea canephora*. In: Loyola-Vargas VM, Vázquez-Flota FA (eds) Plant cell culture protocols. Humana Press, Totowa, NJ, pp 111–117

Reddy MC, Murthy KSR, Pullaiah T (2012) Synthetic seeds: a review in agriculture and forestry. Afr J Biotechnol 11:14254–14275

Redenbaugh K (1993) Synthetic seeds. In: Application of synthetic seeds to crop improvement. CRC Press, Boca Raton. FL, p 481
Redenbaugh K, Nichol J, Kossler ME, Paasch B (1984) Encapsulation of somatic embryos for artificial seed production. In Vitro Cell Dev Biol Plant 20:256–257
Redenbaugh K, Paasch BD, Nichol JW, Kossler ME, Viss PR, Walker KA (1986) Somatic seeds: encapsulation of asexual plant embryos. Biotechnology 4:797–801
Redenbaugh K, Slade D, Viss PR, Fujii J (1987) Encapsulation of somatic embryos in synthetic seed coats. Hort Sci 22:803–809
Rihan HZ, Al-Issawi M, Burchett S, Fuller MP (2011) Encapsulation of cauliflower (*Brassica oleracea* var. *botrytis*) microshoots as artificial seeds and their conversion and growth in commercial substrates. Plant Cell Tissue Organ Cult 107:243–250
Rihan HZ, Kareem F, El-Mahrouk ME, Fuller MP (2017) Artificial seeds (principle, aspects and applications). Agronomy 7(4):71
Roy B, Mandal AB (2008) Development of synthetic seeds involving androgenic and pro-embryos in elite *indica* rice. Indian J Biotechnol 7:515–519
Rslan WM (2018) Sugar beet artificial seeds an overview. Highlights BioSci 1:1–5
Saiprasad G (2001) Artificial seeds and their applications. Resonance 6:39–47
Saunders JW, Tsai CJ (1999) Production of somatic embryos and shoots from sugar beet callus: Effects of abscisic acid, other growth regulators, nitrogen source, sucrose concentration and genotype. In Vitro Cell Dev Biol Plant 35:18–24
Shallal HH, Stanica F, Peticila AG, Butcaru AC, Nicolae CI (2020) The use of artificial seed technology in the production of horticultural plants. Sci Papers Ser B Hortic 64(1):701–711
Sharma S, Shahzad A, da Silva JAT (2013) Synseed technology—a complete synthesis. Biotechnol Adv 31:186–207
Singh D, Pal S, Swapnil SA (2020) Artificial seed/synthetic seed production – brief procedure – advantages and limitations. Res Today 2(6):422–424
Siong PK, Mohajer S, Taha RM (2012) Production of artificial seeds derived from encapsulated in vitro micro shoots of cauliflower. Brassica oleracea var botrytis Romanian Biotechnol Lett 17(4):7549–7556
Srivastava V, Khan SA, Banerjee S (2009) An evaluation of genetic fidelity of encapsulated microshoots of the medicinal plant: *Cineraria maritima* following six months of storage. Plant Cell Tissue Organ Cult 99(2):193–198
Sundararaj SG, Agrawal A, Tyagi RK (2010) Encapsulation for *in vitro* short-term storage and exchange of ginger (*Zingiber officinale* Rosc.) germplasm. Sci Hortic 125:761–766
Tabassum B, Nasir IA, Farooq AM, Rehman Z, Latif Z, Husnain T (2010) Viability assessment of in vitro produced synthetic seeds of cucumber. Afr J Biotechnol Acad J Kenya 9(42):7026–7032
Tian I, Brown DC (2000) Improvement of soybean somatic embryo development and Maturation by abscisic acid steam treatment. Can J Plant Sci 80:721–726
Towill LE (1988) Genetic Considerations for clonal germplasm preservation of materials. Hort Sci 23:91–93
Tsai CJ, Saunders JW (1999) Encapsulation, germination, and conversion of somatic embryos in sugarbeet. J Sugar Beet Res 36(4):11–32
Zhong Z, Smith HG, Thomas TH (1993) In vitro culture of petioles and intact leaves of sugar beet (*Beta vulgaris* L.). Plant Growth Regul 12:59–66

Scope of Cultivation of Sugar Beet Under Indian Subtropical Conditions

9

Simerjeet Kaur, Navjot Singh Brar, and Makhan Singh Bhullar

Abstract

In India, sugar cane is a main crop grown for processing of sugar. As an alternative to sugar crop, sugar beet is a short duration crop, having growth period of 6–7 months as compared to 10–12 months of sugar cane, which results in higher productivity per unit time than sugar cane. Sugar yield of sugar beet is equivalent to that of sugar cane, having more sugar content, recovery and purity and can tolerate adverse conditions like salt and water stress. Sugar beet cultivation results in considerably good yield with use of less irrigation water than sugar cane. Sugar beet crop matures in April–May, when the cane-crushing season is nearly over, thus helps in increasing the operation period of the sugar mills from four to six months in a year. Hence, sugar beet has a potential in sugar industry of the subtropical regions, especially India. For widening the scope of cultivation of sugar beet under subtropical Indian conditions, there is need to select the most appropriate varieties, planting time, planting methods, planting density, sowing depth, adequate crop nutrition, pest management and irrigation scheduling. Further, intensive studies are needed to estimate the economics of its cultivation in comparison to winter crops under cultivation and sugar cane in the region. Also sugar industries need to be upgraded for processing of sugar beet to ensure its marketing at good price for more profit than existing crops.

S. Kaur
Department of Agronomy, Punjab Agricultural University, Ludhiana, Punjab, India

N. S. Brar (✉) · M. S. Bhullar
Farm Science Centre, Guru Angad Dev Veterinary and Animal Sciences University, SAS Nagar, Punjab, India

V. Misra et al. (eds.), *Sugar Beet Cultivation, Management and Processing*,
https://doi.org/10.1007/978-981-19-2730-0_9

Keywords

Constraints · Economic viability · Subtropical conditions · Sugar crop · Sugar productivity · Sustainability

Abbreviation

CPE	Cumulative pan evaporation
IW	Irrigation water

9.1 Introduction

Sugar beet (*Beta vulgaris* L.) is a member of *Chenopodiaceae* family. It has economic importance as a sugar crop is next to sugar cane (*Saccharum officinarum* L.). About 30% of world sugar comes from sugar beet, with Europe at top with regard to sugar beet cultivation. European countries account for 70% of the world's sugar beet production and France is on top in terms of per hectare yield (about 90 t/ha) of sugar beet (Iqbal and Saleem 2015). It is a crop of temperate climate; however, its cultivation is extending to subtropical conditions of India, Pakistan, Bangladesh, etc. In these subtropical countries, it is mainly cultivated in the winter season spanning from October to May months (Brar et al. 2015).

In India, sugar cane is the main crop grown for processing of sugar with an area hovering around five million hectare and the sugar industry encompasses 597 operating sugar mills, 309 distilleries, 213 cogeneration plants and numerous pulp, paper and chemical making units (Solomon et al. 2020). During 2018–2019, sugar production was 33.18 million tonnes in India (Online database 2020). Sugar demand in country is increasing continuously due to stress imposed by population explosion (Singh et al. 2018). This demand cannot be entirely fulfilled from sugar cane crop. Given the physical water scarcity on a global level, especially in subtropical countries, sugar cane being a water guzzling crop cannot be grown on a large area. However, sugar cane by-products such as ethanol are in great demand. Sugar beet can serve as an alternative sugar crop in place of sugar cane for sugar and its by-products. Sugar beet takes 6–7 months for completing its life cycle, whereas sugar cane takes a double time period 10–12 months for completing its life cycle (Mall et al. 2021). Sugar beet, being a short duration crop results in higher yield than sugar cane on unit time basis. Sugar beet has many advantages such as higher yield, short life cycle of 6–7 months, low water use, less fertilizer requirement and withstand abiotic stresses (Rozman et al. 2015).

In subtropical Indian conditions, sugar beet is a short season *Rabi* (Winter) crop (which is planted in October–November and harvested in April–May) and its yields at par to sugar cane with higher sugar content (15–17%), recovery (12–14%) and purity (85–90%) (Sanghera et al. 2016). Further, sugar beet can withstand drought (low water stress) and salinity (high salt stress). In comparison to sugar cane, sugar beet required less water. To produce one kilogram of sugar from sugar beet, about 1.4 cubic metre water is required, whereas 4.0 cubic metre of water is required to produce same quantity of sugar from sugar cane (Sohier and Ouda 2001). Moreover, sugar cane consumes 35–40% more water and fertilizer than sugar beet (Balat and Balat 2009). Thus, sugar beet can play a significant part in lowering production cost without decreasing factor productivity in varied environment conditions (irrigated, drought, saline soils). Sugar beet can assist in sustaining crop productivity in a short time period in comparison to sugar cane.

9.2 Uses of Sugar Beet

Sugar beet is grown for commercial sugar production. It is a biennial crop, and it stores food in underground plant part (root) during first year. For commercial purpose, sugar beet is harvested for its root after first year of its growth cycle, which weighs around 1–2 kg with 15–20% sucrose on dry weight basis. Sugar beet also provides valuable by-products such as green leaves after harvesting, which can be used as green fodder and/or ensilaged for animal feed for cattle. Green beet tops (leaves) may be used as mulch material. After sugar beet processing, along with sugar, molasses are produced as by-product, which can be used in cattle feed and in fermentation industries (Singh et al. 2013). Finkenstadt (2013) reported that sugar beet leaves have almost double crude protein content (15%) on dry weight basis than sugar cane leaves (6–8%) while D-value (Digestibility) is almost similar (55–57). Thus sugar beet can also be used as an alternative for grain in animal feed concentrates/green fodder/silage. Apart from this, ethanol, beer upon fermentation can be produced from impure sugary pulp (molasses).

9.3 Sustainability Through Sugar Beet Cultivation

In India, salt affected soils found in approximate seven million ha area, which have low crop productivity. Sugar beet can be successfully grown in such areas. Sugar beet can yield reasonably well with much less irrigation water than sugar cane. Therefore sugar beet can be successfully grown under water stress conditions. Hence, sugar beet can be explored and exploited for increasing profits and working time span of sugar industries.

Tropical sugar beet is not yet cultivated on large scale in India. In ethanol industry, use of sugar beet as feed stock is in nascent stage (Kulkarni et al. 2013). Research work done so far under Indian subtropical conditions reported that the crop can be grown successfully during winter months in plains of north India for its roots which contains 13–15% sugar (Pathak et al. 2014). Sugar beet cultivation is profitable enterprise amongst various cropping systems of winter season in subtropical conditions. Sugar beet, a new option for farmers, helps to increase profitability as compared to sugar cane especially in saline affected areas. Singh et al. (2018) from

PAU, Ludhiana reported that sugar beet not only helps in diversification of agriculture and but is a viable alternative for paddy–wheat cropping system. Sugar beet cultivation will help in increasing farmer's income along with saving of Rs. 10,000 crores foreign exchange per year and serves as a supplementary crop to augment optimal utilization of operating capacity of sugar industry.

Sugar beet cultivation has a bright scope in subtropical conditions (e.g. India) due to the following four main reasons.

1. Sugar beet is a dependable cash/truck crop of winter season.
2. It performs well under saline conditions and results in its reclamation through improved agronomic practices.
3. Sugar beet has excellent potential to yield reasonably well with much less irrigation water than sugar cane.
4. Sugar beet exhibits tolerance to different climatic and soil stresses. It can be grown on marginal land.

The complex set of interactions of genetic and environmental conditions occurring during growth and development phases of any crop plant decide the crop performance, productivity and its quality. Many environmental and agronomic factors influence sugar beet yield and quality. Both areal (temperature, solar radiation, sunshine hours) and soil (temperature, moisture) environment conditions affect the crop emergence, plant growth and development. Sugar beets emerge fast under temperature range between 15 and 25 °C (Khan 1992; Copeland and McDonald 2001). After emergence, crop growth and development activities are largely influenced by air temperature and crop nutrition. For proper plant growth, development and quality of sugar beet, an average temperature of about 20–22 °C is ideal. Temperatures above 30 °C retard sugar accumulation.

In North India, sugar cane area is almost stagnant year after year due to scarcity of irrigation water. However, sugar beet requires less water and matures within 6–7 months and it is a good substitute of sugar cane (Brar et al. 2015). Working of sugar cane mills is almost over by April to May in northern India. Sugar beet crop cultivation will assist in increasing this operating time of sugar mills by another 4–6 months. This will be an exciting offer for both Indian farmers and sugar industry of northern India (Misra et al. 2017).

However, to widen the scope of cultivation of sugar beet, extensive studies are needed to identify the improved varieties, sowing time, sowing methods, crop geometry, seeding depth, nutrient management, pest management and water management practices suitable for subtropical conditions. Along with this, its economics should be evaluated while comparing it with economics of cultivation of major rabi crops and sugar cane grown in the region. The role of all these crop management practices in successful sugar beet production under subtropical (Indian) conditions is discussed further.

9.4 Scientific Cultivation of Sugar Beet Under Subtropical Conditions

9.4.1 Selection of Varieties

The main aim of plant breeders is to develop varieties which are high yielder and have high sugar recovery, so as to fulfil the deficit in sugar production. Sugar beet can only be possible under subtropical environmental conditions, if varieties to be sown are suitable for these conditions. Although sugar beet is primarily a temperate crop, but some genotypes (called as tropical sugar beet) are suitable for cultivation under subtropical climatic conditions. Under subtropical conditions of Northern India, cultivation of hybrids of tropical sugar beet resulted in average root yield of 60–80 t/ha with 13–15% sucrose content (Anonymous 2020).

Various sugar beet hybrids such as Cauvery, Indus and Shubhra are being grown, however, Cauvery hybrid resulted in maximum root yield (Balakrishnan and Selvakumar 2008). Sugar beet variety Padosa resulted in higher root yield, root/top ratio and sugar recovery than HI 0064. However, two varieties did not differ significantly in sucrose percentage (Bhullar et al. 2009). Ahmad et al. (2012) at Islamabad observed that SD-PAK09/07 resulted in the maximum beet root yield (74.2 t/ha), sugar yield (9.35 t/ha) and highest sugar content (12.60%) among 11 sugar beet hybrids and it was followed by California and Magnolia with sugar yield of 7.08 and 6.99 t/ha, respectively. They reported non-significant difference among varieties for leaf weight, root size and yield.

Bhullar et al. (2014a) at PAU, Ludhiana evaluated four sugar beet genotypes (SV 887 DSO 323, SV 888 DSO324, SV 891 DSO 325, SV 892 DSO 326) and observed that these genotypes performed equally well in terms of root yield and quality under Punjab conditions. In another experiment, Bhullar et al. (2014b) tested six monogerm sugar beet hybrids, *viz.* Cauvery & Shubra (M/s Syngenta India Ltd), Calixta & Magnolia (M/s JK Seeds Ltd) and PAC 60008 & SZ 35 (M/s Sesvanderhave) during 2013–14 (Table 9.1). Sugar beet genotype PAC 60008 resulted in the maximum root yield (87.5 t/ha) and was at par to Cauvery (87.3 t/ha) and Magnolia (80.1 t/ha). When crop was sown on 15 November, PAC 60008 resulted in maximum root yield which was similar to Cauvery but was statistically more than all the other hybrids.

Bhullar et al. (2015) evaluated four hybrids (SV 887, SV 889, SV 891, SV 892) (Table 9.1) from M/s Sesvanderhave in a field study conducted during 2014–15 at Ludhiana, Punjab and observed that at these hybrids recorded root yield varied from 87.9 to 89.0 t/ha with sugar recovery 15.0–15.2% in 150 days. Sanghera et al. (2016) reported that maximum root yield and quality product were obtained from Cauvery genotype, which was seconded by Indus and SV 892 genotypes out of 13 evaluated sugar beet genotypes (Calixta, Cauvery, H10671, H10826, Indus, Magnolia, Shubra, SV 887, SV 889, SV 891, SV 892, SV 893, SV 894).

Table 9.1 Yield and quality of sugar beet genotypes under subtropical Indian conditions

Genotype	Root yield (t/ha)	Sucrose (%) in beet roots	Reference
Cauvery	87.3	14.6	Bhullar et al. (2014b)
Shubra	73.6	14.5	
Calixta	72.9	14.5	
Magnolia	80.1	13.9	
PAC60008	87.5	14.4	
SV 887	88.3	15.1	Bhullar et al. (2015)
SV 889	88.3	15.0	
SV 891	89.0	15.2	
SV 892	87.9	15.0	

9.4.2 Planting Time

The influence of environmental conditions on germination, growth and productivity of any crop mainly depends upon the planting time of crop. Planting time influences the emergence, seedling growth, root yield and sugar recovery of sugar beet. The crop phenology depends upon cumulative growing degree days (sum of heat or temperature units more than threshold or base temperature, below which little growth occur). This threshold or base temperature depends upon season and plant species being grown (Bellin et al. 2007). Seed emergence is function of soil moisture, temperature and aeration. At soil moisture of 20–23% and 15–25 °C soil temperature, emergence of sugar beet seedling is at faster rate (Sroller and Svachula 1990; Khan 1992; Copeland and McDonald 2001; Spaar et al. 2004).

Sowing date of sugar beet in a particular region is fixed to ensure optimum temperature for its faster emergence. Amin et al. (1989) observed non-significant differences in root yield of sugar beet and its quality, when crop was sown on first and 15 October at Mardan, Pakistan. Crop sown later than this period resulted in reduced root yield and sugar recovery. Bhullar et al. (2009) observed non-significant difference for root yield when crop was sown on 25 September and tenth October. Further, sowing on these dates resulted in higher root weight and yield (root and sugar) than crop sown on 25 October. Delay in planting from 10 October to 25 October resulted in substantial reduction (19.4%) in root yield. They further observed that sucrose content is statistically similar in three planting dates.

Bhullar et al. (2014b) at PAU, Ludhiana reported that sowing sugar beet on 15 October resulted in maximum productivity, which was at par to root yield obtained from sowing on 30 October and significantly higher than November sowings. They further reported that Cauvery genotype recorded similar root yield when it was sown between 15 October and 15 November (90.8–98.7 t/ha) but its sowing on 30 November resulted in less yield (61.2 t/ha). Calixta and Magnolia genotypes recorded similar root yield when sowing was done between 15 October and 30 October (82.8–96.3 t/ha) and there was decrease in yield when sowing was done between 15 November and 30th November (54.3–75.1 t/ha, respectively). Shubra and SZ 35 genotypes recorded significantly lower root yield when sowing was done beyond 15 October. PAC 60008 genotype resulted in increased root yield when sowing was delayed from mid-October to mid-November (81.0–115.9 t/ha).

9.4.3 Planting Methods, Density and Depth of Sowing

Method of sowing decides the crop performance and yield. In sugar beet, the economical part is underground root. The soil physical conditions of upper 0–15 cm soil depth decide the growth behaviour of sugar beet roots. Planting method affects the soil physico-chemical properties and microbial activities, which ultimately affect the crop yield. In flat planting method, seed bed is prepared by ploughing and levelling top soil, whereas in ridge or bed planting, top fertile soil is accumulated in a particular shape of raised seed bed above the natural terrain. In ridge and bed sowing, water drains very quickly and sugar beet crop escapes from negative effects of water stagnation.

Sugar beet can be successfully grown on ridges with direct seeding in comparison to its transplanting. The former technique led to establishment of higher number of plants and greater mean weight of individual roots. Sugar yield of crop sown on ridges with direct seeding method is higher than transplanting of seedling on ridges (Garg and Srivastava 1985). Narang and Bains (1987) reported that direct seeding of sugar beet on the southern slope of east-west ridges resulted in higher root yield. Whereas, Bhullar et al. (2009) observed no significant difference for root, forage and sugar yield when sowing was done on flat or ridges on loamy soils.

The quality and quantity of any crop depend upon its plant density. Optimum plant population helps in optimum use of natural resources like water, light and space and therefore increasing the photosynthesis and assimilation of sugars. Optimum plant density is a principal component for achieving higher sugar beet productivity and returns (Freckleton et al. 1999). Bhullar et al. (2010) observed that plant population of 100,000 plants/ha (50 cm × 20 cm) resulted in the highest sugar beet root yield as compared to 83,333 plants/ha (60 cm × 20 cm) and 111,111 plants/ha (60 cm × 15 cm).

Saini and Brar (2017) observed that planting two rows on a bed resulted in maximum sugar beet root yield which was statistically similar to planting two rows per ridge with plant population of 1.23 lakh/ha. Saini and Brar (2018) reported that planting two rows on a bed (2R/Bed) (Fig. 9.1) received maximum interception of the light which resulted in more photosynthesis. This higher photosynthetic efficiency resulted in more dry matter production and more translocation of assimilates to the economic part (roots) ultimately resulted in higher root yield (Table 9.2). Raised bed provided well aerated, friable and well-drained soil conditions conducive for plant growth and root yield of 68.36 t/ha (Behera and Arvadia 2018).

Fig. 9.1 Crop sown on raised bed (two rows/bed) at farmers field in district Tarn Taran (Punjab)

Table 9.2 Root yield, quality and water productivity of sugar beet under different planting methods under subtropical Indian conditions

Planting methods	Root yield (t/ha)	Sucrose (%)	Water productivity (kg m^{-3})	Reference
Flat	55.82	17.9	12.8	Saini and Brar (2018)
	55.87	15.8	–	Behera and Arvadia (2018)
Ridge	57.26	17.7	14.1	Saini and Brar (2018)
	60.15	16.3	–	Behera and Arvadia (2018)
Bed	61.50	18.4	15.0	Saini and Brar (2018)
	68.36	16.5	–	Behera and Arvadia (2018)
Planting two rows on both sides of ridge	67.09	17.8	16.2	Saini and Brar (2018)
Planting two rows per bed	70.67	18.1	17.6	Saini and Brar (2018)

Seedling emergence is affected by soil physical and chemical properties. Among soil physico-chemical properties, soil moisture, temperature and aeration affect the most. Apart from this soil structure and mechanical friction/impedance also decide the emergence (Brar et al. 2015). Seeding at proper depth is essential for good emergence and optimum plant population in a field. Seed of sugar beet is very small, and its emergence is lowered with deeper sowing. The maximum emergence was observed when sugar beet was sown at 1.00 cm–3.00 cm soil depth than its deeper sowing at 3.75 cm–5.00 cm (Yonts et al. 1999; Khan 2013; Saini and Brar 2017).

9.4.4 Nutrient Management

Balanced fertilization is a major factor for achieving higher root yield of sugar beet. Fertilizer addition, especially nitrogen helps in more plant growth, chlorophyll content and higher photosynthesis rate, thus resulting in more dry matter production (Brar et al. 2015).

However, excessive use of nitrogen fertilizer results in more vegetative growth on the expense of root (economical part) growth and its quality (Draycott and Christenson 2003). Under certain conditions, excessive nitrogen application results in increase in root and forage (leaves) yield and reduces the sugar content. Determination of the optimum rate of application of nitrogen, which may produce maximum yield, improve root quality parameters by improving the chlorophyll content of the leaves and increasing root number and size, is of prime importance. In initial crop growth period, nitrogen application results in more dry matter accumulation per unit area, while nitrogen application in later stages of crop growth increases above ground and below ground dry matter production, thus helps in greater sugar production.

Soil organic matter plays a prominent part in natural mineralization, aggregate stability, aeration, favourable soil moisture conditions and retention properties. Soil organic matter is an indicator of inherent nutrient supplying capacity of a soil and decides the availability and relative proportion of different nutrient elements, both macro- and micro-nutrients. Bulky organic manures such as farmyard manure and green manure add humus/organic matter in the soil, and their application helps in nutrient transformation and their availability to the crops (Brar et al. 2015). The release of nutrients from bulky organic manure is very slow, so the maximum sugar beet productivity cannot be achieved with addition of farmyard manure alone. Therefore, bulky organic manures should be used in integration with chemical fertilizers for maximum availability of nutrients to the plants. This integration of chemical fertilizer with organic manures not only helps in nutrients availability but also helps in improvement of soil physico-chemical and biological properties (Kumar and Lokesh 2018).

Balakrishnan and Selvakumar (2008) reported that integrated use of nitrogen through chemical fertilizer, FYM and bio-fertilizer resulted in higher crop growth and yield of sugar beet under clay loam soil (with low available nitrogen status) than chemical fertilizers when used alone. Bhullar et al. (2010) reported that application of 120 kg/ha of nitrogen integrated with 20 t FYM in loamy soils (with high available nitrogen) results in higher sugar beet yield (Table 9.3). Further, this treatment recoded statistically similar root yield with application of 150–180 kg N/ ha. Application of 150 kg/ha of nitrogen fertilizer resulted in maximum sugar beet root and sugar yield per unit area; however, highest values of sugar concentration in roots were recorded with 120 kg N/ha (Barik 2003).

Table 9.3 Sugar yield of sugar beet under different nutrient management practices

Treatments	Root top ratio	Sucrose (%)	Sugar yield (t/ha)
Nitrogen 120 kg/ha	1.80	14.34	9.75
Nitrogen 150 kg/ha	1.70	14.10	10.19
Nitrogen 180 kg/ha	1.52	14.25	10.36
Nitrogen 90 kg/ha + FYM 20 t/ha	1.95	14.54	10.31
Nitrogen 120 kg/ha + FYM 20 t/ha	1.83	14.34	10.36

(Source: Bhullar et al. 2010)

9.4.5 Irrigation Management

Sugar beet has a deep root system, which can effectively extract water from deep soil. Sugar beet yield is lower under both extreme conditions, i.e. water stagnation and drought. Under drought conditions or rainfed farming, water available to the plants is very less which results in poor crop growth and reduced crop yield. If water remains stagnant in field, then it will result in aeration problem and more infestation of disease, which ultimately leads to poor crop growth and reduce yield. Water deficiency in the initial crop growth phase results the maximum reduction in sugar beet yield (Abdollahian-Noghabi 1999). Singh et al. (2018) working in 22 villages of district Amritsar, Gurdaspur and Kapurthala of Punjab reported water productivity 13.98 kg/m^3 in beet crop than 8.17 kg/m^3 in sugar cane. Similarly, Gupta et al. (2013) and Shukla and Awasthi (2013) also reported that sugar beet requires less irrigation than sugar cane. Thus, sugar beet cultivation can be of considerable help in saving precious water.

The intensity and frequency of irrigation affect sugar beet root yield and relative proportion of sugars. Total eight irrigations were given for higher root yield and sugar recovery (Kumar 1993). Further, he observed that sugar beet juice contain less impurity index when irrigated frequently. Irrigation scheduling at IW/CPE of 0.8 resulted in the highest sugar beet root yield, while irrigation scheduling at IW/CPE of 1.0 and 1.2 resulted in lower yield (Saini and Brar 2018). This resulted in significantly higher water productivity under IW/CPE of 0.8 (18.8 kg/m^3) than IW/CPE of 1.0 (14.9 kg/m^3) and 1.2 (11.7 kg/m^3). Sugar beet sowing on beds or ridges (two rows per bed/ridge) resulted in significantly higher water productivity as compared to flat method (Table 9.2) (Saini and Brar 2018; Saini et al. 2020). They further reported that for maximum sugar beet root yield and water productivity, crop should be sown on beds/ridges (with two rows per bed/ridges) keeping plant population of 1.00–1.23 lakh plants/ha and watering should be done at IW/CPE of 0.8.

9.4.6 Pest Management

Sugar beet crop, being short statured is relatively susceptible to the competition of weeds owing to its slow initial growth (Bhadra et al. 2020). Weeds result significant

Fig. 9.2 Crop before and after hand weeding

reduction on yield of sugar beet. Weeds cause maximum damage when these are allowed to grow for initial 60 days (Gerhards et al. 2017). If control measures are not employed during this critical period of crop weed competition, a severe competition occurs which results in full crop damage (Fig. 9.2) (Cioni and Maines 2010; Kropff and Spitters 1991; Salehi et al. 2007). Among 250 weed species infesting sugar beet crop on global level, 60 weed species are detected as major infesting species, out of which approximately 70% are broadleaved and 30% are grass weeds (May and Wilson 2006). Weed competition from dicot weeds is intense as compared to monocot weeds (Roos and Brink 1996; Zoschke and Quadranti 2002). In winter season, *Anagallis arvensis, Chenopodium album, Convolvulus arvensis, Coronopus didymus, Lathyrus aphaca, Malva neglecta, Medicago denticulate, Rumex dentatus* and *Rumex spinosus* are among major broad leaf weeds which infest sugar beet crop under subtropical Indian conditions. Amongst annual grasses, *Avena ludoviciana, Phalaris minor* and *Poa annua* are major ones which infest sugar beet crop.

Weeds result in significant reduction in sugar beet root yield because of intense competition for crop nutrients, water, light and space and may cause complete crop failure if not controlled on time. The most competitive are annual weeds, mostly broadleaved species that emerge with, or shortly after the crop. These weeds attain more height with time and over-shadow the sugar beet plants and develop dense shade (Cioni and Maines 2010).

Mechanical weed management techniques include physical uprooting and mowing of the above ground plant parts is beneficial when weeds are relatively young and annual (Cioni and Maines 2010). In this method, major disadvantage is shifting of dormant weed seeds to the soil's surface, where they may germinate. Use of herbicide for weed management is economical, simple and effective method for keeping the weed population below minimum threshold level. Herbicide use ensures timely control of weeds during critical period of crop weed competition. Bhullar et al. (2013) evaluated different herbicides for weed management at PAU, Ludhiana. Four pre-emergence herbicides, *viz*. pendimethalin at 365 & 562 g, alachlor at 937 & 1250 g, oxadiargyl at 67 & 90 g and oxyfluorfen at 58 & 87 g/ha were evaluated. All herbicides provided effective control of grasses and broadleaves weeds during initial crop growth stage. However, oxyfluorfen 87 g/ha and pendimethalin 562 g/ha were observed phytotoxic for sugar beet crop. All the herbicidal treatments except oxyfluorfen 87 g/ha recorded statistically higher yield as compared weedy check.

Various cutting and feeding insect-pests such as army worm, hairy caterpillar, pod borer, semilooper, cutworms and sucking pests like aphids cause considerable damage to sugar beet. Aphid population decreased with increase in temperature while an infestation of armyworm and pod borer increased with rise in aerial temperature (Sharma et al. 2017).

9.5 Economics of Cultivation

Rice–Wheat is a widely cultivated cropping system in Indo-Gangetic plains and productivity of this system stagnated in the last few years. Farmers are looking for alternative crops which give more returns under existing conditions. Sugar beet crop is a great option to replace wheat crop and results in more returns than rice–wheat cropping system and sugar cane (Brar and Kumar 2019; Iqbal and Saleem 2015). Moreover, the economic viability of sugar industry can be enhanced through increasing the milling period. Sugar beet may provide supplementary/ alternate source of farm produce which can be used as a raw material in sugar industry to increase in its operational time period. Sugar beet cultivation requires significantly less investment than sugar cane and results in more economic returns of Rs. 20–25 thousands on per hectare basis.

Brar and Kumar (2019) did the economic comparison of cultivation of two winter crops, namely wheat and sugar beet (Fig. 9.1) under subtropical region. They reported that sugar beet and wheat resulted in mean yield of 940 q/ha and 47.5 q/ha, respectively. The sale price of sugar beet produce was Rs. 185/quintal (sugar mill situated in district Amritsar) and wheat was sold in local grain market at Rs. 1735/quintal. After making economic comparison, they found that the sugar beet cultivation resulted in more net returns of Rs. 35,945/ha than wheat (Table 9.4). In Punjab, farmers are reaping a harvest of 87.5 tonnes sugar beet per ha and earning gross and net returns of 1.5 lakh and 0.68–0.75 lakh, respectively, from a short

Table 9.4 Comparative yield and economics of sugar beet and wheat

Parameters	Sugar beet	Wheat
Yield (q/ha)	940	47.5
Selling price (Rs/ha)	185	1735
Gross returns (Rs/ha)	173,900	82,413
Cost of cultivation (Rs/ha)	82,612	27,070
Net return (Rs/ha)	91,288	55,343

(Source: Brar and Kumar 2019)

duration sugar crop (Anonymous 2019). Sugar beet crop can replace wheat in the northern India, provided there is assured market for this crop. Saini et al. (2020) observed that sugar beet cultivation is economical if sugar beet is grown on beds with plant population of 1.23 lakh/ha.

9.6 Major Constraints in Cultivation of Sugar Beet in Indo-Gangetic Plains

There are some problems which need to be solved for wider cultivation under Indian subtropical conditions:-

1. Unavailability of processing units for marketing of crop.
2. Huge labour requirement with high wages rate.
3. High cost of cultivation.
4. Market price uncertainty.
5. Poor technical knowledge.

Since it is an industrial crop and additional machinery is required in present-day sugar mills for its processing, which requires huge investment. Sugar cane is principally used to extract white sugar, khandsari and jaggery (Bhatt 2020). Whereas, jaggery cannot be produced from sugar beet, rather it needs additional plant for vacuum-pan sugar production. Sugar beet roots deteriorate faster after harvesting; therefore, it needs immediate transportation from farmer's field to the sugar factories. So its commercial cultivation is possible only around the processing units. Also, there is necessity of incentivization for shifting of area from sugar cane to sugar beet by the farmers and to sugar processor/industries for installing new units required for sugar beet processing (Singh et al. 2018).

The other major limitations in sugar beet cultivation are huge labour requirement with high rate of wages in region, costly seed and absence of label claim of pesticides in sugar beet for controlling different agricultural pests. Presently, production cost of sugar beet is very much high due to costly seed and huge labour demand. Brar and Kumar (2019) reported that sugar beet cultivation requires Rs. 82,612/ha than Rs. 27,070/ha required for wheat cultivation (Table 9.4). Further, there is uncertainty in sale price and sugar beet produce fetches market price of Rs. 190–195/q which is lower than sale price of sugar cane, i.e. Rs. 290–295/q (Saini et al. 2020). Apart from this, the simple farmers could not realize the full yield potential of sugar beet without following complete package of practice.

9.7 Future Prospects

Sugar beet has a vast scope to give more economic returns to farmers and increase economical viability of sugar industry through increasing the milling period. Favourable government policy and/or incentives for upgrading the existing sugar mills can boost the acreage under this crop during October to May months in subtropical region. Further, mechanization solutions must be explored to cut down the labour demand and make it a more viable option for the growers under Indian subtropical conditions.

9.8 Conclusion

Development and identification of suitable sugar beet cultivars for subtropical conditions are of utmost importance for realizing maximum economic returns. Improved cultivars with higher harvest index, higher root and sugar yield should be bred for cultivation and attaining maximum productivity under subtropical conditions. Planting time varies according to genotype selected and yield reduces with delay in sowing from 15 October to 15 November. Sugar beet sowing on ridges/beds with two rows per ridge/bed while maintaining plant population of 100,000 plants per/ha results in maximum yield. On soil test basis, judicious use of chemical fertilizers and organic manures should be followed for higher sugar beet productivity. Water management should be strictly according to climate of the region and field should be well drained to avoid water stagnation. More experimentation is needed to develop different integrated pest management techniques and different pesticide companies should get label claim of their pesticides for this crop.

References

Abdollahian-Noghabi M (1999) Ecophysiology of sugar beet cultivars and weed species subjected to water deficiency stress. Ph.D. Dissertation, University of Reading, UK, p 227

Ahmad S, Zubair M, Iqbal N, Cheema NM, Mahmood K (2012) Evaluation of sugar beet hybrid varieties under Thal-Kumbi soil series of Pakistan. Int J Agric Biol 14:605–608

Amin M, Khan A, Khan D (1989) Effete of date of sowing on yield and quality of sugar beets. Pakistan J Agric Res 10(1):30–33

Anonymous (2019) Annual report 2018–19, ICAR-Indian Institute of Sugarcane Research, Lucknow, (UP) India

Anonymous (2020) Package of practices for crops of Punjab: *Rabi* 2020–21, Punjab Agricultural University, Ludhiana. p 67

Balakrishnan A, Selvakumar T (2008) Integrated nitrogen management for tropical sugar beet hybrids. Sugar Tech 10(2):177–180

Balat M, Balat H (2009) Recent trends in global production and utilization of bio-ethanol fuel. Appl Energy 86(11):2273–2282. https://doi.org/10.1016/j.apenergy.2009.03.015
Barik S (2003) Role of potassium and nitrogen on sugar concentration of sugar beet. African Crop Sci J 11(4):259–268
Behera SD, Arvadia MK (2018) Root yield and quality of tropical sugar beet (*Beta vulgaris* L.) cultivars under land configurations in *vertisol* and impact on soil health. J Crop Weed 14:193–197
Bellin D, Schutz B, Soerensen TR, Salamini F, Schneider K (2007) Transcript profiles at different stages and tap-root zone identity correlated developmental and metabolic pathway of sugar beet. J Exp Bot 58(3):699–715
Bhadra T, Mahapatra CK, Paul SK (2020) Weed management in sugar beet: a review. Fund Appl Agri 5:147–156
Bhatt R (2020) Resources management for sustainable sugarcane production. In: Kumar S et al (eds) Resources use efficiency in agriculture. Springer, pp 647–693. https://doi.org/10.1007/978-981-15-6953-1_18
Bhullar MS, Uppal SK, Kapur ML (2009) Effect of agronomic practices and varieties on productivity of sugar beet (*Beta vulgaris* L.) in semi-arid region of Punjab. J Res Punjab Agric Univ 46: 6–8
Bhullar MS, Uppal SK, Kapur ML (2010) Influence of planting density and nitrogen dose on root and sugar yield of beet (*Beta vulgaris* L.) under sub-tropical semi-arid conditions of Punjab. J Res Punjab Agric Univ 47:14–17
Bhullar MS, Thind KS, Uppal SK, Sharma B, Sandhu SK, Kaur S, Kaur T, Sharma S, Kumar R (2013) Annual project report 2012–13: exploring possibility of introducing sugar beet for enhancing economic viability of sugarmills in Punjab. Department of Agronomy, Punjab Agricultural University, Ludhiana, Punjab (India)
Bhullar MS, Kaur S, Kaur T, Thind KS, Sanghera GS, Sharma S (2014a) Annual project report (2013–14): evaluation of Sesvanderhave sugar beet genotypes for yield and quality under Punjab conditions. Department of Agronomy, Punjab Agricultural University, Ludhiana, Punjab (India)
Bhullar MS, Thind KS, Kaur S, Kaur T, Singh K, Singh O, Sanghera GS, Singh J, Sharma S, Garg T, Randhawa HS, Saini KS, Brar NS (2014b) Annual Project Report 2013–14: exploring possibility of introducing sugar beet for enhancing economic viability of sugarmills in Punjab. Department of Agronomy, Punjab Agricultural University, Ludhiana, Punjab (India)
Bhullar MS, Kaur S, Kaur T, Thind KS, Sanghera GS, Sharma S (2015) Annual project report (2014–15): evaluation of Sesvanderhave sugar beet genotypes for yield and quality under Punjab conditions. Department of Agronomy, Punjab Agricultural University, Ludhiana, Punjab (India)
Brar NS, Kumar B (2019) Innovative cultivation of sugar beet for sweet returns. In: Rana RK, Singh R, Thakur AK, Chahal VP, Singh AK (eds) Contemplating agricultural growth through farmers' frugal innovations. ICAR-ATARI-1, Ludhiana, Punjab, India, pp 131–132
Brar NS, Dhillon BS, Saini KS, Sharma PK (2015) Agronomy of sugar beet cultivation-a review. Agril Rev 36:184–197
Cioni F, Maines G (2010) Weed control in sugar beet. Sugar Tech 12:243–255. https://doi.org/10.1007/s12355-010-0036-2
Copeland LO, McDonald MB (2001) Seed Science and technology. Kluwer Academic Publishers, Boston, Norwell, Massachusetts, pp 72–124. (in English)
Draycott AP, Christenson DR (2003) Nutrients for sugar beet production: soil-plant relationships. CABI Publishing, Wallingford, p 242
Finkenstadt VL (2013) A review on the complete utilization of the sugar beet. Sugar Tech 16:339–346. https://doi.org/10.1007/s12355-013-0285-y
Freckleton RP, Watkinson AR, Webb DJ, Thomas TH (1999) Yield of sugar beet in relation to weather and nutrients. Agric Forest Meteorol 93:39–51

Garg VK, Srivastava SC (1985) Evaluation of techniques suitable for growing sugar beet on sodic soils. J Indian Soc Soil Sci 33:884–887
Gerhards R, Bezhin K, Santel H (2017) Sugar beet yield loss predicted by relative weed cover, weed biomass and weed density. Plant Protect Sci 53:118–125. https://doi.org/10.17221/57/2016
Gupta R, Singh PR, Singh AK (2013) Optimizing irrigation water and land need for sugar production through intercropping sugar beet in autumn planted sugarcane. Souvenir- IISR-Industry Interface on Research and Development Initiatives for Sugar beet in India, 28–29 May, Sugar beet Breeding Outpost of IISR IVRI Campus, Mukteswar-263138, Nainital. Organised by Indian Institute of Sugarcane Research (ICAR) and Association of Sugarcane Technologists of India. p 51–52
Iqbal MA, Saleem AM (2015) Sugar beet potential to beat sugarcane as a sugar crop in Pakistan. Am Eurasian J Agric Environ Sci 15:36–44
Khan AA (1992) Pre-plant physiological seed conditioning. Hortic Rev 13:131–166
Khan M (2013) Sugar beet production guide. NDSU Extension Service, North Dakota State University of Agriculture and Applied Science, and U.S. Department of Agriculture cooperating. Duane Hauck, Director, Fargo, North Dakota, p 9
Kropff M, Spitters C (1991) A simple model of crop loss by weed competition from early observations on relative leaf area of the weeds. Weed Res 31:97–106
Kulkarni VN, Rana DK, Wielandt N (2013) Sugar beet a potential new crop for sugar and ethanol production in India. Souvenir-IISR-Industry Interface on Research and Development Initiatives for Sugar beet in India, 28–29 May, Sugar beet Breeding Outpost of IISR IVRI Campus, Mukteswar-263138, Nainital. Organised by Indian Institute of Sugarcane Research (ICAR) and Association of Sugarcane Technologists of India, p 58–62
Kumar V (1993) Effect of irrigation on yield, sucrose content and impurity levels of sugar beet crop. J Indian Soc Soil Sci 41:26–28
Kumar A, Lokesh B (2018) Integrated nutrient management for maximum economic yield of sugar beet and sustainable soil health. Agrica 7(2):156–163
Mall AK, Misra V, Santeshwari, Pathak AD, Srivastava S (2021) Sugar beet cultivation in India: prospects for bioethanol production and value added co products. Sugar Tech 23:1218–1234
May JM, Wilson RG (2006) Weed and weed control. In: Draycott AP (ed) Sugar beet. London, Blackwell, pp 359–386
Misra V, Sah U, Mall AK, Kishor R, Pathak AD (2017) Identification and evaluation of different sugar beet germplasm for high yield and sucrose content under Indian agro-climatic conditions. Bull Env Pharmacol Life Sci 6:292–295
Narang RS, Bains BS (1987) Techniques for planting sugar beet to advance the harvest date in North Indian conditions. Exp Agri 23:99–103
Online database (2020) Supply and demand position of sugar in India. www.indiastat.com. Accessed 5 Jan 2021
Pathak AD, Kapur R, Solomn S, Kumar R, Srivastava S, Singh RP (2014) Sugar beet: a historical perspective in Indian context. Sugar Tech 16(2):125–132. https://doi.org/10.1007/s12355-014-0304-7
Roos H, Brink A (1996) Optimierung von produkteigenschaften durch formulierungsentwicklung am beispiel rübenherbicide. Mitteilungen aus der Biologischen Bundesanstalt für Land-und Fortwirtschaft 321
Rozman C, Kljajić M, Pažek K (2015) Sugar beet production: a system dynamics model and economic analysis. Organ 48:145–154
Saini KS, Brar NS (2017) Sugar beet (*Beta vulgaris* L.) yield response to varying planting methods, densities and depth of sowing under subtropical conditions. Bioscan 12:1715–1720
Saini KS, Brar NS (2018) Crop and water productivity of sugar beet (*Beta vulgaris*) under different planting methods and irrigation schedules. Agric Res 7:93–97
Saini KS, Brar NS, Walia SS, Sachdev PA (2020) Effect of planting techniques, plant densities and different depths of sowing on production economics, water and sugar productivity of sugar beet. J Crop Weed 16:190–196

Salehi F, Esfandiari H, Rahimian MH (2007) Critical period of weed control in sugar beet in shahrekord region. Iran J Weed Sci 2:1–12
Sanghera GS, Singh RP, Kashyap L, Tyagi V, Sharma B (2016) Evaluation of sugar beet genotypes (*Beta Vulgaris* L.) for root yield and quality traits under subtropical conditions. J Krishi Vigyan 5:67–73
Sharma S, Kooner R, Sandhu SS, Arora R, Kaur T, Kaur S (2017) Seasonal dynamics of insect pests of sugar beet under sub-tropical conditions. J Agrometeorol 19:81–83
Shukla SK, Awasthi SK (2013) Sugar beet: a supplement to sugarcane in non-traditional areas to meet future sugar demand. Souvenir- IISR-Industry Interface on Research and Development Initiatives for Sugar beet in India, 28–29 May, Sugar beet Breeding Outpost of IISR IVRI Campus, Mukteswar-263138, Nainital. Organised by Indian Institute of Sugarcane Research (ICAR) and Association of Sugarcane Technologists of India, p 55–57
Singh RK, Sharma AK, Singh RK, Prakash B (2013) Problems and prospects of sugar beet cultivation as fodder crop in subtropical India. Souvenir- IISR-Industry Interface on Research and Development Initiatives for Sugar beet in India, 28–29 May, Sugar beet Breeding Outpost of IISR IVRI Campus, Mukteswar-263138, Nainital. Organised by Indian Institute of Sugarcane Research (ICAR) and Association of Sugarcane Technologists of India, p 53–54
Singh S, Sidana BK, Kumar S (2018) Water Productivity of sugar beet vs sugarcane cultivation in Punjab. IJIRST 4:61–69
Sohier MM, Ouda (2001) Response of sugar beet to N and K fertilizers levels under sandy soil conditions. Zagazig J Agric Res 28(2):275–297
Solomon S, Rao GP, Swapna M (2020) Impact of COVID-19 in Indian sugar industry. Sugar Tech 22(4):547–551
Spaar D, Dreger D, Zacharenko A (2004) Sugar beet. CUP "Orech", Minsk, p 133–135
Sroller I, Svachula V (1990) Influence of weather on the production and quality of sugar beets. In: Baier J, Bures R, Coufal VI et al (eds) Weather and production. Agropromizdat, Maskva, pp 247–269. (in Russian)
Yonts CD, Smith JA, Wilson RG (1999) Effect of seed type, planter type and depth of planting on sugar beet emergence. J Sugar Beet Res 36:1–9F
Zoschke A, Quadranti M (2002) Integrated weed management: Quo vadis. Weed Bio Manga 2:1–10. https://doi.org/10.1046/j.1445-6664.2002.00039.x

10 Factors Affecting Production Potentials and Adaptability of Sugar Beet Under Subtropical Conditions of Punjab

Gulzar S. Sanghera, Navdeep S. Jamwal, and Prabhjot S. Sanghera

Abstract

Sugar beet (*Beta vulgaris* L.) is the second largest sugar crop in the world, after sugar cane. It is a short-term sugar crop (6–7 months). Of the various factors affecting its production potential under Indian conditions, the lack of information about the effects of cultivars, crop age, and environment on the beet root yield and quality has been the primary factor impacting the sugar industry and processing units. One of the processes about which we do not have a satisfactory understanding is sugar beet crop management with respect to its ripening and the effects of cultivars, crop age, and environment on that. The ripening of sugar beet is completed with the accumulation of sucrose in its roots, which is also influenced by environmental conditions like temperature and soil water deficits. Sugar beet ripening is a complex process, studies of the variables help in choosing better cultivars and time of planting, which is beneficial to farmers, breeders, sugar mills, and the scientific community associated with this crop. In several studies, it has been suggested that sugar beet ripening depends on a complex combination of environmental variables, the genetic potential of cultivars, and crop management practices. In this chapter, attempt is made to discuss the existing understanding of the environmental conditions and its interaction with the ripening of sugar beet, under the influence of cultivar traits and age of crop (time of planting) for its adaptability under subtropical conditions of Punjab.

G. S. Sanghera (✉) · N. S. Jamwal
PAU, Regional Research Station, Kaurthala, Punjab, India
e-mail: sangheragulzar@pau.edu

P. S. Sanghera
Punjab Agricultural University, Ludhiana, Punjab, India

V. Misra et al. (eds.), *Sugar Beet Cultivation, Management and Processing*, https://doi.org/10.1007/978-981-19-2730-0_10

Keywords

Sugar beet · India · Subtropical · Sucrose

Abbreviations

IDM Integrated disease management
LAI Leaf area index

10.1 Introduction

Sugar beet (*Beta vulgaris* L.) is a member of the Chenopodiaceae family that originated in the Middle East, along the Tigris and Euphrates rivers. The wild maritime beet (*B. maritima*) is estimated to have been domesticated approximately 8500 BC by selection (Cooke and Scott 1993) into cultivated forms of wild annual (*B. patellaris, B. Webbiana*, and *B. procumbens*) and perennial species (*B. trigyna, B. lomatogona,* and *B. macrorhiza*) (Cooke and Scott 1993). Based on the use, *Beta vulgaris* is categorized in 'leafy beet' which includes spinach beet and medicinal beet and other is 'root beet' which includes fodder beet, garden beet. In 1747, the pharmacist 'Andreas Markgraf' found sucrose in beets which was around 1.3–1.6% at that time. About 40 years later in 1786, the breeder 'Franz Karl Achard' evaluated 23 local beet varieties from the Halberstadt area in Germany for beet sugar production. Moritz Baron von Koppy selected the local variety 'White Silesian' with sugar content of 6% and this submerged-root variety became the progenitor for all modern sugar beet varieties in the world (Stevanato et al. 2019). Later, with the efforts of breeders, sugar beet varieties with sugar content of 18–20% were developed.

Sugar beet is commercially grown for sugar production, especially in temperate countries. The production of sugar beet in the world during year is about 266.83 million tons, and it is planted in about 4.47 million ha with an average of root yield of 59.60 t/ha (Anonymous 2011). It is contributing around 20% of the world's sugar production, the rest almost entirely from sugar cane (*Saccharum* hybrids L.). Both the crops might be anticipated for increased demand in biofuel production (Jordan et al. 2007). Europe is dominating the world in sugar beet production with yearly 45–50% of total world production. The European Union, United States of America, and Russia are the three largest sugar beet producers in the world. The biggest consumers of beet sugar are Russia, America, European, and middle east countries.

Being a temperate climate crop, sugar beet requires high moisture conditions throughout the growing period and normally grown as spring or early summer crop (Rinaldi and Vonella 2006) in such areas. It is considered as both drought and salinity tolerant species (Francois and Maas 1994). With advancement of genetic and agro-technology, new varieties are developed which can be successfully cultivated in tropical conditions as well. Now, it can be grown successfully in any region with temperature ranging from 12 °C to 45 °C during the crop season (October–May). Around 15 °C is the ideal soil temperature for sugar beet seed germination. Temperatures between 20 and 22 °C are excellent for plant development and sugar accumulation in roots, whereas temperatures above 30 °C slow sugar accumulation. The commercial cultivation of this crop is possible in the plains during winter season. Under tropical conditions, the root yield of sugar beet varied from 60 to 80 t/ha with sucrose percentage 14–19% within its short life span of 5–6 months (Chatin 2004).

India is among top ranks in production as well as consumption of sugar. Most of the sugar are produced from sugar cane (Sanghera and Jamwal 2019a, b, c, 2020). As an alternative sugar crop, sugar beet can be cultivated in higher latitudes of subtropics and to tropics in Maharashtra as an irrigated winter crop. In subtropical India, Punjab, Haryana, Rajasthan, and western Uttar Pradesh are possible winter sugar beet growing areas, whereas Maharashtra, Tamil Nadu, Karnataka, Andhra Pradesh, and Gujarat are potential winter sugar beet growing regions in tropical India. There is also a scope of sugar beet in untraditional areas like Kashmir and salinity affected coastal areas of the country for fodder beet and alcohol production (Anonymous 2011).

Sugar beet cultivation was introduced in India in 1914, and commercial cultivation began in the 1970s in the Sri Ganganagar district of Rajasthan, where it was farmed on around 1000 hectares for sugar production until the mid-1990s (Pathak et al. 2014; Mall et al. 2021). Sugar beet adaptation to tropical environments is a relatively new phenomenon. It wasn't until 1997 that Syngenta began exploring the prospect of creating tropical sugar beet varieties, focusing on disease resistance. In 2007, the first new Syngenta sugar beet varieties were harvested on a commercial scale in tropical India (Goldberg and Alwin 2008). It is now becoming a commercial field crop due to its favourable characteristics, such as its suitability for tropical and subtropical conditions, its shorter duration (5–6 months) compared to sugar cane (10–14 months), its moderate water requirement (600–800 mm) compared to sugar cane (2000 mm), and its higher sugar content (Katerji et al. 1997). Besides, white sugar beet can be utilized for the production of alcohol, ethanol, and pharmaceutical value (Anonymous 2011). The by-products obtained from the sugar beet are—the beet tops can be fed as a fodder for livestock, can also be used as green manure. Sugar beet tops with 10% digestible crude protein can be utilized as forage and silage in feed concentrates. The tops contain the growth stimulant 'saponin'. It also contains 1.4–6.2% carotene along with vitamin C, E and has estrogenic activity.

Sugar cane is the sole sugar crop in India except for a few patches where sugar beet is growing. Continuous sugar cane mono-cropping has led to salinity and reduced yield levels in sugar cane growing belts. Since sugar beet is salinity tolerant, short duration, moderate water requirements, higher brix, as well as sugar recovery coupled with higher yield levels up to 40 t/ha have made farmers think about it (Balakrishnan et al. 2007). In case of sugar cane crop failure due to drought or red rot, sugar beet can also be cultivated as a catch crop. Besides, sugar beet is the second most important source of sugar globally, after sugar cane, providing annually million tonnes of sugar for consumption and beet pulp for animal feed, which is rich in polysaccharides, such as cellulose, hemicellulose, and pectic substances (Fishman et al. 2009; Iqbal and Saleem 2015; Sun and Hughes 1999). Hence, sugar beet could be a best possible alternative to sugar cane in Indian context particularly under subtropical conditions of Punjab (Brar et al. 2015).

The climatic conditions of Punjab are suitable for sugar beet cultivation. Though, it is a short *Rabi* season crop (sown in October–November and harvested during April–May) when sugar cane crop in field is nearly over, its yields are comparable to sugar cane with similar sucrose content (16–22%). Thus, supply of sugar beet can extend the crushing season of sugar mills by nearly 2 months. Additionally, the lowering water table crisis of northern India could also be addressed by replacing the sugar cane area with sugar beet which is short duration and comparably less water requirement than sugar cane. In sugar beet, 1.4 m^3 water is required to produce a kilogram of sugar as opposed to 4.0 m^3 water required for the production of same quantity of sugar from sugar cane (Sohier and Ouda 2001). The short duration nature of sugar beet could better fit in Punjab wheat rice rotation. Farmers who like to take additional crop rather than keeping the ratoon of sugar cane in such conditions sugar beet is best alternative. As a result of its shorter growing period, it produces more sugar and water per unit time than sugar cane. This is due to its short lifespan, greater sucrose and sugar recovery rates as well as its capacity to resist drought and salt (Shrivastava 2006; Misra et al. 2020). Being tolerant to saline conditions, it has the potential to bring saline soil wastelands into cultivation (Pathak et al. 2014). Varieties have an essential role in agricultural crop production, both in terms of yield and quality features. Crop output is influenced by several factors, including cultivar type and environment as well as management approaches. In light of this historical backdrop and economic relevance, the current chapter was aimed to examine the genetic production capability of sugar beet genotypes and variables affecting its adaptability across locations in Punjab.

10.2 Production Needs of Sugar Beet Crop

The successful cultivation of introduced crop like sugar beet in new environment subtropics of Punjab requires understanding of technical knowledge of crop and production technology to support the crop. The adaptability of agricultural crops is influenced by multiple factors such as variety, climate, soil texture, nutrient availability, occurrence of pests and diseases and their interactions. Number of studies demonstrated that the yield potential of sugar beet depends mainly on location and time. The effect of the location can be recognized mainly to its constant characteristics of soil, climate and their interactions. The effect of the time reflects in terms of weather conditions prevailed during the cropping period, which directly influence plant growth, dates of sowing and harvesting and length of the growing season (Singh et al. 2019). In addition to location or environment, date of sowing is also very important variable for determination of appropriate growth period required for proper growth and productivity. Sowing time is a non-monetary input in agriculture but it plays a significant role in increasing the yield of a crop. Thus, planting sugar beet on suitable date according to environmental conditions of region is the best method to maximize sugar beet yield and quality (Curcic et al. 2018). So, it is crucial to find out a genotype specific optimum sowing date with a view to obtain maximum root yield in sugar beet. Nutritional requirement of crop is another factor which directly influenced the yield. Maintaining nutritional level in soil to avoid deficiency as well as toxicity of macro and micronutrient of crop is essential for growth of crop. In order to prevent weeds from interfering with a crop's growth and to avoid issues during harvest, weed management is of paramount importance (Demont and Dillen 2008). The crop stage and weed populations of area are the key deciding factors of controlling treatments to achieve the best crop stand. Insect pest and diseases affect the crop stand and ultimately yield of crop. Sugar beet is affected by sclerotinia root rot and *Cercospora* leaf spot like diseases while army worm, tobacco caterpillar, and cutworm were troublesome insect pests of this crop. Hence, management of insect and prevention of diseases are essential in sugar beet. After successful cultivation of sugar beet, its harvesting and post-harvesting are also essential because it is highly sensitive and it can't be kept in yard as sugar cane for days; hence, its immediate transport to mill and processing is required.

10.2.1 Favourable Climatic Conditions

Climate conditions are the uncontrollable yet important parameters in any crop production. These can be managed by changing the sowing time, spacing, planting depth and with other methods of cultivation. Sugar beet is now cultivated over wide range of climatic conditions (Petkeviciene 2009). Primarily, it was a crop of temperate zone cultivated in the northern hemisphere at latitudes of 30–60°N. The region where day length is longer than 14 hr. is most suitable for sugar beet growth. Normally, under temperate conditions, the most favourable environment for producing a sugar beet requires 90 days after emergence to harvest along with bright and sunny days, day temperature ranged between 18 and 26 °C while night time temperatures of 5–10 °C (Deshpande et al. 2017). But now with advancement in biotechnology and genetics, sugar beet can be cultivated in hotter and more humid environments; however, problems due to insects, disease, and low quality of the crop hinder the success of crop in new geographical areas.

The tropical sugar beet varieties require good sunshine during its growth period. The crop does not prefer high rainfall or high soil moisture even continuous heavy rain may affect development of tuber and sugar synthesis (Elbersen et al. 2010). The optimum temperature for its germination is 20–25 °C, for growth and development 30–35 °C, and for sugar accumulation 25– 35 °C. Optimum temperature for seed germination is 15 °C and for growth and sugar accumulation, it is 21 °C. In contrast, higher temperatures (over 30 °C) slow down sugar build-up while favouring fast growth (Anonymous 2011). Hence, such conditions may coincide with *rabi* season under subtropical conditions where sowing time of sugar beet lied between September and November which favours vegetative growth of crops and it provides the base for better root enlargement. Influence of environment on the performance of variety is a matter of concern for successful crop production.

The interaction of genotype with prevalent environment plays a crucial role for the breeding of improved varieties (Allard and Bradshaw 1964; Eberhart and Russell 1966). When interaction between genotype and environment occurs, the relative ranking of cultivars for yield often differs when genotypes are compared across a series of environments and years. Sugar beet varieties also differ in terms of yield and quality across. Campbell and Kern (1982) determined the presence and magnitude of cultivar x environment interactions for yield and quality data from five locations and 4 years in sugar beet. They reported that cultivar x location interaction for root yield indicated that cultivars tended to rank consistently different in the 4 years of testing at individual locations. Cultivar × location, cultivar × year, and cultivar × location × year variance components for sucrose, sodium, and potassium concentration were relatively small as compared to the cultivar variance component, indicating consistency in the relative expression of these characters. This was confirmed by a regression of cultivar mean yield on environmental mean yield, and by the contributions of various cultivars to the cultivar × environment variance component.

Generally, in plant breeding programs, the potential genotypes are evaluated in environments/locations over years to test the stability of genotype (locations and date of sowings) before selecting desirable genotypes. A genotype and environment interaction (GEI) is the differential genotype response to distinct environmental circumstances. Genotype and environment interaction is of major importance because they provide information about the effects of different environments on cultivar performance and play a key role for assessment of performance stability of the breeding materials (Moldovan et al. 2000). The performance of genetically stable genotypes gives similar yields over years as well as locations (Bjornsson 2002). Number of studies has been reported for evaluating the stability of various sugar beet varieties using the parametric univariate methods (Ebrahimian et al. 2008; Ggyllenspetz 1998; Keshavarz et al. 2001).

Under Punjab conditions, performance of five sugar beet genotypes, *viz*; Calixta, Magnolia, Cauvery, Shubra, and Indus was assessed over nine environments involving three locations, namely Ludhiana, Kapurthala, and Faridkot, Punjab during Rabi season 2014–2015. The effect of locations was significant for number of leaves at maximum growth and Brix (%). Kapurthala location was found earliest to initiate root swelling (DAS) and possessed highest mean value for root length, sucrose (%), juice purity root yield, and sugar yield. The highest plant weight, root fresh weight, and root dry weight were recorded at Faridkot location. Kapurthala and Faridkot were deemed the finest locations for sugar beet farming (Singh et al. 2019).

10.2.2 Ideal Soil Type

Like climate, soil is also an uncontrolled factor which can only managed in cultivation of crops. Sugar beet has deep root system, and light soil with less water holding

Fig. 10.1 Sugar beet at germination stage (3 weeks after sowing)

capacity causes drought like conditions which leads to reduction in root yield. Similarly, heavy soil with less drainage causes wet periods which are also harmful. Because of the high production and, thus, high biological activity, the root system requires large amounts of oxygen. In wet periods, there is often too little oxygen in the soil for maximum growth. The structure of the soil can also negatively affect growth. Sugar beet is relatively tolerant to saline and alkali conditions and performs better within pH of 6.0–8.0. It does not grow well on highly acidic soils. Hence, sugar beet can be cultivated in salt affected soils or marginal areas.

Sugar beet can be grown on a wide range of soils with medium to slightly heavy textured, well-drained soils preferred for good germination and growth (Fig. 10.1). It requires deep root growth system particularly during early growth period for better penetration of roots in soil but soil compaction hinders the root movement and result in formation of sprangled and forked roots. Similar to sugar beet, crust development at the soil surface can adversely influence seed germination, causing poor crop growth. Sugar beet can grow better under irrigated condition with low rainfall. Under Punjab conditions, sugar beet can be grown successfully on well-drained soils, loamy sand to clay loam, saline and alkali soils, though it prefers sandy loam soils (Anonymous 2020).

10.2.2.1 Role of Varieties/Hybrid

Sugar beet is basically a temperate crop but after genetic improvement its cultivation started in tropical region also. But it requires a whole new agronomic production technology for its successful cultivation in the new region. The process starts with seed availability of sugar beet varieties/hybrid for cultivation.

10.2.3 Pure and Healthy Seed

Seed is a key factor in deciding the success of a crop particularly in case of introduced crop. Seed production of sugar beet requires a period of vernalization at the end of the first year before they can flower. Because it is a biennial plant with a thick taproot to store sucrose and having deep and extensive root system. In the first year, rosette of leaves are formed and in the second year, a large inflorescence emerges out. Sometimes, plants flower in their first year instead of swollen root and seeds of such plants persist in the soil which acts as weed in subsequent years. This phenomenon is known as 'bolting' (Sumit 1983; Stevanato et al. 2019). Such plants should be removed from the seed plots because these annual weed beets may pollinate the seed crop. The seed production of sugar beet requires specific climatic parameters. In the transition from vegetative to generative phase, both temperature and day length play a role. Seed size of sugar beet is very small and contains very little perisperm for germination and early growth. This makes seedlings very vulnerable during early growth to competition from weeds and to damage by disease and browsers (Cioni and Maines 2010). In such condition, seed pelleting or priming (Fig. 10.2) can improve the germination and possibility of mechanical sowing of crop using seed dills.

Additionally, seed priming before sowing reduced the effects of salinity on emergence rates which resulted in significantly larger seedlings and recorded higher dry matter production under moderate salinity levels (Stephen and Kurt 2004).

Fig. 10.2 Sugar beet seed pelleting for enhanced germination. Top to bottom: Raw seed, plain pellet, cruiser force, standard pellet treated with fungicides

Unprocessed seeds may be soaked in 0.25% solution of Thiram or 0.1% of Carbendazim overnight to protect seeds from seed and soil borne fungal pathogens (Khan 1992).

Sugar beet has two types of varieties monogerm and multigerm. In seed of the monogerm varieties, one seedling comes out from the glomerule, whereas multigerm varieties give rise to three to four seedlings from a single seed. Generally, monogerm varieties are preferred because singling and thinning operation as is practiced with multigerm varieties are not required and therefore saves labour and time (Elliott and Weston 1993). For sugar beet, seed production in subtropical region of India altitude of 5000 ft. from mean sea level is required. Such station was established by IISR, Lucknow at Mukteshwar located in Kumaon hills (Uttarakhand) (Pathak et al. 2014).

10.3 Indian Perspective of Sugar Beet Varieties

In India, there is limited sugar beet breeding programme which includes evaluation of germplasm for its adaptability and introduction. Under these sugar beet programmes, LS-6, a diploid multigerm open pollinated variety was developed through modified mass selection. Composites, namely IISR Comp.-1, Pant Comp.-3, LKC-2, LKC-95 and synthetics, namely Pant S-10, LKS-10, were also developed. Testing of exotic varieties in Indian conditions was conducted by number of workers, i.e., Deshpande (2013) tested varieties, viz. SZ-35, PAC 6008, and Magnolia in Deccan plateau region of peninsular India and reported Magnolia as higher yielder and PAC 6008 was better in quality. Balakrishnan and Selvakumar (2008) evaluated the tropical sugar beet hybrids of syngenta and found that Cauvery genotype recorded highest root yield while Shubhra was better in brix percentage. In another study, Posada, Dorotea, and HI 0064 are well suited for Belgaum district of Karnataka to obtain higher yields. Dorotea variety which recorded higher brix (16.70%) and top yield was more in HI 0064 (Anonymous 2004).

A successful production of sugar beet under subtropical environmental conditions is not possible without the use of varieties which are highly suitable under these conditions. Amin et al. (1989) at Mardan, Pakistan evaluated three varieties, i.e. Kawe poly, Kawe mira, Kawe terma and reported the superiority of var. Kawe terma for root yield and sugar content. Similarly, Zahoor et al. (2007) also reported that 'Kawe terma' performed better than KWS 1451 variety for sugar beet root yield. Ahmad et al. (2012) evaluated 11 sugar beet varieties, and the results revealed that SD-PAK09/07 attained the highest sugar yield with highest sugar contents and beet root yield followed by California (7.08 t/ha) and Magnolia (6.99 t/ha) with sugar yield. While, the non-significant difference among varieties were observed for the traits like leaf weight, beet root yield and root size.

Under Punjab conditions, plant varieties of tropical sugar beet, viz. Ramonskaya-06, IISR Comp-1, Mezzanopoly, LS-6, Tribel, Plant comp-3, and Pant S-10 were tested at Sugarcane Research Station, Jalandhar (PAU, Ludhiana) and were found to be effectively producing higher yield with better sucrose quality for the region (Toor and Bains 1994). While analysing the sugar beet varieties Padosa and H10064 at

Ludhiana, Punjab, Bhullar et al. (2009) reported the superiority of Padosa for root yield, root top ratio, and sugar yield. However, the two varieties did not differ for sucrose percentage. Sanghera et al. (2016) evaluated 13 sugar beet genotypes/ varieties, viz. Magnolia, Shubra, Calixta, SV 893, H10671, SV 894, SV 887, H10826, SV 892, Cauvery, Indus, SV 889, and SV 891. They reported germination (%) varied from 60% (Calixta) to 90% (Cauvery). The highest root length (31.33 cm) was recorded in Magnolia. The Cauvery was recorded as most promising genotype for both root diameter (46.25 cm) and root volume (1755.78 cm^3). The H10671 was observed to be superior genotype for brix (18%), sucrose (15%), and purity (85%). The root yield of Cauvery was recorded highest followed by Indus and SV 892. The correlation coefficients of root yield with total length (0.416), shoot length (0.475), shoot weight (0.435), root diameter (0.605), root volume (0.538), and germination (%) (0.360) have shown significantly positive associations while root yield was negatively correlated with root length and harvest index. In another study, Singh et al. (2019) evaluated five genotypes of sugar beet, viz. Magnolia, Cauvery, Calixta, Shubra, and Indus at different locations in Punjab and reported highest average dry root weight and sucrose (%) in genotype Calixta while Magnolia variety recorded highest juice purity percentage. Complete package and practices of sugar beet are developed by Punjab Agricultural University for the cultivation of sugar beet crop in the state.

10.4 Agronomic Interventions for Higher Productivity

Agronomic practices required for well growth and development of crop. It started with land preparation till harvesting of crop.

10.4.1 Land Preparation

Selection of suitable field and preparation of land is essential before establishment of the sugar beet crop. The prime objective of this operation is to curtail the soil erosion by improving the soil structure, management of crop residue of previous crop, and most importantly removal of seasonal weeds from the field (Brar et al. 2015).

The requirement of tillage in preparation of land is depending upon type of soil, quantity of previous year crop residue and weeds present, and most importantly the conservation of soil and water in the field. Primary tillage of field agricultural implements like chisel plow, mold board plow, disc harrows, and field cultivators have been used. Sugar beet has also been successfully sown in fields with no tillage or with strip tillage of last year crop residue. In the successful implementation of tillage in the planned field, there is requirement of specialized farm machinery, better planning as well as management (Cattanach et al. 1991). El-Maghraby et al. (2008) reported that sowing of sugar beet at a laser levelled soil along with deep ploughing gave a significant increase in root length, root diameter in comparison to other treatments. Under Punjab conditions, two to three ploughing followed by planking is sufficient for land preparation (Anonymous 2020).

10.4.2 Date of Sowing

Sowing time is also a very important variable for determination of appropriate growth period required for proper photosynthetic activity. It eases out the extreme climatic conditions that can adversely affect the growth and yield of crop, which differ widely from one region to another. It has a vital role in germination, growth, yield, and root quality of sugar beet plants. Sowing dates of sugar beet are considered among most important factors that influenced its growth and productivity (Curcic et al. 2018). Also, planting date is the important factor in organizing and securing work schedule of beet factories. The phenological development in plant species is robustly associated with the uptake of heat above a threshold temperature, under which the plant remain stunted is very poor. This lower threshold temperature varies with plant species (Bellin et al. 2007). Thus, sowing of sugar beet on suitable date according to environmental conditions of region is the best method to maximize sugar beet yield potential and quality. Heath and Cleal (1992) reported 5% higher yield from early transplanting of sugar beet than late planting. Sugar beet emerges at faster when the soil moisture in the seedbed is 20–23% and air and soil temperature ranges between 15 and 25 °C (Khan 1992; Copeland and McDonald 2001; Spaar et al. 2004; Sroller and Svachula 1990). Early planting before onset of winters provides more time for vegetative growth of plants before onset of winter, which checks the growth in favour of sugar accumulation in roots (Metcalfe and Elkins 1980). The crop sown in October reported maximum leaf area index (LAI), root length, root diameter, fresh root and stem weight, sucrose and purity percentages, root yield and sugar yield (Abd EL-Gawad et al. 2000; Ghonema 1998; Leilah et al. 2005). Amin et al. (1989) observed non-significant differences between first and 15 October sown crop w.r.t. yield and quality parameters. Further delay in sowing of crop resulted in reduced yield and quality of sugar beet. Early sowing of sugar beet during October gave higher beet root yield as well as higher sucrose content than late sowing (Balakrishnan and Selvakumar 2009). While Ahmad et al. (2012) found that sugar beet sown in November gave highest root and sugar yield and low incidence of insect pest infestation.

Under Punjab conditions, whole of October to mid-November is recommended as optimum time of sowing of sugar beet crop (Anonymous 2020). Bhullar et al. (2009) analysed the different sowing dates of sugar beet and reported that 25 September and tenth October sown crop had significantly higher single root weight and sugar yield per hectare as compared to 25 October. Delay in planting from 25 September to tenth October, the average root yield was reduced only by 0.6%, while substantial reduction of 19.4% was observed with delaying planting on 25th October. They further observed that the three planting dates (September 25, October 10, and October 25) did not differ significantly in sucrose percentage. In recent study, Singh et al. (2019) reported that the date of sowing significantly influences the traits of agronomic importance, i.e. initiation of root swelling (DAS), biological weight, dry root weight, root yield, and sugar yield. They also found that planting on 30th followed with greatest fresh root weight, whereas one done on 15 November reaped highest sucrose percentage. Sugar beet is best sown in Punjab around the 15 October for early root swelling, better root production, and higher sugar output.

10.4.3 Planting Methods

Planting methods include crop spacing and seed depth, and are the agronomic practices to improve crop yield as well as cultivation efficiency. The sowing methods of sugar beet also influence the performance of the crop. The underground part of sugar beet mainly the root is the main economic yield component. Therefore, the soil physical at the time of sugar beet cultivation affects its root growth. Crop spacing decides the number of plants per unit area which needs to be economical and well calculated based on requirement of the economical part whether root or shoot or fruit, plant canopy and its photosynthesis requirement, weed competition, and climatic conditions. In sugar beet for high root yield, quality narrow row spacing is preferred over wider row spacing because sugar beet competes better with seasonal weeds under narrow row spacing. The most favourable row spacing is 45–60 cm while 55 cm row spacing is most commonly used globally (Cattanach et al. 1991). In case of mechanical harvester and intercropping with other row crops, sugar beet can be planted in 75 cm row spacing also. However, 75 cm row spacing costs the root yield as well as recoverable sugar per acre as compared to 55 cm row spacing of sugar beet, with the same harvest populations (Dean and John 1997). The uniform plant population is easy to establish and maintained on narrow rows, which may ultimately improve higher root yield as well as quality. Under ideal condition, the population of plant in the sugar beet field should be in between 30,000 and 40,000 uniformly spaced plants per acre at the time of harvesting (Cattanach et al. 1991). These populations should produce very good yields of easily harvested high quality sugar beet. El-Kassaby and Leilah (1992) reported that highest root diameter as well as root weight was obtained when sugar beet was sown on the one side of ridge having ridge to ridge distance of 70 cm and plant to plant distance of 30 cm. While the highest root yield and sucrose extraction were obtained at planting distance of 70 cm between ridges and 25 cm between plants.

Under subtropical conditions of Punjab, the estimated sugar yield was higher in crop sown through direct seeding on ridges rather than that transplanting of sugar beet seedling (Garg and Srivastava 1985). Narang and Bains (1987) reported that the influence of direction of sugar beet sowing in the northern side of east-west ridges gave higher root yields (45–50 t/ha) with 12–14% sucrose content. Bhullar et al. (2009) concluded that sowing method (flat and ridge) did not significantly influence the root yield, top yield, root top ratio, sucrose content, and sugar yield on loamy soils indicating that both methods are equally effective. Direct sowing of sugar beet on ridges was more suitable than transplanting seedlings on flat bed or on ridges. Moreover, sugar beet is sensitive to stagnant water, which may be avoided by ridge

Fig. 10.3 Sugar beet planting on flat and ridge methods (ridge method is good for intercultural operation)

planting. The direct sowing technique led to establishment of higher number of plants and greater mean weight of individual roots. Under Punjab conditions, sugar beet sowing is recommended on flat beds or on ridges spaced at 50 cm and plant to plant spacing of 20 cm. The optimum plant population is 40,000 plants per acre. In dibbling method, sowing depth of 2.5 cm is maintained. Maintain only one plant per hill (Anonymous 2020). Bhullar et al. (2010) studied the consequence of three planting densities first 60 × 20 cm, 50 × 20 cm, and 60 cm × 15 cm row to row and plant to plant distance which average plant population of 83,333 plants/ha 100,000 plants/ha and 111,111 plants/ha, respectively, on total root yield and quality. The sugar beet planting density of 100,000 plants/ha (50 cm × 20 cm) produced the highest root yield and sugar recovery. Saini et al. (2020) evaluated the effect of different planting techniques, plant population, and diverse depths of sowing on sugar beet production over years. Flat sowing with 1.23 lakh plants /ha recorded maximum production efficiency (2.98 q ha^{-1} day^{-1} and 2.58 q ha^{-1} day^{-1}) and sugar productivity (9.65 t ha^{-1} and 8.62 t ha^{-1}), which was at par with planting two rows on both side of ridges with planting density of 1.23 lakh plants ha^{-1} (Fig. 10.3). Water efficiency was also recorded significantly higher (11.03 kg m^{-3} and 10.63 kg m^{-3}) under these treatments. Crop sown at sowing depth of 2–3 cm has recorded higher production efficiency, water and sugar productivity than 4–5 cm sowing depth (Saini and Brar 2017, 2018).

10.5 Integrated Nutrient Management

Like other root crops, sugar beet also responds well to fertilizers. N, P, and K must be continuously and adequately supplied in order to produce excellent quality roots. Nutrition requirement of sugar beet varies with location, soil type, soil nutrients status, cultivar, irrigation facility, etc. For profitable sugar beet production, management of growth-limiting factors like soil fertility should be accomplished. The nitrogen (N) requirement of sugar beet should be optimum because its deficiency as well as toxicity adversely affects the yield as well as quality. The nitrogen deficiency leads to poor sugar beet leaf canopy, chlorosis (yellowing of leaves), and ultimately reduced root yield (Stevanato et al. 2019). While excess nitrogen in sugar beet affects its quality particularly sucrose extraction because of increased nitrogen also enhances the impurities and which further lowered sucrose extraction (Cattanach et al. 1991). Nitrogen fertilizer normally increases yield of both roots and sucrose and also may increase impurities and decrease the percent sucrose in the root (Mekdad 2015). Hence, the management of nitrogen in sugar beet is essential for that available nitrogen which is already present in the soil should be determined before sowing of the sugar beet crop and due to mobile nature of nitrogen, its availability should be recorded at regular intervals of growing season (Brar et al. 2015).

To ensure early maximum vegetative growth, adequate nitrogen is required. The effectiveness of nitrogen improves when it is applied in split doses because nitrogen moves to deeper layer after leaching and become unavailable to crop. Hence, nitrogen should be applied at particular plant growth stage. In sugar beet, half dose of nitrogen should be applied at sowing and the rest of nitrogen applied after thinning. The application of excess amount or late application of nitrogen reduces sugar content. Balakrishnan and Selvakumar (2008) recorded higher yield of sugar beet under tropical conditions with integrated nitrogen management, i.e. application of urea, farmyard manure, and bio-fertilizer treatment (150:75:75 kg NPK ha^{-1}). Tropical sugar beet yield (71 and 89 t/ha in 2005 and 2006, respectively) and brix values were considerably higher in 2005 and 2006 (18.2%). In another study, treatment of N and K_2O @ 160 kg/ha with 60 kg P2O5 ha^{-1} resulted in considerably greater root (47.50 t ha^{-1}), top (13.41 t ha^{-1}), and sugar production (7.317 t ha^{-1}) as compared to other fertilizer levels (Deshpande 2013). In another study, Mekdad (2015) reported that the nitrogen had a significant role in increasing the root length, root diameter, fresh weight of root and top, crop yield, level of Na, K, α-amino N, loss sugar percentages, whereas decrease in harvest index. Insignificant effect of nitrogen level on white sugar and purity (%) has been observed.

Potassium plays an important role in photosynthesis, protein synthesis, translocation of assimilates as well as in increasing plant growth and yield. It helps in maintaining the balance of other nutrients to improve the sugar beet root yield as well as quality. Sugar beet is a heavy potassium feeder. It increases the number of leaves per plant along with improved leaf length as well as chlorophyll content compared to nitrogen phosphorus control. In potassium deficient soils, apply 12 kg K_2O (20 kg muriate of potash) per acre at sowing. Barik (2003) reported highest yield of 36 t/ha after applying potassium @ 155 kg/ha, while 35 t/ha yield had been obtained when potassium applied @ 115 kg/ha. Seadh and Farouk (2007) reported that application of 72 kg potassium sulphate recorded improvement of all the traits.

Boron is an essential trace element for sugar beet crop. Its deficiency causes adverse effect on root yield and sucrose quality of roots (Cooke and Scott 1993). Foliar spraying of boron found to be effective in increasing root yield. Root absorbed boric acid and its uptake is depending on soil pH, and boron available in soil (Gerendas and Sattelmatcher 1990). Boron deficiency affects formation of chloroplast and sink limitations (Tersahima and Evans 1988), changes in cell wall, and further led to secondary effects on plant metabolism, development, and growth (Gobarah and Mekki 2005; Loomis and Durst 1992). Boron application increased root fresh weight, sucrose %, root and top yields, root diameter and length, sugar yield, juice by increasing boron levels through both soil application and a foliar spray (Dordas et al. 2007; Gezgin et al. 2001; Hellal et al. 2009; Jaszczolt 1998; Kristek et al. 2009). The purity percentage of sugar beet juice and its sucrose increases after addition of boron at higher concentration. It might be due to decrease in Na and K uptake in root juice. The spray of boron @ 12% improved the root yield as well as sucrose concentration in sugar beet (Armin and Asgharipour 2012). Mekdad (2015) reported that 120 and 150 ppm concentration of boron improved yield of sugar beet and further extraction of gross and white sugar, whereas decrease in concentration of potassium, α-amino nitrogen, sodium, harvest index, loss sugar percentages, and loss sugar yield had been reported.

Under Punjab conditions, the recommended dose of fertilizer applications may be up to 150 kg/ha nitrogen, 50–70 kg/ha phosphorus at planting, and 100–160 kg/ha potassium. Apply 8 t per acre well rotten FYM and mix it well before sowing. In the absence of FYM, apply 60 kg N (135 kg urea) ad 12 kg P_2O_5 (75 kg SSP) per acre. In boron deficient (below 0.5 kg B per acre) soils, apply 400 g boron (4 kg borax) per acre at sowing. Apply 45 kg urea and full phosphorous at the time of sowing and remaining urea in two splits of 45 kg each at 30 and 60 days after sowing. If FYM has been used, reduce the nitrogen dose to 48 kg (105 kg urea) per acre (Anonymous 2020). The soil testing reported high available nitrogen in the soils of Punjab which are high in organic carbon and loamy in texture (Bhullar et al. 2010). The recommended dose of nitrogen is 120 kg/ha along with 20 t/ha of farmyard manure to meet the nutritional requirement of sugar beet. Bhullar et al. (2010) reported that the treatments of soil nutrition with 120 t/ha nitrogen recorded for the root yield were statically at par with the application of 150 and 180 kg N/ha. Similarly, the quality parameter, i.e. brix and sucrose were also non-significant among the different doses of nitrogen and farmyard manure integrated treatments. Kumar and Lokesh (2018) had shown that PAC 60008 had superior impact of FYM than LS 6.

10.6 Irrigation: Role of Micro-Irrigation Systems

Water is an essential component in growth and development of the crop. It may provide from rain or through irrigation facility. The tropical sugar beet requires well distributed rainfall of 300–350 mm across the growing period. The irrigation scheduling is depending on the soil type and rainfall. The highest yield of sugar beet has been observed when irrigation is applied at 50 and 75 mm evaporation. At this irrigation schedule, normally 10–12 irrigations are required for luxuriant growth of sugar beet (Shukla and Awasthi 2013). Sugar beet requires less irrigation, not more than four to five irrigations amounting to 37.5–60 cm (Gupta et al. 2013). However,

Fig. 10.4 Sugar beet performs hardly better in micro-irrigation system (drip) than furrow system save water too

excessive soil moisture or persistent heavy rain might impair tuber growth and sugar synthesis. Sugar beet is very sensitive to over irrigation at all stages of growth. It reduces yield through increased incidence of disease, loss of nutrients from the soil root zone, and reduced oxygen to roots. Over irrigation can also reduce sugar beet quality by lowering the sugar percentage. Similarly, water deficiency during the early growing season ceased water flow in the plant, reduction in nutrients and photosynthates within the plant, potential yield loss in sugar beet production, and increased level of impurities, i.e. α-amino-N, potassium, sodium, and impurity index in sugar beet juice. Kumar (1993) evaluated the effect of irrigation on root yield, sucrose and level of impurities in beet sugar and observed that eight irrigations were required for optimum root production and sugar per unit area. The irrigation to sugar beet should be applied before the available soil water reaches to 60% of its field capacity. Then available soil water should be replenished to field capacity in sugar beet root zone (Fig. 10.4). The maximum root yield was recorded when irrigation applied every 3 weeks followed by every 5 weeks while the minimum root yield was obtained by irrigation every 7 weeks (Abo-Shady et al. 2010; Besheit et al. 1996; Hassanli et al. 2010). Different irrigation methods have been explored in sugar beet. Makrantonaki et al. (2002) reported efficiency of subsurface drip produced a similar root yield by saving 16.6% irrigation water over surface drip. Rajasekaran (2007) used a surface drip method for nitrogen fertigation under tropical condition and found increased yield of sugar beet and substantial quantity of water saving (34.2%). The average sugar beet yield and sugar content were higher in drip irrigation (Fig. 10.4) than with furrow practices (Sharmasarkar et al. 2001).

In case of Punjab, 100% cultivable area is under irrigation. Sugar beet grows well under irrigated condition. For crop establishment, first irrigation is crucial because of sensitivity of seed to water. Therefore, first irrigation should be applied after sowing of seed and given in such a way that water should not flow over the ridges while subsequent irrigation about 2 weeks after sowing. The crop needs irrigations at the interval of 3–4 weeks till February end. After that, intervals should be 10–15 days during March-April due to increase in temperature in the region. The irrigation should be stopped 2 weeks before harvesting (Anonymous 2020).

10.7 Effective Weed Management

The germination and initial vigour of sugar beet seed is slow due to its little perisperm. In comparison to weeds, sugar beet is less vigorous from its emergence until its canopy covers the ground. Due to its slow initial growth, sugar beet leaves take about 2 months to cover the ground. Weeds which emerge even after 4 weeks of sugar beet emergence cause severe damage. Weed competition has been estimated to reduce root yields 6–10%. Weeds have a long period to establish and compete. Competition from annual grasses also suppresses root yields; however, competition from annual grass species is not usually as severe as that from broad leaf weeds because they do not compete for light as effectively as broadleaf weeds (Cioni and Maines 2010). Sugar beet is at two to three leaves stage within 4 weeks which is not enough to contend with weeds in uptaking the sunlight and nutrients. Till 4 weeks, weeds get establish in the field and start covering the space and shadowing the sugar beet which results in severe yield losses. Weed competitiveness, density and length of the time period for which weeds are allowed to compete with the crop highly impacts the yield performance because approximately 70% of weeds found in sugarbeet crops are broadleaved species (Schweizer and May 1993). Broad-leaved weeds become most competitive after they begin shading the crop (Wicks and Wilson 1983). Position of leaf area would be as important as the total area in deciding the competitive outcome between crop and weed (Legere and Schreiber 1989). Sugar beet cultivars may differ in competitiveness with weeds. The yield loss caused by weed competition can be avoided if initial weed population can be checked for at least 4 weeks of sugar beet sowing which give time to sugar beet for its establishment and utilizing soil nutrients. The effective weed control is essential in throughout sugar beet growing season.

The large proportion of production cost is utilized on controlling weeds in sugar beet. Weed management practices are depended on several factors, i.e., weed species present or will be anticipated in the ensuing crop season which determines choice of weed control method and choice of herbicides; availability and cost of hand labour for weeding, availability of equipment which determines choice of herbicide and influences cultivation choices. In sugar beet integrated weed management which includes mechanical, chemical, and cultural methods of weed control needs to be employed for effective weed control (Fig. 10.5). For small fields, hand weeding is an effective method to control weeds. The fields with heavy infestation of weeds should

Fig. 10.5 Integrated weed management practices in sugar beet cultivation

not be avoided for cultivation of sugar beet crop. Careful selection and application of herbicides and planting to stand can reduce the costs considerably. Weed can be controlled effectively with complementary pre-emergence/post-emergence herbicide treatments (Miller and Fornstrom 1989). The advantage of pre-emergence soil applied residual herbicides is that they reduce the number of weeds that emerge with the crop and often sensitize survivors to subsequent post-emergence sprays, provide some flexibility with timing and selection of post-emergence treatments (Cioni and Maines 2010; Duncan et al. 1982). The residual herbicides are applied to the soil surface within 24–48 h. after sowing further delay in spraying may also affect the germination of sugar beet seedlings (May and Hilton 1985). The common pre-emergence herbicides used in sugar beet are cycloate, ethofumesate, lenacil, metamitron, metolachlor, quinmerac, and chloridazon. These herbicides inhibit photosynthesis, lipid synthesis, cell division by a reduction of photosynthesis and respiration and lead to shortening of leaf mid vein, yellowing of leaf vein, stunting, crinkled, fused leaves, yellowing of leaf margin (May and Wilson 2006). There are two categories of post-emergence herbicides first used for broad-leaved weed control and other for the control of grasses. The chemicals used for controlling broad leaves weed in sugar beet are phenmedipham, desmedipham, ethofumesate, chloridazon, clopyralid, endothall, lenacil, metamitron, and triflusulfuron-methyl (May and Wilson 2006). In sugar beet, single chemical is not effective in controlling wide spectrum of weeds and also don't have sufficient residual activity to control weeds for longer duration. Hence, the combinations of two or more herbicides are used to control broad-spectrum weeds (Cioni and Maines 2010). James and Stephen (1990) reported effective dose of pre-emergence herbicide application of cycloate and ethofumesate @ 1.36 (0.68 + 0.68) kg active ingredient /acre, post-emergence application of phenmedipham along with desmedipham @ 0.56 (0.29 + 0.27) kg active ingredient/acre. Application of both pre- and post-emergence herbicides in sugar beet can be significantly reduced the total weed population.

Under subtropical conditions of Punjab, number of *rabi* season weeds, i.e. Sour Dock (*Rumex dentatus*), Blue pimpernel (*Anagallis arvensis*), Little mellow (*Malva parviflora*), etc. (Table 10.1) can compete with sugar beet. There is no recommendation for herbicides use in sugar beet. Weeds can be controlled with cultural practices and two to three weeding (Anonymous 2020).

10.8 Management of Important Pest and Diseases of Sugar beet

Pests are one of the major constraints in the profitable yield of sugar beet in the form of tonnage and sugar content. About 16–20% of the crop is destroyed by diseases every year. The distribution of pests of sugar beet in the region also plays an important role in the beet sugar industry and sugar beet crop in most of the sugar beet growing countries (Diffus and Ruppel 1993). The crop is subject to attack by these diseases or insects from the time of seed-sowing, until the harvest of the crop. All parts of the sugar beet plant (seeds, seedlings, roots, and foliage) are susceptible to attack by a number of diseases and insects which reduce the quantity and quality of roots. The exploitation of genetic resistance against sugar beet diseases has not yielded expected success. Chemical treatments are the important tools for the protection against sugar beet pest and diseases.

10.8.1 Diseases

Sugar beet in a subtropical region has several major pathological issues owing to the high temperatures that are prevalent in the environment. The succulent nature of its foliage and roots are also favourable for quick development, proliferation, and spread of the diseases. The major diseases which affect the sugar beet are mainly caused by fungal infection, i.e. seedling rot, leaf spots, root rot are most destructive. The damage caused by bacteria and viruses is negligible, while fungi and nematodes are proving limiting factors in the profitable cultivation of the crop India. Many diseases like rhizomania and cyst nematodes have not been recorded but may pose serious problems in the future. In India, about 10–15% of the crop is damaged by diseases. About 25 diseases are known to occur in the country out of which 15 are economically important (Srivastava 2004).

10.8.1.1 Seed Borne Diseases

In India, there are many known seed-borne pathogens like *Cercospora beticola, Phoma betae, Rhizopus oryzae, Alternaria alternata,* and *Fusarium* spp. (Singh et al. 1973; Srivastava and Tripathi 1998; 1999). Among these pathogens, the most important seed-borne pathogen (*Phoma betae*) causing disease like damping-off, leaf spot, stem, crown and storage rots has not been yet reported in the subtropical

Table 10.1 Common weeds present during the sugar beet season

S. N.	Botanical name	English name	Common Name
1	*Phalaris minor*	Little seed canary grass	*Gilli Danda*
2	*Avena ludoviciana*	Wild oats	*Jaundhar*
3	*Lolium temulentum*	Poison ryegrass/Ivary	*Rye grass*
4	*Polypogon monspeliensis*	Beard grass	*Loombar gha*
5	*Poa annua*	Sweet grass/Annual blue grass	*Guien/Buien*
6	*Sonchus arvensis*	Perennial sow thistle	*Milk weed*
7	*Carthamus oxyacantha*	Wild safflower	*Pohli*
8	*Cichorium intybus*	Blue daisy	*Kasni*
9	*Rumex dentatus*	Sour Dock	*Jangli palak*
10	*Rumex spinosus*	Dock/ Sorrel	*Kandiali palak*
11	*Euphorbia simplex*	Leafy spurge	*Kaurgandal*
12	*Asphodelus tenuifolius*	Wild onion	*Piazi*
13	*Chenopodium album*	Common Lambs quarter	*Bathu*
14	*Chenopodium murale/ Chenopodiastrum murale*	Nettle leaf	*Karund*
15	*Lathyrus sativus*	Grass pea	*Jangli matar*
16	*Lathyrus aphaca*	Meadow pea	*Dokanni/Pili mattri*
17	*Vicia sativa*	Vetch	*Rari or rewari (broad leaved)*
18	*Vicia hirsute*	Hairy vetch	*Rari or rewari (narrow leaved)*
19	*Medicago denticulata*	Toothed bur clover	*Maina*
20	*Trigonella polycerata*	Wild fenugreek	*Maini*
21	*Melilotus alba*	White sweet clover	*Khandi or wild senji*
22	*Melilotus indica*	Yellow sweet clover	*Khandi or wild senji*
23	*Anagallis arvensis*	Blue pimpernel	*Billi booti*
24	*Spergula arvensis*	Corn spurry	*Jangli dhania*
25	*Stellaria media*	Common Chickweed	–
26	*Saponaria vaccaria*	Cow cockle	*Bara takla*
27	*Silene conoidea*	Forked catchfly	*Chotta takla*
28	*Fumaria parviflora*	Fumatory	*Pitpapra*
29	*Argemone Mexicana*	Mexican poppy	*Satyanasi / Jangli post*
30	*Coronopus didymus*	Garden cress	*Jangli halon*
31	*Sisymbrium irio*	London rocket/Wild Mustard	*Jangli sarson*
32	*Malva parviflora*	Little mellow	*Button weed*
33	*Veronica agrestis*	Green field speedwell	–
34	*Lithospermum arvense*	Stone seed	–
35	*Antirrhinum orontium*	Wild dogflower	–

(continued)

Table 10.1 (continued)

S. N.	Botanical name	English name	Common Name
36	*Gnaphalium purpureum*	Purple cudweed	–
37	*Cannabis sativa*	Indian hemp	*Bhang*
38	*Oenothera laciniata*	Cutleaf evening primrose	–
39	*Galium aparine*	Goosegrass/Coachweed/ Catchweed	–
40	*Arenaria serpyllifolia*	Thyme-leaf sandwort	–
41	*Ranunculus sceleratus*	Cursed buttercup	–

region. These diseases attack the seed as well as seedlings. Sometimes, two fungal pathogens attack successively or simultaneously in seedlings causing damping-off like symptoms (Srivastava 2004). Seed polishing of sugar beet seed by rubbing to remove cortical tissues strikingly reduced the mortality of seedlings in field (Singh et al. 1973; Leach and MacDonald 1976). Seed treatment of sugar beet seeds with various fungicides has been found very effective for the elimination of seed mycoflora and better seedling stand.

10.8.1.2 Soil Borne Diseases

The soil borne fungi, i.e. *Rhizoctonia solani* and *Aphanomyces cochlioides* are the most common seedling pathogen of sugar beet which causes economic loss due to root rot. Further, there are some other fungi which also cause minor damage in sugar beet, i.e. *Fusarium* species, *Pythium aphanidermatum,* and *Erwinia carotovora.* The long survival period of these fungi in soil further worsens the situation leading to minor symptoms and sometimes complete loss of crop. The single control measure can't be effective in controlling these fungi. Hence, the integrated disease management should be followed which includes use of resistant varieties, seed treatment, use of fungicides for soil drenching, and soil solarization. The controlling root rot is not permanent in nature. Hence, continuous as well as effective control measures needs to be applied which are also expensive. The pre-treatment of seed with more than one fungicide is most efficient and widely used method of root rot control (Srivastava 2004).

10.8.1.3 Foliar Diseases

Cercospora leaf spot, *Alternaria* leaf blight, and powdery mildew are the most common sugar beet foliar diseases. Other diseases such as *Ramularia*, Phoma and Colletotrichum leaf spots are of minor importance. Ramularia leaf spot is of rare occurrence and sporadic in nature (Srivastava 2004). *Cercospora* leaf spot (also known as brown spot or leaf blotch), due to *Cercospora beticola*, is one of the most widespread and destructive foliar diseases of sugar beet. The symptoms of the disease start appearing on lower and older leaves. Initially, the disease is characterized by the appearance of minute, translucent spots. Within a few days, the spots turn into discrete circular lesions 3–5 mm in diameter having necrotic grey centres with reddish brown to black margins. By avoiding short rotations of fewer than 3 years, illness can be controlled. The removal of crop waste, the adoption of tolerant cultivars, and the use of foliar fungicides are only a few examples.

Alternaria leaf blight is also an important disease of sugar beet and caused by two species of *Alternaria*, i.e. *A. alternata* and *A. brassicae*. Out of these two, *A. alternata* is more damaging and may destroy up to 30% leaf area (Srivastava 2004). Symptoms appeared on leaves only. *A. alternata* leaf spots are up to 10 mm in diameter, irregularly shaped and black-brown. They are more prevalent on the edges, whereas *A. brassicae* leaf spots are concentric and up to 15 mm in diameter. As the disease progresses, the spots increase in size and become dark brown or black in colour with water-soaked margin. Water-soaked, sub-circular brown spots with necrotic flecks in the centre appear on both surfaces of leaves. The disease is partially managed by spraying of Dithane M-45 (Agnihotri et al. 1972). Out of these, two or three sprays of Dithane M-45 @ 2.5 kg/ha per spray at 15-day intervals before the appearance of the disease give effective control.

Powdery mildew caused by *Erysiphe betae* is also a serious disease of sugar beet. The disease appears first on lower and older leaves and gradually spreads towards the upper and younger leaves. It is characterized by the formation of white, later grey, tan mildew areas on both sides of the leaf. In advanced stage of disease development, mildew patches enlarge and coalesce and leaf looks as if dusted with white powder (Srivastava 2004). An integrated approach involving destruction of crop debris, spraying of fungicides and use resistant varieties should be recommended for managing the disease effectively. Disease control has been exclusively achieved by spraying of fungicides like wettable sulphur and other sulphur formulations (Cicco and Curtis 1993; Karve et al. 1973; Russeel and Mukhopadhyay 1981).

10.8.1.4 Root Diseases

Roots of sugar beet plants are affected by a number of fungal pathogens causing various types of root rots. Among these, Sclerotial root rot (*S. rolfsii*) is the most destructive disease causing about 50% damage of the roots under favourable conditions. Other root rots, of the like dry root rot (*Rhizoctonia solani*) and charcoal root rot (*Rhizoctonia bataticola*) may cause 15–30% destruction. Both yield and sucrose in the root are adversely affected. Neither *Rhizoctonia oryzae* or *Fusarium* root rot (*F. chlamydosporum*) are of major relevance and are sporadic in nature.

Known as 'Southern stem and root rot', sclerotial root rot caused by *Sclerotium rolfsii* has a significant economic impact in the tropical and subtropical regions. Under Indian conditions, the disease appears during March. The symptoms include yellowing and wilting of leaves followed by rotting of roots of affected plants. White cottony mycelium develops on rotted basal portions of roots and causes gradual semi-watery decay. Sclerotia are the means of survival of fungus in soil even in the absence of suitable hosts or conditions favouring its active growth. These are spread via cultivation and irrigation water (Diffus and Ruppel 1993). The integrated disease management (IDM) system involving cultural, chemical, biological, and host resistance may be employed to manage the disease.

Dry root rot caused by *Rhizoctonia solani* also known as 'Rhizoctonia Root or Crown Rot' or 'Dry Root Rot Canker' has been reported from most of the temperate, tropical, and subtropical countries. The disease is characterized by a greyish brown to reddish-brown discolouration of mature roots around the bases of lateral roots. Diseased roots show a woody appearance, and concentric rings develop on the infected portion. The lesions are slightly sunken and beneath there, pockets or deep cankers of dirty brown spongy decayed tissue develop which are sometimes filled with fungal mycelium. The management of the disease through crop rotation (3–5 years) and other cultural practices like deep ploughing and destruction of diseased plants gives substantial protection against dry root rot.

Charcoal root rot (*Rhizoctonia bataticola*) is another important disease prevalent both in tropical and subtropical countries. The disease appears on the upper portions of roots and bases of petioles, and is characterized by a brownish-black discolouration of the crown portion of the root. By soaking the soil with PCNB, the disease's spread can be minimized (Srivastava et al. 1986).

10.8.2 Insects and Pests

Under tropical and subtropical conditions of India, defoliating insects, viz. beet armyworm (*Spodoptera litura* Fabricius), hairy caterpillar (*Diacrisia obliqua* Walker), semilooper (*Plusia orichalcea* Fabricius), cutworm (*Agrotis ipsilon* Rott.) cause appreciable damage to sugar beet at different growth stages (Patil et al. 2007). Manoharan et al. (2010) reported *Spodoptera litura* as the predominant pest on sugar beet. The gram pod borer, *Helicoverpa armigera* (Hubner) is a polyphagous pest of economic importance on many agricultural and horticultural crops (Venette et al. 2003). The other insects of sugar beet which are not present in the region yet may cause problem in future. The root maggot of sugar beet (*Tetanopsmyopae formis*) is a major pest in light-textured soils, and grasshoppers become a big problem during dry conditions. The other minor pest of sugar beet are wireworms, flea beetles, aphids of root, white grubs, and beet webworms. White grubs and wireworm damage germinating seedlings and result in reduced plant stand. Root aphids may cause severe yield loss in dry years. Sugar beet webworm feeds on leaves of sugar beet. In early stages of crop growth, cutworm may infest the crop. It can be controlled by applying chlorpyrifos at 1 kg a.i./ha in soil at the time of sowing. Armyworm seriously damages the crop in tropical as well as subtropical region (Sharma et al. 2017). It appears about 100 DAS. Though this insect has natural enemies, a single spray of quinalphos should be enough to keep it at bay (0.05%). Instead of using pheromone traps, release *Trichogramma chilonis* parasitized eggs at 50,000 eggs/ha using trichocards, create grass piles near sugar beet fields to attract armyworm larvae, and provide bird perches for birds to feed on the larvae (Anonymous 2011). In order to manage aphids, spray 3% neem oil or 2 ml/l dimethoate in water with 0.5 ml of teepol. To control tobacco caterpillar, use 2 ml/l endosulfan or carbaryl in water (Balakrishnan et al. 2007). To control these insect pests, a programme of integrated pest management must be implemented.

Under Punjab conditions, most common diseases of sugar beet are sclerotia root rot, *Cercospora* leaf spot, and heart rot. Army worm, tobacco caterpillar, and cutworm are troublesome insect pests. Sugar beet must be grown in the same field only once in 3 years to prevent pests and diseases (Anonymous 2020).

10.9 Harvesting and Post-Harvesting Constraints

The sugar beet crop matures in about 5–6 months under tropical and subtropical environment. The harvesting of sugar beet started from mid-April to May-end in this region. The indicators of sugar beet maturity are yellowing of the older leaves and around 15–18% brix of root at the time of harvest. The harvested beet tuber should be handled as gently as possible to remove soil and trash to minimize the beet breakage and bruising to get quality beet tuber (Paul et al. 2019). In general, 30–35 tonnes of beet tuber per acre are produced. Webb and Jaggard (1980) observed that planting and harvesting of this crop on different dates influenced sugar yield. Paul et al. (2017) evaluated the effect of two harvesting date, viz. 135 DAS and 155 DAS on root yield and quality of sugar beet. When it came to beet length (24.26 cm), beet girth (25.40 cm), individual beet weight (536.07 g), and beet yield (53.60 t ha^{-1}), the 155 DAS harvest had the greatest results, while the 135 DAS harvest had the highest brix (14.66%). Beet length (25.67 cm), individual beet weight (681.2 g), and beet yield (68.12 tonnes ha^{-1}) were observed in SV 894 with 155 DAS harvest as the highest values in interaction (Curcic et al. 2018).

In developed countries, harvesting is done by expensive machinery such as sugar beet defoliator to remove leaves and sugar beet lifter-loader harvesters to dig out roots from soil and load them in the trucks. Sugar mills have temperature controlled piling ground before processing of sugar beet. The primary role of these piling grounds is to lower down the root respiration which otherwise reduced extracted sugar. In India, harvesting is done with sugar beet harvester/potato digger/cultivator/by manual digging. Herman (2004) compared pinch-wheel beet harvesting machines with spike-wheel harvesters and reported 1–2 tons of beets more per acre in pinch-wheel beet harvesting machines than spike-wheel harvesters due to the method of root extraction. Also, pinch-wheel harvesters can be used to harvest fields a few days earlier after the last irrigation than spike-wheel harvesters. Sugar beet roots must be processed within 48 h after harvesting. The beet leaves should be allowed to remain in the field to serve as green manure or alternately, the leaves can be fed to cattle as forage. The loss of sugar after harvesting is due to unfavourable weather conditions, further reductions in harvested sugar yield occur before the roots are processed in the factory. During storage, sugar is cleaved to provide energy for the life-sustaining processes of the sugar beet and also consume stored sugar in the root (Klotz and Finger 2004). Among other factors such as damage during harvest operations results in subsequent infestation with mould and rots during storage and the genotypic effect on the storability of sugar beet determines the sugar content in sugar beet (Hoffmann and Schnepel 2016; Schnepel and Hoffmann 2016). Therefore, sugar beet roots started deteriorating on rapid rate after harvesting hence its transportation should be ensured to sugar mill preferably within 24 h of harvesting (Fig. 10.6). Sugar beet can't be stored for more than a few hours, therefore a well-coordinated plan for

Fig. 10.6 Optimum root size at harvest in sugar beet requires immediate processing to avoid losses

harvesting and preparing the roots is essential. It is thus only possible to grow it commercially near processing plants.

The processing of sugar beet is different from sugar cane. The availability of efficient sugar processing technology is essential for the successful commercial exploitation of sugar beet. Sugar from sugar beet roots is extracted by adopting diffusion process in special diffusers based on the counter current washing technique. It is not possible to extract juice from sugar beet roots with roller mills. The juice characteristics of sugar beet also necessitate adoption of carbonation process only for clarification, and hence carbonation with diffusers and related accessories is essential to process sugar beet. The use of efficient machinery includes diffusers which have a significant role in utilization of sugar beet by-products and synthesis of high-quality beet molasses. Sugar beet can't be processed for *khandsari* or *gur* like sugar cane. It can only be used for vacuum-pan sugar production (Pathak et al. 2014). Sugar beet doesn't give by-product like bagasse which is generally used as fuel for running sugar mills. However, the by-products of sugar beet, i.e., beet pulp and molasses have a good market and should be expected to fetch extra money to compensate the fuel cost in its processing. The beet molasses can be used as raw material for several special fermentations and also form a rich source of lactic acid, vitamin B, and other pharmaceutical preparations. In India, only few sugar mills have facility to process sugar beet. In Punjab, only one private mill processes sugar beet (Brar et al. 2015).

10.10 Comparative Economics of Sugar beet Cultivation

Sugar beet is an industrial crop, and average economic return from sugar beet has been greater than other crops. Mostly, sugar beet is cultivated on contract basis with sugar mills. Hence, the return from sugar beet cultivation is more or less stable as compared to the other crop. The profit from sugar beet may be high or low but the

Table 10.2 Comparative yield and economic of sugar beet and wheat

S. No.	Parameters	Sugar beet	Wheat
1	Yield (q/ha)	940	47.5
2	Selling Price (/q)	185	1735
3	Gross income	173,900	82,413
4	Cost of cultivation	82,612	27,070
5	Net income	91,288	55,343

(Source: Brar and Kumar 2019)

risk of economic loss from the crop is not high. In USA, cost of cultivation of sugar beet was $ 900 to $ 1800 per hectare and a net output 35% higher than for sugar cane (Elbersen et al. 2010). Sugar beet is considered as high input, as well a high output crop. The costs of seed, tillage, and weed control are high but the benefits as well. Furthermore, the crop often grown under irrigated conditions. In India, according to calculation by Tamil Nadu Agricultural University total cost of cultivation of sugar beet is around Rs.8000–8500 per acre and the income will be Rs.18,000 /acre with a net income of Rs. 10,000/acre (Anonymous 2011).

In the line with the diversification of agriculture in Punjab to bring the farmer out of wheat-paddy cycle and enhancing their income, a project was initiated by sugar mill. The farmers are getting better returns on growing sugar beet than wheat. They have the advantage of an assured market under a contract farming mechanism with the sugar mills and supervised agronomic practices. There is the target of sugar mills to bring around 6000 acres under sugar beet cultivation in Punjab. To achieve the target, the mills need to provide necessary facilities to the farmers to encourage them for sugar beet cultivation. A similar project has been started in the states Karnataka and Maharashtra. The innovative farmers are cultivating sugar beet instead of the usual sugar cane crop due to scarcity of irrigation water. In addition to its lower water requirement, sugar beet matures within 6–7 months and is a good substitute of sugar cane crops. Besides its potential to increase the operation period of the sugar mills from 4 months to 6 months in year, it makes rice–sugar beet cropping system a reality by replacing wheat crop in Punjab for higher returns. Data was recorded to make economic comparison of the cultivation of sugar beet and wheat (Table 10.2). Sugar beet cultivation gave 35,945 rupees per ha additional net returns than wheat (Brar and Kumar 2019). In another study, Singh et al. (2018) compared the cost and return structure of sugar beet and sugar cane cultivation in Punjab. The comparative economics revealed that the total variable cost was four times higher in sugar cane than sugar beet, whereas sugar cane crop gave much higher returns than beet crop (Table 10.3). Per hectare, net returns came out to be Rs. 24,112 for sugar beet, whereas it was Rs.100,513 for sugar cane depicting a difference of Rs. 76,401. The important constraints in the cultivation of this crop in Punjab were labour scarcity, high wages of hired labour, inadequate price, perishability of the crop, and high incidence of insect pest attack.

Table 10.3 Comparative economics of sugar beet and sugar cane in Punjab, 2013–14

S. No.	Particulars	Sugar beet	Sugar cane
1	Human labour	16,569	65,625
2	Machine labour	5653	28,938
3	Seed	1500	24,470
4	Manures & fertilizers	7856	3815
5	Agric chemical/ weedicides	1708	4985
6	Irrigation	–	1600
7	Interest @9% per half of the period of crop on operational cost	1498	5824
8	Total variable cost	34,784	135,257
9	Yield-Main product, qtl/ha	409	813
10	Price-Main product, Rs./qtl	144	290
11	Gross returns, Rs./ha	58,896	235,770
12	Returns over variable cost, Rs./ha	24,112	100,513
13	Benefit-cost ratio	1.69	1.74

(Source: Singh et al. 2018)

10.11 Future Prospects

Though sugar cane is an important agro-industrial crop in Indian scenario, the demand of sugar will be 49Mt by 2025 with exponentially growing rate of population in the country. Projected estimates revealed that with an average sugar recovery of 10.75% and productivity 100 t/ha, about 495 million tonnes of sugar cane will be required which needs 4.75 million hectare area for its cultivation (Pathak et al. 2014) in the country, but with increasing urbanization, fluctuating extremes of various biotic and abiotic stresses and accumulation of salts in the soil, the cultivable land is continuously shrinking. In addition, the increasing pressure on petroleum products forced the policy makers to shift focus on bio-ethanol that will further partition the sugar cane used in sugar making. The long duration of sugar cane crop makes it less productivity per unit area and time. In addition, the lowering of ground water table is also posing serious threat on sustainability of sugar cane cultivation. Additionally, the yield plateau has been achieved in most successful wheat-paddy rotation, now profits are decreasing due to high cost of cultivation. Hence, it is high time to find alternatives which can efficiently cover the lacunas of sugar cane. The cultivation of short duration and high yielding sugar beet could be a best supplementary crop and possible alternative. Its better tolerance towards adverse soil conditions (salinity and alkalinity), short growing period with good economic yields makes it a good choice for the salt affected regions. With improved genetic, biotechnological, and agro-technological interventions/advancements, the cultivation of sugar beet will be more feasible under marginal and fragile areas rendered unfit for other sugar crops. The involvement and deployment of public and private sectors in development of improved varieties, quality seed production and distribution, recent advanced production technologies and mechanization for harvesting and timely transport from field to industry units will further reduce the cost of sugar beet production. For the expansion in cultivation of this important sugar crop, sugar industries may adopt contract farming models for its assured cultivation and by back policies with remunerative prices which will make sugar beet a profitable industrial crop under Indian scenario.

10.12 Conclusions

Sugar beet can realize its high productivity in a short growing season under farmers' field. This makes it sustainable crop in comparison to sugar cane. This crop has a high yield per hectare, making it an effective land use crop. Due its short crop cycle, water requirement is also low. The amount of by-products is limited, although leaves, pulp, and heads of the beets may serve well as seems to make efficient use of inputs such as water and nutrients, associated with high productivity. For a good yield, the crop requires sufficient chemical control of pests and diseases. The crop must be grown in rotation and in this way can contribute to diversification of crops in a cropping system. Sugar beet can be cultivated on marginal areas but due to the high input costs, it is not practically possible. The advancement in sugar beet research and technology will further increase its adaptability and will improve its market in the tropical and subtropical areas. Further, the availability of sugar beet to sugar mills for extra 2–4 months after sugar cane is over makes sugar beet a profitable venture for sugar mills. In area like Punjab where paddy-wheat rotation reached at plateau in terms of productivity and profitability, sugar beet could be a valid option. Moreover, lowering ground water level and increasing cost of production pose a serious threat on sustainability of sugar industry, under such circumstances sugar beet could be valid alternative to long duration sugar cane crop.

References

Abd El-Gawad AA, Hassan HK, Hassany WM (2000) Transplanting technique to adjust plant stand of sugar beet under saline conditions. Response of sugar beet yield to transplanting missed hill by different transplant ages. Proc. 9th Conf. Agron., Minufiya Univ., 1–2 Sept, p 533–548

Abo-Shady KA, Hilal SMM, El-Sheref EEM, Ibrahim MFM (2010) Yield and quality of sugarbeet crop as affected by irrigation interval, cultivars and potassium fertilization in north delta. J Agric Res Kafer ElSheikh Univ 36:361–376

Agnihotri VP, Sen C, Srivastava SN (1972) Chemical control of Alternaria leaf spot of sugarbeet. Annual Report IISR, Lucknow (1971–72), p 64

Ahmad S, Zubair M, Iqbal N, Cheema NM, Mahmood K (2012) Evaluation of sugar beet hybrid varieties under Thal-Kumbi soil series of Pakistan. Int J Agric Biol 14:605–608

Allard RW, Bradshaw AD (1964) Implication of genotype environmental interactions in applied plant breeding. Crop Sci 4:503–507

Amin M, Khan A, Khan D (1989) Effect of date of sowing on yield and quality of sugarbeets. Pak J Agric Res 10(1):30–33

Anonymous (2004) The results from R and D trial obtained Ugar Sugars Ltd. Ugar Khurd, Karnataka. www.syngentaindia.com

Anonymous (2011) Hand book of agriculture. Indian Council of Agricultural Research Publication, pp 1211–1216

Anonymous (2020) Package and practices for crops of Punjab Rabi 2020–21. Punjab Agricultural University, Ludhiana, p 67

Armin M, Asgharipour MR (2012) Effect of time and concentration of boron foliar application on yield and quality of sugarbeet. Am Eurasian J Agric Environ Sci 12(4):444–448

Balakrishnan A, Selvakumar T (2008) Integrated nitrogen management for tropical sugarbeet hybrids. Sugar Tech 10(2):177–180

Balakrishnan A, Selvakumar T (2009) Evaluation of suitable tropical sugarbeet hybrids with optimum time of sowing. Sugar Tech 11(1):65–68

Balakrishnan A, Selvakumar T, Ponnuswamy K (2007) Sugarbeet cultivation in Tamil Nadu for sugar and ethanol production. Training conducted on renewable energy sources through biofuel crops at CAS, Agronomy, Tamil Nadu Agricultural University, Coimbatore, p 12

Barik S (2003) Role of potassium and nitrogen on sugar concentration of sugarbeet. Afr Crop Sci J 11:259–268

Bellin D, Schutz B, Soerensen TR, Salamini F, Schneider K (2007) Transcript profiles at different stages and tap-root zone identity correlated developmental and metabolic pathway of sugar beet. J Exp Bot 58:699–715

Besheit SY, Mekki BB, Beshay MG (1996) Effect of different levels of water supply on sugar beet yield and quality in calcareous soils. J Agric Sci Mansoura Univ 21:3429–3436

Bhullar MS, Uppal SK, Kapur ML (2009) Effect of agronomic practices and varieties on productivity of sugarbeet (*Beta vulgaris* L.) in semi-arid region of Punjab. J Res Punjab Agric Univ 46: 6–8

Bhullar MS, Uppal SK, Kapur ML (2010) Influence of planting density and nitrogen dose on root and sugar yield of beet (*Beta vulgaris* L.) under sub-tropical semi-arid conditions of Punjab. J Res Punjab Agric Univ 47:14–17

Bjornsson J (2002) Stability analysis towards understanding genotype x environment interaction. Plant agriculture department of University of Guelph, Ontorio, Canada, pp 25–26

Brar NS, Kumar B (2019) Innovative cultivation of sugarbeet for sweet returns. In: Rana RK, Singh R, Thakur AK, Chahal VP, Singh AK (eds) Contemplating agricultural growth through farmers' frugal innovations. ICAR-ATARI-1, Ludhiana, Punjab, pp 131–132

Brar NS, Dhillon BS, Saini KS, Sharma PK (2015) Agronomy of sugarbeet cultivation - a review. Agric Rev 36:184–197

Campbell IG, Kern JJ (1982) Cultivar x environment interactions in sugarbeet yield trials. Agron J 22:932–935

Cattanach AW, Dexter AG, Oplinger ES (1991) Sugarbeets. In: Alternative field crops manual. University of Wyoming Cooperative Extension and University of Minnesota Center for Alternative Plant and Animal Products and Minnesota Extension Service. http://www.hort.purdue.edu/newcrop/afcm/sugarbeet.html

Chatin N (2004) Tropical sugar beet: from a business idea to sustaining sugar production in tropical areas. Compiled by Yndgaard, F. in summary of workshop: Globalisation - a great opportunity for sugar beet in developed and developing countries. J Swedish Seed Associat 1–2:12–29

Cicco VDE, Curtis FDE (1993) Powdery mildew of sugarbeet. Informatore Fitopathologico 43:18–20

Cioni F, Maines G (2010) Weed control in sugarbeet. Sugar Tech 12:243–255

Cooke DA, Scott RK (1993) The sugar beet crop: science into practice. Chapman and Hall, London, pp 262–265

Copeland LO, McDonald MB (2001) Seed science and technology. Kluwer Academic Publishers, Boston, Norwell, Massachusetts, pp 72–124

Curcic Z, Ciric M, Nagl N, Taski-Ajdukovic K (2018) Effect of sugarbeet genotype, planting and harvesting dates and their interaction on sugar yield. Front Plant Sci 9:1041

Dean CY, John AS (1997) Effects of plant population and row width on yield of sugarbeet. J Sugar Beet Res 34:21–30

Demont M, Dillen K (2008) Herbicide tolerant sugar beet: the most promising first generation GM crop? Int Sugar J 110:613–617

Deshpande HH (2013) Response of sugarbeet genotypes to nitrogen, potassium, planting methods and dates of sowing in Deccan Plateau of Peninsular India. Ph.D. Thesis, University of Agricultural Sciences, Dharwad, Karnataka (India)

Deshpande HH, Deshmukh HP, Deshpande SH (2017) Agro-techniques for sustainable sugarbeet production. Int J Agric Sci 13:410–418

Diffus JD, Ruppel EG (1993) Diseases. In: Cooke DA, Scott RK (eds) The sugarbeet crop; science into practice. Chapman and Hall, London, p 675

Dordas C, Apostolidesm GE, Goundra O (2007) Boron application affects seed yield and seed quality of sugar beets. J Agric Sci 145:377–384

Duncan DN, Meggitt WF, Penner D (1982) Basis for increased activity from herbicide combinations with ethofumesate applied on sugar beet (*Beta vulgaris*). Weed Sci 30:195–200

Eberhart SA, Russell WA (1966) Stability parameters for comparing varieties. Crop Sci 6:36–40

Ebrahimian HR, Sadeghian SY, Jahadakbar MR, Abasi Z (2008) Study of adaptibilty and stability of sugarbeet monogerm cultivars in different locations of Iran. J Sugar Beet 24:1–13

Elbersen W, Oyen L, Franken YJ (2010) Tropical sugar beet (*Beta vulgaris* L.) potential of tropical sugar beet for bio-ethanol production. FACT Foundation, p 1–6

El-Kassaby AT, Leilah AA (1992) Influence of plant density and nitrogen fertilizer levels on sugar beet productivity. Proc 5th Conf Agron Zagazig 13–15 Sept, 2, p 954–962

Elliott ME, Weston GD (1993) Biology and physiology of the sugar-beet plant. In: Cooke DA, Scott RK (eds) The sugarbeet crop; science into practice. Chapman and Hall, London, pp 37–66

El-Maghraby SS, Gomaa IMA, Rehab FH, Sanad MH, Sanad M (2008) Response of sugar beet to some mechanical management practices, irrigation and plant densities. Sugar Tech 10(3): 219–226

Fishman ML, Hall LF, Hoa K, Chau P, Cooke P, Hotchkiss AT (2009) Physico-chemical characterization of alkaline soluble polysaccharides from sugar beet pulp. Food Hydrocoll 23(6): 1554–1562

Francois LE, Maas EV (1994) Crop response and management on salt-affected soils. In: Pessarakli M (ed) Handbook of plant and crop stress. Marcel Dekker Inc., New York, pp 149–181

Garg VK, Srivastava SC (1985) Evaluation of techniques suitable for growing sugar beet on sodic soils. J Indian Soc Soil Sci 33:884–887

Gerendas J, Sattelmatcher B (1990) Influence of nitrogen form and concentration on growth and ionic balance of tomato (*Lycoperiscon esculentum*) and potato (*Solanum tuberosum*). In: Van Beusichem ML (ed) Plant nutrition physiology and application. Kluwer Academic, Dordrecht, pp 33–37

Gezgin S, Hamurcu M, Apaydin M (2001) Effect of boron application on the yield and quality of sugar beet. Turk J Agric Forestry 25:89–95

Ggyllenspetz U (1998) Genotype × environment interaction and stability of diploid and triploid sugar beet (*Beta vulgaris* L.) varieties. Sveriges Lantbruksuniv, Uppsala (Sweden), p 19

Ghonema MH (1998) Effect of plating dates and harvesting time on yield, yield components and quality of sugar beet (*Beta vulgaris* L.). J Agric Sci Mansoura Univ 23:2971–2979

Gobarah ME, Mekki BB (2005) Influence of boron application on yield and juice quality of some sugar beet cultivars grown under saline soil conditions. J Appl Sci Res 1:373–379

Goldberg RA, Alwin RK (2008) Syngenta international AG: tropical sugar beet. Harvard Business School Case, p 909–404

Gupta R, Singh PR, Singh AK (2013) Optimizing irrigation water and land need for sugar production through intercropping sugarbeet in autumn planted sugarcane. Souvenir- IISR-Industry Interface on Research and Development Initiatives for sugarbeet in India, 28–29 May, sugarbeet breeding outpost of IISR IVRI Campus,Mukteswar-263138, Nainital,

Organised by Indian Institute of Sugarcane Research (ICAR) and Association of Sugarcane Technologists of India, p 51–52
Hassanli AM, Ahmadirad S, Beecham S (2010) Evaluation of the influence of irrigation methods and waterquality on sugar beet yield and water use efficiency. Agri Water Manage 97:357–362
Heath MC, Cleal RA (1992) Transplanting date and pot length for sugar beet. Transplant UK Aspects Appl Biol 32:135–140
Hellal FA, Taalab AS, Safaa AM (2009) Influence of nitrogen and boron nutrition on nutrient balance and Sugar beet yield grown in calcareous Soil. Ozean J Appl Sci 2:1–10
Herman SM (2004) University of California, cooperative extension imperial county guidelines Cir, p 104
Hoffmann CM, Schnepel K (2016) Susceptibility to root tip breakage increases storage losses of sugar beet genotypes. Sugar Tech 141:625–632
Iqbal MA, Saleem AM (2015) Sugar beet potential to beat sugarcane as a sugar crop in Pakistan. Am Eurasian J Agric Environ Sci 15(1):36–44
James KF, Stephen DM (1990) Herbicide and variety effects on sugarbeet stand establishment. J. Sugar Beet Res 27:20–28
Jaszczolt E (1998) Effect of two methods of fertilizing sugar beet with trace elements on the yields of roots and sugar. Gazeta Cukrownicza 106:232–234
Jordan N, Boody G, Broussard W, Glower JD, Keeney D, McCown BH, McIsaac G, Muller M, Murray H, Neal J, Pansing C, Turner RE, Warner R, Wyse D (2007) Sustainable development of the agricultural bio-economy. Science 316:1570–1571
Karve AD, Bhalerao AC, Mhaiskar VG (1973) Powdery mildew of sugarbeet. Proc Ann Convention Deccan Sugar Technol Assoc 25(1):9–15
Katerji N, Van Hoorn JW, Hamdy A, Mastrorilli M, Mou Karzel E (1997) Osmotic adjustment of sugar beets in response to soil salinity and its influence on stomatal conductance, growth and yield. Agric Water Manag 34:57–69
Keshavarz S, Mesbah M, Ranji Z, Amiri R (2001) Study on stability parameters for determining the adaptation of sugar beet commercial varieties in different areas of. Iran J Sugar Beet 17(1): 15–36
Khan AA (1992) Pre-plant physiological seed conditioning. Hortic Rev 13:131–166
Klotz KL, Finger FL (2004) Impact of temperature, length of storage and postharvest disease on sucrose catabolism in sugarbeet. Postharvest Biol Technol 34:1–9
Kristek A, Stojic B, Kristek S (2009) Effect of the foliar boron fertilization on sugar beet root yield and quality. Agricult Sci Prof Rev 12(1):22–26
Kumar V (1993) Effect of irrigation on yield, sucrose content and impurity levels of sugarbeet crop. J Indian Soc Soil Sci 41:26–28
Kumar A, Lokesh B (2018) Integrated nutrient management for maximum economic yield of sugar beet and sustainable soil health. Agrica 7(2):156–163
Leach LD, Macdonald JD (1976) Seed borne Phoma betae as influenced by area of sugarbeet production, seed processing and fungicidal treatments. J Am Soc Sugarbeet Technol 19:4–15
Legere A, Schreiber MM (1989) Competition and canopy architecture as affected by soybean row width and density of red root pigweed. Weed Sci 37:84–92
Leilah AA, Badawi MA, Said EM, Ghonema MH, Abdou MAE (2005) Effect of planting dates, plant population and nitrogen fertilization on sugar beet productivity under the newly reclaimed sandy soil in Egypt. Sci J King Faisal Univ Basic Appl Sci 6:85–110
Loomis WD, Durst RW (1992) Chemistry and biology of boron. Biofactors 3:229–239
Makrantonaki MS, Kalfountzos D, Vyrlas P (2002) Water saving and yield increase of sugar beet with subsurface drip irrigation. Global Nest Int J 4:85–91
Mall AK, Misra V, Santeshwari PAD, Srivastava S (2021) Sugar beet cultivation in India: Prospects for bioethanol production and value added co products. Sugar Tech 23:1218–1234
Manoharan T, Pathma J, Preetha G (2010) Seasonal incidence of sugar beet pests and natural enemies. Indian J Ent 72(1):36–41

May JM, Hilton JG (1985) Reduced rates of pre-emergence herbicides (1982–1984). In: 77th Report of the Norfolk agricultural station, p 14–21

May JM, Wilson RG (2006) Weed and weed control. In: Draycott AP (ed) Sugarbeet. Blackwell, London, pp 359–386

Mekdad AAA (2015) Sugar beet productivity as affected by nitrogen fertilizer and foliar spraying with boron. Int J Curr Microbiol App Sci 4:181–196

Metcalfe DS, Elkins DM (1980) Sugar crops. In: Crop production: principle and practices. Mac Milan Company, New York, pp 511–527

Miller SD, Fornstrom KJ (1989) Weed control and labour requirements in sugarbeets. J Sugar Beet Res 28:1–9

Misra V, Mall AK, Pathak AD (2020) Sugar beet: a sustainable crop for saline environment. In: Hasanuzzaman M (ed) Agronomic crops, vol vol 3. Stress Responses and Tolerance Springer Publications, pp 49–61

Moldovan V, Moldovan M, Kadar R (2000) Item from Romania. SCA Agricultural Res. Stat. Turda, 3350 Str. Agriculturi Jud Chuj Romania, p 27

Narang RS, Bains BS (1987) Techniques for planting sugarbeet to advance the harvest date in north Indian conditions. Exp Agric 23:99–103

Pathak AD, Kapur R, Solomon S, Kumar R, Srivastava S (2014) Sugarbeet: a historical perspective in Indian context. Sugar Tech 16:125–132

Patil AS, Salunkhe AN, Pawar BH, Ghodke PV, Shivankar SB, Zende NB, Shewate SR, Patil DJ (2007) Sugarbeet cultivation in tropical India – a new experience. In: Proceedings of 68th Annual Convocation. STAI, New Delhi, pp 85–111

Paul SK, Paul U, Sarkar MAR (2017) Yield and quality of tropical sugarbeet as influenced by variety, spacing and fertilizer application. Sugar Tech 20:175–181

Paul SK, Joni RA, Sarkar MAR, Hossain MS, Paul SC (2019) Performance of tropical sugar beet (*Beta vulgaris* L.) as influenced by date of harvesting. Arch Agric Environ Sci 4(1):19–26

Petkeviciene B (2009) The effects of climate factors on sugar beet early sowing timing. Agron Res 7:436–443

Rajasekaran M (2007) Effect of drip fertigation on growth, yield and quality of tropical sugarbeet. M.Sc. Agronomy Thesis, Tamil Nadu Agricultural University, Coimbatore, TN

Rinaldi M, Vonella AV (2006) The response of autumn and spring sown sugarbeet (*Beta vulgaris* L.) to irrigation in southern Italy: water and radiation use efficiency. Field Crop Res 95:103–114

Russeel GE, Mukhopadhyay AN (1981) Effect of some systemic and non-systemic fungicides on *Erysiphe betae* and its development on sugarbeet leaves. Phytopathology Z 10:1–6

Saini KS, Brar NS (2017) Sugarbeet (*Beta vulgaris* L.) yield response to varying planting methods, densities and depth of sowing under subtropical conditions. Bioscan 12:1715–1720

Saini KS, Brar NS (2018) Crop and water productivity of sugarbeet (*Beta vulgaris*) under different planting methods and irrigation schedules. Agril Res 7:93–97

Saini KS, Brar NS, Walia SS, Sachdev PA (2020) Effect of planting techniques, plant densities and different depths of sowing on production economics, water and sugar productivity of sugarbeet. J Crop Weed 16:190–196

Sanghera GS, Jamwal NS (2019a) Evaluation of elite sugarcane clones for cane yield and component traits using augmented design. LS: Int J Life Sci 8(3):175–183

Sanghera GS, Jamwal NS (2019b) Identification of potential crosses based on vigour, cane characteristics and HR brix for first clonal selection in sugarcane. Indian J Sugarcane Technol 34:12–16

Sanghera GS, Jamwal NS (2019c) Perspective for genetic amelioration of sugarcane towards water logged conditions. Int J Pure App Biosci 7:484–502

Sanghera GS, Jamwal NS (2020) Unravelling sugar accumulation in sugarcane (*Saccharum* spp.) stalk: recent approaches. LS: Int J Life Sci 9:185–198

Sanghera GS, Singh RP, Kashyap L, Tyagi V, Sharma B (2016) Evaluation of sugarbeet genotypes (*Beta vulgaris*L.) for root yield and quality traits under subtropical conditions. J. Krishi Vigyan 5:67–73

Schnepel K, Hoffmann CM (2016) Genotypic difference in storage losses of sugar beet – causes and indirect criteria for selection. Plant Breed 135:130–137
Schweizer EE, May MJ (1993) Weeds and weed control. In: Cooke DA, Scott RK (eds) The sugar beet crop: science into practice. Chapman & Hall, London, pp 485–519
Seadh SE, Farouk S, El- Abady MI (2007) Response of sugar beet to potassium sulfate under nitrogen fertilizer levels in newly reclaimed soils conditions. Afr Crop Sci Proc 8: 147–153
Sharma S, Kooner R, Sandhu SS, Arora R, Kaur T, Kaur S (2017) Seasonal dynamics of insect pests of sugar beet under sub-tropical conditions. J Agrometeorol 19:81–83
Sharmasarkar EC, Sharmasarkar S, Miller SD, Vance GF, Zhang R (2001) Assessment of drip irrigation and flood irrigation on water and fertilizer use efficiencies for sugarbeet. Agric Water Mgmt 46:241–251
Shrivastava AK (2006) Sugarcane at a glance. IBDC, Lucknow, p 246
Shukla SK, Awasthi SK (2013) Sugarbeet: a supplement to sugarcane in non-traditional areas to meet future sugar demand. Souvenir- IISR-Industry Interface on Research and Development Initiatives for Sugarbeet in India, 28–29 May, Sugarbeet Breeding Outpost of IISR IVRI Campus, Mukteswar-263138, Nainital. Organised by Indian Institute of Sugarcane Research (ICAR) and Association of Sugarcane Technologists of India, p 55–57
Singh K, Srivastava SN, Sen C, Agnihotri VP (1973) Seed mycoflora of sugarbeet (*Beta vulgaris* L.) and its fungicidal control. Proc Indian Natl Sci Acad 39:695–700
Singh S, Sidana BK, Kumar S (2018) Water productivity of sugarbeet vs sugarcane cultivation in Punjab. Int J Innov Res Sci Technol 4:61–69
Singh NS, Thind KS, Sanghera GS (2019) Interactive outcome of dates of sowing and genotypes over environments on root yield and quality attributes in sugarbeet (*Beta vulgaris* L.). Acta Sci Agric 3:164–172
Sohier MM, Ouda (2001) Response of sugar beet to N and K fertilizers levels under sandy soil conditions. Zagazig J Agric Res 28:275–297
Spaar D, Dreger D, Zacharenko A (2004) Sugar beet. CUP Orech, Minsk, pp 133–135
Srivastava SN (2004) Management of sugarbeet diseases. In: Mukerji KG (ed) Disease management of fruits and vegetables, vol Vol 1. Kluwer Academic Publishers, Netherlands, pp 307–355
Srivastava SN, Tripathi RC (1998) Survey, surveillance and monitoring of sugarbeet diseases in root and seed crops. Annual Report IISR, Lucknow (1997–98), p 59
Srivastava SN, Tripathi RC (1999) Seed and seedling pathogens of sugarbeet and their management. National Symposium on Challenges and Prospects of Plant Pathology in the coming Millennium NBRI, Lucknow. December 9–11 (abstract), p 25
Srivastava SN, Misra SR, Singh K (1986) Evaluation of fungi toxicants against *Rhizoctonia bataticola* causing charcoal root rot of sugarbeet. Indian J Sugarcane Technol 3:89–90
Sroller I, Svachula V (1990) Influence of weather on the production and quality of sugar beets. In: Baier J, Bures R, Coufal VI (eds) Weather and production. Agropromizdat, Maskva, pp 247–269
Stephen K, Kurt H (2004) The effects of saline soil, irrigation, and seed treatments on sugarbeet stand establishment. J Sugar Beet Res 41(3):61–72
Stevanato P, Chiodi C, Broccanello C, Concheri G, Biancardi E, Pavli O, Skaracis G (2019) Sustainability of the sugar beet crop. Sugar Tech 21:703–716
Sumit AL (1983) Wageningen Agricultural Research Reports 914 PUDOC; Influence of external factors on growth and development of sugar beet (*Beta vulgaris* L.) Ph D Thesis
Sun RC, Hughes S (1999) Fractional isolation and physico-chemical characterization of alkali-soluble polysaccharides from sugar beet pulp. Carbohydr Polym 38:273–281
Tersahima I, Evans JR (1988) Effect of light and nitrogen nutrition on the organization of the photosynthetic apparatus in Spinach. Plant Cell Physiol 29:143–155
Toor SS, Bains BS (1994) Optimizing nitrogen fertilization for higher yield and quality of sugarbeet. Madras Agril J 81:689–691

Venette RC, Davis EE, Zaspel J, Heisler H, Larson M (2003) Mini risk assessment old world *bollworm, Helicoverpa armigera Hübner (Lepidoptera:Noctuidae)*. Cooperative Agricultural Pest Survey, Animal and Plant Health Inspection Service, US Department of Agriculture, p 20–32

Webb J, Jaggard K (1980) When should you grow the beet crop? Br Sugarbeet Rev 48:7–8

Wicks GA, Wilson RG (1983) Control of weeds in sugarbeets with hand hoeing and herbicides. Weed Sci 31:493–499

Zahoor A, Faridullah PS, Sanaullah B, Kakar KM, Haytham ES, Honna T, Yamamoto Y (2007) Sugar beet (*Beta vulgaris* L.) response to different planting methods and row geometries I: effect on plant growth and yield. Arch Agron Soil Sci 53:49–61

Sugar Beet Crop Production and Management

11

Koç Mehmet Tuğrul

Abstract

Sugar beet (*Beta vulgaris* L.), which is generally regarded as a temperate crop, is grown in winter in countries with subtropical climate characteristics. Its vegetation period is half that of sugar cane; however, the yield is higher and less water is needed in the growing period compared to sugar cane. Environmental and agronomic factors significantly affect the sugar beet yield and quality. In order to improve the quality of sugar beet and to obtain maximum yield from it, it is necessary to select the most appropriate varieties, sowing time, sowing method, sowing density, sowing depth, fertilizer type and amount, and irrigation plan. The most suitable air temperature for the development of sugar beet is between 15 and 25 °C. In conditions other than these temperatures, yield and quality are adversely affected. Therefore, the planting date should be determined to coincide with the given temperature range. The sowing method, density, and depth significantly affect the yield and quality. 11–12 plant m^{-2}, 45–50 cm row distances, 20–25 cm plant distance in row, and 2–3 cm sowing depth are ideal for sugar beet agriculture. Insufficient or excessive irrigation has negative consequences for sugar beet as well as for all agriculture. For an adequate and effective irrigation, the soil moisture level should be monitored at a depth of 0–90 cm and the amount of 600–700 mm of water that is needed and cannot be met by precipitation should be given with irrigation at least 75–80% of the field capacity in each irrigation. Planning the techniques to be applied in agricultural production according to the needs of the plant in sugar beet, as in other crops, is an important issue that ensures an increase in production and quality.

K. M. Tuğrul (✉)
Faculty of Agriculture, Department of Biosystems Engineering, Eskişehir Osmangazi University, Odunpazarı/Eskişehir, Turkey

V. Misra et al. (eds.), *Sugar Beet Cultivation, Management and Processing*,
https://doi.org/10.1007/978-981-19-2730-0_11

Keywords

Crop management · Cultivation · Irrigation · Plant density · Sugar beet

Abbreviations

BBRO	British Beet Research Organization
ET	Evapotranspiration

11.1 Introduction

In addition to being an important source of energy for human food, sugar beet is the most important sugar production source for the world's sugar needs, together with sugar cane as an important commercial commodity and industrial production tool. Currently, approximately 22% of the world's total annual sugar production is produced from sugar beet. The climatic characteristics of the region are an important determining factor in the yield and quality of sugar beet. Sugar beet is a long day plant and requires a lot of light and warmth. Less or too much light affects the sugar formation in sugar beet through photosynthesis. Overcast and cloudy weather adversely affect the photosynthesis in the plant. A total temperature of 2400–2800 °C is needed during the growing season for the development of sugar beet. Especially in June and July, the desire for lighting and warmth reaches its highest level. Sugar beet is a plant that grows well in clear and sunny weather. Long-term low temperatures that occur in the first cultivation period initiate vernalization of the plants and cause low yield.

Vegetation of beets starts at 7 °C in spring and ends at 5 °C in autumn. The length of the vegetation period between these temperature limits should be at least 170 days. During the vegetation period, the highest beet growth and sugar accumulation are observed at temperatures above 15 °C, especially at 20–25 °C (Draycott 2006). At higher temperatures, the growth and sugar accumulation slow down. Sugar beet is very sensitive to low temperatures during the first stages of the growth period. Young seedlings can be damaged by prolonged temperatures as low as 1–4 °C. In the last periods of the growth, on the other hand, the harvest can withstand cold temperatures down to −5 °C and photosynthesis continues. In the dry years, sugar content decreases especially when the temperature is above 30 °C in July and August and the day and night temperatures are close to each other. During the growth period, in addition to temperature and precipitation, the effect of relative humidity is also important, and relative humidity of 60–70% is considered ideal for sugar beet (Petkeviciene 2009).

Crop rotation has a significant impact on production quantity and quality. For this reason, shifting plant production to different areas from year to year will ensure that the nutrients needed are met from different depths. On the other hand, with the cultivation of different agricultural crops, the efficiency of crop-specific diseases and pests will decrease, and an increase in production will be achieved. Crop rotation is one of the most effective ways to reduce the weed population, including disease, pests, and vernalization. To be able to eliminate all these factors that directly affect yield and quality, it is important not to include plants of the same species in sugar beet crop rotation (Koch et al. 2018).

When wheat is planted after sugar beet, it can increase wheat yield by around 15%. If the applied crop rotation interval is not sufficient to prevent the spread of beet cyst nematode and seeded beet, it is necessary to extend the cycle interval to obtain high yield and to produce quality beets. It can be said that legumes, potatoes, and cereals are suitable as the pre-plants for sugar beet cultivation. Forage beet, sunflower, seed beet, rice, hemp, spinach, mustard, rapeseed, carrot, radish, cabbage are not suitable pre-plants for sugar beet as they cause many diseases and pests, especially nematodes (Götzea et al. 2017).

Legumes such as alfalfa, chickpeas, beans, lentils, and vetch can be counted as the best pre-plants in sugar beet due to their deep rooted and extensive root system. Legumes increase the presence of nitrogen and organic matter in the soil by absorbing the free nitrogen of the air through the podocytes (active nitrogen bacteria) in their roots and create a loose soil structure. In determining the pre-plants, the planting and harvesting dates should not be delayed at a level that will adversely affect the development of the main crop, and it should also be suitable for the conditions of the region and the sugar beet.

11.2 Soil Management and Preparation

Sugar beet is a deep-rooted crop and requires deeply cultivated soil. Compacted soil structures and hardened soils are not suitable for sugar beet cultivation. In other words, the taproot does not form a root but creates root bifurcations, cannot grow and develop, and thereby root yield decreases. First of all, no matter what tillage method is used, deep tillage of the soil every few years in autumn according to the planting rotation is an important issue to establish the deep root structure of the beet.

11.3 Autumn Tillage

Autumn tillage is an important process, especially in arid and semi-arid climates, depending on the soil structure, to facilitate the preparation of the seed bed in the spring by making maximum use of the winter precipitation, and to facilitate the conversion of the base fertilizers into a more useful form in the spring. Autumn tillage is generally carried out by chopping the stubble and mixing it into the soil with tillage tools or tool combinations such as moldboard plow, disc harrow, and rototiller (Fig. 11.1).

In sugar beet cultivation, the field is usually plowed at a depth of 25–30 cm in autumn with a moldboard plow or chisel. For the preservation of the soil structure

Fig. 11.1 Stubble-ploughing

Fig. 11.2 Reversible plow and field preparation

and for a sustainable agricultural production, due care must be taken to ensure that the tillage is done in an appropriate soil weathering in all agricultural operations. The fact that soil weathering is suitable for cultivation will prevent the formation of clod and clumping and will also ensure good soil ventilation. When this type of field is prepared, large clods will not form on the land, and an ideal structure will be created for other processes that will support plant production by providing a homogeneous field surface. Improper tillage and wrong sowing time can cause a yield loss of around 30%.

Moldboard plow is the most preferred tillage tool in primary tillage, which is generally applied to remove previous crop residues, improve drainage, and provide a uniform and flat seedbed with medium to fine texture (Fig. 11.2). The first factor in whether the tillage will be done with or without a plow is the soil texture. The parameters related to the soil texture to be taken into consideration regarding the soil cultivation method can be summarized as follows.

Fig. 11.3 Subsoiler

In heavy clay soils (clay loam, silty clay loam, sandy clay loam), it is recommended to make a rough plowing as early as possible in autumn (late October) in order to benefit from the freezing and thawing effect in winter months. Medium-heavy soils (silty loam, sandy loam) can be tilled more easily than heavy soils, but too early plowing will cause the soils to be smeared, especially during the rainy winter months. For this reason, late autumn tillage is recommended. In light soils (loamy sand, sand), it is an important issue to prefer equipment that will prepare it in a single pass with combined tools according to the condition of the soil in spring, instead of plows, which will eliminate the negative effects of erosion (Brown 1999).

Especially in heavy textured soils, it is important to plan the deep cultivation of the soil with a 3–4 year transformation in order to prevent the negative effects of the hard structure created by tillage at the same depth on deep-rooted plants such as sugar beet. The processing performed with subsoiler or chisel will help the root of sugar beet to go deeper, reach water and air more easily, accelerate beet development, increase yield and quality, and prevent salinity to a certain extent (Figs. 11.3 and 11.4). In the context of mitigating the negative effects of this hard layer, subsoiling is also crucial in autumn when the soil is dry enough to effectively eliminate disturbance from tractor tracks and headland turns. The most effective application method is 60–90 cm depth and a width of 1.5–2 times the depth in subsoiling.

Very important benefits of autumn release are known, such as mixing base fertilizers, providing natural processing through winter precipitation and frost, and accelerating the decay of plant residues mixed with the soil (Martindale 2013). For this, although it is not indispensable, it is recommended to apply all of the potassium fertilizer and 2/3 of the phosphorus fertilizer to the soil before autumn cultivation.

Some tillage methods such as minimum tillage, no-till, low-till, non-plowing, eco-tillage are widely applied in light-structured soils such as sandy, sandy-loamy, etc. These methods are mostly applied for erosion control. In these methods, the use of combined tillage tools and herbicides to create a suitable field in sugar beet and to control weeds increases the efficiency. Soil structure and biological activity in the soil are more positive in terms of agriculture in the minimum tillage method. This

Fig. 11.4 Chisel

Fig. 11.5 Minimum tillage practices in the field

provides an advantage at the point of making soil tillage, where the highest energy consumption occurs, more economical (Fig. 11.5).

11.4 Seedbed Preparations

In sugar beet agriculture, a tight and thin seed bed with sufficient air/water ratio, suitable for water retention is desired. Before the seedbed preparation, half of the nitrogen fertilizer and 1/3 of the phosphorus fertilizer left from autumn are given to the soil. If tillage cannot be done in autumn, all phosphorus and potassium fertilizers can be applied in the spring. Since the spring season is usually rainy, there is usually limited time for preparation. Therefore, soil weathering is very important for good preparation. In sugar beet, the seed bed should be loosened as much as the seed sowing depth, the germination path should be short, the water, air, and temperature values should be in a balanced structure. For correct sowing by leaving the seeds at

Fig. 11.6 Spring tine cultivator

Fig. 11.7 Combined tillage tool

the desired depth and ensuring a normal field output, the seed bed should be free of stones, clods, and plant residues, the field should be smooth and the soil should be loose enough in the area to contact the seed. For sugar beet, a 5–8 cm deep seed bed prepared with a harrow, cultivator, or combined tools and a shallow cultivation as much as possible can be suggested as suitable solutions (Figs. 11.6 and 11.7).

Soil preparation should be completed in a single pass as much as possible in order to create a suitable moisture environment in seed bed preparation and to ensure a particle structure of less than 3 mm around the seed. A flat field surface should be created to reduce harvest losses and to ensure a good seed-soil contact. For this, the timing of preparation (i.e., soil condition) is crucial. As a precaution against the possibility of compaction in the soil, settings such as wide tire, double wheel use, and low tire pressure and removing unnecessary weights on the tractor can be solutions (BBRO 2019).

11.5 Soil Requirements of Sugar Beet

The fact that sugar beet is a plant where yield loss is frequently seen due to its physiologically weak soil structure and sensitivity to insufficient drainage conditions requires due attention to the physical, chemical, and biological structure of the soil. When a general definition is made for sugar beet, it is physically not stony, gravelly, sandy, clayey, or heavy in structure, well ventilated, deep groundwater level below 120–150 cm, good drainage, chemical and biological structure neutral-slightly alkaline (pH = 6.4–7.6), soil rich in organic matter with high water retention is defined as an ideal soil structure (Draycott 2006).

Although it is stated that loamy soils are ideal for sugar beet cultivation, soils ranging from sandy loam to clay loam are the soil types where beet cultivation is common. However, it is necessary to consider the possibility that the technological quality of sugar beet will decrease as it moves away from the optimum soil criteria.

Since sugar beet uses high amounts of nitrogen (N), phosphorus (P_2O_5), and potassium (K_2O) during its development, the soil and plant should be supported with the main nutrients determined to be deficient in order to maintain the productivity level.

11.6 Water Use and Irrigation

Although it varies according to the climate and the length of the growing period, an average of 550–750 mm of water is needed in the cultivation of sugar beet without irrigation (FAO 2020). With the distribution of annual precipitation according to months being proportionally equal, it is ideal that half of it occurs seasonally during the growing period and the remaining half as winter precipitation. The need for water in crop production is a concept related to evapotranspiration (ET_o). In the estimation of ET_o in plants, the crop coefficient (k_c) is given as 0.4–0.5 in the emergence period, 0.75–0.85 in the growing period, and 1.05–1.2 in the ripening period. To give an average value, this value can be taken as 0.9–1.0 at the end of the season and 0.6–0.7 at the harvest time (Fig. 11.8).

Irrigation, which is the application of water that is necessary for growth and cannot be provided by natural precipitation, should be applied in an amount that will ensure sufficient moisture in the plant root zone, especially in arid and semi-arid climatic conditions. In addition, irrigation has important benefits such as controlling the temperature in the soil, washing off excess salt, softening the hard layer, helping to take the fertilizers into the plant.

In general, the amount of water retained by the sugar beet is 1% of the total amount of water evaporated during the growing season. In this case, irrigation can be thought of as the completion of water lost through evaporation from plants and soil.

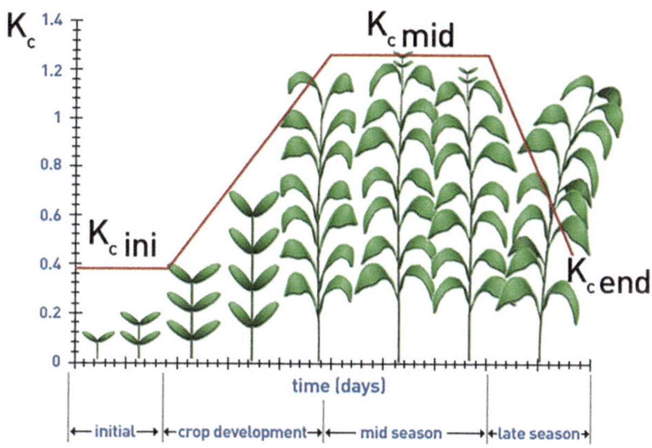

Fig. 11.8 Change in plant water consumption coefficient depending on the beet's development period (FAO 2020)

The sugar beet plant is sensitive to water deficiency, especially when there is not enough rainfall at the time of germination and 3–4 weeks after emergence. In cases where precipitation delay and irrigation cannot be done adequately, it becomes difficult to obtain the necessary plant for optimum efficiency and yield loss may be equivalent to late planting (Hassanli et al. 2010).

Weather has a great impact on agriculture. Sugar beet produces high levels of dry matter in favorable weather conditions, especially in sufficient rainfall and irrigation conditions. The daily water consumption of beets is 1.1–1.5 m^3 in the leaf forming phase, 5.6–8.2 m^3 in the root formation phase, and 6.1–6.8 m^3 in the sugar accumulation period. These amounts increase 2.5–3 m^3 per day as the average daily temperature increases 1 °C (Tortopoğlu 1994; Vazifedousta et al. 2008).

Sugar yield of each 25 mm of irrigation water applied on sandy soils increases by 0.4 t ha^{-1}, this amount being 50% more in dry years (Ober 2004). With the increase in the amount of irrigation water, the sugar content decreases proportionally and the sugar yield increases. For optimum sugar yield, it is sufficient to apply 65–70% of the full water amount in each irrigation (Mahmoodi et al. 2008; Abyaneha et al. 2017; Abbas et al. 2018).

In the traditional method, when the leaves are dark green or the leaves lose their vitality at noon and do not return to their original state in the evening, it is considered that the time has come for irrigation sugar beet. In addition, if the leaves do not break or crack when the leaves are folded transversely, the irrigation is late. One of the factors affecting irrigation planning is soil structure.

Light soils require more frequent irrigation than medium and heavy soils, and higher parts of the field than pits. After knowing about the irrigation time according to the leaves, the moisture presence of the soil should be examined for a definitive decision. If the soils taken from 0 to 60 cm or 0 to 90 cm depth of the soil do not take the shape of a ball and disperse after slightly compacted, it can be thought that it is time to water.

The duration of irrigation is as important as the determination of the irrigation time. The effective root depth of the plant is taken into account in determining the irrigation time. It is known that the roots reach a depth of 90 cm in sugar beet. In a good irrigation planning, monitoring the movement of water during irrigation will be beneficial in terms of preventing unnecessary water loss. Effective root depth can be taken as 60 cm in sugar beet. In this case, water application should be stopped when the irrigation water reaches a depth of 40 cm. When the water is cut off, the topsoil is at the saturation point and it reaches a depth of 60 cm in 1–2 days with the downward movement of the water by gravity.

Water consumption in sugar beet is highest in June, July, and August and it is more sensitive to water deficiency. Irrigation can be started in these months by making thinning-singling and applying the remaining nitrogen fertilizer. Due to the high water holding capacity of heavy textured soils such as clay or clay loam, the number of irrigation is naturally less than that of light textured soils. Daytime irrigation should be avoided to reduce irrigation losses and increase productivity. For high and quality yields, 80 ± 20 mm of water should be given in each irrigation at 10–15 day intervals, depending on the soil structure and the rainfall in July and August (Carlson and Bauder 2020; FAO 2020).

In general, irrigation and nitrogen have mutual effects on plant growth in crop production. Nitrogen must be balanced with the irrigation water needed for a high production. In other words, in case of insufficient nitrogen amount, high production potential cannot be reached even if the amount of water is sufficient. When there is enough nitrogen as needed by the plant, excessive irrigation causes nitrogen to be washed out. This also leads to a possible loss of production. In cases where precipitation is irregular or insufficient, the benefit of nitrogen decreases if irrigation is not performed within 3–4 weeks after planting (Zarski 2020).

11.7 Irrigation Methods

11.7.1 Surface Irrigation

In this method, which is called surface or flood irrigation, water is applied uncontrolled in the direction of the slope or in the direction perpendicular to the slope. Surface irrigation, which is a primitive method, has low investment costs. It is difficult to maintain a homogeneous distribution of water and erosion can occur. This method can generally be applied by small farmers on sloping fields with a smooth field surface and low slopes (Fig. 11.9).

In this method, the problem of ponding is frequently experienced in areas with height difference in the field and thus equal water distribution cannot be achieved. Surface irrigation method, as a method where salinity and drainage problems are seen, is recommended to be applied in irrigation of plants that are resistant to moisture deficiency in the soil, diseases caused by wetting the root collar and are frequently planted.

Fig. 11.9 Surface Irrigation Method

11.7.2 Drip Irrigation

It is a method with high installation cost, where irrigation water is applied to the soil surface in drops with or without fertilizers, under low pressure (Stevanato et al. 2019). However, since it requires less labor forces and allows automatic irrigation, it is especially applied in greenhouses where initial investment costs are high and in the cultivation of plants with high economic value. The basic principle here is to give the required amount of water to the root zone of the plant at frequent intervals in low amounts without creating stress on the plant. Water is given only to the root zone of the plant instead of the entire soil surface in the form of drops by means of drippers. Drops move vertically and laterally in the soil, wetting a larger area under the soil than on the surface and providing sufficient moisture for the plant roots. It is considered as the method with the lowest water loss and the highest irrigation efficiency. The heart of the drip irrigation system is drippers. The drippers are made of plastic and mounted on pipes with a diameter of 12–32 cm and called lateral. Drippers drop the water onto the soil with a flow rate of a few liters per hour. A drip irrigation system consists of four parts apart from the drippers. These are control unit, main pipeline, side main pipeline, and laterals (Fig. 11.10).

11.7.2.1 Advantages of the Drip Irrigation Method

- Water usage efficiency is high, water loss is also very low due to low evaporation.
- The product quality is high, the development of diseases and pests is less.
- Weed growth is lower.
- Since only the root zone of the plant is wetted, soil tillage, spraying, harvesting, and transportation can be done at the same time during irrigation.
- Weather conditions have no effect on irrigation.
- Fertilization and spraying can be done together with irrigation.

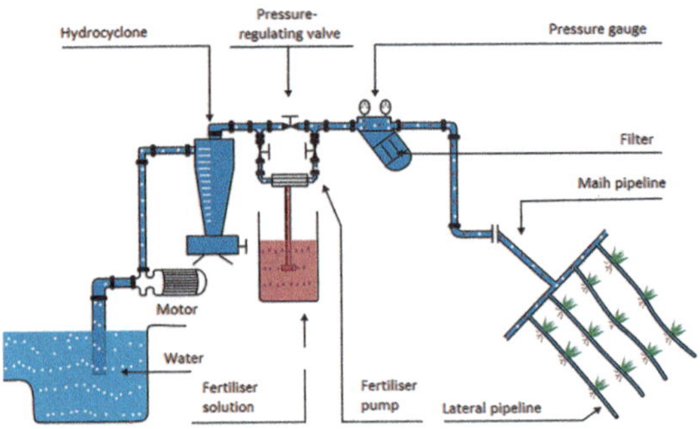

Fig. 11.10 Elements of the drip irrigation system

- It can be applied successfully and easily on sloping lands.
- Labor requirement during irrigation is low.
- It is a suitable method for irrigation automation applications.

11.7.2.2 Disadvantages of the Drip Irrigation Method

- The system is very sensitive to clogging. It is necessary to apply the filtration with precision.
- It can cause salinity problem. For this reason, especially in rainy periods, the system should be operated and the salt should be washed.
- Initial investment costs are high.
- Agricultural practices such as hoeing can damage pipes. Therefore, it is recommended to install the system after the hoeing.
- Not suitable for emergence irrigation.

11.7.3 Sprinkler Irrigation

It is an irrigation system in which water is sprayed in droplets through sprinkler heads under certain pressure. The system works by means of pipes and irrigation heads placed in the field at certain intervals. In order for the water to be supplied under pressure from the nozzles, the water must be taken from a pump or from a high source. At least 2.5 bar is sufficient for the sprinkler system to work effectively. The system generally works as a closed system and consists of the main pipe, side lines, pump, and sprinkler heads (Fig. 11.11).

For an even distribution of water in sprinkler irrigation, a 50–60% overlap is planned in the wetting areas. Since the homogeneity of irrigation deteriorates in windy weather, the headers are placed closer in mandatory situations. For an ideal irrigation in the sprinkler irrigation method, the application should be made according to the physical properties of the soil. Incorrect and irregular irrigation

Fig. 11.11 Sprinkler irrigation system

causes excessive water use, salinity, deterioration of land quality, increased desertification, and economic losses.

11.7.3.1 Elements of the Sprinkler System (Fig. 11.12).

(a) Water source: Stream, lake, caisson well, deep well, pond, dam, irrigation canal are the main sources. The quality of the water is an important factor in irrigation without interruption.

(b) Pump unit: It is the power unit that provides pressure in the system. Centrifugal pumps are common in sources with low suction height, and vertical shaft deep well or submersible pumps in deep wells. The pumps are either powered by fuel or electricity. Electric motor driven pumps have advantages in terms of ease of use and low operating costs.

(c) Pipelines: The transmission of the water taken from the source to the lateral lines is provided by the main pipe, and the transmission of the water taken from the main pipeline to the sprinkler heads is provided by the lateral pipes.

(d) Sprinkler heads: These heads are located in the lateral pipelines. The connection between the lateral pipelines and the sprinkler heads can be adjusted with the riser pipes according to the plant height.

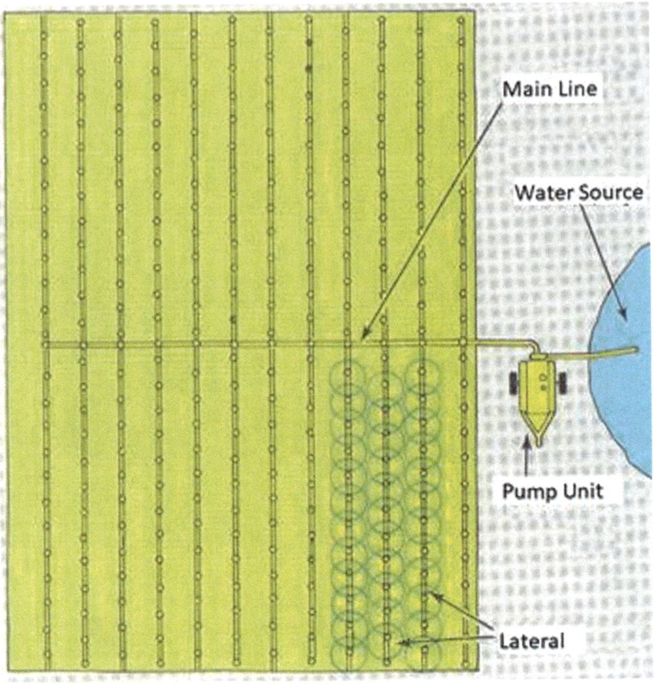

Fig. 11.12 Elements of the sprinkler system

11.7.3.2 Advantages of the Sprinkler Irrigation Method

- High water use efficiency allows effective irrigation even in places where water is scarce.
- It is suitable for irrigation without causing erosion on sloping lands.
- It prevents plant emergence problems caused by slipping.
- Operating cost and labor requirement are low.
- It is the most suitable irrigation system in shallow and permeable soils with low soil depth.
- Salt, dust, and pests in the soil can be washed with sprinklers.
- It has the possibility of controlled irrigation suitable for places with high groundwater and drainage problems.
- The transmission of water through pipes provides easy application opportunity.
- Fertilization can be done with irrigation water.
- It can be used to protect plants from frost and heat.

11.7.3.3 Disadvantages of the Sprinkler Irrigation Method

- The initial installation cost is high, especially in fixed systems (Stevanato et al. 2019).
- Wind has a negative effect on water distribution.
- Power required for pumping increases fuel consumption and operating costs.

11.7.3.4 Installation of Sprinkler Irrigation System

- The main pipes should be parallel to the slope direction.
- Sprinkler lines should be in a vertical position parallel to the leveling curves.
- When the wind speed is high, the lateral lines should be perpendicular to the wind direction.
- Short laterals instead of very long sprinkler laterals reduce labor and provide even water distribution.
- The movement of the laterals on the main line should be arranged in a way that requires the least amount of labor.
- The system should be arranged as square or rectangular as possible so that lateral movements are easy and fewer heads work together.
- The pipe dimensions and layout of the system should be such as to minimize annual costs.
- Having the pump unit in the field is the most economical way of working.

11.7.4 Center and Linear Pivot Irrigation System

Drought, climate variability, and irregularity in precipitation make the sustainability of existing resources even more important. It has become imperative to use limited water resources effectively and to ensure the correct operation of mechanized sprinkler systems in facilities. For these purposes, mobile pivot irrigation systems, whose use is becoming widespread day by day, are divided into two: linear and center pivot irrigation systems.

Center pivot irrigation systems are automatic systems made of galvanized steel pipes, moving from the center, rotating around a fixed tower. The inner surfaces of the pipes are coated with PVC to provide resistance against chemicals. The system moves in a circular rotation on a reinforced concrete platform and performs irrigation automatically. Center pivots can start from 50 m and reach a radius of 1100 m. Thus, one machine is sufficient for 380 hectares of land. In addition, these systems have brought a different and effective usage understanding to irrigation management with their powerful and easy-to-use control panels. The lifespan of this type of systems varies approximately between 25 and 30 years. Irrigation efficiency is as high as 90–95% and it can work on sloping lands such as 15% (Fig. 11.13).

Linear moving irrigation systems have been developed to irrigate geometrically shaped fields. Linear systems can operate effectively on lower slopes such as 4–5%. The systems are 1000 m long and can irrigate up to 98% of the land without leaving any unirrigated areas and operate smoothly. With these systems, it is possible to apply chemical applications together with irrigation, as in sprinkler irrigation systems (Fig. 11.14).

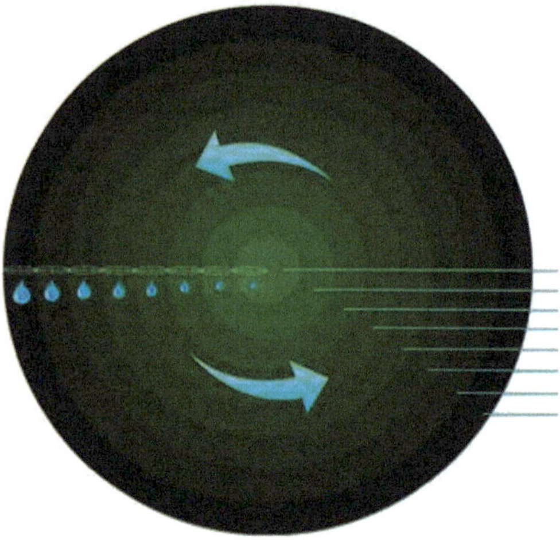

Fig. 11.13 Working principle of center pivot irrigation system

Fig. 11.14 Linear pivot irrigation systems and sprinkler head

11.7.5 Subsurface Drip Irrigation

Subsurface drip irrigation is an irrigation method in which water is given just below the soil surface through point or line source drippers. The materials used in this system, the planning and operation of the method are based on the same design principles as in the surface drip irrigation method. In the method, the laterals are placed under the soil at a depth of 0.02–0.70 m with 0.25–0.50 m intervals. Chemicals are used against clogging due to plant roots (Lamm and Camp 2007). In subsurface applications, no flow or evaporation loss occurs due to the direct application of water to the plant root zone, and the irrigation efficiency exceeds 95%. Within the scope of sustainable water management, subsurface drip irrigation method can be offered as a solution to save water in sugar beet cultivation (Fig. 11.15). In a study comparing linear, sprinkler, surface, and subsurface drip irrigation systems, linear pivot irrigation method gave the best results in terms of

Fig. 11.15 Subsurface drip irrigation applications

beet yield, and subsurface drip irrigation system gave the best results in sugar content and refined sugar yield (Turkseker 2018).

11.8 Crop Establishment

11.8.1 Sowing

Temperature and humidity are the main factors determining planting time. Sowing should be started immediately when soil conditions become suitable for sowing. Sowing should be done in weathered soil at the earliest date when the soil temperature is 5–7 °C, and the possibility of late frost is very low. Accordingly, planting should be started towards the middle of March under favorable conditions in the northern hemisphere and should be completed by April 15 at the latest. In sowing performed after this date, 0.5–0.6 t ha^{-1} daily yield loss occurs (BBRO 2019). In the winter sugar beet cultivation applied in the semi-tropical or tropical climate zone, the most suitable planting time was determined as 1–15 October, and the vegetation period was 195–210 days (Ozgur and Erdal 2002; Gadallah and Tawfik 2017; Gobarah et al. 2019). In winter sugar beet cultivation, the fact that the harvest date coincides with the months of June and July, when the temperature is at its highest, is an important problem and it is necessary to process the beet quickly. Otherwise, high sugar loss in beet silos is inevitable. Tables 11.1 and 11.2 are prepared to assist in selecting row spacing. In the case of planting in rows selected according to field emergence rates in the region shown as dark in the table, the number of plants required per unit area will decrease and the producer will suffer economically. For high profits in beet cultivation, field emergence rates and in-row spacing should be determined correctly.

Table 11.1 Number of plants per area in 45 cm row spacing in case of no double and missing seed (×1000 ha−1)

Seed spacing (cm)	Number of sowing seeds	Field establishment rate (%)				
		40	50	60	70	80
8	278	111	138	166	194	222
10	222	88	111	133	155	177
12	186	**74**	92	111	129	148
14	159	**64**	**79**	95	111	127
15	148	**59**	**74**	88	104	118
16	139	**56**	**69**	83	97	111
17	131	**52**	**65**	**78**	91	104
18	123	**49**	**61**	**74**	86	98
19	117	**46**	**58**	**70**	81	93
20	111	**44**	**55**	**67**	**78**	89
21	106	**42**	**52**	**63**	**74**	84

Table 11.2 Number of plants per area in 50 cm row spacing in case of no double and missing seed (×1000 ha^{-1})

Seed spacing (cm)	Number of sowing seeds	Field establishment rate (%)				
		40	50	60	70	80
8	250	100	125	150	175	200
10	200	80	100	120	140	160
12	167	**67**	84	100	117	134
14	143	**57**	**71**	86	100	114
15	133	**53**	**67**	80	93	107
16	125	**50**	**63**	**75**	88	100
17	118	**47**	**59**	**71**	82	94
18	111	**44**	**56**	**67**	**78**	89
19	105	**42**	**53**	**63**	**74**	84
20	100	**40**	**50**	**60**	**70**	80

In the literature, it has been determined that the optimum plant density for the highest yield and quality is 80,000–100,000 plant ha^{-1} (Jaggard et al. 1995; Ecclestone 2011). Nowadays, in many countries where beet planting is carried out, in order to provide 80,000 ha^{-1} beets, the row space is 45–50 cm, and the distance is 20 cm. Seed germination rates have increased significantly in recent years. Accordingly, if 80% field germination is obtained, 89,000 plants ha^{-1} is provided at a 20 cm row planting distance. Depending on the agricultural technique applied in the period from singling to harvest, 12–20% of the existing beets are lost due to agricultural applications and pest damage. The amount of seed needed in planting should be calculated by taking this into consideration. With plant numbers above 100,000 plant ha^{-1}, high yields can be obtained, but high profits may not be obtained. It is not possible to obtain high yields with plant numbers below 80,000 plant ha^{-1}. The most suitable plant number for high yield and high quality

Fig. 11.16 Mechanical and pneumatic precision sowing machines used in sugar beet cultivation

sugar beet production has been determined as 100,000 plant ha^{-1} by the British Beet Research Organization (BBRO) (BBRO 2019). Seed distances are an important factor that determines the production quality of sugar beet. It is important that the seed is placed on the row properly, which will have an equal living space according to the structure, texture, and soil weathering conditions. Sowing depth is 2.5–3 cm on average (Fig. 11.16). In arid regions and light soils, a roller can be offered as a solution to increase the planting depth by 1–2 cm, to prevent evaporation when necessary and to ensure the contact of the seed with the moist soil (Fig. 11.16) (see Fig. 11.17).

Sowing speed is also important in terms of determining the evenness of seed distribution in row. The optimum sowing speed should be 4–5 km h^{-1}. Before planting, a proper seedbed should be prepared, and a good soil structure should be provided. Especially in clay soils with low organic matter content, too thin soil surface due to improper or excessive processing will reduce field emergence and cause losses. In addition, the formation of a crust layer on the soil after rain or emergence irrigation should be considered as a factor that will cause losses in this context. The shoot that emerges from the seed forms a yellow fold in a way not to exceed the crust layer; therefore, sufficient field emergence cannot be provided and the desired result cannot be reached. In case of dry sowing in fields where soil weathering is not suitable, 15–20 mm exit irrigation should be done immediately by sprinkling. With timely planting and field emergence, sufficient vegetation period will be ensured, sufficient technological maturity will be achieved and finally high quality beet production will be achieved.

Practical applications such as strip processing and direct drilling for sugar beet are becoming widespread for reasons such as ensuring sustainability in agriculture, protecting soils from erosion, and reducing soil processing costs (Fig. 11.18). In strip tillage, time and energy are saved by tilling only the sowing rows. In the direct drilling method, the plant residues on the planting line are crumbled in a way that

Fig. 11.17 Roller application

Fig. 11.18 Field tilled with the strip tillage method

does not affect the germination of the seed, and the furrow is cleaned for sowing (Fig. 11.19).

11.8.2 Plant Population

In the period when the beet has four to six leaves, ten rows are counted in different parts of the field and the average number of plants is found:

50 cm row spacing.

Plants per 20 m row × 1000 = plant population (000 plants ha^{-1}).

45 cm row spacing.

Plants per 22 m row × 1000 = plant population (000 plants ha^{-1}).

Fig. 11.19 Types of direct sowing machines used in sugar beet cultivation

For an ideal structure, 900–1200 cm^2 of living space is sufficient for each plant. This area is regulated by thinning and singling. After the beet emerges to the field surface, plants can be found at a distance of less than 17 cm. Seedlings at distances less than 17 cm should be thinned, and plants closer than 5 cm should be singled out and brought to equal in-row spacing. Thinning is the arrangement of narrow plant spacing on the row, and singling is the reduction of multiple plant shoots to one due to double seed filling in seed holes during sowing. In ideal production conditions, there should be 80,000–100,000 ha^{-1} plants in the land after thinning (Jaggard et al. 1995; Ecclestone 2011). In cases where the plant density falls below 60,000 plant ha^{-1}, if the time is not too late for the beet to complete its growing period, replanting is necessary to increase production. In this case, soil fertility, seasonal conditions, and irrigation possibilities should be considered.

Plant density is a controllable factor that determines the level of production. The low plant density, that is, the distance between the beets in the field, causes an increase in the nutrients that each plant will receive and causes it to grow too much. Overgrowth, non-sugar substances increase in plants, while sufficient sugar accumulation slows down and quality deteriorates. As the spacing in the row increases, the proportion of plant leaves decreases and the water loss in the soil increases. The reason for the sparseness may be the germination power of the seed, pesticide, field emergence, sowing distance in the row, pests, seedbed preparation, frost after planting, improper hoeing, disease, and drought. A good seed bed preparation and suitable weathering increase field yield, and homogeneous plant density is achieved.

11.8.3 Weed Competition and Hoeing

The soil surface remains empty for 1.5–2 months until the development in the initial period is slow and the beet covers the field completely by expanding its living space. Weeds grow very quickly by using air, water, and nutrients quickly, and they can reduce the habitat of other plants and make them unproductive and cause losses. In addition, they create a living environment for pests and increase losses in mechanical harvesting. The most effective method of weed control is hoeing. Although the inter-

Fig. 11.20 Sugar beet hoeing with inter-row hoeing machine

row hoeing machine is preferred against weeds, it is more effective when applied in combination with the in-row tape herbicide application (Fig. 11.20).

Hoeing is an effective tool for breaking the crust layer and controlling weeds. Depending on the weed density and precipitation conditions in the field, two or more hoeing can be done. The first hoeing is carried out when the rows are clear with the start of the seedling emergence. Second hoeing can be done within 15–25 days under suitable conditions after thinning-singling. Weed control with hoeing greatly affects the root yield of the beet. Hoeing has an important role in providing a clean field in order to realize quality production by reducing harvest losses. The combined use of mechanical and chemical weed control methods at certain rates increases the effectiveness of weed control (Kaya and Buzluk 2006).

11.9 Future Prospects

Technological and genetic studies carried out in sugar beet strengthen the thoughts that the increase in production will continue. Developed technologies in tillage, cultivation, and other agricultural practices support increases in crop production. Establishing a deep root structure and meeting high water needs are absolutely essential, among other factors, for high production. Another issue is finding solutions to prevent sugar losses during storage after harvest.

By developing agricultural techniques and applied methods in sugar beet cultivation, 24 t of sugar ha^{-1}has been obtained in field trials. Although it is stated that there are still many unfavorable conditions to achieve this yield in field production, it is known that there is an opportunity to reach this goal agronomically (Hoffman and Kenter 2018). However, the biggest threat to reaching this potential is shown as possible future climate change. It is estimated that the risk of drought, the intensity of diseases and pests will increase with climate change (Kremer et al. 2016). Increased production, prolongation of beet processing times, earlier beet harvest and longer storage periods are the situations where losses are expected to increase (Hoffman and Kenter 2018). In summary, it is foreseen that efforts to accelerate processing times, improve storage properties, reduce processing and storage losses, and increase production will intensify.

11.10 Conclusions

Despite the high energy requirement in its production, sugar beet is an important product with its feature increasing the yield of the next crop and increasing soil fertility in the crop rotation cycle. In addition, by-products of sugar beet constitute an important input used in bioethanol production and animal nutrition. In this respect, it can be said that beet is a strategic energy crop. In sugar beet producing countries, beet yield varies between 60 and 140 t ha^{-1}, and digestion varies between 16 and 18%. These values are not considered sufficient, and higher values are aimed with new agricultural techniques and genetic studies. It is estimated that this target can be achieved if new techniques are developed and sufficient water is provided, in addition to the agricultural techniques practices required for the supply of 100,000 plant ha^{-1}.

References

Abbas MS, Soliman AS, Moustafa ZR, Abd El-Reheem KM (2018) Effect of some soil amendments on yield and quality traits of sugar beet (*Beta vulgaris* L.) under water stress in sandy soil. Egypt J Agron 40(1):75–88

Abyaneha HZ, Jovzi M, Albaji M (2017) Effect of regulated deficit irrigation, partial root drying and N-fertilizer levels on sugar beet crop (*Beta vulgaris* L.). Agric Water Manag 194:13–23

BBRO (2019) Sugar beet reference book. British Beet Research Organisation. https://bbro.co.uk/publications/reference-book/

Brown S (1999) Review of sugar beet cultivation. British Sugar Beet Review 67(2):30–36

Carlson L, Bauder J (2020) Sugarbeet Agronomy 101. Montana State University (MSU), Extension Water Quality Program. https://waterquality.montana.edu/farm-ranch/irrigation/other_crops/sugarbeet.html#:~:text=Sugar%20beets%20require%2022%2D28,per%20day%2C%20on%20the%20average

Draycott AP (2006) Sugarbeet. Blackwell Publishing, Oxford, p 465

Ecclestone P (2011) Sugar beet drills - time to update. British Sugar Beet Review 79(4):3–8

FAO (2020) Sugarbeet. Land & Water, Food and Agriculture Organization of the United Nations. http://www.fao.org/land-water/databases-and-software/crop-information/sugarbeet/en/

Gadallah AFI, Tawfik SF (2017) Effect of harvesting age of some sugar beet varieties grown in a new reclaimed soil in sohag. Alex Sci Exch J 38(4):975–982

Gobarah ME, Hussein MM, Tawfik MM, Aahmed AG, Mohamed MF (2019) Effect of different sowing dates on quantity and quality of some promising sugar beet (*Beta vulgaris* l.) varieties under north delta condition. Egypt J Agron 41(3):343–354

Götzea P, Rücknagel J, Wensch-Dorendorf M, Märländer B, Christena O (2017) Crop rotation effects on yield, technological quality and yield stability of sugar beet after 45 trial years. Eur J Agron 82:50–59
Hassanli M, Ahmadirada S, Beecham S (2010) Evaluation of the influence of irrigation methods and water quality on sugar beet yield and water use efficiency. Agric Water Manag 97(2): 357–362
Hoffman CM, Kenter C (2018) Yield potential of sugar beet - have we hit the ceiling. Front Plant Sci 9:289. https://doi.org/10.3389/fpls.2018.00289
Jaggard KW, Limb M, Proctor GH (1995) Sugar beet: a grower's guide. Sugar Beet Research and Education Committee, London, p 111
Kaya R, Buzluk Ş (2006) Integrated weed control in sugar beet through combinations of tractor hoeing and reduced dosages of a herbicide mixture. Turk J Agric For 30:137–144
Koch HJ, Trimpler K, Jacobs A, Stockfisch N (2018) Crop rotational effects on yield formation in current sugar beet production. Results from a farm survey and field trials. Front Plant Sci 9:231. https://doi.org/10.3389/fpls.2018.00231
Kremer P, Schlüter J, Racca P, Fuchs HJ, Lang C (2016) Possible impact of climate change on the occurrence and the epidemic development of Cercospora leaf spot disease (*Cercospora beticola* sacc.) in sugar beets for Rhineland-Palatinate and the southern part of Hesse. Clim Chang 137: 481–494. https://doi.org/10.1007/s10584-016-1697-y
Lamm FR, Camp RC (2007) Subsurface drip irrigation. In: Microirrigation for crop production, design, operation, and management. Elsevier, Amsterdam, pp 473–551. https://doi.org/10.1016/S0167-4137(07)80016-3
Mahmoodi R, Maralian H, Aghabarati A (2008) Effects of limited irrigation on root yield and quality of sugar beet (*Beta vulgaris* L.). Afr J Biotechnol 7(24):4475–4478
Martindale W (2013) The sustainability of the sugar beet crop - the potential of add value. Br Sugar Beet Rev 81:49–52
Ober E (2004) Irrigation: old stories and new. Br Sugar Beet Rev 72(2):52–54
Ozgur OE, Erdal M (2002, September 10–11) Researches on winter sugar beet cultivation in Turkey. Second National Sugar Beet Production Symposium, Ankara, Turkey. pp. 204–216,
Petkeviciene B (2009) The effects of climate factors on sugar beet early sowing timing. Agron Res 7:436–443
Stevanato P, Chiodi C, Broccanello C, Concheri G, Biancardi E, Pavli O, Skaracis G (2019) Sustainability of the sugar beet crop. Sugar Tech 21:703–716. https://doi.org/10.1007/s12355-019-00734-9
Tortopoğlu AI (1994) Factors of affecting the cost of yield, quality and sugar production in sugar beet cultivation. Turkish Sugar Factories Corporation Publications, Ankara, pp 12–13
Turkseker (2018) Unpublished research results of "Different irrigation systems effect on sugar beet yield and quality". Turkish Sugar Factories Corp. Sugar Institute, Ankara, Turkey
Vazifedousta M, Vandama JC, Feddesa RA, Feizic M (2008) Increasing water productivity of irrigated crops under limited water supply at field scale. Agric Water Manag 95:89–102
Zarski J, Ku'smierek-Tomaszewska R, Dudek S (2020) Impact of irrigation and fertigation on the yield and quality of sugar beet (*Beta vulgaris* L.) in a moderate climate. Agron 2020 10:166. https://doi.org/10.3390/agronomy10020166

Economical Crop Production and Management of Sugar Beet in Serbia and Montenegro

12

Radmila Bojović, Vera Popović, Dragana Popović, Vuk Radojević, Zoran Jovović, Velibor Spalević, and Nataša Ljubičić

Abstract

Sugar beet (*Beta vulgaris* L.) is a member of family *Chenopodiaceae*. It is a crop of area of temperate climate and second of world importance sugar crop, after sugar cane (*Saccharum officinarum* L.). It is mainly produced in Europe and to a lesser extent in Asia and North America. Sugar beet quality is a complex trait conditioned by genetic divergence among cultivars and environmental diverse in the region of growing. One of pertinent agro-technical measures is sowing. Proper sowing (which is during March in the Republic of Serbia and Montenegro) ensures optimal crop density which is important for reaching a high yield. The correct choice of assortment for a certain production area contributes to a larger and more stable production of cultivated plants.

A significant measure of growing technology is properly balanced plant nutrients. NPK mineral nutrients are the main carriers of yield value. Sugar beet

R. Bojović
Faculty of Agriculture, University of Belgrade, Zemun-Belgrade, Serbia

V. Popović (✉)
Institute of Field and Vegetable Crops, Novi Sad, Serbia
e-mail: vera.popovic@ifvcns.ns.ac.rs

D. Popović
Faculty of Economics in Subotica, University of Novi Sad, Novi Sad, Serbia

V. Radojević
Institute of Field and Vegetable Crops, Novi Sad, Serbia

Faculty of Agriculture, University of Novi Sad, Novi Sad, Serbia

Z. Jovović · V. Spalević
Biotechnical Faculty, University of Montenegro, Podgorica, Montenegro

N. Ljubičić
Biosense Institute, University of Novi Sad, Novi Sad, Serbia

V. Misra et al. (eds.), *Sugar Beet Cultivation, Management and Processing*,
https://doi.org/10.1007/978-981-19-2730-0_12

yields are significantly reduced by lack of soil nutrients. The main nutrient elements—nitrogen, phosphorus, and potassium—have major importance in plant nutrition and the greatest impact on increased technological and yield value of root. Ratio of mentioned nutrients and also the secondary and microelements should suit to the needs of plants and to the natural soil fertility. Yield of crystal sugar, as major production value indicator, has statistically very significant dependence of way of plant nutrition, years, genotype, and their mutual interaction. Intensification of plants nutrition with NPK fertilizers significantly increases the total vegetative biomass yield and root sugar content.

Keywords

NPK nutrition · Growing technology · Production · Sugar beet · Yield

Abbreviations

ETP Evapo-Transpiration Parameter
pF Capillary Potential

12.1 Introduction

Sugar beet (*Beta vulgaris L.*) belongs to the family *Chenopodiaceae* and is considered as crop of the temperate region and second, by importance, crop for sugar production worldwide, after sugar cane (*Saccharum officinarum L.*). It is mainly produced in Europe and to a lesser extent in Asia and North America. Sugar beet is a relatively young agricultural crop although it is known as a garden plant as far back as 3000 years ago.

The benefits of its cultivation are multiple and are reflected in its importance for the population nutrition, positive influence on industry, and agro-culture production. The economic value of sugar and other products which are obtained from sugar beet is very high, in comparison with many other field plants. Economically profitable production requires yield higher than 40 t ha^{-1}. For achieving this and more, the production technology must rise to higher level. The technical properties and value of yield of sugar beet depend on: agro-climatic factors, soil, (structure, composition, and cultivation), and cultivation technology (choice of variety, sowing, providing adequate irrigation nutrition, crop protection, and extracting time). Sugar beet quality is a complex parameter, conditioned by genetic differences among grooving sorts and also by different environmental conditions of growing region (Rosso and Candolo 2001; Ćurčić 2014; Bojović et al. 2014; Bojović et al. 2019a, b). It should be emphasized that proper plant nutrition will increase the yield with no negative influence on technological parameters of the roots.

In Serbia and Montenegro, there are favorable soil and agro-ecological conditions for sugar beet production. These conditions are defined by the presence of excellent arable land, relatively favorable conditions of weather and built processing capacities. In Vojvodina, sugar beet is grown mainly on larger agricultural farms, and the reason is using the specific mechanization of production technology and for achievement of better financial results on larger production surface areas (Vlahović et al. 2006).

12.2 Soil Properties and Preparation

12.2.1 Soil Requirements

Sugar beet belongs to the group of plants with the greatest demands from soil. This is understandable considering that its root fully develops in depth of it and for it creates a large organic mass. Soil should be deep, rich in organic matter, good physical properties (structural), loose and fertile. Heavy, compacted, and impermeable soils do not suit the crop, same as too light and sandy soils, which are usually poor in humus. Therefore, best suited soils are: chernozem, alluvial soils, lighter vertisols, and brown forest soil. With modern agricultural techniques (cultivation, fertilization, drainage, irrigation), high yields can also be reached on other soils.

Physical-mechanical properties that affect the water-air-, heat, and biological regime are important for cultivating sugar beet. Plants need a crumbly and stable soil structure. Those soils that quickly lose their structure, increase their compaction, and become unfavorable for production. Excessive compaction soil has negative effect on formation of correct root shape and plant productivity. The optimal air capacity depends on soil moisture and decreases by its increase. The bulk density has variable size and indicates soil compaction. During sowing, at the start of vegetation period, it has lower values; later, the soil subsidence occurs, so it is higher on vegetation termination. On chernozem type soil, the optimal volumetric mass ranges from 1.1–1.3 g cm^{-3}. Only on soils of high fertility, on deep arable land with good physical traits and with neutral to weakly alkaline reaction high yields are obtained. The soil needs to be with good water regime, be permeable, should have the ability to store winter moisture, and groundwater level should not be higher than 1.2 m.

Stanaćev (1979) claims that sugar beet, as a highly productive plant, reacts very much on soil traits. This reaction is manifested in intensity of energy of growth, modification of morphological characteristics, and technological characteristics of its above ground and underground part. Milošević et al. (1989) point out that with favorable climatic conditions, good sugar beet yields can be obtained only on soils of good natural fertility, deep arable layer, favorable mechanical composition, optimal physical-chemical and biological properties.

Sugar beet grows on all continents at different soil types.

For making the right choice of land for cultivation, we must know its requirements from the soil. Stanaćev (1979) emphasizes that the land profile is of special importance. Sugar beet requires a favorable profile structure and its differentiation into genetic horizons or layers. Of special importance is the horizon number, expressiveness, transition from one horizon to another, profile depth, and properties of the parent substrate. Horizon characteristics, such as mechanical composition, porosity, and structure, determine the suitability of the land.

In general, soils with pronounced textural differentiation and compacted layers cannot be considered suitable for it. Soils are considered compacted when the porosity is so low that it reduces soil aeration, interferes with the roots penetration and normal soil drainage. Natural compactness of soil is not only conditioned to their mechanical composition, water regime, or way of origin, but it can also occur as a consequence of using heavy mechanization. Compaction adversely affects water, air, and heat regime of the soil, its biological activity, and the plant in general. Sometimes, anaerobic conditions and water stagnation in the active rhizosphere occur, when, according to Luthin (1966), appear a limited growth of the root system and lot of harmful phenomena that accompany it, lack of oxygen, the appearance of a number of toxic, reduced mineral and organic compounds, anaerobic and slow decomposition of organic matter, with nitrogen remaining organically bound, poor heat regime, the appearance of plant diseases, especially fungal.

Hadzic (1984) states that occurrence of compacted layers, especially if they are in root development zones cause a lot of undesirable consequences. For root then it is difficult to break through, it becomes forked, the thickened root is shorter, and lower yields of poorer quality are achieved.

Factor of fertility is also the soil total and differential porosity. Soils with a total porosity above 45% vol. are suitable for sugar beet growing, while soils with porosity lower than this are considered unhealthy for root system growth. However, differential porosity is also important, i.e., presence of pores of different dimensions which ensure good soil drainage and retention of a certain category and shape of water accessible to plants. According to Kačinski (cit. Živković 1983), the optimal porosity in the arable layer of loamy and clayey soils is 55–65% vol.

We should notice that soil compaction is not always harmful. For example, after sowing, moderate soil compaction ensures better seed-soil contact and increases soil water conductivity. Beside the number and arrangement of layers, their depth must be taken into account for is often the deciding factor for choice of what species to grow. Sugar beet prefers deep soils with a deep humus-accumulative horizon. Skeletal, sandy, or clay soils do not suit. At lowland areas, soils with a profile depth of 100–200 cm are considered deep. It is especially important how deep the physiologically active layer is. For that reason, today, on all those surfaces where the conditions allow, the physiologically active soil layer should be deepened. Impact of mechanical composition of soil can be direct in providing resistance to root penetration and indirect in affecting the soil's water, air, and thermal regime. In sandy and skeletal soils, growing of plants is difficult because of low moisture and not much of nutriments. Extremely clay soils, if they do not have a favorable structure, are compacted, have increased water capacity and unfavorable water-air regime.

Concerning mechanical composition of soil, those with an appropriate ratio of sand, powder, and clay fraction and also a quality of that clay affects the sugar beet. If montmorillonite clay prevails, wide and deep cracks appear at dry period, root system becomes damaged, so when wet period comes, due to drein, anaerobic conditions occur and the plants suffer from lack of oxygen. If the stones and gravel are present in soil, it leads to irregular root growth and its branching.

Soil structure, more precisely its type, is also important. The best are those soils that have a crumbly and granular structure that are also characterized by reduced impact of water dispersion on them and thus provide a continuously favorable water-air regime, while the one-particle structure (in sands) and the coherent structure (in heavy clay soils) are not suitable.

The root system requires normal breathing conditions for normal growth. Lack of oxygen reflects on plant development and can end with decay of plants. The percentage, structure, and rapidity of air renewal in the soil are important. Even before 1927, Kopecky (cit. Baver 1966) pointed out requirement of some plants for air capacity and gave, for sugar beet, limit values of 15–20%, which makes it very demanding for this parameter. In the modern literature, optimal capacity is water–air 1:1. Sugar beet needs good air permeability in arable layer. Soils in which pores less than 0.2 μm predominate is not desirable for air capacity for sugar beet. Beside the amount, the air structure has high importance. Relation between CO_2 and O_2 is variable and depends on root respiration, soil microbiological activity, and the possibility of air renewal. The O_2 content varies from 0.1 to 20% and the CO_2 from 0.1 to 15%. For normal plant growth, the O_2 content has to be more than 10% and growing stops at 5%.

Beside the water-air regime, the soil temperature is of special importance, because it determines the intensity of physical, chemical, and biological soil processes. In addition, each plant species has certain requirements for soil temperatures for optimal development and tolerance limits. Based on their thermal properties, we divide all lands into hot and cold. Cold soils are not suitable for sugar beet growing. Sarić (1971) states the following biological minimum: for the appearance of sprout 2–3 °C; formation of vegetative organs 2–3 °C; formation of generative organs 8–10 °C; fruit formation 10–12 °C. Živković (1983) states that the soil temperature is required for beet germination 3–4 °C while appearance of shoots need 6–7 °C.

Suitable for sugar beet cultvation are medium (3.1–5.0%) to strong (5.01–8.0%) humus soils well provided with calcium. Muckenhausen, cit. Zivkovic (1983). The level of saturation with adsorbed bases, predominantly Ca, should be greater than 90%, while adsorbed sodium or hydrogen is not desirable. Neutral or slightly alkaline soils are suitable (Sarić 1971, pH = 6.5–8; and pH = 6.5–7.2 in nKCl). In addition, the adsorption capacity is of importance and should be higher than 30–40 eq.mol ×/100 g (these have the subtypes and varieties of our chernozem). Zivkovic (1983) emphasizes that redox-potential is important because it affects the soil nutrients. Manojlović et al. (1989) indicate that it should done soil tillage and fertilization that would not lead to damage to chernozem and diminish in its natural fertility (Table 12.1).

Table 12.1 Physical and water–air properties of chernozem in South Banat, Vojvodina

Volume (%)						
Depth	Bulk density g/cm^3	Specific mass gr/cm^3	Total porosity	Max. Water capacity	Water accessible to plants	Air capacity
0–20	1.45	2.67	45.69	44.23	16.15	11.94
30–50	1.45	2.70	48.30	45.15	16.97	11.42
60–80	1.39	2.71	48.71	46.90	19.70	10.11
100–120	1.48	2.75	46.18	45.00	16.93	11.97

Production, beside soil properties, depends on the climate, plants, man, and about improvement of soil characteristics, which is increasingly done in relying on technical capabilities and new knowledge in field of plant growing and science.

12.2.2 Fertility

Fertility of soil is its ability to simultaneously supply plants with water, nutrients, oxygen, and adequate heat throughout the growing season. For normal root growth, soil should be loose and not contain any harmful substance. Fertility is a complex feature of soil that depends on numerous factors as follows:

1. Pedogenetic processes determine the direction and intensity of processes of soil genesis, and in connection with that, they predetermine the characteristics and regimes of the soil.
2. Composition and properties—morphological, physical, chemical, and biological. These include assembly or profile structure, characteristics of individual horizons—layers, profile depth, depth of physiologically active layer, characteristics of individual layers (mechanical composition, structure), specific masses, porosity, water, air, and thermal regime and properties, and physical-mechanical properties. Also, fertility depends on chemical structure of soil solid phase (mineralogical, composition, and properties of humus), characteristics of soil colloids, sorption capacity, acidity and alkalinity, solution, buffer capacity, plant-accessible macro- and microelements, composition of adsorbed cations, presence and absence of toxic substances, soil salinity, i.e., percentage of total and water-soluble salts, content of reduced compounds and gases.
3. Biological factor includes the number and structure of soil microflora and microfauna.
4. Anthropological factor is factor of greatest importance for fertility of arable areas.

12.2.3 Land Tillage

12.2.3.1 Basic Tillage

Soil conditions directly affect the yield during the sugar beet growing because the correct processing provides optimal conditions, longer vegetation, and easier extraction. For that reason, the loosely plowed layer is needed. While growing, a large amount of organic matter is produced and for that production, beside nutrients, enough water is needed. As there is an insufficient or uneven distribution of precipitation in growing season in our climate, soil must retain a larger amount of atmospheric precipitate. This indicates importance of basic tillage. Sugar beet, more than other crops, needs appropriate method, time, depth, quality, and timeliness of tillage. Using heavy mechanization for tillage and transport could damage land composition and a prolonged bad influence in growing subsequent crops can occur (Milošević et al. 1989).

Tillage, if it comes after small grain (which is very usual) could be:

1. Peeling stubble about 15 cm deep.
2. Shallow plowing to a depth of 20–25 cm, with fertilization in early August.
3. Deep plowing to a depth of 30–35 cm (difference is affected by arable layer depth) with fertilization during September.
4. Fertilizing can be one-time.

The benefit of this treatment is multiple. A stable structure of deep loose layer is created, a horizontal and vertical mixing is provided, and if harvest residues are plowed and organic nutrient is added, their layered presence is ensured. Repeated treatment reduces weeds and pest populations and increases population size of beneficial organisms.

Tillage for production intensification was relied on the system of plowing and is doing this way: three plowings with gradual deepening of the arable layer to a depth of 45 cm and with manure applying. Increase of yield by plowing on deepness of 35 cm comparing one on deepness of 25 cm was 10.2%, and for a deepness of 45 cm 15.8%. Stanaćev (1963) proposes three plowings at a depth of 45 cm with the introduction of 1 ton per ha of manure at every 1 cm depth of plowing. Justification of investing in tillage, even if it would lead to an increase in yield is debatable. Therefore, such processing is waived. On some plots manure is used every 30–40 years. By omitting manure, the plowing quality decreases, the appropriate depth is not achieved or the time and interval of performing certain operations do not count. Excessive use of pesticides and herbicides decreases the soil biogenicity. From earlier processing of arable land, only the mixing and overturning of the arable layer remained. By newer understanding, more intensive cultivation is opposite to stability of structural soil aggregates. Therefore, aiming to reduce trampling and deterioration of structural and biogenic soil traits more advanced way of protective tillage was in use. The domain of such processing includes the cultivation of certain plants as the basic crops, intermediate crops for green manure or mulch, and production per system of permanent traces. The system of permanent traces is such that it separates the production surfaces from the surfaces that are trampled by machines. Savić (1989) found that the yield minimization begins with an increase of soil-specific mass over 1.4 t m^{-2}. Trampling soil could be decrease by performing all agro-technical operations at optimal time, related to soil humidity, by aggregating tools, merging of operations, reducing the protection operations, using appropriate pneumatics better organization of extraction, etc.

Lighter soils do not resist to roots permeation, but require more fertilizers, and since they are weaker structures, they do not retain longer the benefits of deep and intensive cultivation. Usually, these soils are in more humid areas, so reduced cultivation is justified.

In Serbia and Montenegro, it is possible to reduce tillage that would correspond to sugar beet cultivation. Instead of this is use of under-digging instead of deep plowing. The under-digger with the feather system does not overturn the soil but shakes it. It is performed at a depth of 45–70 cm. After under-digging, tillage at 30 cm is recommended. On soils of heavier mechanical composition and shallow and uneven shallow arable layer, under-digging enables greater accumulation of water, increases the porosity of deeper layers, and enables deeper penetration of basic root into depth. The achieved conditions enable good yields on less productive lands as well. Rožić (1989) examined the influence of under-digging at a depth of 20, 30, and 40 cm combined with plowing at 20–25, 30–35, and 40–45 cm, respectively. In all variants, the surface treatment was disking. The best results were achieved by plowing at 20–25 cm and under-digging at 40–45 cm. Increasing plowing depth over 35 cm has no significant impact on yield. In 4-year experiments, using under-digging instead of deep plowing led to an increase of yield by 13% on an average.

12.2.4 Pre-Sowing Treatment

For sugar beet sowing, sowing layer preparation is important because it aim is, could provide water contact from deeper layers and be moist to make it ready for sowing. Upper part should be loose and provide aeration and heating. The main goal is to ensure fast and even seed germination and rapid and uniform seed germination. The sowing layer should have three parts: loose, 2–3 cm thick above the seeds and compacted, 3–4 cm thick and then, loose again. Very often, the loose layer is achieved at a greater depth than necessary for sowing seeds. For good preparation of pre-sowing surface is important quality basic processing. At our region, it is often poorly performed for several reasons: previous trampling, untimely harvest, inappropriate plows, bad training of plowmen. After poor plowing, more passes are needed with surface treatment tools. Very often, using of leveler is required. Although this increases production costs, poor plowing causes the surface layer to become excessively crushed during the winter period and turn into dust by frost. At spring, the rain causes the bark occurrence, so germination is very weak. In order for smaller as possible treatment of surface, starting fertilization is increasingly omitted and some of the operations are moved to the period before winter. The land planned for cultivation should enter winter smoothly, without major ridges and depressions. Also, it must not be too finely chopped. The worst thing for the soil is to enter the winter with open furrows because this would require several passages more in the spring with seedbed tiller, and the loss of winter moisture would be inevitable, and the basic aim is to ensure germination and sprouting from winter moisture. Unnecessary trampling with hard machinery should be avoided. One pass more with a seedbed tiller at spring means 5–10% less emergence in field. When surface is left uneven during the winter, at spring the flattened part is undressed and creates a very thick loose layer and at ridges a compacted unfrozen layer exists (Bojović 2014), and so the germination and sprouting becomes uneven.

On sufficiently leveled soil, using the seedbed tillers in one pass is justified because sprouted weeds are destroyed with it, eliminate ascending water flows and create equal conditions for each seed. Ideally, deepness of pre-sowing preparation should be 4–6 cm. On deeply prepared soil, it is impossible to carry out the optimal sowing depth and, in such condition, up to 25% of seeds are sown deeper than 4 cm.

Beside the depth, importance is given to define the correct start of work, for too much of moisture leads to structural change of soil. Starting of surface preparation in our conditions is after the middle of March. With combined crumbs, best results can be achieved.

12.2.5 Nutrient Requirement, Nutrient Deficiency, and Management

12.2.5.1 Sugar Beet Nutrients

Soil fertility for growing a crop is important. It is necessery to known what of soil properties do plant requires.

All fertility factors have equal importance. The deficiency of one factor cannot be recompensed with another, while vegetation period lasts, the fertility of the soil must be at high level, because only in that way can be achieved a high and stabile yields of appropriate quality. Ljubomirovic et al. (2006) followed adding of NPK fertilizers to chernozem for 34 years and found that largest amounts of fertilizers (130,130,130) increased the amount of humus, phosphorus, and potassium; amount of 50, 50, 50 maintained content at initial level, while not using mineral nutrients reduced their level. Long-term use NPK fertilizers affects rising of acidity and decrease of bases amount at adsorption complex, the change in contents of some biogenic elements, but not of accessible microelements (Martinović et al. 1999).

12.2.5.2 Sugar Beet Nutrition

One of the main measures of sugar beet production technology is properly balanced plant nutrition. For high yields accomplishment, not only higher mass of roots and leaves, but also of sugar, plants must be provided with significant amounts of plant assimilates, in a form that is easily accessible to plants. The main nutrients—nitrogen, phosphorus, and potassium, have the greatest importance in plant nutrition and have an impact on yields and increased root technological value, as pointed out by numerous researchers. It is not a unique sugar beet plant nutrition system, because

Table 12.2 Take out in kg ha^{-1}

Parameter	N	P_2O_5	K_2O	Na	Ca	Mg
1 t of root	1.8	0.3	1.5	0.1	0.5	0.4
1 t above ground part	3.2	0.4	4.3	1.6	1.3	0.6
1 t root + above ground part	4.5	0.6	5.2	1.5	1.6	0.8

rise in production was noticed, only if potassium salts without sodium were used (Draycott 1972). Potassium–nitrogen relationship was also examined by Todorčić (1974) and he found that the right chosen ratio of K and N in certain production conditions has a favorable impact on the rise of production per hectare. The top results were at the ratio K: N = 1:1; 1:1.2, all the way to 1:2.

A large number of researchers state that intensification of mineral nutrition has justifications in irrigation conditions because the utilization of plant nutrients is higher in a favorable water regime. Vučić (1992) emphasize the positive effect of nutrition and watering in vegetation period on rise of productive and technological value of roots.

The latest research shows other ways of plant nutrition. Plant nutrition could be enhanced indirectly with non-symbiotic bacteria (Milošević et al. 1999). Hardy and Eagleshaw (1995) point out that using of nitrogen in the world in recent years has increased 27 times and that it is possible to replace it with 20–60 kg ha^{-1} of non-symbiotic nitrogen fixers.

Large amounts of NPK fertilizers lead to crystal sugar yield loss (Bojović et al. 2014; Bojović et al. 2015). Proper crop protection also helps the utilization of assimilates from the soil. Numerous researchers claim that the correct choice of genotype is important for good production. Elements, C, O_2 and H, N, P, K Na, Ca, Mg, and Fe and microelements B, Cu, Mn, and Zn are important for sugar beet. During the entire vegetation period, plants evenly absorb nitrogen and phosphorus, while potassium is absorbed more intensively in the beginning of vegetation.

Real need for nutrients of sugar beet is greater than those that are taken out by harvest, Table 12.2. Twenty-five-year study in Göttingen shows that at an average yield of 50.8 t ha^{-1}, 1.8 kg of N, 0.3 kg of P_2O_5, and 1.5 kg of K_2O on average was taken out from soil per ton of root. Sugar beet uses only a portion of the applied fertilizers for its growth, and the rest is fixed temporarily or permanently in an unsuitable form or washed into deeper layers. Therefore, a detailed analysis of soil is needed. When planning fertilization, beside plants' requirements for nutrients, their presence, planned yield, pre-sowing preparing, etc. must be known. The N-min method enables obtaining information on available N in the soil before sowing and thus determining the required amount of N fertilizer. Phosphorus fertilization is determined on planned yield and the amount of easily accessible phosphorus. Required K is determined in a similar way. Manure has been used very rarely in our production area. Our areas are usually well supplied with microelements and in case of their need pre-sowing or foliar can be introduced.

Sugar beet requires a special way of fertilization for the uneven rhythm of assimilation of nutrients. When using nitrogen fertilizers, a fear of leaching exists, especially in light soils with low adsorption capacity, but also in places with plenty of winter precipitation. If the fertilizer is applied before sowing, plants number decreases, the maturation slows down, and percentage of sugar decreases. Production technology dictates plant protection before sowing, so nitrogen intake is divided into two third during autumn and one third before sowing. If the entry is done correctly, there is no need for additional nutrition. High doses of N nutrients at pre-sowing or top-dressing worsen the value and quality of yield.

P and K fertilizers are easy to use because they are performed completely with basic processing during the fall. The form of using of these fertilizers do not have influence on productive and qualitative properties. Pre-sowing fertilization with these nutrients did not give satisfactory results. Formulations of complex fertilizers are used for fertilizing sugar beet. Foliar treatment with fertilizers has its justification because of sugar beet large leaf area, especially if it is performed correctly and several times. Because of few unresolved issues such as solution concentrations, light intensity, relative humidity, and washing away by precipitation, this form of fertilization is rarely used for macronutrients. It is more acceptable for microelements, growth hormones, and bio-regulators (Bojović 2014). All previous research shows that foliar fertilization are considered a supplementary form of fertilizing.

12.2.5.3 Water Needs and Irrigation

Sugar beet has high requirements for water. In Vojvodina, in Serbia, according to Dragović (1987), sugar beet consumes an average of 1.8 m^3 water at 1 °C at mean daily temperature. Therefore, due attention must be paid to water regime. Favorable are those soils that absorb water well, leak excess water, keep forms of water accessible for plants and ensure its motion from wetter places to the source of consumption—the root system.

Sugar beet contains a high percentage of water in the above- and underground parts, as well as organic matter which is why it has exceptional requirements for water at the vegetation period. The transpiration coefficient shows plant needs for water and the required water quantity necessary to create a unit of dry matter. Sugar beet has a lower transpiration coefficient than lot of crops. Sorghum, corn, and millet are exceptions. The coefficient of transpiration is subject to significant fluctuations, and it is conditioned by external factors and varietal characteristics of plants.

The water needs are different and depend on developmental phase. If enough winter moisture is provided, sugar beet water needs are minimal at starting phases. At the intensive leaf mass growth (from the formation of rows till the beginning of August), large amount of water is needed. This period is critical. The need for water later decreases, but higher than in initial stage.

Sugar beet development requires a significant amount of moisture. Excess moisture could be harmful because then sugar beet contain small amounts of dry matter and sugar. Excess of air moisture can cause a reduction of transpiration and productivity, and at technological maturity phase, it has a negative effect on sugar accumulation.

Weak and short-term drought is well tolerated, especially if it appears at beginning phases. A stronger drought adversely affects and causes great damage to it. According to most authors, a total amount of precipitate of about 600 mm is aplenty for the effectual sugar beet production. Ludecke (1956) points out that 660 mm is needed on loose terrain, with winter precipitate sufficient on colder 220 mm and 250 mm on warmer soils.

Stanaćev (1979) states that in average, for Vojvodina territory, needs about 620 mm of annual precipitate. With the stated amount of precipitate, the needs for winter moisture are 240 mm and from April to October 380 mm of precipitate. However, according to long-term average yields, sugar beet is successfully produced with 500 mm and even 1000 mm of total annual amount of precipitate (in Vojvodina, sometimes, somewhere, fewer than 500 mm). Vojvodina has semi-arid climate, where large deviations are possible both by two neighboring areas and by years.

Numerous researchers state the required amounts of precipitation for certain agro-ecological conditions. However, the Wolthman schedule is best for Balkans, where the total moisture needs are 620 l/m^2. Of that winter humidity (November–March) are 240 l/m^2 and, during vegetation, 380 l/m^2. The needs by months are: in April—40, in May—50, in June—80, in July—80, in August 65, in September—35, and in October—30 l/m^2.

According to some tests and experiences from practice, sugar beet will have good technological and productive parameters, if it is provided with water as follows: with sufficient moisture in winter, it needs relatively dry March, moderately humid April, humid May, and from June to August plenty of rain. Much of precipitation in September and October, a high yield, but a poor root quality can be expected.

Moisture requirement depends on heat. The loss of water by evaporation is the greatest from June to August when, in Serbia, the air is usually warm and dry. Therefore, importance is attached to the distribution of precipitation while vegetation last. There are disagreements among various authors in these quantities, but all this is explained by Ludecke (1961), who connects the quantities and distribution of precipitation with other climate factors. In the southern area of the northern hemisphere, this need is greater because there is higher degree of evaporation in a relatively short time at vegetation period.

When the sum of precipitation during the year is optimum, the production parameters are always high (Ustimenko and Vakomovskij 1979). Spasić (1989) notices a trend of decrease in yield with a downtrend of the amount of precipitation during the autumn of the year before.

Berenyi's research in Hungary shows that beside precipitation, the temperature and duration of insolation in the specified period affected productivity. At maturation phase of sugar beet, big deviation in precipitation quantity by regions of production happens, which affects quality. Vukov (1972) says that August with average precipitation and September without precipitation give a content of 18.3% of sugar in root while each mm of precipitation in September reduces it from 0.03–0.06%. Spasić 1992 states that at first glance relatively small deviations of temperature, for 1.39 C in average, but with precipitates of 146.5 mm at vegetation period, have impact on quite large distinction in the achieved productivity and sugar content.

12.2.5.4 Relative Humidity

Moderate humidity is the best for sugar beet. High humidity causes poorer transpiration, and this has an impact on the yield because if the plant gives away less water, its absorption is lower and so the intake of nutrients is lower. High humidity in July, August, and September causes the spread of the fungal disease *Cercospora*. Low humidity leads to bed production results. The basic role of air humidity consists in its influence on the soil evaporation of moisture and transpiration of plants. At moderate relative humidity, it has normal growth and the course of physiological processes. The wilting process is associated with low relative humidity and high temperatures. The southern parts of Macedonia and Serbia have a pronounced summer air drought, especially in July and August (53–55%). In that period, in those parts, high temperature prevails, which is accompanied by great evaporation and transpiration, which affects sugar beet productive and qualitative parameters. Surplus of moisture in the autumn period of over 80% affects production in Vojvodina and causes appearance and spread of diseases.

12.2.5.5 Hydrothermal Coefficient

Temperature and humidity as inseparable factors are most often expressed together as—hydrothermal coefficient. According to Štojko (1968), Seljanin's ratio of the sums of precipitation and evaporation is use for a complex assessment of the entire vegetation period or its parts. The consumption of moisture for evaporation is approximately equal to the temperature sum reduced by ten times. To rate the favorable weather conditions with a hydrothermal coefficient, the period of vegetation phase or place of production are classified as follows:

- Greater than 1.3—increased humidity, moderate heat.
- 1.0–1.3—moderate humidity and heat.
- 0.8–1.0—relatively dry conditions.
- 0.5–0.8—dry and quite critical period.
- Below 0.5—a very critical period that causes a complete stoppage of physiological processes.

The stated classification according to the hydrothermal coefficient, has practical significance because of its possible application on smaller time intervals, i.e., decades, for establish excesses during the month. The cumulative assessment of climate parameters has an advantage over individual explication of influence of temperature and precipitation.

For favorable evaluation of hydrothermal coefficient, the following values for sugar beet are taken: from germination to closing of rows—1.3, for the period of intensive growth of to the beginning of August, for the period of technological maturity till the end of vegetation. In our climatic conditions, the deviations are significant, depending on the excess of temperature and humidity for a given period.

When the hydrothermal coefficient is lower during the vegetation period, the yield and sugar content in root are lower.

12.2.5.6 Irrigation

For sugar beet production, irrigation is very important, especially in areas of limited humidity. It reduces the adverse climatic effects and enables more uniform yields. Sugar beet presence on irrigated areas is desirable because this provides its production. It irrigates from furrows, by artificial rain or overflow. Overflowing worsens the soil structure, creates a firmer crust, and increases water consumption. It also affects the quality parameters, so that is why its frequent use is less. Irrigation from furrows gives lower yields compared to artificial rain up to 17% on chernozem (Vučić, Miladinović 1964).

Water needs or ETP (evapo-transpiration parameter) are the starting point for irrigation plan. ETP depends on irrigation regime. It is an orientation and not an absolute size, for agro-ecological conditions should be taken into account. Judging by the ETP, the sugar beet is a plant with low water requirements because it needs much less water to produce 1 g of dry matter than many crops. But this is not the case in practice because it requires plenty of water for biomass production. The sugar beet vegetation period can be divided according to Orlovski (1961), at first year, as proportion—1: 9: 3, while by Zolotarev the percentage ratio is 20–25%: 55–65%: 10%. Vojvodina is our largest irrigation area. Considering the average values of ETP (450–500 mm), such amount of precipitation and reserve moisture of 60 mm of easily accessible water in the soil in Vojvodina are rare, it is not real that varietal potential and intensive agricultural techniques can show the highest level, so irrigation is characterized as a need.

The irrigation effect, rational consumption of water, the irrigation economy, and soil protection depend on the correct irrigation regime. Even the simplest irrigation regime, based on certain principles, is better than spontaneous watering. Schemes of periods of plan using water combined with data of precipitation and reserves of water in the soil could serve for establishing the simple irrigation regime. At vegetation start, watering is not necessary, except when spring is very dry and then with a 20 mm watering norm, but in favorable weather conditions, by sprinkling and not by surface watering methods. In the next period, precipitation is insufficient and water consumption is over 4 mm per day. Drought stops the plants growth, reduces the yield, and increases the harmful nitrogen amount, and usually 2–3 watering with a watering norm of 40–60 mm is needed. Although the needs decrease sharply in the last period of vegetation, drought can negatively affect the yield even then. Normal watering is not necessary, but if performed, in mid-August and the drought continues, watering should be repeated in the first half of September. Later, watering has good impact on yield but bad on sugar content in the root. Humidity of 60% is favorable and 40% unfavorable water supply to plants. According to Orlovski (1961), who quotes Zavigordnia, the watering scheme 60-60-60 gives highest root production and amount of sugar and the least of harmful nitrogen, and 40-40-40 is the best of sugar percentage and gives better result than 60-40-40. The trend of soil moisture is possible to monitor, the technical minimum is lento-capillary humidity, which in heavier soil is 65–70% of the field water capacity and in lighter 50% and in heavy 75%. There are procedures for irrigation regime based on physiological indicators of leaf suction force and water concentration. It is not widely used because it depends on other factors (leaf age, fertilization, air saturation with water vapor, temperature). By possibility of dynamic monitoring water using the water of balance method is acceptable. The bioclimatic method that connects the consumption or need for water for the plant biological traits with climate characteristic is the most acceptable method. Based on it, hydrothermal coefficients, in Vojvodina, for sugar beet has been determined as 1.6–1.7 m^3 ha^{-1} (0.16–0.17 mm) of water. Other approaches are possible, but this is the simplest to apply in a wider irrigation area, on the farm or individual plots Vučić (1981).

It is impossible to access irrigation without knowing the basic water-physical properties of soil because without them irrigation norms, rainfall intensity, and furrow length cannot be determined. Knowing of field water capacity is necessary because this indicator expresses quantity of retained water in soil layer after more abundant moisturizing. A larger amount of precipitation drains down and carries nutrients into the deeper layers and is unfavorable except as a meliorating rinsing of salt marshes. The water in the soil is kept by a low power at a PVK of about 300 mb, which corresponds to a value of capillary potential (pF) of 2.7 and is to some extent easily accessible to plants.

On heavier soils, it causes water to lie down, which slows down growth plants. Lento-capillary humidity shows constant in the soil till whose value water moves quickly towards the place of consumption and is easily accessible to plants and it is of great importance. It shows the lower limit of soil moisture in irrigation, and by that participates to define irrigation norm. It is considered to be held in the soil with the strength of 6.2 b which corresponds to a pF value of 3.6–3.7. Further reduction of moisture of soil leads to initial and permanent wilting, mostly accessible water supplies are depleted, and the rest is kept at a power of 15–50 b (pF 4.2–4.5) and the plants are in a critical situation. Important are infiltration data expressed in mm/h (water column in mm which soil absorbed within 1 h). It has variable value, influenced by agro-technical measures and is always the largest in the first hour, except in sandy soils where the decline in time is minimal. The volume of soil mass (a) is also important. It shows compaction of soil, which is important for water-air regime. For Vojvodinas chernozem, p is 1.30–1.35 mg/m^3, for moister, heavy soils less than 1.2 and for fluvisol of sandy composition even p greater than 1.6 make no problem due to porosity.

Several factors determine irrigation: soil properties, pre-irrigation soil moisture, depth of active rhizosphere, method of irrigation, so for that reason is unrealistic to give average values. The depth of the layer that needs to be wet by watering is the active rhizosphere—layer where the plant absorbs 65–70% of water. It is 60 cm for adult plants, while for younger ones it is lesser. With irrigation by artificial rain, the uniformity of wetting is more favorable compared to irrigation in furrows, so the norm can be lower by up to 10%.

However, the conclusion is that irrigation significantly increases root yield. In experiments (Marinković et al. 2001) in some NS hybrids, the percentage in yield rising with irrigation was 10.8 tha^{-1} or 12% up to 14.5 t ha^{-1} or 17% with a variation of 9–23% among hybrids, while in dry farming conditions this difference was much smaller. Pejić (2010) concluded that irrigation has no strong effect on yield rising, based on performed experiments (2000–2006) where mainly it was 8.01 t ha^{-1} with top-dressing.

12.2.6 Crop Establishment

12.2.6.1 Sowing

Its production is influenced by the following factors: agro-climatic conditions, soil (composition, structure, and cultivation) and cultivation technology: choice of varieties, sowing, providing adequate crop nutrition, irrigation schedule, crop protection, (Kosanović et al., 1952; Žeravica, 1965; Shestakova, 1969; Vučić, 1973; Stanaev and Stefanovic, 1974; Draycott et al., 1977; Spasić, 1987; Ming et al., 1989; Stewens et al., 2011; Terzić et al., 2018) and time of extraction. When choosing an assortment, we have to opt for several varieties, which differ from each other by ripening time. Those varieties have to show a higher ability of adaption in different environment area.

Among the major agro-technical measures for sugar beet production there is sowing. Sowing that is done well ensures optimal crop density and high yield. In Serbia and Montenegro, sugar beet is sown after middle of March (Filipović et al. (2009) claim that every day of delayed sowing, 70 kg / ha of sugar is less). Temperature can then be unfavorable and unstable, so we must know the features of seeds and seedlings soil at the time of sowing, germination, and sprouting (Redfearn 1995; Sabovljević et al. 1995). Glamočlija (1986) states that a seed fraction above 4 mm gives higher plants develop with higher sugar yield than a fraction less than 3 mm. Glamočlija (1986) also claim that from larger seeds grow higher plants with longer root and higher sugar yield. Milošević (1989) examined the influence of seed size on germination and found out that the largest fractions of 4.5–5 mm and 4–4.5 mm have the highest value of germination. Dokić et al. (1992) did not find a significant connection of the seed size and productive and quality properties of sugar beet. Quality of seeds are important for it to provide good field germination, good penetrating power, and thus achieve an optimal plant number and even growing at the beginning (Glamočlija 1986). Many researchers have examined the germination and emergence of varieties related to their genetic constitution (Durr 1992; Lovato and Cagalli 1992). In controlled conditions at $t = 5/15$ °C, higher total germination and germination energy have fractions of larger seeds while at $t = 20$ °C this difference is smaller (Bojović 2010). Results of sugar beet seed properties research have found application in the seed technology of other field plants. Properly processed seeds are an important prerequisite for successful sowing and by that, a yield. The sugar beet germ is tender and so it should not be sown deep. Sowing depth is conditioned by soil type and humidity conditions. In mass conditions, it is usually performed at 2–3 cm deep. In dry conditions and on lighter soils, the sowing depth is 4–5 cm. Very deep sowing reduces the access of air to the seeds, which prolongs germination; the path of the germ to the surface is prolonged and this causes excessive elongation of the hypocotyl part of the germ and leads to shortening of the roots. If germinating is from the depths, field germination reduces, germs often decay while the others are depleted, the germ is thin and isn't good base for future growing. High concentrations of salt in the sowing layer (if the pre-sown mineral fertilizer and herbicides are applied shallowly or poorly balanced or there is its stronger presence in the seedling zone) reduce the germinating of plants. Germinating decreases with the bark forming. Sowing too shallow sowing is not right because of the loose and dry layer at the surface, and for that reason seed is at small contact with the soil, the inflow of water is difficult, so germination will be difficult, prolonged, and often depends on precipitation. It needs to be underlined that the sowing depth must be uniform. In general, below 50% of field germination is low germination, 50–65% moderate, 65–80% good, and over 80% very good. Seed mass is small, depends on the size and amounts, per hectare, 1.5–3 kg. The number of seeds for sowing is expressed in sowing units. 1 unit is 100,000 seeds.

$$\text{Seed quantity} = \frac{\text{Row length (m) per 1ha} \times 100}{\text{Seed distance in a row (cm)}}$$

Sugar beet is usually sown between row spacing of 50 cm and in-row spacing 15–19 cm. The best schedule is when sowing at 10 cm with thinning. According to many authors (Smith et al. 1977; Glamočlija 1986; Filipović et al. 2007; Bojović 2014; Bojović et al. 2015), vegetation space weakly affected the increase of yields and this was by extended time of growing. For an earlier extraction period, a smaller vegetation space is better, unlike for a later one, where larger space is better. Stanaćev (1979) found that the best production and quality value were the highest at 1000 cm^2. Lüdecke (1953) pointed 74,000–95,000 as the most suitable density. Glamočlija (1990) states that crop density has high influence on yields. In his researches, few densities (60,000, 80,000, 100,000, 120,000 to 140,000) of two varieties of sugar beet with plant nutrition which levels differed were examined. When density of crop and quantity of mineral nutrients rise, content of sugar gradually falls and "harmful" nitrogen increases. Varieties react differently on crop density and nutrition. Filipović et al. (2007, 2008) concluded that increase of plant density increases sugar beet root yield. Thus, the highest density (120,000 plants per hectare) brings the highest yield, which is on average 5.6% higher than at the lowest density. The highest crystal sugar yield (12.43 t ha^{-1}), was at a highest examined density and the lowermost at a density of 100,000. The tolerance to stress caused by lack of precipitation, high air temperatures and other environmental factors of some genotypes, is advantage in production in dry farming conditions, especially when crop irrigation is impossible. Such varieties have great coefficient of utilization of applied mineral nutrients and in general higher productivity (Čačić et al. 2000; Dobrovnaya et al. 2009; Pejić et al. 2010; Bojović et al. 2014). The assortment choice that fits production territory brings a larger and more stable production. The rhythm of growing and the facility to use a certain vegetation space for the highest possible sugar beet yield is what makes difference between two varieties. They also differ greatly in maturing time for cost-effective sugar processing. Certain varieties have the property of prolonging the growing period and during the autumn they create a significant part of organic matter. When choosing an assortment for growing in a certain production area, producer should choose several varieties, which have different ripening time. We can distinguish three basic types of sugar beet varieties: Z-high-sugar content types, N- normal, and E- high yield types (Bojović 2014). It is profitable that varieties of Z-types are sowed for their earlier extraction. Varieties of the normal type are appropriate for the medium harvesting period, and varieties of E-type are in optimal technological maturity much later, so they should be removed last (Stanaćev 1979). It is economically justified to grow varieties of all three of types. Stanaćev (1976) suggested that ratio should be: 25% of Z-type ,40% of N-type, and 35% of E-type for Vojvodina production territory. The proposed ratio gives the possibility for gradually harvested from the fields and processed, depending of its technological maturation time. It provides good technological properties of root and new sowing of winter cereals on those fields.

The yield per ton depends on the root yield, sugar content in the root, and the purity of the sugar beet juice (Campbell and Kern 1982; Smith et al. 1977). Smith et al. (1977) found that the juice purity depends on genotype, soil type, plant nutrition, and density of crop. Among the mentioned factors which are crucial for quality, nitrogen nutrition is crucial in Beek and Huijrbregts's (1987) opinion. Most scientists agree with some differences between varieties related to uptake and transformation of mineral nutrients, but these differences are not yet so great as to have some relevance to practice. However, when using very large content of N, P, and K, the high-sugar content varieties deteriorated to a lesser extent the quality for processing than the high yield types (Stanaćev 1973).

New varieties, of both domestic and foreign origin are constantly appearing, which by some of their characteristics deserve the attention of producers (Figs. 12.1 and 12.2). For higher results in different climate and soil conditions, varieties must have a higher adaptive ability. At the Institute of Field and Vegetable Crops, Novi Sad, the following varieties were created: Sara, Mara, Lara, Drena, Rama, Ilina, Vera, Irina, Nora, and others. Bojović (2014) states that the year has a greater impact on quality parameters relative to the locality, while the interactions of cultivars × years and cultivars × locality were very similar for yield value.

Sugar yield and root yield are the basic indicators of the value of sugar beet varieties. Basic indicator of technological value of a particular genotype is the sugar content. NS varieties, Vera and Nora, had a high root yield, while the Vera variety had a higher sugar content (16.35%) and belongs to the sugar type (Z). Varieties Vera and Nora have genetic resistance to Rhizomania and good technological characteristics, recommended for early and medium extraction periods in all growing regions in Serbia and Montenegro.

The amount of sugar and non-sugar substances are related with environment and used agro-technics, but these characteristics are features of each variety (Čačić et al. 1997; Čačić et al. 2000; Kolarić et al. 2015). By using plants that improved by biotechnology, production could be higher, using of pesticides and erosion can be reduced, and the groundwater quality can rise (Stojšin 2001). Fundamental science,

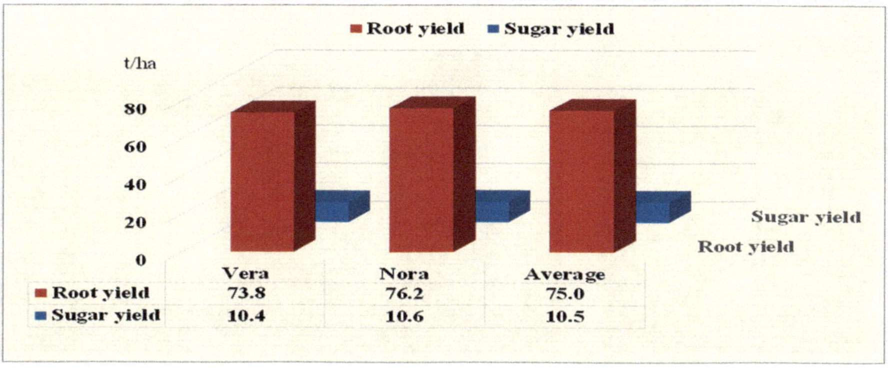

Fig. 12.1 Root yield and sugar yield at NS variety Vera and Nora

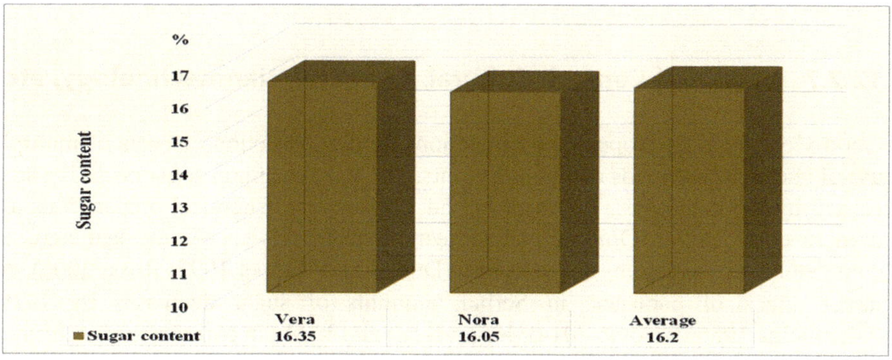

Fig. 12.2 Sugar content at NS variety Vera and Nora

more often than at other cultivars, changes completely in the genetic composition of cultivated varieties of sugar beet several times. Breeding has aim to increase rising of fertility potential and by that, economical and profitability production.

The Institute of Field and Vegetable Crops in Novi Sad mainly effectively followed European and world trends in creating varieties. They did not lag those from multinational companies in terms of their characteristics. In the recent years, the increase in fertility potential is due exclusively to the improved genotype and is 2% per year, which is a great success for geneticists and breeders in Serbia and Montenegro (Kovačev et al. 2006). In the production of sugar beet, the most important is the amount of moisture and nutrients at a depth of 60–90 cm, Kovačev et al. 2011 (Fig. 12.3).

a b

Fig. 12.3 Sugar beet (**a**) and poppy in chernozem soil (**b**) in Bački Petrovac, Serbia

12.2.7 Weed and Control: Cultural, Chemical, Nanotechnology, etc.

Good yield requires proper crop protection. Proper protection consists of multiple use of chemicals on seeds and young plants. The most common sugar beet disease is caused by the pathogens, *Cercospora* and Rhizomania. Chemical preparations are used to control both. Omission of protection treatments causes the leaf mass to completely perish, leading to yield loss (Duffus and Ruppel 1993; Rossi 2000). At larger attack of parasites, in Serbia, amounts of sugar decreases by 1–2% (Glamočlija 1986; Bojović 2014). Kuzevski et al. (2000), in experiments performed 1996–1997 record that in intolerant varieties, have the yield decrease 30.94 t ha^{-1} and the sugar content for 3.31%. The mentioned researchers conclude that by monitoring the sugar beet root juice chemical composition, attack of Rhizomania can be established.

Weed control is an important agro-technical measure during the cultivation of sugar beet. Before the middle of the twentieth century, weed control was performed by mechanical inter-row cultivation or manual hoeing. Discovering selective herbicides contributed that human work is less necessary. Biological control measures, although promising, have almost reached expected level. At South Banat territory, it has been noticed over 160 weed species. Although mostly the same weeds are found in sugar beet growing areas, the composition and number of weed species differ, so no field herbicide used in other countries can be recommended without field trials and inspections in certain conditions and areas. Control by herbicides can be accepted only if high yields are achieved.

Weeds cause the greatest damage to plants after germination and affect the yield of sugar beet to its complete absence. Thus, Wicks and Wilson (1989) state that weeds that germinate up to 8 weeks or within 4 weeks of formation of two pairs of leaves cause the greatest damage. Dawson (1965) found that 132,000/ha of plants *Chenopodium album* reduced sugar beet production by 94%. In experiments by

Brimhall et al (1965), the species *Sinapis arvensis* together with other broadleaf weeds diminishes by 95–100%. The studies of Wicks and Wilson (1983) show that in field experiments that for every 1120 t ha^{-1} of dry matter of weed mass root yield decreases by 10–15 t ha^{-1}.

As it was said, weed control is most often done with herbicides nowadays. By the application time of herbicides, we distinguish three periods as follows:

1. Before sowing—those herbicides that evaporate easily at higher temperatures and those which depend on precipitation. It is introduced with seeders or discs and is recommended for dry areas.
2. After sowing and before germination—pre-emergence, numerous of herbicides are applied. They are adopted through the roots, so their application depends on precipitation, which means that at lack of precipitation, their action is absent also.
3. After germination—post-emergence, used preparations with numerous active substances.

The plant absorbs these herbicides through the leaves and further sends them into other parts of the plant through the conducting bundles. With precipitation, they can be absorbed through the root. One preparation cannot destroy both narrow-leaved and broad-leaved weeds, so we regularly use them in combinations. Lasting of herbicides is problem, so it is better to apply them repeatedly in reduced quantities. Application has to be repeated in the germ leaf phase of weed until the first pair of leaves develop. The effectiveness depends not only on the spectrum and residuals of the herbicide but also on the climatic conditions after application, type, structure, pH, organic matter content, and soil moisture.

We must mention that residues of persistent herbicides can show phytotoxic effects on sugar beet.

We will mention the herbicides used in Serbia:

With active substance S-metolachlor 960 g L^{-1} against annual grass and broad-leaf weeds—before sowing and after germination with 1.4–1.5 L ha^{-1} depending on the percentage of humus once.

Propaquizafop 100 g L^{-1} once when the weeds reach the stage with 3–5 leaves or twice when the weeds are in the stage with more than three leaves in another 15–20 days later.

Cyclosidim 100 g L^{-1}.

Fenoxaprop-P-ethyl when weeds are in phase 2 and before flowering of leaf crops once.

Quizalofop-P-tefuryl once when the weeds are at 3–5 leaf stage or twice within that phase and after 10–15 days in reduced quantities.

Haloxyfop-R-methyl once in the phase of 3–5 leaves or twice at same phase and after 10–15 days in reduced quantities.

Cletodim, after the emergence of crops once.

Desmedipham + phenmedipham once or twice after crop emergence.

12.2.8 Interculturing

12.2.8.1 Cultivation Technology

Crop Rotation

The crop rotation today has much smaller importance, relative to few decades ago, because of agricultural development, especially the mechanization, use of pesticides and mineral fertilizers. This created the possibility that the crop rotation is more conditioned by the market and economic policy than the growing conditions and plant characteristics. However, negative consequences of this have been noticed, especially on keeping of soil biogenicity and its structure.

Sugar beet is a plant that, due to its production requirements, requires production within the crop rotation. It is very sensitive on successive cultivation on the same plot. Cultivation in monoculture and even frequent (to 2–3 years) leads to reduced yields for not providing the soil recovery that cannot be compensated by fertilizers, have good condition for diseases and pest attacks so cultivation in crop rotation is recommended, which provides a return to the same plot only after 4–5 years. Research experiments at Rimski Sancevi, near Novi Sad, conducted a long-term cultivation of sugar beet in monoculture. After the first 10 years of performing experiments, Stanaćev (1973) introduces that there was a decrease in sugar beet yield by 26% compared to five-field crop rotation. Stanaćev et al. (1983), states that after 20 years the yield in monoculture will be 23.5% lower compared to five-field. So, after the initial drop in yield, the yield stabilized during continuous cultivation on the same area. On this experiment fields, there were clear negative changes in soil porosity, microbiological activity, composition, and amount of humus. For proper crop rotation establishment, biological traits of culture and soil and climate conditions must be known. In regions with insufficient rainfall or poor distribution of it, plants with less water demand and soil draining are better as pre-crops. In humid regions, perennial grasses and legumes are desirable. In region with heavier mechanical composition of soil, pre-crops whose root can easily penetrate are desirable, while for those with a lighter mechanical composition, those are crops with a shallower root and lower water requirements. Sugar beet should not be grown behind those crops that have common pests and diseases. In our production conditions, the best pre-crops are small grains. It is usually wheat, because it leaves enough time for adequate preparation of the soil for growing sugar beet. Good pre-crops are also the vegetables, but their representation is small on cultivating area so they are not so important. Potatoes and soybeans are also a good pre-crop, but not in dry years if water consumption is not compensated with winter moisture.

Unsuitable pre-crops for sugar beet are fodder beet, oilseed rape, cabbage, kale, and other crucifers (*Brassicaceae*) because they support the spread of aphids, nematodes, and viruses. The pre-crops impact quality, same as the yield. Legumes leave the soil enriched with nitrogen, so they can reduce sugar utilization by rising of harmful nitrogen amount. Sugar beet, as a pre-crop, is suitable for many cultivated plants (spring barley, wheat, legumes) because the land for its cultivation is well cultivated several times and the harvest residues were plowed. In some localities, yield reducing of corn, sunflower, and hemp was noticed, and of soybeans also, in some more arid areas, if their pre-crop was sugar beet. This is possible for two reasons: Sugar beet gives a high yield and above ground masses, and when it is plowed, it uses so much water to decompose it, so in regions with a lack of precipitation, it is bad as a pre-crop (Glamočlija 1986). Also, if the extraction is performed at increased humidity, its structure is disturbed, it becomes compressed, as a result of which the water-air regime is disturbed, and the biogenic activity is reduced, decreases the yield of crops which are subsequent.

It is considered that the sugar beet participation in crop rotation conditioned on the humidity of the region. Thus, in humid regions, the share of beets can be up to 30%, in regions with uneven distribution of precipitation is 20–25% and in regions with unfavorable humidity up to 20%.

12.2.9 Crop Care Measures

Sugar beet is a highly intensive field crop and for expression of its greatest yield potential, it requires much of work and resources. If these requirements are met, it becomes the most profitable crop. Care begins almost post sowing and includes various operations and procedures: rolling, breaking the bark, cultivating and hoeing, thinning, feeding, irrigation and protection from weeds, diseases and pests.

After sowing, periodically going around is important and sometimes, rolling and harrowing measures are necessary. Rolling doesn't belong to regular measure, but it can be carried out, especially in drier areas or when sowing accompanied dry weather for the surface layer to be compact and come into contact with deeper, wetter soil layers. Crust can occur if heavy rain falls after sowing, thus disrupting aeration and making germination and sprouting more difficult. Then the breaking of the crust is approached, usually by hilling. This measure is not used today. When the crop sprouts, inter-row shallow hoeing (cultivation) is started and if the sowing was not at the final distance, thinning. Hoeing is done for better rooting, better change of gases, uniform absorption of precipitation, faster heating. Hoeing does performed if not all weeds are not destroyed by herbicides and if due to heavy rains, soil structure is severely damaged.

Thinning of plants is the best in the phase of the first pair of permanent leaves and is completed until the appearance of the second pair of permanent leaves (in 8–10-days). The optimal density is about 10^5 plants per hectare (45×25 and 50×20 cm), with $8x10^4$ plants realized during sowing. If the crop is irrigated, the crop density should be slightly lower. From thinning to closing the rows, inter-row cultivation and hoeing are performed. If more hoeing is needed, the first one is at the depth of 4–5 cm and each subsequent one is deeper (up to 10 cm).

Top-dressing is performed with cultivation. It performs not before the sprouting so sprouts would not be damaged. If it is done before closing the rows, it must be at the afternoon, because the leaves are more bent and the damage is less. When soon as the plants develop cotyledon leaves, fed can start. Top-dressing could be done once or twice with 30–60 kg ha^{-1} of nitrogen on soils that are light, shallow, cold, and poor in nitrogen. In fertile soils, this operation is not necessary. If nutrition is done before closing the rows, quality may decrease so it shows that it must be done when a sugar beet has four leaves maximum. Feed can be done with hoeing or superficial. If it is superficial, inter-row cultivation must be immediately done after mixing the soil with fertilizer. This feeding is best performed while the leaf is wet to prevent burns on the leaf. Nitrogen supplementation through leaves has no great justification.

Irrigation is important because, in conditions of limited areas, it reduces the adverse effects of climate and enables more uniform yields. Sugar beet should be present on irrigated areas because that ensures its production.

12.2.10 Intercropping

Cultivated plants grow mainly as monocultures. In agriculture, monoculture has facilitated work operations such as weed removing and harvest. This system maintains crop productivity by using chemical inputs including fertilizers and pesticides but reduces the plant and microorganism diversity (Brooker et al. 2015). Lessening of resources, like arable land, water and energy, high need for devise and practiced new strategies and techniques of production is present, to fulfill the expanding needs for food and forage through sustainable utilization of available inputs (Jabbar et al. 2010). Various environmental and socio-economic reasons have been suggested for explaining the well-known concept of intercropping.

In agriculture, more studies on intercropping have been carried out for evaluation of potential agronomic and economic benefits. Importance of interactions among crop species in shaping the structure and dynamics of plant association is widely acknowledged. Benefit of it is great. Intercropping systems can enhance crop produce. Sometimes productivity is enhanced in intercrops (Fukai and Trenbath 1993), but most of the studies find intercropping plant yields intermediate. Intercropping means that at least two crops species are grown on the same plot of land simultaneously and that results in higher yield. In intercropping systems, both negative interaction (competition) and positive interaction (facilitation) can occur simultaneously. Competition prevents crop growth by sharing the limited resources or allelopathy, whereas facilitation increases crop performance by improving the micro-environment for utilizing resources (Brooker et al. 2015). Intercropping is a widespread agronomic practice in the tropics because it reduces the loss caused by pests, diseases, and weeds and also it guarantees better yield (Andrews 1974). Similarly, Banaszak et al. (1998) carried out experiments with intercropping and found that new varieties of oil radish and white mustard as intercrops had reduced the *H. schachtii* infestation by about 20–40% in sugar beet crop. In order to evaluate the effect of intercropping sugar beet with oilseeds (mustard and canola) and lentils, three sugar beet varieties *viz*., Kaweterma, Aura, and Pamela were tested against four intercropping systems (sugar beet only, sugar beet + mustard, sugar beet + canola, and sugar beet + lentil). Among the sugar beet varieties, Kaweterma had excellent performance in growth, yield, and quality traits. Sugar beet yields and monetary benefits were also maximized in lentil intercropping compared to cereals and oilseeds intercropping. It is recommendation that intercropping of sugar beet variety Kaweterma with lentil should be practiced for higher qualitative, quantitative, and monetary benefits (Usmanikhail et al. 2012).

There was an experiment with sugar beet as monoculture and wheat and intercrop at small or suitable presence of phosphorus. In intercrops, production of dry matter diminished in wheat and increased in sugar beet. The rate of photosynthesis in wheat was lower and in sugar beet it was higher. In intercropping it is fallowed with transpiration rate lowering and higher using of water in the after species. Contents of P, K, and Fe decreased in wheat intercrop and it rose in sugar beet. The similar effect of intercropping was observed in the short-term hydroponic experiment. The three root parameters—length, soluble carbohydrates, and activity of secretory acid phosphatase which are in connection with plants phosphorus-deficiency, raise in both intercropping, regardless of the phosphorus amount. Results show an inter-specific interaction which is above the different capacity nutrient acquisition in the intercropping (Hajiboland et al. 2018).

12.2.11 Crop Management under Abiotic Stress Condition: Drought, Salt, and High Temperature Stress

12.2.11.1 Conditions of Cultivating

Sugar beet is one of major crops grown in temperate climates, especially the northern hemisphere. It is largely distributed because it easily adapts to various agro-ecological conditions. Area of growth is between 30–60°N and 25–35°S latitude. In Serbia conditions, it prefers a moderately warm, sufficiently humid and sunny climate, with sufficiently humid winters, warm and humid May, relatively cool June and July and sufficiently sunny August, and similar September when sugar in sugar beet accumulates, and October sunny and cool. For its production and quality, agro-climatic conditions are not fully defined, but limit values exist.

Rise in global temperature, increased drought, increase in CO_2 levels, and various forms of pollution in different ecological conditions strongly affect all crops (Glamočlija 1986; Glamočlija 1990; Popovic et al. 2016; Popović et al. 2020a, b, c; Terzić et al. 2019; Lakić et al. 2019a, b, Božović et al. 2018; Božović et al. 2020; Rajičić et al. 2020a, b). The weather conditions of an area vary from year to year. Sugar beet, beside ability to adapt, reacts fast on climate changes that greatly affect its productive and qualitative properties. Natural conditions cannot be changed by the will of man but production process and variety can be changed and adjusted. It is possible to regionalize its production, based on plant requirements and production conditions. Quantity and quality are influenced ecological factors: temperature, precipitation, humidity, length of insolation, soil properties, etc. These factors' influence is complex.

12.2.12 Thermal Conditions

Knowledge of production area thermal conditions means a lot for production success. Significant variation of temperature, while vegetation last, have the greatest influence on root quality in Serbia and Montenegro.

Since sugar beet is a plant of early sowing, minimum and maximum temperatures have a great influence on seedlings and shoots. The optimum temperature for germination and sprouting is 25 °C, minimum is 4 °C, and maximum 28–30 °C. Critical temperatures are especially important during sowing and sugar beet germination. The minimum temperature of soil at a depth of 5-6 cm is 6 °C which is optimal for sowing. The minimum temperature for germination of sugar beet is 4–5 °C, which shows that it is resistant to low temperature but then is a fairly long period from germination to sprouting: at $T = 4.4$ °C is 22 days and at 10.3 °C only 9 days. For sowing, the critical T is 4 °C but at T higher than 28 °C the seed can germinate just at high relative humidity. The total sum of T for germination and sprouting is 100–125 °C.

Low, critical temperature affected germination and cotyledon phase until the appearance of the second pair of first leaves. During this period, low T can have a vernalization effect, which, on a greater or lesser extent, at sensitive varieties, lead to appearing of stalks during the summer, which depends on low temperatures lasting. Frost damage depends on moisture of the soil and the previous effect of high temperature on water absorption by plants. Sugar beet is not sensitive to weak frosts in both spring and autumn. However, susceptibility to frost is dependent on the sprouts age. Young, newly sprouted plants can withstand frosts down to −6 °C. In the sprouting (cotyledon outbreak), they are sensitive to frost—it suffers if temperature is less than −3 °C. Critical low temperatures at cotyledons period, which ranges up to −7 °C (in Vojvodina), lead to deterioration of crops state, caused by frost. Adult plants (in autumn) can withstand frost up to −5 °C. Late in autumn, at relatively low T, sugar beet vegetates and increases root yield and sugar content. Sugar beet tolerates low temperature without harmful consequences for processing at autumn if it is in the soil. If the frosts are stronger, and especially if the periods of freezing and melting alternate, it suffers damage and root quality reduces.

Sugar beet has moderate heat requirements. It tolerates both low and high temperatures quite well. Its vegetation period lasts from 120 to 200 days. Most authors agree that the growing season requires temperature between 2400 and 2600 °C. Broadly speaking, the sum of temperatures depends on latitude and is 1900–3600 °C. At the Pannonian plain, it is around 3000 °C (Spasić 1992). Temperature and its sum that sugar beet need are different through vegetation period. Stanaćev (1979) states that the optimal mean temperature for the sugar beet vegetation period is 15.3–16.4 °C. It slightly differs in Europe from North to South. Glamočlija (1986) finds in his experiments that the optimum average temperature is 18 °C, and that high temperatures in July and August reduce the final yield, but, at the end of the season, plant growth does not depend on thermal conditions. The required mean temperature during vegetation differs and is: Part I from germination to row formation (about 60 days), – 10.7 °C ($\Sigma = 650$ °C), Part II from the assembly of rows till beginning of August—18.8 °C (Stanaćev 1979). In Pannonian lowland, it is often warmer 2–3 °C than in Europe. In Part III, which is the maturation and sugar accumulation phase, from the beginning of August until sugar beet extraction, – 16.5 °C (Σ = 1000 °C) is preferred. In Pannonia, temperature oscillations in that period are frequent in certain years, and this leads to quality deviations. The longer vegetation period leads to higher yield.

The role of heat in sugar synthesis process is large depending on the phenological phase. The most intense accumulation of sugar is at $T = 25$ °C and stagnation occurs when the average daily temperature drops below 6 °C. According to Glamočlija (1986), the formation of organic matter is slowing down if temperature is less than 12 °C but also greater than 30 °C. High summer temperatures, especially in the root zone if they are higher than 30 °C, decrease turgor in the leaves, reduce the photosynthesis intensity, and cause leaves drying. The transport of assimilates from the photosynthetic apparatus to root is slowed down and that indirectly effect the absorption of water and nutrients. According to Bojović et al. (2014), the optimal temperature for photosynthesis intensity is 20–25 °C. The tests were conducted under strictly controlled T in the root zone. At the stated T, the highest sugar yield was obtained.

12.2.13 Light

Sugar beet places great demands on light, especially when sugar accumulates most intensively in the root. The sugar percentage is positively correlated with insolation length from August to October. Also, its use of sunlight is poor, only about 2%. It is estimated that lasting of insolation in July, August, and September is more important for the ripening and quality. According to photoperiodic reaction, sugar beet is a long-day plant and for that, at first year, it easily brings a flower and vice versa, in the southern areas with a shorter day, it hardly gets a flowering stem. Since sugar beet intensively makes organic matter, it needs many sunny days during growth. Certain intensity of light is needed for enabling the best assimilation, better sugar content, and largest amount of organic matter. Lack of light decreases the total sugar beet yield and quality, while content of non-sugar substances increase (Bojović 2014). The best flows of assimilates from leaf to root occur during sunny and cloudy days shift. During sunny days, the assimilation takes place intensively, and when it is cloudy, the products of photosynthesis are transported to the root. According to insufficiently checked data, for sugar beet quality, at ripening period (August–October), about 700 h of sunshine are needed. In most production areas and years of production, the stated number of hours is lower. With a smaller number of hours of insolation, sugar content is usually lower.

12.2.14 Relation to CO_2

Content of carbon dioxide greatly affects photosynthesis. The main sources of CO_2 are soil surface and air currents. Sugar beet consumes large amounts of CO_2 and reduces its concentration. Under good growing conditions, photosynthesis is generally limited with amount of CO_2 in the ground layer. It is possible to increase the concentration of CO_2 in the ground layer of the soil by applying organic fertilizers and appropriate tillage and in the same way the yield of sugar beet.

12.2.15 Salt Impact

Sugar beet is relatively resistant to total soil salt. Sarić (1971) states that it thrives at a salt content of 0.8–1% of the soil dry weight, which means, at a strong degree of salinity. A decline in crop yields leads to a reduction in the economical production. Climate change contributes to this, so alternative crops are often introduced into the crop rotation (Popović et al. 2020a, b, c). By present cultivation system, early sowing affects high yield. For achieving of high production results, early sown plants germinate quickly and have speeder row closing in comparison one sown at usual date. Therefore, variety must have a lower temperature of emergence minimum and of leaf formation (<3 °C; Milford et al. 1985), faster growth at $T < 10$ °C (Milford et al. 2010) and lower sensitivity on freezing, to ensure yield formation in a vegetative phase.

12.3 Future Prospects

The knowledge, ability, and skill of the producer are measured by the achieved yield and its quality. The sugar industry has become, in the true sense of the word, a large food industry, whose development is closely connected with the general development of industry in the world, and it is moving in terms of increasing capacity and at the same time in terms of increasing labor productivity. The level of agricultural technology, assortment but also climatic and soil conditions also dictate the yield. That is why we should always strive for new knowledge, and perfect the skill of forming yields, while preserving nature. Climate changes, fast growing world population, increased importance of ecological awareness in industry are factors which affect all aspects of human life, even on the production of sugar beet.

Where is the place of sugar beet in the world of tomorrow? Today sugar beet is the main source of sucrose for the European countries and further (especially the central and southern parts of the Europe) and is involved in agricultural plans of around 50 countries of the temperate regions of the world. Around 20% of world sugar production was made out of sugar beet (FAO 2009). Since sugar is strategical product which is needed in many industries (confectionary industry, beverage and alcohol industry, pastry industry), growing this crop will be continued. The aim of sugar beet growing will be focused on achieving greater white sugar yield (around

15 t/ha) with lower inputs and lower usage of pesticide by using conventional cultivation systems or organic growing in the world and even in Serbia and Montenegro. Due to unstable oil production and limited oil sources, new technologies are oriented towards new generations of biofuels from renewable sources where sugar beet offers interesting solutions using post-harvest sugar beet waste (biomass, sugar beet root residues, etc.).

Sucrose, stored in the sugar beet root, presents an economical source of bioethanol since it can be easily processed by microbes in the fermentation reactions. Also other fuels can be produced from sugar beet such as biomethane, biomethanol, and biohydrogen. In the future, solving the problem of weeds in sugar beet fields using transgenic or GM technologies will also present one of the challenges of sugar beet breeding with different opinions and effects on the environment. Sugar beet is an economically important crop, and the growth of world surfaces is expected in the future in Serbia and Montenegro.

The general tendency of the national economy in our country, in this economic branch is oriented towards increasing the beet processing capacity with the aim of completely freeing the country from sugar imports and with the task of eliminating its deficit despite the planned increase in consumption of this important food. In 2021, the average yield is 55 tons per hectare from 36 thousand hectares. In 2020, the yields were from 35 to 55 tons per hectare, even up to 60, so the yield this year is around the average. Starting from the current situation and our needs, the medium-term plan for the next period envisages a significant increase in the total production of sugar beet. This increase should be achieved to a greater extent by increasing the yield per unit area (hectare) and to a lesser extent by increasing the area under sugar beet.

12.4 Conclusion

In the world and in Serbia and Montenegro, a decline in sugar beet yields leads to a reduction in the economic budget. Climate change significantly contributes to reduced yields. For successful production planning, proper application is necessary for all measures in cultivation technology. Important measures of sugar beet production technology is properly balanced plant nutrition. NPK mineral nutrients are the major carrier of sugar beet productivity. Intensive NPK nutrition significantly increases the vegetative biomass quantity and root sugar content. Lack of soil nutrients significantly reduces yields. The major nutrients—nitrogen, phosphorus, and potassium, play an important role in plant nutrition and greatly impact yield and increase technological value of root. The quantity and ratio of the major, secondary, and microelements should be adjusted to the needs of plants and the natural fertility of the soil. Single plant nutrition system does not exist because the plant needs for certain elements depend on weather conditions, soil, cultivation methods (natural water regime, irrigation), and genotypes used in production. Crystal sugar yield, as major indicator of productive value, has statistically very significant dependence of way of plant nutrition, years, genotype, and interaction of them.

Acknowledgments The authors express their appreciation for providing financial support for this research to Ministry of Education, Science and Technological Development of the Republic of Serbia (Grant No. 451-03-68/2022-14/200032 and 200358) and Bilateral project Serbia and Montenegro 2019-2022: Alternative cereals and oil crops as a source of healthcare food and an important raw material for the production of biofuel.

References

Andrews DJ (1974) Response of sorghum varieties to intercropping. Exp Agric 10:57–63

Banaszak H, Nowakowski M, Szymczak-Nowak J, Ojczyk K (1998) Limiting of Heterodera schachtii Schm. diseases and weeds of sugar beet by tillage system based on mustard or radish intercrops and mulches. J Plant Protec Res 38(1):70–80

Barocka K, Geidel N, und Müller W (1972) Enffluss der Bestandesdichte und N–Dungung auf die Leistung von Zuckerruben. II Die Ainteile von K, Na und α-amino N Zuckerind. 22. Nr 10(3): 556–556

Baver LD (1966) Soil physics, vol 203. John Wiley and Sons, New York

Beek MA, Huijrbregts VD (1987) Internal quality aspects of sugarbeet. Proc Fertilizer Soc 252:20–25

Bojović R (2010) Influence of hybrid combination and temperature on seed germination and germination of sugar beet seedlings, Master's thesis, Faculty of Agriculture, Belgrade

Bojović R (2014) Morphological and production properties of sugar beet genotypes in conditions of intensive plant nutrition. Doctoral dissertation University of Belgrade, Faculty of Agriculture Zemun. Serbia

Bojović R, Glamočlija Đ, Popović V, Popović B, Filipović V, Kuzevski J (2014) Sugarbeet yield parameters on carbon at chernozem soil type. Agric For Podgorica 60(3):41–53

Bojović R, Glamočlija Đ, Popović V, Popović B, Filipović V, Kuzevski J, Ugrenović V (2015) Parameters of yield of sugar beet in conditions of dry farming. J Inst PKB Agroecon Belgrade 21(1–2):45–57

Bojović R, Popović V, Ikanović J, Živanović Lj, Rakaščan N, Popović S, Ugrenović V, Simić D (2019a) Morphological characterization of sweet sorghum genotypes across environments. J Anim Plant Sci 29(3): 721–729. ISSN:1018–7081

Bojović R, Popović V, Janković S, Kolarić L, Ikanović J, Rajičić V, Remiković M, Simić D (2019b) Effect of genotype on morphological and quality features of sugar beet. Agric For 65(2):29–38., Podgorica. ISSN: 0554–5579. https://doi.org/10.17707/AgricultForest.65.2.02

Bornscheuer E (1970) Influence of different crop densities on root and sugar yield in sugar beet cultivation without thinning, translated by Zucker, 23, p 657–662

Božović D, Živanović T, Popović V, Tatic M, Gospavić Z, Miloradović Z, Stankovic G, Đokić M (2018) Assessment stability of maize lines yield by GGE-biplot analysis. Genetika 50(3): 755–770

Božović D, Popović V, Rajičić V, Kostić M, Filipović V, Kolarić L, Ugrenović V, Spalević V (2020) Stability of the expression of the maize productivity parameters by AMMI models and GGE-biplot analysis. Notulae Botanicae Horti Agrobotanici Cluj-Napoca 48(3):1387–1397. https://doi.org/10.15835/nbha48312058

Brandeburg E (1931) Die Herz und Trockenfäule der Rüben als Bormangelerscheinung. Phytopathology 2(3):499–517

Brimhall PB, Chemberlain EW, Alley HP (1965) Competition of annual weeds and sugar beets. Weeds 13:33–35

Brooker RW, Bennett AE, Cong WF, Daniell TJ, George TS, Hallett PD, Hawes C, Iannetta PPM, Jones HG, Karley AJ, Li L, Mc Kenzie BM, Pakeman RJ, Paterson E, Schöb C, Shen J, Squire G, Watson CA, Zhang F, Zhang J, White PJ (2015) Improving intercropping: a synthesis of research in agronomy, plant physiology and ecology. New Phytol 206:107–117

Buchner A (1951) Zur Stickstoffdüngung der Zuckerrüben. Landwirtsch Forsch 3:7–16

Čačić N, Kovačev L, Mezei S, Sklener P (1997) Influence of genotype-environment interaction on production properties of sugar beet. Selection Seed Prod Novi Sad Serbia 4(1–2):127–134
Čačić N, Kovačev L, Mezei S, Sklenar P, Nagl N (2000) The process of maturing different varieties of sugar beet. Selection Seed Prod Novi Sad, Serbia 7(1–2):9–13
Campbell LG, Kern JJ (1982) Cultivar x environment interactions in sugar beet yield trials. Crop Sci 22:932–935
Ćurčić Ž (2014) Genetic divergence and combination abilities of multigerm sugar beet pollinators. Doctoral dissertation. Faculty of Agriculture, University of Novi Sad, Serbia
Dawson JH (1965) Competition between irrigated field beans and annual weeds. Weeds 12:206–208
Dobrovnaya OV, Tishalenko EN, Sakalo VD, Chungunkova TV, Lyalao II (2009) Use of biotechnical methods for increasing sugar content and tolerance of sugar beet to stress factors of environment collected scientific work of Nikit. Botan Gard 202–206
Dokić P, Milošević M, Boćanski J (1992) Biological valuesof sugar beet seed fractions. Proceedings of the Institute of Field and Vegetable Crops, Novi Sad, pp 20–25
Dragović S, Hadžić V, Panić Ž (1987) Sugar beet requirements for void and air in different agroecological production conditions. XXI Seminar of agronomists, Proceedings of the Institute of Field and Vegetable Crops of the Faculty of Agriculture, Novi Sad, p 355–340
Draycott AP (1972) Sugar-beet nutrition. Book. Applied Science Publishers, London
Draycott AP, Webb DP, Wright EM (1977) The effect of time of sowing and harvesting on growth, yield and nitrogen fertilizer requirement of sugar beet. J Agric Sci 81:267–275
Duffus JE, Ruppel EG (1993) Diseases in the sugarbeet crop. Chapman and Hall, London, pp 346–427
Durr I (1992) Analysis of the variability of the sugarbeet growth during the early growth stage II Factors influencing the seedlings-size in field conditions. Agronomic 12:527–535
FAO (2009) Food and Agriculture Organization of the United Nations. FAOSTAT Database. http://faostat.fao.org/. Accessed 12 June 2009
Filipović V, Glamočlija Đ, Jaćimović G (2007) Influence of vegetation space on yield and quality of different varieties of sugar beet. III Symposium with international participation Agro-innovations in Field and Vegetable Production. Belgrade, Serbia. 19–20.10. 2007, Proceedings, p 136–137
Filipović V, Glamočlija Đ, Radivojević S, Jaćimović G (2008) Influence of crop density on yield and quality of different varieties of sugar beet. Arch Agric Sci Belgrade 69(4):33–48
Filipović V, Glamoclija D, Radivojević S, Jaćimović G (2009) The influence of crop density and harvesting time on yield and quality of various sugar beet cultivars. Selekcija & semenarstvo, Belgrade 15:45–53
Fukai I, Trenbath BR (1993) Intercropping for the management of pests and diseases. Field Crop Res 34:381–405
Glamočlija Đ (1986) Influence of nitrogen and sowing density on photosynthesis productivity and sugar beet yield. Doctoral dissertation Faculty of Agriculture Belgrade – Zemun, p 1–24
Glamočlija Đ (1990) Influence of fertilization strength and crop density on yield and quality of sugar beet. INI "PKB Agroekonomik", IV Winter seminar of agronomists - farmers of Serbia", 5–9.02.1990.Vrnjacka Banja, p 148–158
Graf A, Müller HJ (1979) N-Dungung nur so viel wie netwending, Zukerribenbau, informacion. Östereich
Hadžić V (1984) Problems of compaction of agricultural land, Bulletin for control of soil fertility and use of fertilizers, Faculty of Agriculture, Institute of Field and Vegetable Crops, Novi Sad, p 156–163
Hajiboland R, Shekari S, Sadeghzadeh N, Poschenrieder C (2018) Sugar beet profits from intercropping with wheat both under optimum and deficient phosphorus supply. Acta Agric Slovenica 111(1):85–100. https://doi.org/10.14720/aas.2018.111.1.09
Hardy RWF, Eagleshaw ARJ (1995) Ecology and agricultural applications of nitrogen – fixing systems: overview. Kluwer Academic Publishers, Dordrecht, Boston, London

Holmes MR, Devine JR (1976) Nitrogen requirement of sugar beet. J Agric Sci (Camb) 87:559–566
Jabbar A, Ahmad R, Bhatti IH, Rehman A, Virk ZA, Vains SN (2010) Effect of different rice-based intercropping systems on rice grain yield and residual soil fertility. Pak J Bot 42(4):2339–2348
Jaćimović G, Marinković B, Crnobarac J (2006) Yield of refined sugar depending on the level of sugar beet fertilization. Mod Agric Pain 54(3–4):285–289
Jekić M (1974) Influence of various complex NPK fertilizers and the manner of their introduction on yield and percentage of sugar in sugar beet. Arch Agric Sci Belgrade 27(97):17–28
Jelešev RE (1969) The effect of changing certain forms of phosphorus fertilizers for sugar beet. Agricultural News. No. 1/69, 14
Kessel W, Schladen C (1984) Qualitat und Zuckerertrag – die Weichen Werdwn bereits im Fruhjaer gestellt. Die Zuckerrube 33Ig(2):68–73
Kolarić L, Popović V, Paunović J, Živanović L, Ikanović J, Sikora V (2015) Sugar beet yield and quality in the agroecological conditions of central Banat, Serbia. Agric For Podgorica 61(4): 33–41. https://doi.org/10.17707/AgricultForest.61.4.03
Kosanović M (1952) The effect of artificial fertilizers on increasing yields in chernozems. Arch Agric Sci 6:10–20
Kovačev L, Čačić N, Mezei S, Nagl N (2006) Breeding of sugar beet - previous results and further directions. III Symposium of Selection for Breeding the Society of Geneticists of Serbia IV Scientific Symposium on Selection and Seed Production of the Society of Breeders of Serbia. Zlatibor 16–20 May 2006, p 56–62
Kovačev L, Čačić N, Mezei S (2011) New sugar beet hybrid varieties developed at Institute of field and vegetable crops resistant to Rhizomania. Rat Pov Novi Sad, Serbia 48:83–86
Kuzevski J, Krstanović S, Demajo V, Kačarević A, Ivanović M, Krstić B (2000) Changes in the chemical composition of roots as an early indicator of sugar beet rhizomania. Proc Inst PKB Agroekonomik 6(1):85–95
Kuzevski J, Krstanović S, Šurlan-Momirović G, Živanović T, Jeličić Z (2008) The effect of mineral nutrition on the chemical composition of sugar beet root. Conventional and molecular breeding of field and vegetable crops. November 24–27, Novi Sad, p 72–82
Lakić Ž, Stanković S, Pavlović S, Krnjajic S, Popović V (2019a) Genetic variability in quantitative traits of field pea (*Pisum sativum* L.) genotypes. Czech J Genet Plant Breed 55(1):1–7
Lakić Ž, Antić M, Đurđić I, Popović V (2019b) Morphological characteristics of alfalfa genotypes tolerant of low soil pH. Genetika 51(3):907–922. https://doi.org/10.2298/GENSR1903907L
Ljubomirović D, Filipović V, Jovanović B (2006) Influence of different amounts of NPK fertilizers on chernozem properties in the period 1970-2004. year. Contemp Agric 55(5):119–124
Lovato A, Cagalli S (1992) Sugarbeet (*Beta vulgaris* L.) seed vigour compared in laboratory and field tests. Seed Sci Technol 21:61–67
Lüdecke N (1953) Zuckerrübenbau. Ther Ber:62–68
Ludecke N (1956) Zukerrubenbau. Book, Hamburg and Berlin
Ludecke N (1961) Über den Einfluss von klimatischenFaktoren auf das Wachstum und die Qälitat der Zukerrübenunter Berücksihitingung de Düngung. Zuckerindustrie 1
Lüdecke H, Müller A(1965) Die Düngung von Hackfrüchten, Zuckerrübe. HandbuchfürPflanzenernährung und Düngung, Springer – Verlag, 3 Band S. 282
Luthin NJ (1966) Drainage engineering. John Wiley and Sons, New York, pp 2–3
Manojlović S, Kastori R, Hadžić V (1989) Problems of soil, plant nutrition and fertilization in the production of quality and health safe food, SRS SAPV, Provincial Committee for Science and Informatics Novi Sad 26:20–25
Marinković B, Crnoborac J (1993) Technology of sugar beet production in production conditions in 1993. XXX Business Community for Industrial Better Novi Sad, p 44–51
Marinković B, Crnoborac J, Jaćimović G, JankovićS, MD, Polaček M (2001) Significance of organic and mineral fertilizers in sugar beet production. International Symposium Food in the 21st Century. November 14–17, 2001 Subotica. Scientific Institute of Field and Vegetable Crops Novi Sad

Martinović Lj, Stevanović D, Zdravković M (1999): Influence of perennial fertilizer application on chemical soil changes. XIII conference of agronomists, veterinarians and technologists Arandjelovac 5,1,33

Milford GFJ, Pocock TO, Riley J (1985) An analysis of leaf growth in sugar beet. I. Leaf appearance and expansion in relation to temperature under controlled conditions. Ann Appl Biol 106:163–172. https://doi.org/10.1111/j.1744-7348.1985.tb03106.x

Milford GFJ, Jarvis PJ, Walters C (2010) A vernalization-intensity model to predict bolting in sugar beet. J Agric Sci 148:127–137. https://doi.org/10.1017/S0021859609990323

Milošević R, Manojlović S, Jankulovski Ž, Obradović S, Popin J (1989) Optimization of sugar beet fertilization on Pannonian chernozem and similar lands in Vojvodina, Faculty of Agriculture, Institute of Field and Vegetable Crops, Novi Sad, p 20–22

Milošević N, Govedarica M, Kuzevski J, Jarak M, Krstanović S (1999) Application of microbial inoculants in sugar beet production. XIII Conference of Agronomists, Veterinarians and Technologists, Arandjelovac, Prosveta - Nis 5(1):141–142

Milovanović M (1984) Reaction of sugar beet to nitrogen nutrition and root extraction time. Master's thesis. Faculty of Agriculture. Belgrade – Zemun, Serbia

Ming MF (1989) Influence of nitrogen nutrition on yield and quality of sugar beet roots on different soil types. Doctoral dissertation. Faculty of Agriculture, Novi Sad, Serbia

Orlovski NI (1961) Fundamentals of sugar beet biology. Book, Belgrade, Serbia

Pejić B, Maksimović L, Milić S, Rajić M (2010) Influence of irrigation and nitrogen fertilization on yield and productivity of sugar beet. Modern Agric Novi Sad 59(1):45

Popovic V, Tatić M, Sikora V, Ikanovic J, Drazic G, Djukic V, Mihailovic B, Filipovic V, Dozet G, Jovanovic L, Stevanovic P (2016) Variability of yield and chemical composition in soybean genotypes grown under different agroecological conditions of Serbia. Roman Agric Res 33:29–39

Popović V, Vučković S, Jovović Z, Ljubičić N, Kostić M, Rakaščan N, Glamočlija-Mladenović M, Ikanović J (2020a) Genotype by year interaction effects on soybean morpho-productive traits and biogas production. Genetika 52(3):1055–1073

Popović V, Ljubičić N, Kostić M, Radulović M, Blagojević D, Ugrenovic V, Popovic D, Ivosevic B (2020b) Genotype x environment interaction for wheat yield traits suitable for selection in different seed priming conditions. Plan Theory 9(12):1804. https://doi.org/10.3390/plants9121804

Popović V, Jovović Z, Marjanović Jeromela A, Sikora V, Mikić S, Šarčević Todosijević Lj (2020c) Climatic change and Agricultural production, International GEA, Geo Eco Agro Conference Podgorica. http://www.gea.ucg.ac.me, p 160–166

Rajičić V, Popović V, Terzić D, Grčak D, Dugalić M, Mihailović A, Grčak M, Ugrenović V (2020b) Impact of lime and NPK fertilizers on yield and quality of oats on pseudogley soil and their valorisation. Notulae Botanicae Horti Agrobotanici Cluj-Napoca 48(4):1495–1503. https://doi.org/10.15835/nbha48312057

Rajičić V, Popovic V, Perišić V, Biberdžić M, Jovović Z, Gudžić N, Mihailović V, Čolić V, Đurić N, Terzić D (2020a) Impact of nitrogen and phosphorus on grain yield in winter triticale grown on degraded vertisol. Agronomy 10(6):757–767. https://doi.org/10.3390/agronomy10060757

Redfearn M (1995) A biochemical test for seed vigour. 58th Congress IIRB 133-137

Rossi V (2000) Cercospora leaf spot infection and resistance in sugarbeet (*Beta vulgaris* L.). Adv Sugar Beet Res IIRB 2:17–48

Rosso F, Candolo G (2001) Evaluation of sugar beet main quality features through the analysis of the diffusion juices produced by y pilot plant. 64. IIRB Congress, Bruges, Belgium, 26-27 Jun 64:437-442

Rožić R (1989) Some possibilities of rationalization and fertilization of sugar beet. Proc XXIV Conf Agron 16:233–245

Sabovljević R, Petrović S, Sheptovski V, Lekić S (1995) Germination variability of sugarbeet seeds in relation to different temperatures and the leaching procedures. 58th Congress IIRB, p 53–61

Sarić M (1971) Plant physiology, University of Novi Sad, Novi Sad Serbia p 328, 421, 427
Savić B (1989) Influence of basic tillage variants on sugar beet production. Proc Inst Field Vegetable Crops Novi Sad Serbia 16:269–279
Shestakova VA (1969) Influence of mineral fertilizers on some physiological - biochemical processes and sugar beet yield. Agricultural News. No. 4/69, 48
Smith GA, Martin SS, Ash KA (1977) Path coefficient analysis of sugarbeet purity components. Science 17:249–253
Spasić P (1987) Influence of production factors on yield height and quality of sugar beet in Vojvodina in 1986. Faculty of Agriculture, Institute of Field and Vegetable Crops, Novi Sad, Serbia
Spasić (1989) Planning the yield and quality of sugar beet under certain ecological conditions of production. Faculty of Agriculture, Institute of Field and Vegetable Crops, Novi Sad. Serbia, p 16
Spasić P (1992) Sugar beet. Monograph. Belgrade, Serbia
Stanaćev S (1963) Prolonged effect of basic processing for sugar beet on wheat yield in the following year. Contemp Agric 11:68–74
Stanaćev S (1973) Productivity of sugar beet in various two-field crop rotations, Proceedings of the Institute for Agricultural Research. Novi Sad. IX, 9, p 103–114
Stanaćev S (1976) Proposal of the production process for sugar beet in Vojvodina and the results of the latest tests of its basic elements. Improving sugar beet production in Vojvodina. Chamber of Commerce of Vojvodina, Novi Sad, Serbia, p 63–78
Stanaćev S (1979) Sugar beet. Nolit, Belgrade, pp 162–166
Stanaćev S, Milošević R, Stefanović D (1983) Influence of monoculture and various crop rotations on the yield of sugar beet. Proceedings of the scientific conference: "Man and Plant" Novi Sad, p 9–17
Stanaev S, Stefanovic D (1974) Influence of vegetational space and harvesting time on yield of beets and refined varieties sugar beets. Arch Field Sci 30(112):33–52
Stewens NB, Evans RC, Jabro JD, Iversen NM (2011) Sugar beet productivity as influenced by fertilizer band depth and nitrogen rate in strip tillage. J Sugar Beet Res 48(3–4):137
Štojko DA (1968) Orošeniezemledeliena Ukraine. Kiev, p 11–22
Stojšin D (2001) Biotechnology in Agriculture: Benefits and Risks. International Symposium Food in the 21st Century-Proceedings, November 14–17, 2001 Subotica. Scientific Institute of Field and Vegetable Crops Novi Sad. Serbia. p 11
Terzic D, Djekic V, Jevtic S, Popovic V, Jevtic A, Mijajlovic J, Jevtic A (2018) Effect of long-term fertilization on grain yield and yield components in winter triticale. J Anim Plant Sci 28(3): 830–836
Terzić D, Popović V, Malić N, Ikanović J, Rajičić V, Popović S, Lončar M, Lončarević V (2019) Effects of long-term fertilization on yield of side rates and organic matter content of soil in the process of recultivation. J Anim Plant Sci 29(3):790–795
Todorčić B (1974) The ratio of NPK nutrients and the interaction of these ratios with the yield of sugar beet. Economy, Osijek 1:13–20
Troncoso A, Cantos M (1990) Influence of nitrogen fertilization on sugar beet (*Beta vulgaria* L.) quality in an area of southern Spain. Plant nutrition – physio. and appl. Kluwer Acad Publ, pp 581–583
Trzebinski J (1974) Wplyw wysokich dawek azotu na sklad chemycrny burakow cukrowych, Buletyn institutu Hodawli I Aklimatyzaciji Rosein, p 314
Usmanikhail, Jamro GH, Oad FC, Hassan SW, Chachar QD, Khanzada MA, Gandahi AW (2012) Agronomic and economic effect of intercropping sugarbeet with oilseeds and lentil. Pak J Bot 44(6):1983–1988
Ustimenko AV, Vakomovskij I (1979) Vijanije pogodi na urožaj i saharistost svekli, Sah, p 8
Vlahović B, Stevanović S, Tomašević D, Zelenjak M (2006) Agrarian production in the Republic of Serbia. Novi Sad, Serbia, p 34–46

Vučić N (1973) Bioclimatic method for scheduling irrigation: Experiment with maize in Vojvodina, Yugoslavia. In: Ecological studies, analysis and synthesis. Berlin, Heidelberg, New York
Vučić N (1981) Irrigation and two harvests per year. Novi Sad, Serbia
Vučić N (1992) Irrigation of sugar beet. Sugar beet, Monograph, Yugosugar / Jugošećer. Serbia, p 381–398
Vučić N, Miladinović Ž (1964) Study of vegetation space and irrigation regime for sugar beet during furrow irrigation on carbonate chernozem of South Bačka. Agricultural Research, Novi Sad, p 4
Vukov K (1972) Physik und Chemie der Zuckerrube als Grundlage der Verarbeitungsverfahren, Budapest
Wicks GA, Wilson RG (1983) Control of weeds in sugarbeets (*Beta vulgaris*) with hand hoeing and herbicides. Weed Sci 31:493–499
Wicks GA, Wilson RG (1989) Control of weeds in sugarbeets (*Beta vulgaris*) with hand hoeing and herbicides. Weed Sci 31:493–499
Žeravica M (1965) Utilization and dynamics of nitrogen, phosphorus and potassium uptake in sugar beet on chernozem. Doctoral dissertation, Faculty of Agriculture. Novi Sad, Serbia
Živković M (1983) Pedology, Belgrade, University of Belgrade, Faculty of Agriculture, p 197, 282, 399, 525
Zocca A (1982) Results obtained by testing some genetic single-seeded varieties of sugar beet. Translation, Informator. 10

Agronomic Management of Sugar Beet 13

Swapan Kumar Paul

Abstract

Sugar beet (*Beta vulgaris* L.) roots contain a high concentration of sucrose and grown commercially for sugar production next to sugar cane. Sugar productivity from sugar beet is higher than sugar cane per unit of time due to its shorter growth period and higher sucrose concentration in juice. Sugar beet is a temperate crop but cultivating the subtropical environment in the winter season. Beet and sugar yields depend on varietal selection, sowing time, sowing methods, plant spacing, sowing depth, providing adequate crop nutrition, weed control, and irrigation schedule of sugar beet. The time of sowing of sugar beet is decided by the prevailing temperature of the farming areas. Low temperature with adequate soil moisture during sowing enhances seed germination and seedling establishment of sugar beet. November to April appears as the appropriate time for sugar beet growing in Bangladesh. Ridge or bed sowing with 50 cm × 20 cm spacing is suitable for crop growth and root development. Proper growth and beet yields can hamper from both shortage and excess-irrigation. Improvement of growth and yields can be obtained through providing a higher dose of N and K fertilization along with B depending on soil fertility. The collective use of manure and chemical fertilizers can be advantageous for sugar beet production. This chapter discusses agronomic practices of sugar beet notable highlighting on the subtropical environment.

Keywords

Management · Sugar beet · Quality · Yield

S. K. Paul (✉)
Department of Agronomy, Bangladesh Agricultural University, Mymensingh, Bangladesh
e-mail: skpaul@bau.edu.bd

V. Misra et al. (eds.), *Sugar Beet Cultivation, Management and Processing*,
https://doi.org/10.1007/978-981-19-2730-0_13

Abbreviation
TSS Total soluble solids

13.1 Introduction

Sugar beet (*Beta vulgaris* L.) is a member of the family Amaranthaceae (formerly Chenopodiaceae) which is reflected as the second vital sugar crop next to sugar cane and provides 30% of world sugar (Dohm et al. 2014). Sugar beet is generally considered a crop of temperate region and requires vernalization for its flowering (Biancardi et al. 2010). Now, this crop is successfully grown in subtropical countries during the winter season. Top fifteen sugar beet growing countries are Russian Federation, France, Germany, United States of America, Turkey, Poland, China, Egypt, Ukraine, United Kingdom, Iran, Belarus, Netherlands, Italy, and Belgium. Sugar beet is mainly produced in Europe, and, to a lesser extent, in Asia, and North America (Kumar and Pathak 2013). It was introduced in India in the 1950s (Pathak et al. 2014), in Pakistan during the mid-1960s (Zahoor et al. 2007), whereas it has recently been introduced into Bangladesh (Bithy et al. 2020). Sugar beet is a biennial grown from seeds. During the first growing season, it completes the vegetative growth while it requires the second year to produce flowers and seeds. The crop is commercially grown for sugar production because its taproot holds a high quantity of sucrose (15–17%). Sugar beet is a short-duration crop compared to sugar cane and requires 5–6 months for its maturity. Along with sugar, sugar beet offers valuable by-products like palatable foliage and molasses which are of value as livestock feed and bioethanol production. Because of its high chemical quality, beet molasses is a priced item with the potentialities for export. Sugar yield depends on root size and its sugar concentration. Climatic conditions, nutrients in the soil, water management, and to some point variety and spacing of planting principally influence the quick growth of the root and its steady value sugar concentration. This chapter discusses the agronomic management of sugar beet in the subtropical region in distinct reference to the Indian sub-continent and South East Asian countries.

13.2 Soil and Climate

Sugar beet is a versatile crop as it can survive various climatic and soil environments. It can be grown on most soils, except heavy clays with excessively wet, sticky, and very stony soils on which drilling and harvesting are also challenging. Sugar beets do not grow well on highly acidic soils. Soils having pH of 6.0–8.0 are convenient for its optimum growth but also grows well on soils at pH 9.5 or those affected on high salinity (Rahman et al. 2006). Soil compaction in the seedling stage restricts the deep root penetration in the soil and encourages forked roots with lower yields. Organic matter rich well-drained loam to sandy loam soil is influential for sugar beet (Rahman et al. 2006).

Environmental conditions such as temperature, solar radiation, and sunshine hours influence the growth of sugar beet. Crop emergence is a function of soil temperature. Early emergence occurred in sugar beet when air and soil temperature ranges at 15–25 °C (Khan 1992; Copeland and McDonald 2001). For seed germination, the suitable minimal temperature is 7–10 °C. After emergence, the vegetative growth along with root formation is largely influenced by air temperature and crop nutrition. An average temperature of about 20–22 °C is ideal for sugar beet growing and sugar accumulation in roots. During growing period, temperatures above 30 °C retard sugar accumulation (Brar et al. 2015). Sugar formation and transportation depend on climatic factors, viz. light, temperature, moisture, and day length, that act as external stimuli to accelerate sucrose accumulation in roots (Brouwer et al. 1976; Petkeviciene 2009). The crop requires some weeks at low temperature (near 4 °C) to encourage flowering.

13.3 Varietal Performance

The yield and quality performance of sugar beet depend on varietal selection. The yield is a quantitative trait of a cultivated species that is highly dependent on environmental factors and cultivars. Adapting climate change appropriate to variety selection is very crucial to maximizing crop yields. So, the successful farming of sugar beet under subtropical environment would not be possible without suitable variety selection. Due to hereditary characters or various genetic constituent variety shows different potentiality on their growth performance and yield-ability under same environment. Balakrishnan and Selvakumar (2009) stated that among the sugar beet hybrids (viz. Cauvery, Indus, and Shubhra), Cauvery produced higher yield while Shubhra showed higher brix in juice. Kaloi et al. (2019) found the higher beet yield and sugar recovery in SD-12970 and SD-PAK 07/07, respectively, in comparison to additional tested varieties. As sugar beet is a newly introduce crop in Bangladesh, the selection of suitable genotype is essential for its profitable cultivation as a promising sugar crop. Islam et al. (2012) conducted an experiment with fourteen tropical genotypes, viz. SB001, SB002, SB003, SB004, SB005, SB006, SB007, SB008, SB009, SB010, SB011, SB014, SB015, and SB016. Among the genotypes, SB001 (85.30 t ha^{-1}) gave the maximum beet yield which is closely followed by SB006 (84.40 t ha^{-1}) at 165 DAS. Brix (%) and sucrose content (%) were also varied due to genotype. Hossain et al. (2011) depicted that genotype EB-0809 produced the highest beet yield (89.74 t ha^{-1}) compared to other tested genotypes (viz. Shubhra, Cauvery, EB-0616, EB-0626, and EB-0809). Paul et al. (2018a) documented that PAC 60008 showed better performance (yield and% TSS) compared to Shubhra and Cauvery. Varietal performance of tropical sugar beet was stated elsewhere (Rahman et al. 2006; Refay 2010; BSRI (Bangladesh Sugarcane Research Institute) 2010; Hossain et al. 2011; Ahmad et al. 2012; Kaloi et al. 2014, 2019).

In Bangladesh, the feasibility of sugar beet is still investigational stage although some people are growing low sucrose-containing genotype for salad and vegetables.

Thus, suitable genotype selection is a prerequisite to promote sugar beet as a supplementary sugar-based crops scheduling in Bangladesh.

13.4 Sowing Time

The sowing date is an important factor in crop management affecting crop yield and other agronomic traits (Leilah et al. 2005; Ghonema 1998; Karbalaei et al. 2012). Sowing date influences the canopy through growth, numbers, size, and age of green leaves, and thereby affects the light intercepted by the plants throughout the growing period (Rinaldi and Vonella 2006). Many researchers have exposed that all quantity inherited traits can significantly vary depending upon environmental situations and cultivation practices such as sowing time (Caliskan et al. 1999; Ozel and Ozguven 2002; Salmasi et al. 2006). Sugar beet sown in too early and also too late reduce beet and sugar yields (Hocking and Stapper 2001; Robertson et al. 2004; Uzun et al. 2009). Late sowing in many crops resulting in lower yields was documented by previous studies (Hocking and Stapper 2001; Ozer 2003; Robertson et al. 2004; Uzun et al. 2009; Paul et al. 2020). Length of vegetative growth periods in addition to other production factors reflects on the sugar beet quantity and quality parameters (Badawi et al. 1995). If sugar beet sown early that prolong the growth period which is utmost vital influential reasons of its yield variations (Olesen et al. 1990; Karbalaei et al. 2012). Sowing date of tropical sugar beet varies with the environmental status of the area and the genotypes. Therefore, sowing time is the crucial factor affecting the sugar beet yield of a greater extent. Soil moisture status as well as the sowing technique influences the sugar beet sowing time (Romaneckas et al. 2009; Safina et al. 2012). Usually, the planting time of sugar beet at the individual region is decided by the prevailing temperature of the area. Ferdous et al. (2015) noticed that beet yield significantly decreased irrespective of genotypes with the advancement of sowing dates from 1 November onwards. A comparable tendency was depicted by Sohel (2016) who mentioned that early sowing on 30 October showed the best result which was at par with 10 November sowing than delayed sowing. Sugar beet sowing on 10 November gave the best yield than 20 November, 30 November, and 10 December (Hossain et al. 2011). It has been suggested that November to April is the best time for growing tropical sugar beet in Bangladesh.

13.5 Planting Methods

Sowing method influences the quantity and quality traits of sugar beet. Flatbed planting method is likely the seedbed preparation, whereby the topsoil is ploughed and leveled. In the ridge method, the topsoil is scrapped and concentrated in a defined region to deliberately raise the seedbed above the natural rain, which affects physical and chemical properties of soils, and also biological activities, and ultimately the sugar beet yield. A bed planting system has attained popularity for water-saving, and maximum yield in many crops (Connor et al. 2003). Ridge making is expensive and laborious than to sowing on flat and traditional knowledge of famers is enough to create a flatbed through minimum tillage (Zahoor et al. 2007). El-Maghraby et al. (2008) found the substantial increment of root length and diameter when sowing on a laser flattened soil with deep ploughing. Direct sowing of sugar beet on ridges was more suitable than transplanting seedlings on flatbed or on ridges, and the estimated yield of white sugar was also increased when sugar beet grown on ridges by direct seeding compared to transplanting (Garg and Srivastava 1985). El-Kassaby and Leilah (1992) documented that the higher diameter and root weight were recorded in planting beets on the one side of ridges with 70 cm × 30 cm spacing. Planting beets at both sides on ridges with 70 cm × 25 cm spacing produced the higher beet and sugar yields. Whereas, Bhullar et al. (2009) at Ludhiana concluded that the sowing method (flat and ridge) had no significant influence on the yield contributing characters, root and sugar yields of sugar beet in loamy soils indicating that both methods are likewise effective. Zahoor et al. (2007) and Ahmad et al. (2010) reported that the time of emergence, seedling establishment, and root number and yields are significantly influenced by method of planting. The planting pattern also consequently influences weed biomass production and beet yield and quality. Double-row spacing (inter-row spacing 62 cm) greatly reduced weed biomass and increased beet yield (Bayat et al. 2019).

13.6 Sowing Depth

Plant emergence is influenced mostly by soil temperature, moisture, and aeration plus physical impedance from the soil. Physical impedance relates to the sprouts move to emerge through the soil. Soil impedance is directly related to seedling emergence (Hagery and Royle 1978; Limede et al. 2018). Therefore, seed sowing in optimum depth ensures a good percentage of emergences. Romaneckas et al. (2009) and Ririe and Hills (1970) noticed that emergence declined when sowing depth increased, the maximum emergence found at 1 in. seeding depth. For the Rallye 590, as seeding depth increased with sugar beet germination reduced. The maximum emergence happened at seeding depth 1.25 cm, 53.4% to a low of 28.5% at the 5.0 cm seeding depth. For the John Deere 71 planter, sugar beet emergence was maximum and similar to 1.25 and 2.5 cm sowing depths. As sowing depth increased to 3.75 and 5.0 cm, sugar beet emergence decreased by 6.1 and 16.7%, respectively (Yonts et al. 1999). Plant seeds 1.00–1.25 in. deep placement resulted in maximum seedling emergence (Khan 2013).

13.7 Plant Density and Geometry

Planting density is the vital agronomic attribute which influences on light interception during photosynthesis. Plant density determines the amount and superiority of the sugar in roots. Optimum plant population offers adequate amount of nutrients, water, light, and thus increases the photosynthetic activity of leaves which enhance dry matter accumulation and higher root yield (Freckleton et al. 1999). Root yield and sucrose content of sugar beet are considerably affected by the planting density. According to Theurer and Saunders (1995), smooth root and beet yield was boosted if sugar beets were planted at the closer spacing (46 cm × 30 cm) of 71,760 plants ha^{-1}. In addition, sucrose percentage and sugar recovery were linearly declined with the decrease in plant population (Nassar 2001). Ramazan (2002) reported that beet yield and sugar (%) were maximum at planting density 45 cm × 20 cm (103,600 plants ha^{-1}) in comparison to 45 cm × 40 cm (555,000 plants ha^{-1}), 43 cm × 30 cm (73,000 plants ha^{-1}), and 45 cm × 25 cm (88,900 plants ha^{-1}). Sogut and Arioglu (2004) reported that with 45 cm inter-row spacing, narrow plant density either 5, 20, and 25 cm (i.e. 116,000, 95,000, and 81,000 plants ha^{-1}) gave higher root yield than 30 and 35 cm intra-row spacing (71,000, and 58,000 plants^{-1}, respectively). Ismail and Allam (2007) found that sugar beet planting 70,000 and 105,000 plants ha^{-1} gave higher yield and quality. Masri (2008) detected an encouraging result of increasing plant density from 87,500 to 100,000 plants ha^{-1} which significantly increases TSS, sucrose, purity, and sugar yield. Sadre et al. (2012) noted that 12 plants m^{-2} increased root and white sugar yields. Bhullar et al. (2010) noted that planting density of 100,000 plants ha^{-1} (50 cm × 20 cm) produced the highest root and sugar yields than 83,333 plants (60 cm × 20 cm) and 111,111 plants (60 cm × 15 cm) ha^{-1}. A similar observation was documented by Shukla and Awasthi (2013), and Paul et al. (2018a, b) who stated that spacing 50 cm × 20 cm (100,000 plants ha^{-1}) produced the highest beet yield. Sohel (2016) conducted a study with eight spacings, viz. 50 cm × 20 cm, 60 cm × 20 cm, 70 cm × 20 cm, 60 cm × 25 cm, 70 cm × 25 cm, 50 cm × 30 cm, 60 cm × 30 cm, and 70 cm × 30 cm, and noticed that wider spacing produced the maximum values plant^{-1} for all growth and yield attributes while at optimum plant population, the closer spacing (50 cm × 20 cm) produced the highest beet and sugar yields. It is predictable that the lower plant density reduces the sugar content as well as sugar yield because of increased impurity content such as amino nitrogen (Garcia and Bellido 1986).

13.8 Weed Management

Weed control is mandatory to achieve desirable yield and quality of sugar beet, as it is competitive. Most importantly, severe yield penalties can result from a failure to control weeds. The diverse weed flora tremendously declined beet yield, total soluble sugar (TSS %), sucrose (Pol %), and apparent purity (%) in beet juice over control (Seadh et al. 2013). There are 250 weed species recognized in sugar beet fields, and among them, 60 species detected as major damaging where 30% are grass and 70% are broad-leaved weeds (May and Wilson 2006; Bhadra et al. 2020). However, the weed management options in sugar beet depend on geographic location, planting date, weed type, labor cost, weeding equipment, irrigation, etc.

13.9 Cultural Control

13.9.1 Crop Rotation

The weed control through crop rotation schedule is imperative due to its lower cost, higher efficacy, and without or least environmental threat. Weeds are less disquieting if beets are planted succeeding crop rotation. Rotation allows the suppression of weed inhabitants, which may be tough to manage in sugar beets, like velvetleaf. Crop rotation can influence the stability of beet yield with quality through suppressing weed spectrum in the sugar beet field (Winkler et al. 2015; Götze 2017).

13.9.2 Cover Crop or Mulching

Cover crop can help suppress weed growth by altering the soil dynamics, becoming established and growing quicker than weeds, and by smothering weed seedlings. It can suppress weeds by limiting the water, space, light, and nutrients during their growth period. Commonly used cover crop species in sugar beet fields are mustard (*Sinapis alba* L.), Phacelia (*Phacelia tanacetifolia* Benth.), and radish (*Raphanus sativus* var. *oleiformis* (Stokes) Metzg.) (Petersen 2004). These fast growing cover crops reduced light intensity and sometimes added allelopathic properties into soils that can suppress the weeds in crop field (Kelton et al. 2012; Kunz et al. 2016).

13.9.3 Tillage

Tillage has been used since the start of agriculture to prepare the seedbed and reduce weeds that will compete with the crops. Weeds present in any given field will reflect the tillage management systems (Buhler 1995). Weed flora existing at sugar beet farms can be changed by decreasing tilling depth in soils during land preparation. Weed species configuration in sugar beet field differs due to different tillage system (conventional tillage, minimum tillage, and through drilling) of previous crops (Cioni et al. 1998). The consequence of tillage on weed flora configuration was not detected in annual weeds which seems problematic to control in sugar beet, while Polygonaceae, Gramineae, and perennials were preferred by minimum tillage. The minimum tillage could lead to encourage not only perennials and Gramineae but also the Compositeae weed species.

13.10 Mechanical Control

Weeds during their early growth stages have weak shoots and roots as such mechanical control can eliminate them before crop yields suffer (Bayat et al. 2019). Mechanical control removes weeds significantly by uprooting and chopping up the whole plant. The unexpected disseminate of perennial weeds, through splitting up and dispersal roots, rhizomes, stolons, and tubers can produce new ones (Cioni and Maines 2010). The efficacy of machine-driven farming is very imperative for the high efficacy of crop care. There are some tools (finger weeders or a torsion weeder) used in hoeing machines that eliminate weeds from the sugar beet rows (Petersen 2004).

13.11 Chemical Control

Chemical control is efficient and easily pertinent in crop fields which are frequently used to offer a vast range of weed control. The chemical method of weeds control is the dynamic ways of weed control in sugar beet farm. Application of non-selective herbicides before sugar beet emergence can control nearly all the visible weed flora in crop fields. The utmost general active herbicides useful so far for weed management in sugar beet are phenmedipham, metamitron, ethofumesate, desmedipham, triflusulfuron-methyl, lenacil, clopyralid, and chloredazone (May 2001; Deveikyte 2005; Wilson et al. 2005). Triflusulfuron-methyl is selective for broad leaf and grasses with low doses, whereas chloredazone is extensively used only for broad leaf weed control in sugar beet. Field trail showed that weed germination started 30-day after the application of chloredazone @ 1.3 kg ha^{-1} (Majidi et al. 2011). When sugar beets are grown without any weed control measure, then sugar yield decreases up to 95% (Petersen 2003). Herbicidal weed control is economically viable practice in the sugar beet because it augmented the sugar yield while hand weeding is costly methods with lower yields for damaging effect of standing crop. Pendimethalin @ 3.6 kg a.i. ha^{-1} significantly improved sugar yield (0.74 t ha^{-1}) by controlling 82% monocotyledonous and 56% dicotyledonous weeds with minor phytotoxicity on sugar beet crop (Yagoob et al. 2021).

13.12 Biological Weed Control

Biological weed control is an environmentally sound and effective means of suppression weeds using macrobial and microbial organisms. It has gained a particular and attention worldwide as its environmentally friendly behavior (Scavo and Mauromicale 2020). Using insects is conventional biological control organisms but some microorganisms, nematodes, and mites are also used to manage weeds in crop fields. More than 300-insect species have been familiarized for the controlling of weed (Cioni and Maines 2010). Microorganisms like bacteria, fungi, and viruses are also defend weeds in crop fields. Smith (1986) found that fungal pathogens *Colletotrichum gloeosporioides* spp. *aeschynomene* can control *Aeschynomene virginica* in rice and soybean fields. Some fungal pathogens also show potentiality to control *Abutilon theophrasti, Chenopodium album, Datura stramonium, Echinochloa crus-galli,* and *Sorghum halepense* in sugar beet fields (Cioni and Maines 2010). In sugar beet fields to control weeds, currently no biological control or natural phytotoxin statistics is used. However, further research should be taken to test the efficiency of different microorganisms to weed control in sugar beet fields.

13.13 Nutrient Management

Fertilizer provides macro and micronutrients to crop plants. Fertilization, a vital practice is related with the effective use of nutrients for sugar beet cultivation. Macronutrients along with micronutrients especially zinc and boron may involve in the improvement of yield and juice quality of tropical sugar beet (Bairagi et al. 2013; Kashem et al. 2015; Paul et al. 2018a, b). If scarcity of nutrients cannot be amended by the soil applications, foliar spray can be the alternative way of nutrient supplementation (Sarkar et al. 2007). Nitrogen is the highest tested macronutrient for sugar beet due to its direct connection to root yield and quality (Loomis and Conor 1992). Geypens et al. (1998) reported that fertilization of higher nitrogen rate more than the recommended dose (80 kg N ha^{-1}) increased root yield but decreased sugar content, while Paul et al. (2018a) noted that a higher rate of nitrogen increased beet yield without significant hampering TSS (%) in the juice. Phosphorus is a second-most limiting nutrient in comparison to nitrogen for sugar beet cultivation (Malhotra et al. 2018). It is directly involved in the rapid initial root development and uptake of various nutrients in plants. Phosphorus assistances in the transmission of energy inside the plant cells and additionally, it controls the basic integrity of the cell membrane (Ahmad et al. 2017). Madani et al. (2014) noted that phosphorous fertilizations significantly increased beet yield @ 0.98 kg/m^2 when phosphorous applied as $(NH_4)_3PO_4$ @ 375 kg ha^{-1}. Adding phosphorus in soils augmented beet yield by 37 and 47% over the control (Hussain et al. 2014). Ghaly et al. (2019) depicted that P fertilization enhanced beet yield and juice quality, and the optimal P dose for effectual sugar beet husbandry is 48 kg P_2O_5 fed^{-1}. Sugar beet is well recognized as a high potassium-requiring crop (Johanson et al. 1971). Potassium applied @ 100 kg K_2O ha^{-1}augmented yield components, beet and sugar yield over control under various irrigation regimes (Mehrandish et al. 2012). Potassium application @ 180 kg ha^{-1} gave the higher root and sugar yields compared to lower doses (Kashem et al. 2015). The sulfur requirement in sugar beet is relatively low in comparison to cereals and oil-seed crops (Syers et al. 1987) while sulfur helps to proper leaf and root growth (Thomas et al. 2003). The sulfur @ 25 kg ha^{-1}increased beet yield by 25% along with improvement of dry matter addition in plants (Thomas et al. 2003). Sulfur @ 46 kg ha^{-1} along with other regular nutrients endures superior root, and sugar yield was established by Zengin et al. (2009).

Zinc is the vital micronutrient that triggers in various types of enzyme systems involved in not only protein synthesis but also in carbohydrate metabolism (Madani et al. 2014). Piskin (2017) opined that zinc fertilization @ 5 kg ha^{-1}enhanced beet yield (4.62–6.97%), sugar in juice (2.09–5.75%), white sugar (2.60–8.03%), and ultimately sugar yield (7.84–13.06%). Spraying Zn in the foliage of sugar beet plants influencing beet yield was pronounced by many researchers (Nemeat-Alla and El-Geddawy 2001; Enan 2004; Menisy 2009). Sugar beet is generally considered to have comparatively high necessities of boron to other crops (Tlili et al. 2018). Boron is mainly associated with sugar production and translocation from growing parts to developing sugar beet roots (Barker and Pilbeam 2007). Armin and Asgharipour (2012) noted that spraying boron as 12% H_3BO_3 improved beet yield and sucrose (%) in juice. Boron application on foliage of sugar beet significantly responses to vegetative characters and subsequently rise in sugar, juice purity, and crude protein (Soliman et al. 2014). Boron @ 1.5 kg ha^{-1}applied in soil augmented beet yield, sucrose (%) and inhibited crown rot occurrence in tropical sugar beet (Islam et al. 2015). Bithy et al. (2020) noticed that B@150 ppm applied thrice (40, 65, and 90 days after emergence) in foliage is influential for beet yield and juice quality improvement, and suppressive of crown rot prevalence in sugar beet.

13.14 Irrigation

Sugar beet is sensitive to water stress for emergence and seedling growth. Flood irrigation just after seed sowing is beneficial to encourage maximum seed germination and seedling establishment. Light irrigations (20–25 days interval) are beneficial in early growth stage, thus lessen crust development on the soil and decline the salinity of the top soil. The crop requires about 6–10 irrigations depending on the soil moisture level. Excess watering at early growth stage may hinder leaf growth and root development. Water scarcities in the intermediate stage of the growth stage (vegetative and root formation) tend to disturb sugar yields. An ample supply of water in the advanced stage of growth has badly influence on sugar percentage though it can upsurge the root bulk. Water scarcities with nitrogen deficit in the later stage of growth lead to hampered root growth with increased sucrose (%). Generally, top growth toward later growth stage is negatively associated with sugar recovery. Sugar beet crop is commonly necessary to irrigate 100–200 mm water per year (Morillo-Velarde 2011). Water supply should be dropped at least 2–4 weeks before harvest and excess water suggested to be drained out from the field as early as possible.

13.15 Harvesting

Time of harvest is very influential for sugar beet yield and sugar recovery from roots. The root dry matter accumulation upsurges with advancement of growth period of sugar beet and sugar extents by 20–26% during harvesting. The early sowing with prolonged growth period in delayed harvesting influences beet yield (Hemayati et al. 2012). Late harvesting results in higher root and sugar yields compared to early harvesting (Bürcky and Winner 1986; Heidari et al. 2008; Lauer 1995). Jaggard and Scott (1999) and Bürcky and Winner (1986) opined that delay harvesting resulted in better sugar yield under rainfed and cold weather for sugar beet. Although delay harvest improved beet yield in tropical sugar beet but decreased brix (%) content in the juice (Paul et al. 2019). It might be increased soil moisture during the beet harvesting stage in *krarif-1* season because of frequent rain. A similar observation of brix (%) was also reported by BSRI (Bangladesh Sugarcane Research Institute) (2013). Jozefyová et al. (2004) observed the harvesting time on root and sugar yields, and enhancement of root and sugar yields was found by the delay harvest (27 days) than early harvest. Interestingly, Kerr and Leaman (1997) reported an increased yield under the irrigated conditions from first to the last harvest. Sometimes delay harvest later than optimum time leads to decline root yield and total soluble sugar (%) of beet (Islam et al. 2012). The yield reduction in later harvest might be attributed due to leaf dropping and dehydration of the roots. Therefore, harvesting time affects the beet yield and sucrose content that might be depending on some environmental conditions, irrigation facilities, and some unknown factors but optimum maturity is the key factor for better yield of sugar beet.

13.16 Stress Management

Stress in plants refers to external conditions that adversely affect the growth, development, or production of plants. Drought is an important environmental factor with a reverse impact on crop productivity. Drought stress has become the major limiting factor of crop yield worldwide because of environmental changes in dry areas as well as in temperate zones (Yordanov et al. 2012). Plants show a various physiological and biochemical changes at cellular and whole plant under drought stressed. Due to drought stress, often leaf wilting happens for high evaporation although it has deep tap root (Clarke et al. 1993). Succeeding stomata closing can lessen the leaf water potential, thus preserving water uptake, photosynthesis, and possible growth (Clarke et al. 1993). Commonly in dry soils, shoot growth is faster than root, thus increases the dry matter ratio of root-shoot as a reaction to moisture stress (Marschner 1995). Boriŝev et al. (2016) suggested that a nanoparticle fullerenol (FNPs, molecular formula C_{60} $(OH)_{24}$) can support to relieve the drought stress by helping as an extra intercellular water supply. Besides drought, and heat, other abiotic stresses that concern sugar beet farmers worldwide are salinity and cold. Sugar beet is although a salt-tolerant glycophyte crops but as sensitive during germination and establishing stages. Salinity increased seedling emergence time; this outcome was slightly more pronounced at high temperatures and sown under cool conditions could be overcome such stress condition (Mahmoud and Hill 1981). Sugar beet can be able to tolerate salinity by having osmotic regulation system and accumulation of Na^+ and Cl^- in their vacuoles and cytoplasm (Subbarao et al. 2001; Ghoulam et al. 2002). Commonly, crop abiotic stresses can be minimized to some level by alterations in agronomy, suitable varietal selection, and soil and water management strategies.

13.17 Future Prospects

Sugar beet is a new crop in Bangladesh suitable for sugar, brown sugar, and ethanol production. The selection of suitable genotypes and management practices is very important for its commercial cultivation. Compared with sugar cane, farmers will be interested to choose sugar beet cultivation due to its short duration and high sugar. In Bangladesh, agricultural land uses are very poor in coastal areas, because of salinity in the winter season. As sugar beet is a salt-tolerant crop, its cultivation will be expanded in coastal saline-prone areas. This chapter will provide a valuable background for future sugar beet cultivation and expansion as a profitable industrial crop for farmers of Bangladesh as well as other subtropical countries.

13.18 Conclusion

Most of the available sugar beet genotypes can be grown under temperate conditions. However, the selection of suitable genotypes is the vital factor to grow sugar beet in subtropical environments especially in India and Bangladesh. *Sugar beet* ensues faster germination with soil and air *temperature* 15–25 °C, therefore sowing date should be *adjusted accordingly* to *match* the temperature. Considering temperature, November to April is the appropriate time for sugar beet growing under subtropical climate. Ridge or bed sowing and adequate moisture are preferable to get maximum seed germination and proper seedling growth. Among the management practices—planting density, proper fertilization, irrigation, and harvesting time are the critical to yield maximization for sugar beet. Plant population around 100,000 ha^{-1}produced the highest beet yield. Sugar beet requires higher macro- and micronutrients especially nitrogen, potash, and boron. This crop faces mostly water stress which can be managed by frequent irrigation depending on moisture status in crop fields. Therefore, proper agronomic practices should be maintained for sustainable sugar beet production.

References

Ahmad Z, Shah P, Kakar KM, El-Sharkawi H, Gama PBS, Khan EA, Honna T, Yamamoto S (2010) Sugar beet (*Beta vulgaris* L.) response to different planting methods, and row geometries II: effect on plant growth and quality. J Food Agric Environ 8:785–791

Ahmad S, Zubair M, Iqbal N, Cheema NM, Mahmood K (2012) Evaluation of sugar beet hybrid varieties under Thal-Kumbi soil series of Pakistan. Int J Agric Biol 14:605–608

Ahmad I, Ahmad B, Ali S, Kamran M, Fang HQ, Bilegjargal B (2017) Nutrients management strategies to improve yield and quality of sugar beet in semi-arid regions. J Plant Nutr 40(15): 2109–2115

Armin M, Asgharipour M (2012) Effect of time and concentration of boron foliar application on yield and quality of sugar beet. Am-Eurasian J Agric Environ Sci 12:444–448

Badawi MA, El-Agroudy MA, Attia AN (1995) Effect of planting date and N P K fertilization on growth and yield of sugar beet (*Beta vulgaris* L.). J Agric Sci Mansoura Univ 20(6):2683–2689

Bairagi A, Paul SK, Kader MA, Hossain MS (2013) Yield of tropical sugar beet as influenced by variety and rate of fertilizer application. Pakistan Sugar J 28(4):13–20

Balakrishnan A, Selvakumar T (2009) Evaluation of suitable tropical sugar beet hybrids with optimum time of sowing. Sugar Tech 11(1):65–68

Barker AV, Pilbeam DT (2007) Handbook of Plant Nutrition (Books in Soils, Plants and the Environment). CRC Press, Boca Raton, FL, p 773

Bayat M, Kavhiza N, Orujov E, Zargar M, Akhrarov M, Temewei AG (2019) Integrated weed control methods utilizing planting pattern in sugar beet. Res Crops 20(2):413–418

Bhadra T, Mahapatra CK, Paul SK (2020) Weed management in sugar beet: a review. Fund Appl Agric 5(2):147–156

Bhullar MS, Uppal SK, Kapur ML (2009) Effect of agronomic practices, and varieties on productivity of sugar beet (*Beta vulgaris* L.) in semi-arid region of Punjab. J Res Punjab Agric Univ 46: 6–8

Bhullar MS, Uppal SK, Kapur ML (2010) Influence of planting density, and nitrogen dose on root and sugar yield of beet (*Beta vulgaris* L.) under sub-tropical semi-arid conditions of Punjab. J Res 47:14–17

Biancardi E, JM MG, Panella LW, Lewellen RT, Stevanato P (2010) Sugar beet. In: Bradshow J (ed) Handbook of Plant Breeding, Tuber and Root Crops, vol 4. Springer, New York, pp 173–219

Bithy S, Paul SK, Kader MA, Sarkar SK, Mahapatra CK, Islam AKMR (2020) Foliar application of boron boosts the performance of tropical sugar Beet. J Bangladesh Agric Univ 18(3):537–544

Borišev M, Borišev I, Županski M, Arsenov D, Pajević S, Ćurčić Ž, Vasin J, Djordjevic A (2016) Drought impact is alleviated in sugar beets (Beta vulgaris L.) by foliar application of fullerenol nanoparticles. PLoS One 11(11):e0166248

Brar NS, Dhillon BS, Saini KS, Sharma PK (2015) Agronomy of sugarbeet cultivation-a review. Agric Rev 36(3):184–197

Brouwer W, Stählin I, Caesar K (1976) In: Brouwer W (ed) Handbuch des speziellen Pflanzenbaues, vol 2. Verlag Paul Parey, Berlin, Hamburg, pp 188–368

BSRI (Bangladesh Sugarcane Research Institute) (2010) Annual report of 2010. Ishurdi, Pabna, Bangladesh. pp. 1–7

BSRI (Bangladesh Sugarcane Research Institute) (2013) Annual report of 2012, 64–70. Ishurdi, Pabna, Bangladesh

Buhler D (1995) Influence of tillage systems on weed population dynamics, and management in corn, and soybean in the Central USA. Crop Sci 35(5):1247–1258

Bürcky K, Winner C (1986) The effect of plant population on yield and quality of sugar beet at different harvesting date. J Agron Crop Sci 157(4):264–272

Caliskan ME, Isler N, Gunel E, Guler MB (1999) The effects of sowing date, and row spacing on yield, and quality of some sugar beet (*Beta vulgaris* L.) varieties in Hatay ecological conditions. Turk J Agric For 23:1119–1124

Cioni F, Guizelis A, Meriggi P, Vicari A, Villarias JL (1998) Role de la betterave a sucre dans les strategiesde controle des mauvaises herbes dans lesassolements des regions de la mediterranee. In: Proceedings of the 61 IIRB Congress, Brussels

Cioni F, Maines G (2010) Weed control in sugar beet. Sugar Tech 12(3–4):243–255

Clarke NA, Hetschkun H, Jones C, Boswell E, Marfaing H (1993) Identification of stress tolerance traits in sugar beet. In: Jackson MB, Black CR (eds) Interaction stresses on plants in a changing climate. Springer, Berlin, pp 511–524

Connor DJ, Gupta RK, Hobbs PR, Sayre KD (2003) Bed planting in rice-wheat system. In: Addressing resource conservation issues in rice-wheat system of south Asia: a resource book. Rice-wheat consortium for the Indo Gangetic Plains. International Maize, and wheat Improvement Center, New Delhi, India, pp 103–108

Copeland LO, McDonald MB (2001) Seed Science and Technology. Kluwer Academic Publishers, Boston, Norwell, Massachusetts, pp 72–124

Deveikyte I (2005) Sensitivity of *Tripleurospermumperforatum* and *Chenopodium Album* on low rates of phenmedipham, desmedipham, etofumesate, metamitron andchloridazon. LucrariStiintifice, Universitatea de StiinteAgricole Si MedicinaVeterinara "Ion Ionescu de la Brad" Iasi. SeriaAgronomie 48:386–392
Dohm JC, Minochle AE, Holtgrawe D, Capella GS, Zakrzewski F, Tafer H, Rupp O, Sorensen TR, Stracke R, Reinhardt R, Goesmann A, Kraft T, Schulz B, Stadler PF, Schmidt T, Gabaldon T, Lehrach H, Weisshaar B, Himmelbauer H (2014) The genome of the recently domesticated crop plant sugar beet (*Beta vulgaris*). Nature 505(7484):546–549
El-Kassaby AT, Leilah AA (1992) Influence of plant density and nitrogen fertilizer levels on sugar beet productivity. In: Proc 5th Conf Agron, Zagazig, 13–15 September, vol. 2, pp. 954–962
El-Maghraby SS, Gomaa MA, Rehab IF, Hassan HMS (2008) Response of sugar beet to some mechanical management practices, irrigation and plant densities. Sugar Tech 10(3):219–226
Enan SAAM (2004) Effect of transplanting and soil application of boron and zinc on yield, and quality of sugar beet. Ph.D., Al-Azhar University, Egypt
Ferdous HM, Khaliq QA, Karim A (2015) Effect of sowing date on growth and yield of tropical sugar beet. Int J Agron Agric Res 7(1):53–60
Freckleton RP, Watkinson AR, Webb DJ, Thomas TH (1999) Yield of sugar beet in relation to weather and nutrients. Agric For Meteorol 93:39–51
Garcia JEC, Bellido LL (1986) Growth and yield performance of autumn-sown sugar beet: effects of sowing time, plant density and cultivar. Field Crop Res 14:1–14
Garg VK, Srivastava SC (1985) Evaluation of techniques suitable for growing sugar beet on sodic soils. J Indian Soc Soil Sci 33:884–887
Geypens M, Vanongeva JL, Elstamd P, Bries J (1998) Evaluation of nitrogen fertilizer recommendations for sugar beet on the nitrogen-index expert system. Commun Soil Sci Plant Anal 29(11–14):2217–2225
Ghaly FA, Abd-Hady MR, Abd-Elhamied AS, Mosallm MEA (2019) Effect of varieties, phosphorus and boron fertilization on sugar beet yield and its quality. J Soil Sci Agric Eng Mansoura Univ 10(2):115–122
Ghonema MH (1998) Effect of planting dates and harvesting time on yield, yield components and quality of sugar beet (Beta vulgaris, L.). J Agric Sci Mansoura Univ 23(7):2971–2979
Ghoulam C, Foursy A, Fares K (2002) Effects of salt stress on growth, inorganic ions and proline accumulation in relation to osmotic adjustment in five sugar beet cultivars. Environ Exp Bot 47(1):39–50
Götze P (2017) Impact of specialized sugar beet crop rotations on soil fertility parameters and on yield, and yield stability of sugar beet. Ph.D. Thesis, University of Halle Wittenberg, Wittenberg, pp. 9–11
Hagery TW, Royle SM (1978) Soil impedance as a factor reducing seedling emergence and its relation to soil conditions at sowing and to applied water. J Appl Ecol 15:897–904
Heidari G, Sohrabi Y, Esmailpoor B (2008) Influence of harvesting time on yield, and yield components of sugar beet. J Agric For Soc Sci 4(2):69–73
Hemayati SS, Shirzadi M, Aghaeezadeh M, Taleghani DF, Javaheri MA, Aliasghari A (2012) Evaluation of sowing and harvesting date effects on yield and quality of five sugar beet cultivars in Jiroft region (autumn planting). J Sugar Beet 28(1):13–21
Hocking PJ, Stapper M (2001) Effect of sowing time and nitrogen fertiliser on canola and wheat, and nitrogen fertiliser on Indian mustard. I. Dry matter production, grain yield, and yield components. Aust J Agric Res 52:623–634
Hossain MM, Kader MA, Kashem MA (2011) Effect of sowing date and variety on the yield of tropical sugar beet (*Beta vulgaris*). Bangladesh Agron J 14(1 & 2):95–101
Hussain Z, Khattak RA, Irshad M, Mahmood Q (2014) Sugar beet (*Beta vulgaris* L.) response to diammonium phosphate and potassium sulphate under saline–sodic condition. Soil Use Manag 30:320–327
Islam MS, Ahmed S, Uddin MN, Sattar MA (2012) Evaluation of tropical sugar beet (*Beta vulgaris* L) genotypes under Bangladesh condition. Bangladesh J Agric Res 37(4):721–728

Islam MS, Hossain GMA, Islam S, Alam KM, Mitu AS, Rahman MS, Haque MA (2015) Influence of Boron on yield, juice quality and crown rot incidence of tropical sugar beet cultivar 'Shubhra' grown at two AEZs of Bangladesh. Bangladesh J Sugarcane 36:1–8

Ismail AMA, Allam SM (2007) Yield and technological traits of sugar beet as affected by planting density, phosphorus and potassium fertilization. Proc The 3rd Conf of sustain. Agric Develop Fac Agric, Fayoum Univ, pp. 15–28

Jaggard KW, Scott K (1999) The origins of yield improvement in the national sugar beet crop science 1970. Br Sugar Beet Rev 67:7–11

Johanson RT, John TA, Geore ER, George RH (1971) Advances in sugar beet production: principles and practices. The Iowa State University Press, Ames, IA

Jozefyová L, Pulkrábek J, Urban J (2004) Effect of harvest time on sugar beet fertilized with increased nitrogen. J Food Agric Environ 2:232–237

Kaloi GM, Mari AH, Zubair M, Panhwar RN, Bughio N, Junejo S, Unar GS, Bhutto MA (2014) Performance of exotic sugar beet varieties under agro-climatic conditions of lower Sindh. J Anim Plant Sci 24(4):1135–1140

Kaloi GM, Mari AH, Bughio N, Junejo S, Panhwar RN, Bhutto MA, Rajput MA, Arain S (2019) Growth assessment of exotic sugar beet varieties in Southern-zone of Sindh. Pakistan Sugar J 34(2):33–40

Karbalaei S, Mehraban A, Mobasser HR, Bitarafan Z (2012) Sowing date and transplant root size effects on transplanted sugar beet in spring planting. Ann Biol Res 3(7):3474–3478

Kashem MN, Khaliq QA, Karim MA, Karim AJMS, Islam MR (2015) Effect of nitrogen and potassium on dry matter production and yield in tropical sugar beet in Bangladesh. Pakistan Sugar J 30(2):6–14

Kelton J, Price AJ, Mosjidis J (2012) Allelopathic weed suppression through the use of cover crops. In: Price AJ (ed) Weed Control. INTECH Open Access Publisher, Rijeka

Kerr S, Leaman M (1997) To water or not. Br Sugar Beet Rev 65:11–13

Khan AA (1992) Pre-plant physiological seed conditioning. Hortic Rev 13:131–166

Khan M (2013) Sugar beet production guide. NDSU extension service, North Dakota State University of Agriculture and Applied Science, and U.S. Department of Agriculture cooperating. Duane Hauck, Director, Fargo, North Dakota, p. 9

Kumar R, Pathak AD (2013) Recent trend of sugar beet in world. In: Souvenir-IISR-Industry Interface on Research, and Development Initiatives for Sugar beet in India, 28–29 May, Sugar beet Breeding Outpost of IISR IVRI Campus, Mukteswar-263138, Nainital. Organised by Indian Institute of Sugarcane Research (ICAR) and Association of Sugarcane Technologists of India, pp. 46–47

Kunz C, Sturm DJ, Varnholt D, Walker F, Gerhards R (2016) Allelopathic effects and weed suppressive ability of cover crops. Plant Soil Environ 62:60–66

Lauer JG (1995) Plant density and nitrogen rate effects on sugar beet yield, and quality early in harvest. Agron J 87:586–591

Leilah AA, Badawi MA, Said EM, Ghonema MH, Abdou MAE (2005) Effect of planting dates, plant population and nitrogen fertilization on sugar beet productivity under the newly reclaimed sandy soils in Egypt. Sci J King Faisal Univ 6(1):95–110

Limede AC, da Silva Oliveira CE, Zoz A, Zuffo AM, Steiner F, Zoz T (2018) Effects of seed size and sowing depth in the emergence and morphophysiological development of soybean cultivated in sandy texture soil. Aust J Crop Sci 12(01):93–98

Loomis RS, Conor DJ (1992) Nitrogen processes. In: Loomis RS, Conor DJ (eds) Crop Ecology: Productivity and Management in Agricultural Systems. Cambridge University Press, Cambridge, UK, pp 195–223

Madani H, Borji S, Sajedi NA (2014) Effects of Zinc and Phosphorus fertilizers on Sugar Beet (*Beta vulgaris* cv. SBS I005 Crouse) yield in Iranian high Zinc alkaline soil condition. Sci Papers Ser A Agron 57:240–245

Mahmoud EA, Hill MJ (1981) Salt tolerance of sugar beet at various temperatures. N Z J Agric Res 24(98):67–71

Majidi M, Heidari G, Mohammadi K (2011) Management of broad leaved weeds by combination of herbicides in sugar beet production 1: management of broad leaved weeds by combination of herbicides in sugar beet production. Adv Environ Biol 5(10):3302–3306

Malhotra H, Ana V, Sharma S, Pandey R (2018) Phosphorus nutrition: Plant growth in response to deficiency and excess. In: Plant nutrients and abiotic stress tolerance. Springer Nature Singapore Pte Ltd, Singapore, pp 171–190

Marschner H (1995) Mineral nutrition of higher plants, 2nd edn. Academic Press, London, p 889

Masri MI (2008) Effect of nitrogen level and planting density on sugar beet yield and its attributes. Egypt J Agron 30:119–136

May M (2001) Crop protection in sugar beet. Pesticide Outlook 12:188–191

May JM, Wilson RG (2006) Weed and weed control. In: Draycott AP (ed) Sugar beet. Blackwell, London, UK, pp 359–386

Mehrandish M, Moeini MJ, Armin M (2012) Sugar beet (*Beta vulgaris* L.) response to potassium application under full and deficit irrigation. Eur J Exp Biol 2(6):2113–2119

Menisy MGA (2009) Response of sugar beet to nitrogen fertilization rates and foliar application of zinc and boron under newly reclaimed soils at Fayoum. Ph.D. Fac. Agric. Fayoum, Fayoum University, Egypt

Morillo-Velarde R (2011) Water management in sugar beet. Sugar Tech 12(3):299–304

Nassar AM (2001) Effect of plant density on the productivity of some sugar beet varieties. J Agric Sci Mansoura Univ 26(12):7533–7546

Nemeat-Alla EAE, El-Geddawy IHM (2001) Response of sugar beet to foliar spraying time with micronutrients under different levels of nitrogen and phosphorus fertilization. J Agric Res 27(4): 670–681

Olesen JE, Fenhann J, Larsen H, Mackenzie GA, Rasmussen B (1990) Assessing impact of climatic change on crop production using growth models. In: Environmental models: emissions and consequences, pp 139–147

Ozel A, Ozguven M (2002) Effect of different planting times on essential oil components of different mint (*Mentha* spp.) varieties. Turk J Agric For 26:289–294

Ozer H (2003) Sowing date and nitrogen rate effects on growth, yield and yield components of two summer rapeseed cultivars. Eur J Agron 19:453–463

Pathak AD, Kapur R, Solomon S, Kumar R, Srivastava S, Singh PR (2014) Sugar beet: a historical perspective in Indian context. Sugar Tech 16(2):125–132

Paul SK, Paul U, Sarkar MAR, Hossain MS (2018a) Yield and quality of tropical sugar beet as influenced by variety, spacing and fertilizer application. Sugar Tech 20(2):175–181

Paul SK, Paul SC, Sarkar MAR, Hossain MS (2018b) Influence of integrated nutrient management on the growth yield and sugar content of tropical sugar beet (*Beta vulgaris* L.). Pakistan Sugar J 33(2):11–22

Paul SK, Joni RA, Sarkar MAR, Hossain MS, Paul SC (2019) Performance of tropical sugar beet (*Beta vulgaris* L.) as influenced by date of harvesting. Arch Agric Env Sci 4(1):19–26

Paul SK, Nila NY, Sarkar MAR (2020) Grain yield and quality of aromatic *Boro* rice (cv. BRRI dhan50) subject to date of transplanting and nutrient management. Thai J Agric Sci 53(2):85–96

Petersen J (2003) A review on weed control in sugar beet: from tolerance zero to period threshold. In: Derjit (ed) Weed biology and Management. Kluwer Academic Publishers, Dordrecht, pp 467–483

Petersen J (2004) A review on weed control in sugar beet. In: Inderjit (ed) Weed Biology, and Management. Kluwer Academic Publishers, pp 467–483

Petkeviciene B (2009) The effects of climate factors on sugar beet early sowing timing. Agron Res 7:436–443

Piskin A (2017) Effect of Zinc applied together with compound fertilizer on yield, and quality of sugar beet (*Beta vulgaris* L.). J Plant Nutr 40(18):2521–2531

Rahman MK, Kabir LM, Alam JM, Hossain SM, Islam RAKM (2006) Sugar beet cultivation in Bangladesh. Sugarcane Research Institute, Ishurdi, Pabna, Bangladesh. BSRI, Bangladesh, pp 1–11

Ramazan C (2002) Root yield and quality of sugar beet in relation to sowing date, plant population and harvesting date interactions. Turk J Agric For 26:133–139

Refay YA (2010) Root yield and quality traits of three sugar beet (*Beta vulgaris* L.) varieties in relation to sowing date and stand densities. World J Agric Sci 6(5):589–594

Rinaldi M, Vonella AV (2006) The response of autumn and spring sown sugar beet (*Beta vulgaris* L.) to irrigation in Southern Italy: water and radiation use efficiency. Field Crop Res 95:103–114

Ririe D, Hills FJ (1970) Effects of row covering materials and depth of planting on sugarbeet emergence. J Am Soc Sugar Beet Tec 16:168–171

Robertson MJ, Holland JF, Bambach R (2004) Response of canola and Indian mustard to sowing date in the grain belt of north-eastern Australia. Aust J Exp Agric 44:43–52

Romaneckas K, Pilipavicius V, Sarauskis E, Sakalauskas A (2009) Effect of sowing depth on emergence and crop establishment of sugar beet (*Beta vulgaris* L.). Food Agric Environ 7:571–575

Sadre P, Soleymani A, Javanmard HR (2012) Root yield and quality traits of sugar beet (*Beta vulgaris* L.) in relation to nitrogen fertilizer and plant density in Isfahan region. Intl J Agri Crop Sci 4:1504–1507

Safina SA, Hassanin MA, El-Metwally E-MA, Elsherbini NR (2012) Sowing date and plant density influences on yield and quality of some sugar beet varieties grown in sandy soils under drip irrigation system. J Egypt Acad Env Dev 13(2):73–85

Salmasi SZ, Golez KG, Moghbeli S (2006) Effect of sowing date and limited irrigation on the seed yield and quality of dill (*Anethum graveolens* L.). Turk J Agric Forest 30:281–286

Sarkar D, Mandal B, Kundu MC (2007) Increasing use efficiency of boron fertilizers by rescheduling the time, and methods of application for crops in India. Plant Soil 301:77–85

Scavo A, Mauromicale G (2020) Integrated weed management in herbaceous field crops. Agronomy 10(4):466

Seadh SE, Attia AN, Said EM, El-Maghrabby SS, Ibrahim MEM (2013) Productivity and quality of sugar beet as affected by sowing methods, weed control treatments and N fertilizer levels. Pakistan J Biol Sci 16:711–719

Shukla SK, Awasthi SK (2013) Sugarbeet: A supplement to sugarcane in non-traditional areas to meet future sugar demand. Souvenir-IISR-Industry Interface on Research and Development Initiatives for Sugarbeet in India, 28-29 May, Sugarbeet Breeding Outpost of IISR IVRI Campus, Mukteswar-263138, Nainital Organised by Indian Institute of Sugarcane Research (ICAR) and Association of Sugarcane Technologists of India, pp 55-57

Smith RJ (1986) Biological control of northern jointvetch (*Aeschynomene virginica*) in rice (*Oryza sativa*) and soybeans (*Glycine max*)–a researcher's view. Weed Sci 34:17–23

Sogut T, Arioglu H (2004) Plant density and sowing date effects on sugarbeet yield and quality. J Agron 3:215–218

Sohel MAT (2016) Yield of tropical sugarbeet as influenced by variety, spacing, fertilization, date of sowing and intercropping with sugarcane. PhD Thesis, Department of Agronomy, Bangladesh Agril Uni, Mymensingh, Bangladesh

Soliman ME, Hammada MMA, Abd El-Nabi AE (2014) Response of sugar beet plants to different compost types, and boron spraying, and their effect on growth characters, and quality under rassurd conditions. J Soil Sci Agric Eng Mansoura Univ 5(10):1363–1375

Subbarao GV, Wheeler RM, Levine LH, Stutte GW (2001) Glycine betaine accumulation, ionic, and water relations of red beet at contrasting levels of sodium supply. J Plant Physiol 158:767–776

Syers JK, Curtin D, Skinner RJ (1987) Soil and fertilizer sulphur in UK agriculture. In: A proceeding of the international fertilizer society, No. 264. London

Theurer JC, Saunders JW (1995) Row spacing and plant density effects on smooth root sugarbeets. J Sugarbeet Res 32(2&3):69–78

Thomas SG, Hocking TJ, Bilsborrow PE (2003) Effect of sulfur fertilisation on the growth, and metabolism of sugar beet grown on soils of differing sulfur status. Field Crop Res 83(3): 223–235

Tlili A, Dridi I, Fatnassi S, Hamrouni H, Gueddari M (2018) Effects of boron distribution on sugar beet crop yield in two soils of Dour Ismail irrigated perimeter (Beja-Goubellat-North ́ Tunisia). In: Kallel A (ed) Recent Advances in Environmental Science from the Euro-Mediterranean, and Surrounding Regions. Advances in Science, Technology & Innovation, Tunis, Tunisia, pp 417–420

Uzun B, Zengin U, Furat S, Akdesir O (2009) Sowing date effects on growth, flowering, seed yield and oil content of canola cultivars. Asian J Chem 21:1957–1965

Wilson RG, Smith JA, Yonts CD (2005) Repeated reduced rates of broadleaf herbicides in combination with methylated seed oil for post emergence weed control in sugar beet (*Beta vulgaris* L.). Weed Technol 19:855–860

Winkler J, Hledik P, Prochazkova B (2015) Influence of crop rotation on actual weed infestation of sugar beet. Listy Cukrovarnické a Řepařské 131(5):162–166

Yagoob AM, Zaroug MSA, Dafaallah AB (2021) Efficacy and selectivity of pendimethalin for weed control in sugar beet (*Beta vulgaris* L.), Gezira state. Sudan J Res Weed Sci 4(1):16–28

Yonts CD, Smith JA, Wilson RG (1999) Effect of seed type, planter type and depth of planting on sugarbeet emergence. J Sugar Beet Res 36:1–9

Yordanov I, Velikova V, Tsonev T (2012) Plant responses to drought, acclimation and stress tolerance. Photosynthetica 38(2):171–186

Zahoor A, Faridullah SP, Sanaullah B, Kakar KM, El-Sharkawi H, Haytham T, Yamamoto Y (2007) Sugar beet (*Beta vulgaris* L.) response to different planting methods, and row geometries I: effect on plant growth and yield. Arch Agron Soil Sci 53:49–61

Zengin M, Gökmen FF, Yazici MA, Gezgin S (2009) Effects of potassium, magnesium and sulphur containing fertilizers on yield and quality of sugar beets (*Beta vulgaris* L.). Turk J Agric For 33: 495–502

Autumn-Sown Sugar Beet Cultivation in Semiarid Regions

14

Javad Rezaei and Parviz Fasahat

Abstract

The sugar beet has long been grown as a spring crop in relatively cool parts of the temperate zones of the world. However, in recent decades, its role in sugar production has led to its cultivation as an autumn crop in warm regions of South America, Africa (Egypt, Morocco, and Tunisia), the Middle East (Iraq and parts of Iran), and even the south Europe (Spain). Limited water resources in arid and semiarid regions have caused the autumn sown sugar beet to be extended to higher latitudes areas. Coinciding the period of autumn sown sugar beet growth with autumn and winter rainfall leads to the reduction in water consumption and an increase in water use efficiency for sugar production. However, expanding the area under autumn sugar beet cultivation to semiarid regions of higher latitudes faces three main challenges: the occurrence of bolting, increasing the risk of cold and frost stresses, and reducing its yield compared with spring sown sugar beet. Research in this field has shown that by choosing the proper cultivation area as well as the correct planting date, using bolting-resistant cultivars, and accepting the reasonable risk of relative yield reduction, autumn cultivation of sugar beet in arid and semiarid regions can be expanded. Autumn sugar beet cultivation advantages are increase in the sugar factories operation time, sugar production, employment rate, farmers' income, and saving about 50% in irrigation.

J. Rezaei (✉)
Sugar Beet Research Department, Khorasan Razavi Agricultural and Natural Resources Research and Education Center, Agricultural Research, Education and Extension Organization (AREEO), Mashhad, Iran
e-mail: javad.rezaei@alumni.um.ac.ir

P. Fasahat
Sugar Beet Seed Institute, Agricultural Research, Education and Extension Organization (AREEO), Karaj, Iran

V. Misra et al. (eds.), *Sugar Beet Cultivation, Management and Processing*,
https://doi.org/10.1007/978-981-19-2730-0_14

Keywords

Bolting · Freezing stress · Planting date · Water consumption

Abbreviation

GDD Growing degree days

14.1 Introduction

Climate change along with climate diversity, especially in developing countries, has raised concerns about agricultural production for the world's growing population (Parry et al. 2004). These changes have made the need to supply adequate food and its security more important than ever. One of the main strategies for farmers to deal with climate change and increase or at least maintain current crop yields is to adapt the planting date to new conditions (Lauer et al. 1999). Accordingly, the planting date of crops has changed over time for the above-mentioned reasons as well as changes in technology and socio-economic factors (Sacks et al. 2010). Therefore, in order to adapt to the new conditions, it is necessary to review the current crop calendar based on scientific principles and new research.

In the arid and semiarid ecosystems, plants often grow under the effects of drought, salinity, cold, or heat, and each of these stresses, individually or in combination influences plant growth and development and in some cases even cause their death. To ensure proper growth or improve plant survival, these stresses should be managed by adopting appropriate cropping methods or plant breeding to reduce the severity of their effects. In plants in which growing season and yield are limited by these factors, it is necessary to choose a planting date in order to achieve proper yield. In fact, the planting date affects the establishment, vegetative and reproductive growth of the plant by adapting the growth stages of the plant to temperature, day length, rainfall, and other environmental factors, reduces the effect of environmental stresses, and thus improves the quantity and quality of the crop. Therefore, management of the planting time is one of the appropriate agronomic strategies and plays an important role in achieving sustainable crop production in stressful areas.

14.2 Challenges of Spring Planting in Arid and Semiarid Regions

In arid and semiarid regions, spring planting of the crops exposes plants to drought and heat stress, and as a result of the main or interaction effects of these factors, their growth and development are severely affected. Planting sugar beet in the spring, in addition to increase water consumption and damage from summer heat, will also increase the likelihood of damage from early autumn cold. In these areas, due to various reasons such as water shortage and the need to provide sufficient water for autumn cereals, delays in land preparation operations due to rainfall in early spring, reducing the potential risks of late spring frosts and pests, diseases and weed management, late spring planting (or summer planting) is common. In relation to sugar beet, delayed planting causes a significant reduction in its yield due to the shortening of the growth period and the exposure of the critical growth stages of this plant to high summer temperatures. Water scarcity, as the most important and common factor reducing yield in arid and semiarid regions, may occur even in areas with high rainfall (Vamerali et al. 2003). The average annual yield decline due to drought in the world is 17%, which can increase up to more than 70% per year (Nasri et al. 2006). In recent decades, changes related to global warming and heat stress have become one of the serious problems in the production of agricultural products in arid and semiarid regions. Heat affects the physiological and biochemical processes of plants, changes the function of plant organs, and ultimately reduces yield (Hanson et al. 2002).

14.3 Alternative Planting

To deal with the problems of spring planting of crops such as sugar beet in arid and semiarid regions, several strategies have been studied and implemented:

14.3.1 Early Spring Planting

Early planting of plants in spring provides an opportunity for the plant to make proper usage of soil moisture and spring rainfall and provides enough time for the plant to grow. Planting crops early in the spring can also prevent susceptible growth stages from being exposed to hot summer weather. In addition, in the mentioned conditions, due to the earlier emergence of the plant, its growth is ahead of the growth of some spring weeds and has a better chance of using environmental resources (Campbell and Enz 1991). At the same time, with faster establishment and further expansion of leaf area, by increasing the coincidence of larger leaf area index with the peak of solar radiation and lengthening the growing season, it is possible to produce more yield (Scott and Jaggard 1993; Jaggard and Werker 1999).

However, in early spring planting, the percentage of greenness and the establishment of crops may be difficult. Occurrence of low temperatures in early spring sowing will reduce the percentage and speed of seed emergence and will also delay seedling establishment. Campbell and Enz (1991) observed that at 14 °C, it takes 14 days for 50% of sugar beet seeds to germinate, while at 25 °C, 50% of seeds germinate in 9 days. Also, the presence of low temperatures at the beginning of the growing season, by reducing the evaporation of soil moisture and increasing its relative humidity, leads to the vulnerability of seedlings to pathogens (Bolton et al. 2010).

14.3.2 Dormant Seeding

In warm areas where late winter and early spring rains increase the likelihood of delayed land preparation operations in the spring, some crops, such as sugar beet, are planted as expected. In this method, the seeds are sown in late autumn or early winter, when the temperature is lower than the base temperature for germination, and the seeds germinate at the first time in spring. In this case, choosing the right planting time is crucial. Because in early planting, the seedlings will damp off due to winter stress. On the other hand, in case of delay in cultivation, it may not be possible to cultivate the crop due to wet or frozen soil. In general, the expected planting causes the plant to establish itself faster than other plants planted in the spring, surpassing weeds and increasing the plant's chances of using soil moisture storage in early spring, as well as avoiding high temperature and drought in late spring and early summer. In this way, the plant escapes drought stress well and its water use efficiency would improve.

14.3.3 Autumn–Winter Planting

Temperature is an important factor in determining the geographical distribution and production of plant species and even in areas where environmental conditions are ideal for the growth of a particular species, unpredictable changes in temperature may cause damage and reduce the growth and production of the plant (Steponkus et al. 1993). In temperate regions of the world, cold-adapted crops (such as cereals) are usually grown in autumn. Autumn cultivation of crops has the following advantages over spring planting: Production and yield of autumn crops are often higher than spring ones and have more yield stability (Graves 1995). In this regard, it has been stated that the increase in yield and stability of autumn plants are due to the proper establishment of the plant in autumn, and better use of rainfall and escape from heat and drought stress are common in late spring and summer (Nezami and Bagheri 2005). In plants grown in autumn, water use efficiency is often higher than spring ones (Kuschel-Otárola et al. 2020). Autumn grown plants usually provide cover and protection for the soil and prevent soil erosion (McKersie and Leshem 1994). Early harvest of autumn plants provides suitable condition for planting of the second crop (McKersie and Leshem 1994). Due to the mentioned advantages, during the last 120 years, many efforts have been made to produce cold-tolerant plants in order to replace spring planting with autumn planting and achievements have been made.

14.4 Autumn Cultivation of Sugar Beet

Sugar beet is one of the most important crops that is grown in a wide range of climatic conditions, from cold mountain areas to warm plains. In general, in the beet growing regions of the world, sugar beet is cultivated in both spring and autumn. Due to the climatic conditions of different regions, spring sugar beet is usually planted from 25 February to 25 May and its harvest period is from 20 September to 25 December, while autumn sown sugar beet is planted in autumn (20 September to 20 November) and harvested in next spring (20 April to 20 June) (Taleghani et al. 2015).

Autumn cultivation of sugar beet is being developed or studied in different countries of the world, to the extent that the idea of autumn cultivation of sugar beet in northwestern Europe has also been proposed (Rinaldi and Vonella 2006). In the United Kingdom, trials were performed on autumn sugar beet in the 1970s, but there was no increase in yield compared with spring planting (Wood and Scott 1975). Studies performed in Iran (Ahmadi 2012; Rezaei and Haghighat 2019; Sadeghzadeh et al. 2012) on various aspects of agronomy, breeding, plant pathology, economic, quality, and other characteristics of autumn sown sugar beet in Iran during the last two decades showed that sugar beet can be introduced as an important and effective autumn crop in the rotation table of proper areas. Due to the fact that in arid and semiarid regions, water is considered to be the main limiting factor of agriculture, autumn cultivation of sugar beet becomes more important. The main period of autumn sown sugar beet growth takes place in autumn and winter. Since the distribution of rainfall in these areas is generally such that about three-quarters of the annual rainfall occur in the two seasons of autumn and winter, so in the autumn cultivation of sugar beet it is possible that part of the water need is met through rainfall. This ultimately leads to a reduction in water consumption and an increase in its consumption efficiency for sugar production in the autumn sugar beet crop. In Haghayeghi et al. (2015) study, the water use in spring sown sugar beet was 2.3 and in autumn sown sugar beet was 7.4 kg m^{-3}.

14.5 Comparison of Spring and Autumn Sugar Beet Cultivation

The growth period of sugar beet in spring and autumn cultivation is completely different, and this difference has a major effect on the genetic needs of compatible cultivars for planting in these two seasons. In autumn cultivation due to longer growing period and receiving higher degree-day growth, higher root yield would achieve, however due to the coincidence of sugar storage with hot air, the percentage of sugar is lower than spring cultivation. Although the length of the growth period of spring sugar beet (180 days) is less than autumn one (220 days), but after harvesting spring-sown sugar beet there is no opportunity to plant a second crop, while after harvesting autumn-sown sugar beet, the remaining time can be allocated to planting another crop. In the other words, the rate of land productivity in autumn cultivation is much higher than in spring cultivation.

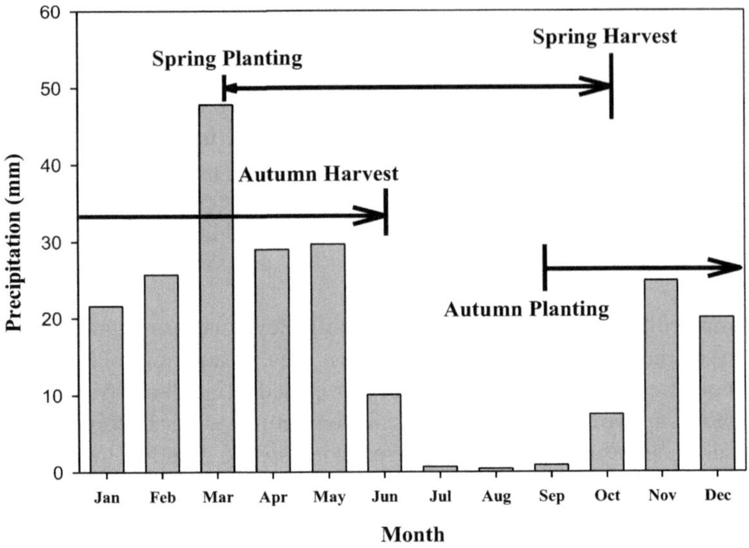

Fig. 14.1 Mean precipitation (mm) per month during autumn and spring sown sugar beet growth period (Region: Mashhad – Iran, 2010–2020)

In semiarid regions, in autumn cultivation, the growing season is completely consistent with seasonal rainfall, while in spring cultivation, most of the growth occurs in summer when there is little or no rainfall. As can be seen in Fig. 14.1, more than 96% of rainfall occurs during the autumn sugar beet growth period and about 50% in spring cultivation. This will meet part of the plant's water needs and use less irrigation water than spring crops.

Research (Haghayeghi et al. 2015) shows that water consumption in autumn sugar beet cultivation is about 30–50% less than spring cultivation. As a result, water use efficiency in autumn cultivation (about 1000 g of sugar per cubic meter of water) is about twice that of spring cultivation (about 580 g of sugar per cubic meter of water). This is in fact the most important advantage of autumn over spring cultivation in arid and semiarid regions. In the current situation and with regard to future climate change, there is no opportunity or possibility for the development of spring crops in arid and semiarid regions, and the only way out of the problem of water shortage in sugar production from sugar beet is development of autumn-sown sugar beet in prone areas especially lower latitudes.

Comparison of spring and autumn sugar beet root compounds shows that there is no difference between them in terms of non-sugar substance percentage, but there is a difference in terms of sugar content, dry matter, and marc. Root quality indices such as dry matter percentage and sugar content in spring sugar beet are better than autumn sugar beet. On the other hand, due to the low level of infection to rhizomania in autumn sown sugar beet fields, it is expected that the amount of sodium content in the root would be less than spring beet (Hoseinpoor and Sadeghzadeh 2019).

14.6 Advantages of Autumn Cultivation of Sugar Beet

The main determinants of root and sugar yield in sugar beet are the length of vegetative growth period and the plant's ability to receive solar radiation, which depends on the plant leaf area index (Rezaei 2014). Sugar beet is basically grown as a spring crop (i.e., it is planted in spring and harvested before the onset of winter cold). Rapid and uniform emergence of seedlings in early spring is a determining factor in the beginning of leaf development as an important factor affecting the final yield. However, due to low temperature at this time of year, the leaves grow slowly and the optimal shade coverage required to receive maximum radiation lasts until June (Hoffmann and Kluge-Severin 2010; Jaggard et al. 2009), as a result, plants cannot utilize about 40% of the total annual radiation. Under this circumstance, the use of autumn crops instead of spring can provide condition to avoid energy loss. Autumn crops develop a significant leaf area index before entering winter period which allows them to receive solar radiation more quickly in the spring of next year. Therefore, autumn planting of sugar beet can also be more attractive. In addition, autumn sown sugar beet has more seedlings emergence than spring ones, and as expected, the leaf area of autumn sown sugar beets is higher. Research has shown that the dry weight of autumn sown sugar beet leaves in December and June was much higher (1–2 and 4–10 t ha^{-1}, respectively) than spring beet in June (2–4 t ha^{-1}). As a result, autumn sown sugar beets receive more radiation during the early stage of their growth (Fig. 14.2).

In arid and semiarid regions such as Iran, the most important factor that can be introduced as a clear indicator for the priority and superiority of autumn sugar beet

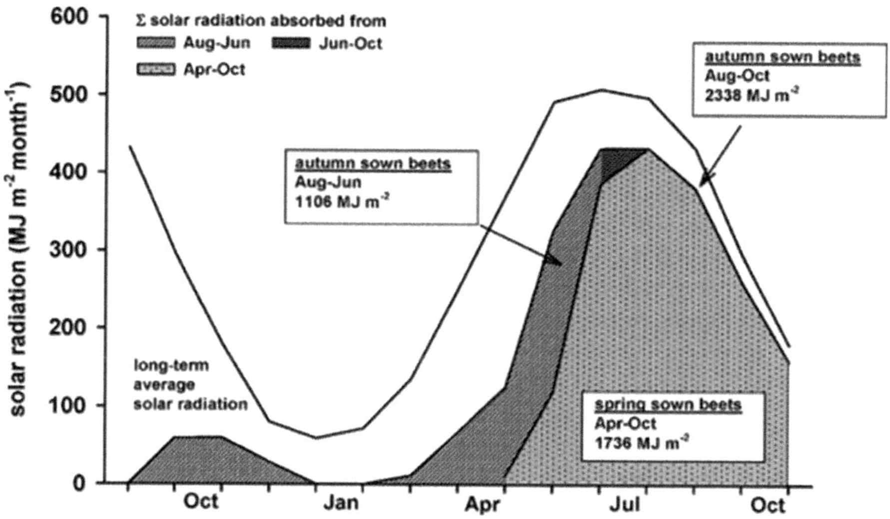

Fig. 14.2 Long-term solar radiation 1952–2007 and mean radiation absorbed by autumn and spring sown beets, 4 trials with 6 sowing and 4 harvest dates in 2005/06 and 2006/07, Göttingen (Hoffmann and Kluge-Severin 2010)

cultivation over spring cultivation is the optimal use of rainfall during the growing season and greater water use efficiency. This issue becomes even more important when water is the main limiting factor for agriculture in these areas; except in special cases where spring rainfall contributes to plant germination and establishment, the main stage of spring sugar beet growth takes place in summer and the plant needs irrigation for almost the entire period of its growth. However, the main period of autumn sugar beet growth takes place in autumn and winter, and part of the plant's water needs is met through rainfall. Since the distribution of rainfall in semiarid regions is generally such that about three-quarters of the annual rainfall occur in the two seasons of autumn and winter, therefore, autumn cultivation of sugar beet has the possibility to meet part of its water needs through rainfall in the region. This ultimately leads to a reduction in water consumption and an increase in its water use efficiency in the production of sugar from autumn sugar beet crop. Autumn and spring sugar beet produce 853 and 532 g of sugar per cubic meter of irrigation water, respectively.

14.7 Challenges of Autumn-Sown Sugar Beet Cultivation in Arid and Semiarid Regions

Expanding the autumn-sown sugar beet cultivation in arid and semiarid regions faces three major challenges including the occurrence of bolting, increasing the risk of cold stress and frost, and reducing its yield compared with spring cultivation due to the shortening of the effective growing season.

14.7.1 Bolting

Bolting is a premature flowering phenomenon in tuberous plants such as sugar beet and carrots that has adverse effects on yield (Alt and Wiebe 2001; Dielen et al. 2005). Sugar beet bolting is influenced by environmental factors and is controlled by a dominant gene (*B*) and several recessive alleles. The emergence of bolting phenomenon in sugar beet depends on agronomic factors, cultivar and environmental conditions (temperature and day length). Occurrence of bolting in sugar beet causes problems such as 0.7% reduction in yield per 1% bolting, seed shattering, growth of sugar beet seedlings in the next crop as a weed (Jaggard et al. 1983) as well as disruption in the harvest process and processing in the sugar factory.

14.7.1.1 Flowering in Sugar Beet

Sugar beet is a long-day plant in which the induction of flowering depends on the vernalization (Lexander 1980). The onset of the transition from vegetative to reproductive is determined by bolting (or longitudinal growth of the stem from the main axis). After bolting, if the sugar beet plants are exposed to suitable environmental conditions (optimal temperature and photoperiod), the flowering is occurred. Bolting and induction of flowering are initiated by a photo thermal sensitive process which demands the sugar beet to be exposed to cold temperatures for a long period of weeks or months (depending on the sugar beet variety) followed by experiencing a critical day length (more than 12–16 h of light). It is natural that if sugar beet is exposed to long days without vernalization, it will show vegetative growth behavior for several years. Also, if sugar beet is exposed to short days after vernalization instead of long days, bolting and flowering will not occur (Mutasa-Göttgens et al. 2010).

In some studies, using gibberellic acid, the onset of bolting has been evaluated. Application of gibberellic acid accelerates bolting and flowering occurrence in vernalized beets. Gibberellic acid can also induce bolting in the absence of vernalization and independently of the photoperiod but cannot increase flowering (Mutasa-Göttgens et al. 2010). Conversely, unlike other plant species—in which gibberellic acids can compensate for the lack of vernalization or photoperiod signaling—this hormone in sugar beet alone cannot have complete control over the flowering process. On the other hand, it has been shown that exposure of sugar beet to long days (with 22 h of light) following vernalization increases bolting and accelerates flowering, indicating a certain threshold of photoperiod time in the transition from bolting to flowering (Pin 2012).

14.7.1.2 Devernalization

There is a difference between bolting and flowering processes in sugar beet. This means that flowering does not necessarily occur after bolting. This phenomenon can occur when vernalized beets are not exposed to short, non-inductive days or too warm temperatures (Van Dijk 2009). Therefore, vernalized sugar beet may lose the ability to start bolting and flowering, which was required during vernalization. This phenomenon is called devernalization. Devernalization can also occur after the onset of bolting, in which bolting halts (leading to dwarfism) and flowering is typically canceled. When the beet is devernalized, it must be re-vernalized to produce flower and seed (Pin 2012).

14.7.1.3 Growth Habits: The Role of the Bolting Gene *B*

The wild ancestor of domestic beet (*B. maritima*) has annual growth habit. Annual beets, when grown under short-day condition, have a habit of continuous vegetative growth and can never enter the bolting phase. However, when exposed to long days, the annual sugar beet will begin to bolt and flower relatively quickly over a period of weeks or months. Increasing the length of the photoperiod can also significantly accelerate bolting in annual beets, as observed in vernalized biennials. Interestingly, bolting does not occur in one-year-old vernalized beets that are subsequently placed under short-day conditions (Mutasa-Göttgens et al. 2010). However, if the plants are exposed to long days, the vernalized annual beets enter the bolting before non-vernalized ones. This suggests that annual beets can respond to vernalization. Genetic studies have shown that annual growth habit prevails over two-years growth and is controlled by a single locus called the bolting *B* gene. Plants carrying dominant gene *B* do not need vernalization and begin bolting and flowering as a direct response to the signal of long days. Resistance to bolting is the main challenge in sugar beet breeding. Breeders seek to improve and produce bolting-resistant cultivars without affecting the flowering and seed growth processes required by breeding and seed production programs. Vernalization, light period, day length, light quality as well as intensity are the most important environmental factors affecting bolting (Milford et al. 2010). The role of vernalization in bolting is very critical.

In the process of sugar beet seed production, bolting is a useful and necessary phenomenon, however, in autumn cultivation of sugar beet, cultivars that do not enter this stage of growth are needed. Accordingly, genotypes with bolting resistance have been developed for environmental conditions with a high probability of bolting occurrence (Ritz et al. 2010).

Over a period of 4–6 weeks, the occurrence of temperatures above freezing up to 12 °C for 3 weeks induces vernalization followed by bolting in sugar beet and increasing the temperature to more than 12.6 °C will cause devernalization (Jaggard et al. 1983). Part or all of the vernalization process may be neutralized under hot temperatures (Fauchere et al. 2003). The temperature at which devernalization occurs is not precisely determined, however, temperatures above 23 °C have the greatest effect on this phenomenon (Longden et al. 1995).

Gaskill (1963) concluded that there is a positive correlation between the age of sugar beet plants at the time of frost occurrence and the incidence of bolting. At germination stage and also early growth stages of sugar beet, susceptibility to low temperatures is weak. In general, it has been found that in late growth stages, cold causes faster bolting induction in sugar beet (Smit 1983). Various statistical methods have been used by researchers to predict the occurrence of bolting. Fauchere et al. (2003) used descriptive statistics while Jaggard et al. (1983) used the logistic regression model. Alt and Wiebe (2001) used sigmoidal and exponential two-stage nonlinear regression models. Ritz et al. (2010) simulated the bolting phenomenon as a function of temperature and time using bioanalysis method. He used the indices of cultivar susceptibility to bolting as well as longitude and latitude in his functions.

One of the effective ways to control bolting phenomenon in autumn cultivation of sugar beet, especially in higher latitudes, is to set the planting date (Fig. 14.3). Many researchers in Iran (Ahmadi 2012; Rezaei and Haghighat 2019; Taleghani et al. 2015) have worked to determine the planting date of autumn sugar beet in different parts of the country. Early planting of autumn sugar beet will lead to greater yield. However, according to the studies, the earlier planting date of sugar beet leads to the higher percentage of bolted plants. In areas where the possibility of bolting is high, plant yield will be reduced if bolting occurs. Therefore, in conditions where the probability of bolting is high, with later planting and less bolting, the yield will practically be higher than early planting (Jaggard et al. 1983). Therefore, in each region, it is possible to determine the planting date for autumn sugar beet by which not only the probability of bolting occurrence is reduced but also the least reduction in plant yield is obtained.

Fig. 14.3 Effect of planting date on bolting occurrence in autumn sown sugar beet, Mashhad-Iran, 2018

Fig. 14.4 Symptoms of freezing stress on autumn sown sugar beet seedling, Torbat-e-Jam, Iran, 2017

14.7.2 Freezing Stress

One of the determinants of planting date for autumn cultivation of sugar beet is frost damage on young seedlings (Draycott 2006) (Fig. 14.4). The occurrence of adverse factors such as temperature has a great effect on the establishment of plants in the field (De Figueiredo et al. 2003) and in this event, germination is the most critical stage of plant growth. In plants such as sugar beet that are not able to withstand cold, the possibility of frostbite is very important (Palva et al. 2002). In a study (Reinsdorf and Koch 2013) conducted in different climatic conditions of Central Europe, it was shown that autumn planting of sugar beet has a 10–35% chance of freezing. The results of Carter et al. (1985) showed that at a temperature of about −4.4 °C, the process of sugar beet photosynthesis was ceased and the transfer of sucrose was disrupted, and if the minimum air temperature reaches −8 °C, the leaves will be dried. In addition, planting autumn sugar beet in colder regions poses a risk of frost and freezing of the cell membrane and leakage of soluble solids (Baker and Rosenqvist 2004). Studies by other researchers (Reinsdorf and Koch 2013) have shown that the lower the temperature below −6 °C, the more damage to the roots will occur.

Although many problems have been reported for the survival of young sugar beet seedlings during the frost period (Kockelmann and Meyer 2006), the temperature effects of snow cover can to some extent protect seedling against freezing (Sokratov and Barry 2002). Autumn sown sugar beet can withstand temperatures up to −23 °C in field under snow cover (Loel and Hoffmann 2014). According to Bürcky (1981), autumn sown sugar beet can withstand temperatures as low as −10 °C for a short time without damage. The freezing point of fully developed sugar beet root tissue is between −2 and −4 °C under in vitro condition (Chelemski 1967). Reinsdorf and Koch (2013) have determined the lethal temperature for root tissue at −6 °C. Nezami et al. (2015) reported the temperature of −7 °C as the onset temperature of electrolyte leakage due to cold under controlled condition for sugar beet cultivars.

However, it should be noted that the growth stage of sugar beet is very important for tolerance to frost temperatures. The results of Jalilian et al. (2005) showed that the flowering and rosette stages are the most susceptible and tolerant stages to frost, respectively. Under greenhouse condition, Reinsdorf and Koch (2013) showed that under long frost period, the temperature of plant tissue is more important than the minimum air temperature. Also, the temperature of the aerial parts of sugar beet plant is correlated with the average daily temperature. They also showed that sugar beet frost tolerance is highest when the root diameter is 25 mm.

The effect of pre-frost plant growth stage and climatic condition during the cold period on the survival of sugar beet seedlings are 46 and 17%, respectively. Therefore, planting date can be effective in sugar beet tolerance to cold. The maximum survival of sugar beet plants is achieved at 600–900 GDD from planting date and plant can tolerate up to −7 °C at this condition (Loel and Hoffmann 2014).

14.7.3 Birds' Damage

In arid and semiarid areas, due to the lack of soil moisture and low rainfall in early autumn, farmlands has little plants for birds. Furthermore, more use of herbicides and the replacement of monogerm seeds with polygerm has increased bird damage to crops, especially to sugar beet seedlings (Dunning 1974). In the autumn cultivation of sugar beet, especially in arid areas, at the beginning of the autumn season, this damage is so great in some fields that it is possible to destroy up to 100% of young sugar beet seedlings (Fig. 14.5). The use of chemical and mechanical methods, especially the use of trap crops, can greatly prevent this damage.

Fig. 14.5 Birds' damage to autumn sown sugar beet, Sarakhs-Iran, 2020

14.8 Future Prospects

In order to expand autumn sugar beet cultivation, important steps must be taken such as farmer training programs and skill-based education in sugar beet fields. In these training programs, farmers are acquainted to the proper planting date and suitable cultivars of autumn-sown sugar beet, breeding of sugar beet crop for bolting and freezing resistance that adapted for arid and semiarid regions, selection of areas for autumn cultivation of sugar beet with lower risk of bolting and freezing and also adoption of farming and chemical methods for birds damage management and control.

14.9 Conclusion

In arid and semiarid regions, water is considered to be the main limiting factor for agriculture, so autumn cultivation of crops is very important. Studies in these regions have shown that sugar beet can be introduced as an important and effective autumn crop in the crop rotation. Autumn cultivation of sugar beet has many advantages. The most important of which for semiarid regions is to reduce water consumption and increase water use efficiency. The spread of autumn cultivation of sugar beet in these regions, especially in higher latitudes, faces some challenges (the most important of which are frost stress and bolting). Therefore, production sustainability for autumn-sown sugar beet is depended to suitable autumn planting date and the use of bolting resistant cultivars.

References

Ahmadi M (2012) Feasibility study of autumn sowing of sugar beet in South Khorasan. SBSI publication no. 41200. Karaj, Iran, 34p

Alt C, Wiebe HJ (2001) Flower formation in celeriac (Apium graveolens L. var. rapaceum) II. Modelling the risk of bolting. Gartenbauwissenschaft 66(1):46–50

Baker NR, Rosenqvist E (2004) Applications of chlorophyll fluorescence can improve crop production strategies: an examination of future possibilities. J Exp Bot 55(403):1607–1621. https://doi.org/10.1093/jxb/erh196

Bolton MD, Panella L, Campbell L, MFR K (2010) Temperature, moisture, and fungicide effects in managing rhizoctonia root and crown rot of sugar beet. Phytopathology 100(7):689–697. https://doi.org/10.1094/PHYTO-100-7-0689

Bürcky K (1981) Gefrieren und Erfrieren von Zuckerrüben – Beobachtungen und Erfahrungen 1980. Zuckindustrie 106:232–236

Campbell LG, Enz JW (1991) Temperature effects on sugarbeet seeding emergence. J Sugarbeet Res 28(3&4):129–140. https://doi.org/10.5274/jsbr.28.3.129

Carter JN, Kemper WD, Traveller DJ (1985) Yield and quality as affected by early and late fall and spring harvest of sugarbeets. J Sugarbeet Res 23(1&2):8–27. https://doi.org/10.5274/jsbr.23.1.8

Chelemski MS (1967) Die Lagerung von Zuckerrüben. National Printing Press, Berlin, p 557

De Figueiredo E Albuquerque, MC, De Carvalho NM (2003) Effect of the type of environmental stress on the emergence of sunflower (Helianthus annus L.), soybean (Glycine max (L.) Merril) and maize (*Zea mays* L.) seeds with different levels of vigor. Seed Sci Technol 31(2):465–479 https://doi.org/10.15258/sst.2003.31.2.23

Dielen V, Notté C, Lutts S, Debavelaere V, Van Herck JC, Kinet JM (2005) Bolting control by low temperatures in root chicory (*Cichorium intybus* var. *sativum*). Field Crops Res 94(1):76–85. https://doi.org/10.1016/j.fcr.2004.11.005

Draycott AP (2006) Sugar beet. Blackwell Publishing, London, p 514

Dunning RA (1974) Bird damage to sugar beet. Ann Appl Biol 76(3):325–335. https://doi.org/10.1111/j.1744-7348.1974.tb01373.x

Fauchere J, Richard-Molard M, Souverain F, Prats S, Pérarnaud V, Decquiedt B (2003) Cartographie des risques de montées en France en relation avec les températures de printemps et d'été: conséquences sur l'expérimentation et le conseil. Proceedings of the 1st Joint IIRB-ASSBT Congress, San Antonio, pp. 189–205

Gaskill JO (1963) Influence of age and supplemental light on flowering of photothermally induced sugar beet seedlings. J Sugarbeet Res 12(6):530–537. https://doi.org/10.5274/jsbr.12.6.530

Graves WR (1995) Stress and stress coping in cultivated plants. Field Crops Res 42(2–3). Springer Science & Business Media. https://doi.org/10.1016/0378-4290(95)90040-3

Haghayeghi SA, Alizadeh A, Ahmadi M, Bannayan M, Ansari H (2015) Effect of irrigation regimes on crop water use efficiency of autumn sugar beets grown on the Mashhad plain. J Agric Eng Res 16(3):15–30

Hanson PM, Chen JT, Kuo G (2002) Gene action and heritability of high-temperature fruit set in tomato line CL5915. HortScience 37(1):172–175. https://doi.org/10.21273/hortsci.37.1.172

Hoffmann CM, Kluge-Severin S (2010) Light absorption and radiation use efficiency of autumn and spring sown sugar beets. Field Crops Res 119(2–3):238–244. https://doi.org/10.1016/j.fcr.2010.07.014

Hoseinpoor M, Sadeghzadeh HS (2019) Autumn Sown Sugar beet in Iran (from research to execution) SBSI publication No. 54072, Karaj, Iran, pp. 112

Jaggard KW, Werker AR (1999) An evaluation of the potential benefits and costs of autumn-sown sugarbeet in NW Europe. J Agric Sci 132(1):91–102. https://doi.org/10.1017/S002185969800608X

Jaggard KW, Wickens R, Webb DJ, Scott RK (1983) Effects of sowing date on plant establishment and bolting and the influence of these factors on yields of sugar beet. J Agric Sci 101(1): 147–161. https://doi.org/10.1017/S0021859600036479

Jaggard KW, Qi A, Ober ES (2009) Capture and use of solar radiation, water, and nitrogen by sugar beet (*Beta vulgaris* L.). J Experimental Bot 60(7):1919–1925. https://doi.org/10.1093/jxb/erp110

Jalilian A, Mazaheri D, Tavakkol Afshari R, Rahimian H, Abdolahian M, Gohari J (2005) Estimation of base temperature and the investigation of germination and field emergence trend of monogerm sugar beet under various temperatures. J Sugar Beet 20(2):97–112

Kockelmann A, Meyer U (2006) Seed production and quality. In: Draycott AP (ed) Sugar beet. Blackwell Publishing, Oxford, pp 89–113

Kuschel-Otárola M, Rivera D, Holzapfel E, Schütze N, Neumann P, Godoy-Faúndez A (2020) Simulation of water-use efficiency of crops under different irrigation strategies. Water 12(10): 2930

Lauer JG, Carter PR, Wood TM, Diezel G, Wiersma DW, Rand RE, Mlynarek MJ (1999) Corn hybrid response to planting date in the northern corn belt. Agron J 91(5):834–839. https://doi.org/10.2134/agronj1999.915834x

Lexander K (1980) Present knowledge of sugar beet bolting mechanisms. The 43rd Winter Congress of the International Institute of Sugar Beet Research, pp. 245–258

Loel J, Hoffmann CM (2014) Importance of growth stage and weather conditions for the winter hardiness of autumn sown sugar beet. Field Crop Res 162:70–76. https://doi.org/10.1016/j.fcr.2014.03.007

Longden PC, Clarke NA, Thomas TH (1995) Control of bolting and flowering in sugar beet. Proceedings of the 58th Winter Congress of the Institut de Reche'rches Sur La Betterave, Brussels, pp. 71–75

McKersie BD, Leshem YY (1994) Stress and stress coping in cultivated plants. In: Stress and stress coping in cultivated plants. Springer, Amsterdam. https://doi.org/10.1007/978-94-017-3093-8

Milford GFJ, Jarvis PJ, Walters C (2010) A vernalization-intensity model to predict bolting in sugar beet. J Agric Sci 148(2):127–137. https://doi.org/10.1017/S0021859609990323

Mutasa-Göttgens ES, Qi A, Zhang W, Schulze-Buxloh G, Jennings A, Hohmann U, Müller AE, Hedden P (2010) Bolting and flowering control in sugar beet: relationships and effects of gibberellin, the bolting gene B and vernalization. AoB Plants 2010. https://doi.org/10.1093/aobpla/plq012

Nasri M, Heidari SH, Shiranirad AH, Majidi H, Zamanizadeh H (2006) Performance of the effect water stress on physiological characters of rapeseed cultivars. J Agric Sci 12(1):127–134. https://www.sid.ir/fa/journal/ViewPaper.aspx?id=63217

Nezami A, Bagheri A (2005) Responsiveness of cold tolerant chickpea characteristics in fall and spring planting: II. Yield and yield components. Iranian J Field Crops Res 3(1):156–170. https://doi.org/10.22067/GSC.V3I1.1300

Nezami A, Khazaei HR, Pouramir F, Saeidnejad AS, Mehrabadi HR (2015) Effect of freezing stress on some of sugar beet (*Beta vulgaris* L.) genotypes electrolyte leakage. J Crop Prod 3(5):63–76

Palva ET, Sari T, Tamminen I, Puhakainen T, Laitinen R, Svensson J, Helenius E, Heino P (2002) Biological mechanisms of low temperature stress response: cold acclimation and development of freezing tolerance in plants. In: Genetic engineering of crop plants for abiotic stress, pp 9–15. JIRCAS Working Report

Parry ML, Rosenzweig C, Iglesias A, Livermore M, Fischer G (2004) Effects of climate change on global food production under SRES emissions and socio-economic scenarios. Glob Environ Chang 14(1):53–67. https://doi.org/10.1016/j.gloenvcha.2003.10.008

Pin PA (2012) Life cycle and flowering time control in beet. Swedish University of Agricultural Sciences, pp 1–66

Reinsdorf E, Koch HJ (2013) Modeling crown temperature of winter sugar beet and its application in risk assessment for frost killing in Central Europe. Agric For Meteorol 182–183:21–30. https://doi.org/10.1016/j.agrformet.2013.08.001

Rezaei J (2014) Modelling the effects of drought, climate change and rhizomania on sugar beet production. Ferdowsi University of Mashhad, Iran

Rezaei J, Haghighat M (2019) Simulation and determination of proper sugar beet autumn sowing date in suitable regions for bolting risk reduction, SBSI publication No. 56965, p 34
Rinaldi M, Vonella AV (2006) The response of autumn and spring sown sugar beet (*Beta vulgaris* L.) to irrigation in southern Italy: water and radiation use efficiency. Field Crops Res 95(2–3): 103–114. https://doi.org/10.1016/j.fcr.2004.12.004
Ritz C, Pipper C, Yndgaard F, Fredlund K, Steinrücken G (2010) Modelling flowering of plants using time-to-event methods. Eur J Agron 32(2):155–161. https://doi.org/10.1016/j.eja.2009.10.002
Sacks WJ, Deryng D, Foley JA, Ramankutty N (2010) Crop planting dates: an analysis of global patterns. Glob Ecol Biogeogr 19(5):607–620. https://doi.org/10.1111/j.1466-8238.2010.00551.x
Sadeghzadeh HS, Shirzadi MH, Aghaeizade M, Taleghani D, Javaheri MA, Aliasgari A (2012) Evaluation of sowing and harvesting date effects on yield and quality of five sugar beet cultivars in Jiroft region (autumn planting). J Sugar Beet 28(1):25–42. https://doi.org/10.22092/JSB.2012.659
Scott RK, Jaggard KW (1993) An analysis of the efficiency of the sugar beet crop in exploiting the environment. J Sugarbeet Res 30(1 & 2):37–56. https://doi.org/10.5274/jsbr.30.1.37
Smit AL (1983) Influence of external factors on growth and development of sugar-beet (*Beta vulgaris* L.). Centre for Agricultural Publishing and Documentation. Wageningen Pudoc, p 106
Sokratov SA, Barry RG (2002) Intraseasonal variation in the thermoinsulation effect of snow cover on soil temperatures and energy balance. J Geophys Res D Atmos 107(9–10):ACL-13. https://doi.org/10.1029/2001JD000489
Steponkus PL, Uemura M, Webb MS (1993) A contrast of the cryostability of the plasma membrane of winter rye and spring oat-two species that widely diVer in their freezing tolerance and plasma membrane lipid composition. In: Advances in low-temperature biology, vol 2. JAI Press, London, pp 211–312
Taleghani D, Mohammadian R, Sadeghzadeh HS (2015) Autumn sown sugar beet planting to harvest. SBSI publication, Karaj, Iran, p 89
Vamerali T, Saccomani M, Bona S, Mosca G, Guarise M, Ganis A (2003) A comparison of root characteristics in relation to nutrient and water stress in two maize hybrids. In: Roots: The Dynamic Interface between Plants and the Earth, pp 157–167. https://doi.org/10.1007/978-94-017-2923-9_15
Van Dijk H (2009) Evolutionary change in flowering phenology in the iteroparous herb *Beta vulgaris* ssp. *maritima*: a search for the underlying mechanisms. J Experimental Bot 60(11): 3143–3155. https://doi.org/10.1093/jxb/erp142
Wood DW, Scott RK (1975) Sowing sugar beet in autumn in England. J Agric Sci 84(1):97–108. https://doi.org/10.1017/S0021859600071926

New Approach to Utilize Nano-Micronutrients in Sugar Beet (*Beta vulgaris* L.)

15

Mohamed S. Abbas, Hemat S. Mohamed, and Mohamed A. Shahba

Abstract

Nano-fertilizers are more efficient and eco-friendly when compared with the other forms of chemical fertilizers. Using nano-fertilizers as a source of needed nutrients has advantages in crop production. Nano-fertilizers have a great impact on crops and soils. They reduce the toxicity of the soil, decrease the frequency of fertilizer application, and increase crop productivity. A clear understanding of sugar beet response to nano-micronutrients may help in programs aiming at yield and quality traits evaluation. In this chapter, we focused on the importance of nano-micronutrients including iron, magnesium, silicon, zinc, copper, and boron in crop production using sugar beet as a case study. In addition, interactive effects of nano-micronutrients as fertilizers are also discussed and reported. Nano-micronutrients application promoted the growth, development, yield, and quality traits of sugar beet and has the potential to improve crop production and plant nutrition. Nano-fertilizations can gradually provide the crops with their essential nutrients and can be of environmental and economic significance in comparison to chemical fertilization.

Keywords

Boron · Iron · Manganese · Micronutrients · Nano-fertilizers · Zinc

M. S. Abbas (✉) · M. A. Shahba
Natural Resources Department, Faculty of African Postgraduate Studies, Cairo University, Giza, Egypt
e-mail: msaelsarawy@cu.edu.eg

H. S. Mohamed
Animal Nutrition Department, Animal Production Research Institute, Agricultural Research Center, Cairo, Egypt

V. Misra et al. (eds.), *Sugar Beet Cultivation, Management and Processing*,
https://doi.org/10.1007/978-981-19-2730-0_15

Abbreviations

B	Boron
Cl	Chlorine
Cu	Copper
EDTA	Ethylene diamine tetra acetic acid
ENPs	Engineered nanoparticles
Fe	İron
Mn	Manganese
Mo	Molybdenum
Ni	Nickel
NMs	Nanomaterials
TSS	Total soluble solids
Zn	Zinc

15.1 Introduction

Nanotechnology is a novel beneficial discovery; it may provide keener solutions for the current problems in the field of agriculture. Nanotechnology concepts can help farmers to know the effects before and input solutions for a better after (Rameshaiah et al. 2015). Nanotechnology is one of the new technologies that entered almost all aspects of our lives and were used in agriculture production. Nanotechnology has the potential to increase food quality, raise global food production, protect plants, detect plant and animal diseases, and monitor plant growth and waste reduction for "sustainable amplification" (Khan and Rizvi 2017). This technology is used in all stages of the production of agricultural products such as processing, packaging, transport, and storage. It is used in the detection and control of diseases. One of the most important uses of nanotechnology is in the field of plant fertilization (Mousavi and Rezaei 2011; Srilatha 2011; Ditta 2012). The aim of the application of nanomaterials in agriculture is to reduce the applied amount of plant protection products, minimize the nutrient loss in fertilization, and increase the yield through optimized nutrient management (Predoi et al. 2020).

Sugar beet (*Beta vulgaris* L.) belongs to Chenopodiaceae family. It is a biennial plant and one of the most important sugar crops in the world (Watson and Dallwitz 1992). Sugar beet ranks as the world's second important sugar crop. The great importance of the sugar beet crop is underlined in its ability to grow in the newly reclaimed areas as economic crop and its ability to produce high sugar yield (Hassnein et al. 2019). Sugar beet plants fertilized with micronutrients achieved the highest root and sugar production (Abd El-Hadi et al. 2002; Ramadan and Nassar 2004; Nemeat-Alla and Mohamed 2005; Nemeat-Alla et al. 2009; Moustafa et al. 2011; Amin et al. 2013; Abbas et al. 2014, 2020; Hassnein et al. 2019). Also, Asadzade et al. (2015) and Mekdad and Rady (2016) mentioned that adding micronutrient mixtures (Fe, Zn, and Mn) has improved yield and other attributes of sugar beet crop. Nemeat-Alla et al. (2014) showed that micronutrients application gave the maximum yield and quality for sugar beet crop. Contrary, deficiency of soil nutrients such as macro- and micronutrients should be added to the rhizosphere according to plant needs and has been known as the major limitations in beet crop production (Abido 2012). Also, the sugar beet exhibits the greatest sensitivity to the deficiency of micronutrients in the soil (Christenson and Draycott 2006). Therefore, the current chapter presents a review of evidence related to the roles of nano-fertilizers in sugar beet production and management.

15.2 Definition of Nano-Fertilization

Nano-fertilizers are known as nanomaterials that can provide nutrients to plants or help to increase the activity of traditional fertilizers without direct contact with crops. Nano-fertilizers are new generation of the synthetic fertilizers which contain readily available nutrients in nano scale range (Janmohammadi et al. 2016), which improves the ability of plants to absorb nutrients (Mousavi and Rezaei 2011; Srilatha 2011; Ditta 2012).These materials have unique properties of very small size ranging from 8 μm to 10 nm (Das et al. 2004), Also, engineered nanoparticles (ENPs) are able to enter plant cells and leaves, and can also transfer DNA and chemicals in plant cells (Galbraith 2007; Torney et al. 2007). Nano-fertilizers are one potential output that could be a major innovation for agriculture; the large surface area and small size of the nanomaterials could allow for enhanced interaction and efficient uptake of nutrients for crop fertilization (DeRosa et al. 2010). The integration of nanotechnology in fertilizer products may improve release profiles and increase uptake efficiency, leading to significant economic and environmental benefits.

Subramanian et al. (2015) indicated that nano-fertilizers are nutrient carriers of nano-dimensions ranging from 30 to 40 nm and capable of holding bountiful of nutrient ions due to their high surface area and release it slowly and steadily that commensurate with crop demand. However, Chhipa and Joshi (2016) reported that nano-fertilizers are divided into three categories. These are macro-nano-fertilizers, micro-nano-fertilizers, and nanoparticulate fertilizers, depending on nutrient requirements of the plants. Also, nanostructured fertilizers in the form of nanocarriers, nano-capsules, and nano-nutrients could be considered as smart fertilizers, which can enhance the efficiency of plant nutrients use, control the nutrients release, and reduce the environmental pollution (Yaseen et al. 2020).

A nano-fertilizer is any product that is made with nanoparticles or uses nanotechnology to improve nutrient efficiency. Nano-fertilizers are being studied as a way to increase nutrient efficiency and improve plant nutrition, compared with traditional fertilizers (Mikkelsen 2018). Current applications of nanotechnology in fertilizer and plant protection can be divided into three categories (Mastronardi et al. 2015):

(a) Nanoscale fertilizer inputs: This category describes examples of a nanosized reformulation of a fertilizer input. The fertilizer or supplement is reduced in size, using mechanical or chemical methods, down to the nanoscale. The input is typically in the form of nanoparticles but may also be in other forms.

(b) Nanoscale additives: This category includes examples where the nanomaterials are added to bulk (>100 nm scale) product. These nanomaterials may be a supplement material added for an ancillary reason, such as water retention or pathogen control in plants or soils.

(c) Nanoscale coatings or host materials for fertilizers: This category describes nano-thin films or nano-porous materials used for the controlled release of the nutrient input. These include, for example, zeolites, other clays, and thin polymer coatings.

15.3 The Importance and Advantages of Nano-Fertilization

Lin and Xing (2007) and Navarro et al. (2008) attributed the high proficiency of the nano-fertilizers to:

- Reactivity of nanomaterials with the other compounds is higher than those of ordinary ones, due to their higher surface area and very less particles size, which provides more sites for plant metabolism.
- Enhancement of nutrients penetration and plant uptake, due to the reduced size of the NPs, that increased its specific surface area and particle numbers per unit, which led to increasing the contact surface between the nano-fertilizers and the plants.

Many studies showed that the use of nano-fertilizers causes an increase in nutrients use efficiency, reduces soil toxicity, minimizes the potential negative effects associated with over dosage, and reduces the frequency of the application. Hence, nanotechnology has a high potential for achieving sustainable agriculture, especially in developing countries (Nadzri and Danesh-Shahraki 2013).

Zulfiqar et al. (2019) mentioned that nano-fertilizers offer benefits in nutrition management through their strong potential to increase nutrient use efficiency. Nutrients, either applied alone or in combination, are bound to nano-dimensional adsorbents, which release nutrients very slowly as compared to conventional fertilizers. This approach not only increases nutrient use efficiency, but also minimizes nutrient leaching into ground water. Furthermore, nano-fertilizers may also be used for enhancing abiotic stress tolerance and used in combination with microorganisms (the so-called nano-biofertilizers) to provide great additional benefits.

Conley et al. (2009) mentioned that the aim of using nanomaterials (NMs) in agriculture is to improve the efficiency and sustainability of agricultural practices by developing fewer inputs and generating less waste in comparison to traditional products and approaches, fertilizers are vital for plant growth and development. Most of the added fertilizers remain unavailable to plants due to several factors such as leaching and degradation by hydrolysis, solubility, and decomposition. The addition of traditional fertilizers at a high- and long-term rate in agriculture has caused major environmental issues around the world.

Nano-fertilizers stand as increasing intelligent materials that enhance nutrients phytoavailability of crops (Jahan 2018). Application of nano-fertilizers may improve solubility and dispersion of insoluble nutrients in soil, reduce nutrient immobilization (soil fixation), and increase the bioavailability (Naderi and Danesh-Shahraki 2013). However, Tavan et al. (2014) reported that the use of nano-fertilizers to precisely control nutrient releases an effective step in achieving sustainable and environmentally friendly agriculture.

Guru et al. (2015), showed that the common features of nano-fertilizers including:

- Delivering the appropriate nutrients for enhancing the plant growth through foliar and soil applications.
- Eco-friendly sources of plant nutrients and of low cost.
- Have high efficiency of fertilization process.
- Have a supplementary role with mineral fertilizers.
- Protect the environment from pollution hazards.
- Nano-fertilizers help us to eliminate the contamination of drinking water and could be considered as emerging alternatives of the conventional fertilizers.

Nano-fertilizers play an important role where the ancient chemical fertilizers are replaced with nano and biofertilizers and preferred largely due to their efficiency and environment friendly nature compared to conventional chemical fertilizers (Janmohammadi et al. 2016). Primary use of adding is fast uptake of nutrients from the soil and giving better, faster yield. The symbiotic exchange between soil and the plant system is very efficient. When the same is applied in slow and efficient way, all the required nutrients are taken up by the plant and restores the required and efficient energy in it for which the yield increases drastically (Rameshaiah et al. 2015).

Recent studies revealed that nano-fertilizers can make strides both on germination of seeds and on development of seedlings. This is attributed to its capacity to enter the seeds effectively and to increment accessibility of diverse supplements into the developing seedlings (Antar and Igor 2018). In addition, Boutchuen et al. (2019) revealed that there is an emerging scientific interest in the use of nanoparticle fertilizers for enhanced agricultural and bioenergy crop production to meet the growing food and energy demands of the world. The objective of designing the nanoparticle fertilizers is to effectively deliver the required nutrients for the plants without adding large quantities of fertilizer to the environment. The use of nanoscale micronutrients conduced to suppressing crop disease and the relationship between nutritional status and plant diseases is investigated. Nanomaterials are capable to penetrate into cells of herbs; they can carry DNA and other chemical compounds in the cells extending the possibility in plant biotechnology to target special gene manipulation (Predoi et al. 2020).

15.4 Using the Nanotechnology Image on Micronutrients Fertilizers

Barker and Pilbeam (2007) revealed that micronutrients are those trace elements which are essential for the normal healthy growth and reproduction of plants and animals. The trace elements essential for plants are boron (B), chlorine (Cl), copper (Cu), iron (Fe), manganese (Mn), molybdenum (Mo), nickel (Ni), and zinc (Zn). Differences in the efficiency with which crop varieties are able to utilize low supplies of B, Cu, Mn, Fe, and Zn have resulted in them being labelled as being either "efficient" or "inefficient" for a specific micronutrient (Alloway 2008).

Micronutrients play an important role in various physiological, metabolic, and cellular processes in plants as these transition metals have unpaired electrons that promote oxidation and reduction reactions (Zargar et al. 2015). Several research studies reported that nanoscale of Fe_2O_3 and ZnO application in different plant. i.e. pumpkins (Zhu et al. 2008), mung beans and chickpeas, (Mahajan et al. 2011) peanuts, (Prasad et al. 2012) tomatoes, (Giordani et al. 2012) soybeans, (Sheykhbaglou et al. 2010; Lopez-Moreno et al. 2010; Ghafariyan et al. 2013; Alidoust and Isoda 2013) cucumbers, (Raliya and Tarafdar 2013) watermelons, (Wang et al. 2013; Li et al. 2013) clover, (Feng et al. 2013) wheat, (Ramesh et al. 2014) rice, (Upadhyaya et al. 2017; Jaksomsak et al. 2018) beans, (Dimkpa et al. 2017) maize, (Subbaiah et al. 2016) cotton (Venkatachalam et al. 2017) and black gram (Raja et al. 2019).

Much recent research on nanoparticles in a number of crops has evidence for enhanced germination and seedling growth, physiological activities, gene expression, and protein level indicating their potential use in crop improvement (Kole et al. 2013). Janmohammadi et al. (2016) said that results revealed that days to a thesis and maturity of barley significantly increased after application of nano-fertilizers. Also, Ali and Al-Juthery (2017) reported that nano-fertilizers enhance growth parameters (plant height, leaf area, number of leaves per plant) dry matter production, chlorophyll production, rate of the photosynthesis which result more production and translocation of photosynthesis to different parts of the plant compared with traditional fertilizers. Nanoparticles enhance crop yield, photosynthetic activity, nutrient use efficiency, grain quality, and nitrogen metabolism (Sekhon 2014).

Use of nano-fertilizers instead of common fertilizers may have properties that are valuable to crops which releases the nutrients requirement, discharge of chemicals fertilizers in a controlled way that standardize plant growth and improve activity of target (Farooqut et al. 2016). Also, Farahat et al. (2007) reported that the nano-micronutrients form an important micronutrient needed in small amounts by crop plants. Its important roles in various metabolic and physiological processes in the plant, where it activates some enzymes, regulate metabolism of carbohydrates and proteins, which are essential for various processes, critical to development and differentiation of plant cells. Researchers reported that using nano-fertilizer leads to impressive reductions in nitrogen related environmental harms and an increase in nitrogen use efficiency, resulting in crops yield increases (Gammon 2017). However,

Kalra et al. (2020) found that the nano-encapsulated micronutrients that facilitate increased nutrient utilization efficiency have several properties:

- They possess large surface area because of very small size of particles, thus providing it more area to ease distinct metabolic process in the plant system resulting in production of more photosynthetic products.
- They have high reactivity with other compounds because of wide surface area and very minute size of particles.
- They are highly soluble in solvent such as water.
- They penetrate more into the plant due to nano-fertilizers' particle size less than 100 nm.
- They improve uptake capacity and efficiency of nutrient utilization due to large surface area of nano-fertilizer.
- Nanoparticles encapsulating fertilizers within themselves will hasten the bioavailability and uptake.

Dey et al. (2018) indicated that the loading of nutrients on the nanoparticles is usually done by:

(a) Absorption on nanoparticles.
(b) Attachment on nanoparticles mediated by ligands.
(c) Encapsulation in nanoparticulate polymeric shell.
(d) Entrapment of polymeric nanoparticles.
(e) Synthesis of nanoparticles composed of the nutrient itself.

15.4.1 Zinc (Zn)

It is absorbed by roots as a cation (Zn^{+2}) and as a component of synthetic and natural organic complexes. Its concentration in plants ranges between 25 and 150 ppm. Zinc is important for the synthesis of tryptophane, a component of some proteins and a compound needed for the production of growth hormones (auxins) such as indole acetic acid and gibberellic acid. It is involved in enzyme systems and metabolic reactions. It is necessary for the production of chlorophyll and carbohydrates (Dey et al. 2018). Zinc is involved in hormone biosynthesis, cytoplasm synthesis, activation and function of different enzymes, protein synthesis (Noaema and Barbara 2018). Zinc is an essential micronutrient that produces growth hormones and chloroplast (Palmer and Guerinot 2009). Also, Zn plays a vital role in plant and the most important biochemical and physiological functions of Zn in plants include: participation in biosynthesis of tryptophan—the precursor of auxins; control of carbonic anhydrase; activation of RNA polymerase; stabilization of cytoplasmic membranes; control of oxidative stress through superoxide dismutase and increased plants resistance to water stress (Khan et al. 2004; Lošák et al. 2011; Hafeez et al. 2013).

Nanotechnology plays an important role for the same nano particles which can be used to coat zinc in order to get diffused and soluble zinc (Milani et al. 2010). The effect on different plants of the foliar exposure to nanomaterials as ZnO conduced to increase in shoot length, root length, increase in chlorophyll, soluble leaf protein or increase in acid phosphatase, alkaline phosphatase and phytase (Khodakovskaya et al. 2011; Tarafdar et al. 2014). Application of zinc nano-fertilizer increased the grain mass up to 6% over to control. The higher 1000 grain mass indicates increased individual grain sink strength. The sink strength can be depicted as the output of sink activity and sink size (Yang et al. 2003). Malik and Kumar (2014) indicated that equal ratios between surface area and size of nano particles should be carefully designed if not, total solubility of the zinc will be affected. This is shown by taking ratio of Nano-ZnO and bulk ZnO available on whole. Tarafdar et al. (2014) who suggested that application of zinc nano-fertilizer on pearl millet (*Pennisetum americanum* L.) significantly improved shoot length, root length, root area, chlorophyll content, total soluble leaf protein, plant dry biomass, and increased the grain yield by 37.7%. Marzouk et al. (2019) indicated that the foliar application of zinc nano-fertilizer increased the studied characteristics for snap bean (the highest values of vegetative growth, fresh pod yield, pod physical quality (length, diameter, and fresh weight), dry weight, and pod nutritional value content expressed as P, K, Zn, Mn, Fe, Cu, crude protein, total soluble solids, and fiber) significantly compared with other nano-micronutrients. Also, the combined effect of Flantino cultivar with zinc nano-fertilizer treatment recorded the highest values of vegetative growth, fresh pod yield, pods physical quality, and nutritional value. Rizwan et al. (2019) said that experimental application of ZnO nanoparticles improved the activities of antioxidant enzymes in leaves and roots. Generally, revealing that ZnO nanoparticles with effect in maize biomass and growth are expressed by accelerated exogenous application of nanoparticles further enhanced with biochar application in combination to nanoparticles.

ZnO NPs is nanomaterial with intense antimicrobial activity that is effective to pathogen control growth, also characterized by a lower toxicity in comparison to Ag and with benefits on soil fertility. The application of ZnO NPs conduced to systemic disruption of cellular function of pathogens as *Botrytis cinerea* or *Penicillium expansum* resulting in hyphal malformation and fungal depth (He et al. 2011). In addition, Raliya and Tarafdar (2013) and Raliya et al. (2015a, b) showed that ZnO NPs increased seed germination and seedling vigor and also increased the stem and root growth. Early germination and establishment of seeds in the soil caused early flowering and promoted leaf chlorophyll content. Foliar treatment of ZnO NPs to *Cyamopsis tetragonoloba* and *Solanum lycopersicum* has shown a positive response in terms of biomass production and the chlorophyll and total soluble leaf protein contents. Shaban et al. (2019), showed that the common bean that is obtained from the plant treated by ZnO NPs has no effect on the lipid parameters as well as the function of the kidney and liver of the rats that feed on this common bean. Zhu et al. (2020) indicated that the main route to cross the wheat leaf epidermis for ZnO NPs is via the stomata; then these nanoparticles accumulate and release Zn ions in the apoplast, and the released Zn ions and ZnO NPs are absorbed by mesophyll cells. Du et al. (2019) represented that the ZnO NPs were more significantly effective on the germination and growth of wheat rather than $ZnSO_4$. In addition, they showed that $ZnSO_4$ was more toxic than ZnO NPs at higher dosages. Applications of ZnO NPs are used as an antibacterial, antifungal, and anticancer drug delivery agent; as a biofertilizer in plant system; and as catalysts, (Husen and Iqbal 2019).

15.4.2 Iron (Fe)

It is absorbed by roots as Fe^{+2} and Fe^{+3}. The sufficiency range of Fe in plant tissue is between 50 and 250 ppm. Fe is a structural component of porphyrin molecules: cytochrome, heme protein, Fe-S protein, and leghaemoglobin. These substances are involved in oxidation-reduction reactions in respiration and photosynthesis. Fe is a catalyst to chlorophyll biosynthesis. It is a constituent of nitrogenase, the enzyme essential for N_2 fixation by N-fixing microorganisms (Dey et al. 2018). Fe is essential for chlorophyll development in cell, without iron photosynthesis it is not possible (Moinuddin et al. 2017). Fe nanoparticles have a potential role in plants as a fertilizer, as it can enhance photosynthesis efficiency and nutrient absorption (Rajabi et al. 2013; Rui et al. 2016; Tombuloglu et al. 2019).

Sheykhbaglou et al. (2010) and Dhoke et al. (2013) showed that iron containing nanoparticles have been used as nano-fertilizer for nutrition of plants. As an example, there was observed a positive effect of nano-FeO and nano-Zn-Cu-Fe oxide on the growth of mung (*Vigna radiata*) seedling, as well a positive influence on leaf and pod dry weight on soybean yield and quality. Azarpour et al. (2013) reported that nano iron fertilizers foliar spraying had significant effects at 1% probability level on fresh flower cover yield of saffron. Also, the foliar and root application of nanoparticles of Fe_2O_3 conduced to the increasing of root elongation of soybean and to the increase of photosynthetic parameters by foliar application (Alidoust and Isoda 2013). When using iron oxide NMs as a nano-fertilizer, Rui et al. 2016 performed a study on the effectiveness of Fe_2O_3 NPs as fertilizer for *Arachis hypogaea* has revealed that the Fe_2O_3 NPs and Fe_2O_3 -EDTA effectively increased the root length and plant height and biomass by regulating the phytohormones and antioxidant enzymes' activity. The Fe_2O_3 NPs were adsorbed onto the soil, increasing easy availability of iron to peanut plants. Likewise, growth parameters of *Solanum lycopersicum* were improved under the influence of Fe_2O_3 NPs (Shankramma et al. 2016). In addition, Sebastian et al. (2018) reported that iron oxide NPs with surface-fabricated phenolics from coconut husk could efficiently adsorb Ca and Cd. However, the interesting attribute of the study was the augmented iron accumulation in rice plants as well as tolerance towards calcium and cadmium stress. Increase of biomass and chlorophyll content attested the plant-growth accelerating action of the iron oxide NPs. Yuan et al. (2018) demonstrated a concentration specific role of Fe NPs in promoting growth in *Capsicum annuum* plants. The Fe NPs increased growth in these plants through reorganization of the leaf, increasing chloroplast per grana stacking, and regulating the vascular tissues within the leaf and stem.

15.4.3 Boron (B)

It plays an active role in protein synthesis during seed and cell wall formation. Boron also helps in water and nutrient transportation from root to shoot (Noaema and Barbara 2018). Boron metabolism and transport of carbohydrates, regulation of, meristematic tissue, cell wall synthesis, lignification growth regulator metabolism, phenol metabolism, integrity of membranes, root elongation, DNA synthesis, pollen formation, and pollination are the functiona of boron (Srivastava and Gupta 1996; Xu et al. 2000; Heckman 2007). In addition, Brown et al. (2002) indicated that boron plays a key role in higher plants by facilitating the short-and long-distance transport of sugar via the formation of borate-sugar complexes. In addition, boron may be of importance for maintaining the structural integrity of plasma plant cells membranes. This function is likely related to stabilization of cell membranes by boron association with some membrane constituents. Boron deficiency is often a problem in grape vines (*Vitis vinifera* L.) and in tree fruits, especially apple (*Malus sylvestris* Mill) and olive (*Olea europaea* L.). In field crops, it affects sunflowers (*Helianthus annuus* L.), sugar beet (*Beta vulgaris* L.), black gram (*Vigna mungo* L.), and oilseed rape (canola) (*Brassica napus* L.) (Rerkasem and Jamjod 2004).

15.4.4 Manganese (Mn)

It is associated with activation of enzymes like decarboxylase, dehydrogenase in photosynthesis (Moinuddin et al. 2017). Also, Mn related with photolysis of water in chloroplasts, regulation of enzyme activities, protection against oxidative damage of membranes (Srivastava and Gupta 1996; Xu et al. 2000; Heckman 2007). The application of Mn in concentration 0.05–1 mg/L on Mung bean roots in a Hoagland culture solution conduced to an increase in shoot and root length, dry and fresh biomass, and rootlet number (Pradhan et al. 2013). Dey et al. (2018) showed that plants absorb Mn^{+2} and low- molecular weight organically complexed Mn. Its concentration in plants typically ranges from 20 to 500 ppm. In Mn Functions, activates several important metabolic reactions, Aids in chlorophyll synthesis in photosynthesis because it is essential to electron transfer through chlorophyll to reduce CO_2 to carbohydrate and produce O_2 from H_2O, accelerates germination and maturity, it activates several enzymes that synthesize several amino acids and phenols important to lignin production and Increases availability of P and Ca. Shebl et al. (2019) said that the result showed that the spraying of manganese oxide nanoparticles on *Cucurbita pepo* plants led to the best vegetative growth characteristics, also, the characteristics of the fruits, yield, and the content of photosynthetic pigments.

15.4.5 Copper (Cu)

It is a constituent of several enzymes, with roles in photosynthesis, respiration, protein and carbohydrate metabolism, lignification, and pollen formation (Srivastava and Gupta 1996; Xu et al. 2000; Heckman 2007). Plants absorb Cu^{+2} and as a component of either natural or synthetic organic complexes. Normal Cu concentration in plant tissue ranges between 5 and 20 ppm. Functions: Lignin is a constituent in cell walls that imparts strength and rigidity, essential for erect stature of plants. Several enzymes (polyphenol oxidase and diamine oxidase) important to synthesis of lignin contain Cu. Copper is part of the enzyme cytochrome oxidase that catalyzes electron transfer in the transfer of electrons in respiration. It is important in carbohydrate and lipid metabolism (Dey et al. 2018). Copper nanoparticles play an important role as an antibacterial and antimicrobial agent in the formation of chlorophyll, enhancing porosity and taking part in some enzyme processes (Abbasifar et al. 2020).

Huo et al. (2014) found that mesoporous aluminosilicates have been noted to use as CuO nanoparticles carriers and thus have the potential for macro and micronutrients delivery to the soil Over the last several years. Guin et al. (2015) found the biologically synthesized CuO NPs to be significantly effective against oxidative stress and less toxic than the precursor material. Duman et al. (2016) reported that the antioxidant and DNA-cleavage properties of CuO NPs biosynthesized with the help of chamomile flower extract and act as a chemical nuclease, can generate DNA-cleavage, and may be useful for preventing cell proliferation.

15.4.6 Titanium (Ti)

Titanium dioxide nanoparticles ($nTiO_2$) are promising as efficient nutrient source for plants to improve biomass production due to enhanced nitrogen assimilation, photoreduction activities of photosystem II and electron transport chain, and scavenging of reactive oxygen species(Morteza et al. 2013; Raliya et al. 2015a, b). However, Zheng et al. (2005) revealed that increase in both germination rate and vigor indexes of aged spinach seeds were observed as a result of seed treatment with 0.25–4% of TiO_2NPs. Moreover, the developing chlorophyll, dry weight of plant, rate of photosynthesis, and the action of ribulose-bisphosphate carboxylase/oxygenase were essentially expanded. Also, Yang et al. (2006) demonstrated that nano-anatase TiO_2 treatment could improve the activities of numerous imperative enzymes including nitrate reductase, glutamine synthase, glutamate dehydrogenase, and glutamic-pyruvic transaminase. The effect of spinach roots exposure to TiO_2 nanoparticles present in soil conduced to an enhanced growth rate and chlorophyll as well as an enhanced rubisco activity and photosynthetic rate (Linglan et al. 2008). In addition, TiO_2 NPs due to their combined photo-catalytic and antimicrobial activity, whereas application of TiO_2 NPs reduced *P. cubensis* infection of cucumber by 91% and increase photosynthetic activity by 30% (Cui et al. 2009). Ahmad and

Rasool (2014) stated that TiO_2 nanomaterials application on different crops, e.g., wheat or soybean has increased the yield and reduced the pathogenic diseases, these effects being based on surface properties of TiO_2 nanoparticles as their photocatalytic characteristics. TiO_2 nanoparticles generally cause positive or non-consequential effects on plant growth for different food crops. For example, in hydroponic conditions, it was observed a significant increase in the root and shoot length of *Brassica juncea* seedling treated with TiO_2 nanoparticles (Garcia-Gomez and Fernandez 2019).

15.4.7 Silicon (Si)

Nano silicon dioxide developed the growth of the plant, net rate photosynthesis, level of transpiration, conductance of stomata, rate electron transport, and photochemical quell (Xie et al. 2011). However, Haghighi et al. (2012) and Siddiqui et al. (2014) showed that lower amounts of nano-SiO_2 increased germination of seeds in tomato, or of *Lycopersicum esculentum* seeds germination in concentration of 8 g L^{-1} nano-SiO_2 for a percentage of 22.16%. The same concentration increased the fresh weight of seedlings by 116.58% and seedlings dry weight by 117.46% compared to control, with an important action upon root and shoot growth. Nano-SiO_2 amplified various factors of the growth and conditions of seedlings, i.e., height, diameter of root collar, main length of roots, seedlings lateral root number as well as induction of chlorophyll synthesis under abiotic stress nanoSiO_2.

15.5 Applications of Nano-Fertilization on Crops

Prasad et al. (2012) used seeds of *Arachis hypogea* to examine the influence of ZnO NPs on their growth and yield parameters. Various doses of ZnO NPs (25 nm) influenced the overall plant-growth response in terms of seed germination, seedling vigor index, root growth, flowering, chlorophyll content, and pod yield. Amirnia et al. (2014) conducted a study of the effect of nano-fertilizers application and maternal corm weight on flowering of some saffron (*Crocus sativus* L.) ecotypes, in Iran. Significant differences between nano-fertilizers levels, saffron ecotypes, maternal corm weight and their interactions in terms of all flowering traits highlighted the importance of the nano-fertilizers on improving saffron yield. In addition, it was also clear that Fe, P, and K nano-fertilizers all had positive effects on the saffron flowering. In this regard, Abdel-Aziz et al. (2016), showed that the foliar application of either normal or nano-fertilizer at different concentrations to wheat plants, induced marked significant variable increases in all growth variables determined at fully vegetative and reproductive growth stages.

To evaluate the effects of foliar spray of micronutrient nano-fertilizer (iron and zinc) and nano-titanium dioxide ($nTiO_2$) solution on grain yield and its components in barley under supplemental irrigation conditions, a field experiment was carried out by Janmohammadi et al. (2016) in the semi-arid highland region of Maragheh, Iran.

A considerable improvement was observed in grain mass, spike length, number of the grains per spike, chlorophyll content, grain yield, and harvest index by application of nano-fertilizer. Foliar application of $nTiO_2$ positively affected some morphophysiological characteristics like as days to anthesis, chlorophyll content, and straw yield.

Aghajani and Soleymani (2017) found that nano fertilization types including (nano Biologic and nano Zn-Fe-Mn) resulted in the highest rate of yield and yield components for bean under water sufficient and deficient conditions. Also, Sabaghnia et al. (2017) performed to study the effects of farmyard manure and nano-fertilizers (Mn, Fe, and Zn) on sunflower. The results of this investigation showed that application of nanoparticles may alleviate the adverse environmental factors and improve the sunflower performance and the integrated application of organic manure and nano-micronutrients is more effective.

Jahan (2018) conducted a field experiment to study the influence of using nano fertilization on growth, physiology, and yield parameters of okra. Nano-fertilizer significantly increased growth and physiological parameters such as leaf numbers, the plant height, chlorophyll content, Chl fluorescences, yield, net photosynthesis rate, photo synthetically active radiation, and relative water content. Moreover, nano-fertilizers significantly increased yield parameters of okra production. Abdelkader et al. (2019) conducted a study that was carried out to evaluate the effect of different phosphorus fertilization rates as P_2O_5, nano-micronutrients concentrations (Fe, Zn, B, Mn, Cu, and Mo) as well as their combinations on growth and production of fennel (*Foeniculum vulgare*, L.). The results showed the importance of the nano-micronutrients on improving fennel growth, fruits, and volatile oil yield compared to control. In general, 45 kg P_2O_5/feddan +500 or 1000 mg/L of nano-micronutrients as foliar spray had significant effects in above-mentioned parameters of fennel plant. Merghany et al. (2019) conducted an experiment to determine the effects of liquid nano-fertilizer on cucumber growth, production and quality of cucumber. The results stated that the nano-fertilizer treatments significantly improved the growth and yield of cucumber compared with control treatment. All treatments of nano-fertilizer led to increase plant height, number of leaves / plants, chlorophyll content, yield, and NPK % in leaves and fruits. It can be concluded that nano-fertilizer improved the plant growth, yield, and fruit quality of cucumber and it can be used as an alternative to mineral fertilizers.

Sajyan et al. (2019) studied that, the effect of nano-fertilizer on vegetative and reproductive growth of salt-stressed tomato plants. Nano-fertilizer increased leaf number and stem diameter in salt-stressed plants regardless of the application dose. Flowering characteristics were also improved by nano-fertilizer application under all salinity levels. Consequently, salt tolerance of tomato was ameliorated by nano-fertilizer application.

Shebl et al. (2019) conducted a field experiment to study the influence of nano-fertilizers on *Cucurbita pepo* L. The result indicated that nano-fertilizers improved the growth and the yield in comparison with untreated plants. Furthermore, the yield of fruit squash was significantly affected with Mn nano-oxide especially when it is used individually or combined with Fe nano-oxide. Also, the content of organic matter, protein, lipids, and energy recorded the higher levels in fruits of squash plants that have been sprayed with Fe nano-oxide.

Ghasemi et al. (2020) investigated the effect of nano-fertilizers (n) on the yield components and antioxidant properties of Dragon's head Balangu (*Lallemantia* sp.), the results indicated that the combination of winter cultivation and NPK-n + Fe-chelated-n fertilizers is the most appropriate treatment to acquire highest qualitative and quantitative yield of Dragon's head. An experiment was undertaken in order to assess the performance of new commercial nano-based water-soluble (Nano- Max NPK foliar spray) foliar fertilizer in comparison to commonly adopted water-soluble foliar fertilizer. Results revealed that the treatment as regards average tomato fruit weight (Panda et al. 2020). Tarafder et al. (2020) indicated that the composition of the proposed hybrid nano-fertilizer was functionally valuable for slow and sustainable release of plant nutrients. The obtained result showed a significant increase of Cu^{2+}, Fe^{2+}, and Zn^{2+} nutrient uptake in *A. esculentus* as a result of slow release from hybrid nano-fertilizer, whereas the slow releasing of hybrid nano-fertilizer is observed during leaching studies and confirmed the availability of Ca^{2+}, PO_4^{3-}, NO_2^-, NO_3^-, Cu^{2+}, Fe^{2+}, and Zn^{2+}.

15.6 Nano-Applications Microelements Fertilizing on Sugar Beet

Nano-fertilizer application promoted the growth, development, yield, and quality traits of sugar beet and has the potential to improve crop production and plant nutrition. These conclusions have been reported by several authors, e.g., Abd El-Hadi et al. (2002), Ramadan and Nassar (2004), Nemeat-Alla and Mohamed (2005), Nemeat-Alla et al. (2009), Moustafa et al. (2011), Amin et al. (2013), Asadzade et al. (2015), Mekdad and Rady (2016), Dewdar et al. (2018), Abbas et al. (2020). However, Liu and Lal (2015) reported that the application of nano particles to sugar beet plants can be beneficial for growth and development due to its ability for greater absorbance and high reactivity.

In western Poland, Barłóg et al. (2015) conducted an experiment to study the effect of zinc band application on sugar beet yield, quality, and nutrient uptake. The significant root and sugar yield increase compared to the control was recorded at a level of 0.5–2.0 Zn kg/ha, the best quality of taproots reflected in biological sugar content was observed at a level of 0.5 kg/ha. Also, Masri and Hamza (2015) conducted an experiment to study the influence of foliar application with micronutrients (zinc (Zn) + Manganese (Mn) + Iron (Fe) + Boron (B)) on productivity of sugar beet. The increasing micronutrients mixture significantly increased root weight, root yield, sugar yield, and quality traits, in terms of total soluble solids (TSS), sucrose%, purity%, and extractable sucrose% were significantly increased by increasing levels of micronutrients. However, Rassam et al. (2015) conducted a field experiment to investigate the effect of foliar application of micronutrients on quality and quantity of sugar beet. The foliar spraying of micronutrients significantly increased root yield, content of sucrose, and refined sugar compared to the control.

Hassnein et al. (2019) recommended that using nano-nitrogen fertilizer (Sissay) and micronutrients (B, Zn, and Mn) with mineral nitrogen fertilizer can save 40% from recommended dose of mineral nitrogen fertilizer without insignificant differences in root and sugar yield per plant of sugar beet plants. Furthermore, El-Sherief et al. (2016) conducted an experiment to study the effect both individual and combined applications of B, Zn, and Mn on juice quality and the content of some macro- and micronutrients of sugar beet. All micronutrients at all levels had significant effect on roots and sucrose yields in sugar beet juice at harvest, also, B, Zn, Mn, and their mixture at the highest levels significantly increased roots and sucrose yields. In context, Dewdar et al. (2018) and Abbas et al. (2020) conducted a field experiment to study the influence of foliar spraying nano-micronutrients (Fe, Mn, Zn, and B) on yield and quality of sugar beet. The findings of the study exhibited that the best results were sugar beet plants treated with nano-microelements 200 mg/L + urea 1% could be ranked as the first favorable treatment, this treatment significantly produced the highest yields with improved quality traits of sugar beet and results in saving the plants' needs from micronutrient and nitrogen fertilizer. Moreover, nano-fertilizers have great impact on the soil, can reduce the toxicity of the soil, and decrease the frequency of fertilizer application. The foliar application of Fe, Mn, and Zn mixture was assessed to improve growth and yields and their qualities in two multigerm cultivars of *Beta vulgaris* L. (Mekdad and Rady 2016).

Qotob et al. (2020) reported that the application of nanotechnology in agriculture as using nano-boron increased the application efficiency, decreased pollution and risk of fertilizers used, and increased sugar beet quality. Increasing nano-boron level under different growth stages increased sugar and white sugar contents, on contrary impurities (Na, K, and α-amino-N) loss and molasses sugar percentage were decreased (Pirzad et al. 2019). Also, Rahimi et al. (2016) indicated that micronutrients (Fe, Zn, B, and Mn) enhanced sugar percentage, amount of K, Na, N, alkalinity, crystallized sugar percentage, sugar yield, and percentage of sugar in molasses.

Matsi et al. (2005) have done a survey conducted in order to estimate micronutrient levels (Cu, Zn, Fe, and Mn) in sugar beet plants and soils. Concentrations of DTPA-extractable Fe and Mn, and plant Zn and Mn, were significantly and negatively correlated with soil pH. Soil pH and DTPA-extractable Fe seemed to have a significant positive impact on root, top, and raw sugar yields. However, in all cases, less than 14% of the variance of the sugar beet parameters was explained by soil characteristics. However, Jakienė et al. (2015) performed an experiment to study the effect of the bio-organic nano-fertilizer on improving sugar beet photosynthesis process and productivity. The results indicated that a single application of the bio-organic fertilizer increased the number of leaves, leaf area, root diameter, canopy dry biomass, root biomass, net photosynthetic productivity, root yield, sucrose content and yield of white sugar in comparison with the control treatment.

15.7 Future Prospects

More efforts are required to investigate the nano-micronutrients application on sugar beet (*Beta vulgaris* L.) under different environmental conditions as well as to apply of nano-micronutrients fertilizations on productivity and quality traits of sugar beet under biotic and abiotic stresses.

15.8 Conclusions

The deficiency of soil nutrients such as micronutrients has been known as the major limitations in beet crop production. Nano-micronutrients application promoted the growth, development, yield, and quality traits of different crops and has the potential to improve crop production and plant nutrition. Nano-fertilizers can be used as an alternative to mineral fertilizers due to the nano-micronutrients facilitating increased nutrient utilization efficiency. The foliar application of nano-zinc (Zn), Manganese (Mn), Iron (Fe), and Boron (B) has an important role to improve growth and yield and quality traits of sugar beet (*Beta vulgaris* L.). Nano-fertilizations can gradually provide crops with their essential nutrients and can be of environmental and economic significance in comparison to the chemical fertilization.

References

Abbas MS, Dewdar MDH, Gaber EI, El-Aleem HAA (2014) Impact of boron foliar application on quantity and quality traits of sugar beet (*Beta vulgaris* L.) in Egypt. Res J Pharm Biol Chem Sci 5(5):143–151

Abbas MS, El-Hassanin AS, Dewdar MDH, Abd Elaleem HA (2020) Impact of nano-micronutrients as foliar fertilization on yield and quality of sugar beet roots. Pak J Biol Sci 23(11):1416–1423

Abbasifar A, Shahrabadi F, ValizadehKaji B (2020) Effects of green synthesized zinc and copper nano-fertilizers on the morphological and biochemical attributes of basil plant. J Plant Nutr 43: 1104–1118

Abd El-Hadi AH, Aly MA, Attat AA, Zidan MA, Zahran F (2002) Response of sugar beet to various forms and rates of nitrogen fertilizer. Egypt J Soil Sci 4:643–658

Abdel-Aziz HMM, Hasaneen MNA, Omer AM (2016) Nano chitosan-NPK fertilizer enhances the growth and productivity of wheat plants grown in sandy soil. Span J Agric Res 14(1):1–9

Abdelkader MAI, Ibrahim FR, Metwaly EE (2019) Growth and productivity of fennel (*Foeniculum vulgare*, mill) plants as affected by phosphorus rate and nano-micronutrients concentration. J Plant Product Mansoura Univ 10(7):483–488

Abido WAE (2012) Sugar beet productivity as affected by foliar spraying with methanol and boron. Int J Agric Sci 4(7):287–292

Aghajani A, Soleymani A (2017) Effects of nano-fertilization on growth and yield of bean (*Phaseolus vulgaris* L.) under water deficit conditions. Curr Nanosci 13:2

Ahmad P, Rasool S (2014) Emerging technologies and management of crop stress tolerance. In: Ahmad P, Rasool S (eds) Biological technique, vol 1. Academic Press, San Diego

Ali NS, Al-Juthery HWA (2017) The application of nanotechnology for micronutrient in agricultural production (review article). Iraqi J Agric Sci 9(48):489–441

Alidoust D, Isoda A (2013) Effect of gamma Fe_2O_3 nanoparticles on photosynthetic characteristics of soybean (*Glycine max* (L.) Merr): foliar spray versus soil amendment. Acta Physiol Plant 35(12):3365–3375

Alloway BJ (2008) Micronutrients and crop production: an introduction. In: Alloway BJ (ed) Micronutrient deficiencies in global crop production. Springer Science + Business Media B. V., New York

Amin GA, Badr EA, Afifi MHM (2013) Root yield and quality of sugar beet (*Beta vulgaris* L.) in response to bio fertilizer and foliar application with micro-nutrients. World Appl Sci J 27(11): 1385–1389

Amirnia R, Bayat M, Tajbakhsh M (2014) Effects of Nano-fertilizer application and maternal corm weight on flowering of some saffron (*Crocus sativus* L.) ecotypes. Turkish J Field Crops 19(2): 158–168

Antar B, Igor S (2018) Effect of nano-fertilizer on seed germination and first stages of bitter almond seedlings' growth under saline conditions. Bio Nano Sci 8:742–751

Asadzade N, Moosavi SG, Seghatoleslami MJ (2015) Effect of low irrigation and Zn and SiO_2 nano-fertilizers and conventional fertilizers on morph physiological traits and seed yield of sunflower. Biol Forum Int J 7:357–364

Azarpour E, Asghari J, Bozorgil HR, Kamalpour G (2013) Foliar spraying of ascophyllum nodosum extract, methanol and iron fertilizers on fresh flower cover yield of saffron plant (*Crocus sativus* L.). Int J Agric Crop Sci 5(17):1854–1862

Barker AV, Pilbeam DJ (2007) Introduction. In: Barker AV, Pilbeam DJ (eds) Handbook of plant nutrition. CRC, Taylor & Francis, Boca Raton, FL, pp 3–18

Barłóg P, Nowacka A, Błaszyk R (2015) Effect of zinc band application on sugar beet yield, quality and nutrient uptake. Plant Soil Environ 62(1):30–35

Boutchuen A, Zimmerman D, Aich N, Masud AM, Arabshahi A, Palchoudhury S (2019) Increased plant growth with hematite nanoparticle fertilizer drop and determining nanoparticle uptake in plants using multimodal approach. J Nanomater, Article ID 6890572, 11p

Brown PH, Bellaloui N, Wimmer MA, Bassil ES, Ruiz J, Hu H, Pfeffer H, Dannel F, Romheld V (2002) Boron in plant biology. Plant Biol 4:205–223

Chhipa H, Joshi P (2016) Nano-fertilizers, nanopesticides and nanosensors in agriculture. In: Ranjan S, Dasgupta N, Lichtfouse E (eds) Nanoscience in food and agriculture. Sustainable agriculture reviews, vol 20, pp 247–282

Christenson DR, Draycott AP (2006) Nutrition – phosphorus, Sulphur, potassium, sodium, calcium, magnesium and micronutrients – liming and nutrient deficiencies. In: Draycott AP (ed) Sugar Beet. Blackwell Publishing Ltd, Bury St, Edmunds, pp 185–219

Conley DJ, Paerl HW, Howarth RW, Boesch DF, Seitzinger SP (2009) Ecology controlling eutrophication: nitrogen and phosphorus. Science 323:1014–1015

Cui H, Zhang P, Gu W, Jiang J (2009) Application of anatase TiO_2 sol derived from peroxotitannic acid in crop diseases control and growth regulation. NSTI-Nanotech 2:286–289

Das R, Kiley PJ, Segal M, Norville J, Yu AA, Wang L, Lebedev N (2004) Integration of photosynthetic protein molecular complexes in solid-state electronic devices. Nano Lett 4(6): 1079–1083

DeRosa MC, Monreal C, Schnitzer M, Walsh R, Sultan Y (2010) Nanotechnology in fertilizers. Nat Nanotechnol 5(2):91–93

Dewdar MDH, Abbas MS, El-Hassanin AS, Abd El-Aleem HA (2018) Effect of nano micronutrients and nitrogen foliar applications on sugar beet (*Beta vulgaris* L.) of quantity and quality traits in marginal soils in Egypt. Int J Curr Microbiol App Sci 7(8):4490–4498

Dey JK, Das S, MawlongInt LG (2018) Nanotechnology and its importance in micronutrient fertilization. J Curr Microbiol App Sci 7(5):2306–2325

Dhoke SK, Mahajan P, Kamble R, Khanna A (2013) Effect of nanoparticles suspension on the growth of mung (*Vigna radiata*) seedlings by foliar spray method. Nanotechnol Dev 3:1–5

Dimkpa CO, Bindraban PS, Fugice J, Agyin-Birikorang S, Singh U, Hellum D (2017) Composite micronutrient nanoparticles and salts decrease drought stress in soybean. Agron Sustain Dev 37(5):1–13

Ditta A (2012) How helpful is nanotechnology in agriculture? Adv Nat Sci Nano Sci and Nanotechnol 3:10

Du W, Yang J, Peng Q, Liang X, Mao H (2019) Comparison study of zinc nanoparticles and zinc sulphate on wheat growth: from toxicity and zinc biofortification. Chemosphere 227:109–116

Duman F, Ismail Ocsoy I, Kup FO (2016) Chamomile flower extract-directed CuO nanoparticle formation for its antioxidant and DNA cleavage properties. Mater Sci Eng C Mater Biol Appl 60:33–338

El-Sherief MAB, Moustafa SMI, Neana SMM (2016) Response of sugar beet yield and quality to some micronutrients under sandy soil. J Soil Sci Agric Eng Mansoura Univ 7(2):97–106

Farahat MM, Ibrahim S, Taha LS, ElQuesni FEM (2007) Response of vegetative growth and some chemical constituents of *Cupressus sempervirens* L. to foliar application of ascorbic acid and zinc at Nubaria. World J Agric Sci 3(4):496–502

Farooqut A, Tabassum H, Ahmad A, Mabood A, Ahmad A, Ahmad IZ (2016) Role of nanoparticles in growth and development of plants: a review. Int J Pharm Bio Sci 7(4):22–37

Feng Y, Cui X, He S, Dong G, Chen M, Wang J, Lin X (2013) The role of metal nanoparticles in influencing arbuscular mycorrhizal fungi effects on plant growth. Environ Sci Technol 47: 9496–9504

Galbraith DW (2007) Nanobiotechnology: silica breaks through in plants. Nat Nanotechnol 2(5): 272–273

Gammon K (2017) Slow-release nitrogen fertilizer could increase crop yields yields. Chem Eng News. http://cen.acs.org/articles/95/web/2017/02/Slowrelease-nitrogen-fertilizer-increase.html

Garcia-Gomez C, Fernandez MD (2019) Impacts of metal oxide nanoparticles on seed germination, plant growth and development. Compr Anal Chem 84. Elsevier:76

Ghafariyan MH, Malakouti MJ, Dadpour MR, Stroeve P, Mahmoudi M (2013) Effects of magnetite nanoparticles on soybean chlorophyll. Environ Sci Technol 47:10645–10652

Ghasemi M, Moghaddam S, Rahimi A, Pourakbar L, Popović-Djordjević J (2020) Winter cultivation and nano fertilizers improve yield components and antioxidant traits of dragon's head (*Lallemantia iberica* (M.B.) Fischer & Meyer). Plants 9:252

Giordani T, Fabrizi A, Guidi L, Natali L, Giunti G, Ravasi F, Cavallini A, Pardossi A (2012) Response of tomato plants exposed to treatment with nanoparticles. EQA Int J Environ Qual 8: 27–38

Guin R, Banu S, Kurian GA (2015) Synthesis of copper oxide nanoparticles using Desmodium gangeticum aqueous root extract. Int J Pharmacol Pharm Sci 7:60–65

Guru T, Veronica N, Thatikunta R, Reddy SN (2015) Crop nutrition management with nano fertilizers. Int J Environ Sci Technol 1(1):4–6

Hafeez B, Khanif YM, Saleem M (2013) Role of zinc in plant nutrition – a review. Am J Exp Agric 3:374–391

Haghighi M, Afifipour Z, Mozafarian M (2012) The effect of N-Si on tomato seed germination under salinity levels. J Biol Environ Sci 6:87–90

Hassnein AM, Azab MA, El-Hawary MA, Darwish NN (2019) Effect of nano fertilization on sugar beet. Al-Azhar J Agric Res 44(2):194–201

He LL, Liu Y, Mustapha A, Lin MS (2011) Antifungal activity of zinc oxide nanoparticles against *Botrytis cinerea* and *Penicillium expansum*. Microbiol Res 166(3):207–215

Heckman JR (2007) Chlorine. In: Barker AV, Pilbeam DJ (eds) Handbook of plant nutrition. CRC Press, Taylor & Francis, Boca Raton, FL, pp 279–291

Huo C, Ouyang J, Yang H (2014) CuO nanoparticles encapsulated inside Al-MCM-41 mesoporous materials via direct synthetic route. Sci Rep 4:3682

Husen A, Iqbal M (2019) Chapter: 7 Natural product-based fabrication of zinc-oxide nanoparticles and their applications. In: Husen A, Iqbal M (eds) Nanomaterials and plant potential, vol 7, pp 193–220

Jahan MS (2018) Nano-fertilizer increased growth, physiology and yield parameters of okra plants. 1st Global Conference on Health, Agriculture and Environmental Sciences. Melbourne, Australia. ISBN: 978-0-6481172-7-8

Jakienė E, Spruogis V, Romaneckas K, Dautartė A, Avižienytė D (2015) The bio-organic nano fertilizer improves sugar beet photosynthesis process and productivity. Zemdirbyste-Agric 102(2):141–146

Jaksomsak P, Tuiwong P, Rerkasem B, Guild G, Palmer L, Stangoulis J (2018) The impact of foliar applied zinc fertilizer on zinc and phytate accumulation in dorsal and ventral grain sections of four Thai rice varieties with different grain zinc. J Cereal Sci 79:6–12

Janmohammadi M, Amanzadeh T, Sabaghnia N, Dashti S (2016) Impact of foliar application of nano micronutrient fertilizers and titanium dioxide nanoparticles on the growth and yield components of barley under supplemental irrigation. Acta Agric Slov 107(2):265–276

Kalra T, Tomar PC, Arora K (2020) Micronutrient encapsulation using nanotechnology: nano-fertilizers. Plant Arch 20(2):1748–1753

Khan MR, Rizvi TF (2017) Application of nanofertilizer and nanopesticides for improvements in crop production and protection. In: Nanoscience and plant–soil systems. Springer International Publishing, New York, pp 405–427

Khan HR, McDonald GK, Rengel Z (2004) Zinc fertilization and water stress affects plant water relations, stomatal conductance and osmotic adjustment in chickpea (*Cicer arientinum* L.). Plant Soil 267:271–284

Khodakovskaya MV, deSilva K, Nedosekin DA, Dervishi E, Biris AS, Shashkov EV, Galanzha EI, Zharov VP (2011) Complex genetic, photothermal, and photoacoustic analysis of nanoparticle-plant interactions. Proc Natl Acad Sci U S A 108(3):1028–1033

Kole C, Kole P, Randunu KM, Choudhary P, Podila R, Ke PC, Rao AM, Marcus RK (2013) Nanobiotechnology can boost crop production and quality: first evidence from increased plant biomass, fruit yield and phytomedicine content in bitter melon (*Momordica charantia*). BMC Biotechnol 13(37):1472–6750

Li J, Chang PR, Huang J, Wang Y, Yuan H, Ren H (2013) Physiological effects of magnetic iron oxide nanoparticles towards watermelon. J Nanosci Nanotechnol 13:5561–5567

Lin D, Xing B (2007) Phytotoxicity of nanoparticles: inhibition of seed germination and root growth. Environ Pollut 50:243–250

Linglan M, Chao L, Chunxiang Q, Sitao Y, Jie L, Fengqing G (2008) Rubisco activase mRNA expression in spinach: modulation by nanoatanase treatment. Biol Trace Elem Res 22(2): 168–178

Liu R, Lal R (2015) Potentials of engineered nanoparticles as fertilizers for increasing agronomic productions. A review. Sci Total Environ 514:131–139

Lopez-Moreno ML, de la Rosa G, Hernandez-Viezcas JA, Castillo-Michel H, Botez CE, Peralta-Videa JR, Gardea-Torresdey JL (2010) Evidence of the differential biotransformation and genotoxicity of ZnO and CeO_2 nanoparticles on soybean (*Glycine max*) plants. Environ Sci Technol 44:7315–7320

Lošák T, Hlušek J, Martinec J, Jandák J, Szostková M, Filipčík R, Maňásek J, Prokeš K, Peterka J, Varga L, Ducsay L, Orosz F, Martensson A (2011) Nitrogen fertilization does not affect micronutrient uptake in grain maize (*Zea mays* L.). Acta Agric Scand B 61:543–550

Mahajan P, Dhoke SK, Khanna AS (2011) Effect of nano-ZnO particle suspension on growth of mung (*Vigna radiata*) and gram (*Cicer arietinum*) seedlings using plant agar method. J Nanotechnol 2011:1–7

Malik S, Kumar A (2014) Approach for nano-particle synthesis: using as nano-fertilizer. Int J Pharm Res Biosci 3(3):519–527

Marzouk NM, Abd-Alrahman HA, AMM EL-T, Mahmoud SH (2019) Impact of foliar spraying of nano micronutrient fertilizers on the growth, yield, physical quality, and nutritional value of two snap bean cultivars in sandy soils. Bull Natl Res Cent 43(84):1–9

Masri MI, Hamza M (2015) Influence of foliar application with micronutrients on productivity of three sugar beet cultivars under drip irrigation in sandy soils. World J Agric Sci 11(2):55–61

Mastronardi E, Tsae P, Zhang X, Monreal C, DeRosa MC (2015) Strategic role of nanotechnology in fertilizers: potential and limitations. In: Rai M, Ribeiro C, Mattoso L, Duran N (eds) Nanotechnologies in food and agriculture. Springer International Publishing, Switzerland, pp 25–67

Matsi T, Maslaris N, Barbayiannis N (2005) Micronutrient levels in sugar beet in soils of Greece. J Plant Nutr 28(12):2093–2099

Mekdad AAA, Rady MM (2016) Response of *Beta vulgaris* L. to nitrogen and micronutrients in dry environment. Plant Soil Environ 62:23–29

Merghany MM, Shahein MM, Sliem MA, Abdelgawad KF, Radwan AF (2019) Effect of nano-fertilizers on cucumber plant growth, fruit yield and it's quality. Plant Arch 19(2):165–172

Mikkelsen R (2018) Nanofertilizer and nanotechnology: a quick look. Better Crops 102:3

Milani N, McLaughlin MJ, Hettiaratchchi GM, Beak DG, Kirby JK, Stacey S (2010) Fate of nanoparticulate zinc oxide fertilisers in soil: solubility, diffusion and solid phase speciation. In: Gilkes RJ, Prakongkep N (eds) 19th World Congress of Soil Science. Soil solutions for a changing world, Brisbane, Australia, 1–6 August 2010, pp 172–174

Moinuddin G, Jash S, Sarkar A, Dasgupta S (2017) Response of potato (*Solanum tuberosum* L.) to foliar application of macro and micronutrients in the red and lateritic zone of West Bengal. J Crop Weed 13(1):185–188

Morteza E, Moaveni P, Farahani HA, Kiyani M (2013) Study of photosynthetic pigments changes of maize (*Zea mays* L.) under nano TiO_2 spraying at various growth stages. Springer Plus 2(1): 1–5

Mousavi SR, Rezaei M (2011) Nanotechnology in agriculture and food production. J Appl Environ Biol Sci 1(10):414–419

Moustafa ZR, Soudi AMK, El-Shenawy MK (2011) Productivity and quality of sugar beet as influenced by nitrogen fertilizer and some micronutrients. Egypt J Agric Res 89(3):1005–1017

Naderi MR, Danesh-Shahraki A (2013) Nanofertilizers and their roles in sustainable agriculture. Int J Agric Crop Sci 5(19):2229–2232

Navarro E, Baun,A, Behra R, Hartmann NB, Filser J, Miao AJ, Quigg A, Santschi PH, Sigg L (2008) Environmental behavior and ecotoxicity of engineered nanoparticles to algae, plants, and fungi. Ecotoxicology 17:372–386

Nemeat-Alla EAE, Mohamed SGA (2005) Agronomical and statistical studies on the response of sugar beet to foliar application with micronutrients under different micronutrients formula. J Agric Mansoura Univ 30(12):7265–7284

Nemeat-Alla EAE, Zalat SS, Bader AI (2009) Sugar beet yield and quality as affected by nitrogen levels and foliar application with micronutrients. J Agric Res Kaferelsheikh Univ 35(4): 995–1012

Nemeat-Alla HEA, Nemeata Alla EAE, Mohamed AAE (2014) Response of sugar beet to micronutrients foliar spray under different nitrogen fertilizer doses. Egypt J Agron 36(2): 165–176

Noaema AH, Barbara S (2018) Using the spray of macro- and micronutrients of fertilizers to increase the productivity of potato tubers (*Solomun tuberosum* L.). VI international congress "agricultural machinery" 25–28 June 2018, Burgas, Bulgaria

Palmer CM, Guerinot ML (2009) Facing the challenges of Cu, Fe and Zn homeostasis in plants. Nat Chem Biol 5:333–340

Panda J, Nandi A, Mishra SP, Pal AK, Pattnaik AK, Jena NK (2020) Effects of nano fertilizer on yield, yield attributes and economics in tomato (*Solanum lycopersicum* L.). Int J Curr Microbiol App Sci 9(5):2583–2591

Pirzad A, Mamyandi MMM, Khalilzadeh R (2019) Physiological and morphological responses of sugar beet (*Beta vulgaris* L.) subjected to nano-boron oxide at different growth stages. Acta Biol Szeged 63(2):103–111

Pradhan S, Patra P, Das S, Chandra S, Mitra S, Dey KK, Akbar S, Palit P, Goswami A (2013) Photochemical modulation of biosafe manganese nanoparticles on *Vigna radiata*: a detailed molecular, biochemical and biophysical study. Environ Sci Technol 47(22):13122–13131

Prasad TNVKV, Sudhakar P, Sreenivasulu Y, Latha P, Munaswamy V, Raja Reddy K, Sreeprasad TS, Sajanlal PR, Pradeep T (2012) Effect of nanoscale zinc oxide particles on the germination, growth and yield of peanut. J Plant Nutr 35:905–927

Predoi D, Ghita RV, Iconaru SL, Cimpeanu CL, Raita SM (2020) Application of nanotechnology solutions in plants fertilization. In: Solankey SS, Akhtar S, Maldonado AIL, Rodriguez-Fuentes-H, Contreras JAV, Reyes JMM (eds) Urban horticulture – necessity of the future [Internet]. IntechOpen, London. [cited 2022 Apr 30]. Available from: https://www.intechopen.com/chapters/71186. https://doi.org/10.5772/intechopen.91240

Qotob MA, Nasef MA, Elhakim HKA, Shaker OG, Habashy NR, Abdelhamid IA (2020) Integrated effect of plant growth regulators with boron sources on some biological parameters of sugar beet. Plant Arch 20(Suppl 1):1783–1793

Rahimi A, Arslan N, Esmzad S (2016) Mikro Elementlerin (Fe, Zn, B ve Mn), Şekerpancarında (*Beta vulgaris* L.) Kalite Kriterleri Üzerine Etkisi. Tarla Bitkileri Merkez Araştırma Enstitüsü Dergisi 25. (Özel sayı-2):71–76

Raja K, Sowmya R, Sudhagar R, Moorthy PS, Govindaraju K, Subramanian KS (2019) Biogenic ZnO and Cu nanoparticles to improve seed germination quality in blackgram (Vigna mungo). Mater Lett 235:164–167

Rajabi HR, Khani O, Shamsipur M, Vatanpour V (2013) High performance pure and Fe^{3+}–ion doped ZnS quantum dots as green nanophotocatalysts for the removal of malachite green under UV-light irradiation. J Hazard Mater 250–251:370–378

Raliya R, Tarafdar JC (2013) ZnO nanoparticle biosynthesis and its effect on phosphorous mobilizing enzyme secretion and gum contents in cluster bean (*Cyamopsis tetragonoloba* L.). Agric Res 2:48–57

Raliya R, Biswas P, Tarafdar JC (2015a) TiO_2 nanoparticle biosynthesis and its physiological effect on mung bean (*Vigna radiata* L.). Biotechnol Rep 5:22–26

Raliya R, Nair R, Chavalmane S, Wang WN, Biswas P (2015b) Mechanistic evaluation of translocation and physiological impact of titanium dioxide and zinc oxide nanoparticles on the tomato (*Solanum lycopersicum* L.) plant. Metallomics 7:1584–1594

Ramadan BSH, Nassar AM (2004) Effect of nitrogen fertilization on yield and quality of some beet varieties. Egypt J Agric Res 82(3):1253–1268

Ramesh M, Palanisamy K, Babu K, Sharma NK (2014) Effects of bulk and nano-titanium dioxide and zinc oxide on physio-morphological changes in *Triticum aestivum* Linn. J Glob Biosci 3: 415–422

Rameshaiah GN, Pallavi J, Shabnam S (2015) Nano fertilizers and nano sensors – an attempt for developing smart agriculture. Int J Eng Res Gen Sci 3(1):314–320

Rassam G, Dashti M, Dadkhah A, Yazdi AK (2015) Root yield and quality of sugar beet in relation to foliar application of micronutrients. Ann West Univ Timişoara Ser Biol 2:87–94

Rerkasem B, Jamjod S (2004) Boron deficiency in wheat: a review. Field Crop Res 89:173–186

Rizwan M, Ali S, Rehhman MZ, Andrees M, Arshad M, Qayyum MF (2019) Alleviation of cadmium accumulation in maize (*Zea mays* L.) by foliar spray of zinc oxide nanoparticles and biochar to contaminated soil. Environ Pollut 248:358–367

Rui M, Ma C, Hao Y, Guo J, Rui Y, Tang X, Zhao Q, Fan X, Zhang Z, Hou T, Zhu S (2016) Iron oxide nanoparticles as a potential iron fertilizer for peanut (*Arachis hypogaea*). Front Plant Sci 7:815

Sabaghnia N, Yousefzadeh S, Janmohammadi M (2017) Treatment by trait biplot analysis of organic manure and nano-fertilizers on sunflower production. Helia 41(69):241–251

Sajyan TK, Naim L, Sebaaly Z, Rizkallah J, Shaban N, Sassine YN (2019) Alleviating the adverse effects of salinity stress on tomato crop (*Solanum lycopersicum*) using nano-fertilizer as foliar application. Acta Hortic 1253:33–40

Sebastian A, Nangia A, Prasad MNV (2018) A green synthetic route to phenolics fabricated magnetite nanoparticles from coconut husk extract: implications to treat metal contaminated water and heavy metal stress in *Oryza sativa* L. J Clean Prod 174:355–366

Sekhon BS (2014) Nanotechnology in agri-food production: an overview. Nanotechnol Sci Appl 7: 31–53

Shaban EE, Elbakry HFH, Ibrahim KS, El Sayed EM, Salama DM, Farrag ARH (2019) The effect of white kidney bean fertilized with nano-zinc on nutritional and biochemical aspects in rats. Biotechnol Rep 23:e00357

Shankramma K, Yallappa S, Shivanna MB, Manjanna J (2016) Fe_2O_3 magnetic nanoparticles to enhance *Solanum lycopersicum* (tomato) plant growth and their bio mineralization. Appl Nanosci 6:983–990

Shebl A, Hassan AA, Salama DM, Abd El-Aziz ME, Abd Elwahed MSA (2019) Green synthesis of nanofertilizers and their application as a foliar for *Cucurbita pepo* L. J Nanomater. Article ID 3476347, 11p

Sheykhbaglou RY, Sedghi M, Shishevan MT, Sharlfi RS (2010) Effects of nano-iron oxide particles on agronomic traits of soybean. Not Sci Biol 2(2):112–113

Siddiqui MH, Al-Whaibi MH, Faisal M, AlSahli AA (2014) Nano-silicon dioxide mitigates the adverse effects of salt stress on *Cucurbita pepo* L. Environ Toxicol Chem 33(11):2429–2437

Srilatha B (2011) Nanotechnology in agriculture. J Nanomed Nanotechnol 2:5–7

Srivastava PC, Gupta UC (1996) Trace elements in crop nutrition. Science Publishers, Lebanon, NH, p 356

Subbaiah LV, Prasad TN, Krishna TG, Sudhakar P, Reddy BR, Pradeep T (2016) Novel effects of nanoparticulate delivery of zinc on growth, productivity, and zinc biofortification in maize (Zea mays L.). J Agric Food Chem 64(19):3778–3788

Subramanian KS, Manikandan A, Thirunavukkarasu M, Rahale CS (2015) Nano-fertilizers for balanced crop nutrition. Nanotechnol Food Agric 3:69–80

Tarafdar JC, Raliya R, Mahawar H, Rathore I (2014) Development of zinc nanofertilizer to enhance crop production in pearl millet (*Pennisetum americanum*). Agric Res 3(3):257–262. https://doi.org/10.1007/s40003-014-0113-y

Tarafder C, Daizy M, Alam M, Ali R, Islam J, Islam R, Ahommed S, Aly MAS, Khan ZH (2020) Formulation of a hybrid nano-fertilizer for slow and sustainable release of micronutrients. Am Chem Soc Omega 5:23960–23966

Tavan T, Niakan M, Norinia AA (2014) Effect of nano-potassium fertilizer on growth factors, photosynthetic system and protein content in wheat (*Triticum aestivum* L. Cv N8019). J Plant Environ Physiol 9(3(35)):61–71

Tombuloglu H, Slimani Y, Tombuloglu G, Almessiere M, Baykal A (2019) Uptake and translocation of magnetite (Fe_3O_4) nanoparticles and its impact on photosynthetic genes in barley (*Hordeum vulgare* L.). Chemosphere 226:110–122

Torney F, Trewyn BG, Lin VS, Wang K (2007) Mesoporous silica nanoparticles deliver DNA and chemicals into plants. Nat Nanotechnol 2:295–300

Upadhyaya H, Roy H, Shome S, Tewari S, Bhattacharya MK, Panda SK (2017) Physiological impact of zinc nanoparticle on germination of rice (*Oryza sativa* L) seed. J Plant Sci Phytopathol 17(1):062–070

Venkatachalam P, Jayaraj M, Manikandan R, Geetha N, Rene ER, Sharma NC, Sahi SV (2017) Zinc oxide nanoparticles (ZnONPs) alleviate heavy metal-induced toxicity in *Leucaena leucocephala* seedlings: a physiochemical analysis. Plant Physiol Biochem 110:59–69

Wang WN, Tarafdar JC, Biswas P (2013) Nanoparticle synthesis and delivery by an aerosol route for watermelon plant foliar uptake. J Nanopart Res 15:1–13

Watson L, Dallwitz M (1992) The families of flowering plants: descriptions, illustrations, identification, and information retrieval. Version Http://biodiversityunoedu/delta/

Xie Y, Li B, Zhang Q, Zhang C, Lu K, Tao G (2011) Effects of nano SiO_2 on photosynthetic characteristics of *Indocalamus barbatus*. J Northeastern Forest Univ 39:22–25

Xu G, Magen H, Tarchitzky J, Kafkaf U (2000) Advances in chloride nutrition of plants. Adv Agron 28:97–150

Yang J, Zhang J, Wang Z, Zhu Q (2003) Hormones in the grains in relation to sink strength and postanthesis development of spikelets in rice. Plant Growth Regul 41(3):185–195

Yang F, Hong FS, You WJ, Liu C, Gao FQ, Wu C, Yang P (2006) Influences of nano-anatase TiO_2 on the nitrogen metabolism of growing spinach. Biol Trace Elem Res 110:179–190
Yaseen R, Ahmed AIS, Omer AM, Agha, MKM, TM (2020) Nano-fertilizers: bio-fabrication, application and biosafety. Novel Res Microbiol J 4(4):884–900
Yuan J, Chen Y, Li H (2018) New insights into the cellular responses to iron nanoparticles in *Capsicum annuum*. Sci Rep 8(1):3228
Zargar SM, Mahajan R, Farhat S, Nazir M, Ahmad Mir R, Nazir M, Salgotra RK, Mallick SA (2015) Understanding the role of iron and zinc in animals and crop plants from genomics perspective. Trends Biotechnol Pharm 9(2):181–196
Zheng L, Hong FS, Lu SP, Liu C (2005) Effect of nano-TiO_2 on strength of naturally and growth aged seeds of spinach. Biol Trace Elem Res 104:83–91
Zhu H, Han J, Xiao JQ, Jin Y (2008) Uptake, translocation, and accumulation of manufactured iron oxide nanoparticles by pumpkin plants. J Environ Monit 10(6):713–717
Zhu J, Li J, Shen Y, Liu S, Zeng N, Xinhua Zhan X, White JC, Gardea-Torresdey J, Xing B (2020) Mechanism of zinc oxide nanoparticle entry into wheat seedling leaves. Environ Sci Nano 7(12):3901–3913
Zulfiqar F, Navarro M, Ashraf M, Akram NA, SergiMunné-Bosch S (2019) Nanofertilizer use for sustainable agriculture: advantages and limitations. Plant Sci 289:110270

Silicon Foliar Application in Sugar Beet Production

16

Arkadiusz Artyszak

Abstract

In Europe, the only raw material for sugar production is sugar beet. In its cultivation, modern and friendly environment sustainable methods for an increase of yield are sought. The implementation of the "Green Deal" in the countries of EU (European Union) assumes limitation of the use of pesticides by half and fertilizers by 20% by the year 2030. This may limit agricultural production. To prevent this, it is a need to search for innovative technologies. One of them can be the foliar usage of products that contain silicon, which increases the root yield, does not deteriorate their technological quality, and thus improves the biological sugar yield and the pure sugar yield.

Keywords

Foliar application · Pure sugar yield · Silicon · Sugar beet

Abbreviations

EU	European Union
Fm′	Maximum fluorescence
Fs	Stationary fluorescence
LAI	Leaf area index
NDVI	Normalized differential vegetation index
PAR	Photosynthetic active radiation
ΦPSII	Effective quantum efficiency PSII

A. Artyszak (✉)
Institute of Agriculture, Warsaw University of Life Sciences—SGGW, Warsaw, Poland
e-mail: arkadiusz_artyszak@sggw.edu.pl

V. Misra et al. (eds.), *Sugar Beet Cultivation, Management and Processing*,
https://doi.org/10.1007/978-981-19-2730-0_16

16.1 Introduction

Sugar beet is of strategic importance for food processing and agriculture in Europe. In 2019, the area of sugar beet cultivation in the European Union (EU) (28 countries) amounted to 1.6 million ha (FAO 2021). The abolition of sugar production limitations and the increasing emphasis on the implementation of environmentally friendly practices resulted in a deterioration of the situation of producers. Therefore, new production methods are sought. In the last 10 years, interest in the use of foliar silicon in various crops, including sugar beet, is dynamically growing in Europe (Artyszak 2018).

Publications on the effects of foliar application of silicon come from the experiments carried out in Poland and the Czech Republic. Because of the regulations in force in the EU, products marketed as foliar fertilizer must contain a macro- or micronutrient addition or be registered as growth stimulants. In the field trials, products containing silicon as a stabilized orthosilicic acid, potassium silicates, and silica nanoparticles, as well as micronized marine calcite were applied (Table 16.1). Sometimes products without silicon were additionally applied in the examined treatments.

Table 16.1 Characteristics of products used in research with foliar usage of silicon in the crop production of sugar beet

Product	Form	Composition, g dm^{-3} (kg^{-1})
Barrier Si-Ca	Calcium foliar fertilizer	Si – 340 g, Ca – 207 g; pH 9.5–11.0
Forte gama	Macronutrient fertilizer with the supplement of boron	N – 96 g; P – 63 g; K – 60 g; B – 6.2 g
Herbagreen basic	Finely ground marine calcite	SiO_2 – 143 g; Ca – 274 g; Fe – 31 g; Mg – 13 g; S – 2 g; K – 0.8 g, Na – 0.4 g, Zn – 0.02 g
Herbagreen Z20	Finely ground marine calcite	SiO_2 – 278 g; Ca – 220 g; Fe – 21 g; Mg – 10 g; K – 5 g; S – 2 g; Na – 2; P – 0.9 g; Zn – 0.02 g
K-gel 175	Macronutrient fertilizer	K – 145 g; S – 58 g
NanoFYTSi	Hydrated SiO_2 nanoparticles	SiO_2 – 230 g; pH 8.0–10.0;
N-Fenol max	Auxiliary plant preparation based on nitrophenols	Data not available
Optysil	Sodium metasilicate and iron chelate	Si – 94 g; Fe – 24 g; neutral reaction
Retafos prim	Macronutrient fertilizer with the supplement of boron	N – 125 g; P – 109 g, K – 208 g, B – 5 g
Sarmin maKSi	Potassium silicate	Si – 150 g, K – 125 g
YaraVita Actisil	Orthosilicic acid stabilized with choline, addition of ca	Si – 6 g, Ca – 20 g, choline, pH 0.1

Source: own elaboration based on information from producers

16.2 Influence of Silicon Foliar Usage on the Yielding and Technological Quality of Roots of Sugar Beet

The final trait in sugar beet production is the yield of pure sugar, which depends on the yield of biological sugar and the content of molasses-forming components that make sugar extraction difficult. The yield of biological sugar is the product of the yield of roots and the content of sugar in the roots. However, the yield of roots results from the plant density and fresh mass of roots. The molasses-forming components include α-amino nitrogen, potassium, and sodium ions. The effect of individual components on the yield of pure sugar is varied, as evidenced by the value of the path coefficients (i.e., standardized regression coefficient based on multiple regression); root yield (+0.84 to −1.00), the content of sugar in the roots (+0.23 to −0.47), α-amino nitrogen (from −0.03 to −0.07), potassium (from −0.02 to −0.07), and sodium (from +0.01 to −0.03) (Artyszak 2012). The aim of sugar beet crop management is therefore a high root yield of the highest technological quality—high sugar content and low molasses—forming components.

The results of the study with silicon foliar application prove that it has a positive effect on the yield of biological sugar and yield of pure sugar, and the effects vary over the years. Most often, better results were obtained when plants were subjected to a strong influence of stress factors, e.g., drought. When the conditions were more favorable, the yield gains were smaller. The increase in the yield of biological sugar obtained in the study amounted to 3.0–24.8%, and the pure sugar yields were 4.8–26.2%. This increase was the result of an increase in the yield of roots which reached 25.1%.

The effect of foliar usage of silicon on the technological quality of the roots was varied. For this treatment, the content of sugar in the roots varied from −0.3 to +1.1 pp., α-amino nitrogen content from −26.3 to 31.8%, potassium from −20.6 to +13.9%, and sodium from −54.4 to +22.6% in comparison to the control. It can therefore be concluded that, in general, foliar application of silicon does not significantly affect the processing quality of beetroots, and it is worth emphasizing that it does not deteriorate it.

16.3 Effect of Silicon Foliar Usage on the Chemical Composition of Sugar Beet Plants

In the available results of other studies, there are only some of them on the silicon content in sugar beet and even less on the impact of foliar application of this element on the plants' chemical composition. In studies conducted in 2013–2014, the use of 6 combinations of foliar nutrition with fertilizers containing macro- and micronutrients without the addition of silicon, the average silicon content in sugar beet leaves during harvest was 1.26 g Si kg^{-1}, and in roots 0.93 g Si kg^{-1}. An average of 9.37 kg Si ha^{-1} was stored in the leaves, and 23.2 kg Si ha^{-1} in the roots (32.57 kg Si ha^{-1} in total). It follows that the leaves stored 29.7% and the roots

70.3% of the taken silicon. It was not found that the silicon content, both in leaves and in roots, was significantly related to pure sugar yield (Artyszak et al. 2019a).

Herbagreen and Optysil were used in another experiment also performed in 2013–2014. Each product was applied once, twice, and three times, and the effects were compared with the control (without foliar application). The application was made in the 4–6 sugar beet leaf stage, 7 and 14 days later. The dose of Herbagreen was 1 kg ha^{-1}, and Optysil was 0.5 dm ha^{-1}. It was found that foliar application had no significant effect on the N, P, K, and Si content in the leaves and P, K, Mg, Ca, and Si in the roots during harvest. On the other hand, it significantly differentiated the dry matter content, Mg and Ca in the leaves, and the content of dry matter and N in the roots. The highest dry matter content (15.6%), Mg (20.1 g kg^{-1}), and Ca (29.3 g kg^{-1}) in leaves were obtained in the treatment with the triple application of Herbagreen. On the other hand, the highest dry matter content (24.9%) and N were found in the variant with a single application of Optysil. The content of silicon in leaves averaged 0.79 g Si kg^{-1}, and in the roots −1.26 g Si kg^{-1}. The silicon accumulation in leaves was 4.59 kg Si ha^{-1}, in roots 32.87 kg Si ha^{-1}, and the total uptake was 37.46 kg Si ha^{-1}. This means that 12.3% of the silicon taken up by plants was accumulated in the leaves, and 87.7% in the roots. The pure sugar yield was significantly positively correlated with the silicon content in leaves (Artyszak et al. 2018).

16.4 Influence of Silicon Foliar Usage on the Morphological Traits of Sugar Beet Plants

Usage of the Herbagreen fertilizer in 2010–2012 had a significant positive effect on the dry matter content of the leaf blades, leaf dry weight and area, and the leaves number per plant. There was also a trend towards the greater accumulation of dry matter in petioles and roots under the influence of foliar application. In all variants, the yield of roots, the yield of biological sugar, and the yield of pure sugar were significantly positively correlated with the roots' dry matter yield, the total dry matter yield (roots + leaves), and the harvest index (HI). On the other hand, the relationship between the root yield, the yield of biological sugar, and the yield of pure sugar with the dry petiole weight and the leaf number per plant was significant only in the variant where Herbagreen was applied twice in the 4–6 leaf stage and 3 weeks later in the doses of 2 kg ha^{-1}. The content of sugar in the roots was significantly positively correlated with the leaf number per single plant (Artyszak et al. 2016b).

In studies with another variety of sugar beet in 2011–2012, the application of Herbagreen fertilizer caused the following tendency to increase dry weight of leaf blades, petioles, root, and whole plant, as well as leaf surface and their number, compared to the control treatment. The yield of roots was significantly positively correlated with the dry weight of petioles and their percentage in plant dry mass and negatively with a dry mass of roots in plant dry mass. Simultaneously, a significant positive correlation was found between the yield of biological sugar and the yield of pure sugar yield with the percentage of petiole dry weight in the dry weight of the total plant, and a negative correlation with the percentage of the dry weight of the root in the total dry weight of the plant (Artyszak et al. 2016d).

16.5 Influence of Silicon Foliar Usage on the Sugar Beet Physiological Parameters

Studies conducted in 2013–2014 assessed the effect of foliar application of Herbagreen Basic

and Optysil used once, twice, and three times on selected physiological traits of sugar beet plants: leaf area index (LAI), absorption of photosynthetic active radiation (PAR); and after leaves adaptation to the light chlorophyll *a* fluorescence parameters: stationary fluorescence (Fs), maximum fluorescence (Fm′), and effective quantum efficiency PSII (ΦPSII). All variants of silicon foliar usage had a positive effect on LAI and PAR absorption. Yields of roots and pure sugar were significantly positively correlated with PAR absorption and LAI. Simultaneously, those yields were negatively correlated with fluorescence parameters like Fs and Fm′ (Artyszak et al. 2016a).

In the studies performed in 2015–2016, foliar usage of silicon in YaraVita Actisil, Herbagreen Z20, and Optysil products modified the leaf area index (LAI), the photosynthetically active radiation absorption (PAR), and the normalized differential vegetation index (NDVI). On average, the mean LAI value assessed 7 days after the first treatment on the variants with foliar application of silicon, increased in comparison to the measurement on the day of the first treatment by the same value as on the control combination – 0.36. After 14 days from the first application, this increase in comparison to the first date of 1.01 (control variant 0.65), and after 21 days – 1.98 (control variant 1.66). In both years of research, the highest LAI value in the date of the fourth measurement was achieved by the use of YaraVita Actisil, then Optysil stimulator, and Herbagreen Z20 fertilizer. The number of foliar fertilization performed with a given product had no major impact on the LAI value assessed at this time of measurement. The exception was the treatment with the triple use of the Optysil stimulator, in which significantly higher LAI values were obtained than with the single and double use of this product.

On average, for both years of the study, PAR absorption assessed 7 days after the first treatment with silicon foliar application variants increased by 15.9 pp. in comparison to the measurement performed 7 days earlier, and by 17.8 pp. in the control treatment. After 14 days from the first treatment, this increase was 35.9 pp. compared to the 14 days earlier (control variant 26.8 pp), and after 21 days from the first 50.5 pp. (control variant 47.8 pp). The highest value of PAR absorption at the last date of measurement was obtained after using the YaraVita Actisil fertilizer. No significant differentiation was found in the value of this trait depending on the number of sprays with this product.

On average, for the years 2015–2016, the NDVI indicator, assessed 7 days after the first treatment, in the variants with foliar application increased by 0.13 compared to the first treatment, and by 0.12 in the control treatment. After the third term, this increase was compared to the first date 0.31 (control variant 0.28), and in relation to the fourth date, 0.46 (control option 0.41). The highest value of the NDVI index during the fourth measurement was found in the case of using the YaraVita Actisil fertilizer and the Optysil stimulator. No significant differences were observed depending on the number of applications made with these products (Artyszak 2017).

16.6 Disease Impact of Silicon Foliar Application in Sugar Production

One of the most dangerous diseases of sugar beet in Europe is a beetroot tassel caused by *Cercospora beticola* Sacc. The experiments performed in Poland since 2010 with the foliar application of silicon was conducted in the area of strong pathogen pressure, which forces the use of up to 4 fungicide sprays, which significantly increases the costs of sugar beet cultivation. In the years of very high risk, it was observed that foliar application of silicon-containing products delayed the onset of infection by several days compared to the control.

In addition, in the years 2019–2020, it was found that 3 times the application of Barrier Si-Ca in the absence of fungicide protection increased the root yield by 17.0%, sugar content by 0.1 pp., the biological sugar yield by 15.1% and the yield of pure sugar by 15.5% compared to the control object. However, in combination with four sprays of fungicides during the growing season, it resulted in an increase in yield by 22.8%, sugar content by 0.2 pp., yield of biological sugar by 25.2%, and pure sugar yield by 26.2% (Artyszak 2019, 2020) (Table 16.2).

16.7 Profitability of Silicon Foliar Usage in Sugar Beet Production

The results of the experiments with foliar application of products containing silicon, regardless of the number of applications and the products used, showed a positive influence on the profitability of sugar beet production. In the experiment carried out in Poland in 2015–2016, the gross production value of sugar beet as a result of foliar application with silicon increased from 1.3 to 22.9%, and the increase in the net production value from 5.5 to 19.0%. For the entire period of the study, the best results were noted with the YaraVita Actisil foliar fertilizer applied two and three times, and three applications of Optysil. The foliar supplementation profitability index ranged from 3.3 to 11.3. It achieved the highest value for a single application of Optysil (Artyszak et al. 2019b).

The nutrition of Herbagreen Basic in the dose of 1 kg ha^{-1} in the 4–6 leaf stage and in the dose of 2 kg ha^{-1} 21 days later in 2010–2012 resulted in an increase in the gross production value of sugar beet by 24.8%, and in the combination with the nutrition of Herbagreen Basic in the dose 2 kg ha^{-1} in the 4–6 leaf stage and 2 kg ha^{-1} 3 weeks later by 25.6% over the control. In an experiment conducted on

Table 16.2 The effect of silicon foliar application on the yield and technological quality of sugar beet roots

Variants	The difference to control (no foliar application)							Source
	%				%			
	Yield of roots	Biological sugar yield	Pure sugar yield	Content sugar, p.p.	Content α-amino nitrogen	Content potassium	Content sodium	
2010–2012 (Poland)								
Herbagreen Basic (1 kg ha^{-1}) in the 4–6 leaf stage + Herbagreen Basic (2 kg ha^{-1}) 3 weeks later	+21.7	+23.9	+24.3	+0.2	+6.5	−1.0	+0.8	Artyszak et al. (2016c)
Herbagreen Basic (2 kg ha^{-1}) in the 4–6 leaf stage + Herbagreen Basic (2 kg ha^{-1}) 3 weeks later	+21.8	+24.8	+25.2	+0.2	+0.5	−3.8	−6.7	
2011–2012 (Poland)								
Herbagreen Basic (1 kg ha^{-1}) in the 4–6 leaf stage + Herbagreen Basic (2 kg ha^{-1}) 3 weeks later	+14.6	+15.9	+17.6	+0.2	−17.9	−8.4	−18.6	Artyszak et al. (2014)
Herbagreen Basic (2 kg ha^{-1}) in the 4–6 leaf stage + Herbagreen Basic (2 kg ha^{-1}) 3 weeks later	+11.6	+15.1	+17.8	+0.6	−26.3	−11.8	−19.8	
2013–2014 (Poland)								
Herbagreen Basic (1.5 kg ha^{-1}) in the 4–6 leaf stage	+16.2	+18.1	+17.7	+0.3	+0.5	0.0	−4.2	Artyszak et al. (2015)
Herbagreen Basic (1.5 kg ha^{-1}) in the 4–6 leaf stage + Herbagreen Basic (1.5 kg ha^{-1}) 1 week later	+11.1	+13.3	+13.6	+0.4	+0.5	−1.3	−8.2	
Herbagreen Basic (1.5 kg ha^{-1}) in the 4–6 leaf stage + Herbagreen Basic (1.5 kg ha^{-1}) 1 week later + Herbagreen Basic (1.5 kg ha^{-1}) 2 weeks later	+10.4	+11.4	+12.2	+0.2	−4.6	−8.2	−32.7	
Optysil (0.5 dm^3 ha^{-1}) in the 4–6 leaf stage	+14.0	+15.1	+15.6	+0.2	−1.4	−2.9	−10.9	

(continued)

Table 16.2 (continued)

Variants	The difference to control (no foliar application)							Source
	%				%			
	Yield of roots	Biological sugar yield	Pure sugar yield	Content sugar, p.p.	Content α-amino nitrogen	Content potassium	Content sodium	
Optysil (0.5 dm^3 ha^{-1}) in the 4–6 leaf stage + Optysil (0.5 dm^3 ha^{-1}) 1 week later	+15.9	+15.7	+14.3	−0.1	+22.5	+2.1	−8.0	
Optysil (0.5 dm^3 ha^{-1}) in the 4–6 leaf stage + Optysil (0.5 dm^3 ha^{-1}) 1 week later + Optysil (0.5 dm^3 ha^{-1}) 2 weeks later	+13.7	+12.7	+12.2	−0.2	+8.7	0.0	−10.1	
2014 (Poland, two locations, production fields)								
Herbagreen Z20 (4 × 2 kg ha^{-1}) every 7 days from the 4–6 leaf stage	0.0	+3.0	+4.9	+0.5	−20.2	−11.8	−15.7	Artyszak and Kucińska (2016)
2014–2015 (Czech Republic)								
NanoFYT Si (0.5 dm^3 ha^{-1}) July 18 + NanoFYT Si (0.5 dm^3 ha^{-1}) August 19	+3.9	+6.4	+7.3	+0.6	−16.7	+4.0	−35.7	Hřivna et al. (2017)
NanoFYT Si (0.5 dm^3 ha^{-1}) August 19	+8.9	+12.1	+13.5	+0.7	−25.0	−2.3	−48.6	
K-gel 175 (5 dm^3 ha^{-1}) August 6 + NanoFYT Si (0.5 dm^3 ha^{-1}) August 19	+12.8	+19.6	+21.3	+1.1	−25.0	−13.9	−54.4	
2015–2016 (Poland)								
YaraVita Actisil (0.5 dm^3 ha^{-1}) in the 6 leaf stage	+11.3	+7.2	+4.8	−0.7	+29.1	+11.3	–	Artyszak (2017)
YaraVita Actisil (0.5 dm^3 ha^{-1}) in the 6 leaf stage + YaraVita Actisil (0.5 dm^3 ha^{-1}) 1 week later	+24.9	+22.6	+21.6	−0.2	+12.3	+5.6	−4.0	
YaraVita Actisil (0.5 dm^3 ha^{-1}) in the 6 leaf stage + YaraVita Actisil (0.5 dm^3 ha^{-1}) 1 week later + YaraVita Actisil (0.5 dm^3 ha^{-1}) 2 weeks later	+25.1	+23.2	+22.2	−0.2	+9.8	+7.8	−8.0	

Herbagreen Z20 (1 kg ha^{-1}) in the 6 leaf stage	+12.0	+8.7	+7.0	−0.6	+20.4	+5.8	+7.5	
Herbagreen Z20 (1 kg ha^{-1}) in the 6 leaf stage + Herbagreen Z20 (1 kg ha^{-1}) 1 week later	+8.3	+8.4	+7.3	0.0	+20.5	+3.3	+6.0	
Herbagreen Z20 (1 kg ha^{-1}) in the 6 leaf stage + Herbagreen Z20 (1 kg ha^{-1}) 1 week later + Herbagreen Z20 (1 kg ha^{-1}) 2 weeks later	+9.0	+9.7	+8.6	+0.1	+19.2	+4.7	+13.6	
Optysil (0.5 dm^3 ha^{-1}) in the 6 leaf stage	+7.0	+7.1	+5.9	0.0	+20.9	+8.9	+2.0	
Optysil (0.5 dm^3 ha^{-1}) in the 6 leaf stage + Optysil (0.5 dm^3 ha^{-1}) 1 week later	+10.8	+8.5	+6.2	−0.4	+31.8	+13.9	+10.6	
Optysil (0.5 dm^3 ha^{-1}) in the 6 leaf stage + Optysil (0.5 dm^3 ha^{-1}) 1 week later + Optysil (0.5 dm^3 ha^{-1}) 2 weeks later	+18.9	+16.4	+14.1	–0.3	+31.5	+11.5	+22.6	
2017 (Czech Republic)								
Forte gama (4 dm^3 ha^{-1}) + N-Fenol max (0.2 dm^3 ha^{-1}) in the 6 leaf stage + K-gel 175 (5 dm^3 ha^{-1}) in the first decade of August + NanoFYT Si (1 dm^3 ha^{-1}) in the second decade of August	+6.5	+6.2	+6.8	−0.1	−3.6	−4.4	−6.1	Urban and Pulkrabek (2018)
2019–2020 (Poland)								
Barrier Si–Ca (1 dm^3 ha^{-1}) In crop cover complete: leaves cover 90% of ground + Barrier Si–Ca (1 dm^3 ha^{-1}) 1 week later + Barrier Si–Ca (1 dm^3 ha^{-1}) 2 weeks later	+22.8	25.2	26.2	+0.2	−1.1	−20.6	−22.3	Artyszak (2019, 2020)

a different variety of sugar beet for the same combinations in 2011–2012, the production value increased by 15.7% and 15.0%. The increase in the net production value in experiment No. 1 was 22.8% and 23.2%, respectively, and in experiment No. 2–13.9% and 12.8%. The profitability index of the foliar nutrition of Herbagreen Basic in experiment No. 1, on average, for the period 2010–2012 was 12.6 in treatment No. 1 and 10.3 in treatment No. 2. In experiment No. 2, the profitability index of foliar fertilization on average for the period 2011–2012 was 9.0 in treatment No. 1. and 6.8 in treatment No. 2 (Litwińczuk-Bis et al. 2019).

16.8 Future Prospects

The implementation of the "Green Deal" in the EU forces farmers to search for new, innovative solutions. One such solution is silicon foliar application. Further studies should provide an answer to the question of the action mechanisms of this element used as foliar application in sugar beet cultivation.

16.9 Conclusion

The foliar usage of products containing silicon has a beneficial effect on the sugar beet yield, especially in unfavorable conditions for plant growth and development, and simultaneously is the application that is safe for the natural environment. It is also a profitable application for the producer.

References

Artyszak A (2012) Agronomic and habitat conditions for sugar beet yielding and health. In: Treatises and monographs, vol 398. Warsaw University of Life Sciences Press, Warsaw, Poland, p 154

Artyszak A (2017) Possibilities of using silicon for foliar fertilization of sugar beet. Wieś Jutra, Warsaw, Poland, p 128

Artyszak A (2018) Effect of silicon fertilization on crop yield quantity and quality—a literature review in Europe. Plan Theory 7:54. https://doi.org/10.3390/plants7030054

Artyszak A (2019) Report from research for Osadkowski joint-stock company. (unpublished)

Artyszak A (2020) Report from research for Osadkowski joint-stock company. (unpublished)

Artyszak A, Kucińska K (2016) Silicon nutrition and crop improvement: recent advances and future perspective. In: Tripathi DK, Singh VP, Ahmad P, Chauhan DK, Prasad SM (eds) Silicon in Plants. CRC Press Taylor & Francis Group Boca Raton, London, UK; New York, NY, pp 297–319

Artyszak A, Gozdowski D, Kucińska K (2014) The effect of foliar fertilization with marine calcite in sugar beet. Plant Soil Environ 60:413–417

Artyszak A, Gozdowski D, Kucińska K (2015) The effect of silicon foliar fertilization in sugar beet—*Beta vulgaris* (L.) ssp. *vulgaris conv. Crassa* (Alef.) prov. *Altissima* (Döll). Turk J Field Crops 20:115–119

Artyszak A, Gozdowski D, Kucińska K (2016a) Effect of foliar fertilization with silicon on the chosen physiological features and yield of sugar beet. Fragm Agron 33:7–14

Artyszak A, Gozdowski D, Kucińska K (2016b) Sugar beet morphological traits after foliar application with marine calcite. Fragm Agron 33:7–17

Artyszak A, Gozdowski D, Kucińska K (2016c) The effect of calcium and silicon foliar fertilization in sugar beet. Sugar Tech 18:109–114

Artyszak A, Gozdowski D, Kucińska K (2016d) Vliv listové výživy s mořským vápencem na morfologické znaky cukrové řepy. Listy Cukrov Řepar 132:176–179

Artyszak A, Gozdowski D, Kucińska K (2018) Vliv hnojení křemíkem a vápníkem či jen křemíkem na chemické složení cukrové řepy. Listy Cukrov Řepar 134:17–25

Artyszak A, Gozdowski D, Kucińska K (2019a) Impact of foliar fertilization on the content of silicon and macronutrients in sugar beet. Plan Theory 8:136. https://doi.org/10.3390/plants8050136

Artyszak A, Klarzyńska E, Litwińczuk-Bis M, Siuda A (2019b) Profitability of sugar beet foliar nutrition with silicon. Ann PAAAE 21:7–13. https://doi.org/10.5604/01.3001.0013.0539

FAO Food and agriculture data (2021). http://www.fao.org/faostat/en/#data/QC (accessed 12 February 2021)

Hřivna L, Hernandez Kong J, Machálková L, Burešová I, Sapáková E, Kučerová J, Šottníková V (2017) Vliv mimokořenové výživy draslíkem a křemíkem na výnos a kvalitu cukrovky v nestandardních povětrnostních podmínkách roku 2014 and 2015. Listy Cukrov Repar 133: 182–187

Litwińczuk-Bis M, Siuda A, Artyszak A (2019) The economic effects of foliar fertilization of sugar beet with marine calcite. Ann PAAAE 21:188–195. https://doi.org/10.5604/01.3001.0013.2073

Urban J, Pulkrabek J (2018) Navyšeni vynosu a jakosti cukrove řepy pomoci listove vyživy a biologicky aktivnich. Listy Cukrov Repar 134:188–194

Mechanization of Weed Management in Sugar Beet

17

A. E. Ghaly and M. M. Ibrahim

Abstract

Sugar is an essential commodity and an integral part of the food chain as the cheapest source of energy. It plays a vital role in the development of taste and texture. Sugar beet ranks second as a sugar-producing crop in the world with a composition of 75% water, 20% sugar, and 5% pulp. Approximately 60 plant species are identified as important weeds in sugar beet fields. About 70% of weeds are broad-leaved species and 30% are grass species. Weeds in sugar beet fields compete with the crop for light, nutrients, and water, thus reducing the yield. Competition from uncontrolled annual weeds that emerge within 8 weeks of sowing or within 4 weeks of the crop reaching the two-leaf stage can reduce yield by 26–100%.

Nonchemical weed control measures provide a significant increase of yield in sugar beet and up to 50% reduction of herbicides use. In recent years, it has become necessary to reduce the use of herbicides in order to protect human and animal health and the environment. However, the use of mechanical weeders combined with herbicidal has been proven an efficient integrated weed control method. This chapter deals primarily with weed problems and weed control in sugar beet with emphasis on weeding machines used for inter-row and intra-row weeding in sugar beet as well as automation of weeding machines and the use of robotic machines.

Keywords

Automation · Inter-row · Robotic machines · Sugar beet · Weeding machines

A. E. Ghaly · M. M. Ibrahim (✉)
Department of Agricultural Engineering, Faculty of Agriculture, Cairo University, Giza, Egypt
e-mail: abdel.ghaly@dal.ca; mahmoud160@cu.edu.eg

V. Misra et al. (eds.), *Sugar Beet Cultivation, Management and Processing*,
https://doi.org/10.1007/978-981-19-2730-0_17

Abbreviations

GA Genetic algorithm
GPS Global positioning system
GPS Global positioning systems
PTO Power take-off
RTK Real-time kinematics

17.1 Introduction

Weeds have long been a concern since humans cultivated plants. Weeds are defined as unwanted plants in production systems such as farms and gardens or in lawns. Therefore, the term weed has no botanical significance, because a plant that is a weed in one context is not a weed when growing in a situation where it is wanted and a valuable crop (Janick 1979; Ashraf et al. 2012). Some native (or non-native) plants are unwanted in a specific location for several reasons: (a) They compete with the desired plants for sunlight, nutrients, water, and space, (b) they reduce crop yield and quality, (c) they delay maturity and hinder harvesting, (d) they provide hosts for plant pathogens, giving them greater opportunity to infect and degrade the quality of the desired plants, (e) they provide food and shelter for seed-eating birds and fruit flies, and (f) they may cause damage to drains and block streams (Baker 1974; DEFRA 2002; Naylor 2002).

Throughout the long history of agriculture practice, farmers have controlled weeds, and weed control is becoming now a highly developed field of knowledge and technology. Weed control methods include (a) hand weeding, (b) hand cultivation with hoes, (c) powered cultivation with cultivators, (d) lethal wilting with high heat, (e) complete burning, and (f) chemical attack with herbicides. However, weed control methods vary according to the growth habit of the weeds and the nature of the crop. For example, the methods of weed control used on a food crop are different from those used on fiber crops because of concern about the health effects of chemicals used on food crops.

Weeds can be categorized by their growing life cycle to annuals or perennials. Annual weeds grow from seeds dropped during the previous growing season, whereas perennial weeds regrow from previously established roots, tubers, rhizomes, or seeds. Therefore, understanding the habit of weeds is important for nonchemical methods of weed control including plowing, scuffling of soil surface, and prevention of seed accumulation in fields (Ashraf et al. 2012).

Sugar beet is a plant whose root contains a high concentration of sucrose and is, therefore, grown commercially for sugar production. The sugar beet plant consists of a conical-white fleshy root with a flat crown and a rosette of leaves as shown in Fig. 17.1 (Pritchard 2020). The root contains 75% water, 20% sugar, and 5% pulp. Sugar is the primary value of sugar beet crop but by-products such as the pulp

Fig. 17.1 Sugar beet plant. Source: Pritchard (2020)

Table 17.1 Top 10 sugar beet producing and exporting countries. Source: FAO (2020)

Producing countries		Exporting countries	
Countries	Tonnes	Countries	%
Russian Federation	46,408,675	Germany	37.5
France	41,516,031	Slovakia	10.6
USA	30,974,859	Belgium	9.5
Ukraine	28,198,893	Slovenia	8.8
Germany	19,821,461	France	8.1
Turkey	14,781,638	Hungary	5.9
Poland	13,887,088	Latvia	4.8
Italy	11,704,879	Austria	3.5
China	11,328,508	Poland	3.2
UK	7,721,789	Denmark	1.9

(mainly composed of cellulose, hemicellulose, lignin, and pectin) which is used in animal feed and molasses can add another 10% to the value of the harvest (FAO 1999). The top 10 producing and exporting countries are shown in Table 17.1 (FAO 2020). Sugar beets accounted for 20% of the world's sugar production (FAO 2009).

Jansen (1972) reported that weeds occur in sugar beet fields every year and approximately 60 plant species have been identified as important weeds in sugar beet fields. A complete list of common weeds and their scientific names was compiled by May and Wilson (2006). About 70% of weeds found in sugar beet fields are broad-leaved species and 30% are grass species. Holm et al. (1977) stated that two perennial weeds (*Elytrigia repens* and *Convolvulus arvensis*) and ten annual weeds comprise the list of major weeds in the sugar beet fields. The annual broad-leaved weeds are *Amaranthus retroflexus, Chenopodium album, Fallopia (Polygonum), Convolvulus, Sinapis arvensis* and *Stellaria media*. The annual grasses are *Echinochloa crus-galli, Poa annua,* and *Setaria viridis. C. album,* species belonging to the same family as sugar beet, is one of the most frequently reported weeds in this crop. Therefore, weed control in sugar beet fields is necessary to ensure high crop yield. However, weed control is one of the most difficult tasks in agriculture and accounts for a considerable share of the cost of crop production (FAO 1999).

Wilson (1987) stated that most agricultural soils contain large reservoirs of weed seeds of many species (4100–137,700 seeds/m^2) that may germinate and emerge to compete with the sugar beet crop. Squire et al. (2003) indicated that total the number and composition of weed seeds in soil vary depending on climate, soil characteristics, tillage methods, and weed control practices. Milton (1943) reported that the depth of seed burial and the intensity of soil cultivation affect seed longevity. Wilson et al. (1985) stated that only a small percentage (14–24%) of the total seed in reserve consists of species that are remnants of past crops while the large percentage (76–86%) can travel by air long distances from other fields. However, only a portion of the weed seeds in the reservoir germinates and produces seedlings each year. Burnside et al. (1996) reported that 28% of *C. album* seeds germinated after 17 years of burial in undisturbed soil. Roberts and Feast (1973) reported that 9% of *C. album* seed germinated after 6 years of burial in cultivated soil and 53% of the seeds germinated after 6 years of burial in undisturbed soil.

Schweizer and Zimdahl (1984) reported that reduced tillage left 50% of the weed seeds in the upper 0–7 cm of the soil, whereas extensively tilled soil distributed weed seeds evenly through the upper 30 cm of soil. With reduced tillage, the seed reservoir moves closer to the soil surface allowing the seeds to be in a better position to germinate and compete with the crop. Thus, the design of sowing and weed control systems in sugar beet that capitalizes on the shallow weed seeds can improve the effectiveness of cropping systems. May and Wilson (2006) stated that good weed management programs must incorporate a combination of crop rotations, herbicides, and tillage practices in order to limit the number and diversity of weed seeds in the soil.

Weeds compete with the sugar beet crop for light, nutrients, and water. Zimdahl (1980) and Werker and Jaggard (1998) indicated that in rain-fed and irrigated regions where water and nutrients are plentiful, light becomes the prime factor affecting weeds. The broad-leaved species grow taller than the crop and produce dense shade that makes the light-limited to sugar beets and thus reduces yield. Schweizer and Dexter (1987) indicated that competition from uncontrolled annual weeds emerging within 8 weeks of sowing or within 4 weeks of the crop reaching the two-leaf stage can reduce root yields by 26–100%. Scott et al. (1979) reported that weeds that emerge 8 weeks after sowing, particularly after the sugar beet plants have eight or more leaves, are less likely to affect yield. Longden (1987) indicated that weed beet at densities of only 1 plant/m^2 can reduce root yields by 11%.

This chapter deals with the principles of weed problems and weed control in sugar beet with emphasis on weeding machines used in sugar beet. However, the biology and ecology of major weeds in sugar beet are dealt with by the excellent publications by May and Wilson [14]. Readers interested in more information on weed biology and weed ecology are referred to the works of Gwynne and Murray (1985) and Radosevich and Holt (1984).

17.2 Weed Control

Weeding is the process of removing unwanted plants (weeds) in the field crops. Dawson (1965) divided weeding in sugar beet into three distinct periods: (a) the first period occurs from sowing to thinning when close-spaced beet are thinned to the final stand, (b) the second period starts from thinning to the last cultivation or the last time a tractor can travel between sugar beet rows without damaging the plants, and (c) the third period starts after last cultivation. The specific date and duration of each period may vary based on local conditions. Dawson (1974) indicated that weeds emerge in all three periods in irrigated fields, but they may not emerge in all three periods in rain-fed areas. Therefore, weed control in one period may not affect the weeds that emerge in other periods, unless the control is by herbicides that persist in the soil for a long time.

During the first period, weed control is most difficult because the small sugar beet seedlings have a low tolerance to herbicides and are easily covered with soil by cultivation. Thus, sugar beets require hand weeding during this period. However, manual weeding may reach more than 150 h/ha according to Miller and Fornstrom (1989).

During the second period, the sugar beet plants are larger and can tolerate mechanical and chemical weed control methods. Schweizer and Westra (1993) compiled a list of the herbicides utilized in sugar beet along with their effect on weed and sugar beet. Although weeds between the rows are easily controlled by cultivation and/or herbicides, weeds within the row constitute a major problem and must be removed by hand. Adetola (2019) stated that mechanical weed control (cultivation) is very effective in killing the weeds, reduces drudgery involved in manual weeding, keeps the soil surface loose, ensures soil aeration, and improves water holding capacity. Bhadra and Paul (2020) reported that chemical weed control is more prominent than manual and mechanical weed control methods, but it has adverse effects on the environment and animal and human health. Schweizer and Westra (1993) reported that some weeds have developed resistance to sugar beet herbicides. Lavaya et al. (2019) and Cloutier et al. (2007) stated that consumers demand high-quality food products and special attention to food safety and thus prefer organically produced food products and mechanical methods of weed control to ensure safe food production.

In the third period, the sugar beet plants become large enough to suppress the growth of newly emerging weeds by limiting the amount of light reaching them. Dawson (1974) indicated that the ability to apply a suitable weed control method at the critical period (when the removal of weeds has the most beneficial effect on crop yield) may be limited. For example, if weed control is to be achieved with hand labor, then farmers are dependent on the availability of laborers. On the other hand, if weed control is to be achieved with herbicides, then post-emergence herbicides must be applied at the proper growth stages of the crop and weeds.

Sullivan and Fischer (1971) stated that the need to ensure freedom from weeds permits efficiently mechanized harvesting and prevents weed seed shed that may affect the following crops requiring rigorous standard of weed control. Perennial weeds such as *Cirsium arvense, Convolvulus arvensis, Sonchus arvensis*, and *Elytrigia repens* infest sugar beet fields and can reduce yields. However, their overall impact in most fields is limited because they are confined to small areas where their growth and development are held in check by cultivations until the crop canopy begins to close.

Several mechanical methods for intra-row weeding in organic sugar beet have been studied in recent years, and the results have shown that early post-emergence cultivation causes severe crop damage by most tools but later cultivation can be made with a wide range of tools when the sugar beet has developed 4–6 true leaves. However, mechanical methods do not form a true solution for effective removal of weeds when the first flushes of seedlings emerge within 2–3 weeks after sowing and manual weeding is the choice (DEFRA 2002).

Schweizer and Dexter (1987) stated that although tractor hoeing and hand labor are still used in many sugar beet production areas, herbicides have been the primary method of weed control. Modern weed control recommendations are still based on the understanding that sugar beet plants need to gain an advantage over weeds early in the season. Therefore, it is important to design weed management programs that limit the renewal of weed seed reservoirs in soil that may germinate and emerge to compete with sugar beet plants. The program must incorporate crop rotations, use of herbicides, and tillage practices and must be employed for 2–4 years where a large seed bank exists.

Davies and Welsh (2002) reported that organic farmers rely on a wide range of preventative and reactive methods to control weeds including crop rotation, the timing of sowing, and mechanical weeding techniques. While spring-tine weeding remains the most common direct method for weed control in organic crops, there are several problems relating to its efficacy and selectivity. They indicated that inter-row hoeing can overcome many of these problems but there are no established agronomic guidelines for its rational implementation which prevented organic farmers from pursuing inter-row hoeing as a method of weed control.

17.3 Manual Weed Control

May and Wilson (2006) stated that the oldest and simplest of all weeding methods is manual weed control. When the first manuals for growing sugar beet were written in Germany by Achard in 1779, the main methods used for weeding were hand pulling and hand hoeing. Cloutier et al. (2007) reported that manual weed control started with farmers using their hands to uproot the weeds and then advanced to using tools such as hand-hoe. Manual weed control includes hand weeding, roguing, and hoeing.

17.3.1 Hand Weeding

May and Wilson (2006) stated that hand pulling is deemed necessary when the crop is small and frail. Hand weeding is still economically viable and essential in the high-value crop. Gianessi and Reigner (2007) reported that manual weeding using human hands provides a very effective weed control but requires substantial human effort and energy. Table 17.2 shows hand weeding work rates (h/ha) for different crops. Schweizer and Dexter (1987) reported that in countries where labor is plentiful and cheap, handwork can still be economically viable. Slaughter et al. (2008) indicated that hand weeding eliminated only 65–85% of the weeds due to workers mistaking weeds for crop plants or missing weeds.

17.3.2 Rogue

Rogue is defined as the removal of plants that are phenotypically different (with undesirable characteristics) from the plants under production, thereby ensuring the purity and quality of the crop produced. Colley (2009) reported that roguing mature weed plants such as weed beet present in low numbers are recommended to prevent seed shedding and keep seed bank levels low in all production systems. Other weeds that can be removed by roguing are blackgrass, wild oats, docks, ragwort (Lainsbury et al. 1999).

Table 17.2 Hand weeding work rates for different types of crops

Crop	Hand weeding (h/ha)
Asparagus	12
Broccoli	50
Carrot	35
Celery	149
Corn	12
Cucumber	74
Dry bean	40
Green bean	30
Green pea	30
Hot pepper	149
Lettuce	94
Mint	45
Onion	158
Peanut	15
Spinach	50
Sugar beet	150*
Sweet potato	59 Miller and Fornstrom (1989)
Tomato	92

Source: Gianessi and Reigner (2007)

Fig. 17.2 A six-person flat bed weeder (DEFRA 2002)

Bond et al. (2007) reported that weed beet occurs as a weed in sugar beet and can originate from wild beet, hybrids between wild and cultivated beet, and from bolters in open-pollinated cultivars. The flowers produce large amounts of windblown pollen that are more self-fertile than normal sugar beet. Weed beet seeds can remain viable in the soil for at least 7 years. Good control is achieved by cutting down bolters 3 times at 2-week intervals, starting 14–28 days after flowering.

Gunn (1982) indicated that bolters and weed beet can become a real problem in sugar beet fields if not dealt with sooner since each plant can produce 1500 viable seeds. Fayed et al. (1997) and Free et al. (1975) reported that the mature weed plants will harbor pests and diseases, compete for water, nutrients, and light which will reduce yield by 11%.

Longden (1982) indicated that hand roguing proved to be the most cost-effective method of controlling weed beet and bolters. The cost of roguing of sugar beet was $82/ha which resulted in reduced herbicide cost. Ward (2021) stated that hand roguing resulted in a decrease in herbicide cost. The total average weed control cost (including chemical control and hand roguing) is $194/ha compared with the herbicide-only approach of up to $369/ha.

Swire (2021) reported on stewardship guidelines aimed at protecting workers while hand pulling bolters and weed beet from sugar beet crops after spraying. The guidelines specify timescale and clothing requirements for workers. Workers should not re-enter the field within 48 h of spraying. Hand pulling can be continued thereafter upto10 days after spraying workers are wearing boots, gloves, and long-waterproof trousers. However, there are no requirements for worker protection after 10 days from spraying.

DEFRA (2002) reported on a flat-bed machine attached to the back of the tractor used for hand roguing (Fig. 17.2). The machine typically holds 4–8 people who lie prostrating to the ground and weed in the crop row. The machine follows an inter-row hoe and thus the entire area is left with the minimum possible weeds.

17.3.3 Hand Hoeing

Longden (1982) reported that hand hoeing was an essential process of sugar beet production until the introduction of monogerm seed in the 1960s were thinning to single sugar beet was unnecessary. Hand hoeing is used after the crop had become firmly established. Monogerm seed allowed the use of long-handled hoes rather than short hoes. However, Gianessi and Reigner (2007) reported that manual weeding using long-handled hoes would damage the crops while also missing some of the weeds. Schweizer and Dexter (1987) indicated that hand hoeing can be economically viable in countries where labor is plentiful and cheap. Hand-held hoes with different designs (Fig. 17.3) are available for this purpose. Longden (1982) indicated that the availability of herbicides and the increased labor cost caused sugar beet growers to rely less on hand hoeing for weed control. However, if the activity of herbicides is reduced so they are not completely effective, then manual weed control is required. Gianessi and Reigner (2007) reported that among vegetables, asparagus required the lowest time for hand weeding (12 h/ha) while onions required the highest time for hand weeding (158 h/ha); and among crops, corn required the lowest time for hand weeding (12 h/ha) while sugar beet required the highest time for hand weeding (150 h/ha). Walz (2004) reported on a national organic farmers' survey in which farmers cited weeds as one of the major causes of reduced profit after weather-related losses due to high labor costs. Earthbound Farms (the Largest Organic Producer in

Fig. 17.3 Hand-held hoes (Longden 1987)

North America) mentioned that weed control was a time consuming and very costly ($1000/ac)part of their operations (EFO 2011).

Gianessi and Reigner (2007) stated that hoeing leads to workers back injuries and reported farmworkers in California suffering permanent back injuries due to extended periods of hoe weeding and as a result the California Industrial Safety Board banned hoe weeding in 1975 which was then extended to hand weeding in 2004 by the California Occupational Safety and Health Standards Board because of concerns for farm laborer health. However, organic crop growers were exempted from these bans.

17.4 Mechanical Weeding

As agriculture became partially mechanized, weeding tools pulled by animals such as buffaloes, cows, and horses were developed. As time progressed, the tractor became the source of a draft, and many types of mechanical weeders became available. According to Gianessi and Sankula (2003), these weeding machines use three physical techniques for controlling weeds: (a) burying weeds, (b) cutting weeds, and (c) uprooting weeds. Burial of weeds is accomplished through the action of tillage tools and is usually done during land preparation when soil conditions are enhanced through tillage which aims at reducing the soil strength, covering plant residue, rearranging aggregates, and removing weeds. Cutting and uprooting weeds are performed by mechanical tearing and breaking the weeds from the soil and are done by mechanical cultivation after the crop is planted and has emerged.

Cloutier et al. (2007) stated that most manufacturers of mechanical weeders produce weeders that are designed to control weeds between rows, or in the inter-row region. There are only a few machines that are designed to perform intra-row weeding. However, the machinery available for weeding can also be divided into three categories depending on where the weeding action takes place: (a) inter-row, (b) intra-row, and (c) broad-spectrum.

17.4.1 Mechanical Inter-row Weeding

The objective of mechanical inter-row weeding is to cultivate the inter-row area without damaging the crop. Cultivation can destroy weeds by completely or partially burying them or uprooting and breaking the weed root contact with the soil. This type of inter-row weed control is used by farmers who do not use herbicides. Cloutier et al. (2007) stated that the limitation of using this method for weed control is that it can only be done during the early crop stages because of limited tractor and cultivator ground clearance and machine-plant contact can potentially damage the crop foliage at later growth stages.

There are many types of machines designed to work between the crop rows. These include cage or basket weeder, rotary hoe, brush weeder, and rotary cultivator (Fig. 17.4). Machines can be front or rear-mounted and can have a second operator to

Fig. 17.4 Inter-row weeders. (a) Cage or basket weeder, (b) Rotary tiller, (c) Brush weeder, (d) Rotary cultivator. Source: DEFRA (2002)

steer the hoes closer to the crop rows (Slaughter et al. 2008). However, automatic vision-guided systems are now replacing the second operator (Lavaya et al. 2019). Also, row protectors (Fig. 17.5) can be fitted to protect the crop from damage and can be removed when the crop is better established to allow some soil to be thrown into the crop row to bury weed seedlings (Adetola 2019; Cloutier et al. 2007).

17.4.1.1 Cage Weeders

The cage (or basket) weeder is an inter-row cultivator that has two horizontal axes on which rotating baskets are mounted. These require a flat working bed and small weed sizes to be effective. The powered rotary hoe is PTO driven and has rotating 'L' shaped blades on a horizontal axis. According to Cloutier et al. (2007), this machine has a more aggressive weeding action and can deal with larger weeds but is more damaging to the soil.

17.4.1.2 Rotary Hoe

The rotary hoes are versatile tillage tools that can be used to accomplish many tasks including weeding, decreasing soil crusting, and enhance crop emergence. It causes little soil compaction and is used for the early cultivation of corn, cotton, soybeans,

Fig. 17.5 Inter-row hoeing using crop protectors. Source: Cloutier et al. (2007)

potatoes, and small grain. It has as many as 12 sections, each mounting several hoe wheels, with the whole machine up to 12 m wide (King and Ball 2012).

17.4.1.3 Brush Weeder

The brush weeder has a series of rotating polypropylene brushes that are PTO-driven. The most common ones rotate on a horizontal axis. This machine is best used in friable soil conditions and can cope with wetter soil than hoes. The narrow crop protection tunnels (6 cm) allowed to work very close to the crop rows (Cloutier et al. 2007).

17.4.1.4 Rotary Cultivator

The rotary cultivator (star/spider hoe) can be used across the whole soil surface but is more commonly adjusted between the crop rows. It is ground-driven with aggressive weeding action and fast working speed. This machine works best on stone-free soil (DEFRA 2002; Cloutier et al. 2007).

17.4.2 Mechanical Intra-row Weeding

The mechanical intra-row weeders control weeds within the crop rows. There are two different approaches depending on the crop density: (a) the first is to use selective machines or add-on tools that can perform weed control close to the crop without damaging the crop and without requiring any sideways movement of the weeder and (b) the second is to use machines that have weeding tools that move sideways to conduct weed around the crop canopy(DEFRA 2002). The machines that have been reported to be effective in weed control in the crop row are finger weeders, torsion weeders, brush weeders, and ECO weeders.

17.4.2.1 Finger Weeder

The finger weeder (Fig. 17.6) is a simple mechanical intra-row weeding machine that uses two sets of steel cone wheels to which rubber spikes (fingers) are affixed and pointing horizontally outwards at a certain angle. These finger weeders operate from the side and beneath the crop row with ground-driven rotary motion. The rubber fingers penetrate the soil below the surface and remove the weeds near the fingers. The finger mechanism performs best in loose soils and poorly in heavily crusted or compacted soils or in soil with long-stemmed residue. This type of weeder is effective against young weed seedlings (up to 25 mm tall) and interacts gently with well-rooted crops. The recommended operating depth and forward speed are 12-is 20 mm and 5–10 km/h, respectively (Cloutier et al. 2007). Alexandrou (2004) evaluated a finger weeder and reported intra-row weed kill efficiency of 61%. Bowman (1997) and Weide et al. (2008) stated that the disadvantage of using the figure weeder is that the tractor must be steered very accurately so that the finger mechanism can work as close as possible to the crop rows.

17.4.2.2 Torsion Weeder

The torsion weeder (Fig. 17.7) is another intra-row weeding machine. Torsion weeders use spring tines connected to a rigid frame and angled downwards and backward to the crop row. The position of tines can be altered depending on the level of aggression required. They vibrate around the crop uprooting and burying small weeds. The coiled spring tines allow the tips to flex with soil contours and around established crops. These weeders reduced the weed density by 60–80% but require very accurate steering with relatively low forward velocities (Cloutier et al. 2007; Bowman 1997; Weide et al. 2008; Melander 2000).

Fig. 17.6 Finger weeder. Cloutier et al. (2007)

Fig. 17.7 Torsion weeder. Cloutier et al. (2007)

(a) (b)

Fig. 17.8 Rotating brush weeders. (a) Horizontal brush weeder, (b) Vertical brush weeder. Source: Cloutier et al. (2007)

17.4.2.3 Brush Weeders

The brush weeders (Fig. 17.8) use flexible brushes made of fiberglass or nylon rotating about vertical or horizontal axes. These weeders uproot, break, and bury weeds. A protective shield or cover can be installed to cover the crop from damage and an operator is required to steer the brushes to cultivate as many weeds as possible without damaging the crop plants (Cloutier et al. 2007). Fogelberg and Gustavsson (1999) investigated the use of a brush weeder for intra-row weed control and found it to be effective at early growth stages (in the 2–4 true leaf stage), uprooting 45–90% of the weeds using a working depth of 15 mm. Kouwenhoven (1997) reported on the use of a brush weeder for intra-row weed control in sugar beet crops and found the best rotational speed was 240–360 rpm with a forward travel speed of 2 km/h. Sugar beet plant damage resulted from steering inaccuracy and the fine soil created by the brushing effect.

Fig. 17.9 ECO weeder. Source: HCC (2011)

17.4.2.4 ECO Weeder

The ECO weeder (Fig. 17.9) is an intra-row weeding machine that is three-point hitch mounted and trails behind a tractor. It uses the tractor's power take-off (PTO) to drive a belt system that powers two discs with tines but does not require any hydraulic power. The minimum tractor size needed to power the ECO weeder is 20 hp. and the PTO speed required is 540 rpm. This weeder requires an operator to move two rotating discs with vertically oriented tines in and out of the crop row. The recommended speed is 1.0–2.5 km/h, and the rotation speed of the weeding element is 150–300 rpm (Kouwenhoven 1997). The ECO weeder can save up to 60% of weeding costs when compared to manual weeding (Univerco 2011).

17.4.3 Broad-spectrum Multipurpose Machines

The weeding action of the broad-spectrum machines is across the whole width of the machine. Examples of these machines are spring or flexi tines, chain or drag harrows, and toppers.

17.4.3.1 Spring Tines

Spring or flexi tines (Fig. 17.10) are coiled loop or spring mounted tines of 6–8 mm diameter mounted on bars. The tines can be raised to allow weeding between crop rows and the aggressiveness of the weeding action of these tines can be adjusted. The

Fig. 17.10 Spring or flexi tines. Source: Zeng et al. (2020)

spring tines have wide usage in many crops including sugar beet, cereals, beans, maize, and some horticultural crops. They work best on small broad-leaved annual weeds (Zeng et al. 2020).

17.4.3.2 Chain Harrow

Chain or drag harrows (Fig. 17.11) have a more rigid construction with steel spikes that stir the soil. The chain harrows are commonly used for knocking down sugar beet ridges and can be more damaging to the crop than flexi tines. The drag harrows kill some weeds but are not very efficient in doing so due to their highly flexible teeth (Jarman 2016).

17.4.3.3 Topper

A general-purpose topper (Fig. 17.12) can be used to control tall weeds such as fat hen in short crops such as beetroot or fodder brassicas. Home built toppers with weed seed collectors have been used successfully to prevent wild oats dropping seeds in cereal crops. It is a highly innovative and compact design and can run in uneven and narrow fields (Pathade et al. 2015).

Fig. 17.11 Chain drag harrows. Jarman (2016)

Fig. 17.12 Mower or topper. Source: Pathade et al. (2015)

17.5 Other Weed Control Machines

There are other types of nonchemical weed control techniques such as flame and pneumatic weeding machines. These techniques require other sources of energy to control weeds.

17.5.1 Thermal Weeders

The flame weeder requires propane gas to produce heat to elevate the temperature of the weed plants and either burns the entire weed biomass or causes weed plant cells to rupture and damage the plant structure. Merfeild (2011) reported that the flam weeder has substantial energy requirements and consumes significant amount of fuel (28–131 L/ha depending on the flam intensity and coverage).

There are two main types of thermal weeder: (a) the flame weeder and (b) the infrared burner (Figs. 17.13 and 17.14). The flame weeder applies heat directly to the soil surface from the combustion of LPG while the infrared burner heats a metal or ceramic surface which then radiates the heat towards the ground. The temperature reaches 100 °C for about one tenth of a second in order to cause plant cells to boil, burst, and desiccate. However, weed seedlings of some species will be easily killed, whereas others (meadow grass or shepherd's purse) are more tolerant to heat.

Fig. 17.13 Flame weeder. Source: Merfeild (2011)

Fig. 17.14 Infrared weeder. Source: Merfeild (2011)

Bond et al. (2003) reported that it is possible to remove weed in rows of sugar beets using computer-controlled propane flames. The flames only hit the weed plants without touching the crop plants. Spagnolo et al. (2018) stated that this approach has the advantage that weed seeds will not be triggered to start growing as is the case with mechanical weeding machines. Thermal control is an acceptable and feasible option for weed management in organic and conventional production systems. The technique was widely used prior to the emergence of agrochemicals and is currently used as an alternative to the use of chemical control of weeds in organic farms (Kanellou et al. 2017; Stepanpvic et al. 2016; Datta and Knezevic 2013; Ulloa et al. 2010, 2012; Sniauka and Pocius 2008; Kang 2001).

17.5.2 Pneumatic Weeders

The pneumatic weeder (Fig. 17.15) requires an air compressor to inject compressed air into the soil to loosen and uproot small weeds in well-anchored crops (Bond et al. 2003). The pneumatic weeder uses substantial power (a 68 hp. tractor) to produce high air pressure to control weeds. This is twice the power required for conventional hoeing (Weide et al. 2008). The pneumatic weeder is commonly used in organic production systems in which chemicals are not used.

Fig. 17.15 Pneumatic weeder. Source: Weide et al. (2008)

17.6 Factors Affecting Mechanical Weeding

The effectiveness of mechanical weeding is influenced by many factors including time and frequency of weeding, depth of soil coverage, type of tool, and type of machine.

17.6.1 Timing and Frequency of Inter-Row Weeding

Studies on weed competition with crop (Cousens et al. 1987; Kropff et al. 1993) concluded that the weeds that emerge with or shortly after the crop pose the most significant threat to crop yield. Several authors (Cousens 1985; Cudney et al. 1989; Wilson and Wright 1990; Kropff et al. 1992) indicated that weed population density is probably the most important among the factors affecting the level of weed competition with crop. Scott et al. (1979) stated that the optimum timing for mechanical weeding operations is at early stage in the lifecycle of the crop which is likely to produce the greatest yield response to weeding. For organic crops, mechanical weeding should commence as early as possible after the crop and weeds have emerged. Welsh et al. (1997) reported that it is necessary to control weeds later in the growing season to prevent them from shedding seeds.

Welsh et al. (1999) stated that inter-row hoeing can be conducted at very early crop growth stages if crop protectors are used. Inter-row hoeing using ducks-foot blades can control a wide range of weeds including both annual broad-leaved and grass species at a wide range of growth stages. Several authors (Bohrnsen 1993; Rasmussen 1993; Welsh 1998) reported reductions of weed density and biomass of up to 99% with inter-row hoeing. However, Bohrnsen (1993) and Hammarstrom et al. (1993) indicated that yield benefits are typically much smaller (4–5%) and only when crop damage is minimal.

17.6.2 Depth of Soil Coverage

Each tillage operation controls weed populations by covering, cutting, and uprooting weeds. Kouwenhoven (1982) reported that the differences observed in root structure and growth habit of weeds suggest that the effectiveness of mechanical weed control will vary depending upon the type and size of weed.

Jones et al. (1996) conducted experiments to investigate the effectiveness of three modes of weed kill on grass and broad-leaved weeds. They selected four species of weeds for their different root and growth habits: (a) chickweed which is a fibrous rooting prostrate broad-leaf weed, (b) field poppy which is a tap rooted broad-leaved rosette forming weed, (c) annual meadow grass which is a prostrate annual grass, and (d) rough-stalked meadow grass *Poa trivialis* L. which is an upright grass. Each treatment was conducted with a soil-based compost under dry and wet conditions and cutting was done at either 1 cm above the surface, at the surface or 1 cm below the surface. Burial was either partial or complete to a depth of 1 cm. Uprooting was done with the roots laid on the surface and with reburial after uprooting. The results showed that for broad-leaved weeds, uprooting (leaving the roots on the surface) and cutting at or below ground level were the most effective treatments, giving approximately 90% reductions in dry weight. The efficacy of these treatments was improved in dry conditions. Uprooting and reburial were also effective in dry conditions but poor (65% reduction) in the wet condition, indicating the importance of ground conditions at or immediately after treatments. The relatively poor results (35–70% reduction) from cutting above ground and stripping indicate the importance of cultivation, as opposed to a mowing operation in controlling these weeds.

Similar results were reported by Jones et al. (1995) for grass weeds and broad-leaved weeds. One exception was that complete burial was always more effective (98–100% reduction) irrespective of moisture. Uprooting grass on the other hand was more sensitive to moisture compared to broad-leaved weeds. Typical reductions of grass weeds were 55% for uprooting in wet conditions compared to 100% in the dry condition.

Terpstra and Kouwenhoven (1981) investigated the depth of soil coverage necessary to kill garden cress (*Lepidium sativum* L.) under laboratory condition. They found that a soil depth of 1.5 cm was lethal for small weeds and 2 cm was required for larger weeds. Increasing the soil coverage depth from 2.5 to 4.0 cm gave only 8% increase in weed kill.

17.6.3 Type of Tool

Adetola (2019) and Cloutier et al. (2007) indicated that the potential for improved weed control depends on selecting the appropriate tool for specific types of weed at particular soil moisture levels. A tool that has a below soil level cutting action may be appropriate to control broad-leaved weeds in dry conditions, but grass weeds in wet conditions may require a tool that will result in a higher proportion of burial.

Kouwenhoven (1982) indicated that the soil-engaging component of the hoe is referred to as the blade which has a wide range of designs as shown in Fig. 17.16. The rake angle (α) is the angle of lift that the hoe blade has from the horizontal. A low rake angle will cause the blade to clean cut the weed with minimal soil disturbance. Increased the rake angle generates more soil movement and mixing of the soil. The sweep angle (γ) is the angle of the cutting facing a line perpendicular to the direction of travel.

Jones et al. (1996) stated that a compromise is needed between self-cleaning, effective cutting, and draught force. A blade with a sweep angle of 30–50° with a low rake angle (2–5°), just enough to prevent scrubbing, is classified as a sweep. When viewed in plan form, sweeps either have 'L' or 'A' blade profiles. Figure 17.17 shows an 'L' blade, illustrating the sweep angle and leg mounting. Often two 'L' blades of opposite hands are used as a pair and staggered with a trailing and leading blade so that trash or stones can flow between them without causing blockages. It also enables the overall width to be adjusted to suit the crop growth stage and soil conditions. An 'A' blade has a centrally mounted leg, with a swept cutting face on each side, with a low rake angle. Variations of the 'A' blade come in the form of the ducks foot (Fig. 17.18). The difference is an increased rake angle (typically 20°) that tends to displace more soil from between rows into the row.

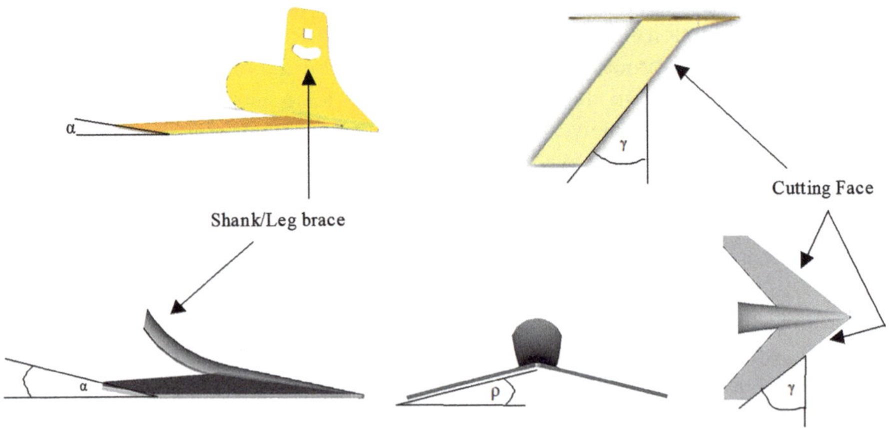

Fig. 17.16 Hoe blade classification. Source: DEFRA (2002)

Fig. 17.17 An L sweep blade. Source: DEFRA (2002)

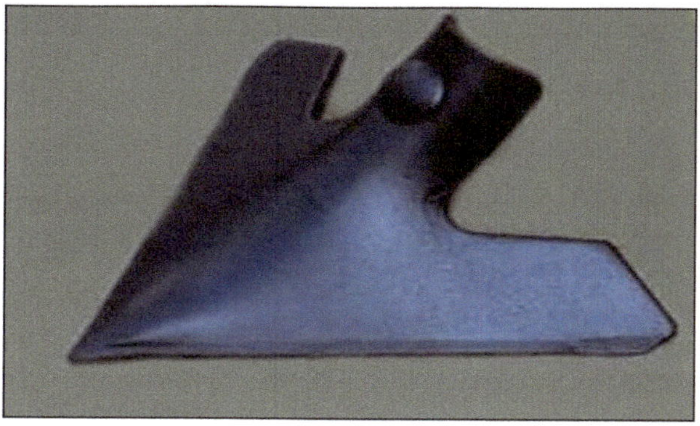

Fig. 17.18 Ducks-foot A blade. Source: DEFRA (2002)

17.6.4 Type of Machine

The tractor hoe (Fig. 17.19) is a generic name given to a tractor and toolbar mounted weeding mechanism (DEFRA 2002). Bowman (1997) indicated that the important aspect of the tractor hoe is the weeding device which is the soil-engaging part of the hoe. Different types of blades can be fitted to the hoe as shown in Table 17.3 (DEFRA 2002).

17.7 Effectiveness of Mechanical Weeding in Sugar Beet

Good weed control in sugar beet could be achieved using any or a combination of several machines (sown in Table 17.3) from the four true leaf stage of sugar beet onwards. Mechanical weed control eliminates weeds physically by uprooting them, chopping up the entire plants or separating weed stems and leaves from their roots and is especially effective for young annual weeds. Fogliatto et al. (2018) stated that mechanical weeding has drawbacks including: (a) bringing dormant weed seeds to the surface of the soil where they may germinate and (b) unintended propagation of

Fig. 17.19 Tractor hoe

Table 17.3 Summary of commercial mechanical weeding equipment (DERFA 2002)

Device	Speed (km/h)	Depth (mm)	Weed control	Mode of action	Weed size
Harrow	7	20–30	Inter-row Intra-row	Uprooting, burial	<50 mm
Brush weeder	<3.5	15–45	Inter-row Intra-row	Uprooting, curial	<25 mm
Split hoe	3	50	Inter-row	Uprooting, burial	<50 mm
Finger weeder	10	12–19	Intra-row	Uprooting	<25 mm
Torsion weeder	<10	25	Intra-row	Uprooting, burial	<25 mm
Hoe ridger	7	25–40	Inter-row Intra-row	Cutting, uprooting, burial	Large
Subsurface tiller	8	100	Inter-row	Cutting	Large
Powered rotary	6	120	Inter-row	Cutting, uprooting, burial	<150
Rotary cultivator	10	20–50	Inter-row	Cutting, mixing	<25 mm
Basket weeder	8	25	Inter-row	Scrubbing, uprooting	<20 mm
Sweep	6	20–40	Inter-row	Cutting, uprooting, burial	Large
Ducks-foot	6	20–40	Inter-row	Cutting, uprooting, burial	Large

perennial weeds by chopping up and spreading root segments, rhizomes, and tubers that will each grow into a new weed plant.

Jones et al. (1996) reported that tractor mounted hoes are very important in most sugar beet producing countries to kill weeds between sugar beet rows. Tractor hoes are used: (a) where herbicides have been sprayed in bands over the rows and weeds between the rows still need to be destroyed, (b) to replace a late herbicide application, especially when weed infestations are low or some weeds are too far advanced to be controlled by the herbicide, or (c) to control difficult weeds such as weed beet and perennials. However, tractor mounted hoes perform better in dry conditions. On the other hand, finger or fine tine weeders tend to work best when the soil is moist.

Blair et al. (2003) reported that in many European countries, the reduced use of herbicides has become a necessity to limit environmental pollution and to safeguard human health. A combination of low dose herbicide spraying and harrow gave a weed reduction of up to 86%. Ogoshi (1987) found the use of intra-row weeding implements resulted in similar weed control as the chemical herbicide application and encouraged the idea of substituting some of the herbicide applications with precision mechanical weeding. Kunz et al. (2016) observed a noteworthy herbicide reduction of up to 80% when mechanical weed control and band-spraying were combined with intra-row and inter-row hoeing treatments. Kruidhof et al. (2009) stated that mechanical weed control aided by precision steering in combination with conservation tillage and cover crops proved to be a good potential for integrated pest management. Cover crop mulches can suppress different weed species in sugar.

A comparison studies (Cornelis et al. 1997; Tugnoli et al. 2002) between mechanical weed control and use of herbicides to control weeds demonstrated that (a) the time and frequency of harrowing can differ from the time and frequency of low dose herbicide spraying, (b) the best weed control was obtained with combinations of finger or star weeder, flexible tine cultivator, split hoe and inter-row rotary tiller, (c) harrowing from the two true leaf stage onwards had a poorer weed control than normal practices, (d) harrowing from the four true leaf stage onwards had a comparable weed control as normal, and one or two low dose herbicide spraying, (e) best weed control was achieved by substituting the last low dose spraying by once or twice harrowing from the six true leaf stage onwards, (f) harrowing onwards the true leaf stage and higher driving frequency had no effect on yield and quality of the beet.

Gummert et al. (2012) pointed out that adverse side effects like soil erosion must be considered when using mechanical weed control. They recommended further studies for the evaluation of the reduced soil coverage with different mechanical weed control measures in relation to soil erosion. Kurstjens and Kropff (2001) reported that mechanical weed control is time consuming and less area efficient compared to chemical weed control. Favorable weather conditions and dry soil conditions are needed for an effective mechanical weed control. Also, the stony and lumpy soils make mechanical weeding less effective.

17.8 Comparison Between Different Weed Control Methods

The various mechanical weed control methods were compared with chemical weed control methods and conventional manual weeding as shown in Table 17.4. The comparison was taken into consideration the work rate, operating speed, operating depth, weed control efficiency, and cost. These costs were determined based on an hourly labor cost of $12 (Gianessi and Reigner 2007) and the estimating farm machinery costs were provided by Edwards (2015). The work rate of manual weeding was based on the work of Gianessi and Reigner (2007). The chemical weeding work rate was based on a 6.1 m boom sprayer operating at a speed of 9.7 km/h. The finger weeder work rate was based on an estimated operating width of 0.76 m. The torsion weeder work rate was based on an estimated operating width of 0.18 m of a single-row torsion weeder. The brush and ECO weeder work rates were based on an estimated operating width of 0.64 m of a twin weeding mechanism for single-row brush weeder and ECO weeder. The flame weeder work rate was based on an estimated operating width of 0.76 m of a tractor mounted flame weeder (Gianessi and Reigner 2007; Edwards 2015).

Manual weeding has the highest cost ($771/ha), followed by flame weeder ($173–222/ha), brush weeder ($183/ha), ECO weeder ($109), finger weeder ($94/ha), torsion weeder ($54/ha), and chemical weeding ($37/ha). Because of this big difference in cost alone, farmers tend to use chemical methods for weed control. In addition, the weed control efficacy of chemical weeding can be as high as 90%. The lowest cost among mechanical methods is the torsion weeder ($54/ha) with a weed control efficacy as high as 80%.

Table 17.4 Comparison between manual, chemical, and mechanical weed control methods

Method	Work rate (ha/h)	Operating speed (km/h)	Operating depth (mm)	Weed control (%)	Cost ($/ha)
Manual weeding	0.01	NA	0–50	65–85	771
Chemical weeding	2.9–5.9	4.8–9.6	Om surface	80–90	37
Torsion weeder	0.1–1.4	6.4–8.1	0–25	60–80	54
Finger weeder	0.3–0.6	4.8–9.6	10–40	55–60	94
ECO weeder	0.05–0.15	0.8–2.4	25–50	60–80	109
Brush weeder	0.1–0.3	1.6–4.8	25–50	60–80	183
Flame weeder	0.1–0.4	1.6–6.4	On surface	80–90	173–222

17.9 Automation in Weeding Machines

Automation is defined as the technique or method of operating and controlling a process or a mechanical device without human intervention using electronic hardware, sensors, actuators, and a software (Chancellor 1981). Weed control is a process that benefits greatly from automation of mechanical weeding machines. Bakker (2009) stated that automation allows a machine to determine and differentiate the crop plants from weed plants and then remove the weed plants with a precisely controlled device.

Precision intra-row weed control can be applied to mechanical, chemical, thermal, or electrical systems. Astrand and Baerveldt (2002) reported that automated weed control machines use mechanical knives that travel in and out of the crop row or use a rotating hoe with height adjustment. Lee et al. (1999) reported that automated chemical weed control such as precision spraying system was developed using independent spray ports for spraying weeds in a spray map generated by vision systems. Lee et al. (1999) and Diprose and Benson (1984) indicated that an electrical weed control system was developed which applies high voltage (15–60 kV) electrical discharge to small weeds using precise probe. Merfield (2011) stated that a precision thermal weed control involves the usage of infrared sensors to detect weeds and automatically opens the flame nozzle to burn the detected weeds.

Slaughter et al. (2008), in reviewing automated weed control systems, identified four core systems needed for automated weed control technology: (a) guidance, (b) detection and identification, (c) precision in-row weed control, and (d) mapping. They indicated that row guidance systems can use machine vision for crop row detection and/or global positioning systems (GPS). Machine vision can identify crop rows at travel speeds in the range of 2.5–10 km/h and produces very small errors (12–27 mm). GPS can provide a lateral positioning accuracy along the row with small error (6–13 cm).

Gonzales et al. (2004) indicated that detection and identification of weeds and crop are a very challenging task to conduct in real time. Weed identification techniques rely on machine vision systems and image processing techniques such as biological morphology, spectral characteristics, and visual structure. Steward and Tian (1999) used environmentally adaptive segmentation algorithm (EASA) to develop real-time machine vision weed detection for outdoor lighting conditions. Tang and Brian (2000) used color image segmentation using a binary-coded genetic algorithm (GA) for outdoor field weed identification under different lighting conditions.

Several researchers (Sabeenian and Palanisamy 2009; Buddha et al. 2019; Rekha and Bhagyalakshmi 2013; Shinde and Shukla 2014; Shanmugam et al. 2020) indicated that weed detection and differentiating between weeds and crops in the field by the robotic weed control systems are based on the size, spectral reflectance, shape, and texture features. However, Noguchi et al. (1999) used a genetic algorithm to optimize decision-making for classifying crops and weeds. It is an intelligent vision system that can also gathers the geographical field information for creating the field map.

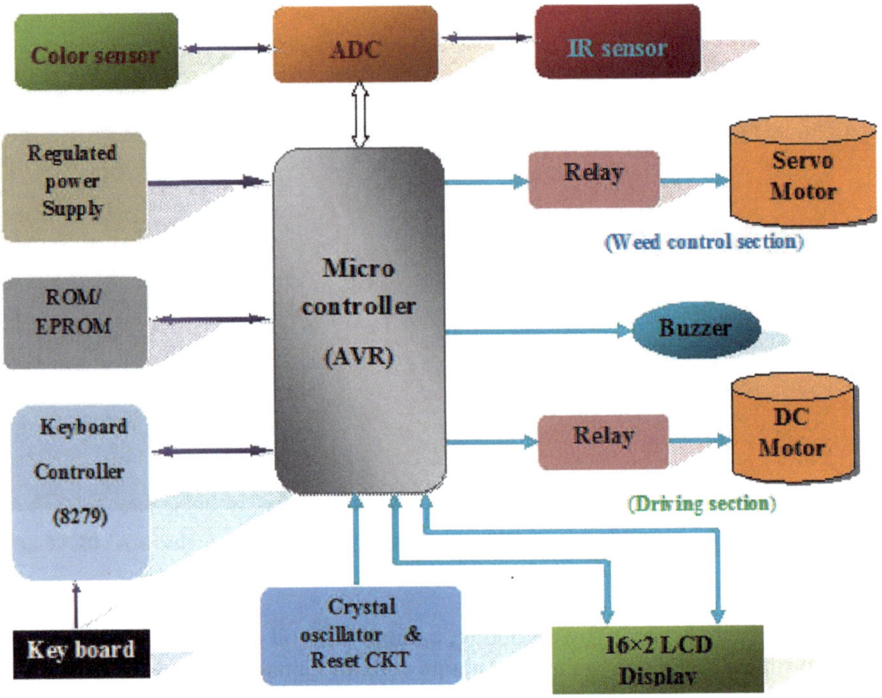

Fig. 17.22 Hardware structure of robotic system. Source: Kulkarni and Deshmukh (2013)

motor is based on simple electromagnetism used to give energy to the wheels of vehicle. When the DC motor starts, the robotic vehicle moves along the columns between the two crop lines. Once the color is sensed by the color sensor, the information is given to the weed control tool present behind the robotic vehicle which then moves down in the soil at particular depth to cut the weed.

Astrand and Baerveldt (2002) developed an agricultural mobile robot with vision-based perception for weed detection and control (Fig. 17.23). This machine had two cameras, one gray-scale camera with a near-infrared filter to obtain high-contrast images (located at the front to identify the crop row location and direction, and a color camera to identify crop plants (located at the center of the machine (facing downwards towards the soil). A weeding tool (a rotating wheel oriented perpendicular to the crop row) was located at the rear of the machine and was lowered using a pneumatic cylinder when gap between crop plants was detected. It provided tilling action in the inter-crop plant area.

Chebrolu et al. (2017) reported on an agricultural field robot BoniRob (Fig. 17.24) with all sensors and JAI camera mounted inside the shroud under the robot chassis and looking downwards. The BoniRob was developed for applications in precision agriculture including mechanical weed control and selective herbicide spraying as well as for plant and soil monitoring. It was equipped with four wheels

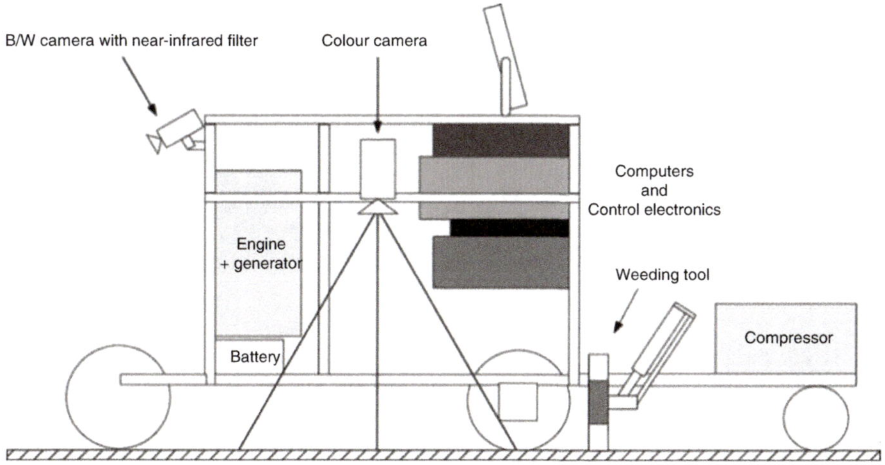

Fig. 17.23 Major components of the mobile robot. Source: Astrand and Baerveldt (2002)

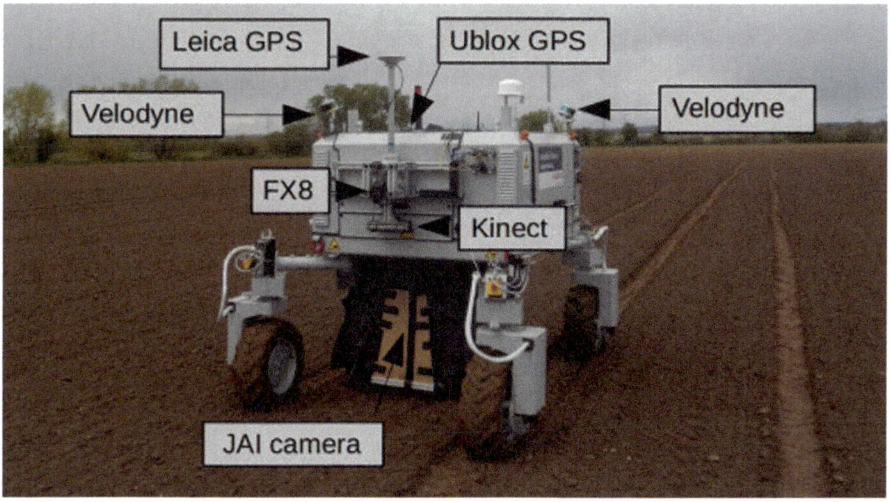

Fig. 17.24 Agricultural field robot BoniRob. Source: Chebrolu et al. (2017)

which can be steered independently, allowing for flexible movements and navigation on rough terrain. Figure 17.25 shows the robot BoniRob operating in the field 3 days after planting and 5 weeks after emergence.

EcoRobotix robot (Fig. 17.26) was the first completely autonomous robot developed by Startup Ticker (2018) and Ecorobotix (2018) for weed control in row crops. It is driven by solar energy (work up to 12 h on battery), does not need a human operator, is controlled and configured with a smartphone, it covers up to 3 ha/d and

(a) (b)

Fig. 17.25 Field robot BoniRob operating on the field. (a) Data acquisition 3 days after planting, (b) data acquisition 5 weeks after emergence. Source: Chebrolu et al. (2017)

Fig. 17.26 The EcoRobotix weeding robot. Source: Startup Ticker (2018)

removes up to 95% of the weeds. It is lightweight (130 kg) and does not have much effect on soil compaction. It carries two tanks of herbicide with 15 L each which is enough for 1 day operation. With the help of cameras, GPS-RTK, and sensors, it can follow rows and detect weeds and use the two robotic arms to apply herbicide on the detected weeds.

Blue River Technologies (2018), which a subsidiary of John Deere, developed a weeding robot called See & Spray Robot (Fig. 17.27) which is considered the next generation of smart agricultural equipment. It uses computer vision and artificial intelligence to see every plant and weed and determine what course of action is best for each one in real time. The robot is driven by a tractor and can eliminate 90% of the herbicide sprayed by farmers. The See & Spray Robot has the advantages of being able to spray a much bigger area and works at a much faster rate than its

Fig. 17.27 The See & Spray robot. Source: Blue River Technology (2018)

Fig. 17.28 The AgBot II. Source: QUT (2018)

smaller competitors. However, being pulled by a tractor it causes soil compaction problems.

Queensland University of Technology developed AgBot 2 (Fig. 17.28), a robot that can apply fertilizer and destroy weeds. It destroys weeds in three ways: (a) mechanically, (b) by applying herbicide, and (c) using microwave destruction methods. Sensors can be added to monitor soil and crop health (Energy Matters 2016; QUT 2018).

Fig. 17.29 The Ladybird robot. Source: The University of Sydney (2014)

Fig. 17.30 The BoniRob. Source: BOSCH (2015)

Ladybird (Fig. 17.29) is a solar-electric powered autonomous agricultural robot developed by the University of Sydney (2014) for the vegetable industry and capable of conducting weeding, farm surveillance, mapping, classification, and detection for a variety of vegetables and automated harvesting using the robotic arm that is to **be** used for weeding.

BOSCH (2015) developed the BoniRob (Fig. 17.30), an agricultural robot with omni-directional drive and adjustable track width. It was built to be a mobile plant laboratory able to monitor crop growth and optimize the amount of fertilizer used. It is completely herbicide free as it uses a rod to smash weeds.

17.11 Future Prospects

The ineffective and costly broadcast-spraying and the problems with soil compaction caused by tractors and heavy weeding machines have led to an increased interest in robot technology in the agricultural industry. The size of agricultural robot market was $817 million in 2013 and has grown to $16.3 billion in 2020, and it expected to grow exponentially in the near future. Farmers have come to realize that they can use lightweight autonomous robots without operator for efficient weeding instead of using heavy machinery and spraying whole fields with herbicides.

17.12 Conclusion

Sugar is an essential commodity and an integral part of the food chain as the cheapest source of energy. It plays a vital role in the development of taste and texture and keeps baked goods soft and moist. Sugar beet ranks second as a sugar-producing crop in the world. The root of the beet contains 75% water, 20% sugar, and 5% pulp. Approximately, 60 plant species have become important weeds in sugar beet production fields. About 70% of weeds found in sugar beet fields are broad-leaved species and 30% are grass species. Weeds in sugar beet fields compete with the crop for light, nutrients, and water and thus reduce yield and make the harvesting and processing difficult. Competition from uncontrolled annual weeds that emerge within 8 weeks of sowing or within 4 weeks of the crop reaching the two-leaf stage can reduce root yields by 26–100%.

Hand pulling is deemed necessary when the crop is small and frail and is still economically viable and essential in high-value crops. Manual weeding using human hands provides a very effective weed control but requires substantial human effort and energy. In countries where labor is plentiful and cheap, hand work can still be economically viable. However, nonchemical weed control provides a significant increase of yield in sugar beet and 50% reduction of herbicides use in the crop. In recent years, it has become necessary to reduce the use of herbicides in order to protect human health and other living organisms. However, during the early growth period, weed control is most difficult because the small sugar beet seedlings have a low tolerance to herbicides and are easily covered with soil by cultivation. Thus, sugar beets require hand weeding during this period. Mechanical weed control combined with herbicidal use has been proven an efficient integrated method.

Weed control equipment include manual tools such as conventional hoes and tractor mounted weeders for inter-row weeding such as cage or basket weeder, rotary hoe, brush weeder, and rotary cultivators as well as for intra-row weeding such as finger weeders, torsion weeders, brush weeders, and ECO weeders. There are other types of nonchemical weed control techniques such as flame and pneumatic weeding machines. The effectiveness of mechanical weeding is influenced by many factors including time and frequency of weeding, depth of soil coverage, type of tool, and type of machine. Some weeding machines are fully automated with GPS, cameras, sensors, and cutting tools. Agricultural field robots with sensors and cameras are currently used in weeding operations, and the use of this technology is expected to increase.

References

Adetola OA (2019) A review on the performance of weeding machines. Futo J Ser 5(2):109–123

Alexandrou A (2004) Evaluation of in-row weed cultivators in organic soybeans and corn. Organic Farming Research Foundation, Minto, South Wales

Amer G, Mudassir SMM, Malik MA (2015) Design and operation of Wi-Fi Agribot integrated system. International Conference on Industrial Instrumentation and Control (ICIC), College of Engineering Pune, Maharashtra, India

Ashraf M, Öztürk M, Ahmad MSA, Aksoy A (2012) Crop production for agricultural improvement. Springer, New York, p 796

Astrand B, Baerveldt A (2002) An agricultural mobile robot with vision-based perception for mechanical weed control. Auton Robots 13:21–35

Baker HG (1974) The evolution of weeds. Annu Rev Ecol Syst 5:1–24

Bakker T (2009) An autonomous robot for weed control – design, navigation and control. Unpublished Ph.D. Thesis, Department of Agricultural Engineering, Wageningen University, Wageningen, The Netherlands, p 149

Bhadra T, Paul SK (2020) Weed management in sugar beet: a review. Fund Appl Agric 5(2): 147–158

Blair AM, Jones PA, Ingle, Tillett ND, Hague T (2003) The integration of herbicides with mechanical weeding for weed control. J Agric 139(4):385–395

Blue River Technology (2018) Optimize every plant. http://www.bluerivertechnology.com/. Accessed 2 Jun 2021

Bohrnsen A (1993) Several years' results about mechanical weed control in cereals. In: Proceedings of the 4th International Conference IFOAM – Nonchemical Weed Control, Dijon, France

Bond W, Turner RJ and Grundy AC (2003) A review of non-chemical weed management. HDRA the Organic Organization, Coventry. https://www.gardenorganic.org.uk/weed-management. Accessed 6 Jun 2021

Bond W, Davies G, Turner P (2007) The biology and non-chemical control of weed beet (*Beta vulgaris* L). HDRS, Rayton Organic Garden, Coventry, UK. https://www.gardenorganic.org.uk/weed-management. Accessed 24 May 2021

BOSCH (2015) Bosch's Giant Robot can Punch weeds to death. A modular agricultural robot from Bosch startup Deepfield Robotics deals with weeds the old fashioned way: violently. https://spectrum.ieee.org/bosch-deepfield-robotics-weed-control

Bowman G (1997) Steel in the field: a farmer's guide to weed management tools. A publication of the Sustainable Agriculture Network, Handbook Series book 2, funding by the Sustainable Agriculture Research and Education Program of CSREES, U.S. Department of Agriculture, Sustainable Agriculture Network Beltsville, Maryland 20705

Buddha K, Nelson HJ, Zermas D, Papanikolopoulos N (2019) Weed detection and classification in high aerial images for robot-based precision agriculture. Mediterranean Conference on Control and automation, Akko, Israel. https://www.researchgate.net/publication/335195710_Weed_Detection_and_Classification_in_High_Altitude_Aerial_Images_for_Robot-Based_Precision_Agriculture. Accessed 22 Jan 2021

Burnside OC, Wilson RJ, Weisberg S, Hubbard KG (1996) Seed longevity of 41 weed species buried 17 years in Eastern and Western Nebraska. Weed Sci 44:74–86

Chancellor WJ (1981) Substituting information for energy in agriculture. Trans ASABE 24(4): 802–807

Chebrolu N, Lottes P, Schaefer A, Winterhalter W, Burgard W, Stachniss C (2017) Agricultural robot dataset for plant classification, localization and mapping on sugar beet fields. Int J Robot Res 36(10):1045–1052

Cloutier DC, van der Weide RY, Peruzzi A, Leblanc ML (2007) Mechanical weed control. In: Upadhyaya NK, Blackshaw RE (eds) No-chemical weed management. CAB International, Wallingford

Colley M (2009) Selection of rogueing in organic seed production. Organic Seed Alliance, Port Townsend, Washington

Cornelis E, Westerdijk CE, van der Weide RY, and Wevers JDA (1997) Integrated weed control (harrowing) in sugar beet. In: Proceedings of the 60th IIRB Congress. Cambridge, UK

Cousens R (1985) An empirical model relating crop yield to weed and crop density and a statistical comparison with other models. J Agric Sci 105:513–521

Cousens R, Brain O, O'Donovon JT, O'Sullivan A (1987) The use of biologically realistic equations to describe the effects of weed density and relative time of emergence on crop yield. Weed Sci 35:720–725

Cudney DW, Lowell SJ, Holt JS, Reints JS (1989) Competitive interactions of wheat (*Triticum aestivum*) and wild oats (*Avena fatua*) grown at different densities. Weed Sci 37:538–543

Datta A, Knezevic SZ (2013) Flaming as an alternative weed control method for conventional and organic agronomic crop production systems: a review. Adv Agron 118:399–428

Davies DHK, Welsh JP (2002) Weed control in organic cereal and pulses. In: Younie D, Taylor BR, Welsh JP, Wilkinson JM (eds) Organic cereals and pulses. Chalcombe Publications, Lincoln

Dawson JH (1965) Competition between irrigated sugar beets and annual weeds. Weeds 13:245–249

Dawson JH (1974) Full season weed control in sugar beets. Weed Sci 22:330–335

DEFRA (2002) Annual weed management in organic system. Leaflet produced by the Department of Environment, Food and Rural Affairs, London, UK

DERFA (2002) A review of knowledge: Inter-row hoeing and its associated agronomy in organic cereal and pulse crops. Final Project Report, Department of Environment, Food and Rural Affairs, London, UK. http://sciencesearch.defra.gov.uk/Default.aspx?Menu=Menu&Module=More&Location=None&Completed=0&ProjectID=10824. Accessed 6 May 2021

Diprose MF, Benson FA (1984) Electrical methods of killing plants. J Agric Eng Res 30(3): 197–209

Ecorobotix (2018) Discover our weeding robot. https://www.ecorobotix.com/en. Accessed 2 Jun 2021

Edwards W (2015) Estimating farm machinery cost. In: Extension and outreach. Department of Economics, College of Agriculture and Life Science, Iowa State University, Ames, Iowa. https://www.extension.iastate.edu/agdm/crops/html/a3-29.html. Accessed 11 May 2021

EFO (2011) Weed control. Earthbound Farm Organic. www.ebfarm.com/learn/organic-101/weed-control. Accessed 1 May 2021

Energy Matters (2016) Weed terminator: solar powered agbot II 2018. https://www.energymatters.com.au/renewable-news/solar-powered-agbot-em5716/. Accessed 2 Jun 2021

FAO (1999) Sugar beets/with sugar. In: Agribusiness handbook, agriculture statistics. Food and Agriculture Organization of the United Nations, Rome, Italy

FAO (2009) Agribusiness handbook. Food and Agriculture Organization of United Nations, Rome, Italy

FAO (2020) Agriculture statistics. Food and Agriculture Organization of the United Nations, Rome, Italy

Fayed MTB, El-Geddawy IH, El-Zeny MM (1997) Influence of weed interference on growth, yield and quality of sugar beet. Egypt J Agric Res 77(3):1239–1249

Fogelberg F, Gustavsson AD (1999) Mechanical damage to annual weeds and carrots by in-row brush weeding. Weed Res 39(6):469–479

Fogliatto S, Milan M, De Palon F, Farrero A, Vidotto F (2018) Effectiveness of mechanical weed control on Italian Flint varieties of maize. Renew Agric Food Syst 34(5):447–459

Free JB, Williams LH, Longden PC, Johnson MG (1975) Insect pollination of sugar-beet (*Beta vulgaris*) seed crops. Ann Appl Biol 81:127–134

Gianessi LP, Reigner NP (2007) The value of herbicides in U.S. crop production. Weed Technol 21(2):559–566

Gianessi LP, Sankula S (2003) The value of herbicides in US crop production. National Centre for Food and Agricultural Policy, Washington, DC

Gonzales RC, Woods RE, Eddins SL (2004) Digital image processing with MATLAB. Pearson/ Prentice Hall Publishing, New Jersey

Griepentrog HW, Norremark M, Nielsen J (2006) Autonomous intra-row rotor weeding based on GPS. In: proceedings of the International Commission of Agricultural and Biosystems Engineering (CIGR) World Congress 7. Bonn, Germany

Gummert A, Ladewig E, Marlander B (2012) Guidelines for integrated pest management in sugar beet cultivation: Weed control. J fürKult 64:105–111

Gunn JS (1982) Population dynamics of weed beet. Proceedings of the British Crop Protection Conference: Weeds, Brighton, UK

Gwynne DC, Murray RB (1985) Weed biology and control in agriculture and horticulture. Batsford Academic and Educational, London

Hammarstrom L, Gillberg B, Pettersson H (1993) Inter-row cultivation in cereals. In: Proceedings of the 4th International Conference IFOAM on Nonchemical Weed Control, Dijon, France

HCC (2011) Specialized cultivators. Hillside Cultivator Company, Lititz, Pennsylvania. http://www.hillsidecultivator.com/?page_id=179. Accessed 4 Jun 2021

Holm LG, Plucknett DL, Pancho JV, Herberger JP (1977) The world's worst weeds: distribution and biology. University Press of Hawaii, Honolulu, Hawaii

Janick J (1979) Horticulture science. WH Freeman Publisher, San Francisco, California, p 508

Jansen LL (1972) Extent and cost of weed control with herbicides and an evaluation of important weeds. Agricultural Research Service, United Sates Department of Agriculture, Washington, DC.

Jarman N (2016) Five reasons for hoeing. Country Small Holding, Britain's Best Selling smallholding Magazine, London, UK https://www.countrysmallholding.com/in-focus/five-reasons-to-harrow-1-4443316. Accessed 11 May 2021

Jiang G, Zhao C (2010) A vision system-based crop rows for agricultural mobile robot. IEEE International Conference on Computer Application and System Modelling, Institute of Electrical Engineering, Piscataway, New Jersey, USA

Jones PA, Blair AM, Orson JH (1995) The effect of different types of physical damage to four weed species. In: Proceedings of the 1995 Brighton Crop Protection Conference: Weeds, Brighton, UK

Jones PA, Blair AM, Orson JH (1996) Mechanical damage to kill weeds. In: Proceedings of the 2nd International Weed Control Congress. Copenhagen, Denmark

Kanellou E, Economou G, Papafotiou M, Ntoulas N, Lyra D, Knezevic S (2017) Flame weeding at archaeological sites of the Mediterranean Region. Weed Sci Soc Am 31(3):396–403

Kang WS (2001) Development of a flame weeder. Trans Am Soc Agric Eng 44(5):1065–1070

King LD, Ball PR (2012) Denitrification in white clover killed by ploughing, rotary hoeing, shallow tillage or herbicide. N Z J Agric Res 35:441–450

Kouwenhoven JK (1982) Weed control by Tillage Tools. In: Proceedings of the 9th Conference of the International Soil Tillage Research Organisation (ISTRO), Osijek, Socialistic Federal Republic of Yugoslavia

Kouwenhoven J (1997) Intra-row mechanical weed control possibilities and problems. Soil Tillage Res 41(1):87–104

Kropff MJ, Weaver SE, Smits MA (1992) Use of ecophysiological models for crop-weed interference: Relations amongst weed density, relative times of weed emergence, relative leaf area, and yield loss. Weed Sci 40:296–301

Kropff MJ, Lotz LAP, Weaver SE (1993) Practical applications. In: Kropff MJ, van Laar HH (eds) Modelling crop-weed interactions. CAB International, Oxford
Kruidhof HM, Bastiaans L, Kropff MJ (2009) Cover crop residue management for optimizing weed control. Plant Soil 318:169–184
Kulkarni VA, Deshmukh AG (2013) Advanced Agriculture robotic weed control system. Int J Adv Res Electr Electron Instrum Eng 2(10):5073–5081
Kunz C, Weber JF and Gerhards R (2016) Comparison of different mechanical weed control strategies in sugar beets. Conference: 27. Deutsche Arbeitsbesprechungüber Fragen der Unkrautbiologie und -bekämpfung, Braunschweig, Germany
Kurstjens DAG, Kropff MJ (2001) The impact of uprooting and soil-covering on the effectiveness of weed harrowing. Weed Res 40:211–228
Lainsbury MA, Hilton JG, Burn A (1999) The incidence of weeds in UK sugar beet crops during autumn 1998. Proceedings of Brighton Crop Protection Conference: Weeds, Brighton, UK
Lavaya NR, Niharika S, Deepika CH, Harini M, Chatane KS (2019) A review on weed detection using image processing. Indian J Sci Technol 12(48):1–3
Lee WS, Slaughter DC, Giles DK (1999) Robotic weed control system for tomatoes. Precis Agric 1: 95–113
Longden PC (1982) Understanding how to control weed beet in sugar beet. Proceedings of British Crop Protection Society, Weeds, Brighton, UK
Longden PC (1987) Weed beet: past, present and future. In: International sugar economic year book and directory 1987. F.O. Licht, pp F5–F16
May MJ, Wilson RG (2006) Weeds and weed control. In: Draycott AP (ed) Sugar beets. Rockwell Publishing Ltd, Arizona
Melander B (2000) Mechanical weed control. In: Transplanted sugar beet. 4th EWRS Workshop on Physical Weed Control, Elspeet, The Netherlands, p 25
Merfeild CM (2011) Physical weeding: flame weeding machine design. Harts Creek Farm, Canterbury. http://www.physicalweeding.com/flameweeding/index.html. Accessed 10 Jun 2021
Merfield CN (2011) Thermal weed management for crop production. http://www.merfield.com/research/2010/thermal-weed-management-for-crop-production-2010.pdf. Accessed 31 May 2021
Miller SD, Fornstrom KJ (1989) Weed control and labor requirement in sugar beet. J Sugarbeet Res 26(3&4):1–9
Milton WE (1943) The buried viable-seed content of a Midland calcareous clay soil. Empire J Exp Agric 11:155–167
Naylor REL (2002) Weed management handbook, 9th edn. British Crop Protection Council, Blackwell Sciences Limited, Oxford, p 423
Noguchi N, Reid JF, Zhang Q, Tian L, Hansen AC (1999) Vision intelligence for mobile agro-robotic system. J Robot Mechatron 11(3):193–199
Ogoshi A (1987) Ecology and pathogenicity of anastomosis and intraspecific groups of *Rhizoctonia solani* Kuhn. Annu Rev Phytopathol 25:125–143
Pathade HP, Shinde P, Magar N, Mundaware S (2015) Multipurpose weeding machine. Int J Multidiscip Res Dev 2(4):402–405
Pingzeng L, Shusheng P, Guansheng Z, Wenshan W, Yushu G, Zhenmin D (2011) Obstacle avoidance system for agricultural robots based on multi-sensor information fusion, IEEE 1181-1185
Pritchard S (2020) Too much of bad things: The use and misuse of UK soil and land to grow sugar. Food, Farming and Countryside Commission, London
QUT (2018) A case study: AgBot II 2018. https://www.qut.edu.au/research/partner-with-us/case-study-agbot-ii. Accessed 2 Jun 2021
Radosevich SR, Holt J (1984) Weed ecology: implications for vegetation management. John Wiley, New York, p 265
Rasmussen J (1993) Can high densities of competitive weeds be controlled efficiently by harrowing or hoeing in agricultural crops? In: Proceedings of the 4th International Conference IFOAM on Non-Chemical Weed Control, Dijon, France

Rekha S, Bhagyalakshmi A (2013) A survey on weed and pest detection system. Int J Adv Res Comput Sci Manage Stud Spec Issue:33–38
Roberts HA, Feast PM (1973) Emergence and longevity of seeds of annual weeds in cultivated and undisturbed soil. J Appl Ecol 10:133–143
Sabeenian RS, Palanisamy V (2009) Texture based weed detection using multi resolution combined statistical and spatial frequency (MRCSF). Int J Comput Electr Auto Control Inform Eng 3(4): 1–5
Schweizer EE, Dexter AG (1987) Weed control in sugar beets (*Beta vulgaris*) in North America. Rev Weed Sci 3:113–133
Schweizer EE, Westra P (1993) Potential for weed to develop resistance to sugar beet herbicides in North America. J Sugar Beet Res 28(1&2):1–22
Schweizer EE, Zimdahl RL (1984) Weed seed decline in irrigated soil after rotation of crops and herbicides. Weed Sci 32:84–89
Scott RK, Wilcockson SL, Moisey FR (1979) The effects of time of weed removal on growth and yield of sugar beet. J Agric Sci 93:693–709
Shanmugam S, Assunção E, Mesquita R, Veiros A, Gaspar PD (2020) Automated weed detection systems: a review. In Proceedings of the International Congress on Engineering: Engineering for Evolution, KnE Engineering 2020:271–284
Shinde AK, Shukla MY (2014) Crop detection by machine vision for weed management. Int J Adv Eng Technol 7(3):818–826
Slaughter D, Giles D, Downey D (2008) Autonomous robotic weed control systems: a review. Comput Electron Agric 61(1):63–78
Sniauka P, Pocius A (2008) Thermal weed control in strawberry. Agron Res 6:359–366
Spagnolo RT, Oldoni A, Custódio TV, dos Reis AV, Lilles A, Machado T (2018) Design specifications of a heat applicator weed controller device for family farms. Ciência Rural Santa Maria 48(2):1–9
Squire GR, Brookes DR, Bohan DA, Champion GT, Daniels RF, Haughton AJ, Hawes C, Heard MS, Hill MO, May MJ, Osborne JL, Perry JN, Roy DB, Woiwod LP, Firbank LG (2003) On the rationale and interpretation of the farm scale evaluations of genetically modified herbicide-tolerant crops. Philos Trans R Soc 358:1779–1799
Startup Ticker (2018) Weeding robots startup ecoRobotix raises CHF 10.6 million. https://www.startupticker.ch/en/news/may-2018/ecorobotix-raises-chf-10-6-million-in-series-b-financing. Accessed 2 Jun 2021
Stepanpvic S, Datta A, Neilson B, Bruening C, Shapiro C, Gogos G, Kmezevic SZ (2016) The effectiveness of flame weeding and cultivation on weed control, yield and yield components of organic soybean as influenced by manure application. Renew Agric Food Syst 31(4):288–299
Steward BL, Tian LF (1999) Real-time weed detection in outdoor field conditions. Proc SPIE Int Soc Opt Eng 3543:266–278
Sullivan EF, Fischer BB (1971) Weed control. In: Johnson RT, Alexander JT, Rush GE, Hawkes GR (eds) Advances in sugar beet production: Principles and practices. Iowa State University Press, Ames, Iowa, pp 66–109
Swire J (2021) New guidelines for sugar beet field workers. Agronomist and Arable Farmer Magazine. Lewis Business Media, River Way, Uckfield. https://www.aafarmer.co.uk/news/new-guidelines-for-sugar-beet-field-workers.html. Accessed 24 May 2021
Tang L, Brian S (2000) Color image segmentation with genetic algorithm for in field weed sensing. Trans ASAE 43(4):1019–1027
Terpstra R, Kouwenhoven JK (1981) Inter-row and intra-row weed control with a hoe ridger. J Agric Eng Res 26:127–134
The University of Sydney (2014) Ladybird developer awarded researcher of the year. http://sydney.edu.au/news/84.html?newsstoryid=13686. Accessed 2 Jun 2021
Tillett ND, Hague T, Grundy AC, Dedousis AP (2008) Mechanical within-row weed control for transplanted crops using computer vision. Biosyst Eng 99(2):171–178

Tugnoli V, Cioni F, Vacchi A, Martelli R, Pezzi F, Baraldi E (2002) Integrated mechanical weed control with reduced herbicide dosages on sugar beet. In: Proceedings of the 65th IIRB Congress. Brussels, Belgium

Ulloa SM, Datta A, Knezevic SZ (2010) Tolerance of selected weed species to broadcast flaming at different growth stages. Crop Prot 29:1381–1388

Ulloa SM, Datta A, Bruening C, Gogos G (2012) Weed control and crop tolerance to propane flaming as influenced by the time of day. Crop Prot 31:1–7

Univerco (2011) ECO weeder. Univerco, Napierville, Quebec. https://univerco.com/en/product/weeder-eco-i/. Accessed 5 Jun 2021

Walz E (2004) Fourth national organic farmers survey: Sustaining organic farms inachanging organic marketplace. Organic Farming Research Foundation, Santa Cruz, California

Ward A (2021) How hand rogueing is winning the battle against blackgrass. ADAMA Agricultural Solutions, Airport City. https://www.adama.com/en/farmtalk/weed-pest-disease/weed-landing/how-hand-rogueing-is-winning-the-battle-against-blackgrass. Accessed 23 May 2021

Weide RYVD, Bleeker PO, Achten VTJM, Lotz LAP, Fogelberg F, Melander B (2008) Innovation in mechanical weed control in crop rows. Weed Res 48(3):215–224

Welsh JP (1998) Developing strategies for weed control in organic arable systems. Unpublished Ph. D. Thesis, Departments of Agricultural Botany and Agriculture, University of Reading, Reading, UK

Welsh JP, Bulson HAJ, Stopes CE, Froud-Williams RJ, Murdoch AJ (1997) Mechanical weed control in organic winter wheat. Asp Appl Biol 50:375–384

Welsh JP, Bulson HAJ, Stopes CE, Froud-Williams RJ, Murdoch AJ (1999) The critical weed-free period in organically grown winter wheat. Ann Appl Biol 134:315–320

Werker AR, Jaggard KW (1998) Dependence of sugar beet yield on light interception and evapotranspiration. Agric For Meteorol 89(3/4):229–240

Wilson RG (1987) Biology of weed seed in the soil. In: Altieri MA, Liebman M (eds) Weed management in agroecosystems: ecological approaches. CRC Press, Boca Raton, Florida, p 354

Wilson BJ, Wright KJ (1990) Predicting the growth and competitive effects of annual weeds in wheat. Weed Res 30:201–211

Wilson RG, Kerr ED, Nelson LA (1985) Potential for using weed seed content in the soil to predict future weed problems. Weed Sci 33:171–175

Winter Green Research (2014) Agricultural Robots: Market Shares, Strategies, and Forecasts Worldwide. http://wintergreenresearch.com/reports/AgriculturalRobots.html. Accessed 30 Oct 2020

Zeng Z, Martin A, Chen Y, Ma X (2020) Weeding performance of a spring-tine harrow as affected by timing and operational parameters. Weed Sci 69(2):247–256

Zimdahl RL (1980) Weed-crop competition: a review. Corvallis, Oregon, International Plant Protection Centre, Oregon State University, p 198

Chemical Strategy for Weed Management in Sugar Beet 18

Mohammad Mehdizadeh, Zoleikha Mehdizadeh, Shahida Anusha Siddiqui, Sepideh Kazemi, Atun Roy Choudhury, Koko Tampubolon, and Mostafa Mehdizadeh

Abstract

The growing population of the world and the needs related to nutrition and food supply for human societies has led farmers and crop producers to increase production and minimize the limiting factors of crop production. Among these, the management of pests, mainly weeds, is of great importance. One of the significant limiting factors in agricultural production systems is the presence of weeds in main crops and especially the sugar beet. Sugar beet, as an inferior competitor, is very sensitive to biotic and abiotic stresses. Despite all the

M. Mehdizadeh (✉)
Department of Agronomy and Plant Breeding, Faculty of Agriculture and Natural Resources, University of Mohaghegh Ardabili, Ardabil, Iran

Ilam Science and Technology Park, Ilam, Iran

Z. Mehdizadeh
Department of Biology, Payame Noor University, Tehran, Iran

S. A. Siddiqui
DIL e.V.–German Institute of Food Technologies, Quakenbrück, Germany

Technical University of Munich Campus Straubing for Biotechnology and Sustainability, Straubing, Germany

S. Kazemi
Department of Horticultural Science, Science and Research University of Tehran, Tehran, Iran

A. R. Choudhury
Fecal Sludge Management Division, Banka BioLoo Limited, Hyderabad, Telangana, India

K. Tampubolon
Program Study of Agrotechnology, Faculty of Agriculture and Animal Husbandry, Universitas Tjut Nyak Dhien, Medan, Indonesia

M. Mehdizadeh
Faculty of Agriculture, University of Guilan, Rasht, Iran

V. Misra et al. (eds.), *Sugar Beet Cultivation, Management and Processing*,
https://doi.org/10.1007/978-981-19-2730-0_18

environmental hazards, impact on human health, and challenges related to herbicide residues in the environment, the chemical weed control method is still considered an effective and promising method for controlling weeds. This chapter is devoted to discuss about the chemical strategy for weed management in sugar beet.

Keywords

Environment · Herbicides · Residues · Sugar beet · Weed management

Abbreviation

MS Mass spectrometry

18.1 Introduction

Sugar beet is an important commercial agricultural crop related to sugar production globally, ranked second in sugar production after sugarcane. One of the most critical factors affecting the yield of sugar beet and the quality of produced sugar is weed management. Weeds are one of the leading causes of damage to crops. According to available data, the damage caused by their existence is not less than for pests and plant diseases. This amount of damage in developed countries, semi-developed countries, and developing countries with traditional agricultural systems are 5%, 10%, and 25%, respectively (Harker and O'Donovan 2013). A 50–100% reduction in sugar beet yield has been reported when weeds were not controlled (Deveikyte and Seibutis 2006). Sugar beet competes poorly with weeds from emergence until the leaves shade the ground. To prevent economic damage and reduced yields, weeds should be entirely controlled within 4 weeks after the emergence of sugar beet plants in the field. Subsequently, the weed management program should be continued throughout the growing season (Gerhards et al. 2017).

Sugar beet is very sensitive to weed competition, especially in the early growth stages (Lobmann et al. 2019). Of all the pests associated with sugar beet, weeds are the most severe and critical pest for this crop (Abouziena and Haggag 2016). From the first stages of sugar beet growth, the competition between this plant and the weeds in the field for water, sunlight, and micronutrients in the soil begins (Bruciene et al. 2021). Other disadvantages of the presence of weeds in crops include reducing the quantity and quality of crops, interference in harvest, hosting some pests and plant diseases, threatening human and animal health and increasing production costs. Soltani et al. (2018) assessed the economic damage in sugar beet crops due to the presence of weeds during 2002–2017. They reported an average yield loss of 70% for this crop with approximately US $1.25 billion in the United States. Manual weeding and mechanical methods for controlling sugar beet weeds are very

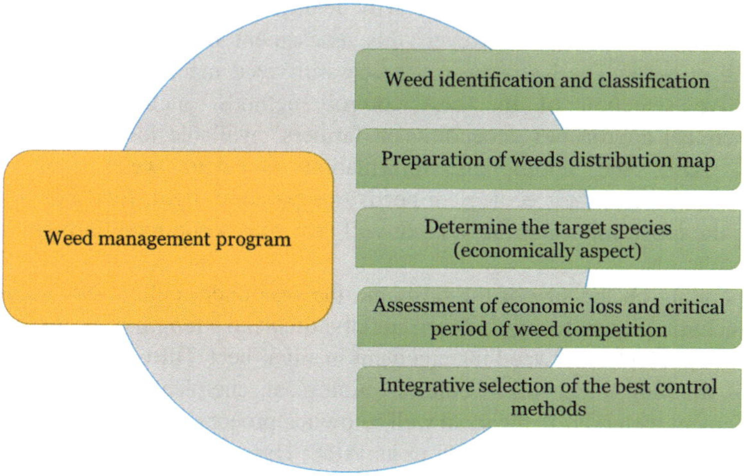

Fig. 18.1 Critical steps to a successful weed management program

expensive and may cause damage to sugar beet seedlings. So, the herbicides application is a more economical practice. Sugar beet cultivated in fields with minor weed infestation and correct agricultural practices only needed post-emergence application of herbicides. However, sugar beet grown in fields with heavy weeds infestation and improper agricultural practices required both the pre-and post-emergence application of herbicides (Cioni and Maines 2010).

Successful weed control in new agricultural systems requires the development of a management plan with considering all the factors affecting the crop and weeds. Critical steps to a successful weed management program are presented in Fig. 18.1.

According to this strategic plan, weed identification and classification is the first step to achieve a successful management plan and needs sufficient knowledge about plant biology and the condition of the field and weed population. Therefore, researchers and farmers should collect, identify, and classify the weeds in the field and evaluate their characteristics, growth cycle, life cycle, biological needs, and competitiveness. The second step involves mapping the distribution of weeds in the field. For this purpose, the field area is divided into smaller sections, and weed density is recorded in each plot. The preparation of this map will help farmers be aware of the status of weeds in the field and how their population changes over the years (Siddiqui et al. 2021; Somerville et al. 2020). Accurate preparation of distribution maps and their monitoring over the growing seasons will help to minimize weed management costs. In the third step, the target species are identified. The target species are weeds whose population is high in the field, or their vegetation structure is the same as the main crop. So, they could cause significant economic damage. Farmers must have a specific plan to control the population of these species and not neglect their existence during the growing season. Assessment of economic loss and critical period of weed competition is one of the other fundamental steps that researchers and farmers should consider (Gantoli et al. 2013). Accurate determination of economic damage will determine the severity of the need for managing weed species. However, this assessment requires scientific data and research projects. The last step in a successful weed management program is the integrative selection of the best control methods, including chemical and non-chemical approaches depending on farmers' available facilities and financial ability. Also, the appropriate information on field weed status, growth characteristics, and their ecological needs (temperature, light, food needs, etc.) are among the most important factors that will critically accompany farmers' management programs.

Depending on the size of the farm and the area under cultivation, access to the latest technologies and new cultivars, and the financial strength of farmers, there are different approaches to weed management in sugar beet. Different methods such as manual weeding, mechanical, cultural, biological, chemical, and integrated weed management methods are the most well-known approaches for weed control in sugar beet and other crops (Mehdizadeh et al. 2018; Hassani et al. 2020). In sugar beet, weed control is necessary to prevent reduced yields, so herbicides are prevalent in all sugar beet farms. It is reported that 70% of the pesticides used in this crop are devoted to herbicides (Marwitz et al. 2012). Due to the risks of excessive use of herbicides for the environment and human health, and the possibility of weed resistance, none of these solo methods is sufficient to control sugar beet weeds. Therefore, it is necessary to implement an integrated weed management system, primarily using reduced doses of herbicides as an effective control method (Kaya and Buzluk 2006). However, reduced amounts of herbicides in sugar beet should be performed when the weeds are most sensitive to herbicides. In most weeds, this stage is the cotyledon stage (Petersen 2004). Nowadays, herbicide splitting, the combination of different herbicides, and integrated weed management are the main methods to reduce the herbicide dose (Cioni and Maines 2010). According to Daneshian et al. (2013), the application of a mixture of Betanal progress AM and sethoxydim herbicides along with manual weeding 100% of sugar beet weeds. Ganbari Birgani et al. (2007) evaluated the control of sugar beet broadleaf weeds in combination with Betanal progress and Safari herbicides and cultivation. They reported that the combination of cultivation and herbicides reduced the density of the weeds by 41% as compared to the solo chemical method. Melander et al. (2005) reported that the application of reduced amounts of herbicides and mechanical methods reduced total herbicide use and increased the yield of sugar beet in Turkey. This chapter is devoted to evaluate the chemical strategy for weed management in sugar beet.

18.2 Problematic Weeds in Sugar Beet

As a short and low-growing crop, sugar beet is highly affected by weeds. For this reason, weeds are the most critical factor limiting the growth, development, and yield of this crop in agricultural systems (MacLaren et al. 2020). Numerous plant species are known as weeds associated with sugar beet production worldwide, the most important of which are presented in Table 18.1. The abundance of broadleaf weeds is higher than other weeds, and they have a more significant share of competition with sugar beet (Soltani et al. 2018). Weeds cause problems for agricultural products in the following cropping seasons due to the production of abundant seeds and the distribution of these seeds in arable soils. On the other hand, weeds pose a severe challenge to the weed management program (Chauhan 2020). A wide range of weeds can be found in sugar beet products that could be classified in different ways. One of the best classification factors is based on plant morphology. Therefore, sugar beet weeds could be divided into broad-leaved and narrow-leaved (grassy) species. However, more than 70% of problematic weeds in sugar beet is devoted to broad-leaved weeds (Lobmann et al. 2019).

18.2.1 Broad Leaf Weeds

These kinds of weeds have wide leaves with netlike veins, and their seedlings emerge with two leaves. More than 70% of the sugar beet weeds are broadleaf weeds (Heidari et al. 2007). As shown in Table 18.1, the most abundant and important broadleaf weeds of sugar beet have belonged to Brassicaceae, Chenopodiaceae, Amaranthaceae, and Asteraceae families. Typically, different types of control methods can be used for these weeds. However, selective herbicides are one of the most effective options to manage these plants in products such as sugar beet (Jhala et al. 2021).

18.2.2 Grasses (Narrow Leaf Weeds)

Although the economic losses associated with narrow-leaved weeds in broadleaf crops such as sugar beet are not significant, several narrow-leaved species of the Poaceae family are found on sugar beet farms. One of the challenges associated with using herbicides to control narrow-leaved weeds is related to the resistance of these plant species to herbicides. Accordingly, the application of other weed management methods with an integrated approach can be practical (Storkey et al. 2021).

18.2.3 Parasitic Weeds

A limited group of weeds called parasitic weeds is found in some agricultural products, such as sugar beet. Due to the severity of economic losses of this group of weeds, their rapid management is of particular importance. Cuscuta spp. is one of the most important parasitic weeds in sugar beet fields (Saric-Krsmanovic et al. 2017). Hoseyni et al. (2018) reported 90.63–100% control of *Cuscuta campestris* in response to application of Propyzamide herbicide in sugar beet fields. In the case of these weeds, the use of herbicides along with a combination of other control methods

Table 18.1 Problematic weeds associated with sugar beet

Scientific name	Family name	Common name	Morphology
Brassica napus	Brassicaceae	Wild buckwheat	Broad-leaved
Chenopodium album	Chenopodiaceae	Common lambsquarters	Broad-leaved
Amaranthus powellii	Amaranthaceae	Powell amaranth	Broad-leaved
Amaranthus retroflexus	Amaranthaceae	Redroot pigweed	Broad-leaved
Kochia scoparia	Amaranthaceae	Kochia	Broad-leaved
Beta vulgaris	Chenopodiaceae	Sea beet	Broad-leaved
Ambrosia artemisiifolia	Asteraceae	Common ragweed	Broad-leaved
Anagallis arvensis	Primulaceae	Ain el-gamal	Broad-leaved
Polygonum lapathifolium	Polygonaceae	Pale persicaria	Broad-leaved
Cirsium arvense	Asteraceae	Canada thistle	Broad-leaved
Convolvulus arvensis	Convolvulaceae	Field bindweed	Broad-leaved
Veronica persica	Plantaginaceae	Persian speedwell	Broad-leaved
Portulaca oleracea	Polygonaceae	Common purslane	Broad-leaved
Galium aparine	Rubiaceae	Goosegrass	Broad-leaved
Helianthus annuus	Asteraceae	Common sunflower	Broad-leaved
Brassica nigra	Brassicaceae	Kaber mustard	Broad-leaved
Chamomilla suaveolens	Asteraceae	Pineappleweed	Broad-leaved
Matricaria chamomilla	Asteraceae	False chamomile	Broad-leaved
Sinapis arvensis	Brassicaceae	Wild mustard	Broad-leaved
Polygonum persicaria	Polygonaceae	Ladysthumb	Broad-leaved
Physalis spp.	Solanaceae	Groundcherries	Broad-leaved
Sonchus arvensis	Asteraceae	Perennial sow-thistle	Broad-leaved
Polygonum aviculare	Polygonaceae	Knotgrass	Broad-leaved
Polygonum spp.	Polygonaceae	Smartweeds	Broad-leaved
Abutilon theophrasti	Malvaceae	Velvet leaf	Broad-leaved
Datura stramonium	Solanaceae	Jimsonweed	Broad-leaved
Lamium purpureum	Lamiaceae	Red dead-nettle	Broad-leaved
Solanum sarachoides	Solanaceae	Hairy nightshade	Broad-leaved
Solanum tuberosum	Solanaceae	Potato	Broad-leaved
Fumaria officinalis	Fumariaceae	Common fumitory	Broad-leaved
Stellaria media	Caryophyllaceae	Common chickweed	Broad-leaved
Viola arvensis	Violaceae	Field pansy	Broad-leaved
Galeopsis tetrahit	Lamiaceae	Common hemp-nettle	Broad-leaved
Matricaria inodora	Asteraceae	Scentless mayweed	Broad-leaved
Thlaspi arvense	Brassicaceae	Field pennycress	Broad-leaved
Vicia sativa	Fabaceae	Common vetch	Broad-leaved
Sisymbrium irio	Brassicaceae	London rocket	Broad-leaved
Helianthus annuus	Asteraceae	Common sunflower	Broad-leaved
Salsola kali	Amaranthaceae	Saltwort	Broad-leaved
Euphorbia helioscopia	Euphorbiaceae	Libbein	Broad-leaved
Cichorium pumilum	Asteraceae	Shikoria	Broad-leaved
Ammi majus	Apiaceae	Common bishop	Broad-leaved
Rumex dentatus	Polygonaceae	Sheep sorrel	Broad-leaved

(continued)

Table 18.1 (continued)

Scientific name	Family name	Common name	Morphology
Avena fatua	Poaceae	Wild-oat	Grassy
Echinochloa crus-galli	Poaceae	Barnyardgrass	Grassy
Poa annua	Poaceae	Annual meadow-grass	Grassy
Agropyron repens	Poaceae	Common couch	Grassy
Setaria glauca	Poaceae	Yellow foxtail	Grassy
Setaria faberi	Poaceae	Giant foxtail	Grassy
Setaria spp.	Poaceae	Foxtail	Grassy
Setaria viridis	Poaceae	Green foxtail	Grassy
Sorghum halepense	Poaceae	Johnsongrass	Grassy

can lead to successful control of these plants and reduce the severity of field contamination in next growing seasons.

18.3 Chemical Weed Management

Herbicides today play a pivotal role in weed management and are widely used due to their high efficiency and economic advantage. One of the most widely used, easily applicable, flexible, and effective weed management methods in most crops is chemical method and the use of herbicides or bioherbicides (Kunz et al. 2016; Mushtaq et al. 2020; Mehdizadeh and Mushtaq 2020). Especially for crops such as sugar beets that have low competitiveness, the use of herbicides to prevent yield loss is critical (Jhala et al. 2021). The success of chemical herbicides in controlling weeds depends mainly on the time of application, application doses, and method of application. According to the herbicide application time, there are three different types of herbicides for controlling weeds (Fig. 18.2). The primary purpose of using herbicides is to reduce production costs and human resources, use labor for more critical farm affairs, increase the product's quantity and quality, and improve weed control and better utilization of agro-ecosystems. One of the other aspects of chemical weed control is the use of biochemical compounds derived from plants with allelopathic properties. Dadkhah (2013) assessed the allelopathic impact of sugar beet on *Portulaca oleracea* and reported that the seedling growth of this weed was significantly affected by the extract of sugar beet.

Accordingly, due to the presence of the main crop, there are more restrictions on the use of appropriate herbicides in the post-emergence application. So that the main crop should not be damaged while controlling weeds.

18.3.1 Herbicides Used in Sugar Beet

Generally, few selective herbicides such as desmedipham, chloridazon, clopyralid, phenmedipham, ethofumesate, and metamitron have been introduced to control

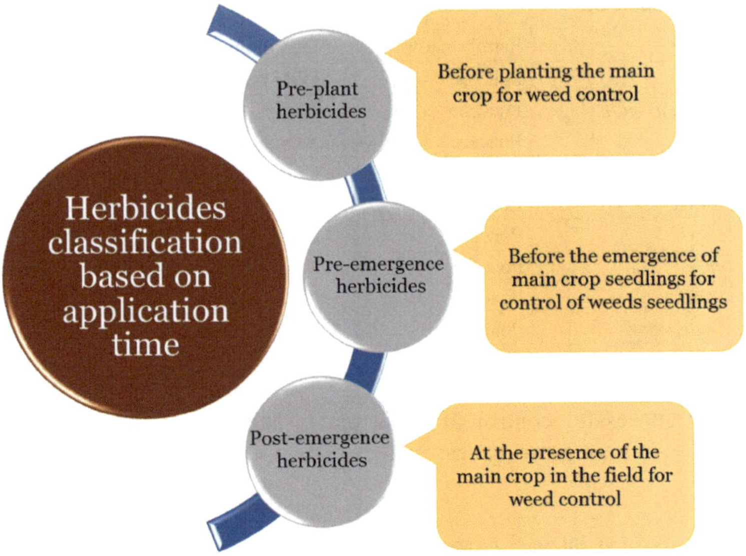

Fig. 18.2 Herbicides classification based on application time

weeds in sugar beet fields (Wilson 1999; Adamczewski et al. 2019). Due to the wide range of weeds related to the sugar beet plant and, on the other hand, the limitations of the selective herbicides for this crop, it is common to use a combination of some herbicides as tank mixes to control sugar beet weeds.

Today's use of pre-plant herbicides is very restricted due to their high persistence and toxicity and the negative impacts on human health and the safety of the agroecosystem (Ayivi et al. 2021; Zimmerman et al. 2021). On the other hand, the use of pre-emergence herbicides in sugar beet fields is only about 10% of the herbicides used in this crop, and in fact, the significant share is related to post-emergence herbicides (Deveikyte et al. 2015). Some circumstances such as rainfall severity and duration, soil moisture, soil physicochemical properties, and microorganism's population in the soil could be affected the efficacy of post-emergence herbicides. Some successful chemical control cases for weed management in sugar beet are presented in Table 18.2.

18.3.1.1 Combination of Herbicides

The combination of herbicides could enhance the control efficacy of a wide range of weed in different crops. Typically, the combination of different herbicides that are compatible in terms of mixing can affect a wide range of weeds in cropping systems due to having several different modes of action. A tank mixture of some different herbicides was successfully performed for weed management in the United States (Carlson et al. 2008). Rasha (2010) reported a significant reduction in weed biomass due to the application of Betanal progress (135 g a.i/fed) combined with Fusilade Super (94.75 g a.i./fed). Rapparini (2008) evaluated the effect of combined

Table 18.2 Herbicides used in sugar beet

Herbicide	Dose	Result	References
Glyphosate	3 L ha^{-1}	100% weed control efficacy	Bezhin et al. (2015)
Phenmedipham	960 g a.i ha^{-1}	Significant weed control	Hamouzová et al. (2013)
Ethofumesate 6.5% + metamitron 28% + phenmedipham 6.5%	2.5 kg/fed	84.5% reduction in total weeds	Abd El-Hamed (2019)
Acetochlor 84% EC	0.75 L/fed	51% reduction in total weeds	Abd El-Hamed (2019)
Betasana-trio	0.9 L fed^{-1}	Completely eliminate the broad-leaved weeds associated with sugar beet	Abd El-Lateef et al. (2021)
Betanal MaxxPro	0.5 L fed^{-1}	Completely eliminate the narrow-leaved weeds associated with sugar beet	Abd El-Lateef et al. (2021)
Desmedipham + phenmedipham + ethofumesate	616.5 g a.i ha^{-1}	90.93% control of *Chenopodium album*	Chitband et al. (2014)
Chloridazon	1300 g a.i ha^{-1}	90.47% control of *Amaranthus retroflexus*	Chitband et al. (2014)
Clopyralid	240 g a.i ha^{-1}	89.67% control of *Portulaca oleracea*	Chitband et al. (2014)
Desmedipham + phenmedipham + triflusulfuron + clopyralid	45 g + 45 g + 4.4 g + 26 g a.i ha^{-1}	78.4% reduction in weed biomass	Armstrong and Sprague (2010)
Glyphosate	840 g a.i ha^{-1}	99.8% reduction in weed biomass	Armstrong and Sprague (2010)

(continued)

Table 18.2 (continued)

Herbicide	Dose	Result	References
Phenmedipham 6.5% + metamitron 28% + ethofumesate 6.5%	2 kg a.i ha^{-1}	96.8 and 59.8% reduction in *Medicago polymorpha* and *Phalaris minor* biomass, respectively	Mahmoud and Soliman (2012)
Acetochlor	0.75 L/fed	51.3 and 47.3% reduction in *Medicago polymorpha* and *Phalaris minor* biomass, respectively	Mahmoud and Soliman (2012)
Desmedipham plus phenmedipham plus ethofumesate	0.23 + 0.23 + 0.23 kg a.i ha^{-1}	Maximum reduction in weed biomass	Abdullahi and Ghadiri (2004)
Goltix + betanal progress	Recommended doses	Best control results for *Chenopodium album* and *Amaranthus retroflexus*	Zargar et al. (2010)
Betanal Expert of	1.7–2.1 L ha^{-1}	Efficient control for *Salvia reflexa*	Chetin et al. (2008)
Ethofumesate + phenmedipham + desmedipham	1.12 g a.i ha^{-1}	45% reduction in Kochia control	Sbatella et al. (2019)
Betanal progress	877 g a.i ha^{-1}	Resulted in lowest weed density and weed dry matter	Anabestani and Armin (2017)
Propyzamide	1500 g a.i ha^{-1}	Significant control of field dodder	Saric-Krsmanovic et al. (2017)
Desmedipham + phenmedipham at	0.045 + 0.045 kg a.i ha^{-1}	Control of *Chenopodium album* and *Amaranthus* spp.	Dale et al. (2006)
Glyphosate	0.84 kg a.i ha^{-1}	Provided 89% weed control in sugar beet	Wilson and Sbatella (2011)
Pendimethalin	3.6–4.8 kg a.i ha^{-1}	Provided 82% grass and 56% broadleaved weed control	Yagoob et al. (2021)

herbicides desmedipham + phenmedipham + ethofumesate and found high efficiency (95% control) of this combination on annual dicotyledonous weeds in sugar beet. Deveikyte and Seibutis (2008) reported significant management of *Chenopodium album* L., *Tripleurospermum perforatum*, *Polygonum aviculare* L., and *Thlaspi arvense* L. due to the application of phenmedipham + desmedipham + ethofumesate mixed herbicides. Significant control of broad-leaved weeds and sugar beet yield improvement was reported by Majidi et al. (2011).

In some cases, considering genetically modified crops or herbicides-tolerant varieties reduces the limitations associated with using a tank mixture of some different herbicides and thus prevents their occurrence of side effects in agricultural ecosystems. The introduction of sugar beet varieties with high tolerant levels to glyphosate herbicide was one of these approaches for effective management of broad-leaved weeds in this crop (Khan 2010). Bezhin et al. (2015) reported 90% weed control efficacy in sugar beet using the tank mixture of pre-emergence application of 1.0 L ha^{-1} Goltix Gold, followed by 2–4 post-emergent applications of 1 L ha^{-1} Goltix Gold +1.5 L ha^{-1} Betanal Expert.

18.3.1.2 Reduced Doses

Given the environmental risks associated with the use of herbicides, it seems necessary to provide practical tactics to reduce these hazards. In general, a significant portion of herbicides used to control weeds reaches places other than the herbicide's site of action. Accordingly, the concentration of the recommended doses is usually considered to be higher than the actual required level. From an environmental point of view, there is no need for maximum weed control to achieve optimal crop yield. So, the recommended and registered doses of herbicides could be shifted to the application of reduced doses. One of these strategies is to reduce the dose of herbicides compared to the recommended doses (Hamill et al. 2004; Benedetti et al. 2020). In other words, by using reduced doses of herbicides, we can prevent the adverse effects of herbicide residues while achieving an acceptable level of weed control (Kudsk 2008). On the other hand, the use of reduced doses of herbicides can play a role in reducing weed resistance (Beckie and Kirkland 2003; Norsworthy et al. 2012). The essential component in applying reduced doses of herbicides is to prevent the reduction of herbicide efficiency in the control of target weeds. Kahramanoglu and Uygur (2010) reported that reducing metribuzin doses from 525 g a.i ha^{-1} (recommended dose) to 183.7 g a.i ha^{-1} was still significantly provided 90% wild mustard control. Bostrom and Fogelfors (2002) reported the satisfactory control of weeds by reducing 50% recommended herbicide doses. The application of reduced doses could achieve acceptable results in weeds control if used in combination with other weed management methods. 70% reduced doses of Atlantis herbicide, and a combination of sunflower and sorghum water extracts resulted in a 90% reduction in weed dry weights (Razzaq et al. 2012).

18.3.2 Herbicide Residues

Monitoring and evaluation of chemical pesticides in the environment are essential components of sustainable agriculture in agro-ecosystems. The issue of herbicide residues should be considered in terms of food security, human health, animal and microorganism's safety, prevention of the damage to non-target crops, etc. (Mehdizadeh et al. 2021). One of the most critical approaches to chemical weed management is maximum weed control without damaging or reducing yield for the main crop. Generally, different plants have different levels of resistance or tolerance to herbicides. Based on this, plants can be divided into resistant, tolerant, and sensitive crops. Resistance level or sensitivity of a plant to a particular herbicide depends on many factors, including the formulation and chemical composition of the herbicide, herbicide application time, herbicide half-life and persistence, herbicide concentration, herbicide mode of action, soil physicochemical properties, plant biology, etc. Sugar beet crops need extensive use of herbicides to control weeds; however, it has a relatively high sensitivity to herbicide residues.

Today, various methods such as instrumental analysis (chromatography (GC, HPLC, TLC), mass spectrometry (MS)), and bioassay methods are used to assess herbicide residues in agricultural ecosystems (Mehdizadeh 2014; Mehdizadeh et al. 2016; Janaki et al. 2018). Crops such as sugar beet, oil seed rape, and tomato, due to their high sensitivity to herbicide residues, have a high potential for selection as biological indicators to track and evaluate the residues of these toxins in agricultural soils (Mehdizadeh 2016, 2019). Matte et al. (2021) evaluated the mobility and persistence of pyroxasulfone herbicide in soil by using some sensitive crops such as lettuce, cucumber, sorghum, sugar beet, and tomato as bioindicators. Very low concentrations of rimsulfuron herbicide residues were successfully assessed using a bioassay method using sugar beet as a sensitive crop (Mehdizadeh and Gholami-Abadan 2018). Mehdizadeh et al. (2017) used a high-performance liquid chromatography along with bioassay methods to evaluate the residues of two sulfonylurea herbicides and reported appropriate results due to the use of HPLC and bioassay for analyzing these herbicides residues in different soils.

18.3.3 Sensitivity of Sugar Beet to Persistent Herbicides

Herbicides with high or moderate persistence in the soil environment could adversely affect sensitive crops in the field or non-target following plants in crop rotation (Greenland 2003). Typically, these kinds of herbicides have a relatively long half-life, and the residues from their degradation can affect plants and microorganisms in the soil (Zaller et al. 2021). There are many factors involved in herbicide residues and their adverse effects on different plants. However, the most important influencing factors are the physiochemical properties and concentration of using the herbicide and the biology of the plant exposed to direct concentrations of the herbicide or its residues over time. Accordingly, it is not unreasonable to expect that different crops show different responses to a particular herbicide. Tandon and

Pal (2021) found no adverse effect of ethofumesate herbicide 2.0 kg ha^{-1} on sugar beet. However, this herbicide with different concentrations could influence the other crops.

Sugar beet is known as one of the most sensitive crops to herbicide residues. Mehdizadeh and Gholami-Abadan (2018) reported the high susceptibility of sugar beet to the trace concentration of rimsulfuron herbicide. According to their study, the root biomass was more sensitive than for shoot. Carneiro et al. (2019) reported a significant reduction in the yield of sugar beet due to the application of tembotrione at the rate of 100.8 g a.i ha^{-1}. The total fresh biomass and carotenoid content of sugar beet were significantly reduced by applying 288 g a.i ha^{-1} mesotrione (Pintar et al. 2020). Dale et al. (2006) evaluated the effects of different herbicides on sugar beet and weed biomass. They reported 44% sugar beet injury due to the application of Desmedipham + phenmedipham + ethofumesate at the rate of 0.03 kg ha^{-1}.

18.4 Future Prospect

Today, crop producers employ a diversity of weed management techniques such as chemical, mechanical, cultural, biological, and integrated weed management (Cheboi et al. 2021). These methods aim to reduce weed damage and to deplete the weed seed bank in crop ecosystems. Given the critical challenges such as ensuring human health and the environment, preventing soil degradation and pollution of water resources, weed resistance, and superweeds creation, the need to review and innovate in weed management methods in the future is absolutely essential (Chauhan et al. 2017). Artificial and robotic control techniques with minimum interference with soil, sensory, computer and information techniques, precision agriculture approaches, expanding the new effective bio-herbicide formulations, genetic engineering, and biotechnology, and considering biological method and allelopathy as environmentally friendly perspective in weed control, could be developed as new prospects for weed management in agricultural systems (Shaner and Beckie 2014; Westwood et al. 2018; Dayan 2019; Mehdizadeh and Mushtaq 2020).

18.5 Conclusion

As discussed before, weeds are among the most critical limiting factors in crop production systems. Damage due to the weed presence in agricultural lands becomes more severe when the crop has lower competitiveness than weeds. Therefore, weed control is one of the most fundamental prerequisites required to achieve acceptable crop yields. Among the various weed management methods, chemical techniques and the application of herbicides play a pivotal role. They are widely used due to their high efficiency, flexibility, easily applicable, and economic advantage. Accordingly, various herbicides have been developed for weed management and are available to farmers worldwide. Despite the relative success of chemical weed management, several challenges such as threatening human health, animals, microorganisms, and the environment, pollution of water and soil resources, and the emergence of weed resistance phenomena have arisen concerning the increasing use of herbicides. Therefore, the side effects of herbicides used in agricultural ecosystems should be evaluated. On the other hand, the researchers should focus on reducing herbicides by using environmentally friendly alternative methods such as robotic, sensory, and computer techniques, expanding precision agriculture and bio-herbicide approaches, biotechnology, and genetic engineering.

References

Abd El-Hamed GM (2019) Performance of some sugar beet (*Beta vulgaris* L.) cultivars under annual weed control condition. J Plant Prod 10(9):739–750. https://doi.org/10.21608/jpp.2019.54481

Abd El-Lateef EM, Mekki BB, Abd El-Salam MS, El-Metwally IM (2021) Effect of different single herbicide doses on sugar beet yield, quality and associated weeds. Bull Natl Res Cent 45:21. https://doi.org/10.1186/s42269-020-00476-9

Abdullahi F, Ghadiri H (2004) Effect of separate and combined applications of herbicides on weed control and yield of sugar beet. Weed Technol 18(4):968–976. https://doi.org/10.1614/WT-03-142R2

Abouziena HF, Haggag WM (2016) Weed control in clean agriculture: a review. Planta Daninha 34 (2):377–392. https://doi.org/10.1590/S0100-83582016340200019

Adamczewski K, Matysiak K, Kierzek R, Kaczmarek S (2019) Significant increase of weed resistance to herbicides in Poland. J Plant Prot Res 59(2):139–150. https://doi.org/10.24425/jppr.2019.129293

Anabestani V, Armin M (2017) Reduction of Phenmedipham + Desmedipham + Ethofumesate herbicides dosage based on application timing in sugar beet. J Crop Ecophysiol 10(4):823–838

Armstrong JJQ, Sprague CL (2010) Weed management in wide- and narrow-row glyphosate-resistant sugarbeet. Weed Technol 24(4):523–528. https://doi.org/10.1614/WT-D-10-00033.1

Ayivi R, Ibrahim S, Colleran H, Silva R, Williams L, Galanakis C, Fidan H, Tomovska J, Siddiqui SA (2021) COVID-19: human immune response and the influence of food ingredients and active compounds. Bioact Compd Health Dis 4(6):100. https://doi.org/10.31989/bchd.v4i6.802

Beckie HJ, Kirkland KJ (2003) Implication of reduced herbicide rates on resistance enrichment in wild oat (*Avena fatua*). Weed Technol 17:138–148. https://doi.org/10.1614/0890-037X(2003)017[0138:IORHRO]2.0.CO;2

Benedetti L, Rangani G, Ebeling Viana V, Carvalho-Moore P, Merotto A Jr, Rabaioli CE, Antonio de AL, Roma-Burgos N (2020) Rapid reduction of herbicide susceptibility in jungle rice by recurrent selection with sublethal dose of herbicides and heat stress. Agronomy 10(11):1761. https://doi.org/10.3390/agronomy10111761

Bezhin K, Santel HJ, Gerhards R (2015) Evaluation of two chemical weed control systems in sugar beet in Germany and the Russian Federation. Plant Soil Environ 61(11):489–495. https://doi.org/10.17221/482/2015-PSE

Bostrom U, Fogelfors H (2002) Response of weeds and crop yield to herbicide dose decision-support guidelines. Weed Sci 50(2):186–195. https://doi.org/10.1614/0043-1745(2002)050[0186:ROWACY]2.0.CO;2

Bruciene I, Aleliunas D, Sarauskis E, Romaneckas K (2021) Influence of mechanical and intelligent robotic weed control methods on energy efficiency and environment in organic sugar beet production. Agriculture 11:449. https://doi.org/10.3390/agriculture11050449

Carlson AL, Luecke JL, Khan MFR, Dexter AG (2008) Survey of weed control and production practices on sugarbeet in Minnesota and eastern North Dakota. North Dakota State University, 2007 Sugarbeet research and extension reports 38:40–63

Carneiro GDOP, Bontempo AF, Guimarães FAR, Rodrigues dos Reis M, Silva DV, Souza MF, Lins HA (2019) Carryover of tembotrione and atrazine in sugar beet. Cienc Investig Agrar 46(3):319–324. https://doi.org/10.7764/rcia.v46i3.2006

Chauhan BS (2020) Grand challenges in Weed Management. Front Agron 1:3. https://doi.org/10.3389/fagro.2019.00003

Chauhan BS, Matloob A, Mahajan G, Aslam F, Florentine SK, Jha P (2017) Emerging challenges and opportunities for education and research in weed science. Front Plant Sci 8:1537. https://doi.org/10.3389/fpls.2017.01537

Cheboi PK, Siddiqui SA, Onyando J, Kiptum CK, Heinz V (2021) Effect of ploughing techniques on water use and yield of rice in Maugo small-holder irrigation scheme. Kenya Agric Eng 3(1):110–117. https://doi.org/10.3390/agriengineering3010007

Chetin AD, Borodin AA, Chmeleva LE (2008) Lanceleaf sage and other problem weeds in sugar beet crops. Sakharnaya Svekla 6:26–27

Chitband AA, Ghorbani R, Rashed MH, Abbaspoor MA, Abbasi R (2014) Evaluation of broadleaf weeds control with selectivity of post-emergence herbicides in sugar beet (*Beta vulgaris* L.). Not Sci Biol 6(4):491–497. https://doi.org/10.1583/nsb649457

Cioni F, Maines G (2010) Weed control in sugarbeet. Sugar Technol 12:243–255. https://doi.org/10.1007/s12355-010-0036-2

Dadkhah A (2013) Allelopathic effect of sugar beet (*Beta vulgaris*) and eucalyptus (*Eucalyptus camaldulensis*) on seed germination and growth of Portulaca oleracea. Russ Agric Sci 39(2):117–123. https://doi.org/10.3103/S1068367413020043

Dale TM, Renner KA, Kravchenko AN (2006) Effect of herbicides on weed control and sugar beet (*Beta vulgaris*) yield and quality. Weed Technol 20(1):150–156. https://doi.org/10.1614/WT-04-278R1.1

Daneshian J, Najari Z, Mavi FL (2013) Investigating of integrated weed management on sugar beet (*Beta vulgaris*) yield in Miandoab. Iran J Weed Sci 8:41–53. (In Persian)

Dayan FE (2019) Current status and future prospects in herbicide discovery. Plan Theory 8:341. https://doi.org/10.3390/plants8090341

Deveikyte I, Seibutis V (2006) Broadleaf weeds and sugar beet response to phenmedipham, desmedipham, ethofumesate and triflusulfuron-methyl. Agron Res 4:159–162

Deveikyte I, Seibutis V (2008) The influence of postemergence herbicides combinations on broad-leaved weeds in sugar beet. Zemdirbyste-Agriculture 53(3):43–49

Deveikyte I, Sarunaite L, Seibutis V (2015) Evaluation of pre- and Postemergence herbicide combinations for broadleaved weeds in sugar beet. In: Price A, Kelton J, Sarunaite L (eds) Herbicides, agronomic crops and weed biology. IntechOpen. https://doi.org/10.5772/61437

Ganbari Birgani D, Hosseinpour M, Shimi P, Noghabi MA (2007) Integrated weed control of sugar beet in Dezful and Boroujerd. Iran J Crop Sci 8:283–299. (In Persian)

Gantoli G, Ayala VR, Gerhards R (2013) Determination of the critical period for weed control in corn. Weed Technol 27(1):63–71. https://doi.org/10.1614/WT-D-12-00059.1

Gerhards R, Bezhin K, Santel H (2017) Sugar beet yield loss predicted by relative weed cover, weed biomass and weed density. Plant Prot Sci 53:118–125. https://doi.org/10.17221/57/2016-PPS

Greenland RG (2003) Injury to vegetable crops from herbicides applied in previous years. Weed Technol 17(1):73–78. https://doi.org/10.1614/0890-037X(2003)017[0073:ITVCFH]2.0.CO;2

Hamill AS, Weaver SE, Sikkema PH, Swanton CJ, Tardif FJ, Ferguson GM (2004) Benefits and risks of economic vs. efficacious approaches to weed management in corn and soybean. Weed Technol 18:723–732. https://doi.org/10.1614/WT-03-166R

Hamouzová K, Jursík M, Zábranský P (2013) Effect of different weather conditions on selectivity of post-emergence herbicides in sugar beet. Listy Cukrovarnicke a Reparske 129(7–8):224–228

Harker KN, O'Donovan JT (2013) Recent weed control, weed management, and integrated weed management. Weed Technol 27(1):1–11. https://doi.org/10.1614/WT-D-12-00109.1

Hassani M, Hemmati Gezaz H, Behfar H, Abdollah Pour S (2020) Construction and evaluation of flame weeding to remove weed in corn and sugar beet rows. J Res Weed Sci 3(4):436–450. https://doi.org/10.26655/JRWEEDSCI.2020.4.3

Heidari GH, Dabbagh Mohammadi Nasab A, Javanshir A, Rahimzadeh Khoie F, Moghaddam M (2007) Influence of redroot pigweed (*Amaranthus retroflexus* L.) emergence time and density on yield and quality of two sugar beet cultivars. J Food Agric Environ 5:261–266. https://doi.org/10.1234/4.2007.1183

Hoseyni SM, Najafi H, Sani B, Mozafari H (2018) Role of new herbicides in dodder (*Cuscuta campestris*) control in sugar beet (*Beta vulgaris*) fields. Appl Ecol Environ Res 16(4):5117–5125. https://doi.org/10.15666/aeer/1604_51175125

Janaki P, Bhuvanadevi S, Dhananivetha M, Murali AP, Chinnusamy C (2018) Persistence of Quizalofop ethyl in soil and safety to ground nut by ultrasonic Bath extraction and HPLC-DAD detection. J Res Weed Sci 1(2):63–74. https://doi.org/10.26655/jrweedsci.2018.9.1

Jhala AJ, Beckie HJ, Peters TJ, Culpepper AS, Norsworthy JK (2021) Interference and management of herbicide-resistant crop volunteers. Weed Sci 69(3):257–273. https://doi.org/10.1017/wsc.2021.3

Kahramanoglu I, Uygur FN (2010) The effects of reduced doses and application timing of metribuzin on redroot pigweed (*Amaranthus retroflexus* L.) and wild mustard (*Sinapis arvensis* L.). Turk J Agric For 34:467–474. https://doi.org/10.3906/tar-0905-17

Kaya R, Buzluk Ş (2006) Integrated weed control in sugar beet through combinations of tractor hoeing and reduced dosages of a herbicide mixture. Turk J Agric For 30:137–144

Khan MFR (2010) Introduction of glyphosate-tolerant sugar beet in the United States. Outlooks Pest Manage 21:38–41. https://doi.org/10.1564/21feb10

Kudsk P (2008) Optimising herbicide dose: a straightforward approach to reduce the risk of side effects of herbicides. Environmentalist 28:49–55. https://doi.org/10.1007/s10669-007-9041-8

Kunz C, Sturm D, Vanrholt D, Walker F, Gerhards R (2016) Allelopathic effects and weed suppressive ability of cover crops. Plant Soil Environ 62:60–66. https://doi.org/10.17221/612/2015-PSE

Lobmann A, Christen O, Petersen J (2019) Development of herbicide resistance in weeds in a crop rotation with acetolactate synthase-tolerant sugar beets under varying selection pressure. Weed Res 59:479–489. https://doi.org/10.1111/wre.12385

MacLaren C, Storkey J, Menegat A, Metcalfe H, Dehnen-Schmutz K (2020) An ecological future for weed science to sustain crop production and the environment. A review. Agron Sustainable Dev 40:24. https://doi.org/10.1007/s13593-020-00631-6

Mahmoud MS, Soliman FS (2012) Chemical weed control in sugar beet. Alex Sci Exchange J 33(4):341–350. https://doi.org/10.21608/asejaiqjsae.2012.3171

Majidi M, Heidari G, Mohammadi K (2011) Management of broad- leaved weeds by combination of herbicides in sugar beet production. Adv Environ Biol 5(10):3302–3306

Marwitz A, Ladewig E, Märländer B (2012) Impact of herbicide application intensity in relation to environment and tillage on earthworm population in sugar beet in Germany. Eur J Agron 39:25–34. https://doi.org/10.1016/j.eja.2012.01.005

Matte WD, Oliveira RS Jr, Constantin R, Mendes RR, Padovese LB, Tseng TM (2021) Bioindicators selection to monitoring pyroxasulfone mobility and persistence in soil. J Res Weed Sci 4(2):142–150. https://doi.org/10.26655/JRWEEDSCI.2021.2.1

Mehdizadeh M (2014) High performance liquid chromatographic determination of an asymmetrical Triazine herbicide from soil. Int J Farming Allied Sci 3:1201–1205

Mehdizadeh M (2016) Effect of pesticide residues on agricultural food production; a case study: sensitivity of oilseed rape to Triasulfuron herbicide soil residue. MOJ Food Process Technol 2: 1–2. https://doi.org/10.15406/mojfpt.2016.02.00054

Mehdizadeh M (2019) Sensitivity of oilseed rape (*Brassica napus* L.) to soil residues of imazethapyr herbicide. Int J Agric Environ Food Sci 3(1):46–49. https://doi.org/10.31015/jaefs.2019.1.10

Mehdizadeh M, Gholami-Abadan F (2018) Negative effects of residual herbicides on sensitive crops: impact of Rimsulfuron herbicide soil residue on sugar beet. J Res Weed Sci 1:1–6. https://doi.org/10.26655/jrweedsci.2018.6.1

Mehdizadeh M, Mushtaq W (2020) Biological control of weeds by allelopathic compounds from different plants: a BioHerbicide approach. In: Egbuna C, Sawicka B (eds) Natural remedies for pest, disease and weed control. Academic Press, Cambridge, pp 107–117. https://doi.org/10.1016/B978-0-12-819304-4.00009-9

Mehdizadeh M, Alebrahim MT, Roushani M, Streibig JC (2016) Evaluation of four different crops' sensitivity to sulfosulfuron and tribenuron methyl soil residues. Acta Agric Scand B-Soil Plant Sci 66:706–713. https://doi.org/10.1080/09064710.2016.1212919

Mehdizadeh M, Alebrahim MT, Roushani M (2017) Determination of two sulfonylurea herbicides residues in soil environment using HPLC and phytotoxicity of these herbicides by lentil bioassay. Bull Environ Contam Toxicol 99:93–99. https://doi.org/10.1007/s00128-017-2076-8

Mehdizadeh M, Asadi GA, Delobel A (2018) Biological control potential of *Spermophagus sericeus* Geoffroy, 1785 (Coleoptera: Chrysomelidae) against field bindweed as the first report from Iran. J Res Weed Sci 1:40–47. https://doi.org/10.26655/jrweedsci.2018.6.5

Mehdizadeh M, Mushtaq W, Siddiqui SA, Ayadi S, Kaur P, Yeboah S, Mazraedoost S, DKA A-T, Tampubolon K (2021) Herbicide residues in agroecosystems: fate, detection, and effect on non-target plants. Rev Agric Sci 9:157–167. https://doi.org/10.7831/ras.9.0_157

Melander B, Rasmussen IA, Bàrberi P (2005) Integrating physical and cultural methods of weed control-examples from European research. Weed Sci 53:369–381. https://doi.org/10.1614/WS-04-136R

Mushtaq W, Mehdizadeh M, Siddiqui MB, Ozturk M, Jabran K, Altay V (2020) Phytotoxicity of above-ground weed residue against some crops and weeds. Pak J Bot 52(3):1–10. https://doi.org/10.30848/PJB2020-3(40)

Norsworthy JK, Ward SM, Shaw DR, Llewellyn RS, Nichols RL, Webster TM, Bradley KW, Frisvold G, Powles SB, Burgos NR, Witt WW, Barrett M (2012) Reducing the risks of herbicide resistance: best management practices and recommendations. Weed Sci 60(1):31–62. https://doi.org/10.1614/WS-D-11-00155.1

Petersen J (2004) A review on weed control in Sugarbeet. In: Inderjit (ed) Weed biology and management. Springer, Dordrecht, pp 467–483. https://doi.org/10.1007/978-94-017-0552-3_23

Pintar A, Stipicevic S, Svecnjak Z, Baric K, Lakic J, Sraka M (2020) Crop sensitivity to mesotrione residues in two soils: field and laboratory bioassays. Chil J Agric Res 80(4):496–504 http://dx.doi.org/https://doi.org/10.4067/S0718-58392020000400496

Rapparini G (2008) New Betarens in the system of postemergence weed control. Sakharnaya Svekla 8:18–20

Rasha GMA (2010) Improving the efficiency of some herbicides in weed control in sugar beet by some adjuvants. M.Sc. Thesis, Fac. Agric. Cairo Univ., Egypt. Saric-Krsmanovic M, Bozic D, Radivojevic L, Gajic Umiljendic J, Vrbnicanin S (2018) response of alfalfa and sugar beet to field dodder (*Cuscuta campestris* Yunck.) parasitism: physiological and anatomical approach. Can J Plant Sci 99:199–209. https://doi.org/10.1139/cjps-2018-0050

Razzaq A, Cheema ZA, Jabran K, Hussain M, Farooq M, Zafar M (2012) Reduced herbicide dosed used together with allelopathic sorghum and sunflower water extracts for weed control in wheat. J Plant Prot Res 52(2):281–285. https://doi.org/10.2478/v10045-012-0045-0

Saric-Krsmanovic MM, Bozic DM, Radivojevic LM, Umiljendic JSG, Vrbnicanin SP (2017) Effect of *Cuscuta campestris* parasitism on the physiological and anatomical changes in untreated and herbicide-treated sugar beet. J Environ Sci Health B 52(11):812–816. https://doi.org/10.1080/03601234.2017.1356167

Sbatella GM, Adjesiwor AT, Kniss AR, Stahlman PW, Westra P, Moechnig M, Wilson RG (2019) Herbicide options for glyphosate-resistant kochia (*Bassia scoparia*) management in the Great Plains. Weed Technol 33(5):658–663. https://doi.org/10.1017/wet.2019.48

Shaner DL, Beckie HJ (2014) The future for weed control and technology. Pest Manag Sci 70(9): 1329–1339. https://doi.org/10.1002/ps.3706

Siddiqui SA, Blinov AV, Serov AV, Gvozdenko AA, Kravtsov AA, Nagdalian AA, Raffa VV, Maglakelidze DG, Blinova AA, Kobina AV, Golik AB, Ibrahim SA (2021) Effect of selenium

nanoparticles on germination of Hordéum Vulgáre barley seeds. Coatings 11(7):862. https://doi.org/10.3390/coatings11070862
Soltani N, Dille JA, Robinson DE, Sprague CL, Morishita DW, Lawrence NC, Kniss AR, Jha P, Felix J, Nurse RE, Sikkema PH (2018) Potential yield loss in sugar beet due to weed interference in the United States and Canada. Weed Technol 32:749–753. https://doi.org/10.1017/wet.2018.88
Somerville GJ, Sønderskov M, Mathiassen SK, Metcalfe H (2020) Spatial modelling of within-field weed populations; a review. Agronomy 10:1044. https://doi.org/10.3390/agronomy10071044
Storkey J, Helps J, Hull R, Milne AE, Metcalfe H (2021) Defining integrated weed management: a novel conceptual framework for models. Agronomy 11:747. https://doi.org/10.3390/agronomy11040747
Tandon S, Pal SL (2021) Residue kinetics of ethofumesate in texturally diverse soils of sugar beet crop under field conditions. J Chem:5566787. https://doi.org/10.1155/2021/5566787
Westwood JH, Charudattan R, Duke SO, Fennimore SA, Marrone P, Slaughter DC, Swanton C, Zollinger R (2018) Weed management in 2050: perspectives on the future of weed science. Weed Sci 66(3):275–285. https://doi.org/10.1017/wsc.2017.78
Wilson RG (1999) Response of nine sugarbeet *(Beta vulgaris)* cultivars to post-emergence herbicide applications. Weed Technol 13:25–29. https://doi.org/10.1017/S0890037X00044857
Wilson RG, Sbatella GM (2011) Late-season weed control in glyphosate-resistant Sugarbeet. Weed Technol 25(3):350–355. https://doi.org/10.1614/WT-D-10-00057.1
Yagoob AM, Zaroug MSA, Dafaallah ABB (2021) Efficacy and selectivity of pendimethalin for weed control in sugar beet (*Beta vulgaris* L.), Gezira state, Sudan. J Res Weed Sci 4(1):16–28. https://doi.org/10.26655/JRWEEDSCI.2021.1.2
Zaller JG, Weber M, Maderthaner M, Gruber E, Takács E, Mörtl M, Klátyik S, Győri J, Römbke J, Leisch F, Spangl B, Székács A (2021) Effects of glyphosate-based herbicides and their active ingredients on earthworms, water infiltration and glyphosate leaching are influenced by soil properties. Environ Sci Eur 33:51. https://doi.org/10.1186/s12302-021-00492-0
Zargar M, Najafi H, Zand E, Mighani F (2010) Study of integrated methods for management of red-root pigweed and lamb-squarters in single- vs. twin-row sugar beet. Proceedings of 3rd Iranian Weed Science Congress. Vol 2: Key papers, weed management and herbicides. Babolsar, Iran, February. pp. 654–657
Zimmerman T, Siddiqui SA, Bischoff W, Ibrahim SA (2021) Tackling airborne virus threats in the food industry: a proactive approach. Int J Environ Res Public Health 18(8):4335. https://doi.org/10.3390/ijerph18084335

Intercropping Sugar Beet with Different Agricultural Crops

19

Mihajlo Ćirić

Abstract

During the recent years, sugar beet production has been declining due to multiple reasons. Low price of the sugar on the world market, high-cost production of sugar beet, changes in the agricultural policy of EU, lower level of subsidies for growers and sugar industry, the rise of the ecological awareness all together lowered its production and changed the crop structure of many areas and growers. However, sugar beet remains the main source of sugar in many temperate climate regions of the world and in countries such as Argentina, Bulgaria, Romania, Russia, Ukraine. Globalization is the importance of ecology in our daily life, and other new trends in the world economy are creating different changes in all the aspect of human life. Intercropping is a relatively old and well-known agricultural technique with the purpose of improving the agricultural systems through principles of biological interactions and plant symbiosis. Intercropping improves plant production in many ways—the rise of biodiversity, land preservation and conservancy of water resources, lower production risks, and higher revenues. In recent years, the interest in intercropping is growing especially in the developing countries. This is common practice in fodder production where certain grass species are grown together with some compatible legume plants. Sugar beet can be grown in the intercropping system with a variety of plant species such as barley, canola, lentil, mustard, onion, rice, bean, soybean, poppy, and many more agricultural plants.

M. Ćirić (✉)
Institute of Field and Vegetable Crops, Novi Sad, Serbia
e-mail: mihajlo.ciric@ifvcns.ns.ac.rs

V. Misra et al. (eds.), *Sugar Beet Cultivation, Management and Processing*,
https://doi.org/10.1007/978-981-19-2730-0_19

Keywords

Beta vulgaris · Intercropping · Intercrops · Multiple cropping · Plant associations sugar beet

19.1 Introduction

Sugar beet presents an important source of sugar in temperate climate regions. Monoculture crop systems productivity is achieved by using high chemical inputs which include fertilizers and pesticides (Brooker et al. 2015). Nitrogen is the carrier of root yield (Bojović 2014; Kolarić et al. 2015; Bojović et al. 2014, 2015, 2019). This causes lower levels of biodiversity in a number of plants, animal species, and microorganisms. Intercropping represents a way of growing two or more plants together in the same field at the same time which produces higher yield or better quality of crops in various soil and environmental factors (Li et al. 2014; Brooker et al. 2015). In this joint life, different effects occur with different relations among the plants such as competition, facilitation, and allelopathy. There are many synonyms of plants used for intercropping systems such as intercrops, companion crops, cover crops, catch crops, or living mulch crops (Romaneckas et al. 2020). Intercrops present components of alternative agriculture to the conventional way of growing plants (Daryanto et al. 2018). The use of intercropping is an additional way for preserving biodiversity, soil fertility, and stability of soil particles (Poeplau et al. 2015; Rücknagel et al. 2016). Through intercropping, water-holding capability and infiltration are increased (Wick et al. 2017) while water erosion is being reduced (Prosdocimi et al. 2016). Applying intercropping brings higher levels of soil biological activity (El-Fakharany et al. 2012; Piotrowska-Długosz and Wilczewski 2012; Alexander et al. 2019) and reduces the density of the pests and weeds in the field (El-Fakharany et al. 2012; Kolota and Adamczewska-Sowinska 2013; Fracchiolla et al. 2020). These interactions among specific plants are present in the space they share above and below the soil surface (Zhang et al. 2001). Many studies show that the subterranean effects are more important for plants' mutual assistance and interaction (Wu et al. 2012). There are many reasons and plans for intercropping utilization such as more efficient usage of mineral fertilizers. Economic effects are also a very important aspect of intercropping since it enables the growers to use their lands more efficiently (Mohammadi and Pankhaniya 2017) and produce more crops on the same field providing more money benefit for them and their families especially in the low input farming systems (Li et al. 2014). Willey (1979) also emphasizes that intercropping is especially suitable for small labor-intensive farms. Intercropping plant crops has many positive and negative effects which appear at the same time. The most important of them is facilitation and completion. Interspecific interactions among plants combine interaction on the ground surface and below the ground surface (Zhang et al. 2001). Some studies suggest that root interactions are a more important aspect of the plant interaction comparing to interactions of shoots and other plant parts (Hajiboland et al. 2018). Sugar beet can be grown with the varieaty of agriculatural crops such as: onion (Badawy and Shalaby 2015), barley (Afshar et al. 2018), sunflower (Sheha et al. 2017) even with sugar cane (Khan and Hussain 2016; Ullah et al. 2018), flax (Hussein and Metwally 2012), faba bean (El-Shamy et al. 2019).

The practice of intercropping is spreading especially in the developing countries with low input agricultural production (Sheha et al. 2017). Intercropping of different crops with its reaction on the pest presence is suggested in specific situations such as a way of positive agricultural practices and as a part of integrated pest management (El-Fakharany et al. 2012). Preventing competition among the main crop and supporting crop and creating conditions for balanced usage of light, water, and mineral nutrients are the most important principles of intercropping (Hollander et al. 2007; Munkholm and Hansen 2012; Wanic et al. 2019; Romaneckas et al. 2020).

Preserving the yield level and quality of the main crop is the main focus of this plant-growing system. Intercropping helps farmers in ways such as higher money benefits, decreases cultivation costs, and enables farmers to use resources and time more efficiently which all together creates more intensified agricultural production (Vishwanatha et al. 2011). Growing sugar beet together with other agricultural plants offers a lot of opportunities and challenges.

One of the first goals for using intercropping was as an insurance in case of failure of crops (Mohammadi and Pankhaniya 2017) On the other hand, sugar beet requires around five times less water quantity than sugar cane (Stevanato et al. 2019). The thoughtful selection of the plant components in the crop system together with the times of establishment is important for efficient usage of space and land resources, where the period of coexistence between the crops determines the productivity of the plant association (Filho et al. 2013). Intercropping has been presented as a more efficient system (Umrani et al. 1984) which is environmentally friendly and more profitable in some cases. This is an interesting process and tool for achieving agricultural intensification and optimization per area unit (Motagally and Metwally 2014).

19.2 The Most Important Relations of the Intercropping Systems

An economically sustainable intercropping system for the most part depends on an adjustment of the planting arrangement and appropriate choice of compatible plants (Seran and Brintha 2009). Choosing the right plants for the intercropping system is related to the speed and rhythm of plant growth, together with water, light, land, and mineral nutrients usage (Brintha and Seran 2008). Andrews (1972) reported that intercrop plants with diverse times of sowing reduce competition between crops and acquire better usage of environmental factors. Various intercropped crops enable higher levels of utilization of plant growth limiting factors (Silwana and Lucas 2002) by developing balance and reducing competition between the main crop and the supporting plant (Romaneckas et al. 2020).

19.3 Crop Maturity

Growing two or more plants together requires that their growing periods do not correspond so that the crops can use environmental factors and resources at different periods. Also intercropping crops with different maturity period have benefits in harvest organization, labor planning together with grain storage schedule and continuous income inflow (Table 19.1).

We can distinguish three basic types of sugar beet varieties: Z- high-sugar content types, N- normal, and E- high yield types (Bojović 2014). This means that period of greatest demands for mineral nutrients, water, light spatial demand of the intercropped plants should not coincide and they represent compatible plant partners. In sugar beet–poppy seed intercropping, sugar beet demands for the water and nutrients are the highest during the beginning of the second three-month interval of the sugar beet vegetation period when the poppy seed is ripe for picking or manual seed pods harvest.

19.4 Crop Compatibility

Selection of compatible plants is the most important part of the intercropping system. The main reason for lower yields in the monocropping systems is competition between the same plant species for mineral nutrients, together with shading and high plant density. Competition can be decreased by an adequate spatial plan. Another way of solving this issue is by using the plants which have the ability of more efficient performance in acquiring mineral nutrients from the soil (Fischer 1977; Willey 1981). Carruthers et al. (1998), Ofori and Stern (1987) suggest that intercropping sugar beet with legumes can present a practical alternative system for achieving greater productivity in mass crop cultivation. Toaima et al. (2001) report the benefits of intercropping sugar beet with onion and garlic. Krall et al. (1996) explored the potential of growing sugar beet and mustard in an intercropping system. Altieri (1994) examined many mixed and relay crops among them intercropping sugar beet and sunflower. Stoyanov et al. (1997) also explored the sugar beet–sunflower combination and its effect on the economical results and soil condition. Singh et al. (1999) explored the economical possibilities of growing sugar beet and cereals together. Mixed intercropping is usual practice in case sugar beet and other root crops are grown together with legumes or cereals and no or little tillage is applied (Agboola 1982). In central and east Europe, there was a common agricultural practice of cropping sugar beet with poppy seed (Figs. 19.1 and 19.2). In subtropical

Table 19.1 Types of sugar beet NS variety

NS variety	Z- high-sugar content types	N- normal	E- high yield types
Early extraction periods	Vera	Lara	Drena
Medium extraction periods	Irina	Nora	Prima
Late extraction periods	Sara	Darija	Neda

Fig. 19.1 Intercropping sugar beet and poppy trials

regions, there are trials about intercropping sugar beet together with sugar cane (Khan and Hussain 2016).

19.5 Plant Density

Decreased number of plants per area together with a lower crop population produces low yields (Seran and Jeyakumaran 2009). However, growing crops in the intercropping system requires adjusting the number of plants per hectare or other surface units. The aim is to decrease the full rate of each crop and avoid overcrowding in order to achieve optimal plant denseness (Seran and Brintha

Fig. 19.2 Intercropping sugar beet and poppy seed trials during poppy flowering

2010). In the case of full-rate plant patterns, both crops would not achieve high yields; therefore, reduced plant patterns enable both crops to develop and yield well in crop mixture (Seran and Brintha 2010). The aim is to find the optimal plant density of both crops so the seeding rate is not reduced to a point where yield is reduced drastically. This can be achieved by pairing rows of specific crops in the mixture or by changing the row orientation of one crop (Sivamaran and Palaniappan 1996). In order to achieve high levels of photosynthesis and greater yield results optimal Leaf Area Index (LAI) is very important to be established (Xiaolei and Zhifeng 2002). Bahadar et al. (2007) conducted the trials where sugar beet and sugar cane were grown together in different spatial and geometrical systems in order to explore the potential of intercropping these two crops, while Vishwanatha et al. (2020) explored the effect of different row proportions on the mutual relations of these two sugar plants in the intercropping way of cultivation.

19.6 Sowing Time

Results of Abou-Elela (2012) showed that sugar beet was under great influence of the sowing date and the intercropping system when is grown with sunflower. Similar studies have been focused on this topic such as Grangeiro et al. (2007), Salama et al. (2016), Badr (2017), Mourad and El-Mehy (2021). Although the main goal of growing sugar beet is for the sugar industry, through intercropping sugar beet can be grown and used in different environments for various purposes such as bioethanol production, forage production, bioremediation, and reclaiming saline soils (Mall et al. 2021; Misra et al. 2020), as a source for new biodegradable and environmentally friendly materials, and for preventing nitrogen leaching (Whitmore and Schröder 2007).

19.7 Positive Aspects of Intercropping

19.7.1 Resource Usage

A different way of using natural resources when plants are grown together compared to classical monocropping system is the most important argument when explaining the higher yields in intercropping systems. Partitioning of the resources among plants is a normal situation that occurs every time plants are grown together (Blade et al. 1997). Biological explanation of intercropping includes complementarity in resources usage by plants (Barhom 2001) which means that a combination of intercrop plants achieves better overall utilization than in the case of separate plant cultivation.

Issues regarding fertility levels of soil are not only agronomic areas but also for economical and social studies. Preserving soil fertility is imperative for every generation and society. In many parts of the world, especially in developing countries, farmers are not prone to risk and cannot invest in expensive mineral fertilizers (Seran and Brintha 2010). New times demand new measures in which all resources in agriculture can be used in order to preserve soil fertility.

Romaneckas et al. (2020) reported results which intercropping sugar beet with Persian clover (*Trifolium resupinatum* L.), white mustard (*Sinapis alba* L.), and spring barley (*Hordeum vulgare* L.) had a positive impact on the content of nitrogen, phosphorus, and potassium in soil than sugar beet grown in mono-crop system. The roots of intercrops produce excretion which solubilizes soil phosphorus and makes it more available for crops. On their roots live the symbiotic bacteria which are specialized for nitrogen fixation.

Various leaf shapes and patterns together with diverse roots can harness more solar energy, more mineral nutrients, and water than in situations when only one plant type appears. In case when only one crop is present between its roots a competition starts to develop, likewise, a similar situation starts among its leaf which have the same developing period and orientation and tends to compete with each other. Waddington et al. (1989) offer an explanation that leaf canopy which consists of different plants can give better light usage. In the tropics, normal agricultural practice tends to favor planting lower plants between the wider rows of higher species. This results in better light absorption and therefore higher yields of lower species. Azam-Ali et al. (1990) also reported that conditions in the intercropping systems are backing up lower and smaller plants. In the situation when two crops with diverse maturity rates are grown, the factor which determines their yield levels is light (Willey 1979).

Water has always been a major factor that determines plant development in the intercropping system. Better water supply implicates higher mineral nutrients uptake and better usage of other environmental factors (Hook and Gascho 1988). The presence of different roots in soil limits water deficiency and raises the level of transpiration and water uptake and establishes colder microenvironments (Innis 1997). Compared with mono-crop system, plants in intercropping take 7% more water (Morris and Garrity 1993).

19.7.2 Weed Management

Intercropping system is also a way of control disease, pests, and weeds. Growing crops as intercrops offers is a way of developing sustainable agriculture with advantages in weed management since it decreases the farmer's dependence on herbicides and other pesticides (Lithourgidis et al. 2006; Fernández-Aparicio et al. 2007; Amini et al. 2020). Comparing to monocropping, growing intercrops gives greater competition against weeds in space or in time and therefore requires lower use of herbicides (Seran and Brintha 2010). The degree and features of competition between the crops and weeds depend mainly on the crop combinations, but also on species, plant density, duration period, growing rhythm and habits of crops, and sowing patterns. Furthermore, crop-weed competition is also under the effect of moisture, fertility capacity, and tillage system in the intercropping. When established expanded leaf cover is helping crops in weakening and slowing down weed (Beets 1990). Growing two or more crops in a mixed intercropping system lowers the weed frequency and spreading (Zuofa et al. 1992). In their short-term investigation, Romaneckas et al. (2020) reported that intercropping sugar beet with Persian clover, white mustard, and spring barley had a positive effect on weed suppression (Adamavičienė et al. 2009; Romaneckas et al. 2009). However, the best effect on weed control reported for Persian clover which although presents a rivalry for the main crop (sugar beet) has the best effectiveness for weed suppression (Marks et al. 2018). The gains of using intercropping for weed management especially in low input farming go to the improvement of weed control and agricultural production.

19.7.3 Pest and Disease Control

During its vegetation, sugar beet can be attacked by many insects and other pests. Intercropping offers promising practices for these threats. There is a general opinion that one of the components in the intercropping system can provide a defense line against spreading insects and other pathogens (Youssef and El-Nagdi 2012). Problems with insects are smaller on agricultural crops grown together in intercropping system comparing to the monocropping way of growing and need fewer chemicals (Singh and Ajeigbe 2002).

19.7.4 Soil Preservation and Erosion Control

Soil erosion and soil silting are presenting important problems in sugar beet cultivation. Lack of plant cover during winter months and the first phases of sugar beet growth are responsible for not providing protection to the surface of the soil from the weather factors (Schmidt 1987). Intercropping way of growing crops reduces soil erosion by leaf cover which prevents direct raindrops from hitting the surface of the soil. Higher plants play the role of a wind barrier for lower crops (Radke and Hagstrom 1976; Reddy and Reddi 2007) and by decreasing wind speed, they also lower the desiccation (Beets 1990). Growing crop mixture enables soil protection through prolonged vegetative growth during critical erosion periods (Siddoway and Barnett 1976; Kolota and Adamczewska-Sowinska 2013). Romaneckas et al. (2020) reported the 3 years to result with intercropping sugar beet with Persian clover, spring barley, and white mustard as living mulch. The results show that soil pH in the field plots with Persian clover was decreased significantly in comparison with control treatments.

19.7.5 Yield Grain

Assessing the value of intercropping yield is the first examination. Plants are grown together in intercropping due to high-level yields and higher economical and biological balance of the system (Francis 1986). El-Nakhlawy and Ismail (2018) found no statistically significant differences between sugar beet fresh root or sucrose yield ha^{-1} under the intercropping systems with clover and sugar beet sole crop. Similar results were reported by Edrees et al. (2019) which also found greater yields of sugar beet root and sucrose yield in the intercropping systems in comparison to sole crop growing.

19.8 Economic Aspects

Intercropping in many cases brings farmers higher money revenue comparing to growing only one agricultural species (Motagally and Metwally 2014; Ward et al. 2016; Ćirić et al. 2016). Through the usage of the intercropping system, greater land areas are being cultivated which in return gives higher financial results (Seran and Brintha 2009). El Dessougi et al. (2003) reported that sugar beet grown with oilseed crops produced higher monetary returns than other companion crops. Lal and

Mukerji (1998) suggested that intercropping system sugar beet-cereals can produce higher money income at small farms. Comparing the result of intercropping sugar beet with lentil and wheat, the analysis showed that intercropping system was considerably more profitable than the monocropping system. The best financial results were received from sugar beet + lentil intercropping (Usmanikhail et al. 2013).

19.9 Intercropping with Sugar Beet

19.9.1 Intercropping Sugar Beet with Cereal Crops and Sugar Cane

Many authors have been exploring the relations between sugar beet and cereal crops such as barley (Ozkan 1971; Usmanikhail et al. 2013; Afshar et al. 2018; Romaneckas et al. 2020), wheat (Abou-Elela 2012; Usmanikhail et al. 2013; El-Dein 2015; Ouda and Zohry 2017; Gomaa et al. 2019; Osman and Haggag 1981), maize (Elmer et al. 1963; Ozkan 1971; Khazaie 2015), and ryegrass (Adamavičienė et al. 2009). In the case of intercropped production of sugar beet and wheat, research shows that the sugar beet had the upper hand and better used the intercropping conditions. Sugar beet roots and shoots had a higher level of making dry matter while the same indicator in wheat has been reduced (Hajiboland et al. 2018). Lal and Mukerji (1998) advocate that intercropping systems like sugar beet with cereal crops can bring the farmer higher financial results. Badraoui et al. (2003) had grown wheat-sugar beet as companion crops in the irrigated fields of Morocco. Sugar beet–wheat plant association showed higher growth of sugar beet while the wheat growth was reduced (Hajiboland et al. 2018). In their study, Hajiboland et al. (2018) suggest that acquired results were made as an effect of underground interaction between sugar beet and wheat plants since all the plants had sufficient light conditions.

Results of Afshar et al. (2018) show that using barley as living mulch contributed to higher sucrose levels in sugar beet root and better-quality traits such as lower content of sodium, potassium, and amino nitrogen. However, Romaneckas et al. (2020) point out that using live mulch in intercropping has undoubtedly negative effect on the sugar beet yield with 20 tonnes/ha lower yields on average compared to control treatment and conventional sugar beet growing. Romaneckas et al. (2020) also point out that all treatments showed a higher level of phosphorus but the highest level was recorded in the treatment where sugar beet was mulched with spring barley. It is determined by Romaneckas et al. (2020) that annual ryegrass has a negative and aggressive reaction to sugar beet when it is grown in intercropping system. Specific intercropping systems can improve environmental factors especially soil conditions (Badawy and Shalaby 2015), raise the soil quality traits and field microclimate so in the end, they raise the level of crop production intensity (Li et al. 2014). Intercropping sugar beet and sugar cane has been an interesting topic for many authors and researchers such as Munir et al. (2008), Khan and Hussain (2016).

19.9.2 Intercropping Sugar Beet with Legumes

There are many studies about intercropping sugar beet with different legumes such as clover (El-Nakhlawy and Ismail 2018; Edrees et al. 2019), Egyptian clover (Osman and Haggag 1981), and Persian clover (*Trifolium resupinatum* L.) (Romaneckas et al. 2020). Sugar beet can be cultivated with various species of legumes especially grain legumes such as soybean in the forms of organic production. Research of Usmanikhail et al. (2013) showed that lentil as the agricultural plant produces the higher money revenue when it is grown together with the sugar beet. Piotrowska-Długosz and Wilczewski (2012) point out that legume biomass stimulates microbial activity in the soil, which results in greater levels of organic matter decomposition. In their 3-year study, Romaneckas et al. (2020) showed that intercropping sugar beet with Persian clover elevated the level of nitrogen, potassium, and magnesium in the soil. Mohammed et al. (2005) point out that sugar beet intercropped with faba bean has lower results in yield and yield components together with slower growth. Using Persian clover as living mulch in sugar beet crop research showed that there was a neutral reaction on the root yield and its quality traits; however, a positive response was achieved in weed control (Romaneckas et al. 2020). In their research, Amini et al. (2020) recommend growing sugar beet intercropped with soybean and Moldavian balm in sustainable production systems. Intercropping patterns sugar beet with this crop could be recommended in sustainable production systems in order to increase crop production per unit area without chemical fertilizer and pesticide application that is consistent with environmentally friendly agriculture.

19.9.3 Intercropping Sugar Beet with Forage Crops

Research about growing sugar beet together with chicory has been made by Czaban et al. (2018). Study of using plants from the *Brassicaceae* family as catch crops and at the same time nematicidal intercrops was made by Curto (2008). In this study was explained the use of fodder radish (*Raphanus. sativus* L. ssp. *oleiformis*) and white mustard (*Sinapis alba* L.) as a measure for sustainable management of cyst nematodes in sugar beet growing areas.

19.9.4 Intercropping Sugar Beet with Oil Crops

Studies of intercropping sugar beet with oil crops have been made by many authors on sunflower (*Helianthus annuus* L.) (Mohammed and Abd El Zaher 2013; Sheha et al. 2017; Mourad and El-Mehy 2021), on white mustard (*Sinapis alba*) (Romaneckas et al. 2020), camelina (Afshar et al. 2018), Chinese mustard—*Brassica juncea* (Motisi et al. 2009), flax—(*Linum usitatissimum* L.) (Hussein and Metwally 2012). Stoyanov et al. (1997) reported that the sugar beet–sunflower intercropping system greatly raises financial results and produces a positive effect on the soil quality and nutrition uptake for the future crops. Lal and Mukerji (1998) suggest that intercropping system like sugar beet with oilseed crops represent a great potential for farmers and their high money returns. Badraoui et al. (2003) have been cultivated wheat-sugar beet or sunflower in the irrigated agricultural areas of Morocco and propose using sugar beet and sunflower as intercrops. Tichy et al. (2001) found that sugar beet–sunflower intercropping enlarged sunflower yield more than 5 tons ha^{-1} and sugar beet/sunflower intercropping was marked as the most successful companion crop with net profits. Ćirić et al. (2016) point out that intercropping sugar beet and poppy seed could be used for more efficient utilization of mineral fertilizers. Large amounts of NPK fertilizers are necessary for high root yield; however, too much of mineral nutrients decrease root quality and amount of sucrose that can be used in the sugar factories (Fig. 19.3).

19.9.5 Intercropping Sugar Beet with Vegetable Plants

Study of growing sugar beet with vegetables has been made with many crops such as bean (Ozkan 1971), faba bean (Elshamy 2016; Ouda and Zohry 2017; El-Shamy et al. 2019; Khalifa et al. 2019; Zohry and Ouda 2019; El-Mehy et al. 2020; El-Refaey et al. 2021), onion (Besheit et al. 2002; Motagally and Metwally 2014; Badawy and Shalaby 2015; Zohry and Ouda 2019; Abd Allah et al. 2020), garlic (Badawy and Shalaby 2015; Hussein and El-Shamy 2017), salad rocket (de Sousa Alves et al. 2020), lentil (Usmanikhail et al. 2013; Afshar et al. 2018), and other vegetable plants such as cucurbits (Sridhar et al. 2002). Sugar beet is very closely related to vegetable plants especially beet (*Beta vulgaris ssp cicla).* In his research where sugar beet was grown together with garlic and onion, the effect of intercropping on insect infestation was studied. Badawy and Shalaby (2015) found that the lowest degree of infestation was recorded in plots where sugar beet was grown together with onion. Research of Toaima et al. (2001) shows that intercropping production of sugar beet with garlic or onion results with higher yield and quality of sugar beet and its yield components.

19.9.6 Intercropping Sugar Beet with Aromatic Plants

Sugar beet can be grown together with many aromatic plants such as Moldavian balm (Amini et al. 2020), fennel (Khafagy et al. 2020), dill (Khafagy et al. 2020), coriander (Khafagy et al. 2020), marjoram (Khafagy et al. 2020). Amini et al. (2020) reported that among intercropping trials with different plants sugar beet achieved the best result when was grown with Moldavian balm (*Dracocephalum moldavica* L.) During the two-year trial period, the highest root yield, sugar content, and sugar yield reported with Moldavian balm or there were no significant differences among the best result and the result where sugar beet was grown with Moldavian balm.

Fig. 19.3 Intercropping sugar beet with poppy seed trials during poppy harvest

19.10 Future Prospects

Sugar beet has a great potential for intercropping since it can be grown in many agricultural areas of the world with a large number of different field and vegetable crops together with fodder plants. Since the human population has been growing very fast during the recent years and decades and soon the number of people on the Earth will reach figure 10 billion together with climate changes, large adjustments of human life and activities are necessary and many countries (India, Pakistan, Egypt, Saudi Arabia) (Bahadar et al. 2007; Salama et al. 2016; El-Nakhlawy and Ismail

2018; Mall et al. 2021) are including sugar beet in their future agricultural plans and developing agendas. Sugar beet has found its place in many aspects of human life. It is used for the sugar industry, dairy farming, and protein production. New times offer new possibilities and there are many possibilities for sugar beet. In close future use of fossil fuels will be forbidden and the production of ecological fuels such as bioethanol will be promoted and sugar beet with its high percentage of sucrose and a large amount of biomass present a potential solution for the production of cleaner energy and fuels. Today many soils are lost due to the road and building construction. Growing sugar beet intercropped with other salt and pollutant tolerant plants for bioremediation and re-cultivation of polluted and saline soils will be necessary for the agriculture of future generations. The discovery of plastic made life easier for many people but today plastic waste presents a global problem because of its slow decomposition and unsatisfactory impact on the environment. With its large amount of cellulose sugar beet and other cellulose-rich plants grown together and could be used for production, new environmentally friendly materials so-called bioplastic would largely help the industries and decrease the amount of human-made waste.

19.11 Conclusion

Sugar beet is an agricultural crop that is an important source of sugar and which is usually produced in mono-crop system. Numerous studies over the past decades have proven that sugar beet can be grown together with many agricultural crops such as cereals, legumes, oil, and aromatic and forage crops, even with sugar cane. During the recent years, practice of growing sugar beet together with other crops has been spreading especially in the developing countries. This phenomenon can be explained by different reasons such as preserving biodiversity, better usage of mineral nutrients, diversification of crop production, and increasing the farmers' profits through production intensification, especially on smaller estates. Through intercropping practice, ideas of sustainable agriculture have been promoted and implemented in crop production. Using intercropping, ecological principles of environmental protection are also advocated since the intercropping system improves soil fertility by using legumes as intercrops; promotes using smaller quantities of pesticides and mineral fertilizers or the use of biological measures in plant protection (utilization of nematocidal plants as a part of cyst nematode management). Protection of soil from erosion caused by wind and rain is also achieved by plant cover produced by intercrops.

Generally, sugar beet intercropping growing technology offers many new perspectives in agricultural production but also requires many adjustments of agricultural machines and more efficient labor planning. Furthermore, investigation on how to minimize the competing relations among sugar beet plants and their intercrops, and how to optimize utilization of mineral nutrition, plant protection measures, soil and climate conditions are also necessary in order to achieve high yields, quality crops, and better financial results.

Acknowledgments The authors express their appreciation for providing financial support for this research to Ministry of Education, Science and Technological Development of the Republic of Serbia, Grant No. 451-03-09/2021-14/200032.

References

Abd Allah AM, El-Mehy AA, Abdel-Baset SH (2020) Effect of intercropping onion with sugar beet on productivity of both crops and root-knot nematodes control under different onion plant densities and slow-release N fertilizer rates. J Plant Prot Res 9(1):61–75

Abou-Elela AM (2012) Effect of intercropping system and sowing dates of wheat intercropped with sugar beet. J Plant Prod 3(12):3101–3116

Adamavičienė A, Romaneckas K, Šarauskis E, Pilipavičius V (2009) Non-chemical weed control in sugar beet crop under an intensive and conservation soil tillage pattern: II. Crop Prod Agron Res 7:143–148

Afshar RK, Chen C, Eckhoff J, Flynn C (2018) Impact of a living mulch cover crop on sugarbeet establishment, root yield and sucrose purity. Field Crops Res 223:150–154

Agboola AA (1982) Crop mixtures in traditional system. In: MacDonald LH (ed) Agroforestry in the African humid tropics. United Nations University, Tokyo, pp 46–48

Alexander J, Venterea RT, Baker JM, Coulter JA (2019) Kura clover living mulch: spring management effects on nitrogen. Agronomy 9(2):69

Altieri MA (1994) The ecological role of biodiversity in agroecosystems. Agric Ecosyst Environ 74:19–31

Amini R, Khoei BC, Nasab ADM, Amirnia R (2020) Effects of intercropping and cultivar on sugar beet (*Beta vulgaris* L.) root and sugar yield. Int J Agric Sci 13(3):159–167

Andrews DJ (1972) Intercropping with sorghum in Nigeria. Exp Agric 8(2):139–150

Azam-Ali SN, Matthews RB, Williams JH, Peacock JM (1990) Light use, water uptake and performance of individual components of a sorghum/groundnut intercrop. Exp Agric 26(4): 413–427

Badawy S, Shalaby G (2015) Effect of intercropping sugar beet with onion and garlic on the insect infestation, sugar beet yield and economics. J Plant Prod Mansoura Univ 6(6):903–914

Badr KS (2017) Integrated crop managements through optimal planting date and nitrogen fertilizer levels in wheat–sugar beet association on competitive relationships and yield advantages. Ann Agric Sci 55(3):511–528

Badraoui M, Agbani M, Bouabid R, El-Gharous M, Karrou M, Zeraouli M (2003) New recommendations for the wheat-sugar beet and sunflower cropping in the Doukkala and Gharb irrigated regions of Morocco. Proc Regional Workshop of the Int Potash Inst, Amman, pp. 113–119

Bahadar K, Sadiq M, Subhan M, Khan AU, Khan P, Khan D (2007) Production potential of sugar beet intercropping with sugarcane under various planting geometry system. Pak Sug J 22(l):76–81

Barhom TIH (2001) Studies on water requirements for some crops under different cropping systems. M.Sc. Thesis. Fac Agric Cairo Univ Egypt

Beets WC (1990) Raising and sustaining productivity of smallholder farming systems in the tropics: a handbook of sustainable agricultural development. AgBe Publishing, Holland, p 754

Besheit SY, Elwafa AA, El-Hamd AA, Bekeet MA (2002) Quality and productivity of sugar beet as affected by intercropping onion in various densities. Al-Azhar J Agric Res 36:87–101

Blade SF, Shetty SVR, Terao T, Singh BB (1997) Recent developments in cowpea cropping systems research. In: Singh BB, Mohan Raj DR, Dashiell KE, Jackei LEN (eds) Advances in cowpea research. International Institute of Tropical Agriculture, Ibadan, pp 114–128

Bojović R (2014) Morphological and production properties of sugar beet genotypes in conditions of intensive plant nutrition. Doctoral dissertation. University of Belgrade, Faculty of Agriculture Zemun. Serbia, pp. 1–100

Umrani NK, Shinde SH, Dhonde PM (1984) Studies on intercropping of pulses in kharif sorghum. Ind J Agron 29(1):27–31

Usmanikhail MU, Tunio SD, Jamro GH, Oad FC, Hassan SWU, Chachar QD, Khanzada MA (2013) Effect of intercropping cereals and lentil in sugar beet on yield and monetary benefits. Pak J Bot 45(2):401–406

Vishwanatha S, Koppalkar BG, Anikumar SN, Desai BK, Naik V (2011) Economics and yield advantages of pigeonpea and sunflower intercropping system influenced by fertilizer management. Res J Agric Sci 2(2):248–251

Vishwanatha S, Shwetha BN, Koppalkar BG, Kavya BM, Hiremath SM, Patil PL (2020) Performance of tropical sugar beet cultivars in intercropping with sugarcane at different row proportions. J Pharmacogn Phytochem SP6:220–226

Waddington SR, Palmer AFE, Edje OT (eds) (1989) Research methods for cereal/legume intercropping: proceedings of a workshop on research methods for cereal/legume intercropping in eastern and southern Africa held at Lilongwe, Malawi, 23–27 January 1989 (no. 17). CIMMYT

Wanic M, Zuk-Gołaszewska K, Orzech K (2019) Catch crops and the soil environment—a review of the literature. J Elem 24:31–45

Ward B, Roe BE, Batte MT (2016) Potential profitability of strip intercropping with corn and soybeans. J ASFMRA:149–166

Whitmore AP, Schröder JJ (2007) Intercropping reduces nitrate leaching from under field crops without loss of yield: a modelling study. Eur J Agron 27(1):81–88

Wick A, Berti M, Lawley Y, Liebig M (2017) Integration of annual and perennial cover crops for improving soil health. In: Soil health and intensification of agroecosytems. Elsevier BV, Amsterdam, pp 127–150

Willey R (1979) Intercropping-its importance and research needs: part 1. Competition and yield advantages. Field Crop Abst 32:1–10

Willey RW (1981) A scientific approach to intercropping research. In Proceedings of the international workshop on intercropping, Hyderabad, India, 10–13 January 1979, pp. 4–379

Wu K, Fullen MA, An T, Fan Z, Zhou F, Xue G, Wu B (2012) Above-and below-ground interspecific interaction in intercropped maize and potato: a field study using the 'target' technique. Field Crop Res 139:63–70

Xiaolei S, Zhifeng W (2002) The optimal leaf area index for cucumber photosynthesis and production in plastic green house, ISHS Acta Hortic (XXVI international horticultural congress. http://www.actahort.org/books/633/633_19.htm.) pp. 633

Youssef MMA, El-Nagdi WM (2012) Antagonistic potential of certain legumes to *Meloidgyne incognita* on sugar beet. Arch Phytopathol Plant Prot 45(14):1724–1731

Zhang FS, Li L, Sun JH (2001) Contribution of above- and below ground interactions to intercropping. In: Horst WJ, Schenk MK, Bürkert A, Claassen N, Flessa H, Frommer WB, Goldbach HE, Olfs HW, Römheld V, Sattelmacher B, Schmidhalter U, Schubert S, von Wirén N, Wittenmayer L (eds) Plant nutrition - food security and sustainability of agro-ecosystems. Kluwer Academic Publishers, Dordrecht, pp 979–980

Zohry A, Ouda S (2019) Intercropping systems for sugar beet to improve its land and water productivity. J Soil Crops 29(2):218–226

Zuofa K, Tariah NM, Isirimah NO (1992) Effects of groundnut, cowpea and melon on weed control and yields of intercropped cassava and maize. Field Crops Res 28(4):309–314

Sugar Beet Production Under Changing Climate: Opportunities and Challenges

20

Aiming Qi

Abstract

Sugar has formed an essential part of human diets for a long time and is an important raw material for the food, beverage and pharmaceutical industries. It is a common name for sucrose and can be extracted from two field crops—sugar beet and sugar cane. Sugar beet (*Beta vulgaris* ssp. *vulgaris*) is mainly grown in countries with temperate climates while sugar cane (*Saccharum officinarum* L.) is cultivated primarily in tropical and subtropical countries. It was demonstrated that sugar beet yield has kept increasing since 1926, but sugar concentration (on fresh weight basis) has not changed much. In the meantime, the improved potential sugar beet yields in the varieties included in the variety trials have been rapidly translated into commercially delivered yields by sugar beet farmers. This can be seen in the increase of farmer-delivered sugar beet yields in parallel with the increase of sugar beet yields in the variety trials. The warming temperature and increasing concentration of CO_2 in the atmosphere due to climate change have benefitted the sugar beet crop in recent decades and will probably create opportunities to further boost sugar crop productivity in the future. However, social and environmental demands to adapt sugar beet production to both less input-intensive and less pesticide-dependent cropping systems to mitigate climate change and to maintain biodiversity friendly environments require sugar beet farmers to balance the trade-offs between maximising the sugar yield and increasing the use efficiencies of inputs such as fertilisers, fungicides, pesticides, herbicides and fuels. Sugar beet breeders and other stakeholders need to breed climate-smart cultivars resistant to diseases and find other effective non-chemical

A. Qi (✉)
Centre for Agriculture, Food and Environmental Management Research, School of Life and Medical Sciences, University of Hertfordshire, Hatfield, Hertfordshire, UK
e-mail: a.qi@herts.ac.uk

V. Misra et al. (eds.), *Sugar Beet Cultivation, Management and Processing*,
https://doi.org/10.1007/978-981-19-2730-0_20

solutions to the reduced availability and/or removal of reliable pesticides in the face of more new pathogens emerging under climate change.

Keywords

Potential crop yields · Crop improvement · Yield gaps · Climate change · Prolonged growing season · Crop health · Plant pathogenic diseases · Pesticide use

Abbreviations

ABF	Associated British Foods plc
AY	Attainable yields
BChV	Beet chlorosis virus
BMYV	Beet mild yellowing virus
BYV	Beet yellows virus
CV%	Coefficient of variation
FY	Farmers' yields
GCM	Global climate model
GHG	Greenhouse gas
PY	Potential yield
PY_{wl}	Water-limited potential sugar beet yields
RCPs	Representative concentration pathways

20.1 Introduction

Sugar beet (*Beta vulgaris* spp. *vulgaris*) is agriculturally important because of its ability to accumulate a large quantity of sucrose (i.e. sugar) in its storage root. Sugar beet and sugar cane are the two crops grown exclusively for extracting sucrose in the world (Draycott 2006). Sugar beet is a biennial temperate root crop while sugar cane is a perennial grass-type crop which grows well in tropical and frost-free warm subtropical areas. Sugar beet is usually sown as a spring crop and harvested in autumn in Europe and Northern America and Japan. However, sugar beet can be sown as an autumn crop in some regions where the climate is Mediterranean such as southern Spain, Algeria, Morocco and Egypt. Europe accounted for about two-thirds while Asian and American countries combined accounted for about 25% of total sugar beet cultivation areas and production in recent years (Table 20.1).

Recent human economic development and related activities have increased concentration of anthropogenic greenhouse gases in the atmosphere and modified the climate of the earth. As a result, the annual atmospheric CO_2 concentration has increased every year since 1959 (NOAA 2021) (Fig. 20.1). It has increased at an annual rate of 2.2 ppm for the past 20 years and reached 414 ppm in 2020. The global

Table 20.1 Annual sugar beet cultivation areas in hectare (ha) and beet productions in metric tonne(t) in the world and in continents of Europe, America, Asia and Africa in years from 2015 to 2019

Region	Item	Year 2015	2016	2017	2018	2019
World	Area	4,215,084	4,591,785	4,989,641	4,797,704	4,609,434
	Production	240,810,018	278,834,790	313,989,402	273,711,753	278,497,980
Europe	Area	2,839,175(67.4)[a]	3,098,131(67.5)	3,440,646(69.0)	3,303,067(68.8)	3,166,508(68.7)
	Production	158,155,276(65.7)[b]	188,276,772(67.5)	219,865,555(70.0)	184,722,390(67.5)	194,460,403(69.8)
America	Area	495,719(11.8)	48,739 (10.6)	480,882(9.6)	486,906(10.1)	425,083(9.2)
	Production	34,718,422(14.4)	36,012,041(12.9)	34,642,231(11.0)	34,005,661(12.4)	28,225,847(10.1)
Asia	Area	584,461 (13.9)	708,397 (15.4)	788,708(15.8)	744,633(15.5)	750,980(16.3)
	Production	31,995,720(13.3)	39,035,256(14.0)	44,798,181(14.3)	40,811,709(14.9)	41,507,477(14.9)
Africa	Area	295,729(7.0)	297,861(6.5)	279,405(5.6)	263,098(5.5)	266,863(5.8)
	Production	15,940,600(6.6)	15,510,721(5.6)	14,683,435(4.7)	14,171,993(5.2)	14,304,253(5.1)

[a]Figures in brackets represent percentage of world sugar beet cultivation area
[b]Figures in brackets represent percentage of world sugar beet production
Data source: http://www.fao.org

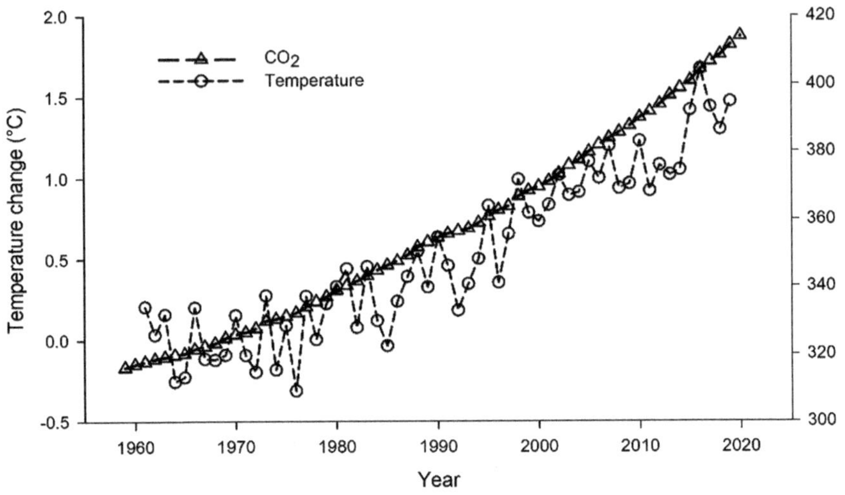

Fig. 20.1 The annual world temperature changes from the long-term mean in 1951 to 1980 from 1961 to 2019 and the annual atmospheric concentration of CO_2 (μmol/mol - ppm) from 1959 to 2020. Temperature data: http://www.fao.org. CO_2 data: NOAA (2021)

mean annual temperature over land has increased, with the six largest temperature increases compared to the long-term mean from 1951 to 1980, ranging from 1.2 to 1.6 °C, occurring in the decade 2010–2019 (Fig. 20.1). Rising temperatures, often combined with more frequent droughts, are not only threatening crop production (Jaggard et al. 2010; Lobell and Gourdji 2012; McKersie 2015) but also encouraging pathogens to move to new regions and exacerbating crop diseases from existing and new races (Rosenzweig et al. 2001; Newbery et al. 2016; Nelson 2020; Juroszek et al. 2020).

Sugar beet productivity is subject to challenges from weather, uncertain efficiency of fertiliser applications and variable control of biotic stressors such as diseases, pests and weeds (Hoffmann et al. 2009; Jaggard et al. 2009). Annual variations in sugar beet yields are caused by the weather in combination with the impact of management decisions (Werker and Jaggard 1997, 1998, Werker et al. 1998; Scott and Jaggard 2000; Richter et al. 2001; Qi et al. 2005; Hoffmann et al. 2009; 2020). For example, cooler than average temperatures during the growing season can be significantly reduced the sugar yield in the UK (Qi and May 2013). As shown in Fig. 20.2, the slower development of the leaf canopy resulted in reduced canopy-intercepted solar radiation and then a 2.2 t/ha reduction in the UK sugar yield in 2012 (9.93 t/ha) compared with 2011 (12.1 t/ha) even though the sowing date was earlier in 2012 (Qi and May 2013).

Sugar beet production has gone through phases from growing multi-germ to monogerm varieties and from being labour-intensive to being highly mechanised. With the ever-increasing world population and the decreasing land and water resources, there is a need to produce more food and to reduce greenhouse gas

Fig. 20.2 Contrast in sugar beet crop canopy cover development due to temperature differences between sowing and 8 June at Broom's Barn, Suffolk, England for crops sown on 26 March 2011 (**a**) or on 24 March 2012 (**b**)

emissions from agriculture to mitigate climate change. The present intensive conventional agriculture is an important source of greenhouse gases (GHG) and produces many unintended negative impacts on the environment. Efforts are needed to change to more sustainable ways of crop production, relying less on synthetic inputs for fertilisation and crop protection and less on fossil-fuel powered machinery. There is an increasing need to adapt to both less input-intensive and less pesticide-dependent sugar beet cultivation to contribute to the net zero GHG emission target by 2050. The challenges for sugar beet breeders, agronomists, farmers and processors to mitigate climate change, to maintain ecological services and to conserve natural resources are increasing.

Analysis of sugar yield in the national variety trials and in the commercial crop between 1976 and 2005 showed that about two-thirds of the increase could be accounted for by changes in climate; other factors such as plant breeding, increased concentration of CO_2 in the atmosphere and agronomic improvement accounted for the rest (Jaggard et al. 2007). Future climate change is predicted to be beneficial to crops whose harvestable part is vegetative organs, such as roots of sugar beet (Jaggard et al. 2010), tubers of potato (Gregory and Marshall 2012) and grassland pastures (Qi et al. 2018). Despite faster development in determinate crops such as wheat and maize, it is projected that crop yields will increase or stay the same under climate change if water supply remains adequate, and crop diseases and pests are adequately and sustainably controlled (Hatfield et al. 2011). However, the interactions between crops and pests/pathogens are complex and poorly understood in the context of climate change (Newbery et al. 2016; Juroszek et al. 2020; IPPC Secretariat 2021; Jeger et al. 2021). Although the increased severity of drought will negatively affect sugar yield, the estimated net impacts of climate change will likely increase sugar beet yields in Europe (Pidgeon et al. 2001; Jones et al. 2003; Richter et al. 2006; Qi and Jaggard 2008, 2012; Okom et al. 2017). However, existing and emerging sugar beet diseases and diseases like virus yellows, powdery mildew (Qi and Jaggard 2008; Qi and Fitt 2014; Dewar and Qi 2021) and Cercospora

(Racca et al. 2015; Kremer et al. 2016; Vogel et al. 2018) will probably increase in severity and frequency.

20.2 Levels of Sugar Beet Yields

Sugar beet production can be grouped into different levels based on growth-limiting factors. Sugar beet farmers want to produce as large yields as possible, but what is achieved and what is the real potential of the crop? The yield levels are illustrated in Fig. 20.3. It is important to recognise these different levels when yield increases are targeted in sugar beet crop improvement programmes (Fig. 20.3).

Potential yields (PY) are the theoretical maximal yield at harvest in one cropping season using well-adapted cultivars grown under optimal conditions. They are determined by the total amount of light energy captured by the crop, its radiation-use efficiency, which is a measure of the efficiency of conversion of that light energy into biomass and its harvest index, which is the proportion of biomass partitioned into sucrose (i.e. sugar) yield. Photosynthetic crop canopy area and its duration over the growing season are factors that moderate PY. Therefore, management practices that speed the full crop canopy cover and increase the duration of light capture can increase PY (Jaggard et al. 2009; Hoffmann et al. 2020). PY are therefore determined by variety, solar radiation and temperature from sowing until harvest. The water-limited potential sugar beet yields (PY_{wl}) are PY that are limited further by the availability of soil moisture under the rain fed cultivation conditions. So, PY and

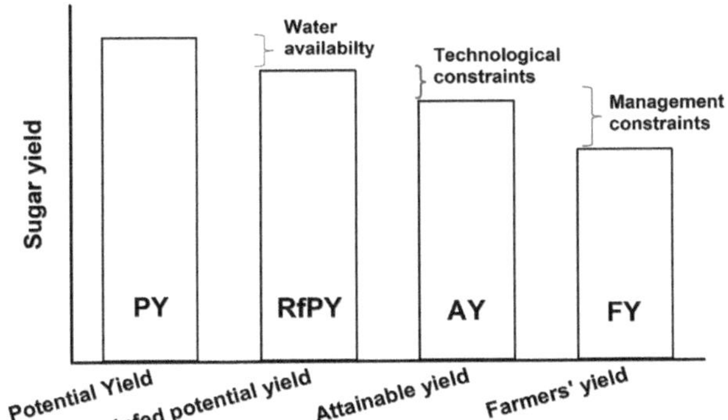

Fig. 20.3 Levels of sugar yield based on yield-limiting factors. The potential yields (PY) are the largest yields possible from a well-established crop grown in the absence of limitations from water, nutrients, pests, diseases and weeds. So, for a given cropping season, PY are determined by variety, solar radiation and temperature from sowing until harvest. Most beet crops are not irrigated, so the water-limited potential yield (PY_{wl}) is PY that is limited by the availability of soil moisture under rainfed conditions. The attainable yields (AY) are those achieved from crops under rainfed conditions using the best management practice with technical constraints. Farmers' yields (FY) are the actual yields that are harvested and delivered to processing factories. Compared with AY, FY may be smaller because of limitations such as poor plant establishment, inadequate nutrient supply, poor management of diseases, pests, weeds and losses during and after harvest

PY_{wl} are the ceilings that indicate the maximum yields that can be achieved under irrigated and rainfed conditions, respectively. The attainable yields (AY) are those achieved from actual crops under rainfed conditions using the best management practice with technical constraints. Farmers' yields (FY) are the actual yields that are harvested and delivered to processing factories. Compared with AY, FY are smaller because of limitations such as non-optimal sowing dates, poor plant establishment, inadequate nutrient supply, imperfect management of diseases, pests, weeds and losses during and after harvest. PY may not change much from year to year but will increase when there is a significant step progress in crop management and high-yielding breeding. However, it is the ratio between FY and AY (i.e. the yield gap) that varies more between countries and from year to year on the farm and between sugar beet farmers (Jaggard et al. 2012; Qi et al. 2012). Yield gap can act as the efficiency of a crop production system. The larger the gap, the poorer the growers' performance. Sugar beet breeders, agronomists and growers should concentrate on finding measures that can increase/maintain the weather-governed PY and reduce the yield gaps using economically feasible inputs. Development of sugar beet crop management decision-making tools can be an integral part of sugar beet production improvement programmes (Fig. 20.4). A growth model can form a core part of these flexible, multi-purpose tools. These crop management tools can then be used by agronomists/growers to assist in identifying factors responsible for yield gaps and by processors to manage their business (Jarvis and Qi 2014). The core crop growth model is overlaid with a series of crop management options, and it can estimate the weather-governed potential sugar yields. The management modules allow users to estimate what factors are limiting yield and facilitate improved crop management decision making, either on a farm or at a field-based level. It can also be used to help plan and validate experimental and strategic developments in the whole chain sugar beet crop production.

20.3 A Route Map to Successful Sugar Beet Production

Although sugar beet is a biennial plant, the crop is grown as an annual, so it is still in vegetative stage at harvest. The harvested parts of the crop are beet roots. The roots start to accumulate sucrose at a very early stage. This implies that (1) the longer the crop grows, the larger the sugar yield will be; (2) the faster the crop grows, the higher the sugar yield will be; (3) any inefficient use of crop canopy intercepted radiation due to biotic and/or abiotic factors on any day during the growing season will lead to potential sugar yield losses; (4) it is vitally important that crop health should be maintained throughout the growing season to achieve the maximum sugar yield. So, the road map for spring-sown beet crops to achieve high and stable sugar yields should be:

- well-prepared seed beds;
- sowing the crops as soon as the soil and temperature allow machinery into the fields;

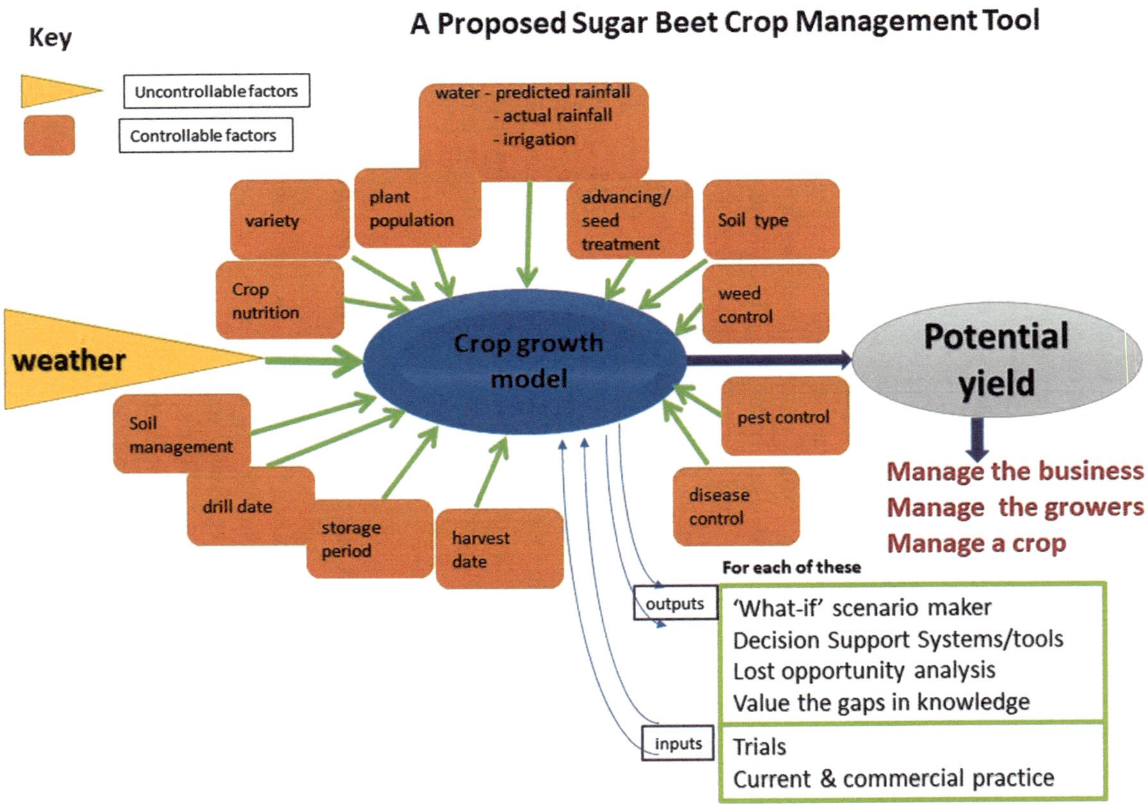

Fig. 20.4 A schematic diagram of a sugar beet crop management tool showing the use of a core sugar beet growth model overlaid with layers of management and agronomic practices that act on the weather-governed potential yield which can be estimated/predicted by the core sugar beet growth model

- good seedling emergence and well and uniformly established crop population at between 80,000 and 100,000 plants per hectare (Jaggard et al. 2011);
- reaching full crop canopy cover as early as possible (i.e. mid- to late-June in the north-hemisphere);
- plenty of sunshine combined with a supply of water that is sufficient to meet the demands of the atmosphere through the reserve in the soil, through rainfall and supplemented by irrigation if necessary;
- well-maintained healthy leaf canopy throughout and making the best use of autumn weather to allow beets to be harvested as planned or before the arrival of freezing temperatures.

20.4 Sugar Yield Increase in the Past

Sugar beet production has gone through phases from growing multi-germ to mono-germ varieties and from being labour-intensive to being highly mechanised. For example, in Europe, the per hectare man-power input was 350–400 man-hours in the mid-1950s but was reduced to 50 man-hours in the early 1980s (Bosemark 2006) and is now about 26 man-hours in the UK (Redman 2021). Along the way, the sugar yield per hectare has increased. For example, the annual commercial sugar beet yields recorded by British Sugar between 1937 and 2017 have increased from about 3 t/ha to more than 13 t/ha. Similarly, in the American Crystal Sugar Company of the United States delivered sugar yields were as low as at 2.0 t/ha in the 1930s, but as high as at 13.51 t/ha in the 2020s (Fig. 20.5). The coefficient of variation (CV%) in sugar yields within these 93 years stood at 44.8% (Fig. 20.5). This figure also shows how the annual sugar concentration has not changed much in the American Crystal area, the mean sugar concentration on fresh weight basis is 16.56% and the CV% is 7.1% (Fig. 20.5).

The stability of the sugar concentration over the decades indicates that yields have increased because total root dry matter has increased, which may be due to the increased capacity of cultivars to produce total biomass and to partition more biomass into root dry matter and sucrose (Jaggard and Qi 2006; Qi et al. 2013; Loel et al. 2014; Hoffmann and Kenter 2018; Hoffmann 2019). Other practices were shown to increase sugar beet yields through conserve soil available water such as transplanting and strip tillage (Deihimfard et al. 2021; Zhu et al. 2020).

However, this steady and continuing yield increase has been the result of many synergistic developments in machinery, breeding, agronomy, control of weed, pest and diseases, seed priming and coating techniques, and more recently in the climate change. For example, the progress made in seed treatment technology and applied in the *Xbeet® enrich 200* by Germains Seed Technology has enabled the sugar beet farmers to establish uniform crop stands with accelerated seedling emergence and increased sugar yields (https://germains.com). The recent warming in temperature resulted in earlier sowing and more rapid development to full crop canopy cover. Rapid growth throughout the autumn has been made possible by effective fungicides

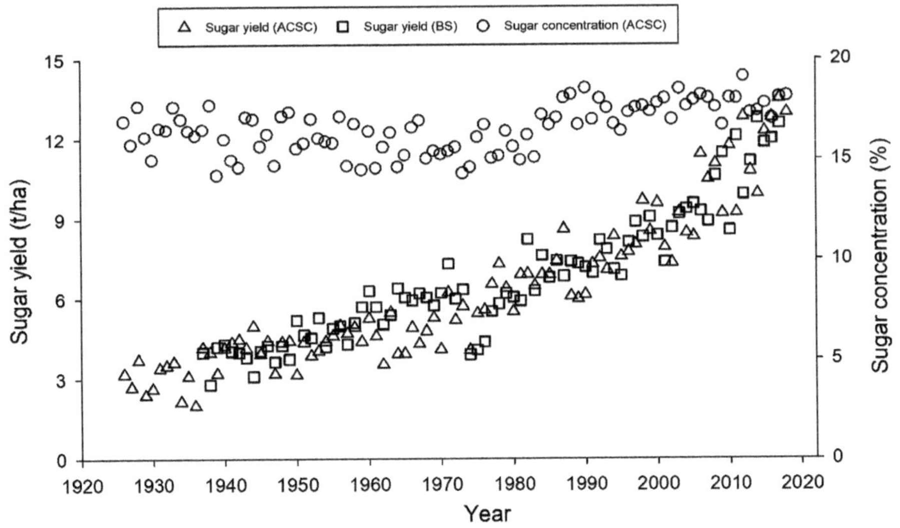

Fig. 20.5 Trend in commercial sugar yields (t/ha) from delivered beets to processing factories operated by American Crystal Sugar Company (ACSC) in 1926–2018 and by British Sugar (BS), a company of Associated British Foods plc (ABF) in 1937–2017. Data from ACSC were supplied by Mr. Tyler Grove, a general agronomist, in American Crystal Sugar Company, Fargo, North Dakota, USA

which have kept the canopy healthy and have contributed the recent yield improvement in the UK (Jaggard et al. 2007).

20.5 Opportunities

In Europe, where most sugar beet is grown (Table 20.1), it was estimated that, relative to those in 2000, summer and winter temperatures are likely to increase by 3.5 °C and 4.7 °C, respectively, by 2050 (Bastin et al. 2019). Recognising the risks and impacts of climate change, the Paris Agreement under the aegis of the United Nations Framework Convention on Climate Change was adopted and signed by many countries in the United Nations to mitigate climate change. The long-term goal within the agreement is to keep the global mean temperature to less than 2 °C above the pre-industrial temperature, but preferably to achieve a temperature increase of less than 1.5 °C by 2100. Climate change is being driven by increases in the carbon dioxide concentration in the atmosphere. It is well-demonstrated that increase in [CO_2] in the atmosphere will enhance photosynthesis of C_3 crops and stimulate the growth of sugar beet (Jaggard et al. 2010; Manderscheid et al. 2010). The estimated yield increase rate was at between 5% and 8% per 100 ppm increase of [CO_2] in the range between 350 and 450 ppm (Ewert et al. 2005; Manderscheid et al. 2010). The warmer temperatures could benefit sugar beet growth by allowing earlier sowing because the risk of vernalisation and bolting will be reduced and warmer weather

Table 20.2 Atmospheric concentration of CO_2 used for the baseline period (1980–2010) and in the projection of climate change scenarios for three representative concentration pathways (RCP) in 2030 (2021–2040), 2050 (2041–2060) and 2090 (2081–2100) when sugar yield was estimated in the UK by the Broom's Barn sugar beet growth model

Representative concentration pathways	Time period	CO_2 (ppm)
Baseline	1980–2010	364
RCP2.6	2021–2040 (2030)	430
	2041–2060 (2050)	442
	2081–2100 (2090)	426
RCP4.5	2021–2040 (2030)	435
	2041–2060 (2050)	487
	2081–2100 (2090)	534
RCP8.5	2021–2040 (2030)	449
	2041–2060 (2050)	541
	2081–2100 (2090)	844

will accelerate growth of the foliage in early summer to increase the amount of solar radiation that is intercepted (Jaggard et al. 2007, 2009, 2010). The recent trend in warming temperatures indeed reduced the vernalisation intensities during the early part of the crop development in spring-drilled sugar beet crops in the UK (Chiurugwi et al. 2013), which contributed to fewer bolters observed in the fields. The breeding of bolting-resistant and frost-tolerant sugar beet varieties could shift sowing-dates even earlier and lead to higher sugar yields (Pin et al. 2010). The potential increases of more than 20% in sugar yield have been shown by the possibility in sowing sugar beet in the autumn in an area where the normal sowing date is in spring (Jaggard and Werker 1999; Hoffmann and Kluge-Severin 2010; Hoffmann and Kluge-Severin 2011; Stephan et al. 2020).

An example of the positive impact of climate change on sugar beet productivity in the UK is given below. The Broom's Barn sugar beet growth model was developed and used to study the contribution of past climate change on the sugar beet yield increase from 1976 to 2004 (Qi et al. 2005; Qi and Jaggard 2008, 2012). The model was then updated with data from crops grown on different soil types at various sites in 2011 (Qi et al. 2013). This updated growth model was used to assess the effects of three representative concentration pathways (RCP) – RCP2.6, RCP4.5 and RCP8.5 in three future time periods—2030 (2021–2040), 2050 (2041–2060) and 2090 (2081–2100) using two global climate models—HadGEM2 and GISS-E2-R-CC (Semenov and Stratonovitch 2010, 2015). The use of baseline and projected CO_2 concentrations is shown in Table 20.2. As the [CO_2] is rising in the atmosphere, temperature is becoming warmer. The sowing date is shifting from the baseline mean sowing date on 19 March to 1 March under projected climate scenario with HadGEM2 in 2090 under RCP8.5 to 9 March under projected climate scenario with GISS-E2-R-CC in 2090 under RCP8.5. As a result, the sugar yield will increase under all three RCPs. The sugar yield benefits more from climate change on the most water-retentive soils such as clay loam and silt loam and in crops that are harvested

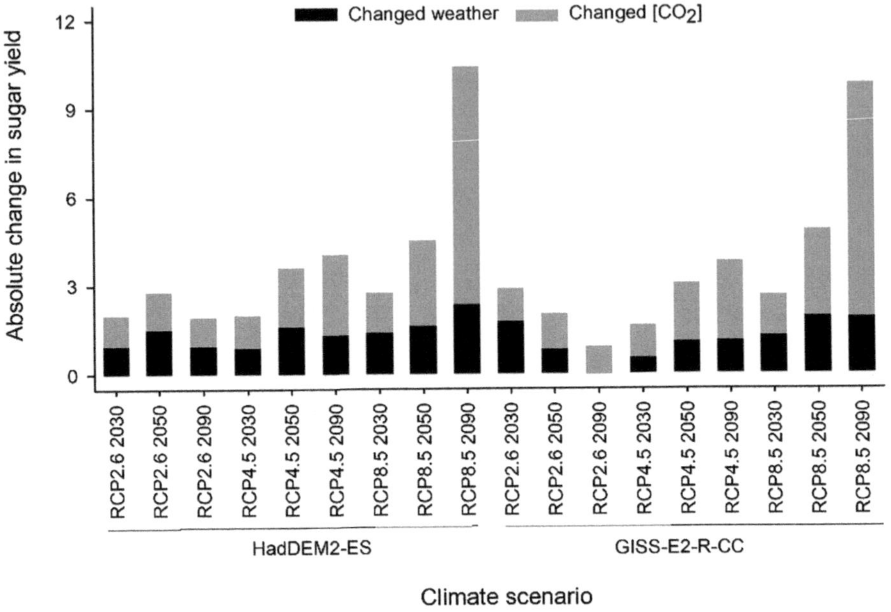

Fig. 20.6 The mean absolute increase of sugar yield (t/ha) on top of the mean sugar yield at 14.64 t/ha under baseline climate (1980–2010) due to changed weather and fertilisation effects of increased atmospheric [CO_2] for beet crops harvested on 31 October in the UK on sandy loam soils in 18 climate change scenarios created using global climate models (GCM)—HadGEM2-ES and GISS-E2-R-CC with three representative concentration pathways (RCPs)—RCP2.6, RCP4.5 and RCP8.5 in three future time periods—2030, 2050 and 2090. For climate scenarios, it is referred to Semenov and Stratonovitch (2010, 2015)

later because of the more extended growth in the autumn under UK production conditions. The likely absolute sugar yield increase for crops grown on sandy loam soils and harvested on 31 October is shown in Fig. 20.6. The yield increase is most likely attributed to the CO_2 fertilisation effects but the effect of the changing climate is still positive. It should be pointed out that the yield increase due to the CO_2 fertilisation effect was assumed to be a linear function of rising [CO_2] without limit up to 800 ppm. This may not be realistic considering the sugar storage capacity of the beet is limited in present sugar beet cultivars (Milford 2006; Hoffman and Kenter 2018). Under the baseline weather (1980–2010), the mean sugar yield was 14.64 t/ha on sandy loam soils. However, the average absolute sugar yield increase will be 2.27, 3.64 and 5.47 t/ha under climate changes projected by HadGEM2 under RCP2.6, RCP 4.5 and RCP8.5 in 2030, 2050 and 2090, respectively. For the climate changes projected by GISS-E2-R-CC, the average absolute sugar yield increase will be 2.39, 3.33 and 4.86 t/ha under RCP2.6, RCP 4.5 and RCP8.5 in 2030, 2050 and 2090, respectively.

20.6 Challenges

Among many biotic and abiotic factors limiting crop productivity, the major limiting factor is disease (Oerke 2006; Savary et al. 2019). The estimated ceiling sugar yield for sugar beet in today's climate can be as high as 24 t/ha (Hoffmann and Kenter 2018). Silva et al. (2020) estimated a yield gap as small as 12% on sugar beet in the Netherland using a crop model approach. If the sugar yield in the variety trials was used as a benchmark, the commercial sugar yield delivered by farmers (i.e. Farmers' yields (FY) referred in Fig. 20.3) only averaged 72% of the benchmark yield in the national variety trials in the UK (Fig. 20.7), ranging from 57 to 85% in years from 1938 to 2017. It is therefore important to close these yield gaps at national levels but also among different famers (Jaggard et al. 2010, 2012; Qi et al. 2012). In general, crop production is struggling to achieve food security for increasing world populations, in the face of climate change, while avoiding further land use change for agriculture coupled with the loss of biodiversity and other ecosystem services. High-input, resource-intensive sugar beet production systems can cause soil depletion and serious greenhouse gas emissions and are not considered sustainable. With the increasing demand to adapt sugar beet production to both less input-intensive and less pesticide-dependent cropping systems to mitigate climate change and to maintain biodiversity friendly environments, the challenge for sugar beet farmers is how to balance the trade-offs between maximising the sugar yield and increasing the use efficiencies of inputs such as fertilisers, fungicides, pesticides, herbicides and fossil fuels (Qi et al. 2010). The beet crop management tool proposed in Fig. 20.4 should facilitate the decision-making in assessing the cost-effectiveness of individual inputs

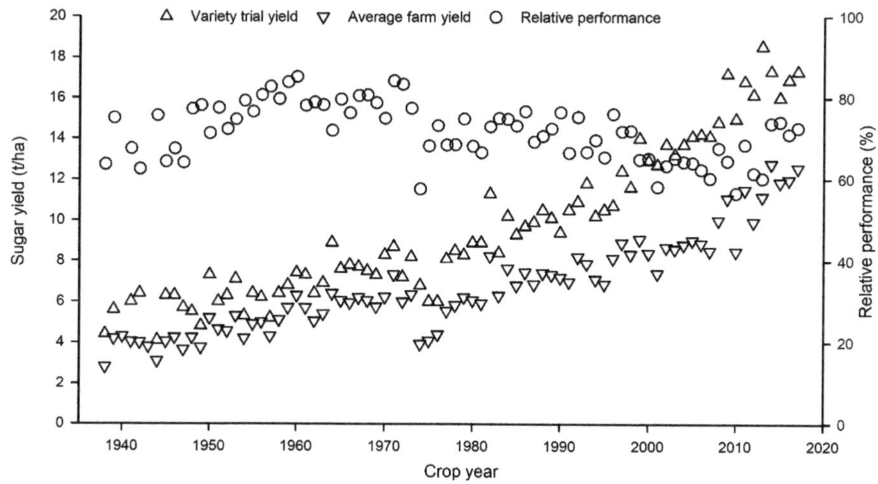

Fig. 20.7 Trends in sugar yields (t/ha) determined in the national variety trials and in commercial crops delivered by all sugar beet farmers in the UK. The ratio of farmer-delivered over variety trial sugar yield expressed as percent acted as a measure of growers' relative performance

since inputs are not always independent of each other and the sequence in which yield is affected by various inputs must be considered in multiple yield-limiting factor situations.

Despite the positive effects of increasing [CO_2] and warming temperatures of recent decades and of climate changes in the future, sugar beet growers still face challenges from increased risk of drought and increasing severity and frequency of diseases of existing and new pathogens (Richter et al. 2006; Qi and Fitt 2014; Kremer et al. 2016; Juroszek et al. 2020; Chaloner et al. 2021; Saunders 2021). For example, Cercospora leaf spot, which causes necrotic lesions and progressive destruction of the crop canopy, in the past was not a problem on sugar beet in the UK but it has started to be an important disease as the temperature is becoming warmer. In Europe, it will appear at an earlier stage of crop development and will require more applications of fungicides in the prolonged growing seasons of the future (Kremer et al. 2016).

Genetic resistance to pathogens is one of the most important elements in non-chemical crop protection. However, the main challenge is to ensure the durability of these genetic resistances, as their effectiveness decreases with time due to the selection of resistant or tolerant races. Meanwhile, due to their detrimental effects on wildlife, some widely used chemicals have been banned, which reduces the number of available pesticides, increases the risk that resistance to the remaining pesticides will develop. This is threatening the yield stability and economic viability of growing sugar beet and increasing the need to develop other solutions (Dewar and Qi 2021). The following example shows the impact of banning the neonicotinoid seed treatments on the control of virus yellows disease on sugar beet in the UK.

Virus yellows have been regarded as one of the worst scourges of sugar beet production in northern Europe (Jaggard et al. 1998; Hossain et al. 2021; Dewar and Qi 2021). It is caused by a complex of three viruses—beet mild yellowing virus (BMYV), beet chlorosis virus (BChV) and beet yellows virus (BYV) (Stevens et al. 2006; Hossain et al. 2021). BYV causes a much higher yield loss (up to 50%) than either BMYV or BChV (up to 25%) (Stevens et al. 2006). All three viruses are transmitted by aphids, of which the peach-potato aphid, *Myzus persicae,* is the most important, and against which control measures are targeted (Qi et al. 2004).

The wide use of seeds treated with neonicotinoids started in 1994 in the UK. As a result of its high efficacy and prolonged duration (protection against virus-carrying aphids until the 12-leaf stage) the virus yellows incidence was kept at about 1% until 2019 (Fig. 20.8). In 2012, the decision to ban neonicotinoid seed treatments in the EU was applied to bee-friendly flowering crops such as oilseed rape, sunflowers and maize and then in 2019, the ban was expanded to include non-flowering crops such as sugar beet and cereals (Dewar and Qi 2021). It was expected that all EU sugar beet industries would follow the banning decision, but 10 counties successfully applied for derogations to allow neonicotinoid seed treatments in sugar beet in 2019.

The UK resumed growing sugar beet without neonicotinoid seed treatments in 2019 after 26 years of successfully controlling virus yellows. That year the weather was unfavourable for aphid activity and there was little carry-over of infection sources from the previous year so national virus yellows incidence was low at

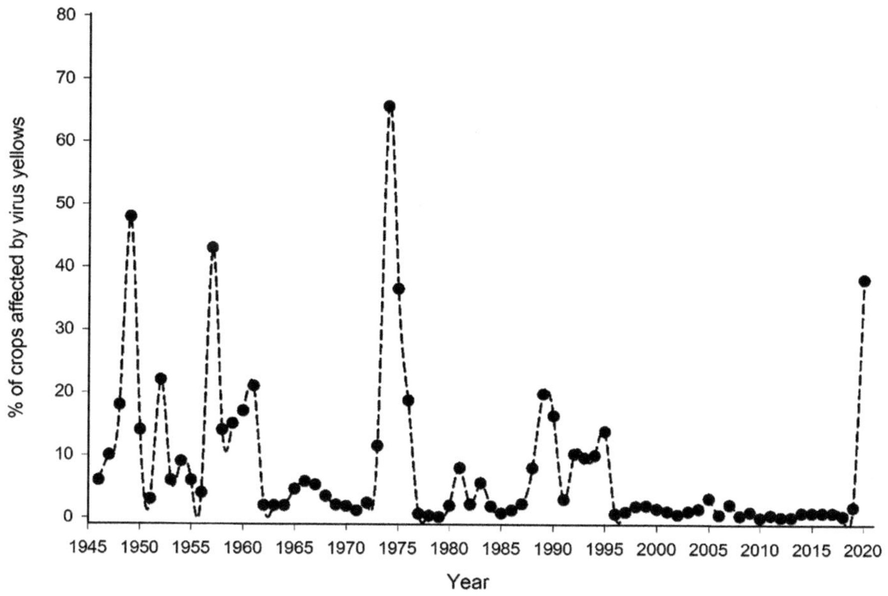

Fig. 20.8 National average incidence of virus yellows recorded in sugar beet crops at the end of August in 1946–2020 in the UK. Use of neonicotinoid sugar beet seed treatments started in 1994 and ended in 2019

1.8% (Fig. 20.8) (Dewar and Qi 2021). However, in the second year without neonicotinoid seed treatments, the beet crop suffered from severe virus yellows infection and the nation-wide incidence averaged 38.1% by the end of August in 2020. The result was a national sugar yield loss at 25% (Dewar and Qi 2021). The high virus yellows incidence was caused by the increased virus sources from the previous autumn, the warmer winter allowing more aphids (i.e. high level of primary inoculum) to overwinter, and the warmer spring allowing early migration of aphids to the sugar beet seedlings. The neonicotinoid ban had allowed virus to spread into host plants in 2019 and allowed overwintered aphids to spread from these sources early in spring where they were not controlled by an effective insecticide (Dewar and Qi 2021). As a consequence, some sugar beet growing countries will continue to allow derogations for the use of neonicotinoid seed treatments, at least if the threat of virus yellows is severe (Sugar Industry 2021). Depending on the economic threshold value that can be tolerated, weather-based virus yellows epidemic models can be used to decide whether the neonicotinoid seed treatment is required in advance (Dewar and Qi 2021).

In an era of global warming, it seems likely that epidemics will become more frequent. Table 20.3 shows the scenarios that might occur in the future as a result of global warming and its influence on virus yellows epidemics in the UK. Some scenarios will become a frequently threat, with virus yellows incidence up to 76% by 2100 if no effective control measures are available. At present, sugar beet

Table 20.3 Mean air temperature in January to February ($T_{Jan–Feb}$), the likely first flight date (first Flight), the likely total aphids caught in the insect suction trap to the end of June and the incidence of beet crops affected by virus yellows (VY%) in the Eastern England, UK, in response to climate change scenarios projected by the UK Met Office Hadley Centre climate model HadGEM2-ES under three representative concentration pathways (RCPs)

RCPs	Time period	$T_{Jan-Feb}$	1st flight	Total aphids	% VY with neonics	% VY without neonics
Baseline	1980–2010	4.2(0.48)	17 May (6)	37(15)	0.6(0.25)	17.6(9.74)
RCP2.6	2021–2040	5.2(0.55)	6 May (7)	94(48)	1.2(0.46)	34.7(14.63)
	2041–2060	5.2(0.51)	5 May (6)	93(35)	1.2(0.44)	36.7(15.42)
	2081–2100	5.4(0.50)	3 May (6)	111(47)	1.4(0.49)	41.9(15.79)
RCP4.5	2021–2040	5.3(0.51)	4 May (6)	104(44)	1.4(0.46)	40.5(14.97)
	2041–2060	5.9(0.58)	27 April (7)	181(92)	1.7(0.51)	48.1(14.80)
	2081–2100	6.5(0.56)	20 April (7)	297(112)	2.3(0.62)	62.1(16.13)
RCP8.5	2021–2040	5.7(0.51)	29 April (6)	153(65)	1.7(0.53)	47.6(16.11)
	2041–2060	6.2(0.50)	24 April (6)	224(98)	1.9(0.50)	51.3(14.42)
	2081–2100	8.2(0.62)	30 March (8)	1560(786)	3.1(0.77)	76.3(13.78)

research institutes and plant breeders across Europe are prioritising virus-carrying aphids and virus yellows research, to produce new varieties with inbuilt resistance or tolerance to the three different viruses. Neonicotinoid seed treatment suppresses virus yellows by controlling the aphid vector. Without the use of neonicotinoid pesticide, the aphids may become problematic.

Figures in brackets are standard deviations. The baseline daily weather (1980–2010) was from Broom's Barn weather station, Suffolk, England and used to derive parameters as inputs to LARS-weather generator to generate 100 years of daily data under climate change scenarios (Semenov and Stratonovitch 2015). The virus yellows model of Qi et al. (2004) was used to calculate the percentage of beet crops affected by the disease. The calculated virus yellows incidence has options for with and without the use of neonicotinoids(neonics) pesticide seed treatments.

20.7 Future Prospects

In countries where sugar beet is well-adapted and it is a main component of the crop rotations, its cultivation had gone from man-powered to horse-powered, then to the present intensive machine-powered stages. The future move is being aimed towards data-driven cultivation and management systems armed with applications of advanced technologies such as genome-editing techniques, robots, Global Positioning System (GPS), unmanned aerial vehicle (UAV) or drones, satellite-derived images and model-based tools for research and decision-making. Natural resources conservation and climate change mitigation will likely demand more use of preventive and pathogen suppression methods like crop rotation and resistant crop cultivars, and more applications of remote sensing technologies to detect early onset of epidemic diseases and of forecasting models to estimate economic thresholds to determine when to apply control measures.

Sugar beet breeders take great pride in breeding the high-yielding varieties with superior agronomic traits. Sugar beet seed companies compete proudly to promote their best-treated and primed heathy seeds to farmers. Sugar beet processors are prepared to pay reasonable prices to purchase and take great interests in securing the beets they need to extract sucrose and produce other by-products. Sugar beet production researchers and agronomists are always determined to innovate and develop most efficient practices to help farmers to manage their crops. These concerted efforts have contributed to the steady sugar yield increase in the past 100 years. In recent decades, global warming due to the climate change has indirectly contributed a significant proportion of the sugar yield improvements. It remains to be seen whether this positive effect will continue from the warming temperatures in the future.

20.8 Conclusion

It has been demonstrated that the rising temperature and increasing concentration of CO_2 in the atmosphere under future climate change will continue to benefit sugar beet crops. The current yield potentials of sugar beet are estimated at 24 t/ha, and they can reach even higher under weather conditions by the mid- and late twenty-first century. For spring-sown sugar beet crops, the sowing date will become earlier with less worry about bolters, the rate of development to full crop canopy cover date can be more rapid, more favourable growing weather conditions in autumn will extend the growth season, all of which signpost to higher sugar beet crop yields. There is also the potential prospect of sowing the sugar beet crop in the autumn with frost-tolerant and bolting-resistant varieties to take advantages of the temperature warming while climate change is taking place since autumn-sown beet crops can yield potentially 26% more than spring-sown beet crops in north-western Europe.

However, the climate change related soil water stress will probably pose an increasing risk to sugar beet production in the future, causing more variability in yield from year to year. The social and environmental demands to adapt sugar beet production to both less input-intensive and less pesticide-dependent cropping systems (e.g. minimum tillage, using cover crops, precision agriculture) to mitigate climate change and to maintain natural resources and to create biodiversity-friendly environments require sugar beet farmers to balance the trade-offs between maximising the sugar yield and increasing the use efficiencies of inputs such as fertilisers, fungicides, pesticides, herbicides and fuels. The use of pathogen-resistant cultivars in sugar beet production should be the most economical and effective way to reduce losses caused by epidemic diseases. Sugar beet breeders and other stakeholders need to breed climate-smart cultivars resistant to pathogenic diseases and find other effective non-chemical solutions (e.g. nature-based, integrated crop management) to the reduced availability and/or removal of reliable pesticides in the face of more severe pathogenic diseases and emerging new pathogens under climate change.

Acknowledgements Thanks are due to Rothamsted Research, Harpenden, UK, for granting access to the meteorological data from the Broom's Barn weather station, Suffolk, UK, to Dr. Mikhail A. Semenov at Rothamsted Research for help of future climate change scenario information and generating the climate change scenarios using his LARS-Weather Generator, to British Sugar of Associated British Foods plc for the virus yellows survey and sugar yield data each year, to Dr. Alan Dewar at Dewar Crop Protection Ltd. for sharing information on the rise and fall of neonicotinoids pesticide use on sugar beet, to Mr. Patrick Jarvis for sharing ideas developing sugar beet management tools, and Mr. Tyler Grove, general agronomist and Ms. Kathy Wang, senior agriculture information analyst in American Crystal Sugar Company, Fargo, USA, for providing the sugar yield and sugar concentration data. Last, but not least, I thank Professor Keith Jaggard at Jaggard Consultancy, Sugar Beet Agronomy Research, Prof. Dr. Christa Hoffmann, Department of Physiology, Institute of Sugar Beet Research, Göttingen, Germany, for giving critical comments and improving this Chapter.

References

Bastin JF, Clark E, Elliott T, Hart S, van den Hoogen J, Hordijki I et al (2019) Understanding climate change from a global analysis of city analogues. PLoS One 14(7):e0217592. https://doi.org/10.1371/journal.pone.0217592

Bosemark NO (2006) Chapter 4- Genetics and breeding. In: Draycott P (ed) Sugar Beet. Blackwell Publishing, Oxford, pp 50–88

Chaloner TM, Gurr SJ, Bebber DP (2021) Plant pathogen infection risk tracks global crop yields under climate change. Nat Clim Chang 11:710–715

Chiurugwi T, Holmes HF, Qi A, Chia TYP, Hedden P, Mutasa-Gottgens ES (2013) Development of new quantitative physiological and molecular breeding parameters based on the sugar-beet vernalization intensity model. J Agric Sci 151:492–505

Deihimfard R, Rahimi-Moghaddamb S, Goudrian J, Damghani AM, Noori O, Nazari S (2021) Can optimizing the transplant of sugar beet by age and date enhance water productivity in arid and semi-arid climates? Field Crop Res 271:108266

Dewar AM, Qi A (2021) The virus yellows epidemic in sugar beet in the UK in 2020 and the adverse effect of the EU ban on neonicotinoids on sugar beet production. Outlooks Pest Manag 32:53–59

Draycott AP (2006) Sugar beet. Blackwell, Oxford, pp. xxi+474

Ewert F, Rounsevell MDA, Reginster I, Metzger MJ, Leemans R (2005) Future scenarios of European agricultural land use: I. Estimating changes in crop productivity. Agric Ecosyst Environ 107:101–116

Gregory PJ, Marshall B (2012) Attribution of climate change: a methodology to estimate the potential contribution to increases in potato yield in Scotland since 1960. Glob Chang Biol 18:1372–1388

Hatfield JL, Boote KJ, Kimball BA, Ziska LH, Izaurralde RC, Ort D, Thomson AM, Wolfe D (2011) Climate impacts on agriculture: implications for crop production. Agron J 103:351–370. https://doi.org/10.2134/agronj2010.0303

Hoffmann CM (2019) Importance of canopy closure and dry matter partitioning for yield formation of sugar beet varieties. Field Crop Res 236:75–84

Hoffmann CM, Kenter C (2018) Yield potential of sugar beet – have we hit the ceiling? Front Plant Sci 9:289. https://doi.org/10.3389/fpls.2018.00289

Hoffmann CM, Kluge-Severin S (2010) Light absorption and radiation us efficiency of autumn and spring sown sugar beets. Field Crop Res 119:238–244. https://doi.org/10.1016/j.fcr.2010.07.014

Hoffmann CM, Kluge-Severin S (2011) Growth analysis of autumn and spring sown sugar beet. Eur J Agron 34:1–9. https://doi.org/10.1016/j.eja.2010.09.001

Hoffmann CM, Huijbregts T, van Swaaij N, Jansen R (2009) Impact of different environments in Europe on yield and quality of sugar beet genotypes. Eur J Agron 30:17–26

Hoffmann CM, Koch H-J, Märländer B (2020) Sugar beet. In: Sadras VO, Calderini D (eds) Crop physiology. Case histories for major crops. Elsevier Inc, San Diego, pp 635–674

Hossain R, Menzel W, Lachmann C, Varrelmann M (2021) New insights into virus yellows distribution in Europe and effects of beet yellows virus, beet mild yellowing virus, and beet chlorosis virus on sugar beet yield following field inoculation. Plant Pathol 70:584–593

IPPC Secretariat (2021) Scientific review of the impact of climate change on plant pests – a global challenge to prevent and mitigate plant pest risks in agriculture, forestry and ecosystems. FAO on behalf of the IPPC Secretariat, Rome. https://doi.org/10.4060/cb4769en

Jaggard KW, Qi A (2006) Chapter 7-Agronomy. In: Draycott P (ed) Sugar Beet. Blackwell Publishing, Oxford, pp 134–168

Jaggard KW, Werker AR (1999) An evaluation of the potential benefits and cost of autumn-sown sugar beet in NW Europe. J Agric Sci 132:91–102

Jaggard KW, Dewar AM, Pidgeon JD (1998) The relative effects of drought stress and virus yellows on the yield of sugar beet in the UK, 1980-95. J Agric Sci 130:337–343

Jaggard KW, Qi A, Semenov MA (2007) The impact of climate change on sugarbeet yield in the UK: 1976-2004. J Agric Sci 145:367–375

Jaggard KW, Qi A, Ober ES (2009) Capture and use of solar radiation, water and nitrogen by sugar beet (*Beta vulgaris* L.). J Exp Bot 60:1919–1925

Jaggard KW, Qi A, Ober ES (2010) Possible changes to arable crop yields by 2050. Philos Trans R Soc B Biol Sci 365:2835–2851

Jaggard KW, Qi A, Milford GFJ, Clark CJA, Ober ES, Walters C, Burks E (2011) Determining the optimal population density of sugar beet crops in England. Int Sugar J 113:12–17

Jaggard KW, Koch H-J, Arroyo Sanz JM, Cattanach A, Duval R, Eigner H, Legrand G, Olsson R, Qi A, Thomsen JN, Van Swaaij N, Minerva N (2012) The yield gap in some sugar beet producing countries. Int Sugar J 114:496–499

Jarvis J, Qi A (2014) Developing and using the AB sugar BeetGro model. Br Sugar Beet Rev 82: 45–48

Jeger M, Beresford R, Bock C, Brown N, Fox A, Newton A, Vicent A, Xu X, Yuen J (2021) Global challenges facing plant pathology: multidisciplinary approaches to meet the food security and environmental challenges in the mid-twenty-first century. CABI Agric Biosci 2. https://doi.org/10.1186/s43170-021-00042-x

Jones PD, Lister DH, Jaggard KW, Pidgeon JD (2003) Future climate change impact on the productivity of sugar beet (*Beta vulgaris* L.) in Europe. Climate Change 58:93–108

Juroszek P, Racca P, Link S, Farhumand J, Kleinhenz B (2020) Overview on the review articles published during the past 30 years relating to the potential climate change effects on plant pathogens and crop disease risks. Plant Pathol 69:179–193

Kremer P, Schlüter J, Racca P, Fuchs HJ, Lang C (2016) Possible impact of climate change on the occurrence and the epidemic development of Cercospora leaf spot disease (*Cercospora beticola* sacc.) in sugar beets for Rhineland-Palatinate and the southern part of Hesse. Climate Change 137:481–494. https://doi.org/10.1007/s10584-016-1697-y

Lobell DB, Gourdji SM (2012) The influence of climate change on global crop productivity. Plant Physiol 160:1686–1697

Loel J, Kenter C, Märländer B, Hoffmann CM (2014) Assessment of breeding progress in sugar beet by testing old and new varieties under greenhouse and field conditions. Eur J Agron 52: 146–156

Manderscheid R, Pacholski A, Weigel H-J (2010) Effect of free air carbon dioxide enrichment combined with two nitrogen levels on growth, yield and yield quality of sugar beet: evidence for a sink limitation of beet growth under elevated CO_2. Eur J Agron 32:228–239
McKersie B (2015) Planning for food security in a changing climate. J Exp Bot 66:3435–3450. https://doi.org/10.1093/jxb/eru547
Milford GFJ (2006) Chapter 3- Plant structure and crop physiology. In: Draycott P (ed) Sugar Beet. Blackwell Publishing, Oxford, pp 30–49
Nelson R (2020) International plant pathology- past and future contributions to global food security. Phytopathology 110:245–253
Newbery F, Qi A, Fitt BDL (2016) Modelling impacts of climate change on arable crop diseases; progress, challenges and applications. Curr Opin Plant Biol 32:101–109
NOAA (2021) NOAA National Center for Environmental information, Climate at a Glance: Global Time Series, published July 2021. https://www.ncdc.noaa.gov/cag. Accessed 14 Jul 2021
Oerke E-C (2006) Crop losses to pests. J Agric Sci 144:31–43
Okom S, Russell A, Chaudhary AJ, Scrimshawa MD, Francis RA (2017) Impacts of projected precipitation changes on sugar beet yield in eastern England. Meteorol Appl 24:52–61
Pidgeon JD, Werker AR, Jaggard KW, Richter GM, Lister DH, Jones PD (2001) Climatic impact on the productivity of sugar beet in Europe, 1961–1995. Agric For Meteorol 109:27–37
Pin PA, Benlloch R, Bonnet D, Wremerth-Weich E, Kraft T, Gielen J, Nilsson O (2010) An antagonistic pair of FT homologs mediates the control of flowering time in sugar beet. Science 330:1397–1400
Qi A, Fitt BDL (2014) Can crop disease control cope with climate change? Outlooks Pest Manag 25:364–368
Qi A, Jaggard KW (2008) Assessing the impact of future climate changes on the UK sugar beet production. In: Aspects of applied biology 88: effects of climate changes on plants: implications for agriculture. Association of Applied Biologists (AAB), Wellesbourne, Warwick, pp 153–159
Qi A, Jaggard KW (2012) Sugar beet yield in England under an extreme climate change scenario. In: In Proceedings of the 73rd IIRB Congress, Brussels, pp. 73–80
Qi A, May M (2013) How the sugar beet crop grew in 2012. Br Sugar Beet Rev 81:3–6
Qi A, Dewar AM, Harrington R (2004) Decision making in controlling virus yellows of sugar beet in the UK. Pest Manag Sci 60:727–732
Qi A, Kenter C, Hoffmann C, Jaggard KW (2005) The Broom's barn sugar beet growth model and its adaptation to soils with varied available water content. Eur J Agron 23:108–122
Qi A, Pettitt I, May M (2010) Assessing the cost-effectiveness of inputs. Br Sugar Beet Rev 78(4): 17–21
Qi A, Eric ES, Jaggard KW (2012) Benchmarking sugar beet yields and the growers' performance. Br Sugar Beet Rev 80:3–6
Qi A, Eric ES, Jaggard KW (2013) Updating the BBRO/Broom's barn sugar beet growth model. Br Sugar Beet Rev 81:29–32
Qi A, Holland RA, Taylor G, Richter GM (2018) Grassland futures in Great Britain – productivity assessment and scenarios for land use change opportunities. Sci Total Environ 634:1108–1118. https://doi.org/10.1016/j.scitotenv.2018.03.395
Racca P, Kakau J, Kleinhenz B, Kuhn C (2015) Impact of climate change on the phenological development of winter wheat, sugar beet and winter oilseed rape in Lower Saxony, Germany. J Plant Dis Prot 122:16–27
Redman G (2021) John Nix pocketbook for farm management 2022, 52nd edn. Melton Mowbray: Agro Business Consultants, Leicestershire, UK
Richter GM, Jaggard KW, Mitchell RAC (2001) Modelling radiation interception and radiation use efficiency for sugar beet under variable climatic stress. Agric For Meteorol 109:13–25
Richter G, Qi A, Semenov M, Jaggard KW (2006) Variability of UK sugar beet yields under climate change and soil use adaptation needs. Soil Use Manag 22:39–47

Rosenzweig C, Iglesias A, Yang XB, Epstein PR, Chivian E (2001) Climate change and extreme weather events. Implications for food production, plant disease, and pests. Global Change Hum Health 2:90–104

Saunders DGO (2021) Will yield gains be lost to disease? Nat Clim Chang 11:648–649

Savary S, Willocquet L, Pethybridge SJ, Esker P, McRoberts N, Nelson A (2019) The global burden of pathogens and pests on major food crops. Nat Ecol Evol 3:430–439

Scott RK, Jaggard KW (2000) Impact of weather, agronomy and breeding on yields of sugarbeet in the UK since 1970. J Agric Sci 134:341–352

Semenov MA, Stratonovitch P (2010) Use of multi-model ensembles from global climate models for assessment of climate change impacts. Clim Res 41:1–14

Semenov MA, Stratonovitch P (2015) Adapting wheat ideotypes for climate change: accounting for uncertainties in CMIP5 climate projections. Clim Res 65:123–139

Silva JV, Tenreiro TR, Spätjens L, Anten NPR, van Ittersum MK, Reidsma P (2020) Can big data explain yield variability and water productivity in intensive cropping systems? Field Crop Res 255:107828

Stephan H, Böttcher U, Sieling K, Kage H (2020) Yield potential of non-bolting winter sugar beet in Germany. Eur J Agron 115:126035

Stevens M, Liu HY, Lemaire O (2006) Chapter 11- Virus diseases. In: Draycott P (ed) Sugar Beet. Blackwell Publishing, Oxford, pp 256–278

Sugar Industry (2021) Emergency authorisations to use neonicotinoids moved westwards. Sugar Ind 146(4)

Vogel J, Kenter C, Holst C, Märländer B (2018) New generation of resistant sugar beet varieties for advanced integrated Management of Cercospora Leaf Spot in Central Europe. Front Plant Sci 9: 222. https://doi.org/10.3389/fpls.2018.00222

Werker AR, Jaggard KW (1997) Modelling asymmetrical growth curves that rise and then fall: applications to foliage dynamics of sugar beet (*Beta vulgaris* L.). Ann Bot 79:657–665

Werker AR, Jaggard KW (1998) Dependence of sugar beet yield on light interception and evapotranspiration. Agric For Meteorol 89:229–240

Werker AR, Jaggard KW, Allison MF (1998) Modelling partitioning between structure and storage in sugar beet: effects of drought and soil nitrogen. Plant Soil 207:97–106

Zhu X, Han B, Song B, Yang J (2020) Effect of extending seedling-raising period on yield of transplanting sugar beet (*Beta vulgaris*) in black soil area of Northeast China. Sugar Tech 22(6). https://doi.org/10.1007/s12355-020-00862-7

Drought Stress Management in Sugar Beet (*Beta vulgaris* L.) Cultivation

21

Abazar Rajabi and Dariush Taleghani

Abstract

Drought stress is one of the main restrictive factors for sugar beet production in arid and semi-arid regions such as Iran and other areas where summer rainfall is significantly lower than potential evapotranspiration. Sugar beet is considered a drought-tolerant crop, but its secondary traits are affected by drought stress. To achieve the maximum potential sugar yield, sugar beet plants should have sufficient moisture available on a daily basis to meet the atmospheric demand so that transpiration, and likewise photosynthesis, can occur without stomatal limitation. However, these proper conditions are often not provided, either because rainfall is insufficient or irrigation water is limited. Sugar forms an integral part of the human diet, and sugar beet is the main source of sugar. Due to the arid and semi-arid climate of Iran, the water shortage has become a key challenge in irrigated agriculture. To address this problem, the development and implementation of water-saving agricultural practices are necessary for providing high yields with low water application. Effects of drought stress on sugar beet could be mitigated by different methods some of which are presented in this chapter.

Keywords

Drought · Irrigation · Mitigation · Planting pattern · Sugar beet · Transplanting · Variety

A. Rajabi (✉) · D. Taleghani
Agricultural Research, Education and Extension Organization, Sugar Beet Seed Institute (SBSI), Karaj, Iran

V. Misra et al. (eds.), *Sugar Beet Cultivation, Management and Processing*,
https://doi.org/10.1007/978-981-19-2730-0_21

Abbreviations

BNYVV	Beet necrosis yellow vein virus
JA	Jasmonic acid
MeJA	Methyl jasmonate
QTLs	Quantitative trait loci

21.1 Introduction

Sugar forms an integral part of the human diet, and sugar beet is the main source of sugar. Drought is a chief constraint on crop productivity worldwide, especially in arid and semi-arid areas. Sufficient moisture is needed to achieve the highest potential yield in sugar beet. Nevertheless, these favorable conditions are often not realized, due to either insufficient rainfall or limited availability of irrigation water (Ober and Rajabi 2010). In arid and semi-arid climates like Iran, the water shortage has become a key challenge in irrigated agriculture. To address this problem, the development and adoption of new water-saving agricultural practices are necessary for providing high yields with low water application. This chapter describes the sugar situation in the world and Iran as well as the influence of drought on this crop. Also, it will present the important measures to mitigate the drought stress effects on sugar beet. These measures include deficit irrigation, use of drought-tolerant varieties, modification of planting patterns, transplanting, and autumn sowing.

21.2 Sugar in the World

Sugar production pattern in the world during 1952–2018 shows an increasing pattern for sugar cane and a relatively stable trend for sugar beet (Licht 2018). The sugar produced from sugar cane amounted to 150.73 MT and that from sugar beet to 43.51 MT in 2018 with 77.56% and 22.44% of total sugar from sugar cane and sugar beet, respectively. The highest quantity of sugar (39.39%) was produced in America and Asia and the least (2.45%) in Oceania in 2018 (Licht 2018). The three main sugar-producing countries in 2017–2018 were Brazil, India, and Thailand with 20.13, 17.45, and 7.48%, respectively (Licht 2018). Iran produced 1.16% of the world's sugar in 2017–2018. There are seven countries where both sugar beet and sugar cane are produced. These are China, USA, Pakistan, Egypt, Iran, Japan, and Morocco.

India is a country where suitable sites for sugar beet production have been recognized but it has not developed at a commercial scale yet. Sugar beet was introduced in India in 1950s and then grown in north India for sugar production. In 1971, a sugarcane-cum-sugar beet sugar factory was developed at Sri Ganganagar which was operational for almost three decades, and sugar beet was well established in that area. However, the area was not expanded because no additional sugar factories were established. Also, the government did not support upgrading the existing sugarcane factories. However, it is possible to produce ethanol from sugar beet. The other reason why sugar beet is not grown commercially in India is that there is not a market for it. Sugar beet is a crop linked to industry; so, seed money or incentives have not been provided to the sugar industry to install the supplementary machinery (Pathak et al. 2014).

21.3 Sugar Industry in Iran

The first beet sugar factory in Iran was founded in 1895, whereas the first cane sugar factory was launched in 1961. More beet sugar factories were then founded in the 1960s. The sugar industry is the key energy user among the food industries. So, the sugar factories were enforced by the government to advance their technologies for optimal use of energy. Thus, natural gas replaced other fossil fuels (esp. Mazut) so that Mazut share reduced from 73% in 2001 to 39% in 2007. The second oldest industry after the textile industry is the sugar industry supported by the government as a cheap energy source. At present, there are 35 beet sugar factories in cold temperate areas of Iran with a total nominal capacity of 75,000 t/d. In addition, seven sugarcane factories with a total nominal capacity of 100,000 t/d are active in the southwest (tropical area) of Iran.

21.4 Drought Stress and Sugar Beet Production

A study conducted by World Resources Institute on water stress in 164 countries showed that 44 countries will face high or extremely high water stress by 2040, of which 17 countries with extremely high water stress including Iran are located in the Middle East (Hofste et al. 2019). Different parts of Iran are influenced by drought stress to various extents. Crop production is mainly constrained by drought stress worldwide (Boyer 1982). Sugar beet is considered a drought-tolerant crop (Vamerali et al. 2009); nevertheless, drought stress influences physiological and biochemical traits of sugar beet (Mohammadian et al. 2003; Choluj et al. 2014; Sattar et al. 2019). Drought stress is also a limiting factor for sugar beet in arid and semi-arid areas like Iran (Noghabi and Williams 2000; Moosavi et al. 2017) and other areas where summer rainfall is significantly lower than potential evapotranspiration (Ober 2006). To realize the highest potential sugar yield, sugar beet should have regular access to sufficient moisture to meet atmospheric demand so that transpiration, and likewise photosynthesis, can occur without stomatal limitation (Ober and Rajabi 2010). However, these proper conditions are often not provided, either because rainfall is insufficient or irrigation water is limited.

In areas where production of sugar beet depends mostly on rainfall, drought-induced yield losses are common. Across Europe, for example, losses ranged from 5% in parts of Northern Europe to 30% in Southern Russia (Pidgeon et al. 2001); whereas in areas where production of sugar beet is normally reliant on irrigation, the crop regularly experiences various degrees of drought stress due to water and energy costs, or limited available water resources (Morillo-Velarde and Ober 2006).

21.5 Mitigation of Drought Stress Effects on Sugar Beet Crop

Sugar beet has been produced in Iran for a century mostly under irrigated conditions. As Iran has an arid and semi-arid climate, limited water availability has become a major challenge in irrigated agriculture (Khozaei et al. 2020). To address this problem, the development and implementation of new water-saving agricultural practices are necessary for providing high yields with low water application. Drought stress effects on sugar beet could be mitigated by different methods some of which are presented below.

21.5.1 Use of Drought-Tolerant Varieties

In arid regions, drought stress could not be mitigated by irrigation because irrigation water is limited. One of the sustainable approaches for mitigating drought stress is to use drought-tolerant varieties, i.e., the varieties which are better able to maintain potential yield when conditions are dry (Ober 2006). There is significant genetic variation for drought tolerance and several morpho-physiological traits associated with drought tolerance (Ober 2006; Rajabi et al. 2008, 2009). Sugar beet Seed Institute of Iran started research on the development of such varieties 20 years ago. It developed and introduced the first monogerm drought-tolerant hybrid variety named PAYA for the disease-free conditions in 2014. The variety produced 7 t/ha sugar by using 8000 m^3 water, whereas the non-tolerant check variety produced 5.5 t ha^{-1} (Orazizadeh et al. 2015).

Development of drought-tolerant varieties by conventional plant breeding methods is a time-consuming process. To speed up this process, molecular markers could be used for genotypic selection. For example, to find the quantitative trait loci (QTLs) of root yield and some drought tolerance-related characteristics in sugar beet, 142 $F_{2:3}$ families were studied under non-stress and water-deficit stress conditions. It was revealed that markers strongly linked with the major QTLs, especially those associated with root yield, leaf senescence, and leaf wilt could be used in marker-assisted selection programs for the selection of superior drought-tolerant lines (Rajabi and Borchardt 2009).

In drought-affected areas of Iran, Rhizomania, the most destructive viral disease of sugar beet caused by beet necrotic yellow vein virus (BNYVV) is also a major threat to sugar beet production. So, breeding drought and rhizomania tolerant varieties can be considered a viable solution for these conditions. Evaluation of 12 monogerm sugar beet hybrids along with drought and rhizomania tolerant checks showed that some hybrids are promising and performing better than the drought-tolerant checks (Paya and IR7) and co-grouped with the rhizomania tolerant check (Mandarin) (Fig. 21.1) (Rajabi 2019).

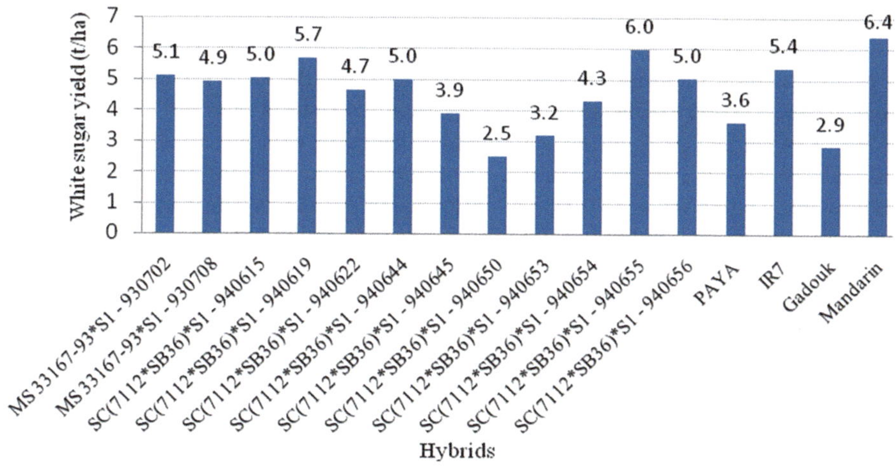

Fig. 21.1 Comparison of white sugar yield of drought and rhizomania tolerant hybrids (The data was compiled by Rajabi 2019)

Table 21.1 Comparison of normal and short-vegetation-period varieties (The data was compiled by Rajabi 2017)

Characteristics	Normal variety	Short-vegetation-period variety	Variation (%)
Sowing date	April	Late June	–
Growing period (d)	180–200	120	−33%
Number of irrigations	10–12	6–7	−41%
Machinery trafficking (times)	7–8	4	−34%
Water use (m^3)	12,873	8558	−27%
Sugar yield (t/ha)	12.8	6.9	−46%

21.5.2 Use of Short-Vegetation-Period Varieties

In some regions of Iran, there is competition for irrigation water between sugar beet and cereals in the early season of sugar beet growth, and priority is given to cereals. Therefore, sugar beet experiences some drought stress which can result in reduced sugar yield (Mohammadian et al. 2005; Monti et al. 2006). Short-vegetation-period varieties are considered a solution for these conditions. Comparison of normal and short-vegetation-period varieties showed sufficient variation among the varieties in late sowing conditions. Several investigations have revealed that short-vegetation-period varieties use 27% less water and need 34% less machinery trafficking but produce 46% less sugar yield than the normal varieties (Table 21.1) (Rajabi 2017). This indicates that short-vegetation-period varieties could only be recommended for situations where a key limitation in the growth period such as competition for irrigation water occurs. There are differences amongst sugar beet genotypes in their responses to shortened growth periods. So, it is feasible to develop varieties with a short growth period.

21.6 Deficit Irrigation

Deficit irrigation in which the water applied is below the crop's water requirement and is a new water-saving practice in arid regions (Oweis et al. 2011; Unlü et al. 2011; Yang et al. 2015). By applying deficit irrigation, excessive vegetation growth is decreased and a higher quantity of water is saved (Padilla-Daz et al. 2016; Hernandez-Santana et al. 2017). The influence of deficit irrigation on the yield and quality of sugar beet depends on different irrigation treatments and methods. When every-other-furrow irrigation method was applied at l0-day intervals on sugar beet, a smaller quantity of irrigation water was used with some yield reduction. Conversely, when frequent every-other-furrow irrigation was applied at 6-day intervals, root yield was similar to that of every-furrow irrigation at the 10-day intervals and saved 23% of water (Sepaskhah and Kamgar-Haghighi 1997).

21.7 Modification of Planting Pattern

Despite the restricted water availability in different parts of Iran, sometimes sugar beet is irrigated more than the plant needs. One alternative in-furrow and tape-drip irrigation method to decrease the over-required irrigation is to change the planting pattern (Taleghani et al. 2004; Mohammadian 2013; Mirzaei et al. 2013; Mohammadian and Sadrqaen 2016). At present, in most sugar beet farms of Iran, the between-row spacing is 50 cm and the plants are irrigated from both sides (Fig. 21.2). However, numerous investigations have revealed that sugar beet can also be irrigated from one side either in-furrow (Fig. 21.2) (Mirzaei et al. 2013) or

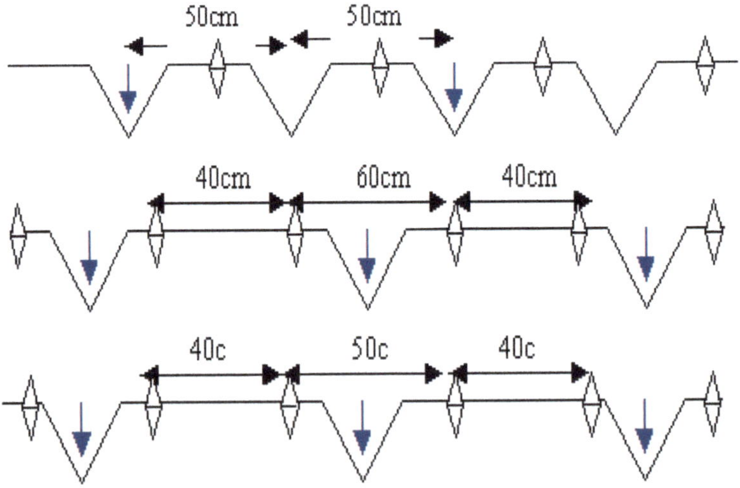

Fig. 21.2 Planting pattern of sugar beet in-furrow irrigation system (after Mirzaei et al. 2013)

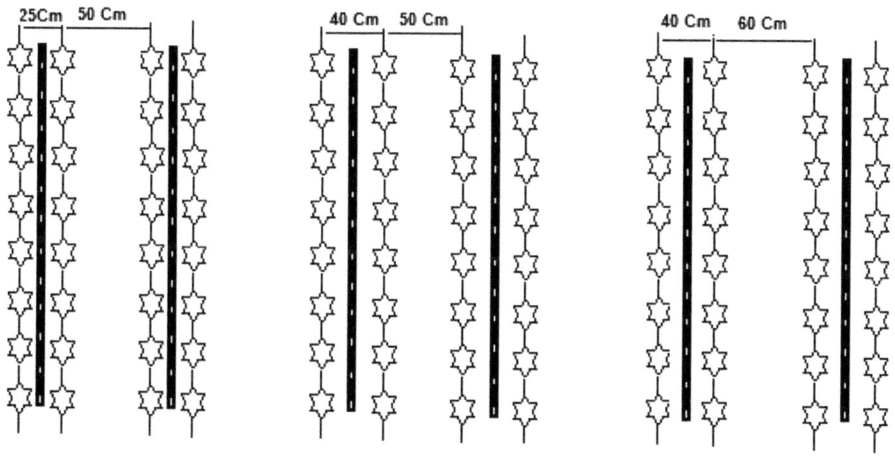

Fig. 21.3 Planting patterns of sugar beet in tape-drip irrigation system (after Mohammadian and Sadrqaen 2016)

tape-drip (Fig. 21.3) irrigation systems (Mohammadian and Sadrqaen 2016). Changing the planting pattern can also increase the effectiveness of using available resources and increase plant density to enhance yield and thus increase resourceful use of water. By decreasing the distance between planting rows and planting two rows on a ridge, while improving plant density and thus increasing yield, water consumption can be decreased which results in increased water use efficiency. The following planting patterns can be suggested.

1. Planting pattern with 100 cm between-row spacing and two planting rows on the ridge (40–60 cm).
2. Planting pattern with 90 cm between-row spacing and two planting rows on the ridge (40–50 cm).
3. Planting pattern with 75 cm between-row spacing and two planting rows on the ridge (25–50 cm).

In the planting pattern of 90 cm between-row spacing with two planting rows on the ridge, water consumption can be reduced by at least 20% without any significant change in yield (Taleghani et al. 2004). Changing the planting pattern in tape-drip irrigation and using tape tubes every other row, while improving plant density, because of the decreased use of tape tubes, reduces the expenses of using tape-drip irrigation method because a major share of the regular annual cost of tape-drip implementation (31%) is associated with the expense of tape tubes. This planting pattern is applicable in all areas where sugar beet is grown and furrow and tape-drip irrigation methods are applied (Mohammadian 2013).

21.8 Transplanting

Realization of the maximum yield in sugar beet production depends, among other factors, on the length of the growth period. Seedling transplantation has been extensively used to decrease the seedling emergence period and increase the emergence rate (Basra et al. 2005). Due to limited water resources in areas where cereal crops and sugar beet compete for irrigation water, sugar beet seeds can be sown early spring in paper pots in the greenhouse and then transfer the young seedlings to the field after cutting off the last irrigation of cereals. Studies have shown that using this method, sugar yield was 10.11 t ha^{-1}, while in direct seeding (sowing after cutting off the last irrigation of cereals) it was 5.66 t ha^{-1}. Where pot is not available to prepare seedlings, it is possible to use bare-root seedlings (potless seedlings) (Yousefabadi 2018). Due to problems associated with preparing paper pots, experimentation was done to examine the possibility of sowing bare-root seedlings and study the influence of root size and transfer date on root yield and quality of sugar beet. Results indicated the success of this sowing method, and the maximum root and white sugar yields (87 and 9.3 t ha^{-1}, respectively) were observed at the first planting date (Yousefabadi et al. 2014). In a similar study in Egypt, the maximum yield was obtained at the first planting date (9th May) with the larger transplant root size (3–4.5 cm root diameter) (Karbalaei et al. 2012).

In transplantation method, 20–40 days of plant development occur in the nursery. This method can help to mitigate the early season water and salinity stresses; it is also suitable for autumn sowing when cold damage is likely. By using this method, a lot of inputs can be saved. For example, irrigation is decreased by 24% (Khozaei et al. 2020), thinning is not needed, application of herbicides and pesticides is diminished by 40–50%, and use of seed is decreased by 65%. Increased number of established plants and harvestable roots are among the other advantages of the transplanting method (Moursy and El-Kady 2019).

The transplanting method can increase irrigation water productivity and white sugar yield water productivity of sugar beet in arid and semi-arid regions. Therefore, a large quantity of irrigation water, which is usually used for seed germination and crop establishment, can be saved using the transplantation technique (Khozaei et al. 2020). Generally, the increased yield observed in the transplanting method can be ascribed to an extended growth period under conditions favorable to rapid plant growth (Anderson et al. 1958).

21.9 Autumn Sowing

Due to water restriction in Iran, autumn sowing is advantageous over spring sowing and is gradually increasing with newly suitable areas being explored. Research on autumn sowing of sugar beet in Iran was started in 1963. Among the advantages of autumn sowing of sugar beet are low water use, low disease pressure, and high root yield (Taleghani et al. 2017). A study conducted in Khuzestan province (the largest autumn sowing area in Iran) demonstrated that on average, autumn sowing has about

Table 21.2 Comparison of characteristics of spring and autumn sowing of sugar beet

	Sugar yield (tha^{-1})				Water use efficiency ($kg\ m^{-3}$)	
Sowing season	Root	Sugar	Sugar content (%)	Water use (m^3)	Root	Sugar
Spring	52	7.25	17	12,000	4.33	0.60
Autumn	70	7.61	14.50	8000	8.75	0.95

20% higher yield, uses 35% less water, and hence uses water more efficiently than spring sowing (Table 21.2) (Taleghani et al. 2017). However, varieties resistant to bolting are a prerequisite for autumn sowing; they can prevent yield and quality reduction caused by the bolting phenomenon. These varieties have already been bred and are available to the farmers in autumn showing regions.

Sugar beet responds differently to drought stress at different growing stages. The life cycle of sugar beet could be divided into four stages: (1) germination to the establishment, (2) establishment to 70–80% canopy development, (3) full canopy development, and (4) beginning of canopy senescence to technological maturity (Farzamnia et al. 2007; Mirzaei and Rezvani 2012). Sugar beet is highly sensitive to drought stress at early growth stages, whereas it demonstrates high-stress tolerance at second and fourth stages and actually a lower amount of water could be given at these two stages without a significant reduction in yield (Kirda 2002; Farzamnia et al. 2007; Mirzaei and Rezvani 2012).

21.9.1 Methyl Jasmonate (MeJA)

Reduction of root and sugar yields in water-limited condition is mainly due to decreased turgor and photosynthesis, and these declines are most detrimental during early development (Clover et al. 1999; Monti et al. 2006). Jasmonates are concerned with plant drought stress responses. Methyl jasmonate delays plant drying out and protects photosynthetic machinery from drought-induced injury. When applied exogenously, MeJA moderates the drought stress effect on sugar beet and may decrease the losses of early season drought stress (Fugate et al. 2018). The effects of mild and severe drought on some physiological traits such as relative water content and photosynthesis were reduced, and drought-induced changes in proline accumulation were altered but transpiration rate, stomatal conductance, or betaine accumulation were not affected by application of MeJA, at 1 and 10 μM (Fugate et al. 2018). In another study conducted in Iran, foliar use of jasmonic acid (JA) enhanced plant water relations mostly because of improvements in osmoregulation (Ghaffari et al. 2020). The positive effects of JA were more evident in stressed sugar beet since it improved chlorophyll content and relative water content in drought stress conditions (Ghaffari et al. 2020). Also, the foliar application of JA significantly mitigated harmful effects of drought stress and improved sugar content and sugar yield of sugar beet under drought stress treatments (Ghaffari et al. 2020).

21.10 Future Prospects

Development and adoption of new water-saving agricultural practices are necessary for providing high yields with low water application. This review has considered some applied methods to cope with drought stress in sugar beet. Due to the limitation of available water in irrigated agriculture of areas like Iran, it is predicted that the spring-sown sugar beet area will not change dramatically, whereas the area of autumn sowing will gradually increase. This needs the establishment of additional sugar factories or further expansion of industrial capacity in newly explored autumn sowing areas. Furthermore, the efficiency of breeding drought-tolerant varieties can be enhanced by using molecular markers which provide a high throughput system to discover the potential traits in large segregating populations (Rajabi and Borchardt 2009). Furthermore, the potential of genomics and proteomics could be integrated with the drought tolerance breeding programs (Navid et al. 2012).

21.11 Conclusions

Because of water limitations and the necessity for sustainable utilization of water resources, the spring sugar beet area will be relatively constant, whereas the cultivation area of autumn sugar beet will gradually increase in Iran. Among the drought mitigation strategies, the use of drought-tolerant varieties, autumn sowing, short-vegetation-period varieties, deficit irrigation, modification of sowing pattern, root transplanting, and jasmonic acid could be suggested as suitable methods to alleviate the effects of drought stress on sugar beet.

References

Anderson DT, Dubetz S, Russell GC (1958) Studies on transplanting sugar beets in southern Alberta. J Am Soc Sugar Beet Technol 10:150–155

Basra SMA, Farooq M, Tabassum R (2005) Physiological and biochemical aspects of seed vigor enhancement treatments in fine rice (*Oryza sativa* L.). Seed Sci Technol 33:623–628

Boyer JS (1982) Plant productivity and environment. Science 218:443–448

Choluj D, Wiśniewska A, Szafrański KM, Cebula J, Gozdowski D, Podlaski S (2014) Assessment of the physiological responses to drought in different sugar beet genotypes in connection with their genetic distance. J Plant Physiol 171:1221–1230

Clover GRG, Smith HG, Azam-Ali SN, Jaggard KW (1999) The effects of drought on sugar beet growth in isolation and in combination with beet yellows virus infection. J Agric Sci 133:251–261

Farzamnia M, Darvishi D, Zareie G, Taleghani D (2007) Optimum irrigation scheduling under deficit irrigation in sugar beet growth stages. Final report of Sugar beet Seed Institute, Karaj, Iran

Fugate KK, Lafta MA, Eide JD, Li G, Lulai EC, Olson LL, Deckard EL, Khan MFR, Finger FL (2018) Methyl jasmonate alleviates drought stress in young sugar beet (*Beta vulgaris* L.) plants. J Agron Crop Sci 204(6):1–11

Ghaffari H, Tadayon MR, Razmjoo J, Bahador M, Karimzadeh Soureshjani H (2020) Impact of jasmonic acid on sugar yield and physiological traits of sugar beet in response to water deficit regimes: using stepwise regression approach. Russ J Plant Physiol 67(3):482–493

Hernandez-Santana V, Fernndez JE, Cuevas MV, Perez-Martin A, Diaz-Espejo A (2017) Photosynthetic limitations by water deficit: effect on fruit and olive oil yield, leaf area and trunk diameter and its potential use to control vegetative growth of super high density olive orchards. Agric Water Manag 184:9–18

Hofste RW, Reig P, Schleifer L (2019) National Water Stress Rankings. World Resources Institute. wri.org/aqueduct

Karbalaei S, Mehraban A, Mobasser HR, Bitarafan Z (2012) Sowing date and transplant root size effects on transplanted sugar beet in spring planting. Ann Biol Res 3(7):3474–3478

Khozaei M, Kamgar Haghighi AA, Zand Parsa SH, Sepaskhah A, Razzaghi F, Yousefabadi V, Emam Y (2020) Evaluation of direct seeding and transplanting in sugar beet for water productivity, yield and quality under different irrigation regimes and planting densities. Agric Water Manag 238:106230

Kirda C (2002) Deficit irrigation scheduling based on plant growth stages showing water stress tolerance. In: Deficit irrigation practices. FAO, Rome, pp 3–10

Licht FO (2018) International highlights. Int Sugar Sweetener Rep 151(16):343

Mirzaei MR, Rezvani SM (2012) Effect of deficit irrigation levels at four growth stages on yield and quality of sugar beet. Iran J Crop Sci 14(2):94–107

Mirzaei MR, Ghadami Firoozabadi A, Mohammadian R, Sadrqaen SH, Ashraf Mansouri GR, Dehghanian SA (2013) Determining the best planting pattern of sugar beet and water use efficiency under tape-drip irrigation system. Final report of Sugar Beet Seed Institute, Karaj, Iran

Mohammadian R (2013) Irrigation of sugar beet. In: Taleghani D, Alimorad I, Sadeghian SY, Mohammadian R, Mahmoudi SB, Abdollahian M, Sadeghzadeh S, Khayamim S, Tohidloo GH, Hosseinpour M, Rezaei V, Aghaeizadeh M, Noshad H, Yousefabadi V, Babaei B, Sharifi H, Matloobi F, Orazizadeh MR, Vahedi S (eds) Development of standards for determining the potential and assessment of damage by managerial and coercive factors at different stages of growth in sugar beet fields. Agricultural Education and Extension Publications, Tehran, Iran, pp 91–120

Mohammadian R, Sadrqaen SH (2016) Determining the most suitable plant density and amount of sugar beet nitrogen fertilizer under tape-drip irrigation conditions. Final report of Sugar Beet Seed Institute, Karaj, Iran

Mohammadian R, Rahimian H, Moghaddam M, Sadeghian SY (2003) The effect of early season drought on chlorophyll a fluorescence in sugar beet (*Beta vulgaris* L.). Pak J Biol Sci 6(20):1763–1769

Mohammadian R, Moghaddam M, Rahimian H, Sadeghian SY (2005) Effect of early season drought stress on growth characteristics of sugar beet genotypes. Turk J Agric For 29:357–368

Monti A, Brugnoli E, Scartazza A, Amaducci MT (2006) The effect of transient and continuous drought on yield, photosynthesis and carbon isotope discrimination in sugar beet (*Beta vulgaris* L.). J Exp Bot 57(6):1253–1262

Moosavi SGR, Ramazani SHR, Sadeghzadeh Hemayati S, Gholizade H (2017) Effect of drought stress on root yield and some morpho-physiological traits in different genotypes of sugar beet (*Beta vulgaris* L.). J Crop Sci Biotechnol 20:167–174

Morillo-Velarde R, Ober ES (2006) Water use and irrigation. In: Draycott AP (ed) Sugar beet. Blackwell Publishing, Oxford, pp 221–255

Moursy MAM, El-Kady MS (2019) Study planting methods to improve water use efficiency and productivity of sugar beet in a newly reclaimed area. Life Sci J 16(12):11–19

Navid A, Khan AA, Rauf S (2012) The potential of breeding okra for water stress tolerance. In: Ashraf A, Ozturk M, Aqeel Ahmad MA, Aksoy A (eds) Crop production for agricultural improvement. Springer Science and Business Media, Berlin, pp 217–235

Noghabi AM, Williams BF (2000) Drought stress and weed competition in sugar beet. Br Sugar Beet Rev 1:47–49

Ober E (2006) Evidence for genotypic differences in water use efficiency (WUE) during drought and how to improve WUE through selection. Proceedings of the 69th IIRB congress. Brussels, Belgium pp. 229–236

Ober E, Rajabi A (2010) Abiotic stress in sugar beet. Sugar Tech 12(3–4):294–298

Orazizadeh MR, Rajabi A, Fatholah Taleghani D, Ahmadi M, Vahedi S, Aghaeizadeh M, Sadeghian-Motahar SY, Chageni MA, Yosefabadi V, Sadeghzadeh-Hemayati S, Abdollahian-Noghabi M, Mohammadian R, Mirzaee MR, Ebrahimi-Koulaie H, Jalilian A, Fathi MR, Khodadadi S, Nouroozi A, Rahnamaian M (2015) Paya, the first sugar beet monogerm variety tolerant to drought in Iran. Res Achieve Field Horticult Crops 4(1):31–42

Oweis TY, Farahani HJ, Hachum AY (2011) Evapotranspiration and water use of full and deficit irrigated cotton in the Mediterranean environment in northern Syria. Agric Water Manag 98: 1239–1248

Padilla-Daz CM, Rodriguez-Dominguez CM, Hernandez-Santana V, Perez-Martin A, Fernndez JE (2016) Scheduling regulated deficit irrigation in a hedgerow olive orchard from leaf turgor pressure related measurements. Agric Water Manag 164:28–37

Pathak AD, Kumar R, Kapur R, Solomon S, Srivastava S, Singh PR (2014) Sugar beet: a historical perspective in Indian context. Sugar Tech 16(2):125–132

Pidgeon JD, Werker AR, Jaggard KW, Richter GM, Lister DH, Jones PD (2001) Climatic impact on the productivity of sugar beet in Europe, 1961–1995. Agric For Meteorol 109:27–37

Rajabi A (2017) Adaptation analysis of sugar beet monogerm hybrids in normal and late sowing conditions in different areas of Iran. Final report of Sugar Beet Seed Institute, Karaj, Iran

Rajabi A (2019) Development and evaluation of the response of sugar beet diploid hybrids to drought stress and Rhizomania. Final report of Sugar Beet Seed Institute, Karaj, Iran

Rajabi A, Borchardt D (2009) QTL mapping for root yield and leaf traits in sugar beet (*Beta vulgaris* L.) under drought stress condition. Iran J Crop Sci 17(1):46–62

Rajabi A, Griffiths H, Ober ES, Kromdijk W, Pidgeon JD (2008) Genetic characteristics of water-use related traits in sugar beet. Euphytica 160:175–187

Rajabi A, Ober ES, Griffiths H (2009) Genotypic variation for water use efficiency, carbon isotope discrimination, and potential surrogate measures in sugar beet. Field Crop Res 112:172–181

Sattar A, Cheema MA, Sher A, Ijaz M, Ul-Allah S, Nawaz A, Abbas T, Ali Q (2019) Physiological and biochemical attributes of bread wheat (*Triticum aestivum* L.) seedlings are influenced by foliar application of silicon and selenium under water deficit. Acta Physiol Plant 41(146)

Sepaskhah AR, Kamgar-Haghighi AA (1997) Water use and yield of sugar beet grown under every-other furrow irrigation with different irrigation intervals. Agric Water Manag 34:71–79

Taleghani D, Tohidloo GH, Gohari J, Habibi D, Sadeghian Y, Mesbah M (2004) Improvement of water productivity in surface irrigation system by changing the plant spacing in sugar beet cultivation. Proceedings of the 4th international crop science congress, Brisbane, Australia

Taleghani D, Mohammadian R, Sadeghzadeh S (2017) Handbook of sowing, husbandry and harvesting of autumn sugar beet. Amoozesh Press, Tehran, Iran, p 92

Unlü M, Kanber R, Ko DL, Tekin S, Kapur B (2011) Effects of deficit irrigation on the yield and yield components of drip irrigated cotton in a Mediterranean environment. Agric Water Manag 98:597–605

Vamerali T, Guariseb M, Ganisb A, Moscab G (2009) Effects of water and nitrogen management on fibrous root distribution and turnover in sugar beet. Eur J Agron 31:69–76

Yang CJ, Luo Y, Sun L, Wu N (2015) Effect of deficit irrigation on the growth, water use characteristics and yield of cotton in arid Northwest China. Pedosphere 25:910–924

Yousefabadi WA (2018) Study of sugar beet seedling transplantation after cutting off the last irrigation of cereals. Final report of Sugar Beet Seed Institute, Karaj, Iran

Yousefabadi W, Taleghani D, Abdullahian Noghabi M, Sadeghzadeh S, Habibkhodaei A (2014) The effect of seedling size and the date of its transfer to the main land on root yield and quality of sugar beet in small farms. Final report of research project, Sugar Beet Seed Institute, Karaj, Iran

Exogenous Putrescine-Mediated Modulation of Drought Stress Tolerance in Sugar Beet: Possible Mechanisms

22

Md Jahirul Islam, Masuma Akter Mou, Md Abdur Razzak, and Young-Seok Lim

Abstract

Drought management is a major challenge under changing climate conditions, and improvement in drought tolerance is the ultimate goal for scientists working in varietal development. Drought stress has a severe effect on the growth and development of sugar beet seedlings, leading to an enormous reduction in total biomass accumulation. Polyamines (PAs) are low-molecular weight positively charged aliphatic polycations, and putrescine (Put) is the central product in PA biosynthesis pathway and also acts as a precursor of Spd and Spm, which are widely used for stress management in various crop. Due to their chemical structure, PAs are polycationic in nature, which modulate the ion balance in cells and bind with polyanionic molecules like DNA, RNA, proteins, or membrane, thus preventing macromolecule and cell degradation membranes under adverse conditions. The binding properties exhibit the ROS scavenging capacity of PAs, which confers the antioxidative role that can prevent the cell from being

M. J. Islam (✉)
Department of Bio-Health Convergence, College of Biomedical Science, Kangwon National University, Chuncheon, Republic of Korea

Physiology and Sugar Chemistry Division, Bangladesh Sugarcrop Research Institute, Pabna, Bangladesh

M. A. Mou
Department of Agronomy, Bangladesh Agricultural University, Mymensingh, Bangladesh

M. A. Razzak
Faculty of Life and Environmental Science, Shimane University, Matsue, Japan
e-mail: razzakma@bau.edu.bd

Y.-S. Lim
Department of Bio-Health Convergence, College of Biomedical Science, Kangwon National University, Chuncheon, Republic of Korea

V. Misra et al. (eds.), *Sugar Beet Cultivation, Management and Processing*,
https://doi.org/10.1007/978-981-19-2730-0_22

damaged by lipid peroxidation and ROS generation. Therefore, the application of Put on sugar beet may be a helpful tool for stress mitigation that increases the yield and quality without any harmful effect on plants and the environment.

Keywords

Drought · Putrescine · Sugar beet · Tolerance

Abbreviations

ADC	Arginine decarboxylase
Agm	Agamitine
AIH	Agmatine iminohydrolase
APX	Ascorbate peroxidase
Arg	Arginine
Cad	Cadaverine
CAT	Catalase
CDC	Citrulline decarboxylase
Chl	Chlorophyll
Cit	Citrulline
DeSAM	*S*-adenosylmethionine
GB	Glycine betaine
H_2O_2	Hydrogen peroxide
LA	Leaf area
LRWC	Leaf relative water content
NCPA	Agm to *N*-carbamoyl put
NCPAH	*N*-carbamoylputrescine amidohydrolase
$O_2^{-\bullet}$	Superoxide
ODC	Ornithine decarboxylase
OH	Hydroxyl radical
Orn	Ornithine
PAL	Phenylalanine ammonia-lyase
PAs	Polyamines
POD	Guaiacol peroxidase
PPO	Polyphenol oxidase
Pro	Proline
PS II	Photosystem II
Put	Putrescine
ROS	Reactive oxygen species
SAM	*S*-adenosyl methionine
SAMA	*S*-adenosyl methionine synthase
SAMDC	*S*-adenosylmethionine decarboxylase
SOD	Superoxide dismutase
Spd	Spermidine

SPDS	Spd synthase
Spm	Spermine
SPMS	Spm synthase
TFC	Total flavonoids
TPC	Total polyphenol
TSC	Total soluble carbohydrate
TSP	Total soluble protein
tSpm	Thermospermine
TSS	Total soluble sugar

22.1 Introduction

Sugar beet (*Beta vulgaris* L.) is the second crucial sugar-producing crop after sugarcane, contributing to about 30% and 40% of world sugar production and world sugar trade, respectively (Mall et al. 2021). Sugar beet production depends on many factors, and drought stress is the most decisive limitation for reducing its yield ranged from 5 to 30% (Chołuj et al. 2014; Hosseini et al. 2019). Drought has a detrimental effect on crop production through its direct negative impact on plant growth and establishment, pigment production, photosynthetic rate, nutrient accumulation, and osmotic adjustment (Farooq et al. 2020; Praba et al. 2009). Water scarcity hampers crop yield by affecting plant's morphological development and quality (Islam et al. 2020; Sohag et al. 2020). Plants grown in water scarcity conditions produce a pool of reactive oxygen species (ROS) through physiological, biochemical, morphological, and molecular changes (Islam et al. 2020), causing an imbalance in component quantities and dysfunction of their typical defensive mechanisms (Zabalza et al. 2008). This interruption of the defensive system provokes the overproduction of ROS consisting of both non-radical (hydrogen peroxide, H_2O_2) and free radical species (superoxide, $O_2^{-\bullet}$; hydroxyl radical, $OH^\bullet$) that are known highly detrimental to plant cells (Bi et al. 2016). The overproduction of ROS triggers lipid peroxidation in cell especially leaf tissue that hampers the chlorophyll accumulation and ultimately reduces the photosynthetic rate and photosynthetic efficiency in plants (Islam et al. 2021a; Seleiman et al. 2021).

Polyamines (PAs) are low-molecular weight positively charged aliphatic polycations containing several amino groups ubiquitously distributed in eukaryotic and prokaryotic cells (de Sousa Araújo et al. 2019). The major PAs abundant in pants are putrescine (Put, 1,4-diaminobutane), spermidine (Spd, *N*-(3-Aminopropyl)-1,4-diaminobutane), spermine (Spm, *N*,*N*′-Bis(3-aminopropyl)-1,4-diaminobutane) with some plants also having thermospermine (tSpm) in place of or along with Spm (Chen et al. 2019; Minocha et al. 2014). Besides this, a diamine named cadaverine (Cad, 1,5-Diaminopentane), less known compared to major PAs, is also identified in the plants belonging to the families Gramineae, Leguminosae, and Solanaceae (de Sousa Araújo et al. 2019). They are synthesized from a group of amino acid precursors like arginine, ornithine, methionine, and lysine, followed by an intricate decarboxylation process (Falahi et al. 2018).

22.2 Polyamines and Their Stress Response in Plants

The principal idea of PAs in the protection of plants comes from their chemical structure. Polyamines are polycationic in nature, which can modulate the ion balance in cells and bind with polyanionic molecules like DNA, RNA, proteins, or membrane lipids by preventing macromolecule degradation and protecting cell membranes from the toxicity produced from the adverse condition. The binding properties exhibit the ROS scavenging capacity of PAs, which confers the antioxidative role that can prevent the cell from damage by ROS generation and lipid peroxidation (Alcázar et al. 2020). The structural simplicity, ubiquitous distribution in all cellular compartments, and potential role in plant's fundamental processes like growth and establishment, senescence, and notably adaptation to abiotic and biotic stresses make PAs an attractive model of metabolites in biological activities. Their high concentration accumulation probably reduces the toxicity of ammonia and modulates the total nitrogen distribution into a diverse pathway by alleviating extra nitrogen from the cell (Bais and Ravishankar 2002; Minocha et al. 2014). Putrescine can alter the plasma membrane of guard cells by controlling the potassium channel and pores to control the pore opening and closing (Chen et al. 2019), thereby regulating evapotranspiration in the plant (Liu et al. 2000). Controlled foliar application of Put can trigger physiological processes and induce osmotic adjustment molecules like proline, total soluble sugars, and amino acids in plants (Chen et al. 2019). It was also reported that PAs affect DNA, RNA, and protein biosynthesis, exacerbate plant growth and establishment, linger aging, and remove ROS from the cell that protect the membrane from oxidative damage in the plants (Hussein et al. 2019).The positive functions of exogenous Put conferring stress tolerance have been well documented in plants (Abd Elbar et al. 2019; Chen et al. 2019; Cui et al. 2020; Shu et al. 2015); however, many aspects of Put-mediated drought stress tolerance remains elusive.

22.3 Putrescine Biosynthesis Pathway

Putrescine is the main product in PA biosynthesis pathway, which acts as a precursor of Spd and Spm and contains two amino groups. Three possible routes of Put biosynthesis were narrated so far in different plants (Fig. 22.1). In the first route, Put is biosynthesized via the arginine decarboxylase (ADC) pathway, which involves the synthesis of agmatine (Agm), intermediate products by removing the No. 8 carbon atom from arginine (Arg) by ADC. Agmatine is further subject to catalyze in two successive steps by agmatine iminohydrolase (AIH) and *N*-carbamoylputrescine amidohydrolase (NCPAH) to synthesize putrescine; First, conversion of Agm to *N*-carbamoyl Put (NCPA) and NH_3 by removing the

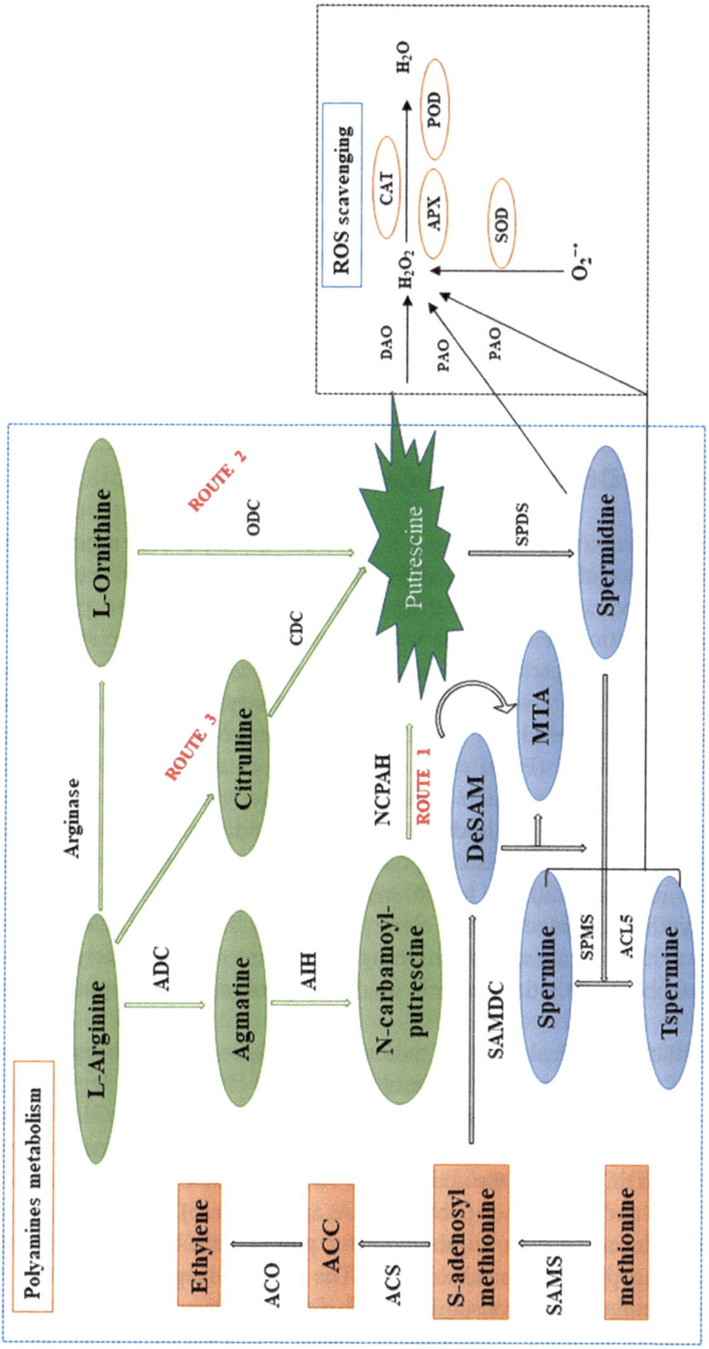

Fig. 22.1 The biosynthesis of PAs and their simple ROS scavenging pathways in plants. The green part denotes the three routes of the putrescine synthesis; the blue part denotes the conversion of putrescine; the orange part denotes the ethylene synthesis pathway (Chen et al. 2019; Yuan et al. 2016)

reduce the growth, development, and economic yield of plants (Mir et al. 2012). Besides, the nutrients like NPK also have the potentiality for their involvement in some morphological and physio-biochemical processes in the cell, such as photosynthetic rate and stomatal movement (AlKahtani et al. 2021). Notably, the concentration of Na^+ and Cl^- was reduced while K^+ level was stimulated by spraying Put in the salinity affected rice seedlings (Prakash and Prathapasenan 1988). It was also described that exogenous Put inhibited Na^+ and Cl^- uptake and accelerated the concentration of K^+, Ca^{2+}, and Mg^{2+} in salt-tolerant rice cultivar (Prakash and Prathapasenan 1988). From these results, it can be assumed that exogenous Put can reform the mineral accumulation and bring the balance among them which indirectly help plants to reduce the toxicity of water scarcity condition.

Secondary metabolites such as total polyphenol (TPC), total flavonoids (TFC), and callose have several important physiological functions in the plants (Chen et al. 2009). These positive physiological roles are directly linked with plant's metabolism and development under both normal and stress conditions. The biosynthesis of secondary metabolites is also regulated by stress conditions which are significantly used in defensive mechanisms of plants (Mundim and Pringle 2018; Ncube and Van Staden 2015; Zhao et al. 2005). In a previous study, both TPC and TFC significantly increased under drought stress in 11 sugar beet cultivars (Islam et al. 2020). On the other hand, total phenolic compounds and the activity of enzymes such as polyphenol oxidase (PPO) and phenylalanine ammonia-lyase (PAL) were subsequently increased under both drought and exogenous Put treatments (Abd Elbar et al. 2019). Generally, phenolic compounds are synthesized in the endoplasmic reticulum and cytoplasm, which play a vital role as secondary protection systems by using them as signaling and ROS scavenging molecules (Hu et al. 2008). Polyphenols scavenge both free radicals and lipid alkoxyl radicals, thus modulate ROS and MDA under adverse conditions. Moreover, flavonoids can be oxidized by peroxidase, thus act as H_2O_2 scavengers in the phenolic/AsA/POD system (Sharma et al. 2012). PAL is the key enzyme responsible for the biosynthesis of plants' secondary compounds, including phenolic compounds, whereas PPO is linked with plant defense response against varietal stress, whose antioxidant properties could help plants in scavenging ROS under hostile conditions (Taiz et al. 2015; Thipyapong et al. 2007). The mode of action of PAs on PPO is not clear but Put increased the activity of PPO in spinach leaves under both normal and salinity stress (Öztürk and Demir 2003). So, it can be hypothesized that under drought conditions, Put may play a positive role in inducing tolerance mechanisms by accumulating secondary metabolites along with antioxidant enzymes in sugar beet.

22.5.4 Changes in Antioxidant Enzymatic Activities and Molecular Events in Drought-Stressed Sugar Beet Plants by Exogenous Put

The activities of antioxidant enzymes have a profound relationship with stress tolerance mechanisms in plants. The potential oxidative damage in plants by excess

ROS under stress can be modulated by the up-regulation of enzymatic activities such as superoxide dismutase (SOD), catalase (CAT), ascorbate peroxidase (APX), and guaiacol peroxidase (POD) (Islam et al. 2020; Sohag et al. 2020). The primary ROS produced as a consequence of stress are superoxide anion ($O_2^{-\bullet}$), hydroxyl radical (OH^-), and H_2O_2, where superoxide is rapidly converted to H_2O_2 by the result of catalysis of $O_2^{-\bullet}$ to H_2O_2 by cellular SOD (Gough and Cotter 2011). In general, drought stress causes oxidative damage, which results from mainly unstable activities of both enzymatic and nonenzymatic antioxidants (DaCosta and Huang 2007). Earlier studies assured a positive relationship between H_2O_2 and membrane damage (lipid peroxidation) in the sugar beet under severe drought stress, where both enzymatic and nonenzymatic antioxidants were species-specific, and their activities either reduced or not significantly increased (AlKahtani et al. 2021; Islam et al. 2020). From these results, it seems that under prolonged drought stress, inconsistencies in the modulation of ROS occur due to an asymmetry of ROS synthesis and its scavenging, which increase lipid peroxidation, membrane permeability, and finally, cellular oxidative damage in plants (Islam et al. 2021a).

Some earlier studies have claimed that exogenous Put application stimulates antioxidant enzyme activity and stress response gene expression of plants under several abiotic stresses (Abd Elbar et al. 2019; Akter et al. 2018; Yuan et al. 2016). Exogenous spraying of Put substantially enhanced the survival capacity with more stress tolerance in *Panax ginseng* by bringing balance in production and scavenging of ROS, which ensured more photosynthetic pigments and photosynthetic efficiency (Islam et al. 2021b). Besides, several potential enzymes are involved in producing key polyamines like Put, Spd, and Spm in plants. For example, during Put biosynthesis process, ornithine and arginine are catalyzed by ODC and ADC, respectively. In another biosynthesis procedure, decarboxylated *S*-adenosylmethionine (DeSAM) (an aminopropyl group) is produced from *S*-adenosylmethionine (SAM) by *S*-adenosylmethionine decarboxylase (SAMDC), which take part in the transformation of Put into Spd and Spm with the help of Spd synthase (SPDS) and Spm synthase (SPMS), respectively (Fig. 22.1). The activity and the transcriptional level of such enzymes were narrated to be induced by abiotic stress. For instance, increased activity of ADC was reported in rice and *Arabidopsis* under salinity and water stress. The expression of SAMDC was also induced by various abiotic stress (Zhou et al. 2020). In a previous study, exogenous Put increased the gene expression of APX, Cu-Zn SOD, Glutathione *S*-transferase in cucumber under salinity stress (Yuan et al. 2016). Exogenous polyamines significantly increased the activities of some antioxidant enzymes and stress-related proteins in bermudagrass, thus enhanced the tolerance to drought and salinity (Shi et al. 2013). Besides, exogenous Spd significantly enhanced endogenous polyamines like Put, Spd, and Spm along with activity and gene expression of some major antioxidant enzymes in saline-stressed *Panax ginseng* (Parvin et al. 2014). Recently, exogenous Put was narrated to subsequently enhance the tolerance mechanisms in ginseng sprouts, where activities of SOD, CAT, APX, and POD enzymes significantly increased under saline stress (Islam et al. 2021b). Moreover, exogenous L-ornithine also significantly increased several osmolytes and enzymes activities, which modulated the ROS and protected the sugar beet plants from drought stress conditions (Hussein et al. 2019). So it is plausible that the modulation of ROS by the enzymatic and nonenzymatic antioxidants along with enhanced endogenous PAs by applying exogenous Put demonstrates a balanced condition that can improve the morpho-physiological characteristics of sugar beet under drought conditions.

22.6 Future Prospects

The effects of Put on sugar beet demands more in-depth study for stress mitigation, increasing the yield and quality without any harmful effect on plants and the environment that might find future practical applications for crop protection against stress.

22.7 Conclusion

Water shortage is a frequent and vital issue in sugar beet cultivation that has a complex impact on plant physiology. Drought creates oxidative stress in sugar beet as a consequence of ROS generation, which induces lipid peroxidation and damage in the cell with reduced pigments and photosynthetic rate results in a retardation of growth and biomass accumulation (Fig. 22.2). Although a decent number of studies have been documented on exogenous Put application or modification of endogenous PAs levels in plants, the precise molecular mechanism underlying improved stress tolerance by the protective effects of PAs is still unknown. The advanced research on Put related to genomics, transcriptomics, and proteomics to sugar beet under stress and their metabolic ameliorations would be an ideal tool by monitoring the up-regulation or down-regulation of genes connected with stress tolerance mechanisms. In recent years, many studies concentrated only on the effect of PAs, including Put on growth and development of some vegetable, fruit, and model crops. From the overall results, plants treated with Put can effectively reduce ROS and damage in the cell by upregulating osmolytes concentration, hormonal and enzymatic activity. The declined ROS by the catabolic activity of Put help to recompensate the photosynthetic pigments, leaf area, and leaf relative water content, which accelerates the plant morpho-physiological mechanisms with higher photosynthate and plant biomass accumulation (Fig. 22.2). However, pathways related to biosynthesis and catabolism of PAs, including Put that regulated at transcriptional, translational, and post-transcriptional levels, were absent or poorly discussed. Sugar beet production is confounded by the decreased performance of photosynthesis, canopy expansion, root dimension, and sucrose accumulation under drought stress (Ober and Rajabi 2010).

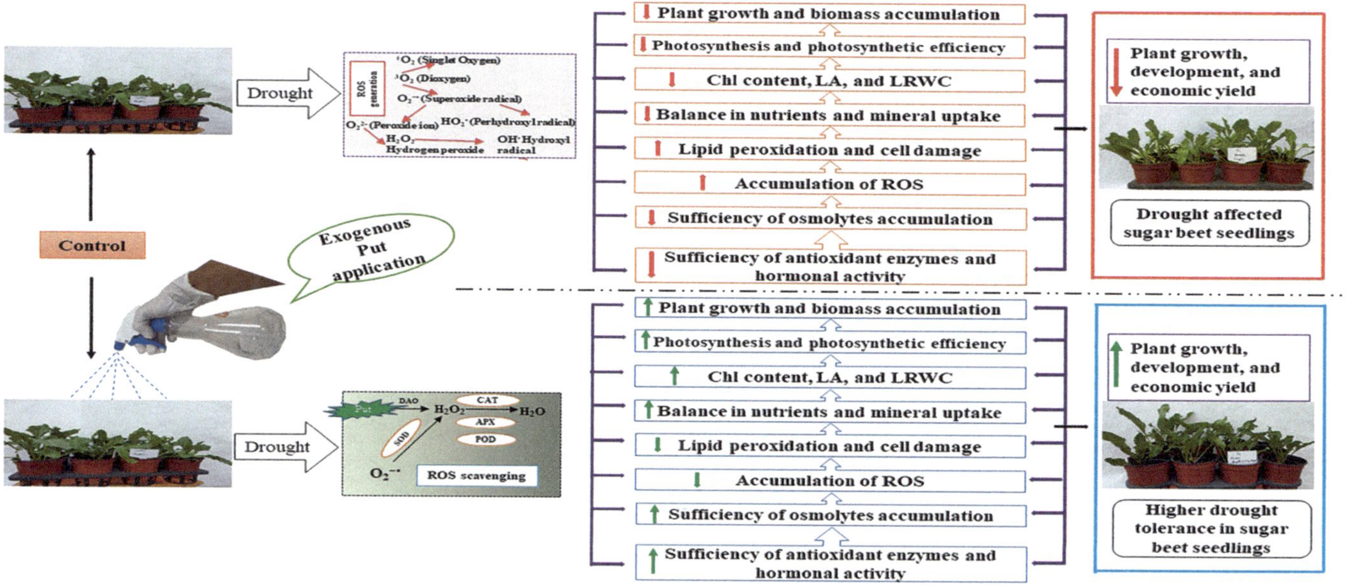

Fig. 22.2 Schematic representation of drought-induced growth inhibition and its recovery by exogenously Put treatment on sugar beet plants. Drought exerts its harmful effects on plants by diminishing the Chl content resulting in reduced photosynthetic rates. Drought increases superoxide, which in turn increases H_2O_2 content. Water scarcity suppresses the levels of osmolytes with lower activity of enzymes such as SOD, CAT, APX, and POD resulting in boosting H_2O_2. Enhanced ROS by drought damages the membrane and promotes lipid peroxidation in the cell. LA and LRWC also decrease under drought, limiting gas exchange and photosynthesis that ultimately hamper plant growth, establishment, and economic yield. Importantly, Put treatment can restore the performance of drought-stressed sugar beet and reduce oxidative damage. Exogenous Put boosts in osmolytes content along with hormonal balance and the activities related to antioxidant enzymes like SOD, CAT, APX, and POD that modulate the amount of ROS. Put attenuates the damage in the cell membrane by reducing lipid peroxidation, therefore increases Chl content, LA, and LRWC in seedlings, thus maintaining higher photosynthesis and biomass accumulation. *APX* ascorbate peroxidase; *CAT* catalase; *Chl* chlorophyll; H_2O_2 hydrogen peroxide; *LA* leaf area; *LRWC* leaf relative water content; *POD* guaiacol peroxidase; *ROS* reactive oxygen species; *SOD* superoxide dismutase

References

Abd Elbar OH, Farag RE, Shehata SA (2019) Effect of putrescine application on some growth, biochemical and anatomical characteristics of *Thymus vulgaris* L. under drought stress. Ann Agric Sci 64:129–137

Abdollahian-Noghabi M, Froud-Williams B (2000) Drought stress and weed competition in sugarbeet. Br Sugar Beet Rev 68:47–49

Akram NA, Shafiq F, Ashraf M (2017) Ascorbic acid-a potential oxidant scavenger and its role in plant development and abiotic stress tolerance. Front Plant Sci 8:613

Akter S, Rasul MG, Zakaria M, Sarker MM, Nila IS, Dutta S, Haque MM, Rohman MM (2018) Effect of polyamine on pigmentation, reactive oxidative species and antioxidant under drought in maize (*Zea mays* L.). Turk J Agric Sci Technol 6:799–811

Alcázar R, Altabella T, Marco F, Bortolotti C, Reymond M, Koncz C, Carrasco P, Tiburcio AF (2010) Polyamines: molecules with regulatory functions in plant abiotic stress tolerance. Planta 231:1237–1249

Alcázar R, Bueno M, Tiburcio AF (2020) Polyamines: small amines with large effects on plant abiotic stress tolerance. Cell 9:2373

AlKahtani MDF, Hafez YM, Attia K, Rashwan E, Al Husnain L, AlGwaiz HIM, Abdelaal KAA (2021) Evaluation of silicon and proline application on the oxidative machinery in drought-stressed sugar beet. Antioxidants 10:398

Anjum SA, Wang LC, Farooq M, Hussain M, Xue LL, Zou M (2011) Brassinolide application improves the drought tolerance in maize through modulation of enzymatic antioxidants and leaf gas exchange. J Agron Crop Sci 197:177–118

Badeck FW, Rizza F (2015) A combined field/laboratory method for assessment of frost tolerance with freezing tests and chlorophyll fluorescence. Agronomy 5:71–88

Bais HP, Ravishankar GA (2002) Role of polyamines in the ontogeny of plants and their biotechnological applications. Plant Cell Tissue Organ Cult 69:1–34

Bhaskara GB, Yang TH, Verslues PE (2015) Dynamic proline metabolism: importance and regulation in water limited environments. Front Plant Sci 6:484

Bi A, Fan J, Hu Z, Wang G, Amombo E, Fu J, Hu T (2016) Differential acclimation of enzymatic antioxidant metabolism and photosystem II photochemistry in tall fescue under drought and heat and the combined stresses. Front Plant Sci 7:453

Blum A (2011) Plant water relations, plant stress and plant production. In: Plant breeding for water-limited environments. Springer, Berlin, pp 11–52

Borišev M, Borišev I, Župunski M, Arsenov D, Pajević S, Ćurčić Ž, Vasin J, Djordjevic A (2016) Drought impact is alleviated in sugar beets (*Beta vulgaris* L.) by foliar application of fullerenol nanoparticles. PLoS One 11:e0166248

Chaves MM, Flexas J, Pinheiro C (2009) Photosynthesis under drought and salt stress: regulation mechanisms from whole plant to cell. Ann Bot 103:551–560

Chen F, Liu CJ, Tschaplinski TJ, Zhao N (2009) Genomics of secondary metabolism in populus: interactions with biotic and abiotic environments. Crit Rev Plant Sci 28:375–392

Chen D, Shao Q, Yin L, Younis A, Zheng B (2019) Polyamine function in plants: metabolism, regulation on development, and roles in abiotic stress responses. Front Plant Sci 9:1945

Chołuj D, Wiśniewska A, Szafrański KM, Cebula J, Gozdowski D, Podlaski S (2014) Assessment of the physiological responses to drought in different sugar beet genotypes in connection with their genetic distance. J Plant Physiol 171:1221–1230

Cui J, Pottosin I, Lamade E, Tcherkez G (2020) What is the role of putrescine accumulated under potassium deficiency? Plant Cell Environ 43:1331–1347

DaCosta M, Huang B (2007) Changes in antioxidant enzyme activities and lipid peroxidation for bent grass species in response to drought stress. J Am Soc Hortic Sci 132:319–326

de Sousa Araújo S, dos Santos ALW, Duque AS (2019) Engineering polyamine metabolic pathways for abiotic stress tolerance in plants. In: Osmoprotectant-mediated abiotic stress tolerance in plants. Springer, pp 287–318

Doneva D, Pál M, Brankova L, Szalai G, Tajti J, Khalil R, Ivanovska B, Velikova V, Misheva S, Janda T (2021) The effects of putrescine pre-treatment on osmotic stress responses in drought tolerant and drought-sensitive wheat seedlings. Physiol Plant 171:200–216

Ebeed HT, Hassan NM, Aljarani AM (2017) Exogenous applications of polyamines modulate drought responses in wheat through osmolytes accumulation, increasing free polyamine levels and regulation of polyamine biosynthetic genes. Plant Physiol Biochem 118:438–448

Efeoğlu B, Ekmekçi Y, Çiçek N (2009) Physiological responses of three maize cultivars to drought stress and recovery. South African J Bot 75:34–42

Falahi H, Sharifi M, Ahmadian Chashmi N, Zare Maivan H (2018) Water stress alleviation by polyamines and phenolic compounds in Scrophularia striata is mediated by NO and H_2O_2. Plant Physiol Biochem 130:139–147. https://doi.org/10.1016/j.plaphy.2018.07.004

Farooq A, Bukhari SA, Akram NA, Ashraf M, Wijaya L, Alyemeni MN, Ahmad P (2020) Exogenously applied ascorbic acid-mediated changes in Osmoprotection and oxidative defense system enhanced water stress tolerance in different cultivars of safflower (*Carthamus tinctorious* L.). Plan Theory 9:104

Forni C, Duca D, Glick BR (2017) Mechanisms of plant response to salt and drought stress and their alteration by rhizobacteria. Plant Soil 410:335–356

Foyer CH (2015) Redox homeostasis: opening up ascorbate transport. Nat Plants 1:1–2

Foyer CH, Lelandais M, Kunert KJ (1994) Photooxidative stress in plants. Physiol Plant 92:696–717

Genard H, Le Saos J, Billard JP, Tremolieres A, Boucaud J (1991) Effect of salinity on lipid composition, glycine betaine content and photosynthetic activity in chloroplasts of Suaeda maritima. Plant Physiol Biochem 29:421–427

Giri J (2011) Glycinebetaine and abiotic stress tolerance in plants. Plant Signal Behav 6:1746–1751

Gough DR, Cotter TG (2011) Hydrogen peroxide: a Jekyll and Hyde signalling molecule. Cell Death Dis 2:e213–e213

Hasegawa PM, Bressan RA, Zhu JK, Bohnert HJ (2000) Plant cellular and molecular responses to high salinity. Annu Rev Plant Biol 51:463–499

Hossain MS, ElSayed AI, Moore M, Dietz KJ (2017) Redox and reactive oxygen species network in acclimation for salinity tolerance in sugar beet. J Exp Bot 68:1283–1298

Hosseini SA, Réthoré E, Pluchon S, Ali N, Billiot B, Yvin JC (2019) Calcium application enhances drought stress tolerance in sugar beet and promotes plant biomass and beetroot sucrose concentration. Int J Mol Sci 20:3777

Hu Y, Burucs Z, Schmidhalter U (2008) Effect of foliar fertilization application on the growth and mineral nutrient content of maize seedlings under drought and salinity. Soil Sci Plant Nutr 54:133–141

Hussein HAA, Mekki BB, El-Sadek MEA, El Lateef EE (2019) Effect of L-ornithine application on improving drought tolerance in sugar beet plants. Heliyon 5:e02631

Islam MJ, Kim JW, Begum MK, Sohel MAT, Lim YS (2020) Physiological and biochemical changes in sugar beet seedlings to confer stress adaptability under drought condition. Plan Theory 9:1511. https://doi.org/10.3390/plants9111511

Islam MJ, Ryu BR, Azad MO, Rahman MH, Cheong EJ, Lim JD, Lim YS (2021a) Cannabinoids accumulation in hemp (*Cannabis sativa* L.) plants under LED light spectra and their discrete role as a stress marker. Biol. https://doi.org/10.3390/biology10080710

Islam MJ, Ryu BR, Azad MOK, Rahman MH, Rana MS, Lim JD, Lim YS (2021b) Exogenous putrescine enhances salt tolerance and Ginsenosides content in Korean ginseng (*Panax ginseng* Meyer) sprouts. Plan Theory 10. https://doi.org/10.3390/plants10071313

Kapoor D, Bhardwaj S, Landi M, Sharma A, Ramakrishnan M, Sharma A (2020) The impact of drought in plant metabolism: how to exploit tolerance mechanisms to increase crop production. Appl Sci 10:5692

Khoshbakht D, Asghari MR, Haghighi M (2018) Influence of foliar application of polyamines on growth, gas-exchange characteristics, and chlorophyll fluorescence in Bakraii citrus under saline conditions. Photosynthetica 56:731–742

Kotakis C, Theodoropoulou E, Tassis K, Oustamanolakis C, Ioannidis NE, Kotzabasis K (2014) Putrescine, a fast-acting switch for tolerance against osmotic stress. J Plant Physiol 171:48–51
Liu K, Fu H, Bei Q, Luan S (2000) Inward potassium channel in guard cells as a target for polyamine regulation of stomatal movements. Plant Physiol 124:1315–1326
Liu M, Chen J, Guo Z, Lu S (2017) Differential responses of polyamines and antioxidants to drought in a centipedegrass mutant in comparison to its wild type plants. Front Plant Sci 8:792
Mall AK, Misra V, Pathak AD, Srivastava S (2021) Sugar beet cultivation in India: prospects for bio-ethanol production and value-added co-products. Sugar Tech 23:1–17
Mansour MMF, Ali EF (2017) Evaluation of proline functions in saline conditions. Phytochemistry 140:52–68
Minocha R, Majumdar R, Minocha SC (2014) Polyamines and abiotic stress in plants: a complex relationship1. Front Plant Sci 5:175
Mir RR, Zaman-Allah M, Sreenivasulu N, Trethowan R, Varshney RK (2012) Integrated genomics, physiology and breeding approaches for improving drought tolerance in crops. Theor Appl Genet 125:625–645
Muhammad F, Abdul W, Lee D (2009) Exogenously applied polyamines increase drought tolerance of rice by improving leaf water status, photosynthesis and membrane properties. Acta Physiol Plant 31:937–945
Mundim FM, Pringle EG (2018) Whole-plant metabolic allocation under water stress. Front Plant Sci 9:852
Ncube B, Van Staden J (2015) Tilting plant metabolism for improved metabolite biosynthesis and enhanced human benefit. Molecules 20:12698–12731
Ober ES, Rajabi A (2010) Abiotic stress in sugar beet. Sugar Tech 12:294–298
Öztürk L, Demir Y (2003) Effects of putrescine and ethephon on some oxidative stress enzyme activities and proline content in salt stressed spinach leaves. Plant Growth Regul 40:89–95
Parvin S, Lee OR, Sathiyaraj G, Khorolragchaa A, Kim YJ, Yang DC (2014) Spermidine alleviates the growth of saline-stressed ginseng seedlings through antioxidative defense system. Gene 537: 70–78
Praba ML, Cairns JE, Babu RC, Lafitte HR (2009) Identification of physiological traits underlying cultivar differences in drought tolerance in rice and wheat. J Agron Crop Sci 195:30–46
Prakash L, Prathapasenan G (1988) Effect of NaCl salinity and putrescine on shoot growth, tissue ion concentration and yield of rice (*Oryza sativa* L. var. GR 3). J Agron Crop Sci 160:325–334
Putnik-Delić M, Nagl N, Maksimović I (2018) Sugar beet tolerance to drought: physiological and molecular aspects. In: Andjelkovic V (ed) Plant, abiotic stress and responses to climate change. IntechOpen, London
Rapacz M, Sasal M, Kalaji HM, Kościelniak J (2015) Is the OJIP test a reliable indicator of winter hardiness and freezing tolerance of common wheat and triticale under variable winter environments? PLoS One 10:e0134820
Ruíz-Sánchez M, Armada E, Muñoz Y, de Salamone IEG, Aroca R, Ruíz-Lozano JM, Azcón R (2011) Azospirillum and arbuscular mycorrhizal colonization enhance rice growth and physiological traits under well-watered and drought conditions. J Plant Physiol 168:1031–1037
Sarker U, Oba S (2018) Drought stress effects on growth, ROS markers, compatible solutes, phenolics, flavonoids, and antioxidant activity in Amaranthus tricolor. Appl Biochem Biotechnol 186:999–1016
Seleiman MF, Al-Suhaibani N, Ali N, Akmal M, Alotaibi M, Refay Y, Dindaroglu T, Abdul-Wajid HH, Battaglia ML (2021) Drought stress impacts on plants and different approaches to alleviate its adverse effects. Plan Theory 10:259
Sharma P, Jha AB, Dubey RS, Pessarakli M (2012) Reactive oxygen species, oxidative damage, and antioxidative defense mechanism in plants under stressful conditions. J Bot 2012:217037
Sharma A, Shahzad B, Kumar V, Kohli SK, Sidhu GPS, Bali AS, Handa N, Kapoor D, Bhardwaj R, Zheng B (2019) Phytohormones regulate accumulation of osmolytes under abiotic stress. Biomol Ther 9:285
Shen J, Wang Y, Shu S, Jahan MS, Zhong M, Wu J, Sun J, Guo S (2019) Exogenous putrescine regulates leaf starch overaccumulation in cucumber under salt stress. Sci Hortic (Amsterdam) 253:99–110

Shi H, Ye T, Chan Z (2013) Comparative proteomic and physiological analyses reveal the protective effect of exogenous polyamines in the bermudagrass (*Cynodon dactylon*) response to salt and drought stresses. J Proteome Res 12:4951–4964

Shu S, Guo S, Sun J, Yuan L (2012) Effects of salt stress on the structure and function of the photosynthetic apparatus in *Cucumis sativus* and its protection by exogenous putrescine. Physiol Plant 146:285–296

Shu S, Yuan LY, Guo SR, Sun J, Yuan YH (2013) Effects of exogenous spermine on chlorophyll fluorescence, antioxidant system and ultrastructure of chloroplasts in *Cucumis sativus* L. under salt stress. Plant Physiol Biochem 63:209–216

Shu S, Yuan Y, Chen J, Sun J, Zhang W, Tang Y, Zhong M, Guo S (2015) The role of putrescine in the regulation of proteins and fatty acids of thylakoid membranes under salt stress. Sci Rep 5: 1–16

Slama I, Abdelly C, Bouchereau A, Flowers T, Savoure A (2015) Diversity, distribution and roles of osmoprotective compounds accumulated in halophytes under abiotic stress. Ann Bot 115: 433–447

Sohag AAM, Tahjib-Ul-Arif M, Polash MAS, Chowdhury MB, Afrin S, Burritt DJ, Murata Y, Hossain MA, Hossain AM (2020) Exogenous glutathione-mediated drought stress tolerance in rice (*Oryza sativa* L.) is associated with lower oxidative damage and favorable ionic homeostasis. Iran J Sci Technol Trans A Sci 44:955–971

Subbarao GV, Wheeler RM, Levine LH, Stutte GW (2001) Glycine betaine accumulation, ionic and water relations of red-beet at contrasting levels of sodium supply. J Plant Physiol 158:767–776

Tahjib-Ul-Arif M, Sohag AAM, Afrin S, Bashar K, Afrin T, Mahamud AGM, Polash MAS, Hossain M, Sohel M, Taher A (2019) Differential response of sugar beet to long-term mild to severe salinity in a soil–pot culture. Agriculture 9:223

Taiz L, Zeiger E, Møller IM, Murphy A (2015) Plant physiology and development, 6th edn. Sinauer Associates, Incorporated, Sunderland, CT

Thipyapong P, Stout MJ, Attajarusit J (2007) Functional analysis of polyphenol oxidases by antisense/sense technology. Molecules 12:1569–1595

Valenzuela-Soto EM, Figueroa-Soto CG (2019) Biosynthesis and degradation of glycine betaine and its potential to control plant growth and development. In: Osmoprotectant-mediated abiotic stress tolerance in plants. Springer, Berlin, pp 123–140

Vuosku J, Karppinen K, Muilu-Mäkelä R, Kusano T, Sagor GHM, Avia K, Alakärppä E, Kestilä J, Suokas M, Nickolov K (2018) Scots pine aminopropyltransferases shed new light on evolution of the polyamine biosynthesis pathway in seed plants. Ann Bot 121:1243–1256

Wang Y, Stevanato P, Yu L, Zhao H, Sun X, Sun F, Li J, Geng G (2017) The physiological and metabolic changes in sugar beet seedlings under different levels of salt stress. J Plant Res 130: 1079–1093

Yuan Y, Zhong M, Shu S, Du N, Sun J, Guo S (2016) Proteomic and physiological analyses reveal putrescine responses in roots of cucumber stressed by NaCl. Front Plant Sci 7:1035

Zabalza A, Gálvez L, Marino D, Royuela M, Arrese-Igor C, González EM (2008) The application of ascorbate or its immediate precursor, galactono-1, 4-lactone, does not affect the response of nitrogen-fixing pea nodules to water stress. J Plant Physiol 165:805–812

Zhang L, Zhang Z, Gao H, Xue Z, Yang C, Meng X, Meng Q (2011) Mitochondrial alternative oxidase pathway protects plants against photoinhibition by alleviating inhibition of the repair of photodamaged PSII through preventing formation of reactive oxygen species in Rumex K1 leaves. Physiol Plant 143:396–407

Zhao J, Davis LC, Verpoorte R (2005) Elicitor signal transduction leading to production of plant secondary metabolites. Biotechnol Adv 23:283–333

Zhou R, Hu Q, Pu Q, Chen M, Zhu X, Gao C, Zhou G, Liu L, Wang Z, Yang J (2020) Spermidine enhanced free polyamine levels and expression of polyamine biosynthesis enzyme gene in rice spikelets under heat tolerance before heading. Sci Rep 10:1–9

Zuckerkandl E, Pauling L (1965) Evolutionary divergence and convergence in proteins. In: Evolving genes and proteins. Elsevier, Amsterdam, pp 97–166

Improving Sugar Beet Production Under Salinity Conditions

23

Samar Khayamim

Abstract

Salinity is one of the most important abiotic stresses especially in dry or semi-dry areas of the world, which are caused by saline land or man-made activities in irrigated fields. Sugar beet is known as a salt-tolerant crop. This crop knowing about its salinity threshold and response to stress would be useful for its management. There are agronomic approaches such as soil leaching, crop rotation, planting pattern, plant density, and use of suitable fertilizers, which would be helpful for sugar beet management in saline areas, which are mentioned in detail through this chapter.

Keywords

Drainage · Nutrition · Rotation · Slope row planting

Abbreviations

CEC	Cation exchange capacity
EC	Electrical conductance
ESP	Exchangeable sodium percentage
PSII	Photosystem II
ROS	Reactive oxygen species

S. Khayamim (✉)
Seed and Plant Improvement Institute, Agricultural Research, Education and Extension Organization, Karaj, Iran
e-mail: skhayamim@spii.ir

V. Misra et al. (eds.), *Sugar Beet Cultivation, Management and Processing*,
https://doi.org/10.1007/978-981-19-2730-0_23

23.1 Introduction

Salt stress is one of the most important stress among abiotic stresses. It is estimated that natural salinization in the world is less than one billion ha, but secondary salinization is about 77 M ha of which 58% is in irrigated areas. It is evaluated that 20% of the irrigated lands of the world are salt affected mostly in intensively cultivated areas of India, Pakistan, China, Iraq, and Iran. Regions at risk of salinity increase are the Mediterranean Basin, Australia, central Asia, the middle east, and northern Africa (Cherlet et al. 2018). In Asia, more than 50% of productive lands will be lost by salinity till 2050 (Nazar et al. 2011). 53 M ha of salt-affected grounds in Asia and the Middle East (Iran, Syria, Egypt, and Tunisia) is caused by man-made activities in irrigated fields (FAO 2000). There are different reasons for natural salinization but in arid and semi-arid regions like Iran, the main factors are dry climate which afforded less precipitation and more transpiration, saline bedrock, and also insufficient drainage of lands which caused more salinization. The secondary causes of salinity in these regions are more utilization of groundwater resources, water lodging, excessive grazing of pastures, frequent use of saltwater, and overuse of fertilizer (Kehl 2006).

Sugar beet is one of the most important industrial crops, which is salt tolerant (Jamil et al. 2006). But knowledge of its salinity threshold and susceptible growth stage would be very useful for its management. The threshold level varies based on the environmental conditions (Abrol et al. 1988). It is reported that germination percentage decreased in 8 dS/m (Jafarzadeh and Aliasgharzad 2007). Also, it is reported that root length decreased but germination didn't at 12 dS/m (Mohammadian et al. 1995) salinity levels. It's the threshold of 50% damage ranged between 13.7 dS m^{-1} (Morillo-Velarde and Ober 2006), 15 (Doorenbaos and Kassam 1979 based on FAO report) to 20 dS/m salinity (Mesbah et al. 1991; Ebrahimian and Ranji 2004a, b), calculated by Abrol et al. (1988) equation. Although seed germination in electrical conductance (EC) = 16 was the same as EC = 20 dS/m under laboratory conditions, 50% seedling losses were observed at EC = 16 dS/m in the greenhouse condition (Khayamim et al. 2011), but the crop has been damaged in less salinity in the field as a result of other soil ions side effect (Duan et al. 2004).

It was previously thought that the sugar beet germination stage was an important phase for susceptibility to salinity (Ghoulam and Fares 2001; Jamil et al. 2006), but it seems that establishment is more susceptible because sugar beet germination decreased to 35% and dead seedlings increased to 80% under salinity stress (EC = 16 dS/m) at the establishment, so it is obvious that salt stress decreased seed germination and then by increasing dead seedlings, establishment and consequently sugar beet yield decreased (Khayamim et al. 2014). Also, it is reported that salt stress didn't affect significantly on sugar beet till 35 (Niazi et al. 2004) and 50 days' growth stages (Delfine et al. 1999). Salinity slowed down the growth and development of seedlings (Rajabi et al. 2014), consequently decreased seedlings establishment (Durrant et al. 1974). Effects of salinity were more on seedling growth reduction than germination in other crops such as chick pea (Okcu et al. 2005) and millet (Al-Taisan 2010), further suitable germination of seeds for seedling development, their survival for the establishment, and achieving good yield under saline condition is necessary (Sadat Noori and McNeilly 2000).

23.2 Agronomic Management Approaches in the Salt-Affected Areas

Generally, salt-affected areas could be managed with modification of environment, crops and hybrid approaches (Singh et al. 2010). There are different methods to change the environment for the normal growth of plants such as leaching of salts from the root zone and drainage practices, also practical agronomic methods including crop rotation, planting pattern, plant density, fertilization, and crop protection are the most useful ways to increase yields under salinity stress.

23.2.1 Leaching and Drainage

Increased soil salinity of agricultural lands, especially in irrigated lands, is inevitable in arid and semi-arid regions. However, the intensity and rate of salinity increase depend on factors that interact with each other, such as the number of soluble salts and the local climate, therefore soil salinity can be managed with proper management of soil moisture, uniformity and efficiency of irrigation system, soil drainage, and proper selection of crops to maintain the fertility of the field (Khorsandi and Nezhad 2019). The existence of sufficient good quality water for leaching the construction of efficient and appropriate drainage systems is essential for the successful and appropriate remediation of saline soils. It is shown that leaching twice of saline soil (EC = 67 dS/m) with EC = 12 dS/m water before planting caused suitable yield (Khosgoftarmanesh and Shariatmadarih 2002). Suitable soil preparation increases water use and leaching efficiency. Plowing in 45–50 cm depth decreases soil salinity 50% and removes more salt from the soil (Karimi 1997). Establish proper drainage in the land despite the high initial cost is highly valued.

23.2.2 Sugar Beet Crop Rotation

Crop rotation is one of the methods to increase soil fertility and crop yield and also reduce environmental pollution. Failure to have proper rotation will cause loss of organic matter and suitable soil conditions, increase soil erosion, and ultimately reduce sugar beet yield in many areas. The existence of restrictions on climatic, soil, agronomic, or social and economic factors has caused the production of conventional single or double cropping, and crop rotation is generally unstable in the conditions of traditional Iranian agriculture (Khayamim 2013).

Different plants are placed in rotation under different conditions, but in choosing a set of crops for a period of rotation in an area, the number of their residues, the

Table 23.1 Some common and suitable rotations of sugar beet

Year	Crop	Description
1	Wheat (barely, rye)	This rotation is the most common rotation used by most farmers. Sugar percentage and yield have increased. The residue must be completely rotted with proper management until the sugar beet is planted. Under these conditions, short-term fallow (during autumn and winter) can be used to completely rot the grain residues. In some studies, the placement of sugar beet after a short fall (half a season) also increases the yield of white sugar.
2	Sugar beet	
1	Corn	Effective in reducing nematodes—usable in fields without a history of root rot disease
2	Sugar beet	
1	Pease (clover)	Decreased population of weeds and fungi causing rhizoctonia root rot
2	Corn (grains)	
3	Sugar beet	
4	Corn (grains)	
1	Alfalfa	Increased soil fertility due to alfalfa and clover, the last forage should be returned to the soil, direct cultivation of sugar beet after alfalfa increases susceptibility to root rot, the presence of corn increases the permeability of sugar beet root, and corn straw should be spread
2	Alfalfa	
3	Alfalfa	
4	Corn (grains)	
5	Sugar beet	
6	Wheat (peas)	

effect of plants on increasing or maintaining soil fertility, water, and soil quality, erosion control, pests, diseases, and weeds, the amount and time of access to water and other inputs, labor, the amount of available agricultural machinery and finally economic efficiency should be considered. For example, when there is alfalfa in the sugar beet rotation program and also the rotation increases from 3 to 6 years, sugar beet production is greatly improved; so that, nematode damage to sugar beet is greatly reduced in 4-to-6-year rotations. Land infected with rhizomania should not be allocated to sugar beet cultivation for up to 15 years (Wilson 2001).

The type of crops that are considered in rotation and also the position of sugar beet in rotation are very notable. Crops have different effects on sugar beet; so, it is important to pay attention to some points in selection of different rotation (Table 23.1).

23.2.3 Planting Pattern

One of the most important factors for suitable plant density and proper use of environmental conditions is the sugar beet planting pattern, which preserves more water resources, increases water use efficiency, and improves productivity in furrow irrigation conditions. Proper cultivation, medium condition, and irrigation management can effectively control soil salinity at critical growth stages. Single-row cultivation is used in most beet-growing areas that do not face salinity constraints.

Fig. 23.1 Sugar beet planting in top (left) and slope of row (right) in saline condition

Salt accumulates at the top of the ridges in the moisture pattern of single-row cultivation. Sowing of seeds in the center of the stack places the seeds exactly where the salt accumulates, so to prevent this problem, methods such as cultivation in the bottom of the furrow (Rhoades et al. 1992) or in the slope of the row (Minhas and Gupta 1993) and two-row planting are suggested because of the facilitated irrigation and lack of salt accumulation at the seedling emergence zone (Jahadakbar 2009) (Fig. 23.1).

23.2.4 Plant Density

Crop yield is the result of internal and external competition for the potential of the environment. Crop management can influence this competition. Maximum performance is achieved when these competitions are minimized. External competition is reduced to a minimum by timely control of weeds and intra-species competition by the uniform distribution of plants at the field level and uniform coverage, competition for light, and the use of water, air, and nutrients are reduced with proper density and uniform distribution of plants at the field level and plants benefit more from light and nutrients.

Choice of the right density for plants using their root crops is very important. The distribution of sugar beet plants on the field is one of the most significant factors affecting the quality of stored roots.

One of the most serious problems of sugar beet cultivation in Iran is low plant density and non-uniformity of field cover. Low plant density reduces the productivity of inputs and the quantitative and qualitative yields of the product. According to research, the quantity and quality of sugar beetroot are directly affected by plant density. So that if the density is higher than usual and recommended, root weight will be low, shoot to root ratio will be high, and root yield will reduce due to intrinsic competition, also below normal and recommended density increases single root weight and impurities and decreases sugar content and extraction coefficient.

Production costs—including the cost of thinning and weeding—are reduced by creating the right density by observing the correct principles of various agronomic and managerial factors, in addition to increasing crop yield per area.

Various agronomic and management factors that affect plant density, establishment, and vegetation at the field level include the type and quality of seeds used, how to prepare the planting bed, type of seed machine, use of chemical fertilizers before planting, soil salinity, temperature, and amount of soil moisture at the time of seed sowing, pest and disease control are the first season and planting arrangement. Maintenance of a proper plant density in sugar beet cultivation has particular importance not only in terms of increasing yield but also in terms of limiting space for weed growth. Finally, creating a suitable density and establishment of sugar beet plant on the field level reduce the competition of weeds with this plant.

The field cover is completed sooner at suitable plant densities and hence the productivity and efficiency of using solar radiation per unit area increase, consequently the quantitative and qualitative root yield increases. According to research studies, the appropriate density in sugar beet cultivation is about 8–11 plants/m^2 (80,000–110,000 plants/ha) at the time of final harvest. Plant density less than this number causes heterogeneous root growth and decreases sugar extraction coefficient and loss of consumption inputs, and higher density increases intra-species competition (between sugar beet plants), and production of smaller roots, shoot to root ratio and finally decreases yield.

23.2.5 Fertilization

One of the most important methods for increasing crop yield is the use of inputs mainly fertilizers. Climate, soil, and management conditions would be affected by the efficiency of fertilizer, especially in saline lands. It has been seen that the use of chemical manures has different effects on crops in salinity conditions. Salinity concentration determines whether fertilizer is useful for plants or not (Maas and Grattan 1999). In low salinity, fertilizer involvement has positive effects while in medium or high salinity it is not completely clear (Hu et al. 1997; Grattan and Grieve 1999). Salinity accumulation around the plant root decreases element uptake based on the type of elements and composition of the soil solution (Maas and Grattan 1999) because of the competition between ions in salt-affected soils (Esmaili et al. 2008). Na^+ in saline soils causes K^+ deficiency in crops under salinity stress which limits plant growth and affects soil physical properties (Maas and Grattan 1999). There are different studies that showed that the application of gypsum would be useful in alkaline soils. Salinity increases soil osmotic pressure, so water absorption and growth of plants would be reduced because of ion toxicity and imbalance in the soil. The scattered salts in the root zone make the plants to absorb fertilizers a bit difficult and this competitive cause imbalances in the nutritional level of the plant. Thus, making plant ability to absorb nutrients at a lesser rate (Salama et al. 2019).

Nitrogen is one of the main nutrition for plants which should be used in the form of fertilizer because nitrate and ammonium forms of nitrogen are rare in the soils

(Khademi et al. 2001). Soil nitrogen usability depends on different factors such as environment, temperature, humidity, crop rotation, soil drainage, pH, and other soil chemical and physical characteristics (Al-Kaisi 2001). Nowadays, planting systems with high nitrogen use efficiency are considered to have better nitrogen management and decrease nitrogen leaching (Mele 2017). Application of nitrogen fertilizers based on soil nitrogen decreases the use of this manure. Firstly, plant requirements for nitrogen to obtain potential yields should be considered, secondly, the amount of soil nitrogen should be noticed for proper fertilizer recommendation (Isfan et al. 1991), so nitrogen fertilizer for sugar beet crop should be used based on soil analysis and determining of soil nitrate in 0–30 cm depth (Noshad 2010). Overuse of nitrogen fertilizer in sugar beet increases root impurity and decreases white sugar content and root quality (Hoffmann 2010). There are different reports on the amount of nitrogen fertilizer in the saline soil but most of the reports recommend 150–250 kg/ha nitrogen from urea or ammonium nitrate sources (Jahadakbar 2005; Shahabi Far 2009; Shahabi 2010; Salama et al. 2019). It is also suggested to use 25% nitrogen in saline soil more than the usual recommendation in normal conditions (Jahadakbar 2005) but recommendation based on soil analysis is the main strategy for using fertilizer in all conditions.

Potassium is one of the most important cations in plants especially sugar beet because of its high absorption and important role in physiological regulation and root quality (Cooke and Scott 2012). Also, it has an important role in the activation of enzymes especially photosynthesis and respiration. Osmotic pressure regulation of cells is one of the most important roles of potassium under environmental conditions. Protein and starch synthesis and stomata guard cell opening and closing are affected by this ion (Hopkins and Huner 2008) as a result it is very important in conserving water in the plant.

Balanced and effective fertilization of potassium in combination with other nutrients such as nitrogen is not only effective in the growth, yield, and sustainable quality of the crop, but also is very effective in plant health and reducing environmental risks (Wang et al. 2013). Sugar translocation and storage in plant roots are affected by potassium, as a result, using potassium fertilizer would be necessary for suitable sugar yield (Cooke and Scott 2012). Potassium fertilizer resources and application methods have different effects on sugar beet yields, for example, it is reported that KCl application was better than K_2SO_4 (Ghani Shayeste et al. 2003). Potassium fertilizer increases root, sugar, and white sugar yields of sugar beet under salinity condition, and it prefers the high amount of potassium about twice as normal condition (Jahadakbar 2005) or about 100 kg/ha (Salama et al. 2019) based on soil analysis.

Solubility of microelements especially Cu, Fe, Mn, Zn is little in saline or sodic soils which causes plants to show symptoms of lack of these elements, application of micronutrition fertilizers such as Zn (Gadallah and Ramadan 1997), Fe (Elfouly et al. 2001) and others were useful in saline land but the efficiency is related to different parameters including plant species, application method. For example, it is reported that the effect of micronutrition fertilizer was better in salt-tolerant crops such as barley than corn or rye (Achmadi et al. 2001). Also the effect of soil application with one-time foliar application of Cu, Fe, Mn, Zn nutrition with an amount of 25% more than normal condition was suggested in saline lands (Jahadakbar 2005).

Chemical fertilizers increase environmental pollution also their cost is too much despite their positive effects on crop yields. Nowadays, the use of organic matter is preferred for sustainable agriculture especially in lands under stress. It is clear that the application of organic matters, biological fertilizers, growth regulators, or stimulators is not enough for suitable crop yield but they decrease the utilization of chemical compounds (Asadi Rahmani et al. 2010). Industrial agriculture, overuse and misuses of chemical compounds decrease soil organic matter and have negative effects on human and animal health as well as in food safety and quality (Amer et al. 2019). So organic matters such as compost and manure are suitable sources of nutrients because of their role in improving soil quality including soil porosity, aggregation, structure, bulk density, water-holding capacity, pH, EC, CEC, ESP, and nutrients (Amer and El-Ramady 2015).

Use of compost affects soil salinity positively based on removing sodium from the root and improving soil physical properties of planting medium (Day et al. 2019). Compost is full of plant macronutrients including N, P, K, Ca, Mg, and S and important microelements (Madeleine et al. 2005). Application of organic matter decreases Na + from the root zone, the ESP, EC and increases water infiltration, water-holding capacity, and aggregate stability (Tejada et al. 2006; Mahdy 2011). It is concluded that the use of compost with mineral nitrogen had the best effect on sugar beet under salinity conditions (Amer et al. 2019).

There are other organic matters which are used recently in agriculture such as biochar (charcoal) which is the solid co-product of pyrolysis usually used as a soil amendment in agriculture. It improves soil pH and nutrition and also increases plant tolerance to biotic and abiotic stresses (Boubakri 2020). It enhances root nitrogen assimilation (Farhangi-Abriz and Torabian 2018), the activity of antioxidant enzymes to remove excess reactive oxygen species (ROS) (Jia et al. 2019), and also photosynthesis and dry matter (Xu et al. 2015). It is reported that the application of biochar organic fertilizer improves growth and yield of sugar beet under salt stress based on improving the activities of nitrogen assimilation and antioxidant enzymes in the root, the synthesis of photosynthetic pigments, PS II (Photosystem II) activity, stomatal opening, and photosynthesis of sugar beet under saline-alkaline conditions (Zhang et al. 2020).

23.3 Future Prospect

Sugar beet is one of the main crops for producing sugar in the world but products such as molasses, ethanol, citric acid, and betaine will be extracted from this crop which would be useful for human, poultry fisheries, food, feed, health, and industry. Environmental stresses such as salinity increase products like betaine which would be useful even for human cancer. So, salinity stress would be excellent opportunity to increase potential efficiency of this crop.

Food security for human beings and maintenance of biodiversity are the main and essential factors in all part of the world. It is not possible to increase yield by increasing agricultural land so it is essential to maximize utility and efficiency of each land. Novel methods in farm management as in breeding approaches are improving daily. Knowledge of applied agronomic factors which improve salinity would still be essential for future because management is more important than breeding in salinity stress.

There are many small farmers especially in arid and semi-arid regions of developing countries who don't know main principle of proper agriculture, their crop yield is minimum and they have to immigrate to big cities. Education and extension of new researches to these farmers are necessary especially to deal with stresses such as drought, salinity, or other biotic and abiotic stresses.

23.4 Conclusion

Salt stress which is caused by saline land or man-made activities is an important stress in the world. It's efficiency on crop yield would be affected based on different factors such as region, crop, management strategies, etc. Sugar beet as an important industrial crop for producing sugar and other by-products is known as salt-tolerant crop in which threshold is about 16–20 dS/m and its most sensitive stage to salinity is established which would affect final yield of crop significantly. There are different methods to improve the normal growth of sugar beet under salinity. The existence of sufficient good quality water for leaching and drainage systems also suitable soil preparation, for example, plowing in 45–50 cm depth is essential for successful and appropriate management of saline soils. Application of proper rotation would also decrease the effects of saline and bad environmental conditions. Crops such as wheat, rye, barley, and corn can be used in rotation with sugar beet but management of their residue should be considered. Cultivation method, for example, planting in the bottom of the furrow or in the slope of the row and two-row planting is suggested because of the facilitated irrigation and lack of salt accumulation at the seedling emergence zone. Sugar beet plant density should be kept in 80–110 (thousand plants in ha) especially in saline land and small farm to have proper yield.

Climate, soil, and management conditions would be affected on the efficiency of fertilizer especially in saline lands. In low salinity, adding fertilizer has positive effects while in medium or high salinity it is not completely clear. There are different studies showed that application of gypsum would be useful in alkaline soils. Nitrogen fertilizer for sugar beet crop should be used based on soil analysis and determining of soil nitrate in 0–30 cm depth. There are different reports on amount of different fertilizer in the saline soil but most of reports recommend 150–250 kg/ha nitrogen from urea or ammonium nitrate sources, 100 kg/ha potassium from KCL, 100 kg/ha P2O5, also foliar application of Cu, Fe Mn, Zn nutrition with amount of 25% more than normal condition is recommended. Use of organic matter such as biological fertilizers, growth regulator or stimulators is not enough for suitable crop yield but they decrease the utilization of chemical compounds. Application of compost with mineral nitrogen had the best effect on sugar beet under salinity condition. Also, other organic matters which are used recently in agriculture such as biochar (charcoal) improve soil pH and nutrition and also increase plant tolerance to biotic and abiotic stresses. It is reported that application of biochar organic fertilizer improves growth and yield of sugar beet under salt stress. Finally, it is important to pay attention than in salinity condition, environmental management is very essential and even more important than breeding strategies for deal with this stress.

Acknowledgments The writer thanks Dr. Hamid Noshad for his technical comments and MS Somaye Rahimi for her English editing on manuscript. These researches were supported by Sugar Beet Seed Research Institute, Agricultural Research, Education and Extension organization, Karaj, Iran.

References

Abrol IP, Yadav JSP, Massoud FI (1988) Crops in saline soils. In: Salt-Affected soils and their management. FAO Soils Bulletin 39. http://www.fao.org/docrep/x5871e/x5871e00.htm#Contents

Achmadi J, Yamad M, Yamada S, Fujiyama H (2001) Salt tolerance of grain crops in relation to ionic balance and ability to absorb microelements. Soil Sci Plant Nutr 47(4):657–664

Al-Kaisi M (2001) Value of crop rotation in nitrogen management, vol 6. Iowa State University, Department of Agronomy IC-486, p 49. https://www.pan-europe.info/old/Campaigns/documents/Agriculture/PAN%20E%20Advantages%20of%20crop%20rotation-1.doc

Al-Taisan WA (2010) Comparative effects of drought and salt stresses on germination and seedling growth of *Pennisetum divisum* (Gmel.) Henr. Am J Appl Sci 7(5):640–646

Amer MM, El-Ramady HR (2015) Alleviation soil salinity and sodicity hazard using some biochemical amendments for production of canola (*Brassica napus* L.) in North Delta region. J Soil Sci Agric Eng Mansoura Univ 6(4):427–444

Amer M, Aiad M, Rashed S, El Ramady H (2019) Sustainable irrigation and fertilization management of successive cultivated sugar beet and cotton under salt-affected soil conditions. Environ Biodiv Soil Secur 3:227–239

Asadi Rahmani H, Khavazi K, Asgharzadeh A, Rejali F, Afshari M (2010) Biofertilizer in Iran: opportunities and problems. First congress of fertilizer problems in Iran: half a century fertilizer application. Sana Publication, pp. 215–228 (In Persianwith English abstract)

Boubakri H (2020) Induced resistance to biotic stress in plants by natural compounds: possible mechanisms. In: Hossain MA, Liu F, Burritt DJ, Fujita M, Huang B (eds) Priming-mediated stress and cross-stress tolerance in crop plants. Elsevier, Amsterdam, pp 79–99

Cherlet M, Hutchinson C, Reynolds J, Hill J, Sommer S, Von Maltitz Q (eds) (2018) World atlas of desertification, 3rd edn. Publication office at European Union Luxembourg, Luxembourg, Catalog Number, KJ-07-17-008-EN-C

Cooke DA, Scott RK (2012) The sugar beet crop. Chapman & Hall, London, p 649

Day SI, Norton JB, Strom CF, Kelleners TJ, Aboukila EF (2019) Gypsum, langbeinite, sulfur, and compost for reclamation of drastically disturbed calcareous saline–sodic soils. Int J Environ Sci Technol 16:295–304. https://doi.org/10.1007/s13762-018-1671-5

Delfine S, Alvino A, ConcettaVilani M, Loreto F (1999) Restriction to carbon dioxide conductance and photosynthesis in spinach leave recovering from salt stress. Plant Physiol 119:1101–1106

Doorenbaos J, Kassam AN (1979) Yield response to water, FAO Irrigation and Drainage. Paper 33, Food and Agriculture Organization of the United Nations, Rome, p 193. ISBN 92-5-100744-6
Duan D, Liu X, Ajmal Khan M, Gul B (2004) Effect of salt and water stress on the germination of *Chenopodium glaucum* L. seed. Pak J Bot 36(4):793–800
Durrant MJ, Draycott AP, Payne PA (1974) Some effect of sodium chloride on germination and seedling growth of sugar beet. Ann Bot 38:1045–1051
Ebrahimian HR, Ranji Z (2004a) Preparing tolerant variety to salt stress with diurnal half sib selection. Final report. Sugar beet seed institute agricultural research, education and extension organization, Karaj, Iran. (In Persian with English abstract)
Ebrahimian HR, Ranji Z (2004b) Comparison of salt tolerant variety 7233–29 * MST with current variety sown in Esfahan. Final report. Sugar Beet Seed Institute Agricultural Research, Education and Extension organization, Karaj, Iran. (In Persian with English abstract)
Elfouly M, Zeniab M, Zeinab AS (2001) Micronutrient spray as a tool to increase tolerance of fababeaen and wheat plants to salinity. XIV International plant nutrition colloquium. Hanver, Germany, pp. 422–423
Esmaili E, Kapourchal SA, Malakouti MJ, Homaee M (2008) Interactive effect of salinity and two nitrogen fertilizers on growth and composition of sorghum. Plant Soil Envrion 54:537–546
FAO AGL (2000) Land and plant nutrition management service: global network on integrated soil management for sustainable use of salt affected soils. http://www.fao.org/ag/agl/agll/spush
Farhangi-Abriz S, Torabian S (2018) Biochar improved nodulation and nitrogen metabolism of soybean under salt stress. Symbiosis 74:215–223
Gadallah MA, Ramadan T (1997) Effects of zinc and salinity on growth and anatomical structure of *Carthamis tinctorius* L. Biol Plant 39:411–418
Ghani Shayeste F, Taibenejad H, Malakouti MJ (2003) Effect of potassium sources and amounts on potassium absorb and sugar beet quality and quantity. The 8th Iranian Soil Science Congress. Sept, Rasht Gilan, pp. 310–312. (In Persian, abstract in English)
Ghoulam C, Fares K (2001) Effect of salinity on seed germination and early seedling growth of sugar beet (*Beta vulgaris* L.). Seed Sci Technol 29:357–364
Grattan S, Grieve C (1999) Mineral nutrient acquisition and response by plants grown in saline environments. Handbook Plant Crop Stress 2:203–229
Hoffmann CM (2010) Root quality of sugar beet. Sugar Technol 12:276–287
Hopkins W, Huner N (2008) Introduction to plant physiology. John Wiley, Hoboken, p 528
Hu Y, Oertli J, Schmidhalter U (1997) Interactive effects of salinity and macronutrient level on wheat. I Growth. J Plant Nutr 20:1155–1167
Isfan D, Cserni A, Tabi M (1991) Genetic variation of the physiological efficiency index of nitrogen in triticale. J Plant Nutr 14:1381–1396
Jafarzadeh AA, Aliasgharzad N (2007) Salinity and salt composition effects on seed germination and root length of four sugar beet cultivars. Biologia Bratislava 62(5):562–564
Jahadakbar MR (2005) Effect of water salinity on N&K efficiency in sugar beet cultivation. Isfahan Agriculture and Natural Resources, Research and Education Center, Agricultural Research, Education and Extension Organization, Iran. (In Persian with English Abstract)
Jahadakbar MR (2009) Sugar beet cultivation in saline lands. Extension BrochureIsfahan Agricultural and Natural Resources, Research and Education Center, Agricultural Research, Education and Extension Organization, Iran. (In Persian)
Jamil M, Lee DB, Jung KY, Ashraf M, Lee SC, Rha ES (2006) Effects of salt (NaCL) stress on germination and early seedling growth of four vegetables species. J Cent Eur Agric 7(2): 273–282
Jia W, Ma C, White JC, Yin M, Cao H, Wang J, Wang C, Sun H, Xing B (2019) Effects of biochar on 2,20, 4, 40, 5, 50 -hexabrominated diphenyl ether (BDE–153) fate in *Amaranthus mangostanus* L.: accumulation, metabolite formation, and physiological response. Sci Total Environ 651:1154–1165
Karimi A (1997) Desalinization of heavy texture soils in Tabriz flood plain by plowing at various depth. Iran J Agric Sci 28(3):27–48

Kehl M (2006) Saline soils of Iran with example from the alluvial plain of Korbal, Zagros Mountains. Proceeding of the international conference. Soil and Desertification- Integrated research for the sustainable management of soils in dry lands. Germant. www.desertnet.de/proceedings/start.htm

Khademi Z, Mohajermilani P, Balali MR, Dorodi MS, Shahbazi K, Malakouti MJ (2001) A comprehensive computer model for fertilizer recommendation towards sustainable agriculture. Soil and Water Research Institute, Tehran, Iran. (In Persian, abstract in English)

Khayamim S (2013) Sugar beet crop rotation. In: Taleghani D, Alimoradi I et al (eds) Development of standards for determining potential and assessing damage by managerial and coercive factors at different stages of growth in sugar beet fields. Sugar Beet Seed Institute Agricultural Research, Education and Extension organization, Karaj, Iran, pp 37–42

Khayamim S, Tavakol Afshari R, Sadeghian Motahar SY, Pustini K (2011) Study of sugar beet germination under different salt stresses in laboratory and green house conditions. Iranian. J Crop Sci 13(1):1–17. (in Persian, abstract in English)

Khayamim S, TavakolAfshari R, SadeghianMotahar SY, Pustini K, Rouzbeh F, Abbasi Z (2014) Seed germination, plant establishment and yield in sugar beet genotypes under salinity stress. JAST 16(2):779–790

Khorsandi FH, Nezhad Y (2019) Guide for agriculture in saline condition. Agricultural Research Education and Extension Organization of Iran, Iran, p 323. (In Persian)

Khosgoftarmanesh AH, Shariatmadarih H (2002) Reclamation of bare saline soils by leaching and barley production. Paper presented at the 17th world congress of soil science (symposium no. 33; paper no. 1828). 14–21 Aug 2002, Bangkok, Thailand

Maas EV, Grattan S (1999) Crop yields as affected by salinity. Agronomy 38:55–110

Madeleine I, Peter S, Tim T, Tom V (2005) Agrodok no. 8: the preparation and use of compost. Agromisa Foundation, Wageningen

Mahdy AM (2011) Comparative effects of different soil amendments on amelioration of saline-sodic soils. Soil Water Res 6(4):205–216

Mele A (2017) Assessing nitrogen use efficiency in sugar beets grown under variable fertilizer and drip irrigation rates. Managing global resources for a secure future. 22–25 Oct, 2017. Tampa FL

Mesbah M, Yavari N, Motahari AR, Arjmand N, Alimoradi I (1991) Results of salt tolerance assessment in different sugar beet lines. Final report. Sugar Beet Seed Institute (SBSI). (In Persian with English abstract)

Minhas PS, Gupta RK (1993) Conjunctive use of saline and non-saline water −1. Response of wheat to initial salinity profile and modes of salinization. Agric Water Manag 23:139–148

Mohammadian R, Ghasemi Golozani K, Moghaddam M, Sadeghian SY (1995) Effects of seed exhausted in germination, seedling and establishment of seven sugar beet lines under salt stress. M.Sc thesis. Tabriz University. (In Persian with English abstract)

Morillo-Velarde R, Ober ES (2006) Water use and irrigation. In: Draycott P (ed) Sugar beet. Blackwell Publishing, Hoboken, pp 50–255

Nazar R, Iqbal N, Masood A, Syeed S, Khan NA (2011) Understanding the significance of sulfur in improving salinity tolerance in plants. Environ Exp Bot 70(2–3):80–87

Niazi BH, Athar M, Rozema J (2004) Salt tolerance in the fodder beet and sea beet: analysis of biochemical relations. Bulg J Plant Physiol 30:78–88

Noshad H (2010) Study and improving nitrogen use efficiency in sugar beet using soil sampling position and soil nitrate and ammonium test. Sugar beet seed institute, agricultural research, education and extension organization. (In Persian with English Abstract)

Okcu G, Kaya MD, Atak M (2005) Effects of salt and drought stresses on germination and seedling growth of pea (*Pisum sativum* L.). Turk J Agric For 29:237–242

Rajabi A, Khayamim S, Abbasi Z, Ober E (2014) Salt stress and sugar beet improvement: challenges and opportunities. In: Ahmad P, Wani MR, Azooz MM, Phan LS (eds) Tran improvement of crops in the era of climatic changes. Springer, London, pp 21–150

Rhoades JD, Kandiah A, Mashali AM (1992) The use of saline water for crop production, Irrigation and Drainage Paper 48. FAO, Rome

Sadat Noori SA, McNeilly T (2000) Assessment of variability in salt tolerance based on seedling growth in Triticum durum Desf Genet Resource. Crop Evol 47:285–291

Salama MA, Al-Azab KF, Hashim ME (2019) Response of sugar beet yield and production to different rates from chemical fertilizers and soil salinity. Sylwan 163(5):36–58

Shahabi AA (2010) Study on the effect of different sources and rates of nitrogen on yield and quality of sugar beet in saline conditions. Isfahan Agriculture and Natural Resources, Research and Education Center, Agricultural Research, Education and Extension Organization, Esfahan, Iran. (In Persian with English Abstract)

Shahabi Far J (2009) Study on effects of nitrogen sources and levels on yield and quality of sugar beet in Qazvin province under salinity situations. Qazvin Agriculture and Natural Resources, Research and Education Center, Agricultural Research, Education and Extension Organization, Qazvin, Iran. (In Persian with English Abstract)

Singh RK, Redona E, Refuerzo L (2010) Varietal improvement for abiotic stress tolerance in crop plants: special reference to salinity in rice. In: Pareek A, Sopory SK, Bohnert HJ, Govindje (eds) Abiotic stress adaptations in plants: physiological, molecular and genomic foundation. Springer Science+Business Media, Berlin, pp 385–415

Tejada M, Garcia C, Gonzalez JL, Hernandez TL (2006) Use of organic amendment as a strategy for saline soil remediation: influence on the physical, chemical and biological properties of soil. Soil Biol Biochem 38(6):1413–1421

Wang M, Zheng Q, Shen Q, Guo S (2013) The critical role of potassium in plant stress response. Int J Mol Sci 14(4):7370–7390

Wilson RG (2001) Crop rotation. In: Wilson RG, Smith JA, SD Miller (2001) Sugar beet production guide. NU Cooperative Extension. EC 01-156. University of Nebraska: Nebraska

Xu CY, Hosseini-Bai S, Hao Y, Rachaputi RCN, Wang H, Xu Z, Wallace H (2015) Effect of biochar amendment on yield and photosynthesis of peanut on two types of soils. Environ Sci Pollut Res 22:6112–6125

Zhang P, Yang F, Zhang H, Liu L, Liu X, Chen J, Wang X, Wang Y, Li C (2020) Beneficial effects of biochar-based organic fertilizer on nitrogen assimilation, antioxidant capacities, and photosynthesis of sugar beet (*Beta vulgaris* L.) under saline-alkaline stress. Agronomy 10(1562). https://doi.org/10.3390/agronomy10101562

Mechanization in Sugar Beet Cultivation 24

Koç Mehmet Tuğrul

Abstract

Mechanization in agricultural production has an important effect on both product quality and yield. It is an important input to agriculture in terms of the timely execution of farm operations. Mechanization has an important role in processes such as timely and low-cost realization of agricultural practices, increasing the productivity of high-cost inputs and soil fertility, improving product quality, and completing time-consuming agricultural works appropriately. It is normal for regional conditions and technological developments to lead to differentiation of mechanization practices. Criteria such as regionally applied mode of production, power source, land size, marketing conditions, purchasing capacity of the farmer are important factors in the selection of mechanization tools for an efficient and economical agriculture. Sugar beet mechanization includes agricultural operations such as leveling, tillage, sowing, fertilizing, hoeing, irrigation, weeding, spraying, harvesting, cleaning, and loading. Timely and appropriate mechanization applications are important for high efficiency, low loss, and high product quality. In order to be useful to all who are interested in sugar beet, the main mechanization steps used in sugar beet farming and some important properties are summarized in this chapter.

Keywords

Agriculture · Sugar beet · Mechanization · Automation · Precision agriculture

K. M. Tuğrul (✉)
Department of Biosystems Engineering, Faculty of Agriculture, Eskişehir Osmangazi University, Eskişehir, Turkey

V. Misra et al. (eds.), *Sugar Beet Cultivation, Management and Processing*,
https://doi.org/10.1007/978-981-19-2730-0_24

24.1 Introduction

Labor costs and low space efficiency increase the need for mechanization and automation in agricultural areas. Due to advanced production techniques, mechanization has become a necessity rather than a requirement for efficient and quality production. Today, engines, hydraulic-pneumatic systems, tractors, tillage, sowing-planting, spraying, fertilizing, irrigation and harvesting machines are the most common types of machines in agricultural areas. Along with technological developments, computer and auto control systems, drone disease monitoring, and product prediction systems are becoming widespread in agricultural production. Machine is a tool that does the defined work by changing the direction and magnitude of the force it receives from the power source. Agricultural machinery refers to all machines and tools that are used in plantation, cultivation, care, and harvesting of agricultural products. Agricultural mechanization is processed based on increasing labor productivity with tools and machines in agricultural operations from field preparation to harvest. There are two main power sources in this process, human and mechanical. In addition, the manufacture, maintenance-repair, management and operation of agricultural tools and machines are important issues for the efficient and effective use of machines. In this section, the mechanization stages used in sugar beet agriculture and the agricultural characteristics of the machines classified below are explained.

1. Soil cultivation machines
2. Sowing, planting machines
3. Hoeing machines
4. Fertilizing machines
5. Harvesting machines
6. Cleaning and loading machines

24.2 Soil Cultivation Machines

Tillage is the process of arranging the land according to the conditions of the product to be grown by using agricultural machinery. Tillage covers the mechanization processes carried out for purposes such as soil tillage, seed bed preparation, weed control, creating a suitable environment for plant growth, reducing erosion, diseases, and pests. Soil tillage practices are generally grouped under four headings:

- To turn upside down of soil (plough),
- To loosen the soil without upside down (harrows and cultivators),
- To till the soil by mixing (rotary cultivators),
- To press on the surface for leveling and compression (land rollers).

Untimely, incorrect and excessive tillage can cause deterioration in soil properties, loss of time and energy. In sugar beet cultivation, the effect of pre-sowing processes on production can reach as high as 70% (Bee et al. 2004).

Some of the most suitable tools and machines for tillage applications are as follows:

- Heavy disc implements
- Wide-board ploughs
- Rotovators, interrow cultivators
- Chisel plough
- Roller

24.2.1 Some Application Guidelines for Tillage Practices

The optimum benefit can be obtained if certain rules are followed in tillage practices. Some of those are as follows:

- Minimize tillage. The more crop residues such as grain and stalks can be reduced, the faster and easier the ideal soil structure can be created. If soil conditions and machines are suitable, it would be better to consider direct planting. In this case, a little more manpower may be needed for weed control.
- Instead of plows and rotavator that overturn and disintegrate the soil, tools such as cultivators or chisels should be preferred.
- The main factor that determines the timing of tillage is soil weathering.
- Unless necessary, tillage deeper than 8–10 cm should be avoided.
- If deep tillage is required, subsoiling, which increases water infiltration and encourages early root growth, should be preferred.

24.3 Soil Tillage Systems

Conventional and conservation tillage methods constitute two main groups.

24.3.1 Conventional Tillage

The plow is the main tool of the conventional (traditional) tillage system and is tilled upside down at a depth of 25–30 cm. By causing intensive and excessive tillage, conventional methods increase soil compaction and erosion. Then seed bed preparation is made with secondary tillage machines. The following practices are mostly used in the traditional tillage system:

- Tillage with moldboard plow
- Crushing with disc plow (1 or 2 times)

- Harrow or cultivator processing (1 or 2 times)
- Planting and fertilization
- Hoeing with a cultivator or interrow cultivator (1 or 2 times)
- Irrigation
- Pulverizing and plant protection applications
- Harvest

Compared to conservation tillage, the traditional method requires higher inputs in machinery investment, maintenance, repair, and labor. In this system, the excessive number of cultivation causes erosion, deterioration of soil structure, and compaction. Studies on the water and energy savings required for sustainable agriculture are not common in traditional tillage. Stubble and plant residues are burned, removed from the soil, or mixed with the soil. In other words, the land remains bare until the next planting period (Fig. 24.1).

24.3.2 Conservational Soil Tillage

Conservation tillage is a system of weed control and seedbed preparation with minimal field traffic. In this system, chisel that makes subsoiling at a certain depth is preferred instead of plow. Chisel breaks down the hard layer that prevents plant growth, and product residues remain on the field surface. The positive effects of conservational system on erosion control have been revealed. As in conventional tillage, basic tillage, seedbed preparation and planting can be applied separately or together (Fig. 24.2).

Conservation tillage provides significant savings in labor, energy consumption, and timeliness. This method has many advantages over traditional tillage. The total power requirement, fuel consumption, operating time, and energy consumption required in the system are significantly reduced. In the soils where this system is applied, the physical structure becomes more stable over time, the carbon-nitrogen balance is provided, and the conversion of plant residues into organic matter occurs faster.

24.3.3 Reduced Tillage

In reduced tillage, which is a subgroup of conservation tillage, the energy requirement is even lower. Chisel or disc tools are used in primary tillage, and disc tools, cultivators, or combined tools are used in secondary tillage and seed bed preparation (Fig. 24.3).

Fig. 24.1 Processing steps of the conventional tillage method

24.3.4 Mulch Tillage

In the mulch method, where the stubble and residues of the previous crop are left on the soil surface, the difference from the direct sowing method is the subsoiling. Soil preparation is done with chisel, disc tools, cultivators, or combinations of tools, and

Fig. 24.2 Processing steps of the conservational soil cultivation method

Fig. 24.3 Processing steps of the reduced tillage method

weed control is done with weeders, herbicides, and/or harrows. The number of applications is limited to ensuring that sufficient residue is left on the surface to provide erosion control (Fig. 24.4).

24.3.5 Strip Tillage

It is an alternative method to no-till agriculture, which is tilled only the area to be planted and leaving the stubble in the remaining area as it is. The strip width is 5–30 cm (Godwin 1990). The strips can be prepared before planting, or it is more suitable to apply fertilizer in one pass with the planting. In areas susceptible to erosion, significant energy savings are achieved compared to full-area tillage in the appropriate method for sustainable agriculture (Fig. 24.5).

As with all no-till methods, the high amount of weeds and pests creates a problem in the strip cultivation method. Nonetheless, it is easier for growers to remove crop residue by strip tillage than full-width tillage. In the method, the correct placement of the seeds on the tilled area depends on the correct matching of the harmony between the planter coulter and the strips.

Fig. 24.4 Mulching in sugar beet

Fig. 24.5 Strip tillage-applied sugar beet field

24.3.6 Direct Sowing (No Tillage, Zero Tillage)

In direct sowing or no-till farming methods, sowing is done without any tillage. In direct sowing, the seeds are sown directly into the line opened by especially designed sowing machines, covered and pressed in accordance with the precision sowing technique. The contact of the seed with the soil is of absolute importance for ideal planting.

There is no traditional plow or disc plowing, and even hoeing is not done during the development and maturation period of the plant (Fig. 24.6). The direct sowing

Fig. 24.6 Direct sowed sugar beet

Fig. 24.7 Some types of direct sowing machines used in sugar beet cultivation

method provides significant time and fuel savings (Köller 2003; Šarauskis et al. 2010); however, the final emergence rate is lower and seedling growth is slower after emergence (Richard et al. 1995; Tuğrul and Dursun 2003; Koch et al. 2009) (Fig. 24.7).

24.4 Sowing, Planting Machines

Sugar beet, like other plants that need large living space, is planted with precise sowing technique. Precision sowing to leave the seeds one by one between the rows, on the row, and at equal depth, and the machines that apply this are called precision sowing machines. It is generally divided into mechanical and pneumatic precision sowing machines according to the way the seed is taken from the seed box and left to the seed bed (Fig. 24.8). The features that the precision sowing machine should have can be listed as follows:

- The double filling and discharging amount of the seeding systems will be low,
- To ensure uniformity in sowing, seed drop height will be maximum 7 cm,
- The adjusted seed distance in a row should not change,

Fig. 24.8 Mechanical sugar beet precision sowing machine

- Sowing units will be able to follow the field surface independently,
- Seed distances should not change during sowing.

The main difference between mechanical and pneumatic precision sowing machines is the way that the seeds are held on the sowing system. In mechanical systems, the sowing wheel located just below the seed box carries the seeds to the seed bed through the holes on its outer surface. In mechanical sowing systems, the hole or slot dimensions and the seed dimensions must match. Sowing wheel, horizontal disc, inclined disc, and banded systems are sowing systems in mechanical precision seed drills (Fig. 24.9).

Pneumatic sowing systems have a sowing disc. The seeds are transported to the seed bed by being held on the holes with a vacuum of 40–50 mbar created by a fan driven by the PTO. While the holes on the sowing wheel differ depending on the seed calibration, the holes on the sowing disc vary depending on the crop type (Fig. 24.10).

There are various scrapers of mechanical and pneumatic (suction and pressure) type to carry out single seed delivery to the seed bed in sowing machines (Fig. 24.11).

Another important part to be considered in sugar beet precision sowing machines is press wheels. Based on the machine type, press wheels can be front-middle-back, front-back, or just back. The front press wheel creates a sowing line of appropriate density to ensure seed-soil contact by pressing the seed furrow. The middle-pressure wheel presses the seed in the furrow providing maximum contact with the soil. The back pressure wheel creates a suitable structure for germination by pressing on the seed covered with soil and creates a structure that will prevent the crusting of the soil. The pressure wheels, for which soil properties are an important factor in the selection, are manufactured from adjustable metal material with a double-sided

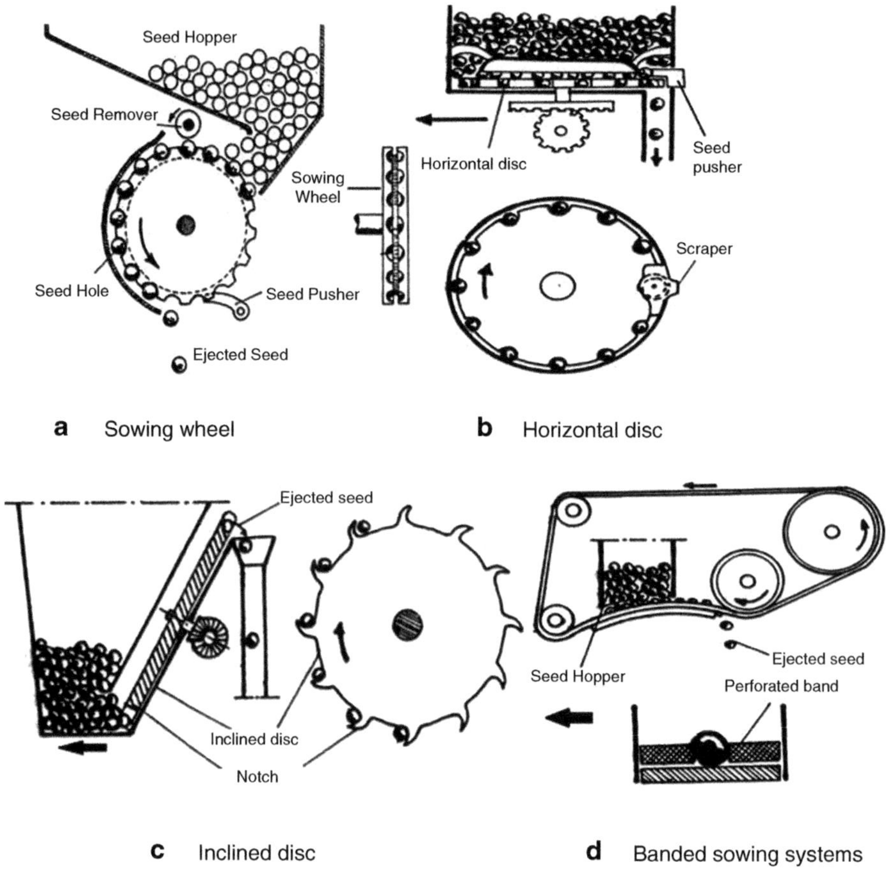

Fig. 24.9 Sowing systems used in mechanical precision sowing machines. (**a**) Sowing wheel, (**b**) Horizontal disc, (**c**) Inclined disc, (**d**) Banded sowing systems

conical structure, rubber filled and zero pressure rubber material, depending on the machine type (Fig. 24.12).

Precision sowing provides much better seed distributions and is commonly used for a wide range of crops such as beets. In precision sowing, the plant distribution on the row is disrupted in cases of poor seedbed (missing or double seeds) and non-germinating seed. The frequency distribution of intervals can usually be multiples of the average range. In such cases, the field emergence rates should be calculated correctly and the seed distance should be adjusted to ensure the ideal number of plants. For this, the sowing range, sowing depth, sowing speed settings of the machine should be checked and correct adjustments should be made. In addition, clod pushers, press wheels, and soil coverers settings and whether there are enough seeds in the drill hoppers should be checked.

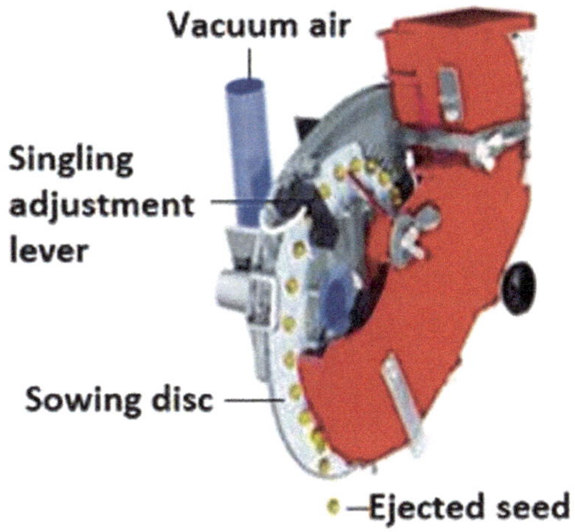

Fig. 24.10 Sowing systems used in pneumatic precision sowing machines

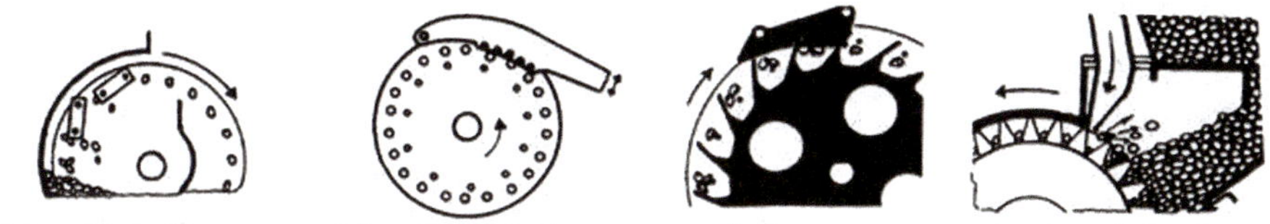

Fig. 24.11 Various types of scrapers using on the sugar beet precision sowing machines

24.5 Hoeing Machines

Hoeing is the process done at 2–6 cm depths for loosening the soil, weeding and mixing the fertilizer. A suitable environment is created for the growth of the plant by performing the throat filling process together with hoeing. Hoeing in the cultivation of the plants sowed with wide row spacing has found a common application area in agricultural operations as it allows to do weed control together with tilling.

Hoes are examined in two categories:

1. Hand hoes
2. Tractor hoes

Tractor hoes are called hoeing machines and are in mounted type. Hoeing machines are also classified according to their attachment to the tractor;

- those attached to the front of the tractor,
- those connected between the axles of the tractor, and
- those mounted on the back of the tractor by towed attachments.

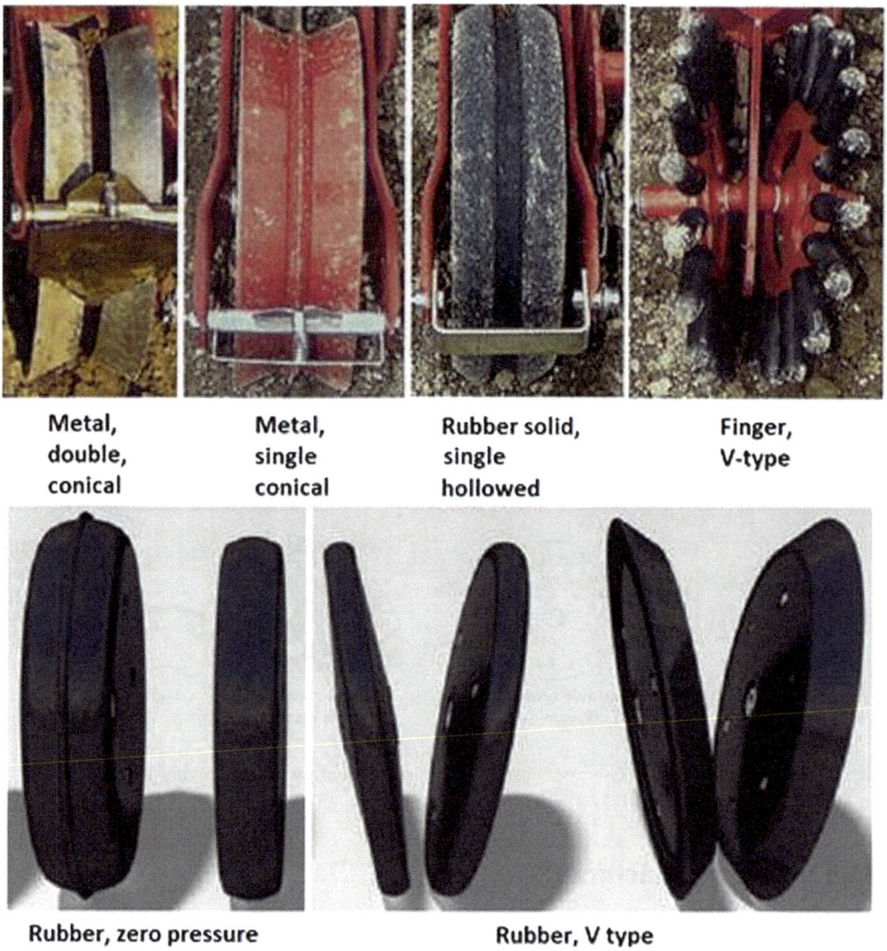

Fig. 24.12 Types of back press wheels used on the sugar beet precision sowing machines

Hoes mounted at the front of the tractor provide easy steering, but the tractor wheels trample the tilled area. Hoes connected between tractors axles are mostly used with tool carrier type tractors. Precise steering is not required. Rear-mounted hoes are the most used machine type. These are generally grouped in two ways depending on foot types:

1. Rigid tine hoes
2. Rotary hoes

The working part in rigid leg hoes is usually a different shaped cultivator attached to a universal chassis. The angle between the cutting edges of duck foot shares is 30–35° and the width is between 7 and 20 cm. Hoe legs can be rigid, semi-spring, or

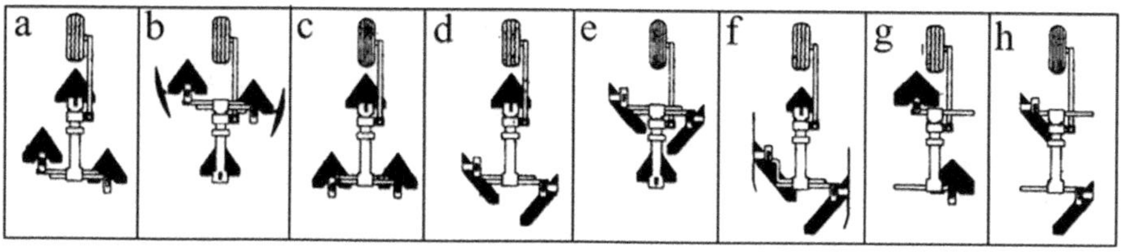

Fig. 24.13 The position of the cultivating legs according to different cultivation features

full spring. The legs are mounted parallel to a universal chassis, allowing independent movement of each unit and achieving a homogeneous hoeing depth.

24.5.1 Rigid Tine Hoes

The compatibility of the width between the rows and the track width of the tractor in hoeing machines is an important issue in preventing plant losses. For this, the units can be mounted symmetrically or asymmetrically. Weed density, plant height, spacing between rows, working speed, and crust condition in the soil are important factors in the selection of different foot combinations and the related hoeing efficiency.

The arrangement of the hoe legs in the form of cultivator primarily affects the entry of feet into the soil and weed. In angled blades, they cause only a very little soil around the plant to move (d, e, h). Plants can be hoed depending on the arrangement of the legs (a, c, g) even at different soil conditions and working speeds. Cultivator leg can be adjusted based on the high weed rates (b and e) and narrow rows (g and h). Areas with high grass density can be controlled by deep hoeing (Fig. 24.13).

24.5.2 Rotary Interrow Cultivator

It is an efficient machine with high success in full area and interrow cultivating. While interrow cultivator removes weeds by cultivating between rows, it also saves labor with fertilizer application. It can work in different widths between 45 and 80 cm row distances. The hoe width on each unit can be increased or decreased by turning the hoe blades inside or outside (Fig. 24.14).

In rotovators and interrow cultivators, the blades on the flange connected perpendicular to the rotor are arranged at different angles relative to the neighboring flanges, creating a helical appearance. This arrangement prevents impact operation when blades entering the soil. Blade types can be in horizontal position such as L, wedge, hoe type, or vertical structure (Fig. 24.15).

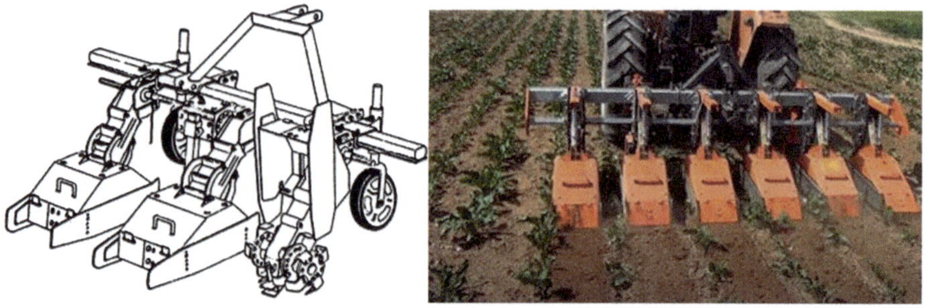

Fig. 24.14 Schematic and general views of a rotary inter row cultivator and its units

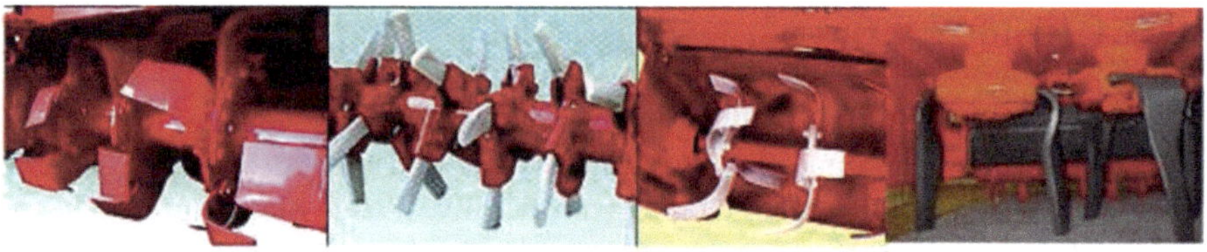

Fig. 24.15 Blade types used in rotovators and rotary interrow cultivators

24.6 Fertilizing Machines

Fertilizer is organic or inorganic substances that can be taken by the leaves or roots of the plant, containing the nutrients that the plant needs for its development. Fertilizers provide the nutrients needed by the plant or the soil, as well as improve the structure of the soil and increase development. Fertilizers are an important tool in increasing agricultural production. Fertilizers generally contain three basic nutrients, nitrogen, phosphorus, and potassium. In a profitable agricultural production, first of all, the nutrient deficiencies in the soil should be determined by soil analysis, and then the type and amount of fertilizer should be determined and applied in a timely manner with the most effective method. Farm manure is an important source for the protection and improvement of the soil with its plant nutrients, humus, and organic substances. Farm manure also has important benefits such as increasing the water holding capacity and aeration in the soil and supporting the formation of humus by improving its physical properties. Fertilizer applications can be divided into four parts as broadcasting, banding, fertigation, and foliar application.

Fig. 24.16 Fertilizer spreader

24.6.1 Broadcasting

A recommended rate of fertilizer is spread over the cultivation area and incorporated into the soil with a cultivator. Broadcasting is a method that is generally applied in large fields, when time or labor is limited (Fig. 24.16).

24.6.2 Banding

Fertilizers are applied to the furrows, which are at about 5–8 cm around the seeds or plants and are 3–5 cm deep from the seed furrows. It provides significant fertilizer savings compared to broadcasting (Fig. 24.17).

24.6.3 Fertigation

Nitrogen and potassium fertilizers are applied to plant production systems with irrigation water at regular intervals. Since phosphorus fertilizers contain insoluble compounds in their structures, they create a risk of clogging, so they are not suitable for application in this way (Fig. 24.18).

24.6.4 Foliar Application

The rapid absorption of nutrients from the leaves by the plant, especially the application of micronutrients together with the main elements such as N, P, and K, has become widespread. Applications can be made with various tools such as sprinkler and mobile irrigation systems and sprayers.

Fig. 24.17 Band fertilizing

24.7 Types of Fertilizing Machines

24.7.1 Organic Fertilizing Machines

24.7.1.1 Liquid Farm Fertilizer Distribution Machine

Liquid manure of farm animals is stored in especially prepared liquid manure wells. Liquid fertilizers taken from the wells are transported and applied in cylindrical shaped tanks (Fig. 24.19).

24.7.1.2 Solid Farm Fertilizer Distribution Machine

They are wheeled vehicles consisting of a conveyor and a distributor. Distribution systems can be located in horizontal or vertical structures, rear or side (Fig. 24.20).

Fig. 24.18 Fertigation

Fig. 24.19 Fertigation

24.7.2 Inorganic Fertilizing Machines: Solid (Mineral) Fertilizer Distribution Machine

24.7.2.1 Tank Type Fertilizer Spreader

In this group, there are systems with a single fertilizer hopper used in grain planting and a united fertilizer hopper in precision planting technique. The amount of fertilizer can be adjusted precisely (Fig. 24.21).

Fig. 24.20 Manure distributer

Fig. 24.21 Tank type fertilizer spreader

24.7.2.2 Centrifugal Fertilizer Spreader

It is used in two types as single and double disc in spread fertilizer application. Fertilizer distributors are divided into two as trailed and mounted according to their capacities (Fig. 24.22).

24.7.2.3 Liquid and Gas Fertilizer Spreader

In these systems, liquid fertilizer is pressurized with a pump and delivered to the desired area. It is possible to apply in combination with sowing machines or field sprayers (Fig. 24.23).

Fig. 24.22 Centrifugal fertilizer spreader

Fig. 24.23 Liquid fertilizing machine

Fig. 24.24 Beet harvesting machine

24.8 Harvesting Machines

Harvesting is the process of taking the mature plant from its growing environment and evaluating it. Conditions for effective harvesting often deteriorate during autumn and the risk of severe frost increases. Therefore, in most soils, beet harvesting should start in mid-September and end in early December. Sugar beet harvest consists of topping, lifting, cleaning, and loading stages, and sugar beet harvesters show structural differences depending on their features at these stages (Fig. 24.24).

24.9 Topping System

Making the beet top straight from the lowest green leaf level is considered a suitable topping. Topping directly determines harvest quality and losses. Insufficient topping causes the green part remaining on the head to continue to grow after harvest. Continuing the growth means consuming the sugar in the root. Conversely, deep heaping causes an economic loss as it causes weight and sugar losses (Fig. 24.25).

Fig. 24.25 Topping system

Fig. 24.26 Lifting system

24.10 Lifting System

It is the unit that removes the roots of the beet from the soil as a single piece without damaging it. Beets damaged during harvesting are exposed to infection and sugar loss increases (Fig. 24.26).

24.11 Cleaning System

It is the unit that cleans the soil that is stuck on the beet and carried together after lifting and leaves it to the field. Otherwise;

Fig. 24.27 Cleaning system

Fig. 24.28 Loading systems of beet harvesting machine

– Taking adherent soil together with beet increases soil loss.
– The soil and straw pieces attached to the beet create a good growth environment for microorganisms in the silo, and decay and sugar loss increase.
– The soil transported with the beet will increase the transportation costs per unit of beet (Fig. 24.27).

24.12 Loading System

It is the system of loading the topped, lifted, and cleaned beets into a vehicle to be sent to reception centers or the factory. Loading is done with the units of harvesting machines or beet cleaning machines (Fig. 24.28).

24.13 Suggestions for an Ideal Harvest

1. Not overtopping and no leaves: All leaves should be cut from the lowest petiole level. In a good harvest, 5% of the harvested beets are allowed to be cut above 2 cm and 5% to be cut deep. Beets with more leaves cause extra sugar loss during storage (Tijink 2010).

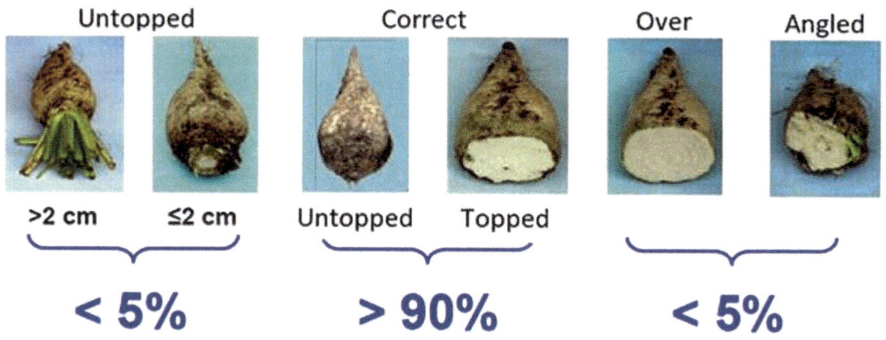

Fig. 24.29 Recommended cutting quality of sugar beet head

2. Good separation of beets and soil: The soil on the beet should be cleaned and there should be no breakage and/or damage to the roots. To achieve this, the lifting shares and the working depth should be adjusted carefully.
3. Beet-friendly cleaning: Direct root breakage losses and storage losses from cleaning should not be allowed, and very high rotation speeds of turbines should be avoided.
4. Cooperation of grower and harvester operator: For high harvesting performance, good cooperation between the producer, contractor, and driver should be ensured. The grower must provide a flat seedbed, homogeneous plant row, and excellent harvesting conditions, while the contractor must provide good machinery and a skilled driver (Fig. 24.29) (Tijink 2010).

If the soil wetness is above the field capacity during harvest, it reduces the bearing capacity of the soil and increases soil compaction. The moisture condition should be controlled during harvest to avoid compactness. Deeper lifting should be done to reduce root tip breakages in dry and hard soil. Excellent harvest is obtained under normal humidity conditions around the field capacity, and subsoil compaction is prevented when the tire pressure is 1.5 bar and below. In very wet soils, it is necessary to wait for the soil to dry sufficiently for harvesting. This waiting period can take several hours in sandy soil and 3–5 days in clay soils (Tijink 2010).

24.14 Machinery Harvesting Systems

Features such as the size of the beet cultivation land, the economic situation, the organization of sugar factories, the habit of using machinery, and the tractor power have led to the emergence of the combined and gradual harvesting system. In the combined harvesting system, topping, lifting, cleaning, and loading operations are performed in one go, and in the gradual harvesting system, each operation is done with a separate machine.

Fig. 24.30 One-row trailed harvester (without bunker)

Fig. 24.31 One-row trailed harvester

The systems in which topping, lifting, cleaning, and loading are done in one step are defined as the combined harvesting system, and the systems in which each operation is done with a separate machine are defined as the multi-stage harvesting system.

24.14.1 Types of Harvesting Machine

24.14.1.1 One-Row Trailed Harvester (without Bunker)

These are the machines that make topping and lifting and leave tha^{-1}beets on the field surface (Fig. 24.30).

24.14.1.2 One-Row Trailed Harvester

These are the harvesting machines that have a hydraulic control system and carry out cleaning, storage, and loading operations in addition to the single row trailed type harvester. Daily capacity (10 h) of this type of machine is 1.8–2.0 t ha^{-1} (Fig. 24.31).

Fig. 24.32 Double-row trailed harvester

Fig. 24.33 One-row self-propelled harvester

Double-row trailed beet harvester: It is a two-row trailed combined harvester. Daily capacity (10 h) of this type of machine is 6–7 t ha^{-1} (Fig. 24.32).

One-row self-propelled beet harvester: They are single row self-propelled combine harvesters. Daily capacity (10 h) of this type of machine is 2.2–2.5 t ha^{-1} (Fig. 24.33).

24.14.1.3 Six-Row Self-Propelled Beet Harvester

It is a six-row self-propelled combine harvester with a capacity of 1.5 ha h^{-1} (Fig. 24.34).

The quality parameters in the beet harvester are as follows:

- Total beet loss
- Topping quality
- Soil and leaf tare
- Beet injury rate

When purchasing or renting a machine, learning the test values from the manufacturer or machine users, if any, provides information in the evaluation of the harvest quality of the machine.

24.15 Cleaning and Loading Machines

They are high-capacity machines that load the beet into a vehicle after it is cleaned from substances such as leaves and soil (Figs. 24.34, 24.35 and 24.36).

The advantages of beet cleaning loading machines can be summarized as follows:

- Facilitating the loading of beets from field silos to vehicles after harvest,
- Reducing low-tare beet delivery and transportation costs with effective cleaning,
- Providing opportunity to producers to start field preparation in a short time for other crops in rotation,

Fig. 24.34 Six-row self-propelled harvester

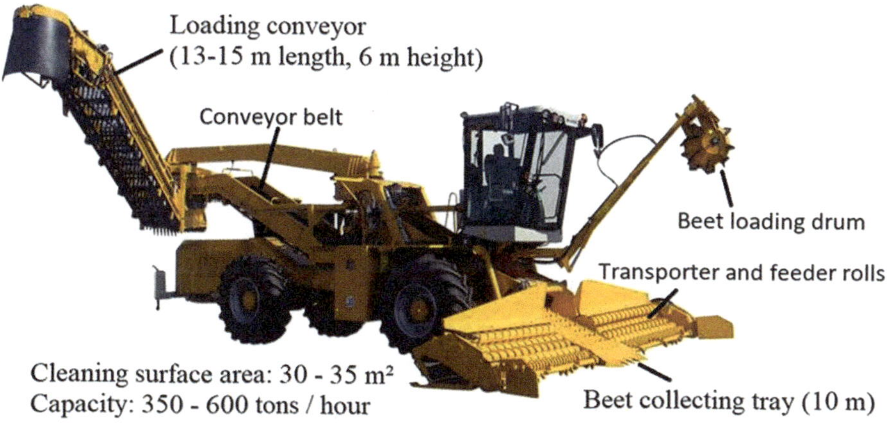

Fig. 24.35 Cleaning and loading machines (mouse)

Fig. 24.36 Cleaning and loading machines (mouse)

Fig. 24.37 Drone application in agriculture

- Preventing soil erosion,
- Effective cleaning of weeds and soil from beets,
- Purchasing low-tare clean beets,
- Reducing business losses.

24.16 Drone Use in Sugar Beet Farming

With the use of drones in agriculture, aerial imaging devices have become widespread. Drones are tools that allow rapid and accurate evaluation in the detection of diseases and pests in plants, water stress, yield/maturity levels detection, weed and flora tracking, irrigation and monitoring of agricultural workers (Figs. 24.36 and 24.37).

The usage purposes of drones in sugar beet agriculture can be summarized as follows:

1. *Monitoring:* Images from drones' built-in cameras allow tracking crop development and identifying areas of poor performance for better crop management.
2. *Easy mapping:* Making field maps facilitates the planning of irrigation, fertilization, and spraying according to the data obtained from the maps.
3. Precision spraying: Drones use advanced sensor combinations to distinguish healthy plants from unhealthy ones. These sensors precisely detect color differences, giving the chance to intervene before diseases spread further. Drones can perform precise spraying at low altitudes, except for equipment scanning the soil.

Crop development in sugar beet is monitored through normalized difference vegetation index (NDVI) data, and even yield estimation can be made with NDVI images taken in autumn (Hoffmann and Blomberg 2004; Bu et al. 2016).

24.17 Future Prospects

The excessive increase in input costs has made economic agriculture impossible in small-scale enterprises. In larger scales, automatic control and smart systems and applications with fast and variable rate instant data are gaining importance. Artificial intelligence methods are taking place in agriculture in the detection of pests and diseases with robotic devices that instantly monitor soil quality and plant growth. Systems that offer or implement fast and effective protection have completed the research phase and have been put into practice. Instead of spraying an entire field, using an agricultural drone that can deliver the required amount of pesticides to the right spot will reduce harmful chemicals and decrease costs. Accurate and fast selection of product types depending on water potential, changing soil properties with soil mapping will contribute to increase agricultural potential. Today, it is thought that the most efficient agriculture can be done with giant agricultural machines on large-scale lands. The heavy machinery used, besides being expensive, causes soil compaction that can last for years. In addition, although the control of the machines has been facilitated, the need for labor in agricultural operations, which are still heavy and tiring, continues. In conclusion, it can be predicted that in the future, small and autonomous robots will be widely used in practice and efficient and economical agriculture will be possible without soil compaction with big data-based applications (King 2017).

24.18 Conclusions

The function and correct use of agricultural machinery have an increasing importance in the sustainability of agricultural production. Therefore, it is necessary to know that the role and characteristics of machinery in agriculture must be well explained in order to be able to plan in the next projections. In addition to having sufficient tractors and equipment in agricultural activities, ensuring the rapid transition from the traditional system to the conversational system is considered important in terms of productivity and sustainability. Individual machines such as moldboard plow, cultivator, disc harrow, and mechanical sowing machines that are still used in agricultural production should be replaced with combined tools/machines that operate in one pass without wasting time.

Because of its deep root structure, sugar beet should be planted in a deep structured field and with a field preparation without compaction. For this, reducing field traffic or planning operations on the track line in large areas is the right solution. Attention should be paid to the timing of agricultural practices and applications should be completed quickly. Early sowing provides a great advantage in sugar beet cultivation. However, the possibility of agricultural frost should be considered in early planting, meteorological data should be followed, and long-term annual statistics should be evaluated at the location of the land. In summary, it can be said that choosing the tools/machines that will perform the agricultural operations in the shortest possible time and performing the operations correctly and quickly with trained operators will guarantee high yield and quality in sugar beet cultivation.

References

Bee P, King J, May M (2004) IIRB congress 2004—report. Br Sugar Beet Rev 72(2):2–8

Bu H, Sharma LK, Denton A, Franzen DW (2016) Sugar beet yield and quality prediction at multiple harvest dates using active optical sensors. Agron J 108(1):273–284

Godwin RJ (1990) Agricultural engineering in development: tillage for crop production in areas of low rainfall. In: FAO agricultural services bulletin no. 83. Food and Agricultural Organization of the United Nations, Rome, p 124. ISBN 92-5-102542-8

Hoffmann CM, Blomberg M (2004) Estimation of leaf area index of *Beta vulgaris* L. based on optical remote sensing data. J Agron Crop Sci 190:197–204

King A (2017) The future of agriculture. Nature 544:521–523. https://doi.org/10.1038/544S21a

Koch HJ, Dieckmann J, Büchse A, Märländer B (2009) Yield decrease in sugar beet caused by reduced tillage and direct drilling. Europ J Agron 30(2009):101–109

Köller K (2003) Conservation tillage-technical, ecological and economic aspects. Proceedings of conservation soil processing and direct sowing workshop, ISBN 975–483–601-9. İzmir

Richard G, Boiffin J, Duval Y (1995) Direct drilling of sugar beet (*Beta vulgaris* L.) into a cover crop: effects on soil physical conditions and crop establishment. Soil Tillage Res 34(3): 169–185

Šarauskis E, Godlinski F, Sakalauskas A, Schlegel M, Kanswohl N, Romaneckas K, Algirdas J, Pilipavicius V (2010) Effects of soil tillage and sowing systems on sugar beet production under the climatic conditions of Lithuania. Landbauforschung Volkenrode 60(2):101–110

Tijink F (2010) Trends and future needs in sugar harvesting. IIRB seminar 2010. https://www.iirb.Org/fileadmin/IIRB/PDF/seminar/2010/1_tijink_IIRB_seminar_2010.PDF

Tuğrul KM, Dursun I (2003) Determination of the efficiency of different soil tillage methods in the sugar beet cultivation. J Agric Sci 9(2):213–221

GPSR Compliance

The European Union's (EU) General Product Safety Regulation (GPSR) is a set of rules that requires consumer products to be safe and our obligations to ensure this.

If you have any concerns about our products, you can contact us on ProductSafety@springernature.com

In case Publisher is established outside the EU, the EU authorized representative is:

Springer Nature Customer Service Center GmbH
Europaplatz 3
69115 Heidelberg, Germany

Batch number: 10755324

Printed by Printforce, the Netherlands

Sugar Beet Cultivation, Management and Processing

Varucha Misra • Santeshwari Srivastava •
Ashutosh Kumar Mall
Editors

Sugar Beet Cultivation, Management and Processing

Volume 2

Springer

Editors
Varucha Misra
ICAR-Indian Institute of Sugarcane Research
Lucknow, India

Santeshwari Srivastava
ICAR-Indian Institute of Sugarcane Research
Lucknow, India

Ashutosh Kumar Mall
ICAR-Indian Institute of Sugarcane Research
Lucknow, India

ISBN 978-981-19-2729-4 ISBN 978-981-19-2730-0 (eBook)
https://doi.org/10.1007/978-981-19-2730-0

This Springer imprint is published by the registered company Springer Nature Singapore Pte Ltd.
The registered company address is: 152 Beach Road, #21-01/04 Gateway East, Singapore 189721, Singapore

To
Late (Mr.) N.K. Misra
(F/O Dr. Varucha Misra)

Foreword

The book entitled *Sugar beet Cultivation, Management and Processing* is a very useful compilation of excellent work on sugar beet that may prove to be the torchbearer in research field of this subject. The editors of this book have cast a broad net to assure how each topic is addressed by some of the world's foremost experts of respective disciplines. Sugar beet is farmed and researched all over the world, as seen by the countries that made a significant contribution in this area. During the last two centuries, sugar seems to have become a staple diet with a yearly intake of 20 kg per person around the globe, and consumption is expected to rise in the near future. Domestic demand for ethanol is also increasing, with projections of 14.5 billion gallons in 2021. Because of its short life span and tremendous potential, sugar beet is increasingly attracting agriculturalists, millers, and researchers for the manufacturing of ethanol. The editors have taken a keen interest in compiling information in the form of chapters on all essential aspects for high and quality production. In a straightforward, crisp, and succinct style, the book summarizes recent achievements and research in this crop. The book's information is simple to comprehend for the reader and will be of great use to those interested in this crop. Furthermore, the book's contributors have provided up-to-date and high-quality information, enhancing the value of this edited volume.

Birsa Agricultural University, Kanke
Ranchi, Jharkhand, India

Onkar Nath Singh

Foreword

Sugar beet (*Beta vulgaris* L.) is cultivated worldwide in many temperate countries like Europe, North America, Asia, South America, and North Africa. It is an annual vegetative crop that has become a key source of sugar production in temperate countries around the world. Sugar beet varieties can adapt to a variety of habitats and growing circumstances. Now its cultivation is being emerged in India after a long halt. It is utilized in diverse industries as raw materials, such as in bioethanol and molasses production and pectin extraction.

Considering the diverse usage and application in sugar and other agro-industries, this crop has been researched meticulously for the benefit of farmers and sugar millers. Several research papers on different aspects are being published, but it is very surprising that only limited information compiled in the form of a book is available in the public domain. I am delighted to hear about this compilation entitled *Sugar beet Cultivation, Management and Processing* by Varucha Misra, Santeshwari Srivastava, and A.K. Mall (eds) devoted specifically to sugar beet which will be published by Springer, Singapore. This book examines sugar beet production, processing, and management from a variety of perspectives, with a particular focus on sugar beet studies in the Indian subcontinent and Southeast Asian countries. This will aid in boosting the long-term development of sugar beets around the world. An array of topics on this crop like crop production and management, mechanization in sugar beet, seed production, weed management, foliar and root diseases, diverse uses of by-products, insect pests, abiotic stress management, and application of artificial intelligence have been covered in this book which is of interest for the readers. One of the highlighting sections in each chapter is of future prospects that give a new route to sugar beet research.

I appreciate the work of the editors who have cast a broad net to guarantee that each topic is addressed by sugar beet researchers worldwide. This edited book is a compilation of 48 chapters by experienced sugar beet experts from continents of Asia, Europe, Africa, North and South America in different fields. This book contains in-depth information on a variety of contemporary sugar beet development and establishment practices and technologies giving a glimpse of recent advancements. Researchers and practitioners interested in understanding the essential dynamics of sugar beet cultivation, processing, and diverse usage for its sustainable development will find the book particularly useful and interesting. This book will serve as a reference book for them. The editor and authors should be commended for compiling so much current material in one spot. Sugar beet will be an important part of the industry's knowledge base and a springboard for further advancement for the next one or two decades, serving practitioners and scholars alike.

I wish this edited book a great success.

Jegor Miladinović

Institute of Field and Vegetable Crops
National Institute of the Republic of Serbia
Novi Sad, Serbia

Preface

Sugar beet is known as an alternative crop for sugar production and is a renowned crop of temperate regions. It runs behind sugarcane in production of sugar contributing to 20% of sugar in the world. It is a biennial and short-duration crop with less water requirement capability compared to sugarcane. Historically, this crop is considered to be a European crop and is known to cultivated in many other countries worldwide.

India is also among such fortunate countries where this crop has been introduced many years ago. Exhaustive research work on this crop has been done in India during 1970s. ICAR-Indian Institute of Sugarcane Research, Lucknow, has always been in the leading role of sugar beet research in India along with other government institutes and sugar industries. The search for the right location for seed production of indigenous and exogenous varieties is a great effort by Indian sugar beet researchers. Successful development of indigenous varieties like LS 6, IISR Comp 1 is another achievement to the determination of the researchers. Agronomic practices for crop production and protection of Indian subtropical and tropical conditions have also been developed and standardized through several attempts. Identification and management of various diseases and insect pests on sugar beet in Indian conditions has also been thoroughly explored. In the current time, germplasm maintenance, seed production, post-harvest quality analysis, drought and saline tolerance, and ethanol production are some of the ongoing research areas. The chapters on seed and crop production, crop adaptability to Indian subtropical climatic conditions, identification of foliar and root diseases, insect pests, bioethanol production, and management measures for diseases and insects have discussed the efforts of Indian research in brief.

Improvement in crop production technologies with the involvement of advanced machines and artificial intelligence has further strengthened the base of this crop in the world. The increased usage of agricultural applications of drones for monitoring and surveillance of diseases, insects, and application of insecticidal sprays has also benefitted this crop. Image processing techniques and deep learning techniques, for disease detection, have great potential benefits for sugar beet protection technologies. The chapter on crop production and management of various countries like Serbia, Turkey, mechanization for this crop, and new technologies for disease monitoring has thrown light on it. As improvement in crop production has always been done for the farmer's benefit, intercropping of sugar beet with other crops has been a good option for doubling farmers' income. The cultivation package and practices have been clearly discussed in the chapter on the intercropping of sugar beet with different agricultural crops.

Climate change is another key ongoing issue that is giving birth to several abiotic stresses. Soil salinity and recurrent droughts are to name a few. The halophytic natural endowment helps sugar beet to not only grow but reclaim saline soil for cultivation, where other crops lag behind. The increasing frequency of water shortage in the world has strongly impacted sugar beet cultivation all over the world. Research have been performed and conclusive results have been obtained. The chapters on Sugar beet production under changing climate: Opportunities and Challenges Improving Sugar beet Production under Salinity Conditions, and Drought stress Management in this crop have covered the worldwide research work.

Furthermore, sugar beet is identified as a promising feedstock for ethanol production. The ethanol blending with petrol in many countries has been the reason behind catching their eyes towards this crop. One ton of sugar beet has been known to attain 100–120 L of ethanol, and this may exceed even through the recent advanced technologies being developed all over the world. Furthermore, this crop has shown its versatile nature by its diverse usage in different fields. The molasses obtained after juice extraction is rich in betaine and is being used as a feed supplement. The non-sucrose dry matter and fleshy leaves are also used as fodder for lactating animals like cattle. The vinasse obtained is being used in the production of petrochemicals. Additionally, sugar beet pectin is now emerging as a natural ingredient for food and allied industries like pharmaceutical, cosmetic, and polymer industries. The chapters based on production and utilization of sugar beet molasses, prospects and scope of this crop as cattle feed, diversified usage of sugar beet pulp and pectin throw complete insight on the expanded application of by-products and recent progress in this field.

This book *Sugar Beet Cultivation, Management and Processing* is a complete package of sugar beet cultivation covering every essential aspect for high production and good yield of sugar beet cultivation, development, and crop management, such as origin, breeding, seed production, physiology, pathology, entomology, biotechnology, and post-harvest technology, in a concise way. This book will give readers an overview of sugar beet crop, beginning with their origins. The book endows with the recent advancements and up-to-date knowledge and research on this crop. We hope this book will serve as a significant reference for students, researchers, scientists, industrialists, and entrepreneurs associated with this field and will emerge out with newer research ideas and work on this issue.

Lucknow, Uttar Pradesh, India

Varucha Misra
Santeshwari Srivastava
Ashutosh Kumar Mall

Acknowledgment

Opportunity is missed by most people
because it is dressed in overalls and looks like work.
—Thomas Edison

A famous quote from Thomas Edison ran into my mind while penning down an acknowledgment note. On behalf of all the editors of this book, I express my gratitude first to the contributors of the chapters who are experts in the sugar beet field of different aspects worldwide. Their experience and research have added value to this book. I would like to thank all of them who in the era of COVID 19 could make up in writing the chapters for this book. Their efforts and cooperation play a vital part in the successful publication of this book.

I would also like to especially thank the people who were always supportive and encouraging behind the curtain. To begin with, I would like to thank my father, Late Mr. N. K. Misra for being there with me when I was down during the compilation of this book. His words of inspiration and motivation have always boosted me up and when I have lost him recently his words made me strong to reinitiate and finish this book. This would not be possible without the strong support from my mother and brother, Er Varun Misra, during this tough time in my life.

I would also like to express my gratitude to Late Dr. T.P. Mall for his blessings during the tenure. Furthermore, I would like to thank the respective family members of editors for their continuous support, motivation, encouragement, and cooperation during the compilation and editing of this book.

Last but not least, I would like to thank God for letting us sail out safely through the hurdles in the path of completion of this book.

I like to conclude the thanksgiving by the renowned line of footballer, Mr. Pele,

Success is no accident.
It is hard work, perseverance, learning, studying, sacrifice and most of all, love of what you are doing or learning to do.

Varucha Misra

Contents

Contents for Volume 1

Editors and Contributors

About the Editors

Varucha Misra is working in ICAR-Indian Institute of Sugarcane Research, Lucknow, as a Research Fellow. She received a Ph.D. degree in Biotechnology from Amity University, Uttar Pradesh, India. Recently, her thesis work was awarded "SAAR Best Thesis Award 2018–2019" for her outstanding work in the field of frontier sciences. Her thesis work was also appreciated and published in the renowned newspaper of *Dainik Jagran*, Uttar Pradesh, India, in 2020. She is a lifetime member of many professional societies such as Biologix Research and Innovation Centre Pvt. Ltd. (BRICPL), India; the Society for Scientific and Social Development (SSSD), Meerut, UP, India; and the Society for Community Mobilization for Sustainable development, India. She has published 34 research papers and 18 review papers in national and international journals of repute. She has also published 15 book chapters with reputed publishers like Springer. She has published three books, two edited books, and three e-bulletins focusing on sugarcane and sugar beet. She has also published several popular articles in *Hindi* language for benefit of farmers in several magazines and has also recently published two articles in *Prabhat Khabar* newspaper, Muzaffarpur, Bihar (India), for the welfare of farmers. She was also appreciated for her reviewing work from Biosciences Biotechnology Research Asia. She is also reviewer of many reputed Journals like Scientific Reports.

In 2021, she received the **AWSAR Award** for Popular Writing from the Department of Science and Technology. In 2020, she got the Plantica Society's **Young Scientist Award**, and in 2018, she received the **Young**

Biochemist Award from Biologix Research and Innovation Centre Pvt. Ltd.. In 2018, she received a **Doctoral Research Fellow Award** from the Society for Scientific and Social Development (SSSD) in Meerut, Uttar Pradesh, India. In 2017, she was awarded **Young Scientist Award** from Samgra Vikas Welfare Society. She was also awarded **Subject Expert** in Microbiology, Agriculture, and Nutrition from the International Consortium for Clinical Research Excellence, Ethics and Education and appreciated the study of "Effect of Sugar Intake towards Human Health" as a noble cause of advancement of the art and science of clinical research in 2017. Further, she received **Prof HS Shrivastava Young Scientist Award (Gold medal)** in National Conference on CCSD-2015 for her oral presentation; **Silver medal** from poster presentation in the International Conference on Impact of Environment on Women's Health organized by Amity University and the National Institute of Environmental Health Sciences, North California, USA, in 2017; and **Best Paper** in the National Conference on Women in Sugarcane Agriculture and Industry in 2013. In addition, she received the **Best Worker Award** in the Research Fellow Category from ICAR-Indian Institute of Sugarcane Research, Lucknow, in 2020 and 2021.

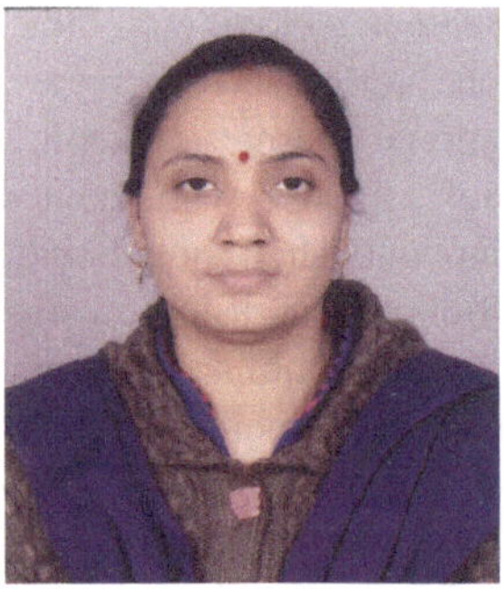

Santeshwari Srivastava is working as Research Fellow at ICAR-Indian Institute of Sugarcane Research, Lucknow and has a M.Sc. and Ph.D. in Zoology. She has an experience of 10 productive years in the research field of entomology as a Research Fellow at the Department of Entomology and Agricultural Zoology, Institute of Agricultural Sciences, Banaras Hindu University, Varanasi. Currently, her research is mainly focused on the economic implications of insect pests on sugar beet and their management. She is actively involved in research in various fields and her research interests include insect taxonomy, insecticidal applications, soil arthropods identification, etc. She has published her 22 research outputs in reputed peer-reviewed journals having citations. Considering her expertise in entomology, she has published 10 research papers in reputed journals. In addition to it, she has 10 popular articles in the field of agriculture. As an editor of books, she has four books to her credit and published more than

12 book chapters with recognized national publishers. She has compiled and edited souvenirs of international conferences during her research tenure. She has presented her studies in various national and international conferences. She has been awarded **Scientist Associate Award** from **Astha Foundation**, Meerut (UP), in 2016 and **Young Achiever Award 2020** from InSc for her excellent contribution in the field of research. For her exemplary work in international conference, she received **Appreciation Award** from the Department of Extension Education, Banaras Hindu University, Varanasi, in 2016. Recently, she was honored **Appreciation Award** from the Society for Agriculture and Allied Research (SAAR) for her marvellous association and inputs in organizing international conference.

Ashutosh Kumar Mall is working as Principal Scientist (Genetics and Plant Breeding), ICAR-Indian Institute of Sugarcane Research, Lucknow, completed his B.Sc. (Ag.) from VBSP University, Jaunpur, UP, in 2001 and both M.Sc. (Ag.) and Ph.D. (Ag.) in Genetics and Plant Breeding from NDUA&T, Ayodhya (UP), during 2003 and 2006, respectively. He has done his post-doctorate from the International Rice Research Institute, Philippines. After serving for 3 years as Scientist with IRRI, Philippines, he served as Senior Scientist at ICAR-Indian Grassland and Fodder Research Institute, Jhansi (UP), and then joined ICAR-IISR, Lucknow. Dr. Mall has research and managerial responsibilities at IISR, Lucknow, as In-charge, IISR Regional Center, Motipur and IISR Sugar beet Outpost, Mukteswar (Nainital). He has developed 15 rice varieties for different ecologies and 13 varieties in different forage crops, namely Guinea grass, Oat, Cowpea, Pearl millet, Tall fescue, Lucerne, and *Sewan* grass, and one variety of sugarcane in collaboration with All India Coordinated Research Project. Through several cross combinations, he has identified 11 germplasm for sugarcane and five germplasm for sugar beet. He has lifetime membership of 27 professional societies.

Dr. Mall significantly contributed to the sugarcane improvement program and is working for seed replacement through Breeder Seed Production under private-public partnership mode in Bihar, which is recognized by task force for improving sugarcane productivity and sugar recovery by the Government of Bihar. He is also working on varietal development of sugar beet for Indian sub-tropical conditions and maintaining nearly 200 germplasm. He has received many prestigious awards. To site a few "Best Scientist Award" by the Society for Scientific and Social Development; "Young Professional Award-2017" by the Society for Community Mobilization for Sustainable Development; and "Dr. Basant Ram Eminent Young Scientist Award-2017" by Uttar Pradesh Academy of Agricultural Sciences, Lucknow. He has published more than 90 research articles in international and national journals of repute, 60 popular articles and manuals, 69 abstracts published in national/international seminar/symposium, published 9 technical bulletins, 3 leaf folders, 3 e-technical bulletins, 3 e-folders, 1 inventory, 3 e-books, 24 e-annual reports and e-proceedings, 10 annuals reports and 7 workshop proceedings compiled and edited to his credit. He has written three books, two edited souvenirs cum abstract books, and written 25 book chapters including publishers like Springer, etc. and handled five externally funded and eight institutional research projects with ongoing one external project and six institutional projects.

Contributors

Mohamed S. Abbas Natural Resources Department, Faculty of African Postgraduate Studies, Cairo University, Giza, Egypt

Mohamed Abdel-Raheem Pests & Plant Protection Department, Agricultural and Biological Research Institute, National Research Centre, Cairo, Egypt

Anuja Rameshchand Adbhai Vasantdada Sugar Institute, Pune, Maharashtra, India

Arkadiusz Artyszak Institute of Agriculture, Warsaw University of Life Sciences—SGGW, Warsaw, Poland

Arun Baitha Division of Crop Protection, ICAR-Indian Institute of Sugarcane Research, Lucknow, Uttar Pradesh, India

Makhan Singh Bhullar Farm Science Centre, Guru Angad Dev Veterinary and Animal Sciences University, SAS Nagar, Punjab, India

Radmila Bojović Faculty of Agriculture, University of Belgrade, Zemun-Belgrade, Serbia

Navjot Singh Brar Farm Science Centre, Guru Angad Dev Veterinary and Animal Sciences University, SAS Nagar, Punjab, India

Atun Roy Choudhury Fecal Sludge Management Division, Banka BioLoo Limited, Hyderabad, Telangana, India

Intiaz Amin Chowdhury Department of Plant Pathology, North Dakota State University, Fargo, ND, USA

Mihajlo Ćirić Institute of Field and Vegetable Crops, Novi Sad, Serbia

Siddhartha Das Department of Plant Pathology, M.S. Swaminathan School of Agriculture, Centurion University of Technology and Management, Paralakhemundi, Odisha, India

Sujaya Dewanjee RRSS, Bidhan Chandra Krishi Viswavidyalaya, Purulia, West Bengal, India

Marijana Djordjević Department of Carbohydrate Food Engineering, Faculty of Technology Novi Sad, University of Novi Sad, Novi Sad, Serbia

Miljana Djordjević Department of Carbohydrate Food Engineering, Faculty of Technology Novi Sad, University of Novi Sad, Novi Sad, Serbia

Ayman Esh Agricultural Research Center, Sugar Crops Research Institute, Giza, Egypt

Parviz Fasahat Sugar Beet Seed Institute (SBSI), Agricultural Research, Education and Extension Organization (AREEO), Karaj, Iran

M. B. Flores-Romero Faculty of Economic and Administrative Sciences, Michoacan University of San Nicolas of Hidalgo, Morelia, Mexico

A. E. Ghaly Department of Agricultural Engineering, Faculty of Agriculture, Cairo University, Giza, Egypt

Meenakshi Goyal Department of Plant Breeding and Genetics, Punjab Agricultural University, Ludhiana, Punjab, India

Md. Alamgir Hossain Department of Crop Botany, Bangladesh Agricultural University, Mymensingh, Bangladesh

Mostafa Hosseinpour Sugar Beet Seed Institute (SBSI), Agricultural Research, Education and Extension Organization (AREEO), Karaj, Iran

M. M. Ibrahim Department of Agricultural Engineering, Faculty of Agriculture, Cairo University, Giza, Egypt

Md. Jahirul Islam Department of Bio-Health Convergence, College of Biomedical Science, Kangwon National University, Chuncheon, Republic of Korea
Physiology and Sugar Chemistry Division, Bangladesh Sugarcrop Research Institute, Pabna, Bangladesh

Navdeep S. Jamwal PAU, Regional Research Station, Kaurthala, Punjab, India

J. D. Janse Institute of Plant Protection NAAS, Kyiv, Ukraine

L. A. Janse Institute of Plant Protection NAAS, Kyiv, Ukraine

D. Jiménez-Islas Renewable Energy Division, Higher Technological Institute of Huichapan, Huichapan, Mexico

Zoran Jovović Biotechnical Faculty, University of Montenegro, Podgorica, Montenegro

Mozhdeh Kakueinezhad Sugar Beet Seed Institute (SBSI), Agricultural Research, Education and Extension Organization (AREEO), Karaj, Iran

K. A. Kalatur Institute of Bioenergy Crops and Sugar Beet NAAS, Kyiv, Ukraine

Nilima Karmakar Department of Soil Science and Agricultural Chemistry, NMCA, Navsari Agricultural University, Navsari, Gujarat, India

Aanchaldeep Kaur Department of Plant Breeding and Genetics, Punjab Agricultural University, Ludhiana, Punjab, India

Simerjeet Kaur Department of Agronomy, Punjab Agricultural University, Ludhiana, Punjab, India

Rıza Kaya Department of Plant Protection, Sugar Institute, Etimesgut, Ankara, Turkey

Sepideh Kazemi Department of Horticultural Science, Science and Research University of Tehran, Tehran, Iran

Mohammed Khan Department of Plant Pathology, North Dakota State University, Fargo, ND, USA

Samar Khayamim Seed and Plant Improvement Institute, Agricultural Research, Education and Extension Organization, Karaj, Iran

Amit Kumar Plant Breeding Section, Division of Crop Science, ICAR-RC-NEH Region, Umiam, Meghalaya, India

Dharmendra Kumar ICAR-CPRI, Shimla, Himachal Pradesh, India

Rajeev Kumar ICAR-IISR, Lucknow, Uttar Pradesh, India

Santosh Kumar ICAR-IARI, Hazaribag, Jharkhand, India

Ram Ji Lal ICAR-Indian Institute of Sugarcane Research, Lucknow, Uttar Pradesh, India

Young-Seok Lim Department of Bio-Health Convergence, College of Biomedical Science, Kangwon National University, Chuncheon, Republic of Korea

Nataša Ljubičić Biosense Institute, University of Novi Sad, Novi Sad, Serbia

A. K. Mall Division of Crop Improvement, ICAR-Indian Institute of Sugarcane Research, Lucknow, Uttar Pradesh, India

Ashutosh Kumar Mall ICAR-Indian Institute of Sugarcane Research, Lucknow, Uttar Pradesh, India

Nikola Maravić Department of Carbohydrate Food Engineering, Faculty of Technology Novi Sad, University of Novi Sad, Novi Sad, Serbia

Mohammad Mehdizadeh Department of Agronomy and Plant Breeding, Faculty of Agriculture and Natural Resources, University of Mohaghegh Ardabili, Ardabil, Iran
Ilam Science and Technology Park, Ilam, Iran

Mostafa Mehdizadeh Faculty of Agriculture, University of Guilan, Rasht, Iran

Zoleikha Mehdizadeh Department of Biology, Payame Noor University, Tehran, Iran

Manish Chandra Mehta Department of Entomology and Agricultural Zoology, Institute of Agricultural Sciences, Banaras Hindu University, Varanasi, India

Anja Milosavljević Research and Development Center Sunoko, Indjija, Serbia

Varucha Misra Division of Crop Improvement, ICAR-Indian Institute of Sugarcane Research, Lucknow, Uttar Pradesh, India

Hemat S. Mohamed Animal Nutrition Department, Animal Production Research Institute, Agricultural Research Center, Cairo, Egypt

Masuma Akter Mou Department of Agronomy, Bangladesh Agricultural University, Mymensingh, Bangladesh

Barbara Magdalena Muir Tereos Corporate Technology Centre, Aalst, Belgium

José M. Mulet Biochemistry and Molecular Biology, IBMCP, Universitat Politècnica de València-CSIC, Valencia, Spain

Mona Nagargade ICAR-IISR, Lucknow, Uttar Pradesh, India

Violeta Oro Institute for Plant Protection and Environment, Belgrade, Serbia

Mehmet Metin Ozguven Department of Biosystems Engineering, Faculty of Agriculture, Tokat Gaziosmanpaşa University, Tokat, Turkey

Kamlesh G. Patel Department of Soil Science and Agricultural Chemistry, NMCA, Navsari Agricultural University, Navsari, Gujarat, India

A. D. Pathak ICAR-Indian Institute of Sugarcane Research, Lucknow, Uttar Pradesh, India

Sanjay Patil Department of Alcohol Technology and Biofuels, Vasantdada Sugar Institute, Pune, Maharashtra, India

Sudeepta Pattanayak Division of Plant Pathology, ICAR—Indian Agricultural Research Institute, New Delhi, India

Swapan Kumar Paul Department of Agronomy, Bangladesh Agricultural University, Mymensingh, Bangladesh

M. E. Pérez-Romero Faculty of Economic and Administrative Sciences, Michoacan University of San Nicolas of Hidalgo, Morelia, Mexico
Business Management Division, Higher Technological Institute of Huichapan, Huichapan, Mexico

W. S. Philanim Plant Breeding Section, Division of Crop Science, ICAR-RC-NEH Region, Umiam, Meghalaya, India

Mohammed Arif Sadik Polash Department of Crop Botany, Khulna Agricultural University, Khulna, Bangladesh

Dragana Popović Faculty of Economics in Subotica, University of Novi Sad, Novi Sad, Serbia

Vera Popović Institute of Field and Vegetable Crops, Novi Sad, Serbia

Aiming Qi Centre for Agriculture, Food and Environmental Management Research, School of Life and Medical Sciences, University of Hertfordshire, Hatfield, Hertfordshire, UK

Vuk Radojević Institute of Field and Vegetable Crops, Novi Sad, Serbia
Faculty of Agriculture, University of Novi Sad, Novi Sad, Serbia

M. Raghuraman Department of Entomology and Agricultural Zoology, Institute of Agricultural Sciences, Banaras Hindu University, Varanasi, India

Abazar Rajabi Agricultural Research, Education and Extension Organization, Sugar Beet Seed Institute (SBSI), Karaj, Iran

Md. Abdur Razzak Faculty of Life and Environmental Science, Shimane University, Matsue, Japan

Javad Rezaei Sugar Beet Research Department, Khorasan Razavi Agricultural and Natural Resources Research and Education Center, Agricultural Research, Education and Extension Organization (AREEO), Mashhad, Iran

Alessandro Romano Council for Agricultural Research and Economics, Plant Protection and Certification Centre, Lonigo, Italy

Md. Arif Sakil Department of Biochemistry and Molecular Biology, Bangladesh Agricultural University, Mymensingh, Bangladesh
Department of Regulatory Biology, Saitama University, Saitama, Japan

Gulzar S. Sanghera PAU, Regional Research Station, Kaurthala, Punjab, India

Prabhjot S. Sanghera Punjab Agricultural University, Ludhiana, Punjab, India

Shahida Sazia Department of Horticulture, Bangladesh Agricultural University, Mymensingh, Bangladesh

Mile Sečanski Maize Research Institute "Zemun Polje", Belgrade, Serbia

Zita Šereš Department of Carbohydrate Food Engineering, Faculty of Technology Novi Sad, University of Novi Sad, Novi Sad, Serbia

Mohamed A. Shahba Natural Resources Department, Faculty of African Postgraduate Studies, Cairo University, Giza, Egypt

Nivedita Shettigar Plant Breeding Section, Division of Crop Science, ICAR-RC-NEH Region, Umiam, Meghalaya, India

A. K. Shrivastava ICAR-Indian Institute of Sugarcane Research, Lucknow, Uttar Pradesh, India

Shahida Anusha Siddiqui DIL e.V.–German Institute of Food Technologies, Quakenbrück, Germany
Technical University of Munich Campus Straubing for Biotechnology and Sustainability, Straubing, Germany

Preeti Singh ICAR-IARI, Hazaribag, Jharkhand, India

Pankaj Kumar Sinha ICAR-IARI, Hazaribag, Jharkhand, India

Archana Siraree Division of Seed Production, Genda Singh Sugarcane Breeding and Research Institute, Kushinagar, Uttar Pradesh, India

Dragana Šoronja-Simović Department of Carbohydrate Food Engineering, Faculty of Technology Novi Sad, University of Novi Sad, Novi Sad, Serbia

Velibor Spalević Biotechnical Faculty, University of Montenegro, Podgorica, Montenegro

Sangeeta Srivastava ICAR-Indian Institute of Sugarcane Research, Lucknow, Uttar Pradesh, India

Santeshwari Srivastava Division of Crop Protection, ICAR-Indian Institute of Sugarcane Research, Lucknow, Uttar Pradesh, India

Marijenka Tabaković Maize Research Institute "Zemun Polje", Belgrade, Serbia

Shadia Taghian Agricultural Research Center, Sugar Crops Research Institute, Giza, Egypt

Dariush Taleghani Agricultural Research, Education and Extension Organization, Sugar Beet Seed Institute (SBSI), Karaj, Iran

Koko Tampubolon Program Study of Agrotechnology, Faculty of Agriculture and Animal Husbandry, Universitas Tjut Nyak Dhien, Medan, Indonesia

Paul Townson Lion Seeds Ltd., Woodham Mortimer, Maldon, Essex, UK

Nenad Trkulja Institute for Plant Protection and Environment, Belgrade, Serbia

Koç Mehmet Tuğrul Department of Biosystems Engineering, Faculty of Agriculture, Eskişehir Osmangazi University, Eskişehir, Turkey

I. Ventura-Cruz Mechatronic Division, Higher Technological Institute of Huichapan, Huichapan, Mexico

Vivek Chandra Verma DIHAR-DRDO, Chandigarh, India

Yusuf Yanar Department of Plant Protection, Faculty of Agriculture, Tokat Gaziosmanpaşa University, Tokat, Turkey

Guiping Yan Department of Plant Pathology, North Dakota State University, Fargo, ND, USA

List of Figures

List of Tables

Part II

Biotic Stress, Post-harvest and Processing Technologies

Etiology, Epidemiology, and Management of Sugar Beet Diseases

25

Ayman Esh and Shadia Taghian

Abstract

Sugar beet (*Beta vulgaris* L.) is commercially cultivated in the northern temperature zone, i.e. between 30° and 60° latitudes north. Sugar beet is the second source of sugar (sucrose) after sugarcane; it is covering about 40% of the world's sugar demands. On the contrary with sugarcane, sugar beet stores sugar in the roots not the stalks as in sugarcane. Like other crops, the plant density in the unit area is one of the most important factors affecting sugar beet production. Various fungal pathogens can decrease the number of cultivated plants and cause substantial economic losses at all plant stages, especially damping-off diseases in the seedling stage and root rot diseases during growth. On the other hand, foliar diseases (leaf spots, rust, powdery mildew, and viral diseases) also affect sugar production as well as the quality of roots. In this chapter, we will discuss the economic root diseases and foliar diseases that affect sugar beet and sucrose production quantitatively and qualitatively. This chapter will discuss foliar and root diseases and the etiology, epidemiology, and management of each disease.

Keywords

Disease control · Epidemiology · Plant pathogens · Sugar beet

Abbreviations

BCLV	Beet leaf curl virus
BCTV	Beet curly top virus

A. Esh (✉) · S. Taghian
Agricultural Research Center, Sugar Crops Research Institute, Giza, Egypt
e-mail: aymanesh@arc.sci.eg

V. Misra et al. (eds.), *Sugar Beet Cultivation, Management and Processing*,
https://doi.org/10.1007/978-981-19-2730-0_25

BDM	Beet distortion mosaic
BMY	Beet mild yellows
BtMV	Beet mosaic virus
BWY	Beet western yellows
BYNV	Beet yellow net virus
BYV	Beet yellows
CLS	*Cercospora* leaf spot
H_2O_2	Hydrogen peroxide
O^{2-}	Superoxide
OH^-	Hydroxyl radical
QoIs	Quinone outside inhibitors
ROS	Reactive oxygen

25.1 Introduction

Plants are surrounded by microorganisms living in the same environment. Different microorganisms can colonize the plant and establish different relationships, mutualistic, neutral, or parasitic, with it (Compant et al. 2010; Raaijmakers et al. 2009). Sugar beet (*Beta vulgaris* subsp. *Maritime* L.) is mainly grown for the production of sugar that accumulated in the roots (Trebbi and McGrath 2004). Like other crops, the plant density in the unit area is one of the most important factors affecting sugar beet production. The ideal number of sugar beet plants is ranged from 72 to 96 thousand plants per hectare to give yield of 60–70 tons roots per hectare (Pervin and Islam 2015). However, various fungal pathogens can decrease the number of cultivated plants and cause substantial economic losses at all plant stages, especially damping-off diseases in the seedling stage. In each plant stage of sugar beet, there are some pathogens affecting plant health. There are three major stages in the growth of sugar beets (Fig. 25.1); the seedling stage (30–45 days from sowing date), growth stage, and repining stage (starts 1 month before harvesting).

In an early stage of plant development, many soil-borne pathogens attack sugar beet seeds and seedlings such as *Rhizoctonia solani, Aphanomyces cochlioides, and*

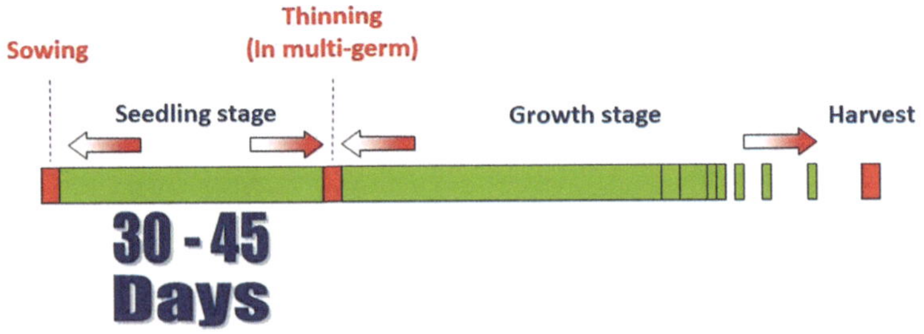

Fig. 25.1 Sugar beet growing stages

Pythium ultimum causing preemergence and postemergence damping-off. In the growth stage, the roots and foliage are attached by numerous fungal, bacterial, and viral pathogens. *Rhizoctonia solani*, *Sclerotium rolfsii Aphanomyces cochlioides*, *Fusarium spp.*, *Pythium ultimum*, and *Pectobacterium carotovorum* attack the roots that cause root rot diseases. *Cercospora beticola*, *Ramularia betae*, *Alternaria tenuis*, *Phoma betae*, *Uromyces betae*, and *Erysiph polygoni* can cause leaf spot diseases that damage the leaves and lead to severe harvest losses during the growth stage (Zachow et al. 2010). In the postharvest stage, *Fusarium* spp. and many other saprophytic fungal and bacterial genera can cause storage root rot and lead to potential losses in root quality and sugar yield (Liebe et al. 2016).

25.2 Foliar Diseases

25.2.1 Leaf Spot Diseases

25.2.1.1 *Cercospora* Leaf Spot Disease (CLS)

Cercospora leaf spot (CLS) disease has been documented for almost a century. The disease is known as the most damaging foliar disease of sugar beets in the world. The disease can result in significant yield and sugar content losses, depending on the severity of the disease. Root yield losses are approximately 20–25 percent and it can reach 42 percent severe attacks (Smith and Ruppel 1973). Shane and Teng (1992) and Byford (1996) reported that the yield was reduced by 3 percent when the spots covered 3 percent of the leaf area. Since then, experts have been attempting to better understand and manage the disease. Because disease damages leaves, it has a negative impact on plants' photosynthetic capacity and reduces yield quantity and quality by increasing impurities in the juice, which reduces sugar extractability.

Causal Agent

Sugar beet leaf spot disease is caused by the pathogenic fungus *Cercospora beticola* Saac. *Cercospora* fungi are very active and damaging diseases that infect a wide range of host plants around the world. *Cercospora* fungi have a far broader host range than many other plant pathogenic fungi, infecting both monocot and dicot angiosperms, gymnosperms, and even some lower plants like ferns. *Cercospora* is an asexual genus, as confirmed by phylogenetic analysis (Goodwin et al. 2001). *Cercospora* has 659 species globally, according to a recent taxonomic research. *Cercospora* fungi infect major crops such as sugar beets, corn, soybeans, coffee, and peanuts, as well as vegetables and ornamental plants. Black Sigatoka, one of the most damaging banana diseases, is caused by the closely similar *Mycosphaerella fijiensis* (anamorph: Pseudocercospora fijiensis) (Ploetz 2001). There was also a lot of morphological, physiological, and genetic variation among *Cercospora beticola* isolates (Groenewald et al. 2006, 2007; Esh and Moghaieb 2011).

Pathogenesis

The fungus persists in the past season's infected leave debris in the fields for almost 2 years and is the main source of infection; on the other hand, the infection can transmit by conidia or stomata from other infected hosts through the wind, rain splashes, or insects (Franc 2010). The appressoria structure germinated from the conidium penetrates the leaf tissue through stomata, then grows and colonizes the intercellular tissue, and after a while the cells surrounding the growing mycelium die due to a photosensitizing compound (photoactivated toxin) Cercosporin and beticolin as well as the hydrolytic enzymes such as cellulases and pectinases produced by the fungus (Rossi et al. 2000). The dead cells of leaf tissue give the appearance of leaf spots which are the characteristic of the infection of *C. beticola* (Weiland and Koch 2004; El-Kholi and Esh 2011). The conidiophores immerge from the stomata in the middle of the necrotic spot carrying new conidia. The immerged new conidia (Fig. 25.2) start to separate from the conidiophore by air and rain insects to other sites of infection on the same leaf or plant or to other plants and repeat the cycle again several times during the growing season in a polycyclic manner (Rossi et al. 2000; Franc 2010; El-Kholi and Esh 2011).

Cercosporin is produced by several species of the genus and it can be isolated from diseased plant tissue and has been demonstrated to be necessary for the

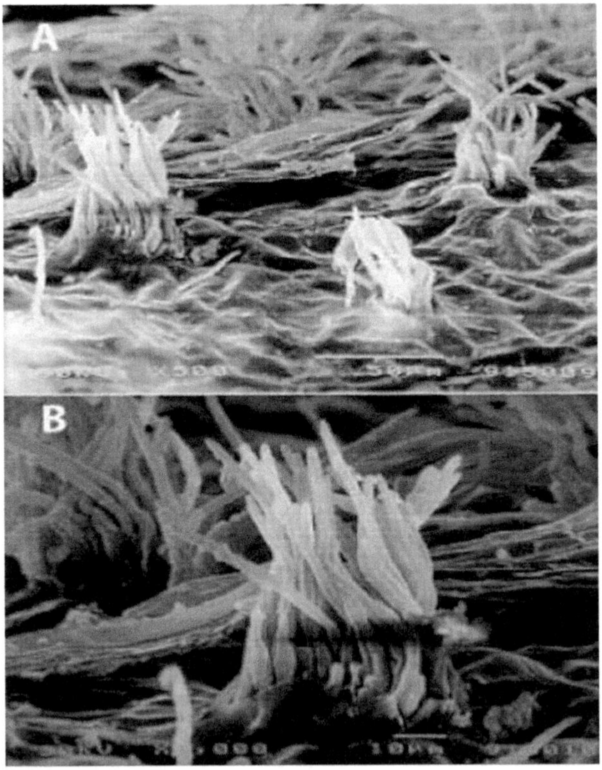

Fig. 25.2 Scanning electron micrographs showing the conidiophores immerge from the stomata in the middle of the necrotic spot carrying new conidia (Source: El-Kholi and Esh 2011)

development of typical disease symptoms in numerous plants. Cercosporin is a reddish-brown substance that is water-insoluble. Cercosporin is a photosensitizer, which absorbs light and converts to an electronically active triplet state that reacts with oxygen via electron transfer (radical) reactions or a reducing substrate to produce reactive oxygen species (ROS) such as superoxide (O^{2-}), hydrogen peroxide (H_2O_2), and the hydroxyl radical (OH) (Girotti 1990). Alternatively, the triplet sensitizer could use an energy transfer mechanism to react directly with oxygen (Spikes 1989). As a result, non-radical yet highly hazardous singlet oxygen ($^1O_2$) is produced. Almost all macromolecules in cells, including lipids, proteins, and DNA, are sensitive to oxidative stress induced by photosensitizers, with the type of cellular damage dictated by where the photosensitizer molecule is found, such as in the membranes, cytoplasm, or nucleus (Moan et al. 1998).

Cercosporin and other photo-activated perylenequinone toxins have a key role in leaf spot disease, according to several lines of evidence. Rotem et al. (1988) and Calpouzos and Stalknecht (1967) reported a strong link between *Cercospora* disease severity in sugar beets and day length as well as high light intensities. The ultrastructural studies on infected sugar beet leaf showed that cell membrane damage is the major characteristic symptom, which is consistent with cercosporin's membrane damaging activity (Steinkamp et al. 1979; El-Kholi and Esh 2011). Studies on cercosporin-deficient mutants of *Cercospora* showed that cercosporin is responsible for disease progression. *C. kikuchii* cercosporin-deficient mutants, for example, show a significant reduction in symptom development on soybeans (Upchurch et al. 1991). In contrast to the wild-type strain's severe necrotic lesions, disruption mutants of a MAP kinase gene required for cercosporin synthesis in *C. zeae-maydis* generated only mild chlorotic lesions on corn. Shim and Dunkle (2003) reported that Cercosporin is the virulence factor for *Cercospora* pathogens.

Symptoms

Cercospora leaf spot symptoms are expressed as the appearance of a separate 0.3–0.5 cm tan-colored (Fig. 25.3), necrotic spots in the middle of a reddish-brown border (Chupp 1953; Mulder and Holliday 1974). The spots may appear zonate and can eventually merge to form a large necrotic area on the leaf. The merged necrotic areas darkened with age; the disease symptom starts on older leaves, but also appears on younger leaves as it progresses. In humid conditions, spots show a grayish-blue fuzzy look as conidiophores and conidia have emerged from the stomata (Fig. 25.4). The fungus may have 4–5 disease cycles during the season, and with each cycle, there is a remarkable increase in the inoculums density. If the infected leaf covered with large necrotic areas, the leaf withers and die, resulting in defoliation (Hudec and Rohacik 2002).

Favorable Conditions

Humid conditions (90–95%) and day temperature (25 and 35 °C) and $\geq$17 °C at night are the favored conditions for *Cercospora beticola* to form conidiophores and conidia (the first appearance of the spot); under favorable conditions, conidia and secondary infections occur within 7–14 days from first infection (Ruppel 1986).

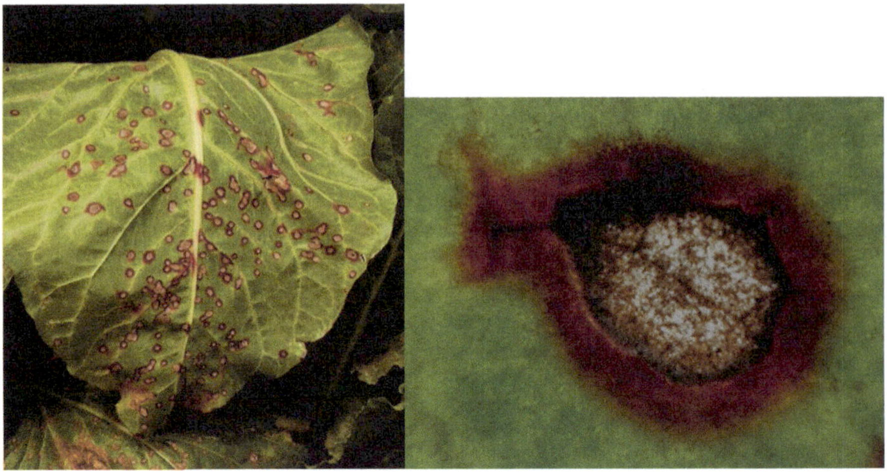

Fig. 25.3 Symptoms of *Cercospora* leaf spot disease (on the right) and a close-up picture of leaf spot (on the left)

Fig. 25.4 Development of *Cercospora* leaf spot disease

Water spray irrigation and rain lead to more sporulation (Bleiholder and Weltzien 1971), due to better release of conidia (Carlson 1967; Meredith 1967) and increased conidial germination and subsequent infection to the host.

Control

Crop rotation and tillage, as well as farming disease-resistant sugar beet varieties, are all utilized to reduce inoculums (Miller et al. 1994). Sugar beets have four to five resistance genes against *Cercospora* leaf spot (Smith and Gaskill 1970). According to Smith and Campbell (1996), the chances of developing a sugar beet cultivar with high yield and disease resistance are limited. As a result, commercial types always contain a moderate level of Cercospora resistance, which needs the use of fungicides to combat the disease (Miller et al. 1994). *Cercospora beticola*, on the other hand, was able to produce fungicide-tolerant strains when applied extensively and regularly (Weiland and Koch 2004). The typical technique for controlling sugar beet *Cercospora* leaf spot disease is to use fungicides. The disease is treated with a variety of systemic and nonsystemic fungicides. Benomyl, bitertanol, cyproconazole, difenoconazole, carbendazim, epoxiconazole, fentin hydroxide, fentin acetate, fluquinconazole, flutriafol, flusilazole, mancozeb, maneb, prochloraz, propiconazole, tetraconazole, and flusilazole are some of the fungicides used to treat *Cercospora* leaf spot (Mukhopadhyay 1992; Esh and Kamhawy 2010; Baltaduonytė et al. 2013).

It is critical to use a variety of fungicides with distinct modes of action to avoid fungicide tolerance. Controlling fungicides for *C. beticola* are classified as follows: (1) Fentin, dithiocarbamates, and nitriles are examples of protective fungicides. These protective fungicides must be sprayed before the first symptoms appear in order to destroy or inhibit the conidia. (2) Systemic fungicides with dual action, such as benzimidazoles, ergosterol inhibitors (DMIs and amines), and quinone outside inhibitors (QoIs), can be used when the disease is already present (Ioanidis and Karaoglanidis 2010). Crop rotation (2–3 year rotation), weed host management, and adjusting nitrogen fertilization and water in irrigated areas all help to prevent and reduce disease incidence and severity (Meriggi et al. 2000; Rangel et al. 2020).

Many studies have been published on the use of biocontrol agents to combat sugar beet leaf spot disease. Hydrolytic enzymes production and antibiotics production, specific colonization and competition for host nutrients, triggering the mechanisms of plant host defense, and interference with pathogenicity elicitors are all examples of biocontrol mechanisms (Punja and Utkhede 2003). *Bacillus subtilis, Bacillus pumilus, Bacillus amyloliquefaciens*, and *Bacillus mycoides* have been reported as biocontrol agents against *C. beticola* on sugar beets (Douglas et al. 2003; Larson 2004; Taghian et al. 2008; Esh et al. 2011a, b). Larson (2004) reported that the effect of a *Bacillus mycoides* isolate was equivalent to the effect of chemical fungicide on cercospora leaf spot.

25.2.1.2 Ramularia Leaf Spot Disease

Ramularia leaf spot disease is found in North America (USA and Canada), Europe (North and East), and Russia, where sugar beets are grown in cool, wet (rainy) climates. The disease can cause yield losses of up to 10% and a 1% reduction in sugar percentage (Baltaduonytė et al. 2013).

Causal Agent

The filamentous fungus *Ramularia beticola* is the causal pathogen for Ramularia leaf spot disease. The fungus belongs to Phylum Ascomycota and Family Mycosphaerellaceae (Asher and Hanson 2006).

Pathogenesis

The pathogenesis of *Ramularia* and *Cercospora* is very similar. The disease can survive in infected beet debris from the previous season; additionally, the spores can survive in the soil for 1 year; also the pathogen can be a seed-borne. The conidium germinates and enters the leaf tissue through the natural leaf openings (stomata) to colonize the leaf intercellular tissue. Leaf spots' appearance on the leaf presents the dead cells of leaf tissue. After a while, the conidiophores of the pathogen start to emerge carrying new conidia from the stomata in the necrotic areas during the growing season. The disease cycle is repeated several times on same plant or near plants by the dispersal of newly produced conidia (Asher and Hanson 2006).

Symptoms

On the contrary, *Cercospora* leaf spot symptoms, *Ramularia beticola*, make larger spots than *C. beticola*, without a red border zone. Smaller greyish-white spots appear in the center of the larger spots. The disease symptoms appear on the older leaves as round light brown leaf spots (4–7 mm diameter). The center of the spots becomes gray to white when the spores immerged from the center; the spots are surrounded with a dark to reddish-brown margin (Fig. 25.5) (Asher and Hanson 2006).

Favorable Conditions

Ramularia leaf spot disease development requires cold temperatures around 17 °C, and weather humidity of more than 95 percent is favorable for disease development. At 17 °C, it takes about 14 days for the disease to develop inside the host before symptoms appear on the leaf, whereas at 25 °C, no symptoms appear (Baltaduonytė et al. 2013).

Control

The same effective fungicides with *Cercospora* leaf spot are effective with *Ramularia* leaf spot. In contrary with *C. beticola, R. beticola* found no developing resistance to fungicides such as strobilurins or triazoles (Thachab et al. 2013).

25.2.1.3 *Alternaria* Leaf Spot Disease

Most *Alternaria* species are common saprophytes and some species are plant pathogens that cause plant diseases, generally foliar diseases like leaf blight or leaf spot on a wide host range. The disease was reported in almost all countries cultivating sugar beet. *Alternaria* diseases can cause a significant loss in economic crops.

Fig. 25.5 *Ramularia* leaf spot disease on sugar beet

Causal Agent

The genus *Alternaria* belongs to the family Pleosporaceae (Pleosporales, Dothideomycetes, Ascomycota) (Lawrence et al. 2016). *Alternaria* leaf spot of sugar beet is caused by different *Alternaria* spp. species of *Alternaria*. Many reports on different species of *Alternaria* were published from all the world's sugar beet growing countries as the causal agent for the disease, *A. tenuis*, *Alternaria alternata*, *Alternaria brassicae*, *A. alternata*, *A. ashwinii* and *A. dilkushana* (Misra et al. 2020; Hudec and Rohacik 2002; El-Kholi et al. 1994; Agnihotri 1990; Mcfarlane et al. 1954).

Pathogenesis

Alternaria conidiospores are spread by the wind or rain and land on the leaf surface, where they germinate and penetrate the leaf tissue, living inter and intracellularly and secreting hydrolytic enzymes and other metabolites that cause an array of dark spots to appear on the infected leaf. Secondary infection cycles can occur when the weather conditions are favorable to the pathogen. The infected leaf will eventually fall, preserving the pathogen for the following season. The disease destroys host tissues by reducing the photosynthesis process's potential. *Alternaria* spp. produces a diverse range of secondary metabolites, including pathogenesis-related toxins (Fujiwara et al. 1988; Wolpert et al. 2002) and mycotoxins (Andersen et al. 2005; Frisvad et al. 2008; Ostry 2008).

Symptom

Different *Alternaria* species cause similar symptoms on sugar beet, despite its taxonomic and pathogenic differences. Irregular or circular dark brown lesions appear as the disease symptoms on the leaves (Fig. 25.6). As the spots or lesions became older, they enlarge to reach around 10 mm in diameter. The lesions then confined within separate parallel veins of the leaves, but gradually dilated and consolidated across veins. Some sugar beet types can have a crimson zone around the borders of the spots. Under long periods of high humidity, a modest production of Alternaria spores can be seen in the center of the spot, which usually appears on the rear surface of the leaf. To create a target spot effect, the spot centers get grey and somewhat zonate. The dead centers may rip and partially fall out afterwards. This necrotic patch may persist until the entire plant's leaves are lost. Spotted leaves become yellow and die (Agnihotri 1990; El-Kholi et al. 1994; Thomma 2003; Misra et al. 2020).

Control

The most effective control is the use of fungicides as seed dressing or spray on plants for the control of the *Alternaria* leaf spot. Effective fungicides for controlling Cercospora leaf spot are effective against the Alternaria leaf spot. Other disease management activities, such as the removal of plant debris from the fields at the end of the season, increase the spaces between plants, and the use of suitable crop rotation for 2–3 years can be carried out.

25.2.1.4 Phoma Leaf Spot Disease

Phoma leaf spot disease is a sugar beet seed-borne and soil-borne disease. By reducing photosynthesis activity, the disease can cause a significant loss in yield quality and quantity, as well as a negative effect on seed germination and seedling emergence.

Fig. 25.6 Symptoms of *Alternaria* spp. on sugar beet

Causal Agent

Phoma betae (Rostr.) causes Phoma leaf spot (Bugbee 1979). The fungus's perfect stage (sexual stage) has recently been identified as Neocamarosporium betae, which belongs to the Phylum Ascomycota, Order Pleosporale, and Family Neocamarosporiaceae (Vaghefi et al. 2019). The fungus attacks the root system of sugar beet plants, including seedlings and foliage. During the growing season, natural wounds are the most common route to the plant. When plants are subjected to abiotic stress, the disease can cause severe damage (Bugbee 1979).

Pathogenesis

The spread of the disease within the field and to other fields caused annually by the sexual stage spores (ascospores) is developed on the old debris of diseased plants from the previous season and on weed hosts. The spores can remain on plant debris for 24 months. However, the conidia of the imperfect stage produced by pycnidia spread to other healthy plants in the fields by rain splash, irrigation, or mechanically through agricultural equipment. The fungal conidia can infect seeds of sugar beet; some reports indicated 40–50% infection in seed lots (Bugbee 1979; Bugbee and Soine 1974). The mechanisms of disease progress and the relative contribution of inoculum sources to epidemics are under research (Pethybridge et al. 2018). Ichihara et al. (1983) reported *Phoma betae* produces seven toxins that are toxic to plant tissues and induced appearance of brown spots on infected leave. These toxins are betanon A, betanon B, betanon C (which is high toxic to plant tissues), and afidicolin, 3-deoxyaffidicholine, afidicholine-17-monoacetic acid, and afidicholine-3, 18-orthoacetic acid.

Symptoms

Symptoms start with occurring of spots on the upper leaf surface (Fig. 25.7). The spots are of brown color and round shape (10–20 mm in diameter) and contain dark concentric circles. The spots then are covered with small black spots (pycnidia). The

Fig. 25.7 Phoma leaf spot disease

older leaves get affected by the disease than the younger ones. As the disease progresses, small, dark, cavernous spots start to appear near the crown, then after a while become water-soaked. The spot's color eventually turns to blackish-brown color and spreads downward into the taproot. A black line is found between diseased and healthy tissues (Pool and McKay 1915; Garibaldi et al. 2007).

Favorable Conditions

The optimum temperatures for infection and disease progress are found between 15 and 20 °C with high humidity of 95% in infected seed fields; late rains before harvest are found to increase the occurrence of infected seeds.

Control

Effective fungicides for controlling *Cercospora* leaf spot are effective against the Phoma leaf spot.

25.2.1.5 Rust Disease

Sugar beet rust disease is also one of the most important diseases, especially when the infection severity is very high. Several reports indicated that the disease decreases the yield by 15% of root weight and 1% of sugar content (O'Sullivan 1996).

Causal Agent

Sugar beet rust disease is caused by *Uromyces betae*. The fungus belongs to Basidiomycota family: Pucciniomycetes.

Pathogenesis

The rust persists on overwintered seed crops, clamped mangolds, ground keeping beet, and mangolds. The viability of teliospores can reach 2 years.

Symptoms

Cinnamon brown pustules are scattered over the lamina (rusty color) under the leaf epidermis (Fig. 25.8). The mature pustules crack the epidermis cover and release the rusty at the center of the spots. The urediniospores spread to other leaves and plants and from one field to another either by air or mechanically. Generally, the disease severity is always low in most beet crops and does not cause any economic damage (Pozhar and Assual 1971; Voegele et al. 2009).

Favorable Conditions

The favorable conditions for disease incidence are when temperature ranged between 15 and 20 °C and moist conditions when the dew presence is long under cloudy weather (Pozhar and Assual 1971).

Control

Many fungicides were used in chemical control of rust of sugar beet (flusilazol, difenoconazol); fenpropimorf has shown good effect (Sorensen and Marcussen

Fig. 25.8 Sugar beet rust disease

1996; O'Sullivan 1996). Biological control agent Bacillus subtilis QST 713 was also used to control the disease; however, chemical control results were better (Kristoffersen et al. 2018).

25.2.1.6 Powdery Mildew

Sugar beet Powdery mildew is a serious disease in the dry climate zone. It can be found in all sugar beet around the world. The disease can decrease the yield by 20–25% in severe infection (Karve et al. 1973).

Causal Agent

The disease caused is by the Ascomycotina fungus *Erysiphe betae* (Vañha) Weltz; the fungus belongs to Family Erysiphaceae. The fungus is an obligate parasite on the Genus *Beta* sp.

Pathogenesis

The infection is caused by sexual spores every year (ascospores). Overwintering debris is infected with fungal resting spores (Chasmothecian). Chasmothecia release asci containing ascospores under favorable environmental conditions (15–25 °C). The wind transports the developed ascospores on the plant leaf's surface. The spores begin to germinate on the leaf's surface, producing a feeding structure called haustoria as well as Asexual spores (conido spores). Secondary infection on the same plant or from one plant to another is caused by conidio spores. Chasmothecium is eventually produced on the surface of the infected organism at the end of the growing season. Chasmothecium is then produced on the surface of the infected tissues (Mukhopadhyay and Russell 1979a; Esh and El-Kholi 2007; Esh and Shalaby 2008).

Symptoms

White, floury patches start to appear on the upper surface of infected old leaves. Under favorable environmental conditions for the fungus, colonies expand gradually and combine to cover large areas of the leaf. The affected leaf tissues turn yellow and the leaves look as if dusted with white powder (Fig. 25.9). The fruiting bodies of the fungus Cleistothecia begin to form and appear most strongly on the infected leaf spots in late summer (Esh and El-Kholi 2007). The cleistothecia of the fungus are small, spherical, and brownish-black bodies (Fig. 25.9). Eventually, the infected leaves collapse and die (Mukhopadhyay and Russell 1979b).

Favorable Conditions

High temperatures (15–28 °C) and low humidity <60% are the most favorable conditions for developing powdery mildew.

Control

Powdery mildews should be prevented and controlled using good cultural techniques. Powdery mildew is most commonly prevented by planting-resistant cultivars, removing the remnants of the previous crop and fertilizing with a balanced fertilizer (Lewellen 2000; Lewellen and Schrandi 2001). Generally, the disease must be controlled by one or two applications of fungicide sprays that start in the beginning of symptoms' appearance. In many sugar beet growing countries,

Fig. 25.9 Powdery mildew symptoms on sugar beet plant. Up-left shows the powdery blotches on the leaf. Up-right shows the Cleistothecia of the fungus; and a severe infection on sugarcane plant in the field

powdery mildew and *Cercospora* leaf spot are treated together as the specific fungicides used to control *Cercospora* found to decrease the incidence and severity of powdery mildew. However, recent reports on the development of sexual stage of the fungus could make its control is difficult.

25.2.1.7 Bacterial Leaf Spot (Bacterial Blight)

Causal Agent

The disease is caused by the gram-negative short rods bacterium *Pseudomonas syringae pv. Aptata.* The pathogen belongs to Phylum: Proteobacteria; Class: Gammaproteobacteria; Family: Pseudomonadaceae.

Pathogenesis

Pseudomonas syringae grows in two stages: inside plant tissues (endophytic phase) and outside (epiphytic phase). The bacteria grow on the phyllosphere, and then penetrate the plant tissue and colonize it intercellularly in order to produce the disease. Stomata and leaf hydathodes are the common entry points to plant tissue or through wounds and lesions on the leaf. After colonizing plant tissues, *P. syringae* start to produce phytotoxin and other metabolites that cause the necrotic spots (Hirano and Upper 2000; Xin et al. 2018).

Symptoms

Stojšin et al. (2015) described the diseases' symptoms as round to irregular necrotic patches (5–20 mm in diameter), each with a black edge and a tan to light brown core (Fig. 25.10) (Rotondo et al. 2020).

Favorable Conditions

The bacteria require high humidity levels. Many studies have been published on the importance of high humidity conditions such as dew, fog, and rain in promoting the population of *P. syringae* on the plant surface, as well as their relationship with

Fig. 25.10 Severe bacterial leaf spot (Blight) on sugar beet

epidemics in the field (Hirano and Upper 1990; Rouse et al. 1985). Temperature, on the other hand, has a crucial influence in the development and pathogenicity of bacterial cells. Wang et al. (2009) found that the warm temperature (about 28 °C) is ideal for development and pathogenicity.

Control

Up to now, there is no chemical control methods reported as an efficient method to control disease. In some situations, increasing the intervals between irrigations of sugar beets fields that are irrigated by spray irrigation may stop or reduce disease speared (Bashan 1997).

25.2.2 Viral Diseases

Sugar beet foliage is affected by many different viral diseases such as Beet curly top virus (BCTV), Beet distortion mosaic, Beet leaf curl virus (BCLV), Beet mild yellows & Beet western yellows (BMY), Beet mosaic virus (BtMV), Beet yellow net virus (BYNV), and Beet yellows (BYV). Here we show the most important economic viral diseases.

25.2.2.1 Beet Curly Top Virus (BCTV)

Beet curly top disease is one of the important sugar beet viral diseases. The disease can cause a dramatic loss in many regions in the tropical and subtropical regions.

Causal Agent

Beet curly top virus is a DNA virus that belongs to genus Curtovirus, family Geminiviridae. It causes curly top disease in several economically important crops. The virus transmits by the leaf hoppers (*Circulifer tenellus*) (Stanley 2008; Horn et al. 2011).

Pathogenesis

The vector of the virus is Leafhoppers that overwinter on annual and perennial weeds. The vectors acquire the disease after feeding on infected plants to be able to transmit the virus for the rest of their lives. Over 300 species in 44 plant families, such as tomatoes, beans, and peppers, are reported as hosts for leafhoppers (Chen et al. 2010).

Symptoms

Many different symptoms of beet curly top virus were reported including vein swelling, leaf curling, yellowing of leaves with purple veins, necrosis and Phloem hyperplasia, and stunting and death of young seedlings (Chen et al. 2010).

Favorable Conditions

The vector is favorable in the warm climates in tropical and subtropical regions. It is also reported in different parts of the world United States, Mexico, South America, the Mediterranean basin, and the Middle East (Chen et al. 2010).

Control

Cultivating-resistant sugar beet cultivars and controlling the weeds that harbor leaf hoppers as well as chemical control of the leaf hoppers are the best ways to prevent the disease in sugar beet fields (Stanley 2008; Strausbaugh et al. 2006).

25.2.2.2 Sugar Beet Yellows Virus (BYV)

The yellows virus is a very important disease that can cause yield reductions of 50% in sugar beet crops.

Causal Agent

Beet yellows is the common yellows virus of three other viruses (Beet mild yellowing virus and beet chlorosis virus). The three viruses make the so-called virus yellows complex. Eight species of aphids can transmit the virus, especially green peach aphids (*Myzus persicae*).

Symptoms

Symptoms appeared as circular yellow areas on leaves and the veins still green, with brittle outer leaves showing yellow between the veins. As the infection develops, it spread throughout the plant (Fig. 25.11).

Control

No control treatment has been developed.

25.2.3 Seedling and Root Rot Diseases

Sugar beet root rot diseases are occurred by several soil born microorganisms during germination, emergence, or juvenile growth as well as post-harvest. Root rot diseases are caused by many fungus species such as *Rhizoctonia solani* (Abadan 1994; Elliger and Halloin 1994; Esh et al. 2004), *Macrophomina* spp. (Koppanyi et al. 1993), *Fusarium* spp. (Bosch and Mirocha 1992; Abadan 1994), *Pythium* spp. (Orlandini and Signorini 1993; Stephens et al. 1993; Whipps et al. 1993; Abadan 1994; Payne et al. 1994), and *Sclerotium rolfsii* Abadan 1994, Esh and El-Kholi 2003). The root rot disease is also caused by *Erwinia carotovora* subsp. *betavasculorum* (Costa and Loper 1994). Sugar beet root rot organisms seriously attacked sugar beet crop in the fields (from seeding to maturity) and through storage (Whitney and Duffus 1986; Ishikuri et al. 1992; Peretyat-ko et al. 1993; Abadan 1994; Cooke and Scott 1995).

Fig. 25.11 Beet yellow virus

25.2.3.1 Seedling Diseases (Damping-off Diseases)

Causal Agent

The disease is caused by different fungi, *Rhizoctonia solani*, *Aphanomyces cochlioides*, *Phoma beta*, and several *Pythium* species (Harveson 2006; Vincelli 2008).

Pathogenesis

The pathogens either attack the seed before germination or attack the seeds after germination. The fungal hyphae penetrate the germinated root and produce hydrolytic degrading enzymes pectinases and cellulases, which hydrolyze the tissues and cells and cause root death or seedling collapse (Esh et al. 2004; El-Kholi et al. 2005).

Symptoms

The fungus induces postemergence damping-off as well as preemergance damping-off. Just below the soil surface, a dark brown cankers extend up the hypocotyl. When the hypocotyl decays, the seedling collapses and dies (Fig. 25.12) (Whitney and Duffus 1986; Ishikuri et al. 1992; Peretyat-ko et al. 1993; Abadan 1994; Cooke and Scott 1995; Esh et al. 2004).

Fig. 25.12 Sugar beet damping-off disease on germinating seeds and seedlings

Favorable Conditions

Warm conditions (25–30 °C) are the most favorable condition for disease incidence (Lamichhane et al. 2017).

Control

Seed treatment with seed coating fungicides and planting in cool soil are the best practices to control the disease (Lamichhane et al. 2017).

25.2.3.2 *Rhizoctonia* Root and Crown Rot

Several root diseases are caused by the fungus *R. solani*. Root and crown rot are considered as the most important sugar beet diseases in all sugar beet countries in the temperate zone.

Causal Agent

The perfect stage of the fungus *R. solani* (imperfect) is *Thanatephorus cucumeris* which belongs to phylum basidiomycetes, order Tulasnellales, and family Ceratobasidiaceae (Parmeter and Whitney 1970).

Pathogenesis

The fungus grows on the root surface, to form a hyphae structure called dome shape or infection court. From the contact area of the dome shape structure and the surface of the root tissue, a penetration tube grows and directly penetrates the root by both mechanical pressure and hydrolytic enzymes that degrade the cell wall. The pathogen hyphae grow inter- and intracellularly in root tissue (Bateman 1970; Ruppel 1973).

Symptoms

The symptoms can be characterized as sudden wilt and chlorosis to plant foliage. At the base of the petioles, dark brown lesions appear. The diseased leaves eventually die and fall, but remain connected to the base, forming a rosette of brown leaves. On the root surface, dark brown to blackish lesions develop, while deep cankers appear

Fig. 25.13 *Rhizoctonia* root rot symptoms. Symptoms on young roots (Right); and symptoms on old roots (left); symptoms on foliage (rosette shape) in sever root infection (down)

on the crown (Fig. 25.13) (Windels and Nabben 1989; Halloin 1994; Esh et al. 2004).

Favorable Conditions

When soil temperatures rise, the disease can infect petioles, crowns, and roots of older plants. The fungus hyphae and sclerotia can survive on plant debris (Boosalis and Scharen 1959; Roberts and Herr 1979), becoming active when soil temperatures reach 25–33 °C (LeClerg 1939).

Control

Generally, as sugar beet root disease is covered under the soil, chemical control of Rhizoctonia root rot is not used in commercial fields. However, Bartholomäus et al. (2017) used a mix of fungicides (such as azoxystrobin and difenoconazole) treatments to control the disease. The disease has a diverse host range and may persist in the soil as a saprophyte for extended periods of time, making the complete control a challenge (Anees et al. 2010). Some agricultural operations, including tillage, crop rotation, plant residue management, reducing soil compaction, and improving drainage efficiency, may assist to reduce the quantity of inoculum in the soil (Buhre et al. 2009). Cultivation of resistant cultivars has been established for *R. solani* control. However, the disease resistance in the available varieties is moderate to Rhizoctonia and gives a low yield compared to other sugar beet commercial varieties (Buddemeyer and Märländer 2005).

25.2.3.3 Wet Rot of Sugar Beet Roots

Causal Agent

Wet rot of sugar beet rots caused by different Oomycota pathogens that belong to Order: Peronosporales; the first two pathogens (*Pythium ultimum* Trow and *P. debaryanum* Hesse) belongs to family Pythiaceae and the third pathogen *Phytophthora drechsleri* Tucker belongs to family Peronosporaceae (Whitney and Duffus 1986)

Pathogenesis

The fungal sporangia in the soil begin to germinate, and the sporangium develops a pronounced beak, after which the sporangial contents are discharged. The sporangium protoplasm begins to cleave, resulting in the formation of biflagellate veniform zoospores that are released into the soil. With the help of hydrolytic enzymes, the zoospores connect to the root epidermis and penetrate the root tissue. The fungal mycelium develops between and inside root tissues, producing the hydrolytic enzymes cellulases and pectinases that induce soft degradation of root tissues (Sutton et al. 2006).

Symptoms

Typically, symptoms include wilting and a deep, dark to blackish wet (watery) rot at the base of the taproot, which spreads upward from the lower section of the root to the crown (Fig. 25.14). Dark-colored, irregular lesions develop on the root surface. Wilted plants may recover at night in the early stages of this illness because infected roots have a "rubbery sensation" (Cooke and Scott 1995)

Fig. 25.14 Symptoms of Wet rot disease on sugar beet

Favorable Conditions

The most favorable conditions for the disease are soil temperatures 28–32 °C and excessive soil moisture for at least 12 h (Vesely 1986; von Bretzel et al. 1988)

Control

Agricultural practices that avoid prolonged periods of high soil moisture are the best management of wet rot (Buhre et al. 2009; Anees et al. 2010). Some reports showed that the use of fungicide seed treatments such as hymexazole and metalaxyl or treating the soil with metalaxyl decrease the disease incidence (Bartholomäus et al. 2017)

25.2.3.4 *Sclerotium* Root Rot

Causal Agent

Sclerotium root rot disease is caused by *Sclerotium rolfsii*; its imperfect stage is soil-borne saprophytic fungus *Sclerotium rolfsii* Sacc. and its perfect stage is *Athelia rolfsii* that belongs to Phylum: Basidiomycota and Class: Agaricomycetes. More than 200 species of plants serve as hosts for the fungus. Sclerotia can be globose, elongate, inflated, or flattened, typically band-like, single or confluent, occasionally covering broad areas, mainly dark colored, commonly black, rigid, especially when dried, and inside brightly colored. Color and cell structure distinguish rind-tissue from the inside (Whitney and Duffus 1986).

Symptoms

A very watery, blackish rot develops in the tap roots, which become covered with thick, ropy strands of cottony hyphae and vast numbers of spherical, white to dark

Fig. 25.15 Symptoms of *sclerotium* root rot on roots and foliage

brown sclerotia (Fig. 25.15), 1–3 mm in diameter (Mukhopadhyay 1987; Esh and El-Kholi 2003). The causal organism of this disease induces permanent wilting. These hyphal strands and sclerotia can be also detected in the soil from diseased roots (Schneider and Whitney 1986, Esh and El-Kholi 2003).

Favorable Conditions

The sclerotia persist in soil for long time and serve as the source of primary inoculums. High soil moisture ($\geq$70% of field capacity) and temperature between 25–35 °C are the favorable conditions for the disease (El-Kholi 1979; Schneider and Whitney 1986; Pinheiro et al. 2010).

Control

The fungus has a host range of more than 200 species making the disease management difficult. Crop rotations that have less susceptible hosts (corn, alfalfa, wheat, or barley) can reduce the fungal inoculum. The use of chemical control as soil applications can provide control of the disease. Dwivedi and Ganesh (2016) reported a list of different fungicides to control the disease such as carbendazim, carboxin, benomyl, sancozeb, dithane M-45, captan, propiconazole, and thiram; they also reported some other plant extracts such as garlic, clove, ginger rhizome, neem leaf and seed oil, and onion bulb. On the other hand, many other researchers reported biological control as an active method to control the disease. Different fungal and bacterial bioagents were reported such as different *Trichoderma sp.*, *Penicillium* sp., *Curvularia* sp., and *Aspergillus niger* as well as bacterial bioagents such as different *Pseudomonas sp.* and *Bacillus subtilis* (Rasu et al. 2013; Babu and Paramageetham 2013; Dwivedi and Ganesh 2016).

25.2.3.5 *Fusarium* Root Rot

Causal Agent

Many *Fusarium* spp. are isolated from rotted roots of sugar beet by plant pathologists (Stanek 1983). Popova et al. (1985) and Guzhova et al. (1988) in the countries of the former USSR isolated *F. moniliforme* Shellden, while Burenin and Timoshenko (1985) isolated unidentified *Fusarium* spp. from sugar beet rotted roots collected from different areas in USSR and Poland. In Egypt, El-Kholi (1979) and Abadan (1994) isolated *F. moniliforme* Sheldon *var. subgulitinans* Wr. and *Reink* and *F. solani* (Mart.) Sacc. from rotted roots of sugar beet. Martyn et al. (1989) and Hanson et al. (2018) stated that *F. oxysporum* is responsible for sugar beet root rot in USA. In Germany, El-Abyad and Abu-Taleb (1991) isolated *F. solani,* while Bosch and Mirocha (1992) isolated *F. moniliforme var. subglutinans*, *F. poae* (Peck) Wolenweber, *F. sporotichioids* Sherbakolf, *F. equiset* (Corda) Saccardo, and *F. gramineum* Corda. The fungus *Fusarium* spp. is the causal organism of *Fusarium* root rot. This fungus belongs to Deuteromycetes, Order Moniliales, and Family Tuberculareaceae (Booth 1977).

Pathogenesis

The remained chlamydospores in the soil from last season germinate to give a macroconidia or hyphae that invade the susceptible plants through natural root wounds (Whitney and Duffus 1986).

Symptoms

Older leaves of plants affected by *Fusarium* show wilting and necrosis and interveinal chlorosis. Eventually, leaves become scorched, dry, and brittle. The root tip shows a black rot with heavy root growth along the tap root. The infected root cross-section shows grayish-brown vascular discoloration. As the disease progresses, leaves become yellow, chlorotic, and have necrotic areas (Martyn et al. 1989).

Favorable Conditions

The disease is initiated in warm soils. The favorable temperature for infection and symptom development is between 25 °C and 30 °C. High soil moisture is also necessary for disease development (Whitney and Duffus 1986).

Control

Agricultural practices such as planting into cool soils, control of irrigation, crop rotation for 2 years with cereals and alfalafa, and controlling the weeds can reduce the inoculum in the soils and the disease incidence (Harveson and Rush 1998).

25.2.3.6 Charcoal Rot

Causal Agent

The disease is caused by *Macrophomina phaseolina* (Tassi) Goid. The fungus is the imperfect stage of *Sclerotium bataticola* Taub. Sclerotia is the pathogen's most obvious sign; they are smooth, black with a spherical to irregular shape, and vary in size from 50 to 150 μm in diameter (Whitney and Duffus 1986).

Symptoms

Wilting of the foliage is the first symptom of infection, which eventually turns brown and dies. Brownish-black, irregular lesions appear externally on the crown and tip of the root; the roots become grayish brown to black with a silverfish reflection (Fig. 25.16). The root tissues turn into sponge like consistency, with colors ranging from lemon yellow and finally brownish to black (Tomkins 1938; Vera et al. 2012).

Favorable Conditions

The fungus attacks sugar beet plants under stress, weakened or injured. High temperatures (optimum of 28–31 °C) with dry conditions are favorable conditions for disease development (Tomkins 1938, Vera et al. 2012).

25.2.4 Root Bacterial Diseases

25.2.4.1 Tuberculosis Disease (Bacterial Pockets)

The disease is sporadically reported in the United States and some other sugar beet growing countries (Harveson et al. 2009; Moliszewska et al. 2016).

Causal Agent

Xanthomona beticola

Fig. 25.16 Symptoms of sugar beet charcoal rot

Symptoms

The symptoms of the disease can be described as the formation of one or more large and irregular tumor-like shapes on the root crown (Fig. 25.17). The disease can decrease the root sucrose contents by 15%. (Moliszewska et al. 2016).

Favorable Conditions

High temperature greater than 28 °C.

Control

No effective chemical control has been developed.

25.2.4.2 Crown-Gall Disease

Sugar beet crown gall disease is not considered as one of the economic diseases; however, there are some recent reports on losses of the disease in other crops.

Causal Agent

Agrobacterium tumefaciens

Fig. 25.17 Sugar beet crown gall symptoms

Symptoms

Formation of galls attached to the side of the root or crown area. The galls size is varied from small to a very large one that sometimes becomes as the same size as the root. The disease causes a metabolic disturbance that leads to stunt the plant and decrease sugar contents (Mafakheri et al. 2016).

Control

Collecting and destroying diseased plants should prevent the spread of infection.

25.2.4.3 Root Soft Rot

Causal Agent

The disease is caused by the short rod, gram-negative motile (peritichous) bacterium *Pectobacterium carotovorum* (*Erwinia carotovora* subsp. *betavasculorum*) (Thomson et al. 1977).

Pathogenesis

The entry of the bacteria to the root is wound that exists naturally or mechanically. The infection starts from crown to the root and other parts of the plant. The bacteria secreting intensive amounts of hydrolytic cell wall degrading enzymes (polygalactrunase, pectinases, and cellulases) make all the root tissues become soft and completely decayed (Thomson et al. 1977; Fassihiani and Nedaeini 2008).

Symptoms

The disease symptoms can be recognized after the root is rotted by wilting of the leaves and become wet dark brown color. The disease can be observed at any time during the growing season when the environmental conditions are suitable (Fassihiani and Nedaeini 2008). When the disease attacks the plants at the end of the season, we notice a dome of bubbles formed from sugar fermentation covering the dead plants.

Favorable Conditions

High temperatures of 25–30 °C and moisture are the favorable conditions for disease development. Excessive irrigation increases the speed of disease development (Thomson et al. 1977, Fassihiani and Nedaeini 2008).

Control

Cultivating resistant varieties is important to prevent the disease incidence and crop rotation is the most effective to decrease the incidence of the disease in infected fields (Thomson et al. 1977).

25.2.4.4 Root Viral Diseases

Beet Necrotic Yellow Vein Virus (Rizomania)

Rhizomania disease can cause great damages to sugar beet crop. The disease reduces root yield, sugar yield as well as sugar extractability.

Causal Agent

The causal virus is beet necrotic yellow vein virus (BNYVV), which belongs to the genus Benyvirus (BNYVV). The virus's vector is the soil-borne protozoa *Polymyxa betae*, which belongs to Plasmodiophoromycetes, and Order: Plasmodiophorales (Koenig and Lennefors 2000).

Pathogenesis

The fungus *Polymyxa betae* lives in the soil as cysts for more than 10 years. When soil temperatures reach a certain level, root exudates from sugar beet roots or other appropriate hosts trigger cyst germination. The germinated cysts produce virulent zoospores, which infect the root cells and spread the virus to the plant (Tamada and Abe 1989; Richards and Tamada 1992).

Symptoms

Sugar beet roots that have been infected are dwarfed (Fig. 25.18). Taproot stimulates rootlet multiplication, resulting in necrotic, numerous, and fragile rootlets. The leaves exhibit vein yellowing, necrosis, and isolated lesions on the foliage. The yellowing of the leaf is followed by necrosis along the veins (Tamada 1975).

Favorable Conditions

The optimum conditions for *P. betae* are high soil moisture and temperature ranging from 15° to 28 °C. The vector of the virus is a protozoan living organism Polymyxabetae, which is a soil born root parasite. In the absence of favorable conditions, the vector and virus complex keeps its infectious potential for more than 20 years. When the soil temperature reaches the optimum and the soil moisture level is high, the dormant spores germinate and produce primary zoospores, which is infectious to sugar beet root (Richards and Tamada 1992).

Control

There is no currently available environmentally safe and economic pesticide effective against *P. betae*; also alternative control strategies such as biological control didn't show a promising effectiveness (Jakubíková et al. 2006).

Breeding sugar beet for rhizomania-resistant is the only strategy for controlling the disease (Scholten and Lange 2000). The development of rhizomania-resistant varieties is the only solution because of high spread of the disease in many sugar beet countries (McGrann et al. 2009).

Fig. 25.18 *Rhizomania* symptoms on foliage

25.3 Conclusion

The nature of the sugar beet crop in terms of its juicy leaves and its high sugar content roots makes it vulnerable to many diseases that affect its quantitative and qualitative production. From our discussion of the diseases that affect the sugar beet, we can conclude that the greatest effect on production comes from diseases of the roots, which cause serious roots damages (root-rot) that dramatically decrease its economic and industrial value. Moreover, there are no economic and effective ways to control these diseases, hence the importance of integrated control of these diseases, starting from the preparation of soil for planting, balanced fertilization, and irrigation, which provide unfavorable conditions for the causal pathogens of these diseases. On the other hand, we find that breeding for disease-resistant varieties is a very important solution to such pathological problems, whether for root diseases or foliage diseases.

References

Abadan KA (1994) Fungi causing damping-off and root rot on sugar beet and their biological control with *Trichoderma harzianum*. Agric Ecosyst Environ 51(3):333–337

Agnihotri VP (1990) Diseases of sugarcane and sugar beet. Oxford and IBH Publishing Co., New Delhi, pp 443–445

Andersen B, Hansen M, Smedsgaard J (2005) Automated and unbiased image analyses as tools in phenotypic classification of small-spored *Alternaria* spp. Phytopathology 95:1021–1029

Anees M, Edel-Hermann V, Steinberg C (2010) Build up of patches caused by *Rhizoctonia solani*. Soil Biol Biochem 42:1661–1672

Asher MJC, Hanson LE (2006) Fungal and bacterial diseases. In: Draycott AP (ed) Sugar beet, 1st edn. Blackwell Publishing Ltd., pp 286–315

Babu GP, Paramageetham CH (2013) Biocontrol of *Sclerotium rolfsii*—a polyphagous plant pathogen by *pseudomonas aeruginosa* isolated from forest litter. Int J Res Plant Sci 3:1–4

Baltaduonytė M, Dabkevičius Z, Brazienė Z, Survilienė E (2013) Dynamics of spread and control of *Cercospora* (*Cercospora beticola* Sacc.) and ramularia (*Ramularia beticola* Fautrey & F. Lamb.) leaf spot in sugar beet crops. Zemdirbyste-Agric 100:401–408

Bartholomäus A, Mittler S, Märländer B, Varrelmann M (2017) Control of *Rhizoctonia solani* in sugar beet and effect of fungicide application and plant cultivar on inoculum potential in the soil. Plant Dis 101(6):941–947

Bashan Y (1997) Alternative strategies for controlling plant diseases caused by pseudomonas syringae. In: Rudolph K, Burr TJ, Mansfield JW, Stead D, Vivian A, von Kietzell J (eds) Pseudomonas Syringae pathovars and related pathogens, Developments in plant pathology, vol 9. Springer, pp 575–583

Bateman DF (1970) Pathogenesis and disease. In: Parmeter JR Jr (ed) *Rhizoctonia solani*: biology and pathology. University of California Press, Berkeley, pp 161–171

Bleiholder H, Weltzien HC (1971) Beitrage zur Epidemiologie von *cercosporabeticolia* an Zuckerrube. III, Ceopathologische Untersuchungen. Phytopathologische Zeitschrift 73:93–114

Boosalis MJ, Scharen AL (1959) Methods for microscopic detection of *Aphanomyceseuteiches* and *Rhizoctoniasolani* and for isolation *Rhizoctoniasolani* associated with plant debris. Phytopathology 49:192–198

Booth C (1977) *Fusarium* laboratory guide to the identification of the major species. Commonwealth Mycological Institute, Kew, Surrey, pp 130–153

Bosch U, Mirocha CJ (1992) Toxin production by *Fusarium* species from sugar beets and natural occurrence of zearalenone in beets and beet fibers. Appl Environ Microbiol:3233–3239

Buddemeyer J, Märländer B (2005) Genotypic reaction of sugar beet to *Rhizoctonia solani* root and crown rot—susceptibility, yield and quality at different levels of infestation. J Plant Dis Prot 112:105–117

Bugbee WM (1979) *Pleospora bjoerlingii* in the USA. Phytopathology 69:277–278

Bugbee WM, Soine OC (1974) Survival of *Phomabetae* in soil. Phytopathology 64:1258–1260

Buhre CC, Kluth Bürcky K, Märländer B, Varrelmann M (2009) Integrated control of root and crown rot in sugar beet: combined effects of cultivar, crop rotation, and soil tillage. Plant Dis 93: 155–161

Burenin VI, Timoshenko ZV (1985) Blackleg resistance in beet. Sobrink Nauchnykh Trudov po Prikladnoi Botanike, Genetikei Selektsii 92:56-62 (c.f. CAB Abstracts 1984-1986)

Byford WJ (1996) A survey of foliar diseases of sugar beet and their control in Europe. In Proceedings of the 59th IIRB conference, 10. Caesar-TonThat, T.C., R.T

Calpouzos L, Stalknecht GF (1967) Symptoms of Cercospora leaf spot of sugar beets influenced by light intensity. Phytopathology 57:799–800

Carlson LW (1967) Relation of water factor to dispersal of conidia of *Cercospora beticola*. J Am Soc Sugar Beet Technol 14:310–323

Chen L-F, Kelly Clark B, Gilbertson RL (2010) Characterization of Curtoviruses associated with curly top disease of tomato in California and monitoring for these viruses in beet leafhoppers. Plant Dis 94:99–108

Chupp CC (1953) A monograph of the fungus genus Cercospora. Cornell University, Ithaca, New York

Compant S, Christophe C, Angela S (2010) Plant growth-promoting bacteria in the rhizo- and endosphere of plants: their role, colonization, mechanisms involved and prospects for utilization. Soil Biol Biochem. http://www.elsevier.com/locate/soilbio

Cooke DA, Scott RK (1995) The sugar beet crop. Chapman & Hall Publisher, New York, London Paris, p 657

Costa JM, Loper JE (1994) Derivation of mutants of *Erwinia carotovora* subsp. beta vasculorum. Appl Environ Microbiol 60(7):2278–2285

Douglas P, Collins A, Barry J, Jacobsen B, Maxwell B (2003) Spatial and temporal population dynamics of a phyllosphere colonizing *Bacillus subtilis* biological control agent of sugar beet cercospora leaf spot. Biol Control 26:224–232

Dwivedi SK, Ganesh P (2016) Integrated management of *sclerotium rolfsii*: an overview. Eur J Biomed Pharm Sci 3:137–146

El-Abyad MS, Abu-Taleb AM (1991) Growth activities of the sugar beet pathogens *Fusarium solani* (Mart.) Sacc., *Rhizoctonia solani* Kuhn and *Sclerotium rolfsii* Sacc. under pyramin stress. Zentralblatt fur Microbiologia 146:419–424. (c.f. CAB abstract, 1992)

El-Kholi MMA (1979) Studies on root-rot of sugar beet in Egypt. M.Sc. Thesis, Fac. Agric., Ain Shams Univ., pp 81

El-Kholi MA, Esh AMH (2011) Comparative structural and biochemical study on calcium effects on cercospora leaf spot disease of sugar beet. Mansoura J Plant Prot Pathol 2:85–97

El-Kholi MM, Ragab MM, Hussein MY (1994) *Alternaria* leaf spot of sugar beet in Egypt. Egypt J Phytopathol 2:179–193

El-Kholi MMA, Aly AZ, Esh AMH (2005) Enzymatic activity and isozymes patterns in mature healthy and infected (*Rhizoctonia solani*) sugar beet roots. Egypt J Agric Res 2:225–241

Elliger CA, Halloin JM (1994) Phenolics induced in Beta vulgaris by Rhizoctonia solani infection. Phytochemistry 37:691–693

Esh AMH, El-Kholi MMA (2003) *Sclerotium rolfsii* saac. Root rot and its effect on sugar beet quality. Proceedings of First Conference on Farm Integrated Pest Management, Fac. Agric, Fayoum Univ. Egypt, pp. 72–80

Esh AMH, El-Kholi MMA (2007) First record of the perfect stage of powdery mildew of sugar beets in Egypt. Zagazig J Agric Res 85:1263

Esh AMH, Kamhawy MA (2010) Carbendazim and Mancozeb combination for controlling Cercospora leaf spot in sugar beet (*Beata vulgaris* L.). Egypt J Appl Sci 25:242–255

Esh AMH, Moghaieb REA (2011) Analysis of morphological, pathological and genotipic diversity in (*Cercospora beticola* sacc.) from different sugar beet cultivation in Egypt. Arab J Biotechnol 14:77–88

Esh AMH, Shalaby MS (2008) Environmentally safe compounds in controlling sugar beet powdery mildew. Egypt J Appl Sci 23:447–461

Esh AMH, El-Kholi MMA, Aly AZ, Shalaby MS (2004) Characterization and diversity of *Rhizoctonia solani* Kuhn infecting sugar beet under Egyptian conditions. Proceeding of the International Conference of Genetic Engineering and its Applications Sharm El-Shaikh, Egypt, pp. 299–316

Esh AMH, Atia MMM, Tohamy MRA, Taghian S (2011a) Systemic resistance in sugar beet eliciated by non-pathogenic, phyllosphere-colonizing *bacillus pumilus* and *b. Subtilus* againstthe pathogen *cercospora beticola sacc*. Mansoura J Plant Prot Pathol 2:67–83

Esh AMH, El-Kholi MAM, Shadia T (2011b) Antagonistic activities of *bacillus amyloliquefaciens* from phyllosphere of sugar beet against *cercospora beticola* sacc. Mansoura J Plant Prot Pathol 2:99–116

Fassihiani A, Nedaeini R (2008) Characterization of Iranian *Pectobacterium carotovorum* strains from sugar beet by phenotypic tests and whole-cell proteins profile. J Phytopathol 156:281–286
Franc GD (2010) Ecology and epidemiology of *Cercospora beticola*. In: Lartey RT, Weiland JJ, Panella L, Crous PW, Windels CE (eds) Cercospora leaf spot of sugar beet and related species. The American Phytopathological Society, St. Paul, MN, pp 7–19
Frisvad JC, Andersen B, Thrane U (2008) The use of secondary metabolite profiling in chemotaxonomy of filamentous fungi. Mycol Res 112:231–240
Fujiwara T, Oda K, Yokota S, Takatsuki A, Ikehara Y (1988) Brefeldin A causes disassembly of the Golgi complex and accumulation of secretory proteins in the endoplasmic reticulum. J Biol Chem 263:18,545–18,552
Garibaldi A, Gilardi G, Bertetti D, Gullino ML (2007) First report of leaf spot and root rot caused by *Phomabetae* on *Beta vulgaris* subsp. *vulgaris* (garden beet group) in Italy. Plant Dis 91:1515
Girotti AW (1990) Photodynamic lipid peroxidation in biological systems. Photochem Photobiol 51:497–509
Goodwin SB, Dunkle LD, Zismann VL (2001) Phylogenetic analysis of *Cercospora* and *Mycosphaerella* based on the internal transcribed spacer region of ribosomal DNA. Phytopathology 91:648–658
Groenewald M, Groenewald JZ, Harrington TC, Abeln ECA, Crous PW (2006) Mating type gene analysis in apparently asexual *Cercospora* species is suggestive of cryptic sex. Fungal Genet Biol 43:813–825
Groenewald M, Groenewald JZ, Linde CC, Crous PW (2007) Development of polymorphic microsatellite and single nucleotide polymorphism markers for *Cercospora beticola* (Mycosphaerellaceae). Mol Ecol Notes 7:890–892
Guzhova NN, Filatov PA, Gorina IN (1988) Permeability of the membranes and resistance to black leg. Sakharnaya Svekla 6:40–41. (c.f. CAB Abstracts, 1987–1989)
Halloin JM (1994) Localization of phenolic compounds in crowns and roots of healthy and Rhizoctonia solani infected sugar beet. Plant Sci Limerick 99(2):223–228
Hanson L, De Lucchi C, Stevanato P, McGrath M, Panella L, Sella L, De Biaggi M, Concheri G (2018) Root rot symptoms in sugar beet lines caused by *Fusarium oxysporum* f. sp. *betae*. Eur J Plant Pathol 150:589–593
Harveson RM (2006) Identifying and distinguishing seedling and root rot diseases of sugar beet. Plant Health Progress. https://doi.org/10.1094/PHP-2006-0915-01-DG
Harveson RM, Rush CM (1998) Characterization of *Fusarium* root rot isolates from sugar beet by growth and virulence at different temperatures and irrigation regimens. Plant Dis 82:1039–1042
Harveson RM, Hanson LE, Hein GL (eds) (2009) Compendium of beet diseases and pests, 2nd edn. APS Press, St. Paul, p 140
Hirano SS, Upper CD (1990) Population biology and epidemiology of *Pseudomonas syringae*. Annu Rev Phytopathol 28:155–177
Hirano SS, Upper CD (2000) Bacteria in the leaf ecosystem with emphasis on *Pseudomonas syringae*-a pathogen, ice nucleus, and epiphyte. Microbiol Mol Biol Rev 64:624–653
Horn J, Lauster S, Krenz B, Kraus J, Frischmuth T, Jeske H (2011) Ambivalent effects of defective DNA in beet curly top virus-infected transgenic sugar beet plants. Virus Res 158:169–178
Hudec K, Rohacik T (2002) *Alternaria alternata* (Fr.) Keissler—new pathogen on sugar beet leaf in Slovakia. Plant Prot Sci 38:81–82
Ichihara A, Oikawa H, Hayashi K, Sakamura S, Furusaki A, Matsumoto T (1983) Structures of betaenones A and B, novel phytotoxins from *Phoma betae* Fr. J Am Chem Soc 105:2907–2908
Ioanidis PM, Karaoglanidis GS (2010) Control of Cercospota leaf spot and powdery mildew of sugar beet with fungicides and tolerant cultivars. In: Lartey RT, Weiland JJ, Panella L, Crous PW, Windels PE (eds) Cercospora leaf spot of sugar beet and related species. The American Phytopathological Society, St. Paul, MN, pp 259–274
Ishikuri S, Uchino H, Kanzawa K (1992) Isolation of *Pseudomonascapacia* D-202: an available biocontrol agent to the storage rot of sugar beet. Ann Phytopathol Soc Jpn 58:456–460

Jakubíková L, Vubíková Š, Nemčovič M, Farkaš V (2006) Selection of natural isolates of *Trichoderma* spp. for biocontrol of *Polymixa betae* as a vector of virus causing rhizomoniain sugar beet. Biology 61:347–351

Karve AD, Bhalerao AC, Mhaiskar VG (1973) Powdery mildew of sugar beet. Proceedings annual convention. Deccan Sugar Technol Assoc 25:9–15

Koenig R, Lennefors BL (2000) Molecular analyses of European A, B and P type sources of beet necrotic yellow vein virus and detection of the rare P type in Kazakhstan. Arch Virol 145:1561–1570

Koppanyi M, Nagy I, Zsembery S, Bodis Z (1993) A Macrophomina wilt and root rot of sugar beet. Cukoripar 46:121–124. (c.f. CAB Abstracts 1995)

Kristoffersen R, Hansen AL, Munk L, Cedergreen N, Jørgensen LN (2018) Management of beet rust in accordance with IPM principles. Crop Prot 111:6–16

Lamichhane JR, Durr C, Schwanok AA, Robin MH, Sarthou JP, Cellier V, Messean A, Auberlot JN (2017) Integrated management of damping-off diseases. A review. Agron Sustain Dev 37:2–25

Larson BJ (2004) Integrated management of Cercospora leaf spot on sugar beet. M.Sc. Montana Univ. USA, pp 121

Lawrence DP, Rotondo F, Gannibal PB (2016) Biodiversity and taxonomy of the pleomorphic genus *Alternaria*. Mycol Progr 15:3

LeClerg EL (1939) Studies on dry-rot canker of sugar beets. Phytopathology 29:793–800

Lewellen RT (2000) Registration of powdery mildew resistant sugar beet germplasms CP 01 and CP 02. Crop Sci 40:1515

Lewellen RT, Schrandi JK (2001) Inheritance of powdery mildew resistance in sugar beet derived from *Beta vulgaris* sub sp. *maritima*. Plant Dis 95:627–631

Liebe S, Wibberg D, Winkler A, Pühler A, Schlüter A, Varrelmann M (2016) Taxonomic analysis of the microbial community in stored sugar beets using high-throughput sequencing of different marker genes. FEMS Microbiol Ecol:92

Mafakheri H, Taghavi SM, Banihashemi Z, Osdaghi E, Lamichhane JR (2016) Pathogenicity, host range and phylogenetic position of *Agrobacterium* species associated with sugar beet crown gall outbreaks in southern Iran. Eur J Plant Pathol:147

Martyn RD, Rush CM, Biles CL, Baker EH (1989) Etiology of a root rot disease of sugar beet in Texas. Plant Dis 73:879–884

Mcfarlane JS, Roy B, Snyder WC (1954) An *Alternaria* leaf spot of the sugar beet. Proceedings of American Society of Sugar Beet Technologists Eighth General Meeting, pp. 240–246

McGrann GR, Grimmer MK, Mutasa-Göttgens ES, Stevens M (2009) Progress towards the understanding and control of sugar beet rhizomania disease. Mol Plant Pathol 10:129–141

Meredith DS (1967) Conidium release and dispersal in *Cercospora beticola*. Phytopathology 57: 889–893

Meriggi P, Rosso F, Ioannides PM, Ayala Garcia J (2000) Fungicide treatments against Cercospora leaf spot in sugar beet (*Beta vulgaris* L.). Adv Sugar Beet Res IIRB 2:77–102

Miller J, Rekoske M, Quinn A (1994) Genetic resistance, fungicide protection and variety approval policies for controlling yield losses from *Cercospora* leaf spot infection. J Sugar Beet Res 31:7–12

Misra V, Mall AK, Kumar M, Srivastava S, Pathak AD (2020) Identification of two new Alternaria isolates on sugar beet (Beta vulgaris L.) plants in Lucknow, India. Arch Phytopathol Plant Protect 54(3–4):164–176. https://doi.org/10.1080/03235408.2020.1824378

Moan J, Berg K, Kvam E, Western A, Malik Z, Ruck A, Schneckenburger H (1998) Intracellular localization of photosensitizers. In: Photosensitizing compounds: their chemistry, biology and clinical use, Ciba foundation symposium. Vol. 146. Wiley, Chichester, pp 95–111

Moliszewska E, Nabrdalik M, Piszczek J (2016) Tubercle disease (*Xanthomonas beticola*) and other gall-malformed diseases of sugar beet roots: a review. J Plant Dis Prot 123:197–203

Mukhopadhyay AN (1987) Handbook on disease of sugar beet, vol I. CRC Press, Bocca Raton, FL, p 196

Mukhopadhyay AN (1992) Cercospora leaf spot of sugar beet. In: Mukhopadhyay AN, Kumar J, Chaubey HS, Singh US (eds) Plant diseases of international importance; diseases of sugar, forest and plantation crops Volume IV. Prentice Hall Inc., pp 62–77

Mukhopadhyay AN, Russell GE (1979a) Light and scanning microscopy of sugar beet powdery mildew. Trans Br Mycol Soc 72:316–319

Mukhopadhyay AN, Russell GE (1979b) Development of *Erysiphe betae* on leaves of four sugar beet varieties. Phytopathology Z 96:15–20

Mulder JL, Holliday P (1974) *Mycosphaerella fijiensis* CMI descriptions of pathogenic fungi and Bacteria n° 413. CMI

O'Sullivan E (1996) Effect of fungicides on the incidence of rust disease, root yield and sugar content in sugar beet. Irish J Agric Food Res 35:159–164

Orlandini L, Signorini E (1993) In vivo screening of beneficial microorganisms in control of Pythium ultimum. Mededelingen van de Faculteit Landbouwwetenschappen Universiteit Gent 58(3b):1297–1305. (c.f. Rev Pl Path (1994) 73:7652)

Ostry V (2008) *Alternaria* mycotoxins: an overview of chemical characterization, producers, toxicity, analysis and occurrence in foodstuffs. World Mycotoxin J 1:175–188

Parmeter JR, Whitney HS (1970) Taxonomy and nomenclature of the imperfect state. In: Parmeter JR (ed) *Rhizoctonia solani*: biology and pathology. University of California, Berkeley, pp 7–19

Payne PA, Asher MJC, Kershaw CD (1994) The incidence of *Pythiumspp.* and Aphanomyces cochlioides associated with the sugar beet growing soils in Britain. Plant Pathol 43:300–308

Peretyat-ko VG, Pokhiton SV, Valovikova GD, Dubich ZI, Man'ko AE, Lunin NK, Kazachenko RF, Kovbasyuk EV, Yakovets VA (1993). Accessions resistant to diseases. Sakharnaya Svekla, No. 6, 28-31 (c.f. CAB Abstracts, 1993-4/95)

Pervin L, Islam MS (2015) System dynamics approach for modeling of sugar beet yield considering the effects of climatic variables. J Sci Food Agric 95:515–521

Pethybridge SJ, Kikkert JR, Hanson LE, Nelson SC (2018) Challenges and prospects for building resilient disease management strategies and tactics for the New York table beet industry. Agronomy (Basel) 8:112

Pinheiro VR, Seixas CDS, Godoy CV, Soares RM, de Oliveira MCN, Almeida ÁMR (2010) Development of *Sclerotium rolfsii,* sclerotia on soybean, corn and wheat straw, under different soil temperature and moisture contents. Pesq agropec Bras 45:332–334

Ploetz R (2001) The most important disease of a most important fruit. Plant Health Instructor. https://doi.org/10.1094/PHI-I-2001-0126-01

Pool VW, McKay MB (1915) *Phoma betae* on the leaves of sugar beet. J Agric Res 4:169–177

Popova IV, Lunin NK , Startseva KN (1985) A new method of selecting beet forms resistant to black leg. Effecktive mery Zashchity Sakh. Svekly ot boleznei priinduster. terknol. ee vozdel, pp. 119–122. (c.f. CAB Abstracts 1987–1989)

Pozhar Z, Assual B (1971) The biology of *Uromyces betae*, the cause of sugar beet rust. Mikolog i Fitopatol 5:166–171

Punja ZK, Utkhede RS (2003) Using fungi and yeasts to manage vegetable crop diseases. Trends Biotechnol 21:400–407

Raaijmakers MJ, Timothy CP, Christian S, Claude AY (2009) The rhizosphere: a playground and battlefield for soilborne pathogens and beneficial microorganisms. Plant Soil 321(1):341–361

Rangel LI, Spanner RE, Eber MK, Pethybridge SJ, Stukenbrock EH, de Jonge R (2020) *Cercospora beticola*: the intoxicating lifestyle of the leaf spot pathogen of sugar beet. Mol Plant Pathol 21: 1020–1041. https://doi.org/10.1111/mpp.12962

Rasu T, Sevugapperuma N, Thiruvengadam R, Ramasamy S (2013) Biological control of sugar beet root rot caused by *Sclerotium rolfsii*. Int J Biol Ecol Environ Sci 2:7–10

Richards KE, Tamada T (1992) Mapping functions on the multipartite genome of beet necrotic yellow vein virus. Annu Rev Phytopathol 30:291–313

Roberts DL, Herr LJ (1979) Soil populations of *Rhizoctonia solani* from areas of healthy and diseased beets within four sugar beet fields differing in soil texture. Can J Microbiol 25:902–910

Rossi V, Meriggi P, Biancardi E, Rosso F (2000) Effect of *Cercospora* leaf spot on sugar beet growth, yield and quality. In: Asher MJC, Holtschulte B, Richard Molard M, Rosso F, Steinruecken G, Beckers R (eds) Advances in sugar beet research Vol. 2: *Cercospora beticola* Sacc. Biology, agronomic influence and control measures in sugar beet. International Institute for Beet Research, Brussels, pp 49–76

Rotem J, Wendt U, Dranz J (1988) The effect of sunlight on symptom expression of *Alternaria alternata* on cotton. Plant Pathol 37:12–15

Rotondo F, Vrisman CM, Rani R, Testen AL, Deblais MSA (2020) First report of *Pseudomonas syringae pv. Aptata* causing bacterial leaf spot on common beet (*Beta vulgaris*) in Ohio. Plant Dis 104:561–561

Rouse DI, Nordheim EV, Hirano SS, Upper CD (1985) A model relating the probability of foliar disease incidence to the population frequencies of bacterial plant pathogens. Phytopathology 75: 505–509

Ruppel EG (1973) Histopathology of resistant and susceptible sugar beet roots inoculated with *Rhizoctoniasolani*. Phytopathology 63:871–873

Ruppel EG (1986) Cercospora leaf spot. In: Whitney ED, Duffus JE (eds) Compendium of beet diseases and insects. American Phytopathological Society, St. Paul, pp 8–9

Schneider CL, Whitney ED (1986) Root diseases caused by fungi. In: Whitney ED, Duffus JE (eds) Compendium of beet diseases and insects. APS Press, St. Paul., MN, pp 17–23

Scholten OE, Lange W (2000) Breeding for resistance to rhizomania in sugar beet: a review. Euphytica 112:219–231

Shane WW, Teng PS (1992) Impact of *Cercospora* leaf spot on root weight, sugar yield, and purity of Beta vulgaris. Plant Dis 76:812–820

Shim WB, Dunkle LD (2003) CZK3, a MAP kinase homolog in *Cercospora zeae-maydis,* regulates Cercosporin biosynthesis, fungal development, and pathogenesis. Mol Plant-Microbe Interact 9: 760–768

Smith GA, Campbell LG (1996) Association between resistance to Cercospora and yield in commercial sugar beet hybrids. Plant Breed 115:28–32

Smith GA, Gaskill JO (1970) Inheritance of resistance to Cercospora leaf spot in sugar beet. J Am Soc Sugar Beet Technol 16:172–180

Smith GA, Ruppel EG (1973) Association of *Cercospora* leaf spot, gross sucrose, percentage sucrose, and root weight in sugar beet. Can J Plant Sci 53:695–696

Sorensen F, Marcussen C (1996) Rust *Uromycesbetae* in Denmark inoculums sources and effect on sugar beet yield. 59th Congress Institute International de Recherche Betteravieres Palais des Congres, Bruxelles, Belgaum 13–17 février, pp. 119–128

Spikes JD (1989) Photosensitization. In: Smith KC (ed) The science of photobiology. Plenum Press, New York, pp 79–110

Stanek M (1983) The possibility of using *Agrobacterium radiobacter* B6 for the biological control of root rot of sugar beet. Sbornik UVTIZ, Ochrana Rostllin. 19: 3, 169–181. (c.f. CAB Abstracts 1984–1986).

Stanley J (2008) Beet curly top virus. In: Brian WJ, Mahy Marc HV, Regenmortel V (eds) Encyclopedia of virology, 3rd edn. Academic Press, pp 301–307

Steinkamp MP, Martin SS, Hoefert LL, Ruppel EG (1979) Ultrastructure of lesions produced in leaves of Beta vulgaris. Physiol Plant Pathol 15:13–16

Stephens PM, Grawley JJ, O'Connel C (1993) Selection of pseudomonad strains inhibiting *Pythium ultimum* on sugar beet seeds in soil. Soil Biol Biochem 25(9):1283–1288

Stojšin VJ, Balaž D, Budakov S, Stanković I, Nikolić Ž, Ivanović PT (2015) First report of *Pseudomonas syringae* pv. *aptata* causing bacterial leaf spot on sugar beet in Serbia. Plant Dis 99:281

Strausbaugh CA, Gillen AM, Gallian JJ, Camp S, Stander JR (2006) Influence of host resistance and insecticide seed treatments on curly top in sugar beets. Plant Dis 90(12):1539–1544

Sutton JC, Sopher CR, Liu TN, Grodzinski B, Hall JC, Benchimol RL (2006) Etiology and epidemiology of Pythium root rot in hydroponic crops: current knowledge and perspectives. Summa Phytopathol 32:307–321

Taghian S, Esh AMH, Aly AZ, Tohamy MRA (2008) *Bacillus subtilis* as bioagent used to control cercospora sugar beet leaf spot disease. Zagazig J Agric Res 35:210–224
Tamada T (1975) Beet necrotic yellow vein virus. CMI/AAB descriptions of plant viruses No. 144. Association of Applied Biologists, Wellesbourne
Tamada T, Abe H (1989) Evidence that beet necrotic yellow vein virus RNA-4 is essential for transmission by the fungus *Polymyxabetae*. J Gen Virol 70:3391–3398
Thachab T, Munkb L, Hansenc AL (2013) Disease variation and chemical control of *Ramularia* leaf spot in sugar beet. Crop Prot 51:68–76
Thomma BPHJ (2003) *Alternaria* spp.: from general saprophyte to specific parasite. Mol Plant Pathol 4:225–236
Thomson SV, Schroth MN, Hills FJ, Whitney ED, Hilbrand DC (1977) Bacterial vascular necrosis and rot of sugar beet: general description and etiology. Phytopathology 67:1183–1189
Tomkins CM (1938) Charcoal rot of sugar beet. Hilgardia 12:75–81
Trebbi D, McGrath MJ (2004) Fluorometric sucrose evaluation for sugar beet. J Agric Food Chem 52(23):6862–6867
Upchurch RG, Walker DC, Rollins JA, Ehrenshaft M, Daub EM (1991) Mutants of *Cercospora kikuchii* altered in cercosporin synthesis and pathogenicity. Appl Environ Microbiol 57:2940–2945
Vaghefi N, Silva A, Koenick LB, Pethybridge SJ (2019) Genome resource for *Neocamarosporium betae* (syn. *Pleospora betae*), the cause of Phoma leaf spot and root rot on Beta vulgaris. Mol Plant-Microbe Interact 32:787–789
Vera S, Dragana B, Ferenc B, Nadežda D, Ranko M (2012) Morphological, cultural and pathogenic characteristics of *Macrophomina phaseolina* isolates from sugar beet. Proceedings of the international symposium on current trends in plant protection, Belgrade, Serbia, pp 251–256
Vesely D (1986) Effect of temperature of the environment and the strain spectrum of the mycoparasite *Pythiumoligandrum* on the protection of sugar beet against blight. Sbornik UVTIZ Ochrana Rostlin 22: 3,189–198. (c.f. CAB Abstracts 1987–1989)
Vincelli P (2008) Seed and seedling diseases of corn: plant pathology fact sheet University of Kentucky College of Agriculture PPFS-AG-C-02
Voegele RT, Hahn M, Mendgen K (2009) The Mycota, 5. Plant relationships, chapter the Uredinales: cytology, biochemistry, and molecular biology, 2nd edn. Springer, pp 69–98
Von Bretzel P, Stanghellini ME, Kronland WC (1988) Epidemiology of Pythium root rot of mature sugar beets. Plant Dis 72:707–709
Wang Y, Bao Z, Zhu Y, Hua J (2009) Analysis of temperature modulation of plant defense against biotrophic microbes. Mol Plant-Microbe Interact 22:498–506
Weiland J, Koch G (2004) Sugar beet leaf spot disease (*Cercosporabeticola* Sacc.). Mol Plant Pathol 5(3):157–166
Whipps JM, Quilken MP, Budge SP (1993) Use of fungal antagonists for biocontrol of damping-off and Sclerotinia diseases. Pestic Sci. Essex. El-Sevier Applied Science Publishers 37(4):309–313. (c.f. Agricola 1/92-6/95)
Whitney ED, Duffus JE (1986) Compendium of beet diseases and insects. American Phytopathological Society, St. Paul, MN
Windels CE, Nabben DJ (1989) Characterization and pathogenicity of anastomosis groups of *Rhizoctoniasolani* isolated from Beta vulgaris. Phytopathology 79:83–88
Wolpert TJ, Dunkle LD, Ciuffetti LM (2002) Host-selective toxins and avirulence determinants: what's in a name? Annu Rev Phytopathol 40:251–285
Xin XF, Kvitko B, He SY (2018) *Pseudomonas syringae*: what it takes to be a pathogen. Nat Rev Microbiol 16:316–328
Zachow C, Fatehi J, Cardinale M, Tilcher R, Berg G (2010) Strain-specific colonization pattern of Rhizoctonia antagonists in the root system of sugar beet. FEMS Microbiol Ecol 74(1):124–135

Foliar Sugar Beet Diseases and Their Management Approaches in India

26

Varucha Misra, Sangeeta Srivastava, Ashutosh Kumar Mall, and Santeshwari Srivastava

Abstract

Diseases in sugar beet are one of the factors on which its growth and development are dependent. It is a limiting constraint to achieve high yield in sugar beet. In India, rate of occurrence of diseases varies from 10% to 15% and this results in influencing the low sugar beet yield. Fifteen sugar beet diseases of economic value have been known in Indian climatic conditions. The juicy content of the leaves acts as a favorable host for several foliar infections causing a strong impact on root yield. Certain foliar infection in sugar beet crop has nonsignificant effect on sucrose content. Management strategies have been adopted in controlling the various foliar diseases worldwide and the application of bioagents has also been recommended to avoid the lucid use of nonsystemic and systemic fungicides. The foliar diseases known in this crop under Indian conditions have been briefly described in this chapter highlighting the approaches adopted to manage them.

Keywords

Sugar beet · Disease · Sucrose · Fungicide · India

26.1 Introduction

Sugar beet was grown and cultivated for vegetable and fodder purposes for many years prior to its importance for its sugar content. Andreas Marggraf first told that sugar beet also produces sugar in experimental lab of Germany in 1747, but the first beet-sugar factory was built in Silesia in 1802. The Napoleons became keen in this crop in 1811 after the Britishers had blocked the French supply of sugar from West

V. Misra (✉) · S. Srivastava · A. K. Mall · S. Srivastava
ICAR-Indian Institute of Sugarcane Research, Lucknow, Uttar Pradesh, India

V. Misra et al. (eds.), *Sugar Beet Cultivation, Management and Processing*,
https://doi.org/10.1007/978-981-19-2730-0_26

Indies. Under the influence of napoleons, 40 factories were recovered to process beet sugar in 1840. Beet sugar production then increased rapidly throughout Europe. Beet sugar now accounts for almost all sugar production in continental Europe and almost one third of total world production. Now, this crop has been introduced in the Indian continent as a secondary sugar-producing crop (Mall et al. 2020). The crop has been established in tropical and subtropical regions of India with respect to the prevailing climatic conditions of these regions (Mall et al. 2021; Misra et al. 2020).

The diseases in this crop cause hindrance in the yield and production with respect to tonnage and sucrose content (Misra et al. 2021). Each year, the crop is known to be harmed by the occurrence of disease and these diseases amount to 16–20% of destruction (Srivastava 2004). The occurrence of disease begins from the very seedling stage till the crop is harvested and affects all parts of the crop, resulting in quality and quantity losses. Mukhopadhyay (1987) reported that more than 40 diseases in sugar beet seem to occur with approximately 20 affecting the economic portion of the crop. In India, sugar beet cultivation is limited due to conditions required for seed production in the country and the occurrence of various diseases. However, the problem of conditions required for seed production has been resolved to a little extent by growing the sugar beet at higher altitudes such as Kumaon hills (Mukteshwar, Uttaranchal), Kalpa hills (Himachal Pradesh), etc. The high temperatures in tropical and subtropical Indian condition are other problems on which occurrence of various diseases depends and so does the variability in sugar beet. The juicy content in the leaves and roots and environmental conditions for crop development cause the diseases to infect and proliferate. Leaf spots diseases and nematode disorders are known to infect sugar beet crop the most in plain regions and hilly areas resulting in limitation in high yield and production. In India, 10–15 per cent of disease infestation has been revealed to infect the crop. Though more than 20 diseases have been reported in this crop, 15 of them hold economic importance and affect the sugar beet crop the most. Among foliar diseases, *Cercospora* leaf spot is the major disease of sugar beet followed by *Alternaria* leaf blight and Powdery mildew. Other diseases, viz., Ramularia and Phoma leaf spot, are of minor importance, a rare occurrence, and sporadic in nature under Indian conditions (Srivastava 2004).

26.2 Major Foliar Diseases of Sugar Beet in India

26.2.1 *Cercospora* Leaf Spot

Cercospora leaf spot (also known as a brown spot or leaf blotch) is the most important destructive sugar beet disease prevalent worldwide. For Indian conditions too, this disease attains the same status in plains and in hills (Mukhopadhyay 1968; Mukhopadhyay et al. 1974). In India, Pantnagar reports its first incidence (Mukhopadhyay 1968a); besides, its incidence has been reported in roots from the plains of Punjab (Sandhu and Bhatti 1969), Delhi (Juneja et al. 1976), Lucknow (Srivastava and Tripathi 1996), Sriganganagar (Rajpurohit and Singh 1992),

Table 26.1 Application of nitrogen and phosphorus fertilizer on *Cercospora* incidence in India

Treatment	Disease's incidence (per cent decrease over mean)
N60P30	14.88 (+)
N60P60	11.33 (+)
N60P90	0.41 (+)
N120P30	1.35 (−)
N120P60	10.09 (−)
N120P90	8.46 (−)
N180P30	3.37 (+)
N180P60	4.48 (−)
N180P90	5.61 (−)
Mean	46.23
SEM	1.63
CD (5%)	3.45

Sundarbans, West Bengal, (Das 1990) and Maharashtra (Pawar et al. 2004), while in seeds, infection has been reported from hills of Jammu and Kashmir and Mukteshwar (Mukhopadhyay et al. 1974; Kaw et al. 1979). This disease causes a loss of 33 per cent in root yield and 44 per cent in sugar production (Mukhopadhyay and Rao 1978). The disease when infested on the seeds causes adverse effect on size and quality during seed production (Mukhopadhyay 1992). The leaf spot incidence is also dependent on the rate of nitrogen and phosphorus applied to the crop (Anonymous 1989–1990) (Table 26.1).

26.2.1.1 Causal Agent

Cercospora beticola is the fungus which causes this disease. On examining surface lesions under a magnifying lens, minute dots (known as pseudostromata) can be easily seen. These pseudostromata are composed of short conidiophores and conidia. The morpho-taxonomic details of *Cercospora* conidia and conidiophore are as follows:

Conidiophores developing from stroma are geniculate and unbranched. Conidia are formed at the tip of conidiophores acrogenously. The size of conidiophores is 80–188 μm long. Conidia are hyaline, multispetate, long, broader at the base 4.5–6.5 μm, and tapering at the apex 1.5–3.2 μm (Agnihotri 1990).

The pathogen is also known to produce nonhost specific toxins such as Cercosporin (Milat and Blein 1995) and beticolins (Milat et al. 1993; Ducrot et al. 1994). Primary infection takes place via mycelia, conidia, and stroma (Sporodochia), which invade into the soil through infected seed or through infected crop debris. Other plants belonging to genera *Chenopodium, Amaranthus,* and *Atriplex* serve as infection sources. Secondary infection occurs via conidia transmitted by winds. Conidia carried over on the seed and infected plant debris had no importance and was even not associated with the recurrence of the disease under Indian conditions (Pundhir 1979). Favorable condition for its infestation and proliferation is warm and wet weather. A higher incidence of this disease has been reported during intermittent rains and high humidity (92–95%) (Rossi et al. 1994; Battilani et al. 1999). The occurrence of this disease varies in the month of February and March and there is heavy infestation during the months of March to April (Waraitch 1985). The disease assumes severe conditions where the crop is extensively cultivated in the same field every year (Waraitch 1985).

26.2.1.2 Symptomatology

Lower/older leaves are infected first. Circular leaves spots range between 0.125 and 0.185 inches (3 and 4 mm) in diameter. Chief characteristic of the disease is the formation of minute translucent spots. These spots are clearly visible only when the infected leaves are held up to sunlight. Within 6–10 days, spots turn into discrete circular lesions of 3–5 mm in diameter having necrotic gray centers with reddish-brown to black margins. Initially, isolated spots on the leaves are being observed which gradually enlarge, coalesce, and form bigger patches. The number of spots per leaf may vary from 150 to 400. Occurrence of leaves spots when numerous in numbers causes quick drying of foliage. This results in premature defoliation causing infected leaves to shrink. The infected spots were light to dark in color with tan centers and possess dark brown to reddish purplish borders. Furthermore, elliptical lesions are also found on leaves particularly on blades, veins, and petioles of leaves. The petioles are mainly seen infected with the diseases when a severe incidence occurs. In flowering stalks, inflorescence, and seeds, the occurrence of circular spots is seen. Reduction in seed size and germination capability has been observed as a result of this infection. Younger leaves usually remain somewhat free and keep on growing throughout the crop season (Waraitch 1985). A disease cycle of this fungus is illustrated in the Fig. 26.1.

26.2.1.3 Management

For management of this disease, following control measures are to be used:

Fungicidal Spray

Spraying of nonsystemic and systemic fungicides has been recommended for controlling this disease. In the case of ***nonsystemic fungicides***, 4–6 sprays of copper compound like Copper oxychloride (at the rate of 2–2.5 kg/ha) or Indofil M-45 (2–2.5 kg/ha) per spray at 15-day intervals were effective in controlling this disease to some extent. Mukhopadhyay and Rao (1973) showed that four to six sprays of Bordeaux mixture, organotins, copper fungicides, and dithiocarbamates (mancozeb and zinceb) at weekly intervals gave fairly good control over this disease and increased average sucrose content of the crop.

In the case of ***systemic fungicides*** like Carbendazim, Thiobendazole, and Thiophanate (200–250 g/ha), per spray showed a positive response in controlling the disease. In India, Mukhopadhyay (1974) revealed the control of this disease with benomyl, duter, cercobin, bavistin (Table 26.2), and brestanol. Spraying of Bavistin appeared to be effective for managing leaf spot of *Cercospora* by reducing the disease incidence and enhancing the root yield (Anonymous 2009–2010). The dosage of duter is given at the rate of 0.75 kg per hectare mixed in 1000 L of

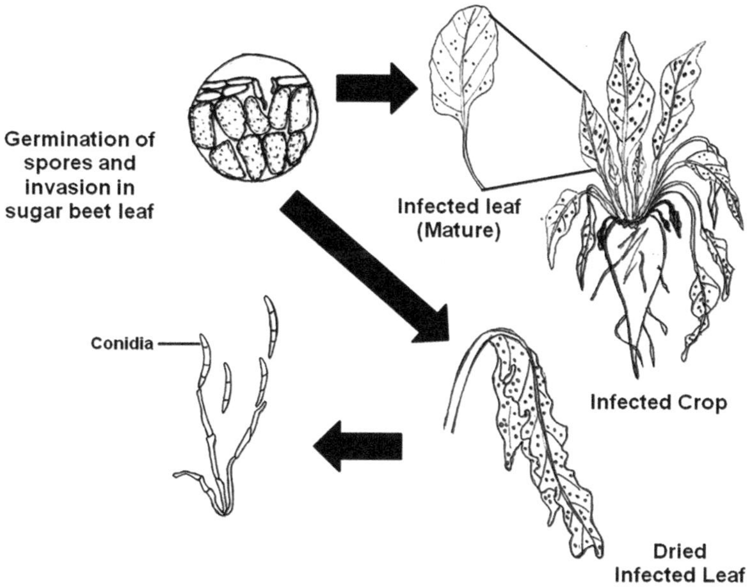

Fig. 26.1 Cercospora leaf spot

water at 10–15 days interval in four sprays. If the disease still persists, then two to three sprays with benlate at the rate of 200 g per hectare mixed in 1000 L of water at 20-day interval have been found to give more effective results than Duter (Waraitch 1985). Fentin acetate, fentin chloride, and fentin hydroxide have also been examined for *Cercospora* management in Indian conditions demonstrated (Mukhopadhyay and Upadhyay 1977; Mukhopadhyay and Thakur 1972). Later benomyl and related compounds, viz., thiophanate methyl and carbendazim, were found much more superior than other protective fungicides (Naidu and Mukhopadhyay 1982; Mukhopadhyay et al. 1974; Mukhopadhyay and Rao 1974). Mukhopadhyay and Bandyopadhyay (1979) reported that one spray of any of these fungicides, viz., benomyl carbendazim and thiophanate methyl at 100 g a.i./ha, gave effective control of the disease with increases in all the yield parameters. Pal and Mukhopadhyay (1983) revealed that during the past several years, the existence of benzimidazole-resistant strains of *C. beticola* has been reported from India. During 2007, at VSI Pune, various fungicides were tested which gave positive results in controlling the *Cercospora* disease (Table 26.3).

Developing Resistant/Tolerant Varieties

Variation in response of varieties towards this disease has been reported by several studies (Mukhopadhyay et al. 1985; Rajpurohit and Singh 1992; Srivastava and Tripathi 1996) affecting sucrose content and root yield (Tables 26.4 and 26.5). During 1993–94, disease incidence was highest reported in CELT (38.03%), Ritma (32.80%), and Ramonskaya 06 (31.83%), while the lowest incidence was in

Table 26.2 Testing of Bavistin for *Cercospora* leaf spot disease in Indian conditions

Treatments	Chemical dose/ha	Disease incidence	Yield (t ha^{-1})			Gain in yield (%) against check		
		0–10 score	Root	Foliage	Sucrose	Root	Foliage	Sucrose
Under uninoculated conditions								
Treated	Sprayed with bavistin and soil drenched with brassicol	<1	±3.62	±11.82	±10.26	7.50	26.82	22.85
Untreated (check)		>7						
		Mean	76.89	36.67	11.56			
Under ephiphytotic conditions								
Treated	Sprayed with bavistin and soil drenched with brassicol	<2	±26.78	±34.55	±28.20	73.17	105.0	78.56
Untreated (check)		>8						
		Mean	58.28	27.5	8.32			

Table 26.3 Evaluation of fungicides in *C. beticola* on different varieties in India

Fungicides	Germination (%)	Plant population (Lakh/ha)	Yield (t/ha)	Disease incidence (%)	Disease intensity (%)	Purity (%)	Pol (%)
Copper oxychloride (0.25%)	68.5	1.10	41.42	0.00	0.00	84.62	17.80
Bayleton (0.1%)	66.67	1.08	37.92	0.00	0.00	84.90	18.18
Bavistin (0.1%)	62.67	1.04	38.5	0.00	0.00	80.76	17.50
Wettable Sulphur (2 kg/ha)	63.17	0.98	33.83	2.57	1.93	87.29	17.77
Mancozeb (0.25%)	66.33	1.03	36.75	0.00	0.00	80.39	17.12
Control	63.42	1.04	32.92	13.93	7.04	77.76	16.89

Table 26.4 Incidence of *C. beticola* in different varieties

Varieties	Germination (%)	Plant population (Lakh/ha)	Yield (t/ha)	Disease Incidence (%)	Disease Intensity (%)	Purity (%)	Pol (%)
Shubhra	60.70	0.72	29.64	2.00	1.10	71.69	9.92
Cauvery	53.53	0.65	28.14	3.49	1.89	77.48	9.17

IISR Comp. 1 (18.13%), Perfo (19.30%), Novantano (19.73%), and Sofie (19.97%). Severity of this disease was not significant in Raspoly, LS 6, LKHY-1, LKC-11, Poly-2, and Freza in farmer's field at West Bengal. During the same year, in advanced varietal trial, M. Pherma had the highest incidence of this disease (48.83%), M. utramono (47.67%), and Ramonskaya 06 (44.77%) with the least disease incidence of M-8603 at West Bengal (Anonymous 1993–1994). At Lucknow, this disease was maximum seen in PP-8 variety (57.86%), while least in Kristall (25.63%) variety; however, at Sriganganagar, the incidence rate of this disease was 4–6 per cent and 2–5 per cent (Anonymous 1996–1997 and 1997–1998). During 1997–98, this disease was also reported in MM Poly and R-06 at farmer's field during Rabi season at Rajasthan (Anonymous 1997–1998), while at Lucknow, the incidence rate of this disease was 4–6% with the highest incidence in LKC LB and Marathon but least in Kristall (25.63%) (Srivastava and Tripathi 1997–1998). A large number of resistant/tolerant varieties have been developed like resistapoly, cercopoly, USH 9B, etc. (Golev et al. 1995; Rossi 1995; Hayashida et al. 1999; Saunders et al. 2000).

Cultural Methods

Several cultural methods are known which help in controlling this disease. Among all, burning of infected crop debris, deep ploughing, field sanitation, crop rotation (at least 3 years with nonhost crops), and the use of disease-free monogerm pelleted seeds help in reducing the occurrence of this disease. The use of certified seeds is also recommended for disease resistance. Early planting of sugar beet in late October in Tarai area of Uttar Pradesh helps to reduce the disease incidence (Waraitch 1985).

26.2.2 *Alternaria* Leaf Blight

Alternaria Leaf spot/blight is a common sugar beet foliar disease. Its first incidence was recorded in India from Pantnagar and Lucknow (Mukhopadhyay 1969; Singh and Srivastava 1969). Infection of leaves by this disease causes reduction up to 30% in leaf area. This disease is more prominently seen in the month of January.

26.2.2.1 Causal Agent

Alternaria spp. is the causal agent. *A. tenuis, A. brassicae, A. ashwinii,* and *A. dilkushana* are the different species of this fungus and are known for occurrence of this disease in sugar beet in Indian conditions. *A. alternata* (=*A. tenuis*) species is

Table 26.5 Variation of *Cercospora* disease incidence rate, loss in sucrose content and root yield in Sugar beet

	Disease incidence (%)										Sucrose (%)			Root yield (m ha^{-1})		
	Kalyani	India			Sriganganagar			Sunderbans	Lucknow		India	Sriganganagar		India	Sriganganagar	
Varieties	1989–90	1990–91	1993–94	1997–98	1989–90	1990–91	1991–92	1990–91	2004–05	2005–06	1990–91	1989–90	1990–91	1990–91	1989–90	1990–91
BR-1	–	–	–	41.79	–	–	–	–	–	–	–	–	–	–	–	–
Cauvery	–	–	–	–	–		–	–	10.42	10.30	–	–	–	–	–	–
CELT	–	–	38.03	–	–	–	–	–	–	–	–	–	–	–	–	–
Denner	–	–	–	–	–		–	59.30	–	–	–	–	–	–	–	–
Donar	–	–	30.08	–	–	–	–	–	–	–	–	–	–	–	–	–
Emma	–	–	–	–	–	–	–	–	–	–	–	–	–	–	–	–
Freeza	–	–	–	–	–	–	–	–	–	–	–	–	–	–	–	–
Hilma	–	–	–	–	–	–	36.66	43.00	–	–	–		–	–		
IISR Comp 1	–	20.00	18.13	43.76	25.90	20.00	–	46.00	29.63	14.93	14.20	15.20	14.20	32.28	63.50	32.28
IISR Comp 2	–	–	–	–	28.50	–	–	–	–	–	–	13.00	–	–	62.60	–
Indus	–	–	–	–	–	–	–	–	16.38	12.09	–	–	–	–	–	–
Kawe Gigapoly	56.23	–	–	–	35.70	–	–	54.10	–	–	–	15.60	–	–	55.10	–
Kawe Mira	50.83	–	–	–	40.00	–	–	51.20	–	–	–	14.80	–	–	53.10	–
Kawe Terma	54.13	–	–	–	31.40	–	–	59.00	–	–	–	11.60	–	–	52.40	–
Kristall	–	–	–	25.63	–	–	–	–	–	–	–	–	–	–	–	–
Laser	–	–	29.60	–	–	–	–	–	–	–	–	–	–	–	–	–
LK 8	–	–	–	35.28	–	–	–	–	–	–	–	–	–	–	–	–
LKC 11	–	–	21.23	–	–	–	–	–	–	–	–	–	–	–	–	–
LKC HB	–	–		54.27	–	–	–	–	–	–	–	–	–	–	–	–
LKC IB	–	–	–	37.83	–	–	–	–	–	–	–	–	–	–	–	–
LKC LB	–	–	–	25.63	–	–	–	–	–	–	–	–	–	–	–	–
LKHY-1	–	–	–	23.33		–	–	–	–	–	–	–	–	–	–	–
LKS 10	–	–	–	–	–	–	29.66	31.70	–	–	–	–	–	–	–	–

(continued)

Table 26.5 (continued)

Varieties	Disease incidence (%)										Sucrose (%)			Root yield (m ha^{-1})		
	Kalyani	India			Sriganganagar			Sunderbans	Lucknow		India	Sriganganagar		India	Sriganganagar	
	1989–90	1990–91	1993–94	1997–98	1989–90	1990–91	1991–92	1990–91	2004–05	2005–06	1990–91	1989–90	1990–91	1990–91	1989–90	1990–91
LKS 11	–	–	–	–	–	–	28.33	41.70	–	–	–	–	–	–	–	–
LS 7	–	–	–	–	40.00	–	–	–	–	–	–	13.20	–	–	68.50	–
LS 6	–	32.00	–	23.37	40.00	32.00	33.33	57.40	28.25	14.63	14.00	14.80	14.00	31.95	60.80	31.95
M Magnapoly	–	–	–	49.88	–	–	–	–	–	–	–	–	–	–	–	–
M. Monova	–	30.00	–	–	–	30.00	30.66	52.80	–	–	14.80	–	14.80	36.60	–	36.60
M-8603	–	–	30.00	–	–	–	–	–	–	–	–	–	–	–	–	–
Magnamono	–	–	–	–	–	–	36.66	54.80	–	–	–	–	–	–	–	–
Marathan	–	–	24.50	32.72	–	–	–	–	–	–	–	–	–	–	–	–
Maribo perma	55.22	35.00	48.83	–	43.40	35.00	33.66	48.80	–	–	15.80	15.00	15.80	42.24	61.20	42.24
Maribo ultramono	59.15	26.50	47.67	–	28.10	26.50	32.00	49.20	–	–	15.60	15.00	15.60	30.29	46.80	30.29
Marita	–	–	33.93	–	–	–	–	–	–	–	–	–	–	–	–	–
Mezzanopoly	55.37	–	–	–	30.70	–	–	57.80	–	–	–	14.60	–	–	61.40	–
MH 2402	–	–	–	–	–	–	–	62.10	–	–	–	–	–	–	–	–
MH 8701	–	–	–	–	–	–	–	63.80	–	–	–	–	–	–	–	–
MHD-2	–		25.00	–	–	–	–	–	–	–	–	–	–	–	–	–
Novanta	–	–	19.73		–	–	–	–	–	–	–	–	–	–	–	–
Pant Comp 3	–	–	–	–	34.30		–	–	–	–	–	15.00	–	–	61.40	–
Pant Comp-6	–	29.00	–	–	33.40	29.00	–	–	–	–	13.60	13.00	13.60	32.28	61.40	32.28
Pant S -10	–	29.50	–	–	36.80	29.50	–	42.60	–	–	13.00	12.60	13.00	28.05	61.80	28.05
Perfo	–	–	19.30	–	–	–	–	–	–	–	–	–	–	–	–	–
Poly-2	–	–	22.33	–	–	–	–	–	–	–	–	–	–	–	–	–

PP 8	–	–	–	57.86	–	–	–	–	–	–	–	–	–	–	–	–
Ramonskaya-06	59.89	30.00	31.83	44.77	27.80	30.00	34.33	60.90	–	–	13.40	13.20	14.00	31.20	60.30	36.52
Raspoly	54.24	–	23.60	–	40.00	–	–	43.70	–	–	–	14.40	–	–	38.70	–
Ritima	–		32.80	–	–	–	–	–	–	–	–	–	–	–	–	–
Shubhra	–	–	–		–		–	–	12.96	11.44	–	–	–	–	–	–
Sofie	–	–	19.97	–	–	–	–	–	–	–	–	–	–	–	–	–
Solid	58.30	30.00	–	–	33.00	30.00	38.00	40.00	–	–	15.00	15.60	15.00	17.01	55.10	17.01
Tribel	48.21	–	–	–	60.00	–	–	48.90	–	–	–	11.20	–	–	39.50	–
Virtus	47.24	20.00	–	–	26.60	20.00	–	48.60	–	–	14.80	14.80	14.80	29.21	43.70	29.21

more destructive than *A. brassicae*. *A. tenuis* could destroy up to 30% of the leaf area and developed rapidly under humid conditions at 20 °C (Agnihotri et al. 1972). Primary infection takes place via air-borne conidia and is transmitted through infected plants while secondary infection occurs via wind-borne conidia. These conidia develop on the debris of infected plants. The infected seeds obtained from diseased crops serve as a basic source of inoculum. Rain, wind, temperature, high humidity, dense mist, fog, and dew play a positive role in developing this disease. The morpho-taxonomic details of the different species of *Alternaria* in this crop are as follows:

A. tenuis: Conidia length with beak ranges from 16.8 to 50.4 μ and width of spore ranges from 8.5 to 15 μ. The length of the beak was 1.5–16.8 μ (Singh and Srivastava 1969; Mukhopadhyay 1969).

A. brassicae: Conidia length with beak ranges from 148 to 184 μ and spore width was 17–24 μ. The length of the beak was 45–65 μ (Singh and Srivastava 1969; Mukhopadhyay 1969).

A. ashwinii: Conidia length with beak ranges from 43.2 to 45.6 μm and the width of the spore was 14.4–16.8 μm. The length of the beak was 7.2 μm and thickness 4.8 μm (Misra et al. 2021).

A. dilkushana: Conidia length with beak ranges from 5.50 to 5.52 μm and the width of the spore was 0.94–0.96 μm. The length of the beak was 4.8 μm and thickness 4.8 μm (Misra et al. 2021).

A disease cycle of this fungus is clearly illustrated in the Fig. 26.2.

26.2.2.2 Symptomatology

Leaves spots of smaller size range up to 10 mm in diameter. These spots are irregular in shape, dark brown to blackish in color, and more commonly seen on margins of leaves. This is the chief feature of *Alternaria alternata*. If leaves spots are found more on lower/older leaves in comparison to newer/younger leaves and leaves spots are concentric and zonate, possessing a size up to 15 mm in diameter, then it is the chief feature of *Alternaria brassicace*. Furthermore, leaves spot may be seen on any portion of leaf lamina. The leaf spots incided by *A. tenuis* are up to 1 cm in diameter, irregular in shape, dark brown to black in color, and are more common on margin, whereas *A. brassicae* form concentrically zonated light to dark brown circular spots up to 1.5 cm in diameter. Leaves spots are subcircular and brown in color with necrosis on the center. The spots are often seen to coalesce. Drying and upward curling of edges of leaves is also seen on margins of leaves. Rarely, small flecks of this disease are seen on petioles of leaves. Advanced stage of infection causes drying off central necrotic lesions which often turn into short holes. Under humid climatic conditions, these necrotic lesions are often covered with blackish fungal growth.

26.2.2.3 Management

For management of disease, following control measures are to be used:

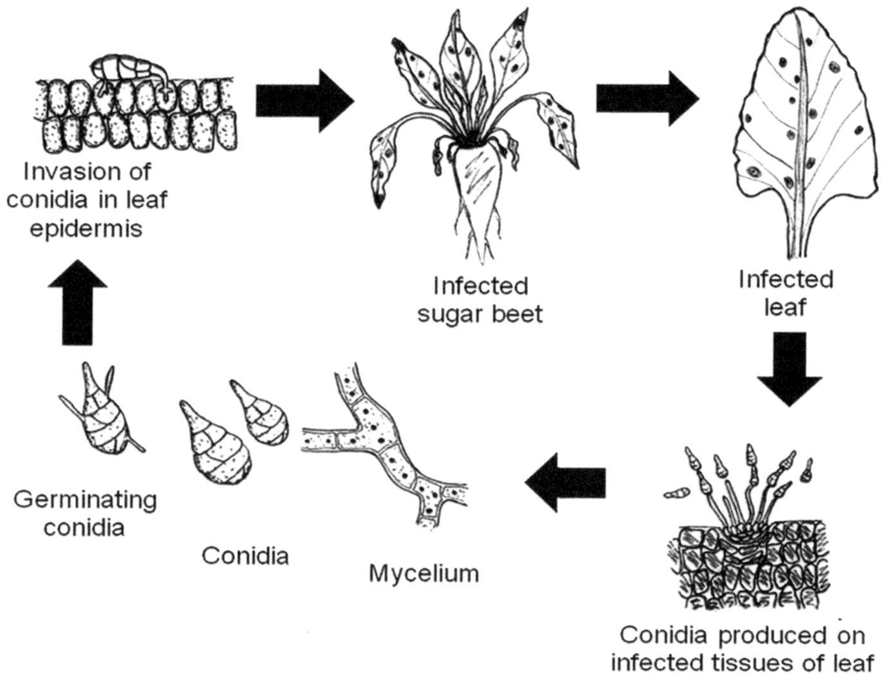

Fig. 26.2 Life cycle of *Alternaria* leaf spot

Control with Nonsystemic Fungicides

The disease is partially managed by the spraying of any nonsystemic fungicides like Dithane M-45 or copper oxychloride or captan @ 2.0–2.5 kg ha^{-1} per spray at 15 day intervals before the appearance or early stage of disease development. Study has been conducted on chemical control of leaf spot caused by *A. tenuis* and showed that the disease incidence could be reduced by the application of Kasumin, Brestanol, and Dithane M 45 (Table 26.6) (Agnihotri et al. 1972). Zineb or Dithane M-45 sprays also had lower incidence rate of this disease to which urea @3% is added at the time of second or third spray for better results.

Developing Resistant/Tolerant Varieties

Breeding for disease resistance has not been found successful due to very wide host range of the pathogen. At Mukteshwar, seeds of IISR Comp. 1 showed black discoloration which later revealed to be infection of *Alternaria* spp. on isolation (Anonymous 1997–1998).

Table 26.6 Chemical control of leaf spot of sugar beet caused by *Alternaria tenuis*

Treatment	Total number of spots	Per cent increase/decrease leaf spot
Benlate	65	(+) 12.07
Brestan	62	(+) 6.90
Brestanol	42	(−) 27.59
Dithane M-45	40	(−) 31.03
Kasumin	37	(−) 36.21
Check	58	

26.2.3 Powdery Mildew

Powdery mildew is prevalent in arid climates of the Middle East, Russia, U.K., U.S. A., and Canada. In India, almost all the sugar beet-growing areas possess an incidence of this disease (Mukhopadhyay 1968b; Singh et al. 1971; Karve 1972). Phalton area of Maharashtra is one among the places in India where its severe form has been observed under warm and dry weather conditions. The occurrence of this disease causes a reduction of 20–25% in root yield. Under the severe conditions, 38% losses occur in gross sugar yield. Besides, losses in purity and sucrose concentrations also occur in this disease (Mukhopadhyay 1968; Mukhopadhyay et al. 1974). Furthermore, this has been one of the main causes of low sugar production in Maharashtra, India (Karve et al. 1973). Pawar et al. (2004) had revealed that this disease has been noticed in tropicalized sugar beet seed of Syngenta varieties at Vasantada Sugar Institute, Pune. The occurrence of this disease is severe where nitrogen levels in soils are low. *Erysiphe betae* Vanha Weltzien was first recorded at Pantnagar (Mukhopadhyay 1968) and then at Lucknow (Singh et al. 1971); this powdery mildew was later observed extensively in Phaltan (Maharashtra), the area having warm and dry weather (Karve 1972). The conditions governing spore production germination and infection have not been worked out.

26.2.3.1 Causal Agent

Erysiphe betae (Syn. *E. polygoni*, *E. communis*, *Oidium erysiphoides*, *Microsphaera betae*) and Ascomycetes fungus. The pathogen is an obligate parasite. Primary infection takes place through ascospores which are produced either on sugar beet plants (Fig. 26.3) or on any alternate hosts, while secondary transmission occurs through conidia produced during primary infection. Infected plants have been reported with higher amounts of sodium and ammonium nitrogen in roots. Favorable condition for this disease is hot and dry weather having cool nights and warm days. Formations of conidia on leaves are mainly seen in the morning, but these are later on released in afternoon. Depending on the temperature, germination of conidia occurs either in the afternoon or in the early evenings. The best temperature for conidial germination is 86 °F while for appressoria formation, it ranges between 59 and 68 °F. The incubation period of this disease is 5 days having a temperature of around 77 °F (Khan 2018). Mukhopadhyay and Russell (1979) were the first to report the critical stages in the development of *E. betae* on sugar beet.

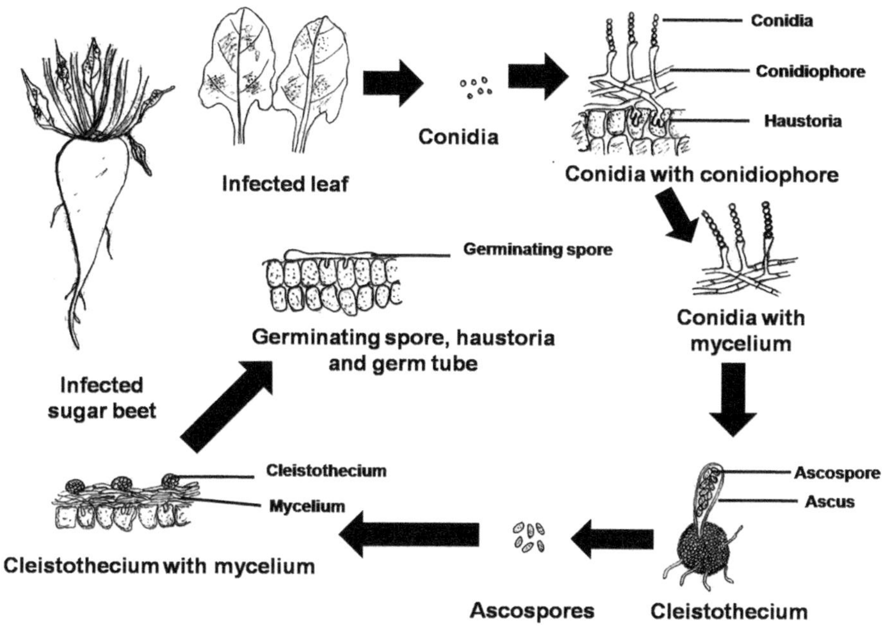

Fig. 26.3 Life cycle of powdery mildew

26.2.3.2 Symptomatology

First appearance is on lower and older leaves as white to gray thread-like filaments that appear as if emerging out from the center point. The movement of appearance of disease starts from lower/older leaves to upper/younger leaves. Chief characteristics of disease are the formation of white powder, which later on turns to gray in color, tan mildew areas on the dorsal and ventral sides of the leaf. However, symptoms are more frequently seen on the upper surface of leaves. **Advanced stage:** Patches become enlarged and get coalesced. The appearance of leaves looks like dusted with white powder. The leaves which were severely affected change their color from green to yellow. These yellow leaves, later on, dry up. In severe infection, symptoms also appear on younger leaves. Furthermore, a characteristic odor alike to musty godown may be noticed where severe infection persists (Khan 2018). Mycelia and spores of pathogen are generally seen in the superficial white mass. At times, leaves' surface gets affected with minute, spherical, orange-brown black fruiting bodies. These fruiting bodies are known as cleistothecia which are embedded in the mycelium of the fungus. This is termed as telomorph stage of the pathogen.

26.2.3.3 Management

For management of disease, following control measures were to be used:

With Systemic and Protective Fungicides

Several systemic and protective fungicides have been reported to possess protection towards this disease (Cicco and Curtis 1993; Asher 2000; Zahradnieck et al. 2002). In this respect, BAY MEB 6447 (Triadmefon) had been revealed to be effective when applied in granular form at the time of planting 3 inches below the seed (@ 1 pound a.i./acre) (Hills et al. 1976). Application of fungicides should be repeated if recurrence of disease takes place. Furthermore, it is important to note that application of fungicides before harvest is of no economic benefit. Theovit was found to be suitable for its control (Arya and Saini 1977). For the control of powdery mildew, spray with Dikar or a mixture of Dithane M-45 and Karathane is recommended. The ratio of dithane and karathane in the mixture is of 16:4. The spray is given thrice at an interval of 2 weeks (Bhatnagar and Pant 1977). The disease can be effectively controlled by dusting sulphur and spraying wettable sulphur, benomyl, or brestan (Weltzien 1968). Karve (1972) recorded good control of this disease in Maharashtra by using oxythioquinox. At the time of initial infection, 20–40 pounds of sulphur dust per acre or 10 pounds of wettable sulphur per acre with 10 gallons of water per acre have been effective against this disease (Hills et al. 1976). However, Karve et al. (1973), Russeel and Mukhopadhyay (1981), and Asher (2000) had recommended dosage of wettable sulphur at the rate of 1.5–2 kg/ha. For managing this disease effectively, two to three sprays have been proved to be beneficial at an interval of 15–20 days. In 2005–06 and 2006–07, wettable sulphur had effectively controlled the disease in HI 0064 and LS 6 at Lucknow (Anonymous 2004–2008). Mosa (2002) had showed that spraying of potassium phosphate (monobasic, dibasic, and tribasic) helps in reducing the incidence of powdery mildew. It also induces systemic resistance against the disease. In case of organic farming, potassium bicarbonate is used as fungicide. Experimental results on powdery mildew management through various chemicals (Copper oxychloride, Carbendazim 0.1%, Tridimefon 0.1%, Wettable sulphur 2.0 kg/ha, and Mancozeb 0.25%) showed that copper oxychloride (0.25% concentration) had effective control by three sprays at regular interval of 12 to 15 days over the disease in HI 0064 and LS 6 at Lucknow, while at ARS Sagauli similar results were seen. The root yield of sugar beet sprayed with Wettable sulphur and Carbendazim (Bavistin) had significantly higher yield over all other treatments and the control. The sucrose content in any of the varieties was not hampered by the fungicidal sprays (Anonymous 2004–2008).

With Bioagent

Trichoderma viridie showed effective results in controlling the powdery mildew disease than Bavistin and thiram application in India (Table 26.7).

Developing Resistant/Tolerant Varieties

Successful approaches have been used for developing resistant/tolerant varieties against this disease (Mukhopadhyay and Russell 1979). Experimental results showed variation in disease incidence from variety to variety (Table 26.8). The mechanism of resistance to powdery mildew was studied on the following four sugar beet varieties: line G, a homozygous diploid line produced by the USDA research

Table 26.7 Biological management of Powdery mildew disease

Treatments	Disease Incidence (%) during 2006–07 to 2007–08
Control	25.42
Rouging + Thiram	10.81
T. viride (sowing)	7.07
T. viride (February)	8.19
T. viride (at sowing and February)	11.65
Bavistin (spraying)	3.13
Bavistin (drenching)	10.10
Thiram (spraying)	6.99
T. Viridie + Bavistin (drenching)	5.45
T. Viridie + Thiram (drenching)	11.17

Table 26.8 Disease incidence of powdery mildew during 2 years

Varieties	Disease incidence (%)				Mean
	Tropical		Subtropical		
	2004–05	2005–06	2004–05	2005–06	
Shubra	6.64	5.71	12.96	11.44	9.19
Cauvery	8.74	9.95	10.42	10.30	9.85
Indus	13.89	14.11	16.38	12.09	14.12
IISR comp 1	11.21	13.04	29.63	14.93	17.20
LS 6	15.54	10.94	28.25	14.63	17.34
Mean	11.20	10.75	19.53	12.68	13.54

Station (USH 7 and USH 8), Salinas, and California (Sharpe Kelin E (SKE)) which is susceptible to powdery mildew in the field (Mukhopadhyay and Russell 1979). Resistance to powdery mildew can be expressed at one or more of the following stages of the development of *E. betae* on sugar beet leaves; germination of conidia, development of haustoria, formation of ESH, and sporulation. SKE was the most resistant variety at each stage. Patrieia, LKS-10, LK-27, and Ramonsakya 06 were some other sugar beet varieties that showed severe incidence of this disease (Anonymous 1996–1997). The leaf disc technique developed by Mukhopadhyay and Russell (1979) could be employed to screen large numbers of individual sugar beet plants for resistance to powdery mildew.

26.2.3.4 Cultural Methods

Destruction of crop debris is the best cultural method adopted for managing this disease. This helps in destroying the surviving structures, especially clestothecia, to some extent.

Table 26.9 Incidence of *Ramularia* leaf spot in different places of India

Places	Disease incidence (%)
Kalyani	5–10
Neempith	2–5
Usti	5–10
Kadwip	2–5

26.2.3.5 Integrated Management

Of all the control measures, integrated management has been recommended as the best management practice which is the combination of crop debris destruction, fungicidal sprays, and growing of resistant varieties.

26.2.4 Minor Foliar Diseases of Sugar Beet in India

26.2.4.1 Ramularia Leaf Spot

In U.K., this disease has been reported to cause sugar yield loss up to 25% when infection is severe. Incidence of this disease has been reported in India in 1989–90 (Table 26.9). Occurrence of this disease has been observed more after the month of September. This disease is often confused with *Cercospora* leaf spot, but it differs in coloration of leaf spot (black in color). This disease causes fall of premature leaves, which in turn causes reduction in root weight, sucrose content, and juice quality (Byford 1975; Nielsen 1991). In Germany, occurrence of this disease causes 10–24% sugar yield loss (Petersen et al. 2001).

26.2.4.2 Causal Agent

Ramularia beticola is the causal agent. Favorable conditions are around 17 °C temperature with humidity of 95% (Ahrens 1987; Hestbjerg et al. 1994). Spores distribution occurs through wind and rain. Higher probability of this fungus has been observed in winter season on residue of crop (Nielsen 1991; Persson and Olsson 2006). A disease cycle of this fungus is illustrated in the Fig. 26.4.

26.2.4.3 Symptomatology

Older leaves are infected with this disease; light brown and fairly large spots ranging between 4 and 7 mm in diameter in size are seen; on maturity, leaves spots turn gray with silvery white center having dark to reddish-brown margin; infected leaves turn yellow from green and later on die.

26.2.4.4 Management

For management of disease, following control measures are to be used:

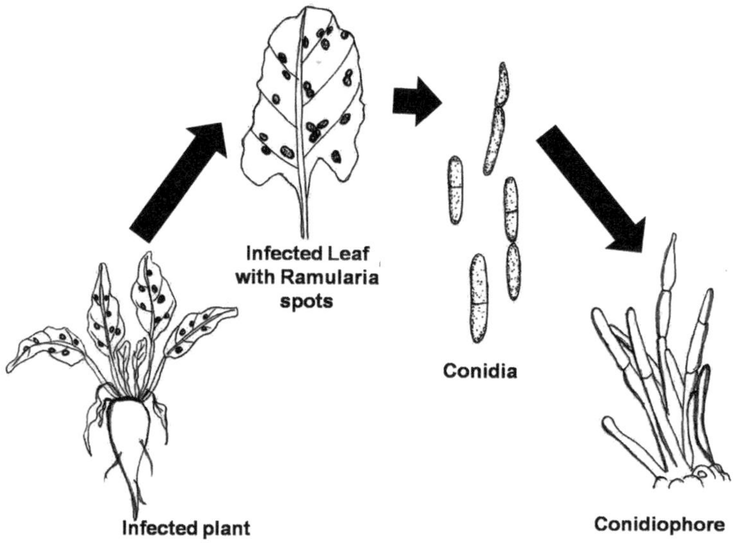

Fig. 26.4 Life cycle of *Ramularia* leaf spot

Control with Triazole Fungicide

Single application of Eminent 125 SL at 13 fl oz/A has been found to be effective. Triazoles (such as epoxiconazole, propiconazole, and difenoconazole (Yoshida and Aoyama 1987, Leroux et al. 2008)) and strobilurins (such as strobilurin pyraclostrobin (Bartlett et al. 2002)) have also found to be effective (Thach et al. 2013).

Rotation of Crop up to 4 Years

26.2.5 Phoma Leaf Spot

This disease is of low economic importance. Occurrence of this disease has caused a loss of 40–50% in seeds. A new leaf spot disease of sugar beet (var. Dobrovicka) caused by *Phoma betae* was observed at Haran Farm Srinagar. The symptoms under field conditions were observed as circular to oval necrotic spots with diffuse margins and with light to dark brown concentric rings (Anonymous 1972). Singh et al. (1973) reported widespread occurrence of leaf spots caused by *P. betae* at Haran in Kashmir Valley. Incidence of spots was more on mature leaves than young foliage. The ecologic significance of this fungus on the foliage is yet to be investigated. During 1997–98, at Mukteshwar, seeds of IISR Comp 1 showed black discoloration of *Phoma betae* with infestation of 72–100% (Srivastava and Tripathi 1997–1998).

26.2.5.1 Causal Agent

Phoma betae (Seed-borne pathogen). Perfect stage of this fungus is *Pleospora bjoerlingii*. Spread of this disease takes place through ascospores, whereas seed gets infected with conidia of asexual stage.

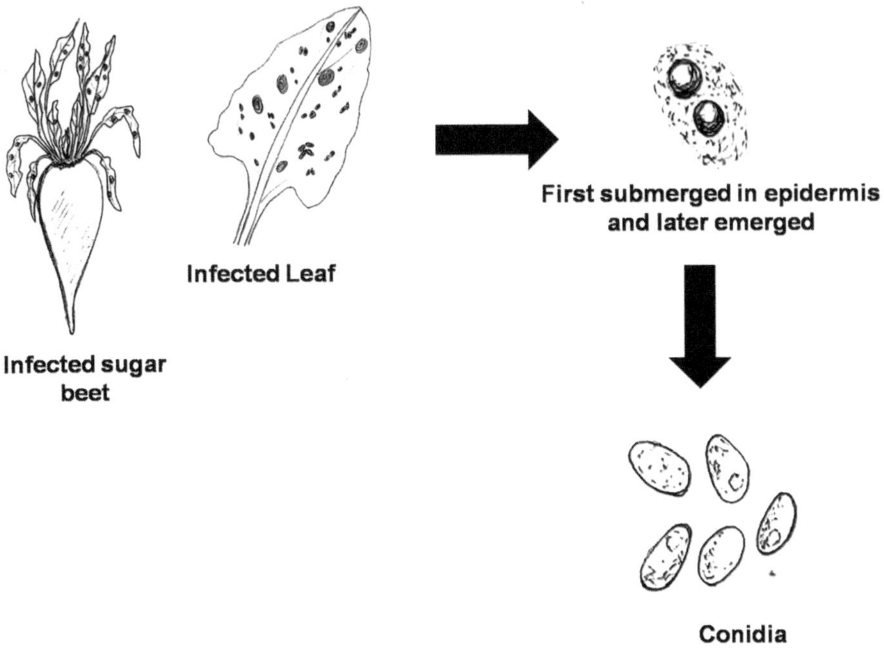

Fig. 26.5 Life cycle of *Phoma betae*

26.2.5.2 Symptomatology

Leaves spots are of 2 cm in diameter, brown in color, and round to oval in shape. Dark concentric rings near the perimeter are also symptoms of this disease. Symptoms are more observed in older, lower leaves in comparison to younger leaves (Fig. 26.5). Small dark pycnidia are originated in concentric rings throughout the spots. Pycnidia are found on seed stalks in dark necrotic streaks with grayish centers; in infection on roots, symptoms first appear near the crown as small, dark, sunken spots which later become soft and water-soaked. These water-soaked spots later on turn dark brown to black in color; older infected tissues become black, dry, shrunken, and somewhat spongy.

26.2.5.3 Management

For management of disease, following control measures were to be used:

Cultural Control

During growing season, proper water and nutrient levels should be maintained. Wounded roots should be avoided.

Control with Fungicides

Application of Captan 4F at 6 fl oz/100 lb seed plus a dye is effective in controlling this disease. Application of systemic fungicide like benlate is also helpful in managing the disease (Gray and Greik 1998). Furthermore, Thiram 50WP at 8 oz/ 100 lb or 42-S Thiram at 8 fl oz/100 lb seed plus a dye is also effective.

Crop Rotation
Crop rotation is also an important management practice for this disease. Crop should be rotated for two or more years for managing the infected leaves debris to decompose completely, making less possibility of reoccurrence of the disease (Gray and Greik 1998).

Seed Cleaning
Seed cleaning method is another way to protect from this disease.

26.3 Future Prospects

Several diseases have been identified in this crop and management practices have been recommended, but at times these diseases are often being ignored causing huge losses to crop. There is a need for proper identification and management of diseases occurring in sugar beet as it has been hampering root yield and production. There is a need for the production of varieties having high genetic resistance against these diseases. Development of such varieties will also lead to avoidance of pathogen resistance to fungicides due to the lucid utilization of chemical treatments. Use of bioagents for management of these diseases will also help in minimizing the rationale application of chemical fungicides.

26.4 Conclusion

Sugar beet is another crop after sugarcane which owes potential for production of sugar and ethanol. Though sugar is being produced all over the world (contribution of 20% in total world's sugar) through this crop, technologies are also being developed for ethanol production. Due to its commercial importance, high root yield is of utmost importance for which crop should be protected from various diseases, insects, and pests. Foliar diseases in sugar beet were often associated with damage to photosynthesis rate, affecting growth and biomass production. Among foliar diseases in India, *Cercospora* leaf spot (*Cercospora beticola*) is the first among the major diseases followed by *Alternaria* and Phoma leaf spot. The root yield and sugar production were reduced to 33% and 44%, respectively, due to *Cercospora* infection, while *Alternaria* infection causes an influence of 30% in leaf area. Powdery mildew disease is a minor disease seen in hilly areas of sugar beet fields, causing reduction in root yield by 20–25%. New species of *Alternaria* have also been reported in sugar beet fields in Lucknow (Misra et al. 2021). Many management approaches through use of systemic and nonsystemic fungicides were tested and found to be effective against the foliar diseases under Indian agroclimatic conditions.

Naidu PH, Mukhopadhyay AN (1982) Relative efficiency of different number of sprays of four systemic funigcides in controlling *Cercospora* leaf spot of sugar beet. Pesticides 16:20

Nielsen S (1991) Experiments with fungicide sprays against ramularia and rust in sugar beets 8. Danske planteværnskonference, Sygdomme og Skadedyr, Statens Planteavlsforsøg, Beretning nr S2109-1990, pp. 327–338

Pal V, Mukhopadhyay AN (1983) Natural occurrence of the strains of Cercospora beticola resistant to systemic fungicides. J Mycol Plant Pathol 13:333

Pawar BH, Vasekar VC, Patil AS, Nerkar YS (2004) Natural occurrence of diseases on sugarbeet in Maharashtra, Personal Communication, pp 1–7

Persson L, Olsson Å (2006) Åtgärder mot förluster av svampangrepp i sockerbetor under odling och Lagring 2003–2005. SBU AB. Slutrapport SLF. Report

Petersen J, Adams H, Schaufele WR, Buttner G (2001) Untersuchungen zur Shadwirkung von Ramularia bet cola in Zuckerruben und Moglichkeiten zur Differenzierung der Sortenanfalligkeit nach kunstlicher Inokulation. Gesunde Pflanzen 53:141–147. (in German)

Pundhir VS (1979) Epidemiology and control of *Cercospora* leaf spot disease of sugar beet. PhD Thesis, GB Pant University of Agriculture and Technology, pp 146

Rajpurohit TS, Singh RD (1992) Field evaluation of sugarbeet varieties against *Cercospora* leaf spot. Indian J Mycol Plant Pathol 22:294–296

Rossi V (1995) Effect of host resistance in decreasing infection rate of *cercospora* leaf spot epidemics in sugarbeet. Phytopathol Mediterr 34:149–156

Rossi V, Racca P, Battilani P (1994) A simulation model for *Cercospora* leaf spot epidemics on sugarbeet. Phytopathol Mediterr 33:105–112

Russeel GE, Mukhopadhyay AN (1981) Effect of some systemic and non-systemic fungicides on *Erysiphe betae* and its development on sugarbeet leaves. Phytopathology Z 10:1–6

Sandhu SS, Bhatti DS (1969) The incidence of sugarbeet diseases in Punjab. J Res Punjab Agric Univ 6:924–926

Saunders JW, Mc Grath JM, Theurer JC, Halloin JM (2000) Registration of "SR 87" sugarbeet germplasm with low tare and resistance to *Cercospora* and *Aphanomyces*. Crop Sci 40:1833–1834

Singh K, Srivastava SN (1969) Annual report IISR Lucknow, pp 54–57

Singh K, Srivastava SN, Misra SR (1971) Diseases of sugarbeet in India. Indian Sugar 21:321–326

Singh K, Sen C, Srivastava SN, Agnihotri VP (1973) Two new diseases of sugarbeet from India. Curr Sci 42:69

Srivastava SN (2004) Management of sugarbeet diseases. In: Mukerji KG (ed) Fruit and vegetable diseases. Disease management of fruits and vegetables, vol 1. Springer, Dordrecht. https://doi.org/10.1007/0-306-48575-3_9

Srivastava SN, Tripathi RC (1996) Reaction of sugarbeet genotypes to *Cercospora beticola* Sacc. Under natural conditions. Indian J Sugarcane Technol 11:180–181

Srivastava SN, Tripathi RC (1997–1998) Annual report of AICRP project on sugarcane and sugar beet network project, pp 8–10

Thach T, Munk L, Hansen AL, Jorgensen LN (2013) Disease variation and chemical control of Ramularia leaf spot in sugar beet. Crop Prot 51:68–76

Waraitch KS (1985) Reaction of sugarbeet genotypes to Cercospora leaf spot and Sclerotium root rot diseases. Indian Phytopathol 38:369–370

Weltzien HC (1968) The effects of triphenyl stannic acetate on the powdery mildew of sugar beets. Zuckerindustrie 21:241–246

Yoshida Y, Aoyama Y (1987) Interaction of azole antifungal agents with cytochrome P-45014DM purified from Saccharomyces cerevisiae microsomes. Biochem Pharmacol 36(2):229–235

Zahradnieck J, Svachula V, Pribyal J, Stipek M (2002) Some findings in the investigation of the effect of impact fungicide on the quality and storage properties of sugarbeet. Listy Cukrova rnicke a Reparske 118(2):52–53, Praha, Czech Republic

Cercospora Leaf Spot Disease 27

Rıza Kaya

Abstract

Cercospora leaf spot, caused by *Cercospora beticola* Sacc., first reported in Italy in 1876, is one of the most devastating and common foliar diseases of sugar beet in the world. The spots of the disease usually appear early in wet and warm areas and are most severe during the vegetation period in case of very early attacks. The disease is common in about 44 percent of sugar beet acreage in the world and the severity of the disease varies between countries and regions in same countries. Because of the disease, beet plants lose their leaves and grow new leaves by using substances stored from roots. During the vegetation season, these activities are repeated. When it cannot cope with the disease, root yield, sugar content, extractable sugar content, and sugar yield decrease up to 26, 13, 18, and 55 percent, respectively. Also, the content of potassium (K), sodium (Na), and alpha-amino nitrogen (α-amino N), having difficulty in getting crystal sugar and reducing sugar production in refining process, increases up to 6, 25, and 40 percent, respectively. Disease is controlled by applying fungicides, besides cultural measures such as planting resistant varieties, crop rotation, use of disease-free seeds, and good agricultural practices. The pathogen forms resistance to fungicides used against it in a very short time. Hence, special combined management strategies must be implemented together safely according to early warning epidemiological models that accurately monitor the onset and progression of the disease.

Keywords

Cercospora beticola Sacc. · *Cercospora* leaf spot · Control · Sugar beet

R. Kaya (✉)
Department of Plant Protection, Sugar Institute, Etimesgut, Ankara, Turkey
e-mail: rkaya@turkseker.gov.tr

V. Misra et al. (eds.), *Sugar Beet Cultivation, Management and Processing*,
https://doi.org/10.1007/978-981-19-2730-0_27

Abbreviation

BmJ	*Bacillus Mycoides*
DF	Dry flowables
DIV	Daily Infection Value
EC	Emulsifiable concentrate
IIRB	International Institute for Beet Research
K	Potassium
ME	Microemulsion concentrate
Na	Sodium
Ps	Pseudostromata
SC	Suspension concentrate
SE	Suspoemulsion
WG	Water dispersible granules
WP	Wettable powder
α-amino N	Alpha-amino nitrogen

27.1 Introduction

Cercospora beticola Sacc. brings about spots on leaves and major pathogen of sugar beet worldwide (Holtschulte 2000). Disease symptoms typically appear after row closure. Sugar beet plants lose the leaves due to disease and grow new leaves by using substances in the roots. In this way, the disease causes continuous leaf damage until harvest (Rossi et al. 2000, Franc 2010). Thus, it reduces the root weight and sugar yield and also increases the substances forming molasses such as sodium, potassium, and alpha-amino nitrogen, leading to sugar losses in refinery (Carruthers and Oldfield 1961, Smith and Martin 1978, Oltmann et al. 1984, Adams and Schaufele 1996). The roots of infected plant in storage are disrupted quickly than healthy plants (Graf 1980, Smith and Ruppel 1971).

The disease can be coped by cultural measures such as crop rotation (Pundhir and Mukhopadhyay 1987), planting resistant varieties (Vogel et al. 2018, Kopisch-Obuch et al. 2020), and good farming practices (Skaracis et al. 2010). In addition, fungicide application is the most effective method (Khan and Khan 2010, Ioannidis and Karaoglanidis 2010). Since pathogen creates resistance to fungicides in a short time (Georgopoulos and Dovas 1973, Ruppel and Scott 1974, D'ambra et al. 1974, Pal and Mukhopadhyay 1985, Weiland and Halloin 2001, Giannopolitis 1978, Cerato and Grassi 1983, Bugbee 1996, Karaoglanidis et al. 2000, Köller 1991, Kirk et al. 2012), fungicides with different effect mechanisms should be selected and their different mixtures should be prepared and applied carefully throughout the season within a program (Ioannidis and Karaoglanidis 2010, Secor et al. 2010). Biological control methods are not to be used in practice due to not being satisfactory (Collins and Jacobsen 2003, Galletti et al. 2008). In this review, the information applied from research results into practice on causal agent, symptoms, distribution, economic importance, epidemiology of the disease, and the management strategies that should be put into effect in accordance with the current conditions are presented.

27.2 Causal Agent

The causal agent of leaf spot disease in sugar beet is *Cercospora beticola* Sacc. The fungus is a member of the class Deuteromycetes (Fungi Imperfecti), order Moniliales, family Dematiaceae, and section Phaeophragmosporae (Barnett and Hunter 1972, Chupp 1953). Hyphae are hyaline to pale olivaceous brown, septate, intercellular, 2–4 μm in diameter. They form pseudostromata in substomatal cavities of the host and conidiophores, 10–100 μm × 3–3.5 μm, unbranched, emerge only from host stomata. There are small conspicuous conidial scars at the geniculations and the apex. Conidia are in dimensions of 36–107 μm × 2–3 μm, straight to slightly curved, hyaline, acicular, 3–14 (sometimes more) septa. Teleomorph stage of *C. beticola* is unavailable (Crous and Braun 2003) (Fig. 27.2).

27.3 Symptoms

Leaf spots created by *C. beticola* are circular, in a diameter of 2–5 mm, tan, pale brown, grey or whitish (Ruppel 1986). First spots develop on the older leaves (Fig. 27.1a–c). At the later stages, elongated lesions grow on the petiole (Fig. 27.5). Sometimes, spots can develop on the beet crown (Giannopolitis 1978). As the disease progresses, individual spots coalesce and parts of the leaf where the spots join together turn brown and necrotic (Figs. 27.1, 27.2, 27.3). Pseudostromata which is minute black dot appears in the middle of mature spots (Fig. 27.2). Conidiophores are formed on the pseudostromata when the weather is humid. After producing conidia, the leaf spots become grey and velvety. Followed by blighted and died leaves, eventually they fall to the ground remaining tied to the head of the root (Figs. 27.4, 27.5). The younger leaves usually get spotted and die later than older leaves (Vereijssen 2004). During the later stages of severe epidemics, leaves can be regrown from the plant surrounded by prostrate (Weiland and Koch 2004) (Fig. 27.6).

27.4 Distribution

Saccardo (1876) described first distribution of the disease on *Beta cicla* in Italy, but to date, it has been determined in all sugar beet growing areas worldwide. *Cercospora* leaf spot in warm and humid regions is most damaging to sugar beet (Lartey et al. 2010). Reichert and Palti (1966) and Weltzien (1967) first started to analyse distribution of *C. beticola* affecting sugar beet worldwide. First general

Fig. 27.1 First spots (**a–c**), increased and coalesced spots (**d**) and the death of leaf tissue (**e**) on beet leaves infected by *Cercospora beticola*

geographical distribution map of *C. beticola* was published in the sugar beet areas of the northern and southern hemisphere by the Commonwealth Mycological Institute (Anonymous 1969). Bleiholder and Weltzien (1972) developed the first detailed map. And then, in the growing zones of sugar beet, Rossi et al. (1995) drew a detailed map of *C. beticola*. The study group including phytopathologists from the International Institute for Beet Research (IIRB), sugar beet breeders, and the staff of seed companies updated the map in 1998. According to the study, a total sugar beet growing area of 6.95 mio ha was estimated and the incidence of *C. beticola* was reported about 44 percent of beet production acreage (Fig. 27.7, Table 27.1).

The disease affects moderately on the average approximately 50% of sugar beet areas in some parts of Belgium, Chile, China, Croatia, Czech Republic, France, Germany, Moldova, Morocco, Poland, Slovakia, Pakistan, Spain, The Netherlands, The Syrian Arab Republic, Ukraine, and USA. A high incidence of the disease in some parts of Austria, Bosnia and Herzegovina, Greece, Italy, Hungary, Japan, Macedonia, Romania, Slovenia, The Cuban Region of The Russian Federation, Turkey, USA, and Yugoslavia has been estimated. *C. beticola* affects on average approximately 63% of sugar beet areas in these countries. Both moderate and high

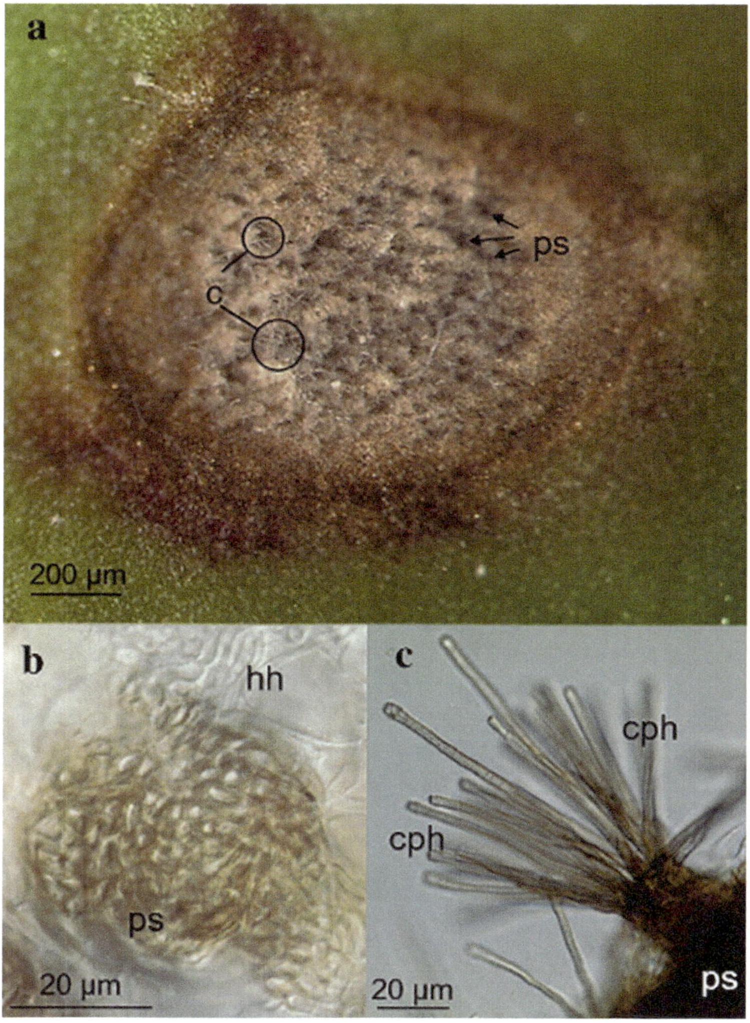

Fig. 27.2 (**a**) Conidia of *Cercospora beticola* on pseudostromata (ps) in the middle of the spot; (**b**) mycelium and pseudostroma; (**c**) conidiophores on the leaf surface (Source: Oerke et al. 2019)

incidence of the disease affecting sugar beet growing areas are more than a third of total acreage worldwide (Holtschulte 2000). The disease occurs severely in Marmora and Black Sea Region in Turkey and it is sometimes seen moderately in the central regions.

Fig. 27.3 Spreading of *Cercospora* leaf spots to neighbouring leaves after initial infection in the field

Fig. 27.4 *Cercospora* leaf spots spreading over all field and killing older leaves

27.5 Epidemiology

Cercospora beticola can infect beet plants between 12–37 °C. Conidia are produced at optimal temperatures between 20–26 °C when the relative humidity prevails in the range of 98–100% (Pool and McKay 1916). Epidemics can severely occur if the relative humidity is above 96% for 10–12 h on a 3–5 succeeding days and the temperature is above 10 °C (Mischke 1960). Although it is rather high temperatures, severe epidemics can develop in Turkey and the Netherlands if the relative humidity is enough. Conidia are disseminated by rain-splash (Pool and McKay 1916, Carlson 1967), wind (McKay and Pool 1918), irrigation water, insects, and mites (McKay and Pool 1918, Meredith 1967). Other potential sources of initial inoculum include the distribution of *C. beticola*-infested plant material via tools or machinery (Knight et al. 2018, Knight et al. 2019) and stromata from other host plants (Khan et al. 2008,

Fig. 27.5 Leaves collapsing and falling to the ground, and regrowth at the head of beet

Fig. 27.6 *Cercospora* disease killing all leaves on the plant and vegetative regrowth

Franc 2010, Skaracis et al. 2010, Tedford et al. 2018, Knight et al. 2020). The most cultivated and wild species of *Beta* are infected by *C. beticola*. The fungus attacks the cultivated plants such as *Spinacia oleracea* (spinach) and *Carthamus tinctorius* (safflower) and weedy species of *Amaranthus, Atriplex, Chenopodium* and *Plantago* (Vestal 1933, Frandsen 1955, El-Kazzaz 1977, Soylu et al. 2003), *Cycloloma, Malva, Limonium*, and Apium (Lartey et al. 2005, Groenewald et al. 2006, Jacobsen and Franc 2009). There have been different races of *C. beticola*, mainly based on cultural and physiological differences in vitro (Schlösser and Koch 1957, Solel and Wahl 1971, Mukhopadhyay and Pal 1981). Conidia of *C. beticola* remain in infected leaf tissues for only 1–4 months (Pool and McKay 1916), but pseudostromata, sources of primary inoculum, can survive for 1–2 years (Pool and McKay 1916, McKay and Pool 1918, Canova 1959b). In the period of 1977–2003, *Cercospora*

Fig. 27.8 The life cycle of *Cercospora beticola* Sacc. (Modified from Jones, Roger K. and Carol E. Windels)

Spain, and 20–40% in Yugoslavia when fungicides were not applied to the disease. Disease incidence in Belgium, Denmark, England, Ireland, and Sweden remained almost negligible (Byford 1996).

The damage of the disease was estimated about 100 and 29 million Euro due to not spraying and wrong fungicide use each year in Northern Italy (Meriggi et al. 1998, Rossi et al. 2000). Without spraying in Italy, Rossi et al. (2000) have reported that 10.1% of root yield, 4.4% of sugar content, 1.3% of the extractable sugar content, and 16.9% of sugar yield have dropped. On the other hand, the contents of potassium (K), sodium (Na), and alpha-amino nitrogen (α-amino N) which consist of molassigenic compounds have increased by 6.4%, 24.7%, and 16.8%, respectively. Root yield, sugar content, extractable sugar content, and sugar yield of beet decreased by 1–26%, 3–13%, 5–18%, and 6–36%, respectively, while potassium (K), sodium (Na), and alpha-amino nitrogen (α-amino N) content increased by 0–5%, 9–20%, and 1–40%, respectively, depending on the severity of infection by years, without spraying in the province of Sakarya in Turkey (Kaya 2012).

27.7 Disease Management

The integrated management of *Cercospora* leaf spot includes cultural practices, host resistance, and then fungicides application (Pool and McKay 1916, Khan et al. 2007). Cultural practices reduce the level of initial inoculum for the following season through rotation with non-host crops. Burying infested plant materials and avoiding planting next to fields previously sown with sugar beets also decrease the inoculum potential of the pathogen. To predict the occurring of the disease and timing of fungicide application, epidemiological models have been established (Rossi and Battilani 1991, Windels et al. 1998, Pitblado and Nichols 2005, Racca and Jörg 2007). Chemicals should be applied prophylactically early to avoid conidia infecting unprotective leaves. Although there have been studies on biocontrol agents including *Trichoderma* and *Bacillus* for *C. beticola* (Collins and Jacobsen 2003, Galletti et al. 2008), they are not to be used in practice.

27.7.1 Cultural Control

The plants which are non-host should be replanted on the same land after at least 3 years. Sugar beet should be sown in the fields in areas at least 300 ft. from last season's plantings. The soil should be plowed deeply to completely bury infected leaf residues. *Cercospora*-free seeds should be sown. Resistant varieties must be sown. Plants should be irrigated during night so as not to keep leaves wet longer.

27.7.2 Crop Rotation

The pseudostromata of the fungus survive in the soil for 2 years. To effectively eliminate inoculum from a field, sugar beets should be planted in a 3-year rotation with non-hosts. The soil should also be plowed to incorporate beet leaf residues. Deep tillage after sugar beet planting will prevent fungus death (Canova 1959a). At least a three-year rotation should be applied to reduce the inoculum potential of *C. beticola* by ensuring the rotting of infected head and leaf residues, which constitute a new source of the disease infection (Pool and McKay 1916, Pundhir and Mukhopadhyay 1987). Spinach, table beet, and chard plants should not be included in rotation and the host weeds should be removed from the field before infection occurs.

27.7.3 Using Disease-Free Seeds

By using seeds that are not contaminated with *Cercospora* spores, the disease is prevented from moving to new planting areas with seeds.

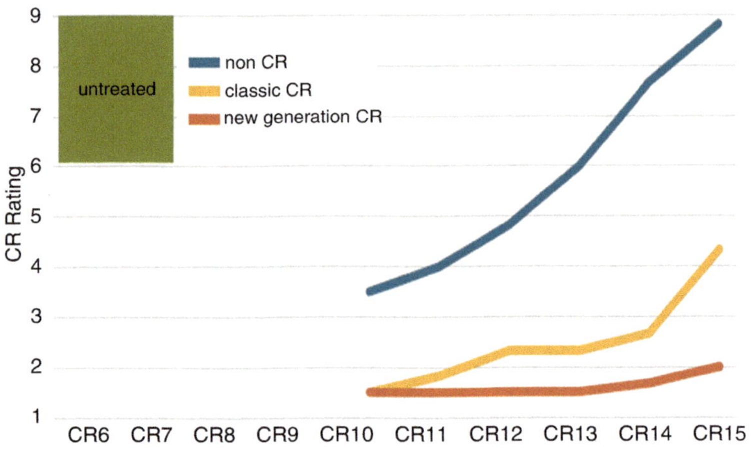

Fig. 27.11 Disease ratings in new generation varieties compared to susceptible and classic resistant ones in untreated plots in Soligenstadt, Frankonia, Germany in 2018 (Source: Kopisch-Obuch et al. 2020)

Fig. 27.12 New generation resistant varieties (right) affected by *C. beticola* compared to susceptible one (left) in untreated plots in lower Bavaria in Germany 2019 (Source: Kopisch-Obuch et al. 2020)

throughout the years (Meriggi et al. 2000). For disease control, a number of fungicides which are protectant and systemic have been registered by different companies and used by the farmers in different countries. Chemical families of available fungicides are as below (Ioannidis and Karaoglanidis 2010):

Table 27.2 Classification of fungicides used in the control of *C. beticola* according to resistance development risk (Source: Ioannidis and Karaoglanidis 2000)

Resistance risk	Fungicides	Chemical group
Low	Maneb Mancozeb	Dithiocarbamats
	Chlorothalonil	Phenolic compounds
	Copper compounds	Copper compounds
Medium	Fentin acetate Fentin hydroxide	Tinned compounds
	Fenpropimorph	Morpholine
	Cyproconazole Difenoconazole Flusilazole Flutriafole	Demethylation inhibitors
High	Benomyl	Benzimidazoles
	Carbendazim	
	Thiophanate-Methyl	

(a) The protective dithiocarbamates, nitriles, and fentin derivatives,
(b) The systemic and curative benzimidazoles,
(c) The systemic, protective, and curative ergosterol inhibitors (DMIs and amines),
(d) The protective, curative, and eradicant quinone outside inhibitors (QoIs) which are relatively new and very effective.

One of the most substantial factors restricting the control of the disease by chemicals is the forming of the resistance to fungicides. Resistance has increased dramatically in the last 40 years. When the same fungicide is used continuously and for many years, fungus *C. beticola* creates resistance. It is the first pathogen to develop resistance to benzimidazole fungicides in some countries, especially Greece, in the early 1970s (Georgopoulos and Dovas 1973, Ruppel and Scott 1974, D'ambra et al. 1974, Pal and Mukhopadhyay 1985, Weiland and Halloin 2001). The pathogen later developed resistance to fentin fungicides (Giannopolitis 1978, Cerato and Grassi 1983, Bugbee 1996). In the 1990s, it developed resistance against demethylation inhibitors (triazoles) in Greece (Karaoglanidis et al. 2000). Since resistance to benzimidazoles is very strong, the efficacy of the fungicide suddenly decreased and disappeared. Resistance to other fungicides developed slowly and was low (Ioannidis and Karaoglanidis 2000) (Table 27.2).

According to the results of the study in Turkey (Maden et al. 2009), *C. beticola* was detected to be resistant to mancozeb and fentin acetate with protecting action and flutriafol with systemic action in all beet growing areas infected by the disease, except Alpullu and Kastamonu regions. The highest resistance to mancozeb was in Susurluk region followed by Adapazarı, Amasya, Kastamonu, and Çarşamba regions. Resistance to fentin acetate was found at the highest rate in the Susurluk factory region, followed by isolates from Amasya and Kastamonu regions.

Table 27.3 Fungicides recently used against *Cercospora* leaf spot

	Fungicides				
Group	The name of active ingredient	Percent of a.i. (%)	Formulation	Dose (kg or L/ha)	
Group I (Systemic curative main fungicides)	Flutriafol	12.5	SC	0.50	L
	Flutriafol	25	SC	0.25	L
	Epoxiconazol + Carbendazim	12.5 + 12.5	SC	0.40	L
	Flusilazol	40	EC	0.20	L
	Difenoconazol	25	EC	0.30	L
	Difenoconazol + Propiconazol	15 + 15	EC	0.30	L
	Prochloraz + Propiconazol	40 + 9	EC	1.25	L
	Tetraconazol	12.5	ME	0.75	L
	Epoxiconazol + Pyraclostrobin	6.25 + 8.5	SE	1.50	L
	Tebuconazole + Trifloxystrobin	50 + 25	WG	0.25	kg
	Epoxiconazol + Fenpropimorph	8.4 + 25	SE	0.75	L
	Azoxistrobin + Difenoconazol	12.5 + 12.5	SC	0.50	L
GroupII (Contactprotectant additional fungicides)	Chlorothalonil	75	WP	0.50	kg
	Maneb	80	WP	1.50	kg
	Mancozeb	80	WP	1.50	kg

WP Wettable powder, *WG* Water dispersible granules, *DF* Dry flowables, *EC* Emulsifiable concentrate, *SC* Suspension concentrate, *SE* Suspoemulsion, *ME* Microemulsion concentrate

15–20 day intervals, depending on the period of fungicides remaining and acting in the leaf, until 3–4 weeks before harvest.

(b) Mathematical early warning prediction model based on climate data collected by means of instruments and computer software. Early warning of *Cercospora* leaf spot disease in sugar beet is predicted by climatic data. The used program widely in the world in this context is the method based on Daily Infection Value (DIV) calculated from the temperature and humidity values around the plants in the field to indicate a spray, developed by Shane and Teng (1985), which is *Cercospora* leaf spot model belonging to the University of Minnesota (Windels et al. 1991). The other is software of risk forming based on incubation and sporulation evaluations according to Bleiholder and Weltzien (1972). Here the model gives the infection directly as mild, moderate, and severe. In DIV evaluations, only the DIV of the day is given. According to the Minnesota DIV, mild disease emergence when DIV6 for a day or a total of 2 days occurs and DIV 7 indicates that severe disease will occur when the total of 2 days is 7 or more in a day.

The transition from thresholds to a climatic-based system is depended upon the intensive amount of labour including field observations needed by using thresholds. Weather- and climate-relative systems developed in the combat of *Cercospora* leaf spot are used in Italy (Rossi 1997), The United States (Windels et al. 1998), and Germany (Jörg and Racca 2000). These models consisted of temperature, humidity, and duration length which are suitable for the germination of *C. beticola* conidia (Vereijssen 2004).

The various models, related to the infection process, developed based on the parameters of climate-environment and damages to plant. Some of them consider only climatic data suitable for disease developing (Shane and Teng 1984, Windels et al. 1998, Khan et al. 2007), while the others the resistance level of the variety (Wolf and Verreet 2002, Racca and Jörg 2007). The disease management for the sustainable sugar beet production might be substantially supported by all the models on condition that they are implemented precisely (Windels et al. 1998, Wolf and Verreet 2002, Khan et al. 2007).

27.9 Integrated Management

One or several of implements including resistant variety, fungicide application, crop rotation, good farming practices, and use of disease-free seeds cannot protect sufficiently beet leaves against *Cercospora* leaf spot. For climatic, efficiency, and economic reasons, these should be cautiously employed altogether to minimize the number of fungicide applications and the possibilities of fungicide resistance, and the increasing in pathogen populations. As a result, integrated disease managements are now adopted and widely guided towards achieving sustainable sugar beet production. Integrated disease management (Fig. 27.13) completely consists of a combination of the practices such as crop rotation, good farming techniques, using

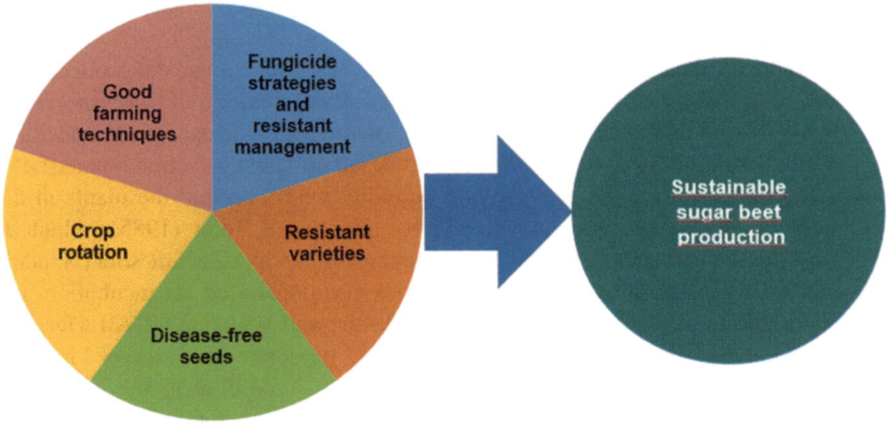

Fig. 27.13 Integrated management of *Cercospora* leaf spot for sustainable sugar beet production

disease-free seeds to decrease inoculum, sowing resistant varieties to make the onset of the disease late and prevent its development, and protecting leaves by fungicides (Jacobsen 2010). The significance of the epidemiological models is that they accurately monitor the onset and progression of the disease. Thus, fungicides are sprayed only when necessary (Skaracis et al. 2010).

27.10 Biocontrol

Biocontrol agents such as *Trichoderma* species and *Bacillus subtilis* were stated for *C. Beticola* in sugar beet (Collins and Jacobsen 2003, Galletti et al. 2008). Unfortunately, they have not been successful in practice. On the other hand, several microbial groups are present together with the disease occurring in the fields of sugar beet and it is supposed that some of microbes may be beneficial to predict the occurrence of disease as biological markers (Kusstatscher et al. 2019). However, biological agents may be used against *Cercospora* leaf spot as a supplementary protection to resistant varieties and fungicides. Bargabus et al. (2002) stated that the systemic resistance caused by *Bacillus mycoides* (BmJ) gave promising result when applied to leaves and also, *Trichoderma* species, a soil-born pathogen (Lartey et al. 2010), can be applied. Galletti et al. (2008) declared that pathogen sporulation and non-competitive or competitive antagonism might be decreased by two *Trichoderma* isolates and also, the incidence of the disease and pathogen sporulation were reduced by repeated sprays of homogenate treated with difenoconazole only once under natural inoculation. Jacobsen (2010) reported that they might contribute to crop protection in times to come. In addition, it was announced that the enzyme laccase gained from a basidiomycete could remove effects of cercosporin and might decrease the cercosporin toxicity when applied to the leaves (Caesar-TonThat et al. 2009). In view of experiments of the troubles owing to the mechanism of resistance, several possible classic and molecular studies to the future improvement are being taken in hand (Skaracis et al. 2010).

27.11 Conclusion

The yield and quality performance of the recent bred varieties resistant to *C. beticola* have reached to that of sensitive ones owing to advances in plant breeding. When disease does not occur, new generation resistant varieties do not cause yield loss and give better performance than sensitive ones. It is supposed that the varieties bred recently will cause a significant reduction in usage of fungicide for an improved integrated pest management. The occurrence of *C. beticola* resistance to the fungicides used should generally be viewed as a big trouble to sustainable sugar beet production. Only obtaining detailed information about the mode of action, method, and time of fungicide usage, the genetics of *C. beticola* and the mechanism of its resistance will identify the risks before fungicide failure. The information gathered will help the resistance management plans and tactics of specific measures for producing sugar beet sustainably, while maintaining yield stability.

Fully comprehending the interaction between *C. beticola* and sugar beet could contribute to new strategies to control disease and thus further reduce yield losses. In this respect, it is the need to research the biology of the pathogen and new developments in its molecular and genetic understanding. Possibly, new biological agents to be discovered in future studies will also contribute to cope with the disease. For advanced integrated *Cercospora* leaf spot management from now onwards, the comprehending of the molecular and genetic characteristics of the pathogen, the properties of the new fungicides to be discovered in detail, the prevention or minimization of resistance improvement, and the exploring of new information on pathogen-beet interaction, especially advances in plant breeding, will allow more competitive and profitable sugar beet production.

References

Adams H, Schaufele WR (1996) Untersuchungen zum Einfluss der *Cercospora*-Blattfleckenkrankheit auf den Alpha-Amino-Gehalt der Zuckerrübe. Ber. 59. IIRB Kongress, Brüssel, pp 129–132

Anonymous (1969) CMI-distribution maps of plant diseases. Ed 4, Map No 96

Bargabus RL, Zidack NK, Sherwood JE, Jacobsen BJ (2002) Characterization of systemic resistance in sugar beet elicited by a non-pathogenic, phyllosphere-colonizing *Bacillus mycoides*, biological control agent. Physiol Mol Plant Pathol 61:289–298

Barnett HL, Hunter BB (1972) Illustrated genera of imperfect fungi, 3rd edn. Burgess, Minneapolis, MN, p 241

Bleiholder H, Weltzien H (1972) Contributions to the epidemiology of *Cercospora beticola* (Sacc.) of sugar beet. III. Geopathological investigations. Phytopathologische Zeitschrift 73:93–114

Bugbee WM (1996) *Cercospora beticola* strains from sugar beet tolerant to triphenyltin hydroxide and resistant to thiophanate methyl. Plant Dis 80:03

Byford WJ (1996) A survey of foliar diseases of sugar beet and their control in Europe. Proceedings of the 59th IIRB Congress, 1–10, Bruxelles

Caesar-TonThat TC, Lartey RT, Solberg-Rodier LL, Caesar AJ (2009) Effects of basidiomycete laccase on cercosporin. J Plant Pathol 91(2):347–355

Canova A (1959a) Ricerche sula biologia e l'epidemiologia della *Cercospora beticola* Sacc. II. Annali della Sperimentazione Agraria 13:157–204

Canova A (1959b) Richerche su la biologia e l'epidemiologia della *Cercospora beticola* Sacc., Parte III. Annali Della Sperimentazione Agraria NS 13:477–479

Carlson LW (1967) Relation of weather factors to dispersal of conidia of *Cercospora beticola* (Sacc.). J Am Soc Sugar Beet Technol 14:319–323

Carruthers A, Oldfield JFT (1961) Methods for assessment of beet quality. Int Sugar J 63:72–74

Cerato C, Grassi G (1983) Tolerance of organo-tin compounds among *Cercospora beticola* isolates. Informatore Fitopatologico 33:67–69

Chupp C (1953) Monograph of the fungus genus *Cercospora*. Ithaca, New York, p 667

Collins DP, Jacobsen BJ (2003) Optimizing a *Bacillus subtilis* isolate for biological control of sugar beet Cercospora leaf spot. Biol Control 26:153–161

Crous PW, Braun U (2003) Mycosphaerella and its anamorphs: I. Names published in Cercospora and Passalora. CBS Biodiversity Series 1. Centraalbureau voor Schimmelcultures, Utrecht, The Netherlands

D'Ambra V, Mutto S, Carula G (1974) Sensibilta e toleranza di isolate di *Cercospora beticola* sensibili e tolerant al benomyl. Industria Saccarifera Italiana 1:11–13

El-Kazzaz MK (1977) Cercospora leaf spot disease of chard in Egypt. Egypt J Phytopathol 9:81–82

EU (2009) Directive 2009/128/EC of the European Parliament and of the Council of 21 October 2009 establishing a framework for Community action achieve the sustainable use of pesticides. Off J Eur Union 309:71–86

Franc G (2010) Ecology and epidemiology of *Cercospora beticola*. In: Lartey R, Weiland J, Panella LW, Crous PW, Windels CE (eds) *Cercospora* leaf spot of sugar beet and related species. American Phytopathological Society, St. Paul, MN, pp 7–19

Frandsen NO (1955) Über den Wirtskreis und die systematische Verwandschaft von *Cercospora beticola*. Arch Mikrobiol 22:145–174

Galletti S, Burzi PL, Cerato C, Marinello S, Sala E (2008) *Trichoderma* as a potential biocontrol agent for Cercospora leaf spot of sugar beet. BioControl 53:917–930

Georgopoulos SG, Dovas C (1973) Occurrence of *Cercospora beticola* strains resistant to benzimidazole fungicides in Northern Greece. Plant Dis Reporter 62:321–324

Giannopolitis CN (1978) Lesions on sugar beet roots caused by *Cercospora beticola*. Plant Dis Reporter 62:424–427

Graf A (1980) Ein einjahrigerLagerversucht mit *Cercospora* und *Erysiphe* befallenen Rüben. Ber. 43. IIRB Wintercongress, pp 221–224

Groenewald M, Groenewald JZ, Braun U, Crous PW (2006) Host range of *Cercospora apii* and *C. beticola*and description of *C. apiicola*, a novel species from celery. Mycologia 98(2): 275–285

Holtschulte B (2000) Cercospora beticola—worldwide distribution and incidence. In: Asher MJC, Holtshulte B, Richard-Molard M, Rosso F, Steinrücken G, Beckers R (eds) *Cercospora beticola* Sacc. biology, agronomic influence and control measures in sugar beet, advances in sugar beet research, vol 2. IIRB, pp 5–16

Ioannidis PM (1994) Fungucide chemicals and techniques in controlling *Cercospora beticola* Sacc. in Greece. Proceedings of Mediterranean Section IIRB Brussels, pp. 89–99

Ioannidis PM, Karaoglanidis GS (2000) Competition between DMIs-sensitive and—resistant strains of *Cercospora beticola* on untreated sugar beet crop. Proceedings of the 63rd IIRB Congress, Interlaken, Switzerland, pp 489–496

Ioannidis PM, Karaoglanidis GS (2010) Control of Cercospora leaf spot and Powdery mildew of sugar beet with fungicides and tolerant cultivars. In: Lartey RT, Weiland JJ, Panella L, Crous

PW, Windels CE (eds) Cercospora leaf spot of sugar beet and related species St. The American Phytopathological Society, Paul, MN, pp 259–274

Jacobsen BJ (2010) Integrated management of Cercospora leaf spot. In: Lartey RT, Weiland JJ, Panella L, Crous PW, Windels CE (eds) Cercospora leaf spot of sugar beet and related species. The American Phytopathological Society, St. Paul, MN, pp 275–284

Jacobsen BJ, Franc GD (2009) Cercospora leaf spot. In: Harveson RM, Hanson LE, Hein GL (eds) Compedium of beet diseases and insects, 2nd edn. American Phytopathological Society, St. Paul, MN, pp 7–10

Jörg E, Racca P (2000) CERCBET 1 2 3–Prognose-modelle zur Simulation von *Cercospora beticola*. Zuckerrübe 49:200–203

Karaoglanidis GS, Ioannidis PM, Thanassoulopoulos CC (2000) Reduced sensitivity of *Cercospora beticola* isolates to sterol demethylation inhibiting fungicides. Plant Pathol 49: 567–572

Kaya R (2012) Cercospora leaf spot (*Cercospora beticola* Sacc.) disease and control strategies in sugar beet. 1st International Anatolian sugar beet symposium, 2022 September, Kayseri, Turkey, pp 13–27

Khan MFR, Khan J (2010) Survival, spore trapping, dispersal, and primary infection site for *Cercospora beticola* in sugar beet. In: Lartey RT, Weiland JJ, Panella L, Crous PW, Windels CE (eds) Cercospora leaf spot of sugar beet and related species. The American Phytopathological Society, St. Paul, MN, pp 67–75

Khan J, del Rio LE, Nelson R, Khan MFR (2007) Improving the Cercospora leaf spot prediction model for sugar beet in Minnesota and North Dakota. Plant Dis 91:1105–1108

Khan J, Rio LD, Nelson R, Rivera-Varas V, Secor G, Khan M (2008) Survival, dispersal, and primary infection site for *Cercospora beticola* in sugar beet. Plant Dis 92:741–745

Kirk WW, Hanson LE, Franc GD, Stump WL, Gachango E, Clark G, Stewart J (2012) First report of strobilurin resistance in *Cercospora beticola* in sugar beet (Beta vulgaris) in Michigan and Nebraska, USA. New Disease Reports 26:3. https://doi.org/10.5197/j.2044-0588.2012.026.003

Knight NL, Vaghefi N, Hansen ZR, Kikkert JR, Pethybridge SJ (2018) Temporal genetic differentiation of *Cercospora beticola* populations in New York table beet fields. Plant Dis 102:2074–2082

Knight NL, Vaghefi N, Kikkert JR, Bolton MD, Secor GA, Rivera VV, Hanson LE, Nelson SC, Pethybridge SJ (2019) Genetic diversity and structure in regional *Cercospora beticola* populations from *Beta vulgaris* subsp. *vulgaris* suggest two clusters of separate origin. Phytopathology 109:1280–1292

Knight N, Koenick L, Sharma S, Pethybridge SJ (2020) Detection of *Cercospora beticola* and *Phoma betae* on table beet seed using quantitative PCR. Phytopathology 110:943–951

Köller W (1991) Fungicide resistance in plant pathogens. In: Pimental D, Hanson AA (eds) Handbook of pest management in agriculture, vol 2, 2nd edn. CRC Press, Boca Raton, FL, pp 679–720

Kopisch-Obuch FJ, Stibbe C, Beyer W, Mechelke W (2020) Keeping sugarbeet competitive with new strategies for Cercospora resistance. 77th IIRB Congress, 11th–12th February, Brussels, Belgium

Kusstatscher P, Cernava T, Harms K, Maier J, Eigner H, Berg G, Zachow C (2019) Disease incidence in sugar beet fields is correlated with microbial diversity and distinct biological markers. Phytobiomes J 3:22–30

Lartey RT, Caesar-TonThat TC, Caesar AJ, Shelver WL, Sol NI, Bergman JW (2005) Safflower: a new host of *Cercospora beticola*. Plant Dis 89:797–801

Lartey RT, Weiland J, Panella L, Crous P, Windels C (2010) Cercospora leaf spot of sugar beet and related species. American Phytopathological Society, St. Paul, MN

Maden S, Katırcıoğlu Z, Demirci F, Kaya R, Ergül A, Esmer B, Özeren P (2009) Şeker Pancarında Cercospora Yaprak Lekesi Hastalığının Entegre Savaşımı. TÜBİTAK Proje No: 105O121, Ankara, 81s

Malandrakis AA, Markoglou AN, Nikou DC, Vontas JG, Ziogas BN (2006) Biological and molecular characterization of laboratory mutants of Cercospora beticola resistant to Qo inhibitors. Eur J Plant Pathol 116:155–166. https://doi.org/10.1007/s10658-006-9052-1

McKay MB, Pool VW (1918) Field studies of *Cercospora beticola*. Phytopathology 8:119–136

Menkissoglou-Spiroudi U, Xanthopoulou NJ, Ioannidis PM (1998) Dissipation of the fungicide tetraconazole from field-sprayed sugarbeets. J Agric Food Chem 46:5342–5346

Meredith DS (1967) Conidium release and dispersal in *Cercospora beticola*. Phytopathology 57: 889–893

Meriggi P, Rosso F (1990) *Cercospora*-Nuovi programmi di difesa per la protezione dele cultivar tolleranti. Agronomica 3:5–13

Meriggi P, Rossi V, Cioni F, Maines G (1998) Vantaggi derivanti dalla ottimizzazione dei programmi di lotta alla cercosporiosi della barbabietola da zucchero. Atti Gionate Fitopatologiche:625–630

Meriggi P, Rosso F, Ioannidis PM, Ayala GJ (2000) Fungicide treatments against Cercospora leaf spot in sugar beet (*Beta vulgaris* L.). In: Asher MJC, Holtshulte B, Richard-Molard M, Rosso F, Steinrücken G, Beckers R (ed) *Cercospora beticola* Sacc. biology, agronomic influence and control measures in sugar beet, advances in sugar beet research, IIRB 2:77–102

Mischke W (1960) Untersuchungen über den Einfluß des Bestandsklimas auf die Entwicklung der Rüben-Blattfleckenkrankheit (*Cercosporabeticola* Sacc.) im Hinblick auf die Einrichting eines Warndienstes. Bayerisches Landwirtschaftliches Jahrbuch 37:197–227

Mukhopadhyay AN, Pal V (1981) Variation among the sugar beet isolates of *Cercosporabeticola* from India. Proceedings of the 3rd international symposium on plant pathology, New Delhi, India. pp. 132–136

Oerke EC, Leucker M, Steiner U (2019) Sensory assessment of *Cercospora beticola* sporulation for phenotyping the partial disease resistance of sugar beet genotypes. Plant Methods 15:133, 13ps. https://doi.org/10.1186/s13007-019-0521-x

Oltmann W, Burba M, Bolz G (1984) Die Qualitat der Zuckerrübe: Bedeutung, Beurteilungskriterien und zückerische Massnahmen zu ihrer Verbesserung. Fortschritte der Planzenzühtung, Heft 12. Verlag Paul Parey, Berlin und Hamburg, p 159

Özgür OE, Kaya R (2002) Einige Fragen der *Cercospora*-Bekämpfung in der Türkei, 65. IIRB Kongreß, février 2002, s. 505–512, Bruxelles (B), pp. 13–14

Pal V, Mukhopadhyay AN (1985) Occurrence of strains of *Cercospora beticola* resistant to carbendazim (MBC) in India. Indian J Mycol Plant Pathol 13:333–334

Parlevliet JE (1979) Components of resistance that reduce the rate of epidemic development. Annu Rev Phytopathol 17:203–222

Piszczek J, Pieczul K, Kiniec A (2018) First report of G143A strobilurin resistance in *Cercospora beticola* in sugar beet (Beta vulgaris) in Poland. J Plant Dis Prot 125:99–101. https://doi.org/10.1007/s41348-017-0119-3

Pitblado R, Nichols I (2005) The implementation of BEETCAST-a weather-timed fungicide spray program for the control of Cercospora leaf spot in Ontario and Michigan. J Sugarbeet Res 42: 53–54

Pool VW, McKay MB (1916) Climatic conditions as related to *Cercosporabeticola*. J Agric Res 6: 21–60

Pundhir VS, Mukhopadhyay AN (1987) Epidemiological studies on Cercospora leaf spot of sugar beet. Plant Pathol 36:185–191

Racca P, Jörg E (2007) CERCBET 3—a forecaster for epidemic development of *Cercospora beticola*. EPPO Bull 37:344–349

Reichert I, Palti J (1966) On the pathogeography of plant diseases in the Mediterranean region. Proc I Congr Medit Phytopat, Union, Bari, Italy

Rosenzweig N, Hanson LE, Mambetova S, Jiang Q, Guza C, Stewart J, Trueman CL, Somohano P (2020) Temporal population monitoring of fungicide sensitivity in *Cercospora beticola* from sugarbeet (*Beta vulgaris*) in the Upper Great Lakes. Can J Plant Pathol 42(4):469–479. https://doi.org/10.1080/07060661.2019.1705914

Rossi V (1995) Effect of host resistance in decreasing infection rate of Cercospora leaf spot epidemics on sugarbeet. Phytopathol Mediterr 34:149–156

Rossi V (1997) Use of a simulation model "CERCODEP" in the control of Cercospora leaf spot on sugarbeet. Proceedings of the 60th IIRB congress, Cambridge, United Kingdom, pp. 355–359

Rossi V (2000) Cercospora leaf spot infection and resistance in sugar beet. in: *Cercosporabeticola* Sacc. Biology, agronomic influence and control measures in sugar beet, advances in sugar beet research, IIRB, vol. 2:17–48

Rossi V, Battilani P (1990) Dinamica delle epidemie di *Cercosporabeticola* Sacc. su barbabietola da zucchero. III. Ruolo della suscettibilita varietale. Phytopthol Mediterr 29:114–119

Rossi V, Racca P, Giosue S (1995) Geophytopathologicalanalysis of Cercospora leaf spot on sugar beet in the Mediter-ranean area. Phytopathologia Mediterranea 34:69–82

Rossi V, Battilani P (1991) CERCOPRI: a forecasting model for primary infections of Cercospora leaf spot of sugarbeet 1. EPPO Bull 21:527–531

Rossi V, Merriggi P, Biancardi E, Rosso F (2000) Effect of Cercospora leaf spot on sugar beet growth, yield and quality. In: Asher MJC, Holtshulte B, Richard Molard M, Rosso F, Steinrücken G, Beckers R (eds) *Cercospora beticola* Sacc. biology, agronomic influence and control measures in sugar beet, advances in sugar beet research, vol 2. IIRB, pp 49–76

Ruppel EG (1986) Foliar diseases caused by fungi. In: Whitney ED, Duffus JE (eds) Compendium of beet diseases and insects. APS Press, St. Paul, MN, pp 8–9

Ruppel EG, Scott PR (1974) Strains of *Cercospora beticola* resistant to benomyl in the U.S.A. Plant Dis Reporter 58:434–436

Saccardo PA (1876) Fungi Veneti novi vel critici. Series V. Nuovo Giornale Bot Italiano 8:162–211

Schlösser LA, Koch F (1957) Rassenbildung bei *Cercosporabeticola*. Zucker 10:489–492

Schürnbrand E (1952) Ein Beitrag zur Frage der Bedeutung der Sameninfektion durch *Cercospora beticola*. Zuckerindustrie 13:295–299

Secor GA, Rivera VV, Khan MFR, Gudmestad NC (2010) Monitoring fungicide sensitivity of *Cercosporabeticola* of sugar beet for disease management decisions. Plant Dis 94(11): 1272–1282

Shane WW, Teng PS (1984) Evaluation and implementation of the Cercospora leafspot prediction model. Sugarbeet Res Ext Rep 15:129–138

Shane WW, Teng PS (1985) Evaluation and implementation of the *Cercospora* leaf spot prediction model. Sugarbeet Res Ext Rep 15:129–138

Shane WW, Teng PS (1992) Impact of Cercospora leaf spot on root weight, sugar yield, and purity of Beta vulgaris. Plant Disease 76:812–820

Skaracis GN, Biancardi E (2000) Breeding for *Cercospora* resistance in sugar beet. In: Asher MJC, Holtschulte B, Molard RM, Rosso F, Steinruecken G, Beckers R (eds) Advances in sugar beet research, *Cercospora beticola* Sacc. biology, agronomic influence and control measures in sugar beet, vol 2. International Institute for Beet Research, Brussels, pp 177–195

Skaracis GN, Ioannidis PM, Ioannidis PI (1996) Integrated management systems against sugar beet Cercospora leaf spot disease. In Proceedings of the 59th IIRB congress, Brussels, pp. 45–54

Skaracis GN, Pavli OI, Biancardi E (2010) Cercospora leaf spot disease of sugar beet. Sugar Tech 12:220–228. https://doi.org/10.1007/s12355-010-0055-z

Smith GA, Ruppel EG (1973) Association of Cercospora leaf spot, gross sugar, percentage sucrose and root weight in sugarbeet. Can. J. Plant Sci. 53:695–696

Smith GA, Martin SS (1978) Differential response of sugarbeet cultivars to Cercospora leaf spot disease. Crop Sci 18:39–42

Smith GA, Ruppel EG (1971) Cercospora leaf spot as a predisposing factor in storage rot of sugarbeet roots. Phytopathology 61:1485–1971

Solel Z, Wahl I (1971) Pathogenic specialization of *Cercosporabeticola*. Phytopathology 61: 10811083

Soylu S, Soylu EM, Kurt S (2003) First report of Cercospora leaf spot on Swiss chard caused by *Cercosporabeticola* Sacc. in Turkey. Plant Pathol:52:804

Tedford SL, Burlakoti RR, Schaafsma AW, Trueman CL (2018) Relationships among airborne *Cercospora beticola* conidia concentration, weather variables and Cercospora leaf spot severity in sugar beet (*Beta vulgaris* L.). Can J Plant Pathol 40:1–10

Vereijssen J (2004) Cercospora leaf spot in sugar beet: Epidemiology, life cycle components and disease management. PhD Thesis Wageningen University, Wageningen, The Netherlands

Vereijssen J, Schneider JHM, Termorshuizen AJ (2005) Root infection of sugar beet by *Cercospora beticola* in a climate chamber and in the field. Eur J Plant Pathol 112:201–210. https://doi.org/10.1007/s10658-004-4172-y

Verreet JA, Wolf PFJ, Weis FJ. (1996) Bekämpfungsschwellen als Grundlage für eine integrierte Bekämpfung von *Cercospora beticola*- Das IPS-Modell Zuckerrübe. Proc 59th IIRB Congress, Brussels pp. 55–69

Vestal EF (1933) Pathogenicity, host response and control of Cercospora leaf-spot of sugar beets. Iowa Agric Res Station Bull 168:43–72

Vogel J, Kenter C, Holst C, Märländer B (2018) New generation of resistant sugar beet varieties for advanced integrated management of *Cercospora* leaf spot in central Europe. Front Plant Sci 9: 222

Wade M (1988) Strategies for preventing or delaying the onset of resistance to fungicides and for managing resistance occurrences. In: Delp CL (ed) Fungicide resistance in North America. APS Press, St. Paul, MN, pp 14–15

Weltzien HC (1967) Geopathologie der Pflanzen. Z. Pflanzenkrankh. 74:175–189

Weiland JJ, Halloin JM (2001) Benzimidazole resistance in *Cercospora beticola* sampled from sugar beet fields in Michigan, USA. Can J Plant Pathol 23:78–82

Weiland J, Koch G (2004) Sugar beet leaf spot disease (*Cercospora beticola* Sacc.). Mol Plant Pathol 5(3):157–166

Windels CE, Lamey HA, Hilde D, Widner J, Knudsen T (1998) A *Cercospora* leaf spot model for sugar beet: in practice by an industry. Plant Dis 82:716–726

Wolf PFJ, Verreet JA (2002) An integrated pest management system in Germany for the control of fungal leaf diseases in sugar beet: the IPM sugar beet model. Plant Dis 86:336–344

Wolf PFJ, Verreet A, Maier J, Köhler R (2000) An integrated Pest Management model (IPM sugar beet model) for threshold-oriented control of *Cercospora beticola* on sugar beet, developed under conditions in Southern Germany. Adv Sugar Beet Res, IIRB 2:103–121

Soil-Borne Pathogen-Mediated Root Rot Diseases of Sugar Beet and Their Management

28

Siddhartha Das and Sudeepta Pattanayak

Abstract

Sugar beet (*Beta vulgaris* L.) is the most important, nutritious, and forage crop globally. World's one fourth of sugar production is dependent only on sugar beet crop. Every year, farmers suffer a havoc production loss due to biotic stresses. Soil-borne pathogen-mediated root rot diseases of sugar beet are considered as a key constrain for beet cultivation. Various soil-borne pathogens like *Rhizoctonia solani*, *Macrophomina phaseolina*, *Sclerotium rolfsii*, *Aphanomyces cochlioides*, *Phytophthora drechsleri*, *Fusarium oxysporum* f.sp. *radicis-betae*, and *Phoma* sp. cause root rot symptoms. Various symptomatic characterizations such as wilting of whole shoot system, brownish-black discoloration at the petiole, and rotting of root with cracked holes on the upper surface are observed in infected plant. Disease progress maximally depends on susceptible epidemiological factors which have a crucial role in pathogenesis. Disease incidence level varies from 25 to 65% in minor to major infected plants, although all the pathogenic attacks may not occur in the same time. Yield loss can be estimated by the production loss which is because of reduced sugar quantity at the farm stage and postharvest losses in storage. Application of disease and stress resistant varieties, biological and cultural practices, chemical method, and soil quality control (water holding capacity, micro-macro nutrient deficiency, soil pH, soil salinity) are the possible management practices against root rot disease.

S. Das
Department of Plant Pathology, M.S. Swaminathan School of Agriculture, Centurion University of Technology and Management, Paralakhemundi, Odisha, India

S. Pattanayak (✉)
Division of Plant Pathology, ICAR—Indian Agricultural Research Institute, New Delhi, India

V. Misra et al. (eds.), *Sugar Beet Cultivation, Management and Processing*,
https://doi.org/10.1007/978-981-19-2730-0_28

Keywords

Epidemiological · Management · Root rot disease · Soil-borne · Sugar beet

Abbreviations

PCNB	Penta Chloro Nitro Benzene
SAM	Shoot apical meristem

28.1 Introduction

Sugar beet (*Beta vulgaris* L) is known as a major nutritious vegetable throughout globe and most in Asian continent. Sugar beet belongs to the family Chenopodiaceae and is considered to be a biennial crop grown mainly in temperate climates. From the ancient historical period, edible sugar beet has been cultivated and consumed for its wide nutraceutical value. Addition in human diet fulfilled calorie deficiency and mitigated societal needs. Sugar beet is also used for multiple purposes such as used as sweetener and preservative in different foods like canned foods, beverages, confectionaries, and pharmaceuticals. Though sugarcane is the major sugar producing crop in India, the varied agroclimatic conditions have driven the Indian agriculture to establish sugar beet cultivation as an alternative in tropical and subtropical regions of the country (Mall et al. 2021). Sugar beet can be the best substitute over the water guzzling sugarcane crop, especially in drought-prone and water-scarce areas. Moreover, sugar beet is preferred for its high value sugar yield and low water use efficiency (6.79 and 10.3 kg root/m^3) (Farag et al. 2017). Sugar production from sugar beet is reported to increase from 20% in 2009 to nearly 30% in 2013 (FAO 2009; Dohm et al. 2014). Biochemical analysis of sugar beet revealed that it contains 75–76% water, 15–20% sugars, 2.6% nonsugars, and 4–6% pulp (Brar et al. 2015). Sugar beets are the biennial crop which required periodical vernalization for flowering, popularly known as "bolting" (OECD 2006). Pulp of sugar beet is used for the production of ethanol and beer. The tops contents approximate 10% digestible consumable protein which is considered to be cheapest substitute of forage-based diet (Karla 2020). The top portion is enriched with saponin. Phytochemical study of sugar beet leaf revealed that leaves contain 1.4–6.2% carotene, vitamin E, vitamin C, and different sugar residue (Brar et al. 2015).

In India, sugar beet is cultivated in tropical and subtropical regions, especially Deccan tracts of Maharashtra, Tamil Nadu, Karnataka, and Andhra Pradesh (Mall et al. 2021). Parts of sugar beet crop are extensively utilized in pharmaceutical industry for the extraction of various pharmaceutical products and vitamin B complex. Although sugar beet is referred as a potential economical crop in India, every year, sugar beet cultivation faces a lot of pest and pathogen-related challenges. Root rot diseases are the major constraints for the sugar beet production. Various soil-borne root rot pathogens like *Sclerotium rolfsii*, *Rhizoctonia solani*, *Macrophomina phasiolina*, *Fusarium oxysporum*, *Phytophthora drechsleri*, and many more played a crucial role in disease development and yield/production loss. Almost all components of sugar beet plant are attacked by the same pathogen, though root is highly susceptible part. In Indian agricultural context, two main hurdles associated with sugar beet production are poor seed production quality and various biotic stress-related challenges (Srivastava 2004). Soil-borne pathogen-mediated root rot diseases are the most probable biotic stresses among them. In Indian context, seed production is carried out in regions of higher altitudes like Srinagar, Darjeeling, Shimla, Auli, and Kalpa (Mall et al. 2021). From global point of view, Indian subcontinent and south East Asian countries are more progressive in sugar beet production. Post-monsoon sugar beet cultivation may be effective and profitable in Karnataka, Haryana, Maharashtra, Rajasthan, Punjab, etc. (Kulkarni et al. 2013). High temperature condition for this temperate crop may be a favorable environmental constrain from the pathological point of view. Succulent nature of the foliage and crop helps to introduce various soil-borne diseases. Fungal infection is more destructive compared to bacterial or viral pathogen attack in sugar beet. Root rot disease incidence on sugar beet varies between 15 and 50% crops damage. In such cases, both yield and sucrose contents are tremendously affected (Srivastava 2004). Root rot symptom of sugar beet appeared as distinctive blackish external decay or rot on the distal tap of the primary tap root and vascular tissue turned to necrotic tissue (Jacobsen 2006). Unfortunately, sometimes before appearance of any symptomatic expression, plants reached to mortality stage. Insufficient amount of knowledge about root rot diseases falls into a havoc amount of yield loss.

28.2 Types of Biotic Stresses

A wide range of pest and pathogen attack the sugar beet production. Among which some are seed-borne and some are foliar phytopathogens, but the key role is played by harmful soil-borne root rot pathogens (Srivastava 2004). Soil-borne pathogens are difficult to control because they do not come into focus until substantial amount of damaged has occurred (Amein 2006). Efforts to manage those pathogens are effective only when predictive techniques are implemented through permanent disease progress block. Soil-borne pathogens can survive in soil for long time and germinate to infect new hosts when favorable weather condition arises (Panth et al. 2020). Germination of spore or resting structure varies from species to species. Though in different times, plant pathologist accepted some predictive techniques which were helpful for estimating potential root rot disease problem and related epidemiological factors, but those techniques are not applicable for all soil-borne pathogen-related diseases (Haque and Parvin 2021). Occurrence of Sclerotial root rot (*S. rolfsii*) disease infestation sometimes reached up to 50% in severe cases like worthy dry root rot (*R. solani*) and charcoal root rot (*R. bataticola*) may cause nearly about 15–30% infestation (Srivastava 2004), etc., but disease infestation varies according to the different agroclimatic regions, soil pH, soil type, soil moisture-temperature, and different epidemiological factors. Rise or decline of any range of those factors turned the disease from moderate to severe form.

28.2.1 Phytophthora Root Rot

The pathogen responsible for *Phytophthora* root rot is *Phytophthora drechsleri.* High soil temperature (28–32 °C) with approximately high soil moisture (80–85%) favored disease development. Generally, Oomycota survives in the soil as chlamydospores (5–15 m) and oospores (20–40 m). Oospores germinate sporangia to produce zoospores, which initiate primary and secondary infection. Symptoms appeared as wilting and a wet rot at the base of the taproot eventually extends upward toward the crown, resulting in meristematic decay. A very fine margin is found between rotted and healthy tissues. Wilted plants can survive if proper protection is taken at the early stages of this disease development. The disease severity increased when wilted beets are irrigated during hot weather condition. In contrast, disease progress will slow down or cease markedly when soil temperatures turn to low. Management by controlling of high soil moisture is an effective strategy for root rot disease (Schneider and Whitney 1986).

28.2.2 *Rhizoctonia* Crown and Root Rot

Rhizoctonia solani, AG 2-2 intraspecific groups IIIB and IV (perfect stage, *Thanatephorus cucumeris*), is the responsible causal organism of *Rhizoctonia* crown and root rot. This disease is the most devastating one and can survive under soil for long time (Windels et al. 1997). In Europe, nearly 10–15% of planted land faced economic loss which is increasing day by day. In USA, this damage has reached 30% (Buttner et al. 2003). In various cases, yield loss estimation was calculated which reached more than 50% in severe disease infestation. Yield reductions take place from harvestable crops which reduced tonnage due to ill roots and less recovery of white sugar. Symptoms appeared as sudden or permanent wilting of leaves and shoots with dark-brown to black discoloration of petiole bases. Taproot lesions are brownish-black in color which appeared near the crown. Lesions are superficial, although differences between diseased and healthy tissues are clearly visible. *Rhizoctonia*-mediated root rot appeared as rough cracked with decayed epidermal and cortical tissue zone. At more severe stage, whole root turned to porous on the ground during harvest. Both *Rhizoctonia. solani* intraspecific groups can also cause damping-off and *Rhizoctonia* crown and root rot. AG 2–2 IIIB can attack various hosts like wheat, maize, rice, and matt rush, which multiply at 35 °C, whereas AG 2–2 IV does not grow at 35 °C temperature or not attack the same host. Both intraspecific groups are found worldwide and widely infect sugar beet cultivation (crop rotation with maize cultivation) in Europian and Asian continents (Jacobsen 2006). *Rhizoctonia solani* interspecific group AG-4 causes damping-off symptoms of different hosts such as sugar beets, *Phaseolus* sp., soybeans, and alfalfa. AG group-specific study is key tool kit for pathobiology and patho-diversity study. Highest yield loss occurs mainly in warm, irrigated cropping areas. Dry land area showed similar production, but persistently, wet areas are affected more. Complex symptomatic expression appears as increasing patchy spots grow towards non-symptomatic to symptomatic region. Pathogen survives through sclerotium. The AG 2–2 IV pathogen grows at 24—35 °C. The AG 2–2 IIIB strain showed similar type of growth orientation AG 2–2 IV. Poorly irrigated land or low lands trigger fast spreading of the disease. Colonization of this major pathogen under the soil turned to be severe which is managed through crop alternation, cleaning of row, early planting, maintaining adequate, balanced fertility, and proper irrigation system. Rush and Winter (1990) and Elmer (1997) examined nitrogen availability under the soil. It was observed that application of chloride salts can lessen the disease incidence on table beet (Elmer 1997). Extreme disease tolerance can be withstood by only resistant varieties. Application of chemicals in the furrow at 4–8 leaves stage and underground soil temperature more than 20–25 °C are beneficial. The effective fungicides against the pathogen are trifloxystrobin, chlorothalonil, pencycuron, tebuconazole, azoxystrobin, and pyraclastrobin, PCNB, etc. Among those fungicides, Azoxystrobin has been identified as the most effective one (Kiewnick et al. 2001; Jacobsen et al. 2005).

28.2.3 *Fusarium* Root Rot

The disease, *Fusarium* root rot, is prevalent and cosmopolitan in almost all countries. Two different strains of *Fusarium* are generally found in India. Those two severe strains are *F. oxysporum* sp. *betae* (Mukhopadhyay and Thakur 1970) and *F. chlymydosporum* (Srivastava et al. 1999). These pathogens also caused wilt of seedling and stalk blight. Primary symptoms developed with interveinal yellowing. With the consecutive increase of symptoms, younger leaves showed yellowing and chlorotic patches were observed in different places. Transverse section of the root showed grayish to reddish brown discoloration (Srivastava 2004). Small lateral roots are the point of penetration of these pathogens. In severe stages, the roots showed wilted, shriveled, distorted, and disintegrated tissue of fibrovascular strands. Profuse growth is sometimes found on the infected part. *Fusarium* produces both micro- and macro-conidia in normal time, but in adverse situation it survives through chlamydospore in root debris for a long period of time (Srivastava 2004). The severity of the casual organism increases due to rise in temperature and moisture conditions. Crop rotation with any nonhost plant like cereals or alpha can be helpful to block the rapid outbreak (Srivastava 2004).

28.2.4 Violet Root Rot

The causal pathogen of violet root rot is *Rhizoctonia crocorum* (Pers. Fr.) De Candolle (perfect stage—*Helicobasidium brebissonii*) and has been appearing sporadically throughout the Europian and Asian continents, wherever sugar beet cultivation takes place (Jacobsen 2006). The disease is reported in some countries like U.S., Spain, Europe, etc. This causal fungus sometimes intimately accompanies with various weed hosts like *Capsella bursa-pastoris* (L), *Lynchnis alba*, and *Senecio vulgaris* L. Infected plants generally showed circular patches. Wilting symptoms of roots are characterized with purple to reddish purple spots with cushiony fungal mycelium on the external part of the root. Sclerotia is generally found on the secondary roots. Crop rotation may be the effective strategy for control of this pathogen. Crop rotation with nonhost plants like sweet potato, alpha-alpha, cabbage, bean, pea, and oil seeds is a significant method to reduce the pathogen spread. Another two possible measures for disease management are early harvesting and increase soil aeration.

28.2.5 *Phoma* Root Rot

The causal organism of *Phoma* rot is *Phoma betae* (Sexual stage *Pleospora bjoerlingii*, Byford). The aforesaid strain is maximally predominant in Asia, North America, Australia, Europe, etc.. The causal fungus consecutively showed the symptoms of root rot, leaf spot, black leg, and damping-off. Phoma root rot symptom is characterized with summative appearance of various symptom complexes such as root rot, damping-off and leaf spot, etc. (Jacobsen 2006). Sometimes wilted part showed dark brown to black shrunken lesions with a watery rot near the branch top. Decaying starts with changes of color from deep brown or blackish with prominent lines. *Phoma* rot is observed after 80 days of harvesting in storage. Root rot starts from the central portion of the crown and increases downward to create a cone-shaped structure. Sometimes white mycelium cushions appeared on the infected surface. In case of this disease, infected seed particles are used as primary source of inoculums where ascospore acts as secondary inoculums (Bugbee and Cole 1981). Seedling damping-off symptoms spreads at 5–20 °C temperature in normal condition. Moist condition is most favorable for conidial germination from pycnidia, which may cause lesions on the seed stalk. Survival seedling from damping-off occasionally showed systemic infection with the causal fungus. The life cycle of *Phoma betae* (Perfect stage *Pleospora bjoerlingii*, Byford) is represented through Fig. 28.1. Properly dried seeds can reduce the chances of pathogen attack. Properly irrigated dry seeds are planted as disease-free propagating material. Sometimes treatments of propagating materials with hot water, benzimidzole, thiram, prochloraz chemicals, or fungicides are potential measures against the pathogen. Root rot at the field condition may be managed well by enhancing nutrient availability and balancing water and soil moisture level. *Phoma* can survive under plant debris up to 30 months and when favorable condition arises, it again germinated.

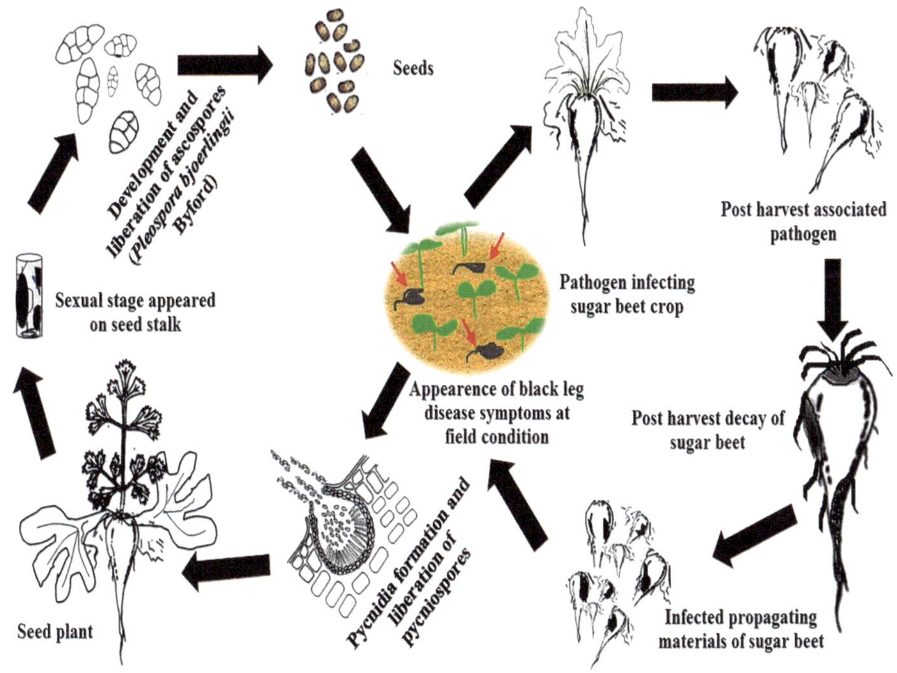

Fig. 28.1 Schematic representation of *Phoma betae* infestation (Perfect stage *Pleospora bjoerlingii*, Byford)

28.2.6 Charcoal Rot

Charcoal rot is one of the harmful diseases of this crop causing a yield loss of nearly 30% (Jacobsen 2006). This disease has minor economic importance as it attacks and damages the plant only under high temperature condition (optimum temperature 31 °C). This disease starts 4 months after sowing, i.e., on March, which may become severe during April in Indian agricultural context (Srivastava 2004). This disease is caused by *Rhizoctonia bataticola* whose pycnidial stage is known as *Macrophomina phaseolina* (Taasi) Goid. It is mostly observed in hot places of California and Greece, Iran, Egypt, India, and Hungary (Jacobsen 2006; Karadimos et al. 2007). The fungus can persist in soil or host tissue for a minimum period of 2 years through microsclerotia. These microsclerotia may be formed in other hosts like beet, strawberry, sunflower, sweet potato, maize, and potato (Su et al. 2001). Pycnidia and pycnidiospores are also produced by this pathogen.

Initially, the plants show significant wilting and browning of the foliage, resulting in plant mortality. Brownish or black, irregular lesions are formed in the external portion of the crown root. When the lesions rupture, masses of charcoal-colored microsclerotia are exposed out which is the characteristic symptoms of charcoal rot. Hence, the disease is named as charcoal rot. In the internal parts of the root, yellow mustard color lesions are observed which later turns to buffy citrine. In later stage of

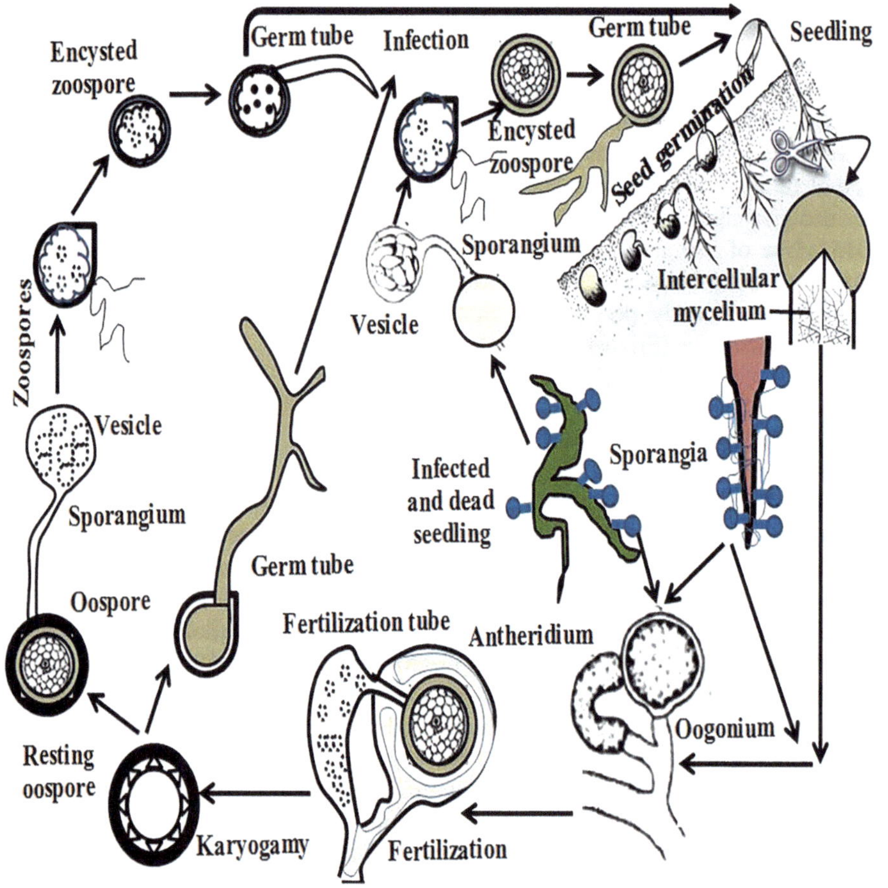

Fig. 28.2 Life cycle of *Pythium*-induced root rot

root tissues, sporangia of the casual organism are the common key player in infection through releasing the biflagellate zoospores. The initiation of secondary infection occurs through zoospores only. The occurrence of several weeds in the cultivated area can favor the existence as well as spread of inoculums in soil.

The disease initiation occurs in two stages, i.e., acute seedling blight and chronic root rot. The initial symptoms appeared as greyish, water-soaked lesions, gradually turns blackish constricting lesions near the collar portion, known as seedling damping-off. Plants look threadlike and fall off from the infected lesion. During the chronic root rot phase, yellow-brown lesion starts initially which further extends to the interior portion of the root. Gradually, the yellowish-brown lesions turn to darkish brown, causing the whole root to rot or may have scab like superficial lesions. In this phase, the root tips become dry and make the root constrict.

Additionally, the shoot system expresses synonymous wilting symptoms in hot times which recover during night. Leaves often show scorching like symptoms.

This pathogen produces numerous oospores in a short period of time. So, alteration of cropping system with nonhost crops is not much effective. The pathogen can be managed by an integrated disease management approach. Cultural methods such as proper soil drainage, soil indexing, early planting, weed control, and use of resistant varieties are the best methods to control the disease (Brantner and Windels 2004). Proper drainage system helps to maintain the soil moisture which reduces the pathogen inoculum. Similarly, weeding of Amaranthus and Chenopodium species can avoid the overwintering of the organism. Treatment of seeds with hymexazole and with bacterial biological compound is also effective against this disease. As infected roots have high respiration and sugar level, early postharvest processing of infected root can decrease the market value (Campell and Klotz 2006).

28.2.10 *Rhizopus* Root Rot

Rhizopus root rot does not have more economic importance due to its sporadic nature. Two pathogens are responsible for this disease, i.e., *Rhizopus arrhizus* and *R. stolonifer*. According to the host, this pathogen acts as both saprophyte and parasite. It is a weak parasite on sugar beet, while saprophyte on organic matter. These pathogens produce sporangia which are airborne. These act as primary inoculum and only infect a host which is damaged mechanically or has excess soil moisture stress. Insect damage of host crop is another factor. *Rhizopus stolonifer* prefers low temperature of 14–16 °C, while *R. arrhizus* favors high temperature of 30–40 °C. This disease is observed in many sugar beet growing areas like California, Italy, Arizona, Iran, France, and Colorado. Excessive soil moisture in addition with atmospheric temperature facilitates the disease.

Initially, the SAM (Shoot apical meristem) portion of shoot system shows wilting. Afterwards, the foliage becomes dry and brittle, resulting in the collapse of the entire plant. The taproots show grey to brown lesions which spread towards downward. In the later stage of the infection, the entire root is covered with the white fungal mycelium which later on becomes dark by producing sporangia. Due to the fungal infestation, both external part and internal parts are affected to a greater extent (Buddemeyer and Märländer 2005). Decaying of internal tissue, browning, and sponging are the common symptoms in the later phase of the infection. The major characteristic symptom observed in the advanced stage is cavities filled with a liquid fluid having vinegar like smell. The secondary roots are not attacked and hampered by this pathogen (Hanson 2010).

The pathogen can be managed by following some cultural methods like proper irrigation and drainage so that excess moisture will not be present in soil. Controlling insect attack can lessen the pathogen infestation to some extent. Avoid mechanical damage by the field implements (Srivastava 2004). Concentrating on the techniques which promote rapid plant growth and healthy plants can surely help to lessen the disease infestation. Combined implementation of soil tillage, resistant varieties, and nonhost crop alternation is very beneficial in managing the disease (Buhre et al. 2009).

28.3 Future Prospects

Root rot diseases are identified as the main constrain for sugar beet (*Beta vulgaris* L.) cultivation. The main challenge comes with sudden occurrence of symptomatic expression with pick mortality turnover rate. Moreover, increasing resistance development in insect pest and strict regulations for approving new varieties are also the major constrains to manage the diseases. To stop the rapid spread of disease severity at the initial level, technical knowledge about pathogen biology-disease cycle, climatic factors, and specific sustainable management strategies is more useful. Another challenge rises in terms of organic management against hazardous toxic chemicals for long sustainable eco-friendly sugar beet production. Reduction of crop or yield loss and quality control can be checked by the cultivation of improved resistant variety for a specific disease or by following integrated disease management practices. New initiatives in the form of Government schemes and policies, laws, and regulations should be proposed and implemented against the enormous and excessive use of toxic agrochemicals. Opportunities and future prospects are associated with the large-scale production of sugar, through establishment of sugar industry as an alternative resource of sugarcane. High yielding, stress-resistant, and root rot disease-resistant improved variety production through molecular tools is one of the overwhelming responses towards a step of sustainable agriculture. Support and proper encouragement of microscale farmers toward beets' production from the Government, ICAR, and NGOs are always welcome and highly appreciated. Engagement of tribal agricultural system into beet production and their allied micro-industry establishment can improve tribal economy in long term.

28.4 Conclusion

Although the sugar beet cultivation is not commercial in most regions of the globe, it has high potential as a supplementary sugar crop. With coming of favorable environmental conditions, *Rhizoctonia* and *Phoma* cause permanent or temporary damage which can be severe. Both the pathogens can infect sugar beets at mature stage also. The aforesaid organisms can survive under the host for long-term basis and occasionally caused black rot. When that pathogen does not infect the roots, it may infect seed stalk. It may be recognized as a significant commercial crop in many regions of the world in near future. Recent study by Liu and Khan (2019) observed that commercial cultivars are more vulnerable and moderately susceptible to resistance against *Rhizoctonia* attack. But maximally those are not effective against Ag 2-2 IIIB strain. Therefore, greater efforts and focus is required to manage the root rot-mediated diseases as these diseases directly affect the quality and quality of the sugar beet crop. In some regions, crop alternation with nonhost crops and some other production practices are followed to control the disease; still total control is a major challenge. Many researchers have given emphasis on fungal root rot diseases of this crop. But more resistance breeding methods should be discovered against both fungal and bacterial root rot by which the production will increase by lessening the damage (Buttner et al. 2004; Strausbaugh and Gillen 2008). The major challenges are to correctly detect the disease with emerging high-throughput phenotyping technologies. Field scale phenotyping in addition with genotyping and genomic approaches is likely to recognize the resistant or susceptible genes. It will be greatly helpful to develop resistant cultivars against several root rot-mediated sugar beet diseases.

References

Abada KA (1994) Fungi causing damping-off and root-rot on sugar-beet and their biological control with *Trichoderma harzianum*. Agric Ecosyst Environ 51(3):333–337

Agribusiness Handbook (2009) Sugar beet white sugar. Food and Agriculture Organization, p 55

Ali ME, Meah MB (2007) Effect of carrier in the formulation of *Trichoderma* based biopesticide for controlling seedling diseases of eggplant. Bangladesh J Plant Pathol 23(1/2):9–16

Amein T (2006) Soil borne pathogens infecting sugar beet in southern Sweden. Plant Pathol J 5(3): 356–361

Brantner JR, Windels CE (2004, June) Sugar beet seedling age and susceptibility to Aphanomyces cochlioides. In Abstr. Phytopathology, vol 94, p S10

Brar NS, Dhillon BS, Saini KS, Sharma PK (2015) Agronomy of sugar beet cultivation—a review. Agric Rev 36(3):184–197

Buddemeyer J, Märländer B (2005) Genotypic reaction of sugar beet to *Rhizoctonia solani* root and crown rot—susceptibility, yield and quality at different levels of infestation. J Plant Dis Prot 112:105–117

Bugbee WM, Cole DF (1981) The effect of seed infected with *Phoma betae* on rot and sucrose yield of stored sugar beet. Phytopathology 71:357–359

Buhre C, Kluth C, Bürcky K, Märländer B, Varrelmann M (2009) Integrated control of root and crown rot in sugar beet: combined effects of cultivar, crop rotation, and soil tillage. Plant Dis 93: 155–161

Buttner G, Pfahler B, Petersen J (2003) *Rhizoctonia* root rot in Europe-incidence, economic importance and concept for integrated control. Proc. 1st joint IIRB-ASSBT congress, 26–1st March 2003, San Anonio (USA)

Buttner G, Pfahler B, Marlander B (2004) Greenhouse and field techniques for testing sugar beet for resistance to Rhizoctonia root and crown rot. Plant Breed 123:158–166

Campell L, Klotz K (2006) Post-harvest storage losses associated with Aphanomyces root rot of sugar beet. J Sugar beet Res 43(4):113–127

Dohm J, Minoche A, Holtgräwe D et al (2014) The genome of the recently domesticated crop plant sugar beet (*Beta vulgaris*). Nature 505:546–549. https://doi.org/10.1038/nature12817

Dyer AT, Szabo LJ, Windels CE (2004) Characterization and spatial distribution of *Aphanomyces cochlioides* in sugarbeet fields. J Sugar Beet Res 41(1–2):1–16

Elmer WH (1997) Influence of chloride and nitrogen form on *Rhizoctonia* root and crown rot of table beet. Plant Dis 81:635–640

Farag AA, Abdrabbo MA, Taher MS, Farag HA, Rasha MA, Radwan HA (2017) Comparative study of the production of sugar crops under limited water. Egypt J Agric Res 95:87–96

Hanson LE (2010) Interaction of *Rhizoctonia solani* and *Rhizopus stolonifer* causing root rot of sugar beet. Plant Dis 94:504–509

Integrated Disease Management in Sugar Beet for Sustainable Productivity

29

Varucha Misra, Ram Ji Lal, Ashutosh Kumar Mall,
Santeshwari Srivastava, and Arun Baitha

Abstract

Integrated disease management is a strategy for management of plant diseases involving all the essential and beneficial methods which a grower needs for obtaining healthy crop. Adapting this technology, the growers are benefitted by coping up with economical losses they face through the occurrence of disease in sugar beet. Extensive researches in the field of pathology are being done for protecting the crop from various diseases by developing new tools and resistant varieties. Disease surveillance and forecasting in the areas where sugar beet are grown is effective and competent method for managing the diseases for a prolonged period. The first and foremost defense line is the development of resistant varieties against various diseases through either conventional or modernized biotechnological means. Biological, chemical, and cultural are also the part of this management strategy. Novel formulations are also being designed for coping up with the problems associated with sugar beet diseases. The antagonistic nature of many beneficial microbes against pathogen-causing diseases has also gained importance considering the environment-friendly aspect. The amalgamation of these strategies will result in improving the sugar beet yield and production.

Keywords

Integrated · Disease management · Pathogens · Sugar beet

V. Misra (✉) · R. J. Lal · A. K. Mall · S. Srivastava · A. Baitha
ICAR-Indian Institute of Sugarcane Research, Lucknow, Uttar Pradesh, India

V. Misra et al. (eds.), *Sugar Beet Cultivation, Management and Processing*,
https://doi.org/10.1007/978-981-19-2730-0_29

Abbreviations

CLEs	Crude lipopeptide extracts
IDM	Integrated disease management
IPM	Integrated pest management
PCNB	Pentachloronitrobenzene
SDHI	Succinate dehydrogenase inhibitor

29.1 Introduction

Sugar beet (*Beta vulgaris* L.) cultivation is well known in many tropical countries of the world. It is being used as an alternative producer of sugar. The healthy production of this crop has been affected by the onslaught of different pernicious plant pathogens. The attack of pathogens on healthy sugar beet crops has been a threatening alarm for the growers and is a concerning problem. The cultivation of sugar beet for years in the same area is a favorable condition for severe disease incidence. There are several root rot diseases, caused by *Rhizoctonia*, *Sclerotium*, and *Fusarium* species, which have caused a strong impact on sugar beet productivity due to which growers suffer huge economical losses (Agnihotri 1990). In addition, foliar diseases also do not lack behind in causing an impact on crop productivity. As the climate is changing, the occurrence of new pathogens as well as species depending on the region is also causing more losses in productivity (Misra et al. 2021). It has been illustrated that when root and seed crops are grown in a nearby area, the chances of disease transmission increase from seed to root crop (Agnihotri 1990).

Management of diseases is an important aspect to maintain the losses from not increasing the disease to above the economic threshold point of injury. It is not necessary that a single pathogen attacks a variety at a time and even the type of pathogen may also vary from fungal, bacterial, and mycoplasmal infection to viral infections. While processing sugar beet in factories, it becomes difficult to treat a single variety at a time and is also a matter of large time consumption. This paved the way for integrated disease management of sugar beet as a better option for controlling the diseases. Plant production through an integrated approach is a new economic method for producing high yield and healthy crop production. The term integrated management (IPM) has been used initially for insect-pests, but later this has also now been used for disease management too. Integrated disease management (IDM) is engaged in the application of pesticides on the basis of plant requirements when the disease incidence surpasses the economic threshold levels. This results in endorsing the use of biocontrol agents. IDM is a strategy for moving towards greener alternatives rather than chemicals and engrosses on the limited application of fungicides. Elimination or reduction of the initial inoculums along with the reduction in its efficiency, delay in disease incidence, and enhancing the resistance capability of the host are some of the objectives of this strategy (Gurjar et al. 2018).

29.2 Components of Integrated Disease Management (IDM) and their Application in Sugar Beet

Generally, there are four main components of IDM. These are host resistance, cultural, biological, and chemical controls.

29.2.1 Host Resistance

In this component, suppression of disease pathogen and its development occurs through the use of resistant genotypes. Growers are always interested in such resistant varieties for their cultivation as it will cause lesser investment during crop protection. In such types of genotypes, disease incidence is slow and the damage caused by the pathogen also appears to be less on the plant. In rhizomania of sugar beet, the use of resistant cultivars can reduce infection to some extent. Studies revealed that genetic resistance is a better option for effectively controlling sugar beet root rot disease, particularly from *Rhizoctonia solani* (Sherf and MacNab 1986; McGrath et al. 2015). However, developing resistant variety against any disease may take a long duration approximately 8–10 years. In the 1950s, the development of *Rhizoctonia*-resistant cultivar was first came into existence with the involvement of multiple resistant genes (Panella and Ruppel 1996; Gaskill 1968; Hecker and Ruppel 1975). Resistant/susceptible varieties have been preferred by the sugar beet growers due to their high yield (Haque and Parvin 2021); however, partial resistant varieties may also be grown to minimize the disease incidence rate (Behn et al. 2012; Brantner and Windels 2009). Bolz and Koch (1983) and Hecht (1989) first time reported the partial resistance varieties, viz., Dora and Lena against rhizomania. Rizor is another such resistant cultivar with superior resistance against this disease (Richard-Molard 1985; De Biaggi 1987).

Jacobsen (2006) revealed that the use of resistant varieties against *Fusarium* root rot disease in sugar beet is a good management strategy. In the case of *Cercospora* leaf spot disease, resistance in cultivar depends on many quantitative aspects (Rossi 1995). Studies reported that measurement of r-reducing resistance could be a good option for assessing the resistance level (Lapwood 1971; Parleviet 1976; Johnson and Wilcoxon 1978). Kawe cercopoly and USH 9B are also resistant to *Cercospora* disease (Agnihotri 1990). For a variety to be resistant against this disease, 3 hydroxytyramine content in the foliage was one of the associated parameters (Agnihotri 1990).

Agnihotri (1990) illustrated some highly tolerant varieties against *Sclerotium* root rot. To name a few, C-W 674, Maribo resistapoly, Kawe Cercopoly, and USH-9 B. Moreover, two resistant lines have also been developed, viz., 75 PI and 7326, and at that time the inclusion was the immediate need at commercial basis for growers. Furthermore, Sharpes Klein E is known to have resistant capability for powdery mildew disease, while some USA varieties such as US-9 and US-10 were susceptible to this pathogen, yet they are commercially being grown over a large area (Agnihotri 1990).

29.2.2 Cultural Practices

In this strategy, the favorable environment for the pathogen is disturbed making the environment less feasible for the pathogen to develop. This is done by either disturbing their reproduction cycles or increasing the growth of natural enemies or by many such means. Intercropping, crop rotation, and shifting in sowing dates for disease escape are also involved under this approach.

29.2.2.1 Use of Healthy Seeds

Healthy seeds are the foundation of a healthy plant. Agnihotri (1990) reported several techniques for protecting the seeds prior to planting. For *Alternaria* leaf spot disease, seed disinfection should be done by 0.25 percent thiram or captan for lowering seed-borne infection. In the case of *Cercospora* leaf spot, treatments with aretan 2 g/kg seed, captan 2.5 g/kg seed, or thiram 2.5 g/kg seed were preferred, but for phoma leaf spot soaking of seeds with thiram solution (0.2% concentration) at 30 °C for 24 h has been reported for seed treatment. Jacobsen (2006) revealed another strategy for the areas where this disease is more prone; seed production for such areas should be done in dry conditions under surface irrigation. In case of diseases in stecklings treatment, thiram or thiophanate methyl (0.1%) was effective for storage rot (Agnihotri 1990).

29.2.2.2 Crop Rotation

Crop rotation may be a more practical approach to the reduction of soil inoculums. Cereals and grasses should be taken in rotation for *Rhizoctonia* root rot (Agnihotri 1990). Buhre et al. (2009) and Koch et al. (2018) reported that in this crop, rotation with other crops, particularly nonhost cereal crops like wheat, should be done at least after 3 years; for instance, in *C. beticola* (Agnihotri 1990). Promising results were illustrated in the number of studies on cover crops (*brassica*) as a controller for *Rhizoctonia* infection and many other soil-borne pathogens in sugar beet (Kundu and Nandi 1985). For *Sclerotium* root rot disease, crop rotation with crops that have less susceptibility helps in reducing the disease potential (Jacobsen 2006; Agnihotri 1990). Leach and Davey (1942) revealed that usage of ample amount of nitrogen as fertilizer and other essential nutrients in the soil in areas prone to this disease will offer strong plant growth causing a reduction in damage by the disease. In violet root rot disease, susceptible crops should not be used for crop rotation like beans, potatoes, peas, etc. (Jacobsen 2006) and so is the case for phoma disease (Agnihotri 1990).

29.2.2.3 Other Miscellaneous Cultural Practices

The removal of crop debris and water management in disease helps in lowering the primary inoculums and disease spread. This is evident more in case of soil-borne diseases such as *Rhizoctonia, Fusarium, Sclerotium*, etc. (Gurjar et al. 2018). Even for *Cercospora* leaf spot, the crop debris remain after harvesting should be buried in the soil by deep plowing (Agnihotri 1990). Early planting/sowing also prevents the onslaught occurrence of diseases. This is seen in rhizomania, *Fusarium* root rot, black root rot, *Cercospora*, and *Sclerotium* root rot diseases (Jacobsen 2006; Agnihotri 1990). In the case of *Sclerotium* root rot, planting date of sugar beet has a strong impact on diseases' incidence (Agnihotri 1990). As per an Indian study, planting of this crop in the submountain areas of Uttar Pradesh by tenth November significantly lowers the rotting disease in sugar beet roots (Thakur and Mukhopadhyay 1972). Agnihotri (1990) revealed that disease incidence rate varies for *Cercospora* with the month of planting. Lower incidence rate was seen when crop was sown in October, while highest when it was sown in December under Indian conditions. Early sowing with optimum dose of fertilizer has been recommended for downy mildew. Seed crop must be separated from the root crop at least by 400 m (Agnihotri 1990). Early plowing of the fields is another way by which the inoculums of pathogens could be reduced. In the case of *Fusarium* root rot, the incidence rate can be controlled by 18% (Maui et al. 2020). Field sanitation by burning of infected crop debris and deep ploughing are recommended. Clean roots without wounds, cuts, or cracks should be stored (Agnihotri 1990). The destruction of sclerotia and hyphae of the fungus *Sclerotium* spp. in soil can be effectively done by deep ploughing. Even burning of infected roots and foliage was also known to be effective for reducing the pathogen of *Sclerotium* root rot. Similarly, in case of *Alternaria* leaf spot, foliage should be destroyed by immersing deep under the soil. Likewise, in case of powdery mildew disease, crop residue annihilation is important as cleistothecia survive in plant residues (Agnihotri 1990).

Fusarium root rot as well as weed growth control in particularly Chenopodiaceae is a must (Jacobsen 2006). The best management for *Pythium* and *Phytopthora* root rots was reported to have reduced moisture content in the soil (Jacobsen 2006; Schneider and Whitney 1986). For the management of black root rot disease, soil drainage is important along with weed control, particularly *Chenopodium* and *Amaranthus,* and rotation with other nonhost crops. The application of oat green manures is a better option for managing this disease (Windels and Bratner 2002). Application of ammonium fertilizer 160 kg N per hectare provides adequate *Sclerotium* root rot control (Agnihotri 1990; Thakur and Mukhopadyay 1972). Soil indexing also plays an important role as it helps in knowing the requirement of resistant varieties for that area where sugar beet is meant to be grown (Windels and Nabben-Schindler 1996). Soil fumigation has also been reported to be efficient in controlling vector of rhizomania (Jacobsen 2006).

29.2.3 Chemical Control

In this method, the application of pesticides on the basis of plant requirements is involved. This approach is adopted in areas where the disease incidence is rapid and severe during the initial stages of crop growth. In the fungicides group, there are two main types of fungicides. These are protectants and eradicates. Protectants are defined as the ones that stick on the surface of the plants and the mode of action is dependent on pathogen contact. It is important that the spraying of chemicals onto the plant should be uniform. On the other hand, eradicates are the ones that are absorbed by the plants and belong to systemic pesticides, implying that only specific fungi can be controlled by their application. In the case of *Erysiphe betae,* application of Bellis 38% WG, Collis 30% SC, and Tilt 25% EC on sugar beet plants had reduced disease incidence with high root weight and total soluble solids (Aly et al. 2020). Zadehdabagh et al. (2020) reported that a combination of azoxystrobin + difenoconazole fungicide (1 l/ha) causes a reduction in *Cercospora* leaf spot disease and better root yield than carbendazim, thus, stating as a better alternative option against carbendazim fungicide. Duter (0.75 kg per hectare) or dithane Z-78 (2.5 kg per hectare) was another effective fungicide in managing this disease provided the prophylactic spray is given before the usual time of appearance of disease. Among systemic, fungicide Bavistin (300 g per hectare) has been found to be very effective. Two to three sprays are required for adequate control of the *Cercospora* leaf spot disease (Agnihotri 1990).

El-Shabrawy and Rabboh Abd (2020) showed that certain chemicals like copper sulfate, zinc sulfate, salicylic acid, ascorbic acid, and potassium silicate had higher effectiveness in controlling powdery mildew disease incidence with high root weight and sucrose content. Agnihotri (1990) reported that application of sulphur (15–20 kg per hectare), wettable sulphur (1.2 kg per hectare), benomyl (0.5 kg per hectare), and brestan (800 g per hectare) were effective in giving promising results towards powdery mildew disease management. Application of rovral, tachigaren seed, and fundazol on sugar beet seeds causes a decrease in root rot infection by 8.1–16.6% with high root yield (an increase of 31.5–50.2 c/ha) (Maui et al. 2020). Jacobsen (2006) demonstrated that thiram, prochloraz hot water, and benzimidazole had a significant reduction in *Phoma* leaf spot disease when sugar beet seeds are priory treated with them; however, *Phythium*, hymexazole, and metalaxyl were effective as seed treatment, while metalaxyl could also be used as a soil treatment. Spraying of dithane Z-78 (2.5 kg per hectare) or brestanol (0.7 kg per hectare) showed significant results when the timely application was given thrice for the control of *Alternaria* leaf spot. It is important that the first spray should coincide with the first secondary infection (Agnihotri 1990).

Quinone outside inhibitors (QoI) fungicides (azoxystrobin and pyraclostrobin) are preferred during sugar beet growth as they help in blocking the electron transfer between cytochrome b and cytochrome c1, resulting in uncertain production of ATP (Markell and Khan 2012–2013). These fungicides were effective in controlling *Rhizoctonia* root rot disease (Balba 2007; Haque and Parvin 2021). Liu and Khan (2016) had reported that penthiopyrad application on sugar beet helps in controlling *R. solani* infection. Penthiopyrad can be used either at planting with a dosage of 210, 280, 420, or 550 g a.i./ha or by soil drenching after 1 month of sowing. Penthiopyrad acts as a good mitigator for developing resistant isolates of *R. solani* (Liu and Khan 2016). Succinate dehydrogenase inhibitor (SDHI) fungicides (sedaxane (0.1 μg/mL), penthiopyrad (0.15 μg/mL), and fluxapyroxad (0.16 μg/mL)) were also effective in controlling *R. solani* infection (Sharma et al. 2021). Carboxin, chloroneb, and certain other fungicides had been found to be used for managing *Sclerotium* root rot disease (Jacobsen 2006). Treatment of hymexaole on seeds proved to be successful in managing the disease (Windels and Branter 2004).

Campbell and Klotz (2005) revealed that a combination of hymexazole and biological control strategy gave promising results against *Aphanomyces* root rot disease. Seed treatment with neonicotinoid in the early time of growth helps in controlling the disease incidence in the crop (Strausbaugh et al. 2010). Application of flutriafol-based fungicide helps in controlling *Cercospora beticola*, *Erysiphe betae,* and *Uromyces betae* by acting as an eradicant and longer persistence (Brown et al. 1986). Poncha beta (insecticide) has also been known to manage the curly top disease with seed treatment under idaho conditions (Strausbaugh and Gillen 2006). A combination of Monocut with pomegranate and black pepper extract was revealed to be effective against sugar beet root rot. Pomegranate methanolic extract showed 93.30% inhibition rate, while similar results were also observed with black pepper methanolic extract against *S. rolfsii* (Osman et al. 2021).

Several fumigants like D-D, Vapam, Chloropicrin, methyl bromide, etc. have been found to reduce the inoculums of *S. rolfsii* appreciably. Fungicide like pentachloronitrobenzene (PCNB) and demosan when applied at the rate of 15–20 kg per hectare provide very effective control of the disease. These fungicides should be applied 10–15 days before the usual appearance of the disease in the field. Application of insecticides like Carbafuran (2 kg a.i. per hectare) also drastically reduces the root rot incidence of sugar beet (Mukhopadhyay and Thakur 1977). Of the various chemicals, only PCNB is widely used in sugar beet growing areas. It is cheap and readily available in India. It gives the cost: benefit ratio of 1:4. PCNB is broken down in soil into two compounds, namely, pentachloroaniline and methyl-thiopentachlorphenyl. Pentachloroaniline is highly fungicidal to *S. rolfsii*. It has been found that light irrigation after PCNB application further enhances the efficacy of the fungicide (Agnihotri 1990). Soil around sugar beet roots should be drenched with brassicol (20 kg/ha) for *R. solani* and *R. bataticola* infection. Although this treatment reduces soil inoculums, it is not cost-effective (Agnihotri 1990).

29.2.4 Biological Control

In this approach, a decrease in pathogen occurrence is known by the application of other living organisms. It is one of the most effective and natural means of coping up with harmful pathogens by the use of beneficial microorganisms. Hyper parasites' application is also one better example of it.

29.2.4.1 Trichoderma Spp. as Biocontrol Agent

Trichoderma harzianum, *T. viridie*, and *T. flavus* are some of the species that are being used for sugar beet disease management. *T. viridie*, *T. harzianum, and Gliocladium virens* have been demonstrated against *Rhizectonia solani* in vitro (Agnihotri 1990). Moussa (2002) observed that *T. harzianum* effectively manages the *R. solani* infection in sugar beet roots. On observing through electron micros-copy, *T. harzianum* was found attaching to the *R. solani* by the hyphal coils. Furthermore, *Trichoderma spp.* formulations (Talc-*T. harzianum* followed by Peat-*T. flavus*, Talc-*T. flavus*) and *Talaromyces* were also found to be potential biocontrolling agents in case of *R. solani*-induced damping-off disease in this crop (Kakvan et al. 2013). The efficiency of *Trichoderma* spp., particularly *T. harzianum*, as biological agent against damping-off and root rot disease in sugar beet resulted in improvement of root weight (Abada 1994). Sawan and Mukhopadhyay (1991) showed that when *T. harzianum* inoculum (17.5 g/m ridge) is used as a soil amendment or when treated with metalaxyl (0.1%), the *Pythium* damping-off in sugar beet was controlled effectively. El-Katatny et al. (2020) demonstrated that combination of mint oil treatment and culture filtrate of *T. harzianum* was effective in reducing the germination of fungal spores causing root rot in sugar beet.

T. harzianum and *T. viridie* are commercially used for management of *Sclerotium* root rot. The application of these fungi was given through irrigation water or as broadcast dosage of 140 kg *Trichoderma* granules per hectare (Mukhopadhyay and Upadhyay 1983; Agnihotri 1990). Upadhyay and Mukhopadhyay (1986) revealed that a combination of pentacholoronitrobenzene (low concentration) with *T. harzianum* causes significant control in *Scleortium* root rot disease by decreasing the incidence rate to 76%. Effective management was also seen when mustard oil cake (25 q per hectare) was applied for *Sclerotium* root rot. The benefit of using this organic amendment was observed in soil improvement where sugar beet was grown as beneficial microbes, like actinomycetes, bacteria, fungi population gets enhanced (Mathur and Sarbhoy 1973). These microbes are known to have antagonistic nature for *S. rolfsii*. Alternate drying and wetting of the field is important for the destruction of *S. rolfsii* (Agnihotri 1990).

29.2.4.2 *Bacillus* Spp. as Biocontrol Agent

Bacillus spp. is being applied as a control measure to many foliar diseases and postharvest diseases in sugar beet. *Cercospora beticola* infection in sugar beet has been known to be controlled with the application of Bac B which provides resistance to the crop against this disease (Collins and Jacobsen 2003). Another bacterial isolate, Bac J from *Bacillus mycoides,* had reported having decrease in the incidence rate of *Cercospora* leaf spot by 38–91% (Bargabus et al. 2002). Further, crude lipopeptide extracts (CLEs) of *Bacillus amyloliquefaciens* strains (SS-12.6) had reduced foliar disease incidence rate (Nikolic et al. 2019). Bargabus et al. (2004) showed that two strains of *B. pumilus,* viz., 203-6 and 2037, caused the decline in *Cercospora* leaf spot symptoms by 70%. Kodiak, prepared from *Bacillus subtilis,* has also shown effective results in decreasing the *R. solani* AG 2-2 IIIB infection (Kiewnick et al. 2001). The production of bacteriocin from *B. subtilis* plays an important role in antagonistic mechanisms against pathogens, resulting in pore formation, cell disintegration, and other processes (Caulier et al. 2019). MSU-127, bacillus strain, with Azoxystrobin (low concentration) was helpful in improving the sugar beet yield by 16%; however, when the fungicide was sprayed after 1 month of sowing, root yield was enhanced by 17% as a result of suppression of diseases (Kiewnick et al. 2001).

Rhizobacteria were even efficient in controlling the density of *R. solani in this crop* (Homma 1996). *Application* of *Pseudomonas putida* 40 RNF on pelleted seeds had reduced the incidence of *Pythium* damping-off disease in sugar beet

(Shah-Smith and Burns 1996). Errakhhi et al. (2007) reported that *S. rolfsii* damping-off disease had significantly reduced incidence rate when J-2 isolate of *Streptomyces* was used. Furthermore, two other isolates of the same bacteria (S2 and C) had shown reduction in *Rhizoctonia solani* infection by the formation of siderophore and chitinase (Sadeghi et al. 2006).

29.2.4.3 Other Miscellaneous Fungi as Biocontrol Agent

Mycofumigation is another approach by which sugar beet diseases can be managed. *Muscodor albusitalic* and *M. roseus* application on sugar beet had reduced disease rigorousness against *R. solani*, *Pythium ultimum*, and *Aphanomyces cochliodies*. Furthermore, *Fusarium* wilt disease of this crop was even manageable by mycofumigation (Stinson et al. 2003). Shawki et al. (2020) reported that treatment of seeds with nicotinic acid (5 mM) acts as a protective agent against *F. moniliforme* pathogen. El-Tarabily (2004) illustrated that isolates of *Candida valida, Rhodotorula glutinis,* and *Trichosporon asahii* act as protectants for seedling and mature plants against *R. solani* diseases in sugar beet. These microorganisms have the capability of root colonization. Spores of fungi like *Aureobasidium pullulans, Sporobolomyces papraroseus*, *Torulopsis candidus* or *Cladisporium cladosporioides* have been mixed with spores of *Phoma betae* and were sprayed on the plant for curtailing the development of the lesions in plants (Agnihotri 1990).

29.3 Benefits of Integrated Disease Management Approaches

IDM strategy is an amalgamation of preventive and manageable methods which shows promising results in controlling the pathogen from causing severe and strong impact on the sugar beet crop with a lowest human hazardous risk. The benefits of IDM are as follows:

1. Encourages healthy sugar beet crop
2. Encourages disease management through bio-based alternatives
3. Lowers risk related to environment due to disease management
4. Lowers the need of insecticides and pesticides usage and problems associated with pesticide residues.
5. Lowers soil and water pollution through use of environment-friendly products.

29.4 Future Prospects

The changing climate has caused the occurrence of many new sugar beet diseases, pathogens, and insect-pests. This has shown the importance of disease surveillance and forecasting for further managing new diseases in sugar beet. Though researches have been focused on developing resistant varieties particularly for soil-borne diseases that affect the beet root yield (the economical part), there is a need to strengthen the identification of such cultivars and their resistant sources through the

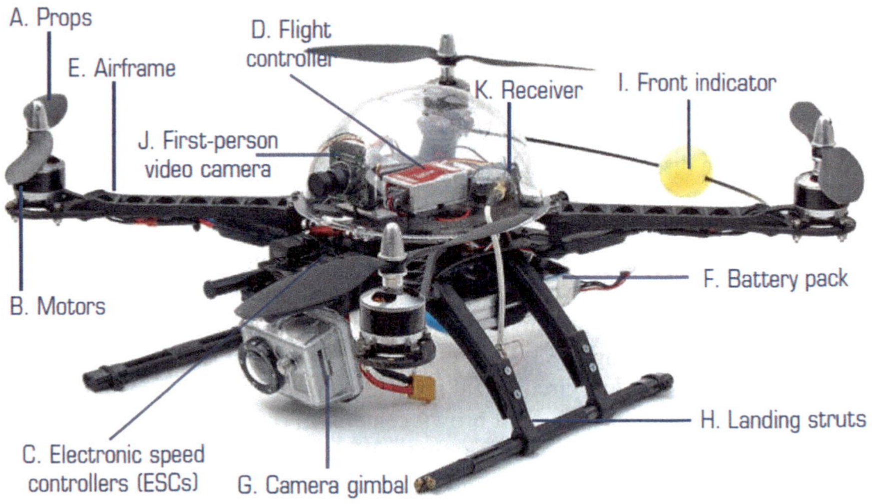

Fig. 30.1 Components of drone (Baichtal 2016)

system installation of UAVs. In drones, however, drone might be used immediately after purchasing drone together with the apparatus without the need for any other costs. The drones are preferred in agricultural applications because of their ease of use and lower cost (Özgüven 2018). The most preferred drones are quadrotor drones shown in Fig. 30.1 (Ozguven 2018). The examples of different forms of rotors are tricopters, hexacopters, and octocopters. The number of optional rotors might be increased, thereby increasing the capacity of the drone.

The quadrotor has an advantage over other rotors because of its highest maneuverability, such as vertical takeoff and landing capability in hazardous areas. However, a quadrotor cannot fly for long periods because of high power consumption (Merç and Bayılmış 2011).

Key attributes that enable UAV operation are as follows (Clarke 2014):

1. Ability to go and return to the target position within the operation area,
2. A set of controls and maneuverability over the UAV attitude, direction, and speed of movement,
3. Remote data streams to maintain timely awareness of movement, attitude, and location, and,
4. Sufficient power to sustain movement, to perform the controls, and to run sensors and data streams, for the period of the flight,
5. Situational awareness through tracking operational space,
6. Ability to avoid collisions and navigate through obstacles,
7. Robustness to withstand various dangerous situations, such as bird strike, stroke of lightning, turbulence, and wind-shear,
8. Ability to fly in all atmospheric conditions.

30.3.1 Components of Drone

Although drones are designed with different technical features in accordance with their usage areas, their basic components are given below (Fig. 30.1) (Akyüz 2013; Johnson 2015; Baichtal 2016; Szabó et al. 2018):

(a) Propellers: Propellers used by mechanically connecting to motors for spatial movements of drones are generally made of carbon-fiber material. For example, the propellers of a quadcopter typically consist of two standard and two pusher propellers rotating in opposite directions.

(b) Motors: Although DC or AC motors might be used, electric motors are used most commonly, brushed or brushless direct current motors. So as to do the same amount of work on all rotors, the same kind of motor is used. Compared to brushed motors, brushless motors are more widely used owing to their advantages such as quiet operation, long life, much more efficiency and less wearing parts, no electrical noise, no regular maintenance, and ability to run at higher speed and high torque in a lower voltage range.

(c) Electronic speed controllers (ESCs): ESCs convert DC to AC for brushless motors and also trigger the motors' power supply. One is used for each engine. ESCs' firmware might be changed to create different motor behaviors. For instance, ESCs are often configured to slow down the motor rather than stopping abruptly.

(d) Flight controller: The flight controller controls the entire electromechanical system of the drone. It assists manual flight with certain autonomous functions. For instance, many flight controllers have an accelerometer sensor that keeps the drone level. The flight controller provides movement and stabilization in the desirable direction by changing speed of the motors according to the data from the sensors. Thus, the drone can stabilize itself even if the engines are given different pusher.

(e) Airframe: The airframe consists of components, such as including motor booms alongside an enclosure or platform for housing the electronics. The drone ought be light and thin enough to lift off and tough enough not to break in a minor accident. Carbon-fiber, plastic, wood, and aluminum alloy materials are generally preferred in the construction of the main body.

(f) Battery pack: Usually a LiPoly battery, the drone's battery pack keeps the propellers turning while also powering whatever electronics are onboard.

(g) Gimbal: Gimbal is a rotating platform on which a camera is placed. Servomotors allow the operator to turn and angle the camera during flight.

(h) Landing struts: Landing struts are used to prevent damage to the camera or other protuberance under the drone. Drones without cameras, on the other hand, do not have landing struts and the drone lands with its entire airframe.

(i) Front indicator: Especially the front side of the drone must be known by the operator. For this, different colored lights, LEDs, reflective materials, or colored balls are used.

Fig. 30.4 Pixel tagging outputs

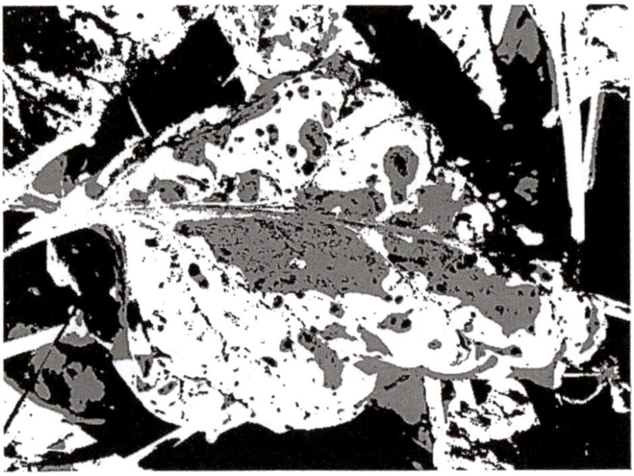

Fig. 30.5 Segmentation outputs: segments of (**a**) black, (**b**) green and (**c**) brown

Using the pixel labels, the pixels in the image were colored as seen in Fig. 30.5 and three images (i.e., $K = 3$) were obtained. The disease image was selected from among three clusters.

Contrast enhancement of color images is done by converting one of the components of the image to a color space with image brightness, for example L*a*b* color space. Therefore, each image was converted from RGB into L*a*b* color space and then, the brightness L layer of the image was worked. The brightness layer was changed with the processed data and the image was reverted to RGB (Fig. 30.6).

The severity of the disease was calculated as the ratio of diseased area to total area.

$$Ak = \sum_{x=1}^{m} \sum_{y=1}^{n} B(x, y) \tag{30.1}$$

$$B(x, y) = \left\{ \begin{array}{ll} 1 & \text{if} \quad B(x, y) \in k \\ 0 & \text{if} \quad B(x, y) \notin k \end{array} \right\} \tag{30.2}$$

Fig. 30.6 Contrast enhancement

Table 30.2 Comparison of image processing methods and visual evaluation results

Image	Image processing (%)	Visual evaluation (%)	Difference (%)
a	100	100	0
b	48	50	−2
c	42	45	−3
d	21	20	+1
e	80	80	0
f	28	30	−2
g	74	75	−1
h	47	50	−3
i	29	30	−1
j	46	50	−4
k	20	20	0
m	51	50	−1

$$\text{Disease severity } (\%) = \frac{\text{Ak}}{\text{Total Area}} \tag{30.3}$$

where Ak = diseased area, $B(x,y)$ = value at given xth row, yth column of the image identified.

In their study, Altas et al. (2018) used the image processing toolbox module of MATLAB program to examine the presence of leaf spot disease on sugar beet leaves and to assess disease severity in 12 images, indicating the different developmental levels of the disease. The results achieved were given in Table 30.2.

When Table 30.2 is analyzed, it is seen that the results were very close to each other. It was reported that the assessment results acquired by visual evaluation were approximate integer values, the image processing methods results, given the exact value of the diseased area with a sensitivity that cannot be acquired by observation and the research was achieved successfully.

30.4.2 Deep Learning Technique

Deep learning technique is a subcategory of machine learning. Machine Learning is a subcategory of artificial intelligence. In machine learning, various algorithms and methods are utilized to look at historical data, a mathematical model that will determine the complex pattern between the data is determined, and then predictions are made about what is desired to be estimated from the data. In machine learning, processing is done in an only layer, while deep learning processes in many layers at once. The difference of deep learning algorithms from machine learning algorithms is that there is a very great amount of labeled data, and owing to the complicated structure of the data, they need GPU-based computers and hardware with very high computational power to process these data. In machine learning, the relevant features are manually extracted from the images and these features are utilized to create a model that categorizes the objects in images. In deep learning, the related features are automatically ejected from the images and it is learned how to perform a task like classification automatically (Özgüven 2019). Furthermore, using many nonlinear processing layers for feature eject and conversion from a great amount of labeled training data, each successive layer uses the output of the previous layer as an input (Deng and Yu 2014). Deep learning is based upon learning from the representation of the data. In the representation of an image, some features better represent the data such as a vector of per-pixel intensity values or edge clusters and certain shapes (Song and Lee 2013). Commonly used deep learning models, which consist of a set of algorithms and models running on neural networks with multiple layers, are listed below (Özgüven 2019):

- Convolutional Neural Networks (CNN),
- Auto-encoders,
- Recurrent Neural Networks (SRN),
- Deep Belief Network (DBN).

Plant disease observations of experts may sometimes be misleading due to exhaustion of decrement of concentration experienced by the experts. Therefore, visual ratings of the samples gathered from the field should be reassessed later. In addition, there is a requirement for standard field schemes for expert assessments (Bock et al. 2010). Deep learning method offers conveniences for more effective plant protection through diagnosis of plant diseases and owing to monitoring the plant development (Ozguven 2020). The study of the automatic classification and diagnosis of leaf spot disease on sugar beet by the deep learning technique by Ozguven and Adem (2019) is summarized and the deep learning method is explained in practice.

A new 1–3 scale was developed by researchers to determine sugar beet leaf spot disease with the help of expert systems. In the new scale, 0 the whole plant is healthy, 1 low severity of disease, 2 a severe disease, and 3 a low and severe of disease. The 24-bit 1024 $\times$ 576 resolution images received from sugar beet leaves images dataset in the research were performed to determine and classify the disease

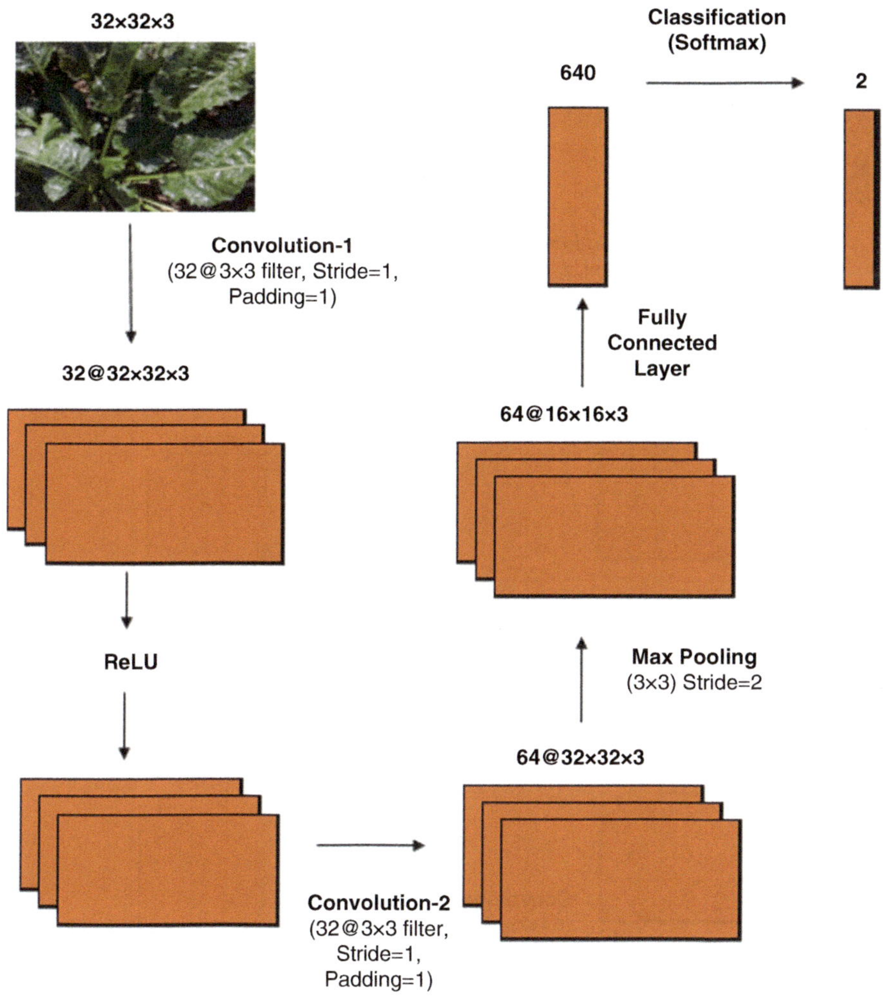

Fig. 30.7 Faster R-CNN architectures

severity as healthy, mild disease, severe disease, or mild and severe mixed disease. The dataset consists of 155 sugar beet leaves images, including 38 healthy, 20 mild diseased, 35 severe diseased, and 62 mild and severe diseased. The Faster R-CNN model was preferred in the study to better determine and classify highly complicated objects. The Faster R-CNN and Updated Faster R-CNN architectures in the research are seen in Figs. 30.7 and 30.8.

The input layer takes the raw data from the network (Figs. 30.7 and 30.8). The raw image was taken as $32 \times 32 \times 3$. The convolution layer is used for feature extraction of the leaf images. The first convolution layer uses 32 different 3×3 filters with 1 stride and 1 padding. The rectified linear unit layer (ReLU) is the

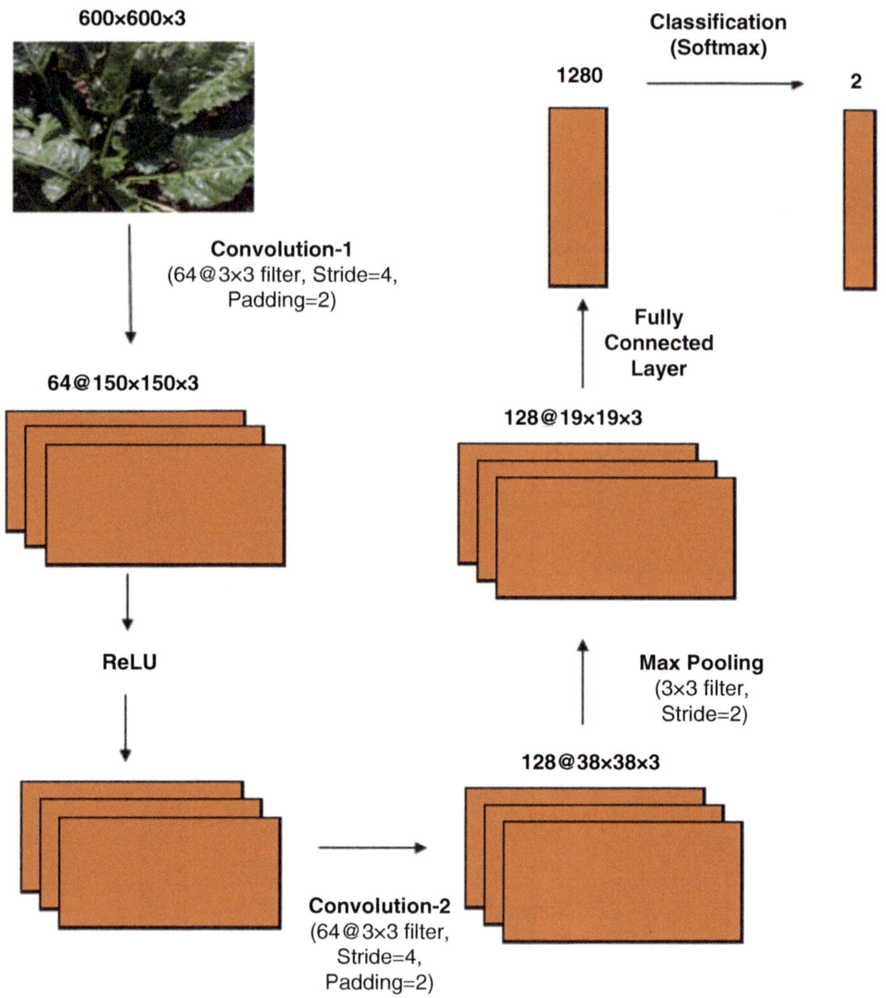

Fig. 30.8 Updated Faster R-CNN architectures

usually used rectifier unit for neuron outputs. There is max pooling to decrease the input size (width × height) for the next convolution layer after the ReLu layer. After the max pooling layer comes a 64-node fully connected layer. This layer connects to all parts of the previous layers. Then there is the classification layer where the classification is done. The last layer is the softmax classifier. This layer is a generalization of the logistic function that could be used for multiclass classification. It gives estimated probability of each class.

In the updated Faster R-CNN model, the input image size is 600 × 600 × 3. In this model, by setting the input image size higher than in the Faster R-CNN model, it was possible to determine insensible diseased areas. Updated Faster R-CNN convolution layer uses 64 different 3 × 3 filters with 4 stride and 2 padding. In addition, the number of filters has been increased to 64 to get comprehensive information about the image. Thus, the number of stride has been increased to 4. For better diagnosis of the diseased areas in the corners of leaves, the padding size was also increased to 2. Accuracy of the classification process was developed by aid of 16-node fully connected layer. Weights were optimized through trainings performed by aid of a heap size of 64, a momentum of 0.85, and a weight reduction of 0.001.

Ozguven and Adem (2019) in their research provided the results of applying deep learning methods to sugar beet leaf images which are given in Fig. 30.9. It was seen that healthy areas were misclassified due to shadows in some images, and diseased areas could not be detected in some images due to reflections. However, disease detection was better with Updated Faster R-CNN architecture. This shows that it was crucial to adjust the parameters in the CNN architecture according to images to which the Faster R-CNN model has been applied. Table 30.3 shows the confusion matrix and Table 30.4 shows the sensitivity, specificity, and accuracy values.

As can be seen in Table 30.3, 111 out of the 117 sugar beet leaves images containing the disease were correctly classified by the proposed model. There was only 1 incorrect classification in 38 images without disease. This demonstrates that the specificity values of Updated Faster R-CNN approach were higher than the sensitivity values as seen in Table 30.4. Updated Faster R-CNN model applied to present dataset yielded an efficiency of 95.48% in detection of sugar beet leaf spot disease. As known, to apply the deep learning methods successfully, diversity of samples and the number should be large. Thus, disease might be classified more effectively. However, with the parameter changes we made in the proposed approach, similar success rates were achieved with fewer images. It is thought that the model performance will be better when the number of images is increased.

30.5 Conclusion

Early identification of plant diseases and timely interventions are of major importance for reducing crop losses. Especially in wide areas of production, the application of technological methods that are faster, practical, and eliminate the margin of human error in the diagnosis of various plant diseases, identifying the severity and change of diseases, offers very important advantages. Thus, cameras mounted on drones are a cost-effective option for capturing images covering disease areas. Computer vision applications, especially image processing techniques and deep learning techniques, for disease detection have great potential benefits to assure that plant protection applications are realized more effectively. The images captured during the flight of the drone can be processed with the developed algorithms to identify diseases in real-time and automatically.

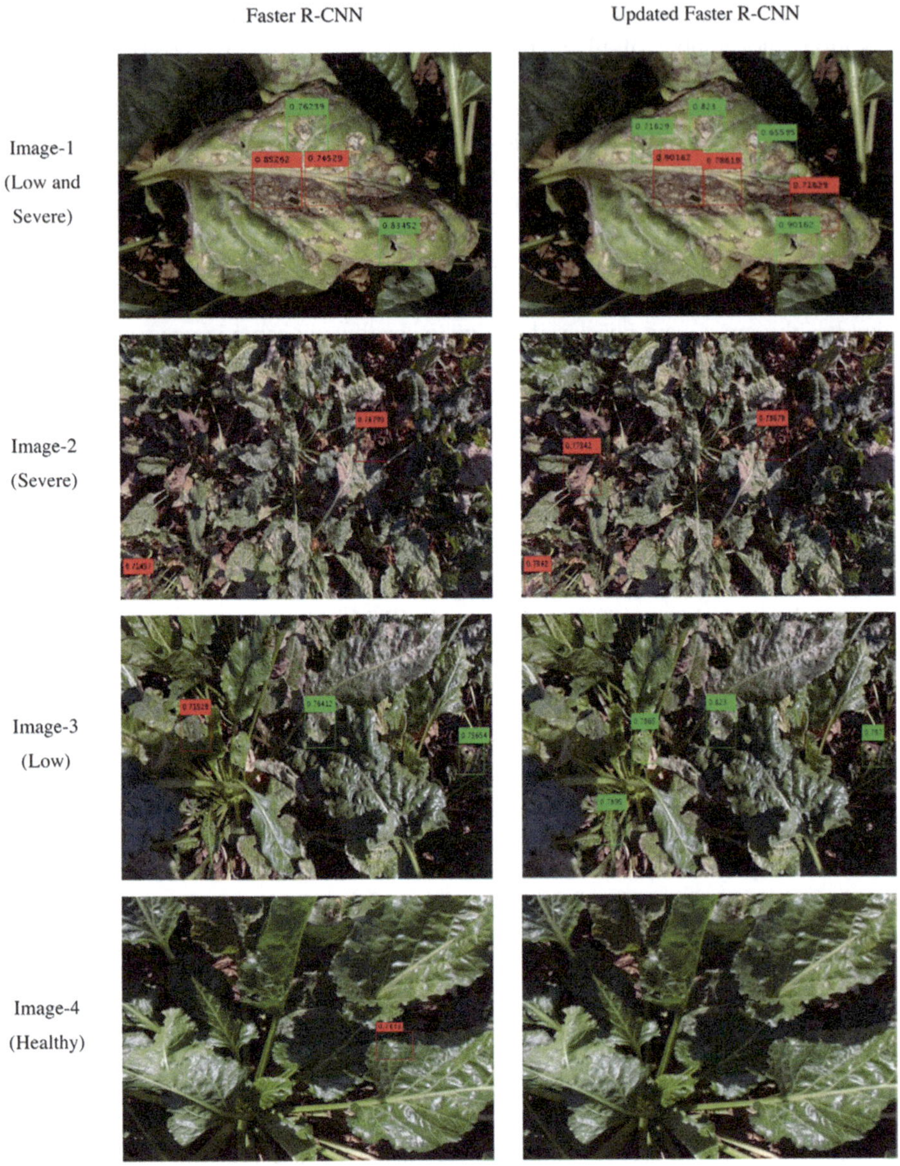

Fig. 30.9 Results of disease detection with deep learning techniques

Table 30.3 Confusion matrix of the updated Faster R-CNN architecture proposed in the study

		Predict			
		0 (healthy)	1 (low)	2 (severe)	3 (low and severe)
Actual	0 (healthy)	37	1	0	0
	1 (low)	0	19	1	0
	2 (severe)	0	1	33	1
	3 (low and severe)	0	1	2	59

Table 30.4 Assessment of the success of the updated Faster R-CNN architecture proposed in the study

	Sensitivity	Specificity	Accuracy
0 (healthy)	97,37	100	99,36
1 (low)	95	97,84	97,48
2 (severe)	94,28	97,6	96,87
3 (low and severe)	95,16	98,97	97,48
Overall	95,48	95,48	95,48

30.6 Future Prospect

Image processing techniques and machine learning techniques have been extensively studied over the last years for the determination of diseases in plants. There are many different diseases seen in plants. To detect these diseases promptly and correctly, methods such as image processing, K-means clustering, ANN, SVM, SR, and CNN might be used together. Platforms or robots might be improved to raise the image quality captured. So, it will contribute to the increment of model performances. As the model performance increases, the success in the determination and diagnosis of plant diseases will increase and so the suitable disease management program can be applied effectively.

In the future, it is expected that like disease detection processes, expert systems that automatically perform spraying operations without human intervention will be established. Robots and drones that will roam autonomously in the field or garden will identify diseases, then send expert system spraying drones or robots to spray the designated areas. In fact, spraying might be done at a variable rate, that is, only the diseased and damaged areas will be sprayed with pesticides to the extent necessary. In addition, drones, robots and other smart machines will be able to real-time communicate with each other and perform their tasks together in coordination, cooperation, or collaboration. In this way, working together will be possible with real-time communication, and with drones, robots, and smart machines, knowing where each other is and what they are doing.

Savary S, Willocquet L (2014) Simulation modeling in botanical epidemiology and crop loss analysis. The Plant Health Instructor. 173 p
Schmittgen S (2014) Effects of cercospora leaf spot disease on sugar beet genotypes with contrasting disease susceptibility. Forschungszentrum Jülich. Energie&Umwelt/Energy& Environment Band. Vol. 244. ISBN: 978-3 95806 021-0
Schoofs H, Delalieux S, Deckers T, Bylemans D (2020) Fire blight monitoring in pear orchards by unmanned airborne vehicles (UAV) systems carrying spectral sensors. Agronomy 10(5):615. https://doi.org/10.3390/agronomy10050615
Shane WW, Teng PS (1983) Sugarbeet yield losses due to Cercospora leafspot. Sugarbeet Res Exten Rep 23:193–198
Skaracis G, Pavli O, Biancardi E (2010) Cercospora leaf spot disease of sugar beet. Sugar Tech 12: 220–228
Song HA, Lee S-Y (2013) Hierarchical representation using NMF. International conference on neural information processing, pp 466–473
Szabó G, Bertalan L, Barkóczi N, Kovács Z, Burai P, Lénárt C (2018) Zooming on aerial survey (Chapter 4). In: Casagrande G, Sik A, Szabó G (eds) Small flying drones, applications for geographic observation. Springer International Publishing AG. ISBN: 978-3-319-66576-4
Tamada T, Asher MJC (2016) The Plasmodiophorid protist Polymyxa betae. In: Biancardi E, Tamada T (eds) Rhizomania. Springer, Cham, pp 135–153
Tan M, Özgüven MM, Tarhan S (2015) Drone sistemlerin hassas tarımda kullanımı, 29. Tarımsal Mekanizasyon Kongresi ve Enerji Kongresi, 2-5 Eylül Diyarbakır, S:543-547
Townsend GK, Heuberger JW (1943) Methods for estimating losses caused by diseases in fungicide experiments. Plant Dis Reporter 27:340–343
Turner D, Lucieer A, Malenovský Z, King DH, Robinson SA (2014) Spatial co-registration of ultra-high resolution visible, multispectral and thermal images acquired with a micro-UAV over antarctic moss beds. Remote Sens 6(5):4003–4024. https://doi.org/10.3390/rs6054003
Vereijssen J, Schneider JHM, Termorshuizen AJ, Jeger MJ (2003) Comparison of two disease assessment keys to assess cercospora beticola in sugar beet. Crop Prot 1:201–209
Walker R (2002) Development of bacterial seed treatments for the control of *Aphanomyces cochlioides* on sugar beet. Ph.D Thesis, University of Nottingham, UK
Ward L, Koenig R, Budge G, Garrido C, McGrath C, Stubbley H, Boonham N (2007) Occurrence of two different types of RNA-5-containing beet necrotic yellow vein virus in the UK. Arch Virol 152:59–73
Wolf PFJ, Kraft R, Verreet JA (1998) Schadrelevanz von *Cercospora beticola* (Sacc.) in Zuckerrüben als Grundlage einer Verlustprognose. J. Plant Dis Prot 105:462–474

Insect-Pests of Sugar Beet and Their Integrated Management

31

Arun Baitha, Santeshwari Srivastava, and Varucha Misra

Abstract

The sugar beet agro-ecosystem has a few key or primary pests that may actually limit production under certain conditions. A couple of these are available all over the world. In addition to the primary pests, there are numerous species that cause periodic losses to sugar beet, while a few species have such a low population rate that no serious damage occurs. Sugar beet crops could be considered a long-term cropping system from the time of planting to harvest. Knowledge of the species complex and their roles in the ecosystem can be essential for deciding whether or not to use pesticides. The indiscriminate use of pesticides may lead to outbreaks of leaf-feeding and sucking pests. There is a need for more emphasis on the augmentation and conservation of natural enemies. Farmers are advised for practicing integrated insect pest management strategies for controlling the insect-pests and their damage and encourage the natural build-up of parasitoids and predators. The chapter provides a comprehensive overview of various insect-pests that affect sugar beet crops, with a focus on Indian conditions.

Keywords

Insect-pest · IPM · Sugar beet

A. Baitha (✉) · S. Srivastava
Division of Crop Protection, ICAR-Indian Institute of Sugarcane Research, Lucknow, Uttar Pradesh, India
e-mail: arun.baitha@icar.gov.in

V. Misra
Division of Crop Improvement, ICAR-Indian Institute of Sugarcane Research, Lucknow, Uttar Pradesh, India

V. Misra et al. (eds.), *Sugar Beet Cultivation, Management and Processing*,
https://doi.org/10.1007/978-981-19-2730-0_31

31.1 Introduction

Sugar beet, *Beta vulgaris* Linnaeus, is a valuable commercial crop since it is a pure source of sucrose and its cultivation is more cost-effective than sugarcane, yielding 30% more sugar in a shorter period of time, providing an ideal opportunity to boost sugar productivity (Lange 1987; Sharma et al. 2017). The roots of sugar beet contain a high concentration of sucrose and commercially it is grown for sugar production (Rashid 1999; Misra et al. 2020), but nowadays provides many by-products such as beverages, feed for livestock, baker yeast, alternative energy production from molasses (producing ethanol and butanol), and molasses used for the de-icing road during heavy snowfall (BSRI 2005; Pathak et al. 2017; Gangwar et al. 2013; Srivastava et al. 2013; Mall et al. 2021).

It is primarily a crop of the temperate region and contributes about 22–28 per cent of the world's sugar production (Solomon 2013), but advances in genetics and agro-technology have extended its scope to subtropics where it can be cultivated during the winter season. The sugar beet growing was found to be profitable compared to the existing cropping systems in the post rainy season in Rajasthan, Punjab, Haryana, Maharashtra, and North Karnataka. Farmers are willing to experiment with new crop options looking to the profitability of the crop, especially in sugarcane growing areas and saline-affected areas of south Maharashtra. It can be profitably produced in highly saline conditions or saline water irrigations as found from the experiment in the Kachh area of Gujarat and Kolhapur area of Maharashtra (Kulkarni et al. 2013).

Sugar beets are attacked by more than 150 species of insect-pests and mites and 40–50 of these species can cause damage either directly or indirectly to the tap root, often causing a great yield loss and quality decrease throughout growth stages (Evaristo 1983; Lange 1987; Bassyouny 1993). Like many agro-ecosystems, sugar beets have a few keys or primary pests that may actually limit production under certain conditions (Blickenstaff 1976; Dunning and Byford 1982; Jones and Dunningg 1972; Lange 1971; Lange and Suh 1980; Reed 1964). In addition to the major pests, there are a number of species with worldwide distribution that inflict periodic losses to sugar beet, as well as a few species that exist at such low population levels that no serious harm is caused. In Egyptian sugar beet ecosystems, commonly known insect-pests are cotton leaf worm *Spodoptera littoralis* (Boisd.), *S. exigua* Hubner, sugar beet fly *Pegomya mixta* Vill., sugar beet beetle, *Cassida vittata* Vill., and sugar beet moth, *Scrobipalpa ocellatella* Boyd (Badawy and Shalaby 2015). Beet army worm, *Spodoptera exigua*; the clover cutworm, *Scotogramma trifolii*; the salt marsh caterpillar, *Estimene acrea*; and the western yellow—striped army worm, *Spodoptera praefica* are the few insect-pests that inflict significant damage to sugar beet (Open blade species) in California (Bisabri-Ershadi and Ehler 1981). Many studies reported that defoliating insects, viz., beet army-worm, *Spodoptera litura* Fabricius; hairy caterpillar, *Diacrisia obliqua* Walker; Semilooper, *Plusia orichalcea* Fabricius; cut worm, and *Agrotis ypsilon* Rott caused appreciable damage to sugar beet at different growth stages in India (Khan and Sharma 1971; Avasthy and Srivastava 1972; Singh et al. 1980; Tewari et al. 1980;

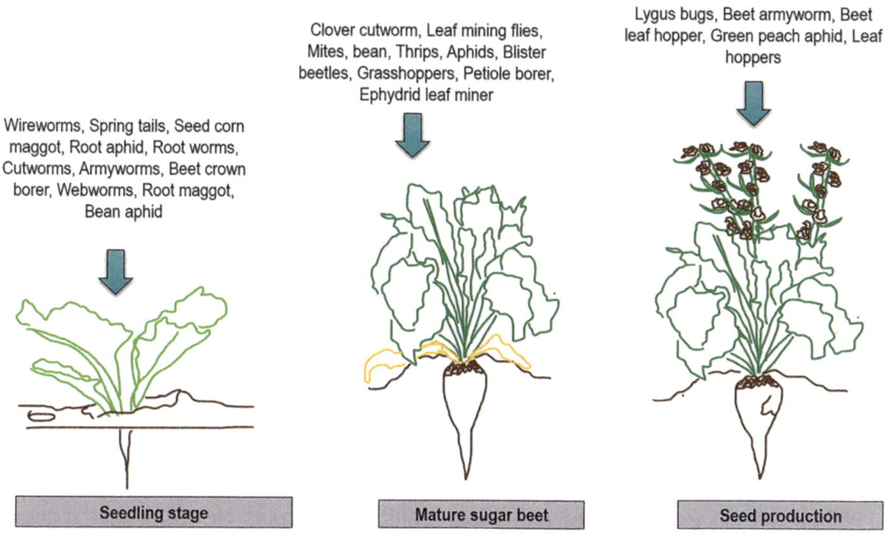

Fig. 31.1 Some major insect-pests in different stages of sugar beet development

Patil et al. 2007). Wire worms, springtails, seed–corn maggots, root worms, soil mites, cutworms, army worms, and other pests are frequently present at the time of sowing and feeding on germinating seeds or young seedlings (Baker and Dunning 1975; Jones and Dunningg 1972; Lange 1971; Rimsa 1979).

A large number of insect-pests have been reported in sugar beet at different stages of its growth. Some of the major insect-pests have been mentioned in Fig. 31.1.

31.2 Insect-Pests of Sugar Beet

Sugar beet insect-pests are divided into three categories, viz., leaf and crown feeders or defoliators, root feeders, and sucking pests.

31.2.1 Leaf and Crown Feeders or Defoliators

These pests cause severe damage to the sugar beet crop at various phases of development by either devouring the entire leaf portion leaving behind only midrib or sometimes eat away only the green portion, leaving behind the network of veins (Avasthy and Srivastava 1972). The most important defoliating insects are sugar beet crown borer, *Hulstia undulatella* Clents; web worm, *Spoladea recurvalis* Fab.; cut worm, viz., black cut worm, *Agrotis ipsilon* Hufnagel, cutworm, *Euxon auxiliaries* Grote; Army worms, *Spodoptera litura, S. exigua* Hub, and grass hoppers, *Melonoplus differentialis* Thomos (Whitney and Duffus 1993). In India, defoliating insects, viz., *S. litura* Hub, *Diacrisia obliqua* Walker, *Plusia orichalcea*

Fabricius, *and Agrotis ipsilon* Rott have caused appreciable damage to the crop at different growth stages (Khan and Sharma 1971; Avasthy and Srivastava 1972; Singh et al. 1980; Patil et al. 2007; Santeshwari et al. 2021).

31.2.1.1 Sugar Beet Moth, *Scrobipalpa ocellatella* Boyd. (Lepidoptera: Gelechiidae)

It was found on sugar beet plants in Portugal, the Canary Islands, North Africa, Europe, Middle East Iran, Russia, Caucasus, Ukraine, Moldova, Georgia, and Turkmenistan (Minoranskii 1987). The newly hatched larvae peel the leaves, hide in the petioles, central leaves, and damage older leaves and petioles with high numbers of larvae in late-planted (October) sugar beet (Renou et al. 1980; Khalifa 2017). The initial incidence of beet moth appeared at the end of December and increased gradually towards the end of the sugar beet growing season (Amin et al. 2008; Khalifa 2018; Youssef 1994; Ad El-Ghany 1995). The infestation negatively affects the root weight and severe infestation caused a significant reduction of 38.20% and 52.40% in root weight and sugar content, respectively (Abo-Saied 1987). Larvae devour leaves and roots, causing the roots to decay and reducing root yield and sugar content (Valic et al. 2005).

31.2.1.2 Sugar Beet Army Worm, *Spodoptera litura* Fabr. (Lepidoptera: Noctuidae)

Sugar beet army worm, *Spodoptera litura,* is one of the important major leaf feeder and polyphagous pests of sugar beet (Holloway 1989; Manoharan et al. 2010). It is almost cosmopolitan in distribution and the most important pest of the Asian continent. In India, it acts as a disastrous pest that caused an outbreak of epidemic in Maharashtra during 2008–09 (Vennila et al. 2016). Young larvae gregariously feed on leaves and skeletonize them (Fig. 31.2). The older larvae, though seen alone, eat up the foliage in a short period of time, leaving leaves with large irregular holes. Incidence appears from the last week of January and reaches its peak during the fourth week of March in Indian conditions after which its populations start to decline and its first incidence was observed after 110 days of sowing and remained active up to 140 days at IISR, Lucknow, and population density of this pest ranged 10.8–15.6 m^2 (Santeshwari et al. 2020).

It shows huge potential to invade into new areas and adapt to climatic and ecological situations. Kapur et al. (2008) found damage in the sugar beet crop by this insect. Complete defoliation is seen when young larvae occur in the beet field and cause 40–60% damage to the foliage. The economic losses vary between 20 and 30 per cent alone. This insect is not primarily reliant on beet leaves, but has also been observed feeding on roots, particularly mature larvae that hide in the soil during the day (Kumar and Regupathy 2000; Santeshwari et al. 2020).

31.2.1.3 Black Cutworm, *Agrotis ipsilon* (Lepidoptera: Noctuidae)

It is polyphagous pest of sporadic importance on a wide range of crops. The early instars generally remain on the foliage of the host plants for a week or two, and then the caterpillars mature, they travel deeper into the soil, and adopt cutworm behavior.

Fig. 31.2 *Spodoptera litura* in sugar beet. (**a**) Egg with newly hatched first instar larvae. (**b**) Second instar larvae in clusters. (**c**) Third instar larvae on leaf. (**d**) Pupa. (**e**) Adult. (**f**) Infected sugar beet. (**g**) Larvae skeletonizing the leaves. (**h**) Skeletonization of leaves due to *S. litura* feeding

the suppression of soil-borne, polyphagous insects such as larvae of white grub, wireworm, and a few species of cutworm. Deep ploughing injures larvae and exposes them to predators, birds.

31.3.2 Resistant Variety

Growing of tolerant varieties such as Calixta, Magnolia, Sandrina, 7KO1, and HI 0064 (Shubhra) against leaf-eating caterpillar (*Spodoptera* sp.) (Shivankar and Patil 2013; Kulkarni et al. 2013).

31.3.3 Biological Control and Use of Pheromone

The egg parasitoids, i.e. *Trichogramma* spp. and *Telenomus* spp., play a significant role in the management of leaf feeding insects in the sugar beet ecosystem. The release of *Trichogramma chilonis* @10,0000/ha in two instalments (50,000 adults/ release and two spraying of Sl NPV @600 mL/ha at 15 days interval in the winter season and 500 mL/ha in summer month) was very effective for reducing the incidence of *Spodoptera litura* (Shivankar and Patil 2013). The pest densities (aphid, tortoise beetle, sugar beet moth, sugar beet fly pests) decreased significantly on sugar beet plants on which green lace wing larvae (first and second instars) had been released (Tauber et al. 2000; Solangi et al. 2013; Youssef et al. 2020). Installation of pheromone traps @25/ha after 4 months sowing (second fortnight of February) in the winter season and 1 month after sowing during summer month has been found effective against *S. litura.*

31.3.4 Insecticides

Application of heptachlor granules or dust @1.0 kg a.i/ha significantly reduced the incidence of subterranean pests in subtropical India, whereas spraying of Malathion or endosulfan 35EC @ 500 mL/ha or Lannate 40SP @25 g/ha or quinalphos 25EC@0.05% suppressed the severe incidence of armyworm (Motiwale et al. 1991; Shivankar and Patil 2013). Systemic granules of phorate and aldicarb, applied in the seed rows or side-dressed into the beds, have been used successfully in many parts of the world (Blickenstaff 1976) against insect-pests of beet. The insecticides have proven successful for controlling the aphids vectoring the yellows viruses (Bryan 1979; Dunning and Winder 1976; Lange 1971). *S. litura* has a long history of exposure and acquired resistance to many insecticides, viz., endosulfan, cypermethrin, fenvalerate, and monocrotophos (Radhika and Subbaratnam 2006).

31.4 Future Prospects

The indiscriminate use of pesticides may lead to outbreaks of leaf-feeding and sucking pests. There is a need for more emphasis on the augmentation and conservation of natural enemies. Natural enemies are found to manage the aphid effectively, and farmers to be advised not to apply insecticides and encourage the natural build-up of parasitoids and predators. The efforts should be made to mass multiply natural enemies (*Chrysoperla carnea*, *Trichogramma* spp. *Cotesia flavipes*, and *Tetrastichus howardi*) in the laboratory, conserve it in situ in the field, and also redistribute it to new areas of infestation. The crop rotation influences future pest management practices and plays an important role.

The development of resistant sugar beets would curtail the need for excessive use of pesticides. New and safer pesticides evaluated in different geographical areas against pest complex, pheromone trap, repellents, and antifeedants may replace conventional types of chemicals. A more concerted effort is needed to work on the genetics of insect-pests in their geographical areas, host plant resistance, and habitat manipulation exploitation of bio-agents with proper identification and evaluation of newer pesticides that are safer to bio-agents and the environment.

There is scope for R&D on habitat manipulation, biological control, and varietal resistance towards being included in the development of IPM of sugar beet insect-pests. Locally prevalent key mortality factors are also to be identified for better management by exploiting the weak links in the life cycle of pests. There is a need to study density-dependent and population-regulating factors of the pests which can also be exploited. Large-scale validation of the impact of entomopathogen on these pests is the need of the hour.

31.5 Conclusion

Sugar beet is being cultivated in more than 40 countries of the world for sugar production. Due to its rich sucrose content and soft juicy leaves and roots, it is attacked by more than 150 different species of insects worldwide. Out of these, 27–34% of insects are known to cause economical loss to the crop. Insect attack is seen from the seedling to its maturity stage or to seed production. *Spilosoma obliqua*, *Agrotis ipsilion*, and *Spodoptera litura* are the few insects which majorly attack this crop worldwide. Occurrence of weeds also plays an effective role in insect incidence rate as often the insects harbour from weeds to sugar beets. The positive relation between insect and weed in sugar beet crop lies with weeds such as Chenopodium, Amaranthus spp., etc. Identical species of aphids, thrips, cutworms, leafhoppers, and spider mites are often seen in both weeds and sugar beet. The production of innovative transgenic sugar beet lines has resulted from advancements in the biotechnological sector combined with traditional approaches. Genes that provide resistance to a variety of damaging insect-pests have been identified, and these genes have served as the foundation for the generation of newer transgenic lines.

Acknowledgments The authors like to thank Dr. SN Sushil, Principal Scientist, ICAR Indian Institute of Sugarcane Research, Lucknow, for taking photographs of Insects.

References

Abo-Saied AMB (1987) Studies on the insects of sugar beet in Kafr EI-Sheik Governorate, Egypt. Ph D Thesis. Fac. of Agric Tanta: Univ Tanta, Egypt, pp 152

Ad EI-Ghany MMA (1995) Studies on the beet moth, *Scrobipalpa ocellatella* Boud at Kafr EI-Sheikh region. M.Sc. Thesis, Faculty of Agriculture Tanta University, Egypt, pp. 116

Amin AH, Helmi A, EI-Serwy SA (2008) Ecological studies on sugar beet insects at Kafr EI-Sheikh Governorate. Egypt J Agric Res 86(6):2129–2139

Anonymous (2008–2009) Annual report of IISR Lucknow, pp. 48

Avasthy PN, Srivastava TN (1972) Pest problem in sugar beet cultivation. India Farmers Digest 10(5):27–29

Badawy S, Shalaby G (2015) Effect of intercropping of sugar beet with onion and garlic on insect infestation, sugar beet yield and economics. J Plant Production 6(6):903–914

Baker AN, Dunning RA (1975) Association of populations of onychiurid Collembola with damage to sugar beet seedlings. Plant Pathol 24:150–154

Bassyouny AM (1993) Studies on preferability and injury level of some main insects to certain sugar beet varieties. Egypt J Appl Sci 8(1):213–219

Bazok R, Drmic Z, Cacija M, Mrganic M, Gasparic HV, Lemic D (2018) Moths of economic importance in the maize and sugar beet production. IntechOpen. https://www.intechopen.com/chapters/62094

Bereś P (2011) Occurrence and harmfulness of cutworms (Agrotinae) on maize (Zea mays L.) in south-East Poland in 2004–2010. Progr Plant Prot 51(2):593–598

Bisabri-Ershadi B, Ehler L (1981) Natural biological control of western yellow-striped armyworm, Spodoptera praefica (Grote), in hay alfalfa in northern California. Hilgardia 49(5):1–23. https://doi.org/10.3733/hilg.v49n05p023

Blickenstaff CC (1976) Sugar beet insects: how to control them. USDA Farmers's Bull 2219:1–20

Bryan KMG (1979) Control of spoil pests of sugar beet with granular formulation of bendocard. In Proc Br Crop Prot Conf, pp 231–237

BSRI (2005) *Sugar beet cultivation in Bangladesh.* Bangladesh Sugarcane Research Institute, Ishurdi, p 10

Čamprag D, Jovanić M (2005) Cutworms (Lepidoptera: Noctuidae)-pests of agricultural crops. Poljoprivredni fakultet, Serbia

Dunning RA, Byford WJ (1982) Pests diseases and disorders of the sugar beet. Deleplanque & Cie, Paris, p 167

Dunning RA, Winder GH (1976) Seed–furrow application of granular pesticides and their biological efficiency on sugar beet. Br Crop Prot Council Monogr 18:37–45

Elliott WM (1973) A method of predicting short term populations trends of the green peach aphid, *Myzus persicae* (Homoptera: Aphididae) on potatoes. Can Entomol 105:11–20

Evaristo FN (1983) Studies on the insect fauna of sugar beet in Portugal. Bio Soc Port Entomol 2: 77–94

Farage AMI, Magada KM, Nadia HH (1998) Survey of mites inhabiting cucurbitaceous and leguminous vegetables in Kaliobia and Giza governorates. J Agric Res 76(1):63–68

Gangwar LS, Sah AK, Hasan SS (2013) Current status of sugar beet production in India: problems and future scenario. In: Kumar S, Singh PK, Swapna M, Pathak AD (eds) Souvenir IISR-industry interface on research and development initiatives for sugar beet in India May 28–29, 2013. Sugar beet Breeding Outpost IISR, IVRI Campus, Mukteshwar, Nainital, p 50

Gu H, Edwards RO, Hardy AT, Fitt GP (2008) Host plant resistance in grain crops and prospects for invertebrate pest management in Australia: an overview. Aust J Exp Agric 48:1543–1548

Hamdany MKH, Aassar MREI (2017) Effect of intercropping three faba bean varieties with sugar beet plants on piercing sucking insect pests and associated natural enemies under ridge space and seedling rates in relation crop yield. Egypt Acad J Biol Sci 10(6):57–77

Holloway JD (1989) The moths of Borneo: family Noctuidae, trifine subfamilies: Noctuidae, Heliothinae, Hadeninae, Acronictinae, Amphipyrinae, Agaristinae. Malayan Nat J 42(2–3): 57–228

Iverson KL, Bromel MC, Anderson AW, Freeman TP (1984) Bacterial symbionts in the sugar beet root maggot, *Tetanops myopaeformis* (von RÖder). Appl Environ Microbiol 47(1):22–27

Jakubowsk M, Walczak F (2009) Current problems of sugar beet plants against turnip moth and heart and dart moth. Progr Plant Prot 49(1):116–121

Jones FGW, Dunningg RA (1972) Sugar beet pests. Minist Agric Fish Food Bull 162 London, pp 133

Kapur R, Pathak AD, Srivastava SN, Yadav A (2008) Varietal response of sugar beet to north Indian growing conditions. Proceedings of international symposium on meeting the challenges of sugar crops & integrated industries in developing countries. Al Arish, Egypt. pp. 237–240

Kennedy JS, Day MF, Eastop VF (1962) A conspectus of aphids as vectors of plant viruses. Commonwealth Institute of Entomology, London, p 114

Khalifa AA (2017) Population dynamics of insect pests and their associated predators at different plantations of sugar beet. J Plant Prot Pathol Mansoura Uni 8(12):651–656

Khalifa AA (2018) Natural enemies of certain insect pests attacking sugar beet plants at Kafr El-Sheikh Governorate. Plant Prot Pathol Mansoura Uni 9(8):507–510

Khan RM, Sharma SK (1971) Some new pests recorded on sugar beet in India. Indian J Ent 33(1): 105–106

Kulkarni VN, Rana DK, Wielandt N (2013) Sugar beet a potential new crop for sugar and ethanol production in India. In: Kumar S, Singh PK, Swapna M, Pathak AD (eds) Souvenir IISR-industry interface on research and development initiatives for sugar beet in India May 28–29, 2013. Sugar beet Breeding Outpost IISR, IVRI Campus, Mukteshwar, Nainital, pp 58–62

Kumar BVN, Regupathy A (2000) Generating base line data for insecticide resistance monitoring in *Spodoptera litura* (Fabricius). Pesticide Res J 12(2):232–237

Lal RJ (2013) Sugar beet diseases and their management. In: Kumar S, Singh PK, Swapna M, Pathak AD (eds) Souvenir IISR-industry interface on research and development initiatives for sugar beet in India May 28–29, 2013. Sugar beet Breeding Outpost IISR, IVRI Campus, Mukteshwar, Nainital, p 64

Lange WH (1971) Insects and mites of sugar beet and their control. In: Johnson RT, Alexander JT, Rush GE, Hawkes GR (eds) Advances in sugar beet production: principles and practices. Iowa State Uni Press, Ames, IA, pp 287–333

Lange WH (1987) Insect pests of sugar beet. Annu Rev Entomol 32:341–360

Lange WH, Suh JB (1980) Leaf feeding and other caterpillars of sugar beet. Calif Sugar Beet:35–36

Lušin V (1971) Izveštajna služba o pojavi i suzbijanju štetnika i bolesti bilja u Sr. Hrvatskoj u 1970. godini. Biljna Zaštita 1:5–29

Maceljski M (1970) O ovogodišnjim štetama od sovica pozemljuša. Biljna Zaštita 6:150–152

Mahrt CG, Blickenstaff CC (1979) Host plants of the sugar beet root maggot, *Tetanops myopaeformis*. Ann Entomol Soc Am 72:627–631

Mall AK, Misra V, Santeshwari, Pathak AD, Srivastava S (2021) Sugar beet cultivation in India: prospects for bio-ethanol production and value added co-products. Sugar Tech 23:1218–1234. https://doi.org/10.1007/s12355-021-01007-0

Manoharan T, Pathma J, Preetha G (2010) Seasonal incidence of sugar beet and natural enemies. Indian J Entomol 72(1):36–41

Minoranskii VA (1987) Formation of noxious entomofauna on beet plantations. Zashchita rastenii 11:32–34

Misra V, Mall AK, Pathak AD (2020) Sugar beet: a sustainable crop for salt stress conditions. In: Hasaanzuman M (ed) Agronomic crops. Springer Nature Singapore Pte Ltd. Publications, Singapore, pp 40–62

Motiwale M.P. Chauhan R.S and Agnihotri VP (1991) Sugar beet cultivation (Eds.) Bulletin No. 32 IISR, Lucknow

Mousa KM, Ueno T (2019) Intercropping potato with citrus trees as ecologically-based insect pest management. J Fac Agric Kyushu Univ 64:71–78

Noori H, Forouzan M, Marzban R, Amin AM (2019) Biological control of the turnip moth, Agrotis segetum Denis & Schiffermuller (Lep.: Noctuidae) by Bacillus thuringiensis (Bt) semisolid bait on sugar beet. J Sugar Beet 35(1):81–89

Pathak AD, Misra V, Mall AK (2017) Prospects of sugar beet in India. In: Souvenir of international symposium on sugarcane research since Co 205: 100 years and beyond, pp 90–97

Patil AS, Salunkhe AN, Pawar BH, Ghodke PV, Shivankar SB, Zende NB, Shewate SR, Patil DJ (2007) Sugar beet cultivation in tropical India—a new experience. In: Proceedings 68th annual convention, 22–24 August, New Delhi

Petrikova C (1952) Zapažanja o pojavi štetočina i bolesti na šećernoj repi u Vojvodini obzirom na klimu i tip zemljišta. Zaštita Bilja 9:27–30

Radhika P, Subbaratnam GV (2006) Insecticide resistance management in cotton-Indian Scenario—a review. Agric Rev 27(3):157–169

Rashid MM (1999) SabjiBiggan. 2nd ed. Rashid Publishing House, Bangla Academy, Dhaka, Bangladesh. (in Bengali). pp. 387-390

Reed JL (1964) The influence of certain physical factors on the relative abundance of the green peach aphid, *Myzus persicae* (Sulzer) (Homoptera: Aphidae). PhD Thesis, University of California, Davis, pp 151

Renou M, Dascoins C, Lallemand JY, Priesner E, Lettere M, Gallois M (1980) Lacetoxy-1 dodecene 3 E, composant principal de la betterave: *Scrobipalpa ocellatella* Boyd. (Lepidoptere :Gelechiidae). Zangew Entomol 90:275–289

Rimsa V (1979) Protection of emerging sugar beet in Czechoslovakia. In Proc Br Crop Prot Conf, pp 245–250

Santeshwari, Misra V, Mall AK, Kumar D (2020) Problems and integrated pest management strategy for *Spodoptera litura* in sugar beet in India. J Exp Zool 23(2):1887–1890

Santeshwari, Shyamrao ID, Misra V, Mall A (2021) Bio-intensive integrated pest management of *Spodoptera litura* in sugar beet. In: Kumar A, Shyamrao ID, Tamang S (eds) Biointensive IPM for commercial crops. Agrobios, pp 99–109

Scherr S, McNeely JA (2008) Biodiversity conservation and agricultural sustainability: towards a new paradigm of ecoagriculture landscapes. Phil Trans R Soc B 33:477–494

Sharma S, Kooner R, Sandhu SS, Arora R, Kaur T, Kaur S (2017) Seasonal dynamics of insect pests of sugar beet under sub-tropical conditions. J Agrometeorol 19(1):81–83

Shivankar SB, Patil AS (2013) Integrated management of *Spodoptera litura* in sugar beet. In: Kumar S, Singh PK, Swapna M, Pathak AD (eds) Souvenir IISR-industry interface on research and development initiatives for sugar beet in India May 28–29, 2013. Sugar Beet Breeding Outpost IISR, IVRI Campus, Mukteshwar, Nainital

Singh KN, Sachan GC, Chibber RC (1980) Effects of some insecticides on sugar beet defoliation by lepidopterous larvae. Indian Sugar:157–160

Solangi AW, Lanjar AG, Baloch N, Rais MUL, Khuhro SA (2013) Population, host preference and feeding potential of *Chrysoperla carnea* (Stephens) on different insect hosts in cotton and mustard crops. Sindh Univ Res Sci Ser 45:213–218

Solomon S (2013) Sugar beet as an energy crop. In: Kumar S, Singh PK, Swapna M, Pathak AD (eds) Souvenir IISR-industry interface on research and development initiatives for sugar beet in India May 28–29, 2013. Sugar Beet Breeding Outpost IISR, IVRI Campus, Mukteshwar, Nainital, pp 1–7

Srivastava AK, Shukla SP, Sawnani A (2013) Useful products from sugar beet. In: Kumar S, Singh PK, Swapna M, Pathak AD (eds) Souvenir IISR-industry interface on research and development initiatives for sugar beet in India May 28–29, 2013. Sugar Beet Breeding Outpost IISR, IVRI Campus, Mukteshwar, Nainital, pp 43–45

Tauber MJ, Tauber CA, Dane KM, Hagen KS (2000) Commercialization of predators: recent lessons from green lacewings (Neuroptera: Chrysopidae: Chrysoperla). Am Entomol 46:26–38

Tewari RK, Mehrotra AK, Bajpai PK, Avasthy PN (1980) Assessment of losses caused by defoliating pests to sugar beet crop (*Beta vulgaris* Linn.). Co-operative Sugar 17(12):943–945

Thielemann R, Nagi A (1977) Further contributions to the incidence of 2 yellow viruses on sugar beets in West Germany with reference to their vectors and yield losses. Z Pflanzenkr Pflanzenschutz 84:257–269

Ueno T (2006) Current status of insect pests attacking green bunching onion in central and southern Vietnam. J Fac Agric Kyushu Univ 51:275–283

Ueno T, Trans DH (2015) *Neochrysocharis okazakii* (Hymenopytera: Eulophidae) as major parasitoid was of stone leek leafminer *Liriomyza chinensis* (Diptera: Agromyzidae) in Central Vietnam. Psyche:179560

Valic N, Vucajnk F, Ferencak B, Mlinaric M, Trdan S (2005) Monitoring of sugar beet moth (*Scrobipalpa ocellatella* Boyd, Lepidoptera,Gelechiidae) in Slovenia using Pheromone traps. In Lectures and papers presented at the 7th Slovenian Conference on Plant Protection, Zrece, Slovenia,8–10 March 2005. Drustvo za varstvo rastlin Slovenije, pp 454–458

Van Emden HF, Eastop V, Hughes RD, Way MJ (1969) The ecology of *Myzus persicae*. Annu Rev Entomol 14:197–270

Vasant TS (2014) Introduction of white grub. International E – Publication, pp 1–23. ISBN: 978-93-84648-38-1. http://www.isca.co.in/BIO_SCI/book/ISBN%20978-93-84648-38-1.pdf

Vennila S, Lokare R, Singh N, Ghadge SM, Chattopadhyay C (2016) Crop pest surveillance and advisory project of Maharashtra—a role model for an e-pest surveillance and area wide implementation of integrated pest management in India. ICAR-National Research Centre for Integrated Pest Management, New Delhi, p 56

Walczak F, Jakubowska M (2001) Wzrost szkodliwości rolnic (Agrotinae) w Polsce. Progr Plant Prot 41(2):386–390

Wallis RL, Turner JE (1969) Burning weeds in drainage ditches to suppress populations of green peach aphid and the incidence of beet western yellows disease in sugar beets. J Econ Entomol 60:307–310

Whitfield GH, Weiss MJ, Howard SMA (1984) A bibliography of the sugar beet root maggot, *Tetanops myopaeformis* (Roder) (Diptera: Otitidae). J ASSBT 3 & 4:268–277

Whitney ED, Duffus JE (1993) Biotic diseases and disorders part-I. In: Compendium of beet disease and insects. APS Press, pp 40–50

Youssef AE (1994) Studies on certain insects attacking sugar beet. Ph.D. Thesis, Faculty of Agriculture Kafr EI-Sheikh, Tanta Univ.

Youssef AE, Ibrahim AS, Bazazo KG, Khattab HM, Ueno T, Mousa KM (2020) Micronutrients foliar fertilization and releasing green lacewing *Chrysoperla carnea* (Stephens) could efficiently suppress sugar beet insect pests. J Fac Agric Kyushu Univ 65(2):269–275

Biological Control of Sugar Beet Insect-Pests

32

Mohamed Abdel-Raheem and Santeshwari Srivastava

Abstract

Biological control is based on the beneficial activities of parasites, viruses, and predators in minimizing pests and their repercussions. Biological control can be used to control a wide range of pests, including animals, plant pathogens, weeds, and insects; however, the methods and agents used differ according to the pest. These live organisms' "natural enemies" biocontrol is particularly imperative for lowering pest insect and mite populations. Entomopathogenic fungi (EPFs) have played a distinctive role in the management of insect-pests since last six decades, and their use has been continuously increasing. Biological pest control is recognized for human and environmental safety; because of this an increasing number of fungi are sold each year around the world to alleviate concerns about the harmful effects of conventional pesticides. *Beauveria bassiana*, *M. anisopliae,* and *V. lecanii* are EPFs, soil-borne, and widely distributed. These have been documented in more than 750 species of host insects and are isolated from soil samples using the Galleria bait technique based on Zimmermann's method. EPFs can grow and maintain on peptone medium, Potato dextrose agar, and Rice Grains. *B. bassiana* showed fastest effect against wax moth, *G. mellonella*, and tortoise beetle, *C. vittata,* followed by *M. anisopliae* then *V. lacanii.*

M. Abdel-Raheem (✉)
Pests & Plant Protection Department, Agricultural and Biological Research Institute, National Research Centre, Cairo, Egypt
e-mail: ma.abdel-raheem@nrc.sci.eg; abdelraheem_nrc@homail.com

S. Srivastava
Division of Crop Improvement, ICAR-Indian Institute of Sugarcane Research, Lucknow, India

V. Misra et al. (eds.), *Sugar Beet Cultivation, Management and Processing*,
https://doi.org/10.1007/978-981-19-2730-0_32

Keywords

Sugar beet · Insect · Entomopathogenic fungi (EPFs) · Isolation · Control

32.1 Introduction

Sugar-beet, *Beta vulgaris,* is an important economic crop of Egypt and almost many countries of the world. It plays a vital role in Europe, Africa, North and South America, Australia, and Asian countries. During recent years, sugar beet has received special attention in Egypt due to its great value in extracting sugar (15–20% sucrose) from the roots. In order to meet the growing demand for local sugar production, scientists, researchers, and farmers need to boost the sugar beet output. Sugar beet cultivation is being performed after 3 years of rotation during the winter season from September till mid-November in Egypt. The most suitable locations for its cultivation are the newly reclaimed lands of Kafr El-Sheikh, Nobaria, Dakahlia and Behira Governorates in Delta, and Fayoum of Middle Egypt. This crop withstands many insects' attack from seeding till harvesting. Some of these insects are common pests that inflict substantial harm to their host plants, resulting in lower agricultural yields in both quantity and quality. Among key pests, tortoise beetle, *Cassida* spp., received a great deal of attention followed by beet moth, *Scrobipalpa ocellatella*, and Sugar beet fly, *Pegomyia mixta*. Larvae and adults of the *Cassida vittata* have been reported as serious pest infesting sugar beet and other host plants belonging to Chenopodiaceae. These insects feed on the plant foliage in different regions of the world. In Egypt, Willcocks (1922) reported an infestation of *C. vittata* adults feeding on leaves of sugar beet for the first time. Steiner (1937) stated that *C. vittata* was of minor important sugar beet pest of Turkey, while this species induced serious damage and caused noticeable losses to crop in United Kingdom (Edward and Heath 1964).

In Bulgaria (Slavechev 1976), southern Europe (Byford et al. 1982), Morocco (Laraichi et al. 1984), and Egypt (Abdel-Raheem 2000, 2005), intensive damage to the sugar beet crop has been reported. Fourth and fifth larval instars are reported to consume up to 85–100% of the total crop. The initial pest has been described as starting on the outer leaves of sugar beet plants at the field's edge. Adult infestations were limited to leaf tissues proximal to the midrib and lateral veins. Injured areas turn brown and become dry and brittle after a few days of insect feeding. According to Guirguis (1985), the average amount of crop consumed up to larval stages of successive five instars is 5.832 mm^2. In case of adults, the rate of infested plants ranged between 11 and 39% estimated the theoretical economic injury level of C. *vittata* in sugar beet in both laboratory and the field conditions (Mostafa et al. 1992). High infestation levels reduced sugar amounts in root of sugar beet (Ali et al. 1993).

To keep the pest population below the economic threshold and to prevent loss of productivity and yield, many insecticides and pesticides are applied. These are employed in agricultural farms to boost crop yield also. However, because it is toxic in nature, it shows negative consequences on human health, including death sometimes. Farmers are seen using pesticides indiscriminately to improve agricultural productivity; as a result, these food residues remain to persist in agricultural goods and cause health concerns in humans after ingestion. These residues are more effective in children than in adults. The most crucial factors in agriculture are environmental stewardship and food security. Misuse of insecticides has resulted in population comeback, chemical residues, and pest resistance in several circumstances. Biopesticides are widely available, and because of the utilization of fewer opportunities, this notion is proving to be incredibly effective. There are also pesticides that are made from natural living elements such as bacteria, minerals, animals, and plants, which control the pest population below damaging levels. These are divided into three categories, viz., predators, parasitoids, and pathogens (Altieri et al. 2005, Mahr et al. 2008).

(i) **Predators:** Predators attack many kinds of insects and eat them. Many staphylinid beetles, ladybird beetles, big-eyed bugs, lacewings, carabid beetles, syrphid (hover) flies, nabid bugs, minute pirate bugs, and spiders are recorded as common predatory arthropods.

(ii) **Parasitoids:** Parasitoids are also known as parasites. These do not eat their hosts directly. These insects lay their eggs inside or on the body of other arthropods. When the eggs hatch, the immature parasitoid comes out and feeds on the victim, called the host, finally killing it. Developing parasitoid mainly kills only one host in his whole life cycle.

(iii) **Pathogens:** The organism that causes a disease is called pathogen. Insect, diseases, and entomopathogens (Insect-parasite) are microorganisms that attack insects and contain nematodes, viruses, fungi, and bacteria. The warm-blooded animals are some exceptions, which cannot be affected by the disease attacked by arthropods. To control weeds, plant disease agents and insects are used (Smith and Capinera 2017).

That is why, biopesticide, which is also known as natural enemy or biological control, is an alternative approach of pest management that is both ecologically benign and effective. It is often known as biocontrol. The biological management of insects and similar species will be the emphasis of this publication. The use of EPFs was the first to be recognized as microbial diseases in insects (Ainsworth 1956). Some of the EPFs are reported, which play a unique role in the history of insect-pest control through microbial processes. *Beauveria bassiana* and *Metarhizium anisopliae* are also well-known EPFs, commonly known as white muscardine fungus and green muscardine fungus attacks on many species of insects and arthropods. Both of the EPFs are soil-borne and widely distributed. These fungi have been documented to occur naturally in approximately 750 species of host insects (Saleh et al. 2016). Soil is considered to be the natural environment of these fungi because they deposit their infectious spores there and remain in the soil for some duration of their life cycle. Therefore, it was determined that soil is the most appropriate place to determine their occurrence (Medo and Cagan 2011).

The occurrence and distribution of insect pathogenic fungi in the agricultural field have been extensively investigated in previous studies by Klingen et al. (2002) and Meyling and Eilenberg (2006). Nevertheless, there are several studies on the isolation of these fungi from insect cadavers (Abdel-Raheem 2005). Wild isolates are still of great importance due to their potential unique characteristics in biological control of insect-pests. The presence of certain entomopathogenic fungal species can be considered as an indicator of their ability to survive in that environment.

In this chapter, amount of isolation and abundance, mass production, and field applications of EPFs will be covered.

32.2 Amount of Isolation and Abundance of the Entomopathogenic Fungi (EPFs)

Abdel-Raheem et al. (2016a, b, c) had isolated 25.03% highest white muscardine fungus *B. bassiana* of total collected samples followed by the green muscardine fungus, *M. anisopliae,* which was 17.76%, and then the metallic pink fungus *V. lecanii* which was 14.49%. Lacey (1997) and Salem et al. (2015) stated that the percentages of infected *G. mellonella* larvae samples ranged from 1.11 to 17.78%. Asensio et al. (2003) described a study conducted on positive soil samples containing *B. bassiana* or *M. anisopliae* that comprised less than 2% of total samples collected from El Behaira, Kafr Elshaikh, or Aswan. Bing-Da and Xing-Zhong (2008) reported from Spain that *B. bassiana* was the most frequent entomopathogenic fungus in the soils followed by *M. anisopliae.* In China, *B. bassiana* was more abundant than *M. anisopliae* in soils (Charnley 1997). *M. anisopliae* had two periods of occurrence; the first was from January to March while the second was from mid-April to May with a peak of 50% positive samples and a minimum of 10%. Abdel-Raheem et al. (2016a, b, c) reported that the percentage of *G. mellonella* larvae infected with *M. anisopliae* was 1–12.2%. Relationship between fungus and plant cover *B. bassiana* was found in the soil under all kinds of fruit trees and recorded 10–40% of collected samples. This fungus was found mostly under Mango trees (50% of samples) than under Pomegranates (40%). Its occurrence was between 10 and 30% for the rest of fruit kinds. *M. anisopliae* was found under seven fruit kinds. Abdel-Raheem et al. (2016a, b, c) found that collected samples of *V. lecanii* were fewer than six kinds of fruits with no clear relationship between the fungus type and the fruit kind. Charnley (1997) stated that there was no clear relationship between the distribution of EPFs and the kind of fruit tree; he also stated that the organic content and temperature are among the factors affecting fungal abundance and activity in the soil. The abundance of *Beauveria bassiana, Metarhizium anisopliae,* and *Verticillium lecanii* in different crops has been described in Table 32.1.

Table 32.1 Abundance of *Beauveria bassiana*, *Metarhizium anisopliae*, and Verticillium lecanii in different crops

	Abundance			
Crop	*Beauveria bassiana*	*Metarhizium anisopliae*	*Verticillium lecanii*	References
Citrus	+ +	+	+ +	Saleh et al. (2016)
Pomegranates	+ + + +	+	+	
Figs	+ +	+	+ +	
Grapes	+ +	–	+ +	
Palm	+ +	–	–	
Annona	+ +	–	+	
Olives	+ + +	+ +	+	
Mango	+ + + + +	+	–	
Guava	+ + +	+ + +	–	
Pears	+ +	+ +	–	
Sugar beet	+ +	+	–	Abdel-Raheem et al. (2016c)

– Absent, + present, each +/– indicates intensity of entomopathogenic fungi (EPFs)

32.3 Process of Isolation from Insects Cadavers

Sample of insect cadavers was collected from different regions of Kafr El-Sheikh Governorate, Egypt. The process of isolation begins from keeping these in petri-dishes at 24 °C temperature. Daily observance for pure growth of the fungus is important for confirmation of the disease cycle through Koch's postulates. For usage in subsequent experiments, storage of fungal growth on slant of PDA artificial media at 4 °C was preferred (Abdel-Raheem 2005). The new fungal generation was isolated from the surface of insect cadaver that was kept on wetted filter paper in a petri-dish and incubated at 24 ± 1 °C for 7 days. Later on, it was cultured in PDA medium in petri-dishes. Fungal cultures were purified weekly until pure cultures were obtained. *B. bassiana* and *M. anisopliae* were isolated from infected cadavers of *C. vittata* and *S. ocellatella* (Abdel-Raheem 2005; Saleh et al. 2016; Abdel-Raheem et al. 2016b).

Abundance ratio varies from crop to crop in *Beauveria bassiana, Metarhizium anisopliae,* and *Verticillium lecanii* that has been shown by various studies (Table 32.1).

32.4 Mass Production of Entomopathogenic Fungi (EPFs)

As per Abdel-Raheem et al. (2020) for mass production, EPFs were grown and maintained mainly by two mediums, i.e., peptone medium and potato dextrose agar medium.

32.4.1 Peptone Medium

For the preparation of peptone medium, 10 g peptone, 40 g dextrose, 2 g yeast extract, 15 g agar, and 500 mL chloramphenicol should be compiled in 1 L of distilled water; after that, autoclave this mixture at 120 °C for 20 min. The autoclaved mixture should be poured in 9 cm diameter and 1.5 cm height petri-dish. The prepared petri-dish was inoculated with the EPFs at 25 ± 2 °C and 85 ± 5 R.H. After 15–30 days, the fungal isolates were re-cultured and stored at 4 °C.

32.4.2 Potato Dextrose Agar Medium

250 g potatoes and 20 g agar should be stirred in 1 L of distilled water and autoclaved at 120 °C for 20 min to prepare potato dextrose agar medium. This medium was poured in 9 cm diameter and 1.5 cm height petri-dish. After 14–30 days, the fungal isolates were re-cultured and stored at 4 °C. The isolates were transferred through their natural host or wax moth larvae *G. mellonella* to restore their pathogenicity.

Isolates of *B. bassiana* and *M. anisopliae* were cultured on wetted rice for isolation. Two kilograms of wet rice were washed for 10 min in boiling water before being placed in thermal bags. These thermal bags were autoclaved for 20 min at 120 °C, then infected with isolates and cultured at 25 ± 2 °C for 15 days. Tween 80 percent was applied after the Conidia were harvested with distilled water and filtered through cheese cloth to remove mycelium clumps (Humber 1996).

32.5 Field Applications of EPFs Against Insect-Pest Control

For field applications of EPFs against some insect-pest, i.e., Tortoise beetle (*Cassida* vittata), Beet moth (*Scrobipalpa ocellatella*), and Sugar beet fly, *Pegomyia mixta* were studied by Abdel-Raheem (2005), Abdel-Raheem and Ragab (2010), Zaki and Abdel-Raheem (2010), Abdel-Raheem et al. (2020).

32.5.1 Tortoise Beetle (*Cassida vittata*)

More information on the tortoise beetle, *C.* vittata, were given by many authors in different countries (Menozzi 1931). *C. vittata* is described as a serious pest of sugar beet and its adult appears in the second half of April and oviposited 20–30 eggs/female in 15–20 days. Abdel-Raheem (2000) mentioned that the eggs are usually deposited in groups (2–12); eggs are arranged in one or two rows. Most of them are laid on the petioles, while few on the leaf surfaces or on the root collar. The average incubation period was 4.39 days. The percentage of hatchability was 93.33%. The average number of deposited eggs was 49.43 egg/female. The larvae start feeding on leaves; this stage passes through four instars. The first instar larva bores its tunnel into midrib and extends to the roots. About three larvae may be found in one tunnel. The duration of each larval instar was 3–5, 2–3, 2–4, and 4–5 days for the first, second, third, and fourth instars, respectively. The whole larval duration ranged between 13–16 days and pupated after some time (Abdel-Raheem 2000). The average duration of pre-pupal and pupal stages ranged between 1–2 and 5–7 days, respectively. The life span of male and female varied, but not significantly according to locality and laboratory conditions. Abdel-Raheem and Ragab (2010) had described the effect of entomopathogenic fungus *B. bassiana* and *M. anisopliae* on *C. vittata* at different concentrations.

32.5.1.1 Effect on Mortality of *C. vittata*

Effect of *B. bassiana* and *M. anisopliae* at different concentrations on mortality (%) of fourth instar and adult of *C. vittata* had been studied (Tables 32.2 and 32.3). The result showed that entomopathogenic fungus *B. bassiana* was found most effective to decrease the total number of *C. vittata* fourth instar larvae and adults at concentration of 10^6, 5×10^6, 10^7, 5×10^7, 10^8 at temperature 24 ± 1 °C and relative humidity (R.H.)100%.

32.5.1.2 Mortality Variation of *C. vittata*

Mortality variation of *C. vittata* at different life stages (Adults, Pupae, Larvae, and Eggs) treated with entomopathogenic fungus *B. bassiana* and *M. anisopliae* at the rate of par 25 plants had been represented in Table 32.4. The field application treatment had done for subsequently 2 weeks. After fourth spray, *B. bassiana* was recorded most effective followed by *M. anisopliae*.

Table 32.2 Effect of *B. bassiana* concentration on mortality (%) of *C. vittata*

	10^6		5×10^6		10^7		5×10^7		10^8	
Days	FI	A	FI	A	FI	A	FI	A	FI	A
3	5	0	10	0	10	0	15	0	20	10
5	10	10	20	10	20	10	20	15	25	20
7	46	40	45	40	45	40	50	50	70	50
9	80	50	80	60	85	65	80	75	100	90

FI Fourth instar, *A* Adult

Table 32.3 Effect of *M. anisopliae* concentration on mortality (%) of *C. vittata*

	10^6		5×10^6		10^7		5×10^7		10^8	
Days	FI	A	FI	A	FI	A	FI	A	FI	A
3	0	0	0	0	0	0	0	0	0	10
5	10	15	15	15	20	15	15	25	20	35
7	20	20	35	20	35	25	30	30	35	35
9	40	30	50	40	50	30	40	60	60	80

FI Fourth instar, *A* Adult

Table 32.4 Mortality variation of *C. vittata* at different stages treated with entomopathogenic fungus *B. bassiana* and *M. anisopliae*

	Average number of *C. vittata*/25 plants							
	Adults		Pupae		Larvae		Eggs	
Experimental time	B	M	B	M	B	M	B	M
Before first spray	0.80	0.00	0.20	0.00	0.80	0.16	0.80	0.16
Before second spray	4.40	2.00	1.44	0.16	8.80	3.84	3.12	2.64
Before third spray	2.60	4.96	0.92	1.08	12.40	7.28	3.36	3.20
After fourth spray	10.40	2.48	8.00	1.76	9.72	9.76	3.68	0.16

B *B. bassiana*, M *M. anisopliae*

32.5.2 Sugar Beet Mining Moth/Beet Moth (*Scrobipalpa ocellatella*)

In Egypt, *Scrobipalpa ocellatella* is becoming a significant pest of sugar beet leaves and roots (Abdel Rahman 2018). It favors hot, dry weather and attacks near the boundaries of fields at first. Depending on temperature fluctuations, geographical region, and sowing date, the sugar beet moth shows three to six generations in a single vegetative season (Kheiri 1991). Fertilized females laid 15–80 eggs singly or in groups on the leaves. The majority of *S. ocellatella* eggs are laid on the center bud and at the place of root collar. The eggs are pale yellow in color and oval in shape. The eggs are orange in color during the pre-hatching period (Kheiri 1991). The egg stage ends after 4 days; newly hatched larvae pierce the petioles and tunneled down into the roots. The first and second larvae instars feed on leaves and the leaves converted in curly, stained black and discolored leaves. In May–June, the larval stage lasted 15–16 days, and in June–July, it lasted 14–15 days. The larvae pupated inside their tunnels and lived 7–8-day pupal stage. Adults live 4–7 days. It takes about 30–35 days to complete its life cycle. Iskander (1982) reported that the larvae of this pest were serious rib minor and also described the infestation levels at different sowing periods. Abdel-Raheem and Ragab (2010), Abdel-Raheem et al. (2020) reported the treatment of *B. bassiana* and *M. anisopliae* at different concentrations.

32.5.2.1 Effect on Mortality of *S. ocellatella*

Effect of *B. bassiana* and *M. anisopliae* at different concentrations on mortality (%) of fourth instar and pupae of *S. ocellatella* had been studied (Tables 32.5 and 32.6). The result showed *M. anisopliae* was found most effective to decrease the total number of *C. vittata* fourth instar larvae at 10^6 concentration, but in pupae stage both represented same mortality percentage. At 5×10^6, 10^7, 5×10^7, and 10^8 concentration, *B. bassiana* was most effective, but in pupae stage entomopathogenic fungus represented same mortality percentage at temperature 24 ± 1 °C and relative humidity (R.H.) 100% (Tables 32.5 and 32.6).

Table 32.5 Effect of *B. bassiana* concentration on mortality (%) of *S. ocellatella*

	10^6		5×10^6		10^7		5×10^7		10^8	
Days	FI	P	FI	P	FI	P	FI	P	FI	P
3	0	0	5	0	5	0	20	0	30	0
5	30	30	30	40	25	60	30	90	40	100
7	30	70	30	90	25	95	30	100	50	100
9	40	75	65	90	40	95	45	100	55	100

FI Fourth instar, *P* Pupae

Table 32.6 Effect of *M. anisopliae* concentration on mortality (%) of *S. ocellatella*

	10^6		5×10^6		10^7		5×10^7		10^8	
Days	FI	P	FI	P	FI	P	FI	P	FI	P
3	25	0	10	0	10	0	5	0	5	0
5	35	30	20	40	20	60	15	90	15	100
7	50	70	25	90	25	95	20	100	20	100
9	55	75	35	90	35	95	40	100	35	100

FI Fourth instar, *P* Pupae

32.5.2.2 Insect Susceptibility Against EPFs

Susceptibility of different stages of *S. ocellatella* had been studied against EPFs *B. bassiana* and *M. anisopliae at* different concentrations (Table 32.7). *B. bassiana* showed highest mortality rate of fourth instar larvae at 10^7 concentration, whereas pupae showed increase in mortality rate with increase in concentration. *M. anisopliae* showed increase in mortality rate with increased concentration. Highest mortality rate of fourth instar larvae and pupae had been observed at 10^8 concentration. The least calculated LC_{50} (0.4×0^5) value of *B. bassiana* represents its high toxic ability of fourth instar larvae, whereas high LC50 (19.2 × 107) value of *M. anisopliae* represents low toxic ability of pupae. Fiducial limits calculated lower in pupae treated with *B. bassiana* and higher in pupae treated with *M. anisopliae* are often used as the limits in control charting when the parameters of the underlying distribution are unknown. Slope ± SE were calculated 0.04 ± 0.14 in fourth instar larvae and 1.20 ± 0.19 in pupae treated with *B. bassiana* and 0.52 ± 0.09 in fourth instar larvae and 0.88 ± 0.18 in pupae treated with *M. anisopliae* (Table 32.7).

32.5.3 Sugar Beet Fly, *Pegomyia mixta*

The sugar beet fly, *Pegomyia mixta,* is considered to be the most serious insect-pest attacking sugar beet all over the world. This leaf miner lays eggs on the underside of the leave of small beets. Abdel-Moneim et al. (2014) and Abdel-Raheem et al. (2016a, b, c) described the egg, larvae, and pupae of *P. mixta*. The adults of *P. mixta* emerge between October and June and were most abundant in November, and reached their peak in December. The eggs are usually deposited in groups, each

Table 32.7 Susceptibility of different stages of *S. ocellatella* against EPFs i *B. bassiana* and *M. anisopliae* concentrations

Treated stages	% Mortality at different concentrations															
	10^6		5×10^6		10^7		5×10^7		10^8		LC_{50}		Fiducial limits 95%		Slope ± SE	
Days	B	M	B	M	B	M	B	M	B	M	B	M	B	M	B	M
4th instar Larva	0	0	60	70	75	70	60	75	60	90	0.4×10^5	2.27×10^5	–	6.14×10^5–6.53×10^5	0.04 ± 0.14	0.52 ± 0.09
Pupa	0	0	75	20	90	20	95	35	100	70	2.99×10^5	19.2×10^7	0.9×10^5–5.63×10^5	6.14×10^6–9.6×10^7	1.20 ± 0.19	0.88 ± 0.18

B *B. bassiana*, M *M. anisopliae*

of the single layers consisting of 3–8 eggs arranged in one or two rows. Eggs are laid on the leaf surfaces. The newly deposited egg is oval-shaped, white in color, and darkens to yellowish green shortly before hatching. The incubation period of the egg ranged between 4–5 days with an average of 4.41 ± 0.57. The percentage of hatchability was 90.56%. Hatched larvae enter into the leaf and mine out its inner portions. Several biological control techniques have been carried out in different countries (Abdel-Raheem et al. 2014, 2015, 2016a, b, c). One of these methods is the use of EPFs *B. bassiana, M. anisopliae,* and the bacteria *Bacillus thuringiensis* (Dipel 2×). Dipel 2× eliminated 15.82% of *P. mixta*, while *B. bassiana* suppressed 35.53 of the population (Sabbour and Abdel-Raheem 2015), isolated some EPFs and used it to control insect-pests in sugar beet and other crops in Egypt.

32.5.3.1 Effect on Mortality of *Pegomyia mixta*

Mortality percentage of *Pegomyia mixta* eggs and larvae, collected from Kafr El-Shik, was studied.

Mortality % of *Pegomyia mixta* Eggs

The percent mortality of *Pegomyia mixta* eggs reached 100% mortality rate sixth days after treated with *B. bassiana* at C_3 concentration, whereas 73% after treated with *M. anisopliae* at C_3 concentration (Table 32.8).

Mortality % of *Pegomyia mixta* Larvae

The percent mortality of *Pegomyia mixta* larvae, treated with *B. bassiana,* reached to 100% mortality at C_3 concentration after tenth day, and when treated with *M. anisopliae,* reached to 100% mortality at C_3 concentration after 11th day (Table 32.9).

Mortality % of *Pegomyia mixta* Pupae

The percent mortality of *Pegomyia mixta* pupae, treated with *B. bassiana,* reached to 100% mortality at C_3 concentration after 12th day, and when treated with *M. anisopliae,* reached to 100% mortality at C_3 concentration after 14th day (Table 32.10).

Table 32.8 Mortality % of *Pegomyia mixta* eggs treated with *B. bassiana* and *M. anisopliae*

	% of Mortality						
		B. bassiana KB1			*M. anisopliae* KM1		
Days after treatment	Control	C_1	C_2	C_3	C_1	C_2	C_3
2nd	0.0	0.0	0.0	0.0	0.0	0.0	0.0
3rd	0.0	0.0	0.0	0.0	0.0	0.0	0.0
4th	0.0	10	13	20	8	11	14
5th	0.0	45	55	60	25	33	50
6th	0.0	60	75	100	45	60	73

$C_1 = 2 \times 10^2$, $C_2 = 2 \times 10^3$ spores/mL, $C_3 = 2 \times 10^4$ spores/mL

Table 32.9 Mortality % of *Pegomyia mixta* larvae treated with *B. bassiana* and *M. anisopliae*

	% of Mortality						
		B. bassiana KB1			*M. anisopliae* KM1		
Days after treatment	control	C_1[a]	C_2	C_3	C_1	C_2	C_3
2nd	0.0	0.0	0.0	0.0	0.0	0.0	0.0
3rd	0.0	0.0	0.0	0.0	0.0	0.0	0.0
4th	0.0	13	15	18	10	14	16
5th	0.0	32	33	44	17	32	44
6th	0.0	40	43	60	40	41	44
7th	0.0	70	75	85	55	55	66
8th	0.0	73	77	87	55	60	77
9th	0.0	80	83	97	60	60	77
10th	0.0	90	92	100	66	65	75
11th	0.0	100	100	100	73	80	100

[a]$C_1 = 2 \times 10^2$, $C_2 = 2 \times 10^3$ spores/mL, $C_3 = 2 \times 10^4$ spores/mL

Table 32.10 Mortality % of *Pegomyia mixta* pupae treated with *B. bassiana* and *M. anisopliae*

	% of Mortality						
		B. bassiana KB1			*M. anisopliae* KM1		
Days after treatment	Control	C_1[a]	C_2	C_3	C_1	C_2	C_3
2nd	0.0	0.0	0.0	0.0	0.0	0.0	0.0
3rd	0.0	0.0	0.0	0.0	0.0	0.0	0.0
4th	0.0	0.0	0.0	0.0	0.0	0.0	0.0
5th	0.0	0.0	0.0	0.0	0.0	0.0	0.0
6th	0.0	0.0	0.0	0.0	0.0	0.0	0.0
7th	0.0	10	13	15	8	11	15
8th	0.0	18	19	27	11	18	33
9th	0.0	20	26	45	13	23	35
10th	0.0	40	43	60	40	45	44
11th	0.0	70	75	85	55	70	73
12th	0.0	92	97	100	60	78	80
13th	0.0	100	100	100	72	82	96
14th	0.0	100	100	100	85	86	100
15th	0.0	100	100	100	100	100	100

[a]$C_1 = 2 \times 10^2$, $C_2 = 2 \times 10^3$ spores/mL, $C_3 = 2 \times 10^4$ spores/mL

32.6 Field Experiments

For the study of field application with *B bassiana* and *M anisopliae* on *Pegomyia mixta*, an area of about 2100 m^2 was chosen in Abo-Ghalab region, Kafr El-Shikh Governorate. Sugar beet plants were cultivated in mid-September 2014. Infestations of *P. mixta* started to appear during the second week of January 2015. In mid-February, the applications were carried out using two EPFs, *B. bassiana*, and *M. anisopliae*.

The data pertaining to infected plants clearly indicated that the first treatment with *B. bassiana* and *M. anisopliae* was insignificant. After second treatment, numbers of infected plants were gradually decreased. After third treatment with *B. bassiana, it* got significant effect followed by *M. anisopliae* (Table 32.11).

32.7 Future Prospects

EPFs are one of the most promising agents for the biological control of insect's population, where it permits the cost of production and preservation of public health. Insects are infected by a wide range of fungal species from several different groups. In order to kill target insect-pests, these biopesticides do not need to be consumed; only its physical touch is sufficient. As the fungus *B. bassiana* successfully used against different insects as well as the fungus *M. anisopliae* and especially the rank of sheaths wings also *Verticillium lecanii* against insects sucking mouth parts. The production of EPFs in various liquid and powder forms, as well as in the forms of nanoparticles through laboratories and factories, can be very useful for human health in near future. The use of this method in Integrated Pest Management (IPM) programmes appears to be promising. Increasing farmer's awareness about importance of biological pesticides control methods is also needful because EPFs are one of the most promising agents for the biological control of insects' population, where it permits the cost of production and preservation of public health. Insects are infected by a wide range of fungal species from several different groups. In order to kill target insect-pests, these biopesticides do not need to be consumed; only its physical touch is sufficient. As the *B. bassiana, M. anisopliae* and *Verticillium lecanii* EPFs are successfully used against different insects, especially in the rank of sheathed wings and sucking mouth parts. It is produced in various liquids and powders as well as in the form of nanoparticles through laboratories and factories. The use of this method under Integrated Pest Management (IPM) programmes appears to be promising. It is essential to spread awareness of biological pest management among the farmers to control the insect-pests because it is developed as eco-friendly mycopesticide. Aside from insect-pest population control, EPFs play major role to colonize green plants and perform additional tasks such as plant growth promoter, stress tolerance, and water absorption enhancer that opens up new possibilities for their use in plant management. As a result, scientists are paying more attention to EPFs because of their potential application in the near future as a replacement to chemical pesticides and inorganic fertilizers. The secondary product of EPFs serves as a rich reservoir of biologically active compounds in agriculture and medicine industry. They can produce a broad range of secondary metabolites that are necessary for insecticidal, antibacterial, and antifungal applications. Integrated Pest Management (IPM), disease pathogen resistance, and host plant nutrition had been discovered to be dependent on mutual interactions between

Table 32.11 Field application with *B bassiana* and *M anisopliae* on *Pegomyia mixta* at Kafr El-Shikh (2014–2016)

Treatment	Virulence of *B. bassiana* and *M. anisopliae*							
	No of infested/10 plants		No of infested leaves		No of infested/egg patch/plant		No of infested larvae/plant	
	2014–2015	2015–2016	2014–2015	2015–2016	2014–2015	2015–2016	2014–2015	2015–2016
1st week of March								
Control	7	6	12	14	10	12	20	25
B. bassiana	7	6	11	13	13	12	22	25
M. anisopliae	8	7	12	13	12	12	21	23
2nd week of March								
Control	7	6	13	14	11	13	21	25
B. bassiana	1	2	4	3	5	4	4	12
M. anisopliae	2	3	6	6	6	6	15	14
3rd week of March								
Control	6	7	12	13	10	13	20	24
B. bassiana	0	0	0	0	0	0	0	0
M. anisopliae	1	0	2	0	3	0	0	0

EPFs and their colonized hosts. EPFs that dwell in the host tissues give sustenance and protection to the host, as well as being directly involved in food intake and biotic and abiotic stress protection. FEPs have a variety of consequences on the health, development, and growth of their hosts. EPFs can have an impact on plant communities, ecological functioning, and population dynamics. The majority of studies are prejudiced and adversely focused only on few significant *Beauveria*, *Metarhizium*, and *Lecanicillium* sp. species. It's critical to isolate and identify a number of new fungal isolates with more efficacy and potential for controlling pests, diseases, and even weeds. Al-Ani et al. (2021) advised that new fungal-derived secondary metabolites be developed in order to improve the efficacy of mycotoxins.

32.8 Conclusion

Beauveria bassiana, commonly known as white muscardine fungus, attacks a wide range of immature and adult insects. *M. anisopliae,* a green muscardine fungus, is reported to infect 200 species of insects and arthropods. Both of these EPFs are soil-borne and widely distributed. These fungi have been documented to occur naturally in over 750 species of host insects. The EPFs were isolated from soil samples using the Galleria bait technique based on Zimmermann's method (Zimmermann 1986). The EPFs were identified morphologically based on the morphological characteristics of reproductive structures according to Salem's method (Salem et al. 2015). The infected insects which were covered with fungal mycelium were collected from the field and carried to the laboratory. Fungi were grown and maintained on peptone medium, Potato dextrose agar and Rice Grains. *B. bassiana* had the fastest effect against larvae of *G. mellonella*, followed by *M. anisopliae,* then *V. lacanii.* The fungus *B. bassiana* was the most effective to control the tortoise beetle, *C. vittata,* than the Fungus *M. anisopliae*.

References

Abdel Rahman IE (2018) Biological and ecological studies on sugar—beet moth, *Scrobipalpa ocellatella*. Int J ChemTech Res 11(11):274–277. https://doi.org/10.20902/IJCTR.2018.111128

Abdel-Moneim et al (2014) Biological and ecological studies on the sugar -beet fly, *Pegomyia mixta* Vill. (Diptera: Anthomyiidae) on sugar-beet in Egypt. Arch Phytopathol Plant Protect 47(13): 1557–1562. https://doi.org/10.1080/03235408.2013.85078

Abdel-Raheem MA (2000) Ecological and biological aspects of certain key pests of sugar-beet in Egypt, M.Sc. Faculty of Agriculture, Cairo University, Cairo, 73 pp

Abdel-Raheem MA (2005) Possibility of using the entomopathogenic fungi *Beauveria bassiana* and *Metarhizium anisopliae* for controlling the sugar-beet insects *Cassida vittata* Vill. and *Scrobipalpa ocellatella* Boh. in Egypt. Ph.D. Faculty of Agriculture, Cairo University, Cairo, 86 pp

Abdel-Raheem MA, Ragab ZA (2010) Using entomopathogenic fungi to control the cotton aphid, Aphis *Gossypii glover* on sugar beet. Bull NRC 35:57–64

Abdel-Raheem MA, Abdel-Rhman IE, Reyad NF (2014) Tomato leaf miner, *Tuta absoluta* (Meyrick) (Lepidoptera: Gelechiidae) on tomato crop, biology and morphology. Bull NRC 39:101–112

Abdel-Raheem MA, Ismail IA, Abdel-Rahman RS, Farag NA, Abdel-Rhman IE (2015) Entomopathogenic fungi, *Beauveria bassiana* (Bals.) and *Metarhizium anisopliae* (Metsch.) as biological control agents on some stored product insects. J Entomol Zool Stud 3:329–333

Abdel-Raheem et al (2016a) Evaluation of some isolates of entomopathogenic fungi on some insect pests infesting potato crop in Egypt. Int J Chem Tech Res 9(8):479–485

Abdel-Raheem et al (2016b) Nano entomopathogenic fungi as biological control agents on cabbage worm, *Pieris rapae* L. (Lepidoptera: Pieridae). Der Pharma Chemica 8(16):93–97

Abdel-Raheem et al (2016c) Isolates, virulence of two entomopathogenic fungi as biological control agent on sugar beet fly, *Pegomyia mixta* in Egypt. Der Pharma Chemica 8(18): 132–138. http://derpharmachemica.com/vol8-iss18/DPC-2016-8-18-132-138.pdf

Abdel-Raheem MA, Reyad N, Alghamdi H (2020) Virulence of nano – particle preparation of entomopathogenic fungi and entomopathogenic bacteria against red palm weevil *Rhynchophorus ferrugineus* (Olivier) (Coleoptera: Curculionidae). Roman Biotechnol Lett 25: 1151–1159

Ainsworth (1956) *Agostino Bassi*, 1773–1856. Nature 177:255–257

Al-Ani LKT, Aguilar-Marcelino L, Salazar-Vidal VE, Becerra AG, Raza W (2021) Role of useful fungi in agriculture sustainability. In: Yadav AN (ed) Recent trends in mycological research. Springer, Berlin, Heidelberg, Germany, pp 1–44

Ali et al (1993) Survey and geographical distribution of sugar—beet insect pests in Egypt. J Agric Sci Mansoura Univ 18:1793–1804

Altieri M, Nichols CI, Fritz MA (2005) Manage insects on your farm: a guide to ecological strategies. Sustainable agriculture network handbook series book 7. (Available online at: http://www.sare.org/Learning-Center/Books/Manage-Insects-on-Your-Farm) (Verified 25 April 2011)

Asensio et al (2003) Entomopathogenic fungi in soils from Alicante province, Spanish. J Agric Res 1(3):37–45

Bing-Da, Xing-Zhong (2008) Occurrence and diversity of insect-associated fungi in natural soils in China. Appl Soil Ecol:100–108

Byford W, Dunning A, Lejealle F, Broom's Barn Experimental Station (1982) Pests, diseases and disorders of the sugarbeet, Maisons Lafitte: Deleplanque Cie, 1982 Pests, diseases and disorders of the sugar beet (Book, 1982) [WorldCat.org]

Charnley (1997) Entomopathogenic fungi and their role in pest control. In: Wicklow DT, Soderstrom B (eds) The mycota IV. Environmental and microbial relationships. Springer, Berlin, Heidelberg, Germany, pp 185–201

Edward, Heath (1964) The principles of agriculture entomology. Chapman Hall, Ltd., London, 418 pp

Guirguis (1985) studies on certain insects, attacking sugar—beet in Western desert Egypt. Ph. D. Thesis Fac of Agric, Monofya Univ, 277 pp

Humber (1996) Fungal pathogen of the Chrysomelidae and prospects for their use in biological control. In: Jolivet PH, Cox ML, Hsiao TH (eds) Biology of chrysomilidae, vol IV. SPB Academic Publishing, The Hague

Iskander (1982) Studies on certain sugar-beet insects [Msc thesis]. Egypt: Fac of Agric, Cairo Univ, 165 pp

Kheiri M (1991) Important pests of sugar beet and their control. Ministry of Agriculture, Agricultural Research and Education organization. Kalameh Publication Institute, Tehran, Iran, 126 pp (in Persian)

Klingen et al (2002) Effects of farming system, field margins and bait insect on the occurrence of insect pathogenic fungi in soils. Agric Ecosyst Environ 91:191–198

Lacey (1997) Manual of techniques in insect pathology. Academic Press (Elsevier), New York, p 409

Laraichi et al (1984) Biological study of the small green tortoise beetle *Cassida vittata* (Vill.) in the Gharb region. Actes Inst AgroVit 4(1):87–92
Mahr DLP, Whitaker, Ridgway NM (2008) Biological control of insects and mites: An introduction to beneficial natural enemies and their use in pest management. University of Wisconsin Cooperative Extension, No. A3842
Medo, Cagan (2011) Factors affecting the occurrence of entomopathogenic fungi in soils of Slovakia as revealed using two methods. Biol Control 59:200–208
Menozzi (1931) Informations on the damage caused by insects to beet during the sugar – beet season of 1930 and on their control. Indust Saccarif Ital 24(2):7
Meyling, Eilenberg (2006) Isolation and characterization of Beauveria bassiana isolates from phylloplanes of hedgerow vegetation. Mycol Res 110:188–195
Mostafa FF, Salama RAK, Hemida LA, Elnagar S (1992) Foliage consumption by tortoise beetle, *Cassida vittata* Vill. And economic injury level to sugar – beet yield. Fayoum J Agric Res Dev 6(1):14–24
Sabbour, Abdel-Raheem (2015) Toxicity of the fungus *Beauveria bassiana* and three oils extract against *Sitophilus granaries* under laboratory and store condition. Am J Innov Res Appl Sci 1(7):251–256
Saleh et al (2016) Natural abundance of Entomopathogenic fungi in fruit orchards and their virulence against *galleria mellonella* larvae. Egypt J Biol Pest Control 26(2):203–207. http://search.proquest.com/openview/900310a09c58a08cc6a3e69403576742/1?pq-origsite=gscholar&cbl=886351
Salem et al (2015) Lab-field evaluation of some Egyptian isolates of entomopathogenic fungi *Metarhizium anisopliae* and *Beauveria bassiana* against sugar beet beetle *Cassida vittata* Vill. Swift J Agric Res 1(2):9–14
Slavechev (1976) *Cassida nebulosa* L and *Cassida nobilis* L and measures for their control. Rastitelna Zashchita 24(5):23–26
Smith HA, Capinera JL (2017) Natural enemies and biological control [Internet], pp 1. http://edis.ifas.ufl.edu/in120
Steiner (1937) Contributions to the knowledge of the pests in Asia, minor pests of sugar – beet in Turkey. Z Ange Ent 24(1):1–24
Willcocks (1922) A survey of the more important and economic insects and mites of Egypt. Bull Sult Agric Tech Sec, (1), Cairo, 482 pp
Zaki, Abdel-Raheem (2010) Using of Entomopathogenic fungi and insecticide against some insect pests attacking peanuts and sugar beet. Arch Phytopathol Plant Protect 43(18):1819–1828. https://www.academia.edu/6684478/Use_of_entomopathogenic_fungi_and_insecticide_against_some_insect_pests_attacking_peanuts_and_sugarbeet_in_Egypt
Zimmermann (1986) The galleria bait method for detection of entomopathogenic fungi in soil. J Appl Entomol 102(2):213–215

Biology, Pest Status and Management of Armyworm *Spodoptera litura* and Cutworm *Agrotis ipsilon* (Noctuidae: Lepidoptera) on Sugar Beet

33

Manish Chandra Mehta, Santeshwari Srivastava, A. K. Mall, and M. Raghuraman

Abstract

Sugar beet (*Beta vulgaris* L.: Chenopodiaceae), a temperate, biennial crop, is infested by more than 500 insects and non-insect pests throughout the world. Among all the defoliating pests, beet armyworm (*Spodoptera litura*) and cut worm (*Agrotis ipsilon*) are of major concern in India. In this chapter, the spectrum of various pests, pest status, and the economic losses caused by them are discussed. The biology and distribution of armyworm and cutworm, which are the major pests of the sugar beet crop in the country, are illustrated with larval, pupal, and adult periods. Suitable integrated management practices including cultural, biological, and chemical interventions and future prospects are also discussed.

Keywords

Sugar beet · Armyworm · Cut worm · Insect pests

33.1 Introduction

Sugar beet (*Beta vulgaris* L.) is a temperate, biennial crop that belongs to the family Chenopodiaceae. Processed sugar is extracted mainly from two crops, sugarcane (*Saccharum officinarum* L.) and sugar beet (*Beta vulgaris* L.). Sugarcane accounts for nearly 80% of the world's sugar production and the rest is extracted from the

M. C. Mehta · M. Raghuraman
Department of Entomology and Agricultural Zoology, Institute of Agricultural Sciences, Banaras Hindu University, Varanasi, India

S. Srivastava (✉) · A. K. Mall
Division of Crop Improvement, ICAR-Indian Institute of Sugarcane Research, Lucknow, India

V. Misra et al. (eds.), *Sugar Beet Cultivation, Management and Processing*,
https://doi.org/10.1007/978-981-19-2730-0_33

roots of sugar beet crop. The by-products of sugar beet (molasses and pulps) are used as fodder and biofuel production. The rest of the material left after biofuel production can be used as organic manure. This benefit provides surplus value to the crop. Sugar beet yields 14–21% of sucrose, among which 90% is root derived and the rest is obtained from hypocotyl (Shrivastava et al. 2013).

Presently 43 countries produce sugar from sugar beet and 9 countries process it from both sugar cane and sugar beet (FAO 2009). Marzo et al. (2019) reported that, during 2014, sugar beet production was 269,714,066 tons worldwide and the estimated cultivated area was about 4,471,580 ha. Russia is the largest producer of sugar beet (54,350,115 tons), followed by France (38,024,390 tons), Germany (29,728,300 tons), USA (25,945,480 tons), Turkey (18,085,528 tons), Italy (1,779,130 tons), Poland (13,836,620 tons), Ukraine (10,204,530 tons), Britain (7,450,000 tons), and Spain (2,752,710 tons) (FAOSTAT 2019). Among the neighbouring countries of India, China ranks ahead with 12,272,900 tons of production, followed by Pakistan (38,620 tons) (FAOSTAT 2019). The countries having the maximum yield are Spain (912,097 kg/ha), Chile (1,016,617 kg/ha), and France (851,419 kg/ha) (FAOSTAT 2019).

The sugar beet crop is not a much-popularized crop in India even though it has many advantages compared to sugarcane. The government of India should take efforts to popularize this crop because it is a short-duration crop (6–7 months), while sugarcane takes 10–12 months from the date of establishment to harvesting. Therefore, the farmers can harvest the previous crop within 5 months and go for another lucrative crop production on the same land. It also has a sugar content of 12–20% (Marzo et al. 2019), while sugarcane has only 13–15% sugar in the form of sucrose. Sugar beet has a higher sugar recovery (12–15%) than sugarcane and a purity of 85–90%. It is tolerant against frost and salinity and can withstand a multitude of climatic conditions (Yang et al. 2012, Wedeking et al. 2017), with a tolerant capacity of up to 9.5 m mhos/cm (Shrivastava et al. 2013). The farmers can save water with the cultivation of sugar beet. The sugar factories can run throughout the year if the cultivation of sugarcane is concomitant with sugar beet. The intercropping of sugarcane and sugar beet can give financial support to the farmers, and fulfil the sugar and bioethanol needs of the country.

In the 90s, many researches had already been done in India and being a new crop numerous insect pests had been observed to harm the crop. Such problems can be surmounted by meticulous and perspicuous study of the ecology (biology, habitat, behaviour, and population density of the insects damaging the crop and its natural enemies), to anticipate its pest status and frame scrupulous plant protection measures in the form of cultural, biological, and chemical control. Motiwale et al. (1991) reported the challenges of managing this new crop from pests, shifting of their hosts, and migrating to attack on sugar beet crop, non-insect pests, and diseases. This becomes a conundrum mainly during favourable seasons, such as rainy and summer (Patil et al. 2007). Data suggest that more than 500 insects and non-insect pests are known to infest the sugar beet crop throughout the world. Among all the defoliating pests, beet armyworm (*Spodoptera litura*) (Santeshwari et al. 2021) and cut worm (*Agrotis ipsilon*) are of major concern (Patil et al. 2007).

The problem of pests can be surmounted by meticulous and perspicuous study of the ecology (biology, habitat, behaviour, and population density of the insects damaging the crop and its natural enemies) to anticipate its pest status and by framing scrupulous plant protection measures in the form of cultural, biological, and chemical control.

33.2 Sugar Beet Pest Status in India

India is blessed with a smorgasbord of climatic conditions extending from the southern tropical to temperate and alpine in the Northern Himalayan region of the country, which assists greatly in seed production of sugar beet (Pathak and Srivastava 2013). Sugar beet is composed of 15–20% sugar, 2.6% non-sugars, and 4–6% pulp, and is rich in amino acids and minerals, such as nitrogen, phosphorus, potassium, calcium, magnesium, iron, manganese, and zinc (Bichsel et al. 1991). These essential nutrients are required by the pests for their survival, growth, reproduction, and metabolism (Chapman 2000; Busch and Phelan 1999; Joern et al. 2012). This plant can be stratified into three parts, viz., "leaf", "top" or "crown", and "root". Owing to such a lavish nutritional profile, most of the sugar beet has been suffering from multiple insect pest infestation since their introduction in India. Infestation by different insect pests on sugar beet crop has been reported from time to time by several workers (Khan and Sharma 1971; Avasthy and Shrivastava 1972; Singh et al. 1980; Tewari et al. 1986; Venette et al. 2003; Nathan and Kalaivani 2005; Patil et al. 2007; Manoharan et al. 2010; Santeshwari et al. 2020, 2021).

Anonymous (1967–1968) reported, from IISR, Lucknow, of the periodic occurrences of cut worm (Agrotis sp.) and hairy caterpillar (Diacrisia oblique Wkr.). Leaves at various stages of growth of the crop were observed to be damaged by cut worms. In the initial stages, the incidence ranged from 50 to 80% on plant basis but declined subsequently. A species of Coccinellid beetle (*Chilomanes* sp.) had been recorded feeding on the eggs of the cut worm. During the period from January to April, when the temperature continuously increases day by day, the crop suffers from serious attack by the hairy caterpillar. Leaves of the infested clumps were almost completely eaten away by this pest. A coccinellid beetle, *Brumus auturalis,* had been recorded feeding on the eggs of the hairy caterpillar.

Kalra and Srivastava (1964) had reported three species of parasites: two dipterous, viz., *Carcelia corvincides* Wulp and *Strobliomyia orbata* Wied., and one Hymenopterous, *Ananteles* sp. glomeratus, group were recorded from the larvae of the hairy caterpillar. Besides these, several insect pests, viz. *Eretmocera impactella* Wkr., *Scoliophthalmus micans* Lamb., *Psilopa* sp., Epyris sp., *Euchal cidiasp, Omalus timidus* Nurse, *Enicospilus* sp., *Circulifer tenellus* Baker, *Empoasca* sp., *Aethus* sp., *Geocoris* sp., *Tricentrus* sp., *Prostemma carduelis* Dohrn, *Eysarcoris* sp., *Nezara antennata* Scott, *Atractomorpha crenulate* Fabr., *Crenulate* Fabr., *Eyprepocnemis plorans* Charp., *Pyrgomorpha bispinosa* Walk., had also been recorded. Mehrotra et al. (1981) reported some other coleopteran insects (*Monolepla* sp., *Epilachna* sp. and *Aulacophora* sp.) from sugar beet crop.

Khan and Sharma (1971) reported important pests of sugar beet, viz. *Spodoptera* (Laphygma) *exigua* (Hb.), *Plusia orichalcea* (F.) larvae of *Agrotis ipsilon* (Hfn.), *A. segetum* (Schiff.), *Autographa (Plusia) nigrisigna* (Wlk.), and *Euxoa intracta* (Wlk.), from Sri Ganganagar (Rajasthan). He also stated that larvae of *Mythimna separata* (Wlk.), *Mythimna (Pseudaletia) unipuncta* (Haw.), and unidentified aphids caused sporadic damage. Zhang et al. (2011) stated that leaf-feeding insects such as *Spodoptera exigua* and cabbage armyworm *Mamestra brassicae* cause serious damage, particularly in the Asian region, including India and China.

In view of the involvement of a large number of insect species in different incidences on sugar beet crop, there was a need to have complete information of various insect species involved in causing multiple injuries to sugar beet. Sharma et al. (2017) reported from Ludhiana, Punjab that aphids and *Myzus persicae* had a minor damage-causing effect on sugar beet crop, whereas *Spodoptera litura* and *Helicoverpa armigera* (Hubner) showed major damage-causing effects on sugar beet crop.

Santeshwari et al. (2020) have listed several insect species, mostly from indigenous and adjacent countries, that caused a damage of approximately 26–30 percent to the sugar beet crop. Severe attack by defoliating insect pests of sugar beet tobacco cutworm (*Spodoptera litura* Fabricius) was reported by Manoharan et al. (2010) and Santeshwari et al. (2020), which causes appreciable damage in India, and occurs in epidemic form in many states during the winter and summer seasons. Besides, semilooper (*Plusia orichalecea* Fabricius), hairy caterpillar (*Diacrisia obliqua* Walker), and cutworm (*Agrotis ypsilon* Rott.) also cause higher damage and affect sugar beet yield (Patil et al. 2007). Young larvae of *Spodoptera* sp. skeletonize the leaves; however, the older ones eat the entire lamina and are able to defoliate the crop completely in a very short period (Cooke 1993; Patil et al. 2007; Santeshwari et al. 2021).

33.2.1 Insect Pest Spectrum on Sugar Beet

Several insect species so far have been recorded by different workers at various locations as infesting sugar beet in India. These insects are represented by 35 species belonging to 33 genuses, 18 families, and 6 orders (Table 33.1). They consist of all categories of pests, including the three main groups, i.e. leaf and crown feeders, root feeders, and sucking pests. Information on sugar beet insects recorded in India is updated with recent findings as well as past records, to make a ready reckoner for the use of researchers and field workers.

Table 33.1 Identified insect pest on sugar beet crop in India

Order	Family	Genus	Species	Damaging state	Nature of damage
Coleoptera	Chrysomelidae	*Monolepta*	*Monolepta* sp.	Leaf and crown feeder	Larvae feeds on leaves and flowers as well as underground on the roots.
		Aulacophora	*Aulacophora* sp.	Leaf and crown feeder	Adult feeds on the foliage and flowers. The beetles feed between the veins, often cutting and removing circular discs, which they then eat. The larvae tunnel into the roots, which become swollen, discoloured, and distorted, and the plant may die.
	Coccinellidae	*Epilachna*	*Epilachna* sp.	Leaf and crown feeder	Both adults and larvae feed on the leaves and leave a fine net of veins. Damaged leaves shrivel soon and dry up.
Lepidoptera	Noctuidae	*Agrotis*	*A. ipsilon* (Hfn.).	Leaf and crown feeder	Cuts the plants at or just below the soil surface larvae destroy much foliage in a short period of time and cause severe damage in spring by first-generation cut worm
			A. segetum (Schiff.)		
			A. ypsilon (Rott.)		
		Autographa	*A. (Plusia) nigrisigna* (Wlk.)		
		Plusia	*P. orichalcea* (Fabricius)		
		Euxoa	*E. intracta* (Wlk.)		
		Spilosoma or Diacrisia	*S. oblique/ D. oblique* (Walker)		
		Helicoverpa	*H. armigera*		
		Mamestra	*M. brassicae*		
		Mythimna	*M. (Pseudaletia) unipuncta* (Haw.)		
			M. separate (Wlk.)		
		Spodoptera	*S.* (Laphygma) *exigua* (Hb.)		
			S. litura Fabricius		

(continued)

Table 33.1 (continued)

Order	Family	Genus	Species	Damaging state	Nature of damage
	Scythrididae	*Eretmocera*	*E. impactella Wkr.*		
Diptera	Chloropidae	*Scoliophthalmus*	*S. micans Lamb.*	Shoot feeder	Larvae develop in the terminal shoots of reeds and damage them
	Ephydridae	*Psilopa*	*Psilopa* sp.	Leaf and shoot	Makes narrow serpentine mines, and patches of leaf tissue around the mines. Sometimes death from an unknown cause.
Hymenoptera	*Bethylidae*	*Parasierola*	*Parasierola* sp.	–	Parasitoid on the larvae of lepidopteron
		Epyris	*Epyris sp.*	–	
	Chalcididae	*Euchalcidia*	*Euchalcidia* sp.	–	–
	Chrysididae	*Omalus*	*O. timidus Nurse*	–	–
	Ichneumonide	*Enicospilus*	*Enicospilus* sp.	–	–
Hemiptera	Aphididae	*Myzus*	*M. persicae*	Root feeder	The nymph and adult both reduce the size and quality of the beet root by sucking the sap. They secrete a white waxy material on the roots, due to which wilting of the plant is observed and it shrinks in size. Aphids reduce the sugar content, juice purity, and root yield, and also transmit the viral diseases.
	Cicadellidae	*Circulifer*	*C. tenellus* Baker	Sucking pest	The nymph and adult both cause damages to the plant's tissue by sucking juices. Heavily infested plants become yellow, wilted, deformed, or stunted, and may eventually die.
		Empoasca	*Empoasca* sp.		
	Cydnidae	*Aethus*	*Aethus* sp.	–	–
	Lygaeidae	*Geocoris*	*Geocoris* sp.	–	–
	Membracidae	*Tricentrus*	*Tricentrus* sp.	–	–
	Nabidae	*Prostemma*	*Prostemma carduelis* Dohrn	–	–

	Pentatomidae	*Eysarcoris*	*Eysarcoris sp.*	–	–
		Nezara	*Nezara antennata* Scott	–	–
Orthoptera	Acrididae	*Atractomorpha*	*Atractomorpha crenulata* Fabr.	–	–
		Crenulate	*Crenulate* Fabr.	–	–
		Eyprepocnemis	*Eyprepocnem isplorans* Charp.	–	–

33.3 Economic Losses of Sugar Beet Due to Pests

Assessment of crop losses was investigated for many pests and diseases by several workers (Hull 1953; Hills et al. 1980; Campbell et al. 1998). There are several methods for estimating the crop losses by insect pests. In the direct method, crop loss is estimated through the relation between insect densities/damage symptoms and yield index (Walker 1991). It is the most precise method to estimate crop losses. Ocete et al. (1994) reported that the larvae of sugar beet weevil, *Lixusin canescens*, can cause up to 75% root weight loss in Iran and other countries like south of Ukraine, south east of Russia, Caucasia, Kazakhstan, Turkmenistan, and Turkey (Davatchi and Kheyri 1960). The limited data available indicate that arthropods may be destroying an estimated 18–20% of the annual crop production worldwide, estimated at the value of more than US$ 470 billion (Sharma et al. 2017). Yield losses in sugar beet due to plant pathogens and all group of pests accounted to be as much as an estimated, in general, 26% with and more than 80% without crop protection (Oerke and Dehne 2004). Jones et al. (1955) reported a significant reduction in yield due to the loss of half or more of the leaf area. Loss of stand was more serious than defoliation, but no significant effect on yield was noted until half or more of the plant had been destroyed. Defoliation tended to cause a decrease in plant size, while removal of plants resulted in increased size of remaining plants.

The loss of half the leaf area in the 4- or 8-leaf stage caused an average decrease in root yield of only 5%, and for complete defoliation, not more than 30%. Similarly, loss of half the initial plant population caused an average reduction in root yield of 10%, while loss of three quarters gave a yield slightly superior to that obtained by re-sowing (Anonymous, 1978–1979). Most pest attacks are light, and few cause more than 50% defoliation or 50% loss of stand. From such attacks a very good recovery can be expected. Roebuck (1932) described the effect of defoliation and destruction in sugar beet and mangold upon the yield. The root crops show losses of half of the leaf surface by the end of May and cause a crop loss of 17% and 25%, respectively (Anonymous 1980). Hanse et al. (2011) stated that despite crop protection measures, the calculated yield losses due to pests and diseases for the top growers were 30.2% and 13.1% and for average growers were 37.1% and 16.7% on sandy and clay soils, respectively. Therefore, pest and disease infestation levels partly explained the differences in sugar yield between the top and average growers analyzed.

Losses due to insect pests in Indian agriculture have also been estimated from time to time by several workers. Mehrotra et al. (1981) described the assessment of losses caused by defoliating insects under field conditions on Ramonskaya—06. Variations in the population of the defoliating caterpillars in the different blocks were created using insecticides. The populations of different species of defoliators were recorded from nine plants selected at random along the two diagonals in each plot. It was observed that the activities of *Plusia* sp. and *Spodoptera* sp. were more pronounced compared to the other defoliators.

At harvest, data on root yield, foliage weight and gross sugar, pest incidence, and intensity per hectare were recorded. Incidence and the intensity of damage based on the number of leaves defoliated and the extent of defoliation ranged between 3.53–56.18% and 5.0–67.25% of the lamina. The extent of losses in root yield and gross sugar per hectare varied from 2.20 to 9.65 tonnes and 1.07 to 2.66 tonnes, respectively. The extent of defoliation increased with the incidence of defoliators. The relationship between the two values was positive ($r = 0.978$). The maximum loss due to defoliators was estimated to be 9.65 t/ha of roots.

Mehrotra et al. (1981) reported the crop variety Ramonskaya creating variations in the population of the defoliating caterpillars; the crop was given insecticidal treatment at pre-determined intervals. Insecticide quinalphos, known for its efficacy against defoliators in general and Bihar hairy caterpillar *Diacrisia oblique* in particular, was sprayed according to the programme.

The population of different species of defoliators was recorded, and it was observed that the activity of *Plusia* sp. was more pronounced than that of the others. The harvested root yield, foliage weight, and gross sugar per hectare were recorded along with pest incidence and intensity values. Incidence and intensity of damage values based on the number of leaves defoliated and the amount of defoliation were observed to range between 4.0–62.0% and 6.6–72.0%, respectively. The extent of losses in root yield and gross sugar per hectare varied from 10.95 to 18.10 ton and from 2.33 to 3.74 ton, respectively.

It was observed that the intensity of defoliation increased with the increase in incidence. The relationship between two values, losses in yield, and sugar per hectare were positive. The maximum loss that had been recorded due to defoliation was approximately 18.14 ton per hectare.

33.3.1 Assessment of Losses by Artificial Defoliation

Assessment of losses by artificial defoliation had been done by Tewari et al. (1986). Defoliation was carried out at 25, 50, 75, and 100% once, twice, thrice, and on the basis of no defoliation, respectively. The defoliation obtained by three rounds and two rounds was 25% and 50%, respectively. In one round, the defoliation was 75%, but the root yield and quality were unaffected. Insect pests were not damaging to sugar beet to a certain extent in the studied fields. This can be ascribed to the insecticide treatment of the seeds in those areas where insect pests can cause yield losses (Heijbroek and Huijbregts 1995).

Mehrotra et al. (1981) carried out 25, 50, 75, and 100% defoliation once, twice, thrice, and on the basis of no defoliation, respectively. Root and foliage yield, length, girth, and pol per cent of the harvested plant were also observed. The result obtained was that beet yield, pol per cent in roots, and gross sugar (calculated) decreased significantly as per cent defoliation increased. However, 25% single as well as double defoliation and 50% single defoliation were as good as no defoliation in respect of the above characters.

33.4 Biology of the Armyworm and Cutworm Infesting Sugar Beet

33.4.1 Armyworm *Spodoptera litura* Hübner, 1808 (Lepidoptera: Noctuidae)

Spodoptera litura is a polyphagous pest (Holloway 1989) with plenteous of host, which has earned several names, such as taro caterpillar, cotton leafworm, rice cutworm, tobacco budworm, Indian leafworm, cluster caterpillar, tobacco caterpillar or tobacco leaf caterpillar, tobacco cutworm, and common cutworm. It is also known as armyworm because of its field-to-field migratory habit in groups (Nagoshi et al. 2011, 2012), destroying the verdant as they march and munch together. The larva feeds during the night and rests at daytime.

It is listed as a pest of quarantine significance by the European and Mediterranean Plant Protection Organization (EPPO), Caribbean Plant Protection Commission (CPPC), North American Plant Protection Organization (NAPPO), and Organismo Internacional Regional de Sanidad Agropecuaria (OIRSA). The European Food Safety Authority (EFSA) panel on Plant Health (PLHP) considers *S. litura* to be a potential quarantine pest of European Union (Bargard et al. 2019).

33.4.2 Distribution

S.litura is a major pest of tropical and temperate Asia, Australia, and Pacific islands (Feakin 1973; Kranz et al. 1977). Some localized occurrence is found in Russia (EPPO 2020), African countries, viz., Central African Republic (Cauquil et al. 1986), Ghana (Obeng-Ofori and Sackey 2003), and the Reunion (EPPO 2020). In the European countries it occurs in France (Cocquempot and Ramel 2008) and Portugal, while it has been eradicated from Germany (EPPO 2020) and Netherlands (NPPO 2013). In the United States it is found in Florida and Hawaii (EPPO 2020).

33.4.3 Habitat

As it is polyphagous in nature (Brown and Dewhurst 1975; Holloway 1989), it can invade a wide range of terrestrial ecosystems such as agricultural lands, greenhouse or glasshouse, forests, orchards, grasslands, riverbanks, and wetlands.

33.4.4 Host

Armyworm attacks 120 species of host crops and apart from sugar beet, it feeds on maize, rice, potato, sweet potato, capsicum, tomato, taro, melon, cotton, flax, jute, soybeans, tea, tobacco, sweet pepper, and ornamental plants.

33.4.5 Nature of Damage

Larva is the main damaging stage of this pest and it chiefly feeds on the leaves of the sugar beet plant. They are voracious feeders, and they can completely skeletonize the leaves. Apart from the leaves they also feed on the terminal portion of beet root. They march from field to field and feed gregariously on the leaves of the sugar beet, causing skeletonization or complete defoliation of the field. Consequently, the field wears a grazed-up appearance just like by cattle movement. It is a major pest of temperate sugar beet (Manoharan et al. 2010).

33.4.6 Biology of the Pest

33.4.6.1 Egg

The female lays a fluffy egg mass on the abaxial surface of the leaf and usually has a fecundity of 2000–2600 eggs (Ahamad et al. 2013; Shekhawat et al. 2018). The egg mass contains about 200–300 eggs in 3 to 4 layers (Hely et al. 1982; Hill 1983), guarded by the abdominal hairs of the female, and hatches in about 2 to 5 days in summer or 11–12 days in winter. Fand et al. (2015) observed a longer incubation period of 14 days at 15 °C. The eggs are orange-brown or pink in colour and the shape is spherical, or somewhat flattened.

33.4.6.2 Larva

Fand et al. (2015) reported larval development of 27 days at 20 °C. The larval developmental period is about 8–22 days. The pest passes through five to seven larval instars (Ranga Rao et al. 1989). The instars can be contemplated based on the head capsule width. The early instar larvae are translucent light green with a dark thorax and feed the veins and leaf ribs (Gupta et al. 2015). The later instars are dark green or greyish in colour and have a pattern of broad red, yellow, and green stripes on the sub-dorsal and lateral sides of the body along with wavy lines running down the back. The larvae are characterized by the presence of a conspicuous mid-dorsal line accompanied by a series of black and yellow dots along the sides of each abdominal segment. The later instars feed during night and hide in the cracks and crevices during the day.

33.4.6.3 Pupa

Pupation occurs in the soil, at 12.7–25.4 mm of depth in an earthen chamber made from soil and trash. The pupation period is about 5–8 days. Gupta et al. (2015) reported the pupal period to be of 12 days at 25 °C. The pupa is reddish-brown with two small spines at the tip of the abdomen.

33.4.6.4 Adult

The adult moths are rusty brown in colour, replete with eclectic scratch marking from wavy to straight, in the white or canary yellow background of the forewings, along with arbitrary black spots. The hindwings are whitish, with a darker

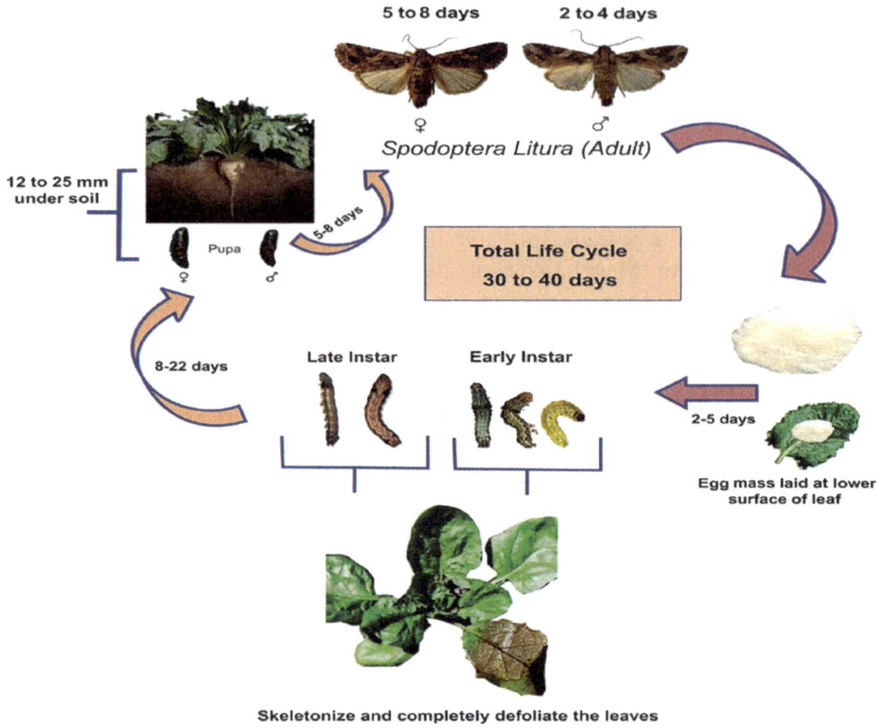

Fig. 33.1 Life cycle and nature of damage of *Spodoptera litura* on sugar beet

circumference and veins (Hamza and Norsyazwina 2019). Adult male longevity varies between 2 and 4 days and that of an adult female is about 5–8 days. Post first night of emergence, the heterosexuals copulate and oviposit (Ahmed et al. 1979). The total life cycle of the beet armyworm takes about 30–40 days. None of the stages of this pest was seen to have entered diapause (Miyashita 1971). This is subjected to variation based on the food sources and climatic conditions (Fig. 33.1).

33.5 Cutworm *Agrotis ipsilon* Hufnagel, 1766 (Lepidoptera: Noctuidae)

Agrotis ipsilon is a highly polyphagous pest and cosmopolitan in nature (Ram et al. 2001; Napiorkowska and Gawowska 2004). This pest is commonly known as black cutworm, potato cutworm, greasy cutworm, floodplain cutworm, ipsilon dart, dark sword-grass, or thread caterpillar. The moth is named cutworm, as the late instars cut the stem of the tender and young plants or seedlings near the ground. The species name "*ipsilon*" is based on the "Y"-shaped black marking on the forewings resembling the Greek letter "upsilon". It is a serious agricultural pest.

33.5.1 Distribution

Though its origin is ambiguous, it has a wide geographical distribution (Vendramim et al. 1982). It is an obnoxious pest ubiquitously present in the continents of Africa, Asia, Europe, North America, and South America (UK CAB International 1969). In India, it was reported from the Southern states such as Andhra Pradesh (Murthy et al. 1982), Karnataka (Mutalikdesai et al. 1973), and Tamil Nadu (Abraham et al. 1972); North-Eastern states such as Assam (Borah et al. 1982) and Tripura (Das and Ram 1988); Eastern states such as Odisha, West Bengal (Pramanik and Basu 1971), and Bihar (Das and Ram 1988); Western states such as Gujrat (Chari and Patel 1972) Rajasthan (Khan and Sharma 1971), and Maharashtra (Patil and Pokharkar 1979); Northern parts such as Haryana, Punjab (Gill 1987), Himachal Pradesh (Verma and Verma 2002), Jammu and Kashmir (CAB International 1969), Uttarakhand (Bisht et al. 2005), and Uttar Pradesh (Nag and Nath 1990); Central parts such as Madhya Pradesh (Meshram and Pathak 1990), and in the Union Territory of Delhi (Prasad et al. 1983; Paul et al. 2016).

33.5.2 Habitat

A. ipsilon has a terrestrial habitat and is an underground pest, which usually dotes on wet fields or soil that is flooded. Apparently, because of its fondness for the wet soil it is also known as overflow worm in the USA. Agricultural fields, pasturelands, grasslands, or any terrestrial ground rich in fauna are obliterated by this pest. They are also found in hilly areas with low to mid-elevation (Pathania 2010).

33.5.3 Host

This pest feeds voraciously on the root and shoot of smorgasbord of the host plant. Almost 100 plant species are attacked by *A. ipsilon* (Liu et al. 2009). Oil seeds such as groundnut, mustard, linseed, rapeseed, safflower, sunflower, and castor; vegetables such as potato, sweet potato, onion, beans, sugar beet, cabbage, okra, celery, asparagus, cauliflower, broccoli, Chinese cabbage, turnip, carrot, radish, and cucumber; fruits namely watermelon, citrus fruits, banana, apple, plum, peach, and grape; plantation crops like tea and coffee; spice crops including ginger, cloves, and black pepper; cereal crops such as barley, maize, wheat, and sorghum and fibre crops such as cotton and flax; legumes like cowpea, beans, lentil, peas, grams, and greenhouse plants such as strawberries are the host plants of black cut worm.

33.5.4 Nature of Damage

They cut off the plant near the soil surface and feed on the crown of the plant on the newest leaves and stems. The damage incurred by the pest depends on the stages of larva. Early instar (first and 2nd) usually feeds on the epidermis of the tender foliage and causes pin holes on them. Late instar (third to 7th) exhibits negative phototaxis; takes shelter in the cracks and clods at daytime, and during the night, feeds on the leaves or cuts the stem near to the ground to feed on the tender plant parts (Bhattacharyya et al. 2014). Occasionally, they store the cut stem in their shelter, to be fed later in the day. One larva can mangle several seedlings and indubitably, they are the most damaging to the crop (Showers 1997).

This obliteration causes reduced vigour and suppressed yield; partial cutting can cause wilting of the plant. Eventually, the pin holes or shot holes in the leaves, wilting, and toppling of the plant, are the characteristic symptoms of damage by the cutworm.

33.5.5 Biology of the Pest

The female lays a single egg, or a group of eggs latched onto the substrate. The female prefers dense verdant areas located in dewy or dampish, low-lying regions. The larva emerges in 3–6 days and migrates towards the soil to take shelter in the day and ascend to feed the plants at night. There are six to nine larval instars (Capinera 2019) but most commonly six to seven, and the total larval period varies between 25 to 35 days. Pupation occurs inside the soil below 2–10 cm and the pupal period is about 10–17 days. The adult longevity is about 6–10 days and preoviposition period is 7–10 days.

33.5.5.1 Egg

The egg is spherical in shape with a smooth plane base, ribbed (35–40 ribs) and sculptured, white in colour, which later turns brown. The height and width of the egg are 0.43–0.50 mm and 0.51–0.58, respectively. The female lays about 1900 eggs throughout its life history, clustered under the side of the leaves.

33.5.5.2 Larva

The larva is characterized by a light brownish dorsum, without bands, and the ventral surface is portrayed with a lighter shade. The cuticle of the larva is fraught with supernumerary black coarse granules. The head of the larva is brownish, replete with dark spots. Fourth to seventh instar onwards, the larva becomes negatively phototoxic for the leaves of the plant and takes shelter in the soil by burrowing, during the scorching hours of daylight. The larva exhibits cannibalism. The head widths of the larva from the first instar to fourth are nearly similar. Head capsule widths from one to eight instars are about 0.26–0.35, 0.45–0.53, 0.61–0.72, 0.90–1.60, 2.1–2.8, 3.2–3.5, 3.6–4.3, and 3.7–4.1 mm, respectively. The length of the first, second, third, fourth, fifth, sixth, seventh, and eighth instars is 3.5, 5.3–6.2, 7, 10, 20–30, 30–45, and 50 mm, respectively.

33.5.5.3 Pupa

The pupa is deep brown in colour, and it is found at a depth of 3–12 cm. The length of the pupa is 17–22 mm and width is 5–6 mm.

33.5.5.4 Adult

The remigium portion of the forewing is dark brown, while the distal portion is characterized by light-coloured bands and the presence of bean-shaped spots extending distally. The hind wings are relatively lighter in colour or greyish. The adult female lays eggs on the underside of the broad-leaved host plant (Fig. 33.2).

33.6 Management

33.6.1 Management of Armyworm

33.6.1.1 Cultural Control

As the larva is conspicuous, it can be handpicked and destroyed. Since the pupa of the beet armyworm is sub-terrestrial and found during April–May, deep ploughing of the field makes the overwintering pupae vulnerable to scorching sunlight and

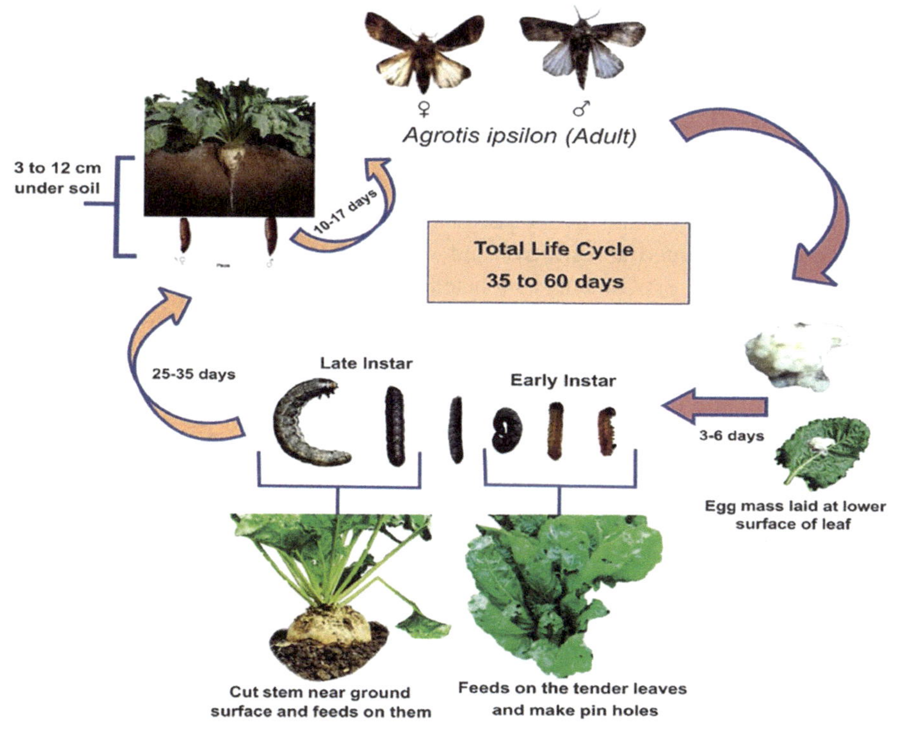

Fig. 33.2 Life cycle and nature of damage of *Agrotis ipsilon* on sugar beet

entomophagous birds. The alternate hosts may cajole the overwintering pests; hence their removal and destruction are indispensable.

33.6.1.2 Biological Control

A documentation of 100 parasitoids, 50 predators, and more than 12 entomopathogens for *S. litura* is presently available in different countries. The families of parasitoids that act against the genus *Spodoptera* are Braconidae, Eulophidae, Ichneumonidae, and Tachinidae (Rios-Velasco et al. 2011). The microorganisms infecting them are viruses belonging to the family Baculoviridae and bacteria, fungus, protozoa, and nematodes.

The most common parasitoids against *S. litura* are *Trichogramma australicum, Trichogramma chilonis, Chelonus helipae, Telenomus remus, Apanteles* sp., *Microplitis sp., Bracon brevicorni, Ichneumon* sp., *Brachymeria* sp., and *Diadegma argenteopilos*. The parasitoids are classified into egg, egg-larval, larval, larval-pupal, pupal, adult, and nymphal parasitoids, based on the stage that they parasitize.

Predators such as *Liposcelis sp., Chrysopa crassinervis, Chrysopa* sp., *Coccinella* sp., *Epilachna* sp., *Andrallus spinidens, Harpactor costalis, Rhinocoris squalis, Rhynocoris fuscipes, Polistes stigma,* and *Ropalidia* sp. are actively used in the biological management of the beet armyworm.

Among pathogens, fungus (*Aspergillus flavus, Beauveria* sp., *Beauveria bassiana*, and *Nomuraea rileyi*), viruses (Granulosis virus and Nuclear polyhedrosis virus), bacteria (*Bacillus cereus, Bacillus thuringinsis, Metarhizium anisopliae*, and *Micrococcus* sp.), nematodes (*Hexamermis* spp., *Neoplectana carpocapsae, Ovomermis albicans*, and *Pentatomimermis* spp.), and protozoans (*Nosema sp.*) are exploited to keep the population of this pest in check.

Shivankar et al. (2008) observed that the augmentative release of egg parasitoids, *T. chilonis* @ 50,000/ha, in conjugation with spraying of Azadiractin 3000 ppm. (5 mL/L) gave 89.71 and 89.39 per cent reduction of the larval population of *Spodoptera*, and found an effective and safe treatment to combat the sugar beet pests. In another experiment conducted by Park et al. (2001), the authors contemplated a 100% mortality in the second and third instar larvae of *S. litura* caused by *Steinernema carpocapsae, Heterorhabditis bacteriophora*, and *Steinernema longicaudum* after 20 h of inoculation. The fourth instar larva exhibited 0% mortality after 20 h of inoculation by the entomopathogenic nematodes. Sneh et al. (1983) observed that a collaborative effect of *Bacillus thuringiensis* and *Bracon hebtor* caused 70% mortality in *S. litura* larvae. Whitlock et al. (1990) found two strains of *Bacillus thuringiensis* (K-2074 and K-2178) very effective and consistent in controlling *S. litura* in laboratory assays. This bioassay revealed a low growth rate of larvae and intercepts in successful pupation.

The combined effect of NPV (4×10^9 POB ml^{-1}) and emamectin benzoate (0.1 ppm) caused 100% mortality (Yasin et al. 2020). The same result was obtained by El-Helay et al. (2020). Ayyub et al. (2019) tested different concentrations of the NPV isolate of *S. litura* (V-*Splt*NPV) against second, third, and fourth instar larvae and recorded the LC_{50} value. 88.08% mortality rate was observed in the early instar, while 65.52% mortality was observed in the fourth instar larva. The authors concluded that the mortality of the larva is directly proportional to the LC_{50} value. Hernández et al. (1989) obtained 78–100% mortality in *S.litura* by releasing 5000 *Telenomus remus* in maize during three consecutive weeks; Linares (1998) obtained a successful control of *S. litura* on 87.5 ha farm land by releasing 4000–6000 wasps per hectare of maize.

33.6.1.3 Chemical Control

The practitioners of modern agriculture and the savant of insect toxicology, unambiguously, know the importance of synthetic insecticides. Although quick production boom is just a tinsel aspect of its use, it is an essential enemy that requires reconciliation with other such components of IPM for a blockbuster pest management. For controlling third and fourth instar larvae, application of Indoxacarb 14.5 SC @75 g per ha or Flubendiamide 480 SC @ 48–60 g a.i. per ha is recommended. Shivankar et al. (2008) obtained a percentage population reduction of 97.27%, 80.24%, and 94.17% by spraying of Quinolphos 25 EC @ 0.05%, Imidacloprid 17.8 SL @ 0.008%, and Chlorpyriphos 20 EC @ 0.1%, respectively. Venkataiah et al. (2015) observed a higher rate of larval reduction in this pest, on groundnut crop, with the spray application of Chlorantraniliprole 20 SC @ 125 mL/ha and Novaluron 10 EC @ 500 mL/ha.

Spraying of Flubendiamide 480 SC @ 150 mL/ha resulted in the lowest mean number of larvae, and percent defoliation was only 9.67 per 10 plants. A bio-efficacy experiment conducted by Waykule et al. (2020), revealed that Chlorantraniliprole @ 0.0185% was most efficacious against *S.litura* along with Indoxacarb @ 0.01% and Emamectin Benzoate @ 0.002%. Shaila and Rao (2013) undertook an application trial of different concentrations of abamectin, emamectin benzoate, novaluron, lufenuron, mancozeb, carbendazim, and chlorothalonil on the larva of *S. litura* and observed that the respective concentrations of 350 ppm, 180 ppm, 600 ppm, and 900 ppm of abamectin, emamectin benzoate, novaluron, and lufenuron were effective in controlling the pest. In contradiction to insecticides, fungicides such as Mancozeb, Carbendazim, and Cholorthalonil proved to be non-toxic to the pest.

33.6.2 Management of cut Worm

33.6.2.1 Cultural Control

Similar to *S.litura*, the pupa of *A.ipsilon* is found in earthen cocoon, and deep summer ploughing helps in digging them out and in the collection and destruction of the larva. Application of well-decomposed organic manure helps in reducing the weeds that could be an alternate host for the pest as well as suppress the pest and its inoculum. Destruction of weed 10–14 days before planting crop on the field and intercropping with wheat or linseed or mustard greatly reduces infestation. Adoption of crop rotation, growing marigold as trap crop on bunds, avoiding planting of sugar beet on agricultural field already infested with cutworm, and flooding of the fallow field for a few days help to kill the caterpillars of cutworm in the soil.

33.6.2.2 Biological Control

A. ipsilon is cosmopolitan in nature and is attacked by a gamut of natural enemies, among which *Telenomus nawaii, Telenomus remus, Trichogramma dendrolimi*, and *Trichogramma evanescens* parasitize its eggs; *Apanteles bourquini, Cotesia marginiventris, Microplits feltiae*, and *Campolestis flavicincta* parasitize the larva, and *Netelia fuscicornis* and *Archytas marmoratus* parasitize the pupa of this pest. There are numerous predators present that run amok in the natural ecosystem to forfend the population of pest from reaching EIL. Amid many such predators, *Abacidus permundus*, *Carabidae*, *Chrysoperla carnea*, *Lapidura riparia*, *Stelopolybia pallipes*, *Zelus tetracanthus* are the most common.

The pathogens that actively retard the development and reproduction, and intimidate the survival of this pest are bacteria (*Bacillus thuringiensis* sp.), fungus (*Beauveria bassiana*, *Metarhizium anisopliae*, *Nomuraea rileyi*), virus (Nucleopolyhedrosis virus (*Agip*MNPV) and Granulosis virus), and nematodes (*Hexamermis arvalis, Noctuidonema guyanense*).

Conservation and augmentative release of parasitoids such as *Braconids, Microgaster* sp., *Bracon kitcheneri, Fileanta ruficanada*, and parasitic wasps like Cotesia ruficrus are very effective in controlling the black cutworm (Joshi et al. 2020). Arrangement of bird perches and birdbaths on the field would attract the predatory birds, which in turn controls the population of this pest. Baculoviruses are very effective in controlling the larvae of this pest (Goodman et al. 2001). Hussaini (2003) suggested that endemic nematodes and alginate formulation of entomopathogenic nematodes proved to be panacea against *A. ipsilon* larva. Yuskel and Canhilal (2018), in a laboratory experiment, checked the pathogenicity of four isolates of EPNs against the fourth instar larva of *A. ipsilon*; the authors found out that the highest mortality (90%) in the larvae was obtained by *Steinernema carpocapse* (E76-S isolate) at a concentration of 100IJs/larva/Petri dish. *A.ipsilon* nucleopolyhedrovirus (*Agip*MNPV) provided 76–86% mortality of the early instar larvae as observed by Bixby-Brosi (2011).

Along with synthetic insecticides, natural or botanical insecticides such as neem products (Viji and Bhagat 2001), powdered *Nerium oleander* leaves in the form of pills, methanol extract of *Melia azedarach* fruits (Schmidt et al. 1997), extract of *Bassia muricata* against first instar larva of *A.ipsilon* (El-Sayed et al. 1998), extract from the leaves of Lantana, Parthenium, Ipomoea, and Hyptis (Ramesh-Chandra 2004), and root extract of *Rumex nepalensis* (Thakur 1997) are utilized in controlling this pest.

33.6.2.3 Chemical Control

Many researchers have undertaken trials with synthetic insecticides in successfully controlling A. ipsilon. Tripathi et al. (2003) suggested that diazinon 20 EC, Quinalphos 25 EC, Fenitrothion 50 EC, Deltamethrin 2.8 EC, and Malathion 5% dust work effectively against the black cutworm. According to Mishra (2002), Chlorpyriphos, Quinolphos, Cypermethrin, Phosalone, and Carbaryl give good results against *A. ipsilon*. Andersch and Schwarz (2003) proposed the use of a new synthetic insecticide, Clothianidin, for seed treatment against *A. ipsilon*.

Broadcasting or surface application of Bt. Mixed bait (*Bacillus thuringiensis* + 1 kg wheat barn) @ 10 kg bait/ha 1 week before planting of crop is efficacious. In case of severe infestation, insecticides such as Polytrin C 44 EC @ 1000 mL/ha diluted in 500–600 L of water is recommended. In an experiment, Bhattacharyya et al. (2014) claimed the effectiveness of soil application of Imidacloprid 200 SL@ 48 g a.i. ha^{-1} +NSKE 5 mL L^{-1} +Gram bait in controlling black cutworm. Drenching of the soil with Dursban 10% G@ 20 kg/ha is usually advocated before planting to prevent damage by the pest.

33.7 Future Prospects

The milestone researches and achievements attained in this crop under Indian climatic conditions are not enough as the population of India is on the rise and is projected to reach a benchmark of 1.4 billion by 2025, and will require 49 million tonnes of sweeteners to meet the consumption demand (National Commission on Agriculture 1976). According to the report of "OECD-FAO Agricultural Outlook 2019–2028", the Asian and African countries will have a higher rate of sugar consumption with the trajectory of 60% and 25%, respectively. For the future demands, India needs systematic and structured sugar-processing industries, well equipped with modern technology, and the farmers should be provided with an efficient plant protection technique to manage the pests and diseases of sugar beet. This is an opportunity and economical leverage for sugar industries to take an initiative to supply the domestic consumption and contribute to international trade and commerce.

33.8 Conclusion

Unequivocally, sugar beet crop surpasses sugar cane in all aspects. In 2013, the contribution of sugar beet to the world's sugar production was 21.8%, which mushroomed to 40% by the end of 2017, as India, China, Brazil, Thailand, and the European Union started sugar production from sugar beet. The contribution from these countries was 20%, 11%, 11%, 9%, and 5%, respectively (OECD-FAO 2018). By the end of 2013, the number of sugar beet-producing countries increased to 57 (Kumar and Pathak 2013), and the contemporary period is marked by the presence of 110 countries processing sugar either from sugarcane or from sugar beet, and 8 countries producing sugar from both (ISO 2019). This incendiary increase in the production of sugar beet revived the moribund production and is the proof of the rising popularity of this crop among the farmers. This drastic shift in the cultivation of sugar beet occurred due to the realization of the economic benefits of this crop.

Amid benefits, pest management is one of the hurdles in its cultivation. The insect pests cause nearly 26–30 percent loss of sugar beet. Among the pests, *Spodoptera litura* alone inflicts about 80–100 percent damage to the crop, and beet armyworm and sugar beet cutworm both cause huge economic losses (Meagher et al. 2008). These pests have developed resistance against many insecticides such as endosulfan, cypermethrin, fenvalerate, and monocrotophos (Radhika and Subbaratnam 2006). So, to counteract this situation, an indomitable IPM module and plant protection technique has to be developed, which must include biopesticides (neem products), natural enemies, green insecticides, novel insecticides, and their synergistic combination along with effective cultural practices by adhering to an effective forecast model (i.e. statistical or mechanistic models) for major pests of sugar beet.

References

Abraham EV, Padmanaban MD, Mohandoss A (1972) Control of potato cutworms. Indian J Agric Sci 42(5):418–420

Ahamad M, Ghaffar A, Rafiq M (2013) Host plants of leaf worm *Spodoptera litura* (Fabricius) (Lepidoptera:Noctuidae) in Pakistan. Asian J Agric Biol

Ahmed AM, Etman AM, Hooper HS (1979) Development and reproductive biology of *Spodoptera litura* (F.) (Lepidoptera: Noctuidae). Australian. J Entomol 18(4):363–372

Andersch W, Schwarz M (2003) Clothianidin seed treatment (Poncho Reg.): the new technology for control of corn rootworms and secondary pests in US-corn production. Pflanzenschutz Nachrichten Bayer 56:147–172

Anonymous (1967–1968) Annual report, Indian Institute of Sugarcane Research, Lucknow, India, 76. Star Press, Gautam BuddhaMarg, Lucknow

Anonymous (1978–1979) Annual Report. Indian institute of sugarcane research, Lucknow

Anonymous (1980) Annual report, Indian Institute of Sugarcane Research, Lucknow, India, 1–165. Star Press, Gautam Buddha Marg, Lucknow

Avasthy PN, Shrivastava TN (1972) Pest problem in sugarbeet cultivation. Indian Farmers Digest 10(5):27–29

Ayyub MB, Nawaz A, Arif MJ, Luqman A (2019) Individual and combined impact of nuclear polyhedrosis virus and Spinosad to control the tropical armyworm, *Spodoptera litura* (Fabricius) (Lepidoptera:Noctuidae), in cotton in Pakistan. Egypt J Biol Pest Control. https://doi.org/10.1186/s41938-019-0170-4

Bargard C, Dehnen-Schmutz K, Serio FD, Gonthier P, Jacques MA, Jaques Miret JA, Justsen AF, Mafnusson CS, Milonas P, Navas-Cortes JA, Stephen P, Potting R, Reignault PL, Thulke HH, Van der Werf W, Civera AV, Yuen J, Zappala L, Malumphy C, Czwinenczek E, MacLeod A (2019) Pest categorisation of *Spodoptera litura*. EFSA J 17(7):e0565. https://doi.org/10.2903/j.efsa.2019.5765

Bhattacharyya B, Pujari D, Bhuyan U, Baruah AALH (2014) Management of potato cutworm, *Agrotis ipsilon* (Hufnagel) in Assam. Pestic Res J 26:82–85

Bichsel SE, Cleary MF, Olson RF, Dak N (1991) Mineral enriched sugar beet. US5,043,171 American Crystal Sugar Company, 8 June 1990. Google patents. https://patents.google.com/patent/US5043171A/en

Bisht RS, Tripathi DM, Mishra PN (2005) Seasonal abundance of cutworms (Agrotis spp: Noctuidae: Lepidoptera) caught at light trap in Garhwal Himalaya. Indian J Entomol 67(3): 247–251

Bixby-Brosi AJ (2011) Biological control of the black cutworm, *Agrotis ipsilon* (Lepidoptera: Noctuidae), and endophyte mediated tritrophic interactions in turfgrass. University of Kentucky Doctoral Dissertations. 183. https://uknowledge.uky.edu/gradschool_diss/183

Borah B, Sharifullah M, Borah DC (1982) Efficacy of some of the granular insecticides and malathion dust for the management of soil pests of potato. J Soil Biol Ecol 2(1):47–51

Brown ES, Dewhurst CF (1975) The genus Spodoptera (Lepidoptera, Noctuidae) in Africa and the near east. Bull Entomol Res 65(2):221–262
Busch JW, Phelan PL (1999) Mixture models of soybean growth and herbivore performance in response to nitrogen-Sulphur phosphorous nutrient interactions. Ecol Entomol 24(2):132–145
CAB International (1969) *Agrotis ipsilon.* [distribution map]. In Distribution maps of plant pests, Wallington, UK
Campbell LG, Anderson AW, Dregseth R, Smith LJ (1998) Association between sugar beet root yield and sugar beet root maggot (Diptera: Otitidae) damage. J Econ Entomol 91(2):522–527
Capinera JL (2019) Black cutworm, *Agrotis ipsilon* (Hufnagel) (insects: Lepidoptera: Noctuidae). EENY-395, Florida
Cauquil J, Girardot B, Vincens P (1986) Cotton pest incidence in the Central African Republic: defining the control means. Coton et Fibres Tropicales 41(1):5–19
Chapman RF (2000) The insects: structure and function. University Press, New York
Chari MS, Patel NG (1972) Record of a new host plant of *Agrotis ipsilon* Rott. Indian J Entomol 34(2):175
Cocquempot C, Ramel JM (2008) Is cotton leafworm becoming sedentary in France? (La noctuelle africaine du coton en voie de sédentarisation en France?). PHM Revue Horticole 506:33–36
Cooke DA (1993) Pests. In: Cooke DA, Scott RK (eds) The sugar beet crop. Chapman & Hall Publ., pp 428–465
Das BB, Ram G (1988) Incidence, damage and carry-over of cutworm (*Agrotis ipsilon*) attacking potato (*Solanum tuberosum*) crop in Bihar. Indian J Agric Sci 58(8):650–651
Davatchi A, Kheyri M (1960) The most important pest of sugar beet and their control. The Plan Organization, Tehran, Iran
El-Helay AA, Sayed AA, Bendary HM (2020) Impact of emamectin benzoate on nucleopolyhedrosis virus infectivity of Spodoptera litoralis (Boisd.) (Lepidoptera: Noctuidae). Egypt J Biol Pest Control 30:111. https://doi.org/10.1186/s41938-020-00314-0
El-Sayed NH, Mogahed MI, Haron AA, Mabry TJ (1998) Flavonoids and other constituents from Bassia muricata and their insecticidal activities. Revista Latinoamericana de Química 26(3): 81–85
EPPO (2020) EPPO Global database. In: EPPO Global database. EPPO, Paris, France. https://gd.eppo.int/
Fand BB, Sul NT, Bal SK, Minhas PS (2015) Temperature impacts the development and survival of common cutworm (*Spodoptera litura*): simulation and visualization of potential population growth in India under warmer temperatures through life cycle modelling and spatial mapping. PLoS One 10:e0124682
FAO (2009) Sugar beet/white sugar, Agribusiness handbook pp 1–55
FAOSTAT (2019) Food andAgriculture Organization of the United Nations, production: Crops http://faostat.fao.org
Feakin SD (1973) Pest control in groundnuts, PANS manual no 2, 3rd edn. UK Centre for Overseas Pest Research, London, p 197
Gill KS (1987) Insect pests of linseed. In: Linseed. Indian Council of Agricultural Research (ICAR), New Delhi, India, pp 342–355
Goodman CL, El-Sayed GN, McIntosh AH, Grasela JJ, Stiles B (2001) Establishment and characterization of insect cell lines from 10 lepidopteran species. In vitro Cell Dev Biol Anim 37(6):374–379
Gupta M, Tara JS, Sharma S, Bala A (2015) Biology and morphometry of *Spodoptera litura* Fabricus, a serious defoliator of mango (Mangifera indica) in Jammu region (J & K). Munis Entomol Zool 10:215–221
Hamza AA, Norsyazwina M (2019) Crops damaged by armyworm (Spodoptera litura F.) in Malaysia and control tactics. Paper presented at Meleka International Conference on Social Sciences, Science and Technology (MIC3ST), Universiti Teknologi MARA, 7–8 August 2019
Hanse B, Schneider JHM, Termorshuizen AJ, Varrelmann M (2011) Pests and diseases contribute to sugar beet yield difference between op and averagely managed farms. Crop Prot 30:671–678

Heijbroek W, Huijbregts AWM (1995) Fungicides and insecticides applied to pelleted sugar-beet seeds. 3. Control of insects in soil. Crop Prot 14:367–373

Hely PC, Pasfield G, Gellatley JG (1982) Insect pests of fruit and vegetables in NSW. Inkata Press, Sydney

Hernández D, Ferrer F, Linares B (1989) Introducción de *Telenomus remus* Nixon (Hym.: Scelionidae) para controlar *Spodoptera frugiperda* (Lep.: Noctuidae) en Yaritagua-Venezuela. Agronomía Trop 39:199–205

Hill DS (1983) Agricultural insect pests of the tropics and their control, 2nd edn. Cambridge University Press, London, p 746

Hills FJ, Chiarappa L, Geng S (1980) Powdery mildew of sugar beet: disease and crop loss assessment. Phytopathology 70(7):680–682

Holloway JD (1989) The moths of Borneo: family Noctuidae, trifine subfamilies: Noctuidae, Heliothinae, Hadeninae, Acronictinae, Amphipyrinae, Agaristinae. Malayan Nat J 42:57–226

Hull R (1953) Assessments of losses in sugar beet due to virus yellows in Great Britain, 1942-52. Plant Pathol 2(2):39–43

Hussaini SS (2003) Progress of research work on entomopathogenic nematodes in India. In: Hussaini SS, Rabindra RJ, Nagesh M (eds) Current status of research entomopathogenic nematodes in India. PdbC, Publication, Bangalore, India, pp 27–26

ISO (2019) About sugar. http://www.isosugar.org. Accessed 23 Oct 2021

Joern A, Provin T, Behmer ST (2012) Not just the usual suspects: insect herbivore populations and communities are associated with multiple plant nutrients. Ecology 93:1002–1015. https://doi.org/10.1890/11-1142.1

Jones FGW, Dunning RA, Humphrifs KP (1955) The effects of defoliation and loss of stand upon yield of sugar beet. Ann Appl Biol 43(1):63–70

Joshi MK, Rana A, Varadharasu PR, Kaushal S, Inamadar AG, Verma KS, Chandel RS (2020) The potency of chemical insecticides in management of cutworm, Agrotis ipsilon Hufnagel (Noctuidae: Lepidoptera): a review. J Entomol Zool Stud 8:307–311

Kalra AN, Srivastava RC (1964) Biology of Apanteles flavipes cam. (Braconidae: hymenoptera), a parasite of moth borers of sugar-cane, under micro-ecological conditions. Proc. all India conf. sugar cane res dev workers, Vol 5, pp 605–609

Khan RM, Sharma SK (1971) Some new pest records on sugarbeet in India. Indian J Entomol 33(1): 105–106

Kranz J, Schumutterer H, Koch W (1977) Diseases pests and weeds in tropical crops. Hamburg, Verlag Paul Parley, Berlin

Kumar R, Pathak AD (2013) Recent trend of sugarbeet in world. Paper presented at the IISR-industry interface on research and development initiatives for sugarbeet in India, sugarbeet breeding outpost of IISR IVRI campus, Mukteswar, Nainital. 28–29 May 2013

Linares B (1998) Farm family rearing of egg parasites in Venezuela. Biocontrol News Information 19:76N

Liu X, XianQuan X, Ge Y, Cheng SW (2009) Studies on the main species and occurrence characteristics of tobacco pests in field growth phase in Sichuan province. Southwest China J Agric Sci 22(1):87–91

Manoharan T, Pathma J, Preetha G (2010) Seasonal incidence of sugarbeet pests natural enemies. Indian J Entomol 72(1):36–41

Marzo C, Díaz AB, Caro I, Blandino A (2019) Status and perspectives in Biethanol production form sugar beet. In: Ray RC, Ramchandran S (eds) Bioethanol production form food crops sustainable sources, interventions, and challenges. Elsevier, Amsterdam

Meagher RL, Brambila J, Hung E (2008) Monitoring for exotic Spodoptera species (Lepidoptera: Noctuidae) in Florida. Fla Entomol 91:517–522

Mehrotra AK, Tiwari RK, Avasthy PN (1981) New record of sugar beet seed pests. Sci Cult 47(12): 434

Meshram PB, Pathak SC, Jamaluddin (1990) Population dynamics and seasonal abundance of some forest insect pests (nursery stage) through light trap. Indian Forester 116(6):494–503

Mishra DN (2002) Chemical control of cut worm, *Agrotis ipsilon* on potato in mid hill conditions. Ann Plant Prot Sci 10(1):151–153

Miyashita K (1971) Effects of constant and alternating temperatures on the development of Spodoptera litura F. (Lepidoptera: Noctuidae). Appl Entomol Zool 6:105–111

Motiwale MP, Chauhan RS, Agnihotri VP (1991) Sugarbeet cultivation, bulletin no. 32. Indian Institute of Sugarcane Research, Lucknow

Murthy KSRK, Lakshminarayana K, Rao BHK (1982) Dissipation of certain commonly used insecticides from cauliflower (*Brassica oleracpa* L. var. capitata). Pestology 6(8):21–26

Mutalikdesai KS, Panchabhavi KS, Malipatil MB, Bankapur VM, Thimmiah G (1973) Effect of different levels of fertilizers and spacing on the extent of damage caused to potato tubers by cutworms. Mysore J Agric Sci 7(1):38–40

Nag A, Nath P (1990) Spatial distribution behaviour of cutworm, Agrotis ipsilon (Hufn.) (Lepidoptera: Noctuidae) in gram, Cicer arietinum L. (Leguminosae) field. Arq Biol Tecnol 33(2): 457–468

Nagoshi RN, Brambila J, Meagher RL (2011) Use of DNA barcodes to identify invasive armyworm *Spodoptera* species in Florida. J Insect Sci 11:154

Nagoshi RN, Meagher RL, Hay-Roe M (2012) Inferring the annual migration patterns of fall armyworm (Lepidoptera: Noctuidae) in the United States from mitochondrial haplotypes. Ecol Evol 2:1458–1467

Napiorkowska KJ, Gawowska J (2004) Increase of harmfulness of caterpillars (Hadeninae and Noctuinae, Lepidoptera: Noctuidae) on cabbage and other cole crops. Progr Plant Prot 44:978–980

Nathan S, Kalaivani K (2005) Efficacy of nucleopolyhedrovirus and azadirachtin on Spodoptera litura Fabricius (Lepidoptera: Noctuidae). Biol Control 34:93–98

National Commission on Agriculture-Report-India (1976) Government of India. https://indianculture.gov.in/report-national-commission-agriculture-1976-5. pp. 1–751

NPPO (2013) National Plant Protection Organization. http://www.ppqs.gov.in

Obeng-Ofori D, Sackey J (2003) Field evaluation of non-synthetic insecticides for the management of insect pests of okra *Abelmoschus esculentus* (L.) Moench in Ghana. Sinet. Ethiopian J Sci 26(2):145–150

Ocete R, Ocete ME, Perez-Izquierdo MA, Rubio IM (1994) Approximation to the phenology of Lixus junci Boh. (Col.: Curculionidae) in La Rioja Alta: estimate of the damage it causes. Boln San Veg Plagas 20:611–616

OECD/FAO (2018) OECD-FAO agricultural outlook 2018–2027, Rome, Italy

Oerke EC, Dehne HW (2004) Safeguarding production- losses in major crops and role of crop protection. Crop Prot 23:275–285

Park SH, Ys Y, Park JS, Choo HY, Do Soon B, Nam MH (2001) Biological control of tobacco cutworm *Spodoptera litura* Fabricius with entomopathogenic nematodes. Biotechnol Bioprocess Eng 6:139–143

Pathak AD, Srivastava S (2013) Six decades of crop improvement in sugarcane and sugarbeet: an overview. In: Srivastava TK (ed) Souvenir six decades of research at IISR towards excellence. IISR, Lucknow, pp 5–16

Pathania V (2010) Distribution of biology of cutworms infesting potato in Himachal Pradesh. Dissertation, CSK Himachal Pradesh Krishi Vishwavidyalaya, Palampur

Patil SP, Pokharkar RN (1979) Some new records of insect pests infesting cruciferous vegetable crops in Maharashtra state. J Maharashtra Agric Univ 4(2):222–223

Patil AS, Salunkhe AN, Pawar BH, Ghodke PV, Shivankar SB, Zende NB, Shewate SR, Patil DJ (2007) Sugar beet cultivation in tropical India—a new experience. In: Proc 68 Annual Conv, STAI, New Delhi, 22–24 August 2007

Paul M, Das SK, Singh R, Shashank PR (2016) Moth (Lepidoptera: Heterocera) Fauna of Delhi with notes on their roles as potential agricultural pests. J Entomol Zool Stud 4(2):435–438

Pramanik LM, Basu AC (1971) Record of *Agrotis ipsilon* Rottenburg and *Gryllus mitratus* Burmeister causing serious damage to cotton. Indian J Entomol 33(3):362–363

Prasad D, Singh KM, Katiyar RN (1983) Succession of insect pests in early maturing high yielding variety of pea, Pisum sativum Linn. Indian J Entomol 45(4):451–455

Radhika P, Subbaratnam GV (2006) Insecticide resistance management in cotton-Indian scenario-a review. Agric Rev 27(3):157–169

Ram G, Misra SS, Dhamayanthi KPM (2001) Relative susceptibility of advanced hybrids and promising cultivars of potato, Solanum tuberosum L. to greasy cutworm, Agrotis ipsilon (Hufnagel) in north-eastern plains. J Entomol Res 25:183–187

Ramesh-Chandra (2004) Status of medicinal plants with respect to infestation of insect pests in and around Chitrakoot, district-Satna (M.P.). Flora Fauna Jhansi 10(2):88–92

Ranga Rao GV, Wightman JA, Ranga Rao DV (1989) Threshold temperatures and thermal requirements for the development of Spodoptera litura. Environ Entomol 18:548–551

Rios-Velasco C, Gallegos-Morales G, Cambero-Campos J, Cerma-Chavez E, Del Rincon-Castro M, Valenzuela-Garcia R (2011) Natural enemies of the fall armyworm *Spodoptera frugiperda* (Lepidoptera: Noctuidae) in Coahuila, México. Florida Entomol 94(3):723–726

Roebuck A (1932) Measurements in nature: or the toll of insect and other animal pests on the farms of Lindsey during 1932. Presidential address to the Lincolnshire Naturalists' Union. November 1932

Santeshwari, Misra V, Mall AK, Kumar D (2020) Problems and integrated pest management strategy for *Spodoptera litura* in sugar beet in India. J Exp Zool India 23(2):1887–1890

Santeshwari, Shyamrao ID, Mishra V, Mall A (2021) Bio-intensive integrated pest management of Spodoptera litura in sugarbeet. In: Kumar A, Shyamrao ID, Tamang S (eds) Biointensive IPM for commercial crops. Agrobios, Jodhpur, pp 99–109

Schmidt GH, Ahmed AAI, Breuer M (1997) Effect of Melia azedarach extract on larval development and reproduction parameters of Spodoptera littoralis (Boisd.) and Agrotis ipsilon (Hufn.) (Lep Noctuidae) Anzeiger fuumlautr Schaumlautdlingskunde. Pflanzenschutz, Umweltschutz 70(1):4–12

Shaila O, Rao SRK (2013) Efficacy of Avermectins, chitin synthesis inhibitor and fungicides against *Spodoptera litura* and *Aspergillus flavus*. Central European Journal of Experimental Biology 2(4):1–6. http://scholarsresearchlibrary.com/archive.html

Sharma S, Kooner R, Arora R (2017) Insect pests and crop losses. In: Arora R, Sandhu S (eds) Breeding insect resistant crops for sustainable agriculture. Springer Nature Singapore Pte Ltd., pp 45–66. https://doi.org/10.1007/978-981-10-6056-4_2

Shekhawat SS, Shafiq Ansari M, Basri M (2018) Effect of host plants on life table parameters of Spodoptera litura. Int J Pure Appl Biosci 6:324–332

Shivankar SB, Magar SB, Shinde VD, Yadav RG, Patil AS (2008) Field bio-efficacy of chemical, botanical, and bio-pesticides against *Spodoptera litura* fab. In sugarbeet. Ann Plant Prot Sci 16(2):312–315

Showers WB (1997) Migratory ecology of the black cutworm. Annu Rev Entomol 42:393–425

Shrivastava AK, Sawnani A, Shukla, SP, Solomon S (2013) Unique features of sugarbeet and its comparison with sugarcane. Paper presented at the IISR-industry interface on research and development initiatives for sugar beet, Sugarbeet Breeding Outpost of IISR IVRI Campus, Mukteswar, Nainital. 28–29 May 2013

Singh KN, Sachan GC, Chhibber RC (1980) Effects of some insecticides on sugarbeet defoliation by lepidopterous larvae. Indian Sugar:157–160

Sneh B, Schuster S, Gross S (1983) Improvement of the insecticidal activity of *Bacillus thuringiensis var. entomocidus* on larvae of *Spodoptera littoralis* (Lepidoptera, Noctuidae) by addition of chitinolytic bacteria, a phagostimulant and a UV-protectant. Z Angew Entomol 96:77–83

Tewari RK, Mehrotra AK, Bajpai PK, Avasthy PN (1986) Assessment of losses caused by defoliating pests to sugarbeet crop (Beta vulgaris Linn.). Co-operative Sugar 17(12):943–945

Thakur SS (1997) Rumex nepalensis—a biopesticide against black cutworm, Agrotis ypsilon (Rottenburg). Insect Environ 3(2):468–473

Tripathi DM, Bisht RS, Mishra PN (2003) Bio efficacy of some synthetic insecticides and bio pesticides against black cutworm *Agrotis ipsilon* infesting potato (*Solanum tuberosum*) in Garhwal Himalaya. Indian J Entomol 65(4):468–473

UK CAB International (1969) Agrotis ipsilon. [distribution map]. In: Distribution maps of plant pests. CAB International. Map 261, Wallingford, UK. https://doi.org/10.1079/DMPP/20056600261

Vendramim JD, Ferraz MCVD, Parra JRP (1982) Biologia comparada de Agrotis subterranea (F) em meios natural e artificial 74(1):76–80

Venette RC, Davis EE, Zaspel J, Heisler LM (2003) Mini risk assessment old world bollworm, Helicoverpa armigera Hübner (Lepidoptera: Noctuidae). Cooperative Agricultural Pest Survey, Animal and Plant Health Inspection Service, US Department of Agriculture

Venkataiah M, Anil Kumar B, Chauhan S (2015) Efficacy of newer insecticides against *Spodoptera litura* in groundnut (*Arachis hypogea* L.). J Oilseeds Res 32(2):152–154

Verma KS, Verma AK (2002) Cutworm species associated with different crops in Himachal Pradesh. Insect Environ 8(1):23

Viji CP, Bhagat RM (2001) Bioefficacy of some plant products, synthetic insecticides and entomopathogenic fungi against black cutworm, *Agrotis ipsilon* larvae on maize. Indian J Entomol 63(1):26–32

Walker PT (1991) Quantifying the relationship between insect populations, damage, yield and economic thresholds. In: Teng PS (ed) Crop loss phytopathological society, St. Paul, MN.

Waykule PK, Mutkule DS, Jadhav AS, Dhormare AP, Bankar DR (2020) Bio-efficacy of different insecticides against *Spodoptera litura* on groundnut. Int J Curr Microbiol App Sci 9(12): 1697–1708

Wedeking R, Mahlein AK, Steiner U, Oerke EC, Goldbach HE, Wimmer MA (2017) Osmotic adjustment of young sugar beets (*Beta vulgaris*) under progressive drought stress and subsequent rewatering assessed by metabolite analysis and infrared thermography. Funct Plant Biol 44:119–133

Whitlock VH, Lo MC, Kuo MH, Soong TS (1990) Two new isolates of *bacillus thuringiensis* pathogenic to *Spodoptera litura*. J Invertebr Pathol 58:33–39

Yang L, Ma C, Wang L, Chen S, Li H (2012) Salt stress induced proteome and transcriptome changes in sugar beet monosomic addition line M14. J Plant Physiol 169(9):839–850. https://doi.org/10.1016/j.jplph.2012.01.023

Yasin M, Quazi MS, Wakil W, Qayyum MA (2020) Evaluation of nuclear polyhedrosis virus (NPV) and emamectin benzoate against *Spodoptera litura* (F.) (Lepidoptera:Noctuidae). Egypt J Biol Pest Control 30(1):1–6

Yuskel E, Canhilal R (2018) Evaluation of local isolates of entomopathogenic nematodes for the management of black cutworm, *Agrotis ipsilon* Hufnagel (Lepidoptera: Noctuidae). Egypt J Biol Pest Control 28(1):82–88. https://doi.org/10.1186/s41938-018-0087-3

Zhang LW, Liu YJ, Yao J, Wang B, Huang B, Li ZZ, Sun JH (2011) Evaluation of *Beauveria bassiana* (Hyphomycetes) isolates as potential agents for control of *Dendroctonus valens*. Insect Sci 18:209–216

Natural Enemies of Sugar Beet

34

Arun Baitha, Santeshwari Srivastava, and Varucha Misra

Abstract

Natural enemies have recently gained much interest because of the problems encountered by the use of pesticides and environmental concerns. Because sugar beet crop is a food crop, it is wise to avoid or minimize the use of pesticides. The indiscriminate use of pesticides may lead to outbreaks of leaf-feeding and sucking pests. There is a need for more emphasis on the augmentation and conservation of natural enemies. Natural enemies are found to manage the aphids effectively, and farmers are to be advised not to apply insecticides and to encourage the natural build-up of parasitoids and predators. Using of natural enemies is very economical, ecologically sound, and capable of giving more or less permanent control. The only expenditure involved is at the development stage, i.e. the study of natural enemies to understand their life history, seasonal cycle, ecological adaptations, population dynamics, and the development of suitable methods to produce them in sufficient numbers for release against insect-pests.

Keywords

Sugar beet · Natural enemies · Augmentation · Conservation

A. Baitha (✉) · S. Srivastava
Division of Crop Protection, ICAR-Indian Institute of Sugarcane Research, Lucknow, Uttar Pradesh, India
e-mail: arun.baitha@icar.gov.in

V. Misra
Division of Crop Improvement, ICAR-Indian Institute of Sugarcane Research, Lucknow, Uttar Pradesh, India

V. Misra et al. (eds.), *Sugar Beet Cultivation, Management and Processing*,
https://doi.org/10.1007/978-981-19-2730-0_34

34.1 Introduction

The use of natural enemies to reduce the impact of insect-pests has a long history. The ancient Chinese, observing that ants were effective predators of many citrus pests, augmented their populations by taking their nests from their surrounding habitats and placing them in their orchards. Today's insectaries and airfreight delivery of natural enemies across the country or around the world are simply modern adaptations of these original ideas. It is an important biological tool widely used around the world for the biocontrol of insect-pests (Sampaio et al. 2008; Kalyanasundaram and Kamala 2016).

The ill effects of synthesized chemical pesticides on the naturally occurring natural enemies, development of pesticide resistance, and environmental hazardousness have been documented (Ueno 2006; Ueno and Tran 2015). The release of natural enemies and their conservation are advantageous to minimize insect-pest overrun and for the sustainable use of biodiversity (Gu et al. 2008; Scherr and McNeely 2008; Mousa and Ueno 2019).

The suppression of pest populations can, hopefully, be achieved to significantly low levels by using their natural enemies including parasitoids, predators, and pathogens, which form an integral part of the ecosystem. Using of natural enemies is very economical, ecologically sound, and capable of giving more or less permanent control. The only expenditure involved is at the development stage, i.e. the study of natural enemies to understand their life history, seasonal cycle, ecological adaptations, population dynamics, and development of suitable methods to produce them in sufficient numbers for release against insect-pests.

34.2 Aspects of Natural Enemies

The use of natural enemies generally involves four important aspects, viz, basic studies, importation, augmentation, and conservation of natural enemies (Gautam 2003).

34.2.1 Basic Studies

The first phase of the study includes researches on taxonomy, biology, physiology, genetics, ecology and demography, behaviour, multiplication methods, and nutrition. The second phase of basic studies, i.e. the ecological approach to pest control, includes ecological field studies designed to evaluate the relative importance of bio-agents with respect to other factors in the regulation of host population densities.

34.2.2 Importation

The importation of natural enemies covers all phases of the importation and establishment of exotic natural enemies is usually based on the fact that many agricultural pests have been accidentally introduced into the area concerned, while their indigenous natural enemies have been left behind.

34.2.3 Augmentation

The augmentation of natural enemies deals with the manipulation of natural enemies themselves in order to make densities. Many cases are known of bio-agents being periodically decimated by extreme summer heat or winter cold. This can be achieved by the following two means.

34.2.3.1 Periodic Colonization

The natural enemies are to recognize the following adverse periods, so that a satisfactory balance between the host and natural enemies could be rapidly established.

34.2.3.2 Selective Breeding

It involves the development of new strains through selective breeding of the natural enemies, which could survive in adverse conditions.

34.2.4 Conservation

The enhancement of biological control would modify the environment in such a way that any adverse environmental effects would be eliminated or mitigated or simply alter the environment better to suit certain needs or responses of the natural enemies which were previously unsatisfied.

34.3 Status of Natural Enemies in the Sugar Beet Ecosystem

A number of natural enemies were tried for the suppression of sugar beet insect-pests with varying levels of success. Since sugar beet is an important industrial crop, there is active interest in coordinating the work for use of natural enemies against major pests like armyworms, leaf and crown feeders, root feeders, webworms, cutworms, and aphids by extension agencies and the Government with the cooperation of the farmers. Natural enemies have recently gained much interest in the management of insect-pests of sugar beet (it is a food crop) because of the problems encountered by the use of pesticides and environmental concerns.

34.3.1 Parasitoids and Predators

Natural enemies have long been sought for control of insect pests. The use of two egg parasitoids, *Aphelinoideo plutella* and *Abbella subflava*, was attempted against beet leaf hopper in Idaho (Henderson 1955). The five species of introduced egg parasitoids (Trichogrammatidae and mymarid) that were tried to colonize the beet leaf have helped make surveys of natural egg parasites for possible releases (Huffaker et al. 1954; Flock et al. 1962).

The natural enemies, i.e. *Monorthochaeta nigra*; *Agathis* sp. *Diadegma originators*; *Praon flavinode*; *Diaeretiella rapae*; *Aphelinus* sp.; and *Pachycrepoideus vindemmiae*, were found to parasitize on different stages of insect-pests of sugar beet (Youssef 1994; Abd El-Ghany 1995; Shalaby 2001; Bazazo 2010; Khalifa 2018; Hamdany and El-Assar 2017; Bazazo and Besheit 2020). The natural parasitization of tortoise beetle, *Cassida vittata*, by egg parasitoid, *Monorthocheata nigra*, was recorded as 6.97–8.46% (Khalifa 2018). In the Egyptian sugar beet ecosystem, *Monorthochaeta nigra* Blood and *Tetrastichus* sp. have been identified as two parasitoid species, where *Monorthochaeta nigra* Blood was used as egg parasitoid, while *Tetrastichus* sp. was used as pupal parasitoid (Bazazo 2010; Hawila 2021). Hegacy and El-Sheikh (2021) had shown that *Cassida vittata* Vill. parasitoids and ecdysone agonists proved to be an efficient tool in integrated pest management programs of crucial insects. Bazazo (2010) investigated 38 parasitoid species from 20 Hymenoptera groups.

The larval parasitoid, *Cotesia marginiventri*, was recorded to parasitize the larvae of black cutworm, spotted beet worm, and beet armyworm, and a new larval pupal parasitoid, *Diadegma oranginator* Aubert (Hymenoptera: Ichneumonidae), was observed on beet moth first time in Egypt; their percentages of parasitization ranged from 55.17 to 68.91 (Sourakov and Mitchell 2000; Bazazo and Ibrahim 2019). The insect growth regulators support the presence of the parasitoid *D. oranginator* in the field, whereas pesticides do harm (Bazazo and Ibrahim 2019).

The natural parasitization of *Cassida vittata* by the larval-pupal parasitoid *Aprostocetus* sp. was recorded as 5.69–6.61%, whereas that of *Opius nitidulator* on *Pengomyia mixta* varied from 7.76 to 8.87% (Khalifa 2018). The occurrence of *Diadegma pusio* (Holgren) was found for the first time in Iran as a parasitoid of *Scrobipalpa ocellatella* Boyd, while *Bracon intercessor* Nees and *Microchelonus subcontractus* Abdinbekova were also identified as larval-pupal parasitoids in the sugar beet ecosystem (Abbasipour et al. 2012; Mahmoudi et al. 2012). El-sheikh et al. (2022) revealed that *Enicospilus repentinus* (Hol) as a parasitoid in sugar beet helps in reducing the incidence of beet moth, *S. ocellatella*, in Egypt.

The predators that were observed on insect-pests of beet crops in different months were *Chrysoperla carnea* Steph.; *Syrphus corollae* Fabricius and *S. syriacus; Coccinella undecimpunctata* L. and *Paederus alferii* (El-Dessouki et al. 2014; Hamdany and El-Assar 2017; Khalifa 2018; Bazazo and Besheit 2020). The majority of coccinellid beetles (ladybirds) are aphid predators in sugar beet (Heathcote 1978). Heathcote (1963) had shown that coccinellids help in prey searching on this crop much prior to the incidence attack of aphids on sugar beet. Furthermore, these predators may help in the reduction of the population rate of aphids. Adult coccinellids have a significant function in the destruction of *Myzus persicae* on either weeds or sugar beet or any other supportive host (Heathcote 1978). Hasan (2021) found that the role of *Paederus alfierii* (L.) as a predator in sugar beet has proved to be effective against many insect pests like *C. vittata* larvae and *P. mixta* larvae.

34.4 Inundative Releases of Natural Enemies

The egg parasitoids, i.e. *Trichogramma* spp. and *Telenomus* spp., play a significant role in the management of leaf-feeding insects in the sugar beet ecosystem. The release of *Trichogramma chilonis* @100,000/ha in two instalments (50,000 adults/release and two sprayings of SlNPV @600 mL/ha at 15 days interval in the winter season and 500 mL/ha in summer month) was very effective for reducing the incidence of *Spodoptera litura* in India (Shivshankar and Patil 2013). The pest densities (aphid, tortoise beetle, sugar beet moth, and sugar beet fly pests) decreased significantly on sugar beet plants on which green lace wing larvae (first and second instars) had been released (Tauber et al. 2000; Solangi et al. 2013; Youssef et al. 2020).

34.5 Conservations of Natural Enemies

Modern agriculture has often caused the simplification of biological and environmental structures in the agroecosystem mainly through intensive cropping practices (Altieria 1999). One of the methods of enhancing the population of natural enemies is by enriching the field neighbourhood with flowering plants and intercropping systems to create more favourable conditions for natural enemies and reduce insect infestations (Ruppert and Mollhan 1991; Al-Beltagy 2015). Wnuk and Wojciechowicz-Zytko (2007) found that when *Phacelia tanacetifolia* Benth was intercropped with Faba bean, the population of *Aphis fabae* was reduced because of the synergistic effect of *P. tanacetifolia* pollens and nectars on the predatory Syrphids that feed upon aphids. The rates of infestations by *Pegomyia mixta* and *Cassida vittata* were less in sugar beet plants intercropped with faba bean as compared with their numbers in sole sugar beet (El-Fakharany et al. 2012). The adoption of intercropping is to create more favourable conditions for natural enemy species and inhibit pest infestations.

The Faba bean planted within and around (on borders) sugar beet fields has significantly increased a number of natural enemies as compared with sole sugar beet (Bazazo and Besheit 2020). The results indicated the reduction of sugar beet insect infestation in the sugar beet+ faba bean field due to the high populations of various natural enemies in the field in comparison with sole sugar beet ones.

34.6 Future Prospects

Efforts should be made to significantly multiply field-collected natural enemies (*Chrysoperla carnea*, *Trichogramma* spp., *Cotesia flavipes Tetrastichus howardi*, and ear bug) in the laboratory and also to redistribute it to new areas of infestation. There is scope for R&D on locally prevalent key mortality factors (parasites, predators, and pathogens) with respect to their identification for better management by exploiting the weak links in the life cycle of sugar beet insect-pests.

34.7 Conclusion

There is only limited research done on habitat diversity in natural enemies. The biodiversity of the natural enemies of insect-pests could be conserved and this reservoir could be utilized later for control of insect-pests. This facility will also ensure good genetic traits in the natural enemies used for their inundative release. There is an increasing amount of experimental data reported in the literature that documents the effects of plant diversity on the regulation of insect herbivore populations by favoring the abundance and efficacy of associated natural enemies.

References

Abbasipour H, Mahmoud M, Basji M, Lozan A (2012) Report of the parasitoid wasps, *Microchelonus* (Chelonus) subcontractus and *Bracon intercessor* (Hym.; Braconidae) from Iran. J Entomol Soc Iran 32(1):89–92

Abd El, Ghany MA (1995) Studies on the beet moth, Scrobipalpa ocellatella Boud at Kafr El-Sheikh region, M.Sc. Thesis. Fac. Agric. Tanta Univ., Egypt, 116 pp

Al-Beltagy FA (2015) Biological control of some insect pests infesting sugar beet at Kafr ElSheikh Governorate. M.Sc., Thesis, Fac. of Agric. Tanta University

Altieria M (1999) The ecological role of biodiversity in agroecosystems. Agric Ecosyst Eniviron 74:19–31

Bazazo KGI (2010) Studies on some insect pests and natural enemies in sugar beet fields at Kafr ElSheikh region. Ph.D. Thesis, Fac. Agric., Tanta Univ., 139 pp.

Bazazo KGI, Besheit RS (2020) Role of faba bean planted around and within sugar beet fields on insect infestations. Egypt J Plant Protect Res Inst 3(1):196–203

Bazazo KGI, Ibrahim ASM (2019) New record of *Diadegma oranginator* Aubert as a parasitoid of *Scrobipalpa ocellatella* Boyd. In Egyptian sugar beet fields. Egypt J Exp Biol (Zool) 15(2): 289–294

El-Dessouki AA, El-awady SM, Elkhawass KAMH, El-Dessouki WAA (2014) Popilation fluctuation of some insect pests infesting sugar beet and the associated predatory insects at Kafr El-Sheikh governorate. Ann Agric Sci 59(1):119–123

El-Fakharany S, Samy M, Ahmed S, Khattab M (2012) Effect of intercropping of maize, bean, cabbage and toxicants on the population levels of some insect pests and associated predators in sugar beet plantations. J Basic Appl Zool 65(1):21–28

El-Sheikh MF, Eldefrawy BM, Mashall REF (2022) The larval pupal parasitoid and resistant varieties as bioagents for regulating population of *Scrobipalpa ocellatella* Boyd. In the Egyptian sugar beet fields. Egypt Acad J Biol Sci F Toxicol Pest Control Egypt Acad J Biol Sci 14(1): 55–62

Flock RA, Doutt RL, Dickson RC, Laird EF (1962) A survey of beet leafhopper egg parasites in the Imperial Valley, California. J Econ Entomol 55:277–281

Gautam RD (2003) Classical biological control of pink hibiscus mealy bug, *Maconellicoccus hirsutus* (green) in the Caribbean. Plant Prot Bull 55:1–8

Gu H, Edwards RO, Hardy AT, Fitt GP (2008) Host plant resistance in grain crops and prospects for invertebrate pest management in Australia: an overview. Aust J Exp Agric 48:1543–1548

Hamdany M, El-Assar M (2017) Effect of intercropping three faba bean varieties with sugar beet plants on piercing sucking insect pests and associated natural enemies under ridge space and seedling rates in relation crop yield. Egypt Acad J Biol Sci 10(6):57–77

Hasan HM (2021) Role of the rove beetle *Paederus memnonius* (Erichson) (Coeloptera Staphylinidae) in controlling certain sugar beet pests for the first time in Egypt. J Plant Prot Pathol Mansoura Univ 12(2):117–120

Hawila HZ (2021) Ecological and biological studies on the main insect pests infesting sugar beet and their associated natural enemies. Mansoura Univ., Ph.D. Thesis, Fac. Agric., 140 pp

Heathcote GD (1963) The effect of coccinellids on aphids infesting insecticide-treated sugar beet. Plant Pathol 12:80–83

Heathcote GD (1978) Coccinellid beetles on sugar beet in eastern England 1961-75. Plant Pathol 27:103–109

Hegacy FH, El-Sheikh MF (2021) Role of parasitoids, conventional insecticides and ecdysone agonists in regulating population of *Cassida vittate* Vill in Egyptian sugar beet fields. J Plant Prot Pathol Mansoura Univ 12(9):569–573

Henderson CF (1955) Parasitisation of the beet leafhopper in relation to its dissemination in southern Idaho. US Department of Agriculture, Washington DC, Circular No. 968, pp 16

Huffaker CB, Holloway JK, Doutt RL, Finney GL (1954) Introduction of egg parasites of the beet leafhopper. J Econ Entomol 47(5):785–789

Kalyanasundaram M, Kamala IM (2016) Parasitoids. In: Omkar (ed) Ecofriendly pest management for food security. Academic Press, London, pp 109–138

Khalifa AA (2018) Natural enemies of certain insect pests attacking sugar beet plants at Kafr El-Sheikh Governorate. J Plant Prot Pathol Mansoura Univ 9(8):507–510

Mahmoudi J, Askarianzadeh A, Karimi J, Abbasipour H (2012) Introduction of two parasitoids of braconid wasps on the sugar beet moth, *Scrobipalpa ocellatella* Boyd. (Lep.: Gelechidae) from Khorasan-e-Razavi province, Iran. J Sugar Beet 28(2):103–106

Mousa KM, Ueno T (2019) Intercropping potato with citrus trees as ecologically–based insect pest management. J Fac Agric Kyushu Univ 64:71–78

Ruppert V, Mollhan J (1991) Augmentation of aphid antagonists by field margins rish in flowering plants. Behavior and impact of aphidophaga. SPB Academic Publishing, the Hague, the Nertherlands, pp 243–247

Sampaio MV, Bueno VHP, Silveira LP, Auad AM (2008) Biological control of insect pests in the tropics. Encyclopedia of life support systems (EOLSS). Unesco, Eolss Publishers, Oxford, UK, pp 28–70

Scherr SJ, McNeely JA (2008) Biodiversity conservation and agricultural sustainability: towards a new paradigm of "ecoagriculture" landscapes. Philos Trans R Soc B 363:477–494

Shalaby GA (2001) Ecological studies on some important sugar beet pests and natural enemies and their control. Ph.D. Thesis, Fac. Agric., Tanta Univ. Egypt, pp. 141

Shivshankar SB, Patil AS (2013) Integrated management of Spodoptera litura in sugar beet. In: Kumar S, Singh PK, Swapna M, Pathak AD (eds) Souvenir on IISR industry interface on research and development initiatives for sugar beet in India, pp 24–35

Solangi AW, Lanjar AG, Baloch N, Rais MUL, Khuhro SA (2013) Population, host preference and feeding potential of Chrysoperla carnea (Stephens) on different insect hosts in cotton and mustard crops. Sindh Univ Res J Sci Ser 45:213–218

Sourakov A, Mitchell ER (2000) Featured creatures. USDA http://creatures.ifas.edu/wasps/diadegma_insulare.html

Tauber MJ, Tauber CA, Daane KM, Hagen KS (2000) Commercialization of predators: recent lessons from green lacewings (Neuroptera: Chrysopidae: Chrysoperla). Am Entomol 46:26–38

Ueno T (2006) Current status of insect pests attacking green bunching onion in central and southern Vietnam. J Fac Agric Kyushu Univ 51:275–283

Ueno T, Tran DH (2015) *Neochrysocharis okazakii* (Hymenoptera: Eulophidae) as a major parasitoid wasp of stone leek leaf miner *Liriomyza chinensis* (Diptera: Agromyzidae) in Central Vietnam. Psyche:179560

Wnuk A, Wojciechowicz-Zytko E (2007) Effect of intercropping of broad bean (*Vicia faba* L.) with tancy phaceloia (*Phacelia tanacetifolia* benth) on the occurrence of *Aphis fabae* Scop. and predatory Syrphidae. Aphids Other Hemipterous Insects 13:211–217

Youssef AE (1994) Studies on certain insects attacking sugar beet, Ph.D. Thesis. Fac. Agric., Kafr El-Sheikh, Tanta Univ., Egypt, 130 pp

Youssef AE, Ibrahim AS, Bazazo KG, Khattab HM, Ueno T, Mousa K (2020) Micronutrients foliar fertilization and releasing green lacewing *Chrysoperla carnea* (stephens) could efficiently suppress sugar beet insects. J Fac Agric Kyushu Univ 65(2):269–275

Sugar Beet Nematodes: Their Occurrence, Epidemiology, and Management in Ukraine 35

K. A. Kalatur, J. D. Janse, and L. A. Janse

Abstract

This chapter describes plant parasitic nematode species of sugar beet, their occurrence, biology, and harmfulness, and presents a system of integrated nematode prevention and control methods for sugar beet under Ukrainian cultivation conditions. It is shown that the beet cyst nematode (*H. schachtii*) is by far the most important one in terms of spread all over the country (18 regions), plant damage, economic losses, and research and monitoring activities. It is followed by the stem nematode (*Ditylenchus dipsaci*) and the needle nematode (*Longidorus elongatus*). Root-knot nematodes (*Meloidogyne* spp.) and stubby root nematodes (*Trichodorus* and *Paratrichodorus* spp.) are of lesser importance and their occurrence in Ukraine less documented. The false root-knot nematode (*Nacobbus aberrans*) is not known to occur in Ukraine. The spread of these nematodes is facilitated by the reuse of sugar factory waste (sludge, washing water, heads, leaves, and root tips), especially when returned to the farm land or used as cattle fodder. Furthermore, wind, animals, man, and his machinery play a role. A system of prevention and control adapted to the Ukrainian cultivation conditions has been developed over the years and is presented here.

Keywords

Beet cyst nematode · Control · Docking disorder · Occurrence · Sugar beet

K. A. Kalatur (✉)
Institute of Bioenergy Crops and Sugar Beet NAAS, Kyiv, Ukraine
e-mail: kkalatur@meta.ua

J. D. Janse · L. A. Janse
Institute of Plant Protection NAAS, Kyiv, Ukraine

V. Misra et al. (eds.), *Sugar Beet Cultivation, Management and Processing*,
https://doi.org/10.1007/978-981-19-2730-0_35

Abbreviation

PEBV	Early browning virus of pea
PepRSV	Ring spot virus of pepper
RRV	Raspberry ringspot virus
TBRV	Tomato black ring virus
TRV	Tobacco rattlesnake virus

35.1 Introduction

Sugar beet (*Beta vulgaris* subsp. *vulgaris* convar. *Vulgaris* var. *altissima*) is in Ukraine, as in many countries, the only crop for the production of sugar, which is an important food product for the humans. By the end of the twentieth century, the cultivated area of sugar beet had reached 1.6 million hectares. There were 195 operating sugar factories in Ukraine, and the average annual production volume of beet sugar exceeded five million tons, with a sugar beet root production volume of 45–50 million tons. There are favourable conditions for sugar beet cultivation in the right-bank (Dnipro river) forest-steppe area (Vinnytsia, Cherkasy, Kyiv, Khmelnytskyi, Ternopil, Chernivtsi, Lviv, Rivne, Volyn, and Zhytomyr regions), as well as in the left-bank (Dnipro river) forest-steppe area (Poltava, Sumy, Kharkiv, and Chernihiv regions). Small acreages are cultivated in the northern steppe and the southern Polissia (Roik 2001; Prymak et al. 2009). The experience of many farms demonstrated that sustained adherence to good agricultural practice allows obtaining stable yields of sugar beet at the level of 55–70 t/ha and higher (Roik 2001; Prymak et al. 2009). However, in the last decade (2010–2020), the total cultivated area did not exceed 300,000 hectares, and in many regions of the country there was a significant decrease in the crop yield (Table 35.1) (Trybel and Stryhun 2012).

Table 35.1 Cultivated area and productivity of sugar beet in Ukraine[a]

Year	Cultivated area (1000 hectares)	Yield (t/ha)	Sugar yield (t/ha)
2011	515.8	36.3	4.51
2012	448.9	41.1	4.77
2013	270.5	39.9	4.67
2014	329.6	47.7	6.23
2015	238.9	43.6	6.19
2016	292.4	48.2	6.97
2017	318.0	47.5	6.87
2018	279.1	50.9	6.63
2019	220.6	46.1	6.69
2020	218.9	42.4	5.32

[a]Data from State Statistics Service of Ukraine (http://www.ukrstat.gov.ua/)

According to Trybel and Stryhun (2012), the main factors of a low root yield are insufficient fertilisation, violation of cultivation technology, and unsatisfactory levels of pest and pathogen control. Among the latter, the most important are the plant parasitic nematode (eelworm) species that live in the soil and feed on living plant tissues that they use as food and, often, also as an environment for reproduction and development (Decker 1969; Perry et al. 2018; Turner and Rowe 2006; Kazachenko and Muhina 2013; Borzykh et al. 2017; Sigareva et al. 2017). The infestation of sugar beet crops by these microscopic plant parasitic organisms not only causes yield reduction, but may even result in a complete yield loss (Cooke 1993; Marić and Čamprag 1982; Kalatur 2008a; Sigareva and Kalatur 2014; Kalatur et al. 2015; Pylypenko et al. 2016).

This chapter is a full update of an earlier review of the nematodes in sugar beet published before (Pylypenko et al. 2016), and presents the last survey and monitoring data as well as new insights into the taxonomy, biology, and management of the nematode diseases of sugar beet under Ukrainian growing conditions.

35.2 The Species of Plant Parasitic Nematodes in the Sugar Beet-Cultivated Areas in Ukraine, Their Occurrence, Host Plants, Biology, Symptoms, and Harmfulness

Sugar beet is affected by a large number of plant parasitic nematode species; the most damaging ones are the beet cyst nematode (*Heterodera schachtii*), the root-knot nematode (*Meloidogyne* spp.), the false root-knot nematode (*Nacobbus aberrans*), the stem nematode (*Ditylenchus dipsaci*), the needle nematode (*Longidorus elongatus*), the and stubby root nematode (*Trichodorus* and *Paratrichodorus* spp.) (Decker 1969, Marić and Čamprag 1982, Cooke 1993. Turner and Rowe 2006, Kazachenko and Muhina 2013, Manzanilla-López et al. 2002).

For Ukraine, the beet cyst nematode (*H. schachtii*), the stem nematode (*D. dipsaci*), and the needle nematode (*L. elongatus*) are the most important in terms of economic loss (Korab 1924; Korab 1929; Korab 1961; Shcherbak 1973; Kitsno 1984; Sigareva and Fylenko 1983; Sigareva et al. 1996; Kalatur et al. 2015; Pylypenko et al. 2016; Borzykh et al. 2017).

35.2.1 Beet Cyst Nematode *Heterodera schachtii* (Nematoda, Order Rhabditida, Family Heteroderidae)

35.2.1.1 Occurrence

The beet cyst nematode occurs in most European countries (Albania, Austria, Azerbaijan, Belgium, Croatia, Czech Republic, Denmark, Estonia, Finland, France, Germany, Greece, Hungary, Ireland, Italy, Latvia, Russia, and Ukraine), some African countries (Algeria, Cape Verde, Gambia, Libya, Morocco, Senegal, and South Africa), in Asia (China, India, Iran, Iraq, Israel, Jordan, Kazakhstan, Korea,

Kyrgyzstan, Pakistan, and Syria), in North and South America (Canada, Chile, Mexico, Peru, USA, and Uruguay), as well as in Oceania (Australia and New Zealand) (EPPO GD, https://gd.eppo.int).

In Ukraine, the beet cyst nematode was first reported in 1923 in the fields of the Pii State Sugar Farm in the Kyiv region (Korab 1924). Since then, monitoring of the areas of sugar beet cultivation for soil infestation by this parasite is done almost yearly. In the 1980s, *H. schachtii* was detected in 16 regions (Linnik 1978) and in the early 2000s, in 17 regions of Ukraine (Sosenko 1998).

Continuing the systematic monitoring of soils and sugar beet sowings during vegetation over the past 20 years has demonstrated the existence of new foci of *H. schachtii* in the traditional beet-growing areas in Ukraine. Thus, during the years 2000–2004, monitoring was conducted in the Vinnytsia, Zhytomyr, Kyiv, Kirovograd, Poltava, and Khmelnytskyi regions on a total area of 18969.7 ha, of which 11310.7 ha (59.6%) was found to be infested with the beet cyst nematode (Sigareva and Pylypenko 2001; Sigareva et al. 2004a).

Monitoring conducted by the Institute of Bioenergy Crops and Sugar Beet NAAS during 2010–2020 in Kyiv, Chernihiv, Cherkasy, Khmelnytskyi, Kirovograd, Vinnytsia, and the Ternopil regions on a total area of 13271.9 ha detected the presence of the beet cyst nematode on 3471 ha (26.2%) of the total surveyed area.

Therefore, the results of nematological monitoring conducted over the past 40 years confirmed the presence of *H. schachtii* in 18 regions of Ukraine: Kyiv, Cherkasy, Vinnytsia, Sumy, Zhytomyr, Chernihiv, Khmelnytskyi, Ternopil, Rivne, Volyn, Lviv, Ivano-Frankivsk, Chernivtsi, Kharkiv, Poltava, Kirovograd, Dnipro, and Donetsk (Kalatur et al. 2015; Pylypenko et al. 2016). Given such a significant distribution of the beet cyst nematode, we can assume its presence in some of the remaining sugar beet-growing areas, for one reason or another not yet covered by nematological surveys. It is also necessary to point out that we found the beet cyst nematode not only in the production fields but also in small private vegetable gardens, where it parasitises on fodder and table beets (Fig. 35.1) (Pylypenko et al. 2016).

We consider that the main reasons for such a wide distribution of *H. schachtii* in the Ukraine area are (a) neglecting preventive measures (including the lack of systematic nematological survey and monitoring programs; poor sanitation of farm machinery leading to *H. schachtii* distribution, with the soil infested; application of infested sugar beet waste from the industry back to the field); (b) a large share (more than 20%) of nematode host plants other than sugar beet in crop rotations; (c) lack of availability of effective nematicides on the market (Korab 1961; Kitsno 1984; Sigareva and Pylypenko 2001; Sigareva et al. 2004a; Kalatur et al. 2015; Pylypenko et al. 2016).

35.2.1.2 Host Plants

Today, a wide range of host plants are known to us (Steel 1965). Host crops are all kinds of beet roots (*Beta* spp.), cabbage (*Brassica* spp.), rapeseed (*B. napus*), turnip (*Brassica rapa* subsp. *rapa*), radish (*Raphanus sativus*), mustard (*Sinapsis alba*,

A B

Fig. 35.1 Symptoms of beet cyst nematode infestation of fodder (**a**) and Table (**b**) beets (wilting, yellowing, stunting and complete necroses of the lower leaves, resulting in dead plants (Kyiv region, Ukraine. Photo: K. Kalatur)

B. juncea, B. nigra), and spinach (*Spinacia oleracea*) (Korab 1929; Korab 1961; Decker 1969; Kalatur et al. 2015; Borzykh et al. 2017).

About 235 species of weed hosts are known, of which almost 70% belong to 6 families: Brassicaceae (67 species), Chenopodiaceae (37 species), Fabaceae (23 species), Caryophyllaceae (19 species), Polygonaceae (12 species), and Asteraceae (12 species) (Kalatur and Pylypenko 2017). It has been found that weed infestation increases the number of nematodes in the ploughed soil layer. Thus, studies conducted in Ukraine (Babich 2004) showed that weed infestation of sugar beet crops with goosefoot (*Chenopodium album*) and wild turnip (*Barbarea vulgaris*) (8–10 plants/m^2) increased the number of *H. schachtii* in the soil 7.3–7.6 times within a period of 4 years. The presence of weed species in crops such as shepherd's purse (*Capsella bursa-pastoris*), wild radish (*Raphanus raphanistrum*), and goosefoot (*Chenopodium album*) increased the initial density of the nematode population by 50%, 56%, and 90%, respectively, over 4 years (Babich 2004). Such results show that host plants of both weeds and crops are important biological factors that contribute to the maintenance and preservation of the beet cyst nematode population in the soil at a high level (Kalatur and Pylypenko 2017).

Meanwhile, some crops do not promote the penetration, feeding, and/or reproduction of this parasite on their roots. These include alfalfa (*Medicago sativa*), asparagus (*Asparagus officinale*), barley (*Hordeum vulgare*), buckwheat (*Fagopyrum esculentum*), carrot (*Daucus carota*), chicory (*Cichorium intybus*), clover (*Trifolium* spp.), esparcet (*Onobrychus viciifolia*), flax (*Linum usitatissimum*), hairy vetch (*Vicia villosa*), hemp (*Cannabis sativa*), kidney bean (*Phaseolus* spp.), cucumber (*Cucumis sativus*), lettuce (*Lactuca sativa*), lupin (*Lupinus* spp.), maize (*Zea mays*), melon (*Cucumis melo*), millet (*Pennisetum glaucum*), oat (*Avena sativa*), onion (*Allium cepa*), pea (*Pisum sativum*), poppy (*Papaver* spp.), potato (*Solanum tuberosum*), rye (*Secale cereale*), tobacco (*Nicotiana spp.*), tomato (*Solanum lycopersicum*), sunflower (*Helianthus annuus*), watermelon (*Citrullus lanatus*), andwheat (*Triticum* spp.) (Korab 1929; Korab 1961; Decker 1969; Kitsno 1984; Kalatur et al. 2015; Borzykh et al. 2017). They are therefore useful alternatives in crop rotation.

35.2.1.3 Biology

The development cycle of beet cyst nematode includes six stages: eggs, four juvenile stages (invasive—second stage, parasitic—third and fourth stages), and adults (males: vermiform, i.e., worm-shaped, 1.3–1.6 mm long; females: lemon-shaped, 0.5–1.3 mm long and cysts, also lemon-shaped, which are dead females with eggs and juveniles inside) (Decker 1969; Turner and Rowe 2006; Sigareva et al. 2017).

In the beet-cultivated areas in Ukraine, hatching of the second-stage juveniles from cysts is observed during the spring at a soil temperature of 8–10°C (Ladygina 1961). The juveniles infest the roots and feed on the sap of living cells, moult twice, and then turn into adult females and males. The juveniles, which turn into males, keep the wormlike shape and leave the root of the plant where they have been developing. The body of the juveniles, which turn into females, rapidly increases in size, takes a lemon-like shape, and begins to press hard on the surface tissues of the root. Under this pressure, the root cover breaks and a mature female appears on its surface, where it is fertilised by a male that lives freely in the soil. After fertilisation, the female lays eggs in a mucous egg sac that is situated at the back of the body. The female produces 10–650 eggs with an average of 200–300 eggs. Subsequently, when the whole body of the female is filled with eggs, its internals die, and the white shell darkens from light to dark brown (Fig. 35.2). The mature cyst measures on average 1 mm in length, eventually drops from the root into the soil, and can stay there for up to 10 years without the eggs losing their viability. Depending on the environmental conditions (temperature, humidity, etc.) the development of one generation of the beet cyst nematode takes 42–67 days. In Ukraine, up to 2–3 generations of this parasite can develop during one growing season depending on the weather conditions and the availability of host plants (Korab 1929; Korab 1961; Kitsno 1984; Babich 1990; Sosenko 1998; Kalatur et al. 2015; Sigareva et al. 2017).

The main source of the long-distance spread of the beet cyst nematode is via wastes from sugar factories derived from washing and cutting sugar beet roots that is dumped untreated into the surface water or brought back to the fields, e.g., leaf and upper root material, soil, and root tips (tare). Another source is the mature sugar beet roots collected from the nematode-infested fields. Nematode cysts can furthermore be distributed with machines, transport vehicles, tillage tools, rain water, wind, animals (including birds), and humans (Korab 1961; Kalatur et al. 2015; Pylypenko et al. 2016).

Fig. 35.2 Mature cysts of the beet cyst nematode, *Heterodera schachtii*, c. 1 mm in length (photo: K. Kalatur, magnification 8× (МБС-10 Stereomicroscope)

35.2.1.4 Symptoms

The juveniles of beet cyst nematode penetrate the root system of beets in the early spring, but the symptoms of infestation become noticeable only in late June–early July (after the development of the first-stage generation) and can be found until the end of vegetation (Decker 1969; Kalatur et al. 2015; Pylypenko et al. 2016).

The degree of damage to beets depends on the pre-sowing density of the nematode population (juveniles + eggs) in the soil and a complex of abiotic factors, primarily temperature and humidity. In particular, at low (<200 juveniles and eggs/ 100 cm^3) and medium (201–600 juveniles and eggs/100 cm^3) numbers of nematodes in the soil, the affected plants will not look different from the healthy ones (Kalatur et al. 2015, Pylypenko et al. 2016). At daytime, however, when the air temperature reaches 20 °C and more, their leaves wither and eventually drop (Fig. 35.3). At a high level of soil infestation (>600 juveniles and eggs/100 cm^3), a vast majority of plants lag behind in growth and development; initially their leaves become pale green, then the outer leaves turn yellow and die. If such plants are dug up, the root would have a 'bearded' appearance, due to a large number of newly formed lateral roots, on which white, swollen female nematodes are clearly visible (Figs. 35.4 and 35.5). Sometimes there is a complete loss of crop plants in the foci of the infected area, resulting in the formation of patches of poor growth (roughly circular foci of stunting, yellowing plants) in the field (Fig. 35.6) (Korab 1961; Decker 1969; Marić and Čamprag 1982; Kitsno 1984; Cooke 1993; Turner and Rowe 2006; Kalatur et al. 2015; Pylypenko et al. 2016).

Fig. 35.3 Sugar beets infested by the beet cyst nematode, showing typical yellowing, necrosis, stunting, and wilting (Vinnytsia region, Ukraine, photo: K. Kalatur)

35.2.1.5 Harmfulness

Parasitism of the beet cyst nematode in beet roots, first of all, impairs the plant transport function and therefore it does not receive the necessary minerals and water from the soil. This in turn leads to pathological changes in a number of important physiological processes in the plant body: decreasing number of leaves and leaf area, the content of green pigments, carotenoids, phosphorus, nitrogen compounds, and potassium, slowing down the intensity of photosynthesis and respiration, and hormone balance (growth regulation) impairment (Kitsno 1984).

Sigareva et al. (2007) and Kalatur et al. (2020) noted that the penetration, feeding, and development of the beet cyst nematode juveniles inside the plant cells not only worsens the physiological state of plants but also 'opens the gate' for fungal pathogens that damage the root system, such as *Rhizoctonia solani*, *Aphanomyces*

Fig. 35.4 White, swollen females of beet cyst nematode on sugar beet (lateral) roots (Kyiv region, Vinnytsia region, Ukraine, photo: K. Kalatur, O. Polovynchuk)

Fig. 35.5 'Bearded' roots (newly formed lateral roots under the influence of nematode attack) of sugar beet (Kyiv region and Vinnytsia region, Ukraine, photo: K. Kalatur, O. Polovynchuk)

Fig. 35.6 Poor growth patches in a sugar beet field as a result of infestation by beet cyst nematodein Vinnytsia region, Ukraine (Source: Pylypenko et al. 2016)

cochlioides, Fusarium sp., *Pythium* sp., and *Phoma* sp. Sigareva et al. (2007) demonstrated that along with an increase in the soil infestation by the beet cyst nematode, the number of plants affected by *Pythium* and the degree of the infestation increased too. In a field plot, where the number of nematodes in the soil did not exceed 20 juveniles and eggs/100 cm^3, the incidence of *Pythium*-infected plants was 48.6% and the disease severity was 24.1%. Increasing the nematode population density from 500 to 1000 juveniles and eggs/100 cm^3 contributed to an increase in disease incidence of60.9–62.4% with a disease severity of up to 33.2%. Examination of the plants in the cotyledon stage showed that in a field plot where only soil fungi were present, 3.8% of the plants died, while in the treatment with fungi and high nematode invasion (1000 juveniles and eggs/100 cm^3) 10.5% of the plants died. The infestation of sugar beets by both the nematode and *Pythium* also negatively affected the weight of the plants. In the two-leaf stage, the weight loss of plants due to soil infestation by the beet cyst nematode alone was 12.5–15.0%, while co-infestation by fungi along with the beet cyst nematode led to 27.5–30.0% weight loss (Sigareva et al. 2007).

Negative physiological changes that occur in the beet plants due to the impact of the beet cyst nematode, along with a strong infestation by various fungal pathogens, lead to a significant reduction in root yield, reduced sugar content, and sometimes complete loss of plants. Sugar beet seed-bearing plants are particularly sensitive to infestation by *H. schachtii* (Korab 1961; Kitsno 1984; Sigareva et al. 2004a; Kalatur et al. 2015; Pylypenko et al. 2016).

Results of the studies conducted in Ukraine (Babich 1990) showed that beet cyst nematodes ranging from 210 to 280 juveniles and eggs/100 cm^3 caused a reduction of about 5–10% in the root yield. Along with the increase in population density to 500 juveniles and eggs/100 cm^3, the reduction reached 20% at 850 juveniles and eggs/100 cm^3 and 30% at 1550–2600 juveniles and eggs/100 cm^3–40 – 50%. The reduction in the yield of sugar beet seed-bearing plants at the above-mentioned levels of nematode infestation in the soil reached 7–14%, 29%, 42%, and 57–70%, respectively. A statistically significant decrease in the sugar content of roots occurred only at a high degree of soil infestation by beet cyst nematodes and ranged from 0.8 to 2% (Babich 1990).

The vast majority of fields infested by *H. schachtii* in Ukraine have an average to high level of beet cyst nematode in the soil, ranging from 200 juveniles and eggs/ 100 cm^3 to more than 600 juveniles and eggs/100 cm^3. However, in some foci, the population density of the nematode was 142,000 juveniles and eggs/100 cm^3, which led to the death of plants during vegetation (Pylypenko et al. 2016). The mortality rate in sugar beet seed-bearing plants in Kyiv, Zhytomyr, Cherkasy, and Ivano-Frankivsk regions of Ukraine was 40–100% (Sigareva et al. 2004a; Kalatur et al. 2015).

35.2.2 Rootknot Nematodes, *Meloidogyne* Spp. (Nematoda, Order Rhabditida, Family Meloidogynidae, *Meloidogyne*)

35.2.2.1 Species Composition and Occurrence

To date (2021), about 98 species of root-knot nematodes have been described and they are found world-wide (Subbotin et al. 2021). The most extensive damage to sugar beet crops, reported over the years in Greece, Italy, and the United States, is by the Javanese root-knot nematode (*Meloidogyne javanica*), south root-knot nematode (*Meloidogyne incognita*), and peanut root-knot nematode (*Meloidogyne arenaria*). In Kyrgyzstan, the former Yugoslavia, and Japan, the harmful nematode was reported to be the northern nematode (*Meloidogyne hapla*); furthermore, Japan and the United States also reported that the barley root-knot nematode (*Meloidogyne naasi*) was very harmful (Decker 1969; Matiashov 1971; Maas and Maenhout 1978; Marić and Čamprag 1982; Cooke 1993; Perry et al. 2018; Kazachenko and Muhina 2013; Subbotin et al. 2021).

In Ukraine, to date four species of root-knot nematodes have been identified. They are not found in sugar beet crops; however, they cause significant damage to vegetables, flowers, and ornamental crops in greenhouses. These four species include the Javanese root-knot nematode (*M. javanica*), which is found only in greenhouses in the Crimea, southern root-knot nematode (*M. incognita*), which is found all over the country, northern root-knot nematode (*M. hapla*), which is found in Zakarpattia, Lviv, Kyiv, Zhytomyr regions, and in the Crimea), and peanut root-knot nematode (*M. arenaria*), which is found in Kyiv and Kharkiv regions (Borzykh et al. 2017; Sigareva et al. 2017).

35.2.2.2 Host Plants

Root-knot nematodes affect about 4000 plant species in both open and protected cultivation. Field crop hosts (apart from sugar beet) include alfalfa, barley, clover, cotton (*Gossypium hirsutum*), maize, rye, oat, soybean (*Glycine max*), tobacco, pea, potato, rice, and sugar cane (*Saccharum officinarum*); vegetables such as cabbage, carrot, celery (*Apium graveolens*), cucumber, lettuce, pepper (*Capsicum annuum*), okra (*Abelmoschus esculentus*), parsley (*Petroselinum crispum*), pumpkin (*Cucurbita moschata*), spinach (*Spinacia oleracea*), tomato, and zucchini

(*Cucurbita pepo)*; many flowers, ornamental plants, trees and bush species; and many weed species (Decker 1969; Perry et al. 2018; Rich et al. 2009; Kazachenko and Muhina 2013; Borzykh et al. 2017; Kalatur and Pylypenko 2017; Sigareva et al. 2017; Subbotin et al. 2021).

35.2.2.3 Biology

The morphology of the root-knot nematodes differs significantly from the other groups of plant parasitic nematodes. The juveniles and males of the second stage are colourless and are worm shaped. Juveniles of the third and fourth stages and mature females have a pear-like or globose shape; they are white, rarely slightly greyish, with a protruding head end. The developmental cycle of the root-knot nematodes starts with the release of juveniles of the second stage, which have a length of 0.4–0.5 mm. Once in the soil, they actively migrate in both horizontal and vertical directions and penetrate into the root near its tip. After a short migration in the root bark, the juveniles orient themselves parallel to the longitudinal axis of the root, become immobile, and then begin to feed on the contents of the cells. As a result, hypertrophy of the root bark cells occurs, which leads to the formation of root-knots. After that, the juveniles moult, passing the third and the fourth stages. After the third moult, the juveniles that turn into males elongate, and after the fourth stage, they acquire a needle shape, 1.0–1.4 mm long and 30–40 μm wide. The juveniles that turn into females acquire a pear shape, 0.5–1.0 mm long and 0.4–0.5 mm wide. Males leave the roots and enter the soil, where they fertilise the females. Mature females secrete a gelatinous substance, in which they lay about 400–800 eggs (Decker 1969; Perry et al. 2018; Kazachenko and Muhina 2013; Subbotin et al. 2021). Depending on the humidity and temperature of the soil, the development of one generation of the root-knot nematode in Ukraine lasts from 21 to 56 days (Borzykh et al. 2017, Sigareva et al. 2017).

35.2.2.4 Symptoms

A characteristic feature of the plant infestation by root-knot nematodes is the formation of galls (outgrowths) on the root system, i.e., root-knots. Usually, at a low number of these parasites in the soil and single root-knots on the root system, the affected plant will not look different from a healthy one. There is a high density of populations of root-knot nematodes in the soil and the formation of a large number of root-knots. As a result, the plant's underground part does not receive enough water and minerals. This affects the appearance of plants: they begin to lag behind in growth and development, lose turgor, and wither in the heat. They have small, pale-green leaves that gradually turn yellow and dry up (Decker 1969, Marić and Čamprag 1982, Cooke 1993, Kazachenko and Muhina 2013, Borzykh et al. 2017, Sigareva et al. 2017, Subbotin et al. 2021).

35.2.2.5 Harmfulness

Matiashov (1971), Marić and Čamprag (1982), and Cooke (1993) reported that in the fields infected with root-knot nematodes, sugar beet yield reduction can reach >30%. It is also noted that the parasitism of the southern root-knot nematode

(*M. incognita*) inside the root system of the beets contributes to the infestation of plants by fungal pathogens, in particular *Pythium ultimum* and *Rhizoctonia solani* (Pandey 1984; Kalatur et al. 2020).

35.2.3 False Root-Knot Nematode *Nacobbus Aberrans* (Nematoda, Order Rhabditida, Family Pratylenchidae)

The false root-knot nematode (*N. aberrans*) causes significant damage to sugar beet crops in the temperate and subtropical latitudes of North and South America, including Mexico, the United States (Arkansas, Colorado, Kansas, Nebraska, Montana, South Dakota, Wyoming, and Utah), Argentina, Bolivia, Chile, Ecuador, and Peru (EPPO GD, https://gd.eppo.int; CABI Crop Protection Compendium, https://www.cabi.org).

Infestation of beet crops with *N. aberrans* inhibits the growth and development of leaves and roots throughout the vegetation. In hot weather, the plants wither and turn yellow. The most characteristic symptoms of the infestation include the formation of irregularly shaped root-knots and the formation of numerous lateral roots (Marić and Čamprag 1982; Cooke 1993; Harveson 2014). Studies showed that *N. aberrans* can cause a root yield reduction of 10–20% (Harveson 2014).

In Ukraine, the false root-knot nematode *N. aberrans* has not been found until now (2021) in sugar beet crops, although there is no active monitoring of this nematode, and its symptoms are easily mistaken for those caused by *Meloidogyne* spp. (hence its common name). This nematode can only be diagnosed microscopically in a specialised laboratory (EPPO PM7/5 (2) 2009). It is included in the Ukrainian A-1list, whichcontains regulated pests that have the status of a quarantine organism, not found in the country.

35.2.4 Stem Nematode, *Ditylenchus dipsaci* (Nematoda, Order Tylenchida, Family Anguinidae, *Ditylenchus*)

35.2.4.1 Occurrence

The stem nematode, *D. dipsaci*, occurs in many European countries with temperate climates, as well as in Africa, Asia, South and North America, and Oceania (EPPO GD, https://gd.eppo.int; CABI Crop Protection Compendium, https://www.cabi.org). In Ukraine, the stem nematode, *D. dipsaci,* is placed in the list of regulated non-quarantine harmful organisms.

35.2.4.2 Host Plants

D. dipsaci is known to affect about 450 plant species, including beets, celery, garlic (*Allium sativum*), onion (*Allium cepa*), pea, pumpkin, rhubarb (*Rheum rhaponticum*), and strawberry (*Fragaria x ananassa*); ornamental bulb species (hyacinth, narcissus, and tulip), oat, rye, and many weed species (Decker 1969, Gubina 1982, Sigareva et al. 2017, EPPO GD, https://gd.eppo.int/taxon/DITYDI/

hosts; CABI datasheet; https://www.cabi.org/isc/datasheet/19287#tohostsOrSpeciesAffected). There are more than 30 physiological races of the stem nematode, many of which are named after the main host crop (for example, onion race, oat race, and beet race) (Decker 1969; Gubina 1982).

35.2.4.3 Biology

D. dipsaci is a migrating obligate endoparasite that develops and reproduces in the living tissues of the host plant. Unlike the cyst-forming and root-knot species nematodes, males, females, and juveniles of the stem nematodes have a worm-like shape, with a body length of 1.0–1.3 mm. After penetrating the plant, the juveniles moult several times and turn into adult males or females. Once fertilised, the female lays an average of 207–408 eggs, from which the next generation of nematodes develops. Depending on the conditions of the environment, the development of one generation lasts from 19 to 23 days (Decker 1969; Gubina 1982; Marić and Čamprag 1982; Cooke 1993; Borzykh et al. 2017; Sigareva et al. 2017). Stem nematodes can be viable for a long time on dry plant residues in a state of anabiosis and come to life under favourable temperature and humidity conditions (Decker 1969, Gubina 1982).

35.2.4.4 Symptoms

Symptoms of sugar beet infestation with stem nematodes appear from the emergence of seedlings to the end of the growing season. In young plants, swellings may be formed on the leaves. Necrosis appears on the petioles and the base of the cotyledons. They begin to rot at high humidity. At the end of vegetation, necrotic zones or cracks are observed on the surface of the crown-root. Then they spread deep inside the root, forming the cavities (Decker 1969; Shcherbak 1973; Marić and Čamprag 1982; Cooke 1993). The affected roots may rot as a result of fungal and bacterial pathogens entering the wounded tissues (Hillnhütter et al. 2011).

35.2.4.5 Harmfulness

According to the research performed by Shcherbak (1973) in the Zaporizhzhia region of Ukraine in the year 1973, stem nematode can cause a sugar beet yield reduction of up to 54.6%. It was also noted that, in addition to yield reduction, the sugar content of roots decreased by 1–2% and dry matter by 2.5%, while the substances undesirable in sugar production were ash and nitrogen (Graf and Meyer 1973; Kuthe 1974).

35.2.5 Needle Nematode, *Longidorus elongatus* (Nematode, Order Dorylaimida, Family Longidoridae)

35.2.5.1 Occurrence

The needle nematode (*L. elongatus*) is common in most countries of Europe, Asia (India, Kazakhstan, Pakistan, Tajikistan, Uzbekistan, and Vietnam), North America (Canada and the USA), South Africa, and New Zealand (CABI database; https://

www.cabi.org). In Ukraine, *L. elongatus* is found in Sumy and Kharkiv regions (Sigareva and Fylenko 1983; Sigareva et al. 1996).

35.2.5.2 Host Plants

Host plants of *L. elongatus* include (apart from sugar beet) field crops such as cotton and maize, fruit and berry crops such as apple (*Malus* spp.), black currant (*Ribes nigrum*), cherry (*Prunus avium*), grape (*Vitis* spp.), peach (*Prunus persica*), pear (*Pyrus communis*), plum (*Prunus domestica*), raspberry (*Rubus idaeus*), strawberry, and forest tree species (Decker 1969; Sigareva et al. 2017).

35.2.5.3 Biology

The needle nematode (*L. elongatus*) is a large (5–10 mm) migrating, ectoparasitic nematode, which inhabits mainly deep (30–60 cm) soil layers. All stages of nematodes are vermiform. The female lays eggs in the soil, from which juveniles of the second stage emerge after 9–12 days of development. They quickly find the young roots of a plant, use their long stylet to pierce the epidermal cells, and begin to feed on their contents. After the fourth moult, the juveniles turn into adult individuals, males or females. Cool rainy spring and summer seasons promote the reproduction of this nematode species (Marić and Čamprag 1982; Cooke 1993; Sigareva and Kalatur 2014; Sigareva et al. 2017).

35.2.5.4 Symptoms

The infestation of plants with *L. elongatus* leads to the formation of slight swellings and galls at or just behind the tips of the lateral roots, as a result of which the main root dies, while lateral roots with a large number of minor roots are formed. Plants are delayed in growth and development (stunted) and have small, narrow leaves. The lower leaves may develop red discolouration at the edges (Decker 1969; Marić and Čamprag 1982; Cooke 1993; Sigareva and Kalatur 2014; Sigareva et al. 2017).

35.2.5.5 Harmfulness

It was found that at the number of *L. elongatus* in soil ranging between 65 and 100 individuals/100 cm^3 root, the yield reduction can reach 60% (Sigareva and Fylenko 1983). In addition to the negative impact on the yield, *L. elongatus* can transmit the tomato black ring virus (TBRV) (a Scottish strain of this virus causes ring spots on beets, see Cadman and Harrison 1960) and raspberry ringspot virus (RRV) (Harrison et al. 1961; Kalatur et al. 2016). This causes additional crop losses.

35.2.6 Stubby Root Nematodes, *Trichodorus* Spp. (Nematoda, Order Dorylaimida, Family Trichodoridae) and *Paratrichodorus* Spp. (Nematoda, Order Dorylaimida, Family Trichodoridae)

35.2.6.1 Occurrence

Nematodes of *Trichodorus* spp. and *Paratrichodorus* spp. are found in most countries of Europe, Africa, Asia, South and North America, and Oceania (CABI database; https://www.cabi.org).

35.2.6.2 Host Plants

Host plants of *Trichodorus* spp. and *Paratrichodorus* spp. include (apart from sugar beet) field crops (e.g., cotton, pea, potato, and tobacco), and many vegetables, flowers and ornamental crops (CABI database; https://www.cabi.org; Decker 1969).

35.2.6.3 Biology

Trichodorus spp. and *Paratrichodorus* spp. are ectoparasitic vermiform nematodes, 0.3–2.0 mm long. They feed externally on the tips of beet roots. The cycle of their development lasts for 21–22 days at a temperature of 22 °C, and for 16–17 days at a temperature of 30 °C (Decker 1969; Cooke 1993).

35.2.6.4 Symptoms of Infestation

Nematodes of these species mainly affect the cells at the tip of the main root or the cells behind it. As a result, the cells stop dividing and the root stops growing in length, and then dies. At the same time, a large number of lateral roots, growing in horizontal direction, are formed. On light sandy soils, these nematodes cause the so-called docking disorder, with symptoms of irregularly stunted plants with many side roots, giving a fangy appearance. Many of the stunted plants cannot be harvested, leading to important yield losses, especially after the heavy rainfall in spring (Decker 1969; Marić and Čamprag 1982; Cooke 1993; Sigareva and Kalatur 2014).

35.2.6.5 Harmfulness

It was found that when the number of *Trichodorus* spp. in the soil exceeds 500 individuals/kg, the yield reduction in sugar beets cultivated on sandy and light soils can reach 50% (Decker 1969; Marić and Čamprag 1982; Cooke 1993; Sigareva and Kalatur 2014). *Trichodorus* and *Paratrichodorus* spp. can transmit plant viruses, in particular, tobacco rattlesnake virus (TRV), early browning virus of pea (PEBV), and the ring spot virus of pepper (PepRSV) (Brown and Trudgill 1998; Kalatur et al. 2016).

35.2.7 Other Nematode Species

In the course of the studies conducted in Ukraine, several other parasitic nematode species have been detected and identified in the rhizosphere of sugar beet, such as

Paratylenchus nanus, the meadow nematode (*Pratylenchus pratensis*), *Helicotylenchus dihystera*, and *Tylenchorhynchus dubius* (Kalatur 1998, 2008a; Galagan and Hryhoriev 2004; Sigareva and Kalatur 2014). These nematode species did not cause any significant damage to sugar beet crops at a low soil density, but a higher number of their population in the soil (1200 individuals/100 cm^3) at the beginning of the growing season had a negative effect on the plant weight (losses reach 35–55%), and contributed to stronger damage to the seedlings by *Pythium* pathogens (Sigareva and Sosenko 2001; Sigareva et al. 2004b; Kalatur 2008b).

35.3 Measures to Prevent and Control Parasitic Nematode Species in Sugar Beet Cultivation

From the above-mentioned research results, it is clear that the beet cyst nematode is the most common and harmful parasitic nematode on sugar beet in Ukraine (Kalatur et al. 2015; Pylypenko et al. 2016). To reduce its harmfulness in sugar beet crops, a system of integrated preventive and control measures, useful under the Ukrainian growing conditions, has been developed, which includes agrotechnical, chemical, and biological methods, and a laboratory analysis before sowing (Korab 1929, Korab 1961, Kitsno 1984, Kalatur 2008b, Kalatur and Polovynchuk 2012, 2013, Kalatur et al. 2015, Hauer et al. 2016, Pylypenko et al. 2016, Borzykh et al. 2017). For a review of the integrated control methods for the beet cyst nematode and use of trap plants, see Held et al. (2000), and Matthias (2020). For the preference of use of resistant varieties over trap plants, see Hauer et al. (2016). Chemical nematicides are not registered in Ukraine. For possibilities in this field from the past, see Cooke (1989); for a more recent urgent plea of the British sugar beet industry for the use of a pesticide on the basis of oxamil (with nematicide and insecticide activity), see https://www.nfuonline.com/sectors/nfu-sugar/nfu-sugar-news/british-sugar-beet-industry-applies-for-an-emergency-authorisation-for-vydate/. For the use of fungicides and insecticides that showed a positive effect on the yield under beet cyst nematode infections in southern Europe, see Sasanelli et al. 2021.

The present control system for nematode infested soil in Ukraine consists of the following:

(1) adequate field intelligence (soil sampling for nematode detection and identification to aid decision making in sugar beet production); (2) measures to prevent the spread of cysts into other fields together with tillage tools, sugar factory waste (sludge, washing water, heads, leaves, and root tips), etc. for which hygiene practice should be administered (Korab 1961; Kalatur and Polovynchuk 2012; Kalatur et al. 2015; Pylypenko et al. 2016); and (3) application of a crop rotation that includes crops that adversely affect the development and reproduction of beet cyst nematodes, such as trap crops, and resistant and/or tolerant varieties/hybrids (Korab 1961; Kitsno 1984; Babich 1990; Kalatur and Polovynchuk 2012; Kalatur et al. 2015; Pylypenko et al. 2016).

35.3.1 Soil Sampling

Soil sampling for lab analysis should occur in autumn or spring, before sowing. This not only enables detection of the fields infested with the beet cyst nematode, but also helps predict any future yield loss due to the infestation level determined. As a rule, detection of *H. schachtii* by external symptoms of the infestation on plants during sugar beet vegetation is late and does not allow prevention of its distribution and thereby mitigation of its negative impact on the yield (Kalatur et al. 2015, Pylypenko et al. 2016).

35.3.2 Crop Rotation Schemes

1. In the fields where the population density of *H. schachtii* reaches a medium or high level (from and above 600 juveniles and eggs in 100 cm^3), it is necessary to exploit a ten-field crop rotation with a two-field share of sugar beet; and the following crop alternation: maize for green fodder–winter wheat–sugar beet–barley–perennial grasses/pea–winter wheat–sugar beet–pea–winter wheat–maize.
   Other possible combinations (depending on farm specialisation and field availability) may include:
   (a) sugar beet–oat with alfalfa–alfalfa–alfalfa–potato–winter wheat–winter barley–rye for green fodder + silage maize;
   (b) sugar beet–pea–winter wheat with alfalfa–alfalfa–alfalfa–alfalfa–potato–rye;
   (c) rye + vetch + maize for green fodder–potato–sugar beet–barley with clover–clover–clover–winter wheat;
   (d) winter/ spring wheat–sugar beet–barley–potato/ chicory–maize–bare fallow;
   (e) alfalfa–alfalfa–alfalfa–alfalfa–potato–sugar beet–barley;
   (f) alfalfa–alfalfa–alfalfa–alfalfa–potato–barley–sugar beet.
2. To use the so-called 'cleaning' crop rotations, i.e. rotations without the main host plant sugar beet):
   (a) rye with vetch–maize for green fodder–winter wheat–chicory–barley with clover–clover/alfalfa–alfalfa–alfalfa;
   (b) alfalfa–alfalfa–alfalfa–chicory–barley/spring wheat–rye with vetch–maize.

In practice, often the following short rotations are performed:

1. maize for green fodder/pea–rye for green fodder/for grain;
2. maize for green fodder/pea–rye for green fodder/for grain–rye for green fodder/ for grain;
3. pea–maize for green fodder–rye;

Table 35.2 Cultivated area of sugar beet and rapeseed in Ukraine[a]

Year	Cultivated area (1000 hectares)	
	Sugar beet	Rapeseed (winter and spring varieties)
2011	515.8	870
2012	448.9	566
2013	270.5	1017
2014	329.6	881.6
2015	238.9	684.4
2016	292.4	456
2017	318	789.1
2018	279.1	1042.4
2019	220.6	1285.4
2020	218.9	1115.2

a Data from State Statistics Service of Ukraine (http://www.ukrstat.gov.ua/)

4. barley with clover–clover–rye;
5. alfalfa–alfalfa–wheat

(Korab 1961, Kitsno 1984, Babich 1990, Kalatur and Polovynchuk 2012, Kalatur et al. 2015; Pylypenko et al. 2016).

Based on the monitored results in various regions of Ukraine in the recent years (Sigareva and Pylypenko 2001; Pylypenko et al. 2016), it was determined that for a number of economic reasons, over the past decade, some farmers have reduced the area under sugar beet (or not sown at all) and increased the cultivated area of rapeseed, which is the host plant of the beet cyst nematode (Table 35.2). Thus, the cultivated area of sugar beet in Ukraine for the period from 2011 to 2020 decreased almost 2.4 times, from 515.8 thousand hectares in 2011 to 218.9 thousand hectares in 2020. At the same time, the cultivated area under rapeseed gradually increased, reaching 1042.4–1285.4 thousand hectares in the years 2018–2020.

The favourite conditions for expanding the cultivated area of winter and spring rapeseed are provided in Vinnytsia, Volyn, Zhytomyr, Ivano-Frankivsk, Kyiv, Lviv, Rivne, Ternopil, Khmelnytskyi, and the Chernivtsi regions. For the cultivation of spring rapeseed only, Kropyvnytskyi, Poltava, Sumy, and most southern regions have favourable conditions. However, in some of these regions, outbreaks of the beet cyst nematode with a high number of cysts in the soil have already been detected (Kalatur and Polovynchuk 2013), particularly in some farms, where the interval between sugar beet and rapeseed was just a year. The shortening of the intervals between sowing the host plants of *H. schachtii* became one of the main reasons for the significant spread of beet cyst nematodes in Ukraine (Sigareva and Pylypenko 2001; Kalatur et al. 2015; Pylypenko et al. 2016). Thus, in farms growing the rapeseed, it is necessary to introduce dedicated rapeseed crop rotations with the share of the rapeseed not exceeding 20–25%, and with the maximum share of grain crops such as:

(a) winter rapeseed–winter wheat–winter rye–bare fallow;
(b) winter rapeseed–winter wheat–bare fallow–spring barley;
(c) spring rapeseed–spring wheat–bare fallow–spring barley.

In case such crop rotation is not an option, rapeseed should be placed in crop rotations in such a way that it returns to the previous place not earlier than 4–5 years.

The interval between rapeseed and sugar beet should also be at least 4–5 years, for example:

(a) perennial grasses–winter rapeseed–winter wheat–maize for silage–pea–winter wheat–spring rapeseed (sugar beet)–barley with perennial grasses;
(b) perennial grasses–winter wheat–spring rapeseed–maize for grain–pea–maize for silage–winter wheat–sugar beet–barley with perennial grasses.

Crop rotation is considered as one of the most effective methods to control the beet cyst nematode and other nematode species in sugar beet and rapeseed crops. Moreover, it is economically profitable and the safest method for the environment and humans (Kalatur et al. 2015, Pylypenko et al. 2016).

35.3.3 General Agricultural Practices

To adhere to the correct agricultural practices of sugar beet cultivation, steps such as timely and high-quality tillage and seedbed preparation, application of balanced rates of organo-minerals and microfertilizers, optimal timing of sowing, and systematic (chemical) weed control in all fields of crop rotation should be taken (Trybel and Stryhun 2012; Kalatur et al. 2015; Pylypenko et al. 2016).

35.3.4 Trap Crops

A good practice is a short-term cultivation of the so-called trap crops (mustard, oil seed radish, and rapeseed), which stimulate the hatching of nematodes. These crops should be sown in August or September after harvesting pea, winter wheat, and/or other early grain crops; then, after 40–45 days, they should be mowed and incorporated into the soil. A decrease in the number of nematodes in the soil (up to 50–60%) will be due to the release of juveniles from cysts, their penetration into the roots of plants, and death during ploughing (Babich 1990; Sigareva and Sosenko 1997; Kalatur et al. 2015; Pylypenko et al. 2016).

35.3.5 Resistant and Tolerant Varieties of Sugar Beet

Growing sugar beet hybrids that are resistant/tolerant to the beet cyst nematode is the best control method in the long term. Despite the fact that plants of tolerant hybrids are still invaded by *H. schachtii*, their yield in fields infested with the parasite is higher compared to the conventional hybrids. Such hybrids are already available in the portfolio of most sugar beet seed producers. In Ukraine, research work on the development of sugar beet hybrids that are tolerant to beet cyst nematodes began in the 1980s (Pylypenko and Kalatur 2015). About 556 breeding genotypes and hybrids of sugar beet have been tested for nematode resistance in the years 1982–1990 and 2013–2020, and this work continues to date. Two domestic sugar beet hybrids, 'Bilotserkivskyi Odnonasinnyi 45' and 'Yuvileinyi', are recognized as tolerant to the beet cyst nematode (Pylypenko and Kalatur 2015). A few resistant/tolerant varieties of foreign origin are also registered on the national variety list (Attack, Balu, Bison and Federica).

35.3.6 Resistant and Tolerant Varieties of Oil Seed Radish and Yellow Mustard

Growing nematode-resistant/tolerant varieties of oil seed radish and yellow mustard is also advised. In Ukraine, trials to determine the tolerance/resistance of local varieties are still ongoing and are in the stage of breeding lines (Sigareva and Pylypenko 2001).

35.3.7 Use of Pesticides

When seeds were treated with insecticides (in the absence of any allowed nematicides) with active compounds carbofuran, thiamethoxam, imidacloprid, and bifenthrin, there was a 40% reduction in the number of the first-generation beet cyst nematodes under Ukrainian growing conditions (Babich 1990; Kalatur 2007). By the end of the sugar beet-growing season, however, due to the development of the second and subsequent generations, this parasite reproduced, and in some places even exceeded its initial number in the soil, even when insecticides were applied (Kalatur et al. 2015, Pylypenko et al. 2016).

35.4 Future Prospects

Studies conducted abroad have shown that the application of the biological control formulation Clariva pn (containing the mycelial and endospore-forming bacterium *Pasteuria nishizawae*, see https://www.syngenta-us.com/seed-treatment/clariva-pn and Perry et al. 2018) together with an insecticide based on thiamethoxam as seed treatment may greatly improve the yield of the beet cyst nematode tolerant varieties. Clariva pn, however, has only been registered for sugar beet in the USA and Brazil. The hyperparasitic fungus, *Hyalorbilia* sp. strain DoUCR50, proved to be effective in reducing the yield loss in tolerant varieties in studies performed in Germany (Eberlein et al. 2020).

35.5 Conclusion

The proposed measures to control the beet cyst nematode in sugar beet crops will also be effective against the other nematode species mentioned in this chapter. In particular, to limit the harmfulness of the root knot and migrating root nematodes, it is necessary to collect and destroy nematode-infested roots. Also, it is necessary to adhere to the correct agricultural techniques of growing crops (Decker 1969; Cooke 1989; Cooke 1993).

References

Babich AG (1990) Вредоносность свекловичной нематоды и пути ее снижения в Правобережной Лесостепи Украинской ССР. [Beet nematode harmfulness and the ways of its mitigation in the Right-Bank Forest-Steppe of the Ukrainian SSR]. (In Russian) Abstract of the Cand Sc (Agr) thesis Ukrainian Agricultural Academy, Kyiv, pp 17

Babich AG (2004) Бур'яни-живителі бурякової нематоди. [Weed host species of beet nematode]. (in Ukrainian). In: Issues of weeds and ways to reduce weed infestation of arable land: Scientific Conference of the Ukrainian Scientific Society of Herbologists: Abstracts, 2004, Kyiv, Ukraine, pp 84–86

Borzykh OI, Sigareva DD, Pylypenko LA, Kovtun AM (2017) Найбільш небезпечні нематодози рослин та системи захиснихзаходів. [The most dangerous plant nematode diseases and the systems of control measures]. (In Ukrainian). Interservis, Kyiv. pp 140. ISBN: 978-617-696-569-5

Brown DJF, Trudgill DL (1998) Nematode transmission of plant viruses—a 30-year perspective. Host pathogen interactions & crop protection. Ann Rep Scottish Crop Res Inst (SCRI):121–125

Cooke DA (1989) Damage to sugar-beet crops by ectoparasitic nematodes, and its control by soil-applied granular pesticides. Crop Prot 8:63–70. https://doi.org/10.1016/0261-2194(89)90101-4

Cooke DA (1993) Pests. In: Cooke DA, Scott RK (eds) The sugar beet crop: science into practice. Chapman & Hall, London, pp 428–483. ISBN: 0–412–25130–2

Decker H (1969) Phytonematologie. Biologie und Bekämpfung pflanzenparasitärer Nematoden. VEB Deutscher Landwirtschaftsverlag, Berlin 526 s

Eberlein C, Heuer H, Westphal A (2020) Biological suppression of populations of *Heterodera schachtii* adapted to different host genotypes of sugar beet. Front Plant Sci 11:812. https://doi.org/10.3389/fpls.2020.00812

EPPO PM 7/5 (2) (2009) *Nacobbus aberrans* sensu lato. Bull OEPP/EPPO 39:376–381

Galagan TO, Hryhoriev VM (2004) Комплекси фітонематод бурякового агроценозу. [Phytonematode complexes of beet agrocenosis]. (In Ukrainian) Plant protection and quarantine 50:184–189

Graf A, Meyer H (1973) Importance of sugar beet stem eelworm in Switzerland and possibilities for control. J Int Inst Sugar Beet Res (IIRB) 6(3):117–126

Gubina VG (ed) (1982) Нематоды растений и почвы. [Plant and soil nematodes]. (In Russian) Genus *Ditylenchus*. Nauka, Moscow, pp 248

Harrison C (1960) Studies on the behavior in soils of tomato black ring, raspberry ringspot, and arabis mosaic viruses. Virology 10:1–20

Harrison BD, Mowat WP, Taylor CE (1961) Transmission of a strain of tomato black ring virus by *Longidorus elongatus* (Nematoda). Virology 14(4):480–485

Harveson RM (2014) The false root-knot nematode: a unique plant pathogen native to the Western hemisphere. APS Features. https://doi.org/10.1094/APSFeature-2014-06

Hauer M, Koch H-J, Krüssel S, Mittler S, Märländer B (2016) Integrated control of *Heterodera schachtii* Schmidt in Central Europe by trap crop cultivation, sugar beet variety choice and nematicide application. Appl Soil Ecol 99:62–69. https://doi.org/10.1016/j.apsoil.2015.11.017

Held LJ, Jennings JW, Koch DW, Gray FA (2000) Economics of trap cropping for sugar beet nematode control. J Sugar Beet Res 37:45–55

Hillnhütter C, Albersmeier A, Sikora RA (2011) Synergistic damage by interaction of *Ditylenchus dipsaci* and *Rhizoctonia solani* (AG 2-2IIIB) on sugar beet. J Plant Dis Prot 118:127–133

Kalatur KA (1998) Видовий склад, екологічні угруповання, статус домінування фітонематод в ризосфері цукрових буряків [Species composition, ecological groups, and dominance status of phytonematodes in the rhizosphere of sugar beet]. (in Ukrainian). Scientific bases of the production of sugar beet and other crops of beet crop rotation in contemporary economic and ecological conditions: Scientific papers of the Institute of Sugar Beet UAAS 2:15–19

Kalatur KA (2007) Вплив інсектицидів при обробці ними насіння цукрових буряків на чисельність бурякової нематоди *Heterodera schachtii* Schmidt. [The effect of insecticides in the treatment of sugar beet seeds on the number of beet nematode *Heterodera schachtii* Schmidt]. Sugar beet 5:7–8. (in Ukrainian)

Kalatur KA (2008a) Шкідливість паразитичних нематод у посівах цукрових буряків. [Harmfulness of parasitic nematodes in sugar beet sowings]. Scientific Papers of the Institute of Sugar Beet 10:313–317. (in Ukrainian)

Kalatur KA (2008b) Влияние паразитических нематод на поражение корнеедом и продуктивность сахарной свеклы. [The effect of parasitic nematodes on *Pythium* disease and productivity of sugar beet]. Plant Prot Quarantine 10:36–38. (in Russian)

Kalatur KA, Polovynchuk OYU (2012) Методы ограничения вредоносности свекловичной нематоды. [Methods of limiting the harmfulness of beet nematode]. Sugar beet 6:31–34. (in Russian)

Kalatur KA, Polovynchuk OYU (2013) Свекловичная цистообразующая нематода на рапсе. [Beet cyst nematode in rapeseed crops]. Plant Prot Quarantine 10:14–16. (in Russian)

Kalatur KA, Pylypenko LA (2017) Бур'яни—резерватори популяцій паразитичних видів фітонематод. [Weeds: reserves of parasitic phytonematode species populations]. Adv Agritechnol 5. (in Ukrainian). https://doi.org/10.21498/na.5.2017.122232

Kalatur KA, Suslyk LO, Pylypenko LA (2015) Захист посівів цукрових буряків від бурякової нематоди : рекомендації. [Control of beet nematode in sugar beet crops: recommendations] (in Ukrainian) IBCSB, Kyiv, 22 p

Kalatur KA, Pylypenko LA, Boiko AL (2016) Роль фітонематод родин Longidoridae і Trichodoridae у векторному перенесенні збудників вірусних хвороб рослин. [Role of Longidoridae and Trichodoridae phytonematodes in vectorial transfer of viral plant pathogens]. Scientific Papers of the Institute of Bioenergy Crops and Sugar Beet 24:100–111 (in Ukrainian)

Kalatur KA, Borzykh OI, Sigareva DD, Janse LA (2020) Взаємовідносини між фітонематодами та іншими патогенними організмами : монографія. [Relationships between phytonematodes and other pathogens: A monograph]. (in Ukrainian). Lazuryt-Polihraf LLC, Kyiv. 192 p. ISBN: 978-966-1543-58-3

Kazachenko IP, Muhina TI (2013) Корневые галловые нематоды рода *Meloidogyne* Goeldi (Tylenchida: Meloidogynidae) мировой фауны. [Root gall nematodes of the genus *Meloidogyne* Goeldi (Tylenchida: Meloidogynidae) of the world fauna]. (In Russian). Dalnauka, Vladivostok, pp 307. ISBN: 978-5-8044-1366-9

Kitsno LV (1984) Биологическое обоснование мер борьбы со свекловичной нематодой в условиях Лесостепи Украинской ССР. [Biological substantiation of the measures to control beet nematode in the Forest-steppe conditions of the Ukrainian SSR] (in Russian) Abstract of the Cand Sc (Biology) thesis Ukrainian Research Institute of Plant Protection, Kyiv, 23 p

Korab II (1924) Sugar beet nematode (*Heterodera schachtii* Schm.) in Ukraine. Bull Kyiv Station of Plant Prot 2:4–7

Korab II (1929) Материалы по изучению свекловичной нематоды (*Heterodera schachtii* Schm.) по данным работ нематодной лаборатории Белоцерковской селекционной станции ССУ. [Materials of the study on beet nematode (*Heterodera schachtii* Schm.) based on the data from the nematode Laboratory of the Bila Tserkva Breeding Station of the BSD]. Scientific Papers of the Beet-Sugar Directorate 8(16):29–67 (in Russian)

Korab II (1961) Главнейшие меры борьбы со свекловичной нематодой [The main measures of beet nematode control]. (In Russian). In: Skriabin KI, Turlygina ES (eds) Issues of phytohelminthology. Helminths and helminthiasis of agricultural crops and the control measures. Publishing House of the Academy of Sciences of the USSR, Moscow, pp 84–95. 248 p

Kuthe K (1974) The effect of sugar beet nematode (*Ditylenchus dipsaci* Filipjev) infestation on the impurity percentage, sugar content and yield of sugar beet. Gesunde Pflanz 26(3):48–50, 52–57

Ladygina NM (1961) Реакция свекловичной нематоды на температуру и влажность. [Reaction of sugar beet nematode to temperature and humidity]. (In Russian). In: Skriabin KI, Turlygina ES (eds) Issues of phytohelminthology. Helminths and helminthiasis of agricultural crops and the control measures. Publishing House of the Academy of Sciences of the USSR, Moscow, pp. 129–141

Linnik LI (1978) Свекловичная гетеродера на Украине. [Beet heterodera in Ukraine]. Plant Prot 5:40–41. (in Russian)

Maas PWT, Maenhout CAAA (1978) Het graswortelknobbelaaltje (*Meloidogyne naasi*) bij suikerbieten. [The cereal root knot nematode (*Meloidogyne naasi*) in sugar beet]. (In Dutch) Gewasbescherming 9(6):159–166

Manzanilla-López RH, Costilla MA, Doucet M, Franco J, Inserra RN, Lehman PS, del Prado C, Vera I, Souza RM, Evans K (2002) The genus *Nacobbus* Thorne & Allen, 1944 (Nematoda: Pratylenchidae): systematics, distribution, biology and management. Nematropica 32(2): 149–227

Marić A, Čamprag D (1982) Štetočine i bolesti šećerne repe. [Pest and diseases of sugar beet] (in Serbo-Croatian) Nolit, Beograd. 168 p. ISBN 2-904263-00-4

Matiashov VD (1971) Галловая нематода *Meloidogyne hapla* Chitwood, 1949 на сахарной свекле в Киргизии. [Gall nematode *Meloidogyne hapla* Chitwood, 1949 on sugar beet crops in Kyrgyzstan]. In: Helminthological research in Kyrgyzstan. Ilim, Frunze, pp 63–70. (in Russian)

Matthias D (2020) Effect of winter oilseed rape cropping on the development of the sugar beet cyst nematode, *Heterodera schachtii*, and control of volunteer plants as a trap crop method. Agronomy 10:355, 14 pp. https://doi.org/10.3390/agronomy10030355

Pandey S (1984) Associative effects of *Meloidogyne incognita, Pythium ultimum* and *Rhizoctonia solani* on sugar beetseedlings. Indian Phytopathol 37(3):462–465

Perry RN, Moens M, Starr JL (eds) (2018) Root-knot nematodes. CABI Publishing, Wallingford, UK. pp. 488. ISBN-10: 9781786390837

Prymak ID, Fedorenko VP, Kozak LA, Horodetskyi OS, Lapa OM (2009) Буряківництво: підручник. [Beet growing: Textbook]. Kolobih, Kyiv. pp. 464 (in Ukrainian). ISBN: 978–966–8610–43–1

Pylypenko LA, Kalatur KA, Hallmann J (2016) Sugar beet nematode *Heterodera schachtii* distribution and harmfulness in Ukraine. Agric Sci Pract 3(3):3–11. https://doi.org/10.15407/agrisp3.03.003

Pylypenko LA, Kalatur KA (2015) Breeding and usage of sugar beet cultivars and hybrids resistant to sugar beet nematode *Heterodera schachtii*. Agric Sci Pract 2(1):12–22. https://doi.org/10.15407/agrisp2.01.012

Rich JR, Brito JA, Kaur R, Ferrell JA (2009) Weed species as hosts of *Meloidogyne*: a review. Nematropica 39:157–185

Roik M (2001) Буряки [Beets]. (in Ukrainian). Publishing House XXI—RIA Trud-Kyiv, Kyiv. pp 320. ISBN: 966–95049–0–2

Sasanelli N, Konrat A, Migunova V, Toderas I, Iurcu-Straistaru E, Rusu S, Bivol A, Andoni C, Veronico P (2021) Review on control methods against plant parasitic nematodes applied in southern member states (C zone) of the European Union. Agriculture 11:602. https://doi.org/10.3390/agriculture11070602

Shcherbak PD (1973) Распространение и вредоносность стеблевой нематоды свеклы в Запорожской области. [Occurrence and harmfulness of beet stem nematode in Zaporizhzhia region]. Bulletin of the All-Union Skriabin Institute of Helminthology 11:95–97. (in Russian)

Sigareva DD, Fylenko VL (1983) Новое заболевание сахарной свеклы, вызываемое эктопаразитическими нематодами *Longidorus elongatus.* [A new disease of sugar beet caused by ectoparasitic nematodes *Longidorus elongatus*]. (In Russian). In: Stem nematodes in agricultural crops and control measures. Proceedings of the All-Union symposium on plant stem nematodes, September 27–29, 1983, Voronezh, Russia, pp 152–155, pp 167

Sigareva DD, Kalatur KA (2014) Видовий склад та шкідливість паразитичних нематод в агроценозах цукрових буряків. [Species composition and harmfulness of parasitic nematodes in sugar beet agrocenoses]. Plant Prot Quarantine 60:303–317. (in Ukrainian)

Sigareva DD, Pylypenko LA (2001) Бурякова нематода в бурякових та ріпакових сівозмінах: моніторинг та шляхи зниження чисельності. [Beet nematode in beet and rapeseed crop rotations: monitoring and ways to reduce the population density]. Plant Prot 4:11–12. (in Ukrainian)

Sigareva DD, Sosenko EB (1997) Пожнивные посевы крестоцветных культур как средство защиты от свекловичной нематоды (в условиях Лесостепной зоны Украины). Производство экологически безопасной продукции растениеводства: региональные рекомендации. [Cruciferous catch crops as a method to control beet nematode (in the Forest-Steppe zone of Ukraine). Environment-friendly plant crop production: Regional recommendations]. 3:148–151 (in Russian)

Sigareva DD, Sosenko EB (2001) Вредоносность комплекса паразитических нематод на сахарной свекле. [Harmfulness of a complex of parasitic nematodes on sugar beet]. In Issues of zoo engineering and veterinary medicine: scientific papers of the Kharkiv State Zooveterinary Academy. Proceedings of the V congress of parasitocenologists of Ukraine, April 5–6, Kharkiv, Ukraine, pp 270–271 (in Russian)

Sigareva DD, Fylenko VL, Sosenko EB (1996) Защита посевов сахарной свеклы от лонгидороза (в условиях Лесостепной зоныУкраины). Производство экологически безопасной продукции растениеводства: региональные рекомендации. [Control of longidorosis in sugar beet crops (in the Forest-Steppe zone of Ukraine). Environment-friendly plant crop production: Regional recommendations]. 2:186–187 (in Russian)

Sigareva DD, Shvets HD, Pylypenko LA (2004a) Нематологічні обстеження насінницьких господарств Вінницької області. [Nematological surveys in seed farms of Vinnytsia region]. In Contemporary science: areas of research, status and prospects: proceedings of the fourth interuniversity scientific and practical conference for postgraduates, April 5–7, Vinnytsia, Ukraine, pp 49–51 (in Ukrainian)

Sigareva DD, Pylypenko LA, Svinar OP (2004b) Фитонематоды сахарной свеклы: определение вредоносности. [Sugar beet phytonematodes: definition of harmfulness]. Bulletin of Zoology. Issues of parasitology. Proceedings of the Ukrainian Scientific Society of Parasitologists 18:131–133 (in Russian)

Sigareva DD, Kalatur KA, Hryhoriev VM (2007) Вплив бурякової нематоди (*Heterodera schachtii* Schmidt) на розвиток хвороб цукрових буряків. [The effect of beet nematode (*Heterodera schachtii* Schmidt) on the development of sugar beet diseases]. Plant Protection and Quarantine 53:174–180. (in Ukrainian)

Sigareva DD, Pylypenko LA, Borzykh OI, Kovtun AM (2017) Сільськогосподарська нематологія. [Agricultural nematology]. (In Ukrainian) Ahrarna nauka, Kyiv 340 p. ISBN: 978-966-540-465-1

Sosenko OB (1998) Поширення бурякової нематоди в Україні. [Occurrence of beet nematode in Ukraine]. Bull Agric Sci 5:83. (in Ukrainian)

Steel AE (1965) The host range of the sugar beet nematode, *Heterodera schachtii* Schmidt. J Sugar Beet Res 13:573–603

Subbotin SA, Palomares-Rius JE, Castillo P (2021) Systematics of root-knot nematodes (Nematoda: Meloidogynidae). Series: nematology monographs and perspectives 14. Brill, Leiden, The Netherlands

Trybel SO, Stryhun OO (2012) Динаміка вирощування цукрових буряків в Україні та фітосанітарний стан посівів. [Dynamics of sugar beet cultivation in Ukraine and phytosanitary condition of crops]. Scientific Papers of the Institute of Bioenergy Crops and Sugar Beet 14:217–222. (in Ukrainian)

Turner SJ, Rowe JA (2006) Cyst nematodes. In: Perry RN, Moens M (eds) Plant nematology. CAB International, Wallingford, Oxfordshire, pp 9B130–9B122

Diseases Caused by Nematodes on the Sugar Beet

36

Intiaz Amin Chowdhury, Guiping Yan, and Mohammed Khan

Abstract

Nematodes are one of the most abundant multicellular organisms in the world. A group of these soft-bodied roundworms, called plant-parasitic nematodes (PPN), can parasitize plants and cause serious damage. Most PPN are soil-borne pests and can cause yield loss by feeding on the root tissue of host plants and depriving them of nutrients and water. Annual worldwide crop losses due to PPN have been estimated to range from 8.8 to 14.6%. However, the severity of yield losses caused by PPN can vary with the species present in the fields. Among the numerous species of PPN that infest sugar beet fields, sugar beet cyst nematode (SBCN) and stubby root nematode (SRN) are two of the more economically important groups for sugar beet growers. The above-ground symptoms caused by these PPN mimic the symptoms caused by other biotic diseases and abiotic stresses. Below the ground, tiny lemon-shaped white to yellow adult SBCN females can be seen on SBCN-infected roots. Feeding of SRN on taproot and lateral root tips causes the roots to get swollen, giving the roots a stubby-ended appearance. Effective management of PPN relies on an integrated approach that focuses on preventing the introduction of PPN into an un-infested field and reducing their reproduction in infested fields using strategies such as crop rotation and host resistance. Since effective management tactics can vary with the species present in the fields, sampling and diagnosis are critical. Through such proactive management strategies, yield losses caused by PPN can be reduced.

I. A. Chowdhury · G. Yan (✉) · M. Khan
Department of Plant Pathology, North Dakota State University, Fargo, ND, USA
e-mail: guiping.yan@ndsu.edu

V. Misra et al. (eds.), *Sugar Beet Cultivation, Management and Processing*,
https://doi.org/10.1007/978-981-19-2730-0_36

Keywords

Heterodera schachtii · Nematode management · Nematode symptoms · *Paratrichodorus* · Plant-parasitic nematodes · Sugar beet · Sugar beet cyst nematode · Stubby root nematode · *Trichodorus*

Abbreviations

PPN	Plant-parasitic nematodes
SBCN	Sugar beet cyst nematode
SRN	Stubby root nematodes

36.1 Introduction

Nematodes have existed on the earth for more than a billion years, making them one of the most ancient and diverse groups of animals (Lambert and Bekal 2002). Today, these soft-bodied roundworms have evolved to occupy almost every ecological niche from the deserts to the snowy mountains of the world. Moreover, they are one of the most abundant groups of multi-cellular organisms on the earth. In a single 2-acre field of any soil type, more than 7.5 billion nematode individuals can be found in the top 20 cm of soil (Hartman et al. 2015). A group of these nematodes, called plant-parasitic nematodes (PPN), survive in these fields by feeding on plant roots. Plant-parasitic nematodes can be distinguished from other nematodes by the presence of a specialized hollow spear-like feeding structure near their mouth, called a stylet, that is visible when the nematode is viewed under a microscope. However, shapes and sizes of the stylet can vary with the genera and species of PPN. Usually, the worm-like bodies of PPN are less than 2 mm long and 0.1 mm in diameter (Mai et al. 1996). Thus, a magnification of at least 10× or higher is required for identification. However, a few nematodes, such as the cyst nematodes in the adult stage, can be seen without the aid of a microscope since the females enlarge as they produce eggs inside their body.

Plant-parasitic nematodes can cause serious damage to the host plant's root system, reducing the plant's ability to absorb nutrients and water. Annual worldwide crop losses due to PPN have been estimated to range from 8.8 to 14.6%, causing economic losses worth 100 to 157 billion US$ (Singh et al. 2013). Additionally, yield losses caused by PPN often go unnoticed or are attributed to other causes because the symptoms caused by PPN mimic the symptoms caused by other biotic diseases and abiotic stresses (Norton 1978). In PPN-infested sugar beet fields, yield losses due to nematodes have been estimated to range between 10 and 80% (Hafez 1998). However, the severity of damage caused by PPN can vary with the species present in the field and their population density. More than 37 species of PPN have been reported to infest sugar beet fields; however, only a few PPN species have been reported to cause serious damage to the sugar beet crop (Harveson et al. 2009; Karegar 2006). Among them, the sugar beet cyst nematode and the stubby root nematode are two of the most economically important PPN pests of the sugar beet.

36.2 Sugar Beet Cyst Nematode

Sugar beet cyst nematode (SBCN; *Heterodera schachtii*) is considered the most devastating threat to sugar beet production worldwide. These cyst-forming nematodes were first discovered near Halle, Germany in 1859. Within a few years of its discovery, SBCN was determined to be a major cause of "beet weariness" disease that resulted in the closure of 24 sugar beet factories in Germany (Harveson and Jackson 2008). Again, during the latter half of the nineteenth century SBCN infestation devastated sugar beet production in several European countries including Germany. Since then this nematode has spread to almost all the major sugar beet-producing regions of the world, including Europe, the United States, and Canada (Harveson et al. 2009).

36.2.1 Symptoms and Sign

Above the ground, symptoms of SBCN infestation in sugar beet fields can appear as circular to oval areas of pale-yellow and stunted plants (Khan et al. 2016a). In the afternoons during warm sunny days, the outer leaves of SBCN-infected plants usually appear wilted with pronounced yellowing (Fig. 36.1). Since SBCN can attack sugar beet plants at any growth stage, the seedlings and young plants can be seriously injured or killed when sugar beet is planted in SBCN-infested fields. The infected plants that survive until the adult stage often remain stunted and unreproductive (Harveson and Jackson 2008).

Below the ground, the invasion and feeding of SBCN on the roots of host plants results in the fibrous growth of secondary lateral roots, while the taproot becomes stunted. Thus, the root system of SBCN-infected plants can appear as "bearded" or "hairy" (Bridge and Starr 2019; Harveson et al. 2009).

One of the most important ways to confirm SBCN infestation is to check for the presence of white to yellow lemon-shaped females attached to the feeder roots (Fig. 36.2). However, when the adult SBCN females mature and die, they become lemon-shaped brown cysts (Fig. 36.3) that can be found attached to the roots or in the soil around the root system (Bridge and Starr 2019). Although SBCN adult females and cysts are visible to the naked eye, a magnifying glass is useful for easier detection.

Fig. 36.1 Sugar beet cultivars susceptible to sugar beet cyst nematode (SBCN) planted in the middle of the SBCN-infested field and in the foreground. Infected plants becoming pale yellow and wilted in contrast with the healthy green plants can be observed. (Courtesy of Dr. Steve Poindexter, Michigan State University)

Fig. 36.2 White lemon-shaped females of sugar beet cyst nematode attached to sugar beet roots. (Courtesy of Dr. Steve Poindexter, Michigan State University)

Fig. 36.3 Yellow-brown cysts of sugar beet cyst nematode (SBCN) collected from a SBCN-infected plant (Courtesy of Dr. Guiping Yan, North Dakota State University)

36.2.2 Host Range

Although a majority of cyst-forming nematodes have a very narrow host range, SBCN can parasitize a relatively larger group of host plants (greater than 200 plant species), including all other types of beets (Hafez 1998). This nematode species has been reported to be an economically important pest of the table beet. Other economically important hosts of SBCN include members of the Brassicaceae family such as broccoli, brussels sprout, cabbage, canola, cauliflower, mustard, radish, and turnip. Additionally, several weed species, including lambs quarters, pigweed purslane, and shepherd's purse, have been reported to be good hosts of SBCN (Harveson et al. 2009).

36.2.3 Life Cycle and Survival

The life cycle of SBCN consists of an egg stage, four juvenile stages, and an adult stage. Inside the egg, the embryo of SBCN develops into the first-stage juvenile (J1). Stimulated by the exudates of nearby roots, the J1 then molts into the second-stage juvenile (J2) as it hatches and emerges from the eggshell (Fig. 36.4). The J2 is the infective stage of SBCN that migrates through the soil to the host root system, following the chemical gradients created by the root exudates. The J2 then uses its stylet to infect the elongation zone of the roots and penetrate into the root cortex (Hafez 1998).

After entering the root cortex, the J2 travels intracellularly towards the vascular tissue to find a suitable feeding site. Once it finds a suitable feeding site, it causes physiological changes to the nearby host cells to become a nutrient sink called a syncytium (Thurau et al. 2003). The nematode then starts to enlarge and molt two more times to become an adult. At this stage, the males become vermiform

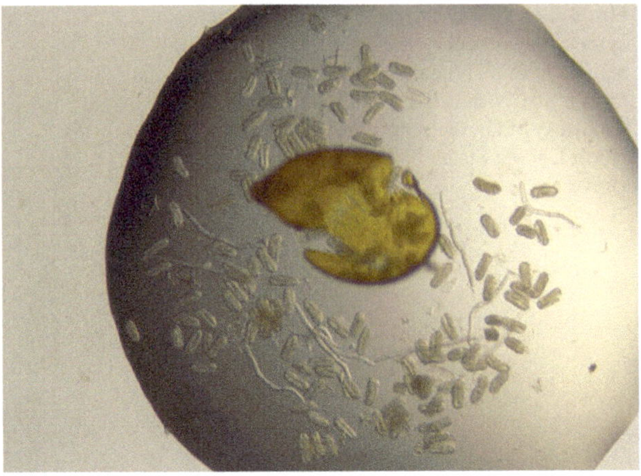

Fig. 36.4 Eggs and juveniles released from the cysts of sugar beet cyst nematode after crushing. (Courtesy of Dr. Guiping Yan, North Dakota State University)

(worm-like) and mobile again, while the females remain sedentary and continue to enlarge into their characteristic lemon shape, producing eggs inside their body. The vermiform males then fertilize the females to produce the eggs. As the females enlarge, their posterior end bursts through the root tissue, becoming exposed to the exterior of the roots, while the head and neck remain embedded in the root tissue. As the female matures and dies, its body wall thickens and undergoes a tanning process, becoming a brown cyst containing up to 500 eggs inside its body (Hafez 1998). On an average, 200 to 300 eggs can be found inside each cyst (Khan et al. 2016a). The cysts can detach from the roots and remain free in the soil, and the eggs can survive inside the cysts for many years, not hatching until stimulated by the favorable root exudates and the right environmental conditions. The life cycle of SBCN normally takes 4–6 weeks to complete, depending on the soil moisture and soil temperature (Hafez 1998).

36.2.4 Epidemiology and Spread

The optimal temperature for SBCN reproduction is between 21 and 27 °C; however, SBCN can reproduce at any temperature between 10 and 32 °C. The temperature most conducive to SBCN egg hatching is 25 °C, while the maximum J2 activity and parasitism is favored by a temperature of 24 °C and a low to moderate soil moisture level (Harveson et al. 2009). Moreover, sandy loam soil texture is optimal for SBCN movement, reproduction, and survival. Anything that moves the soil can move SBCN, such as animals, people, harvested beets equipment, tare soil, and water. In the soil profile, SBCN cysts can be found between 1 and 60 cm below the soil surface; however, the maximum number of cysts is usually found between 5 and 25 cm deep in the root zone (Khan et al. 2016a).

36.2.5 Management

Effective management of SBCN relies on an integrated approach that focuses on preventing the introduction of SBCN intoan uninfested field and reducing SBCN reproduction in an infested field (Khan et al. 2016a). Prevention of SBCN infestation requires strict sanitation measures that include cleaning and sterilizing equipment and other tools that can transfer the soil after being used in an infested field. Tare soil from infested fields contains concentrated amounts of SBCN cysts. Thus, tare soil should not be dumped into fields where sugar beet may be planted in the future (Harveson et al. 2009). Composting tare soil can also reduce the nematode population in it (Hafez 1998).

In the infested fields, reducing SBCN reproduction and the subsequent population decline can be achieved by implementing a combination of management practices that include rotation with non-host crops, planting tolerant cultivars, planting early, controlling weed hosts, using trap crops, and the use of chemical and biological agents that control nematode populations (Khan et al. 2016a). Crop rotation with non-host crops is the most economical and easiest way to reduce SBCN related yield loss. Non-host crops can reduce the initial SBCN population by 40–60% in a single year (Khan et al. 2016a). Crops such as alfalfa, wheat, barley, bean, corn, mint, onion, and potato are good examples of crops that do not support SBCN reproduction (Hafez 1998; Khan et al. 2016a). A minimum of 3–4 years of crop rotation with non-host crops should be utilized in the fields heavily infested with SBCN.

Planting SBCN-tolerant or -resistant cultivars is another way to reduce sugar beet yield loss due to SBCN. Although the genes that confer resistance to SBCN were reported to lack in cultivated *Beta* species (Cai et al. 1997), a SBCN resistance gene, $Hs1^{pro\text{-}1}$ has been cloned with wild relatives of sugar beets such as *Patellifolia procumbens* (Cai et al. 1997; Ghaemi et al. 2020). Consequently, several SBCN-resistant commercial cultivars (eg. Nemakill, Evasion, and Nematop) have been developed by crossing *Beta vulgaris* with *P. procumbens* (Ghaemi et al. 2020; Pylypenko and Kalatur 2015). Moreover, several cultivars with lower levels of resistance (tolerance) have been developed and implemented in the United States (Harveson et al. 2009). Examples of such tolerant cultivars approved for different production regions of the United States include Crystal 932NT, Crystal A702NT, Crystal A404NT, BTS 437N, BTS 188N, BTS 380N, SV099N, and SV1686N. In Europe, cultivars such as Daphna and Cantona KWS are available as SBCN-tolerant cultivars. However, it is critical to be aware of such SBCN-resistant/tolerant cultivars' susceptibility to other diseases such as Cercospora and Rhizoctonia root rot (Khan et al. 2016a). It is also important to manage weed hosts effectively since they can support SBCN reproduction even if non-host crops or resistant/tolerant cultivars are planted (Harveson and Jackson 2008).

The older the sugar beet plants are when SBCN infection occurs, the lesser the damage caused by the nematode. Thus, planting early in the season when temperatures are not favorable for SBCN activity can reduce yield loss at harvest (Harveson and Jackson 2008). Trap crops are also effective against SBCN, since they can trigger hatching of eggs and can attract SBCN juveniles, but do not let them develop and reproduce, thus preventing them from completing their life cycle and reducing their population densities (Khan et al. 2016a). In Europe, some SBCN-resistant commercial cultivars of mustard and radish have been effectively used to reduce SBCN population levels in infested fields (Harveson et al. 2009). Some nematicides such as organo-carbamates and fumigants such as 1,3-dichloropropene and isothiocyanates are effective in controlling the nematode population. However, they are difficult to apply and uneconomical in a large-scale sugar beet production. Moreover, they have significant negative environmental effects. Thus, the use of such chemicals for nematode management is not commonly recommended (Hafez 1998; Harveson et al. 2009; Khan et al. 2016a). On the other hand, biological seed treatments such as *Pasteuria nishizawae* spores may help to manage SBCN on tolerant sugar beet cultivars (Khan et al. 2016a).

36.3 Stubby Root Nematode

Stubby root nematodes (SRN) are another economically important group of pests that can parasitize the sugar beet. These nematodes belong to the *Trichodorus* and *Paratrichodorus* genuses (Khan et al. 2016b). Unlike SBCN, SRN is a migratory ectoparasite, meaning they remain vermiform throughout their life cycle and their body remains outside of the root tissue feeding on it using their stylet. This strategy allows them to migrate between plants and graze over multiple roots during their lifetime. Yield losses due to SRN have been reported to be as high as 50% under favorable conditions (Khan et al. 2016b). Some SRN species are also good vectors of viruses that infect other economically important crops such as potatoes (Harveson et al. 2009). Although SRN has a worldwide distribution, they are considered economically important on sugar beet in the European countries such as the United Kingdom and in major sugar beet-growing regions of the United States such as California and Idaho (Bridge and Starr 2019; Khan et al. 2016b).

36.3.1 Symptoms

The above-ground symptoms caused by SRN resemble the symptoms caused by the other nematodes, which include mild yellowing of leaves and interveinal chlorosis (Hafez 1998). SRN infection deprives the host plants of absorbance of nutrients and water, reducing plant growth and causing plants to get stunted. In infested fields, especially in hot spots, patches of stunted yellowed plants can be seen.

Below the ground, SRN feeds on the taproot as well as the lateral root tips, causing the roots to get swollen and giving the roots a stubby-ended appearance (Hafez 1998), which is the namesake of SRN. SRN feeding on the tip of the roots can also cause them to die resulting in the surviving roots becoming branched and distorted. Stubby root nematode infection at the seedling stage can also destroy the tip of the taproots, thereby reducing their size, and the lateral roots become thick and brown (Khan et al. 2016b).

36.3.2 Causal Organism

Species of SRN that can infect the sugar beet include *P. anemones*, *P. allius*, *P. christiei*, *P. pachydermus*, *P. teres*, *T. cylindricus*, *T. primitivus*, and *T. viruliferus* (Bridge and Starr 2019; Harveson et al. 2009; Ashmit 2019; Yan et al. 2016). The SRN comprise generally small vermiform nematodes that are 0.4–1.8 mm in length at their adult stage (Harveson et al. 2009). Thus, they require nematode extraction from the soil and a microscope for visual identification. Under the microscope, SRN can be distinguished from other nematodes by their characteristic curved onchio stylet and rounded (blunt) head and tail regions.

36.3.3 Host Range

Stubby root nematodes have a wide host range. They have been reported to cause yield losses on multiple economically important crops. *Paratrichodorus allius*, a SRN species, is considered a serious pest of the potato because of its ability to vector *Tobacco rattle virus* that causes corky ringspot disease in the potato (Mojtahedi and Santo 1999). Other economically important hosts of SRN include apple, avocado, cereals, corn, grapevine, onion, and sugar beet (Khan et al. 2016b).

36.3.4 Life Cycle and Survival

Similar to SBCN, the life cycle of SRN has an egg stage, four juvenile stages, and an adult stage. However, unlike SBCN, SRN do not become sedentary at any stage during their life cycle and they remain in the soil throughout their life cycle (Khan et al. 2016b). The juveniles of SRN have the same appearance as the adults; however, the juveniles are smaller than the adults. Some SRN species can reproduce parthenogenetically (e.g., *P. teres*); therefore, males are not required for reproduction. On the other hand, for some species (e.g., *P. pachydermus*) sexual reproduction is more frequent (Harveson et al. 2009). The population of SRN in an infested field can rise rapidly in the presence of a host; however, in the absence of a host, the population can decline rapidly. It takes 3–7 weeks for SRN to complete their life cycle depending on the environmental conditions. Thus, multiple generations can develop in a single growing season. These nematodes, being mobile (vermiform) throughout their life cycle, can move vertically in the soil column and can be found up to 60 cm below the soil surface. Thus, they can overwinter by migrating deep into the soil (Khan et al. 2016b).

36.3.5 Epidemiology and Movement

The optimum soil temperature for SRN development and reproduction ranges from 21 to 24 °C (Khan et al. 2016b), while the yield losses due to SRN can be most devastating in light sandy soil as it is conducive to nematode movement and activity (Harveson et al. 2009). Similar to SBCN, SRN can be spread by anything that moves the soil, including farm machinery, animals, workers, floodwater, and wind.

36.3.6 Management

Management of SRN, in infested fields, can be difficult because of their wide host range (Khan et al. 2016b). Hence, crop rotation may not be effective in reducing SRN population levels. Previous studies have shown that sugar beet cultivars can vary in their ability to host SRN species such as *P. allius* (Ashmit 2019). However, cultivars that are designated as poor hosts to one species of SRN may not be poor hosts for the other SRN species. Additionally, there is a lack of information in previous literature about the sugar beet's hosting ability to many different SRN species. Thus, preventing or delaying the introduction of SRN into an area where it has not been reported is important for SRN management. Since anything that moves the soil can also move SRN, using proper sanitation/sterilization measures on farm machinery and equipment is critical (Khan et al. 2016b). Historically, nematicides such as granular carbamates have been effectively used to control SRN in countries like the United Kingdom (Harveson et al. 2009). However, nematicides can be difficult to apply, uneconomical, and carry significant environmental risks (Khan et al. 2016b).

36.4 Other Nematode Parasites of the Sugar Beet

Although root-knot nematodes (*Meloidogyne* spp.) are considered one of the more devastating groups of PPN, only a few species can parasitize the sugar beet and even fewer can cause a substantial economic damage. False root-knot nematodes (*Nacobbus* spp.), on the other hand, have been reported to cause serious damage to the sugar beet, causing gall-like symptoms (Harveson et al. 2009). However, they are mostly found in some areas of the western United States (Bridge and Starr 2019). The oat and onion race of stem and bulb nematode (*Ditylenchus dipsaci*) is another species of PPN that can parasitize the sugar beet, but it has a limited geographical distribution. Thus, it is considered a serious pest of the sugar beet only in the temperate regions of Europe (Bridge and Starr 2019; Harveson et al. 2009). Needle nematodes (*Longidorus* spp.) are the largest plant-parasitic nematodes in terms of length, and they have been reported to cause damage to the sugar beet in the United Kingdom. Like SRN, needle nematodes cause the most damage in light sandy loam soil, and the feeding of needle nematodes causes necrosis in the site of stylet penetration in the roots due to severe damage. Management measures effective against SRN are also effective against needle nematodes (Bridge and Starr 2019, Harveson et al. 2009).

36.5 Nematode Sampling

The strategies used to manage PPN can vary with the species of PPN present in a field and their abundance. Hence, sampling for nematodes is essential for minimizing nematode-related yield loss. Since nematode populations can be aggregated in most nematode-infested agricultural fields, it is important to follow the recommended protocols for sampling. The density of PPN can vary with the time of year when sampling is conducted. Sampling is usually conducted prior to planting or at harvest, but can be done at any time of the year. Sampling can be focused on high-risk areas such as field entrances, along fence lines, previously flooded areas, low spots, areas with variations in crop growth, and areas with high pH, or the entire field can be divided into subdivisions representing different sections of the field. Samples should then be collected from multiple sampling points from each sampling area or division. A soil probe can be used to collect soil cores from depths of 1–30 cm below the soil surface. The soil cores should then be composited and mixed prior to their transport and storage. Exposure of samples to direct sunlight should be avoided during transport to prevent nematode death due to heat. Samples can be stored at temperatures ranging from 4 to 10 °C (Khan et al. 2016b).

36.6 Future Prospects

Biological agents that have shown efficacy in their ability to inhibit PPN population levels have been developed and released, and more are in the developmental process. However, their long-term impact on the soil ecology is still not well known. Studies are also being conducted by researchers worldwide to identify the sources of resistance against PPN, such as SBCN, and different crops are being evaluated for their hosting ability and trapping ability against nematodes on the sugar beet. Recent advances in molecular technology have the potential to aid in the development of nematode-resistant/tolerant cultivars. Moreover, molecular tools facilitate the identification of the species of nematodes. Conventional microscopic identification of nematode species can be tedious, time consuming, and require substantial expertise. Hence, molecular identification and quantification through methods such as polymerase chain reaction (PCR) and real-time quantitative PCR with species-specific primers can reduce the time and effort required for nematode species detection and population-level estimation (Huang et al. 2017a, b). Moreover, new molecular diagnostic techniques such as recombinase polymerase assays have the potential to further streamline the nematode detection process. However, it is important to note that there are numerous different species of PPN that can infect sugar beets but molecular diagnostic techniques have not been developed for all of them. In such cases, DNA sequencing technology can be useful for nematode species identification.

36.7 Conclusion

Plant-parasitic nematodes pose a substantial threat to sugar beet production worldwide. Sugar beet cyst nematodes and SRN are two important groups of PPN that can cause economic losses to sugar beet growers. However, accurate diagnosis of the PPN species present in sugar beet fields and subsequent proactive management can substantially reduce yield losses caused by these pests. Since the above ground symptoms caused by PPN can mimic the symptoms caused by other biotic diseases and abiotic stresses, it is important to examine the below-ground symptoms and signs. In uninfested fields, the most effective management strategy against PPN is to utilize proper sanitation measures to prevent the introduction of PPN, such as SBCN and SRN, as PPN can be spread by any means that moves the soil. In infested fields, management strategies such as crop rotation, host resistance, trap crops, and chemical and biological control agents can be used to manage the PPN populations. However, the host range and resistance/tolerance of sugar beet cultivars can vary between different species of PPN. Hence, soil sampling and subsequent identification of the PPN species present in infested fields are critical.

References

Ashmit KC (2019) Plant-parasitic nematodes on sugar beet in North Dakota and Minnesota. Thesis. North Dakota State University, Fargo

Bridge J, Starr J (2019) Plant nematodes of agricultural importance: a color handbook. Academic, Burlington

Cai D, Kleine M, Kifle S, Harloff HJ, Sandal NN, Marcker KA, Klein-Lankhorst RM, Salentijn EM, Lange W, Stiekema WJ, Wyss U (1997) Positional cloning of a gene for nematode resistance in sugar beet. Science 275:832–834

Ghaemi R, Pourjam E, Safaie N, Verstraeten B, Mahmoudi SB, Mehrabi R, De Meyer T, Kyndt T (2020) Molecular insights into the compatible and incompatible interactions between sugar beet and the beet cyst nematode. BMC Plant Biol 20:1–16

Hafez S (1998) Sugar beet nematodes in Idaho and eastern Oregon. Extension bulletin CIS 1072. University of Idaho, Moscow

Hartman GL, Rupe JC, Sikora EJ, Domier LL, Davis JA, Steffey KL (2015) Compendium of soybean diseases and pests. The American Phytopathological Society, St. Paul

Harveson RM, Jackson TA (2008) Sugar beet cyst nematode. Extension bulletin G1906. University of Nebraska, Lincoln

Harveson RM, Hanson LE, Hein GL (2009) Compendium of beet diseases and pests. American Phytopathological Society, St. Paul

Huang D, Yan GP, Gudmestad N, Skantar A (2017a) Quantification of *Paratrichodorus allius* in DNA extracted from soil using TaqMan probe and SYBR green real-time PCR assays. Nematology 19:987–1001

Huang D, Yan GP, Skantar AM (2017b) Development of real-time and conventional PCR assays for identifying stubby root nematode *Paratrichodorus allius*. Plant Dis 101:964–972

Karegar A (2006) Identification of plant-parasitic nematodes associated with sugar beet fields and their distribution in Hamadan Province, Iran. Iranian J Plant Pathol 42:1

Khan MF, Arabiat S, Chanda AK, Yan GP (2016a) Sugar beet cyst nematode. Extension bulletin PP1788. North Dakota State University, Fargo

Khan M, Arabiat S, Yan GP, Chanda AK (2016b) Stubby root nematode and sampling in sugar beet. Extension bulletin A1821. North Dakota State University, Fargo
Lambert K, Bekal S (2002) Introduction to plant-parasitic nematodes. Plant Health Instr. https://doi.org/10.1094/PHI-I-2002-1218-01
Mai WF, Mullin PG, Lyon HH, Loeffler K (1996) Plant-parasitic nematodes: a pictorial key to genera. Cornell University Press, Ithaca
Mojtahedi H, Santo GS (1999) Ecology of *Paratrichodorus allius* and its relationship to the corky ringspot disease of potato in the Pacific Northwest. Am J Potato Res 76:273–280
Norton DC (1978) Ecology of plant-parasitic nematodes. Wiley-Interscience, New York
Pylypenko L, Kalatur K (2015) Breeding and usage of sugar beet cultivars and hybrids resistant to sugar beet nematode *Heterodera schachtii*. Agri Science Practice 2:12–22
Singh SK, Hodda M, Ash GJ (2013) Plant-parasitic nematodes of potential phytosanitary importance, their main hosts and reported yield losses. Eppo Bulletin 43:334–374
Thurau T, Kifle S, Jung C, Cai D (2003) The promoter of the nematode resistance gene $Hs1^{pro-1}$ activates a nematode-responsive and feeding site-specific gene expression in sugar beet (*Beta vulgaris* L.) and *Arabidopsis thaliana*. Plant Mol Biol 52:643–660
Yan GP, Khan M, Huang D, Lai X, Handoo ZA (2016) First report of the stubby root nematode *Paratrichodorus allius* on sugar beet in Minnesota. Plant Dis 100:1022

Sugar Beet Cyst Nematode (*Heterodera schachtii* Schmidt): Identification and Antagonists

37

Violeta Oro, Nenad Trkulja, Anja Milosavljević, Mile Sečanski, and Marijenka Tabaković

Abstract

Sugar beet is listed within the top ten most important crops in the world. The paleobotanic data suggest that the sea beet was grown in ancient times, while the beets with swollen roots were cultivated in the Middle Ages in Europe. Sugar beet cyst nematode, *Heterodera schachtii*, is an invasive organism causing high economic loss to sugar beets worldwide. The fundamental steps in the control of harmful organisms in plant protection and food safety are grounded on rapid detection of the causative agent and its proper identification. Prompt reaction before obvious symptoms occur can prevent devastating consequences. To confirm the identity of an invasive organism, the process demands a combination of identification techniques, such as morphology and molecular characterization. The phylogeography of available *H. schachtii* populations, based on matching historical data with phylogenetic analyses of the ITS rRNA region, pinpointed a possible place of origin of the European *H. schachtii* populations. Due to the long persistence of the parasite in soil, cysts harbor a large number of bacteria and fungi, the presence of which can lead to cyst death and population decline. Bacteria, fungi, and other antagonists, being an inevitable part of the soil ecosystem, are also part of those mechanisms in nature that limit the excessive number of invasive organisms and return the ecological system to its stable equilibrium.

V. Oro (✉) · N. Trkulja
Institute for Plant Protection and Environment, Belgrade, Serbia

A. Milosavljević
Research and Development Center Sunoko, Indjija, Serbia

M. Sečanski · M. Tabaković
Maize Research Institute "Zemun Polje", Belgrade, Serbia

V. Misra et al. (eds.), *Sugar Beet Cultivation, Management and Processing*,
https://doi.org/10.1007/978-981-19-2730-0_37

Keywords

Antagonists · Bacteria · Fungi · Identification · Sugar beet cyst nematode

Abbreviations

BI	Bayesian inference
GTR	General time reversible model
ITS	Internal transcribed spacer
ML	Maximum likelihood
SEM	Scanning electron microscopy

37.1 Introduction

Sugar beet (*Beta vulgaris* ssp. *vulgaris*) is an economically important crop classified within the top ten ranks of the world in importance (Biancardi et al. 2012) that provides almost one-third of the world's annual sugar production and is a source of bioethanol and animal feed. Leafy beets have been cultivated since ancient times, but the sugar beet is one of the most recently domesticated crops (Dohm et al. 2014). The genus *Beta* is divided into several sections, such as the section Beta including *B. vulgaris* ssp. *vulgaris, B. vulgaris* ssp. *maritima, B. vulgaris* ssp. *adanensis, B. patula,* and *B. macrocarpa,* the section Corollinae with the species *B. macrorhiza, B. corolliflora, B. lomatogona, B. intermedia*, and *B. trigyna,* the section Procumbentes including *B. procumbens, B. webbiana, B. patellaris,* and the section Nanae with the only species *B. nana* (OECD 2001). Based on molecular studies, the section Procumbentes was transformed into the section Patellares and the species within this section were transferred to a new genus, *Patellifolia* (Kadereit et al. 2006). The wild relatives *Patellifolia patellaris*, *P. procumbens,* and *P. webbiana* are sources of resistance to the sugar beet cyst nematode *Heterodera schachtii* (Biancardi et al. 2012). Taxa of section Beta are widely distributed, occupying the littoral zone of Europe, the Middle East, and the Indian subcontinent (Letschert 1993).

The sea beet *Beta vulgaris* ssp. *maritima* is considered the wild ancestor of all cultivated beets (Leys et al. 2014). The sea beet was known in the prehistoric age. Tybrind Vig, a late Danish Mesolithic archeological site from the sixth millennium BC, revealed the charred fragments from the roots of the sea beet (Kubiak-Martens 1999). In the Late Neolithic site of the North Holland (third millennium BC), the drift deposits contained the sea beet and the herbaceous seepweed, *Suaeda maritima* (Kubiak-Martens et al. 2015). The sea beet is indigenous to the European coast, particularly the Mediterranean. In Europe, *B. vulgaris* species with distinctly swollen roots were cultivated in the Middle Ages. Central European types are supposed to originate from those used in Arabian horticulture in Spain. These plants were taken to the Netherlands, where they were cultivated in the early sixteenth century, later spreading throughout Germany. The crop was introduced into the USA in the nineteenth century, where it became known as a garden beet (OECD 2001, loc. cit. Mansfeld 1986).

The seed balls of *Beta* are resistant to saltwater so that ocean currents can move seeds over relatively long distances. Above the high water line, strong winds distribute them over the shoreline, and sometimes even inland (OECD 2001, loc. cit. Smart 1992). The sugar content has increased 10% (from 8 to 18%) in today's cultivars during the last 200 years of sugar beet breeding. Breeding has also been selected for traits like resistance to plant pathogens and parasites (viral and fungal diseases, nematodes), improved yield, monogamy of the seed, and bolting resistance (Dohm et al. 2014).

Since cyst nematodes can follow their host crops for centuries (Oro et al. 2014), the most economically important nematode parasite of the sugar beet is the cyst nematode, *Heterodera schachtii*. At the present time, its distribution is recorded in almost all sugar beet-growing areas. According to CABI (2019) data, the nematode is present in the following countries in Africa: Algeria, Cabo Verde, Egypt, Morocco, and Tunisia. In Gambia and Senegal it is sporadically present and the nematode is widespread in S. Africa. In Asia, the nematode is present in Azerbaijan, China, Iran, Iraq, Israel, Japan, Jordan, Kazakhstan, Kyrgyzstan, Pakistan, Syria, and Turkey, but is absent in India. In Europe, the sugar beet nematode is found in Albania, Bulgaria, Croatia, Czech Republic, Estonia, Finland, Germany, Greece, Hungary, Ireland, Latvia, Moldova, Poland, Portugal, Romania, Serbia, Slovakia, Switzerland, UK, and Ukraine, and is widespread in Austria, Belgium, Denmark, France, Italy, Netherlands, Russia, Spain, and Sweden.

In N. America the nematode is widespread in Canada and the United States, while in S. America it is present in Chile, Peru, and Uruguay. The sugar beet cyst nematode is present in Australia and widespread in New Zealand. The relatively late introduction and growth of the sugar beet (Pathak et al. 2014; Mall et al. 2021) may be the reason for the nematode absence in India.

The biological cycle of Heteroderidae is similar among different genera and *Globodera* spp. are among the best-studied species. The biological cycle begins with the stage of the egg being in the cyst, protected from external influences and with long-lasting vitality. In the egg, a larva of the first stage (J1) is formed; it does not feed, and has a partially formed stylet and a thick cuticle. It transforms into an invasive larva (juvenile) of the second stage (J2) that hatches from the egg, migrates to the soil, and burrows into the sugar beet root. The invasive larva experiences changes during its biological cycle in order to eventually form a cyst. After burrowing into the root, J2 is stationed at the place where a kind of metabolic reservoir syncytium is formed, and there the larva transforms into immobile larva of the third (J3) and fourth (J4) stage. From J4, either a female or a male develops. After fertilization females transform into cysts and produce a large number of eggs. Hatching means the exit of the larva from the cyst, although it is preceded by the hatching from the egg, so it actually includes both processes (Oro 2011). The main factor of hatching is the presence of exudates of sugar beet roots that contain various chemical components—the hatching agents. The hatched larva, attracted by various attractants, penetrates the epidermis of the root, making repeated stings with a stylet (Goverse et al. 2000). In order to be attracted to the chemical attractants of the host plant, there must be chemoreceptors or sensory organs for receiving chemical stimuli. Such chemoreceptors are amphids, which are necessary for locating the host plant, forming the initial syncytium cell, as well as locating females by males (Jones and Perry 2004). The cell perforation is also aided by the secretion of a number of degradable enzymes that are secreted through the stylet cavity (Smant et al. 1997). Some enzymes were thought to be produced exclusively by plants. These are pectin lyases (Popeijus et al. 2000) and expansins (Qin et al. 2004). Enzymes are secreted by the dorsal and two subventral pharyngeal glands. The subventral glands of J2 are responsible for the enzymes that break down the cell wall, such as amylase, pectin lyase, and xylanase (Bakker 2002). In addition, there are other proteins whose function has not been sufficiently elucidated. Chorismate mutase is an enzyme produced by both the subventral and dorsal glands, which inhibits the production of salicylic acid and phenolic phytoalexins, i.e., inhibits plant defense responses. Protein secretions of the subventral glands are responsible for entering the root tissue. Dorsal gland enzymes are responsible for the formation and maintenance of syncytia (Kudla 2006). Syncytium is a multinuclear tissue formed by the degradation of plant cell walls and the subsequent fusion of the protoplasm of surrounding cells (Jones et al. 2003).

In the list of the top ten plant-parasitic nematodes of scientific and economic importance, cyst nematodes took second place, behind the root-knot nematodes (Jones et al. 2013). The exact field data on sugar beet yield loss due to *H. schachtii* are scarce. In Serbia, the yield loss was more than 60% when a field of sugar beet had an average number of cysts of 4–68/100 g soil (Grujicic 1958). The average yield loss in the province of L'Aquila in Italy was 21%, with an infestation level of 87% (Greco et al. 1993). In Ukraine, the yield loss was up to 70% when sugar beet was grown in narrow crop rotation (Pylypenko and Kalatur 2015).

The above-ground symptoms caused by *H. schachtii* are stunted growth, reduced chlorophyll, and wilting (Hillnhütter et al. 2012a), while the below-ground symptoms of sugar beet consist of the development of secondary roots and beet deformity (Hillnhütter et al. 2012b).

37.2 Identification

Until the expansion of molecular techniques, morphology was the main and mostly the only way of nematode identification. Nowadays, these two methods complement each other and give the most reliable results regarding nematode species identification.

37.2.1 Morphology

Heterodera schachtii was first discovered and studied by Schacht in 1859. A detailed morphological description of the nematode was first given by Strubell 1881, who carried out his observations on materials obtained from the sugar beet (Triffitt 1928). Since its recognition as a plant parasite, other researchers have supplemented Strubell's description, apparently describing different species as strains of *H. schachtii*. *Heterodera* species belong to the family Heteroderidae and the subfamily Heteroderinae, which include nematodes with swollen bodies called cysts containing eggs and larvae. Some eggs and larvae may be laid in a viscous medium attached to a cyst.

Cysts of *Heterodera* species are oval, pear, or lemon-shaped with a short neck. The cyst can vary in color from yellow, light brown to dark brown. The vulva and anus are situated within the vulval cone. One of the key characters for identification is the type of fenestrae. The fenestrae can possess one opening-circumfenestral type, two openings with a narrow vulval bridge-ambifenestral type, or two openings with a wide vulval bridge-bifenestral type (Turner and Rowe 2006). Based on the ITS sequence region, Subbotin et al. (2001) differentiated *Heterodera* spp. within Heteroderinae into six groups, namely Avenae, Cyperi, Goettingiana, Humuli, Sachari, and Schachtii group, with the exclusion of *H. bifenestra* linked to the Cyperi group but not designated as a separate group. The Schachtii group comprises *H. mediterranea, H. glycines, H. trifolii, H. cajani,* and *H. medicaginis* as well as *H. schachtii* as the type species (Subbotin et al. 2001; Ma et al. 2008).

37.2.1.1 Description of *Heterodera schachtii* Schmidt, 1871 (After Turner and Rowe 2006)

Female

White, body enlarged, lemon-shaped with a tapering posterior end, ambifenestral type. Secretory-excretory pore near the base of the neck. The anus is normally situated dorsally, sub-terminally on the cone. The head is weakly sclerotized and the stylet is slight with small basal knobs, pharyngeal bulb prominent and spherical. The glands overlap the intestine lateroventrally. Ovaries are paired, coiled and long; some eggs are laid into an egg sac but most are stored in the body.

Cyst

The cyst protects the eggs; there are approximately 500 eggs/cyst of *H. schachtii*. The vulval slit is long at 70 μm, the longest within the *Heterodera* groups. The nematode cyst is ambifenestrate with a robust underbridge that attaches itself to the walls of the cone. It is dark brown in color and has a thickening in the center, which is a remnant of the virginal. The size range is about the same as adult females; fenestral length 38.7 μm, fenestral width slightly less; vulva–anus distance 77 μm; cyst: 550–950 μm.

Male

Length 900 μm; Width 28 μm; stylet 28 μm; spicules 34 μm bidentate; gubernaculum 11 μm. Head shape is hemispherical, bearing 3–5 annules. Four lateral lines extend around the tail. The secretory–excretory pore is located midway between the pharyngeal gland and the posterior margin of the sub-ventral pharyngeal glands.

Second Stage Juvenile (J2)

Head offset with four head annules, hexaradiate, amphidial apertures small in the lateral region near the mouth. Length 470 μm; width 21 μm; stylet 25 μm; stylet shape heavy and hooked in shape; annulations 1.41 μm at mid-body; tail length 60.3 μm; hyaline tail region 36.4 μm; tail shape short tapering abruptly, rather blunt; phasmids visible on the tail.

37.2.1.2 Light Microscopy

The bright field microscopy combined with low magnification reveals gross morphology or some critical moments in the nematode life cycle, such as hatching, eclosion, etc. A population of *H. schachtii* from Serbia is found in the sugar beet-growing region Kula, and its morphology is given. Figure 37.1 shows lemon-shaped cysts of *H. schachtii* with a long cone, comprising sometimes 1/5 of the cyst length without the posterior part.

The dark field microscopy reveals the natural color of cyst specimens that is golden brown, shown in Fig. 37.2.

Nomarski contrast or differential interference contrast visualizes three-dimensional appearances of specimens, enabling monitoring from different focal planes. In addition, the Nomarski interference contrast generates an image with "shadow" effect, particularly useful in the nematode diagnostics. Figure 37.3 shows the vulval area with semi-fenestrae and a vulval slit situated on the top of the cone, as well as a three-dimensional cone with ridges.

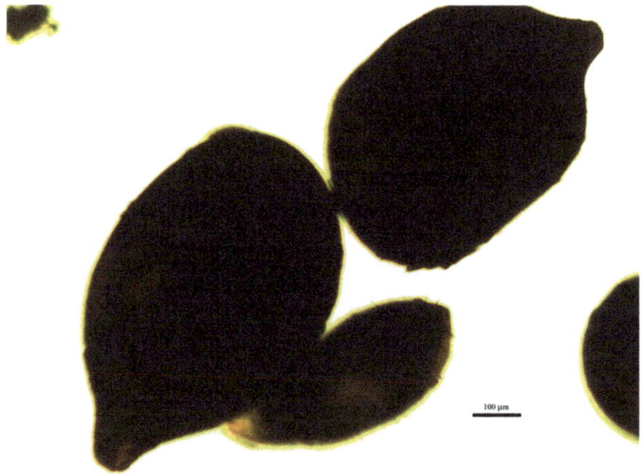

Fig. 37.1 Bright field microscopy of *H. schachtii* cysts

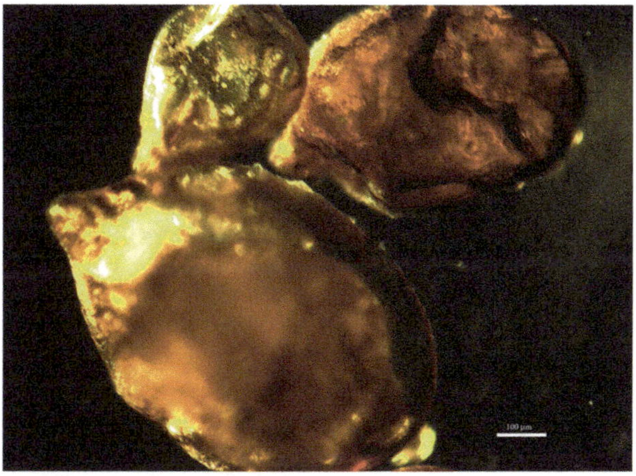

Fig. 37.2 Dark field microscopy of *H. schachtii* cysts

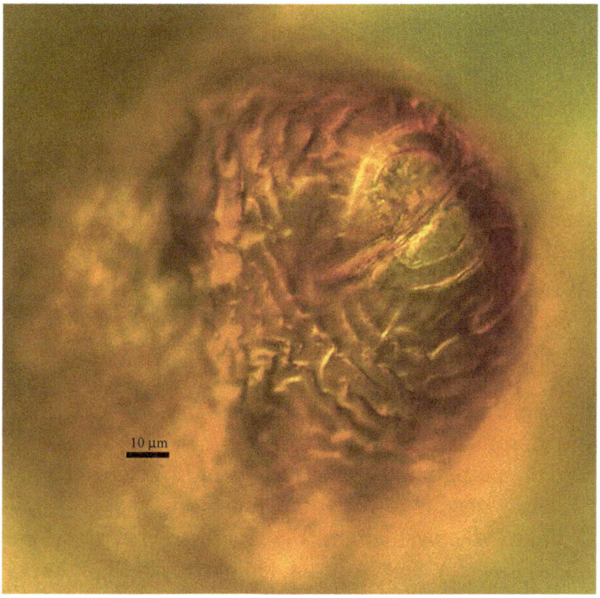

Fig. 37.3 Nomarski contrast showing the semifenestrae of *H. schachtii* cyst

Figure 37.4 shows the underbridge and spherical bodies called bullae.

The second-stage juveniles, as a difference from females, are vermiform and in a migrating stage, in search for the roots of the sugar beet and other host plants. The anterior and posterior ends of an invasive juvenile (larva) under Nomarski contrast are given in Figs. 37.5 and 37.6.

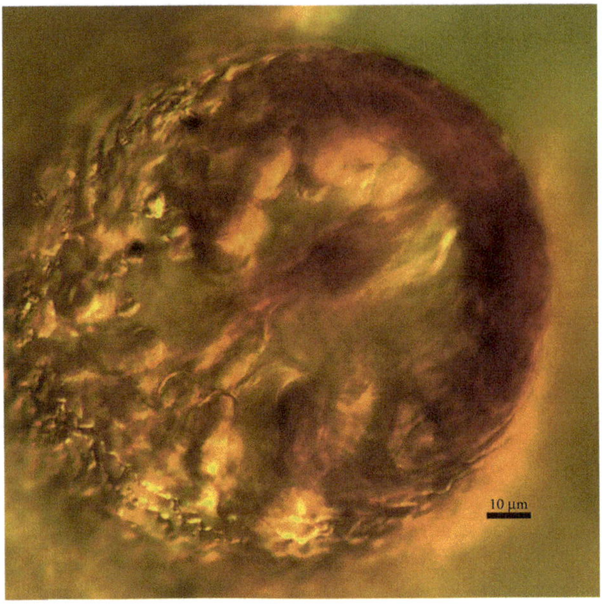

Fig. 37.4 Nomarski contrast showing the underbridge and bullae of *H. schachtii* cyst

37.2.1.3 Scanning Electron Microscopy

Scanning electron microscopy (SEM) with its high resolution and a large depth of field reveals minutious details with dimensions even in nanometers. Figures 37.7 and 37.8 present the SEM of a cyst and the vulval area of *H. schachtii.*

37.2.1.4 Morphometrics

Aside from the strictly morphological description of the sugar beet cyst nematode, there are morphometric data regarding cyst and J2 dimensions, which supplement the morphology (Table 37.1).

The morphometrics of *H. schachtii* from Kula describe a population of the sugar beet cyst nematode with an average dimension of cysts around half of a millimeter (500 μm); the cyst length is greater than the cyst width by 50%. The same ratio may be applied for the fenestral length and width. The vulval slit is longer than the semifenestral width, while the underbridge is around 100 μm. Regarding the larval dimensions, J2 is less than half of a millimeter, with the mean stylet length of 25 μm. The larval tail is almost 4 times longer than the width at the anus level. Comparison of mean morphometric values among different world populations, namely, Kula (this study), Emilia Romagna and Abruzzo, Italy (Ambrogioni and Irdani 2001), Yellowstone Valley, the United States (Nelson et al. 2012), Aleppo, Syria (Haidar et al. 2016), Jeongseon, South Korea (Mwamula et al. 2018), Tepeaca Valley, Mexico (Escobar-Avila et al. 2019), and Hara Village, Japan (Sekimoto et al. 2017), is presented in Table 37.2.

Comparison of different world populations of the sugar beet cyst nematode revealed slight differences among them. Regarding the cyst size, the smallest cysts

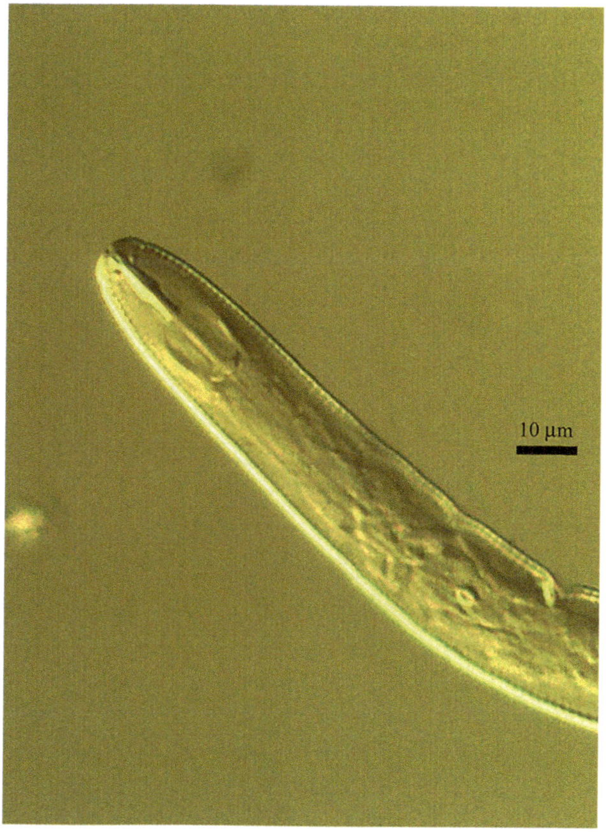

Fig. 37.5 Nomarski contrast showing the anterior part of *H. schachtii* J2

were from the Serbian population Kula, while the largest cysts were from the Italian populations. The cyst length varied from 679 to 815 µm, while the cyst width ranged from 454 to 529 µm. The cyst length/width ratio varied from 1.5 to 1.6. The shortest juveniles were from Kula (405.7 µm), while the longest J2 (472 µm) were from Hara Village. It was noticed that juvenile length was in agreement with the cyst size. The shortest juveniles emerged from the smallest cysts and vice versa. The given description of Turner and Rowe (2006) reported the vulval slit length of 70 µm, while in the investigated populations the values vary from 37.4 to 48.7 µm. In addition, the tail length of the second-stage juvenile was reported to be 60 µm, whereas J_2 tail length varied from 43.2 to 49.9 µm in the investigated populations.

37.2.2 Molecular Identification

The molecular identification based on DNA is a sensitive technique, in which small amounts such as nanograms of DNA are used and the technique is independent of the nematode developmental stage or environment. The main DNA regions used in

Fig. 37.6 Nomarski contrast showing the posterior part of *H. schachtii* J2

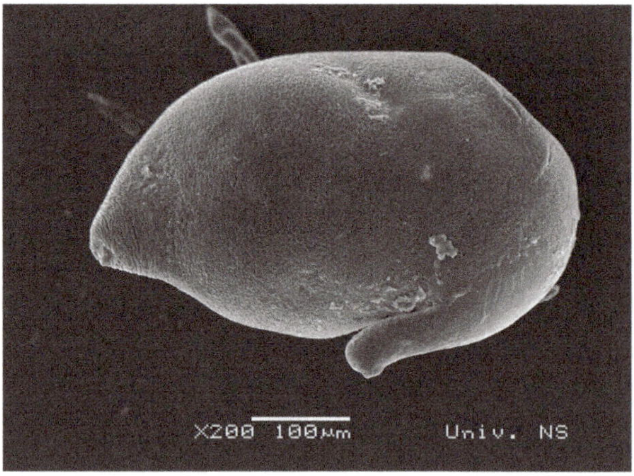

Fig. 37.7 SEM of *H. schachtii* cyst

Fig. 37.8 SEM of the vulval area of *H. schachtii*

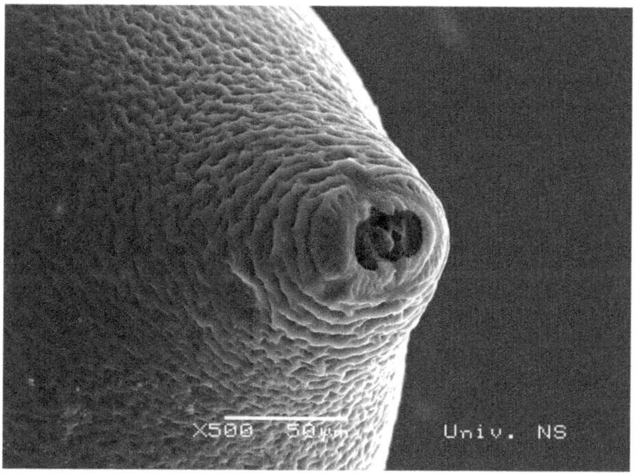

Table 37.1 Morphometrics of the population Kula of *Heterodera schachtii* from Serbia

	Cysts			Juveniles (J_2)		
Characters (in µm, except for ratios)	$\bar{x}$	sd	min–max	$\bar{x}$	sd	min–max
Cyst length	679	103	471–926	–	–	–
Cyst width	454	100	297–739	–	–	–
Cyst length/width	1.5	0.2	1.2–1.9	–	–	–
Fenestral length	43	6	31–52	–	–	–
Semifenestral width	27	4	20–33	–	–	–
Vulval bridge width	6	0.8	5–7	–	–	–
Vulval slit	42	5	32–52	–	–	–
Underbridge length	106	20	75–139	–	–	–
J_2 length	–	–	–	405.7	14	374–436
J_2 width	–	–	–	18.5	0.7	17.1–19.8
J_2 stylet	–	–	–	25	1.3	22.3–26.9
J_2 anterior end to median bulb valve	–	–	–	62.6	7.2	45.3–82.9
J_2 tail length	–	–	–	45.9	4.4	36.6–54.0
J_2 tail width–anus level	–	–	–	12.2	1.1	9.3–14.0
J_2 hyaline length	–	–	–	25.2	2.5	18.6–30.3
a	–	–	–	21.9	1.0	20.4–23.6
b	–	–	–	5.2	0.4	4.2–5.9
c	–	–	–	8.9	0.8	7.8–11.3
c′	–	–	–	3.8	0.3	3.2–4.7

$\bar{x}$ mean, *sd* standard deviation, *min–max* range

diagnostics of cyst nematodes are nuclear ribosomal RNA genes comprising 18S, 28S rRNA genes and internal transcribed spacer (ITS1 and ITS2) regions (Subbotin et al. 2010). Amiri et al. (2001) designed the specific primer SGR1, which in combination with the universal TW81 primer amplified 850 bp fragment used for identification of species from *H. schachtii* sensu stricto group. The amplification of

Table 37.2 Comparison of the mean morphometric characteristics of *Heterodera schachtii* among different populations

Characters (in μm, except for ratios)	Kula	Emilia R.	Abruzzo	Aleppo	Jeong-seon	Yellow-stone V.	Tepeaca Valley	Hara Village
Cyst length	679	768	815	–	795.8	701.2	740.1	779
Cyst width	454	529	512	–	510.8	469.2	517.8	489
Cyst length/width	1.5	1.5	1.6	–	1.6	1.5	–	1.6
Fenestral length	43.0	35.1	38.7	92.3	42.2	–	33.3	32.0
Semifenestral width	27.0	27.7	31.1	44.5	28.2	–	26.0	27.1
Vulval bridge width	6.0	4.8	6.6	7.8	4.5	–	–	–
Vulval slit	42.0	40.7	44.1	48.7	37.4	–	–	42.4
Underbridge length	106.0	107.8	109.9	–	102.3	–	–	93.9
J_2 length	405.7	436	444.5	501.5	484.2	437.1	436.4	472
J_2 width	18.5	19.8	20.2	–	20.3	–	23.6	–
J_2 stylet	25.0	25.5	26.0	25.2	22.1	25.0	23.1	25.6
J_2 anterior end—med. bulb	62.6	71.0	73.6	–	76.4	–	–	–
J_2 tail length	45.9	44.9	47.3	43.2	49.9	46.6	47.4	48.7
J_2 tail width–anus level	12.2	12.6	13.2	–	–	–	–	–
J_2 hyaline length	25.2	24.4	25.4	26.4	29.2	27.3	25.3	28.4
a	21.9	22.1	22.0	–	23.8	–	–	–
b	5.2	5.0	4.9	–	4.2	–	–	–
c	8.9	9.8	9.5	11.6	9.6	–	–	–
c′	3.8	–	–	–	3.5	–	–	–

the D2-D3 expansion segments of 28S rRNA yielded a single fragment of *ca* 700 bp, distinguishing *H. schachtii* from other cyst nematodes but not differentiating *H. trifolii* from *H. betae* (Sekimoto et al. 2017). The sugar beet cyst nematode was identified with the help of nearly full-length SSU rDNA sequences among 339 nematode taxa, the representatives of the entire phylum (Holterman et al. 2006). Cytochrome oxidase I mitochondrial DNA gene sequencing could distinguish *H. schachtii* from *H. trifolii* using COI-F4a-Het and COI-R10b-Het primers (Powers et al. 2019). The Real-Time PCR assay for the rapid discovery of *H. schachtii* was developed using SH6Mod, SH4 primers, and SYBR green I dye (Madani et al. 2005). Phylogenetic relationships based on beta-tubulin DNA sequence data within the cyst nematodes were more informative at higher taxonomic levels (Sabo and Ferris 2004).

In the recent study of Oro and Tabakovic (2020), the phylogeography of the European populations of *H. schachtii* based on ITS rRNA region using Maximum likelihood (ML) and Bayesian inference (BI) analyses was investigated. The sequences of *H. betae* were also included to test the sequence similarity between the two sister species, which frequently occur together. By matching the current nematode molecular data and historical dispersal routes of the sugar beet, an effort was made to find a possible center of origin of the sugar beet nematode in Europe in the light of the host–parasite relationship. The cysts were found in sugar beet-growing areas in Nova Crvenka and Kula in Serbia. DNA was extracted from individual cysts using the Dneasy blood & tissue kit (Qiagen), following the manufacturer's instructions. PCR was done with primers for direct sequencing: TW81 (5′-GTTTCCGTAGGTGAACCTGC-3′) and AB28 (5-′-ATATGCTTAAGTTCAGCGGGT-3′) as per Skantar et al. (2007). The ITS sequences of *H. schachtii* from Nova Crvenka, Kula 1, and Kula 2 were deposited in the NCBI nucleotide database (United States), under accession numbers MF975709, MF975710, and MF975711, respectively. The sequences were aligned with ClustalW within MEGA 4 (Tamura et al. 2007). Phylogenetic analyses were performed with the available sequences of *H. schachtii* and *H. betae* from GenBank using PhyML 3.1 (Guindon and Gascuel 2003) and MrBayes 3.1.2 (Huelsenbeck and Ronquist 2005) computer programs. The Maximum likelihood (ML) and Bayesian phylogenetic trees were obtained with the General Time Reversible model (GTR), invariable sites, and gamma distribution (GTR + I + G). The consensus dendrogram with 50% majority rule obtained by Bayesian inference was created by 3.2×10^6 generations of MCMC (Markov Chain Monte Carlo), with the sample frequency of 100 and a burnin function of 20%. *Heterodera avenae* and *H. filipjevi* were selected as outgroups.

The ITS region of *H. schachtii* was composed of the partial ITS1 region (1st–520th nucleotide), 5.8S rRNA gene (521st–678th nucleotide) and the partial ITS2 region (679th–730th nucleotide). The content of guanine and thymine was higher in the ITS region of the sugar beet nematode, compared to the contents of cytosine and adenine, and likewise, to the content of the same bases of the potato cyst nematodes (Oro and Oro-Radovanovic 2012). The obtained circle phylogenetic trees (both ML

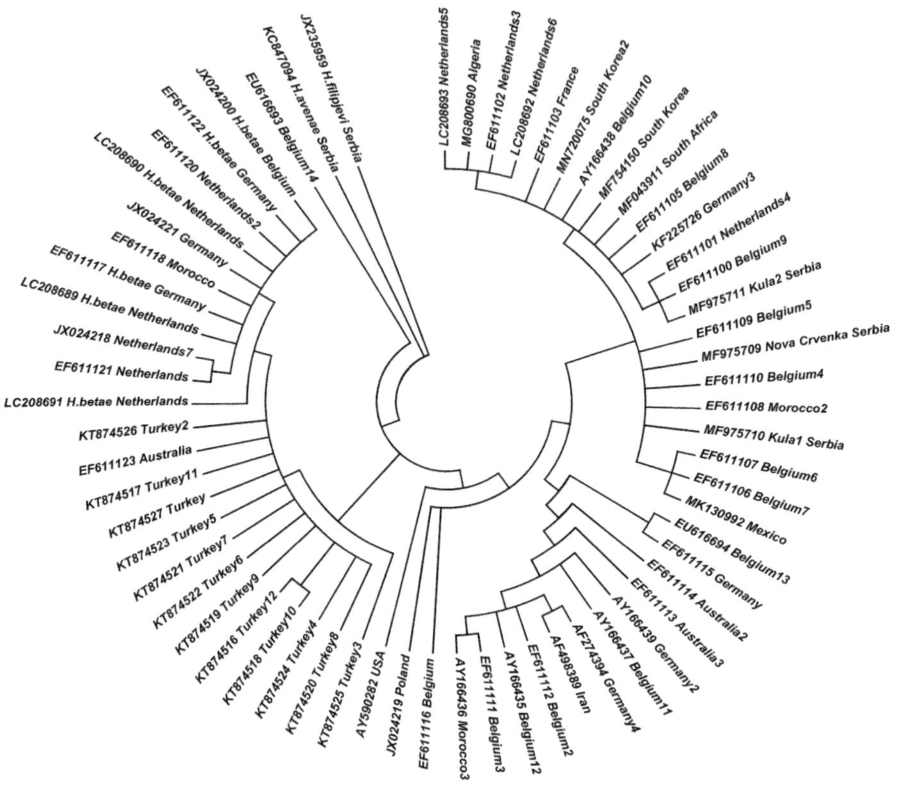

Fig. 37.9 ML circle dendrogram of *H. schachtii* and *H. betae* populations

and BI) presented the same relationships among the examined species (Figs. 37.9 and 37.10).

Three distinct groups were created. Four populations from the Netherlands, seven populations from Belgium, and populations from France, Germany, and Serbia were placed in the first group. Populations from Algeria, South Korea, S. Africa, Morocco, and Mexico belonged to the same group. The second group comprised populations from Iran, Morocco, and Australia. The last group, genetically the most divergent, encompassed some European populations, the clones of a Turkish population, and the populations of *H. betae*. The placements of the populations from Belgium, Poland, and the USA were not resolved.

Since the historical data consider Europe the ancestral region of sugar beet domestication, the geographic positions of the European populations of *H. schachtii* have been summarized. The populations were grouped towards the Dutch–Belgian direction starting from the coastal zone with the population Borsel (or probably Borssele), near the North Sea.

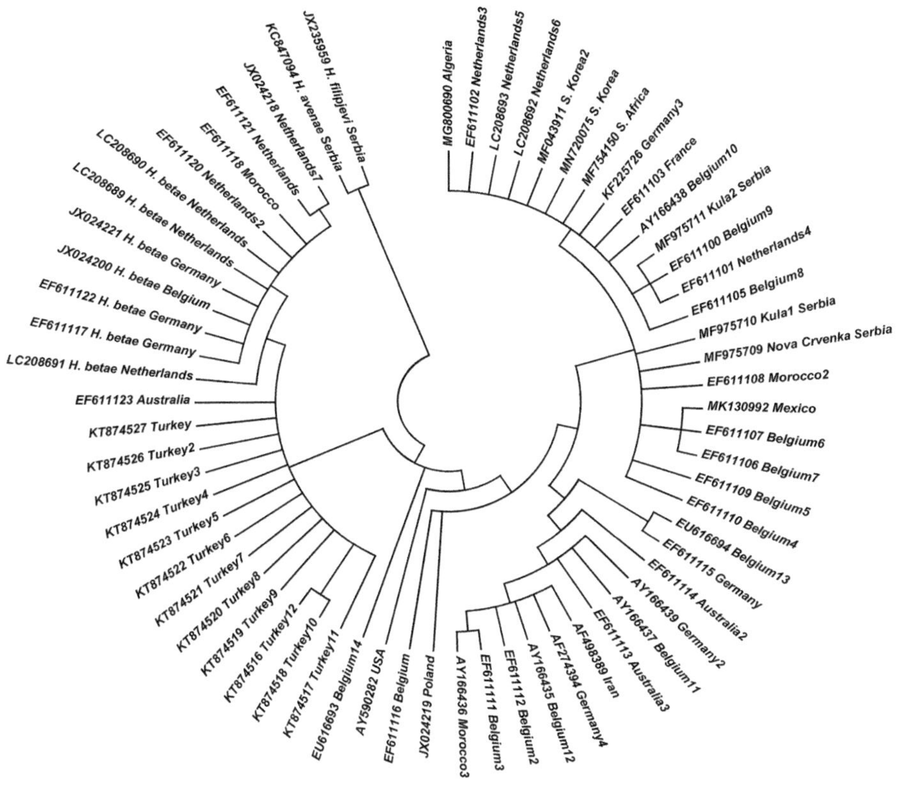

Fig. 37.10 BI circle dendrogram of *H. schachtii* and *H. betae* populations

In the sixteenth century, as mentioned, sugar beet from Spain was introduced to the Netherlands and Borssele presumably by the medieval maritime ships. Even the sea current may be involved in cyst dissemination together with the seeds of the sugar beet. The subsequent spread followed a radial direction towards the nearby countries, such as Belgium, Germany, etc. The majority of the populations were localized near the borders, presumably as a result of human transport activities. The Belgian population Momalle was located close to the Dutch border, while the German population Muenster (Münster) was positioned near the Netherlands, but on the opposite side. The French population Aisne was located near the Belgian border. The population from Molenbaix was situated near the French border.

Regarding the countries from other continents, their populations were always clustered with either Dutch or Belgian populations. A population from Algeria was genetically identical to the Dutch populations 3 (Borssele) and 6 (an unknown population). The South African and South Korean nematode populations were identical to the Belgian population 10 (Ohain). The study of Escobar-Avila et al. (2019) showed that the Mexican population was identical to the one from Belgium, which was congruent with our dendrograms. The Moroccan population 2 was the most similar to the Belgian populations Momalle, Molenbaix, Gingelom, and the Serbian populations Kula 1 and Nova Crvenka. The remaining populations with the highest level of divergence were grouped into the *H. betae* clade.

The evolution of domestication is associated with the rise of many diseases related to an agricultural origin. The increased population of people, domestic animals, and plants in the same area, where diseases are intensively transmitted, has created the basis for new diseases. Such facilitated disease transmission in new environments and new hosts has resulted in an increased virulence of invasive organisms. This process has long been recognized as an origin of diseases in the human population, but it has become clear that the origin of many plant diseases is more recent (Smith et al. 2014). Tracking the historical distribution of the sugar beet seems to be an interesting approach in the search for the ancestral crop population of *H. schachtii*, indicating that its place of origin could be the area across the Dutch-Belgian coastal region, starting with the Borssele population and its further distribution to the nearby countries. Studies of host–parasite relationships are strongly associated with the understanding that hosts and their parasites interact over both relatively long evolutionary and relatively short ecological times. The historical biogeography of host–parasite relationships helps us to understand the present and future ecology of both hosts and parasites (Morand and Krasnov 2010).

37.3 Antagonists

Nematode cysts remain in the soil for a long time exposed to different environmental conditions and host various bacteria and fungi. The antagonistic microorganisms within a cyst may cause cyst destruction and population decline (Oro et al. 2020). A cyst of *H. glycines* revealed the existence of almost 300,000 bacteria (Nour et al. 2003). Such microorganisms are potential candidates for use in the biocontrol of cyst nematodes.

37.3.1 Bacterial Antagonists

Many soil bacteria produce nematode toxins. The actinomycete *Streptomyces avermitilis*, used as a commercial biocontrol agent, has anthelmintic properties (Kerry 2000). Ryan and Jones (2004) found that *Bacillus, Arthrobacter, Acinetobacter,* and *Staphylococcus* isolates reduce the hatching of potato cyst nematodes. The mechanisms of action of bacteria are diverse and are still not fully understood. About 1/20 of the rhizobacteria, when reinoculated on plants, have a direct and indirect stimulatory effect on plant growth. Directly, they produce stimulants and phytohormones, reduce the content of ethylene, improve the nutrition, and stimulate induced resistance. Indirectly, they stimulate the growth of other beneficial symbionts or degrade xenobiotics (Antoun and Prevost 2006). Reitz et al. (2000) found that lipopolysaccharides of *Rhizobium etli* G12 induce systemic resistance of potatoes to *Globodera pallida*. Induction of systemic resistance in the presence of *Bacillus sphaericus* has also been reported by Hasky-Günther et al. (1998). *Rhizobium* species can produce rhizobitoxins and an antibiotic, bacteriocin. Rhizobacteria, including *Pseudomonas fluorescens*, can modify root exudates and reduce J2 hatching or can bind to lectins on the root surface, and thus reduce larval invasion (Kerry 2000). Long-term climate factors create conditions for the development of certain species. Isolation of bacterial microbiota from the cysts of the potato cyst nematode, *G. rostochiensis*, inhabiting two different soil samples revealed the dominance of the order Bacillales. The lower values of air temperatures, insolation, and precipitation and the higher values of relative humidity and cloudiness created conditions for the development of psychrophilic species. *Bacillus frigioritolerans* and a *Psychrobacillus* sp. were the representatives of the indigenous microbiota of a cooler microclimate (Oro et al. 2020). *Bacillus* spp. are well-known nematode antagonists. *Bacillus pumilus* was effective against *Meloidogyne arenaria*, causing 93% larval mortality after 3 days of exposure to a 10% solution of bacteria (Lee and Kim 2016). *Bacillus subtilis* and *B. pumilus* reduced the population of *M. incognita* on *Vigna unguiculata* by 82% (Padgham and Sikora 2007). An isolate of *B. megaterium* decreased the invasion of *M. graminicola* up to 60% compared to a control sample of rice (Abd-El-Khair et al. 2019). *Psychrobacillus soli* was able to decompose almost 3/4 of the oil components in the concentration of 0.15% (Pham et al. 2015). *Psychrobacillus insolitus* and *Curtobacterium oceanosedimentum* exhibited the highest anticandidal effect against *Candida albicans* and *C. glabrata*, among other endophytic bacterial organisms (Das et al. 2017), while two strains of *P. insolitus* were inhibitory towards food strains of *Staphylococcus* sp. (Oliveira et al. 2012). The results of Huang et al. (2021) demonstrated that root colonization by the strain of *B. firmus* significantly protected *Arabidopsis thaliana* from *H. schachtii*. The bacterium also negatively affected the nematode parasitic cycle and juvenile development.

37.3.2 Fungal Antagonists

Fungi have a significant place among nematode antagonists, demonstrating a vast ability in biocontrol. Both internal and external sides of a cyst contain various organic compounds that can be used by microorganisms as nutriments. The starting points for fungal invasion are cyst orifices such as the vulva, positioned in a cone (Oro et al. 2021).

Fungal antagonists mainly consist of four groups: predacious fungi, endoparasites, cysts and egg parasites, and fungi-producing toxins. Fungal spores have adapted and specialized in capturing and penetrating the nematode cuticle. Spores of *Catenaria anguillulae* mostly reside around the natural openings of the nematode body (Mankau 1980). *Arthrobotris* species form traps having a variety of shapes, including mucose cells, adhesive nets, or rings. In the nematode proximity, the fungal ring spreads rapidly, crushing prey, and then digest it in a few hours. *Nematoctonus* spp. produce cellulases and ligninases, the main enzymes produced by wood-degrading fungi (Barron 2003). *Meria coniospora* is a nematode endoparasite. Its conidium enters through the mouth of *Panagrellus redivivus* (Jansson et al. 1984). *Nematophthora gynophila* was destructive for the oat cyst nematode in less than 7 days (Kerry 1980). *Verticillium chlamydosporium, Acremonium strictum,* and *F. oxysporum* are the most common parasites of *H. schachtii* (Dackman and Nordbring-Hertz 1985). *Fusarium oxysporum* produces nematotoxic filtrates. The filtrates decreased the motility of *M. incognita* larvae within 10 min, while 24 h of exposure resulted in 100% mortality (Hallmann and Sikora 1996). The combined use of *Purpureocillium lilacinus* and *Monacrosporium lysipagum* decreased almost 2/3 of the *H. avenae* cysts (Khan et al. 2006). Flavipin, the key metabolite of *Chaetomium globosum*, is responsible for nematode antagonistic activities (Nitao et al. 2002). *Pleurotus ostreatus* and *Conocibe lactea* have small appendages on hyphae that secrete toxins (Barron 2003). Fungal endophytes have recently attracted much attention owing to their great biological potential as participants in constitutive and induced plant defense responses (Gao et al. 2011), as producers of enzyme complexes that enable bioconversion of biomass to biofuels (Cabezas et al. 2012), and as producers of bioactive compounds (Su et al. 2014). Fungal endophytes are prospective bioremediators of environments contaminated with heavy metals (Soldi et al. 2020), petroleum hydrocarbons (Marin et al. 2018), textile dyes (Henagamage 2019), greenhouse gases, especially methane and carbon dioxide (Stepniewska and Kuzniar 2013), pesticides, and radionuclides (Krishnamurthy and Naik 2017). The recent study of Oro et al. (2021) on mycobiota isolated from cysts of *H. filipjevi* showed the presence of diverse fungi from Ascomycota, Basidiomycota, and Mucoromycota. *Heterodera filipjevi* and *H. schachtii* are two congeneric species that are frequently found together in the same field (Oro and Tabakovic 2020) and share the same mycobiota. The representatives of Ascomycota were *Pochonia chlamydosporia, Sarocladium* (syn. *Acremonium*) *kiliense, Fusarium avenaceum,* and *Setophoma terrestris. Pochonia chlamydosporia* is a common parasite of nematode eggs in suppressive soils. It can survive as a saprotroph in soil. The fungus has been found to be an egg parasite of the sugar beet cyst nematode (Ayatollahy et al. 2008). *Pochonia* spp. are endophytes of some species of Poaceae and Solanaceae (Manzanilla-Lopez et al. 2013). *Sarocladium kiliense* has antagonistic properties toward root-knot nematodes. The methanol extract and filtrate of *S. kiliense* had a lethal effect on *M. incognita* juveniles up to 37% (Gamboa-Angulo et al. 2015). *Fusarium* is a large genus of filamentous fungi (Ohara et al. 2004) that are widespread as saprotrophs in soil and organic matter. *Fusarium avenaceum* produces numerous metabolites, such as moniliformin, beauvericin, enniatin, chlamidosporol, chrysogin, acetamido-butenolide, antibiotic I, fusarin, aurofusarin, etc. (Uhlig et al. 2007). *Setophoma terrestris* decreased 60% of the larval hatching of *H. glycines,* the cyst nematode parasite of soybean (Chen et al. 1996).

Crops that host pathogenic fungi of the phylum Ascomycota indicate that fungal presence is recent. Basidiomycota was represented by *Bjerkandera adusta* and *B. albocinerea, Burgoa* sp., *Cerrena unicolor, Phlebia/Mycoacia* spp., *Phlebiopsis* spp., and *Trametes hirsuta*. Ligninases are the main enzymes in the process of biodegradation of benzopyrene by *B. adusta* SM46 (Andriani et al. 2016). A strain of *B. adusta* was not efficient toward the enthomopathogen *Steinernema carpocapsae* (Balaes and Tanase 2016). *Burgoa* species are in symbiosis with lichens and can damage frescoes and gravestones (Kiyuna et al. 2015). *Cerrena unicolor*'s laccase reduces oxygen and oxidizes phenolic substrates, showing good bioremediation capacity (Gianfreda et al. 1998). *Cerrena maxima* can decompose atrazine, the pesticide used in weed control (Gorbatova et al. 2006). *Phlebia* sp. can produce ethyl alcohol from cellulose, glucose, and xylose and is a prospective bio-fermenter (Kamei et al. 2020). *Phlebiopsis gigantea* attacks trees and breaks down the resin and other wood components, showing the great potential in the wood industry (Behrendt and Blanchette 1997). *Phlebia* and *Phlebiopsis* could not infect *Aphelenchoides* spp. (Tzean and Liou 1993). A strain of *T. hirsuta* can degrade ferulic acid that is considered an environmental pollutant (Patil and Yadav 2018). *Trametes trogii* was not pathogenic to *S. carpocapsae* larvae (Balaes and Tanase 2016). Birch, poplar, and willow were the preferred hosts of *Bjerkandera adusta, Cerrena unicolor, Phlebia* spp., and *T. hirsuta* (Park et al. 2020), and found as endophytes on multiple hosts (Martin et al. 2015). There is no nitrogen available in the tree so that nematophagous fungi, that are also wood decomposers, meet their nitrogen needs by catching nematodes (Barron 2003). The presence of wood-degrading fungi in the soil indicates that the plants before crops were trees. Basidiomycota found in agricultural soils were associated to deciduous trees, indicating that deforestation occurred during a long time period, turning forests into agricultural land. Mucoromycota was represented by *Linnemannia*, i.e. *Mortierella elongata*. Isolates of *M. elongata* have multiple beneficial properties and act as plant growth promoters, biodegrading agent of complex organic compounds and environmental toxicants (Zhang et al. 2020; Horel and Schiewer 2020), and plant resistance inducer (Li et al. 2018). *Mortierella globalpina* was found to be a predator of *Meloidogyine chitwoodi* by fastening to the nematode's cuticle and subsequently consuming larvae (DiLegge et al. 2019). Similar fungi have been identified in sugar beet cyst nematodes in the United States and Germany. Chen et al. (2020) detected hyperparasitic fungi, namely *Hyalorbilia oviparasitica, Pochonia chlamydosporia*, certain *Fusarium* spp. etc., that could be biologically suppressing cyst nematodes below a damaging threshold in California's Central coast soils. *Pyrenochaeta* sp., *Pochonia chlamydosporia*, and *Exophiala* sp. were found in infected cysts of *H. schachti* in Germany. The fungi could re-infect the cysts and colonize the eggs of *H. schachtii* in vitro. In greenhouse trials, the sugar beet yield was significantly higher in substrates inoculated with both nematodes and fungi compared to plants inoculated with nematodes only (Haj Nuaima et al. 2021), confirming the antagonism between fungi and nematodes.

37.3.3 Other Antagonists

Aside from bacteria and fungi, some other organisms can inhabit cyst nematodes. In the study of Fosu-Nyarko et al. (2016), it was suggested that a novel viral genome was discovered in *H. schachtii*. Some insects, mites, and spiders can feed on cysts.

During the study of potato cyst nematodes, Oro (2011) found predatory Dorylaimids in cysts. All these organisms are potential biocontrol agents for cyst nematodes, enabling the maintenance of the ecological equilibrium in nature without toxic residues. In addition, there are birds that feed on cysts and thus reduce their number, but also transport them over large areas. Radice and Myers (1984) suggested that cyst nematodes (*Punctodera punctata*) could be transmitted over long distances by migratory birds, such as the Canadian geese (*Branta canadensis).* SEM images of *Globodera* revealed unknown invertebrates that were not detected by an optical microscope. It has been shown that a group of organisms with strong jaws colonizes a cyst. The organisms were between 5 and 20 μm in size and resembled mites (the order Acarina) as well as water bears (the phylum Tardigrada), meaning the list of potential nematode biocontrol agents should be extended with novel microscopic invertebrates with properties that should be explored in the future.

37.4 Future Prospects

The fundamental steps in the control of harmful organisms in plant protection and food safety are based on rapid detection of the causative agent and its proper identification. The prompt reaction before obvious symptoms occur can prevent devastating consequences and save finances for the eradication process. In the future, we can expect that video monitoring covering large areas, e.g., the use of drones equipped with cameras linked to devices and programs that will alert diagnosticians in real time, will facilitate rapid detection of plant symptoms and harmful organisms (Martinez et al. 2020). Confirming the identity of an invasive organism demands a combination of identification techniques. The identification to the species level is very complex due to nematode-species resemblance and morphometric overlap. The process of identification is further complicated in the case of similar molecular identities between different species, which demands an experienced investigator. In the coming years, the machine-learning technologies based on nematode phenotypic characters will enable us to discern a specific nematode species in a complex background (Akintayo et al. 2018; Hakim et al. 2018).

The next step is the determination of the nematode's natural antagonists. Bacteria and fungi are inevitable parts of soil microbiota and also part of those mechanisms in nature that limit the excessive number of invasive organisms such as pests, pathogens, and parasites. The extensive use of pesticides resulted in the disappearance of many beneficial organisms, which together with toxic pollutants create long-term threats for human health and existence. The phylum Firmicutes, especially the order Bacillales, are ubiquitous organisms, some of which have been proven to be nematode antagonists and also environmental detoxifiers. Regarding fungi, it has been demonstrated that some species of Basidiomycota, with their multi-purpose activities, are promising agents in both biocontrol and bioremediation. Returning to nature something that has been previously taken from it enables maintaining the ecological balance, which is in accordance with Regulation (EC) no. 1107/2009, in which the application of non-chemical and natural alternatives should be the first choice in plant protection and integrated pest management.

Acknowledgements The study was supported by The Ministry of Education, Science and Technological Development of The Republic of Serbia.

References

Abd-El-Khair H, Wafaa MA, El-Nagdi WMA, Mahmoud MA, Youssef MMA, Abd-Elgawad MMM, Dawood MG (2019) Protective effect of *Bacillus subtilis, B. pumilus,* and *Pseudomonas fluorescens* isolates against root knot nematode *Meloidogyne incognita* on cowpea. Bull Natl Res Cent 43:64. https://doi.org/10.1186/s42269-019-0108-8

Akintayo A, Tylka GL, Singh AK, Ganapathysubramanian B, Singh A, Sarkar S (2018) A deep learning framework to discern and count microscopic nematode eggs. Sci Rep 8:1–11. https://doi.org/10.1038/s41598-018-27272-w

Ambrogioni L, Irdani T (2001) Identification of *Heterodera schachtii* group species in Italy by morphometrics and RAPD-PCR. Nematol Mediterr 29:159–168

Amiri S, Subbotin S, Moens M (2001) An efficient method for identification of the *Heterodera schachtii* sensu stricto group using PCR with specific primers. Nematol Mediterr 29:241–246

Andriani A, Tachibana S, Itoh K (2016) Effects of saline-alkaline stress on benzo[a]pyrene biotransformation and ligninolytic enzyme expression by *Bjerkandera adusta* SM46. World J Microbiol Biotechnol 32:39. https://doi.org/10.1007/s11274-015-2001-4

Antoun H, Prevost D (2006) Ecology of plant growth promoting rhizobacteria. In: Siddiqui ZA (ed) PGPR: biocontrol and biofertilization. Springer, The Netherlands, pp 1–39

Ayatollahy E, Fatemy S, Etebarian HR (2008) Potential for biological control of *Heterodera schachtii* by *Pochonia* chlamydosporia var. *chlamydosporia* on sugar beet. Biocontrol Sci Technol 18:157–167. https://doi.org/10.1080/09583150701824079

Bakker J (2002) Durability of resistance against potato cyst nematodes. Euphytica 124:157–162

Balaes T, Tanase C (2016) Basidiomycetes as potential biocontrol agents against nematodes. Rom Biotechnol Lett 21:11185–11193

Barron GL (2003) Predatory fungi, wood decay, and the carbon cycle. Biodiversity 4:3–9. https://doi.org/10.1080/14888386.2003.9712621

Behrendt CJ, Blanchette RA (1997) Biological processing of pine logs for pulp and paper production with *Phlebiopsis gigantea.* Appl Environ Microbiol 63:1995–2000. https://doi.org/10.1128/aem.63.5.1995-2000.1997

Biancardi E, Panella LW, Lewellen RT (2012) Beta maritima: the origin of beets. Springer, New York, p 250. https://doi.org/10.1007/978-1-4614-0842-0

Cabezas L, Calderon C, Medina LM, Bahamon I, Cardenas M, Berna AJ, Gonzalez A, Restrepo S (2012) Characterization of cellulases of fungal endophytes isolated from *Espeletia* spp. J Microbiol 50:1009–1013

CABI (2019) Heterodera schachtii. Invasive species compendium. Distribution table. https://www.cabi.org/isc/datasheet/27036#todistributionDatabaseTable

Chen SY, Dickson DW, Mitchell DJ (1996) Pathogenicity of fungi to eggs of *Heterodera glycines.* J Nematol 28:148–158

Chen Y-Y, Koike ST, Logan GD, Drozd C, De Oliveira Silva J, Colindres NB, Peacock BB, Smith Becker J, Loffredo A, Wu H, Ruegger PM, Becker JO, Borneman J (2020) Detection of nematophagous fungi from *Heterodera schachtii* females using a baiting experiment with soils cropped to *Brassica* species from California's central coast. PhytoFrontiers 1:4–12

Dackman C, Nordbring-Hertz B (1985) Fungal parasites of the cereal cyst nematode *Heterodera avenae* in southern Sweden. J Nematol 17:50–55

Das G, Patra JK, Choi J, Baek K-H (2017) Anticandidal effect of endophytic bacteria isolated from *Equisetum arvense L.* against *Candida albicans* and *Candida glabrata*. Braz Arch Biol Technol 60:e17160433. https://doi.org/10.1590/1678-4324-2017160433

DiLegge MJ, Manter DK, Vivanco JM (2019) A novel approach to determine generalist nematophagous microbes reveals *Mortierella globalpina* as a new biocontrol agent against *Meloidogyne* spp. nematodes. Sci Rep 9:7521. https://doi.org/10.1038/s41598-019-44010-y

Dohm J, Minoche A, Holtgräwe D, Capella Gutiérrez S, Zakrzewski F, Tafer H, Rupp O, Rosleff Sörensen T, Stracke R, Reinhardt R (2014) The genome of sugar beet (*Beta vulgaris*) a recently domesticated crop. Nature 505:546–549. https://doi.org/10.1038/nature12817

Escobar-Avila I, Cruz-Alvarado Y, Tovar-Soto A, Subbotin S (2019) First report of sugar beet cyst nematode, *Heterodera schachtii* on beetroot and broccoli in Mexico. Plant Dis 103:1434–1434. https://doi.org/10.1094/PDIS-11-18-2000-PDN

Fosu-Nyarko J, Nicol P, Naz F, Gill R, Jones MG (2016) Analysis of the transcriptome of the infective stage of the beet cyst nematode, *H. schachtii*. PLoS One 11:e0147511. https://doi.org/10.1371/journal.pone.0147511

Gamboa-Angulo M, Moreno-Escobar JA, Herrera-Parra E, Perez-Cruz J, Cristóbal-Alejo J, Heredia-Abarca G (2015) Toxicidad in vitro de micromicetos del tropico Mexicano en juveniles infectivos de *Meloidogyne incognita*. Mex J Phytopathol 34:100–109. https://doi.org/10.18781/r.mex.fit.1507-3

Gao Y, Zhao JT, Zu YG, Fu YJ, Wang W, Luo M, Efferth T (2011) Characterization of five fungal endophytes producing cajaninstilbene acid isolated from pigeon pea (*Cajanus cajan* (L.) Millsp.). PLoS One 6:e27589. https://doi.org/10.1371/journal.pone.0027589

Gianfreda L, Sannino F, Filazzola MT, Leonowicz A (1998) Catalytic behavior and detoxifying ability of a laccase from the fungal strain *Cerrena unicolor*. J Mol Catal B Enzym 4:13–23

Gorbatova O, Koroleva O, Landesman E, Stepanova E, Zherdev A (2006) Increase of the detoxification potential of basidiomycetes by induction of laccase biosynthesis. Appl Biochem Microbiol 42:414–419. https://doi.org/10.1134/S0003683806040132

Goverse A, De Engler JA, Verhees J, Van Der Krol S, Helder J, Gheysen G (2000) Cell cycle activation by plant parasitic nematodes. Plant Mol Biol 43:747–761. https://doi.org/10.1023/a:1006367126077

Greco N, D'Addabbo T, Brandonisio A, Elia F (1993) Damage to Italian crops caused by cyst-forming nematodes. J Nematol 25:836

Grujicic G (1958) *Heterodera schachtii* Shmidt a beet nematode in our country. Plant Prot 49(50): 167–174

Guindon S, Gascuel O (2003) A simple, fast, and accurate algorithm to estimate large phylogenies by maximum likelihood. Syst Biol 52:696–704. https://doi.org/10.1080/10635150390235520

Haidar AM, Al-Assas KM, Dawabah AA (2016) Prevalence, distribution and intraspecific variation of *Heterodera schachtii* populations from semiarid environment. Saudi J Biol Sci 23:293–299

Haj Nuaima R, Ashrafi S, Maier W, Heuer H (2021) Fungi isolated from cysts of the beet cyst nematode parasitized its eggs and counterbalanced root damages. J Pest Sci 94:563–572. https://doi.org/10.1007/s10340-020-01254-2

Hakim A, Mor Y, Toker IA, Neuhof M, Markovitz Y, Rechavi O (2018) WorMachine: machine learning-based Levine phenotypic analysis tool for worms. BMC Biol 16:1–11. https://doi.org/10.1186/s12915-017-0477-0

Hallmann J, Sikora RA (1996) Toxicity of fungal endophyte secondary metabolites to plant parasitic nematodes and soil-borne plant pathogenic fungi. Eur J Plant Pathol 102:155–162

Hasky-Günther K, Hoffman-Hergarten S, Sikora RA (1998) Resistance against the potato cyst nematode *Globodera pallida* systemically induced by the rhizobacteria *Agrobacterium radiobacter* (G12) and *Bacillus sphaericus* (B43). Fundam Appl Nematol 21:511–517

Henagamage AP (2019) Bioremediation of textile dyes by fungal-bacterial biofilms. IJEAB 4:3. https://doi.org/10.22161/ijeab/4.3.7

Hillnhütter C, Mahlein AK, Sikora RA, Oerke EC (2012a) Use of imaging spectroscopy to discriminate symptoms caused by *Heterodera schachtii* and *Rhizoctonia solani* on sugar beet. Precision Agric 13:17–32. https://doi.org/10.1007/s11119-011-9237-2

Hillnhütter C, Sikora RA, Oerke EC, van Dusschoten D (2012b) Nuclear magnetic resonance: a tool for imaging belowground damage caused by *Heterodera schachtii* and *Rhizoctonia solani* on sugar beet. J Exp Bot 63:319–327. https://doi.org/10.1093/jxb/err273

Holterman M, Van Der Wurff A, Van Den Elsen S, Van Megen H, Bongers T, Holovachov O, Bakker J, Helder J (2006) Phylum-wide analysis of SSU rDNA reveals deep phylogenetic relationships among nematodes and accelerated evolution toward crown clades. Mol Biol Evol 23:1792–1800. https://doi.org/10.1093/molbev/msl044

Horel A, Schiewer S (2020) Microbial degradation of different hydrocarbon fuels with mycoremediation of volatiles. Microorganisms 8:163. https://doi.org/10.3390/microorganisms8020163

Huang M, Bulut A, Shrestha B, Matera C, Grundler FMW, Scleke ASS (2021) *Bacillus firmus* I-1582 promotes plant growth and impairs infection and development of the cyst nematode *Heterodera schachtii* over two generations. Sci Rep 11:14114

Huelsenbeck JP, Ronquist F (2005) Bayesian analysis of molecular evolution using MrBayes. In: Statistics for biology and health part II, pp 183–226

Jansson H-B, Von Hofsten A, Von Mecklenburg C (1984) Life cycle of the endoparasitic nematophagous fungus *Meria coniospora*: a light and electron microscopic study. Antonie Van Leeuwenhoek 50:321–327. https://doi.org/10.1007/bf00394645

Jones J, Perry R (2004) Plant-parasitic nematodes-small animals, big impact. Biology 51:155–159

Jones JT, Furlanetto C, Bakker E, Banks B, Blok V, Chen Q, Phillips M, Prior A (2003) Characterization of a chorismate mutase from the potato cyst nematode *Globodera pallida*. Mol Plant Pathol 4:43–50

Jones JT, Haegeman A, Danchin EG, Gaur HS, Helder J, Jones MG, Kikuchi T, Manzanilla-López R, Palomares-Rius JE, Wesemael WM (2013) Top 10 plant-parasitic nematodes in molecular plant pathology. Mol Plant Pathol 14:946–961

Kadereit G, Hohmann S, Kadereit JW (2006) A synopsis of Chenopodiaceae subfam. Betoideae and notes on the taxonomy of Beta. Willdenowia 36(Special Issue):9–19. https://doi.org/10.3372/wi.36.36101

Kamei I, Uchida K, Ardianti V (2020) Conservation of xylose fermentability in *Phlebia* species and direct fermentation of xylan by selected fungi. Appl Biochem Biotechnol 192:895–909. https://doi.org/10.1007/s12010-020-03375-x

Kerry B (1980) Biocontrol: fungal parasites of female cyst nematode. J Nematol 12:253–259

Kerry BR (2000) Rhizosphere interactions and the exploitation of microbial agents for the biological control of plant-parasitic nematodes. Annu Rev Phytopathol 38:423–441

Khan A, Williams KL, Nevalainen HKM (2006) Infection of plant-parasitic nematodes by *Paecilomyces lilacinus* and *Monacrosporium lysipagum*. BioControl 51:659–678. https://doi.org/10.1007/s10526-005-4242-x

Kiyuna T, An K-D, Kigawa R, Sano C, Miura S, Sugiyama J (2015) "Black particles," the major colonizers on the ceiling stone of the stone chamber interior of the Kitora Tumulus, Japan, are the bulbilliferous basidiomycete fungus *Burgoa anomala*. Mycoscience 56:293–300. https://doi.org/10.1016/j.myc.2014.08.005

Krishnamurthy YL, Naik BS (2017) Endophytic fungi bioremediation. In: Maheshwari DK, Annapurna K (eds) Endophytes: crop productivity and protection, sustainable development and biodiversity. Springer International Publishing AG, Cham, pp 47–60. https://doi.org/10.1007/978-3-319-66544-3_3

Kubiak-Martens L (1999) The plant food component of the diet at the late Mesolithic (*Ertebolle*) settlement at Tybrind Vig, Denmark. Veg Hist Archaeobot 8:117–127

Kubiak-Martens L, Brinkkemper O, Oudemans TF (2015) What's for dinner? Processed food in the coastal area of the northern Netherlands in the late Neolithic. Veg Hist Archaeobot 24:47–62. https://doi.org/10.1007/s00334-014-0485-8

Kudla U (2006) The role of cell wall-modifying proteins in plant penetration and feeding site proliferation by the potato cyst nematode *Globodera rostochiensis*. PhD thesis, Wageningen University, Wageningen

Lee YS, Kim KY (2016) Antagonistic potential of *Bacillus pumilus* L1 against root-knot nematode, *Meloidogyne arenaria*. J Phytopathol 164:29–33. https://doi.org/10.1111/jph.12421

Letschert JPW (1993) Beta section Beta: biogeographical patterns of variation and taxonomy. PhD thesis, University of Wageningen, Wageningen

Leys M, Petit EJ, El-Bahloul Y, Liso C, Fournet S, Arnaud JF (2014) Spatial genetic structure in *Beta vulgaris* subsp. maritima and *Beta macrocarpa* reveals the effect of contrasting mating system, influence of marine currents, and footprints of postglacial recolonization routes. Ecol Evol 4:1828–1852. https://doi.org/10.1002/ece3.1061

Li F, Chen L, Redmile-Gordon M, Zhang J, Zhang C, Ning Q, Li W (2018) Mortierella elongata's roles in organic agriculture and crop growth promotion in a mineral soil. Land Degrad Dev 29: 1642–1651. https://doi.org/10.1002/ldr.2965

Ma H, Overstreet RM, Subbotin SA (2008) ITS2 secondary structure and phylogeny of cyst-forming nematodes of the genus *Heterodera* (*Tylenchida: Heteroderidae*). Org Divers Evol 8: 182–193

Madani M, Subbotin SA, Moens M (2005) Quantitative detection of the potato cyst nematode, *Globodera pallida*, and the beet cyst nematode, *Heterodera schachtii*, using real-time PCR with SYBR green I dye. Mol Cell Probes 19:81–86

Mall AK, Misra V, Santeshwari PAD, Srivastava S (2021) Sugar beet cultivation in India: prospects for bioethanol production and value added coproducts. Sugar Tech 23:1218–1234. https://doi.org/10.1007/s12355-021-01007-0

Mankau R (1980) Biocontrol: fungi as nematode control agents. J Nematol 12:244–252

Manzanilla-Lopez RH, Esteves I, Finetti-Sialer MM, Hirsch PR, Ward E, Jean Devonshire J, Hidalgo-Diaz H (2013) Pochonia chlamydosporia: advances and challenges to improve its performance as a biological control agent of sedentary endo-parasitic nematodes. J Nematol 45:1–7

Marin F, Navarrete H, Narvaez-Trujillo A (2018) Total petroleum hydrocarbon degradation by endophytic fungi from the Ecuadorian Amazon. Adv Microbiol 8:1029–1053. https://doi.org/10.4236/aim.2018.812070

Martin R, Gazis R, Skaltsas D, Chaverri P, Hibbett D (2015) Unexpected diversity of basidiomycetous endophytes in sapwood and leaves of *Hevea*. Mycologia 107:284–297

Martinez B, Reaser JK, Dehgan A, Zamft B, Baisch D, McCormick C, Giordano AJ, Aicher R, Selbe S (2020) Technology innovation: advancing capacities for the early detection of and rapid response to invasive species. Biol Invasions 22:75–100. https://doi.org/10.1007/s10530-019-02146-y

Morand S, Krasnov BR (2010) Introduction. In: Morand S, Krasnov BR (eds) The biogeography of host-parasite interaction. Oxford University Press, New York, pp 1–5

Mwamula AO, Ko HR, Kim Y, Kim YH, Lee JK, Lee DW (2018) Morphological and molecular characterization of *Heteroderaschachtii* and the newly recorded cyst nematode, *H. trifolii* associated with Chinese cabbage in Korea. Plant Pathol J 34:297. https://doi.org/10.5423/PPJ.OA.12.2017.0262

Nelson BD, Bolton MD, Lopez-Nicora HD, Niblack TL, Del Rio Mendoza L (2012) First confirmed report of sugar beet cyst nematode, *Heterodera schachtii*, in North Dakota. Plant Dis 96:772. https://doi.org/10.1094/PDIS-02-12-0112-PDN

Nitao J, Meyer S, Chitwood D, Schmidt W, Oliver J (2002) Isolation of flavipin, a fungus compound antagonistic to plant-parasitic nematodes. Nematology 4:55–63. https://doi.org/10.1163/156854102760082203

Nour SM, Lawrence JR, Zhu H, Swerhone GDW, Welsh M, Welacky TW, Topp E (2003) Bacteria associated with cysts of the soybean cyst nematode (*Heterodera glycines*). Appl Environ Microbiol 69:607–615. https://doi.org/10.1128/AEM.69.1.607-615.2003

OECD (2001) Consensus document on the biology of Beta vulgaris L. (Sugar beet). Series on harmonization of regulatory oversight in biotechnology OECD no: 18. https://www.oecd.org/env/ehs/biotrack/46815688.pdf

Ohara T, Inoue I, Namiki F, Kunoh H, Tsuge T (2004) *REN1* is required for development of microconidia and macroconidia, but not of chlamydospores, in the plant pathogenic fungus *Fusarium oxysporum*. Genetics 166:113–124

Oliveira VF, Abreu YJL, Fleming LR, Nascimento JS (2012) Anti-staphylococcal and antifungal substances produced by endospore-forming bacilli. J Appl Pharm Sci 2:154–157. https://doi.org/10.7324/JAPS.2012.2428

Oro V (2011) Potato cyst nematodes–morphology, molecular characterization and antagonists. PhD thesis, Faculty of Biofarming

Oro V, Oro-Radovanovic V (2012) Molecular characterization of PCN populations from Serbia. Genetika 44:189–200. https://doi.org/10.2298/GENSR1201189

Oro V, Tabakovic M (2020) Phylogeography of some European populations of the sugar beet cyst nematode. Russ J Nematol 28:91–98. https://doi.org/10.24411/0869-6918-2020-10009

Oro V, Nikolic B, Josic D (2014) The "potato road" and biogeographic history of potato cyst nematode populations from different continents. Genetika 46:895–904. https://doi.org/10.2298/GENSR1403895O

Oro V, Knezevic M, Dinic Z, Delic D (2020) Bacterial microbiota isolated from cysts of *Globodera rostochiensis* (Nematoda: *Heteroderidae*). Plants 9:1146. https://doi.org/10.3390/plants9091146

Oro V, Stanisavljevic R, Nikolic B, Tabakovic M, Secanski M, Tosi S (2021) Diversity of Mycobiota associated with the cereal cyst nematode *Heterodera filipjevi* originating from some localities of the Pannonian plain in Serbia. Biology 10:283. https://doi.org/10.3390/biology10040283

Padgham JL, Sikora RA (2007) Biological control potential and modes of action of *Bacillus megaterium* against *Meloidogyne graminicola* on rice. J Crop Prot 26:971–977. https://doi.org/10.1016/j.cropro.2006.09.004

Park J-H, Pavlov IN, Kim M-J, Park MS, Oh S-Y, Park KH, Fong JJ, Lim YW (2020) Investigating wood decaying fungi diversity in Central Siberia, Russia using ITS sequence analysis and interaction with host trees. Sustainability 12:2535. https://doi.org/10.3390/su12062535

Pathak AD, Kapur R, Solomon S, Kumar R, Srivastava S, Singh PR (2014) Sugar beet: a historical perspective in Indian context. Sugar Tech 16:125–132

Patil PD, Yadav GD (2018) Comparative studies of white-rot fungal strains (*Trametes hirsuta MTCC-1171* and *Phanerochaete chrysosporium NCIM-1106*) for effective degradation and bioconversion of ferulic acid. ACS Omega 3:14858–14868. https://doi.org/10.1021/acsomega.8b01614

Pham VHT, Jeong S-W, Kim J (2015) *Psychrobacillus soli* sp. nov., capable of degrading oil, isolated from oil-contaminated soil. Int J Syst Evol Microbiol 65:3046–3052. https://doi.org/10.1099/ijs.0.000375

Popeijus H, Overmars H, Jones J, Blok V, Goverse A, Helder J, Schots A, Bakker J, Smant G (2000) Degradation of plant cell walls by a nematode. Nature 406:36–37

Powers T, Skantar A, Harris T, Higgins R, Mullin P, Hafez S, Handoo Z, Todd T, Powers K (2019) DNA barcoding evidence for the North American presence of alfalfa cyst nematode, *Heterodera medicaginis*. J Nematol 51:11–17. https://doi.org/10.21307/jofnem-2019-016

Pylypenko LA, Kalatur KA (2015) Breeding and usage of sugar beet cultivars and hybrids resistant to sugar beet nematode *Heterodera schachtii*. Agric Sci Pract 2:12–22

Qin L, Kudla U, Roze EH, Goverse A, Popeijus H, Nieuwland J, Overmars H, Jones JT, Schots A, Smant G (2004) A nematode expansin acting on plants. Nature 427:30

Radice AR, Myers RF (1984) Distribution and dissemination of *Punctodera punctata* on turfgrasses in New Jersey. Plant Dis 68:256–257

Reitz M, Rudolph K, Schroder I, Hoffmann-Hergarten S, Hallman J, Sikora RA (2000) Lipopolysaccharides of *Rhizobium etli* strain G12 act in potato roots as inducing agent of systemic resistance to infection by the cyst nematode *Globodera pallida*. Appl Environ Microbiol 66:3515–3518

Ryan NA, Jones P (2004) The ability of rhizosphere bacteria isolated from nematode host and non-host plants to influence the hatch in vitro of the two potato cyst nematode species *Globodera rostochiensis* and *G. pallida*. Nematology 6:375–387

Sabo A, Ferris V (2004) ß-tubulin paralogs provide a qualitative test for a phylogeny of cyst nematodes. J Nematol 36:440

Sekimoto S, Uehara T, Mizukubo T (2017) Characterisation of populations of *Heterodera trifolii* Goffart, 1932 (Nematoda: *Heteroderidae*) in Japan and their phylogenetic relationships with closely related species. Nematology 19:543–558. https://doi.org/10.1163/15685411-00003067

Skantar AM, Handoo ZA, Carta LK, Chitwood DJ (2007) Morphological and molecular identification of *Globodera pallida* associated with potato in Idaho. J Nematol 39:133–144

Smant G, Goverse A, Stokkermans JP, De Boer JM, Pomp H, Zilverentant JF, Overmars HA, Helder J, Schots A, Bakker J (1997) Potato root diffusate-induced secretion of soluble, basic proteins originating from the subventral esophageal glands of potato cyst nematodes. Phytopathology 87:839–845

Smith O, Clapham A, Rose P, Liu Y, Wang J, Allaby RG (2014) A complete ancient RNA genome: identification, reconstruction and evolutionary history of archaeological barley stripe mosaic virus. Sci Rep 4:4003. https://doi.org/10.1038/srep04003

Soldi E, Casey C, Brian R, Murphy BR, Hodkinson TR (2020) Fungal endophytes for grass based bioremediation: An endophytic consortium isolated from *Agrostis stolonifera* stimulates the growth of *Festuca arundinacea* in lead contaminated soil. J Fungi 6:254. https://doi.org/10.3390/jof6040254

Stepniewska Z, Kuzniar A (2013) Endophytic microorganisms—promising applications in bioremediation of greenhouse gases. Appl Microbiol Biotech 97:9589–9596. https://doi.org/10.1007/s00253-013-5235-9

Su H, Kang J-C, Cao J-J, Mo L, Hyde KD (2014) Medicinal plant endophytes produce analogous bioactive compounds. Chiang Mai J Sci 41:1–13

Subbotin SA, Vierstraete A, De Ley P, Rowe J, Waeyenberge L, Moens M, Vanfleteren JR (2001) Phylogenetic relationships within the cyst-forming nematodes (Nematoda, *Heteroderidae*) based on analysis of sequences from the ITS regions of ribosomal DNA. Mol Phylogenet Evol 21:1–16. https://doi.org/10.1006/mpev.2001.0998

Subbotin SA, Mundo-Ocampo M, Baldwin JG (2010) Systematics of cyst nematodes (Nematoda: Heteroderinae). In: Hunt DJ, Perry RN (eds) Nematology monographs & perspectives 8BA, Brill, Leiden-Boston, pp 86–101

Tamura K, Dudley J, Nei M, Kumar S (2007) MEGA4: molecular evolutionary genetics analysis (MEGA) software version 4.0. Mol Biol Evol 24:1596–1599

Triffitt MJ (1928) On the morphology of *Heterodera schachtii* with special reference to the potato-strain. J Helmintol 6:39–50. https://doi.org/10.1017/S0022149X00029783

Turner SJ, Rowe JA (2006) Cyst nematodes. In: Perry RN, Moens M (eds) Plant nematology. CABI, Wallingford, pp 91–120. https://doi.org/10.1079/9781845930561.0091

Tzean SS, Liou JY (1993) Nematophagous resupinate basidiomycetous fungi. Phytopathology 83: 1015–1020

Uhlig S, Jestoi M, Parikka P (2007) *Fusarium avenaceum*—the North European situation. Int J Food Microbiol 119:17–24. https://doi.org/10.1016/j.ijfoodmicro.2007.07.021

Zhang K, Bonito G, Hsu C-M, Hameed K, Vilgalys R, Liao H-L (2020) *Mortierella elongata* increases plant biomass among non-leguminous crop species. Agronomy 10:754. https://doi.org/10.3390/agronomy10050754

Endophytes for Sustainable Sugar Beet Production

38

Preeti Singh, Santosh Kumar, Santeshwari Srivastava, Mona Nagargade, Dharmendra Kumar, and Pankaj Kumar Sinha

Abstract

Endophytes live inside the plant in a non-pathogenic way and develop a mutualistic relationship. Endophytes help in the adaptation of plants against biotic and abiotic stresses. They produce secondary metabolites, which protect plants against pathogens and pests. Some endophytes produce growth hormones like IAA, ethylene, and cytokinin. These microbes can be used in sustainable agricultural practices as this improves the crop yield without harming the environment. Manipulating bacterial populations in soils and within crops will be crucial if endophytes are to be utilized in crop production systems. However, their role in plant-stress tolerance and nutrient accumulation is not much explored. Hence, the study of endophytes and their compounds in crop production and protection is needed in the present scenario of environmental pollution and climate change. Different parts of the sugar beet are host to an abundance of endophytes. Both seeds and soil provide specific beneficial bacteria for rhizosphere assembly and microbiota-mediated pathogen tolerance. This can be translated into microbiome management strategies for the sugar beet and ecosystem health. This chapter explains the endophytes, their analysis, factors affecting their growth in general, entry mechanism, growth-promoting abilities, and diversity of endophytes with respect to the sugar beet crop.

P. Singh (✉) · S. Kumar · P. K. Sinha
ICAR-IARI, Hazaribag, Jharkhand, India

S. Srivastava · M. Nagargade
ICAR-IISR, Lucknow, Uttar Pradesh, India

D. Kumar
ICAR-CPRI, Shimla, Himachal Pradesh, India

V. Misra et al. (eds.), *Sugar Beet Cultivation, Management and Processing*,
https://doi.org/10.1007/978-981-19-2730-0_38

Keywords

Endophytes · Growth hormones · Nutrient accumulation pathogen protection · Rhizosphere

Abbreviations

APX	Ascorbate or thiol-dependent peroxidases
CAT	Catalases
CPA	Cellulose-proline agar
DHAR	Dehydroascorbate reductases
GR	Glutathione reductases
IAA	Indole-3-acetic acid
ISP	International Streptomyces Project
MDHAR	Mono-dehydroascorbate reductases
NAP	Naphthalene
PAHs	Polycyclic aromatic hydrocarbons
PGPE	Plant growth-promoting endophytes
POX	Peroxidase
PPO	Polyphenol oxidases
PY	Peptone yeast agar
ROS	Reactive oxygen species
SCA	Starch casein agar
SGN	Starch-glycerol-nitrate agar
SOD	Superoxide dismutases
SREZ	Secondary root emergence zones
YA	Yeast extract agar
YMA	Yeast-mannitol agar

38.1 Introduction

The green revolution in India has been primarily achieved through high-yielding varieties and chemical fertilizers (Foley et al. 2005). However, the indiscriminate use of manufactured inputs and improper use of agricultural practices negatively affect soil ecology. Therefore, new eco-friendly approaches have to be employed to maintain sustainable agricultural production and manage unfavorable conditions that can cause a reduction in crop yield, including plant stresses associated with unfavorable environmental conditions, such as water stress, high/low temperature, or soil salinity, as well as biotic stress induced by plant pathogens. In this context, harnessing the contribution of beneficial endophytes for agricultural management in general and, more particularly, for nutrients and pest management now became essential (Singh et al. 2011). Endophytes are isolated from surface-sterilized plant tissues, and they do not harm the host during their life cycle in the plant. They have been isolated from plants growing in various ecosystems, i.e., temperate to tropical ecosystems.

Initially, De Bary, in the year 1866, used the term endophyte in the nineteenth century for fungus residing inside the plants (de Bary 1866). Endophytes produce secondary metabolites, which act as plant protectants or induce plant immunity. In general, an endophyte genome contains 5–15 terpenoid synthase, 8–21 nonribosomal, and 7–29 polyketide synthase genes required for bioactive compounds diversity in endophytes (Wang et al. 2015). Different microbes, including archaea, bacteria, fungi, and unicellular eukaryotes, have been reported as endophytes (Tremouillaux-Guiller et al. 2002). Ascomycetes and fungi imperfectly grouped among endophytic microorganisms represent the largest endophytic fungal groups containing as many as 106 species, and are sources of untapped biologically active small molecular natural products. In fungi, Glomeromycota is the dominant division in endophytic fungi, followed by Ascomycota, Basidiomycota, Zygomycota, and unidentified phyla. There are several beneficial effects attributed to endophytic microorganisms to improve plant growth, disease and pest management, biological nitrogen fixation, systemic resistance induction, production of siderophore, and antibiotics.

Sugar beet (*Beta vulgaris* L.) is an important root crop and the main source of sugar in temperate climates. Sugar beet is also an interesting model crop for microbiome studies (Zachow et al. 2008; Mendes et al. 2012, Kusstatscher et al. 2019a, b), known for their genome and breeding history (Würschum et al. 2013; Dohm et al. 2014). The presence of endophytic bacteria within healthy sugar beet (*Beta vulgaris* L.) roots has been demonstrated (Jacobs et al. 1985). The endophytic bacteria could boost sugar beet growth and photosynthesis, and increase sugar content, due to increased chlorophyll, leading to a consequent higher carbohydrate synthesis (Shi et al. 2009, 2010, 2011). These endosymbionts boost plant-nutrient absorption, resulting in better vegetative development of the host plant. Endophytes with growth-promoting characteristics are desirable not only for the agronomic development of sugar beet, but also for their subsequent impact on boosting tolerance to diseases via growth enhancement. A beneficial endophyte–host interaction in commercial cultivars would minimize the use of agricultural inputs, such as fertilizers and pesticides, thus saving expenditures and decreasing pollutants to the environment.

38.2 Major Types of Endophytes

38.2.1 Bacterial

Endophytic microorganisms play a multifunctional role in ecosystems and plant physiology, and these bacteria colonize the intercellular and intracellular spaces of the inner tissue. The endophytic habitat provides a conducive environment for their colonization. The bacteria genera of *Bacillus* and *Pseudomonas* are identified as frequently occurring in crops (de Souza Leite et al. 2013). The diversity of endophytes in host plants depends upon plant species, type of bacteria, and environmental factors. The endophytic species mostly encountered are a, b, and c-proteobacteria subgroups closely related to epiphytic species (Kuklinsky-Sobral et al. 2004). The c-proteobacteria group is the most diverse and dominant. Most Gram-negative endophytic bacteria function as biocontrol agents (Kobayashi and Palumbo 2000), whereas the dominant Gram-positive endophytic bacteria (*Bacillus* species) act as a growth promoter by improving nutrient uptake. Shi et al. (2009) identified the endophytic bacteria *B. pumilus* 2-1, *C. indologene* 2-2, and *A. johnsonii* 3-1 *Pseudomonas fluorescens*, *Bacillus flexus*, *Pseudomonas fulva*, *Bacillus pumilus*, *Paeniba-cillus polymyxa*, *Chryseobacterium indologene,* and *Enterococcus faecalis* from the field-cultivated sugar beet leaves.

38.2.2 Fungal

Most endophytic fungi were .isolated from Ascomycota and Basidiomycetes division (Rungjindamai et al. 2008). Endophytic fungi, as stated earlier, have been categorized into two major groups based on phylogenetic traits: clavicipitaceous endophytes, which colonize grasses, and non-clavicipitaceous endophytes, which colonize nonvascular plants, ferns and allies, conifers, and angiosperms (Rodriguez et al. 2009), i.e., *Acremonium terricola*, *Monodictys Castanea*, *Penicillium glandicola*, *Phoma tropica*, and *Tetraploa aristata* were isolated for the first time (Bezerra et al. 2012). Nonclavicipitaceous endophytes have three major groups based on colonization and transmission in host plants, plant biodiversity, and plant growth traits deliberated to hosts, while the clavicipitaceous group has just one class. *A. alternata*, *F. oxysporum*, and *Pythium aphanidermatum* were the fungi most frequently isolated from sugar beet (Shi et al. 2009).

38.3 Colonization of Plants by the Endophytes

Endophytes follow a similar pattern as the pathogenic microorganism to enter into plants, and the secretion of defence-related molecules from the plant is less for endophytes than for pathogens. Endophytes can colonize in different seed parts, embryos, and vegetatively propagated materials. Endophytes' colonization goes through several critical stages, i.e., host finding, recognition, colonization on the plant surface, and entrance into internal plant tissues. These endophytes likely mobilize and grow in the developing seedlings during germination and early seedling growth (Nelson 2018). As seedlings emerge and plant growth begins, interactions between the roots and the soil microbiome commence. Secretion of root exudates promotes microbial activities in the rhizosphere, which helps in the attachment and entry of endophytes into the plant roots. Eventually, specific endophytes initiate colonization of tissues beyond the roots, such as the stems and leaves, and ultimately throughout the plant endosphere.

Moreover, endophytes passed on to seeds resumed endophytic activity after the seeds were planted (Doty 2017). The complex process of endophytic colonization usually starts from the roots, and recognition of specific compounds in the root exudates by the endophytic microorganism (Rosenblueth and Martinez-Romero 2006). Plant root releases these compounds to interact with the beneficial microorganism for their ecological advantage (Hallmann et al. 1997).

38.3.1 Rhizosphere Colonization by the Endophytic Microorganism

The rhizosphere colonization is a highly complex process and a very competitive task for the endophytic microorganism to occupy spaces and get nutrients. Bacterial traits like motility and polysaccharide production are essential in the colonization of the plant rhizosphere. Bacterial detection systems based on *different* immune markers have revealed that after inoculation in the soil, the bacterial cells first colonize the rhizosphere after being inoculated into the soil (Gamalero et al. 2003). The bacterial cells then attach to the rhizoplane, forming microcolonies. The bacteria then colonize the entire root surface and some rhizodermal cells, leading to the formation of biofilms by the bacteria (Benizri et al. 2001). For improving plant growth, the bacteria have to colonize the plant rhizosphere competently. They also have to compete with various microorganisms present in the soil during colonization. The microorganism's colonization in the root of the host plant is not uniform.

38.3.2 Root Colonization by the Endophytic Microorganism

After establishing the rhizosphere and rhizoplane, endophytesenter inside the plant root and colonize themselves. This requires bacterial adhesionto cell-surface structures of the host plant, which is mediated by polysaccharides, pili, and bacterial adhesions(Hori and Matsumoto 2010). After the bacteria have established themselves on the rhizoplane, they penetrate the root interior using specialized mechanisms. Once on the root surface, the endophytes reach the root entry sites, like lateral root emergence and wounds. Nevertheless, every endophytic microorganism has its distinct colonization pattern and colonization site preferences.

The process of penetration into the host can be passive or active. Passive penetration can occur at wounds caused by abrasion with soil particles, abiotic mechanical injury, root tips, or those made by pathogenic microorganisms (Hardoim et al. 2008). Active penetration by the endophytes is done by the process of attachment and proliferation. This involves the presence of lipopolysaccharides, flagella, pili, twitching motility, and quorum sensing, which can affect endophytic colonization and bacterial movement inside the host plants (Böhm et al. 2007). The microbe mainly produces pectinases and cellulases enzymes to penetrate the cell wall, which helps in bacterial penetration and spreading in the plant tissues (Elbeltagy et al. 2000).

38.3.3 Colonization of Above-Ground Plant Parts by the Endophytes

After traveling from soil to the roots, the endophytes can colonize systemically in above-ground tissues. Nevertheless, only a few microorganisms can colonize aerial-plant parts due to the physiological requirements needed to occupy these plant niches. Thus, the microorganisms that migrate to the above-ground plant tissue are well adapted to this particular endophytic niche. The bacterial movement inside the plant is supported by bacterial flagella and the plant transpirationstream. For bacterial movements inside the plant, a tissue requires enzymes that are responsible for cell wall degradation. However, movement through xylem elements occurs through perforated plates that allow the movement of bacteria through large pores, without requiring cell-wall-degrading enzymes (Sapers et al. 2005). These endophytes ultimately reach leaf tissues for further colonization. Endophytic microorganisms mostly colonize the leaf tissues from plant roots, but like phytopathogenic bacteria, endophytic microorganisms can gain entry into the leaves from the phyllosphere *via* leaf stomata (Senthilkumar et al. 2011).

38.4 Factors Affecting Endophytic Bacterial Diversity

Environmental and plant factors determine the endophyte's ability to colonize the host throughout its life cycle. Endophyte diversity primarily depends on its types and host plant genotypes, growth stages, and health status.

38.4.1 Environmental

Endophyte diversity in host plants is influenced by environmental factors like weather, altitude, spatial coordinates, and soil factors (Chiellini et al. 2014). Among various soil factors, nitrogen content has shown a more significant relationship with bacterial diversity. Microbial diversity increased with an increase in nitrogen content. Change in environmental conditions and the maintenance of a shifting and diverse endophytic community may form part of the physiological strategy that plants adapt to their environment.

38.4.2 Plant Factors

Endophytes show different associations, either parasitism or mutualism, depending on the type of plant tissues, plant growth condition, and the type of host plant. Endophyte diversity depends on several factors like physiological structures, metabolites, and different growth patterns of plants (Kawaguchi and Minamisawa 2010). Plant health affects endophyte diversity: the bacterial community of *Paullinia cupana* with asymptomatic anthracnose comprised mainly Firmicutes, whereas plants with symptomatic anthracnose comprised mainly Acidobacteria (Bogas et al. 2015). In a PCR-based molecular study, it was found that cultivable bacteria, *Echinacea angustifolia* (DC.) and *Echinacea purpurea* (L.), isolated from different parts of the plant possessed a different type of bacterial diversity/community structure, suggesting the intense selective pressure and a low degree of strain sharing in plant tissue (Chiellini et al. 2014).

38.5 Mechanisms of Sugar Beet Plant Growth Promotion

Endophytic microorganisms have been shown to impart several beneficial effects on their host plant, directly or indirectly. The direct benefits are improving nutrient uptake, nutrient use efficiency, and plant growth by modulating growth-related hormones, which can help the plants grow better under normal and stressed conditions (Ma et al. 2016). However, indirect ones are an improvement in plant growth by ceasing pathogen activities using mechanisms like antibiotic and lytic enzyme production, nutrient unavailability for the pathogens, and priming plant defense mechanisms, thereby increasing the plants' tolerance to the pathogen (Miliute et al. 2015). Shi et al. (2010) investigated the growth-promoting effects of endophytes *B. pumilus* 2-1, *C. indologene* 2-2, and *A. johnsonii* 3-1 on host sugar beet seedlings. Endophyte infection resulted in a significant increase in chlorophyll content and carbon assimilation in the sugar beet (Shi et al. 2010). In the sugar beet, electron transport in chlorophyll has been found to accelerate, for example, in the thylakoids, by a higher ATP usage for carbon fixation. As light increased, the electron transport system could meet the increased demand for NADPH and ATP, resulting in enhanced carbon assimilation. Endophyte-infected plants had a total glucose content of approximately 1.5 times that of uninfected plants. Sucrose content was greater than fructose content in infected plants, indicating an increased sucrose synthesis as a result of increased electron transport ability. Additionally, *B. pumilus* 2-1, *C. indologene* 2-2, and *A. johnsonii* 3-1 were able to synthesize IAA in vitro from various precursors. The results indicated that when the L-tryptophan content increased, the growth of seedlings inoculated with the three bacterial endophyte species also increased. This indicates that IAA production by bacteria via tryptophan-dependent pathways has an effect on the sugar beet seedling growth. Tryptophan is naturally secreted in the root exudates of beet plants, and the majority of auxin observed in the rhizosphere is considered to be the result of microbial production (Kamilova et al. 2006). Exogenous sources of IAA, such as those produced by microbes, have been shown to alter the shape of root systems, hence affecting plant nutrient intake (Arteca 1996).

Szymańska et al. (2020) studied the salinity effect on the endophyte-infected sugar beet and found bioaugmentation of *B. vulgaris* with the selected plant growth-promoting endophytes (PGPE) strains. The presence of PGPE reduced the negative effects of salinity, resulting in enhanced plant growth and a decrease in proline and hydrogen peroxide concentrations in plant organs. The results of this study revealed that *K. marisflavi* CSE9, characterized by higher salinity tolerance, is effective in reducing salt stress in sugar beet. Gao et al. (2010) found that application of genes of

1, 3-glucanase, present in a strain of *Lysobacter enzymogenes,* has the biocontrol activity toward the damping-off disease of the sugar beet caused by *Pythium* and tall fescue leafspot disease.

Müller et al. (2013) identified the endophyte *Pseudomonas poae* RE*1-1-14, which is a member of the group of pseudomonads that interact beneficially with plants. They have the ability to use sucrose (Behrendt et al. 2003). These endophytes extensively colonize the developing roots, a primary requirement for the effective suppression of root pathogens (Zachow et al. 2010). In field studies performed over six consecutive years, this isolate was proved to inhibit late root rot caused by *R. solani*. Shi et al. (2009) isolated endophytic bacteria from sugar beet plants in Changji County, Xinjiang Province. Three endophytic bacterial strains, *Paenibacillus polymyxa*, *Bacillus flexus*, and *Stenotrophomonas sp.*, exhibited moderate antagonistic activity against *Cercospora* leaf spot (*Cercospora beticola*). Numerous experiment trials demonstrated that endophytic bacteria could help lower sugar beet disease incidence (Shi et al. 2009). The control efficiency ranged between 67.6 and 80.2%, demonstrating that biocontrol, with endophytic bacteria, was a viable alternative strategy for sugar beet fungal disease control. Abudureheman (2012), in his greenhouse screening experiment, discovered that 12 strains of *Gluconacetobacter spp.* boosted the sugar beet growth to varied extents. *G. diazotrophicus* colonizes the sugar beet and fixes nitrogen when the nitrogen content in the soil is less. Root tips, root hairs, and lateral root junctions were the infection sites of *G. diazotrophicus*. Microscopic inspection of samples showed that root tips, root hairs, and junctions of lateral root emergence were the most potential infectious sites for bacteria. Carrión et al. (2019), in his work on the sugar beet, identified and cultivated several bacteria whose abundance inside plant roots was increased upon inoculation of a plant with the fungal pathogen *Rhizoctonia solani,* in suppressive soil. They discovered important bacterial genes whose expression increased in response to plant infection in suppressive soil, elucidating pathogen suppression mechanisms and showing the way for the establishment of robust consortia, conferring the suppressive phenotype. These genes include carbohydrate-active enzymes (CAZymes, which degrade carbohydrates), particularly those that may be active against fungal cell walls, and biosynthetic gene clusters (BGCs), which are responsible for the production of specific metabolites. Certain *Paraburkholderia*, *Pseudomonas*, and *Streptomyces* species are found in greater abundance near the roots (rhizosphere) of plants cultivated in these *R. solani*-suppressive soils, and some isolates of these genera can confer suppressive behavior (Mendes et al. 2011; Carrión et al. 2018). This activity has been linked to production of a chlorinated lipopeptide (thanamycin) by *Pseudomonas* and of sulfurous volatile compounds by *Paraburkholderia* (Fig. 38.1). According to Carrión et al. (2019) these rhizosphere microorganisms may serve as the initial line of defense against soilborne *Rhizoctonia*; subsequent pathogen attacks and colonization of plant roots induce the plant to mobilize a second line of defense by bacteria within the root. However, there is more to be learned about how these defenses are coordinated and what features of suppressive soils are important.

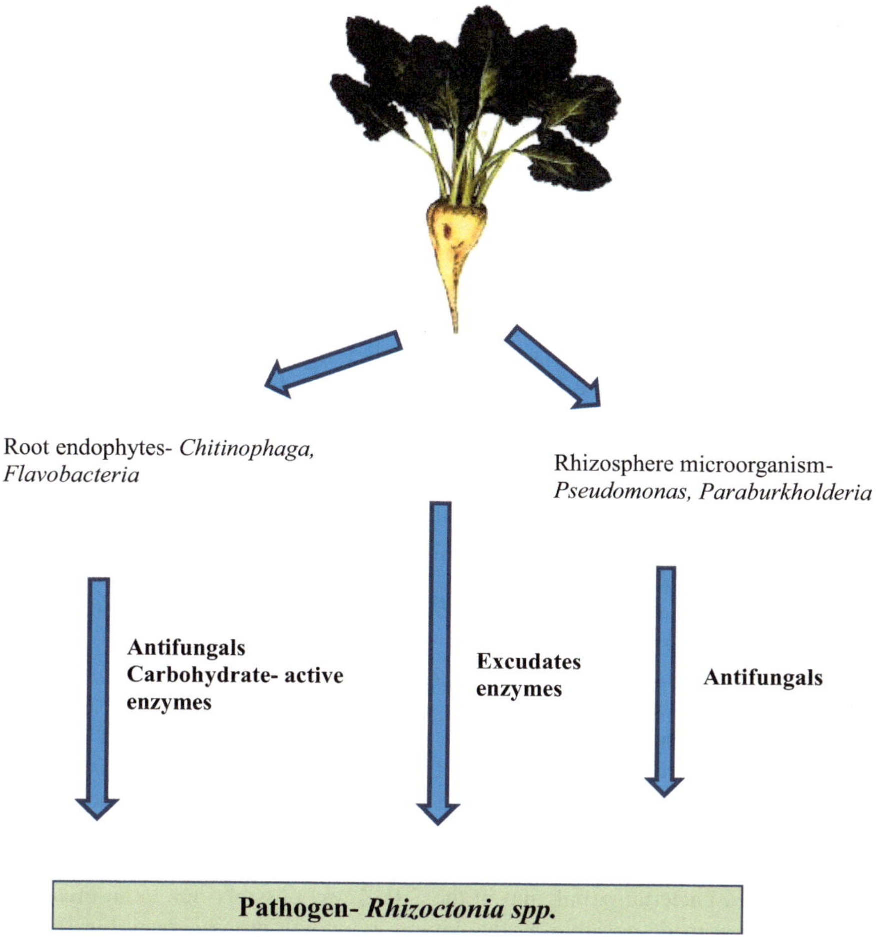

Fig. 38.1 Plant–pathogen–microbiota interactions. The plant (for example, sugar beet), the root (endophytic) microbiota, and the rhizosphere microbiota defend the plant against pathogen (for example, *Rhizoctonia solani*) attack

38.6 Endophytic Bacterial Diversity of Sugar Beet

Sugar beet is an exciting model crop for microbiome studies (Zachow et al. 2008), known for its genome and breeding history (Würschum et al. 2013). Larran et al. (2000) extracted fungal endophytes from the sugar beet and observed that the species most frequently isolated from beet leaves (decreasing order of frequency) were *Alternaria alternata*, *Pleospora herbarum*, *Stemphylium sp.*, and *Epicoccum nigrum*. Yeasts were primarily isolated from leaves and are present in large quantities. Other species of fungi such as *Chaetomium sp.*, *Phomopsis sp.*,

Cladosporium spp., Colletotrichum dematium, C. gloeosporioides (Penz.) Sacc., Phoma betae Frank, and *Phomopsis sp.* were isolated and are present in lower quantities. They further found that most taxa are recovered only sporadically, possibly due to environmental factors influencing their spatial distribution. More competitive endophytes have already achieved significant colonization of the host tissues simply because of the sampling and isolation techniques used (Petrini et al. 1992). With age, the increase of *Alternaria alternata* and *Pleospora herbarium* infection could result from increased exposure time to the propagules. According to Miller and Roy (1982), such an increase may be related to an alteration in the nutritional content of the leaf and the secretion of leachates that enhance germination and growth of propagules. Cabral et al. (1993) point out that *Alternaria alternata* are solely found in the sub-stomata area, where they may benefit from some nutrient leakage from the host or are afforded protection from desiccation or mycophagous invertebrates. Other fungi that were isolated included both *Fusarium oxysporum* and *Penicillium* species. These two genera of endophytes have also been found in other plants. The species isolated in this work may be classified into three groups: (1) well-known and economically significant pathogens of the beet, that is, *F. oxysporum*; (2) commonly abundant phylloplane fungi considered primary saprophytes and minor pathogens (Zillinsky 1984), that is, *A. alternata*; and (3) species that are sometimes found in beet, that is, *Bacillus sp.* There were significant differences between microorganisms, beet growth stages, and growth stages. *A. alternate, F. oxysporum, and Pythium aphanidermatum* were the fungi most frequently found in the sugar beet. The other microorganisms were present in lower quantities. There are two secondary root emergence zones (SREZ) on opposite sides of a sugar beetroot. The emergence of roots from these SREZ causes wounding, which may provide a natural entry path for bacteria. They found that counts of root bacteria from the SREZ increased 100- to 1000-fold over the 7 weeks. No significant increase in bacterial numbers was found in the peripheral or core tissues during the same period. The increased bacterial population in the SREZ was possibly due to an increase in tissue invasion made possible by the natural wounding process, resulting from the emergence of secondary roots. This was the only period during which a change in the bacterial population was observed.

Shi et al. (2009) cultured 360 sugar beet roots and 60 sugar beet leaf segments and isolated 221 bacteria, 34 fungal, and 5 actinomycetes isolates. Seven bacterial species, six fungal species, and two actinomycete species were identified from all the isolates. *Pseudomonas fluorescens*, *Bacillus flexus*, *Pseudomonas fulva*, *Bacillus pumilus*, *Paenibacillus polymyxa*, *Chryseobacterium indologene*, and *Enterococcus faecalis* were the seven bacteria isolated. The majority of isolated endophytic fungi are anamorphs of Deuteromycotina, which includes various hyphomycetes, but Ascomycotina was very sparse. The study of Miao et al. (2020) found that the bacterial community structures and compositions in the sugar beet-cultivated soil had undergone some changes before and after continuous cropping. The effects of continuous cropping on endophytic bacteria of sugar beet were not statistically significant.

Sphingomonas, *Pseudarthrobacter*, *Paracoccus*, *Planococcus*, *Novosphingobium*, *Nesterenkoni*, *Nocardioides*, *Acinetobacter*, *Bacillus*, and *Halomonas* were identified in the non-continuous soil sample. Under continuous cropping conditions, the bacteria found included *Acinetobacter*, *Bacillus*, *Halomonas*, *Nesterenkonia*, *Nocardioides*, *Paracoccus*, *Planococcus*, *Pseudarthrobacter*, *Sphingomonas*, and *Terribacillus*. Endophytic bacteria genera vary with cropping conditions, i.e., under continuous cropping and non-continuous cropping. There are some differences in the diversity and compositions of the microbial communities in the samples of both continuous and non-continuous sampling. The endophytic bacterial groups found in sugar beet samples were *Pseudarthrobacter, Bacillus, Achromobacter, Pantoea, Pseudomonas, Sphingomonas, Novosphingobium Stenotrophomonas, Terribacillus, Paracoccus, Nesterenkonia, Weissella, Leuconostoc,* and *Nocardioides.*

According to Aeini et al. (2018), three bacterial species are particularly well adapted to colonize the inner plant tissues of the sugar beet. These species occur most frequently and in huge numbers, and could be considered as the dominating leaf endophytes. *A. calcoaceticus* was discovered as one of the most prevalent bacteria. The bacterium *A. calcoaceticus* appears to be extremely prevalent in the leaf endophytic communities that have been examined so far. This species contributes positively to plant growth promotion and the formation of physiologically active metabolites (Indiragandhi et al. 2008; Kang et al. 2009). *P. aeruginosa* was found as the second dominating bacteria. Due to *P. aeruginosa* biocontrol potential, certain strains have been recommended for use in integrated pest management programmes (Kumar et al. 2013). *S. maltophilia*, the last major endophytic bacteria to be discovered, was previously identified as a rhizosphere resident in western Iran (Aeini and Khodakaramian 2017). According to the study, rhizosphere bacteria colonize the roots first and then spread to the plant's upper part via xylem vessels (Compant et al. 2011). Additionally, this research adds to our understanding of the dominating phyla that live as endophytes in sugar beet leaves. One of the more notable findings from this study is the identification of *A. calcoaceticus*, *P. aeruginosa*, and *S. maltophilia* as endophytes in sugar beet leaves for the first time. Jacobs et al. (1985) in their experiment found that the most common bacterial isolates from fresh sugar beet root tissues are *Bacillus subtilis, Corynebacterium sp., Erwinia herbicola, Lactobacillus sp., Pseudomonas aeruginosa, Xanthomonas sp.,* and *Pseudomonas fluorescens.* Wolfgang et al. (2020) analyzed microbial communities in seeds, roots, and the corresponding soil to investigate sugar beet microbiota assembly and composition. They reported that the seeds of all sugar beet cultivars were highly colonized by bacteria and included a core of the sugar beet microbiome that contributed considerably to rhizosphere assembly. Sugar beet seeds contain a significant number of *Proteobacteria*, *Actinobacteria*, *Firmicutes*, and *Bacteroidetes*, which is quite typical for seed microbiota (reviewed by Nelson 2018). However, *Enterobacteriaceae*, in particular, has been recognized as a significant component of seeds. *Pantoea*, another important taxon discovered in sugar beet seeds, was similarly dominating. *Pantoea* encompasses a diverse range of lifestyles, including plant pathogens, growth stimulants, and strains commercially generated for phytopathogen biocontrol. Thus, *Pantoea* serves as a model group for adaptations to specific niches (Walterson and Stavrinides 2015). Additionally, the prevalence of many *Enterobacteriaceae* (*Kosakonia* and *Enterobacter*) and *Paenibacillus* is correlated with *Rhizoctonia* tolerance in sugar beet seeds. *Paenibacillus* is reported to have species that promote plant growth and act as a pathogen antagonist (reviewed by Rybakova et al. 2016). Similarly, several *Enterobacter* species are hostile toward *Rhizoctonia* (e.g. Abdeljalil et al. 2016).

38.7 Future Prospects

Endophytes are untapped microbes and help the plants in their function. Endophytes isolated from the sugar beet can promote the growth of sugar beet plantlets. These isolates are amenable to artificial inoculation, and their non-host specificity enabled them to infect and colonize new host plants. The beneficial association of endophytes with other crops may be extended to the sugar beet. There is also a need to identify the mechanisms of growth promotion and optimize the conditions for endophyte application so that the endophytes serve as growth promoters and as an approach for increasing sugar production for the sugar beet. However, over time, researchers should focus on endophyte biology, procedures to confirm endophytes, and differentiate endophytes from epiphytes. In the upcoming time, the endophytes formulation-based biofertilizers may increase soil fertility and crop yield.

38.8 Conclusion

Endophytes are non-pathogenic, naturally colonized, bacteria or fungi with various beneficial traits for the plants. They are involved in different metabolic activities and induced tolerance under detrimental conditions such as drought, heat, high salinity, poor nutrient availability, and various biotic stresses. Endophytes and their secondary metabolites proved to be a great source of bioactive compounds and potentially be used under detrimental climatic conditions.

Sugar beet is an exciting model crop for microbiome studies, known for its genome and breeding history. Isolation of 221 bacteria, 34 fungal, and 5 actinomycetes isolates from cultured 360 sugar beet roots and 60 sugar beet leaf segments has been reported. Fungal endophytes have also been extracted from the sugar beet. The most frequently isolated species from beet leaves (decreasing order of frequency) were *Alternaria alternata*, *Pleospora herbarum*, *Stemphylium sp.*, and *Epicoccum nigrum*. Three endophytic bacterial strains, *Paenibacillus polymyxa*, *Bacillus flexus*, and *Stenotrophomonas sp.*, exhibited moderate antagonistic activity against *Cercospora* leaf spot (*Cercospora beticola*). Twelve strains of *Gluconacetobacter spp.* have also been identified that boosted sugar beet growth to varied extents. *G. diazotrophicus* colonizes the sugar beet and fixes nitrogen when the nitrogen content in soil is less. Endophytic bacteria genera vary with cropping conditions. In continuous cropping condition of the sugar beet, *Acinetobacter*,

Bacillus, Halomonas, Nesterenknia, Nocardioides, Paracoccus, Planococcus, Pseudarthrobacter, Sphingomonas, and Terribacillus bacteria were majorly seen. However, *Bacillus subtilis, Corynebacterium sp., Erwinia herbicola, Lactobacillus sp., Pseudomonas aeruginosa, Xanthomonas sp.*, and *Pseudomonas fluorescens* are the bacterial isolates frequently seen in fresh sugar beet root. Sugar beet endophytic and soil-microorganism research is vital for boosting its cultivation levels as well as the progress of sugar beet industries.

References

Abdeljalil NOB, Vallance J, Gerbore J, Bruez E, Martins G, Rey P et al (2016) Biocontrol of *Rhizoctonia* root rot in tomato and enhancement of plant growth using rhizobacteria naturally associated to tomato. J Plant Pathol Microbiol 7:356. https://doi.org/10.4172/2157-7471.1000356

Abudureheman A (2012) Improving sugar beet (*Beta vulgaris* L.) productivity by inoculation with *Gluconacetobacter* spp.

Aeini M, Khodakaramian G (2017) Rhizosphere bacterial composition of the sugar beet using SDS-PAGE methodology

Aeini M, Khodakaramian G, Mirzaei Najafgholi H (2018) Sugar beet leaf culturable endophytic bacterial composition from the major sugar beet growing areas in the west of Iran. J Genet Resour 4(2):105–113

Arteca RN (1996) Plant growth substances: principles and applications. Chapman and Hall Press, New York, p 332

Behrendt U, Ulrich A, Schumann P (2003) Fluorescent pseudomonads associated with the phyllosphere of grasses; Pseudomonas trivialis sp. nov., Pseudomonas poae sp. nov. and Pseudomonas congelans sp. nov. Int J Syst Evol Microbiol 53(5):1461–1469

Benizri E, Baudoin E, Guckert A (2001) Root colonization by inoculated plant growth-promoting rhizobacteria. Biocontrol Sci Technol 11:557–574

Bezerra JDP, Santos MGS, Svedese VM (2012) Richness of endophytic fungi isolated from Opuntia ficus-indica Mill. (Cactaceae) and preliminary screening for enzyme production. World J Microbiol Biotechnol 28(5):1989–1995

Bogas AC, Ferreira AJ, Araújo WL, Astolfi-Filho S, Kitajima EW, Lacava PT (2015) Endophytic bacterial diversity in the phyllosphere of Amazon Paullinia cupana associated with asymptomatic and symptomatic anthracnose. Springerplus 4:258

Böhm M, Hurek T, Reinhold-Hurek B (2007) Twitching motility is essential for endophytic rice colonization by the N2-fixing endophyte *Azoarcus* sp. strain BH72. Mol Plant Microbe Interact 20:526–533

Cabral D, Stone JK, Carroll GC (1993) The internal mycobiota of Juncus spp.: microscopic and cultural observations of infection patterns. Mycol Res 97(3):367–376

Carrión VJ, Cordovez V, Tyc O, Etalo DW, de Bruijn I, de Jager VC, Medema MH, Eberl L, Raaijmakers JM (2018) Involvement of Burkholderiaceae and sulfurous volatiles in disease-suppressive soils. ISME J 12(9):2307–2321

Carrión VJ, Perez-Jaramillo J, Cordovez V, Tracanna V, de Hollander M, Ruiz-Buck D, Mendes LW, van Ijcken WFJ, Gomez-Exposito R, Elsayed SS, Mohanraju P, Arifah A, van der Oost J, Paulson JN, Mendes R, van Wezel GP, Medema MH, Raaijmakers JM (2019) Pathogen-induced activation of disease-suppressive functions in the endophytic root microbiome. Science 366(6465):606–612

Chiellini C, Maida I, Emiliani G, Mengoni A, Mocali S, Fabiani A, Biffi S, Maggini V, Gori L, Vannacci A, Gallo E, Firenzuoli F, Fani R (2014) Endophytic and rhizospheric bacterial communities isolated from the medicinal plants Echinacea purpurea and Echinacea angustifolia. Int Microbiol 17(3):165–174. https://doi.org/10.2436/20.1501.01.219

Compant S, Mitter B, Colli-Mull JG, Gangl H, Sessitsch A (2011) Endophytes of grapevine flowers berries and seeds: identification of cultivable bacteria comparison with other plant parts and visualization of niches of colonization. Microb Ecol 62(1):188–197. https://doi.org/10.1007/s00248-011-9883-y

de Bary A (1866) Morphologie und Physiologie der Pilze, Flechten, und Myxomyceten. Hofmeister's Handbook of Physiological Botany, Volume II. Engelmann, Leipzig, Germany

de Souza Leite T, Cnossen-Fassoni A, Pereira OL, Mizubuti ESG, de Araújo EF, de Queiroz MV (2013) Novel and highly diverse fungal endophytes in soybean revealed by the consortium of two different techniques. J Microbiol 51(1):56–69. https://doi.org/10.1007/s12275-013-2356-x

Dohm JC, Minoche AE, Holtgräwe D, Capella-Gutiérrez S, Zakrzewski F, Tafer H, Rupp O, Sorensen TR, Stracke R, Reinhardt R, Goesmann A, Karft T, Schulz B, Stadler PF, Schmidt T, Gabaldon T, Lehranch H, Weissharr B, Himmelbauer H (2014) The genome of the recently domesticated crop plant sugar beet (*Beta vulgaris*). Nature 505(7484):546–549. https://doi.org/10.1038/nature12817

Doty SL (2017) Endophytic N-fixation: controversy and a path forward. In: Doty SL (ed) Functional importance of the plant microbiome. Springer International Publishing, New York, pp 17–58

Elbeltagy A, Nishioka K, Suzuki H, Sato T, Sato YI, Morisaki H, Mitsui H, Minamisawa K (2000) Isolation and characterization of endophytic bacteria from wild and traditionally cultivated rice varieties. Soil Sci Plant Nutr 46(3):617–629

Foley JA, Defries R, Asner GP, Barford C, Bonan G, Carpenter SR, Chapin FS, Coe MT, Daily GC, Gibbs HK, Helkowski JH, Holloway T, Howard EA, Kucharik CJ, Monfreda C, Patz JA, Prentice IC, Ramankutty N, Snyder PK (2005) Global consequences of land use. Science 309(5734):570–574

Gamalero E, Lingua G, Berta G, Lemanceau P (2003) Lemanceau methods for studying root colonization by introduced beneficial bacteria. Agronomy 23:407–441

Gao FK, Dai CC, Liu XZ (2010) Mechanisms of fungal endophytes in plant protection against pathogens. Afr J Microbiol Res 4:1346–1351

Hallmann J, Quadt-Hallmann A, Mahaffee WF (1997) Bacterial endophytes in agricultural crops. Can J Microbiol 43:895–914

Hardoim PR, van Overbeek LS, van Elsas JD (2008) Properties of bacterial endophytes and their proposed role in plant growth. Trends Microbiol 16:463–471

Hori K, Matsumoto S (2010) Bacterial adhesion: from mechanism to control. Biochem Eng J 48: 424–434

Indiragandhi P, Anandham R, Madhaiyan M, Sa TM (2008) Characterization of plant growth-promoting traits of bacteria isolated from larval guts of Diamondback moth Plutella xylostella (Lepidoptera: Plutellidae). Curr Microbiol 56:327–333

Jacobs MJ, Bugbee WM, Gabrielson DA (1985) Enumeration, location, and characterization of endophytic bacteria within sugar beet roots. Can J Bot 63:1262–1265

Kamilova F, Kravchenko LV, Shaposhnikov AI, Azarova T, Makarova N, Lugtenberg B (2006) Organic acids, sugars, and L-tryptophan in exudates of vegetables growing on stone wool and their effects on activities of rhizosphere bacteria. Mol Plant Microbe Interact 19:250–256

Kang SM, Joo GJ, Hamayun M, Na CI, Shin DH, Kim YK, Hong JK, Lee IJ (2009) Gibberellin production and phosphate solubilization by newly isolated strain of Acinetobacter calcoaceticus and its effect on plant growth. Biotechnol Lett 31:277–281

Kawaguchi M, Minamisawa K (2010) Plant-microbe communications for symbiosis. Plant Cell Physiol 51:1377–1380

Kobayashi DY, Palumbo JD (2000) Bacterial endophytes and their effects on plants and uses in agriculture. In: Bacon CW, White JF (eds) Microbial endophytes. Marcel Dekker, Inc, New York, pp 199–233

Kuklinsky-Sobral K, Araujo WL, Mendonca C, Geran LC, Piskala A, Azevedo JL (2004) Isolation and characterization of soybean-associated bacteria and their potential for plant growth promotion. Environ Microbiol 6:1244–1251

Kumar A, Munder A, Aravind R, Eapen SJ, Tümmler B, Raaijmakers JM (2013) Friend or foe: genetic and functional characterization of plant endophytic Pseudomonas aeruginosa. Environ Microbiol 15:764–779

Kusstatscher P, Cernava T, Harms K, Maier J, Eigner H, Berg G, Zachow C (2019a) Disease incidence in sugar beet fields is correlated with microbial diversity and distinct biological markers. Phytobiomes J 3:22–30. https://doi.org/10.1094/PBIOMES-01-19-0008-R

Kusstatscher P, Zachow C, Harms K, Maier J, Eigner H, Berg G, Cernava T (2019b) Microbiome-driven identification of microbial indicators for postharvest diseases of sugar beets. Microbiome 7:1–12. https://doi.org/10.1186/s40168-019-0728-0

Larran S, Monaco C, Alippi HE (2000) Endophytic fungi in beet (*Beta Vulgaris* L.,) leaves. Adv Hortic Sci 14:193–196

Ma L, Cao YH, Cheng MH, Huang Y, Mo MH, Wang Y (2016) Phylogenetic diversity of bacterial endophytes of *Panax notoginseng* with antagonistic characteristics towards pathogens of root-rot disease complex. Antonie Van Leeuwenhoek 103:299–312

Mendes R, Kruijt M, De Bruijn I, Dekkers E, van der Voort M, Schneider JH, Piceno YM, DeSantis TZ, Andersen GL, Bakker PA, Raaijmakers JM (2011) Deciphering the rhizosphere microbiome for disease-suppressive bacteria. Science 332(6033):1097–1100

Mendes R, Kruijt M, de Bruijn I, Dekkers E, van der Voort M, Schneider JHM, Piceno YM, DeSantis TZ, Andersen GL, Bakker PAHM, Raajimakers JM (2012) Deciphering the rhizosphere microbiome for disease-suppressive bacteria. Science 332:1097–1100. https://doi.org/10.1126/science.1203980

Miao L, Yang F, Wu X, Liu Y (2020) Effects of continuous cropping of sugar beet (*Beta vulgaris* L.) on its endophytic and soil bacterial community by high-throughput sequencing. Ann Microbiol:70:39. https://doi.org/10.1186/s13213-020-01583-8

Miliute I, Buzaite O, Baniulis D, Stanys V (2015) Bacterial endophytes in agricultural crops and their role in stress tolerance: a review. Zemdirbyste-Agriculture 102:465–478

Miller WA, Roy KW (1982) Mycoflora of soybean leaves, pods and seeds in Mississippi. Can J Bot 60:2716–2723

Müller H, Zachow C, Alavi M, Tilcher R, Krempl PM, Thallinger GG, Berg G (2013) Complete genome sequence of the sugar beet endophyte *Pseudomonas poae* RE* 1-1-14, a disease-suppressive bacterium. Genome Announc 1(2):e0002013

Nelson EB (2018) The seed microbiome: origins, interactions, and impacts. Plant Soil 422:7–34. https://doi.org/10.1007/s11104-017-3289-7

Petrini O, Sieber TN, Toti L, Viret O (1992) Ecology metabolite production, and substrate utilization in endophytic fungi. Nat Toxins 1:185–196

Rodriguez RJ, White JF Jr, Arnold AE, Redman RS (2009) Fungal endophytes: diversity and functional roles. New Phytol 182(2):314–330

Rosenblueth M, Martinez-Romero E (2006) Bacterial endophytes and their interactions with hosts. Mol Plant Microbe Interact 19(8):827–837

Rungjindamai N, Pinruan U, Choeyklin R, Hattori T, Jones EBG (2008) Molecular characterization of basidiomycetous endophytes isolated from leaves, rachis and petioles of the oil palm, Elaeis guineensis, in Thailand. Fungal Divers 33:139–161

Rybakova D, Cernava T, Köberl M, Liebminger S, Etemadi M, Berg G (2016) Endophytes-assisted biocontrol: novel insights in ecology and the mode of action of *Paenibacillus*. Plant Soil 405:125–140. https://doi.org/10.1007/s11104-015-2526-1

Sapers GM, Gorny JR, Yousef AE (2005) Microbiology of fruits and vegetables Gamalero E, Lingua G, Berta G, Clemenceau. Methods for studying root colonization by introduced beneficial bacteria. Agronomy 23:407–418

Senthilkumar M, Anandham R, Madhaiyan M, Venkateswaran V, Sa T (2011) Endophytic bacteria: perspectives and applications in agricultural crop production. In: Bacteria in agrobiology: crop ecosystems. Springer, Berlin, pp 61–96
Shi Y, Lou K, Li C (2009) Isolation, quantity distribution and characterization of endophytic microorganisms within sugar beet. Afr J Biotechnol 8(5)
Shi Y, Lou K, Li C (2010) Growth and photosynthetic efficiency promotion of sugar beet (*Beta vulgaris* L.) by endophytic bacteria. Photosynth Res 105(1):5–13
Shi Y, Lou K, Li C (2011) Growth promotion effects of the endophyte *Acinetobacter johnsonii* strain 3-1 on sugar beet. Symbiosis 54(3):159–166
Singh SK, Strobel GA, Knighton B, Geary B et al (2011) An endophytic *Phomopsis* sp. possessing bioactivity and fuel potential with its volatile organic compounds. Microb Ecol 61:729–739
Szymańska S, Tyburski J, Piernik A, Sikora M, Mazur J, Katarzyna H (2020) Raising beet tolerance to salinity through bioaugmentation with halotolerant endophytes. Agronomy 10(10):1571
Tremouillaux-Guiller J, Rohr T, Rohr R, Huss VAR (2002) Discovery of an endophytic alga in Ginkgo biloba. Am J Bot 89:727–733
Walterson AM, Stavrinides J (2015) Pantoea: insights into a highly versatile and diverse genus within the Enterobacteriaceae. FEMS Microbiol Rev 39(6):968–984
Wang Y, Gao BL, Li XX, Zhang ZB, Yan RM, Yang HL, Zhu D (2015) Phylogenetic diversity of culturable endophytic fungi in Dongxiang wild rice (*Oryza rufipogon* Griff), detection of polyketide synthase gene and their antagonistic activity analysis. Fungal Biol 119(11):1032–1045
Wolfgang A, Zachow C, Müller H, Grand A, Temme N, Tilcher R, Berg G (2020) Understanding the impact of cultivar, seed origin, and substrate on bacterial diversity of the sugar beet rhizosphere and suppression of soil-borne pathogens. Front Plant Sci 11:1450
Würschum T, Reif JC, Kraft T, Janssen G, Zhao Y (2013) Genomic selection in sugar beet breeding populations. BMC Genet 14:85. https://doi.org/10.1186/1471-2156-14-85
Zachow C, Tilcher R, Berg G (2008) Sugar beet-associated bacterial and fungal communities show a high indigenous antagonistic potential against plant pathogens. Microb Ecol 55:119–129
Zachow C, Fatehi J, Cardinale M, Tilcher R, Berg G (2010) Strain specific colonization pattern of Rhizoctonia antagonists in the root system of sugar beet. FEMS Microbiol Ecol 74:124–135
Zillinsky FJ (1984) Enfermedades comunes de los cereales degranos pequeñ os. Una guí a para su clasificacion CIMMYT, El Batan, Mexico

Rhizoctonia Disease and Its Management 39

Nenad Trkulja, Anja Milosavljević, and Violeta Oro

Abstract

Sugar beet (*Beta vulgaris* L.) has been grown for centuries as one of the most important agricultural plants worldwide for the purpose of white crystal sugar production. A large number of pathogens are present on the sugar beet, causing a large number of various symptoms and damaging production in a big scale. Plant diseases and pests are present on leaves and roots, causing damage to the above- and below-ground parts. *Rhizoctonia solani* Kuhn. is one such disease that attacks the sugar beet and is widespread throughout the world. *R. solani* is a soil-borne pathogen that survives in soil, where its life cycle is maintained for a long time period. It grows in the temperate and continental climate zone, making it a very important pathogen with respect to this crop due to its favourable condition. It overwinters in the soil for many years in the form of sclerotia or mycelium. The first symptoms appear on the upper part of sugar beet plants; usually symptomatic plants appear in bordered zones, often present in certain parts of the field with a clearly bordered belt. The infected plants have lost their turgor, and most of the leaf rosette gets laid on the ground. In the cross section, the central part of the root is initially healthy, while the disease is primarily observed on the rim of the root. The most important and only fully effective approach in controlling *R. solani* is the application of fungicides from SDHI and QoI groups at the appropriate time of application.

N. Trkulja (✉) · V. Oro
Institute for Plant Protection and Environment, Belgrade, Serbia

A. Milosavljević
Research and Development Center Sunoko, Indjija, Serbia

V. Misra et al. (eds.), *Sugar Beet Cultivation, Management and Processing*, https://doi.org/10.1007/978-981-19-2730-0_39

Keywords

Disease management · Rhizoctonia solani · Root rot · Sugar beet

Abbreviations

AGs	Anastamosis groups
IGs	Intraspecific groups
QoIs	Qinone outside inhibitors
SDHI	Succinate dehydrogenase fungicides inhibitor

39.1 Introduction

Sugar beet (*Beta vulgaris* L.) has been grown for centuries for the purpose of white-crystal sugar production and represents, strategically, an important agricultural plant in the world (Schnieder et al. 2002). In the beginning sugar was available only for the rich people of society, but later the situation changed. With passing time sugar became cheaper and available to all layers of the society. At present, sugar positions as one of the most important part of the food chain, especially for the poorer sections of the society (Coons 1949). Areas with temperate, continental, and colder climates are the main distribution areas in Europe, Russia, USA, Turkey, Japan, and China. The production of sugar beet is very technologically demanding. Significant problems in its production are caused by different pests and diseases; when the crop attains favourable conditions for growth, the same conditions are also beneficial for pests and diseases. Sugar beet is a crop that is susceptible to a lot of plant pathogens and different kinds of pests, causing a large number of various symptoms and damaging production in a big scale. Plant diseases and pests can be divided into those that cause damage to the above-ground parts, primarily the leaf mass, and those that damage the root. The most significant cause of leaf decay is the pathogenic fungus *Cercospora beticola* Sacc., which can cause a significant reduction in yield up to 50%, especially in recent years when fungicide resistance has been detected as leading to a reduction in fungicide effects (Trkulja et al. 2015, 2017; Karaoglanidis and Ioannidis 2010). Plant disease agents that attack sugar beet roots mass carry out their life cycle in the soil, causing root damage and maintain their fruiting bodies in the soil waiting for new sugar beet or plants that are its temporary hosts to start the new reproduction cycle and to ensure the survival of the fungus. Damage caused by root rot is often difficult to notice during the growing season and it is often very difficult to assess their presence in the field. Once the root gets infected, it should be processed in a short period of time. It can cause additional losses because the root rot process accelerates after extraction. In certain cases, it is necessary to hire additional labor because the diseased roots from the piles of harvested beets have to be removed before loading on trucks and reach the factory, which affects the increase in production cost together with losses in yield due to root rot itself. The most significant disease that causes root rot is *Rhizoctonia solani* Kuhn., and it, along with *C. beticola* as the most important leaf disease (Trkulja et al. 2013, 2014), is the second most significant disease of sugar beet worldwide. In addition to *R. solani*, significant damage to the root of sugar beet is caused by other root diseases too, such as *Aphanomyces cochlioides*, *Fusarium oxysporum*, *Fusarium seecory* and *Macrophomina phaseolina* (Holmquist 2018).

39.2 Prevalence of *Rhizoctonia solani* and Symptomatology

Rhizoctonia solani Kuhn. [Telemorph: *Thanatephorus cucumeris* (A.B. Frank) Donk] is a disease that attacks sugar beet and other agricultural crops besides being widespread (Harveson et al. 2009). The presence of *R. solani* has been reported in major sugar beet cultivation areas in Europe and is estimated to be about 5–10% of the area (Harveson et al. 2009), while in America its presence was recorded in 2015 (Mukhopadhyay 1987), and its occurrence has been recorded in about 25% of fields in the main cultivation areas of North Dakota and Minnesota (Khan et al. 2009; Harveson et al. 2009). Symptoms caused by the presence of *R. solani* are at first hard to recognize since the main damage is the underground portion, i.e., root. The first signs on the upper part of sugar beet plants are the decay of leaf mass, which is more intense when warm days start after the plant roots have been infected by the rot. Usually symptomatic plants appear in zones that are often present in certain spots in the field with a clearly bordered belt, the plants have lost their turgor, and the leaf rosette has been laid on the ground. The leaf stalks at the base of the rosette may take on a dark brown color, and dark brown to purple mycelium of the fungus can be seen on the infected root. In cross section, the central part of the root is initially healthy, while the disease is primarily observed on the rim of the root. As a result of a stronger attack after the plants wither, the leaf mass dries out and the underground parts are completely affected by the rot, and the root mass is all brown coloured. These roots are soon infected by secondary pathogens that accelerate the rot process further and in drastic cases lead to complete devastation of the whole plant. These drastic phenomena are mainly related to those places in the field where moisture content is high, usually during heavy spring rains. When the wet conditions of the field persist for a long period, the plants are exposed to severe infection, resulting in such drastic cases of decay in the lowest parts of the plots, where water retains the longest (Figs. 39.1 and 39.2). If the infection is not much severe, then these plants usually cause changes only on the outer part of the roots. These plants survive until the next rainy period after which they start to decay and the remains of these plants can be found later in the vegetation. There are also examples when the plant roots epidermis tissue of the plants becomes infected with *R. solani* during wet conditions, but because of the disappearance of surface moisture a higher percentage of these plants can close the wounds and no further rot of these roots appears. Such plants can survive and continue their development until the end of vegetation. Based on what stage of sugar beet plant *R. solani* conducts the

a b

Fig. 39.1 *Rhizoctonia solani* symptoms in the field. (**a**) Typical symptom of leaf damping after infection and collapse of root. (**b**) Place in the field where water lay for a longer period of time, causing mass infection of plants in the early stages, which leads to total loss and appearance of bold spots (N. Trkulja)

Fig. 39.2 Damage by *Rhizoctonia solani. Rhizoctonia solani* total damage on bigger plants in field; they disappear from the crop but usually are not noticeable until the harvest stage (N. Trkulja, A. Milosavljevic)

infection, it can cause the decay of young plants, and so called damping-off or root rot of grown plants (Fig. 39.3).

Damping-off of young plants occurs in the initial stages of development and such plants decay very quickly and disappear from the field, which appears as bald spots in the field. Root rot occurs at a somewhat later stage, when the plant has already developed a main root. It is observed that dark lesions that appear first at the outer

Fig. 39.3 Symptoms of *Rhizoctonia solani* after infection due to the rainy season. (**a**) Mature roots. (**b**) Infected younger plants with a healthy plant as comparison (A. Milosavljevic)

Fig. 39.4 *R. solani* complete decay of plants that disappear from crops but are not noticeable up to the harvest time (N. Trkulja)

part of the root can be positioned anywhere at the root itself but can also be seen near the root hairs (Fig. 39.4). The root rot can cause cracks and deformations at the root, while the transition between the healthy and the diseased part is clearly visible, until infection with secondary pathogens occurs. If the infection originated from the crown, the rot spreads from the head to the lower part of the root (Fig. 39.5) and can be expected as a consequence of applying soil and crop residues infected with *R. solani* to the leaf rosette and root head during inter-row cultivation.

Fig. 39.5 *R. solani* infection on the crown of the sugar beet root. (**a**) Infected root. (**b**) Cross view of the crown infected

39.3 Conditions for the Development of *Rhizoctonia solani*

Conditions conducive to the development of *R. solani* are the heavy soils that are not well drained with higher moisture content and longer retention period (Bolton et al. 2010), typically soils with a low possibility of breathing and releasing water. In terms of temperature, *R. solani* best develops and performs its vital functions at temperatures from 20 to 35 °C; at the same time, to infect the sugar beet root it is necessary that the soil moisture is at least 25%, and as the humidity increases, the greater is the chance of infection. High soil moisture with a temperature of 20–25 °C and the presence of *R. solani* in the soil will certainly lead to infection of both young and advanced stages of plants. On the other hand, if there are conditions in terms of humidity, and the temperature is lower than 15 °C, the disease will not progress but will be in a latent phase waiting. Following these conditions, as soon as the temperature rises to over 16 °C for few days, the pathogen will activate from latent to active mode, the infection will be achieved, and symptoms will start to appear on the root surface causing damage in yield and sugar content in the roots.

39.3.1 Life Cycle of *R. solani*

R. solani is a pathogen that carries out its life cycle in the soil; it is a soil-borne basidiomycete and a pathogen that infects a large number of hosts in the temperate and continental climate zone, making it the most important pathogen of sugar beet after *Cercospora* leaf spot (Harveson et al. 2009; Trkulja et al. 2012).

The sexual stage for *Thanatephorus cucumeris* is very rare in nature and can rarely be found; the reason for this probably lies in the fact that basidiospores are very difficult to germinate and also very often less virulent. Basidiospores can be carried by the wind and can serve to spread infection and as an inoculum for disease maintenance, but scientific findings confirm that they are not the primary source of inoculum (Cubeta and Vilgalys 1997; Holmquist 2018).

Since the teleomorph stage is very rare, the most significant spread of the disease and maintenance in nature takes place within the anamorph stage of *R. solani*. The fruiting organs responsible for the spread of disease are mycelia and sclerotia, through which the pathogen persists in the soil and debris-harvest residues for many years. It is believed that asexual spores and conidia do not participate in the spread of the disease (Cubeta and Vilgalys 1997). In order to determine the pathogen, *R. solani* can be isolated from the plant tissue affected by the disease by cutting a piece of tissue (5 mm in size) from the healthy/diseased border zone of the root tissue, washed, sterilized, and placed on a nutrient medium for incubation (Fig. 39.6a). After 24–48 h, the mycelium will develop, which can be observed morphologically in order to determine the presence of *R. solani* in root tissue.

Also, further molecular analyses can be performed for the purpose of identification of the anastomosis groups of pathogen, which cannot be done by simply isolating the pathogen on nutrient media since no morphological differences are determined for different anastomosis groups of *R. solani* yet. Apart from the plant tissue, *R. solani* can also be isolated from infected soil via selective media, where a number of methods can be used for this purpose. Some of those methods involve the use of test plants or specific plant parts or autoclaved seeds sensitive to *R. solani*.

The vegetative mycelium of *R. solani* is colorless to light whitish in the initial stages of development (Fig. 39.6b). After few days in a controlled environment it acquires a brownish, somewhat orange color (Figs. 39.6c and 39.7a). The mycelium is built of characteristic hyphae, which are limited by the septum that has septal pores. Through those septa, the pathogen communicates in time of the development of infection in plant tissue and exchange the cytoplasm, mitochondria, and nuclei from cell to cell. Hyphae are coarse, and pale to dark brown in color. They grow in branches or branch at the distal septum. The angle of branching is usually the right angle of 90°. Individual hyphae cells are multinucleate and most often have more than three nuclei and can have up to 14 nuclei (Sneh et al. 1996).

Colony morphology colors varied from dark brown to light brown and light tan; variations between isolates are also present in the formation of sclerotia and the presence of aerial mycelium hyphae (Das et al. 2020) (Fig. 39.7). Regardless of the morphological variations of the *R. solani* colonies, they are characteristic in their appearance and can be clearly distinguished from the colonies of other

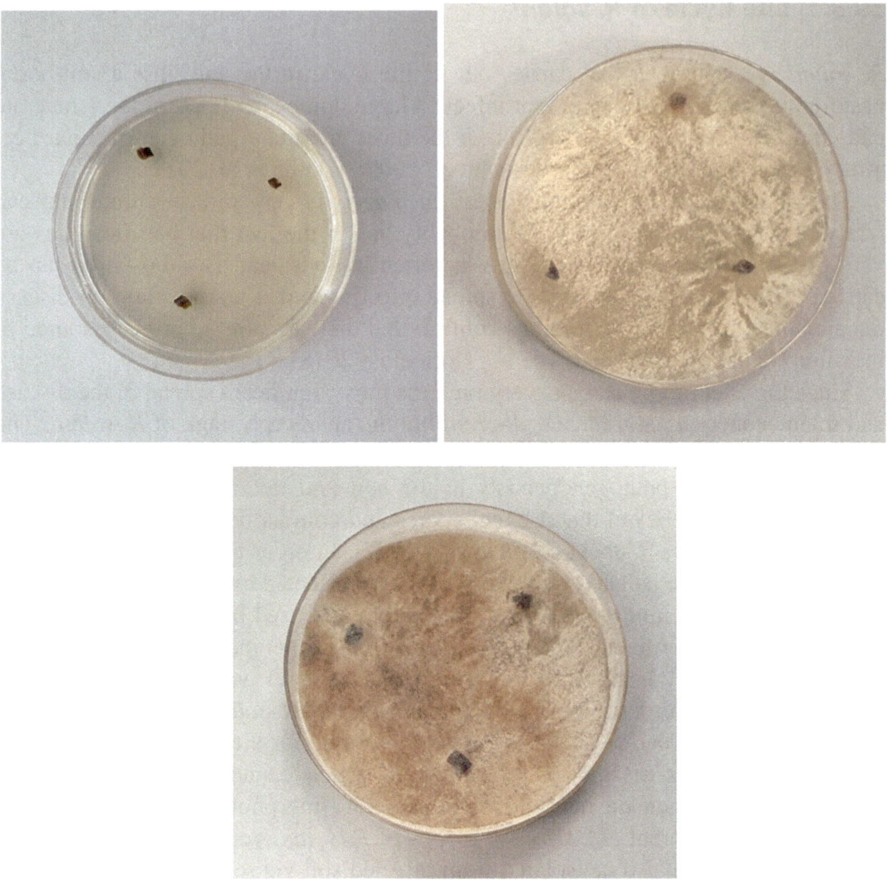

Fig. 39.6 *R. solani* isolation. (**a**) from plant tissue on a nutrient medium. (**b**) vegetative mycelium of *R. solani* in the initial stages of development. (**c**) *R. solani* after few days in controlled environment (Courtesy A. Milosavljevic)

phytopathogenic fungi. However, isolates of *R. solani* can vary greatly depending on the conditions and substrate from which they are originally isolated, and it is reflected through their growth characteristics, pathogenicity, and virulence to different plant species and varieties within the same species (Carling 1996; Carling et al. 2002; Ogoshi 2003; Dubey et al. 2012).

39.3.2 *Rhizoctonia solani* Infection Realization

R. solani overwinters in the soil for many years in the form of sclerotia or mycelium, where the sclerotia are spherical in shape, usually 1–3 mm in diameter, and brown to black. Sclerotia can be maintained in water and plant tissue also. *R. solani* can also

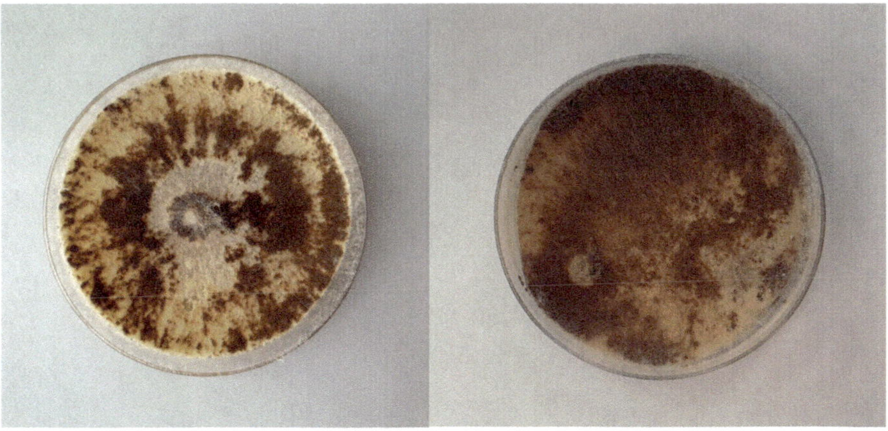

Fig. 39.7 (**a**) *Rhizoctonia solani* after few more days in controlled environment. (**b**) Colony morphology colors varied from light brown and light tan to dark brown; variations between isolates are also present in the formation of sclerotia and the presence of aerial mycelium, hyphae (A. Milosavljevic)

overwinter as mycelium that colonizes the plant debris in the soil and survives as a saprophyte. When new plants start to grow in the field, sclerotia and mycelium, when found near this new plant tissue, become chemically attracted to secretions from the roots and germinate into a vegetative hypha. This way, the pathogen infects the roots and after infection the mycelium is maintained on the outer part of the plant tissue. The infection process takes place further by enzyme secretion that break down the cellulose (extracellular enzymes) and degrade various components of plant cell walls (e.g., cellulose, cutin, and pectin). As the fungus destroys the plant cell, the hyphae continue to grow and colonize the dead plant tissue and eventually form sclerotia that await a new host and begin a new cycle.

The realization of the infection is directly dependent on the humidity and temperature of the soil. The development of *R. solani* is favored by high soil moisture, 75–100%, which has been proven to be the most ideal for infection and disease development. No matter if conditions are favorable in the field, it is evident that the disease can occur at 25% humidity, but in this case it is without significant impact on plants. It shows that the presence of *R. solani* in the soil is constant and it only waits for appropriate conditions in order for the pathogen to start development and make great losses in sugar beet production. Temperatures at which infection can occur range from 20 to 35 °C; however, the ideal temperature that leads to severe symptoms and, at the end, plant decay is a temperature of 20–25 °C. If after a humid and warm spring, at the beginning of the summer period the temperature rises rapidly to 30 °C, already-made infections will not progress and there won't be further infections. Further problems with the infections that have already been realized is that in those spots it can cause infections with secondary pathogens because *R. solani*

Fig. 39.8 Secondary pathogens on roots already damaged with *R. solani* lead to root rot and complete decay of the plants (A. Milosavljevic)

made open the door to saprophytes to enter the plant tissue and make more damage than *R. solani* itself. Depending on the degree of infection, either such plants will heal and remain in the field with the outer part of the root damaged by the initial infections, or secondary pathogens will lead to root rot and complete decay of the plants (Fig. 39.8).

39.3.3 Anastamosis Groups of *Rhizoctonia solani*

Strains of *R. solani* are classified into so-called anastamosis groups (AGs) based on anastamosis hyphal reaction (Ajayi-Oyetunde and Bradley 2018), although they are genetically very similar; in addition, they can be further divided into intraspecific groups (IGs), which significantly affect the possibility of characterization in relation to infected crops (Ogoshi 1987; Grosh et al. 2004; Hietala et al. 1994). There are currently 13 recognized and described anastomosis groups; further division into subgroups was made based on the host range, cultural morphology, biochemical characteristic, and molecular difference (Ogoshi 1987) (Table 39.1). *R. solani* is a very complex group of phytopathogenic fungi that are capable of infecting a large number of plant species such as soybeans, potato, beans, etc., besides the sugar beet, and are considered to be regular, common inhabitants of the soil (Windels et al. 1997). The sugar beet can be infected by *R. solani* strains from groups AG1 to AG5, but the most significant are strains from groups AG2 and AG4 that can cause root rot (Herr 1996). In previous years, *R. solani* AG11 infection was detected, which caused a sugar beet infection in Poland. This finding represents the first occurrence of AG11 strain infection in the sugar beet. Until this finding, strain AG11 was known mainly as a pathogen of soybeans and rice (Moliszewska et al. 2020). The *R. solani* that is responsible for crown and root rot is AG 2-2, which has two intraspecific groups, AG

Table 39.1 Host range of *Rhizoctonia solani* and *Rhizoctonia* diseases arranged by anastomosis groups (Source: Sneh et al. 1991)

Anastomosis group	Diseases	Host
AG 1-IA	Sheath blight, sheath spot	Rice
	Sclerotial disease, leaf blight, banded leaf	Corn
	Leaf blight, banded leaf	Sorghum
	Leaf blight	Bean
	Leaf blight	Soybean
	Summer blight	Crimson clover
	Southern blight	Camphor seedlings
	Brown patch	Turfgrass
AG 1-IB	Web blight	Bean
		Rice
	Rot	Soybean
		Figs
		Leguminosous Woody plants
		Hortensia
	Bottom rot	Cabbage
		Lettuce
AG 1-IC	Damping off	Buckwheat
	Damping off, crown root rot	Carrot
	Damping off	Soybean
		Flax
		Pine
AG 2-1	Damping off	Crucifers
	Bud rot	Strawberry
	Leaf blight	Tulip
	Root rot	Japanese radish and
		Subterranean clover
AG 2-2 IIIB	False sheath blight	Rice
	Sheath blight	Mat rush
		Ginger
		Gladiolos
	Black scurf	Edible burdock
	Brown patch	Turf grass
	Crown rot, brace rot	Corn
	Damping off	Sugar beet
		Tree seedlings
		Chrysanthemum
	Root rot	Konjak
		Chinese yam
AG 2-2 IV	Root rot, leaf blight	Sugar beet
	Large patch	Turfgrass

(continued)

Table 39.1 (continued)

Anastomosis group	Diseases	Host
AG 3	Black scurf, stem/stolon cankers	Potatoes
	Target spot	Tobacco
	Leaf blight	Tomato
	Brown spot	Egg Plant
AG 4 (HG I, HG II and HG III)	Fruit rot	Tomato
	Stem rot	Pea
	Damping off, stem canker	Potato
	Damping off, root rots	Soybean
		Lobolly pine seedlings
		Onion
		Stevia
		Pea
		Snap bean
		Cotton
		Peanuts
		Slash
	Pod rot	Snap bean
AG 5	Black scurf	Potato
	Brown patch	Turf grass
	Root rot	Beans
		Soybeans
		Adzuki beans
AG 6	Nonpathogenic group	–
AG 7	Nonpathogenic	–
AG 8	Bare patches	Cereals
AG 9	Weak pathogen	Crucifers
		Potatoes
AG 10	Nonpathogenic	–
AG 11		Wheat
AG BI	Nonpathogenic	–

2-2 IIIB and AG 2-2 IV, and they can also be categorized into IGs (Bolton et al. 2010). Differences between those are connected to a possibility to make infection on the sugar beet. It has been experimentally proven that only anastamosis groups AG 2-2 IIIB and AG 2-2 IV can infect and cause symptoms on sugar beet plants at 10 weeks of age, while other *R. solani* groups can infect only seedlings, but do not cause more serious disease of plants that can lead to more significant plant decay (Bolton et al. 2010). There are also meaningful differences in the realization of infection between AG 2-2 IIIB and AG 2-2 IV connected to other crops in the crop rotation. Strains AG 2-2 IIIB and AG 2-2 IV can cause symptoms of damping-off, root rot, and crown rot in infected soybeans and beans, besides the sugar beet. However, the difference between these two strains is that only AG 2-2 IIIB can infect maize, wheat, and rice, and is notably more resistant to high temperatures, so it can be developed in temperatures up to 35 °C. AG 2-2 IV strain cannot develop successfully at higher levels of temperatures and is not able to carry out the infection at a temperature of 35 °C. In addition, there is a significant difference in crops that this strain infects because it cannot infect wheat and maize. This difference is very important because it significantly reduces the likelihood of infection with strain AG 2-2 IV as it cannot infect wheat and corn, which are very often in crop rotation with sugar beet, and often direct precursors. The knowledge of which anastamosis groups we have in which region can affect the control of the disease in great scale, because careful selection of crops in the crop rotation can majorly affect the reduction of inoculum and reduce the possibility for the next crop infestation with the pathogen. Previous studies indicated that the anastamosis group AG 2-2 IIIB is more present in Europe compared to AG 2-2 IV, which is crucial for further monitoring and control of *R. solani* because with the reduction of maize and wheat cultivation in areas where strain AG 2-2 IIIB appear we can reduce disease occurrence in the sugar beet.

39.4 Management of *Rhizoctonia solani*

Since *R. solani* is the major disease of sugar beet root and causes damping-off, root, and crown rot, its control is given considerable attention, and research conducted in the past indicates that there are several possibilities to control this disease. The appearance of *R. solani* and its intensity depend on several factors like the aggressiveness of the AG strain, the crop itself, the variety used, and, most importantly, the environment in which the infection occurs. If the environment is with high humidity and a temperature of around 20–25 °C, we need to implement control measures through the use of tolerant varieties, crop rotation, and biological and chemical control.

39.4.1 Management of *R. solani* by Tolerant Variety

The commercial varieties used are susceptible to *R. solani* on higher or lower levels, so there is small possibility of selecting tolerant varieties, because tolerance is controlled by multiple genes. Also, if the selection leads to complete tolerance to *R. solani*, a negative side effect is a drastic reduction in yield. In addition to this, problems may arise in the increase in susceptibility of the given varieties to other diseases too that may affect sustainable sugar beet production (Jacobsen et al. 2004). Experiments established that there are certain differences in commercially available varieties, but these differences are not sufficient to avoid more radical control measures. Disease control still needs to be done with the use of fungicides in infected fields (Wigg and Goldman 2020). Experiments done under controlled conditions in a greenhouse and those done in the field indicate that data obtained indoors can be used for extrapolation in field conditions (Buttner et al. 2004; Campbell and Altman 1976; Schneider et al. 1982; Scholten et al. 2001; Weiland et al. 1999; Wigg and

Goldman 2020). Experiments with isolates of different anastomosis groups and different sugar beet varieties showed that there are certain differences considering their interactions, but further selection is equally necessary so as to reach an adequate solution to the problem (Yassin 2013). Due to reduced yields, growers are not willing to use tolerant varieties; they use commercial varieties that provide high and stable yields, but in regions where *R. solani* activity is present, they apply fungicides, preventive before infection (Bolton et al. 2010). However, use of moderately tolerant varieties with a not-so-great decline in yield could be a way to select new highly tolerant varieties that would have high yields like other conventional varieties (Behn et al. 2012).

39.4.2 Management of *Rhizoctonia solani* by Crop Rotation

Crop rotation is a very significant control measure as it reduces the disease incidence. For the needs of sugar beet cultivation, crop rotation has to be with sugar beets not grown in the same field for 3 years. Additionally, the choice of pre-crops is important, so sowing of soybeans and corn should be avoided in the crop rotation because it is established that they can be important in the maintenance of *R. solani* such as AG 2-2 IIIB, and as pre-crops should be used cereals such as wheat, barley, and rye, which are not host plants for this pathogen (Koch et al. 2018; Buhre et al. 2009). If the disease is present, additional analysis identifying the strain will help in a possible choice of crop rotation, leading to reduction in disease incidence. If the AG 2-2 IIIB strain appears in the field, reducing the growth of maize in the crop rotation can lower the maintenance of the infectious potential of the strain, which would result in lower disease pressure in successive crops (Sumner and Bell 1982; Sumner 1999). It is confirmed that cover crops can significantly reduce the *R. solani* population and substantiation of sugar beet crop infection. Plants from the Brassica family adversely affect the maintenance and viability of *R. solani*, but also decrease some other soil pathogens and phytopathogenic fungi such as *Fusarium*, *Pythium*, *Sclerotinia*, and *Phytohpthora* (Kundu and Nandi 1985). Prevention of disease needs to include improvement of the soil structure by aeration and drainage so that the plants get more air with less presence of free water in the soil. However, the only fully effective approach in controlling of *R. solani* in the field, among other measures, is fungicide application.

39.4.3 Management of *R. solani* by Chemical Control

There are several effective approaches in controlling *R. solani* through fungicide application. Studies performed in conditions with natural source of inoculum resulted in the conclusion that the sugar beet can be preventively protected from this disease appearance. Disease control can be achieved in several ways, and the first is treating the seeds with succinate dehydrogenase inhibitor (SDHI) fungicides, sedaxane, fluxapyroxad, and penthiopyrad. SDHI fungicides exert their action in mitochondrial complex II by disrupting the electron transfer process, resulting in respiratory arrest (Keon et al. 1991; Hagerhall 1997). Isolates of phytopathogenic fungi with reduced susceptibility were found in a large number of populations (FRAC Code List 2021). When it comes to resistance-risk development, they are classified into the group of fungicides with medium risk, and cross-resistance between individual active substances from this group depends on the specificity of mutations (Amiri et al. 2014; Hu et al. 2016). Penthiopyrad applied to the seed had high efficacy in controlling *R. solani* and provided early season control by protecting plant populations compared with the nontreated control. However, penthiopyrad in this way does not provide adequate crop protection in the later stages, and that is why, in regions with a stronger occurrence of *R. solani*, we need to mix it with other fungicides that have an effect on this disease. When penthiopyrad is applied earlier in plant development and disease appearance, and reinforced with azoxystrobin afterwards, it gives a higher percentage of survival for plant (Liu et al. 2021). In addition, the yield of recoverable sucrose is significantly higher in comparison to control (untreated) areas. A very effective way to protect sugar beet crops from *R. solani* is application of fungicides from the strobilurins group (Qinone Outside Inhibitors, QoIs), such as azoxystrobin and pyraclostrobin, which inhibit electron transport in mitochondrial complex III, leading to interruption of the respiratory process (Khan et al. 2009). This also reduces the possibility of resistance development in populations to these two groups of fungicides (Khan et al. 2009). Azoxystrobin usage in disease management in endangered areas is very important and it can be used in several stages. Primarily, the application can be made in furrows at planting and then the treatment can be performed later, after 4–6 true leaves develop. Also, different studies point to the time of application, where we can use it only when plants have 4–6 leaves, but there is a slightly smaller effect than when it is used combined with in-furrow application (Kirk et al. 2008; Khan et al. 2010, 2017). QoIs are characterized as a fungicide group with a high risk of resistance development. Resistance is known in different kind of fungi, but also there is cross-resistance noted with active substances from within this group (Brent and Hollomon 1998). If resistance occurs, anti-resistance measures are necessary, so application of active substances from other modes of action is a good strategy. The use of penthiopyrad in seed treatment and azoxystrobin later in the season, when the plant is more developed, is very effective.

Good effects in the control of *R. solani* have been given by other fungicides such as polyoxin-d and flutolanil, and they are able to lower the disease severity by nearly half relative to water control, additionally to azoxystrobin (Bolton et al. 2010). However, these fungicides are not significantly used to control *R. solani* in sugar beet crops, but their effect is known in other cultures (Paulitz and Reinertsen 2005). The best effects so far are achieved by the system in which the azoxystrobin use is positioned at the phase of 4–10 leaves and at a depth of 10 cm when the average temperature reaches 16 °C. This treatment must be positioned before the rainy season to lower the fungicide entry into the root zone (Khan et al. 2010, 2017). In this way, the application of fungicides is timely, i.e., in the phase when the infection has not yet occurred, which ensures a better fungicide efficacy and a significantly lower number of infections, which is manifested by a larger number of plants per hectare and higher recoverable sucrose.

39.5 Future Prospects

Development of tolerant varieties further would have a satisfactory sugar yield and would significantly facilitate the fight against this disease. Further research could suggest which crops should be planned as a cover crop after the sugar beet or any crops that are also hosts of *R. solani*.

39.6 Conclusion

R. solani is a very important pathogen that causes root disease with a large distribution area and a wide range of host plants. As *R. solani* infects the root, i.e., economical part causing root rot, it became one of the most significant diseases of the sugar beet in the world. In favorable conditions of temperature and humidity, the soil is a regular source of production and therefore special care must be paid to its monitoring and control. Crop rotation is an extremely important measure for decreasing the pressure of *R. solani*, but other control methods like fungicide use are necessary in critical areas. The established control technology with SDHI fungicides in treatments on the seed and preventive treatments with QoI fungicides significantly facilitate production in infected sugar beet-growing areas. Research and some other control measures like use of biological agents would make a significant contribution in environmental terms by decreasing chemical compound use. Use of cover crops also gives a significant result, but this control measure is still little applied. Tillage in regions with *R. solani* presence must be adjusted to the given conditions to keep from retention of free water in the soil. Basic tillage must increase air concentration in the soil, so this can change environmental conditions in favor of the sugar beet development and lower the disease appearance. Finally, it is important to keep records of disease presence in the fields so that disease control in the sugar beet crop can be performed in a planned manner.

References

Ajayi-Oyetunde OO, Bradley CA (2018) Rhizoctonia solani: taxonomy, population biology and management of rhizoctonia seedling disease of soybean. Plant Pathol 67:3–17

Amiri A, Heath SM, Peres NA (2014) Resistance to fluopyram, fluxapyroxad, and penthiopyrad in Botrytis cinerea from strawberry. Plant Dis 98:532–539

Behn A, Ladewig E, Manthey R (2012) Resistance testing of sugar beet varieties against *Rhizoctonia solani*. Sugar Industry 137:49–57

Bolton MD, Panella L, Campbell L, Khan MFR (2010) Temperature, moisture, and fungicide effects in managing Rhizoctonia root and crown rot of sugar beet. Phytopathology 100:689–697

Brent K, Hollomon D (1998) Fungicide resistance: the assessment of risk FRAC monograph no. 2. GCPF, Brussels

Buhre C, Kluth C, Buercky K (2009) Integrated control of root and crown rot in sugar beet: combined effects of cultivar, crop rotation, and soil tillage. Plant Dis 93:155–161

Buttner G, Pfahler B, Marlander B (2004) Greenhouse and field techniques for testing sugar beet for resistance to Rhizoctonia root and crown rot. Plant Breed 123:158–166

Campbell CL, Altman J (1976) Rapid laboratory screening of sugar beet cultivars for resistance to Rhizoctonia solani. Phytopathology 66:1373–1137

Carling DE (1996) Grouping in Rhizoctonia Solani by hyphal anastomosis reaction. In: Rhizoctonia species: taxonomy, molecular biology, ecology, pathology and disease control. Springer Netherlands, Dordrecht, pp 37–47

Carling DE, Kuninaga S, Brainard KA (2002) Hyphal anastomosis reactions, rDNA internal transcribed spacer sequences, and virulence levels among subsets of Rhizoctonia solani anastomosis Group-2 (AG-2) and AG-BI. Phytopathology 92:43–50

Coons GH (1949) The sugar beet. Sci Monthly 68(3):149–164

Cubeta MA, Vilgalys R (1997) Population biology of the Rhizoctonia solani complex. Phytopathology 87:480–484

Das S, Plyler-Harveson T, Santra DK, Maharjan B, Nielson KA, Harveson RM (2020) A longitudinal study on morpho-genetic diversity of pathogenic Rhizoctonia solani from sugar beet and dry beans of western Nebraska. BMC Microbiol 20:354

Dubey SC, Tripathi A, Upadhyay BK (2012) Molecular diversity analysis of Rhizoctonia solani isolates infecting various pulse crops in different agroecological regions of India. Folia Microbiol (Praha) 57:513–524

FRAC Code List (2021) Fungal control agents sorted by cross resistance pattern and mode of action, pp 1–8

Grosh R, Schneider JHM, Kofoet A (2004) Characterization of Rhizoctonia solani anastomosis groups causing bottom rot in field-grown lettuce in Germany. Eur J Plant Pathol 110:53–62

Hagerhall C (1997) Succinate: quinone oxidoreductases. Variations on a conserved theme. Biochim Biophys Acta 1320:107–141

Harveson RM, Hanson LE, Hein GL (2009) Compendium of beet diseases and pests. The American Phytopathological Society, St. Paul

Herr LJ (1996) Sugarbeet diseases incited by Rhizoctonia spp. In: Sneh B, Jabaji-Here S, Neate S, Dijst G (eds) Rhizoctonia species: taxonomy, molecular biology, ecology, pathology and disease control. Kluwer Academic Publishers, Dordrecht, pp 35–38

Hietala AM, Sen R, Lilja A (1994) Anamorphic and telomorphic characteristic of a uninucleate Rhizoctonia sp. isolated from the roots nursery grown conifer seedlings. Mycol Res 98:1044–1050

Holmquist L (2018) Rhizoctonia solani and sugar beet responses. Genomic and molecular analysis. Doctoral thesis Swedish University of Agricultural Sciences, Uppsala, pp 1–52

Hu MJ, Fernandez-Ortuño D, Schnabel G (2016) Monitoring resistance to SDHI fungicides in *Botrytis cinerea* from strawberry fields. Plant Dis 100:959–965

Jacobsen B, Kephart K, Zidack N, Johnston M, Ansley J (2004) Effect of fungicide and fungicide application timing on reducing yield loss to Rhizoctonia crown and root rot. Sugarbeet Res Ext Rep 35:224–226

Karaoglanidis GS, Ioannidis PM (2010) Fungicide resistance of Cercospora beticola in Europe. In: Lartey RT, Weiland JJ, Panella L, Crous PW, Windels CE (eds) Cercospora leaf spot of sugar beet and related species. American Phytopathological Society, St. Paul, pp 189–211

Keon JPR, White GA, Hargreaves JA (1991) Isolation, characterization and sequence of a gene conferring resistance to the systemic fungicide carboxin from the maize smut pathogen, Ustilago maydis. Curr Genet 19:475–481

Khan AF, Liu Y, Khan MFR (2009) Efficacy and safety of generic azoxystrobin at controlling Rhizoctonia solani in sugar beet. Crop Prot 93:77–81

Khan MFR, Khan J, Bradley CA (2010) Effect of azoxystrobin applications based on soil temperature on Rhizoctonia root and crown rot of sugar beet. Int Sugar J 112(1342):557–560

Khan AF, Liu YX, Khan MFR (2017) Efficacy and safety of generic azoxystrobin at controlling Rhizoctonia solani in sugar beet. Crop Prot 93:77–81

Kirk WW, Wharton PS, Schafer RL, Tumbalam P, Poindexter S, Guza C, Fogg R, Schlatter T, Stewart J, Hubbell L, Ruppal D (2008) Optimizing fungicide timing for the control of Rhizoctonia crown and root rot of sugar beet using soil temperature and plant growth stages. Plant Dis 92:1091–1098

Koch HJ, Trimpler K, Jacobs A (2018) Crop rotational effects on yield formation in current sugar beet production—results from a farm survey and field trials. Front Plant Sci 9:231

Kundu PK, Nandi B (1985) Control of Rhizoctonia disease of cauliflower by competitive inhibition of the pathogen using organic amendments in soil. Plant Soil 83:357–362

Liu Y, Qi A, Haque ME, Bhuiyan MZR, Khan MFR (2021) Combining penthiopyrad with azoxystrobin is an effective alternative to control seedling damping-off caused by Rhizoctonia solani on sugar beet. Crop Prot 139:105374

Moliszewska E, Nabrdalik M, Ziembik Z (2020) Rhizoctonia solani AG 11 isolated for the first time from sugar beet in Poland. Saudi J Biol Sci 27:1863–1870

Mukhopadhyay AN (1987) Handbook on diseases of sugar beet. CRC Press, Boca Raton

Ogoshi A (1987) Ecology and pathogenicity of anastomosis and intraspecific groups of Rhizoctonia solani Kühn. Annu Rev Phytopathol 25:125–143

Ogoshi A (2003) Ecology and pathogenicity of anastomosis and intraspecific groups of Rhizoctonia Solani Kuhn. Annu Rev Phytopathol 25:125–143

Paulitz T, Reinertsen S (2005) In-furrow application of Quadris and Moncut for control of Rhizoctonia root rot in spring barley 2004. In: Plant disease management reports. The American Phytopathological Society, St. Paul, 59:CF011

Schneider CL, Ruppel EG, Hecker RJ, Hogaboam GJ (1982) Effect of soil deposition in crowns on development of *Rhizoctonia* root rot in sugar beet. Plant Dis 66:408–410

Schnieder K, Schafer-Pregl R, Borchardt C, Salamini F (2002) Mapping QTLs for sucrose content, yield and quality in a sugar beet population fingerprinted by EST-related markers. Theor Appl Genet 104:1107–1113

Scholten OE, Panella LW, De Bock TSM, Lange W (2001) A greenhouse test for screening sugar beet (*Beta vulgaris*) for resistance to *Rhizoctonia solani*. Eur J Plant Pathol 107:161–166

Sneh B, Burpee L, Ogoshi A (1991) Identification of Rhizoctonia species. APS Press, St. Paul, p 135

Sneh B, Jabaji-Hare S, Neate S, Dijst G (1996) Rhizoctonia species: taxonomy, molecular biology, ecology, pathology and disease control. Kluwer Academic Publishers, Dordrecht, p 578

Sumner DR (1999) Rhizocotnia crown and brace root rot. In: White DG (ed) Compendium of corn diseases, 3rd edn. American Phytopathological Society, APS Press, St Paul, pp 12–13

Sumner DR, Bell DK (1982) Root diseases induced in corn by *Rhizoctonia solani* and *Rhizoctonia zeae*. Phytopathology 72:86–91

Trkulja N, Blagojević J, Ivanović Ž, Milosavljević A, Popović T, Kuzmanović S, Bošković J (2012) Morfološke i odgajivačke karakteristike izolata Cercospora beticola. Zaštita Bilja, 63(1) 279: 45–52

Trkulja N, Ivanović Ž, Pfaf-Dolovac E, Dolovac N, Mitrović M, Toševski I, Jović J (2013) Characterisation of benzimidazole resistance of *Cercospora beticola* in Serbia using PCR-based detection of resistance-associated mutations of the β-tubulin gene. Eur J Plant Pathol 1358:89–902

Trkulja N, Dolovac N, Mitrović M, Milosavljević A, Pfaf-Dolovac E (2014) The sensitivity of the Serbian populations of *Cercospora beticola* Sacc. to benzimidazole and triazole fungicides. In: Dehne HW, Deising HB, Fraaije B, Gisi U, Hermann D, Mehl A, Oerke EC, Russell PE, Stammler G, Kuck KH, Lyr H (eds) Modern fungicides and antifungal compounds, vol 7, pp 281–282

Trkulja N, Milosavljević A, Stanisavljević R, Mitrović M, Jović J, Toševski I, Bošković J (2015) Occurrence of *Cercospora beticola* populations resistant to benzimidazoles and demethylation-inhibiting fungicides in Serbia and their impact on disease management. Crop Prot 75:80–87

Trkulja N, Milosavljević A, Mitrović M, Jović J, Toševski I, Khan M, Secor G (2017) Molecular and experimental evidence of multi-resistance of Cercospora beticola field populations to MBC, DMI and QoI fungicides. Eur J Plant Pathol 149:895–910

Weiland JJ, Smith GA, Panella L (1999) Greenhouse assay for the evaluation of sugarbeet resistance to *Rhizoctonia* root rot. J Sugar Beet Res 30:211–212

Wigg KS, Goldman IL (2020) Variability in reaction to root and crown rot caused by *Rhizoctonia solani* among table beet cultivars, breeding lines, and plant introductions in controlled environment conditions. HortScience 55(9):1482–1494

Windels CE, Kuznia RA, Call J (1997) Characterization and pathogenicity of *Thanatephorus cucumeris* from sugar beet in Minnesota. Plant Dis 81:245–249

Yassin MA (2013) Interaction of *Rhizoctonia solani* anastomosis groups and sugar beet cultivars. J Pure Appl Microbiol 7(3):1869–1876

Post-harvest Sucrose Deterioration in Sugar Beet

40

Varucha Misra, Ashutosh Kumar Mall, and Santeshwari Srivastava

Abstract

In terms of sugar beet quality, post-harvest sucrose degradation is a major concern. After harvest, sugar beet is one of many quickly perishable crops. The primary and secondary losses are, in essence, the principal causes of this loss. These losses after harvest are also caused by pre-harvest circumstances. Harvesting to slicing duration, loading pattern, beet size, physical damage, and disease prevalence are only a few of the elements that contribute to sucrose deterioration after harvest in this crop. All of these variables add up to a significant financial loss for growers and millers. Growers frequently lose a significant amount of money due to a lack of knowledge about the nature and reasons for these losses, adequate preservation procedures, and transportation and marketing techniques. However, by employing appropriate cultural procedures, such as careful handling and packaging, this can be greatly decreased. The chapter highlights the causes, post-harvest sugar degradation issues, and types of deterioration (physical, physiological, and microbiological) for a better understanding of the sucrose losses after harvest in sugar beet.

Keywords

Damage · Deterioration · Mechanical · Microbial · Physical · Storage · Sugar

V. Misra (✉) · A. K. Mall · S. Srivastava
ICAR—Indian Institute of Sugarcane Research, Lucknow, Uttar Pradesh, India

V. Misra et al. (eds.), *Sugar Beet Cultivation, Management and Processing*,
https://doi.org/10.1007/978-981-19-2730-0_40

40.1 Introduction

Sucrose deterioration in sugar beet is one of the most important and crucial issues. Sugar beet degradation begins at a rapid pace immediately after harvest. In Western Europe, sugar beet-roots are stored for about 2 months as the processing period has been increased till mid-January (Huijbregts et al. 2013). Sucrose losses in stored sugar beet are roughly 0.02% each day (Jaggard et al. 1997), which is around 0.1% of the sucrose content per day. The stored beets deteriorate within the first 40 days of storage under cold and high ambient temperatures, resulting in one-half to one-third of the entire sugar loss after harvest. Wyse and Dexter (1971) had found that when harvested beetroots were stored for 130 days at 3 °C, there was an equal loss in recoverable sugars due to reduction in sucrose content and presence of impurities. In the case of direct sucrose loss in stored beets, sugar recovery is influenced throughout the slicing process, where sugar recovery is lowered as slice rates are reduced, resulting in an increase in sugar cost of production. Loss of sucrose during storage of sugar beets had been reported to cause losses of more than $30 million in North America. These stale sugar beets are supplied in mills for sugar and ethanol recovery. Variation in sugar losses along with the production of undesirable products depends on various factors such as varieties, storage duration, beet condition, method of storage, and environmental conditions. Jaggard et al. (1997) reported that loss in sucrose after sugar beet harvest is due to the enzymatic breakdown of sucrose, which supplies energy for wound reactions. However, Wyse et al. (1978) had shown lengthy storage and increased tissue respiration as primary reasons behind these losses. Hexoses like glucose and fructose are also built during sucrose mobilization, leading to lowering of the quality of sugar beet taproots. With respect to the heavy losses in sugar recovery, this chapter highlights the issues, causes, and factors of post-harvest deterioration in sugar beet.

40.2 Post-harvest Sugar Degradation Issues

There are several issues that emerged due to the post-harvest sugar degradation, which need to be looked into:

1. The period between digging and processing has a negative impact on the quality of the processing in sugar beet. To begin with, the physical effect, in which the root's outer layer darkens in colour, loses compactness, and shrinks with pulpy properties due to moisture loss. As a result, chopping/shredding/cossetting becomes more difficult, and the ideal size and width of cossette for processing (length 5 cm and width 4 mm) is not achieved, resulting in partial sugar extraction. Consequently, the sugar content of the beet pulp decreases.
2. Harvested sugar beetroot is a living organism, and has a physiological effect. To stay alive, it respires or burns sugar, just like all living tissues. Sugar is lost as a result of respiration. Sugar is converted to carbon dioxide and some contaminants during respiration. Temperatures influence the intensity of the reaction and the

amount of sugar lost. At low temperatures, the losses are smaller than at higher temperatures.

3. The chemical reaction that causes sugar inversion as well as a surge in non-sugar impurities including alpha amino-N, potassium, sodium, colloidal sugar dextrin, and saponins. The crystallization process is hampered by invert sugar. The presence of alpha-amino acids darkens the colour of the juice, while sodium and potassium are strongly molassgenic.
4. Sugar beet is a perishable crop, so its storage quality is poor. As a result of the time lag between digging and harvesting, the beet root begins to rot and convert sugar into dextrin, which is very colloidal, reducing the filtration process' efficiency. Saponin levels in deteriorated roots are higher, resulting in enhanced diffuser foaming. Furthermore, if the de-topping and cleaning of the beet roots are not done properly, it will affect the sugar processing. The roots are inadequately de-topped and filthy, which makes factory processing difficult and reduces sugar recovery. If the beets' leaves or petioles are left on due to faulty de-topping, the natural circulation of air in the heap is hampered, and new growth may begin at the expense of the beet's sugar-stored roots. Beets should not be stored unless the bud has been removed by scalping them by at least one inch.

40.3 Causes of Sugar Beet Deterioration

After harvest, there are two basic reasons for sugar beet degradation, *viz.*, primary and secondary losses. Mechanical loss, microbiological action, environmental variables, and other causes are among the primary losses. Secondary losses, on the other hand, are caused by insufficient harvesting, transportation, storage, and marketing facilities, which create conditions conducive to secondary sources of loss. Pre-harvest conditions also have an impact on root quality after harvest. Irrigation frequency, fertilizer application, insect control, growth regulators, and climatic conditions (wet and windy weather, natural climates such as hailing, high wind velocity, heavy rainfall) are among them. Some of the summarized reasons for sugar beet post-harvest deterioration are:

- ***Weather condition at the time of harvesting:*** Time of harvesting is one of the causes for post-harvest losses in sugar beet. Morning sugar beet harvesting is generally preferred as it prevents further losses due to high temperatures and sunburn.
- ***Delay and improper transportation:*** There are two stages in which transportation may take place: from the field to the distribution centres, and then from the distribution centres to the mills. Transportation for small-scale farmers in the mill area is reasonably safe because the beets are taken to the nearby mill simply transported on carts or bicycles rather than vehicles. However, transportation of beets for medium-scale farmers or groups of farmers is more difficult, and the product is more subject to mechanical and heat damage. Mechanical damage (fatigue) happens during transportation as a result of vibrations experienced while

traveling long distances on unpaved roads. The time lag of harvesting to slicing is the major cause for post-harvest sucrose losses in sugar beet. After the harvest of sugar beet, the plant needs to go through the different processes (leaf stripping, turning, and loading into trucks) prior to reaching sugar processing in mills, which is the reason behind the heavy losses due to time lapse from harvest to processing (Alami et al. 2021). This in turn causes either rate of respiration to occur at a faster pace or convert sucrose to non-sucrose carbohydrates by enzymes (Fugate and Campbell 2009). Longer shipment and distribution periods cause heavy losses in sugar beet.

- ***Pattern of loading:*** The loading pattern is also important for ensuring fresh produce quality. The stacking height, gap, and arrangement of harvested beets in stacks or piles are referred to as the loading pattern. Another significant factor to consider while loading is the interlocking of the piles. The quality of the root is directly affected by the loading and unloading of harvested beets. It can be done manually or with the help of clampers. Exposure to the sun while waiting to be loaded at a farmer's field, a storage facility, or for transit can severely diminish root quality. The exposed area turns black or brown and begins to rot.
- ***Immature/preharvest/overmature harvest:*** Sugar beet roots that have developed properly should be harvested at the right time. They must be picked at the appropriate developmental stage, as determined by physiological maturity. The quality of sugar beet root, as well as the storage potential and the appearance of numerous storage illnesses, is always influenced by maturity. Sugar beet quality and post-harvest life potential are heavily influenced by harvest maturity.
- ***Inadequate storage condition:*** The unhygienic environment of the storehouses and mill yards causes an increase in respiration, transpiration rates, and other biochemical reactions. Poor air circulation in the storage rooms causes increased loss after harvest. Temperature and relative humidity conditions also contribute to an increase in post-harvest sucrose losses of sugar beet. Improper storage facilities also pave the way for invasion and proliferation of microorganisms in harvested sugar beets, adding more to loss in quality.
- ***Size of sugar beet roots:*** According to Augustinussen et al. (1995), the size of the sugar beet even plays a significant role in post-harvest degradation. Because huge beets have more surface area, little beets have substantially higher respiration losses than large beets.
- ***Fungal Rots and insect attack incidence:*** High incidence of fungal rotting and insect attack causes heavy degradation to the sucrose accumulated in roots.
- ***Improper handling of harvested products:*** Careless handling and improper harvesting method causes bruising, splitting, and skin breaks in sugar beet.
- ***Weather alterations***
- ***Inadequate modern technologies and skills***

During the different stages of harvesting and processing (from field to mill), causes of post-harvest sucrose losses in sugar beet have been illustrated in Fig. 40.1.

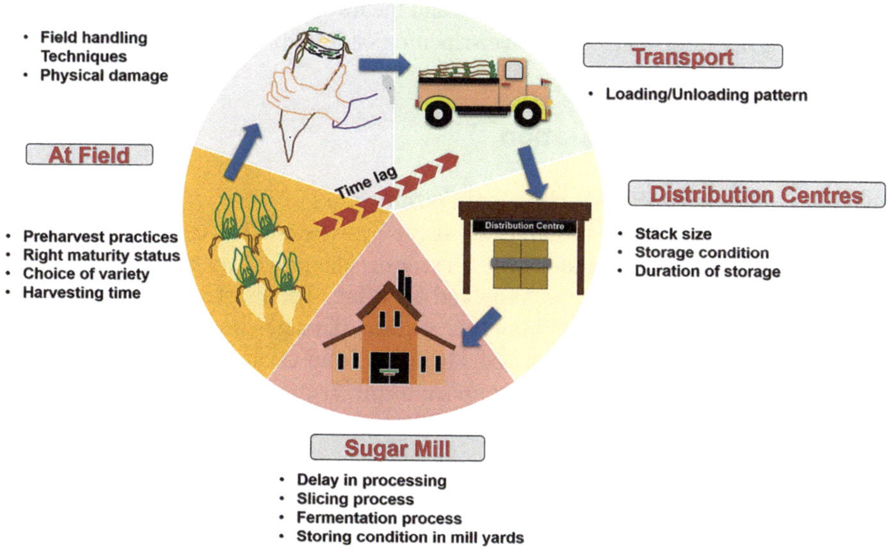

Fig. 40.1 Post-harvest sucrose losses chain in sugar beet

40.4 Types of Deterioration in Sugar Beet

40.4.1 Physiological Deterioration

Conditions that accelerate natural deterioration, such as high temperature, low air humidity, and physical injury, produce an increase in the rate of loss due to normal physiological processes. When the fresh product is exposed to extremes of temperature, atmospheric alteration, or contamination, abnormal physiological degradation ensues. Careless handling of fresh food causes internal bruising, which leads to aberrant physiological damage or splitting, as well as skin breaks, causing rapid water loss and physiological breakdown. Infection by disease organisms that cause deterioration can also be spread through skin breaches.

40.4.1.1 Temperature

The first and most important element that determines the rate of respiration, which contributes to post-harvest sucrose losses, is temperature. According to Houghton and Hopkinson (1998) the temperature rises in clamps of stored sugar beet. This is attributed to an increase in respiration rate, which increases sugar losses by 0.1% per day in the pile beet clamp. Respiration in storage is the process whereby the root converts sucrose into energy to maintain its physiological integrity. The rate of respiration doubles for every 10 °C increase in temperature. The deterioration of beets is accelerated by high temperatures and the development of gases that promote enzyme activity (and hence cause over-ripening or softening) and microbial activity.

40.4.2 Physical Deterioration

40.4.2.1 Physical and Mechanical Damage

Injury or physical damage to harvested sugar beets is another condition that has a significant impact on respiratory rate. The higher the rate of respiration, the greater the degree of harm caused by topping and handling. The most visible and certainly most economically significant injury is surface damage, but interior bruising is equally significant. Root tip breakage, abrasions from shear forces, splits and other surface holes from impacts, and cuts made to the beet in handling—notably in the removal of leaf material, what is known as "topping" procedures—are all examples of surface damage. Marco and Elisabetta (1995) examined cleaner loader performance, including work rate and cleaning efficiency, but gave damage a low priority, focusing mainly on breakage losses and not the level of bruising. Bruising becomes more severe during harvesting and cleaning as breakage losses increase, demonstrating that good breakage prevention practices also reduce bruising (Brown and Pilbrow 1996; Brown 1998). Steensen (1996) counted how many roots had bruising on the sides and in the affected area. According to Peterson et al. (1981), the damage caused by stacking at factory intake was around two-thirds of the damage caused by harvest. Sucrose loss is accelerated by bruising. Imura et al. (1986) found a 19% sucrose loss after mechanical harvest and handling over a four-month storage period, compared to a 2% sucrose loss for hand-harvested beet. According to Hopkinson and Houghton (1998), bruising during harvest causes a loss of roughly 63 kg sugar per hectare, as measured by washing immediately after injury. If there is time for wound healing before leaching episodes, commercial losses from bruising followed by leaching may be lower. Vukov (1977) states that a delay of roughly 2 days between damage and washing significantly reduced sucrose losses compared to washing immediately after dropping, and that this could be explained by the rate of suberization of the wounded surfaces. Suberization appears to occur exclusively in small surface wounds, according to Vukov (1977), whilst large wounds remain mostly unprotected. Although the exact mechanisms at work are unknown, one working concept is that if the periderm is torn, washing can leach away sucrose that has flowed from ruptured cell vacuoles into the apoplast of the underlying tissue. When plant tissues are damaged, cells around the site of the injury reinforce cell walls and seal any damage with glycoproteins, lignin, or suberin secretion (Satoh et al. 1992). Following sugar beet damage, this wound-healing mechanism appears to be significant, sealing the wound and preventing the entry of water and microorganisms, as well as the release of sucrose. However, it is possible that some of the apoplastic sucrose retained by suberization is lost as a result of conversion to reducing sugars by plant or microbial invertase activity, and that reducing sugars could be taken up into intact cells and then sucrose resynthesized and stored in the vacuole by the same mechanism that occurs after phloem unloading.

40.4.2.2 Microbial Deterioration

Microbial infection is connected to sucrose loss after mechanical or physical damage. Because of the changing physiological state of the harvested beets, inappropriate handling, packaging, storage, and transportation may result in decay and the creation of microorganisms. The most critical factor influencing the severity of post-harvest root infection is root damage prior to storage (Mumford and Wyse 1976). Injury makes it easier for the soil's extensive microbial inoculum to infect the roots (Wyse 1980). Microbial infection of bruised beet in the clamp, on the other hand, is likely to result in higher sucrose losses. Unprocessed beets have a high water content (76%) and sugar content (18%), which create ideal conditions for microbial colonization (Jaggard et al. 1997), especially when cracks, root tip breaking, and new wounds on the surface give simple entrance routes (Sebastian et al. 2016).

Sucrose is also lost by hydrolysis, which produces the reducing of sugars glucose and fructose, which are either used in respiration, stored in the root, or devoured by microorganisms. Fungal infection increases the reduction of sugar content, resulting in a reduction in the size of the sucrose pool. Roots infected to 15% of their surface area had a three-fold rise in reducing sugar concentration, with the greatest increase in the infection area, but some increase throughout the root, notably in severe infections (Mumford and Wyse 1976; Wyse 1980). Sucrose can also be lost from injured roots due to leaching, which happens after freeze-thaw cycles and mechanical injury. This can happen during storage if the beets aren't protected from rain and other damp conditions, but it can also happen in the factory's water transport system (Houghton and Armstrong 1994). During sugar beet storage and processing, a rise in microbial activity causes higher sucrose losses. The production of slimy microbial polysaccharides, which cause serious processing and quality issues, is one of the reasons (Sidebotham 1974; Atkins and McCowage 1984; Barfoed and Mollgaard 1987; Clarke et al. 1997; De Lucca et al. 1992; Greenfield and Geronimos 1982).

Bacteria can considerably decrease the processing quality of sugar beet by producing polysaccharides (Augustinussen and Smed 1990), thereby causing difficulty in processing (Hein et al. 2012). *Pseudomonas fluorescens* and *Corynebacterium beticola* are the two important bacteria that have been identified from both intact and degraded sugar beets (Schneider et al. 1968, 1969a, b). In degraded sugar beets, *E. amnigenus* or *Rahnella* sp., both belonging to the Enterobacteriaceae family, have also been found (Tallgren et al. 1999). Microorganisms linked with decaying sugar beets, such as *Penicillium* (saprophytic fungi) and post-harvest pathogens have been found (Liebe et al. 2016; Bugbee 1975; Snowdon 1990). Microbial colonization, primarily by pathogenic or saprophytic fungus, including *Fusarium*, *Penicillium*, and *Botrytis* spp., reduces sugar output significantly. Microbial inversion of sucrose into undesired glucose and fructose molecules is a major finding (Klotz and Finger 2004). Sugar losses of up to 50–60% can occur during storage due to a combination of microbial degradation, beet root respiration, raffinose production, and other factors (Hoffmann 2012; Kenter and Hoffmann 2009).

40.5 Future Prospects

To minimize or entirely eliminate or control quality losses in sugar beet, post-harvest quality deterioration must always be managed. Although there have been studies on sucrose breakdown in harvested sugar beets, more research is needed to understand the molecular underpinnings of this component. This will aid in the identification of potential genes capable of reducing post-harvest sucrose loss. Furthermore, management solutions that will aid in lessening the rotting problem caused by fungus invasion in harvested sugar beet held for longer periods of time are required. As a result, there will be less quality decline following harvest.

40.6 Conclusion

Quality decline in harvested sugar beets has been a major issue for sugar beet workers. After sugar beet harvesting, the sucrose degradation begins at a rapid pace. Harvesting beets and storing them for a lengthy period of time is a common approach all over the world where this crop is grown. After harvest, about 3% of total sucrose is degraded in sugar beet. Sugar beet quality decline is mostly caused by primary and secondary losses. The time between digging and processing, root damage, and poor storage conditions are just a few of the reasons. The physical damage to the roots facilitates root infection by the soil's large microbial inoculum. On the other hand, microbial infection of damaged beet in the clamp is responsible for increased sucrose losses. With the passage of time after harvest, respiration, transpiration, and other biochemical activities rise, contributing to significant losses in sucrose degradation. The greater the degree of harm produced by topping and handling, the higher the rate of respiration. The sucrose losses are further hampered by a two-day delay between damage and washing during mill processing. Another important contribution is the problem of fungal rotting in harvested sugar beets. *Fusarium, Penicillium,* and *Botrytis spp.* drastically limit sugar yield. In the race to lose the sugar, bacterial invasion is not far behind. To mention a few, *Pseudomonas fluorescens, Corynebacterium beticola*, and *Leuconostoc* spp. The understanding and awareness of these losses to growers and millers will definitely boost the quality of harvested sugar beet.

References

Alami L, Terouzi W, Otmani M, Abdekkhalek O, Salmaoui S, Mbarki M (2021) Effect of sugar beet harvest date on its technological quality parameters by exploratory analysis. J Food Qual. https://doi.org/10.1155/2021/6639612

Atkins PJ, McCowage RJ (1984) Proceedings of the 1984 sugar processing research conference. Dextran—an overview of the Australian experience. Sugar Processing Research Institute, New Orleans, pp 108–140

Augustinussen E, Smed E (1990) Quality of sugar beet after frost and thawing during the storage period. Tidsskrift for Planteavl (Denmark) 94(3):249–255

Augustinussen E, Smed E, Steensen JK (1995) Sugar loss in damaged sugar beet, p 39. Losses in field, during storage and through washing. SP Rapport-Statens Planteavlsforsog. No. 7
Barfoed S, Mollgaard A (1987) Dextranase löst Dextranprobleme in DDS-Zuckerfabriken. Zuckerindustrie 112:391–395
Brown SP (1998) Quality harvesting of sugar beet. In: Aspects of applied bioloa, vol. 52, Protection and production of sugar beet and potatoes, pp 179–184
Brown S, Pilbrow J (1996) Effective harvesting: the way to cut sugar beet losses. Br Sugar Beet Rev 64(3):8–10
Bugbee WM (1975) Penicillium claviforme and Penicillium variabile: pathogens of stored sugar beets. Phytopathology 65(8):926. https://doi.org/10.1094/Phyto-65-926
Clarke MA, Edye LA, Cole F, Kitchar JL (1997) Sugarcane factory trials with dextranase enzyme (from *Chaetomium gracile*). Sugar J 59:20–22
De Lucca AJ, Kitchen RA, Clarke MA, Goynes WR (1992) Mesophilic and thermophilic bacteria in a cane sugar refinery. Zuckerindustrie 117:237–240
Fugate KK, Campbell LG (2009) Postharvest deterioration of sugar beet. In: Harveson RM, Hanson LE, Hein GL (eds) Compendium of beet diseases and pests, 2nd edn. The American Phytopathological Society, St. Paul, MN, pp 92–94
Greenfield PF, Geronimos GL (1982) Effect of dextrans on the viscosity of sugar solutions and molasses. Int Sugar J 80:67–72
Hein W, Bauer H, Emerstorfer F (2012) Processing of long-stored sugar beet. Zuckerindustrie Sugar Ind 137(1):25–32
Hoffmann C (2012) Lagerfähigkeit geköpfter und entblätterter Rüben. Sugar Ind 137:458–467
Houghton B, Armstrong M (1994) Fragile. Handle with care. Br Sugar Beet Rev 62(3):13–15
Hopkinson IP, Houghton BJ (1998) Electronic sugar beet measurements: assessments of beet handling and harvesting sugar losses. In: Aspects of applied biology, vol. 52, Protection and production of sugar beet and potatoes, pp 169–172
Houghton BJ, Hopkinson IP (1998) Sugar beet technology. Sugar for Industry, British Sugar plc. Agriculture research and development. Asp Appl Biol 52:191–198
Huijbregts T, Legrand G, Hoffmann C, Olsson R, Olsson Å (2013) Long-term storage of sugar beet in North-West Europe. COBRI (Report No. 1–2013)
Imura E, Hayasaka M, Saito H, Kanzawa K (1986) Relation between mechanical damage and storability in sugar beets. 1. Influence of root damage given in the harvesting and piling processes on the quality of stored sugar beets. Proceedings of the Sugar Beet Research Association, Japan, vol. 28, pp 108–114
Jaggard KW, Clark CJA, May MJ, McCullagh S, Draycott AP (1997) Changes in the weight and quality of sugarbeet (*Beta vulgaris*) roots in storage clamps on farms. J Agric Sci 129:287–301
Kenter C, Hoffmann CM (2009) Changes in the processing quality of sugar beet (*Beta vulgaris* L.) during long-term storage under controlled conditions. Int J Food Sci Technol 44:910–917
Klotz KL, Finger FL (2004) Impact of temperature, length of storage and postharvest disease on sucrose catabolism in sugarbeet. Postharvest Biol Technol 34:1–9
Liebe S, Christ DSM, Ehricht R, Varrelmann M (2016) Development of a DNA microarray-based assay for the detection of sugar beet root rot pathogens. Phytopathology 106:76–86
Marco B, Elisabetta B (1995) Post-raccolta della barbabietola da zucchero: problematiche e operativita della macchine puliscicaricatrici. L'Industria Saccarifera Italiana 88(2):55–65
Mumford DL, Wyse RE (1976) Effect of fungus infection on respiration and reducing sugar accumulation of sugarbeet roots and use of fungicides to reduce infection. J Am Soc Sugar Beet Technol 19:157–162
Peterson CL, Muller ER, Hall MC, Thompson JC (1981) An analysis of root damage during piling. J Am Soc Sugar Beet Technol 21:136–149
Satoh S, Sturm A, Fujii T, Chrispeels MJ (1992) cDNA cloning of an extracellular dermal glycoprotein of carrot and its expression in response to wounding. Planta 188:432–438
Schneider F, Hoffmann-Wahlbeck HP, Abdou MA-F (1968) Über Polysaccharidbildner in der Zuckerfabrikation, II. Mitteilung Zucker 17:465–473

Schneider F, Hoffmann-Wahlbeck HP, Abdou MA-F (1969a) Über Polysaccharidbildner in der Zuckerfabrikation, I. Mitteilung Zucker 23:652–657

Schneider F, Hoffmann-Wahlbeck HP, Abdou MA-F (1969b) Über Polysaccharidbildner in der Zuckerfabrikation, III. Mitteilung Zucker 20:561–566

Sebastian L, Daniel W, Anika W, Alfred P, Andreas S, Mark V (2016) Taxonomic analysis of the microbial community in stored sugar beets using high-throughput sequencing of different marker genes. FEMS Microbiol Ecol 92(2):fiw004

Sidebotham RL (1974) Dextrans. Adv Carbohydr Chem Biochem 30:371–444

Snowdon AL (1990) A colour atlas of post-harvest diseases and disorders of fruits and vegetables. In: General introduction and fruits, vol 1. Wolfe Scientific Ltd, London

Steensen JK (1996) Root injuries in sugar beets as affected step wise by lifting, dumping and cleaning. Proceedings of the 59th lIRB congress, February 1996, Brussells, pp 525–532

Tallgren AH, Airaksinen U, Weissenberg RV, Ojamo H, Kuusisto (1999) Exopolysaccharide producing bacteria from sugar beets. Appl Environ Microbiol 65(2):862–862

Vukov K (1977) Physics and chemistry of sugar-beet in sugar manufacture, English Edition. Elsevier Scientific Publishing Company, Amsterdam

Wyse RE, Dexter ST (1971) Effect of agronomic and storage practices on raffinose, reducing sugar and amino acid content of sugar beet varieties. J Am Soc Sugar Beet Technol 16:369–383

Wyse RE, Theurer JC, Doney DL (1978) Genetic variability in post-harvest respiration rates of sugar beet roots. Crop Sci 18:264–266

Wyse RE (1980) Injury and mold growth as determinants of storage life. Proceedings of the 43rd IIRL3 congress, February 1980, Brussels, pp 5–11

Management Strategies for Reducing Post-harvest Deterioration of Sugar Beet (*Beta vulgaris* L.) 41

Mohammed Arif Sadik Polash, Md. Arif Sakil, Shahida Sazia, and Md. Alamgir Hossain

Abstract

Post-harvest losses of sugar beets are gaining remarkable concern in sugar-producing countries of Southeast Asia. As the extent of post-harvest losses of beets attains about 30–50%, nowadays, reduction of losses has become a prominent issue for maintaining the production and supply chain of sugar beets. Post-harvest losses of beets occur during various post-harvest handling stages, such as harvesting, cooling, washing, followed by disinfecting, grading, packing, and storage. So, appropriate post-harvest handling techniques and technologies along with post-harvest treatments are crucial for reducing the extent of losses and enhancing storage life. Therefore, this recent work bestows a clear concept about appropriate post-harvest handling management and treatments for quality standards and prolonging the storage life of sugar beets.

Mohammed Arif Sadik Polash and Md. Arif Sakil contributed equally with all other contributors.

M. A. S. Polash (✉)
Department of Crop Botany, Khulna Agricultural University, Khulna, Bangladesh
e-mail: arifsadik@kau.edu.bd

M. A. Sakil
Department of Biochemistry and Molecular Biology, Bangladesh Agricultural University, Mymensingh, Bangladesh

Department of Regulatory Biology, Saitama University, Saitama, Japan
e-mail: arifsakilbmb@bau.edu.bd

S. Sazia
Department of Horticulture, Bangladesh Agricultural University, Mymensingh, Bangladesh

M. A. Hossain
Department of Crop Botany, Bangladesh Agricultural University, Mymensingh, Bangladesh
e-mail: alamgir.cbot@bau.edu.bd

V. Misra et al. (eds.), *Sugar Beet Cultivation, Management and Processing*,
https://doi.org/10.1007/978-981-19-2730-0_41

Keywords

Post-harvest loss · Post-harvest management · Shelf life · Sugar beets · Quality

Abbreviations

1-MCP	1-methyl cyclopropene
AVG	Amenoethoxyvinylgycine
$CaCl_2$	Calcium chloride
HDP	High density polyethylene
LDP	Low density polyethylene
MAP	Modified atmosphere packaging
PET	Polyethylene terephthalate
PVC	Polyvinyl chloride

41.1 Introduction

Sugar beet (*Beta vulgaris* L.) belongs to the Amaranthaceae family, is commercially cultivated for sugar production next to sugarcane, as it contains 20% sugar of its weight (Draycott 2008; Biancardi et al. 2012; Cooke and Scott 2012). Although, from time immemorial, it has been commercially cultivated in the USA, Russia, France, Germany, and Turkey; but in Southeast Asia, its cultivation has gained popularity during the last few decades. But being a perishable crop, there is a considerable gap between the gross production and net availability due to post-harvest loss at any moment during the post-harvest chain (Campbell et al. 2008; FAO 2016; Porat et al. 2018).

Almost 30–50% of cultivated sugar beets are lost before reaching the plate due to poor pre-production and improper post-harvest management (Fig. 41.1). These losses have several adverse impacts on farmer income prices and nutritional quality of the produce (Atanda et al. 2011). So, post-harvest losses of sugar beets are one of the alarming problems of the sugar industry and have pulled in far-flung attention. In this context, scientific post-harvest managements are a practical way to eliminate undesirable elements, improve product appearance and establish quality standards. Reduction in post-harvest losses plays an important role in sustainably feeding the citizens in the future (Gustavsson et al. 2011). Therefore, appropriate post-harvest practices are crucial for reducing post-harvest losses and extending storage life so that quality produce is really consumed.

Thus, proper and convenient methods are necessary to be suggested, which could lead to preserving the attribute of harvested sugar beets until consumption. Keeping the above facts in view, the current study was designed to look at scientific

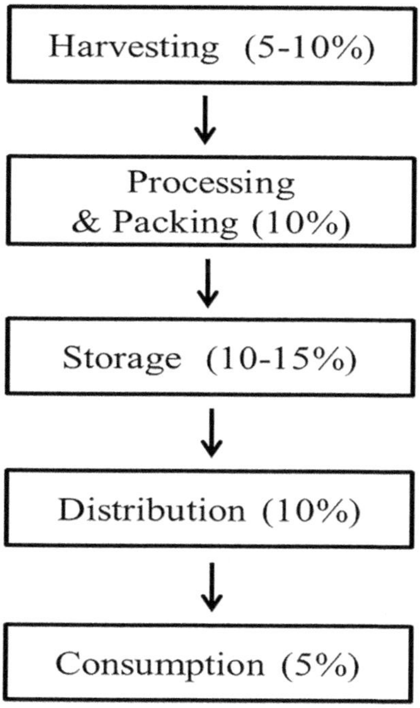

Fig. 41.1 Estimated loss at post-harvest supply chain

management strategies for reducing post-harvest losses and improving the storage life of sugar beets.

41.2 Strategies of Reducing Post-harvest Deterioration

Fresh sugar beet contains around 75% water of its weight, which is subjected to desiccation and mechanical injury from harvesting. Various researchers have estimated 30–50% losses prior to reaching the plate. Therefore, this perishable product needs very sensible handling till consumption to restrict post-harvest deterioration and increase post-harvest shelf life. Certain post-harvest management practices, such as harvesting, transportation, cooling, sorting, cleaning, grading, packaging, and storage, are reviewed below.

41.2.1 Proper Harvesting Practices

Harvesting is an important unit operation that decides the quality as well as storage life and helps in preventing the losses of sugar beets (Beckles 2012; Elik et al. 2019). The goal of good harvesting is to maximize crop yield, to minimize crop losses and quality deterioration, and to be able to keep the harvested produce in good condition until it is consumed or sold. Sometimes farmers harvest sugar beets too early due to market deficiency or the desperate need for cash, which leads to a loss in nutritional value and may get wasted if it is not suitable for consumption and, on the other hand, late harvesting reduces quality as well as yield. Therefore, it is important to harvest sugar beets at the proper maturity stage.

41.2.1.1 Harvest at Correct Stage of Maturity

Maturity at harvest is the most important factor that determines post-harvest life and final qualities (Table 41.1), such as appearance, texture, flavor, nutritive value. Harvesting before maturity is subjected to shriveling and improper ripening, whereas harvested after maturity cause fibrous feel and shorter storage life (El-Ramady et al. 2015). Usually, tropical sugar beet mature 5–6 months from planting, and harvesting is done in the month of April and May (Aly 2012; Paul et al. 2019). Yellowing of lower leaf whirls and root brix reading at 15–18% indicate proper maturation for harvesting.

41.2.1.2 Harvesting at the Proper Time of the Day

Harvesting at the proper time of the day is also essential for reducing post-harvest losses of the produce. Harvesting should be done during the coolest time of the day, either early or late hours of the day, and should be kept shaded in the field to avoid excessive field heat generation and mechanical injury.

41.2.1.3 Correct Harvesting Method

Inappropriate and defectively designed harvesting tools may cause post-harvest losses (Kiaya 2014). Farmers in this subcontinent practice hand defoliation at an early stage and this removal of the photo-synthetically active leaves cause the reduction of 2–3% sugar content. The proper harvesting method includes defoliation of green leaves from the top at maturity to reduce respiratory losses. It can be done manually, but the use of a defoliator machine may lessen the post-harvest losses due to mechanical injury. Then sugar beets are pulled from the soil by a pinch wheel harvester or sugar beet puller.

41.2.2 Post-harvest Handling

Suitable post-harvest handling practices are essential for producing high-quality sugar beets for dinner tables or factories. Post-harvest handling involves transportation, dumping, pre-cooling, sorting, washing and cleaning, grading, post-harvest

Table 41.1 Challenges of both early and late harvesting of sugar beets

Challenges during early harvest	Challenges during late harvest
1. Drop in sugar content	1. Uprooting becomes difficult
2. Weight loss	2. Microbial infections
3. Deterioration of quality	3. Formation of grassy flavor
4. Impedes consumers' choice	4. Not saleable or low price

treatments, packing, and storage. The importance of post-harvest handling is (1) Facilitates a cooling system that avoids moisture loss; (2) slows down unwanted chemical changes within the produce; (3) Preserves the quality of the produce; (4) Escapes physical/mechanical injury; and (5) reduces loss/spoilage till consumption.

41.2.2.1 Transportation to Packing House

In Southeast Asia, a large distance is present between the cultivation lands and packing or manufacturing houses. This considerable distance and unfavorable road networks provide great post-harvest losses during transportation (Adepoju 2014). So, transportation in refrigerated/frozen vehicles is highly suggested, which is not only appropriate but also effective in conserving the attribute of sugar beets.

41.2.2.2 Packing House Operations

It is very essential to reduce mechanical damages at packhouse by avoiding dropping, uneven handling, and bruising. The packing house operations are as follows:

Packing House Dumping

Dumping must be done smoothly either using wet or dry dumping aimed to remove dust and dirt from the produce. There are two damping methods commonly used in packing houses such as wet dumping and dry dumping. Wet dumping is done by dipping the sugar beets into water. As water is more pleasant to the produce, it reduces mechanical injury, abrasions, and bruising on sugar beets. On our subcontinent, wet dumping is popular because of the availability of water and ease to conduct. Another dumping method is dry dumping, which is done by soft brushes fitted on the inclined ramp or moving belts.

Pre-cooling in Packing House

Pre-cooling (or preliminary cooling) indicates rapid exclusion of heat from freshly harvested sugar beets with the intention to provide quality products to the consumers. Heat may come in two ways, (1) from the surrounding air, generally from the radiation of the sun, and (2) from metabolic heat (reactions within the produce). This excessive heat gives rise to an unwanted condition that accelerates the quality of deterioration and decreases shelf life (Bachmann and Earles 2000). Pre-cooling helps to ease the undesirable conditions along with the reduction in metabolic activity, microbial activity, and ethylene biosynthesis (Shahi et al. 2012). Besides, pre-cooling minimizes bruise damage during transit, water loss, ripening rate, and decay, thereby offering good quality beets with prolonged shelf life. Generally, there are five ways for pre-cooling freshly harvested sugar beets, such as hydro cooling, ice cooling, room cooling, forced air cooling, and vacuum cooling, as are practiced in Southeast Asia (Tables 41.2 and 41.3).

Table 41.2 Different types of pre-cooling techniques for sugar beets

No.	Name of the methods	Characteristics
1	Hydro cooling	Cooling of warm produce by water (either spraying or dropping into water) is known as hydrocooling. In this technique the water is mixed with thiabendazole and sodium hypochlorite as surface disinfectant (Genanew 2013). It is a very cheap and competent mode of cooling and becoming an appropriate and smart method for large-scale growers of sugar beets on this subcontinent. Hydro cooling not only prevents water loss but may also help to restore water in slightly wilted produce (El-Ramady et al. 2015)
2	Ice cooling	Here ice is used as a cooling material. Though ice cooling helps to remove heat rapidly, it is inefficient, as about half of the cooling outcome is lost to heat exchange with the atmosphere instead of effectively cooling the produce
3	Room cooling	This process comprises placing the sugar beets in a cool room where cold air through a fan is used to lower the temperature of produce. It is a traditional method for cooling beets as it costs low, but it takes hours to days to adjust the temperature at room set point
4	Forced air cooling	Forced air or 'pressure cooling' involves advanced air pressure with the aim to reduce produce temperature. This is the fastest approach and may reduce cooling times by ten-fold compared to room cooling. As it cools the produce fast, moisture loss is not a problem. Forced air cooling time depends on three basic issues such as (1) variation of temperature between cold air and the produce; (2) the diameter of produce; and (3) airflow
5	Vacuum cooling	Vacuum cooling is another rapid cooling system that involves the evaporation of water from the produce at very low air pressure. In this technique, every 6 °C reduction of temperature causes 1% produce weight loss. Therefore vacuum cooling is not an adequate technique to the growers on the Indian subcontinent

Table 41.3 Comparison of pre-cooling techniques

Variable	Cooling techniques				
	Ice	Hydro	Vacuum	Forced air	Room
Cooling time (h)	0.1–0.3	0.1–1.0	0.3–2.0	1.0–10	20–100
Water contact with the produce	Yes	Yes	No	No	No
Produce moisture loss (%)	0–0.5	0–0.5	2.0–4.0	0.1–2.0	0.1–2.0
Efficacy of energy	Low	High	High	Low	Low
Cost	High	Low	Medium	Low	Low

Table obtained from Kader and Rolle (2004), with the permission from FAO

Pre-sorting

Splitting the different sorts of products from a lot is known as sorting. It is required to subtract injured, damaged, diseased, insect cutting, and misshapen sugar beets from the fresh ones. Pre-sorting not only prevents the spread of infectious microorganisms from decayed produce to other healthy produce during post-harvest management but also saves money and energy as misshapen produce is removed from the lot.

Washing and Cleaning

Cyclospora, hepatitis, salmonella, cryptosporidium, etc. harmful microorganisms may reach consumers via fresh produce of sugar beets. Therefore, appropriate sanitation is a must not only for post-harvest management but also for the avoidance of foodborne ailments. On this subcontinent, chlorine solution (100–150 ppm), with a pH range between 6.5 and 7.5, is frequently used to eliminate soil dust, debris, microorganisms from sugar beets (El-Ramady et al. 2015).

Grading and Sizing

Grading and sizing engages the scrutiny, assessment, and categorization of produce regarding color, size, quality, freshness, maturity state, and ripening level that permits farmers and/or handlers for comfortable handling (Dhatt and Mahajan 2007). For example, grading on the course of maturity or ripening stage assists to exclude overripe sugar beets that will remarkably produce ethylene to accelerate the ripening process in the entire lot. Proper grading and sizing are very crucial as they fetch high income to farmers and/or handler. Besides, it reduces handling losses and assists in the improvement of the packing, storage, and promotion system. A few decades ago, grading was done manually, which is pricey and time-consuming, and even that was affected due to the deficiency in labor during the beet harvesting period. Therefore, lack of labor and overall consistency resulted in the innovation of automatic grading lines, which grade accordingly and preserve the quality till consumption.

41.2.2.3 Post-harvest Treatments

Least post-harvest alterations of sugar beets may be desirable, but utmost are not to the customers. Although post-harvest alterations cannot be stopped, these can be slowed down through proper post-harvest treatments. Post-harvest treatments of sugar beets before storage lessen stresses, minimize deterioration, boost shelf life, and improve quality and marketability. Some popular post-harvest treatments are considered below.

Chemical Treatments

Ethylene Inhibitor

Ethylene is a gaseous compound naturally synthesized by plants which have various physiological and developmental effects on plants. Besides, it also affects some important traits of horticultural products such as senescence, abscission, post-harvest physiology, and ripening, so the action of ethylene often requires to be hindered to enhance the attribute and shelf life of sugar beets. Ethylene inhibitors, such as 1-methyl cyclopropene (1-MCP), amenoethoxy vinylglycine (AVG), etc. help to slow down various metabolic actions (Table 41.4) related to the ripening process and thus prolong shelf life of sugar beets (Watkins 2008; El-Ramady et al. 2015).

Table 41.4 Some ethylene inhibitors and their mode of action

Name of inhibitors	Abbreviation	Site(s) of action	Mode of application
2-Aminoethyoxyvinylglycine	AVG	ACC synthase	Liquid
2-Aminooxyacetic acid	AOA		Liquid
Aminoisobutyric acid	AIB	ACC oxidase	Liquid
Cobalt ions	Co^{2+}		Liquid
1-Methylcyclopropene	1-MCP	Ethylene receptors	Gas
Trans-cyclooctene	TCO		Gas
2,5-Norbornadiene	NBD		Gas
Silver ions	Ag^{+}		Liquid
Silver nitrate	$AgNO_3$		Liquid
Silver thiosulfate	STS		Liquid

Table obtained from Schaller and Binder (2017), with permission from Springer Nature

Application of Calcium Chloride

Post-harvest treatment with minerals, especially calcium chloride ($CaCl_2$), have gained significant consideration for prolonging storage life, freshness, and quality of sugar beets (Senevirathna and Daundasekera 2010). Post-harvest application of calcium chloride on beets assists in slowing down senescence, ripening, physiological disorders, and metabolic activities without detrimental effects on consumers' health. Besides, $CaCl_2$ prevents the softening of beets and fungal attacks. The reasonable cost of the $CaCl_2$ and the quite easy formulation and application mark it as a promising choice to growers/handlers for decreasing post-harvest losses on this subcontinent. There are two appropriate methods for the application of calcium chloride: (1) dipping-washing and (2) impregnation process, of which the first one is commonly practiced for freshly harvested beets in Southeast Asia.

Thermal Treatment

Post-harvest thermal treatment of sugar beets not only helps to control disease and insect infestation but also improves peel color and post-harvest quality (Paull and Chen 2000). In this process, sugar beets are dipped into water at 40–45 °C for 5 min to kill the larva, pupa, and adult insects and harmful microorganisms.

Irradiation

Irradiation is another useful post-harvest technique for destroying microorganisms and controlling insect and parasite infestation, which minimizes food spoilage and prolongs shelf life (Khademi et al. 2013). As it is a costly procedure, it is not commonly used on the Indian subcontinent.

Curing

Curing is a reliable post-harvest technique for root vegetables to lower water loss during post-harvest handling. This process develops a periderm over the cut area that helps in wound healing, reduces water loss, and averts microorganisms from causing decay in the produce. Generally, it is done with 32–40 °C temperature and 90% relative humidity for 1–7 days.

Waxing

External appearance is the foremost attribute that charms the buyers. Therefore, waxing (placing a fine film of edible wax to the external surface of the produce) of beets is a common post-harvest technique. Waxing helps in reducing moisture loss, shriveling during storage, and helps in sealing scratches and tiny injuries on beet's surface. Besides, waxing inhibits sprouting and that improves storage life.

41.2.2.4 Packing

Packing is the act or process of placing the produce into bags or containers with the help of packing materials. Packing is very important in the whole post-harvest chain (Table 41.5) as it reduces food waste and prolongs shelf life (Idah et al. 2007). Packing is done for the fulfillment of three fundamental objectives such as, (1) it encloses produce and simplifies handling and marketing; (2) it shields produce from unfavorable environmental situations (high temperature and low relative humidity) and injuries during storage, transport, and marketing; and (3) it provides information about variety, quality, quantity, and weight. It also provides IDs of the producer and region of the source.

In Southeast Asia, farmers mostly use paper (rigid and printable), polythene (durable; impermeable to water, gases, and odors; resistance to chemicals), jute bag (eco-friendly and biodegradable), and aluminum foil (splendid appearance with low permeability to water, odors, and gases) to wrap the products. Besides, nowadays, cellulose film, cellulose acetate, polyvinyl chloride, vinylidene chloride, rubber hydrochloride, and polyethylene terephthalate are gaining popularity as packing materials on this subcontinent as these are (1) non-toxic and consistent

Table 41.5 Potential packing solution for reducing post-harvest spoilage of produce

Reasons for post-harvest losses	Potential packaging recommendation
Bruising	1. Use of shallow and smooth surface containers 2. Decrease in mass of the beets in a container
Vibration injury during transportation	1. Use of restrainers, individual wrapping and cushioning
Impact injury	1. Use of rigid containers with cushioning of each product
Puncture injury	1. Use of rigid containers with proper handling equipment
Water loss or wilting	1. Use of advanced packaging techniques such as MAP and active packaging
Microbial growth	1. Use of strong, rigid, and microbial-resistant packing, with combined treatment of MAP and irradiation
Inadequate ventilation	1. Use of packaging materials allowed respiration
Consumer behavior (excessive purchase)	1. Use of brilliant appearance packing material

Table obtained from Elik et al. (2019), with permission from Dr. Aysel Elik with the beets; (2) moisture protectant; (3) size, shape, and weight limitations; (4) easily disposable; and (5) inexpensive (Simson and Straus 2010).

The importance of packing is that it (1) protects the sugar beets from microbial infection and deterioration; (2) provides protection from mechanical injury and damage; (3) reduces moisture and weight loss; (4) provides ventilation for gaseous exchange and maintains the freshness of the produce; (5) controls the rate of metabolic activities and ethylene concentration, which prolongs shelf life.

41.2.2.5 Storage

Storage is an indispensable phase in post-harvest handling that helps in trouble-free transportation, operating a proper supply chain, and prolonging the consumption period. Proper storage extends the spell of the processing season and helps to deliver products throughout the seasons. There are many factors governing this aspect. These are (1) post-harvest microbial infections; (2) impairment of beets due to overfilling; (3) collapse of the bottom layer of packet due to overloading and moisture; (4) storage without pre-cooling and chemical treatments. Various storage methods that are commonly practiced on the Indian subcontinent are discussed below.

In situ or Natural Storage Method

In this method, sugar beets are harvested when required and it is suitable for root crops such as beets, cassava, etc.

Sand Method

The sand method is normally practiced on the Indian subcontinent to store different root crops for a long time. In this method, farmers bury the sugar beets with sand underground.

Cellars Method

Cellars refer to underground or partly underground rooms that have been conventionally used as a storeroom. Cellars act as an insulator that keeps the temperature cool in summer and, on the other hand, shields the beets from chilling temperatures in winter.

Night Ventilation

The night ventilation system utilizes the oscillation of temperatures between day and night to cool the storeroom. Here an adjuster fan is switched on once the temperature outside the storeroom drops then the temperature inside the storeroom and switched off after attaining the equilibrium condition in both out and inside of the storeroom.

Natural Ventilation

Natural ventilation is the easiest technique for storage of sugar beets among varied sorts of storage systems. In this process, natural airflow is used to exclude the heat and humidity of the atmosphere and is produced by respiration. Although it is a naïve and widely practiced method, there is a risk of diseases and insect invasion (as it allows storage in natural conditions) which deteriorates the quality of the produce.

Forced-Air Ventilation

In forced-air ventilation system, air is forcefully passed through a perforated floor from air duct. This technique is more efficient for bulk storage, but attention should be given to even circulation of air throughout the stored produce.

Refrigeration

Controlling temperature is the fundamental phenomenon for prolonging post-harvest shelf life because low temperature helps to slacken the metabolic activities inside the product and gives protection against microbial infections. In this process, an airtight and thermal insulated refrigerated room is used, where the capacity of refrigeration equipment should be sufficient to chill the produce. The perfect temperature for a refrigerated room ranges between 1.7 and 4 °C.

Modified Atmosphere Packaging (MAP)

Modified atmosphere packing involves the modification of the atmosphere inside the package in order to expand the shelf/storage life of the produce (Benyathiar et al. 2020). The higher the oxygen level inside the package the higher the metabolic activities (oxidation, respiration rate) and microbial growth which shortens the shelf life of sugar beets. So, drop in oxygen level and its replacement with other gases (such as nitrogen or carbon dioxide) may degrade oxidation reactions and microbiological deterioration. Flexible films such as low density polyethylene (LDP), high-density polyethylene (HDP), polyethylene terephthalate (PET), polyvinyl chloride (PVC), and polypropylene are commonly used packing material as they facilitate proper permeability of gases and water vapor until a steady atmosphere is attained between outside and inside the packet or container (Kang et al. 2008).

41.3 Future Prospects

Not only quantitative but also qualitative deterioration occurs in sugar beets during post-harvest handling tenure. As per estimation, approximately one-third to half of the produce is never consumed by consumers. During the last two decades, several signs of progress have been made in point of view of biological and environmental factors that influence the post-harvest loss. However, much more research for the understanding of physiological and biochemical mechanisms involved in post-harvest management is still indispensable for further enhancements of the attribute as well as storage/shelf life of sugar beets.

41.4 Conclusion

Post-harvest deterioration of sugar beets is considered a foremost problem in Southeast Asia, minimization of which is a potent approach to confirm both food security and quality. The major reasons for post-harvest injury are bad handling and inadequate management of the produce. Therefore, appropriate post-harvest handling and management along with proper use of post-harvest inventions and technologies play a significant role for maintaining quality and improving shelf life. Besides, suitable post-harvest treatment, viz., heat treatment, application of 1-methyl cyclopropene (1-MCP), amenoethoxy vinylgycine (AVG), calcium chloride ($CaCl_2$), irradiation, curing, and waxing are also indispensable for quality standards and prolongation of storage/shelf life. Thus, this chapter concluded that the quality, worth, and storage/shelf life of beets can be conserved easily by practicing appropriate post-harvest handling techniques and treatments, as otherwise post-harvest deterioration of quality and storage/shelf life will be continued and become a key challenge to sugar beet growers/handlers in Southeast Asia.

41.5 Recommendations

1. The agricultural extension department should provide knowledge on post-harvest losses along with appropriate post-harvest handling and management practices with the aim to overcome the post-harvest deterioration of produce.
2. Government and voluntary organizations should cooperate more for financial support in purpose to purchase appropriate storage material.

References

Adepoju AO (2014) Post-harvest losses and welfare of tomato farmers in Ogbomosho, Osun state, Nigeria. J Stored Prod Postharvest Res 5(2):8–13

Aly MSM (2012) Performance study of some sugar beet varieties under sowing and harvesting dates. J Plant Prod 3(9):2439–2449

Atanda SA, Pessu PO, Agoda S, Ilond IU, Ikotun I (2011) The concepts and problems of post-harvest food losses in perishable crop. Afi J Food Sci 5(11):603–613

Bachmann J, Earles R (2000) Postharvest handling of fruits and vegetables. ATTRA:1–19

Beckles DM (2012) Factors affecting the postharvest soluble solids and sugar content of tomato (*Solanum lycopersicum* L.) fruit. Postharvest Biol Technol 63(1):129–140

Benyathiar P, Harte B, Harte J (2020) Shelf life extension of fresh asparagus using modified atmosphere packaging and vacuum skin packaging in microwavable tray systems. Packaging Technol Sci 33(10):407–415

Biancardi E, Panella LW, Lewellen RT (2012) Beta maritime. Springer, New York, pp 85–136

Campbell LG, Klotz KL, Smith LJ (2008) Postharvest storage losses associated with rhizomania in sugar beet. Plant Dis 92(4):575–580

Cooke DA, Scott JE (2012) The sugar beet crop. Springer Science & Business Media, p 704

Dhatt AS, Mahajan BVC (2007) Harvesting, handling and storage of horticultural crops, pp 1–30

Draycott AP (ed) (2008) Sugar beet. Wiley-Blackwell, p 496

Elik A, Yanik DK, Istanbullu Y, Guzelsoy NA, Yavuz A, Gogus F (2019) Strategies to reduce post-harvest losses for fruits and vegetables. Strategies 5(3):29–39
El-Ramady HR, Domokos-Szabolcsy É, Abdalla NA, Taha HS, Fári M (2015) Postharvest management of fruits and vegetables storage. Sustain Agric Rev 65–152
FAO (2016) The state of food and agriculture: climate change, agriculture and food security. Food and Agriculture Organization of the United Nations. Accessed 23 Oct 2016
Genanew T (2013) Effect of post-harvest treatments on storage behavior and quality of tomato fruits. World J Agric Sci 9(1):29–37
Gustavsson J, Cederberg C, Sonesson U, Otterdijk R, Meybeck A (2011) Global food losses and food waste, food and agriculture organization of the united nation (FAO). Swedish Institute for Food and Biotechnology, Düsseldorf
Idah PA, Ajisegiri ESA, Yisa MG (2007) Fruits and vegetables handling and transportation in Nigeria. AU J Tech 10(3):175–183
Kader AA, Rolle RS (2004) The role of post-harvest management in assuring the quality and safety of horticultural produce. Food & Agriculture Organization (FAO), p 152
Kang HM, Choi IL, Kim IS (2008) Comparison packing materials for proper MAP of mixed sprout vegetables. Protect Hortic Plant Fac 17(3):226–230
Khademi O, Zamani Z, Poor Ahmadi E, Kalantari S (2013) Effect of UV-C radiation on postharvest physiology of persimmon fruit (*Diospyros kaki* Thunb.) cv. Karaj during storage at cold temperature. Int Food Res J 20(1):247–253
Kiaya V (2014) Post-harvest losses and strategies to reduce them. Technical paper on postharvest losses. Action Contre la Faim (ACF), p 25
Paul SK, Joni RA, Sarkar MAR, Hossain MS, Paul SC (2019) Performance of tropical sugar beet (Beta vulgaris L.) as influenced by date of harvesting. Arch Agric Environ Sci 4(1):19–26
Paull RE, Chen NJ (2000) Heat treatment and fruit ripening. Postharvest Biol Technol 21(1):21–37
Porat R, Lichter A, Terry LA, Harker R, Buzby J (2018) Postharvest losses of fruit and vegetables during retail and in consumers' homes: quantifications, causes, and means of prevention. Postharvest Biol Technol 139:135–149
Schaller GE, Binder BM (2017) Inhibitors of ethylene biosynthesis and signaling. Methods Mol Biol 1573:223–235
Senevirathna PAWANK, Daundasekera WAM (2010) Effect of postharvest calcium chloride vacuum infiltration on the shelf life and quality of tomato (cv. Thilina). Ceylon J Sci Biol Sci 39(1):35–44
Shahi NC, Lohani UC, Chand K, Singh A (2012) Effect of pre-cooling treatments on shelf life of tomato in ambient condition. Int J Food Agric Vet Sci 2(3):50–56
Simson SP, Straus MC (2010) Post-harvest technology of horticultural crops. Oxford Book Company/Mehra Offset Press, Delhi
Watkins CB (2008) Overview of 1-methylcyclopropene trials and uses for edible horticultural crops. HortSci 43(1):86–94

Sugar Beet Processing to Sugars

42

Barbara Magdalena Muir

Abstract

The beet sugar factory is really a sugar extraction facility where liquid or crystalline sugar, water, animal feeds, fertiliser, molasses, and other by-products are produced. Sugar beet accounts for about one-third of the sugar production in the world and most of this is done in Europe followed by the Russian Federation and the United States of America. Sugar beet entering the factory is cleaned and sliced into thin strips to afford maximum extraction of sucrose while, at the same time, minimising the extraction of non-sucrose. The pulp remaining after extraction is dried and used as animal feed. There may be an option to burn this pulp as fuel or produce biogas via methanisation, but the well-established current feed markets feed would first need to be negotiated. The sugar juice is subjected to a double carbonation process which removes a large portion of the non-sucrose. Factories often operate a lime kiln on-site to provide the active lime and carbon dioxide needed for the carbonation process. Residual soluble calcium can be removed by ion exchange. After evaporation, the white sugar product is obtained through a final purification by crystallisation. The sugar crystals are dried, cooled, and conditioned to produce a free-flowing, mature sugar product ready for packaging or distribution. The run-off syrups from crystallisation are exhausted through further crystallisation steps. These sugars and syrups are recycled and all by-products are valorised.

Keywords

Sugar beet · Diffusion · Clarification · Carbonation · Decalcification · Evaporation · Crystallisation

B. M. Muir (✉)
Tereos Corporate Technology Centre, Aalst, Belgium
e-mail: barbara.muir@tereos.com

V. Misra et al. (eds.), *Sugar Beet Cultivation, Management and Processing*,
https://doi.org/10.1007/978-981-19-2730-0_42

Abbreviations

ICUMSA International Commission for Uniform Methods of Sugar Analysis

42.1 Introduction

The process to produce crystallised sugar from sugar beet was developed by German chemist, Franz Carl Achard, who also built the first beet sugar factory in Europe in early 1801. Achard is thus considered the 'father of the beet sugar industry' (Müller 2021). Initially, raw sugar was produced which was, similar to cane raw sugar, refined to produce white sugar. However, the development of raw house carbonation purification in the 1960s allowed the production of direct white beet sugar (Honig 1965; Van der Poel et al. 1998), which is today the standard for the industry. In fact, the current process for the production of beet sugar has not changed fundamentally since then. Focus on optimisation, energy reduction, and valorisation of by-products has only confirmed that carbonation technology is one of the most efficient and still by far the most cost-effective way of refining beet juice on a large scale. Unfortunately, the carbon footprint due to the use of coal/anthracite is not negotiable without integral changes to the kiln and this might ultimately cause a swing towards physical rather than chemical purification. For good reviews on alternative processes such as membrane filtration and resin technology, see Kochergin (2009) and Johnsonn et al. (2019).

The beet sugar factory can be divided into a few distinct Unit Operations with specific sub-departments (Fig. 42.1). The beet raw material is considered part of the factory as soon as it enters the weighbridge gate, where it is weighed, sampled, and analysed. The beet is then stored on a designated flat pad. This is the first department called **The Yard**. The next operation is that of **Cleaning**, destoning, and de-weeding with concomitant separation of small broken beet particles (tails). Next, the beet is sliced in the **Slicing Station** and prepared for the extraction process. After **Extraction,** the solid plant material (**pulp**) is pressed to a suitable dry substance and valorised further (e.g. by drying and palletisation to produce animal feeds). The beet raw juice is clarified (**Purification**) into a clear juice ready for **Evaporation**. Some factories have large tanks for the storage of evaporated thick juice.

The crystallisation department is the final purification stage of the beet syrup and is conventionally arranged into a three-stage cascading batch or continuous crystallisers and centrifuges to produce white sugar and final molasses. White sugar is finally conditioned by drying, cooling, and conditioning to produce a stable, flowable white crystal product.

This chapter will consider the current mainstream unit operations of a beet sugar production facility.

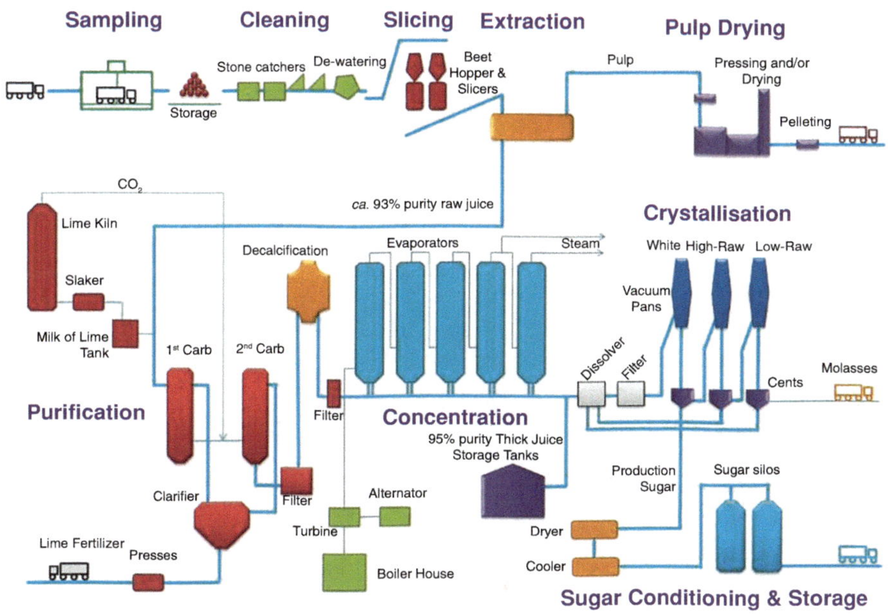

Fig. 42.1 Overview of the beet sugar manufacturing process

42.2 Beet Harvesting Campaign

Sugar beet is planted from seeds at the time of the spring rains. The beet is then harvested after about 8 months when the sucrose levels peak and just as autumn starts, giving the crop plenty of summer sun and rain during the growing period to thrive and accumulate sugar. Today, sugar beet is only harvested mechanically and is directly transported to the factory for processing to avoid deterioration. In colder countries, storage is often considered where there is a risk of prolonged frost as freezing ground cannot be penetrated, potentially locking the beet in the soil until after winter. In such cases, beet must be harvested before the winter sets in and stored in piles, either in the field or at the factory. For example, beet is commonly stored in western Europe from mid-December onwards (Fig. 42.2). As a consequence, the beet harvesting campaigns tend to be highly intensive and short; only about 3–-4 months. Much work has been done to understand and control sucrose losses during storage (Akeson and Stout 1978; Huijbregts 2009; Van Swaaij and Huijbregts 2010, Hoffmann et al. 2017).

Fig. 42.2 Beet stored in a pile next to the field and covered

42.3 Beet Reception

Beet trucks are weighed at a weighbridge as they enter the factory and the trucks are again weighed after offloading of the beet. The sampled beet is cleaned with high pressure water, often in combination with tumbling equipment to remove the soil and as much of the skin as possible. The beet is then inspected visually to look for root damage such as bruises or obvious infections (Fig. 42.3). The beet can be rated in terms of their physical condition or the infected areas can be separated and weighed to produce a comparative measurable value. Mostly, factories will have contracted agreements with the farmers about the quality of the beet that is delivered.

The cleaned beet sample is grated to a fine pulp in preparation for chemical analyses using a solvent and/or reagent and then filtered with the help of filter aids. Water is almost exclusively used as solvent with clarifying reagents, including basic lead subacetate, aluminum sulphate, or other proprietary clarifying agents such as Octapol. Most factories are now shying away from the use of hazardous substances such as lead compounds.

The filtrate is analysed for sucrose content using polarimetry (ICUMSA Method GS6-1 1994 or ICUMSA Method GS6-3 1994), potassium using flame photometry (ICUMSA Method GS6-7 2007), and α-amino nitrogen using UV spectroscopy (ICUMSA Method GS6-5 2007). Sometimes glucose is also measured as routine (ICUMSA Method GS8/4/6-4 2007). The sucrose content is used for payment purposes, although various payment formulas exist within private contracts between

Fig. 42.3 Beet inspection conveyor in the factory tarehouse

farmers and factories/companies, which can include levels of nitrogen as well as potassium to encourage farmers to improve field health. Most often, the potassium and nitrogen values are used to give feedback to the farmers about soil fertilisation as well as in combination with glucose to give the factory a measure of the technical quality of the beet, i.e. how well the beet will behave in the factory. See Van der Poel et al. (1998) and Vermeulen (2015) for more information on formulas used to describe this technical quality.

42.4 Yard Operations

Once in the yard, the truck will offload the beet into a designated area where feeding into the factory will be managed on a first-in-first-out basis (Fig. 42.4). Typically, beet can spend between 12 and 24 h on this pad, depending on the opening hours of the weighbridge. Some factories also have controlled beet silos that allow for longer storage and to simplify logistics. In colder parts, beet can be stored and even sliced while frozen to lend exceptional preservation potential to the beet.

In the yard, extra care is needed to control excessive damage to the beet. Front loaders and the sides of conveyor belts are typically equipped with rubber buffering to ensure smooth movement of the beets. Beet is pushed onto conveyor belts or water troughs located under the level of the flat pad to facilitate continuous feeding of the factory. Since the sucrose losses in the yard is extremely difficult to assess or even estimate, yard operations must be very well managed (Akeson and Stout 1978).

Fig. 42.4 The beet yard and storage flatpad

42.5 Cleaning

Sugar beets grow under the ground and, therefore, soil adheres to the beet after harvesting. Despite developments in harvesting techniques to minimise the levels of soil arriving at the factory, typical dirt levels of 4–8% are still common, depending on the nature of the soil in each area. Soil causes damage to equipment, contains a lot of microorganisms that feeds on sucrose, and can be a cause of high ash levels in products and co-products. The first step after arrival and offloading at the sugar factory is, therefore, cleaning. In sandy soil areas, such as some parts of the USA and the north of Spain, this can be done dry—however, most beet cleaning is done with water along with mechanical separation from stones, weeds, and beet fragments (Fig. 42.5).

Fig. 42.5 Cleaned beet on a conveyor belt

Since beet has a density slightly lower than 1 kg/L, water can be used to wash and at the same time transport the floating beets. This water can be reused for a period of weeks, but care is needed to ensure a reasonable level of soil is maintained and to ensure the pH drop of the water does not cause undue erosion of the equipment used in the cleaning station. Sucrose will inevitably leach into the water, where it will decompose chemically or biologically to organic acids which will cause a reduction in the pH. If this is not managed, the chemical damage to equipment can become quite expensive.

The soil, potentially highly fertile, is recovered from the water in settling bonds, while the stones and organic material can be valorised within the local community.

42.6 Slicing

Beet is sliced into long, V-shaped strips known as *cossettes* (Fig. 42.6). The cossette shape is one of the great innovations of the 1930s and is carefully produced by using two sets of knives iteratively arranged to give a long strip of beet with a notch on the inside to preserve the integrity of the beet structure (Van der Poel et al. 1989). The shape and size of the cossettes are measured and expressed in a number of different ways to enable consistent optimisation of the slicer settings while, at the same time, allowing changes based on the quality of the beet. For example, sucrose from thicker cossettes tend to diffuse slower so that thinner cossettes are preferred as long as structural integrity can be maintained. On the other hand, deteriorated beet benefits from thicker strips.

In the factory, cossette quality can be determined in three ways (Asadi 2007). The **Silin number** expresses the length (in meters) of 100 g of cossettes laid out in a row. The longer the row, the thinner the cossettes. Any pieces shorter than 10 mm (0.4

Fig. 42.6 Sugar beet cossettes

inches) or thinner than 5 mm (0.2 inches) are collected separately as rejects. Good cossettes have a Silin number of 10–18 m (30–54 ft). The mass of rejects should not be more than 5 g (not over 5%).

Secondly, the **Swedish number** expresses the permeability of the cossettes and is measured as the ratio of mass of cossettes longer than 50 mm to those shorter than 10 mm in 100 g of cossettes. The higher the ratio, the better with a required minimum ratio of 10. Lastly, the **Mush content** is expressed as the ratio of the mass of cossettes less than 10 mm long to the total cossette mass (100 g). A mush content of less than 5% is required for good operation.

These characteristics can be used individually or in combination and should be carefully monitored to provide a compromised balance between maximising extraction of sucrose into water (maximising opening of the sucrose containing beet cells) and the subsequent pressing of the exhausted cossettes into manageable pulp fragments (Asadi 2007; Van der Poel et al. 1998; Prati and Maniscalco 2013).

42.7 Extraction

Since the beet is stored outside, the cossettes are relatively cold even after slicing (or can still be frozen in some regions). The cossettes are therefore often pre-heated in a counter current mixer with diffusion juice in a liquid temperature gradient from 73 °C to around 45 °C, to prevent thermal shock of the fibres. The sucrose is then extracted from the beet cossettes in a counter-current diffusion process. A number of different diffuser types and designs exist with specific operational considerations depending on the design (Fig. 42.7).

The temperature of the supply water is kept high to denature the sucrose containing beet cell walls which so become permeable and to inactivate any natural

Fig. 42.7 The RT4 beet diffuser

enzymes and mesophilic and thermophilic bacterial. In addition, biocide is used either continuously or as periodic shock dosages or interchangeably with steam, to control other bacteria. Diffusion can be run completely sterile, but more often, a controlled fermentation is allowed to benefit from the lactic acid effect on dewatering of the beet pulp after diffusion. A temperature of 73 °C is normally targeted. At higher temperatures, the beet cell wall structure starts to break down into small, degraded particles of protein, causing lower juice purity, foaming, slow settling and/or filtration problems. Once again, a balance is maintained to extract between 96.0 and 98.5% of the sucrose as diffusion juice. While beet contains all of the water needed for extraction, some hot diffusion supply water (usually from the pulp pressing station) is supplemented to obtain a ratio of beet to juice of 1.0:1.1, called the draft ratio.

Depending on the equipment and capacity, the diffusion process takes around 40 min, after which the beet and juice are separated via screens. Juice is used to heat the incoming cossettes (counter current mixer) or simply forwarded to the juice heaters for purification.

42.8 Pressed Pulp

After diffusion, the cossettes are considered exhausted and are mechanically pressed to remove as much water as possible to a dry substance of about 30%. Dewatering aids can be added to obtain a higher dry substance; the most prominent pressing aid is calcium sulphate (Gypsum). Low levels of lactic acid (around 150 ppm) are known to also have good dewatering characteristics, but the implied loss of sucrose to form lactic acid needs to be weighed up against the benefit in dry substances.

Sugar beet pulp is used to produce a range of sought-after animal feed products. The pulp is rich in fibre, easily digestible, and high in energy for ruminants, particularly milk cows (Kelly 1983). At the pressed pulp stage, the product is not microbially stable for more than a day or two and should therefore be used or dispensed to customers immediately. Where this is not possible, the pulp can be stored in encapsulated silos where lactic acid from bacterial action will actually serve to conserve the pulp for an extended period of time. The pressed pulp can also be further dried to a maximum of 88% dry substance to afford an additional saving in transport. However, since drying operations are often dependent on fossil fuels, a lot of innovation is seen in this area. For example, drying can be done by low-pressure steam (Deur and Yacine 2015) or in warmer climates with solar energy (Anon 2022).

Alternatively, the pulp can serve as co-feed into an anaerobic digestor for the production of biogas (Maurus et al. 2018) and it has been demonstrated that pulp can be used as biofuel to produce steam and energy for the factory after drying to 50% dry substance (Jensen 2016). However, these innovations will need to take into consideration the primary use of beet pulp as animal feed, an essential step in our current food chain.

42.9 Purification

The resulting diffusion juice is purified by alkaline and thermal treatment, rendering a large portion of the impurities insoluble, followed by either decantation or filtering. The most common purification process used in the beet sugar industry is carbonation—a method that involves treatment of the juice with an excess of milk of lime (calcium hydroxide) to denature, precipitate, and/or convert various impurities, followed by crystallisation of the excess calcium with carbon dioxide as calcium carbonate. The resulting carbonate crystals have a large surface area and will absorb impurities to form a thick sludge that can be separated by filtration or by settling with the aid of a flocculant.

The objectives of purification are to remove enough of the non-sucrose components to be able to make white sugar of good quality at a sufficient rate and yield and to stabilise the juice for the evaporation stage. The target, therefore, is to reduce the levels of impurities that have a detrimental effect on the resulting sugar quality (for example, colourants, ash, odorants, and flavourants), on the crystallisation rate (e.g. oligosaccharides), and exhaustibility of molasses (α-amino nitrogens). Note that impurities do not need to be removed completely, but only sufficiently, and this must be managed carefully to get the right compromise of obtaining the highest possible sucrose recovery at the lowest possible cost (McGinnis and Moroney 1951).

The reader is referred to Van der Poel et al. (1998) for the history of the development of purification processes and also for a number of variations that are still used today. Here, only the double carbonation process will be discussed, using active lime and carbon dioxide (dissolved as carbonic acid) in steps of preliming, main liming, and a two-step carbonation, a process that was finally developed around the 1940s in Germany (Briones 2005). Many variations exist, for example, the use of cold liming in the USA (to make use of the characteristic anomaly of calcium salts to be more soluble at colder temperatures), intermediate liming between first and second carbonation, or addition of either acid or base as required in different parts of the process. Regardless of the exact process used, it is imperative to monitor the nature and quality of the beet to determine the specific target setpoints (such as pH or alkalinity) on a weekly or monthly basis.

42.9.1 Production of Milk of Lime and Carbon Dioxide

Most beet sugar factories today operate their own lime kiln to produce both active calcium hydroxide and carbon dioxide from limestone and either coal or anthracite as fuel (Fig. 42.8). For a discussion on other types of lime kilns, refer to Asadi (2007). Understanding of the kiln process can be simplified by considering the main chemical equations for both production and use of the various components.

Limestones (calcium carbonate, $CaCO_3$) of typically 200 mm diameter are introduced at the top of an operating lime kiln together with pure anthracite or

Fig. 42.8 A multiple shaft mixed-feed kiln and slaker

coal (C) in a fixed ratio (see Eq. 42.1). The process is called calcination, in which calcium carbonate is thermally converted to calcium oxide (quick or active lime, CaO) and carbon dioxide (CO_2). The coal acts purely as fuel and produces some additional carbon dioxide through a reaction with oxygen from the air. Stoichiometry in Eq. 42.1 is purely illustrative (assuming 7.5% coal on limestone). Some carbon monoxide, methane as well as oxygen, nitrogen, and additional carbon dioxide (from the air) will be present in the runoff kiln gas. Typically, this gas will be between 32 and 40% CO_2 and can be scrubbed or dried before use, if needed.

$$CaCO_3 + 0.32C + 0.32O_2 \leftrightarrow CaO + 1.32CO_2 \tag{42.1}$$

The resulting quick lime powder is collected at the bottom of the kiln. As the name implies, this product is very active as well as hygroscopic and is mixed with water from the factory in an adjacent slaker to produce milk of lime (calcium hydroxide, $Ca(OH)_2$), in a highly exothermic reaction (Eq. 42.2). While most calcium salts are sparingly soluble in water, the slaking process produces calcium hydroxide which is partially soluble and presents as a colloidal suspension (milk). Water with low levels of sucrose is often used to also produce some calcium saccharide, which improves the solubility and behaviour of the lime (Rogé 2007).

$$CaO + H_2O \leftrightarrow Ca(OH)_2 \tag{42.2}$$

42.9.2 Liming of Diffusion Juice

While pH is often used as an easy measure to determine setpoints and controls in the factory, the success of the liming process still often depends on the skills and observations of the operators. Supported, of course, by key analyses performed in the laboratory; analyses which still cannot yet be fully automated for integration into advanced control systems. One of the difficulties is the changing nature of the incoming beet: the technical quality as determined in the reception area. A quick response to any changes is needed to maintain a balanced operation.

One of the most useful measurements is that of alkalinity. Alkalinity is defined as the total amount of hydroxide, carbonate, and bicarbonate in a solution measured by titration with an acid. It described the buffering capacity of the juice and is expressed as the equivalent mass of CaO. It can be used to both set up the factory and evaluate the performance, but is most useful in aiding the chemist to balance the factory in terms of cations and anions (Van der Poel et al. 1990; Roten and Schulze 2019).

The chemical reactions that are associated with lime purification are described by Van der Poel et al. (1998) as follows:

1. Precipitation and flocculant of colloidal substances (proteins, pectins, hemicellulose, saccharides, colourants)

Table 42.1 pH and alkalinity profile from a hot pre-liming factory

	pH	Alkalinity (g CaO/L juice)	Temperature (°C)	Averaged laser particle size D [4,3] (μm)
Pre-limed juice	11.4	1.500	72	15
Limed juice[a]	n/a	6.000	83	25
1st carbonation juice	11.2	0.650	83	35
2nd carbonation juice	9.2	0.065	90	25
Pre-evaporator juice	9.2	0.010	80	0

[a] pH reading is no longer sensible for these high alkalinities (a maximum 12.5 is obtained)

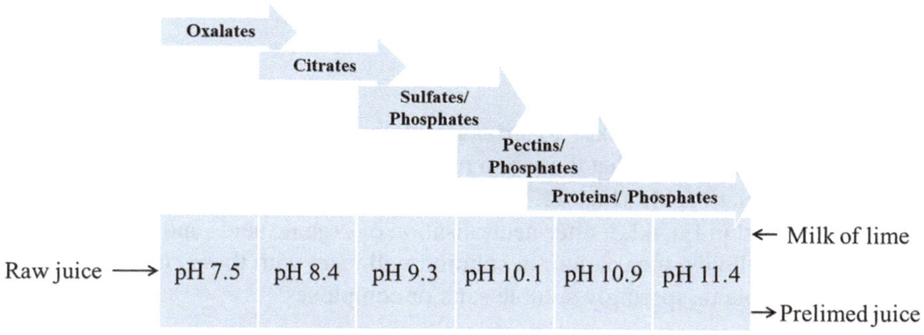

Fig. 42.9 Precipitation reactions in the pre-limer

2. Precipitation of anions forming insoluble or sparingly soluble salts with calcium (phosphate, sulphate, organic acids)
3. Alkaline conversion of fructose, glucose, and amides to organic acids
4. Formation of cations such as ammonia (NH^{4+})
5. Precipitation of cations such as magnesium
6. Adsorption of other juice components on the precipitate and carbonate crystals.

Table 42.1 shows typical pH, alkalinity, and particle size data from a factory in western Europe.

42.9.3 Pre-liming

Milk of lime is introduced to heated diffusion juice in a counter-current pre-liming system with five or six compartments to afford residence time. Each compartment has a target pH and alkalinity to enable the various precipitation reactions to occur at their optimum pH (Fig. 42.9). The temperature of pre-liming between factories can vary widely according to preference.

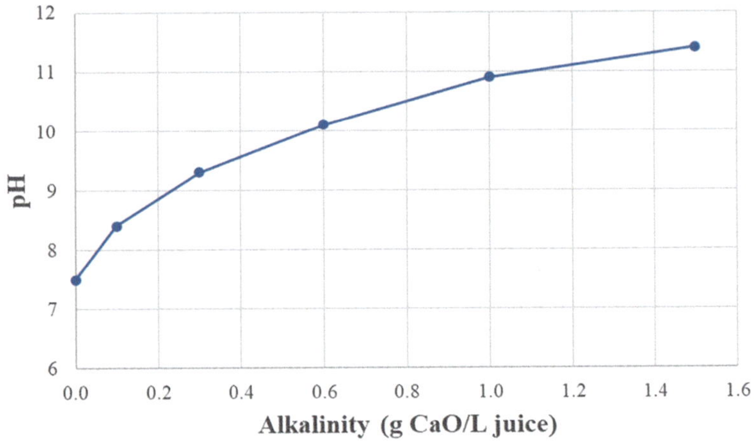

Fig. 42.10 Alkalinity vs pH in pre-liming

While pH is much quicker to measure in the laboratory than alkalinity, it is important to bear in mind that these two parameters do not have a linear, but rather a second order relationship (Fig. 42.10).

As illustrated in Eq. 42.3, after neutralisation of organic acids and stabilisation of proteins and colloids, the calcium in solution will react with these compounds (X^-) to form insoluble or sparingly soluble salts or complexes.

$$X^- + Ca(OH)_2 \leftrightarrow CaX_2 + 2OH^- \tag{42.3}$$

The precipitate from the first carbonation process is often recycled to one of the stages in the pre-limer to act as a seed and aid with adsorption of impurities on the calcium carbonate sludge. The pH at the point of addition needs to be high enough to prevent redissolving of the carbonate. Note that these reactions are all reversible.

42.9.4 Main Liming

Finally, an excess of lime is added to the pre-limed juice, about two to three times the stoichiometric ratio required for the necessary chemical reactions to occur, followed by a period of maturation (5–20 min). The target lime content of the limed juice is in the range of 1.7–2.5% CaO on beet and varies significantly between regions as a function of the weather conditions during growing and the technical beet quality.

42.9.5 Carbonation 1st and 2nd

Finally, once all liming reactions have been allowed to proceed to completion, the excess calcium in the solution is crystallised out with carbon dioxide gas from the

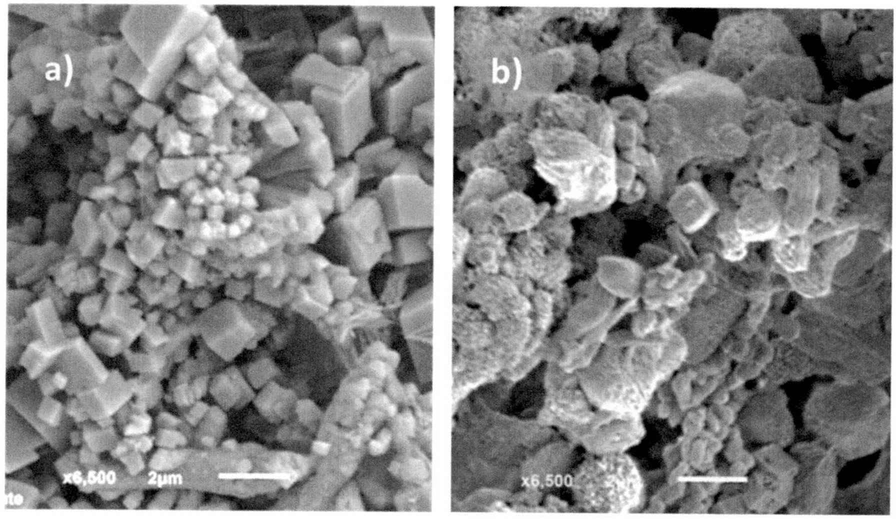

Fig. 42.11 Precipitated calcium carbonate (**a**) as is and (**b**) in the presence of limed beet juice magnified 6500 times

lime kiln (Eq. 42.5). Bear in mind that the carbon dioxide will only react with calcium if it is in solution as carbonic acid (H_2CO_3) which forms on the dissolution of carbon dioxide in water (Eq. 42.4). This could potentially be the rate-limiting step and carbonation vessels are carefully designed to enable the desired dissolution of carbon dioxide in the feed gas mixture.

$$CO_2 + H_2O \leftrightarrow H_2CO_3 \tag{42.4}$$

$$Ca(OH)_2 + H_2CO_3 \leftrightarrow CaCO_3 + 2H_2O \tag{42.5}$$

The crystallised calcium carbonate has a high and highly absorbent surface area which aids in settling and agglomeration of the solids and precipitants from the purification process. Figure 42.11 shows Stereo Electron Microscopy images of calcium carbonate precipitated from the reaction of milk of lime with carbon dioxide (a) on its own as the typical calcite crystal structure and (b) in the presence of limed juice (first carbonation precipitate). A typical averaged particle size of the first carbonation juice is 35 μm (using laser measurement) (Table 42.1). It is essential to obtain the correct particle size and consistency of carbonate precipitate as this will determine the settling and filtration behaviour of this material (Šárka et al. 2008).

Depending on the cation/anion balance of the juice, it is sometimes necessary to add sodium or potassium alkali to ensure sufficient soluble cations to allow for maximum calcium precipitation (Van der Poel et al. 1990; Roten and Schulze 2019).

The solids, which are removed either in a settling clarifier or by bag filtration, can be press-dried to form a filter cake of up to 70% dry substance, or separated by rotary vacuum drum dryers to 50% dry substance. In a settling clarifier, anionic flocculant is often added as a matter of course.

The carbonate filter cake is extensively used as a soil pH-conditioning fertiliser and for improvement of the soil structure. It contains a high level of available calcium, nutrients such as magnesium, phosphates, and nitrogen and is suited to a variety of agricultural spreading techniques (Draycot 2006).

42.9.6 Decalcification

Prior to evaporation, the clarified juice can be decalcified further using, e.g. polystyrene-type cation exchange resins. Under normally good operating conditions, the residual soluble calcium in clarified juice should be around 50 mg CaO/L juice, depending on factors such as beet quality (variety, farming practices, growing conditions, weather impacts), time of the season (early, mid, late) and operations, especially carbonation. In cases where the residual calcium is still too high to obtain the desired sugar quality or to protect the evaporators from scaling, decalcification can be done using strong or weak cation exchange resins. The calcium in the solution is exchanged for sodium (Roten and Schulze 2019) on the resins. Resin operations need to be well controlled and spent resin is regenerated using, for example, sodium hydroxide at least once or twice a day. Spent regenerate is recycled back into the factory. In the Gryllus process, low purity run-off syrups from crystallisation are used as regenerant instead of sodium hydroxide (Borroughs 2007; Van der Poel et al. 1998).

42.10 Sulphitation

Sulphitation via treatment with sulphur dioxide or soluble sulphate compounds (e.g. ammonium bisulphite) often follows the decalcification process. Sulphur treatment will both reduce colour by up to 30% and inhibit the Maillard reaction which is responsible for colour formation in evaporation and crystallisation. Sulphur dioxide is either produced by the factory in a sulphur stove or purchased as a liquid (Asadi 2007).

Since sulphur dioxide is an allergen, the residual levels in sugar are the only sugar quality parameter that is regulated. Allowable levels vary from region to region. The Codex Alimentarius of the Food and Agriculture Organisation of the United Nations (Anon 1999) specifies a maximum allowable level of 15 ppm on sugar.

42.11 Evaporation

The juice prior to evaporation is free from turbid material and stable in terms of pH (around 9.2), so the application of heat should not have a significant effect on the pH during the process. During evaporation, water is removed in a multiple-effect

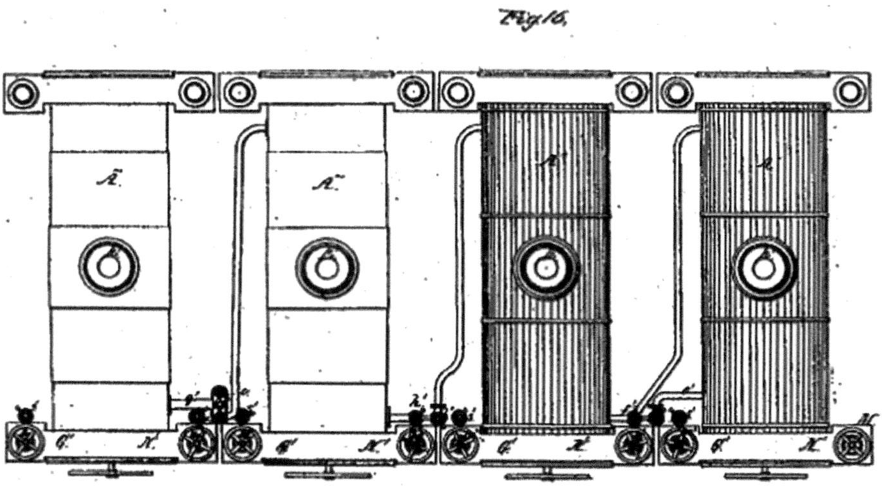

Fig. 42.12 Rillieux's illustration from US Patent US4879A

evaporation system using five or six stages to produce an evaporation syrup called thick juice, condensate, and different grades of steam for supply to the rest of the factory. Interestingly, multiple-effect evaporation under vacuum was developed specifically for the sugar process in New Orleans by French chemical engineer, Norbert Rillieux (Rilleux 1846) (Fig. 42.12).

Steam is produced in large boilers in the factory using gas (natural or biogas) or still sometimes coal or oil (Jensen and Morin 2015). The high-pressure steam from gas is often first fed into alternator turbines to produce electricity which will also regrade the steam to a suitable quality that can be used in the factory. This factory steam is used to heat the first set of evaporators. The steam that is produced from the evaporated water in each evaporator is in turn used to heat the next evaporator all the way through the five or six evaporator sets. Through each subsequent set of evaporators, the temperature and therefore pressure of the steam is reduced due to heat loss and vapour bleed and the pressure inside each evaporator is therefore decreased, allowing the juice to boil at the lower temperatures provided in each subsequent evaporator. This prevents excessive colour formation due to thermal reactions of the juice and allows the evaporator station to produce steam of different grades that can be used throughout the factory in other heating operations as needed. The whole factory is therefore optimised to produce and use just the right amount of steam of each type at any given time. For more information on evaporation and steam economy, see Asadi (2007) or Van der Poel et al. (1998 h).

42.12 Crystallisation

Sucrose crystallisation is the final purification step in the sugar factory. It is also used to optimise the economic recovery of sucrose from the syrup typically over two or three cascading crystallisation steps with full recycling of the raw and final sugars. A simple 3-boiling scheme with recycling options is shown in Fig. 42.13.

The sugar beet industry uses both batch and continuous crystallisers and can operate either batch pans or mother-daughter pans in which a seed massecuite is produced and then used as a footing for further crystallisation pans.

The Brix and purity profiles associated with the scheme in Fig. 42.13 is unique to the beet sugar factory due to the high purity of evaporator syrup (ca. 95%). In addition to a much more thorough purification process (carbonation vs defecation), the type of colourants typically associated with the beet sugar factory will not readily crystallise with the sucrose (Godshall and Baunsgaard 2000). Compared to the cane sugar factories, where crystallisation colour elimination ratios of around 10 is normal, the typical beet sugar factory achieves a colour elimination ratio of 100 quite easily. Table 42.2 shows typical targeted values in beet juice crystallisation.

For a great compendium of the sucrose crystallisation process, see Ziegler (2022). Once further crystallisation is no longer economically feasible, the run-off syrup is considered sufficiently exhausted, though the dry substance will still contain a lot of sucrose. This syrup is called molasses and can be used as animal feed supplement or as feedstock for chromatographic or fermentation processes. Sugar beet molasses

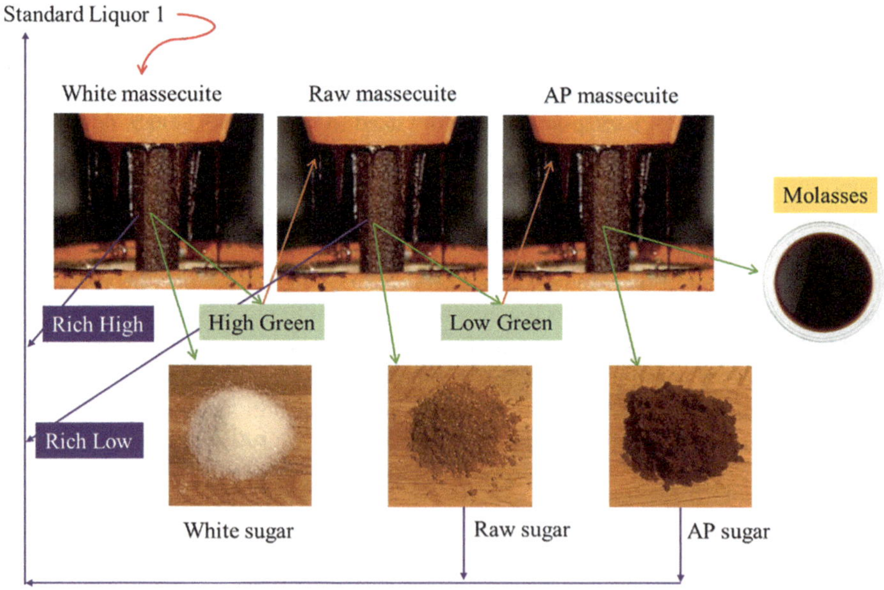

Fig. 42.13 3-boiling crystallisation scheme in a beet sugar factory

Table 42.2 Typical targets in the beet sugar crystallisation process

Parameter	Standard liquor 1	High green	Low green	Molasses
Purity (%)	96	86	77	66
Sucrose content (%)	72	65	58	50
Colour (IU)	2500	5000	11,000	n/m

nm not measured

Table 42.3 Moisture levels in beet sugar from centrifuge to silo

Sugar	Free moisture (%)	Bound moisture (%)	Inherent moisture (%)	Total moisture (%)
(a) Wet	1.20	0.01	0.02	1.23
(b) Dried	0.07	0.01	0.02	0.73
(c) Conditioned	0.02	0.01	0.02	0.05

differs substantially in composition from its sugar cane counterpart and typically has a higher sugar content (65% in beet versus 35% in cane) in part due to a much lower level of invert sugars (i.e. glucose and fructose) (Olbrich 1963; Kelly 1983; Schiweck 1994; DeCloux 2000).

With around 10–15% of the beet sucrose ending up in molasses, it is not surprising that the recovery of sucrose from molasses has always been of interest. Desugarisation by chromatographic fractionation was established by the 1960s with sucrose as the main product although multicomponent separations of invert sugars, betaine, inositol, organic acids, amino acid mixtures, and individual amino acids have also been achieved commercially (Paananen and Kuisma 2000; Hongisto 1977; Šárka et al. 2013). Most beet sugar companies in the USA operate chromatographic molasses desugarisation in some shape or form (Johnsonn et al. 2019).

42.13 Sugar Drying, Cooling, and Conditioning

The sugar crystals are separated from the massecuite in centrifuges and washed with either syrup or hot water. Since most of the colour and ash in the sugar resides on the surface of the crystals, a little increase in washing will markedly improve the quality (Godshall and Baunsgaard 2000). On the other hand, some sugar is dissolved and every ton of water used will need to be evaporated again, establishing an important payoff between energy utilisation and sugar quality.

The wet sugar leaves the centrifuges with between 0.5 and 1.5% moisture content (Table 42.3). The level of moisture is a function of viscosity, particle size, and spin time in the centrifuge. There are three types of moisture in a sugar crystal: inherent moisture, bound moisture, and free moisture. Free moisture exists as a sucrose syrup on the surface of the crystal as sucrose would readily dissolve (Fig. 42.14a). If the crystals are not dried and conditioned, the free moisture on the surface of the crystals will migrate according to temperature and/or humidity gradients in the surrounding

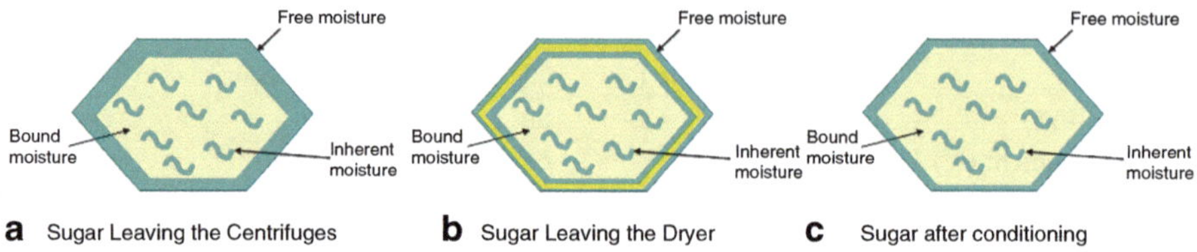

Fig. 42.14 Sugar crystal with moisture during drying and conditioning

area, causing the dissolved sucrose to form a permeable amorphous layer on the surface of the crystals. The moisture migration would therefore cause adjacent crystals to fuse together through this unstable amorphous layer on the surface and will eventually form one big, conglomerated sugar lump if the process is not interfered with. Sugar, therefore, needs to be dried.

However, drying on its own is often not enough. As can be seen from Fig. 42.14, the removal of the free moisture will result in an amorphous sugar layer which would trap some of the free moisture. This layer is still permeable and some time, between 12 and 48 h, is needed for this conditioning to proceed in a controlled relative humidity environment, after which the sugar will be matured. As a general rule, large temperature and relative humidity differences between the sugar and dryer, cooler and conditioning silo should be avoided. Once the sugar is mature, humidity exchange with the environment will be limited to the small amount of moisture that remains on the surface, and moisture and temperature variations in direct contact with the sugar should no longer cause any problems with lumping (De Bruijn 1999; Rogé and Mathlouthi 2003; Starzak and Mathlouthi 2010).

The rotary drum dryer with an integrated cooler, often referred to as a granulator, will dry and cool the sugar by lifting and dropping it through a fast-moving airstream. Another popular design is the rotary louvre dryer, where the air is passed through a moving bed of sugar. Other designs exist of which the fluidised bed drier is of interest as it is the only dryer that will also remove fines. All other dryers are followed by a screen for fines removal. The air for the dryer does not need to be heated except in mid-winter, as the sugar itself will supply the heat necessary for drying. The relative humidity of the air in the dryer and cooler parts are controlled separately for optimum control (Starzak and Mathlouthi 2010; Van der Poel et al. 1998).

After cooling (and screening), the sugar will enter the silo where conditioning takes place. Particle size distribution and especially fines content are key parameters in determining how well the sugar will mature; in other words, how long the sugar will take to reach stability in terms of humidity exchange with the environment. Inside the silo, conditioned air with carefully controlled humidity is forced through the bed of sugar to afford conditioning. In addition to the temperature of the sugar, ambient conditions, including the temperature of the silo walls, play a major role, especially during cold spells when the conditions could easily be near the dew point within the silo (Schindler 2021).

42.14 Thick Juice Campaign

Thick juice after evaporation can either be used immediately for crystallisation or stored for processing outside of the beet harvesting season. Compared to harvested beet that, if not frozen, are preferably stored under cold conditions for a few weeks or months, thick juice can be stored for much longer periods of time. This allows factories to extend the sugar production beyond the 5-month harvesting season.

42.14.1 Thick Juice Storage

After evaporation, the thick juice should be sterile and free from any microbial organisms (Justé et al. 2008). The juice is stored very close to the sucrose saturation point to limit water activity which is necessary for most microbes to become active. pH of stored syrup is above 9.2 where chemical degradation of both invert and sucrose occurs slowly and syrups are cooled to below 15 °C prior to storage to promote conservation. Some infections can still occur mainly starting as deterioration on the surface where the syrup is in contact with the air and therefore exposed to opportunistic microbes such as common yeast and molds. At first, the osmotic pressure of the high Brix syrups will prevent any such activity. However, the atmosphere inside a closed tank is dynamic and condensation will occur on the sides and ceiling that can dilute the surface and thus increase the water activity.

At this stage, the high pH will inhibit microbial growth so that the sucrose destruction will be very slow, yet some molds and yeasts from the air could be attracted to the sucrose and establish themselves on the surface. Gradually, the formation of organic acid products will reduce the pH so that conditions are slowly becoming more favourable for microbial activity and thus sucrose losses. Below pH 8.3, acid-catalysed sucrose inversion begins and becomes more and more significant with pH drop. Further destruction of inverts to lactic and other organic acids will contribute to the drop in pH. At the same time, the lower pH will support the activity of more and more microorganisms. Finally, the degradation can no longer remain just on the surface but could rapidly spread to the body of the tank (Eggleston and Amorim 2006, Eggleston and Vercellotti 2000, Muir et al. 2018). An example of this phenomenon is shown in Fig. 42.15.

42.14.2 Syrup Processing

Processing of the stored syrup for the production of sugar and molasses will occur during the summer months. Since the syrup is already purified, the variation in quality is much reduced compared to the beet harvesting campaign and the processibility of each tank can usually be estimated to a large degree based on the main quality criteria of the juice. Apart from capital utilisation, one of the main advantages of operating a syrup campaign is that the post-evaporation section of the factory does not need to match the size and throughput of the beet operations, making cooperation

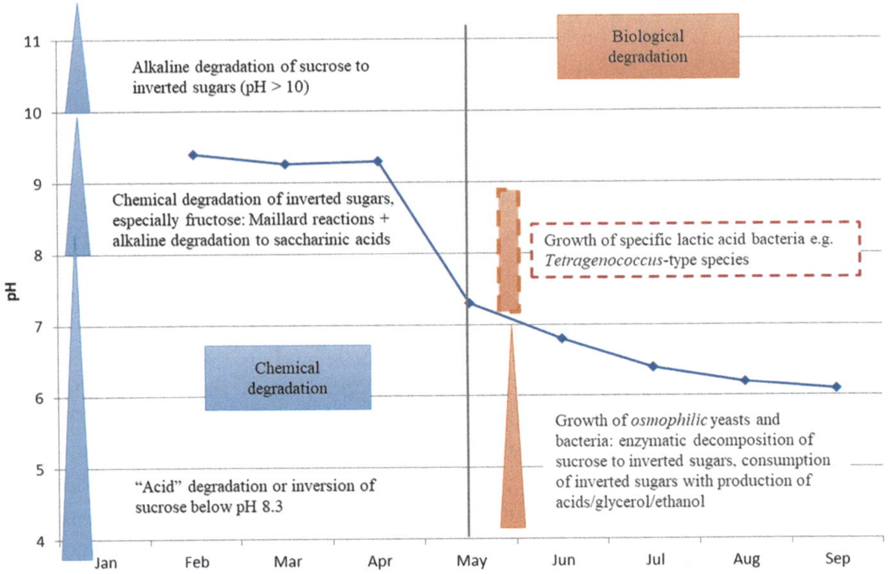

Fig. 42.15 Example of biological and chemical degradation mechanisms of thick juice during storage

during the beet campaign between these two interacting parts highly flexible. This is particularly interesting where a significant increase in the throughput of a factory is considered.

A challenge of the syrup campaign is the complexity of energy management when operating only the back part of the factory, since the full evaporation station, which is both the main user and the provider of different types of steam to the rest of the factory, are not in operation.

42.15 Sucrose Losses

Sucrose is readily hydrolysed at pH below 8.3 in the sugar factory into its constituent molecules, glucose, and fructose (Eggleston and Vercellotti 2000). The sugar factory, or rather sucrose extraction plant, has therefore been designed with one of the primary goals to keep sucrose hydrolysis (be it thermal, chemical, or microbial) as low as possible. The understanding, management, and reduction of sugar losses are one of the main focus points of the process chemist in a factory and substantial monetary savings could be unlocked through continuous improvement in this area (El Shahaby et al. 2014). The main levers are to carefully control the temperature, pH, and residence time in each phase of the process (De Bruijn 2012).

42.16 Future Prospects

There are a number of new technologies with exciting prospects in the beet sugar industry. The application of pulsed-electric fields during either extraction of sucrose or dewatering of pulp is showing great promise in both sucrose extraction and energy reduction (Vidal 2014; Almohammed et al. 2017). Membrane and ion exchange applications are patiently waiting for the technologies to mature so that it would become economically viable on such a large scale (Johnsonn et al. 2019). Furthermore, decarbonation of the industry has started with a whole range of new technologies up for grabs (Rademaker and Marsidi 2019).

One of the limitations of growing beet in colder climates has been the short harvesting season to get the beet lifted before the ground freezes. Areas where beet can be grown and harvested year-round therefore have a distinct advantage (Duraisam et al. 2017).

Over the last few decades, the focus on renewable resources has intensified and a switch from fossil-based essential chemicals and bio-based equivalents are now not far off. The SucroChemistry initiatives that started in the 1970s in the United States of America (Hickson 1977) has not been concluded and the stage is now being set for many of these innovations to see the light despite the more expensive raw materials (Muir and Anderson 2021). Within the fermentation-based industry, sugar cane and especially sugar beet has been identified as two of the most interesting crops for the biobased economy due to both high crop and high hydrocarbon yields (Ragauskas et al. 2006; Anon 2014; E4tech, RE-CORD, WUR 2015).

Recently, the European Commission sponsored an international program called Towards a Sustainable Sugar Industry in Europe (TOSSIE) lead by the Warsaw University of Technology in Poland. The program identified a number of high-level research topics of relevance in the fields of sugar manufacturing, applications of biotechnology and biorefinery processing, sugar beet breeding and growing, and horizontal issues such as funding, training, and best practice sharing (Brühns et al. 2010).

42.17 Conclusion

The beet sugar process has evolved over the past 221 years into massive multistage operations that paint the background of the beet growing areas in the world with steam, silos, kilns, and ponds. The process is, on the one hand, simple enough to be able to produce an edible white sugar crystal quite easily without much more than typical Industry 2.0 level operations and, on the other, intricately balanced and sensitive to a whole range of (often uncontrolled) internal and external factors, the understanding of which should enable the production of a really good quality product in high yield with valorised by-products and optimised energy and chemical consumption.

With the latest industrial environmental targets and the drive towards renewable resources, the stage of the beet sugar industry will be changing, since beet has been identified as one of the most promising crops for the biobased chemical industry (Anon 2014, E4tech, RE-CORD, WUR 2015). However, the main process as described in this chapter will remain intact in the foreseeable future.

References

Akeson WR, Stout EL (1978) Effect of impact damage on sucrose loss in sugarbeets during storage. J Am Soc Sugar Beet Technol 20(2):167–172

Almohammed D, Mhemdi H, Vorobiev E (2017) Purification of juices obtained with innovative pulsed electric field and alkaline pressing of sugar beet tissue. Sep Purif Technol 173:156–164

Anon (1999) Codex standard for sugars—CODEX STAN, 212. https://www.fao.org/input/download/standards/338/CXS_212e_u.pdf

Anon (2014) Opportunities for the fermentation-based chemical industry. An analysis of the market potential and competitiveness of North-West Europe. Deloitte Market Analysis, Amsterdam, 85 pp. https://www2.deloitte.com/content/dam/Deloitte/global/Documents/Manufacturing/gx-mfg-opportunities-for-the-fermentation-based-chemical-industry-2014.pdf. Accessed June 2020

Anon (2022) Pulp drying in Azucarera, Spain. AB Sugar. https://vimeo.com/193865168. Accessed 1 Feb

Asadi M (2007) Sugar-beet handbook. Wiley, New Jersey. a: 143–151; b: 180–183; c: 202–205; d: 333–336; e: 297–302

Borroughs P (2007) Juice decalcification and evaporation protection. European Society for Sugar Technology, pp 107–114

Briones L (2005) Fundamentals of beet juice carbonation. Sugar Indus Technol, p 6

Brühns M, Glavič P, Jensen AS, Narodoslawsky M, Pezzi G, Urbaniec K, Vaccari G (2010) Research for a sustainable European sugar sector. Sugar Ind 135(8):487–495

De Bruijn JM (1999) Activité de l'eau dans les cristaux de sucre. Influence des conditions d'extraction et application au stockage du sucre. Spécial Sucres, pp 23–30

De Bruijn JM (2012) The fascinating sweet world of sugar technology, never a dull moment. Sugar Ind 137(11):697–706

DeCloux M (2000) Literature survey on molasses exhaustion. In: SPRI Conference on Sugar Processing Research Sugar Process Research Institute, pp 322–376

Deur O, Yacine C (2015) Designing a drying process using superheated steam al low pressure. European Society for Sugar Technology 73 (abstract only)

Draycot AP (2006) Sugar beet. Blackwell Publishing, Oxford, pp 1–29

Duraisam R, Salelgn K, Kerebo A (2017) Production of beet sugar and bio-ethanol from sugar beet and it bagasse: a review. Int J Eng Trends Technol 43(4):222–233

E4tech, RE-CORD, WUR (2015) From the sugar platform to biofuels and biochemicals. Final Report, Brussels: European Commission, 183 pp

Eggleston G, Amorim H (2006) Reasons for the chemical destruction of sugars during the processing of sugarcane for raw sugar and fuel alcohol production. Int Sugar J 108:271–282

Eggleston G, Vercellotti JR (2000) Degradation of sucrose, glucose and fructose in concentrated aqueous solutions under constant pH conditions at elevated temperature. J Carbohydr Chem 19(9):1305–1318

El Shahaby OA, Zohri AA, Mohmed MA, Hafez Al Sayed E, Yousef MM (2014) Determination of sucrose losses in beet sugar manufacturing at Dakahlia sugar company, Egypt. Egyptian Sugar J 7:28–50

Godshall MA, Baunsgaard D (2000) The nature of colourant. In: SPRI conference on sugar processing research, pp 122–137

Hickson JL (ed) (1977) Sucrochemistry. ACS symposium series, vol 41. Americal Chemical Society, Washington, DC, p 381

Hoffmann CM, Engelhardt M, Gallmeier M, Gruber M, Märländer B (2017) Importance of harvesting system and variety for storage losses of sugar beet. Sugar Ind 143:74–484
Hongisto HJ (1977) Chromatographic separation of sugar solutions. Int Sugar J 79(100–104): 131–134
Honig P (ed) (1965) Principles of sugar technology, vol 3. Elsevier Publishing Company, Amsterdam, pp 454–454
Huijbregts T (2009) Sugar beet quality during long-term storage in clamp and field. European Society for Sugar Technology, pp 23–32
ICUMSA Method GS6-1 (1994) The determination of the polarisation of sugar beet by the macerator or cold aqueous digestion method using lead acetate as clarifying agent – official, p 3
ICUMSA Method GS6-3 (1994) The determination of the polarisation of sugar beet by the macerator or cold aqueous digestion method using aluminium sulphate as clarifying agent—official: 3 pp
ICUMSA Method GS6-5 (2007) The determination of α-amino nitrogen in sugar beet by the copper method ('blue number')—after defecation with basic lead acetate—official—after defecation with aluminium sulphate—official: 3 pp
ICUMSA Method GS6-7 (2007) The determination of potassium and sodium in sugar beet by flame photometry—official, p 2
ICUMSA Method GS8/4/6-4 (2007) The determination of glucose and fructose in beet juices and processing products by an enzymatic method—accepted: 4 pp
Jensen AS (2016) Environmental and energy saving large pressurized bulk driers. In: The 20th International Drying Symposium (IDS 2016), p 9
Jensen AS, Morin B (2015) Energy and the environment in beet sugar production. European Society for Sugar Technology, pp 75–88
Johnsonn E, Morrow M, Peacock S, Kochergin V (2019) Molasses desugarization in the U.S. beet sugar industry: recent update. Int Sugar J 121(1449):668–681
Justé A, Lievens B, Frans I, Klingeberg M, Michiels CW, Willems KA (2008) Present knowledge of the bacterial microflora in the extreme environment of sugar thick juice. Food Microbiol 25: 831–836
Kelly P (1983) Sugar beet pulp—a review. Anita Feed Sci Technol 8:1–18
Kochergin V (2009) Affinity based separation technologies and their role in the current and future sugar industry. European Society for Sugar Technology, pp 127–147
Maurus K, Ahmed S, Getz W, Kazda M (2018) Sugar beet silage as highly flexible feedstock for on demand biogas production. Sugar Ind 143(12):691–698
McGinnis RA, Moroney CJ (1951) Chapter 7. Juice purification. Fundamental chemistry. In: Beet sugar technology, pp 117–213
Muir BM, Anderson AR (2021) Development and diversification of sugar beet in Europe. Sugar Tech, 27 pp
Muir BM, Parmentier E, De Smet B (2018) Conservation of beet intermediate syrups to extend the sugar producing season. Adv Sugar Crop Proc Conv 2:17 pp
Müller H (2021) The "father of the beet sugar industry." On the 200th anniversary of the death of Franz Carl Achard. Sugar Ind 146(4):216–221
Olbrich H (1963) The molasses. 2006. Biotechnologie-Kempe GmbH, Berlin, p 131
Paananen H, Kuisma J (2000) Chromatographic separation of molasses components. Zuckerindustrie 125(12):978–981
Prati E, Maniscalco M (2013) Recommendations on how to increase the sugar beet pulp press efficiency. Int Sugar J 115(1374):10
Rademaker K, Marsidi M (2019) Decarbonisation options for the Dutch sugar industry *PBL Netherlands Environmental Assessment Agency*. The Hague, PBL publication number: 3481:40
Ragauskas AJ, Williams CK, Davison BH, Britovse G, Cairney J, Eckert CA, Frederick WJ Jr et al (2006) The path forward for biofuels and biomaterials. Science 27:484–489. www.sciencemag.org
Rilleux N (1846) Improvement in sugar-making. US Patent US4879A, granted December 10, 1846
Rogé B (2007) Effect of calcium on white sugar turbidity. Zuckerindustrie 132(3):170–174
Rogé B, Mathlouthi M (2003) Caking of white crystalline sugar. Int Sugar J 105(1251):128–136

Roten CD, Schulze BC (2019) Alkalization pretreatment of thin juice supplied to weak acid cation (WAC) thin juice softening. Agribusiness Intelligence Int Sugar J (5):338–345
Šárka E, Bubnik Z, Kadlec P, Vesela-Trilcova A (2008) The particle size of carbonation mud, and possibilities for influencing it. J Food Eng 87(1):45–50
Šárka E, Bubnik Z, Hinkova A, Gebler J, Kadlec P (2013) Beet molasses—desugarisation, composition, properties and application possibilities. Sugar Ind 138(2):105–114
Schindler H (2021) Storage of white sugar in large-capacity silos. Sugar Ind 146(8):457–475
Schiweck H (1994) Composition of sugar beet molasses. Zuckerindustrie 119(4S):272–282
Starzak M, Mathlouthi M (2010) Formation of amorphous sugar in the syrup film—a key factor in modelling of industrial sugar drying. Food Chem 122:394–409
Van der Poel PW, De Bruijn JM, de Visser NHM, Konings J (1990) Balance of cations and anions in sugar beet processing. Zuckerindustrie 115(11):943–949
Van der Poel PW, Schiweck H, Schwartz T (1998) Sugar technology—beet and cane sugar manufacture. Verlag Dr Albert Bartens KG, Berlin. a: 42–44; b: 216–225; c: 312–313 d: 391–292; e: 484–485; f: 500–523; g: 517–519; h 602–613; i: 874–880
Van Swaaij N, Huijbregts T (2010) Long-term storability of different sugarbeet genotypes—Results of a joint IIRB study. Sugar Ind 135(11):661–667
Vermeulen DP (2015) Development of the sugar beet quality in the Netherlands since 1980 and introduction of invert sugar as a new parameter for beet quality assessment. Sugar Ind 140(2): 95–103
Vidal O (2014) First pulsed electric field (PEF) application at industrial scale in beet sugar industry. Sugar Ind 139(1):37–39
Ziegler J (2022) https://www.zieglerassociates.com/upload/624183/documents/7120006BAE9F7E9B.pdf. Accessed 1 Feb

43 Bioethanol: Technologies, Trends, and Prospects

Sanjay Patil

Abstract

India is the fourth largest ethanol-producing country in the world and produces first generation ethanol from molasses (C molasses and B heavy molasses), sugarcane juice or syrup, and different grains. Present mandate of Government of India is to achieve 10% blending of ethanol in petrol up to 2022 and 20% up to 2025. In the year 2022–2023, Oil Marketing Companies (OMCs) in India are expecting to achieve 10.0% blending for which they have offered attractive differential pricing for ethanol to be produced from sugarcane juice or syrup or sugar, B heavy molasses, C molasses, rice, and other grains. The Ministry of Food and Civil Supplies, Department of Food and Public Distribution (DFPD) has also offered financial support through interest subvention scheme for setting-up of new distilleries or for modernization/expansion of existing distilleries. In last three decades, technologies for fermentation of sugary raw materials have shifted from batch fermentation to continuous fermentation to fed-batch fermentation. For starchy raw materials, the fermentation is carried out typically in batch fermentation using high alcohol tolerant yeast strains. Distillation technology to produce alcohol has also shifted from atmospheric to multi-pressure distillation systems with inbuilt heat integration/recovery concepts. The removal of water from rectified spirit to produce absolute alcohol has shifted from Azeotropic distillation to more efficient Molecular Sieve-based Dehydration (MSDH) technique. Ethanol or distillery plants in India are required to achieve Zero Liquid Discharge (ZLD), which has resulted in development of different downstream effluent treatment technologies and combination of such technologies are to be used to achieve ZLD.

S. Patil (✉)
Department of Alcohol Technology and Biofuels, Vasantdada Sugar Institute, Pune, India
e-mail: sv.patil@vsisugar.org.in

V. Misra et al. (eds.), *Sugar Beet Cultivation, Management and Processing*,
https://doi.org/10.1007/978-981-19-2730-0_43

Feedstock used in ethanol production is also an equally important point that affects ultimately the final cost of ethanol and therefore, considerable research is being carried-out to minimize the feedstock cost. In such circumstances, sugar beet can play a significant role in offering additional and cheaper feedstock for ethanol production and, at the same time, offers an opportunity to enhance the capacity utilization of sugar mills with attached distilleries. The above-mentioned aspects of bioethanol production in India are discussed in this chapter. It explains the technologies useful for ethanol production from sugar beet and how bioethanol from sugar beet can help Indian sugar industry to improve its techno-economic performance.

Keywords

Bioethanol · Biofuel · Effluent Treatment Plant · Sugar beet

Abbreviations

AA	Absolute alcohol
CJ	Cane juice
CPU	Condensate polishing unit
CSTR	Continuous stirred tank reactor
ENA	Extra neutral alcohol
MEE	Multiple effect evaporation
MSDH	Molecular sieve-based dehydration
RS	Rectified spirit
SW	Spent wash
UASBR	Up-flow anaerobic sludge blanket reactor
ZLD	Zero liquid discharge

43.1 Introduction

As the worldwide requirement for liquid biofuels are increasing, future targets and investment plans are suggesting strong growth in near future. Total alcohol production of the world during 2019 was 128,400 million liters, while the alcohol production in India during 2019 was 3488 million liters (FO Licht's World Ethanol and Biofuels Report 2019). India is the second largest producer of sugarcane and sugar in the world after Brazil. There are 564 installed sugar mills in India. The sugar production has increased from 18.9 million MT in 2009–2010 to 33.0 million MT in 2020–2021. Sugar production has increased mainly because of use of improved sugarcane varieties used and better processes for sugar recovery.

Ethanol can be produced using all feedstock which contain mono, oligo, as well as other polysaccharides (Lin and Tanaka 2006). Alcohol can be manufactured from various biomass materials, but the potential for its use as feedstock depends on the availability, cost, sugar content, and the way by which they can be fermented into alcohol (Ogbonna 2004). The production of first generation (1G) ethanol mainly produced using sugarcane juice and sugarcane molasses in Brazil and India, ethanol using corn as substrate especially in the USA, and oilseed biodiesel in Germany are characterized by the commercial market with developed technologies. The production of second generation ethanol (2G) is mainly based on the low-cost crop, forest residues, and organic portion of municipal solid waste after proper chemical and or enzymatic pre-treatments. 2G ethanol is somewhat recently developed, and hence will require more time for optimization of the cost economics with major improvements required in pre-treatments and fermentation processes.

Nearly 61% of the world's alcohol production is from sugar crops (Christoph and Licht 2004). Sugarcane molasses is mainly used as substrate for alcohol production in tropical regions like Colombia, Brazil, and India. Corn is the chief feedstock used in the European Union, the United States, and in China (Vohra et al. 2014). Sugarcane molasses contains about 50% total sugars, of which 30–33% is sucrose and rest is reducing sugar. Ethanol production in distilleries consists of major steps such as preparation of feed of substrate, fermentation, distillation, storage, and sale (Satyawali and Balakrishnan 2008). Molasses is suitably diluted to get proper sucrose level and then supplemented with nitrogen source like urea or ammonium sulfate and phosphate, if required. Fermentation is conducted by using active culture of *Saccharomyces cerevisiae* at 32 °C. After fermentation, around 8–10% (v/v) alcohol accumulates in the wash. Afterward, the fermented wash is distilled to recover the alcohol (Pathade 1999).

Several other feedstocks like corn, potatoes, grain (wheat, barley and rye), sugar beet, sugarcane, and vegetable residues can also be used for ethanol production through fermentation (Icoz et al. 2009). Several firms have pilot scale plants on 2G ethanol. But the time required for pilot scale to full commercialization of cellulosic ethanol may be still more. Therefore, sugarcane and sugar beet are conventionally considered for advanced ethanol production. Cane and beet have almost similar sugar contents (typically, 15% vs. 18%, respectively) while they are different in terms of their non-sugars (non-sucroses) and fibers contents (Rajaeifar et al. 2019). Sugar beet has been recognized as the promising substrate option for ethanol fermentation due to its high land-use efficiency and sucrose content (Alexiades et al. 2018). Beet molasses contains higher level of sugar than sugar beets, and therefore results in a high rate of ethanol production as well as plant efficiency (Maung and Gustafson 2011). Hence, in the current chapter, the author has tried to show the potential of sugar beet and its intermediate processing products for production of ethanol.

43.2 Fuel Ethanol Prospects in India

More than 400 molasses-based distilleries have been installed in India with an installed capacity of 4200 million liters. Capacity utilization of these distilleries has now improved to above 75%. In India, more than 110 grain-based distilleries are installed having capacity of 1800–2000 million liters per annum. Therefore, the total installed capacity of distilleries in India is 6800–7000 million liters with an expectation of 2000 million liters of additional capacity in the next 1–2 years. Several types of alcohol can be produced on an industrial scale in distilleries, such as Rectified Spirit (RS), Extra Neutral Alcohol (ENA), and Absolute Alcohol (AA)/Fuel ethanol. RS contains 94–96% of ethanol along with impurities, such as aldehyde, ketone, acetal, diacetyl, methanol, several higher alcohols, acetic acid, and furfural. Using RS, ENA is produced with further fractionation and contains ethanol content of 94–96%, in which impurity levels are very less in comparison with RS. Majority of impurities are removed during the recovery of ethanol by distillation. Hence ENA is mostly used for potable purposes. The ethanol of commerce (RS and ENA) contains about 4–5% of water, which is also termed as hydrous (water-containing) alcohol. Anhydrous or absolute alcohol (water free) is produced with the removal of the residual water.

Alcohol is mostly used in industries for the manufacture of downstream chemicals, as solvent, used in perfumery industry, manufacture of alcoholic beverages (Country liquor and Indian Made Foreign Liquor), and fuel ethanol for blending with petrol and diesel (Bailey 2018). Though the energy content of ethanol is lower as compared to petrol, it has certain other advantages compared to petrol, like lesser vapor pressure and flammability, no gum formation associated with ethanol, anti-oxidants and detergent additives are not required, and it improves the octane number. Ethanol blending program in India was launched in January, 2003. The first phase this program was started in nine States as well as four Union Territories with supply of 5% ethanol-mixed petrol. The aim of the program was to promote the utilization of alternative and environmentally friendly fuels, as well as to trim down import dependency for energy requirements. The Ministry of Petroleum and Natural Gas directed the Oil Marketing Companies (OMCs) to sell 5% ethanol mixed petrol subject to commercial viability as per Bureau of Indian Standards specifications in 20 States along with four UTs with effect from 1 November, 2006. OMCs are selling petrol blended with ethanol up to 10% throughout India except Union Territories of Andaman Nicobar and Lakshadweep islands with effect from 1 April, 2019. According to the new biofuels policy, Government of India (GoI) aims to achieve a goal of 20% of ethanol mixed with petrol by 2025 and reduce oil imports by 10% up to 2022 and 20% up to 2025. OMCs have offered differential pricing to ethanol depending upon the feedstock used for the production of ethanol. Current prices offered by OMCs per liter are Rs. 46.66 for the ethanol produced by using C-molasses, Rs. 59.08 for the ethanol produced by using B heavy molasses, Rs. 52.92 for the ethanol produced by using grain, Rs. 56.87 for the ethanol produced by using rice as substrate, and Rs. 63.45 for the ethanol produced by using cane juice (CJ) or sugar.

An extensive analysis of energy demand in the transport sector under the business-as-usual (BAU) scenario projected that gasoline demand will increase from 14.2 billion liters in 2010 to 45.6 billion liters in 2030 (Shukla and Dhar 2016). To achieve 20% blending targets, the country will need to produce 9.1 billion liters of ethanol by 2025. Similarly, diesel demand will increase up to 163 million ton (196,400 million liters) by 2030. Therefore, if we go for 5% blending of ethanol in diesel, the country will need to produce 9.8 billion liters of ethanol up to 2025. This indicates that there is going to be excellent demand for fuel ethanol in the near future.

Recently, in 2020, Department of Food and Public Distribution (DFPD), GoI has announced the scheme of soft loans to mills for diverting the excess sugarcane towards the production ethanol with the intention of improving the sustainability of the sugar sector as well as to encourage sugar mills. Soft loans of about Rs. 18,600 Crores are being provided through banks to 362 projects (349 sugar mills and 13 distilleries using molasses as substrate) for increasing and expansion of ethanol production capacity, for which an interest subvention amount of Rs. 4045 Crore for a period of 5 years is going to be borne by the Government.

43.3 Alcoholic Fermentation

Yeast is termed as the heart of alcoholic fermentations. Various yeast species were screened and developed for alcohol fermentation using different substrates, but their efficiency and productivity is low in comparison with *S. cerevisiae*. *Schizosaccharomyces pombe* is also used in some Indian distilleries in Biostil continuous fermentation process as it has high osmotolerance and results in less effluent generation (Dhamija et al. 1996). Yeasts are defined as ascomycetous or basidiomycetous fungi and reproduces mostly by budding or fission (Boekhout and Kurtzman 1996). There are approximately 1500 known yeast species. The total numbers of yeast species on earth are expected to be around 150,000 (Barriga et al. 2011). The length of some yeast cells are only 2–3 μm, while the other species having length of 20–50 μm with a width of 1–10 μm (Hough et al. 1982; Phaff and Stammer 1987). Several yeast species, including *Saccharomyces* spp., are ellipsoidal or ovoid in shape and produce creamy colored colonies on the solid media (Walker 1998; Walker and White 2017).

In anaerobic condition, yeast always takes the fermentative route to utilize glucose. In India, molasses is the chief raw material used for the fermentation of alcohol. Molasses consists of about 50% total sugars. Out of the total sugars, 30–33% is cane sugar (sucrose) and the remaining are reducing sugars (glucose and fructose). The following equation represents the chemical transformation of sucrose, glucose, and fructose to alcohol:

$$\underset{\text{Sucrose}}{C_{12}H_{22}O_{11} + H_2O} \underset{\text{Invertase}}{\rightarrow} \underset{\text{Glucose + Fructose}}{2C_6H_{12}O_6} \quad (43.1)$$

$$\underset{\substack{180\\ \text{Glucose/Fructose}}}{C_6H_{12}O_6} \overset{\text{Yeast}}{\rightarrow} \underset{\substack{2\times 46 + 2\times 44\\ \text{Ethyl alcohol + Carbon dioxide}}}{2C_2H_5OH + 2CO_2 \uparrow} \quad (43.2)$$

Thus, theoretically, 180 g of sugars on reaction gives 92 g of alcohol. Therefore, 1 MT of sugar gives 511.1 kg of alcohol. With 0.7934 specific gravity of alcohol, 511.1 kg of alcohol is equivalent to 644.19 liter of alcohol. Some sugar is required for cell maintenance, and by-products like glycerin, succinic acids, etc. are also generated during fermentation. Therefore, about 94.5% of total fermentable sugars are available for alcohol fermentation. Thus, theoretically, from one MT of fermentable sugar, maximum 608.6 liters of alcohol can be produced under ideal conditions.

Propagation of yeast is usually conducted in several stages. The sugarcane or beet molasses (with lesser concentration of fermentable sugar-based media) based media with additional sources of nutrients such as urea, di-ammonium phosphate, and magnesium sulfate having pH adjusted to 4.5–5.0 is used. Initially, yeast is developed on a laboratory scale using the pure culture of yeast (*S. cerevisiae or S. pombe)* and incubated aerobically at 32.5 °C.

Yeast is propagated in a series of steps starting form 10 mL to 10 L scale medium. Propagation of yeast up to 10 L requires 36–40 h (Gomez-Pastor et al. 2011). At the propagation section in the distillery, there are generally three stages. They are 100 L, 500 L, and 5000 L. All these equipment are designed to sterilize molasses solution, cooling it to 32 °C and inoculation with yeast culture in an aseptic manner. Afterwards, yeast is transferred in well-designed tanks with cooling and recirculation arrangements, i.e., pre-fermenter that requires almost 8 h in order to generate required viable yeast count. Finally, the content of pre-fermenter is transferred in an empty pre-cleaned fermenter. Diluted molasses solution is allowed for filling up to the working volume of main fermenters.

During alcohol fermentations, yeast cells suffer from various stresses. It includes nutrient deficiency, high temperature, and contamination. Other stresses such as accumulation of alcohol and its inhibition on yeast cell growth and ethanol production are related to the metabolism of the yeast. Along with alcohol and CO_2 various by-products are formed. It includes formation of organic acids and higher alcohols. Hence it is very essential to screen and develop yeast strains having tolerance to stress and inhibitors and to produce minimum by-products as well. It will be more industrially important if the yeast strains are developed with high sugar tolerance (more than 25%, w/v), high alcohol tolerance (above 12%, v/v), and high temperature tolerance (more than 40 °C).

43.3.1 Feedstock for Alcohol Fermentation

Sugar mills in India produce C molasses (containing 45–54% total reducing sugar) by the conventional route (three boiling system) and this C molasses is stored and afterwards used in the distillery for ethanol production. B heavy molasses (containing 55–65% total reducing sugar) can be produced by sugar mills by two boiling system and this B heavy molasses is stored and then used in the distillery for ethanol production. Sugar mills can divert a quantity of sugarcane juice/syrup required for the distillery, and from the remaining sugarcane crushing, production of C molasses or B heavy molasses can be carried out which can be stored and used in distillery for alcohol fermentation during the sugarcane off-season. Characteristic of different substrates and yields of alcohol are highlighted in Table 43.1. Sugarcane juice/syrup or C molasses or B heavy molasses are currently being used as feedstock for alcohol production. These routes are desired to fulfill the increasing demand of ethanol and to reduce the increasing sugar stocks of the country. Production of B heavy molasses or diversion of syrup will also reduce steam and power consumption in sugar mills and can reduce the working days of sugar mills due to the additional crushing capacity. It is also possible to increase export of power from the cogeneration unit. These routes can further help in the minimization of effluent treatment expenses of the distillery.

Table 43.1 Typical characteristics of different feedstock and yields of alcohol

S. no.	Parameters	C molasses	B heavy molasses	Cane juice
1.	pH	5.01	5.41	4.74
2.	°Brix	88.0	86.0	57.00
3.	Total reducing sugars, %	50.08	61.00	52.58
4.	Unfermentable sugars, %	5.01	2.60	0.67
5.	Fermentable sugars,%	45.07	58.40	51.91
6.	Carbonated ash,%	10.0	9.8	1.0
7.	Sulphated ash,%	13.0	11.5	2.5
8.	F/N	1.05	2.2	6.4
9.	Volatiles acidity (ppm)	5000	2000	1000
10.	Sp. gravity	1.40	1.35	1.19
11.	Total microbial count (CFU/g)	5.6×10^3	8.8×10^1	7.5×10^1
12.	Shelf life	1–2 years	1–2 years	Perishable
13.	Yields of alcohol (L/MT)	257	295	67

43.4 Fermentation Types

43.4.1 Batch Fermentation

In case of batch fermentation, the substrate along with necessary nutrients are added at start of the fermentation or soon after inoculation, and the fermenter is allowed to progress under controlled parameters until maximum end product concentration is achieved. In this technique, 8–10 fermenters are mostly used for the alcoholic fermentation process. Diluted molasses of about 15% sugar and Baker's yeast cake or propagated pure culture of *S. cerevisiae* is used for the alcohol production. The pH of the wash is usually maintained at 4.5–4.7 and temperature at 32–33 °C (Dombek and Ingram 1987). After around 28–30 h of fermentation, 8.0–8.5% (v/v) alcohol content in the wash is achieved. The majority of the distilleries are now replacing old batch fermentation by cascade continuous or fed-batch fermentation.

43.4.2 Cascade Continuous Fermentation

It is the latest technology in comparison with the old batch fermentation technology (Gyamerah and Glover 1996). In this technique, propagation of yeast is conducted separately and transferred in the first fermenter having required diluted molasses and other supplementary nutrients which is permitted to overflow to the next fermenter. The fermented wash of the second fermenter is subsequently transferred to the third and then to the fourth fermenter which forms up to 8.5–9.5% (v/v) alcohol. Water ring air blowers are provided to the first and second fermenters for the supply of the necessary oxygen essential for the yeast growth. Carbon dioxide formed in the first and second fermenter is collected and transferred into the third fermenter for proper mixing of the fermented broth, while some component of the carbon dioxide formed in the last fermenter is collected and passes to the CO_2 scrubber. The wash coming out from fourth fermenter is settled in the wash settling tank. The sludge content of the wash is allowed to separate from the bottom side of the sludge settling tank, while top supernatant wash is transferred in the wash holding tank through overflow. This wash is then fed to the primary column for its distillation to separate alcohol.

43.4.3 Fed-Batch Fermentation

C-molasses is of inferior quality with high volatile acids and less fermentable sugars, it became difficult for the distilleries to run the continuous mode of fermentation with proper yield and efficiency. In fed-batch fermentation, the single fermenter is fed with substrate at a particular period to achieve the product concentration. From the viewpoint of achieving zero liquid discharge (ZLD), it has now become essential to minimize the spent wash (SW) generation to the lowest possible level. Considering these difficulties, substrate feeding rate may be manipulated (Li et al. 2012). In

Table 43.2 Comparison of different fermentation systems (with use of C-molasses)

S. no.	Parameters	Batch fermentation	Fed-batch fermentation	Cascade continuous fermentation
1.	Fermentation efficiency	87–88	88–90	89–91
2.	Alcohol % in wash (v/v)	8.0–8.5	9.5–10.5	8.5–9.5
3.	Possible yield, L/MT	250	275	280
4.	Molasses quality	Can work with poor quality of molasses	Can work with poor quality of molasses	Requires good quality molasses
5.	Retention time, h	28–30	28–30	22—24
6.	Spent wash generation, L./L. of alcohol	14–15	8–9	9–10
7.	Susceptibility to contamination	Not highly susceptible	Not highly susceptible	Highly susceptible

fed-batch fermentation, the sugar concentration, alcohol concentration, cell count, etc. also vary along with time. In this fermentation mode, yeast is exposed gradually to high alcohol concentration with sufficient retention time of 24–30 h. The final alcohol concentration achieved is about 9.5–10.5% (v/v) with 88.0–89.0% of fermentation efficiency. A comparison of these fermentation processes and different parameters is shown in Table 43.2.

43.4.4 Recovery of Alcohol

After fermentation and removal of yeast sludge, around 8–10% (v/v) alcohol containing fermented broth is distilled, fractionated, and rectified to produce RS or ENA (Pathade 1999). Several types of distillation technologies are available and used for the recovery of alcohol from fermented wash.

43.4.5 Atmospheric Distillation

This system works under atmospheric pressure. In the atmospheric distillation system, the fermented wash is preheated in beer heater, then goes to degasifying column, and afterwards reaches to the top plate of the wash column. The steam is passed through the steam sparger located at the bottom of the column. As the steam goes in upward direction, the wash descending from the top side to the bottom of the column gets heated and reaches to bottom plate. The vapors coming from the wash column consist of about 50% (v/v) alcohol and water along with various impurities. A fraction of these vapors enters into Pre-rectifier column, where impurities having lower boiling point are separated. The other part of the vapors enters the rectifying column that also removes the fusel oil. The vapors coming from wash column flows to the top of rectifying column, then the alcohol content goes on increasing to up to 95.5% (v/v) alcohol (Robinson and Gilliland 1950; Van Baelen et al. 2010).

43.4.6 Multi-pressure Distillation

In this system, various columns operate at different pressures such as some at normal pressure of atmosphere, some at under vacuum, and some are under pressure with good amount of heat recovery that results in steam saving and improved separation of impurities. In this type of distillation, the fermented wash is preheated in pre-heater and fed to the top side of the analyzer column which is fitted with thermosyphon reboiler. The liquid from the bottom side of the analyzer column is heated with rectifier column vapors coming from the top side of the column and the condensate is again fed to the top of the rectified column as a reflux. The rest of the fermented broth which comes out from the analyzer column is termed as SW. Rectifier exhaust column is operated under pressure and bottom liquid is heated with steam through reboiler. The spent lees are then drained out from the bottom part of exhaust column. The resultant alcohol vapors are enriched at the top and are removed as rectified spirit. Fusel oil build-up is controlled in the rectifier column by tapping streams of fusel oil which is transferred to fusel oil concentration column, from where the fusel oil is transferred to decanter for improved separation. The fusel oil wash water is recycled back to the rectifier column (Madson 2003). The comparison of atmospheric and multi-pressure distillation techniques is given in Table 43.3.

43.4.7 Dehydration with Entrainer Process (Azeotropic Distillation)

Here the scheme consists of two to three columns such as dehydration column followed by recovery column. The rectified spirit and cyclohexane (as entrainer) are fed to the dehydration column. Vapor of ethanol, water, and cyclohexane close to its azeotropic concentration are collected from the top, whereas absolute alcohol is collected from the bottom side of the column. When RS is mixed with entrainer like cyclohexane and distilled, a ternary azeotrope is formed. This ternary mixture of cyclohexane, water, and ethanol after condensation is sent for decantation, which forms two layers.

The upper layer is loaded with cyclohexane which is refluxed back, whereas the bottom layer is water rich which is sent to the recovery column. Thus, water is collected from the bottom side of the recovery column whereas ternary mixture of cyclohexane, water, and ethanol comes out of the top, which is condensed and partially sent to dehydration column.

Table 43.3 Comparison of atmospheric and multi-pressure distillation techniques

S. no.	Particulars	Atmospheric distillation	Multi-pressure distillation
1.	Distillation efficiency	97–98%	98.50%
2.	Steam consumption	3.0 kg/L Of R.S. production	2.4 kg/L of R.S. production
3.	Impure spirit production	10–15%	5–6%
4.	Down time	Very frequent due to scaling problems in wash boiling column	Rare shutdown is required and for a very short duration
5.	Plant operation	Manual	PLC/SCADA-based control
6.	Spent wash generation	12–15 L/L of alcohol production	9–12 L/L of alcohol production
7.	Reuse of steam condensate	Nil	80% condensate can be used as boiler feed water
8.	Finished products	Configured to produce one at a time	Two products can be produced, depending on requirement
9.	Quality of R. S./ENA	As per I.S. specifications	Better resolution of impurities. Matches with international specifications
10.	Yield of alcohol per MT of molasses	10–15 L less as compared to continuous fermentation and MPR distillation plant	10–15 L more as compared to batch fermentation and atmospheric distillation plant
11.	Consumer base	Mostly from industries and country liquor manufacturers	Product can be sold in the country and exported worldwide
12.	Selling price	Lower than same alcohol produced in MPR distillation plants	Higher than same alcohol produced in atmospheric distillation plants

43.4.8 Dehydration with Molecular Sieve Process

In this dehydration technology, the RS from the rectifier is superheated with steam in super heater. Superheated hydrous alcohol from super heater is allowed to enter a bed of molecular sieve (3 A type) for some minutes. After a specific interval of time, the flow of superheated RS vapor is switched to the alternate pair of beds. A part of the fuel ethanol vapor leaving the fresh adsorption bed is used to regenerate the loaded bed. A reasonable vacuum is applied by vacuum pump after condensation of the regenerated mixture of ethanol water. This condensate is collected in a condensate collection tank and transferred to the rectified column in the hydrous distillation plant via a recycle pump. The fuel ethanol draw is condensed in product condenser and passed to product storage (Al-Asheh et al. 2004). The lifespan of a molecular sieve may be around 10 years. However, the operating cost is considerably low as compared to azeotropic distillation.

43.5 Effluent Treatment Options

Distilleries using molasses as substrate consume significant quantities of fresh water and generate huge quantities of SW having very high pollution load (Khandekar and Shinkar 2020; Shinde et al. 2020). As per the type of technology used and characteristics of molasses, the SW generation can vary in the range of 7.0–12.0 L per liter of alcohol produced (Qazi 2014). The general characteristics of SW generated through distilleries are specified in the Table 43.4. Commonly, two routes of SW disposal technologies are implemented for achieving ZLD in Indian distilleries, i.e., route 1-Raw SW to biomethanation to evaporation to bio-composting and route 2-Raw SW to evaporation to incineration (Fig. 43.1). As mentioned previously, it is mandatory to achieve ZLD in distilleries.

43.5.1 Raw Spent Wash to Biomethanation to Evaporation to Bio-Composting

In this route, primary SW treatment is biomethanation followed by evaporation for concentration of biomethanated SW up to 30% total solids as a secondary SW treatment system. Concentrated biomethanated SW and press mud cake (PMC) are used for making bio-compost as a tertiary SW disposal system and to achieve zero liquid discharge. More than 70% of the Indian distilleries have implemented biomethanation of distillery SW as the primary treatment method. The major advantages include recovery of energy in the form of biogas produced and neutralization of acidic SW (Mohana et al. 2009). To concentrate the biomethanated SW

Table 43.4 Characteristics of spent wash from different routes

S. no.	Parameters	C molasses	B heavy molasses	Cane syrup	Cane syrup after recycle
1.	pH	4.0	4.2	4.5	4.5
2.	Color	Dark Brown	Yellowish brown	Pale yellow	Pale yellow
3.	Quantity (L/L alcohol)	10	8	6	3
4.	°Brix	12	8	2.5	4.5
5.	COD (mg/L)	120,000	80,000	25,000	45,000
6.	BOD (mg/L)	60,000	40,000	12,000	22,000
7.	Dissolved solids (mg/L)	60,000	50,000	18,000	24,000
8.	Suspended solids (mg/L)	30,000	20,000	3000	6000
9.	Total solids (mg/L)	90,000	70,000	21,000	30,000
10.	Nitrogen (mg/L)	1000	700	400	600
11.	Phosphorus (mg/L)	300	200	100	150
12.	Potassium (mg/L)	10,000	4000	1000	1500

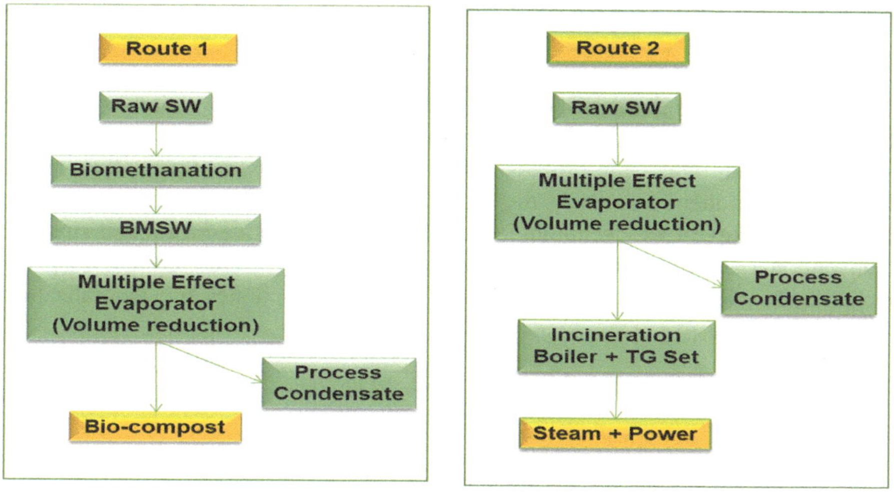

Fig. 43.1 Two routes to achieve ZLD in distilleries

before bio-composting, evaporation plants are used in distilleries. The biomethanated SW is concentrated from 6% total solids to about 30% total solids by standalone multiple effect evaporation (MEE) plant. Types of MEE's used in distillery industry are: Falling film evaporators, Forced circulation evaporators, Combination of falling film and forced recirculation evaporators, and Integrated evaporators. The mixing of SW and PMC (1.6:1) has to be conducted in surface windrows using an aerotiller machine (Self-propelling) for spraying, mixing, turning, and aeration of compost material. Addition of special blend of cultures or cow dung provides microbial inoculum essential for bio-composting process. The total composting cycle period is 60 days.

43.5.2 Raw SW Concentration by Multiple Effect Evaporation Followed by Incineration

In this route, raw SW is concentrated from 12% total solids to about 60% total solids by the multiple effect SW evaporation plant. The concentrated SW is incinerated in a modern incineration boiler to achieve zero liquid discharge. Raw SW (12% total solids) is fed to MEE for concentrating it to 60% total solids. Concentrated SW at about 60% solids is usually fired as a source of energy using a specially designed boiler with use of supporting fuel (coal or bagasse or rice husk, etc). Steam generated is utilized to run a steam turbine to generate electricity and exhaust steam is utilized in the distillery as well as evaporation plant operation. The organics in SW are totally burnt and inorganics are transformed into ash.

43.5.3 Treatment of Process Condensate and Distillation Spent Lees Through Condensate Polishing Unit (CPU)

To achieve ZLD and for disposal of effluents such as evaporation condensate (generated during concentration of biomethanated SW or raw SW by evaporation), spent lees and other non-process effluents, CPU is used. Spent lees are a low-strength effluent formed in the recovery columns of the distillation process. There are multiple CPU technologies available to treat above-mentioned streams. Conventional CPU involves pretreatment, secondary, and tertiary disposal technologies (Figs. 43.2 and 43.3). Pretreatment incorporates the neutralization

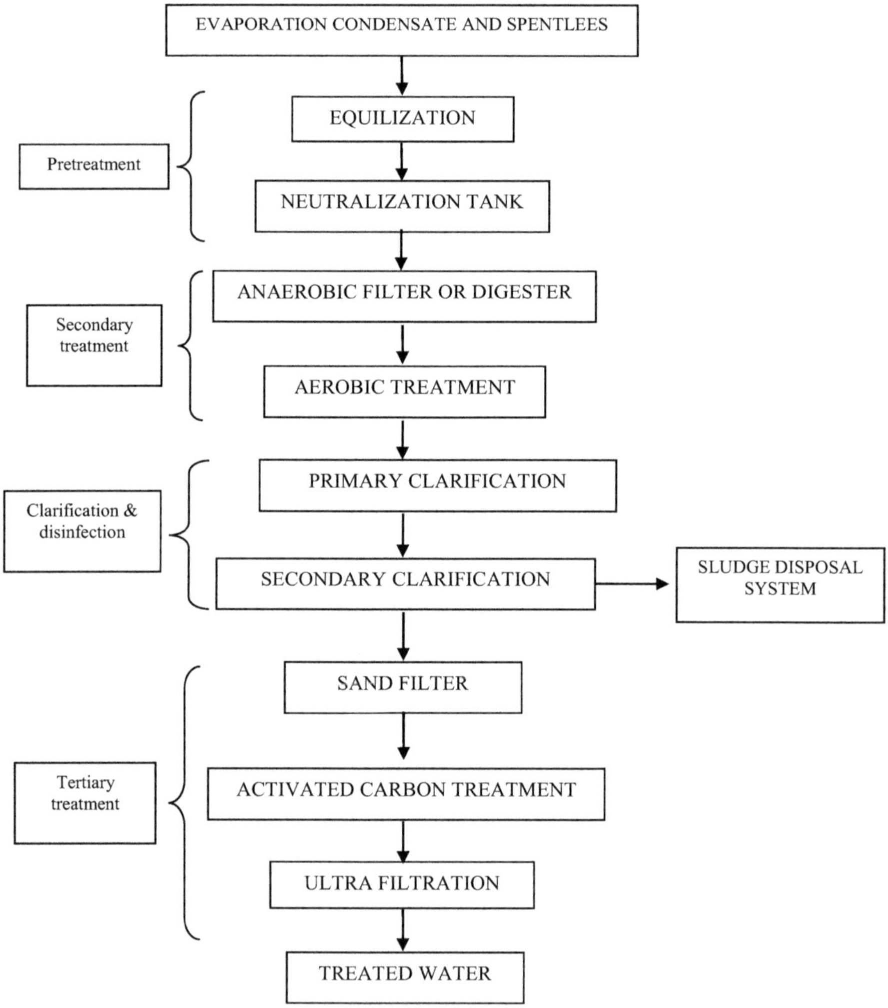

Fig. 43.2 Typical process flow diagram for CPU

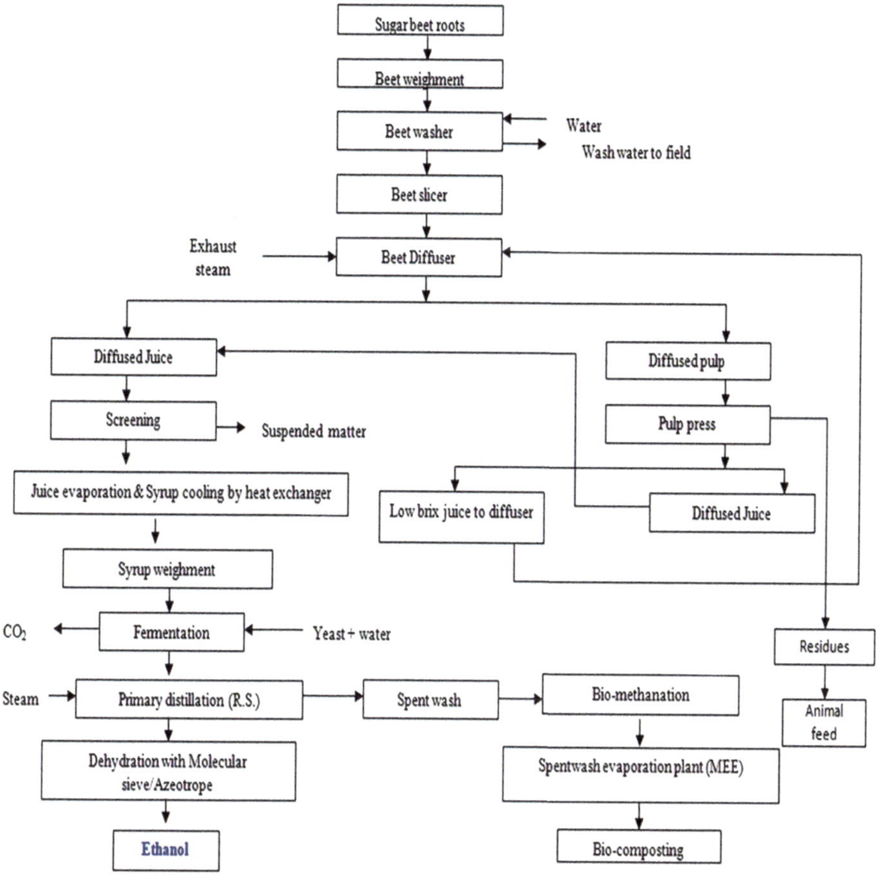

Fig. 43.3 Process flow diagram of ethanol production from sugar beet juice/syrup and animal feed from sugar beet pulp

system which is provided to neutralize the effluent using lime slurry (10%) or soda ash. Secondary treatment involves anaerobic treatment using up-flow anaerobic sludge blanket reactor (UASBR) or continuous stirred tank reactor (CSTR). The anaerobic waste water treatment method is an effective method for the handling of many organic wastes. In Primary clarifier, effluent coming from first aeration tank along with biomass (MLSS) gets settled. In secondary clarifier, effluent coming from second aeration tank along with biomass (MLSS) gets settled. During tertiary treatment, the raw water is first transferred through a multigrade sand filter to minimize the suspended solids present in the treated water. Activated carbon filter is used for minimization of undesired color and odor. Finally, ultrafiltration is employed to get the desired quality of processed water.

43.5.4 Sugar Beet to Ethanol

In India, sugarcane is the main crop grown for processing of sugar. However, sugar beet has a vital role in decreasing the manufacturing cost, reducing crop period, and sustaining higher crop productivity under water and salt stresses. This is mainly due to short growth cycle (around 5 months) in comparison with sugarcane (10–12 months), lower water requirement (about 1/3 to 1/2 to grow than sugar cane), more sugar concentration (15–17%), high sugar recovery (12–14%), high purity (85–90%), and ability to withstand drought and tolerance to salinity (Zheng et al. 2013; Finkenstadt 2014; Rezbova et al. 2013). Its average biomass yield is found to be ranging from 40 to 100 MT per hectare (Panella and Kaffka 2010). Sugar beet is a temperate region crop, but due to the availability of new resistant varieties, it is now emerging as the potential cash crop for tropics and subtropics (Gumienna et al. 2016). Several varieties of sugar beet are available, but varieties such as HI0064, Dorotea, and Posada are suitable for cultivation in India. Various varieties of sugar beet and yield are shown in Table 43.5 (VSI experimental data). The major sugar in sugar beet is sucrose with little amounts of other carbohydrates and pectin (Saulnier and Thibault 1999). Sugar concentration in beet molasses is about 50% and the major constituent is sucrose (Dong et al. 2008). The typical composition of raw sugar beet juice is given in Table 43.6. For obtaining highest yield of sugar beet,

Table 43.5 Varieties and yields of sugar beet (VSI experimental data)

S. no.	Varieties	Root weight (MT/ha)	MT/Acre
1.	H10064	110.70	44.30
2.	Dorotea	107.64	43.05
3.	Posada	106.11	42.44

Table 43.6 Composition of raw sugar beet juice

S. no.	Particulars	Composition % (wet basis)
1.	Water	75.0
2.	Sucrose	17.5
3.	Nitrogenous (amino acids, betaine, other)	1.1
4.	Non nitrogenous (Glucose, fructose, raffinose, other)	0.9
5.	Minerals (K, Na, Ca, Mg, SO_4, PO_4, others)	0.3
6.	Other	0.2
7.	Insoluble solids (Pectin, Celluolose, Hemicellulose, Protein, Saponins, Minerals)	5.0
8.	Pectin	2.4
9.	Cellulose	1.2
10.	Hemicellulose	1.1
11.	Protein	0.1
12.	Saponins	0.1
13.	Minerals	0.1

factors such as breeding, variety of the roots, and the levels of phosphorous, potassium, and sodium in the soil are important (Gumienna et al. 2016).

Sugar from sugar beet is extracted through different processes like leaching, boiling, pressing, and crystallization (Fares et al. 2003; Belitz et al. 2009). The root is processed for obtaining raw juice. It includes several other substances which obstruct the crystallization process of sugar and reduces the crystal's purity that needs more purification steps before its crystallization (Pezzi 2011). Juice purification process removes only a part of non-sugar contents of the sugar juice (proteins, pectins, inorganic salts, and coloring substances). The usual purification management technique of raw juice is called calco-carbonic technique (Jarski et al. 2012). This process involves addition of concentrated lime ($Ca[OH]_2$) water suspension which increases the juice pH and precipitates several organic acids as calcium salts. At this alkaline pH, a major portion of proteins and pectin are also precipitated (Minarovicova et al. 2007; Sarka et al. 2015). A clear juice free from impurities called thin juice containing 14–16% of dry matter is obtained after the carbonation process. For thick juice production, it is concentrated up to 60% (w/w) sugar content using multiple effect evaporators (Cubero et al. 2004). Using multi-effect evaporation technique, the thin juice containing 14–16% of dry matter is concentrated to get thick juice having 60–75% of dry matter. In comparison to thin juice, it is highly stable and can be stored (at lower temperature) for prolonged periods. The final step of crystallization is attempted in crystallizers, which is induced by seeding very small sugar crystals into the thick juice, while the excess quantity of water is removed under vacuum (Vaccari et al. 1996, 2002; Rezbova et al. 2013; Lu et al. 2017). The residual liquid part is called molasses which contains about 50% (w/w) sugars, other oligosaccharides, organic acid salts. One ton of sugar beet gives about 160 kg of sugar, 500 kg of wet pulp along with 38 kg of molasses (FAO 2009).

The raw sugar beet juice as well as molasses is used as feedstock for alcohol production, and pulp residue can be useful in the form of an animal feed (Marzo et al. 2019). The pH of the raw sugar beet juice can be adjusted and directly fermented exclusive of any purification step as well as without any nutrients additions, by adding yeast for the production of bioethanol and other bioproducts (Dodic et al. 2009; Vargas-Ramirez et al. 2013). Sugar beet pulp is the major by-product of sugar refining industry which is sold as animal feed at a somewhat low cost (Gumienna et al. 2016; Zheng et al. 2013). Drying, pelletizing, and transporting are the various processes to be carried out on beet pulp (Rorick et al. 2009). The residue generated after pressing step of beet pulp is processed to make animal feed. It is also dehydrated as well as granulated in the form of pellets and then sold as animal feed. The process flow diagram of the production of ethanol using sugar beet juice/syrup as well as animal feed production from sugar beet pulp is given in Fig. 4. Pulp is also to be converted into either fuel ethanol (Rezic et al. 2013; Bellido et al. 2015) or biogas production (Ziemínski et al. 2014*). The average composition of pulp of sugar beet is given in Table 43.7. Sugar content in sugar beet pulp is available in complex polysaccharide structures, hence either chemical or enzymatic breakdown is necessary (Bonnin et al. 2000; Kuhnel et al. 2011). The pulp of sugar beet contains 20–25% cellulose, 25–36% hemicellulose (mainly arabinans), 20–25% pectin,

Table 43.7 Composition of sugar beet pulp

S. no.	Components	Dry weight %
1.	Carbohydrates	68
2.	Glucose	22
3.	Arabinose	18
4.	Uronic acids	18
5.	Galactose	5
6.	Rhamnose	2
7.	Xylose	2
8.	Mannose	1
9.	Saccharose (Residual)	4
10.	Ester-linked substituents of polysaccharides	0.5
11.	Ferulic acid	1.6
12.	Acetic acid	0.4
13.	Methanol	8.0

10–15% protein, and 1–2% lignin (dry weight basis) (Bellido et al. 2015). However, the main disadvantage of using pulp to bioethanol process is the requirement of higher concentration of dose of hydrolytic enzymes required to convert cellulose, hemicelluloses, and pectin into their constituent sugars, adding a significant manufacturing cost towards the production of ethanol. After hydrolysis of hemicellulose, an important content of pentoses (especially arabinose and xylose) are produced, which are not metabolized directly by *S. cerevisiae* (Diaz et al. 2017; Zhong et al. 2015).

The raw sugar beet produced using a beet washing machine, cossette maker, diffuser, etc. can be clarified in the existing sugarcane juice clarification system and further concentrated to syrup in the existing sugar mill boiling house. The concentrated syrup can be used in the existing distillery unit for the production of alcohol. This will allow increasing the capacity utilization of the existing sugar mill with attached distillery unit.

43.6 Future Prospects

As per the current mandate of the Government of India, blending of ethanol in petrol has become the talk of the town and targets of ethanol blending of 10% up to 2022 with a further rise of 10% up to 2025 in petrol. Oil Marketing Companies (OMCs) in India intend to achieve 10.0% blending in the year 2022, for which they have provided attractive differential pricing for ethanol made from sugarcane juice, syrup, or sugar, B heavy molasses, C molasses, rice, and other cereals. In this respect, sugarcane crop has been the most exploited due to higher ethanol production. However, the application of sugar beet for ethanol production purposes will turn the table to the other side due to its higher ethanol content and other benefits compared to sugarcane (short life span and fewer water needs). The Department of

Food and Public Distribution (DFPD) of the Ministry of Food and Civil Supplies has also offered financial assistance through an interest subvention program for the establishment of new distilleries or the modernization/expansion of existing distilleries. Under the Indian context, sugar beet can be considered to increase the crushing period of the existing sugar mill with an attached distillery. Government interest and financial support through different schemes will help in the sugar beet foundation in India.

43.7 Conclusion

At present, no fuel ethanol is commercially produced using sugar beet in India. The idea behind this chapter is to evaluate the feasibility of producing ethanol from sugar beet using the existing sugar mill with an attached distillery and SW disposal system. It can be considered as an additional crop in line with sugarcane for the production of sugar as well as ethanol. The ethanol production using sugar beet and various types of intermediates as well as molasses can be an alternative solution for sugar factories having interest in the production of a combination of sugar and bioethanol. Using intermediate products of sugar beet processing could also be attractive for distilleries as the raw materials for bioethanol production, which would definitely minimize high transportation costs. The use of appropriate fermentation techniques, such as either fed-batch or repeated batch, can significantly increase the yield of ethanol. It will ultimately improve the performance as well as the economics of the whole process. If sugar beet juice can be used for the production of ethanol, there will be also a reduction in effluent generation. Overall, it will ultimately improve the sugar industry's economics. Thus, due to its close proximity to the sugar beet supply market, cool climate, and already existing processing infrastructure, India has good prospects to produce sugar beet-based fuel ethanol.

References

Al-Asheh S, Banat F, Al-Lagtah N (2004) Separation of ethanol-water mixtures using molecular sieves and biobased adsorbents. Chem Eng Res Des 82(7):855–864

Alexiades A, Kendall A, Winans KS, Kaffka SR (2018) Sugar beet ethanol (*Beta vulgaris* L.): a promising low-carbon pathway for ethanol production in California. J Clean Prod 172:3907–3917

Bailey BK (2018) Performance of ethanol as a transportation fuel. In: Handbook on bioethanol. Routledge, Abingdon, pp 37–60

Barriga EJC, Libkind D, Briones AI, Iranzo JU, Portero P, Roberts I, James S, Morais PB, Rosa CA (2011) Yeasts biodiversity and its significance: case studies in natural and human-related environments, ex situ preservation, applications and challenges. In: Grillo O, Venora G (eds) Changing diversity in changing environment, pp 55–86

Belitz HD, Grosch W, Schieberle P (2009) Springer food chemistry 4th revised and extended edition. Annu Rev Biochem 79:655–681

Bellido C, Infante C, Coca M, González-Benito G, Lucas S, García-Cubero MT (2015) Efficient acetone–butanol–ethanol production by *Clostridium beijerinckii* from sugar beet pulp. Bioresour Technol 190:332–338

Boekhout T, Kurtzman CP (1996) Principles and methods used in yeast classification, and an overview of currently accepted yeast genera. In: Nonconventional yeasts in biotechnology. Springer, Berlin, pp 1–81

Bonnin E, Grange H, Lesage-Meessen L, Asther M, Thibault JF (2000) Enzymic release of cellobiose from sugar beet pulp, and its use to favour vanillin production in Pycnoporus cinnabarinus from vanillic acid. Carbohydr Polym 41(2):143–151

Christoph B, Licht FO (2004) World fuel ethanol analysis and outlook. FO Licht, England

Cubero MG, Sanz MC, Benito GG, Puerta MGD (2004) Distribution of colorants in the evaporation and first crystallization stages. Zuckerindustrie 129(9):660–666

Dhamija SS, Bhaskar D, Gera R (1996) Thermotolerance and ethanol production are at variance in mutants of *Schizosaccharomyces pombe*. Biotechnol Lett 18(11):1341–1344

Diaz AB, Marzo C, Caro I, De Ory I, Blandino A (2017) Valorization of exhausted sugar beet cossettes by successive hydrolysis and two fermentations for the production of bio-products. Bioresour Technol 225:225–233

Dodic S, Popov S, Dodic J, Rankovic J, Zavargo Z, Mucibabic RJ (2009) Bioethanol production from thick juice as intermediate of sugar beet processing. Biomass Bioenergy 33(5):822–827

Dombek KM, Ingram LO (1987) Ethanol production during batch fermentation with Saccharomyces cerevisiae: changes in glycolytic enzymes and internal pH. Appl Environ Microbiol 53(6): 1286–1291

Dong AJ, Zhang YC, Yang X, Zhang H, Ma Y (2008) Determination of glucose, sucrose and raffinose in beet molasses by high performance liquid chromatography. Sci Technol Food Indus 29:245–247

FAO, Sugar Beet/White Sugar (2009) FAOSTAT. Food and Agriculture Organization of the United Nations

Fares K, Rzina Q, El Amrani M, Crepeau MJ, Thibault JF (2003) Extraction and degradation of polysaccharides from sugar beets under factory conditions. Zuckerindustrie 128(4):243–249

Finkenstadt VL (2014) A review on the complete utilization of the sugar beet. Sugar Tech 16(4): 339–346

FO Licht's World Ethanol and Biofuels Report, 18(4) 2019

Gomez-Pastor R, Perez-Torrado R, Garre E, Matallana E (2011) Recent advances in yeast biomass production. In: Biomass—detection, production and usage. InTech, Rijeka

Gumienna M, Szwengiel A, Szczepańska-Alvarez A, Szambelan K, Lasik-Kurdyś M, Czarnecki Z, Sitarski A (2016) The impact of sugar beet varieties and cultivation conditions on ethanol productivity. Biomass Bioenergy 85:228–234

Gyamerah M, Glover J (1996) Production of ethanol by continuous fermentation and liquid–liquid extraction. J Chem Technol Biotechnol 66(2):145–152

Hough JS, Briggs DE, Stevens R, Young TW (1982) Malting and brewing science: Hopped Wort and Beer. Springer, New York, p 528

Icoz E, Tugrul KM, Saral A, Icoz E (2009) Research on ethanol production and use from sugar beet in Turkey. Biomass Bioenergy 33(1):1–7

Jarski HR, McGillivray TD, Groom DR (2012) A comparison of anion concentrations across carbonation at the five valley factories of American Crystal Sugar Company. Int Sugar J 114(1359):156

Khandekar YS, Shinkar NP (2020) An overview of distillery spent wash treatment techniques. J Innov Res Adv Eng 7:248–251

Kuhnel S, Schols HA, Gruppen H (2011) Aiming for the complete utilization of sugar-beet pulp: examination of the effects of mild acid and hydrothermal pretreatment followed by enzymatic digestion. Biotechnol Biofuels 4(1):14

Li Z, Dewan A, Karim MN (2012) Optimization of bioethanol ethanol production in fed-batch fermentation. IFAC Proceedings 45(15):816–821

Lin Y, Tanaka S (2006) Ethanol fermentation from biomass resources: current state and prospects. Appl Microbiol Biotechnol 69(6):627–642
Lu Y, Thomas L, Schmidt S (2017) Differences in the thermal behavior of beet and cane sucrose sources. J Food Eng 201:57–70
Madson PW (2003) Ethanol distillation: the fundamentals. The alcohol textbook, p 4
Marzo C, Díaz AB, Caro I, Blandino A (2019) Status and perspectives in bioethanol production from sugar beet. In: Ray RC, Ramachandran S (eds) Bioethanol production from food crops: sustainable sources, interventions, and challenges. Academic Press, pp 61–79
Maung TA, Gustafson CR (2011) The economic feasibility of sugar beet biofuel production in central North Dakota. Biomass Bioenergy 35(9):3737–3747
Minarovicova L, Dandar A, Jancurova M, Schneiderova R (2007) Reduction of lime in the process of purification. Listy Cukrovarnicke A Reparske 123(1):27–30
Mohana S, Acharya BK, Madamwar D (2009) Distillery spent wash: treatment technologies and potential applications. J Hazard Mater 163(1):12–25
Ogbonna JC (2004) Alcoholic fermentation. Fuel ethanol production from renewable biomass resources. Concise encyclopedia of bioresource technology, pp 346–362
Panella L, Kaffka SR (2010) Sugar beet (*Beta vulgaris* L) as a biofuel feedstock in the United States. In: Sustainability of the sugar and sugar—Ethanol industries. American Chemical Society, pp 163–175
Pathade GR (1999) A review of current technologies for distillery wastewater treatment. In: Advances in industrial wastewater treatment. Technoscience Publications, Rajasthan, pp 180–239
Pezzi G (2011) Considerations of the technological quality assessment of sugar beet. Sugar Industry 136(2):85–89
Phaff HJ, Stammer WT (1987) Yeasts associated with plants, insects and soil. In: Rose AH, Harrison JS (eds) The yeasts, vol 1. Academic Press, pp 123–179
Qazi GN (2014) Opportunities for green chemistry initiatives: molasses-based distilleries
Rajaeifar MA, Hemayati SS, Tabatabaei M, Aghbashlo M, Mahmoudi SB (2019) A review on beet sugar industry with a focus on implementation of waste-to-energy strategy for power supply. Renew Sust Energ Rev 103:423–442
Rezbova H, Belova A, Skubna O (2013) Sugar beet production in the European Union and their future trends. Agris on-line Papers in Economics and Informatics 5(665–2016-44967):165–178
Rezic T, Oros D, Markovic I, Kracher D, Ludwig R, Santek B (2013) Integrated hydrolyzation and fermentation of sugar beet pulp to bioethanol. J Microbiol Biotechnol 23(9):1244–1252
Robinson CS, Gilliland ER (1950) Elements of fractional distillation. McGraw Hill Book, p 492
Rorick RE, Nahar N, Pryor S (2009) Enzymatic hydrolysis and fermentation of sugar beet pulp. In 2009 Reno, Nevada, June 21–June 24, 2009. American Society of Agricultural and Biological Engineers, p 1
Sarka E, Bubnik Z, Pour V (2015) Carbonatation precipitate-origin, separation and uses. Sugar Industry 140(7):434–442
Satyawali Y, Balakrishnan M (2008) Wastewater treatment in molasses-based alcohol distilleries for COD and color removal: a review. J Environ Manag 86(3):481–497
Saulnier L, Thibault JF (1999) Ferulic acid and diferulic acids as components of sugarbeet pectins and maize bran heteroxylans. J Sci Food Agric 79(3):396–402
Shinde PA, Ukarde TM, Pandey PH, Pawar HS (2020) Distillery spent wash: an emerging chemical pool for next generation sustainable distilleries. J Water Process Engineering 36:101353
Shukla PR, Dhar S (2016) India's GHG emission reduction and sustainable development. In: Enabling Asia to stabilize the climate. Springer, Singapore, pp 41–54
Vaccari G, Mantovani G, Sgualdino G, Dosi E (1996) Cooling crystallization of raw juice: laboratory investigations of sucrose crystal growth kinetics. Zuckerindustrie 121(2):111–117
Vaccari G, Wawro P, Tamburini E, Sgualdino G, Bernardi T (2002) Cooling crystallization of microfiltered raw juice and of traditional thick juice: a comparison. Sug Ind 127(1):22–28

Van Baelen G, Vreysen S, Gerbaud V, Rodríguez-Donis I, Geens J, Janssens B (2010) Isopropyl alcohol recovery by heteroazeotropic batch distillation. Process technology and sustainability. In: European Meeting on Chemical Industry and Environment EMChiE, 17–19 May 2010, Mechelen, Belgium, pp 979–986

Vargas-Ramirez JM, Haagenson DM, Pryor SW, Wiesenborn DP (2013) Determination of suitable storage conditions to preserve fermentable sugars in raw thick beet juice for ethanol production. Biomass Bioenergy 59:362–369

Vohra M, Manwar J, Manmode R, Padgilwar S, Patil S (2014) Bioethanol production: feedstock and current technologies. J Environ Chem Eng 2(1):573–584

Walker GM (1998) Yeast physiology and biotechnology. Wiley, Hoboken, p 362

Walker GM, White NA (2017) Introduction to fungal physiology. Fungi: biology and applications, pp 1–35

Zheng Y, Lee C, Yu C, Cheng YS, Zhang R, Jenkins BM, VanderGheynst JS (2013) Dilute acid pretreatment and fermentation of sugar beet pulp to ethanol. Appl Energy 105:1–7

Zhong C, Wang C, Huang F, Wang F, Jia H, Zhou H, Wei P (2015) Selective hydrolysis of hemicellulose from wheat straw by a nanoscale solid acid catalyst. Carbohydr Polym 131:384–391

Sugar Beet Molasses Production and Utilization

44

Anuja Rameshchand Adbhai, Sujaya Dewanjee, Kamlesh G. Patel, and Nilima Karmakar

Abstract

Molasses is mostly produced from sugarcane or sugar beet in the world. The sugar content of molasses mainly comprises sucrose and inverted sugar. The inverted sugars in sugar beet molasses do not exceed 1%. Basically, there are two types of molasses, i.e., soft and hard molasses, depending upon the concentration of the soluble salts of Ca^{++} and Mg^{++} which impart the hardness to the molasses. Beet molasses is proven to be very nutritious as it contains a pretty good amount of essential amino acids, different minerals and betaine. Industrial production of sugar beet molasses follows the steps like extraction, beet juice purification, concentration, and crystallization. Molasses desugaring by chromatographic process (MDC) is a recent method for the separation of sugar from the molasses. Beet molasses can be utilized in several ways, like human food, beverage, pharmaceutical, etc., but the major use of beet molasses lies as cattle feed. Future research and development programs should focus on finding alternative use of sugarcane as well as making government policies and incentives for sugar beet industries.

Keywords

Amino acids · Betaine · Cattle feed · Desugaring · Molasses

A. R. Adbhai
Vasantdada Sugar Institute, Pune, Maharashtra, India

S. Dewanjee
RRSS, Bidhan Chandra Krishi Viswavidyalaya, Purulia, West Bengal, India

K. G. Patel · N. Karmakar (✉)
Department of Soil Science and Agricultural Chemistry, NMCA, Navsari Agricultural University, Navsari, Gujarat, India

V. Misra et al. (eds.), *Sugar Beet Cultivation, Management and Processing*, https://doi.org/10.1007/978-981-19-2730-0_44

Abbreviations

CMS	Concentrated molasses solids
CSB	Concentrated separated by-product
MDC	Molasses desugaring by chromatographic process
PHA	Polyhydroxyalkanoates

44.1 Introduction

The principal source of molasses is either sugarcane or sugar beet. Sugar beet molasses can be defined as non-crystallized syrup, which can be obtained as the by-product of sugar. There are various grades of molasses that may be obtained clarifying, extracting, and/or concentrating the raw sugar from beet juice in the sugar factory. The sugar beet molasses is a dark colored viscous substance, which has a caramel flavor and sweet taste. It is variously used as all types of livestock feed starting from cattle to poultry. This molasses is a great appetizer and a source of energy as well. It functions as a binder in compound feeds and is also rich in non-protein nitrogen like urea (CNC 2002). Beet molasses has a significant role in production of ethanol (EU Beet Sugar Sustainability Partnership 2019). The countries of the European Union exploit 70% of sugar beet molasses to serve the purpose of ethanol production. In the USA, this molasses is used along with chloride salt for de-icing the roads as it is a more environment-friendly process (Midwest Agri 2019). The leftover beet pulp after the production of sugar followed by molasses consists of a significant amount of cell wall materials, which is predominantly polysaccharide in nature, including pectin and dietary fibers that are processed by dehydration and are sold as animal feed in the market. The more the use of co-products, the less is waste reduction and value is added to the product as well. The sugar content of molasses mainly comprises sucrose and inverted sugar. Compared to cane molasses, the beet molasses contains a higher amount of sucrose and a lesser amount of inverted sugar not exceeding 1%. The protein content is almost double in sugar beet molasses compared to cane molasses. Molasses is usually heavier than water as their specific gravity ranged between 1.35 and 1.45 and, under normal conditions, molasses can be stored for longer periods as the sugar content in it leads to high osmotic pressure and hinders the development of bacteria and other microorganisms.

44.2 Quality of Beet Molasses

Beet molasses must contain 48% and or more sugar, which mainly consists of sucrose and inverted sugar. Market available sugar beet molasses also contains 79.5% Dry solid.

In beet sugar factories, the desugaring of beet molasses is generally performed by chromatographic method, by which they produce different by-products, namely, Raffinate (also known as concentrated molasses solids or CMS) and Betaine. Raffinate is basically a by-product containing mineral along with a lower amount of sucrose having a density of 1300 kg/m^3. Raffinate is mainly used as a component of the animal feed industry. On the other hand, Betaine is one of the nitrogenous compounds which contributes 5% of the DS (dry substances) and density with 1250 kg/m^3 in beet molasses. During the desugaring of molasses, a pretty good amount of it is recovered and is sold in the form of liquid feed supplement for the animals. Betaine is a product that has a multipurpose use in the pharmaceuticals, cosmetic, and fermentation industries.

The mother liquor or molasses, which may be called runoff syrup, is obtained from the final stage of crystallization. This must contain the following characteristics:

1. Is produced at about 5% on beet
2. Has a high concentration (about 80%)
3. Has a high amount of sugar (about 50%)
4. Has about 15% of sugar coming with the beets
5. Has about 80% of sugar loss of the beet-sugar factory
6. Has as its most valuable components sugar, betaine, and amino acids
7. Can be desugared by the MDC process to recover sugar, betaine, and minerals
8. Can be used in animal feed, yeast, citric acid, alcohol, and pharmaceutical industries.

However non-food grade beet molasses (Fig. 44.1) contains sucrose (46–52%), Ash (10–12%), protein (8.0–10.0%), betaine (4–6%), water (18–20%) (NOVUS 1996).

44.2.1 Soft Molasses and Hard Molasses

Soft molasses are basically devoid of soluble salts of Ca++ and Mg++, lime salts which contribute to the hardness. Technically, it can be said that the hardness content of soft molasses is <3 *mEq/100 DS* (milliequivalent per 100 g of dry substance). Soft molasses can be prepared by two different methods, namely, thin-juice softening and molasses softening. Hard molasses has a lime salt content of more than 3 *mEq/100 DS*. Usually, hard molasses has 5–40 *mEq/100 DS* hardness in the thin juice.

44.3 Chemical Components of Sugar Beet Molasses

The major sugar content in beet molasses is sucrose as mentioned, but along with these, various types of sugars are present in the beet molasses like, glucose, fructose, raffinose, and some other oligo- or polysaccharides as minor sugars. The

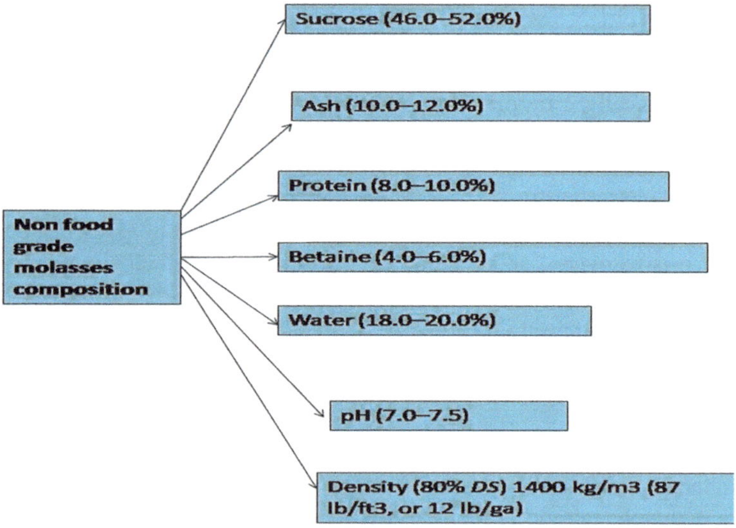

Fig. 44.1 Quality standards for nonfood-grade molasses (approximate)

concentration of these minor sugars is lesser than 1% and mostly depends on the manufacturing and refinery process. In case of major chemical composition (Fig. 44.2), sugar beet contains crude ash (6.6–10%), crude protein (6.6–11%), and the main sugar, i.e., sucrose (43–50.5%) (NOVUS 1996). Different types of amino acids are also present in beet molasses (Fig. 44.3), of which the predominant is lysine (46–52%) (NOVUS 1996). Betaine is the main nitrogenous component (Fig. 44.4) in beet molasses (NOVUS 1996). Potassium is the major mineral content present in molasses, followed by sodium, calcium, and magnesium (Fig. 44.5). The mineral content in the molasses basically depends on the soil type and moisture content prevailing during sugar beet cultivation. Moreover, the amount of calcium

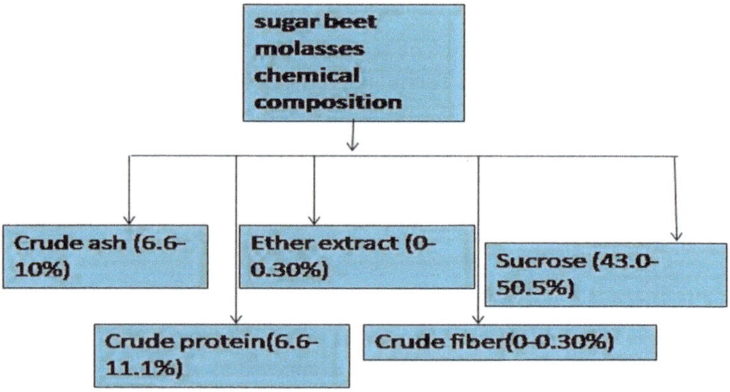

Fig. 44.2 Chemical composition of sugar beet molasses

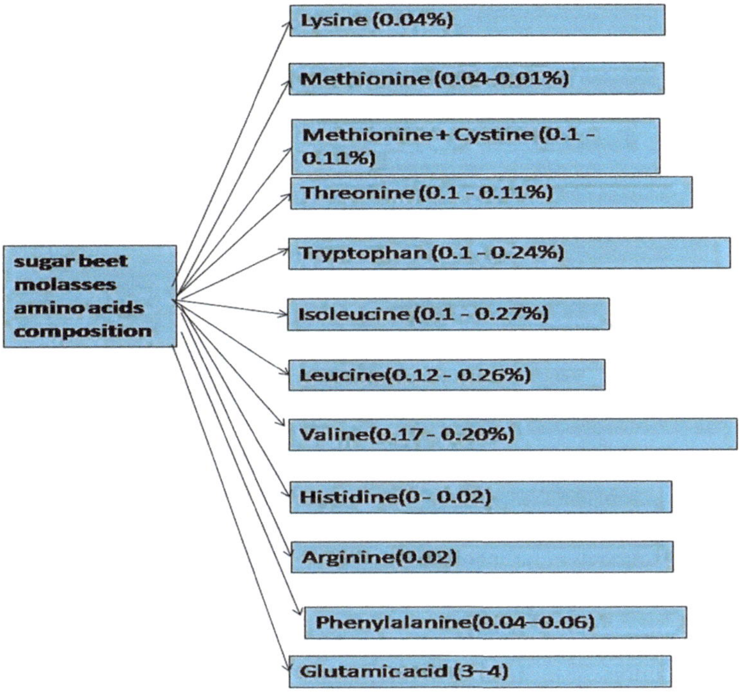

Fig. 44.3 Amino acid composition in sugar beet molasses

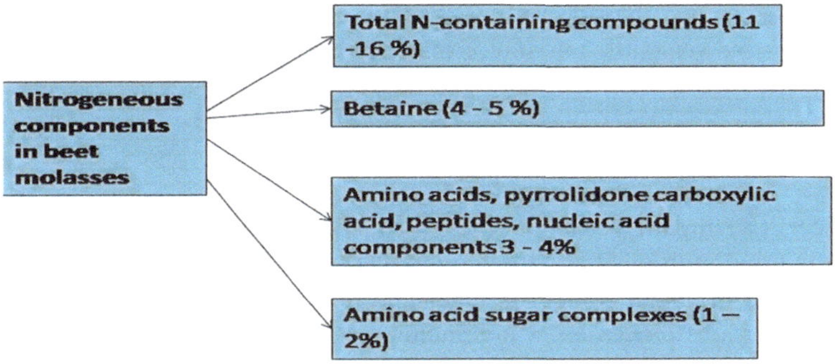

Fig. 44.4 Contents of nitrogen-containing organic compounds in beet molasses

and sodium content in molasses is partly contributed by the processing practices. The non-sucrose organic matter is about 20% of the total biomass, which mainly consists of betaine. Besides these substances, beet molasses contains free and conjugated amino acids and pyrrolidone carboxylic acid, which are basically a conversion product of the amino acid glutamine. Alteration of most of the amino acids during processing leads to lowering of the amino acid amount in the molasses

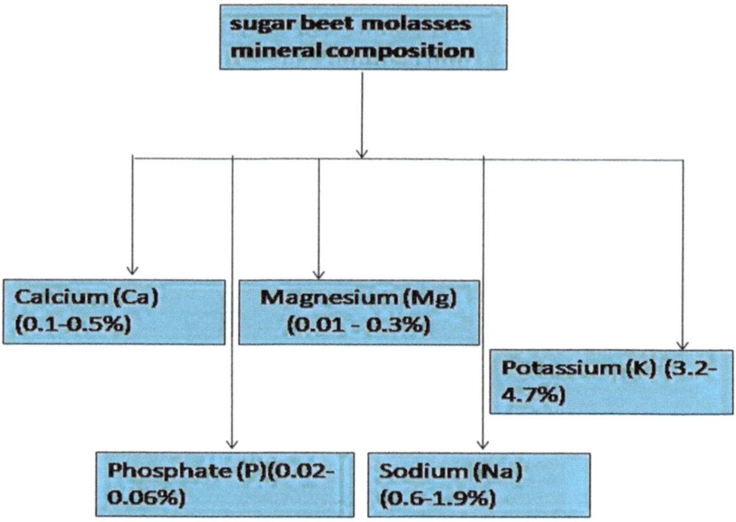

Fig. 44.5 Major minerals in sugar beet molasses

compared to beet juice (Reinefeld et al. 1982a, b; Schiweck et al. 1993). Chloride is the chief anion present in sugar beet molasses (NOVUS 1996). The most prevalent organic acid is lactic acid (4%) which comes from the degradation of invert sugar (1.7%) present in beet juice during processing. The organic acids present in little amounts besides lactic is malic, citric, fumaric, and oxalic acids. The anions present in beet molasses are chloride, sulfate, nitrate, and traces of phosphate and nitrite, while it contains very little amount of trace elements but without iron.

44.3.1 Production of Sugar Beet Molasses

Sugar beet is a temperate crop, but there are some varieties developed for resistance which can grow in tropical and tropic regions as a cash crop. Beetroots of the sugar beet crops are the main source of sugar and this sugar can be extracted in liquid solution by various processes, which include leaching, boiling of beetroot juice, or pressing from pulp. After these processes, sugar can be converted into the crystalline form or during boiling process. After the sugar is extracted, it is tested in food labs mainly for sucrose content (Belitz et al. 2014).

Sugar beet *(Beta vulgaris)* is grown commercially or as an industrial crop for its main constituent, which is sucrose present in the root part. Beetroot is not only useful for sugar production but also gives various by-products such as pulp residues, molasses, and greens as animal feed or for the production of fibers or alcohol production (Bonnina et al. 2012).

One of the important by-products from sugar extraction is molasses; it is a dark brown to black colored, viscous running syrup which is the leftover after extraction and crystallization of sugar from the raw pulp. Molasses was the first form of sugar or say sweetener used for consumption by humans, used by the poor population because of its cheap price as compared to honey or refined sugars. It is mainly used for the production of ethanol (recently for biofuel production) by fermentation technology, or as animal feed. Other than this, molasses can be useful in various biotechnological processes such as an additive in veterinary, as an activator in order to eliminate toxic microbial products from waste water or soil, as a raw material for the production of nitrogen containing compounds like amino acids, betaine, etc. In the traditional sugar extraction method where ethanol is not produced through fermentation, sugar recovery should be minimal in molasses (Sarka et al. 2012).

44.3.1.1 Industrial Process Processing of Sugar Beet

Sugar beets are harvested from the field and then transported to the sugar processing plant for the production of sugar by using preexisting procedures (Rezbova et al. 2015).

Extraction

Washing of beetroots to remove soil and then cutting into slices. From sugar beet slices, raw juice and pulp are obtained and this raw juice remains thermally stable up to the temperature of around 85 °C. After extraction of juice, the remaining pulp of beetroot can be used for cattle feed or it can be used to extract fibers for human consumption.

Beet Juice Purification

Raw juice of beetroot contains non-sucrose fraction too, which gets removed during the purification process, yielding high recovery of crystalline sucrose with improved quality like color of sugar, odor, and taste. Raw juice contains sucrose, non-sucrose compounds like proteins, pectins, inorganic and organic coloring substances, and water. So, to remove unwanted material, the most commonly used purification method of raw juice is the addition of lime and CO_2. Sugar beets are first diffused in hot water to get raw juice, followed by first carbonation, in which raw juice is heated to high temperature, then, at high temperature, water with milk of lime and calcium oxide suspension is added. During this process, lime removes non-sucrose compounds and colors or it can be removed by absorption by calcium carbonate. CaO and CaO_2 are produced by heating calcium carbonate, i.e., lime rock at high temperatures. In the first carbonation method, CaO and CaO2 recombine, forming calcium carbonate mud. This mud is removed either by appropriate filters or by setting clarifiers. Again, second carbonation step is followed in order to remove remaining traces of calcium precipitates (lime salts) which can hinder the evaporation process. Second carbonated juice needs to be filtered to remove these precipitations of lime salt.

In different industries, different methods of carbonation and liming are being used for the purification, some of them are as follows:

(a) Pre-liming—In this process, controlled addition of lime occurs before carbonation process where some of the non-sucrose is precipitated, which could be removed in the subsequent process.
(b) Main liming—This process involves addition of lime juice at high temperature and high pH, destroying some non-sucrose and making the juice more thermostable. Main liming is conducted before first carbonation.
(c) Defeco-carbonation—This process involves the continuous addition of CO_2 and lime.
(d) Adjustable processes—This process focuses on the amount of lime to be added and temperature. Also, some other chemicals are added (sulfur dioxide or soda ash) in order to make the process more adaptable to changing conditions (Anonymous 1995).

Beet Juice Concentration
This treated juice is known as "thin juice" containing non-removable non-sucrose (because of their different chemical nature), of which only 20–30% of the non-sucrose gets removed by carbonation treatment. Thin juice contains about 10–15% of the solids, so it is further processed by multi-effect evaporators for the concentration process, in which solid content increases by 60–70%, and this is known as "thick juice."

Crystallization
It is the process of converting thick juice in to crystalline form which is the marketable product and referred to as "sugar." Evaporation of water from thick juice leads to the formation of crystals; these crystals are separated from syrup by centrifugation. During the crystallization process, not all sucrose from thick juice is converted into sugar and it gets lost in the form of molasses. And this is due to the carbonation and liming process in which not all non-sucrose is removed, which further hinders the crystallization of recovery of sugar. This is the actual step where molasses are produced. Molasses are the end by-product in the process of sugar production by repetitive evaporation, crystallization, and centrifugation of beet juice. Non-crystallized syrup here is known as "beet molasses." It is a dark colored syrup from which sugar (sucrose) can't be recovered in crystal form; it consists of about 50% of sugar by weight (Bonnina et al. 2012).

Molasses are further fermented to produce alcohol and the remaining solid extract is a rich source of nitrogen, so it can be used either as fertilizer or as animal feed. The remaining parts of beet pulp are rich with cell wall polysaccharides like pectin and dietary fibers, so it is pressed and used to feed animals. In Fig. 44.6, procedure of sugar production is given, with molasses being produced as a by-product.

44.3.2 Prospect of Production of Beet Molasses

Beet molasses is the by-product in the form of syrup yielded after the crystallization of sugar from thick juice extracted from the roots of sugar beets (*Beta vulgaris* L.).

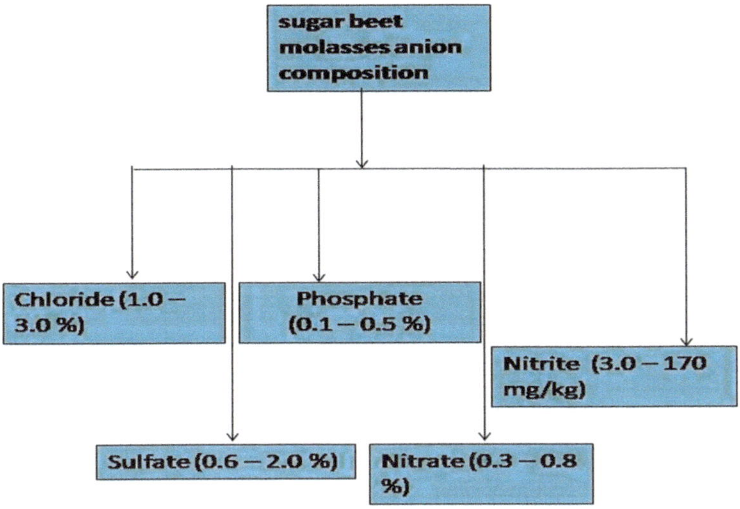

Fig. 44.6 Contents of major anions in beet molasses

Molasses is viscous, dark, sweet, sugar-rich, with a caramel flavor. The production process of Beet sugar molasses and its desugaring chromatographic process are depicted in Figs. 44.7 and 44.8. Just like the sugarcane molasses, it is very palatable to livestock. It is used as a major feed ingredient for all types of livestock, including poultry. Beet molasses is used as an appetizer, as a binder in various compound feeds and energy sources. Beet molasses are a rich source of various polysaccharides, nitrogen, and some part of sucrose also, so they can be used in the production of various products.

1. In the USA, beet molasses are added to chloride salt used for de-icing roads, making the process more environmentally friendly (Heuze et al. 2019).
2. Nikulina et al. (2021) have studied the prospects of using beet molasses in the production of synthetic rubbers. In order to reduce pressure on the environment, they have used beet molasses together with sulfuric acid to improve the coagulation system for the separation of rubber from latex. By using beet molasses, they have produced styrene-butadiene rubber with the brand of SKS-30 ARK, in which they have reduced consumption of coagulating agents by 2–3 times, which saves energy and is also a less harmful process to the environment.
3. Beet molasses can also be useful in the production of the first and second generations of bioethanol. Dilute acid pretreated waste distillery stillage and beet molasses with the cellulose enzyme in the fermentation processes can be successfully used for the production of biofuel. Beet molasses with distillery stillage has shown to increase production of aldehydes, methanol, l-propanol, and l-butanol, but has reduced higher alcohols such as 2-methyl-l-bytanol, isobutanol,

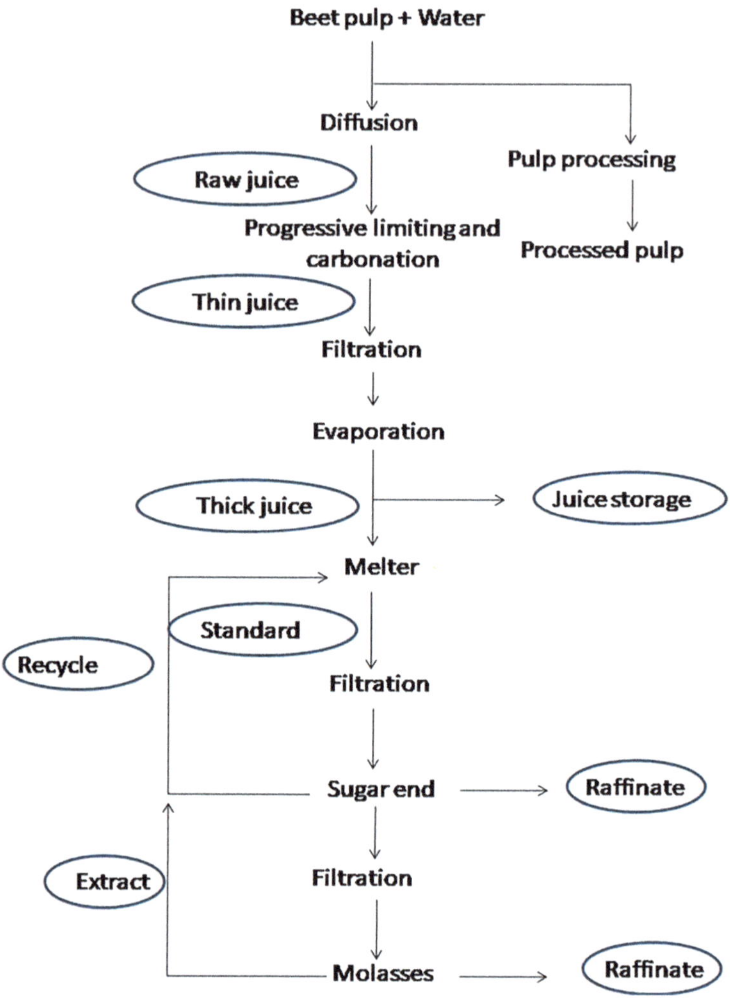

Fig. 44.7 Production process of beet sugar

etc., concluding that enzymatic hydrolysis of the raw material needs to be optimized (Mikulski and Klosowski 2021).

4. Beet molasses has great scope in increasing the production of bacterial cellulose. Beet molasses contains some sugars which are considered as a cheap and renewable source of energy; also beet molasses contains some proteins, organic nitrogen, and some amount of sulfur. So, due to the presence of these compounds, beet molasses has been shown to increase the yield of pellicle (bacterial cellulose productivity and degree of polymerization) by *Gluconacetobacter xylinus* when HS media was added with molasses (Sherif et al. 2006).

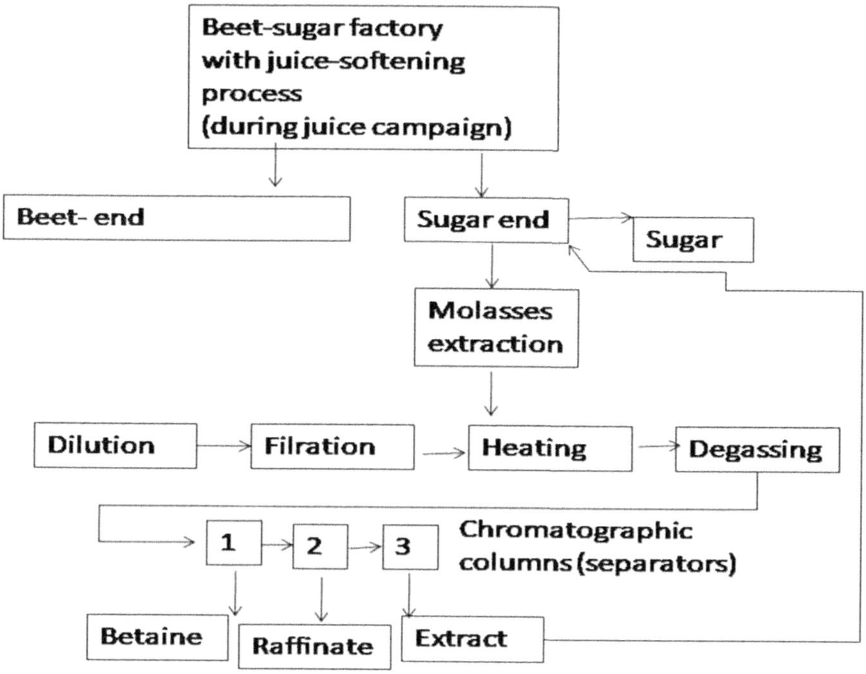

Fig. 44.8 Beet molasses desugaring by chromatographic process

5. Molasses contains some amount of betaine which could be used in diverse processes such as in sports nutrition, as an emulsifier and stabilizers in cosmetic and food industries.
6. Molasses after processing can be used in the production of polyester granules such as polyhydroxyalkanoates (PHA), and since PHA is biodegradable plastic, it can control the degradation and hence could be useful in various medical processes such as surgical sutures, wound dressings, drug delivery, and tissue engineering (Kumar et al. 2013).

44.3.3 The Advantages and Disadvantages of Beet Molasses Production

44.3.3.1 Advantages

Beet molasses have wide application or uses in various food and nonfood industries because of its carbohydrate, nitrogen content, and sweet taste. As we know, traditionally it is used in the production of alcohol through fermentation and also in animal feed; besides these, it has other applications as given below.

1. As a substrate for enzyme or co-enzyme bioproduction
2. It is a low-cost carbon source (used in European countries)
3. Molasses are useful in the production of amino acids and their salts because of nitrogen content
4. As a substrate for biopolymer production
5. In de-icing of roads using molasses-based materials
6. As a carbon source, it can be useful in various processes such as in denitrification of wastewater, biological metal-leaching, remediation of contaminant soil, and biodegradation of pollutants in water, etc.
7. In the production of vitamins (B12) and antibiotics
8. In the production of biogas.

44.3.3.2 Disadvantages

1. Though molasses can be consumed as an alternative to sugars, consuming in excess (for that matter, any added sugar) can cause problems, mainly to people having diabetes.
2. It can cause digestive problems like loose stool or diarrhea.
3. Molasses are also used in making liquid feed for animals. So main disadvantage is that due to its water content, it causes problems in storage, conservation, and transport (Mordenti et al. 2021).
4. Although molasses have various benefits, if it is not used in time, it can lead to environmental pollution.

44.3.4 Desugaring of Beet Molasses

Usually, the conventional sugar factory instruments are unable to separate the leftover sugar in molasses, which needs additional processing for the desugaring. Molasses desugaring by chromatographic process (MDC process) is a recent method for recovery of remaining molasses. This process is able to recover up to 90% sugar from the beet sugar. Ion exchange-based chromatography was first used for desugaring in corn syrup in the US and, later on, the technique was employed in the beet sugar industry. It is a multi-component process that not only separates the sucrose fraction (extract) from the non-sucrose fraction (raffinate) but also separates other valuable compounds like betaine, as the by-product of MDC contains about 50% of the said compound. The MDC process is a cost-effective and eco-friendly process. This technology has helped improve the beet sugar industry a lot.

44.3.5 Utilization of Beet Molasses

Beet molasses is the syrupy by-product yielded after the crystallization of sugar from concentrated sugar juice extracted from the roots of sugar beets *(Beta vulgaris* L.) Syndicat National des Fabricants de Sucre 2015; Crawshaw 2004). It is viscous, dark, sweet, sugar-rich, with a caramel flavor. Like sugarcane molasses, it is very palatable to livestock. It is a major feed ingredient for all types of livestock, including poultry. It is used as an energy source, an appetizer, a binder in compound feeds, and as a carrier for other ingredients such as sources of non-protein nitrogen (urea) (CNC 2002).

Beet molasses is produced worldwide and available wherever beetroots are extracted for sugar production. The yield of beet molasses is about 36–40 kg per t of sugar beet root processed, which represents about 285 g of molasses per kg of sugar produced (CNC 2002). In 2013, 231 million t of sugar beet were processed worldwide (out of a total of 247 million t) (Heuze et al. 2019), resulting approximately in nine million t of beet molasses. In EU, more than 70% beet molasses go to alcohol and yeast production (EU Beet Sugar Sustainability Partnership 2019). In 2004, it was estimated that 13% of the molasses (sugarcane and beet) produced worldwide were used as feed (Rolet 2005). Assuming the same ratio, the amount of beet molasses used for feed should be about one million tonnes.

Beet molasses, like sugarcane molasses, is primarily an energy source with a very high sugar content (60–65% DM), most of this sugar being saccharose (CNC 2002). Its protein content of 12–16% DM is between two and three times higher than that of sugarcane molasses. The protein fraction of beet molasses is completely soluble: it contains 50% amino acids with a low amount of essential amino acid and 50% NPN. Beet molasses is rich in betaine, an N compound that provides methyl groups in metabolic reactions and can reduce animal requirements for choline (FEDNA 2012). Beet molasses has the same feed value as sugarcane molasses, but tends to act more as laxative: the amount of beet molasses fed should be less than those recommended for sugarcane molasses (CNC 2002).

Molasses is a relatively stable product because its sugar concentration results in a high osmotic pressure that does not allow bacteria and other microorganisms to develop. The viscosity of beet molasses, particularly at low temperatures, tends to hinder its transfer from the storage tanks to feed mixers (Mavromichalis 2013). In France, it is recommended to store it in a heated tank when the temperature goes below 10 °C and to heat molasses when it will be used quickly, as a temperature over 50 °C can alter the quality of the product (CNC 2002). In the US, heated tanks are recommended in all circumstances (Midwest Agri 2019. Tubing should be wide enough (7–8 cm in diameter) to prevent clogging (CNC 2002; Midwest Agri 2019).

A common way to mitigate viscosity problems is to dilute beet molasses to 72–75% DM (Harland et al. 2006). Dilution should be done by putting water first to facilitate the mixing since molasses is heavier than water (Bernard et al. 1991). However, diluted molasses has a lower osmotic pressure, and may become a fermentation medium. Fermented molasses has a strong alcoholic smell and its surface is liquefied. Molasses containing more than 5% added water should be consumed quickly, within 8 days if the temperature is high (Midwest Agri 2019; Harland et al. 2006; CNC 2002).

Beet molasses is an energy feed due to its high sugar content, and thus a valuable feed ingredient for all livestock species. The stickiness of beet molasses makes it useful as a binder for making pellets, as it allows the feed granules to stick together during pelletization, resulting in pellets that are less likely to break down during transportation and passage through feeding equipment. The addition of molasses also reduces dustiness in the dietary mixture (Blair 2008). Beet molasses is used as an additive for ensiling as it provides quickly fermentable sugar (Crawshaw 2004). Beet molasses is also used to make molasses-urea mineral blocks (FAO 2011; Beames 1963).

Molasses is used for ethanol production and for yeast and fermentation industry (EU Beet Sugar Sustainability Partnership 2019). In the EU, 70% of beet molasses is used in ethanol production (EU Beet Sugar Sustainability Partnership 2019). In the USA, it is added to chloride salt and used for de-icing roads to make the process more environmentally friendly (Midwest Agri 2019).

There are many ways to distribute beet molasses to livestock. It can be given directly to the animals, in troughs or sprayed on the roughage, in dry blocks (with urea, for instance), or mixed with feed ingredients (CNC 2002; Bernard et al. 1991). The amount of molasses that can be absorbed depends on the substrate, on its particle size (finely ground materials absorb molasses better than coarsely ground ones), on its humidity (dry materials absorb molasses better than humid ones), and on the temperature of the molasses (heated molasses is less viscous and better absorbed) (Bernard et al. 1991).

Heavy snowfall makes travel by road very difficult. Salt is used for de-icing roads during heavy snowfall. Desugared sugar beet molasses is mixed with salts in equal parts for de-icing or anti-icing products in winter control operations; and it has reduced the amount of salt on the roads as much as 30%. Sugar beet molasses, in combination with salts, is rather more effective than the road salt used alone, as it reduces corrosion to some extent and also lowers the freezing point of the de-icing mixture remains, and it remains more effective under such conditions. Additionally, use of this mixture reduces the bounce and scatter of the rock salt used and decreases the time for the snow to melt.

Sugar beets are used primarily for the production of sucrose, a highly energy-pure food, and which is the principal use for production of sugar in sugar manufacturing industries. Sugar beets contain 13–20% sucrose which is the maximum among the already known alternative resources of sugar. High fiber sugar beet pulp is used for the manufacturing of biofuels. In addition to sugar beet pulp, molasses are obtained as processing by-products that is widely used as feed supplements for the livestock, energy generation, biofuel production, environmental, and for pharmaceutical inputs. Cultivation and adaptation of sugar beet enhances not only the sugar manufacturing industry but also supports the local people through creating job opportunity, useful animal feed provision, renewable energy supply, and overall productivity. Generally, sugar beet is the number one highly efficient input for sugar industries compared to some other raw materials. So that introducing, cultivating, and using sugar beet is a step-up in economic development to the national and in multiple directions.

44.3.5.1 Utilization for Consumption of Beet Molasses

Sugar beet molasses, on dry weight basis, contains about 50% sugar. The latter contains predominantly sucrose, and also glucose and fructose. Its non-sugar components are salts like oxalate and chloride of calcium and potassium. Molasses also contains betaine and a tri-saccharide, raffinose (Shrivastava et al. 2013). These make beet molasses unpalatable for human beings. But it is mainly used as an additive to animal feed and also as a feedstock for fermentation. In comparison to sugarcane molasses, the beet molasses lacks biotin (vitamin H or B7). Besides extraction of sugar, sugar beet, as well as pulp and molasses obtained from its processing, have been utilized for preparing some of the useful products for the benefit of mankind. Some of these are given below.

44.3.5.2 Human Food

Sucrose from sugar beets is the principal use for sugar beets in the United States. Sugar beets contain from 13 to 22% sucrose. Sucrose obtained from processing sugar beets is widely used as a high energy food as also a palatable food additive. High fiber dietary food additives have been manufactured from sugar beet pulp. In the United States, these dietary supplements have been recently introduced in breakfast items. A sugary syrup is also produced by cooking shredded sugar beet for several hours. After pressing and filtering, a honey-like dark syrup is obtained. In some parts of Germany, this syrup (called *Zuckerruben-Sirup* or *Zapp*) is used as a spread on sandwiches, to sweeten sauces, cakes, and desserts. This syrup is also hydrolyzed to a product akin to high fructose corn syrup.

44.3.5.3 Beverages

In many countries, molasses from sugar beet is used to make a rum-flavored hard liquor like *Tuzemak* in the Czech Republic and Slovakia, *Kobba Libre* in the Aland Islands, and rectified spirit and *vodka* in the Czech Republic and Germany.

44.3.5.4 Feed for Livestock

Once the juice has been extracted, pressed, or dehydrated, beet pulp provides an ideal foodstuff for cattle. Pulp can also be used to produce industrial pectin or dietary fiber for use with foods enriched with fiber. Processing by-products of sugar beet, the beet pulp and molasses are also widely used as feed supplements for livestock. It contributes to fiber in the feed and adds to its palatability. In France, sugar beet molasses is used as cattle fodder supplement.

44.3.5.5 Pharmaceuticals

Molasses, a by-product from sugar beet processing, is widely used in producing alcohols (ethanol and butanol), other pharmaceuticals, and also producing baker's yeast. A tonne of molasses yields approximately 300 liters of alcohol. Alcohol derived from sugar beet is suitable for human consumption (in spirits, perfume, vinegar, pharmaceutical products, etc.) and ideal for use in household products (cleaning fluids, methylated spirits, etc.) and some other useful chemicals (solvents, etc). For ethanol fermentation from sugar beet molasses, a new alginate-maize stem tissue matrix has been developed as a carrier for the yeast, *Saccharomyces cerevisiae*. The latter led to an ethanol production of 2.51 g/L/h. Fermenting beet molasses with potassium ferro- or ferricyanide with *Aspergillus niger* produced citric acid (approximately 50% of the available sugar as sucrose). *Betaine* can be isolated from molasses, a byproduct of sugar beet processing using a chromatographic technique called Souvenir _ 45 "simulated moving bed." This chemical is used as osmo-protectant for commercial crops under drought conditions. *Uridine* is also isolated from sugar beet molasses. In combination with omega-3 fatty acids, use of this compound overcomes depression in rats.

44.3.5.6 Bio Fuel/Fuel Additives

Alcohol produced from sugar beet molasses is also used as fuels (or mixed with petrol and diesel) for automobiles in many countries. Therefore, it takes 6.22 kg of sugar beet to produce 1 kg of ethanol (approximately 1.27 L).

44.3.5.7 Industrial Utilization of Beet Molasses: Experiment on Industrial Utilization of Beet Molasses

Fermentation of the industrial vitamin B12 by *Pseudomonas denitrificans* usually utilizes sucrose or maltose as the sole carbon source, which results in increased medium costs. In order to decrease the fermentation cost, it is crucial and essential to employ a low-cost and convenient raw material as an alternative medium substrate for industrial vitamin B12 production. The results obtained in chemically defined medium showed that glutamate and sucrose were favorable for cell growth and vitamin B12 biosynthesis of *P. denitrificans. Due* to containing a mass of ingredients such as sucrose, glutamate, and betaine, beet molasses was consequently chosen as the main medium substrate for industrial *P. denitrificans* fermentation in a 120,000 L fermenter. Vitamin B12 production reached 181.75 mg L^{-1}. Beet molasses, a by-product of the sugar industry, was successfully used as an efficient and low-cost substrate for vitamin B12 production by *Pseudomonas denitrificans* in a 120,000 L fermenter. As a result, 181.75 mg L^{-1} of vitamin B12 was obtained at the end of fermentation, and fermentation costs were significantly reduced (Li et al. 2013).

Another study had been performed on comparing the potentiality of sugar beet molasses and thick juice as raw materials for bioethanol production, as renewable and sustainable energy sources. Ethanol fermentation of a wide range of initial sugar concentrations (100–300 g/L) was performed using either free or immobilized *Saccharomyces cerevisiae* in calcium alginate beads in the absence of any added nutrients. In general, immobilized cells showed better fermentative performance, enhanced ethanol productivity, stability, and cell viability compared with free cells under the same fermentation conditions. The high concentration of non-sugar components contained in molasses affected yeast fermentation performance and viability. Maximum ethanol concentration in fermented media of 84.6 and 109.5 g/L were obtained by immobilized cells for initial sugar concentrations of 200 and 250 g/L for molasses and thick juice, respectively. However, the highest ethanol yields of 31.7 L per 100 kg of molasses and 37.6 L per 100 kg of thick juice were obtained by immobilized cells at an initial sugar concentration of 175 g/L. In the high gravity fermentation process, thick juice resulted in a higher ethanol yield per mass of raw material compared with molasses. This study shows the advantage of immobilized yeast for the efficient production of high gravity bioethanol from thick juice, which was a more favorable raw material than molasses. The presented results have industrial relevance as they indicate a more convenient substrate and biocatalyst for efficient bioethanol production (Vesna et al. 2018).

The present study was conducted to observe the optimization of an industrial medium from molasses for bioethanol production using the Taguchi statistical experimental-design method. First, the growth rate of yeast cells and the amount of ethanol produced by the *Saccharomyces cerevisiae* strain sah and 101 were investigated in aerobic and aerobic–anaerobic conditions. The yeast strain produced 8% (v/v) bioethanol in a medium containing molasses with 18% Brix in aerobic–anaerobic conditions. The main factors of the medium, including molasses, ammonium sulfate, urea, and pH, were optimized for the increase of bioethanol production by the Taguchi method. Bioethanol production reached 10% (v/v) after optimization of the medium in flask culture. The yeast strain produced 11% (v/v) bioethanol in the bioreactor culture containing the optimized medium, which is an acceptable amount of bioethanol produced from molasses at the industrial scale. The results showed that the Taguchi method is an effective method for the design of experiments aiming to optimize the medium for bioethanol production by reducing the number of experiments and time (Darvishi and Moghaddami 2019).

In the present study, sugar beet industry wastewater without and with beet molasses was used as potential low-cost substrate for production of the biopolymer Polyhydroxyalkanoates (PHA) by a local bacterial strain. This strain was selected after screening of 30 bacterial isolates for PHA production and was identified according to 16S rRNA gene sequencing as *Bacillus megaterium* AUMC b 272. On the other hand, the bacterial strain produced negligible levels of PHA when grown on the same medium constituents under the same conditions with replaced distilled water and molasses by sugar beet wastewater. It is worthy to mention that the COD in the sugar beet wastewater medium at the end of fermentation period was removed by 69%. Characterization of the PHA was performed by using Fourier transform-infra-red spectroscopy (FT-IR) and gas chromatograph mass spectrometry (GCMS). Accumulation of considerable level of PHA as well as high levels of COD reduction from sugar beet wastewater strongly introduced this biotechnological process as a valuable and economic method for production of PHA as biodegradable biopolymer from sugar beet industry wastewater in presence of beet molasses as potential low-cost substrates and, at the same time, for biological treatment of industrial wastewater (Zohri et al. 2019).

High sugar content in sugar beet molasses enables its use for fermentations, while Sugar beet pulp represents an interesting cheap raw material source for enzyme production (Donkoh 2012). Alternative fuels can be used to reduce our dependence on foreign oil. Bioethanol could be a substitute for fossil fuel. Presently, most of the bioethanol produced in developed countries like the United States of America is derived from corn. In order not to disturb the food security and for ethanol production from lignocellulosic material, it is advantageous because this material is abundantly available with minimum cost (Ramesh et al. 2017). Hence, sugar beet pulp is an attractive feedstock for ethanol production, because it is a co-product from the table sugar industry (Hansen et al. 2012; Hamley-Bennetta et al. 2016). Bio-ethanol production, with sugar beet used as raw material, hopes to provide a new way for producing bioenergy. Fuel alcohol can be obtained after a multi-tower pressure distillation, and products are isolated from sugar-beet bagasse and can be used as feed or degradable materials as well. Economic benefits of ethanol plants primarily depend on materials, energy and resources, labor cost, and investment (Ramesh et al. 2017).

44.3.5.8 Methane Production

Sugar production from sugar beet generates a co-product stream called raffinate, desugarized molasses, or concentrated separated by-product (CSB). About 0.25 tonnes of raffinate is generated for every tonne of sugar produced. Bio-gasification of raffinate at 55 °C produced methane gas. Every metric tonne of raffinate can generate 4300 MJ of thermal energy from combustion of methane or about 300 KWH of electricity. Anaerobic digestion method for whole beet/ensilaged beet to produce bio-methane is also an important and viable technology.

44.4 Future Scope in Production and Utilization of Beet Molasses

Efforts should be made to evolve a production technology for high yield of quality sugar beet molasses. As sugar beet molasses is important for its chemical characteristics, care should be taken that the sugar producing ability of the roots is not affected by cultural practices. The contemporary and up-to-date sugar processing technology is of prime need to our country for successful commercial exploitation of sugar beet. Efforts should be given for systematic studies which will lead to definite information, enabling selection of the most ideal diffusers, production of beet molasses of high chemical quality, and efficient utilization of the by-products.

The researcher is highly attracted towards production of sugar from sugar beet due to its composition and uses. The scope of research to be focused on preparation of sugar from sugar beet, finding of renewable energy resource and pollutant therapy in sugar industrial outlet. Economic progress and sugar scarcity can be overcome by generating new research ideas and can also contribute to the upliftment of the homeland.

A crucial study and keen observation on various aspects are needed to further confirm regarding the profit-making viability of the beet molasses in the country and economic gain to farm and industry would be of great value.

44.4.1 Future Research and Development Could Focus on the Following Areas

1. Detecting other uses of sugar beet molasses such as jaggery, jam, etc.
2. Encouraging sugar beet molasses for cattle feed

3. For ethanol production, technology and economics of sugar beet molasses
4. Government policies and incentives for the sugar beet Industry.

44.5 Conclusion

In a nutshell, beet molasses obtained as a by-product syrup after sugar crystallization mainly comprises sucrose and inverted sugar. Depending on different concentrations of soluble salts, the molasses has been classified as hard or soft molasses. Beet molasses is nutritionally rich with good amounts of essential amino acids, minerals, and betaine. Separation of sugar from molasses needs involvement of chromatographic techniques. Molasses is used as human food, cattle feed, beverages and in the pharmaceutical industry. Diversification of utilization of sugar beet molasses should be promoted in food industry (jam or jelly preparation). Adoption of new government policies and giving incentives to promote the sugar beet industries may be worthwhile.

References

Anonymous (1995). https://patentscope.wipo.int/search/en/detail.jsf?docId=WO1995016794

Beames RM (1963) Provision of urea to cattle in a salt/urea/molasses block. Queensland J Agric Sci 20:213–230

Belitz HD et al (2014) Food chemistry. 4th revised and extended edition. Springer

Bernard M, Chapoutot P, Chatelet M, Gueroult M, Jubert M, Morel d'Arleux F, Taccard M, Mariani M, Tierny M (1991) Synthèse sur la mélasse. Comité des sous produits—RNED Bovins, Juillet: 19 p

Blair R (2008) Nutrition and feeding of organic poultry. Cabi series. CABI, Wallingford

Bonnina E et al (2012) Enzymic release of cellobiose from sugar beet pulp, and its use to favour vanillin production in *Pycnoporuscinnabarinus* from vanillic acid. Carbohydr Polym 41(2000): 143–151

CNC (2002) Mélasse de betterave et de canne. Comité National des Coproduits, ADEME, Idele, Fiche N° 8—Coproduits de la betterave

Crawshaw R (2004) Co-product feeds: animal feeds from the food and drinks industries. Nottingham University Press, Nottingham

Darvishi F, Moghaddami NA (2019) Optimization of an industrial medium from molasses for bioethanol production using the Taguchi statistical experimental-design method. Fermentation 5:14. https://doi.org/10.3390/fermentation5010014

Donkoh E (2012) Optimization of enzymatic hydrolysis of dilute acid pretreated sugar beet pulp using response surface design. J Sugar Beet Res 49(1& 2):26–37

Duraisam R, Salelgn K, Berekete AK (2017) Production of beet sugar and bio-ethanol from sugar beet and it bagasse: a review. Int J Eng Trends Technol 43(4):222–233

EU Beet Sugar Sustainability Partnership (2019) Keep molasses for bioenergy and bio-based products. EU Beet Sugar Sustainability Partnership

FAO (2011) Successes and failures with animal nutrition practices and technologies in developing countries. Proc. FAO Electronic Conference, 1–30 September 2010, Rome, Italy

FEDNA (2012) FEDNA (Fundacion Espanola para el Desarrollo de la Nutricion Animal). Fundacion FEDNA.

Hamley-Bennetta C, Lye GJ, Leak DJ (2016) Selective fractionation of sugar beet pulp for release of fermentation and chemical feed stocks; Optimizations of thermo-chemical pretreatment. Bioresour Technol 209:259–264

Hansen MEZ et al (2012) Enzymatic depolymerization of sugar beet pulp: production and characterization of pectin and pectic-oligosaccharides as a potential source for functional carbohydrates. Chem Eng J 192:29–36

Harland JI, Jones CK, Hufford C (2006) Co-products. In: Draycott AP (ed) Sugar beet. Blackwell Publishing, Hoboken, pp 443–463

Heuze V, Tran G, Lebas F (2019) Beet molasses. Feedipedia, a programme by INRAE, CIRAD, AFZ and FAO. https://www.feedipedia.org/node/711

Kumar S, Singh PK, Swapna M, Pathak AD (2013) In souvenir: IISR-industry interface on research and development initiatives for sugar beet in India. Nainital, IVRI, p 2

Kun-tai Li, Wei-fu Peng, Jia Zhou, Sai-jin Wei, Xin Cheng (2013) (wileyonlinelibrary.com). https://doi.org/10.1002/jctb.4025

Mavromichalis I (2013) Molasses in poultry diets—an unusual ingredient with potential. Animal nutrition views. WATTAgNet.com

Midwest Agri (2019) Beet molasses. Midwest Agri-Commodities Company, San Rafael

Mikulski D, Klosowski G (2021) Integration of first- and second-generation bioethanol production from beet molasses and distillery stillage after dilute sulfuric acid pretreatment. Bioenergy research molasses production data

Mordenti AL, Parazza P, Formigoni A (2021) A review regarding the use of molasses in animal nutrition. Animals 11:115. https://doi.org/10.3390/ani11010115

Nikulina NS, Bulatetskaya TM, Nikulin SS, Pugacheva IN, Rudyka EA (2021) Prospect of using beet molasses in the production of synthetic rubbers. IOP Conf Series Earth Environ Sci 640: 052010

NOVUS (1996) Intern.: raw material compendium, 2nd edn, Brussels

Reinefeld E, Bliesener K-M, Schulze J (1982a) Beitrag zur Kenntnis des Verhaltens der Aminosäuren imVerlauf der Zuckergewinnung. 1. Mitteilung: Gaschromatographische Untersuchungen anZuckerfabrikprodukten auf freie und durch Säurehydrolyse freisetzbare Aminosäuren. Zuckerind 107:283–292

Reinefeld E, Bliesener K-M, Schulze J (1982b) Beitrag zur Kenntnis des Verhaltens der Aminosäuren imVerlauf der Zuckergewinnung. 2. Mitteilung: Untersuchungen über die Auswirkung vonProzeßparametern der Saftreinigung. Zuckerind 107:1111–1120

Rezbova H et al (2015) Sugar beet production in the European Un ion and their future trends, Agris. Economics Informatics 5(4):165–178

Rolet V (2005) Table ronde usages industriels des produits agricoles. Journée matières premières AFTAA

Schiweck H, Jeanteur-DeBeukelaer C, Vogel M (1993) The behavior of nitrogen containing non-sugar substances of beet during the sugar recovery process. Zuckerind 118:15–23

Sarka E, Bubnik Z, Hinkova A, Gebler J, Kadlec P (2012) Molasses as a by-product of sugar crystallization and a perspective raw material. 20th international congress of chemical and process engineering CHISA, Prague, Czech Republic. Procedia Eng 42(2012):1219–1228

Sherif MAS, Keshk Taha MA, Razek KS (2006) Bacterial cellulose production from beet molasses. Afr J Biotechnol 5(17):1519–1523

Shrivastava AK, Shukla SP, Sawnani A (2013) Useful products from sugar beet. Souvenir. Sugar beet breeding outpost of IISR. IVRI Campus, Nainital

Syndicat National des Fabricants de Sucre (2015) Synd. Nat. Fab. Sucre, Paris, France

Vučurović VM, Puškaš VS, Miljić UD (2018) Bioethanol production from sugar beet molasses and thick juice by free and immobilised Saccharomyces cerevisiae. J Inst Brew. https://doi.org/10.1002/jib.536

Zohri Adb El-Naser A, Kamal El-Dean AM, Abuo-Dobara MI, Ali MI, Bakr MMN, Khashaba RA (2019) Production of Polyhydroxyalkanoate by local strain of Bacillus megaterium AUMC b 272 utilizing sugar beet wastewater and molasses. Egypt Sug J 13:45–70

Bioethanol Production from Sugar Beet Juices and Molasses for Economic and Environmental Perspectives

45

Rajeev Kumar, Vivek Chandra Verma, Ashutosh Kumar Mall, and A. D. Pathak

Abstract

In the present scenario, the demand for fossil fuel has become a fundamental issue for mankind across the globe. To render adequate the energy demand for transport, the mixing of bioethanol with gasoline has been a promising aspect in India as well as other developing and developed countries. The potential of co-products and transitional products of sugar beet processing as raw material for bioethanol production has a tremendous scope in view of the demand for ethanol as an alternative for fossil fuel. Molasses is one of the important by-products of sugar beet or sugarcane refining industries which can be utilized as a raw material in the fermentation industry, such as the production of feed yeasts, baker's yeast, antibiotics, citric acid, amino acids, acetone/butanol, organic acids, and enzymes. Sugar beet molasses are enriched with different minerals and vitamins used as a potent medium to enhance the shelf life of fruits and vegetables through osmotic dehydration. Evaluation of molasses for their industrial application cannot be based on their chemical composition and origin as various benchmarks are established for their use in different processes. The utilization of molasses as the sole carbon source in a particular process, pre-treatment of molasses, and removal of inhibitor should be prerequisites. Calcium carbonate is used as a pre-treatment agent for the neutralization of the molasses during yeast and methanol production. However, for various other processes, they are boiled in an acidic or alkaline medium and separated out from the precipitate. For the citric acid, production molasses are boiled with potassium ferrocyanide and generally fermented together with precipitate. Currently, in India, sugarcane molasses is

R. Kumar (✉) · A. K. Mall · A. D. Pathak
ICAR-IISR, Lucknow, Uttar Pradesh, India

V. C. Verma
DIHAR-DRDO, Chandigarh, India

V. Misra et al. (eds.), *Sugar Beet Cultivation, Management and Processing*,
https://doi.org/10.1007/978-981-19-2730-0_45

being used for the production of bioethanol, but cannot fulfil the demand for bioethanol. Therefore, the crop residues such as sugar beet molasses may be explored for biofuel production to meet the demand for alternative and renewable energy sources. In the rapid urbanization and industrial development, bioethanol production from agricultural wastes provides economic as well as environmental benefits. The present status of bioethanol production in India can be encouraged by the development of new low-cost technology for the bioconversion of agricultural wastes which might be helpful for economic and environmental insights.

Keywords

Sugar beet · Molasses · Bioethanol production · Crop residue · Market demand · Low-cost technology

Abbreviations

GRAS	Generally recognized as safe
PHA	Polyhydroxyalkanoates
SBP	Sugar beet pulp

45.1 Introduction

Henry Ford said ethanol is "the fuel of the future." He later stated, "The fuel of the future will come from agriculture commodities, waste or its by-product". Today, Henry Ford's futuristic vision seems to be realistic (Chandel et al. 2007). The continuous depletion of limited fossil fuel stock leads to the search for sustainable, economical, ecologically, and environmentally friendly fuel sources. The rapid utilization of fossil fuel in the present scenario seems to be producing an environmental emergency worldwide. The burning of fossil fuel produces hazardous greenhouse gas emissions like carbon dioxide (CO_2), methane (CH_4), and the bulk of nitrous oxide (N_2O) resulting in global warming (Saini et al. 2015; Van Fan et al. 2018). These harmful gases are formed due to partial combustion of fossil fuel; since bioethanol consists of 35% oxygen, that helps in more complete combustion of fuel and thus decreases hazardous emissions (Chandel et al. 2007). Countries across the world have initiated and followed state policies towards the increased and economic application of biomass for fulfilling their upcoming energy demands in order to achieve an objective to reduce carbon dioxide (CO_2) emission as specified in the Kyoto Protocol and decreases dependence on the supply of fossil fuels (Sarkar et al. 2012). Bioethanol can be used as a huge source of transport, as well as produces both power and heat, through combustion. Many countries around the world are moving toward renewable sources for power production because of diminishing crude oil reserves. Therefore, bioethanol has been drawing broad interest worldwide. The widespread market for bioethanol has acquired a phase of fast and transitional growth. However, the cost of bioethanol production is more compared to fossil fuels (RFA 2017).

The generation of bioethanol from biomass resources is one of the favoured ways by which the transportation zone can be made "eco-friendly" (Kumar et al. 2019). Amongst the aforementioned biofuels, the world production capacity of bioethanol is higher and is the most preferred choice in terms of policies adopted by the government on biofuel blend with gasoline (Niphadkar et al. 2018). Sugars, starch, and lignocellulosic materials are utilized for the production of bioethanol, which has a larger octane number and higher heat of vaporization due to which it can be efficiently mixed with the gasoline (Singh et al. 2016). Conventionally, starchy grains and sugar crops are widely accepted for bioethanol production due to lesser input energy.

The chemical composition of sugar beetroots makes this raw material an attractive feedstock for ethanol fermentation. During the industrial process of sugar extraction from sugar beet crop, various intermediates, transitional products, and wastes are generated which can be used for the production of energy fuel and other value-added products (Marzo et al. 2019).

Keeping in view of all these beneficial effects of bioethanol production in present and future scenarios, bioethanol production technologies should immediately achieve momentum and the hurdles imposed should be removed successfully for the production of bioethanol at the commercial level.

45.2 Global Scenario of Bioethanol Production

Sugarcane, sugar beet, corn, and starch are the cheaply available raw material for bioethanol production, and these materials are primarily used in the world for bioethanol production, although sugar beet is a relevant and simple material as compared to the starchy material, and it does not require additional steps. Sugar beetroots are one of the important feedstocks for bioethanol production due to their chemical constituents, which make it an attractive raw material (Mall et al. 2021).

Extraction, purification, evaporation, and crystallization carried out in the sugar factory produces three major by-products such as sugar beet pulp (SBP), molasses, and lime sludge. Following the sucrose extraction from sugar beet, 830 kg of sugar juice and 170 kg of wet sugar beet pulp can be obtained from sugar beet on per ton basis (Rezic et al. 2013). Globally, sugar beet pulp in the form of pellets can be utilized for animal feed, which can be processed by dehydration and granulation methods in sugar mills. Both dehydration and granulation consumes 30–40% of the energy cost, while the conversion of pulp to bioethanol will be cost reliable. Hence, bioethanol production from pulp is more economically sound than pulp processing for animal feed.

Molasses produced from sugar beet contains 47–48% sugar, 23–26% water, 9–14% minerals, and 8–12% nitrogenous compounds such as amino acid and protein (Taskin et al. 2016). Therefore, it can be utilized as the low cost raw material for diverse sets of value-added products such as enzymes, lipids, acids, and ethanol in India and other developed/developing countries (Razmovski and Vucurovic 2012).

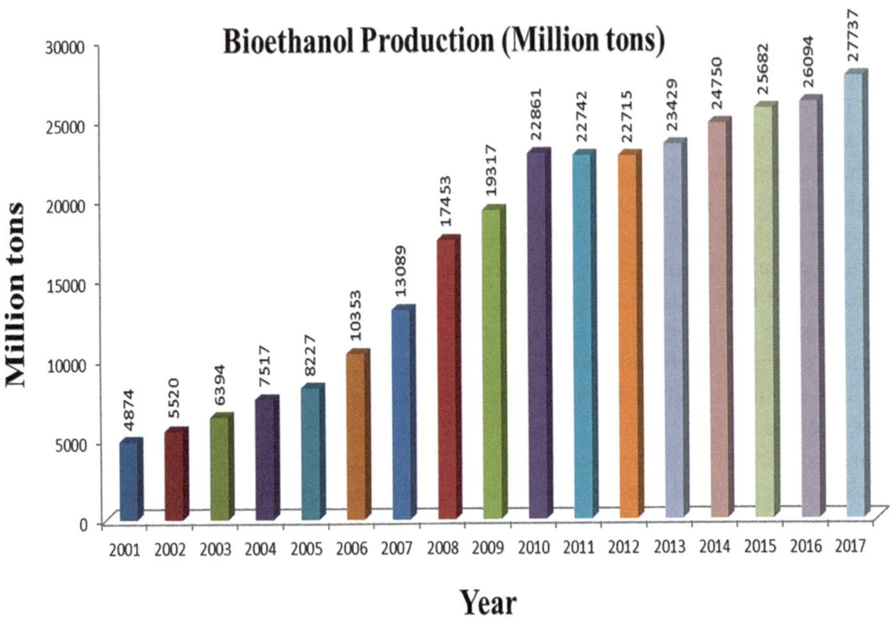

Fig. 45.1 World bioethanol production scenario in million tons from the year 2001–2017

Kim and Dale (2004) reported the world bioethanol production was 31 billion litres in 2001. The bioethanol production was recorded as 39 billion litres in 2006 and is expected to reach 100 billion litres in 2015 (Taherzadeh and Karimi 2007). In 2016, the United States and Brazil produced about 15.25 billion gal (~57.7 billion L) and 7.3 billion gal (~27.6 billion L) of fuel ethanol from starchy and sugar-rich feedstock's (Mohanty and Swain 2019). In the United States and Brazil, corn starch, sugarcane juice, and molasses, respectively, are primarily utilized for bioethanol production and these two countries contributed to 89% of world ethanol production (Singh et al. 2016). The USA, Brazil, and China produced 25,754 million gal of bioethanol in 2016 (GRFA 2017).

Except top leading countries in bioethanol production, other countries such as India, France, Germany, and Australia produce about 1 billion L, 1 billion L, 750 million L, and 500 million L, respectively, primarily from sugarcane, molasses, and sugar beet (RFA 2017). The world bioethanol production is given in Fig. 45.1. In India, sugarcane molasses is principally utilized for bioethanol production and has reached about 330 distilleries' annual production with capacity of over 4.0 billion litres (Venkatesh 2012).

45.3 Feedstocks for Bioethanol Production

Sucrose-containing feedstocks are chiefly used as substrate for bioethanol production. These feedstocks are the commonly preferred choice because the transformation of sucrose into ethanol is easily accessible compared with starchy and

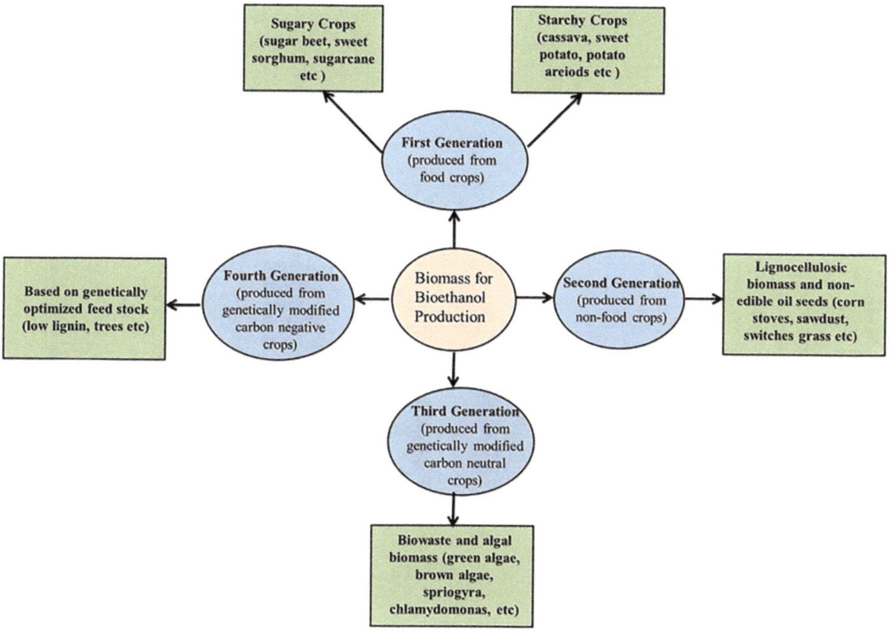

Fig. 45.2 Biomass for different generation's bioethanol production

Table 45.1 Major feedstock used in bioethanol production

S. no.	Feedstock	Ethanol yield (L/ha)
1	Sugar beet	5000
2	Cane bagasse	5882
3	Corn stover	1050–1400
4	Corn	3460–4020
5	Forest residues	3000–5000
6	Sawdust	1500–3000
7	Paddy rice	2250
8	Wheat	2590
9	Cassava	3310
10	Sweet sorghum	3050–4070
11	Switch grass	10,760
12	Algae	46,760–140,290

lignocellulosic biomass. Different feedstocks have been utilized for the production of bioethanol as shown in Fig. 45.2. First generation bioethanol production comprises food crops like grains, sugarcane, and tuber crops. Some of the major feedstocks utilized for the production of bioethanol and their yields are mentioned in Table 45.1. The first generation feedstocks are important for bioethanol production. Corn-based or sugarcane-based bioethanol production cannot compensate the one trillion gallons of fossil fuel which is presently consumed globally every year. The

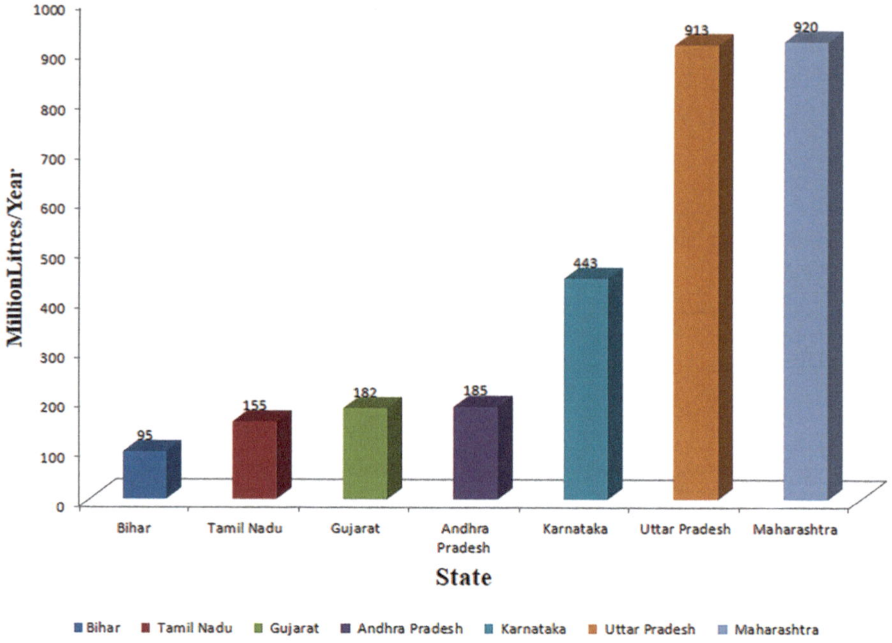

Fig. 45.3 Bioethanol production in different states of India

Utilization of edible crops for bioethanol generation impacts on rise in food price, hunger index and food insecurity. So, there is an urgent demand to produce bioethanol from some other resources that do not put pressure on edible crops. As a result, this led to the bioethanol generation from inedible potential feedstocks like agricultural waste or its co-product.

More advanced technologies for bioethanol generation from lignocellulosic biomass (second generation feedstocks) have been focused. Currently, usage of lignocellulosic biomass for bioethanol generation is the better option as it does not compete with the food crops (Singh et al. 2014). Second generation feedstocks such as bagasse, straw, stover, stems, leaves, de-oiled seed residues, and grass biomass have achieved much concern in the past two decades (Mohapatra et al. 2017).

Recently, algae are regarded as the third generation feedstock and very potential biomass for bioethanol production due to their multisided important aspects, like faster-growing rates as compared to terrestrial plants, high availability, and ability to withstand harsh conditions (Khan et al. 2017). In India, bioethanol generation mainly depends on molasses, which in turn depends on the sugar industry. The two states—Maharashtra and Uttar Pradesh, followed by Karnataka and Tamil Nadu, produce more sugar and contribute to about 60% of the total sugar production (Fig. 45.3). As a result, these states are the only bioethanol suppliers of India, and approximately three billion litres of bioethanol is produced annually (Singh et al. 2017).

45.4 Sugar Beet for Bioethanol Production

The agricultural crops that are well suited for bioethanol production comprises maize, sweet sorghum, sugarcane, cassava, and sugar beet (Ray et al. 2019). In India, the sugar beet seeds are sown in September to November and harvested during March and May. Sugar beet cultivation requires approximately four times less water than sugarcane. Also, sugar content of sugar beet is about 25% higher than that found in sugarcane (Marzo et al. 2019). Sugar beets have comparative advantages over others feedstocks with respect to yield of ethanol in less energy input, better acclimatization in harsh climatic conditions, soils, and could be grown on marginal land (Finkenstadt 2014). The sugar beetroot called taproot is composed of 75% water and 25% dry matter. The dry matter comprises about 5% pulp and 75% sugar of the total dry matter. The pulp is water insoluble and made up of cellulose, hemicellulose, lignin, and pectin. The sugar content in sugar beet can vary, ranging from 12 to 20%. The chemical composition of sugar beet makes this raw material very feasible feedstock for bioethanol production (Gumienna et al. 2016).

Sugar beet contributes about 20% of the worldwide demand for sugar. However, the United States contributes about 11% of the world's supply of sugar beet (Biancardi et al. 2010). In addition, the whole beet with its co-products, such as greens leaves, molasses, and pulp residue, could be used as animal feed or feedstock for alcohol generation. The constituent of sugar beet pulp could be a promising resource for bioethanol production with approximately 68% of carbohydrates (Table 45.2). It is reported that aqueous sugars extracted from 1 kg sugar beet can give an ethanol recovery of 0.07 kg under optimum conditions (Šantek et al. 2010).

The sucrose based feedstocks are mostly grown in Brazil, Germany, France, and India are leading countries for bioethanol production. They mostly cultivated sugarcane, sugar beet, and sweet sorghum with yields of 62–74 tons/ha, 54–111 tons/ha,

Table 45.2 Fermentable sugars in sugar beet pulp

S. no.	Components	Dry weight (%)
1.	Carbohydrate	68.0
a	Glucose	22.0
b	Arabinose	18.0
c	Uronic acids	18.0
d	Galactose	5.0
e	Rhamnose	2.0
f	Xylose	2.0
g	Mannose	1.0
h	Saccharose (residual)	4.0
2.	Ester-linked substituents of polysaccharides	
a	Ferulic acid	0.5
b	Acetic acid	1.6
c	Methanol	0.4
3.	Protein	8.0

and 50–62 tons/ha, respectively. Bioethanol production from sugar beet is the convenient process since the extracted sugar juices can be directly fermented to produce bioethanol. Bioethanol production from second generation feedstocks is more expensive in terms of energy input to soften cellulosic materials such as by acidic hydrolysis after the pulverization and steaming before the saccharification process. From an economic point of view, sugar beet and its intermediates are very precise raw materials for alcohol production due to their good content of fermentable sugars, which can be directly used for fermentation (Hattori and Morita 2010).

45.5 Technology for Bioethanol Production

The step-wise illustration of industrial process of bioethanol production (Fig. 45.4) from first generation feedstocks are explained in the following section (Mohapatra et al. 2019).

- Harvesting and transportation
- Beet washing and cleaning
- Beet slicing
- Diffusion (juice extraction)

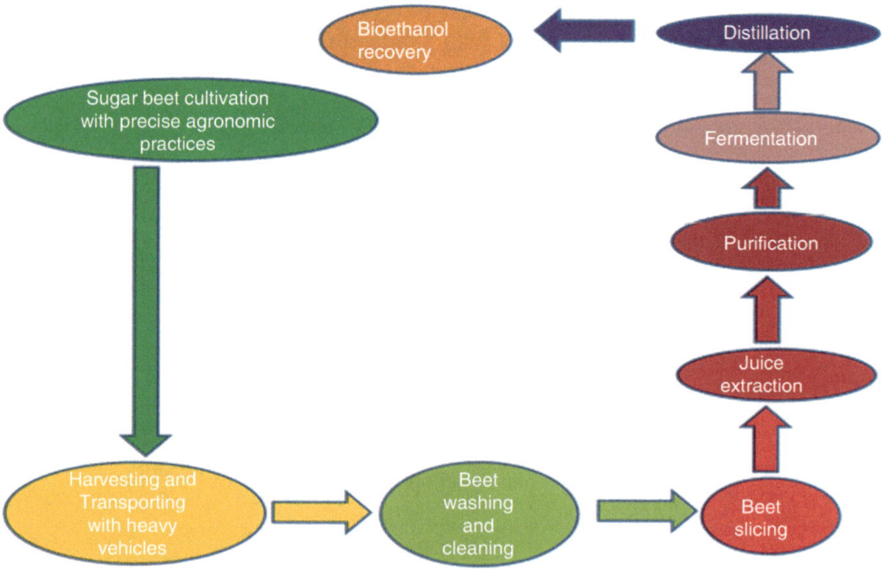

Fig. 45.4 Industrial process of bioethanol production from sugar beet juice

45.5.1 Harvesting and Transportation

During harvesting, sugar beetroots are uprooted mechanically and the bulbs are taken out from the soil and the leaves are ripped off at field. The detached stalks and leaves are generally kept on the field for animal feed and the roots are rapidly transported to the storage unit by several transport services such as truck, train, or ship, depending on the distance (Biancardi et al. 2010).

45.5.2 Beet Washing and Cleaning

At storage units, there are many other units for pre-treatment which includes dry-cleaning, conveying, and stone-trash separation units. Dry-cleaning is usually done when the sugar beets are transported from the storage unit to the main processing line of the factory, with the target of removing tar material by pre-cleaning of the roots. Sugar beetroots are transported from storage area to stone trash by conveying unit. In the stone trash stage, leaves and weeds are separated. Further, remaining sand, dust, and leaves are removed by pressurized water jet. The washing sludge is poured out in order to reuse the water. After completely draining off washing water, sugar beetroots are carried inside the processing plant by conveyer belt. Arm washer, drum washer, and spray washer are the preferred choices as beet washers used at industrial scale. After cleaning, the feedstocks are transferred to the slicing station (Asadi 2006).

45.5.3 Beet Slicing

In this stage, the cleaned beets are cut into long, thin strips called "cossettes." This could promote and augment further sucrose extraction during the stage of diffusion. Beet slicing unit comprises slicers, knife maintenance shop, conveyor, and beet hopper. Beet slicing is the crucial stage in the process of sugar production, since the quality of slicing process would affect the purity of the extracted juice.

45.5.4 Diffusion (Juice Extraction)

The resultant cossettes from the slicing stage are transported to the rotating or tower drum diffuser where the juice is extracted from the beet cells by a counter current diffusion process. More precisely, when the cossettes are kept in nearly 70 °C in the diffuser, the beet cells are decomposed and this phenomenon would facilitate the movement of sucrose and non-sucrose from the cossettes, thereby generating a concentrated solution of impure sucrose commonly known as raw juice. Generally, raw juice obtained during diffusion stage consists of 14–15% dry substance and 80–85% of purity. Sugar beet pulp could be used as a feedstock for producing fuel ethanol by chemical, enzymatic treatment, and several other processes. During the extraction of juice, the three major co-products that are recovered are beet pulp, lime sludge, and molasses. Pulp is obtained after the beet strips diffusion operation and recovery of sugar juice and it is mainly composed of pectin (38–62% by dry matter), hemicellulose (24–32%) and cellulose (22–30%) and low lignin content (around 1%) (Kamzon et al. 2016).

45.6 Pre-treatment of Molasses and Other Lignocellulosic Materials

Residues or by-products such as molasses and SBP obtained from the sugar beet after sugar extraction are lignocellulosic in nature. Different types of polymers is linked with each other such as cellulose, hemicellulose, and lignin in varying concentrations and numerous arrangements based on the kind of the by-products (Kim et al. 2015). "Pre-treatment" is the process popularly used to liberate the cellulose, hemicellulose, and lignin from their complex native form and arrange them in such a way that can be easily targeted for the enzymatic hydrolysis (Manzanares et al. 2020). Perfect pre-treatment process should have the ability to render the lignocellulosic mass absolutely on the exposure to celluloses as per the above description. Pre-treatments can be carried out in different ways (Fig. 45.5); however, initial rendering of the biomass to their constituent monomer (sugars) is the major confined access.

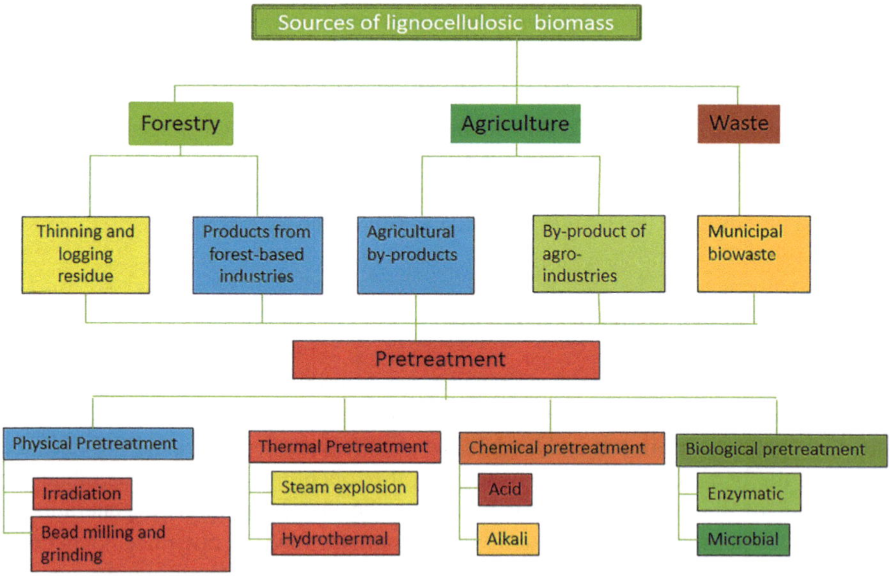

Fig. 45.5 Mode of pre-treatment for the bioethanol production from lignocellulosic mass

Removal of the lignin and hemicellulose of the substrate also needs the pre-treatment that can lead to easy access of the cellulose for the microorganism for their conversion into sugars (van der Pol et al. 2015). During the process of pre-treatment, hemicellulose and lignin degraded, result in the surface area gets increased, leading to the sorting out of the three major constituent polymers and exposure of the cellulose surface. Numerous physicochemical processes have been developed and they are in practice for molasses, SBP, and sugarcane bagasse pre-treatment. These encompass steam explosion, gamma rays irradiation, acid and alkali treatment, hydrogen peroxide treatment and implementation of various ionic solvents, etc. Weak acid treatment such as HCl or H_2SO_4 are used to hydrolyse the hemicellulose fractions efficiently (Agi et al. 2019), following either chemical or enzymatic treatment for the depolymerization of the resultant solid fractions.

Different classes of hydrolytic enzymes commonly produced by the various filamentous fungi have been exploited for enzymatic treatment. The combined action of three enzymes on different substrates involved to obtain the glucose by the breakdown of cellulose which comprises the endo-β-1,4-glucanases (EC 3.2.1.4), β-glucosidases (β-D-glucosidic glucohydrolases, EC 3.2.1.21), and exo-cellobiohydrolases (EC 3.2.1.91) (Benoliel et al. 2013). Pre-treatment of numerous lignocellulosic materials with the aid of dilute H_2SO_4 has become a significant technique since it promotes direct conversion of hemicellulose into fermentable monomer of sugar and enhances the cellulose accessibility for the hydrolysis catalysed by exo- and endoglucanases (Canilha et al. 2011). Whereas, alkali pre-treatment is acknowledged as more effective as well as economical for the lignin solubilization, while drawback of the process is its inability to breakdown the complete hemicellulosic materials. It can be executed at relatively low temperature and pressures.

Treatment with lime ($Ca[OH_2]$) is acknowledged as robust, efficient, economical, and highly recoverable since it can be utilized for the pre-treatment of the sugarcane bagasse (Melendez-Hernande et al. 2019). Lignocellulosic biomass can be also degraded with biological treatment. Numerous groups of fungi like white, brown, and soft rot fungi can be utilized to degrade lignin as well as hemicellulose polymer from lignocellulosic material.

White rot fungi-mediated biological pre-treatment can be used to degrade lignin and seems promising since the pre-treatment method involves less energy requirement, hence it causes minimal damage to the environment (Sindhu et al. 2016).

45.7 Biochemical Compound in Molasses and Their Importance

Difference in the composition of initial raw material, variation in the technology implemented at the juice purification stage, and the process involved in sucrose crystallization are the factors which indicate the wide variation in the composition of the molasses and can be used as a polycomponent system (Higginbotham and McCarthy 1998). Fermentable sugars such as sucrose, glucose, and fructose as well as non-sugar materials generated during the purification stage from the compounds which do not precipitate, chemical as well as enzymatic transformed substances such as D- and L- lactic acid, short-chain fatty acid are the main constituents of the molasses (Higginbotham and McCarthy 1998). Beet molasses have notable contents of potassium (around 3.6%) (Higginbotham and McCarthy 1998). Minerals found in the molasses in dissolved forms, hence they can easily be absorbed in the organism (Susic and sinobad 1989). Higher content of potassium makes molasses attractive for the use in human nutrition and better scope for their diversification through nutraceutical approaches. Apart from these facts, beet molasses has remarkable antioxidative properties, which can be further exploited on the mass scale as source of antioxidants as well as one of the ingredients in the functional foods (Chen et al. 2015; Chou 2003). Phenolic compounds and their derivatives, melanin, melanoidins, and products of the sugar caramelization presence in molasses contribute to the antioxidant capacity of the molasses. (Filipcev et al. 2016). Several compounds such as pantothenic acid, trace elements, inositol, and biotin in a minute concentration are found in molasses; these compounds can promote or inhibit the microbial population. Hence, these compound can be exploited as substrates for the biochemical transformations (Higginbotham and McCarthy 1998). Sugar beet molasses can be used as a substrate at the largescale production of baker's and brewer's yeast, monosodium glutamate (MSG), ethanol, citric acid, and lysine (Filipcev and Levic 2014).

45.8 Fermentation Technology

Micro-organisms are very crucial for fermentation industries since it helps in the production of bioethanol by fermentation of sugar. Micro-organisms play two main applications during the action of ethanol fermentation. In the first application, micro-organisms convert fermentable substrates into ethanol, and in the second application micro-organisms produce the enzymes which catalyse conversion of complex carbohydrates into simpler sugars. Sugar acts as a substrate for several bacteria, yeasts, and fungi which, on fermentation, produces bioethanol (Thatoi et al. 2014). Like any other fermentation technique, the fermentation of tuber crops for bioethanol generation can be accomplished by three processes, viz. batch, fed batch, and continuous fermentation.

45.8.1 Batch Fermentation

In batch fermentation technique, fermentation occurs in a closed-loop system. Fermenter is filled with the prepared mash of raw materials to be fermented. Microbial fermentation occurs in controlled pH and temperature conditions and, at different intervals, nutritive supplements are added as per requirements. The mash is steam-sterilized and the inoculum is added to the fermenter from a separate culture vessel. There are four phases in batch culture, namely, lag phase, exponential or logarithmic phase, stationary phase, and death phase. The batch system includes tanks that are designed on the basis of fermentation tank capacity and holding time. The tank is furnished with heat exchangers, agitators, mixing impellers, and a motor for optimum performance. Batch culture is useful where the shelf life of the end product is less and is mainly used for the product generated only at the stationary phase (Zhao and Bai 2009). The bioethanol productions from intermediates of sugar beet processing in batch culture by free *S. cerevisiae* cell were reported extensively (Grahovac et al. 2012). The ethanol yield from the media based on intermediates of sugar beet reported up to 490 g/kg. Bioethanol production from industrial sweet potatoes involved liquefaction, saccharification, and fermentation using α-amylase and glucoamylase for the production of ethanol by batch fermentation technique. Batch systems can be replaced by fed-batch cultures for lower manufacturing costs and better productivity.

45.8.2 Fed-Batch Fermentation

The technique which is between batch and continuous fermentation is called fed-batch fermentation. A proper feed rate with the correct component constitution is maintained during the process. Fed-batch process is a more effective cultivation strategy as compared to batch process in which micro-organisms perform at low substrate concentration with an increasing ethanol concentration. The yield of fed-batch cultures is more as compared to batch culture and it prevents contaminations (Wang et al. 2007).

45.8.3 Continuous Fermentation

In continuous fermentation, the substrate is added continuously to the fermenter at a fixed rate. This keeps the micro-organisms in the logarithmic growth phase. The fermentation products are removed continuously. The benefits of continuous fermentation technique over batch and fed-batch fermentation are that the yield of the last product in continuous system is much higher than batch and fed-batch culture. The ethanol yield reached a level of 83.53% theoretically with a maximum total substrate (reducing sugars) conversion of 92.68%. These fermentation techniques have a vital role on the bioethanol production from the tuber crops, and the type of fermentation technique can be modified depending on the type of substrate and micro-organisms used (Thatoi et al. 2014).

45.9 Impact of Different Factors on Fermentation Ethanol Production

Several components such as temperature, pH, fermentation time, agitation rate, initial sugar concentration, and inoculum size have an important role on the fermentation process as well as ethanol yield.

45.9.1 Temperature and pH

The ideal fermentation temperature for ethanol production ranges between 20 and 35 °C. The application of immobilized yeast cell in sweet sorghum juice yields 75.59% of ethanol at 28 °C, while at 37 °C, the ethanol yield was 89.89% (Liu and Shen 2008). High temperature causes stress to microorganisms, which is not suitable for their optimum growth and they produce heat-shock proteins in response to the high temperature. In addition, microbial activity and different enzymes used during fermentation process are also sensitive to high temperature. Enhanced ethanol production can be obtained by maintaining pH of the broth because pH has direct influence on organisms as well as on their cellular processes. The optimum pH range for *S. cerevisiae* used in fermentation for ethanol production is 4.0–5.0 (Lin et al. 2012). Depending on feedstocks, different optimum pH range was also reported such as 2.8–3.4 for sugarcane juice and 4.0–4.5 for sucrose (Isono and Hoshino 2000).

45.9.2 Fermentation Time and Agitation Rate

Shorter time in fermentation leads to sufficient development of micro-organisms. On the other hand, higher fermentation time leads to toxic effect on microbial growth mainly in batch mode due to the high concentration of metabolites produced in the fermented broth. Agitation plays an important role in obtaining maximum yield of ethanol during fermentation by increasing the movements of nutrients from the fermentation broth to inside the cells. Agitation also increases the sugar consumption and reduces the inhibition of ethanol on cells. The agitation rate is 150–200 rpm for yeast cells in fermentation. Liu and Shen (2008) reported the maximum ethanol yield (85.73%) at 200 rpm of agitation.

45.9.3 Sugar Concentration

Initial sugar concentration is very crucial as it has a direct impact on fermentation rate and microbial cells. The complexity between initial sugar content and the fermentation rate is more complex as the fermentation rate increases with the increase in sugar concentration up to a certain level.

45.9.4 Distillation and Bioethanol Recovery

During bioethanol production, separation process is very important and expensive as it consumes the highest energy in the process. Different novel separation techniques have been established to separate and clarify ethanol more efficiently. The distillation column is widely accepted as the crucial method for separation due to its performance and accuracy (Amornraksa et al. 2020). In the ethanol industry, two-column distillation separation system was widely implemented to separate ethanol from the fermentation broth. The distillation column method is an energy-intensive process, which could increase the cost of bioethanol recovery from the fermentation broth. In this context, several separation methods have been implemented such as molecular sieve by adsorption, solvent extraction, pervaporation (Membrane distillation), gas stripping, and vacuum fermentation.

45.9.4.1 Molecular Sieve by the Adsorption

The molecular sieve by adsorption process has been commercialized as a conventional method for the separation of bioethanol from broth (Tgarguifa and Abderafi 2016). The difference in molecular size between water and ethanol is the basis of molecular sieve. In this technique, small molecules that pass through the pores are adsorbed, while the larger molecules are not. The pore diameter of molecular sieve is typically 3A° for ethanol dehydration. Therefore, molecular sieve is capable of adsorbing water of diameter of 2.5–2.8 A° but not ethanol that has a diameter of 4–4.4A° (Kumar et al. 2010). Thus, this technique helps in the separation of bioethanol from broth.

45.9.4.2 Solvent Extraction

Apart from the conventional molecular sieve, extractive distillation method can be widely utilized for the production of anhydrous ethanol. The extractive distillation consists of two columns, in which one is extractive distillation column and the other is recovery column. In solvent extraction technique, ethylene glycol is used to change the relative volatilities of the components as well as used for ethyl alcohol dehydration. The experimental and simulation data revealed that extractive distillation could be widely accepted in industry to attain high purity of ethanol with low energy expenditure (Bastidas et al. 2010).

45.9.4.3 Pervaporation (Membrane Distillation)

Pervaporation is a promising membrane separation technique that can be utilized to produce anhydrous ethanol. Basically, in this technique, a liquid feed is separated into two streams, namely, permeate and retentate. The water passes through the membrane as vapour called permeate, while the ethanol remains in the liquid phase as retentate. The differential pressure on both sides of the layer is controlled by temperature difference in membrane distillation technique. The pressure difference is caused due to permeate and retentate (Jaimes et al. 2014). Therefore, membrane distillation is adopted as more efficient, easy to operate, and with low energy consumption.

45.9.4.4 Gas Stripping

Gas stripping separation techniques remove components by forming mixture into a gas passing through the fermentation broth. In this technique, carbon dioxide or nitrogen is used to evaporate the bioethanol which will be collected from the gas stream via a condenser, while carbon dioxide or nitrogen is recycled through the bioreactor for next cycle. In the gas stripping separation technique, the increase in stripping factor results in an increase in cell concentration, substrate utilization, enhancement in ethanol productivity, and to check the ethanol inhibitory effect. Air, CO_2, and N_2 are the most commonly used stripping gases for ethanol recovery in industry (Zentou et al. 2019).

45.9.4.5 Vacuum Fermentation

In this process, application of vacuum pressure in fermentation broth leads to the evaporation of ethanol which is subsequently condensed by condensation cooling system or chilling water. In this technique, ethanol concentration can be regulated at a low level that reduces the inhibitory action of ethanol on yeast metabolism and fermentation process (Huang et al. 2015).

45.10 Utilization of Molasses

Molasses obtained during the sugar production can be further exploited to get several value-added products and enhances the remuneration generation from the crop. It can be utilized as raw material is several industries such as animal husbandry, food processing, plastic and composite manufacturing, etc., to get the desirable end product are discussed in the following section.

45.11 Livestock Feed Material

The molasses obtained during the sugar processing is generally used to obtain bioethanol through fermentation. The rest of the molasses, generally rich in nitrogen, can be utilized either as animal feed or fertilizer. However, the remainder of the beet pulp is pressed as well as dehydrated and further, it can used to feed the animals (Razmovski and Vucurovic 2012). A significant fraction of the structural polysaccharide such as pectin and dietary fibre are also present in the sugar beet pulp. Judicious exploitation of the co-products can reduce the wastes and further increase the value of the crop.

45.12 Food Quality Enhancer

Value addition of the various food products can be carried out with incorporation of molasses without affecting their palatability. Food products endowed with beet molasses exhibit increased mineral as well as antioxidant profile (Filipcev et al. 2016). It can be utilized to augment wheat bread up to 5–10% (Flour basis), semi-sweet cookies at up to 25%, while in the formulation of ginger-based bread-type biscuit, it can be added up to 50% as a honey replacer (Filipcev et al. 2012). Extraction of fibre from sugar beet pulp is a relatively simple process and the obtained fibre from SBP has been generally recognized as safe (GRAS) (Nordic Sugar 2012). Nutritive value of the commercial products of the beet fibre entails the presence of protein (8%) on dry weight basis and carbohydrates mainly hemicellulose 28%, cellulose 19%, and pectin 18% (Michel et al. 1988).

Fibre produced from sugar beet has a broad spectrum of human health-promoting effects (Ralet et al. 2009) since, fibre obtained from sugar beet can be either in the form of pulp or an unadulterated pectic substance such as arabinan (Goodban and Owens 1956). Utilization of sugar beet fibre in the processed foods is restricted by their texture and taste. Generally, it is utilized in the preparation of meat patties, bakery based products, cereals, and assorted products which requires the thickening or bulking promoter (Dhingra et al. 2012). Pectic oligosaccharides such as arabinan in SBP have the prebiotic properties for the human gut (Tamimi et al. 2006). Prebiotic substances have the ability to modulate the microbial population of the human gut and it favours the growth of the beneficial bacteria with respect to human health. Sugar beet molasses can be utilized as a substrate (Sugar/carbon) for the fermentation of xanthan carried out by the microorganisms (Moosavi and Karbassi 2010). Food industry extensively uses the xanthan as a thickener.

45.13 Production of Pectin

Pectin, being fibrous in nature, is a structural heteropolysaccharide primarily composed of the galactouronic acid, derived from the galactose, while rhamnose is attached in the side chains in varying fractions. Pectin has the established role for their gelling properties in the fruit products (Norsker et al. 2000). Emulsifying properties of the sugar beet pectin is better over other known sources of pectin. Pectinolytic enzymes and treatments with mild organic acid can be used to get attractive pectin yields from the SBP (Concha-Olmos and Zuniga-Hansen 2012). Stabilization properties of the pectin derived from sugar beet in emulsions are contributed by the attachment of protein (Fishman et al. 2013). Several low-weight molecules such as arabinan can be isolated from SBP, which can be further exploited in the synthesis of adhesive, emulsion stabilizer (Fishman et al. 2009), and the suspension agent utilized in cosmetics as well as pharmaceuticals (Goodban and Owens 1956).

45.14 Plastics and Composites Manufacturing

Most of the plastics found in the global market are generally petroleum driven; many research efforts in the public domain are available to replace the petro-based plastic with bioplastics generated through renewable resources. In most of the cases, polymers need to be extracted from the plant before their use, while in some of the cases, polymers are synthesizes from small molecules which are also obtained from the plants. Thermoplastic films can be obtained from the sugar beet pulp after their processing in a twin-screw extruder with the aid of plasticizer (Liu et al. 2011). Finally, composite product derived from the SBP can be visualized as the network of suspended cellulose microfibrils woven in a pectin matrix. Productions of urethanes from SBP as a polyol source are also used in the practice (Pavier and Gandini 2000). Biobased polymer, such as polylactic acid, can be blended with sugar beet pulp and form a composite polymer with comparable tensile strength as found in the commodity plastics (Chen et al. 2008). Polyesters like polyhydroxyalkanoates (PHA) derived from either plant or microbial source are gaining popularity in the market for the plastic substitutes. Sugar beet juice as a sugar substrate can be utilized in PHA production (poly hydroxybutyrate) (Wang et al. 2013). PHB is one of the important polymers having plastic properties like polypropylene that can be obtained from biological source. Being biodegradable in nature, they can be compostable and eco-friendly. Cost of the PHA production based on the carbon source used during the process and it can be determined based on the sugar beet molasses as an exclusive feedstock (Castilho et al. 2009).

45.15 Transformation to Platform Chemicals

Exploration for an alternative source of fuel which can be renewable in nature led to the technology-driven utilization of complete biomass in s sustainable way (Hood et al. 2013). The advancement in technology to utilize uses fermentable sugar and their further conversion to the broad range of energy-rich chemicals such as ethanol, which is the current choice of the hour, most of these techniques were developed keeping in view the lignocellulosic grasses. Due to their low lignin content and better carbohydrate digestibility, SBP can be utilized as feedstock material for biorefineries. Breakdown of the complex networks of cell wall and their building blocks such as pectin and cellulose in SBP is a prerequisite for the fermentation. Yield of the fermentable sugar obtained from the SBP was influenced due to the level and severity of the pre-treatment as well as use of the different enzymes (Kuhnel et al. 2011). SBP can be utilized to produce the vanillin naturally through bioconversion with the aid of the fungal enzymes using both the ferulic and cellobiose in small and large amounts (Bonnin et al. 2000). Food as well as drug industries exploit the ferulic acid as a precursor (Kroon et al. 1996). Vinasse, obtained during the ethanol production from sugar beet, has the betaine contents (15%) that can be used for production of the amphoteric surfactants, which can be further used for the formulation of personal care products. Proteomics and pharmaceutical studies utilizes the Galactinol dehydrate and myo-inositol obtained from the sugar beet syrup (Mccready et al. 1965). Sugar beet molasses was used to produce oxalic acid in the presence of vanadium as a catalyst with 75% yield (Guru et al. 2001).

45.16 Source of Carbon to Remove the Contaminants

Growing concerns towards the emission of greenhouse gases and their impacts on the atmospheric changes in the environment leads to the adoption of a renewable and sustainable energy approach which can utilize biomass as raw material and the process involved is known as carbon-neutral. Water bodies contamination with

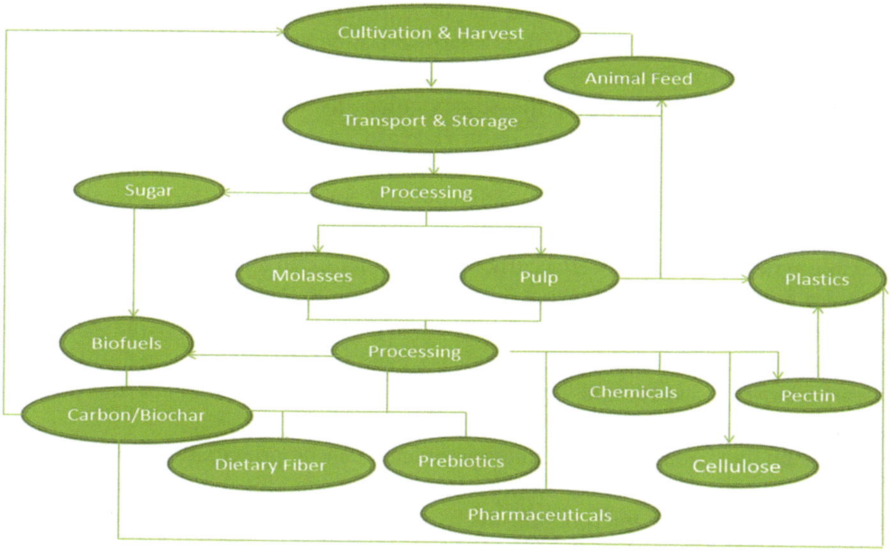

Fig. 45.6 Diagrammatic presentation of value-added products obtained from sugar beet

toxic material is a current issue in context of environmental and health perspective. Utilization of agriculture-based materials can be advantageous over the conventional processes such as their low cost, ability to regenerate biosorbent, and efficient recovery of the heavy metals (Kolodynska et al. 2012). Binding potential of SBP can be enhanced as an ion-exchanger (Dronnet et al. 1998) that can lead to their increased value and demand in the market.

45.17 Production of Cellulose

Cellulose can be isolated from SBP (Togrul and Arslan 2003) (Fig. 45.6) that can be further used for the preservation of fruits and vegetables with maintenance of their freshness during their transportation phase and prolonged storage (Togrul and Arslan 2004). Internal bond energy in the paper products can be enhanced by the use of SBP with their extended network of cellulose microfibrils in the gelatinous pectin matrix (Gigac et al. 2008).

45.18 Future Perspective

Basically, fermentation stage is targeted to enhance the bioethanol production from juice and by-products of the sugar beet. Since targeting of other stages for their modification in the bioethanol production is tedious for various reasons: firstly, production of juices (Sugar beet juice extraction) is the integral part of the sugar industry which is a highly optimized and beneficial process. Accordingly, introduction of changes into the process flow chart cannot be easy without influencing the economic accomplishment of the sugar extraction plant. Therefore, bioethanol industry should have to utilize the sugary juices as well as by-products as it is produced by sugar processing unit (Hahn-Hagerdal et al. 2006). Second, with respect to manipulation at the separation stage (bioethanol distillation), nowadays most of the technology practised for fermentative ethanol concentration (Extractive distillation, inverse osmosis, vacuum distillation, reactive distillation, etc.) already attained maximum development as well as efficiency, further extensive improvements in this field are not expected in the short run (Shihadeh et al. 2014). Therefore, focus on fermentation stage for improvement in bioethanol production can be achieved in two possible ways: (1) development and enhancement of fermentation capability of the microorganism (Caspeta and Nielsen 2015); and (2) improvement in the process productivity (Fakruddin et al. 2012).

Fermentative capacity of the microorganisms can be principally improved with the aid of genetic engineering. Approaches implemented for this purpose based on the development of the microorganisms which can be utilized in different classes of sugars simultaneously as a substrate and ferment them to yield ethanol, elimination of inhibitors, ability to adapt in adverse environment (temperature extremes, pH, high salt concentration, lack of nutrients) without affecting the yield. However, conventional evolutionary approach can be one of the technical options for attainment of those goals (Fadel et al. 2013). Conventional evolutionary approaches strategy requires more efforts as well as time to get significant results, while genetic engineering strategy can achieve it within the timeframe with maximum chances of success. Utilization of genetically modified organism in energy sector is not as rejected as in the food and agriculture sectors. Co-culturing of various microorganisms can be utilized to increase the fermentation capacity, in which highly specific strains are exploited in specific sugar progressively or simultaneously for the production of ethanol (Hickert et al. 2013). In view of the implementation of fermentative process, two approaches for the improvement: (1) immobilization of the microorganism for continuous mode fermentation (Razmovski and Vucurovic 2012), (2) while, in case of solid by-products such as sugar beet pulp, solid state fermentation can be implemented. Different types of substrates such as alginate polymers (Duvernay et al. 2013), naturally occurring solid polymers (Kirdponpattara and Phisalaphong 2013), or agricultural milled residue (Pacheco et al. 2010); and unique strain of the immobilized microorganisms or various types of coimmobilized bacteria and fungi (Guo et al. 2010) can be utilized for the microorganism immobilization to carry out continuous fermentation. Bioethanol can be produced from exploitation of the sugar beet pulp. However, the main constituent of the waste product, polysaccharide, must be hydrolysed into simple sugar, which can be further fermented and ethanol obtained. Optimization of this process can be attained by the use of unique microorganism, which has the ability to carry out both saccharification and fermentation following the solid state fermentation (Moukamnerd et al. 2013).

45.19 Conclusion

Extensive consumption of fossil fuel leads to decline in their reservoir and that necessitates to explore the economically, ecologically, and environmentally viable and sustainable alternative source of energy. Bioethanol-based energy approach can lead to lesser emission of gaseous pollutants. Worldwide initiatives are taken to formulate policies for the utilization of the biomass to meet future energy demand and their strict implication to achieve their target for the reduction of CO_2 emission. Sugar beetroots are one of the alternative substrates for bioethanol production due to their chemical composition. Numerous by-products, intermediates, and garbage are produced from the sugar beet after extraction of sugar that can be further exploited to meet the energy demand and other value-added products. Nowadays, technology-driven approaches are utilized and focused to generate ethanol from lignocellulosic mass (Second generation feedstock) due to their non-competitiveness with food crops (First generation feedstock). Various kinds of polymers found in the by-products such as molasses and SBP are linked with each other in varying amounts and different arrangements. Separation of such types of polymers can be achieved with pre-treatment to release their sugar constituent. Various types of biochemicals such as minerals, antioxidant compounds, vitamins present in molasses enhance their scope for application in nutraceuticals and functional foods. Micro-organisms are a limiting factor for the fermentation industries due to their ability to ferment the sugary substrate into bioethanol. Generally, fermentation stages are targeted to enhance the bioethanol production from various types of feedstocks. The utilization of advanced tools and techniques of genetic engineering for the improvement of fermentative capacity of the microorganism can improve the bioethanol production rather than targeting the other stages involved in the process. Development of microorganisms which can be able to utilize different classes of sugars as substrates and ferment them simultaneously for a better yield of ethanol, high adaptability in harsh environment without affecting the yield, ability to neutralize the inhibitors. Adoption of conventional evolutionary approach can be one of the options to reach those goals. However, extensive efforts and time involved in conventional approach to get better output while, it can be achieved within the short timeframe with maximum chances to get better result.

References

Agi A, Junin R, Gbadamosi A, Abbas A, Azli NB, Oseh J (2019) Influence of nanoprecipitation on crystalline starch nanoparticle formed by ultrasonic assisted weak-acid hydrolysis of cassava starch and the rheology of their solutions. Chem Eng Process 142:107556. https://doi.org/10.1016/j.cep.2019.107556

Amornraksa S, Subsaipin I, Simasatitkul L, Assabumrungrat S (2020) Systematic design of separation process for bioethanol production from corn stover. BMC Chem Eng 2(1):1–16

Asadi M (2006) Beet-sugar handbook. Wiley, Hoboken

Bastidas PA, Gil ID, Rodríguez G (2010) Comparison of the main ethanol dehydration technologies through process simulation. In 20th European symposium on computer aided process engineering

Benoliel B, Torres FAG, de Moraes LMP (2013) A novel promising *Trichoderma harzianum* strain for the production of a cellulolytic complex using sugarcane bagasse in natura. Springerplus 2(1):656. https://doi.org/10.1186/2193-1801-2-656

Biancardi E, McGrath JM, Panella LW, Lewellen RT, Stevanato P (2010) Sugar beet. In: Root and tuber crops. Springer, New York, pp 173–121

Bonnin E, Grange H, Lesage-Meessen L, Asther M, Thibault JF (2000) Enzymic release of cellobiose from sugar beet pulp, and its use to favour vanillin production in Pycnoporus cinnabarinus from vanillic acid. Carbohydr Polym 41:143–151

Canilha L, Santos VT, Rocha GJ, e Silva JBA, Giulietti M, Silva SS, Felipe MG, Ferraz A, Milagres AM, Carvalho W (2011) A study on the pretreatment of a sugarcane bagasse sample with dilute sulfuric acid. J Ind Microbiol Biotechnol 38(9):1467–1475. https://doi.org/10.1007/s10295-010-0931-2

Caspeta L, Nielsen J (2015) Thermotolerant yeast strains adapted by laboratory evolution show trade-off at ancestral temperatures and pre-adaptation to other stresses. mBio 6:e00431–15

Castilho LR, Mitchell DA, Freire DMG (2009) Production of polyhydroxyalkanoates (PHAs) from waste materials and byproducts by submerged and solid-state fermentation. Bioresour Technol 100:5996–6009

Chandel AK, Chan ES, Rudravaram R, Narasu ML, Rao LV, Ravindra P (2007) Economics and environmental impact of bioethanol production technologies: an appraisal. Biotechnol Mol Biol Rev 2(1):14–32

Chen F, Liu L, Cooke PH, Hicks KB, Zhang J (2008) Performance enhancement of poly(lactic acid) and sugar beet pulp composites by improving interfacial adhesion and penetration. Ind Eng Chem Res 47:8667–8675

Chen M, Zhao Y, Yu S (2015) Optimisation of ultrasonic-assisted extraction of phenolic compounds, antioxidants, and anthocyanins from sugar beet molasses. Food Chem 172:543–550

Chou CC (2003) Preparation antioxidants enriched functional food products from sugar cane and beet. United States patent applications #127141

Concha-Olmos J, Zuniga-Hansen ME (2012) Enzymatic depolymerization of sugar beet pulp: production and characterization of pectin and pectic-oligosaccharides as a potential source for functional carbohydrates. Chem Eng J 192:29–36

Dhingra D, Michael M, Hradesh R, Patil RP (2012) Dietary fibre in foods: a review. J Food Sci Technol 49:255–266

Dronnet VM, Axelos MAV, Renard CMGC, Thibault JF (1998) Improvement of the binding capacity of metal cations by sugar-beet pulp. 1. Impact of cross-linking treatments on composition, hydration and binding properties. Carbohydr Polym 35:29–37

Duvernay WH, Chinn MS, Yencho GC (2013) Hydrolysis and fermentation of sweet potatoes for production of fermentable sugars and ethanol. Ind Crop Prod 42:527–537

Fadel M, Keera AA, Mouafi F, Kahil T (2013) High level ethanol from sugar cane molasses by a new thermotolerant *Saccharomyces cerevisiae* strain in industrial scale. Biotechnol Res Int 2013:6

Fakruddin M, Abdul Quayum M, Ahmed MM, Choudhury N (2012) Analysis of key factors affecting ethanol production by *Saccharomyces cerevisiae* IFST- 072011. Biotechnology 11: 248–252

Filipcev B, Levic LJ (2014) Primena melase secerne repe i poluproizvoda od osmotski dehidriranog voca - povrca u melasi kao nutritivno vrednih dodataka za obogacivanje pekarskih i finih pekarskih proizvoda, Monografija, Univerzitet u Novom Sadu, Naucni institut za prehrambene tehnologije u Novom Sadu, Novi Sad

Filipcev B, Bodroza-Solarov M, Simurina O, Cvetkovic B (2012) Use of sugar beet molasses in processing of gingerbread type biscuits: effect on quality characteristics, nutritional profile, and bioavailability of calcium and iron. Acta Alimentaria Hungarica 41(4):494–505

Filipcev B, Misan A, Saric B, Simurina O (2016) Sugar beet molasses as an ingredient to enhance the nutritional and functional properties of gluten-free cookies. Int J Food Sci Nutr 67:249–256

Finkenstadt VL (2014) A review on the complete utilization of the sugarbeet. Sugar Tech 16(4): 339–346

Fishman ML, Chau HK, Cooke PH, Yadav MP, Hotchkiss AT (2009) Physico-chemical characterization of alkaline soluble polysaccharides from sugar beet pulp. Food Hydrocoll 23:1554–1562

Fishman ML, Chau HK, Qi PX, Hotchkiss J, Yadav MP (2013) Physico-chemical characterization of protein-associated polysaccharides extracted from sugar beet pulp. Carbohydr Polym 92: 2257–2266

Gigac J, Fsierova M, Rosenberg M (2008) Improvement of paper strength via surface application of sugar beet pectin. Chem Pap 62:509–515

Goodban AF, Owens HS (1956) Isolation and properties of sugar beet araban. J Sugar Beet Res IX:129–132

Grahovac JA, Dodić JM, Dodić SN, Popov SD, Vučurović DG, Jokić AI (2012) Future trends of bioethanol co-production in Serbian sugar plants. Renew Sust Energ Rev 16(5):3270–3274

GRFA (Global Renewable Fuel Alliance) Reports (2017). http://globalrfa.org/news-media/2017-forecast-global-ethanol-production-stable

Gumienna M, Szwengiel A, Szczepańska-Alvarez A, Szambelan K, Lasik-Kurdyś M, Czarnecki Z, Sitarski A (2016) The impact of sugar beet varieties and cultivation conditions on ethanol productivity. Biomass Bioenergy 85:228–234

Guo X, Zhou J, Xiao D (2010) Improved ethanol production by mixed immobilized cells of *Kluyveromyces marxianus* and *Saccharomyces cerevisiae* from cheese whey powder solution fermentation. Appl Biochem Biotechnol 160:532–538

Guru M, Bilgesu AY, Pamuk V (2001) Production of oxalic acid from sugar beet molasses by formed nitrogen oxides. Bioresour Technol 77:81–86

Hahn-Hagerdal B, Galbe M, Gorwa-Grauslund MF, Linden G, Zacchi G (2006) Bioethanol: the fuel of tomorrow from the residues of today. Trends Biotechnol 24:549–556

Hattori T, Morita S (2010) Energy crops for sustainable bioethanol production; which, where and how? Plant Prod Sci 13(3):221–234

Hickert LR, Da Cunha-Pereira F, De Souza-Cruz PB, Rosa CA, Ayub MAZ (2013) Ethanogenic fermentation of co-cultures of *Candida shehatae* HM 52.2 and *Saccharomyces cerevisiae* ICVD254 in synthetic medium and rice hull hydrolysate. Bioresour Technol 131:508–514

Higginbotham JD, McCarthy J (1998) Quality and storage of molasses. In van der Poel PW, Schiweck H, Schwartz T, Verlag, Albert Bartens KG (eds) Sugar technology-beet and cane manufacture. Berlin, pp 973–992

Hood E, Teoh K, Devaiah S, Vicuna Requesens D (2013) Biomass crops for biofuels and bio-based products. In: Christou P, Savin R, Costa-Pierce B, Misztal I, Whitelaw CB (eds) Sustainable food production. Springer, New York, pp 250–279

Huang H, Qureshi N, Chen MH, Liu W, Singh V (2015) Ethanol production from food waste at high solids content with vacuum recovery technology. J Agric Food Chem 63(10):2760–2766

Isono Y, Hoshino A (2000) Production of ethanol using granulated yeast cells prepared by a spray dryer. J Gen Appl Microbiol 46(4):231–234

Jaimes JHB, Alvarez ME, Rojas JV, Maciel Filho R (2014) Pervaporation: promissory method for the bioethanol separation of fermentation. Chem Eng 38:139–144

Kamzon MA, Abderafi S, Bounahmidi T (2016) Promising bioethanol processes for developing a biorefinery in the Moroccan sugar industry. Int J Hydrog Energy 41(45):20880–20896

Khan MI, Lee MG, Shin JH, Kim JD (2017) Pretreatment optimization of the biomass of microcystis aeruginosa for efficient bioethanol production. AMB Express 7(1):1–9

Kim S, Dale BE (2004) Global potential bioethanol production from wasted crops and crop residues. Biomass Bioenergy 26(4):361–375

Kim Y, Kreke T, Ko JK, Ladisch MR (2015) Hydrolysis determining substrate characteristics in liquid hot water pretreated hardwood. Biotechnol Bioeng 112(4):677–687. https://doi.org/10.1002/bit.25465

Kirdponpattara S, Phisalaphong M (2013) Bacterial cellulose alginate composite sponge as a yeast cell carrier for ethanol production. Biochem Eng J 77:103–109

Kolodynska D, Wnetrzak R, Leahy JJ, Hayes MHB, Kwapinski W, Hubicki Z (2012) Kinetic and adsorptive characterization of biochar in metal ions removal. Chem Eng J 197:295–305

Kroon PA, Faulds CB, Brezillon C, Williamson G (1996) Enzymic release of ferulic acid from sugar beet pulp using a specific esterase from *Aspergillus niger*. In: Visser J (ed) Progress in biotechnology pectins and pectinases proceedings of an international symposium. Elsevier, Amsterdam, pp 761–768

Kuhnel S, Schols HA, Gruppen H (2011) Aiming for the complete utilization of sugar-beet pulp: examination of the effects of mild acid and hydrothermal pretreatment followed by enzymatic digestion. Biotechnol Biofuels 4:14

Kumar S, Singh N, Prasad R (2010) Anhydrous ethanol: a renewable source of energy. Renew Sust Energ Rev 14(7):1830–1844

Kumar P, Varkolu M, Mailaram S, Kunamalla A, Mait SK (2019) Biorefinery polyutilization systems: production of green transportation fuels from biomass. In: Polygeneration with polystorage for chemical and energy hubs. Academic, pp 373–407

Lin Y, Zhang W, Li C, Sakakibar K, Tanaka S, Kong H (2012) Factors affecting ethanol fermentation using *Saccharomyces cerevisiae* BY4742. Biomass Bioenergy 47:395–401

Liu R, Shen F (2008) Impacts of main factors on bioethanol fermentation from stalk juice of sweet sorghum by immobilized *Saccharomyces cerevisiae* (CICC 1308). Bioresour Technol 99(4): 847–854

Liu B, Zhang J, Liu L, Hotchkiss AT (2011) Preparation and properties of water and glycerol-plasticized sugar beet pulp plastics. J Polym Environ 19:559–567

Mall AK, Misa V, Santeshwari PAD, Srivastava S (2021) Sugar beet cultivation in India: prospects for bioethanol production and value added coproducts. Sugar Tech 23:1218–1234

Manzanares P, Ballesteros I, Negro M, Gonzalez A, Oliva J, Ballesteros M (2020) Processing of extracted olive oil pomace residue by hydrothermal or dilute acid pretreatment and enzymatic hydrolysis in a biorefinery context. Renew Energy 145:1235–1245. https://doi.org/10.1016/j.renene.2019.06.120

Marzo C, Díaz AB, Caro I, Blandino A (2019) Status and perspectives in bioethanol production from sugar beet. In: Ray RC, Ramachandran S (eds) Bioethanol production from food crops: sustainable sources, interventions, and challenges. Academic Press, pp 61–79

McCready RM, Stark JB, Goodban AE (1965) Preparation of galactinol and myoinositol from sugar beet sirup by chromatography on a cation exchange resin. J Sugar Beet Res 14:127–132

Melendez-Hernande PA, Hernandez-Beltran, JU, Hernandez-Guzman A, Morales-Rodriguez R, Torres-Guzman J, Hernandez-Escoto H (2019) Comparative of alkaline hydrogen peroxide pretreatment using NaOH and $Ca(OH)_2$ and their effects on enzymatic hydrolysis and fermentation steps. Biomass Convers Biorefin 1–11. https://doi.org/10.1007/s13399-019-00574-3

Michel F, Thibault JF, Barry JL, de Baynast R (1988) Preparation and characterisation of dietary fibre from sugar beet pulp. J Sci Food Agric 42:77–85

Mohanty SK, Swain MR (2019) Bioethanol production from corn and wheat: food, fuel, and future. In: Bioethanol production from food crops. Academic, pp 45–59

Mohapatra S, Pattathil S, Thatoi H (2017) Structural and functional characterization of two Pennisetum sp. biomass during ultrasono-assisted alkali pretreatment and enzymatic hydrolysis for understanding the mechanism of targeted delignification and enhanced saccharification. ACS Sustain Chem Eng 5(8):6486–6497

Mohapatra, S, Ray, R, Ramachandran S (2019) Bioethanol from biorenewable feedstocks: technology, economics, and challenges. In: Bioethanol production from food crops. Academic, pp 3–27

Moosavi A, Karbassi A (2010) Bioconversion of sugar-beet molasses into xanthan gum. J Food Process Preserv 34:316–322

Moukamnerd C, Kawahara H, Katakura Y (2013) Feasibility study of ethanol production from food wastes by consolidated continuous solid-state fermentation. J Sust Bioenergy Syst 3:143–148
Niphadkar S, Bagade P, Ahmed S (2018) Bioethanol production: insight into past, present and future perspectives. Biofuels 9(2):229–238
Nordic Sugar (2012) GRAS notification for sugar beet fiber. http://www.fda.gov/Food/IngredientsPackagingLabeling/GRAS/NoticeInventory/ucm347789.htm. Accessed 20 Oct 2013
Norsker M, Jensen M, Adler-Nissen J (2000) Enzymatic gelation of sugar beet pectin in food products. Food Hydrocoll 14:237–243
Pacheco AM, Gondim DR, Gonçalves LRB (2010) Ethanol production by fermentation using immobilized cells of *Saccharomyces cerevisiae* in cashew apple bagasse. Appl Biochem Biotechnol 161:209–217
Pavier C, Gandini A (2000) Oxypropylation of sugar beet pulp. 1. Optimisation of the reaction. Ind Crop Prod 12:1–8
Ralet MC, Guillon F, Renard C, Thibault JF (2009) Sugar beet fiber: production, characteristics, food applications and physiological benefits. In: Cho SS, Samuel P (eds) Fiber ingredients: food applications and health benefits. Taylor & Francis, Florence
Ray RC, Uppuluri KB, Trilokesh C, Lareo C (2019) Sweet sorghum for bioethanol production: scope, technology, and economics. In: Bioethanol production from food crops. Academic, pp 81–100
Razmovski R, Vucurovic V (2012) Bioethanol production from sugar beet molasses and thick juice using *Saccharomyces cerevisiae* immobilized on maize stem ground tissue. Fuel 92:1–8
Rezic T, Oros D, Markovi I, Kracher D, Ludwig R, Santek B (2013) Integrated hydrolyzation and fermentation of sugar beet pulp to bioethanol. J Microbiol Biotechnol 23:1244–1252
RFA (2017) Ethanol industry outlook. Renewable Fuels Association, Washington, DC. www.ethanolrfa.org/wp-content/uploads/2017/02/Ethanol-Industry-Outlook-2017.pdf
Saini JK, Saini R, Tewari L (2015) Lignocellulosic agriculture wastes as biomass feedstocks for second-generation bioethanol production: concepts and recent developments. 3 Biotech 5(4): 337–353
Šantek B, Gwehenberger G, Šantek MI, Narodoslawsky M, Horvat P (2010) Evaluation of energy demand and the sustainability of different bioethanol production processes from sugar beet. Resour Conserv Recycl 54(11):872–877
Sarkar N, Ghosh SK, Bannerjee S, Aikat K (2012) Bioethanol production from agricultural wastes: an overview. Renew Energy 37(1):19–27
Shihadeh JK, Huang H, Rausch KD, Tumbleson ME, Singh V (2014) Vacuum stripping of ethanol during high solids fermentation of corn. Appl Biochem Biotechnol 173(2):486–500
Sindhu R, Binod P, Pandey A (2016) Biological pretreatment of lignocellulosic biomass–an overview. Bioresour Technol 199:76–82. https://doi.org/10.1016/j.biortech.2015.08.030
Singh R, Shukla A, Tiwari S, Srivastava M (2014) A review on delignification of lignocellulosic biomass for enhancement of ethanol production potential. Renew Sust Energ Rev 32:713–728
Singh R, Srivastava M, Shukla A (2016) Environmental sustainability of bioethanol production from rice straw in India: a review. Renew Sust Energ Rev 54:202–216
Singh S, Adak A, Saritha M, Sharma S, Tiwari R, Rana S, Nain L (2017) Bioethanol production scenario in India: potential and policy perspective. In: Sustainable biofuels development in India. Springer, Cham, pp 21–37
Susic S, Sinobad V (1989) Ispitivanja u cilju unapređenja industrije secera Jugoslavije. Hemijska Industrija 43(1–2):10–21
Taherzadeh MJ, Karimi K (2007) Acid-based hydrolysis processes for ethanol from lignocellulosic materials: a review. Bioresources 2(3):472–499
Tamimi MA, Palframan RJ, Cooper JM, Gibson GR, Rastall RA (2006) *In vitro* fermentation of sugar beet arabinan and arabino-oligosaccharides by the human gut microflora. J Appl Microbiol 100:407–414

Taskin M, Ortucu S, Aydogan MN, Arslan NP (2016) Lipid production from sugar beet molasses under non-aseptic culture conditions using the oleaginous yeast *Rhodotorula glutinis* TR29. Renew Energy 99:198–204

Tgarguifa A, Abderafi S (2016) A comparative study of separation processes for bioethanol production. In: 2016 international renewable and sustainable energy conference (IRSEC). IEEE, pp 890--895

Thatoi H, Dash PK, Mohapatra S, Swain MR (2014) Bioethanol production from tuber crops using fermentation technology: a review. Int J Sustain Energy 35(5):443–468

Togrul H, Arslan N (2003) Flow properties of sugar beet pulp cellulose and intrinsic viscosity—molecular weight relationship. Carbohydr Polym 54:63–71

Togrul H, Arslan N (2004) Extending shelf-life of peach and pear by using CMC from sugar beet pulp cellulose as a hydrophilic polymer in emulsions. Food Hydrocoll 18:215–226

van der Pol E, Bakker R, van Zeeland A, Garcia DS, Punt A, Eggink G (2015) Analysis of by-product formation and sugar monomerization in sugarcane bagasse pretreated at pilot plant scale: differences between autohydrolysis, alkaline and acid pretreatment. Bioresour Technol 181:114–123. https://doi.org/10.1016/j.biortech.2015.01.033

Van Fan Y, Perry S, Klemeš JJ, Lee CT (2018) A review on air emissions assessment: transportation. J Clean Prod 194:673–684

Venkatesh V (2012) Biofuels in India: potential, policy and emerging paradigms. Indian J Agric Econom 67(2):268

Wang FQ, Gao CJ, Yang CY, Xu P (2007) Optimization of an ethanol production medium in very high gravity fermentation. Biotechnol Lett 29(2):233–236

Wang B, Sharma-Shivappa RR, Olson JW, Khan SA (2013) Production of polyhydroxybutyrate (PHB) by Alcaligenes latus using sugarbeet juice. Ind Crop Prod 43:802–811

Zentou H, Abidin ZZ, Yunus R, Awang Biak DR, Korelskiy D (2019) Overview of alternative ethanol removal techniques for enhancing bioethanol recovery from fermentation broth. Processes 7(7):458

Zhao XQ, Bai FW (2009) Yeast flocculation: new story in fuel ethanol production. Biotechnol Adv 27(6):849–856

Sugar Beet as Cattle Feed: Scope and Prospects

46

Meenakshi Goyal and Aanchaldeep Kaur

Abstract

Forages are valuable in maintaining the cattle system and sustaining the production of milk products worldwide. Sugar beet is one of the major sugar producing crops and, nowadays, extending its scope as cattle feed. In the search for sustainability and economic value, the complete utilization of the crop is necessary. The roots and beet tops along with its value-added by-products are the efficient sources for cattle feeding. The by-products such as beet pulp and molasses can be included as alternative ingredients in cattle rations. They can surpass the need of grains and also help in waste disposal produced during sugar extraction. This chapter explores the nutritional and anti-nutritional aspects of sugar beet. Sugar beet contains different quantities of carbohydrates, proteins, minerals and vitamins and also vary in the proportion of their tissue that can be digested by the cattle. The highly digestible dietary fibre fraction is responsible for higher acetate to propionate ratio. The high acetate level helps in increasing the milk fat and its yield in cattle. In addition to nutrients, the anti-nutrients such as nitrate and oxalate are also present in sugar beet. Nitrate and oxalate, though playing an essential role in cattle, can become toxic beyond a certain limit.

Keywords

Feed beets · Nutritional constituents · Dietary fibre · Digestibility · Antinutritional constituents

M. Goyal (✉) · A. Kaur
Department of Plant Breeding and Genetics, Punjab Agricultural University, Ludhiana, India
e-mail: meenakshigoyal@pau.edu

V. Misra et al. (eds.), *Sugar Beet Cultivation, Management and Processing*,
https://doi.org/10.1007/978-981-19-2730-0_46

Abbreviations

ADF	Acid detergent fibre
ADS	Acid detergent solution
Fd	Ferredoxin
Hb	Haemoglobin
LAB	Lactic acid bacteria
Met	Methaemoglobin
N	Nitrogen
NDF	Neutral detergent fibre
NDS	Neutral detergent solution
NFC	Non-fibre carbohydrates
NH_3	Ammonia
NiR	Nitrite reductase
NO_2^-	Nitrite
NO_3^-	Nitrate
VFA	Volatile fatty acids
WSC	Water soluble carbohydrates
WSV	Water soluble vitamins

46.1 Introduction

Forages play a key role in agricultural scenario and contribute heavily towards the livestock sector. Forage grasslands are currently occupying 26% of land area and 70% of agricultural area (FAO 2019). Forages have major impact on the economic efficiency of milk and meat production. In developing countries, the scarcity of fodder is declining the performance of cattle. The global population is rising rapidly along with the demand for dairy products. This can only be fulfilled when cattle are fed with sufficient number of forages. Therefore, the development of cheap and high quality forage species is necessary to meet the future demands for cattle production.

Sugar beet (*Beta vulgaris* L.) is emerging as a cash crop all through the world and belongs to the amaranth family. It is mostly cultivated for sugar production and contributes substantially in cattle nutrition. It thrives in the regions of temperate climate and, nowadays, is extending its scope towards the subtropics. In temperate regions, it is sown in the spring season and harvested in the autumn season. In subtropical climate, it can be grown as winter crop and sown in the autumn season. On an average, it requires temperature of 15–21 °C, rainfall of 460 mm, sunlight of longer duration and low wind speed. It can be grown in wide varieties of soil, but the best suited is sandy loam having pH 6.0–8.0. Globally, it yields about 58.2 tonnes per hectare. The leading producer of sugar beet is Russia followed by France and the United States with 42.0, 39.5 and 30 million tonnes production, respectively, in

2018. In India, the major beet producing states are Maharashtra, Karnataka, Lucknow (Uttar Pradesh) and Tamil Nadu (FAO 2018).

Sugar beet is a crop known for its agricultural multifunctionality. It is the world's second sugar producing crop and accounts for 20% of sugar production worldwide (Anonymous 2013). It is a 'biorefinery' crop, meaning all its by-products can be utilized for a variety of purposes. The molasses is used for ethanol production, factory lime for soil improvement and the remnant tops and pulp as cattle feed. With the paucity of fodder, the usage of sugar beet as forage crop is rising day by day. Sugar beet tops are the preferred source of feeding cattle (Tawab et al. 2020). It can be consumed by the livestock as green/dry roughage or silage. Additionally, the pulp remaining after sugar extraction is another alternative for cattle feeding (Singh and Garg 2013). Feeding cattle with agro-industrial by-products is a good initiative towards waste disposal (Venkataswarlu et al. 2012). It can also improve the sustainability in terms of dairy production. Dairy cattle is often fed with grains along with forage to enhance the milk production. Sugar beet having high dry matter (DM) and energy can surpass the need of grains. It can reduce the feed cost and minimize the need of concentrate feeds used for animal feeding. Sugar beet used for cattle feeding is often known as 'feed beets' to distinguish it from fodder beet (Evans and Messerschmidt 2017).

Quality of any crop resides in the nutrients it holds. The ratio of protein, lipid and carbohydrate determines the nutritional status of the crop. The digestibility of the crop also relies on the ratio of these nutrients. Feed beet is an excellent source of these nutrients, providing complete nutrition to cattle. Most of the DM in sugar beet consists of carbohydrates, predominantly sucrose and structural saccharides. Cellulose and hemicellulose are the structural saccharides forming the dietary fibrous part of beets. Feed beets are low in sucrose and other non-structural carbohydrates, but high in dietary fibre (Filipovic et al. 2007). Lignin, a polyphenolic compound, has a dramatic impact on animal digestibility. Its high concentration can interfere with the digestion of the fibrous part of the crop. Lignin is generally low ($<5\%$) in feed beets (Ozboy et al. 1998). Lipid part mostly consists of polyunsaturated fatty acids such as α-linolenic acid (Hatfield et al. 2007). Nitrogen is a key element forming the protein portion of sugar beet. It is required for the optimum growth and yield of feed beet. Ash represents the inorganic mineral fractions and is considerably important for cattle health. Feed beet mainly comprises 0.5% P and 0.8% Ca (Habeeb et al. 2017).

Along with the nutritional constituents, feed beets also contain anti-nutrients. Nitrate and oxalate are the main anti-nutrients present in feed beets. These anti-nutrients have a specific limit up to which these are safe for cattle feeding. The permissible range for nitrate-N and oxalate is 0.2% and 2–10%, above which they can cause poisoning in animals (Kaur and Goyal 2016; Rahman et al. 2013). This chapter would be focussing on the nutritional and anti-nutritional aspects of sugar beet in regard to cattle feed.

46.2 Parts of Sugar Beet and Its By-products Used as Cattle Feed

Sugar beet consists of conical fleshy root and a flat crown known as beet tops. The root is sliced into long strips called 'cossettes' and treated with hot water at the processing point. The extracted sugar mixture and the hot water is further processed into bagged or bulk sugar. The desugared cossette obtained after this process is called beet pulp. Beet pulp can be fed to the cattle as wet (pressed shreds) or dry by-product (pellets). The wet beet pulp is beneficial for the lactating cows and reduces the danger of bloating and digestive disturbance. The cossettes are dried with the help of pulp press and pulp dryer. The dried beet pulp is high in fibrous residue and increases the digestibility and palatability of feed. Molasses is another alternative for cattle feeding and separated through the process of centrifugation from beet juice. Further, it is refined through molecular exclusion chromatography to produce desugared molasses. It is also called condensed separator by-product of feed beet. Beet molasses binds with dried beet pulp and are successfully included in cattle rations. Beet molasses are the carrier of minerals, vitamins and non-protein nitrogen compounds. The extraction process of beet is summarized in Fig. 46.1.

The small beets, damaged or broken beets along with beet tops are called beet tailings. These are the rejected material at the processing point, i.e. not suitable for the production of sugar. Beet tops are the rosette of leaves with high nutritive value suitable for cattle feeding. The small and broken beets can be made into silage and fed with the requirement of cattle. Silage is a type of fodder made from green forage with fermentation activity of microbes. Silage can be packaged into bags, silo pits, tubes and bales. Whole beet is also sometimes utilized as a feeding source. It is broken into small pieces and spread on stalk fields using manure spreaders. Silage is made through ensiling process. High moisture level sometimes becomes a problem, but addition of some dry ingredients facilitates proper ensiling. Dry ingredient may be grain screenings, chopped forages and other by-products.

46.3 Nutritional Aspects of Sugar Beet

46.3.1 Carbohydrate Composition

Carbohydrate fraction of feed beets is classified into two broad categories, i.e. structural and non-structural carbohydrates. The structural carbohydrates can be also named as fibre fractions of feed beets. Feed beets contain high dietary fibre ($>75\%$) content and can be further classified into insoluble and soluble dietary fibre. Insoluble dietary fibre is principally composed of cellulose, hemicellulose and the phenolic macromolecule lignin (Filipovic et al. 2007). These components reside in the cell wall of plants and provide mechanical strength to the plant. Cellulose is the most abundant carbohydrate on earth and resides in the cell walls of all plants. It is a polymer of many glucose units held in ß acetal linkage and is a linear polysaccharide. Ruminants such as cattle, sheep have symbiotic association with bacteria.

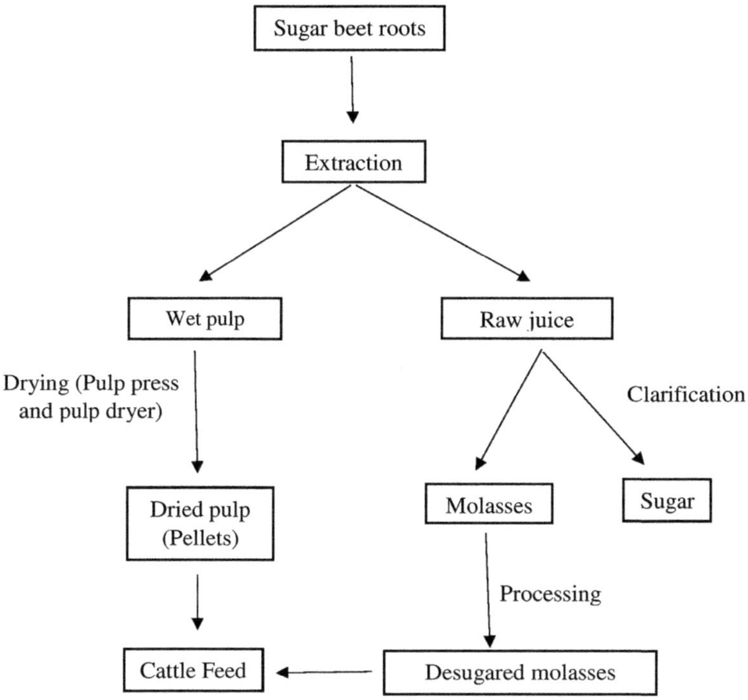

Fig. 46.1 Sugar beet extraction and formation of by-products used as cattle feed

These symbiotic bacteria having the necessary enzymes for the degradation of cellulose resides in the rumen of cattle. Hemicellulose is a group of complex polysaccharide such as xylan, mannan, xyloglucan and mixed linked ß-glucan. The sugar monomers of hemicellulose include xylose, arabinose, mannose, galactose and deoxy sugar rhamnose. It, along with pectin, surrounds the cellulose forming a network of cross-linked fibres. Lignin is not a carbohydrate but is a part of dietary fibre fraction. Feed beets are high in cellulose but low in hemicellulose and lignin. The 20% of glucose present in fibrous residue of feed beet is of cellulosic origin. Feed beets are low in mannose and xylose, the chief components of hemicellulose. The chemical linkages formed by lignin is structurally complex and difficult to be degraded by cattle. Since the fibrous fraction in feed beets are only rich in cellulosic part, it is beneficial for the cattle to utilize it as the feeding source. The carbohydrate profile of feed beet is summarized in Fig. 46.2.

The soluble dietary fractions include the pectins and glucans (Fadel et al. 2000). These are also known as neutral detergent soluble fibres. Pectin is a cell wall heteropolysaccharide made up of galacturonic acid, galactose, arabinose and fucose (Fig. 46.3). The pectin content in beet cell wall is high with 20% arabinose and galacturonic acid (Levigne et al. 2002). The fermentation of pectin maintains the rumen pH and results in high acetate to propionate ratio (Hall et al. 1998). Acetate is

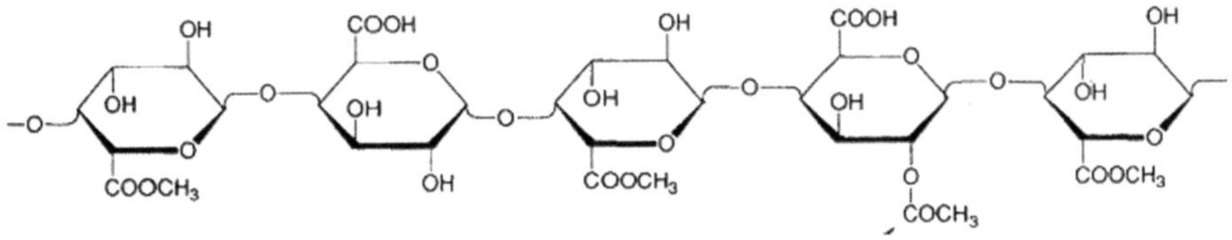

Fig. 46.2 Structure of pectin

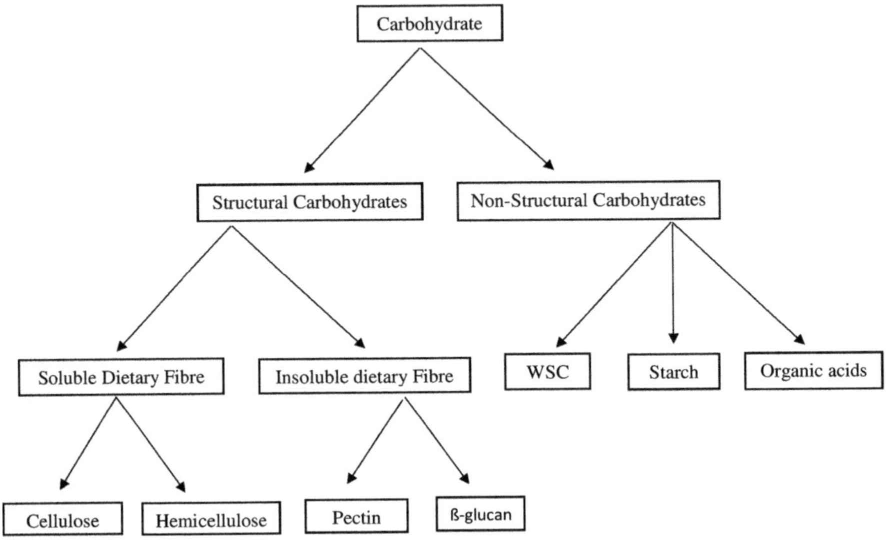

Fig. 46.3 Carbohydrate profile of feed beets

lipogenic, meaning it can result in high fat content in the milk of cattle (Smith and Johnson 2014). The propionate in the liver is metabolized into glucose. Acetate and propionate both are directed towards fat deposition in cattle. Acetate can lead to subcutaneous (back) fat deposits and glucose formed from propionate preferentially resulted in intramuscular fat (marbling) accumulation in young cattle. As the cattle matures, the glucose is substituted with acetate for marbling process (Choi et al. 2014).

In general, acid detergent fibre (ADF) and neutral detergent fibre (NDF) are the chief variables used for the estimation of fibre fraction of forages. ADF and NDF fractions are the insoluble portions remaining after the treatment with acid detergent solution (ADS) and neutral detergent solution (NDS). ADS dissolves the cell contents and the leftover is cellulose, lignin and silica which represents ADF (Van Soest 1994). Similarly, NDS dissolves the cell contents and the remaining portion of cellulose, hemicellulose, lignin and silica represents NDF. The cell content fraction dissolved in ADS and NDS includes the proteins, lipids, nucleic acid, nutrient ions plus non-structural carbohydrates and some structural carbohydrates (Hall et al.

1999). The ADF content in beet tops is 23.5–30.7% (lamina portion) and 17.7–22.3% (petioles), while in beet pulp it is 13.1–16% (Goyal et al. 2015). Similarly, the NDF content is 40.2–46.7% in leaf lamina, 29.7–36.8% in leaf petiole and 17.7–22.5% in beet pulp.

The non-structural carbohydrates, also named as non-fibre carbohydrates (NFC) consist of organic acids, water soluble carbohydrates and starch. The NFC accounts for about 30% in feed beets. The NFC can provide more energy to cattle and is readily fermented in rumen in comparison with the fibre fractions (Favarola et al. 2016). Starch is a polysaccharide having glucose monomer units in α-1, 4 linkage. Water soluble carbohydrates (WSC) include glucose, fructose, fructans and sucrose. Sucrose is about 16–20% in sugar beet, but generally desugared by-products are fed to the animals. Starch and WSC reduce the pH in rumen and also slow down the fermentation of pectin. Generally, starch and WSC are low in feed beets and account for about 1% and 10% on dry matter basis, respectively. It is often said that pectin rich diet is more advantageous than starch rich diet because of its high digestibility in cattle (Munnich et al. 2018).

46.3.2 Other Nutritional Constituents

In addition to the carbohydrate content, there are numerous other agronomic metrics that reflect feed beet nutritional quality. Metrics range from the estimation of crude protein to the analysis of fat content. The nitrogen demand is fulfilled through forage proteins in cattle. Crude protein represents the quantity of nitrogen (N) present in the plant. Crude protein is a total sum of true protein and non-protein nitrogen. True protein represents the nitrogen present in the protein portion of the plant. Non-protein nitrogen refers collectively to the compounds such as amides, amines, ammonia, urea and biuret. These compounds are not proteins but could be a source of nitrogen to ruminants. It includes all nitrogen sources that are not a part of polypeptide. CP content is generally low in feed beets and sometimes supplemental protein is required to fulfil the protein demand (Hartnell et al. 2005). Sugar beet comprises about 6.8% protein, which is low as compared to corn (8.5%) and barley grain (12.8%) (Schafer and Larder 2008 and NRC 2000). The majority of total N fraction in feed beets is of true protein having a range between 54 and 77% of total protein (Castle et al. 1981). The total amino acid content in the roots portion of beet ranged from 0.3 to 0.62%, with 0.10–0.20% essential amino acids (Hu et al. 2018). These amino acids help in improving the growth, development and reproduction of cattle. Betaine is a basic nitrogenous compound and the methyl derivative of glycine. The root portion of beets contains about 0.14–1.09% betaine (Hu et al. 2018). It has a similar role as that of methionine for reducing fatty liver problems in cattle. Betaine is also rich in beet molasses (3–8% on dry solids basis) and serves as methyl donor in transmethylation reactions (Filipcev et al. 2015).

Ash is the inorganic mineral matter required for the bone formation of cattle. Ash content in feed beets range from 3 to 8% (Ozboy et al. 1998). The mineral elements have variable composition in feed beets and perform numerous functions in cattle.

The root of beets contains 3.41–4.79 mmol 100 g^{-1} potassium and 4.08–11.54 mmol 100 g^{-1} sodium (Hu et al. 2018). Beet molasses are abundant in calcium, magnesium, sodium and potassium and iron. It comprises 100–500 mg/100 g calcium, 10–300 mg/100 g magnesium, 600–900 mg/100 g sodium and 3–11.7 mg/100 g iron (NOVUS 1996; Susic and Sinobad 1989; Curtin 1983). Beet pulp is rich in calcium but low in phosphorus. It constitutes about 0.72–0.77% calcium and 0.09–0.20% phosphorus (Mustafa et al. 2009; Castle et al. 1981). Beet pulp also contains magnesium (≈1.2 g/kg), iron (≈20 mg/kg), copper (≈40 mg/kg), manganese (≈50 mg/kg), zinc (≈80 mg/kg), selenium (≈0.4 mg/kg) and iodine (≈0.8 mg/kg) (Lardy and Schafer 2008). Calcium is necessary for blood clotting, contraction of muscles and accomplishes various biochemical reactions in cattle. Phosphorous is the principal component of energy currency (ATP) of the cell. Many biochemical reactions in the cattle system need energy in the form of ATP. Potassium and zinc maintain the acidic conditions in body fluids and are also required for several enzyme reactions of protein and carbohydrate synthesis. Magnesium helps in bone growth and iron is the prime component of haemoglobin, the oxygen carrier in the blood. Manganese is necessary for the utilization of carbohydrates. Copper and selenium maintain the fertility of cattle. Iodine is necessary for the thyroid gland.

Along with the minerals, vitamins also play a chief role in cattle body. Feed beets contain both water soluble as well as fat soluble vitamins. Beet molasses contains some water soluble vitamins (WSV) such as niacin (2.9 mg/100 g), pantothenic acid (≈4.58 mg/kg), riboflavin (0.08–0.14 mg/100 g) and biotin (≈0.7 mg/kg). The cereals barley, wheat, sorghum are poor in biotin and they can be compensated by beet molasses in cattle diet (Mordenti et al. 2021). Beet molasses are deficient in thiamine which is destroyed during heat treatment at the processing point. They are also poor in fat soluble vitamins (Piccioni 1989). They can be used as an additive in silage. Legume forages such as clover and alfalfa lack in sugar content and also have high buffering capacity. The addition of molasses in legume silage helps in lowering the pH required for fermentation. It can produce higher amount of lactic acid with low levels of organic acids and ammonia. Overall, it can raise the organoleptic characteristics of legume silage (Rooke et al. 1985). The WSV ranged from 16 to 100 mg/kg in beet pulp (Lardy and Schafer 2008). The WSV present in beet pulp are vitamin B1, vitamin B2, vitamin B3, vitamin B6, vitamin B7, vitamin B9 and vitamin C. These WSV help in the production of milk components in cattle. They act as co-factors for the enzymes involved in the metabolism of biomolecules. It also contains fat soluble vitamins in detectable amount. It contains approximately 1600 IU/kg vitamin A, 1200 IU/kg vitamin D3, 400 mg/kg and 4 mg/kg vitamin K. Choline is a nutrient found in feed beets having similarity with vitamin B and is necessary for optimum growth and functioning of cattle. It is approximately 40 mg/kg in beet pulp. It improves the milk fat and its yield and also reduces the cholesterol level and serum non-esterified fatty acids in dairy animals (Lardy and Schafer 2008).

Ether extract or crude fat content is another significant metrics important for dairy animals. It acts as the reserve energy and forms the structural tissues of the body. Fat content is <2% in feed beet, which is low in compared with other forages (2–4%). Though fat is low in feed beets, the intake of feed beets by the cattle stimulates the milk fat concentration. The primary reason of it is higher fibre intake and greater acetate level in rumen. Beet top silage can be substituted with corn silage as it has highly digestible dietary fibre, vitamins, minerals and some other nutrients. It is estimated that 15% of beet tops silage can replace about 50% of corn silage (Tawab et al. 2020).

46.4 Digestibility

The digestibility (D-value) mainly refers to the extent up to which the feed is digested by an animal. The calculation of D-value is an important criterion that defines the in vitro digestibility and fermentation kinetics of feed beet in cattle system. The nutritional composition of sugar beet (described in Sect. 46.3) determines its digestibility. The process of digestion differs with the type of biomolecule (carbohydrates, protein and lipids). Monogastric animals have single compartment stomach, whereas polygastric animals have multi-compartment stomach. Cattle is a polygastric animal (ruminant) having a well-developed four chambered stomach. The four chambers are rumen, the reticulum, the omasum and the abomasum. The majority of digestion takes place in rumen which is the home of a vast array of microbes.

46.4.1 Digestibility of Carbohydrates

Different amounts of carbohydrates within different parts of feed beet alter downstream digestibility of cattle. Sugar beet contains 68% of carbohydrates mostly consisting of complex polysaccharides. These carbohydrates are cleaved into simple monosaccharides through the cleavage of glycosidic bonds. The cattle cannot secrete endogenous enzymes in their small intestine for the cleavage of carbohydrates. The cleavage is done either by chewing action of cattle or through microbes present in the rumen. The rate of digestion depends upon the type of carbohydrate. The non-structural and soluble carbohydrates (soluble dietary fibre) are readily fermented by the ruminal microbes. Fructans is a non-structural carbohydrate found in sugar beet having high D-value (Chalmers et al. 2005). At the same time, the structural carbohydrate (insoluble dietary fibre) fraction requires much time for its digestibility. These cell wall components contain lignin that can form complex with sugars and proteins, restraining their digestion. Since the fraction of lignin is found low in feed beet, its fibre fraction is easily digestible. The main dietary fibre fraction comprises cellulose and pectin in feed beet. The cellulolytic enzymes produced by microbes such as cellulase causes the breakdown of cellulose. Beet pulp can modify the rumen's physical and chemical features to avoid digestive disturbance, depressed appetite and acidosis. In general, the assimilation of soluble carbohydrates is 100 times faster than the non-structural carbohydrates. The assimilation of non-structural carbohydrates is five times more than cell wall carbohydrates. The digestion of carbohydrates is summarized in Fig. 46.4.

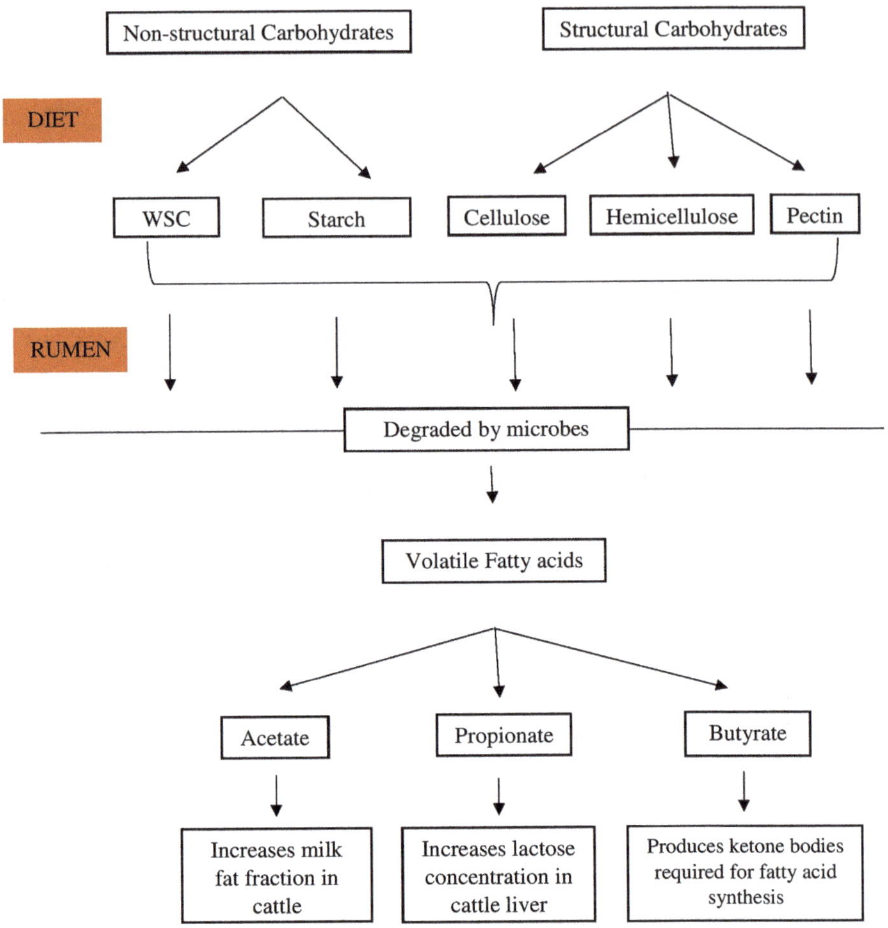

Fig. 46.4 Digestion of carbohydrates

The microbial action generally results in simple sugars. These sugars are metabolized to pyruvate, which is further changed into volatile fatty acids (VFA). The three major VFA produced after carbohydrate fermentation include acetate, propionate and butyrate. The proportion in which these are produced determines the protein and fat fraction in cattle milk. The propionate is generally formed in starch-rich diets and acetate is formed in pectin-rich diets. The feed beet is high in pectin but low in starch. Hence, the most dominating VFA produced after the fermentation of carbohydrates is acetate. Acetate is the major precursor of acetyl CoA and synthesizes lipids. Acetate is mainly responsible for increasing the fat content in milk of cattle. The VFA formed in the rumen can furnish 70% of energy requirement in animals. Cattle absorb VFAs through the rumen into the bloodstream and the rate of absorption depends upon ratio of individual VFA, rumen pH and the absorptive area in the ruminal lining. Another end product of microbial fermentation of carbohydrates includes gases such as methane and carbon dioxide (CO_2). They can be removed through the rumen wall or expelled by belching (eructation). The cattle as well as intestinal microbes both use the carbon dioxide to maintain the fraction of bicarbonate ions in saliva.

46.4.2 Digestibility of Proteins

The dietary proteins are classified into rumen-degradable proteins and rumen-undegradable protein. The digestion of rumen-degradable proteins takes place in the rumen. The microbes present in the rumen produce proteolytic enzymes such as proteases and peptidases. The action of these enzymes leads to the degradation of long peptide chains forming amino acids. The amino acids eventually release ammonia and C-skeleton through deamination reaction. Ammonia acts as the nitrogen source for microbial growth. In addition, ammonia is metabolized into urea in the liver. The final fate of urea is excretion through urine or it can also be recycled back into the rumen as non-protein nitrogen source for microbial protein synthesis. The microbes present in the rumen are the major source of protein in cattle diet. The microbes are washed from the rumen through the omasum to the abomasum where they are killed and digested by the cow. The amino acids formed in this process are absorbed through the small intestine. The digestion of the remaining dietary protein (Rumen-undegradable protein) that has escaped microbial digestion takes place in the abomasum and small intestine.

The consumption of feed beet and its by-products helps in the production of microbial protein. The addition of dried beet pellets in cattle diet can increase the total nitrogen and amino acid fraction in duodenal digesta. The total digestible nutrients (TDN) present in feed beet varied according to the plant part and its by-product. Whole beet contains 75–81% of TDN, beet pulp (wet and dry) contains 72%, beet tops contain 58% and molasses contain 75% (Lardy 2018). The digestion of proteins is summarized in Fig. 46.5.

46.4.3 Digestion of Fats

Like proteins, fats can be categorized into rumen-degradable fats and rumen-protected fats. The fats present in feed beet are triglycerides (made of glycerol and three fatty acids) and glycolipids (one fatty acid is replaced by sugar). The fat digestion starts in the rumen where bacterial action splits off the triglyceride molecule into three fatty acids and a glycerol moiety through hydrolysis. Similarly, the glycolipids can be broken into glycerol, two fatty acids and sugar moiety. Further, the process of bio-hydrogenation converts unsaturated fatty acids into saturated ones. The released fatty acids enter the small intestine and form micelles with the help of bile and pancreatic secretions. These micelles are then absorbed through the gut wall where they are converted back into triacylgylcerols (TAGs) and

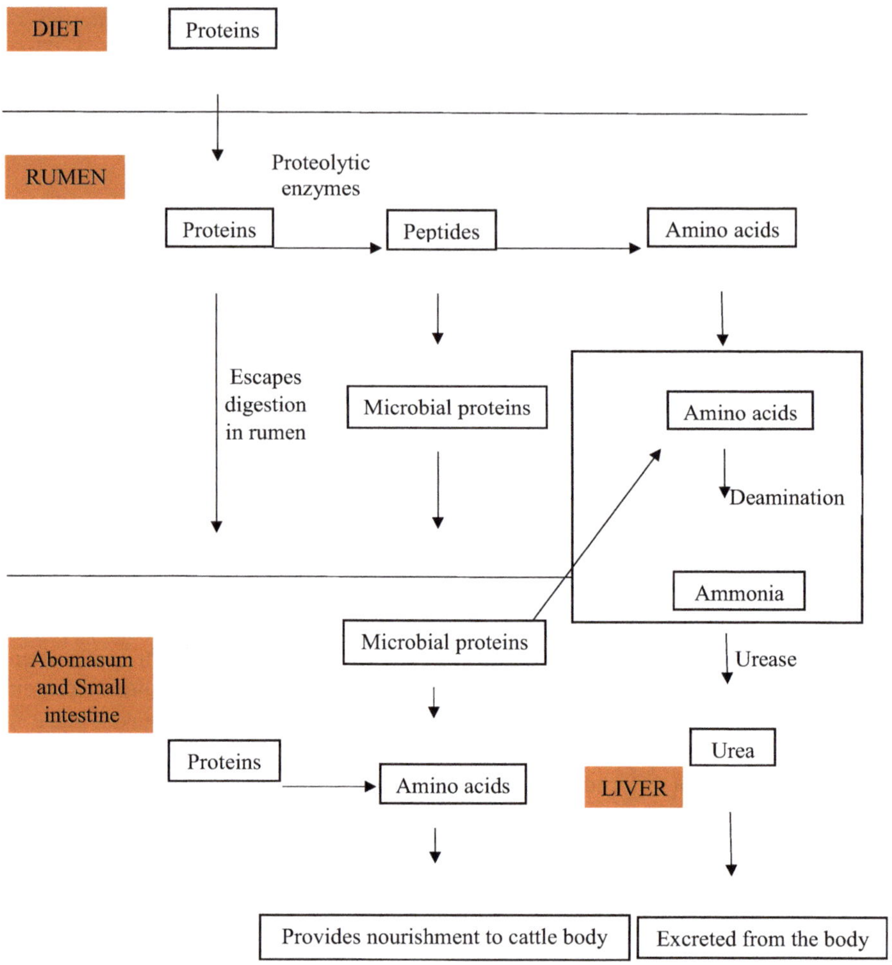

Fig. 46.5 Digestion of proteins

packaged into lipoproteins and chylomicrons. Then they enter the lymphatic system and are finally delivered to the tissues. Rumen-protected fats escape rumen digestion and are absorbed in the jejunum portion of the small intestine. High amounts of rumen-degradable fats sometimes causes digestive upsets in cattle. Sugar beet have very low amount of fat content (<2%), so it prevents the danger of digestive problems and bloating in the cattle (Castle et al. 1981).

The digestibility can be estimated by both in vitro and in vivo methods. The in vitro dry matter digestibility (IVDMD) varied with the plant portion of feed beet. The IVDMD is 69–81% in leaf lamina, 79–85% in leaf petiole, 86–93% in root and 91–94% in desugared cossettes (Goyal et al. 2015). The IVDMD of beet tops can range from 80 to 93.6% (Sandhu et al. 2015). The IVDMD in dried beet pulp is

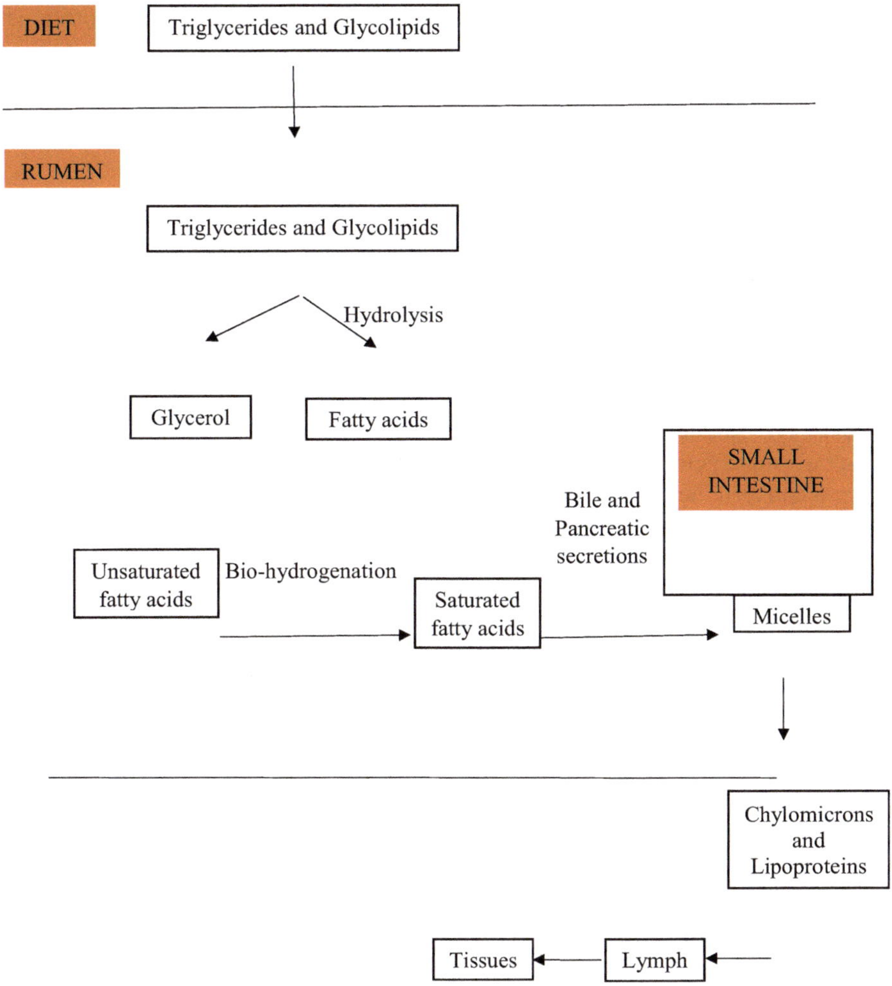

Fig. 46.6 Digestion of fats

92.9% and in beet tails (including root tips, hairs) is 87.3% (Rodríguez et al. 2019). The digestion of fats is summarized in Fig. 46.6.

46.5 Anti-nutritional Aspects of Sugar Beet

Anti-nutritional components are those which interfere with the absorption of nutrients such as proteins and minerals. They can bind with these nutrients, making them unavailable for assimilation. Poisoning incidents in cattle can be seen when anti-nutrients are consumed above a certain limit. They can cause severe health problems and even cause death of the cattle. The two main anti-nutritional components present in feed beet are nitrate and oxalate.

46.5.1 Nitrate Content

Nitrate is an inorganic nitrogen source present in the feed beet and is important for its physiological regulation. The sugar beetroot absorbs the nitrate from the soil and transports to the beet tops via xylem vessels. The nitrate is converted into nitrite via nitrate reductase and nitrite is further reduced to ammonia via nitrite reductase. The final fate of ammonia is proteins formed with the ammonia-assimilating enzymes. The nitrate generally accumulates in feed beet only under certain conditions.

1. **Excessive nitrogen fertilization**—Nitrogen fertilizers are often added in the soil to raise the efficiency of the crop. It is essential to fulfil the nitrogen demand of the crop and also increases yield of the crop. But sometimes, excessive use of it may raise the nitrate level because the absorption of nitrate from soil exceeds the assimilation rate (Ugrinovic et al. 2012).
2. **Low temperature**—Sugar beet requires a temperature of 15–21 °C for its growth and emergence. Low temperature stress can accumulate nitrate in it. Nitrate reductase (NR) consists of five conserved domains; Mo-MPT domain, a cytochrome b domain, a dimer interface domain, NADH domain that combines with FAD domain (Campbell 1999). These domains constitute an electron transport chain through which electrons are transferred from NADH to nitrate (Fig. 46.7). Low temperature stress interrupts the electron transfer between the heme and MoCo domain (Aydin and Nalbantoglu 2010). As a result, the enzyme activity gets inhibited and nitrate accumulates in the feed beet.
3. **Low light intensity—**The sunlight also has a profound effect on nitrate accumulation. Like NR, nitrite reductase (NiR) also constitutes an electron (e^-) transport chain (Fig. 46.8). The photosystem 1 converts oxidized ferredoxin (Fd) to reduced Fd which acts as an e^- donor and transfers six e^- to nitrite (NO_2^-) via the electron transport chain of nitrite reductase. The PS1 is activated in the presence of optimum sunlight and low light intensity inhibits the above process. The requirement of reduced Fd in dark could also be met through NADP reduction by pentose phosphate pathway (Ali 2020). However, this route of

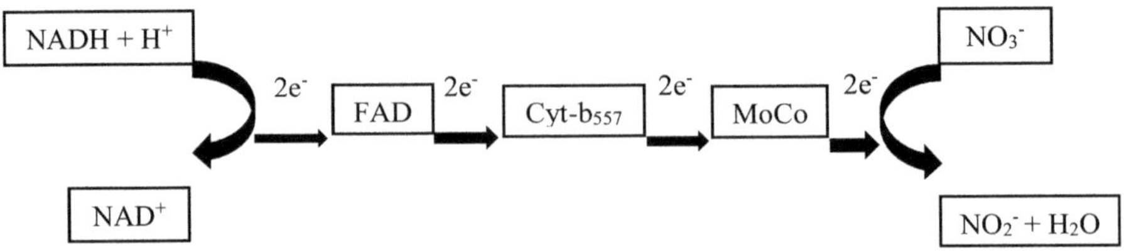

Fig. 46.7 Electron transfer during nitrate reduction (Cyt b: Cytochrome b; MoCo: Molybednum Coenzyme; NO_2^-: Nitrite' NO_3^-: Nitrate)

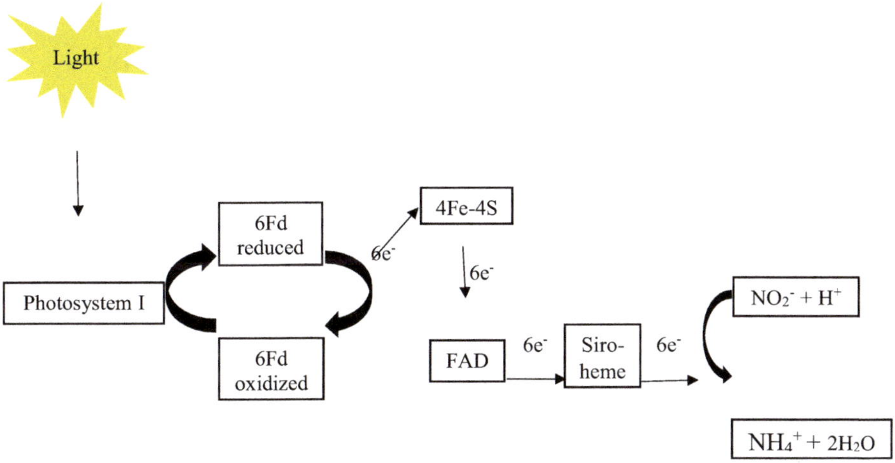

Fig. 46.8 Electron transfer during nitrite reduction

ferredoxin formation is limited to some extent and at some point the nitrite starts accumulating in the cell. Hence, low light intensity limits the NiR activity. Due to less NiR activity, the NR has to be slowed down to avoid the accumulation of nitrite, which is ten times more toxic than nitrate. Sugar beet requires sunlight of longer duration for its proper growth and productivity. Low light intensity may lead to nitrate accumulation in feed beet.

4. **Extraction process**—The nitrate can also get accumulated in feed beet by-products. During the extraction of beet, the nitrate reduction to nitrite takes place with the action of thermophilic bacteria. The reaction takes place when cossettes are treated in the extractor and cossette scalder (Emerstorfer et al. 2014). But this thermophilic bacterium inhibits the gram-positive bacterial activity that helps in the active transport of amino acids and sugars across the plasma membrane (Bianco et al. 2007). Therefore, formalin (a biocide) is sometimes added to suppress the activity of thermophilic bacteria. In this course, the nitrate reduction also gets inhibited. As a result, the nitrate pass as such during the purification process and accumulates in beet molasses.

The consumption of feed beet incorporates nitrate in the cattle body. Nitrate has two major roles in animal nutrition. Firstly, it reduces the production of methane in the rumen, and secondly, it synthesizes microbial protein in the rumen by acting as the superior non-protein nitrogen source. Methane is formed in the rumen through hydrogenotrophic route in which hydrogen and carbon dioxide act as substrates. It is a greenhouse gas responsible for global warming. It is 28 times more potent than carbon dioxide in trapping heat in the atmosphere (Pachauri et al. 2014). Nitrate can reduce methane production in cattle by directing hydrogen away from

Fig. 46.9 Nitrate reduction chelates hydrogen away from methanogenesis

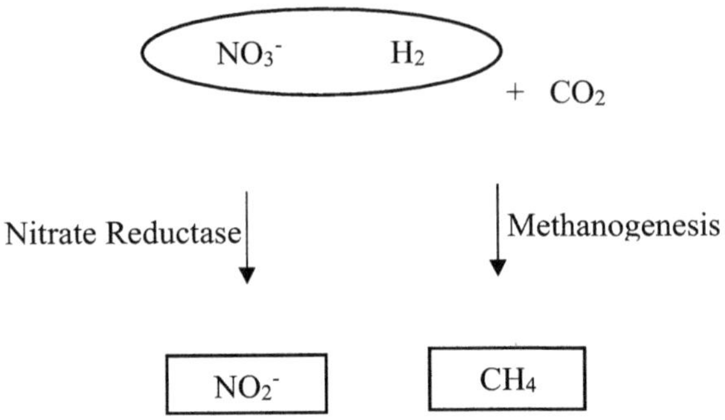

methanogenesis. The thermodynamically favourable reaction of nitrate $(NO3_{-})$ reduction is responsible for this action (Fig. 46.9).

The normal metabolism of nitrate in cattle system (rumen) follows the route in which it is converted to nitrite by rumen microbes. The nitrite is ultimately converted to ammonia, the main N source for microbial protein synthesis. Thus, below the toxic level, nitrate is a necessary requirement for the cattle body. When the nitrate-N level reaches 2000 ppm or above in feed beet, nitrate starts acting as an anti-nutritional factor. Under these conditions, the frequency of nitrate reduction exceeds the nitrite reduction in rumen. The excess of nitrite gets absorbed across the rumen wall and forms complex with haemoglobin in the red blood cells (Fig. 46.10). Haemoglobin is then changed to its dysfunctional form called methaemoglobin that cannot transport oxygen. The shortage of oxygen causes difficulty in breathing, leading to a condition known as tissue hypoxemia. The cattle consuming nitrate rich diets show clinical signs when 20% of haemoglobin is changed into methaemoglobin and death results when it reaches to 60–80% (Qudah et al. 2009). The symptoms of nitrate toxicity include tachypnea, brown mucosa, staggering gait, bloating, frequent urination, lateral recumbency (Gontijo et al. 2017). Radositits et al. (2007) observed chocolate coloured blood and intensely red coloured skeletal musculature of ruminants suffering from nitrate poisoning. Another study reported brownish brain and lungs of cattle with nitrate intoxication due to oat and ryegrass (Jonck et al. 2013). Hence, cattle consuming nitrate-rich diet (>2000 ppm) risk illness or even death from methaemoglobinaemia. Sugar beet that contained high nitrate content can be diluted with grains in the diet or with other forages low in nitrate and then can be fed safely. Intravenous dose of methylene blue helps to treat methaemoglobinaemia. Mineral oil may be given orally to protect the mucous membrane of the cattle.

Sugar beet nitrate-N content differs in relation to the part and its by-product. The roots contain highest nitrate-N (955–2853 ppm) content because the parts nearer to the ground often have high nitrate content. The sugar beet tops comprises 200–605 ppm (leaf lamina) and 600–1500 ppm (leaf petiole) nitrate-N. The beet pulp has 499–1699 ppm nitrate-N (Goyal et al. 2015). The different ranges of nitrate and its impact on livestock are summarized in Table 46.1.

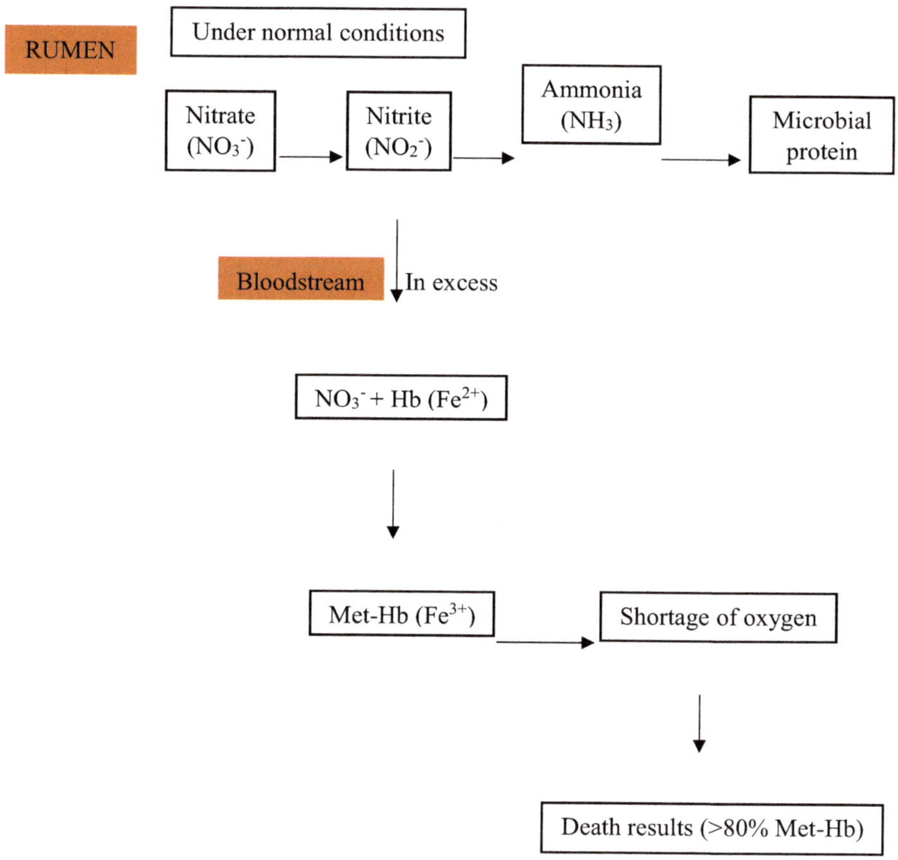

Fig. 46.10 Effect of high nitrate concentration in cattle

Table 46.1 Nitrate level (ppm, DM basis) in forages and its influence on livestock

Nitrate (NO_3) content	Effect on livestock
<1000	Safe to feed under all conditions
1000–1500	Safe to feed to non-pregnant animals under all the conditions. It may be best to limit its use to pregnant animals to 50% of the total ration on a dry matter (DM) basis
1500–2000	Feeds are fed safely if limited to 50% of rations total dry matter
2000–3500	Feeds should be limited to 35–40% of total DM in the ration. Feeds containing over 2000 ppm nitrate-N should not be used for pregnant animals
3500–4000	Feeds should be limited to 25% of total DM in ration. Do not use for pregnant animals
>4000	Feeds containing over 4000 ppm are potentially toxic. Do not feed

46.5.2 Oxalate

Oxalate is a dicarboxylic acid having multifunctional role in plants. It helps in quenching oxidative burst under pathogenic attack, involved in programmed cell death and can cope with metal toxicity. The oxalate production occurs by many routes such as photorespiratory glyoxylate oxidation, hydrolysis of oxaloacetate and cleavage of ascorbate and isocitrate (Nakata 2003). Sugar beet contains oxalate that exists in two forms, i.e. soluble and insoluble. The soluble form can bind with monovalent ions (sodium, potassium and ammonium ions), while the insoluble form binds with divalent ions (calcium, magnesium and iron ions) (Savage et al. 2000). Oxalate is safe for the cattle when consumed below 2% (Rahman et al. 2013). Afterwards, it starts showing harmful effects and causes illness in the cattle. Soluble oxalate exerts its harmful effects by binding to divalent counterions. The resultant disturbed calcium and phosphorus metabolism leads to excessive mobilization of bone mineral. The beet tops rich in oxalate, when fed to cattle, may result in hypocalcaemia and hypomagnesemia (El-Khodery et al. 2008). Oxalate has antimicrobial activity against rumen microbes that results in rumen dysfunction in cattle. It also reduces the nutrient digestibility and feed intake of cattle. Additional effects include kidney failure because of insoluble salt formation in the blood.

The oxalate metabolism in cattle system takes place by the following four routes. First, soluble oxalate is degraded by the rumen microbiota (Allison et al. 1977). Second, when soluble oxalate is high in dietary feed along with calcium, it can form insoluble oxalate crystals by binding with calcium (Ca^{2+}) and magnesium (Mg^{2+}) ions in rumen. These crystals cannot be absorbed by the cattle system and are excreted from the body. Third, when the calcium concentration is low, the soluble oxalate is absorbed from the intestine into the bloodstream. When the oxalate levels reaches high in the blood, it can combine with Ca^{2+} and Mg^{2+} ions. The resultant insoluble oxalate crystals can block urine flow and even cause kidney failure (Blaney et al. 1982). Fourth, the insoluble oxalate can pass directly without causing any problem to the cattle system.

Oxalate intoxication from feed beet is dependent upon numerous factors such as oxalate chemical form, portion of beet and its by-product, animal's age, quantity of oxalate consumed (Radositits et al. 2007). Another major consequence of oxalate accumulation in sugar beet takes place during its storage. In some countries, beet is harvested in advance and stored to prevent the problem of frozen lands after which harvesting becomes impossible. After storing the beets for a passage of time, these are processed to form thick juice. The juice made in large quantities are then stored in tanks. During the storage period, the chelated calcium is released and can form precipitates with anions (such as oxalate, carbonate, etc). These precipitates can be filtered at the processing point, but calcium oxalate is hard to filter and is a limited step. So, the molasses obtained when fed to cattle may cause oxalate poisoning in them. However, addition of anti-scaling agents and decalcification can avoid this problem, but low temperature during storage may also sometimes limit these antidotes (Muir et al. 2019).

Cattle is a polygastric animal so it has higher tolerance capacity of oxalate than other animals like horses and pigs. The bacterium named *Oxalobacter formigens* residing in the rumen of cattle is capable of oxalate degradation. This slow growing bacteria cannot utilize other types of substrates and depends only upon oxalate. When oxalate is present in smaller amounts, it can be easily degraded by the rumen bacteria. Under its high level, the rumen is overwhelmed and unable to metabolize it, thereby resulting in its poisoning. Oxalate-rich diets diminish animal performance and also the production of milk. The calcium and fat concentration in milk is mostly affected. The oxalate fraction varied with different parts of feed beet and also with its by-product. It varied from beet tops to roots and also from beet pulp to molasses. The variation in oxalate content in feed beet is mentioned in Table 46.2.

Ensiling can reduce oxalate concentration in feed beet and is done using lactic acid bacteria, LAB (Tawab et al. 2020). Many species of LAB can be employed, for example, *Lactobacillus plantarum, Lactobacillus gasseri, Bifidobacterium breve* and *Leuconostoc mesenteroides.* These bacteria use oxalate as a C source and have oxalate-degrading activity (Miller et al. 2014). It secretes oxalyl-CoA decarboxylase enzymes that participate in catabolism of oxalate. LAB can decrease the oxalic acid concentration in beet tops silage and ultimately increases the accessibility of calcium ions. Ensiling silage (beet tops) with LAB reduced about 56% of oxalate (Tawab et al. 2020). Oxalate is not only present in feed beet but also in other forage species like napier bajra hybrid, pearl millet, bathu, guinea grass, setaria, sorghum, kikuyu grass and baffle grass. The pearl millet, guinea grass, sorghum, kikuyu grass and baffle grass have lesser oxalate content in comparison with sugar beet. The oxalate content ranged from 0.4 to 2.4% in these forage species. The remaining species had comparable oxalate content (2.58–5.98%) with feed beet. The anti-nutrition composition in feed beet is summarized in Table 46.2.

46.6 Conclusion

Sugar beet in cattle diet is extremely beneficial from a nutritional point of view. It comprises highly digestible dietary fibre with low acid detergent fibre (ADF) and neutral detergent fibre (NDF). The insoluble dietary fibre is mainly of cellulosic origin. The cellulose can be digested with the enzymes secreted by the rumen microbes. This dietary fibre acts as the source of roughage in cattle. Another major dietary fibre present in it is pectin, which resulted in high acetate to propionate ratio. The large quantity of acetate produced during fermentation in cattle helps in increasing the fat content in milk and also its yield. Secondly, feed beets recycle the waste remaining after sugar extraction. The agro-industrial by-products and their disposal is a serious environmental issue since they are potential pollutants. The cattle feeding with these by-products is a good initiative for utilizing the residual waste. This not only helps in solving the environmental problem but also increases the cattle performance. The pulp and molasses are produced as by-products during sugar extraction. These by-products are of paramount importance because these can substitute the grains in the diet. Thirdly, the feed beet and its value-added

Table 46.2 Anti-nutritional composition in feed beet and its by-product

Plant material	Oxalate (% DW basis)	Reference
Beet tops	3.50–4.12	El-Khodery et al. (2008)
Beet tops (silage)	27	Tawab et al. (2020)
Beet tops (silage treated with LAB)	11.9	
Root	0.783–0.988	Ugrinovic et al. (2012)
Leaf lamina	4.3–5.51	
Leaf petiole	1.87–5.02	
Beet tops	3.3–4.89	Bendary et al. (1992)
Beet tops	4.9	Mostafa et al. (2003)
Root	0.367	
Beet pulp	0.237	
Plant material	**Oxalate (% FW basis)**	**Reference**
Beet tops (silage)	0.967	Tawab et al. (2017)
Beet tops (silage treated with LAB)	0.431	
Plant material	**Nitrate (% DW basis)**	**Reference**
Leaf lamina	0.001–0.018	Ugrinovic et al. (2012)
Leaf petiole	0.023–0.27	
Root	0.063–0.32	
Leaf lamina	0.02–0.61	Goyal et al. (2015)
Leaf petiole	0.06–0.15	
Root	0.096–0.285	
Desugared cossettes	0.050–0.170	
Beet tops	0.69–0.94	El-Khodery et al. (2008)
Beet leaves	0.09–0.18	Sandhu et al. (2015)

by-products can be made into silage and stored for long periods of time. It can furnish the need of cattle during the scarcity of forages like in winter. Despite its several advantages, it has the disadvantage of having a large quantity of anti-nutrients present in it. The nitrate and oxalate are the two anti-nutrients found in feed beet. These can diminish the accessibility of other nutrients by forming complexes with them. These are proven hazardous to cattle when taken above the permissible range. Hence, sugar beet cannot be fed solely to the cattle but mixed along with other forages low in these anti-nutrients.

References

Ali A (2020) Nitrate assimilation pathway in higher plants: critical role in nitrogen signalling and utilization. Plant Sci Today 7:182–192

Allison MJ, Littledike ET, James LF (1977) Changes in ruminal oxalate degradation rates associated with adaptation of oxalate ingestion. J Anim Sci 45:1173–1179

Anonymous (2013). Handbook of agriculture. Indian Council of Agricultural Research, New Delhi

Aydin B, Nalbantoglu B (2010) Effects of cold and salicylic acid treatments on nitrate reductase activity in spinach leaves. Turk J Biol 35:443–448

Bendary MM, Mahmoud AM, Koriet IS, Abdel-Raouf EM, Awadalla SA (1992) Nutritional studies on using sugar beet tops in animal feeding-minerals content of different varieties of sugar beet tops and some minerals balances. Vet Med J 10:89–99
Bianco CA, Rojas A, Nimbubona D (2007) Effects of acidity and molecular size on bacteriostatic properties of beer hop derivates. Trends Food Sci Technol 18:144–149
Blaney BJ, Gartner RJW, Head TA (1982) The effects of oxalate in tropical grasses on calcium, phosphorus and magnesium availability to cattle. J Agric Sci 99:533–539
Campbell WH (1999) Nitrate reductase structure, function and regulation: bridging the gap between biochemistry and physiology. Annu Rev Plant Physiol Plant Mol Biol 50:277–303
Castle ME, Gill MS, Watson JN (1981) Silage and milk production: a comparison between barley and dried sugar beet pulp as silage supplements. Grass Forage Sci 36:319–324
Chalmers J, Lidgett A, Cummings N, Cao Y, Forster J, Spangenberg G (2005) Molecular genetics of fructan metabolism in perennial ryegrass. Plant Biotechnol J 3:459–474. https://doi.org/10.1111/j.1467-7652.2005.00148.x
Choi SH, Silvey DT, Johnson BJ, Doumit ME, Chung KY, Sawyer JE (2014) Conjugated linoleic acid (t-10, c-12) reduces fatty acid synthesis de novo, but not expression of gene for lipid metabolism in bovine adipose tissue ex vivo. Lipids 49:15–24. https://doi.org/10.1007/s11745-013-3869-0
Curtin LV (1983) Molasses-general considerations. In: Molasses in animal nutrition. National Feed Ingredients Association, Des Moines, pp 1–11
El-Khodery S, El-Boshy M, Gaafar K, Elmashad A (2008) Hypocalcaemia in Ossimi sheep associated with feeding on beet tops (Beta vulgaris). Turk J Vet Anim Sci 32:199–205
Emerstorfer F, Bergwall C, Hein W, Bengtsson M, Jensen JP (2014) Investigation of nitrite pathways in sugar beet extraction. Sugar Ind 137:88–94
Evans E, Messerschmidt U (2017) Review: sugar beets as a substitute for grain for lactating dairy cattle. J Anim Sci Biotechnol. https://doi.org/10.1186/s40104-017-0154-8
Fadel JG, DePeters EJ, Arosemena A (2000) Composition and digestibility of beet pulp with and without molasses and dried using three methods. Anim Feed Sci Technol 85(12):121–129
FAO (2018) FAOSTAT database. Food and Agriculture Organization. Archived from the original on 2011-07-13
FAO (2019) FAOSTAT database. Food and Agriculture Organization of the United Nations. http://faostat.fao.org/. Accessed 30 Nov 2019
Favarola VR, Ezequiel JMB, Almeida MTC, D'Aurea AP, Paschoaloto JR, van Cleef EHCB (2016) Carcass traits and meat quality of Nellore cattle fed different non-fiber carbohydrates sources associated with crude glycerin. Animal 10:1402–1408. https://doi.org/10.1017/S1751731116000094
Filipcev B, Simurina O, Handnadev TD, Jevtic-Mucibabic R, Filipovic V, Lonacar B (2015) Effect of liquid (Native) and dry molasses originating from sugar beet on physical and textural properties of gluten-free biscuit and biscuit dough. J Texture Stud 46:353–364
Filipovic N, Djuric M, Gyura J (2007) The effect of the type and quantity of sugar-beet fibers on bread characteristics. J Food Eng 78:1047
Gontijo DA, Borges A, Wouters F (2017) Nitrate/nitrite poisoning in dairy cattle from the Midwestern Minas Gerais, Brazil. Ciência Rural, Santa Maria. https://doi.org/10.1590/0103-8478cr20170373
Goyal M, Tiwana US, Bhullar MS (2015) Nutritional evaluation of sugar beet (*Beta vulgaris* L.) genotypes as fodder. Indian J Anim Nutr 32(2):237–241
Habeeb AAM, Gad AE, El-Tarabany AA, Mustafa MM, Atta MAA (2017) Using of sugar beet pulp by-product in farm animals feeding. IJSRST 2(3):107–120
Hall MB, Pell AN, Chase LE (1998) Characteristics of neutral detergent-soluble fiber fermentation by mixed ruminal microbes. Anim Feed Sci Technol 70:23–39
Hall MB, Hoover WH, Jennings JP, Webster TK (1999) A method for partitioning neutral detergent-soluble carbohydrates. J Sci Food Agric 79:2079–2086. https://doi.org/10.1002/(SICI)1097-0010(199912)79:15<2079::AID-JSFA502>3.0.CO;2-Z

Hartnell GF, Hvelplund T, Weisbjerg MR (2005) Nutrient digestibility in sheep fed diets containing roundup ready or conventional fodder beet, sugar beet and beet pulp. Anim Sci 83:400–407

Hatfield RD, Jung HJG, Broderick G, Jenkins TC (2007) Nutritional chemistry of forages. In: Barnes RF, Nelson CJ, Moore KJ, Collins M (eds) Forages: the science of grassland agriculture, 6th edn. Blackwell Publishing), Ames, IA, pp 467–485

Hu ZH, Zhou JC, Yang HZ (2018) Comprehensive evaluation of different sugar beet varieties by using principal component and cluster analyses. J Phy Conf Ser. https://doi.org/10.1088/1742-6569/1176/4/042021

Jonck F, Gava A, Traverso SD, Joelma L, Fernando HF, Gueller E (2013) Intoxicacao espontancea e experimental por nitrato/nitrito em bovinos alimentados com *Avena sativa* (aveia) e/ou Lolium spp. (azevem). Pesq Vet Bras 33(9):1062–1070

Kaur G, Goyal M (2016) Nitrogen assimilation potential in relation to nitrate-N toxicity at different growth stages and N inputs in oats (*Avena sativa* L.). Indian J Biochem Biophys 53:126–134

Lardy GP (2018) Feeding sugar beet byproducts to cattle. Technical report. https://www.researchgate.net/publication/325654319

Lardy G, Schafer R (2008) Feeding sugar beet byproducts to cattle. AS-1365, NDSU. Extension Service, North Dakota State University, Fargo

Levigne S, Ralet MC, Thibault JF (2002) Characterization of pectins extracted from fresh sugar beet under different conditions using an experimental design. Carbohydr Polym 49:145

Miller AW, Kohl KD, Dearing MD (2014) The gastrointestinal tract of the white-throated woodrat (Neotoma albigula) harbors distinct consortia of oxalate-degrading bacteria. Appl Environ Microbiol 80(5):1595–1601

Mordenti AL, Giaretta E, Campidonico L, Parazza P, Formigoni A (2021) A review regarding the use of molasses in animal nutrition. Animals 11(1):115. https://doi.org/10.3390/ani11010115

Mostafa MRM, Salama R, Lashin ME, Abodo AA (2003) Utilization of olive cake in fattening lambs. Egyptian J Nutr Feeds 6(special issue):811–820

Muir BM, Schopf SD, Meade C (2019) Filtration of stored thick juice for the syrup campaign. Technology 9:522–528

Munnich M, Klevenhusen F, Zebeli Q (2018) Feeding of molassed sugar beet pulp instead of maize enhances net food production of high-producing Simmental cows without impairing metabolic health. Anim Feed Sci Technol 241:75–83. https://doi.org/10.1016/j.anifeedsci.2018.04.018

Mustafa MMM, Saleh HM, El-Fouly HA (2009) Effect of olive by-product inclusion in lambs diet on productive performance and some blood parameters. Egypt J Nutr Feeds 12(3 special issue):253–262

Nakata PA (2003) Advances in our understanding of calcium oxalate crystal formation and function in plants. Plant Sci 164:901–909

NOVUS (ed) (1996) International, raw material compendium, 2nd edn. Novus Europe, Brussels, pp 1–26

NRC (2000) Nutrient requirements of beef cattle. 7th rev. ed. The National Academies Press, Washington, DC

Ozboy O, Sahbaz F, Koksel H (1998) Chemical and physical characterization of sugar beet fiber. Acta Aliment 27:137

Pachauri RK, Allen M, Barros V, Broome J, Cramer W, Christ R (2014) Climate contribution of working groups I, II and III to the fifth assessment report of the intergovernmental panel on climate change. Climate change 2014: synthesis report. IPCC, Geneva

Piccioni M (1989) Dizionario Degli Alimenti Per Il Bestiame, 5th ed. Edagricole, Bologna, pp 587–597

Qudah KM, Rousan LM, Ereifej KI (2009) Nitrate/nitrite poisoning in dairy cattle associated with consumption of forages irrigated with municipally treated wastewater. Toxicol Environ Chem 91:163–170

Radostits OM, Gay CC, Hinchcliff KW, Constable PD (2007) Veterinary medicine, 10th edn. Saunders- Elsevier, London

Rahman MM, Abdullah RB, Khadijah WEW (2013) A review of oxalate poisoning in domestic animals: tolerance and performance aspects. J Anim Physiol Anim Nutr 97:605–614

Rodríguez JG, Ranilla MJ, France J, Moretón HA, Carro MD, López S (2019) Chemical composition, in vitro digestibility and rumen fermentation kinetics of agro-industrial by-products. Animals 9(11):861. https://doi.org/10.3390/ani9110861
Rooke J, Brett P, Overend M, Armstrong D (1985) The energetic efficiency of rumen microbial protein synthesis in cattle given silage-based diets. Anim Feed Sci Technol 13:255–267
Sandhu SK, Sarao NK, Goyal M, Uppal SK, Singh P, Kaur S, Kaur J (2015) Profiling of sugar beet genotypes for agronomical, sugar quality and forage traits and their genetic diversity analysis using SSR markers. Electron J Plant Breed 7(2):253–266
Savage GP, Vanhanen L, Mason SM, Ross AB (2000) Effect of cooking on the soluble and insoluble oxalate content of some New Zealand foods. J Food Comp Analysis 13:201–206
Singh D, Garg AK (2013) Evaluation of beet varieties for forage yield and quality parameters. Range Mgmt Agroforest 34(2):182–185
Smith SB, Johnson BJ (2014) Marbling: management of cattle to maximize the deposition of intramuscular adipose tissue. In: 2014 beef research white paper on product quality. Cattlemen's Beef Board and National Cattlemen's Beef Association, Centennial
Susic S, Sinobad V (1989) Researches aimed at advancing the sugar industry of Yugoslavia-about chemical compositions of bee honey and sugar beet molasses and their biological values in human diet. Hemijska Industrija 43(Suppl. 1–2):10–21
Tawab AMAE, Hassan AAM, Khattab MSAE, Matloup OH, Farahat ESA, Khalel MS, Morsy TA, Fouad MT (2017) Research article productive performance of lactating Frisian cows fed sugar beet leaves silage treated with lactic acid bacteria. Int J Zool Res 13(2):74–82
Tawab AMAE, Kholif AE, Hassan AM, Matloup OH, El-Nor SAA, Olafadehan OA, Khattab MSA (2020) Feed utilization and lactational performance of Friesian cows fed beet tops silage treated with lactic acid bacteria as a replacement for corn silage. Anim Biotechnol 31(6):473–482. https://doi.org/10.1080/10495398.2019.1622556
Ugrinovic K, Kmecl V, Custic MH, Znidarcic D (2012) Contents of oxalic acid, nitrate and reduced nitrogen in different parts of beetroot (*Beta vulgaris* var. *conditiva* Alef.) at different rates of nitrogen fertilization. Afr J Agric Res 7(20):3066–3072
Van Soest PJ (1994) Nutritional ecology of the ruminant. Cornell University Press, Ithaca. https://doi.org/10.7591/9781501732355
Venkataswarlu M, Reddy YR, Nagakakshami D, Mahender M, Reddy AR (2012) Effect of unconventional creep feed on growth and economics in early weaned Nellore ram lambs. Indian J Anim Nutr 29:360–365

Sugar Beet Pulp and Research Efforts to Diversify Its Use

47

D. Jiménez-Islas, M. E. Pérez-Romero, I. Ventura-Cruz, and M. B. Flores-Romero

Abstract

Sugar beet (*Beta vulgaris* L) is one of the most important sugar crops in the world. The sugar beet pulp (SBP) is the main by-product obtained from the extraction of sugar from the sugar beet root. The majority of SBP is used for animal feeding; however, in recent years, a series of worldwide studies have tried to add value to the by-product through the diversification into new products that improve the overall profitability of the raw material. Various uses for SBP have been evaluated based on its content of cellulose, hemicellulose, lignin, pectin, and residual sugars. Many researchers identified that the SBP can be used as feed-stock, including the production of methane, biohydrogen, bioethanol, bio-polyols, fertilizers, particleboard, prebiotics, and alternative materials for the paper industry. This chapter describes the general processes for the use of sugar beet pulp and the conditions for obtaining products derived from the processing of the raw material.

D. Jiménez-Islas (✉)
Renewable Energy Division, Higher Technological Institute of Huichapan, Huichapan, Mexico
e-mail: djimenez@iteshu.edu.mx

M. E. Pérez-Romero
Faculty of Economic and Administrative Sciences, Michoacan University of San Nicolas of Hidalgo, Morelia, Mexico

Business Management Division, Higher Technological Institute of Huichapan, Huichapan, Mexico

I. Ventura-Cruz
Mechatronic Division, Higher Technological Institute of Huichapan, Huichapan, Mexico

M. B. Flores-Romero
Faculty of Economic and Administrative Sciences, Michoacan University of San Nicolas of Hidalgo, Morelia, Mexico

V. Misra et al. (eds.), *Sugar Beet Cultivation, Management and Processing*,
https://doi.org/10.1007/978-981-19-2730-0_47

Keywords

Beet-particle board · Beet pulp · Bioenergy · Polyols · Prebiotics

Abbreviations

AD	Anaerobic digestion
BP	Beet pulp
CE	Cellulose
PU	Polyurethane foam
RPUF	Rigid bio-based polyurethane foam
SBP	Sugar beet pulp
SBR	Sugar beet root

47.1 Introduction

Sugar beets are a valuable crop used to make sugar all over the world. After the sugar is extracted, a large amount of sugar beet pulp (SBP) residue is left, which conventionally is utilized as low value animal feed using cost-intensive drying and pelletizing process (Abou-Elseoud et al. 2021; Li et al. 2020). Waste in the food industry cause environmental and economic problems if they are not treated, thus it has constantly been sought to find others uses of the by-products to obtain other high-value products (Simić et al. 2021). Among the alternatives for the comprehensive use of waste from the sugar industry, there is the development of co-products of industrial interest that represent a commercial complement for increased profitability in the sugar industry. Figure 47.1 shows the areas in which

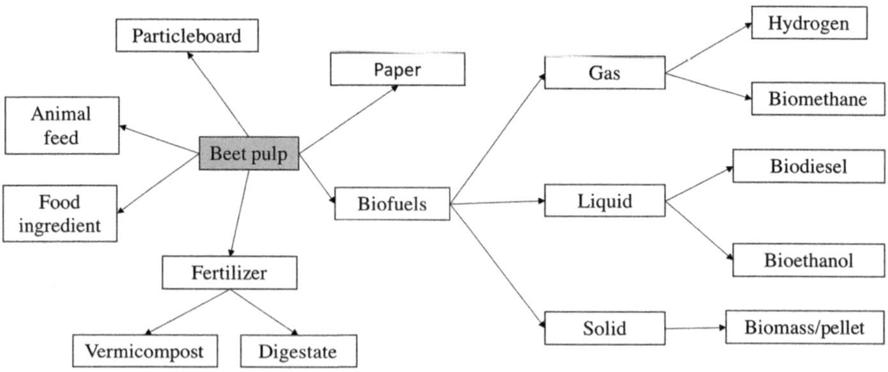

Fig. 47.1 Sugar beet pulp as a raw material to generate high value products

SBP is used as feedstock to generate products with commercial value, experimentally and commercially.

At present, SBP is mainly sold as animal feed at a very low price, due to its relatively low protein content compared to the requirements of ruminants; however, alternative uses contemplate the generation of renewable energy (biofuels), polymers, and a source of pectin (Brachi et al. 2017). In addition, the by-products generated from SBP treatment can also have additional uses, such as digestate. The alternatives for the use of SBP are of interest to the sugar industry in the search to improve energy, environmental and regulatory financing of each country. In this context, it is necessary to update the various uses that are developed to take advantage of the raw material in other products.

47.2 Sugar Beet Process and By-products

More than 25% of the world sugar requirement is met from sugar beet. The beet sugar industry is well established in 45 countries spread over four continents. Sugar beet is a man-made crop with its origin in the nineteenth century from table and fodder beets (Pathak et al. 2014). Figure 47.2 shows the general process for obtaining sugar from sugar beet root (SBR) from the cultivation field. The processing efficiency of SBR can vary depending on factory equipment and how it is operated (Joanna et al. 2018). The SBR is washed through rotating units to remove impurities that affect the extraction of sugar, later it is transferred to slicers where

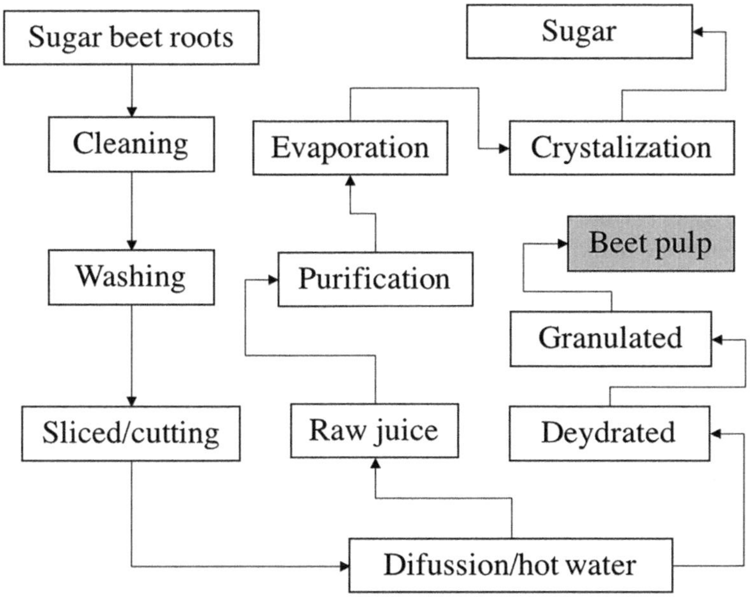

Fig. 47.2 Main operations for sugar beet processing

they are cut into cossettes (Ladakis et al. 2020). Later, cossettes are mixed with hot water (55–75 °C) in tanks called diffusers (Marzo et al. 2019), producing raw juice stream and residual wet pulp stream (Zicari et al. 2019). Then, the raw juice is purified and concentrated in multiple steps to reach an average sugar content of 67% (Vučurović and Razmovski 2012). After, the thick juice is vacuum-crystallized.

In a different part of the process, the sugar beet pulp is the plant material left from sucrose extraction (Cárdenas-Fernández et al. 2017); the chemical composition of the sugar beet pulp (SBP) makes it an attractive raw material for different bio-reactions. During the process of generating sugar from sugar beet, a large amount of SBP is stored in huge piles and to avoid material degradation processes it is necessary to subject them to a drying process.

47.3 Sugar Beet Pulp Applications

47.3.1 Sugar Beet Pulp as a Raw Material for Particleboard

Particleboards are one of the most widely used wood products in the world (Borysiuk et al. 2019). Traditionally, particleboard (for furniture) is made up of 7% resin and 93% wood; however, as the population grows, the demand for furniture grows, and the industry tries to reduce production costs due to intense competition (Pinkl et al. 2020). Various materials have been evaluated in the search for alternatives to wood in the furniture industry, including: (a) sugar beet left fiber into which different assays can be included to improve the physico-mechanical properties of particleboard with a possible commercial value (Das and Chanda 2020); (b) a research was conducted using sugar beet pulp as material with the core layer made from a mixture with industrial wood; in this process, the results show that the standards for furniture boards are met by materials with a 30% SBP addition and the particleboards production process is not required; on the other hand, the density profile alignment is also affected by increasing sugar beet pulp concentration in the core. Additionally, a search of Scopus database was conducted to find research papers on the topic of "particleboard" and "sugar beet pulp"; however, only two papers were found, indicating the topic has been in development in recent years.

47.3.2 Sugar Beet Pulp in the Paper Industry

Because the paper industry uses cellulose from trees primarily, it is regarded as a viable approach to investigate raw resources such as sugarcane bagasse and sugar beet pulp, which have qualities that help to reduce environmental effect (Vaccari et al. 2005). Various pretreatments are used in this case to keep the desired qualities on the paper; although the paper will never be as white as pure cellulose paper, it may be used for printing, photocopying, and other similar tasks.

The SBP may be effectively utilized in the making of paper as a partial substitute for wood fibers and as a natural material that strengthens the water resistance of the paper, according to the findings, a 15% beaten sugar beet pulp level in the combination is still acceptable in terms of paper machine run-ability and paper qualities (Fišerová et al. 2007). In the commercial sense, it has been reported that the company Crown Van Gelder has become the first in the world to produce sugar beet paper on a large scale, the new product line called "Crown Native" reduces the environmental impact by 16% through the usage of less wood fibers compared to traditional paper. In addition, Cosun Beet Company and the paper manufacturer Crown van Gelder are collaborating to develop paper and packaging materials from sugar beet fibers. As you can see, the need for natural fibers to make paper and cardboard has allowed sugar beet pulp to have commercial applications.

47.3.3 Compost of Sugar Beet Pulp

Composting is currently at the core of environmental issues and the circular economy as one of the greatest ways to recycle organic waste and produce a high-value final product (Haouas et al. 2021). Vermicomposting may have been an alternate process for converting sugar beet waste into valuable fertilizer (Bhat et al. 2015). The combination of SBP and paper waste has been evaluated under the process of composting, the results show that the quality compost increased with the combination of 25% SBP and 10% of paper waste around 20 days (Zhang and Sun 2018).

Anaerobic digestion produces biogas and digestate, a material that may be used to improve the qualities of agricultural soil. The application of the residues of anaerobic digestion of SBP has been applied in the fertilization of the sugar beet crop and it has been determined that the experimental plots met the quality criteria, which could reduce the amount of mineral fertilizer (Baryga et al. 2016). It has been shown that SBP by-products (digestate) can be an alternative for improving soil for the production of corn in rotation with sugar beet crops (Baryga et al. 2021). Also, the application of SBP pulp digestate can be treated as an effective method of environmental treatment (Baryga et al. 2020).

47.4 Sugar Beet Pulp as Energy Source

The high population growth rate in the world and the demand for energy are a linked phenomenon that has an adverse impact on the environment. Various raw materials have been evaluated in the production of biofuels, among which the use of fruit juice, lignocellulosic biomass such as sugarcane bagasse and sugar beet stand out, the latter from leaf fibers, stillage, molasses and pulp. The biofuels generated from different SBP treatment processes are bioethanol, hydrogen, biogas (methane), and dry biomass for burning (Fig. 47.3).

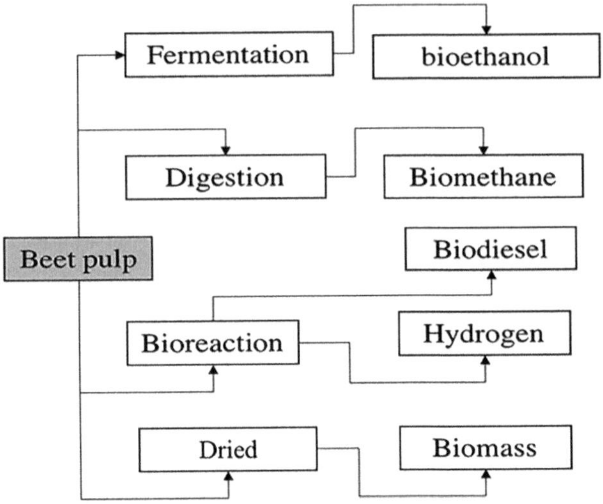

Fig. 47.3 Biofuels obtained from sugar beet pulp feedstock

47.4.1 Biogas

Huge volumes of organic wastes are created throughout various industrial operations, which must be handled or disposed of. This is particularly true in the food and beverage sectors. During the procedure, energy in the form of heat or power is also required (Bochmann et al. 2020). The need for electricity and heat is determined by the manufacturing process, the area, the feedstock, and the technologies used in the manufacturing process itself. Anaerobic digestion (AD) might be a cost-effective way to generate renewable energy from these high material volumes (Suhartini et al. 2019). Sugar beet pulp may be digested anaerobically to create biogas. This is a biological process in which organic matter decomposes in the absence of oxygen, usually by putting a certain amount of biomass in a specifically built reactor for many days (Tomaszewska et al. 2018). The degradations of SBP under co-culture fermentation system was developed using consortium of bacterium *Clostridium cellulovorans* and microbial flora for the methane production, the yield was 34 L of CH_4/kg SBP d/w (Tomita et al. 2019). The SBP has been subjected to treatments to increase the biogas yield during anaerobic digestion, the use of mills up to 2.5 mm improve in yield up to 29% with respect to the SBP without treatment (Ziemiński and Kowalska-Wentel 2016).

47.4.2 Bioethanol

Sugar beet pulp contains (dry weight basis) 75–80% polysaccharides, consisting of 22–24% cellulose microfibers, 30% hemicellulose, and 25% pectin (Vučurović and Razmovski 2012), in addition to having the advantage of having a low lignin content

(1–2%), making the saccharification process relatively easy by some hydrolysis methods (Hamley-Bennett et al. 2016). Four key steps are considered in obtaining fuel ethanol from sugar beet pulp: the treatment of raw materials, fermentation by microorganisms, separation of ethanol, and purification to a specific purity.

Bioethanol is produced by microorganisms through the alcoholic fermentation of sucrose, starch, and biomass that has been subjected to chemical, thermochemical, or enzymatic pretreatments. During the pretreatment of the biomass, some inhibitory compounds like furfural and 5-hydroxymethylfurfural are generated and affect the growth of fermentative microorganisms; to prevent this, it is required to determine the severity of the treatment and the detoxification methods. In this context, the detoxification of hydrolisates of sugar beet pulp increases the production of ethanol by *Pichia stipitis*; the final ethanol concentration was 10.8–12.2 g/L; moreover, increasing the inoculate concentration increases the overall performance (Günan Yücel and Aksu 2015).

Several studies have focused on the monomers produced by the hydrolysis processes (released of pentoses and hexoses); nevertheless, the primary bottleneck with the utilization of sugars is the restricted number of commercial microbes capable of metabolizing both substrates simultaneously. In this case, different options have been developed, such as media fermented with *S. cerevisiae* and *Sch. stipitis* yeast strains produced the most ethanol (12.6 g/L), particularly, in briquetted sugar beet pulp-based hydrolysates; also, lower ethanol yields were obtained during fermentation with *S. cerevisiae* and *K. marxianus* NCYC179 (Berlowska et al. 2017).

When sugar beet pulp hydrolysis is carried out, not all sugars obtained are fermented by commercial yeasts such as *S. cerevisiae*; in this case, the hydrolysis conditions have been evaluated with microorganisms such as *Escherichia coli* KO11, *Klebsiella oxytoca* strain P2, and *Erwinia chrysanthemi* EC 16 pLOI 555, in which the results showed the effectiveness of *E. coli* KO11 in the conversion of galacturonic acid to ethanol (Doran et al. 2000). Figure 47.4 shows three phases of the process of ethanol production: the sugar beet pulp is hydrolyzed using some treatment to release the fermentable sugars; the microorganism utilized can uptake sugars during the fermentation process to produce bioethanol; separation and purification of bioethanol using the distillation.

From an environmental and economic perspective, bioethanol is generated from different materials such as sugar beet pulp; in the process, the incorporation of pretreatments is necessary to avail sugars during fermentation (Gumienna et al. 2014). After pretreatment, a key factor for the fermentation success is to consider the chemical composition of the hydrolysates to define the microorganism to be used, which can be commercially or genetically modified. Finally, after fermentation, the distillation process requires the incorporation of energy, a key factor in the economic viability of the process.

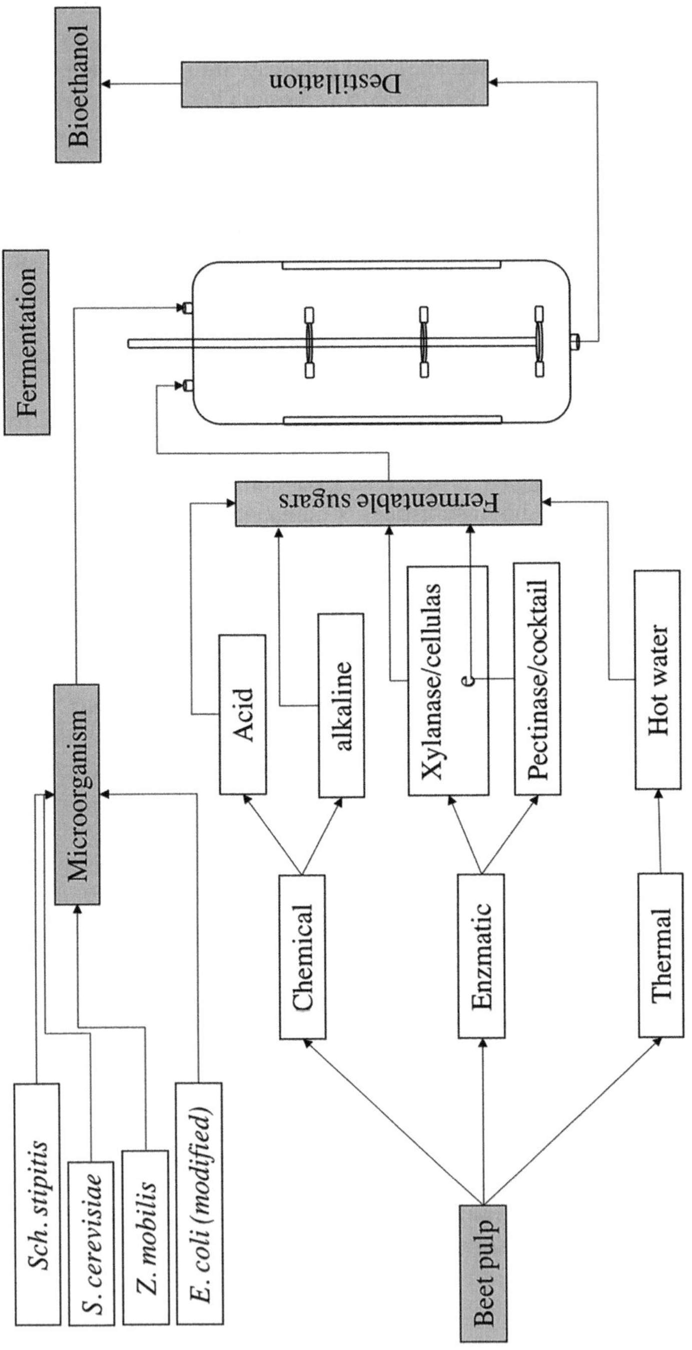

Fig. 47.4 General process for ethanol production using pretreatment of sugar beet pulp

47.4.3 Hydrogen

Because hydrogen combustion produces just water vapor rather than greenhouse gases such as CO_2, it is a clean and ecologically beneficial fuel (Ozkan et al. 2011). Hydrogen is projected to play a significant role in meeting future energy needs. Sugar beet is one of the most attractive raw materials for hydrogen fermentation. This, just like other forms of biomass, must be processed before it can be used as a fermentable feedstock (Grabarczyk et al. 2011). Due to the structural characteristics of SBP, pretreatment is necessary before being used for the biological processes for obtaining hydrogen. In addition, the addition of Fe_2O_2 improves the hydrogen yield from SBP hydrolysates (Cieciura-Włoch et al. 2020).

47.4.4 Biodiesel

The use of biodiesel as a renewable source of energy offers sustainability advantages and that the CO_2 released in combustion will be used by the crops that gave rise to it. Vegetable oils, animal fats, and waste cooking oils are the most common feedstocks for biodiesel manufacturing (ZulqarnainAyoub et al. 2021). Due to their rapid production rates and scale-up for industrial processing, microbial lipids, also known as single-cell oils, are being investigated as viable biodiesel alternatives to vegetable oils (Mhlongo et al. 2021).

As alternative substrates for the generation, increasing the content of cellular lipid in *Lipomyces starkeyi* has been proposed for SBP and molasses, which are the main by-products the process of sugar beet; the authors showed that when SBP is used as raw material, the oil content is 19.2%, but when a mixture of SBP with molasses is used, the oil content increases to 49.2% (Martani et al. 2020). When *Mucor circinelloides* and *A. oryzae* were evaluated as single-cell oil producers from SBP as raw material, the results show 25% oil content in *M. cirninelloides*; although the yield is low compared to other substrates, the production from SBP is an alternative due to its availability and low cost (Ozsoy et al. 2015).

47.5 Animal Feed from Sugar Beet Pulp

Traditionally, SBP is used to feed goats, horses, cows, and pigs. In recent years, the possibility of applying them to a wider range of animals has been explored in the search to improve the nutritional characteristics of food. Fibrous substances such as beet pulp (BP) and cellulose (CE) are widely utilized by pet food manufacturers in diets aimed at weight loss and intestinal health (Donadelli et al. 2021).

The incorporation of SBP in the diet of laying hens improves the performance and quality traits of the egg and improves the health of the hens (Selim and Hussein 2020). When SBP is incorporated into the diet of Gimmizah laying hens in 20%, the results obtained showed poor performance (Emam and Abdel Wahed 2020).

47.6 Food and Aggregated of Sugar Beet Pulp

Pectin has been widely used as food additive (Huang et al. 2017). Sugar beet pectic polysaccharides consist of different structural elements; although pectin extracted from sugar beet pulp is known to have poor gelling property, it has excellent emulsification property (Abou-Elseoud et al. 2021). Different extraction and purification processes have been proposed in order to obtain pectin for different industrial applications from SBP.

47.7 Bio-products of Sugar Beet Pulp

47.7.1 Lactic Acid

Lactic acid is a molecule of industrial interest due to its diverse applications; the hydrolysates of sugar beet pulp are used as a substrate in continuous cultures of *Bacillus coagulans*; on an average, 79% of the sugars of sugar beet pulp are transformed into lactic acid (Alves de Oliveira et al. 2020). Other raw material treatment processes have been developed to favor the fermentation process; under simultaneous saccharification and fermentation, the exhausted sugar beet pulp pellets were used as a raw material for lactic acid production by *Lactobacillus casei* 2246, with an increase in the productivity (27 g/L) compared to separated hydrolysis and fermentation (Díaz et al. 2020). *Lactobacillus plantarum* and exhausted sugar beet pulp pellets were used to produce lactic acid through enzymatic hydrolysis, the production of the metabolite was 30 g/L (Marzo et al. 2021). In the search to reduce the presence of by-products that affect fermentation, commercial enzymes have been used, as reported with Viscozyme® and Ul-traflo® Max, of which the results found are lower processing costs and higher productivity due to saccharification and simultaneous fermentation of sugar beet pulp, which yields around 30 g/L and requires fewer enzymatic loads (Berlowska et al. 2018).

47.7.2 Biopolyols

Polyurethane (PU) foam is a polymer obtained by chemical reaction between polyol polyether and polyisocyanate; however, the need for renewable sources as a material to generate it has allowed the development of various methods of preparing of PU with different physico-mechanical properties (Zheng et al. 2016). Some previous studies have shown that SBP is a source of biopolyols; the polyurethane foam was reinforced with SBP impregnated with Aminopropylisobutyl-polyhedral, increasing the physico-mechanical properties with 1–2% SBP filler (Strąkowska et al. 2020). Rigid bio-based polyurethane foam (RPUF) composites that incorporate sugar beet pulp (SBP) particles as reactive fillers have advantages for thermal insulation engineering materials (Akdogan and Erdem 2021).

47.7.3 Prebiotics

SBP might be used to make prebiotics, which could be a novel and cost-effective use. The market for new-generation animal feeds remains untapped (Berlowska et al. 2018). Different technologies were applied to SBP extract rich in pectin to obtain prebiotics (pectin or saccharides) using *Lactobacillus* species (Prandi et al. 2018). The extraction of pectin oligosaccharides was evaluated for its prebiotic potential using fecal inocula (Gómez et al. 2016).

47.8 Keywords Associated to Beet Pulp

Through an analysis of the keywords associated with SBP, four clusters were identified (Fig. 47.5). The clusters show the relationship of the keywords with the description of the research, the green cluster shows sugar beet traditionally associated with animal nutrition. The red cluster associated with the development of bioenergy. The blue cluster considers the treatments and bio-reactions involved. The yellow cluster shows the characterization in food. The keywords were obtained with the word "beet pulp" in the Scopus database.

In the four clusters, the development of bio-products requires SBP pretreatment. When SBP treatment processes are implemented, the availability of sugars is improved, likewise the structural changes that contribute to improving the surface area for the incorporation of enzymes or microorganisms. From a general approach,

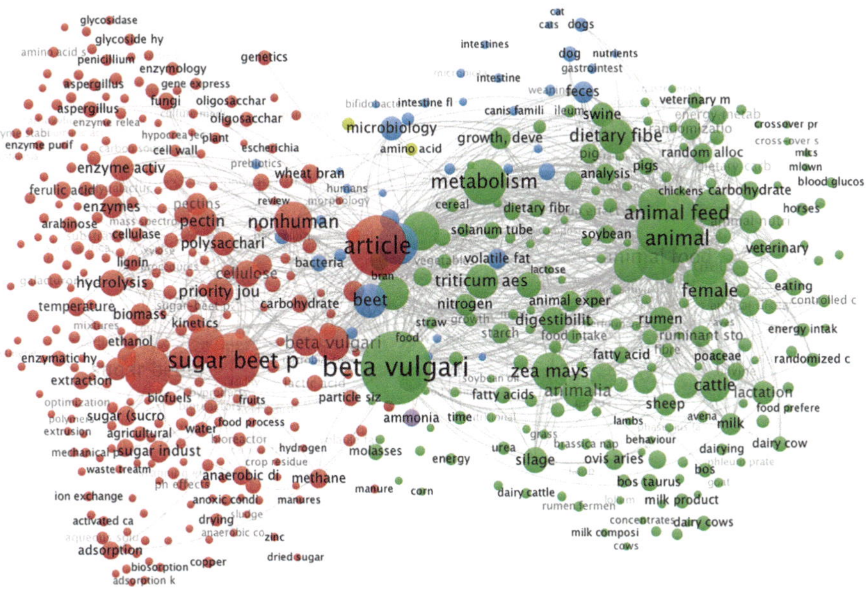

Fig. 47.5 Cluster of keywords associated to sugar beet pulp

the SBP should be treated prior to its use in different processes that generate value-added products.

47.9 Future Prospects

The application of new treatment methods for waste from the sugar industry have generated interest from an economic and environmental perspective. The use of SBP as a feedstock for the generation of products through the biorefinery concept is an attractive way to improve the competitiveness of companies. For the development of products based on SBP, it is necessary for companies to know the various developments that researchers carry out, in addition to improving the link between companies and universities (or research centers) to carry out the development of specific solutions to take advantage of the comprehensive form of the SBP.

From a business perspective, there are other industries such as food, furniture, and paper that can benefit from advances in basic science developed with the SBP. In this sense, the characteristics of each industry must be known to determine the input conditions of the feedstock.

Also, from the research point of view, it is necessary to characterize the by-products generated during each SBP treatment to determine if it is possible to apply it in the development of new products; here it is convenient to build multidisciplinary working groups which help to find the scaling conditions of products up to the industrial scale.

Finally, from the energy point of view, international regulations require reducing greenhouse gas emissions in order to do an exploration in the diversification of energy needs through the implementation of SBP to generate biofuels that are adjusted to the production of the sugar industry and net energy needs can be carried out. The generation of biogas from SBP is an option; however, it can be diversified to improve the profitability of the companies.

47.10 Conclusion

Different investigations worldwide continually seek to take advantage of SBP as a feedstock for the generation of different products that allow the incorporation of value. From a technical perspective, the requirement for raw material processing increases the usage of SBP, furthermore the fact that the feedstock can be fully used. By the same token, the possible generation of energy (renewable) benefits the global process of the sugar industry; however, the dysfunctionality of the feedstock and the technological conditions of the production plants must be taken into consideration. Some researches have been carried out with the goal of developing the fundamental science of the processes; however, all technologies are not yet mature and evaluations are required within the research groups.

References

Abou-Elseoud WS, Hassan EA, Hassan ML (2021) Extraction of pectin from sugar beet pulp by enzymatic and ultrasound-assisted treatments. Carbohydr Polym Technol Appl 2:100042. https://doi.org/10.1016/j.carpta.2021.100042

Akdogan E, Erdem M (2021) Improvement in physico-mechanical and structural properties of rigid polyurethane foam composites by the addition of sugar beet pulp as a reactive filler. J Polym Res 28(3):1–17. https://doi.org/10.1007/s10965-021-02445-w

Alves de Oliveira R, Schneider R, Hoss Lunelli B, VazRossell CE, MacielFilho R, Venus J (2020) A simple biorefinery concept to produce 2G-lactic acid from sugar beet pulp (SBP): a high-value target approach to valorize a waste stream. Molecules 25:2113. https://doi.org/10.3390/molecules25092113

Baryga A, Polec C, Skibniewska KA, Seciu E, Grabara J (2016) Utilisation of residual waste from sugar beet pulp fermentation as fertiliser in sustainable agriculture. J Environ Prot Ecol 17: 1048–1057

Baryga A, Połeć B, Klasa A (2020) Quality of sugar beets under the effects of digestate application to the soil. Processes 8:1402. https://doi.org/10.3390/pr8111402

Baryga A, Połeć B, Klasa A, Olejnik TP (2021) Application of sugar beet pulp digestate as a soil amendment in the production of energy maize. Processes 9:765. https://doi.org/10.3390/pr9050765

Berlowska J, Pielech-Przybylska K, Balcerek M, Cieciura W, Borowski S, Kregiel D (2017) Integrated bioethanol fermentation/anaerobic digestion for valorization of sugar beet pulp. Energies 10:1255. https://doi.org/10.3390/en10091255

Berlowska J, Cieciura-Wloch W, Kalinowska H, Kregiel D, Borowski S, Pawlikowska A, Binczarski M, Witonska I (2018) Enzymatic conversion of sugar beet pulp: a comparison of simultaneous saccharification and fermentation and separate hydrolysis and fermentation for lactic acid production. Food Technol Biotechnol 56:188–196. https://doi.org/10.17113/ftb.56.02.18.5390

Bhat SA, Singh J, Vig AP (2015) Vermistabilization of sugar beet (*Beta vulgaris* L) waste produced from sugar factory using earthworm *Eiseniafetida*: genotoxic assessment by *Allium cepa* test. Environ Sci Pollut Res 22:11236–11254. https://doi.org/10.1007/s11356-015-4302-4

Bochmann G, Pesta G, Rachbauer L, Gabauer W (2020) Anaerobic digestion of pretreated industrial residues and their energetic process integration. Front Bioeng Biotechnol 8:1–11. https://doi.org/10.3389/fbioe.2020.00487

Borysiuk P, Jenczyk-Tolloczko I, Auriga R, Kordzikowski M (2019) Sugar beet pulp as raw material for particleboard production. Ind Crop Prod 141:111829. https://doi.org/10.1016/j.indcrop.2019.111829

Brachi P, Riianova E, Miccio M, Miccio F, Ruoppolo G, Chirone R (2017) Valorization of sugar beet pulp via torrefaction with a focus on the effect of the preliminary extraction of pectins. Energy Fuel 31:9595–9604. https://doi.org/10.1021/acs.energyfuels.7b01766

Cárdenas-Fernández M, Bawn M, Hamley-Bennett C, Bharat PKV, Subrizi F, Suhaili N, Ward D, Bourdin S, Dalby PA, Hailes HC, Hewitson P, Ignatova S, Kontoravdi C, Leak DJ, Shah N, Sheppard TD, Ward JM, Lye GJ (2017) An integrated biorefinery concept for conversion of sugar beet pulp into value-added chemicals and pharmaceutical intermediates. Faraday Discuss 202:415–431. https://doi.org/10.1039/c7fd00094d

Cieciura-Włoch W, Borowski S, Domański J (2020) Dark fermentative hydrogen production from hydrolyzed sugar beet pulp improved by iron addition. Bioresour Technol 314:123713. https://doi.org/10.1016/j.biortech.2020.123713

Das S, Chanda S (2020) Particle board from leaf fibre residues of sugar beet. J Sci Indus Res 59: 306–308

Díaz AB, González C, Marzo C, Caro I, Blandino A (2020) Feasibility of exhausted sugar beet pulp as raw material for lactic acid production. J Sci Food Agric 100:3036–3045. https://doi.org/10.1002/jsfa.10334

Donadelli RA, Dogan H, Aldrich G (2021) The effects of fiber source on extrusion parameters and kibble structure of dry dog foods. Anim Feed Sci Technol 274:114884. https://doi.org/10.1016/j.anifeedsci.2021.114884

Doran JB, Cripe J, Sutton M, Foster B (2000) Fermentations of pectin-rich biomass with recombinant bacteria to produce fuel ethanol. Appl Biochem Biotechnol 84:141–152. https://doi.org/10.1385/ABAB:84-86:1-9:141

Emam RMS, Abdel Wahed HM (2020) Effect of inclusion of sugar beet pulp in the diets on the performance and egg quality of Gimmizah laying hens. Egypt J Nutr Feeds 23:305–319

Fišerová M, Gigac J, Boháček S (2007) Application of pre-treated sugar beet pulp in paper manufacture. Cellul Chem Technol 41:283–289

Gómez B, Gullón B, Yáñez R, Schols H, Alonso JL (2016) Prebiotic potential of pectins and pectic oligosaccharides derived from lemon peel wastes and sugar beet pulp: a comparative evaluation. J Funct Foods 20:108–121. https://doi.org/10.1016/j.jff.2015.10.029

Grabarczyk R, Urbaniec K, Koukios E, Bakker R, Vaccari G (2011) Options of sugar beet pretreatment for hydrogen fermentation. Sugar Industry 136:784–790. https://doi.org/10.36961/si12360

Gumienna M, Szambelan K, Jeleń H, Czarnecki Z (2014) Evaluation of ethanol fermentation parameters for bioethanol production from sugar beet pulp and juice. J Inst Brew 120:543–549. https://doi.org/10.1002/jib.181

Günan Yücel H, Aksu Z (2015) Ethanol fermentation characteristics of *Pichia stipitis* yeast from sugar beet pulp hydrolysate: use of new detoxification methods. Fuel 158:793–799. https://doi.org/10.1016/j.fuel.2015.06.016

Hamley-Bennett C, Lye GJ, Leak DJ (2016) Selective fractionation of sugar beet pulp for release of fermentation and chemical feedstocks; optimisation of thermo-chemical pre-treatment. Bioresour Technol 209:259–264. https://doi.org/10.1016/j.biortech.2016.02.131

Haouas A, El Modafar C, Douira A, Ibnsouda-Koraichi S, Filali-Maltouf A, Moukhli A, Amir S (2021) Evaluation of the nutrients cycle, humification process, and agronomic efficiency of organic wastes composting enriched with phosphate sludge. J Clean Prod 302:127051. https://doi.org/10.1016/j.jclepro.2021.127051

Huang X, Li D, Wang L (2017) Characterization of pectin extracted from sugar beet pulp under different drying conditions. J Food Eng 211:1–6. https://doi.org/10.1016/j.jfoodeng.2017.04.022

Joanna B, Michal B, Piotr D, Agnieszka W, Dorota K, Izabela W (2018) Sugar beet pulp as a source of valuable biotechnological products. In: Advances in biotechnology for food industry, pp 359–392. https://doi.org/10.1016/b978-0-12-811443-8.00013-x

Ladakis D, Papapostolou H, Vlysidis A, Koutinas A (2020) Inventory of food processing side streams in European Union and prospects for biorefinery development. In: Food industry wastes, pp 181–199. https://doi.org/10.1016/b978-0-12-817121-9.00009-7

Li G, Zhang Y, Zhao C, Xue H, Yuan L (2020) Chemical variation in cell wall of sugar beet pulp caused by aqueous ammonia pretreatment influence enzymatic digestibility of cellulose. Ind Crop Prod 155:112786. https://doi.org/10.1016/j.indcrop.2020.112786

Martani F, Maestroni L, Torchio M, Ami D, Natalello A, Lotti M, Porro D, Branduardi P (2020) Conversion of sugar beet residues into lipids by *Lipomycesstarkeyi* for biodiesel production. Microb Cell Factories 19:1–13. https://doi.org/10.1186/s12934-020-01467-1

Marzo C, Díaz AB, Caro I, Blandino A (2019) Status and perspectives in bioethanol production from sugar beet. In: Ray RC, Ramachandran S (eds) Bioethanol production from food crops: sustainable sources, interventions, and challenges. Academic Press, pp 61–79. https://doi.org/10.1016/b978-0-12-813766-6.00004-7

Marzo C, Díaz AB, Caro I, Blandino A (2021) Valorisation of fungal hydrolysates of exhausted sugar beet pulp for lactic acid production. J Sci Food Agric 101:4108–4117. https://doi.org/10.1002/jsfa.11046

Mhlongo SI, Ezeokoli OT, Roopnarain A, Ndaba B, Sekoai PT, Habimana O, Pohl CH (2021) The potential of single-cell oils derived from filamentous fungi as alternative feedstock sources for biodiesel production. Front Microbiol 12:1–27. https://doi.org/10.3389/fmicb.2021.637381

Ozkan L, Erguder TH, Demirer GN (2011) Effects of pretreatment methods on solubilization of beet-pulp and bio-hydrogen production yield. Int J Hydrog Energy 36:382–389. https://doi.org/10.1016/j.ijhydene.2010.10.006

Ozsoy HD, Arikan EB, Cinkir C, Eryilmaz G, Kucuk D, Van Leeuwen J (2015) Fungal oil production from oleaginous fungi *Mucor circinelloides* and *Aspergillus oryzae* cultivated on sugar beet pulp. ISITES 2015, Valencia

Pathak AD, Kapur R, Solomon S, Kumar R, Srivastava S, Singh PR (2014) Sugar beet: a historical perspective in Indian context. Sugar Tech 16:125–132. https://doi.org/10.1007/s12355-014-0304-7

Pinkl S, Veigel S, Gindl-Altmutter W, van Herwijnen H, Riegler M (2020) Effects of fibrillar cellulosic additives on particleboard production and properties. In: Wood material science & engineering, pp 1–7. https://doi.org/10.1080/17480272.2020.1786722

Prandi B, Baldassarre S, Babbar N, Bancalari E, Vandezande P, Hermans D, Bruggeman G, Gatti M, Elst K, Sforza S (2018) Pectin oligosaccharides from sugar beet pulp: molecular characterization and potential prebiotic activity. Food Funct 9:1557–1569. https://doi.org/10.1039/c7fo01182b

Selim S, Hussein E (2020) Production performance, egg quality, blood biochemical constituents, egg yolk lipid profile and lipid peroxidation of laying hens fed sugar beet pulp. Food Chem 310:125864. https://doi.org/10.1016/j.foodchem.2019.125864

Simić S, Jovana P, Dušan R, Biljana P, Ivana L, Antun J, Aleksandar F, Sanja N, Marijana B, Borislav M (2021) The influence of extruded sugar beet pulp on cookies' nutritional, physical and sensory characteristics. Sustainability 13:5317. https://doi.org/10.3390/su13095317

Strąkowska A, Członka S, Kairytė A (2020) Rigid polyurethane foams reinforced with POSS-impregnated sugar beet pulp filler. Materials 13(23):5493. https://doi.org/10.3390/ma13235493

Suhartini S, Heaven S, Zhang Y, Banks CJ (2019) Antifoam, dilution and trace element addition as foaming control strategies in mesophilic anaerobic digestion of sugar beet pulp. Int Biodeter Biodegrad 145:104812. https://doi.org/10.1016/j.ibiod.2019.104812

Tomaszewska J, Bieliński D, Binczarski M, Berlowska J, Dziugan P, Piotrowski J, Stanishevsky A, Witońska IA (2018) Products of sugar beet processing as raw materials for chemicals and biodegradable polymers. RSC Adv 8:3161–3177. https://doi.org/10.1039/c7ra12782k

Tomita H, Okazaki F, Tamaru Y (2019) Biomethane production from sugar beet pulp under cocultivation with *Clostridium cellulovorans* and methanogens. AMB Express 9:2–10. https://doi.org/10.1186/s13568-019-0752-2

Vaccari G, Tamburini E, Sgualdino G, Urbaniec K, Klemeš J (2005) Overview of the environmental problems in beet sugar processing: possible solutions. J Clean Prod 13:499–507. https://doi.org/10.1016/j.jclepro.2003.09.008

Vučurović VM, Razmovski RN (2012) Sugar beet pulp as support for *Saccharomyces cerevisiae* immobilization in bioethanol production. Ind Crop Prod 39:128–134. https://doi.org/10.1016/j.indcrop.2012.02.002

Zhang L, Sun X (2018) Influence of sugar beet pulp and paper waste as bulking agents on physical, chemical, and microbial properties during green waste composting. Bioresour Technol 267:182–191. https://doi.org/10.1016/j.biortech.2018.07.040

Zheng Z, Wang L, Li D, Huang Z, Adhikari B, Chen XD (2016) Mechanical and thermal properties of polyurethane foams from liquefied sugar beet pulp. Int J Food Eng 12:911–919. https://doi.org/10.1515/ijfe-2016-0166

Zicari S, Zhang R, Kaffka S (2019) Sugar beet. In: Integrated processing technologies for food and agricultural by-products, pp 331–351. https://doi.org/10.1016/b978-0-12-814138-0.00013-7

Ziemiński K, Kowalska-Wentel M (2016) Effect of different sugar beet pulp pretreatments on biogas production efficiency. Appl Biochem Biotechnol 181:1211–1227. https://doi.org/10.1007/s12010-016-2279-1

Zulqarnain Ayoub M, Yusoff MHM, Nazir MH, Zahid I, Ameen M, Sher F, Floresyona D, Budi Nursanto E (2021) A comprehensive review on oil extraction and biodiesel production technologies. Sustainability 13:788. https://doi.org/10.3390/su13020788

Sugar Beet Pectin and Its Diverse Uses

48

Miljana Djordjević, Marijana Djordjević, Nikola Maravić,
Dragana Šoronja-Simović, and Zita Šereš

Abstract

Growing relevance of sugar beet pectin as a natural ingredient enables gradual venture into diverse study fields enveloping food, pharmaceutical, cosmetic, and polymer industries. The presented chapter provides a bottom-up overview of the sugar beet pectin starting from the elucidation of its structural elements and their conformation across used extraction methods and specific properties towards further sugar beet pectin application in diverse fields. After sugar beet pectin structure disclosure, a close-up of emerging extraction methods as well as conventional methods for sugar beet pectin isolation is presented focusing on induced differences in structural features and introduction of tailoring possibilities. Particular emphasis on sugar beet pectin viscosity, emulsifying, gelling, and probiotic properties in the light of interdependence on structure and extraction method are described in this chapter. Finally, a comprehensive review on studies addressing sugar beet pectin role in emulsification, encapsulation, preparation of edible films and coatings, likewise heavy metals removal is elucidated.

Keywords

Edible films · Encapsulation · Extraction · Properties · Structure · Sugar beet pectin

M. Djordjević (✉) · M. Djordjević · N. Maravić · D. Šoronja-Simović · Z. Šereš
Department of Carbohydrate Food Engineering, Faculty of Technology Novi Sad, University of Novi Sad, Novi Sad, Serbia
e-mail: miljana.djordjevic@uns.ac.rs

V. Misra et al. (eds.), *Sugar Beet Cultivation, Management and Processing*,
https://doi.org/10.1007/978-981-19-2730-0_48

Abbreviations

AE	Acid extraction
AFM	Atomic force microscopy
APW	Alcohol precipitation with washing
Ara	Arabinose
AraOS	Arabinooligosaccharides
DA	Degree of acetylation
D-Dha	2-keto-3-deoxy-D-lyxo heptulosaric acid
DM	Degree of methylation
EAE	Enzyme-assisted extraction
Fru	Fructose
Fuc	Fucose
Gal	Galactose
GalA	α-(1,4)-linked D-galacturonic acid
GalOS	Galactooligosaccharides
GOS	Glucooligosaccharides
GRAS	Generally recognized as safe
HG	Homogalacturonan
HPE	High pressure extraction
KDO	Keto-3-deoxy-D-manno octulosonic acid
MAE	Microwave-assisted extraction
Man	Mannose
OGalA	Oligogalacturonides
RG-I	Rhamnogalacturonan-I
RG-II	Rhamnogalacturonan-II
Rha	L-Rhamnose
SE	Sequences
SWE	Subcritical water extraction
UAE	Ultrasound-assisted extraction
UMAAE	Ultrasound-microwave assisted acid extraction

48.1 Introduction

"Natural" food ingredients are widely gaining relevance on the market with consumers embracing the idea of clean label products alongside increased health awareness. Pectin represents a vast natural biopolymer present in plants which is generally recognized as safe (GRAS) (FDA 2013), thus matching the clean label criteria. Besides citrus and apple, sugar beet and sugar beet pulp as a by-product of the sugar production process, are recognized as pectin sources of industrial relevance due to season-independent availability and high pectin content (15–30% dry basis) (Lv et al. 2013), which remains unchanged regardless of fruit maturity. Additionally, utilization of sugar beet pulp as a source for pectin production approaches the sustainable development goals and circular economy (Reichembach and de Oliveira Petkowicz 2021). Producers engaged in the commercial production of sugar beet pectin with diverse trade names include CP Kelco, Herbstreith, Fox, and UNIPECTIN Ingredients AG.

Sugar beet pectin differs significantly from other commercial pectins in its structural characteristics, consequently reflecting on both physico-chemical and functional properties which govern its application. The branched structure, many side chains, high quantity of acetyl and methyl groups, likewise the quantity of ferulic acid and proteins, distinguish this type of pectin from other plant pectins such as citrus or apple. Structural differences introduce aggravated gel formation but, in contrast, enhanced emulsifying activity and emulsion stabilizing effect of sugar beet pectin. Furthermore, manipulation with extraction methods and conditions represents a crucial step for sugar beet pectin isolation as well as obtention of tailored structure suitable for exhibiting the desired property (Alba and Kontogiorgos 2017).

Chapter conceptualization is aimed to delve into sugar beet pectin structural features, summarize conventional and emerging extraction methods applied for isolation with emphasis on structure outcome, elucidate properties dependence on structure, and disclose sugar beet pectin application in encapsulation, food packaging, and heavy metals removal enabled by the corresponding properties.

48.2 Sugar Beet Pectin Structure

Sugar beet pectin macromolecular structure was the subject extensively explored by multiple researchers (Bohn 1998; Chen et al. 2016b; Funami et al. 2011). However, a definite conclusion regarding its structure is still vague.

Sugar beet pectin represents a complex carbohydrate chain comprising the following carbohydrates presented in descending order: galacturonic acid (GalA), galactose (Gal), arabinose (Ara), rhamnose (Rha), glucose (Glc), and xylose (Xyl) (Chen et al. 2016b; Dronnet et al. 1996; Guo et al. 2017). Nevertheless, the corresponding order can be altered primarily concerning the quantity of Gal, Ara, Rha, and Glc resulting from used sugar beet varieties as well as applied extraction conditions (pH and extraction time), as shown by more recent studies (Funami et al. 2007; Lara-Espinoza et al. 2021; Li et al. 2015; Ma et al. 2013). Furthermore, negligible amounts of fructose (Fru), fucose (Fuc) as well as traces of mannose (Man) were also reported as sugar beet pectin constituents (Babbar et al. 2016; Bohn 1998; Dronnet et al. 1996; Lara-Espinoza et al. 2021; Li et al. 2015). Along with carbohydrates, existence of ferulic acid (Voragen et al. 2009) and proteinaceous moieties (Funami et al. 2007) was also established in the sugar beet pectin structure, which distinguishes the corresponding pectin from conventional pectins and introduces differences in its physico-chemical properties.

The corresponding carbohydrates are further organized in substructures regarded as pectin structural elements or domains (Voragen et al. 2009).

- Homogalacturonan (HG),
- Rhamnogalacturonan I (RG-I)
- Rhamnogalacturonan II (RG-II)
- Arabinan
- Galactan or arabinogalactan I and
- Arabinogalactan II.

48.2.1 Sugar Beet Pectin Structural Elements

Homogalacturonan (HG) represents a linear homopolymer whose backbone is built up from α-(1,4)-linked D-galacturonic acid (GalA) units. The GalA units are usually partly methyl and acetyl-esterified at *C*-6 and at positions *O*-2 and *O*-3, respectively (Remoroza et al. 2014). HG was estimated to constitute nearly 60% of the pectin macromolecule (Ochoa-Villarreal et al. 2012). However, for sugar beet pectin, reported HG share is lower as well as its length (Buchholt et al. 2004), which was estimated at 72–100 GalA units (Voragen et al. 2009). The number of methyl esters and acetyl groups per 100 GalA units is referred to as degree of methylation (DM) and degree of acetylation (DA), respectively (Lara-Espinoza et al. 2018). Most of the published studies conducted on sugar beet pectin ascertain the DM range of 50.8–58% (Dronnet et al. 1996; Lara-Espinoza et al. 2021), yet fewer detected DM above 60% in commercial sugar beet pectin (up to 67%) (Chen et al. 2016b) as well as pectin obtained in a laboratory (Buchholt et al. 2004; Remoroza et al. 2014). Moreover, at laboratory scale extraction, DM up to 76%, and even 81%, for sugar beet pectin isolated by different extraction methods and conditions suggests a strong influence of applied isolation technique (Michel et al. 1985; Peng et al. 2016). Accordingly, sugar beet pectin may be referred to as high-methyl-esterified pectin since over 50% GalA units are methyl (or methoxy) esterified at the *C*-6 position (BeMiller 2019), although opposite claims were also reported (Zhang et al. 2021). The reported sugar beet pectin DA commonly ranges from 14 to 27% (Chen et al. 2016b; Dronnet et al. 1996; Lara-Espinoza et al. 2021), while greater values were also reported (Chen et al. 2015b; Buchholt et al. 2004; Remoroza et al. 2014; Peng et al. 2016). Compared to conventional apple and citrus pectins, sugar beet pectin tends to exhibit a higher DA (Chen et al. 2016b; BeMiller 2019). Previous research also elucidated approximately the same acetyl groups' presence at *O*-2 and *O*-3 in sugar beet pectin and confirmed the truancy of 2,3-di-*O*-acetylation, likewise the paucity of simultaneously methyl- and acetyl-esterified GalA units (Ralet et al. 2005).

Rhamnogalacturonan-I (RG-I) represents branched heteropolymer with a backbone composed of alternating α-(1,2)-linked L-rhamnose (Rha) units and α-(1,4)-linked D-galacturonic acid (GalA) units (Ochoa-Villarreal et al. 2012). While methyl-esterification of RG-I GalA units was rarely reported, *O*-2 or *O*-3 positions of GalA units are reported to be partially acetyl-esterified, even at greater extent compared to HG (Ropartz and Ralet 2020; Morris et al. 2010). Furthermore, Rha units are partially substituted at *O*-4 (mainly) and/or *O*-3 (scarcely) positions with polymeric side chains of neutral sugars, namely, branched arabinans and linear galactans (Yapo 2011; Oosterveld et al. 2000). The length of the RG-I for sugar beet pectin was estimated to be 120 Rha-GalA repeating units (Ropartz and Ralet 2020).

The RG-I arabinan represents a highly branched polymer whose backbone comprises α-(1,5)-linked arabinose (Ara), single or double substituted with α-(1,2)-linked and/or α-(1,3)-linked arabinose as monomeric or oligomeric side chains susceptible to further branching (Westphal et al. 2010). Additionally, ferulic acid is usually ester-linked to Ara in arabinan backbone at the *O*-2 and *O*-5 positions as typical feature of sugar beet pectin (Ropartz and Ralet 2020; Westphal et al. 2010).

The RG-I arabinogalactan I represents a polymer whose backbone is built up from α-(1,4)-linked Gal units which also contain ester-linked ferulic acid at *O*-6 position (Oosterveld et al. 2000; Ropartz and Ralet 2020). Ferulic acid residues' existence and their capability to create dehydrodiferulic acid dimers by in vivo oxidative coupling reactions (Lara-Espinoza et al. 2018) enables cross-linking of the RG-I heteropolymer side chains of sugar beet pectin (Ropartz and Ralet 2020; Oosterveld et al. 2000). Another component suspected to be a constituent of the RG-I is arabinogalactan II, generally linked to the cell wall structural proteins (Voragen et al. 2009).

Two classes of hydroxyproline-rich glycoproteins (HGPS), namely, arabinogalactan proteins (AGP) and extensin were found to be associated with sugar beet pectin (Funami et al. 2007; Kirby et al. 2006; Nuñez et al. 2009), but the origin of the corresponding association is still not fully revealed. Based on the conducted enzymatic digestion of sugar beet pectin extract using proteases, it was suggested that AGP represents a covalently linked integrated constituent of the RG-I arabinogalactans (Funami et al. 2007, 2011; Ngouémazong et al. 2015). Furthermore, proteomics approach enveloping digestion with trypsin was applied for the recognition of extensin, but with no precise evidence on the nature of its linkage to sugar beet pectin (Nuñez et al. 2009). In addition, the establishment of the pectin-protein complexes' existence was acquired through atomic force microscopy (AFM). The AFM revealed appearance of the pectin-protein complexes in a structure termed tadpole where globular proteinaceous moiety is attached to one end of the polysaccharide chain (Liu et al. 2019; Kirby et al. 2006; Kirby et al. 2008). Additionally, other forms of tadpoles were depicted by AFM where globular proteinaceous moiety is partially (loosely) or entirely (tightly) enveloped by the network of pectin structural elements (Kirby et al. 2008).

Despite the establishment of the pectin-protein complexes, a clear insight into the bonding nature and precise globular proteinaceous moiety position within the pectin structural elements network still remained unanswered.

Rhamnogalacturonan-II (RG-II) represents the most divergent sugar beet pectin structural element present in minor amounts. Its backbone comprises of seven to nine GalA units and four different side chains which contain a variety of monosaccharides, some of which are very rare (2-O-methyl xylose, 2-O-methyl fucose, aceric acid, 2-keto-3-deoxy-D-lyxo heptulosaric acid [D-Dha] and 2-keto-

3-deoxy-D-manno octulosonic acid [KDO]) (Ochoa-Villarreal et al. 2012; Voragen et al. 2009). The corresponding structural elements are further arranged to give sugar beet pectin macromolecule. The potential structural elements' arrangement was depicted by diverse structural models.

48.2.2 Sugar Beet Pectin Structural Models

Familiarity with the pectin structural elements does not necessarily reveal how they are organized into a macromolecular structure. Hence, the extensive research regarding the pectin structural elements arrangement in past years issued several pectin structural models known nowadays:

- "Smooth and hairy regions" model (Schols and Voragen 1996)
- "Side chain model" or RG-I backbone model (Vincken et al. 2003)
- "Combined side chain-hairy regions" model (Ralet and Thibault 2009; Schols et al. 2009).

In the "smooth and hairy regions model," the pectin backbone consists of linear HG referred to as "smooth" region alternated with branched RG-I referred as "hairy" region. In the "side chain model" or RG-I backbone model, RG-I represents the pectin backbone at which linear HG and branched RG-II are attached as side chains (Voragen et al. 2009). In the "combined side chain-hairy regions" model, pectin backbone constituents are alternately placed HG and RG-I, while HG also appears as side chain of RG-I (Ralet and Thibault 2009, Schols et al. 2009). Nevertheless, the definite pectin structural elements' arrangement still remains unknown as well as the exact proteinaceous moiety position within the complex macromolecular pectin structure. Toward the established pectin structural models, HG and RG-I pectin structural elements are regarded as backbone substructures with attached arabinan, arabinogalactan I, arabinogalactan II and RG-II as side chain substructures.

48.3 Sugar Beet Pectin Extraction Methods

The complete process of pectin isolation comprises the following stages (Adetunji et al. 2017):

- Pre-treatment of raw material-particle size reduction, depigmentation (blanching)
- Extraction
- Post-extraction stage-pectin precipitation from the extraction solvent, drying.

Factors affecting the extraction and hence extracted pectin quality and quantity are classified into following groups (Adetunji et al. 2017):

- Operating parameters (solid–liquid ratio, temperature, pressure, extraction time)
- Internal factors regarding raw material condition (moisture, particle size)
- External factors regarding properties of the applied solvent (pH, polarity, volatility, molecular weight, toxicity).

To date, diverse agents (from acid to base) were employed for pectin extraction from sugar beet and pulp under conventional heating (Li et al. 2015). In addition, the evolution in innovative green extraction methods and understanding of their principles raised aspiration toward their usage for sugar beet pectin isolation. Their potential in overcoming conventional extraction issues such as solvent consumption, wastewater generation, long extraction time, and requirement for high temperatures will also result in enhanced quality and more consistent properties of pectin. However, the green extraction methods' way to industrial integration is still vague and requires the acquisition of adequate conditions particular for each method (Adetunji et al. 2017).

The following sections give an overview of the applied sugar beet pectin extraction methods, from conventional extraction methods (including primarily acid extraction (AE)) to emerging green extraction methods including enzyme-assisted extraction (EAE), microwave-assisted extraction (MAE), ultrasound-assisted extraction (UAE), subcritical water extraction (SWE), and high pressure extraction (HPE).

48.3.1 Conventional Extraction Methods

Conventional methods of sugar beet pectin extraction envelop the use of water acidified with mineral (HCl or HNO_3, scarcely H_2SO_4), likewise organic (citric, malic, lactic) acids and usage of chelating agents (EDTA) subjected to conventional heating. A summary of the applied conventional extraction methods for sugar beet pectin isolation is presented within Table 48.1.

The most oft-reported acid extraction conditions for sugar beet pectin are pH 1.5–2.0, temperature 70–100 °C, solid to liquid ratio 1:10–1:20, and extraction time 1–4 h. Corresponding conditions' effects regarding pectin yield, DM, DA, and protein content were previously elucidated (Li et al. 2015; Lv et al. 2013). Stronger acidic environment, as well as stronger acid application, prolonged extraction time and higher applied temperatures, induced increase in sugar beet pectin yield (Ma et al. 2013; Levigne et al. 2002; Li et al. 2015; Lv et al. 2013; Yapo et al. 2007a). Conversely, pectin's DM, DA, and protein content are negatively influenced by the corresponding conditions (Yapo et al. 2007a; Li et al. 2015).

Sequential sugar beet pectin extraction conducted in two stages with ammonium oxalate and sulfuric acid, respectively, was also investigated (Liu et al. 2019). Variations in the ferulic acid and proteins content as well as significantly different amino acid profiles were observed for the oxalate-extracted fractions compared to acid-extracted fractions, which reflected negatively on their emulsifying efficiency (Liu et al. 2019). Same conclusion was drawn by the researchers elucidating the

Table 48.1 Summary of the conventional extraction methods applied for sugar beet pectin extraction

Source	Extraction method	Pectin yield [%]	DM/DE	DA	Protein content [%]	References
Dried sugar beet roots	AE, HCl	5–6.3	50.8–57.4	19.64–26.1	8.6–10.3	Lara-Espinoza et al. (2021)
Dried sugar beet pulp		6.16–24.96	–	–	–	Lv et al. (2013)
Fresh sugar beet		2.5–36	34–94	15–43	–	Levigne et al. (2002)
Sugar beet pulp		5.7–22.4	58.92	–	5.20	Mesbahi et al. (2005)
Protein-free sugar beet pulp		11.62–16.94	66.66–77.51	14.33–14.46	–	Sun and Hughes (1998)
Sugar beet pulp		15.81–20.50	–	–	–	Huang et al. (2018)
Sugar beet pulp		42.6–53.4	–	–	–	Almohammed et al. (2017)
Sugar beet pulp		6.7–24.6	–	–	0.5–6	Chen et al. (2018)
Dried sugar beet pulp		20.75	57	14.3	–	Jafarzadeh-Moghaddam et al. (2021)
Fresh sugar beet	AE, HNO_3	2.3–32.9	42–84	6–42	–	Levigne et al. (2002)
Sugar beet pellets		11.1–23.7	27–76	–	–	Michel et al. (1985)
Pressed sugar beet pulp		13.60	49.29	–	4.3	Pacheco et al. (2019)
Ensiled sugar beet pulp		18.94	50.14	–	3.4	Pacheco et al. (2019)
Dried sugar beet pulp		16.72	48.39	–	4.1	Pacheco et al. (2019)
Dried sugar beet pulp	AE, H_2SO_4	4.1–16.2	14.4–65.6	3.1–29.2	0.9–8.6	Yapo et al. (2007a)
Dried sugar beet pulp		5.23–22.24	28.13–43.02	–	1.56–3.01	Li et al. (2019)
Dried sugar beet pulp	AE, Citric acid	6.3–23	–	–	1.5–4.5	Li et al. (2015)
Dried sugar beet pulp		8.7–21.6	37.9–48.0	14.9–24.3	5.3–8.5	Guo et al. (2017)
Dried sugar beet pulp		25.83	59	18.75	–	Adiletta et al. (2020)

Dried sugar beet pulp		6.79–18.25	23.56–43.78	–	1.69–3.09	Li et al. (2019)
Protein-free sugar beet pulp	EDTA	6.70–14.12	61.27–61.54	12.90–14.06	–	Sun and Hughes (1998)
Protein-free sugar beet pulp	Ammonium oxalate	10.90	78.89	11.18	–	Sun and Hughes (1998)
Dried sugar beet pulp	SE, Ammonium oxalate and H_2SO_4	0.6–5.1	22.5–32.3	15.9–22.8	4.0–9.8	Liu et al. (2019)
Protein-free sugar beet pulp	Sodium hexametaphosphate	14.81	64.95	10.80	–	Sun and Hughes (1998)

DE/DM degree of esterification/methylation, *DA* degree of acetylation, *AE* acid extraction, *EDTA* ethylenediamine tetraacetic acid, *SE* sequential extraction

extraction solvent type and extraction time as indisputable factors affecting sugar beet pectin structure and consequently its properties.

In addition, studied pre-treatments that precede the AE provided enhanced sugar beet pectin yield and properties. AE of sugar beet pectin from the superfine ground sugar beet pulp micronized by multidimensional swing high-energy nano-impact-milling increased the pectin yield by 30% due to cell wall breaking effect occurrence (Huang et al. 2018). Furthermore, high-voltage electrical discharges (voltage 40 kV, number of pulses 100, energy input Qe = 76.2 kJ/kg) prove to be a beneficial pre-treatment for achieving 25.3% higher sugar beet pectin yield with lower applied acid quantity (Almohammed et al. 2017).

As regards post-extraction handling of sugar beet pectin, recovery and purification alongside drying were explored. Alcohol precipitation with washing (APW) is a standard method used for sugar beet pectin recovery from aqueous extract. Nevertheless, because of its limited selectivity (precipitation of co-solubilized protein moieties alongside pectin), membrane ultrafiltration with diafiltration was (10 kD molecular weight cut-off) employed in yielding higher GalA amount (Yapo et al. 2007b). Different drying methods and conditions, namely, hot air-drying (40, 50, 60 °C), vacuum drying (40, 50, 60 °C), freeze-drying, spray-drying (160, 190, 220 °C) and their influence on sugar beet pectin structure and emulsifying properties were determined (Huang et al. 2017). Pectin structure remained unchanged regardless of the applied drying method, while drying conditions affected sugar beet pectin's apparent viscosity and molecular weight (Huang et al. 2017).

AE methods typically deliver high pectin yield, considerable protein content, and approximate range of DM and DA, but, at the same time, are economically and energy consuming as well as limited regarding achieving environmental protection demands.

48.3.2 Emerging Extraction Methods

Emerging extraction methods applied in the sugar beet pectin isolation are listed within Table 48.2. In order to further expedite the pectin extraction process and increase yield, emerging methods were also combined, resulting in hybrid extraction methods such as ultrasound-microwave assisted acid extraction (Peng et al. 2015).

By utilizing enzymes capable of catalysing the hydrolysis of pectin from the cell wall matrix, a high level of selectivity is ensured. This introduces numerous benefits compared to AE such as exclusion of pre-treatment, acidic environment and high temperatures, reduction in solvent amount, and wastewater generation leading to decrease in overall extraction time, higher extraction yield, and preservation of pectin structural and functional features, thus enhanced pectin quality. However, the main obstacles for the achievement of the promising advantages in industrial scale-up are high enzyme costs as well as difficulties in monitoring and controlling of different enzymes' response to changes in extraction conditions, which reflect in extraction time and pectin yield (Adetunji et al. 2017). Enzymes by various strains of *Bacillus polymyxa* were among the first ones applied for sugar beet pectin extraction

Table 48.2 Summary of the emerging and hybrid extraction methods applied for sugar beet pectin extraction

Source	Extraction method	Pectin yield [%]	DM/DE	DA	Protein content [%]	References
Pressed sugar beet pulp	EAE, Celluclast®	3.91	47.08	–	1.6	Pacheco et al. (2019)
Ensiled sugar beet pulp		13.40	48.36	–	2.0	Pacheco et al. (2019)
Dried sugar beet pulp		7.50	45.21	–	2.8	Pacheco et al. (2019)
Dried sugar beet pulp	EAE, *Bacillus polymyxa*	4–7	37	–	–	Matora et al. (1995)
Dried sugar beet pulp	EAE, Cellulase	4.24–8.39	–	–	–	Abou-Elseoud et al. (2021)
Dried sugar beet pulp	EAE, Xylanase	4.37–5.97	–	–	–	Abou-Elseoud et al. (2021)
Dried sugar beet pulp	EAE, Mixture cellulase/xylanase	22.41–28.84	64.5–67.7	–	5.6–6.7	Abou-Elseoud et al. (2021)
Sugar beet pulp	SE, Water-EDTA-EAE, Amyloglucosidase-pectinases	~14–33.7	56	–	–	Concha Olmos and Zúñiga Hansen (2012)
Dried sugar beet pulp	UAE + EAE, Mixture cellulase/xylanase	13.3–33.5	52.6–65.9	–	5.29–6.81	Abou-Elseoud et al. (2021)
Dried sugar beet pulp	MAE, HCl	8.91–16.80	71.5–113.3	36.7–67.6	–	Fishman et al. (2008)
Micronized sugar beet pulp		5.8–22.6	–	–	–	Mao et al. (2019)
Sugar beet pulp	MAE, H_2SO_4	5.2–32.4	–	–	–	Li et al. (2012)
Micronized sugar beet pulp	MAE, sodium hexametaphosphate	7.5–11.0	–	–	–	Mao et al. (2019)
Micronized sugar beet pulp	MAE, H_2O	5.6	–	–	–	Mao et al. (2019)

(continued)

Table 48.2 (continued)

Source	Extraction method	Pectin yield [%]	DM/DE	DA	Protein content [%]	References
Micronized sugar beet pulp	MAE, NaOH	2.0–23.4	–	–	–	Mao et al. (2019)
Dried sugar beet pulp	UAE, HCl	20.85	63	13.2	–	Jafarzadeh-Moghaddam et al. (2021)
Dried sugar beet pulp	UAE, Lactic acid	5.3–12.2	54–71	–	4.5–5.4	Ma et al. (2013)
Dried sugar beet pulp	UAE, Citric acid	7.5–17.3	42–65	–	3.3–4.5	Ma et al. (2013)
Dried sugar beet pulp	UAE, Malic acid	6.1–16.1	45–62	–	3.5–4.6	Ma et al. (2013)
Dried sugar beet pulp	SWE	13.63	15.3–56.2	12.8–26.8	0.18–0.32	Pińkowska et al. (2021)
Dried sugar beet pulp		20.65	84.19	25.96	–	Peighambardoust et al. (2021)
Dried sugar beet pulp	UAE + SWE	24.63	55.20	26.74	–	Chen et al. (2015a)
Dried sugar beet pulp	UAE – SWE	0.4–29.1	25–90	24–80	–	Chen et al. (2015b)
Dried sugar beet pulp	HPE, HCl	3.5–12.09	~33	–	–	Kaya et al. (2021)
Dried sugar beet pulp	UMAAE	19.45–25.31	82.03	36.46	8.31	Peng et al. (2015)

DE/DM degree of esterification/methylation, *DA* degree of acetylation, *EAE* enzyme-assisted extraction, *SE* sequential extraction, *UAE* ultrasound-assisted extraction, *MAE* microwave-assisted extraction, *UMAAE* ultrasound-microwave assisted acid extraction, + indicates pre-treatment, – indicates extraction performed in sequences (SE) with obtained pectin yield of 5.5–7% (Matora et al. 1995). Recently, similar values were reported after Celluclast® application, but higher for treated ensiled sugar beet pulp (Pacheco et al. 2019). Nevertheless, the highest sugar beet pectin yield (22.41–28.84%) was noted after applying different ratios of xylanase and cellulose mixture (Abou-Elseoud et al. 2021). Sequential sugar beet pectin extraction consisted of water, EDTA, and enzymes application (Concha Olmos and Zúñiga Hansen 2012), as well as UAE pre-treatment followed by EAE (Abou-Elseoud et al. 2021) (Table 48.2).

Water coupled with acid is the most favourable solvent for MAE, with high energy absorption capacity originating from the water's fast dipole rotation and acid's high ionic conductivity (Kumar et al. 2020). Besides solvent properties, the pectin extraction efficiency by MAE further depends on applied microwave power, temperature, time, and material moisture. The temperature gradient absence and homogeneous temperature distribution in the material are regarded as MAE advantages. Furthermore, lower solvent and energy consumption bringing higher yields in a reduced extraction time are also prominent advantages (Kumar et al. 2020). Still, generation of a certain amount of wastewater and equipment maintenance due to problems induced by acid usage are to be resolved. However, several advantages favour MAE industrial implementation compared to other emerging extraction methods. The obtained yields and characteristics of sugar beet pectin isolated by MAE and different applied agents are listed in Table 48.2.

UAE was applied as pre-treatment in the sugar beet pectin extraction (Abou-Elseoud et al. 2021; Chen et al. 2015a), substitution to AE conventional heating (Jafarzadeh-Moghaddam et al. 2021) as well as a constituent of hybrid extraction methods(Chen et al. 2015b; Peng et al. 2015) (Table 48.2).

Although readily available, cheap and GRAS solvent with simple equipment requirements, the upscaling of the SWE is still challenging in terms of process control considering possible thermal degradation and pectin hydrolysis (Adetunji et al. 2017; Kumar et al. 2020). SWE of sugar beet pectin delivered pectin yield of 13–20% (Pińkowska et al. 2021; Peighambardoust et al. 2021) and even higher yields were reached (>20%) when UAE preceded SWE (Chen et al. 2015a) or upon SWE coupling with other emerging extraction methods (Chen et al. 2015b; Peng et al. 2015) (Table 48.2).

Faster extraction under ambient temperature using easily operative equipment resulting in extracts containing native compounds are labelled as main HPE advantages (Kumar et al. 2020). Unlike other green extraction methods, sugar beet pectin isolation by HPE is still at the beginning of research, with promising results regarding sugar beet pectin emulsifying properties (Kaya et al. 2021).

48.4 Sugar Beet Pectin Properties

48.4.1 Emulsifying Properties

Considering that sugar beet pectin is generally recognized as a hydrophilic molecule, its emulsifying activity on to the oil–water interface is primarily assigned to the hydrophobic proteinaceous moiety (Funami et al. 2007; Leroux et al. 2003) and, in a certain extent, to acetyl and ferulic acid esters (Siew and Williams 2008; Siew et al. 2008; Williams et al. 2005). However, the emulsion stabilizing effect of sugar beet pectin is associated with carbohydrate moiety structural characteristics (presence of neutral sugar side chains) and conformation (Funami et al. 2011; Leroux et al. 2003; Nakauma et al. 2008).

Factors affecting sugar beet pectin emulsifying potential in oil–water emulsions can be classified as intrinsic (linked to structural characteristics) and external (linked to environmental conditions) (Ngouémazong et al. 2015). Paramount structural features of sugar beet pectin influencing its interfacial activity include the proteinaceous moiety, ferulic acid content, DM and DA, molecular weight, and sidechains branching degree (Alba and Kontogiorgos 2017). Sugar beet pectin concentration, solution pH, and ionic strength are external, i.e. environmental factors affecting its emulsifying performance (Ngouémazong et al. 2015). However, the assessment of dominance or contribution of aforementioned factors, alone or in combination, on sugar beet pectin emulsifying potential described in the subsequent section is not unambiguous and is still debatable.

Amount of protein in sugar beet pectin is highly dependent on isolation conditions and detection methods with an obtained value up to 10.3% (Tables 48.1 and 48.2) compared to citrus and apple pectin with protein quantity of approximately 3% and 1%, respectively (Alba and Kontogiorgos 2017; Lara-Espinoza et al. 2021; Liu et al. 2019; Schmidt et al. 2015). Dominant role of sugar beet pectin proteinaceous moiety in the adsorption onto oil–water interface was elucidated by determining the adsorbed sugar beet pectin fractions, and its subjection to enzymatic modification, aiming to partially or completely remove protein (Funami et al. 2007, 2011; Leroux et al. 2003; Siew and Williams 2008; Siew et al. 2008; Williams et al. 2005). Performed studies reveal that sugar beet pectin fractions adsorbed at the oil–water interface had higher protein content than the original pectin sample (7.9–21.2%), particularly at low pectin concentration used (Leroux et al. 2003; Siew and Williams 2008; Siew et al. 2008). Furthermore, alteration of sugar beet pectin emulsifying activity and stability upon enzymatic modification is manifested through enlargement in interfacial tension and droplet mean diameter, broader droplet size distribution, creaming and phase separation occurrence (Funami et al. 2007, 2011). Conversely, preservation of sugar beet pectin proteinaceous moiety and thus emulsifying potential upon enzymatic modification observed by Siew et al. (2008) was ascribed to steric inaccessibility of protein, likewise nature of the amino acid sequence. However, greater protein quantity in sugar beet pectin fractions after fractionating does not ensure better emulsifying properties as evidenced by Williams et al. (2005). As concluded by Chen et al. (2016a, 2018), optimal interfacial activity of sugar beet pectin is attained when protein concentration is approximately 3%. Additionally to the already mentioned disagreements, sequenced enzymatic modification of sugar beet pectin conducted in a recent study disclose more prominent effect of covalently linked ferulic acid-arabinogalactan-protein complex on interfacial activity and emulsifying properties compared to protein fraction alone (Chen et al. 2016b). All these observations imply ambiguous aspects of the complex relationship among sugar beet pectin structural elements and its emulsifying potential, which is not determined by a single factor.

Apart from proteinaceous moiety, the existence of ferulic acid, acetyl and methyl groups in sugar beet pectin structure positively affect its emulsifying properties (Chen et al. 2016b; Leroux et al. 2003; Siew and Williams 2008; Siew et al. 2008; Williams et al. 2005). Investigation of the sugar beet pectin fraction adsorbed at the oil–water interface confirmed higher ferulic acid (0.5–2.16%) and acetyl contents (1–3.9%) compared to non-adsorbed fraction (Leroux et al. 2003, Siew and Williams 2008, Siew et al. 2008, Williams et al. 2005), proving the contribution of these hydrophobic groups to overall emulsifying potential, especially at low protein contents (Chen et al. 2016b). Nevertheless, sugar beet pectin emulsifying properties were considerably less influenced by methyl groups content as evidenced upon methyl-esterase application (Chen et al. 2016b).

It is considered that sugar beet pectin molecular weight determinates its conformation in solution and consequently the protein, ferulic acids, and acetyl groups' accessibility, although molecular weight effect on emulsifying properties is not unambiguous (Alba and Kontogiorgos 2017; Ngouémazong et al. 2015). Sugar beet pectin with lower molecular weight ($153–282 \times 10^3$ g mol^{-1}) yielded emulsions with initially smaller droplet diameters stable over time compared to those stabilized with its high molecular weight counterparts ($306–562 \times 10^3$ g mol^{-1}) (Williams et al. 2005). Conversely, reduction of sugar beet pectin molecular weight (from 517 to 254×10^3 g mol^{-1}) after enzymatic removal of protein resulted in coarser emulsions (Funami et al. 2007). This inconsistency in results addressing the impact of sugar beet pectin molecular weight on its emulsifying properties can be fully understood in further research only when it is viewed along with the aforementioned structural characteristics (ferulic acid, acetyl and methyl groups).

Proposed preferential adsorption of neutral sugar-rich sugar beet pectin fractions onto oil droplets was linked to superior emulsifying properties of pectin containing RG-I compared to those having linear backbone (Siew et al. 2008). This link was further confirmed upon sugar beet pectin enzymatic modification by polygalacturonase (Funami et al. 2011), arabinase, and galactase (Chen et al. 2016b), resulting in less stable emulsions with coarser droplets prone to coalescence.

Considering that only a small part of sugar beet pectin adsorbs onto oil droplets during emulsification, concentration of sugar beet pectin in solution represents a crucial external factor affecting its emulsifying activity (Ngouémazong et al. 2015). Increment of sugar beet pectin concentration in solution up to 2% leads to enlargement of the adsorbed sugar beet pectin and creation of protective layer covering the oil droplets (Siew and Williams 2008; Williams et al. 2005) and reducing their diameter (Leroux et al. 2003; Nakauma et al. 2008; Siew and Williams 2008;

Williams et al. 2005). With further increase in sugar beet pectin concentration in solution the share of adsorbed sugar beet pectin remains constant implying that 2% represents its threshold concentration for achieving good emulsifying properties attained with considerably higher concentrations of gum Arabic or soy soluble polysaccharide (10–30 w/w% and 5–10 w/w%, respectively) (Nakauma et al. 2008). Moreover, at identical emulsifier concentrations, sugar beet pectin displayed superior emulsifying properties when compared with corn fibre gum (0.1–2 w/w%) (Bai et al. 2017), octenyl succinate modified maltodextrin, and sugar beet fibre (0.5–1 w/w%) (Maravić et al. 2019), but inferior to chicory root pulp pectin (0.5–2 w/w%) (Pi et al. 2019). In addition, with increasing sugar beet pectin concentrations, increase in emulsion viscosity (Lv et al. 2013) should be accounted as an additional factor influencing emulsion stabilization (Ngouémazong et al. 2015). Sugar beet pectin viscosity, as one of its properties, was discussed in Sect. 48.4.3.

Solution pH and ionic strength are other external factors altering sugar beet pectin emulsifying properties. It is hypothesized that the affinity of sugar beet proteinaceous moieties towards oil surface adsorption is pH-independent, implying that changes in carbohydrate moiety structure determine its emulsifying properties at various pH (Nakauma et al. 2008). Sugar beet pectin was able to stabilize oil–water interfaces at pH 3.7 due to decreased ionization of carboxylic groups from GalA residues which enables electrostatic complexation with positively charged proteinaceous moieties among the sugar beet pectin chains and formation of thick interfacial layers (107 nm) (Alba and Kontogiorgos 2020; Siew et al. 2008). With increasing pH, sugar beet pectin chains adopt more extended conformation as a result of electrostatic repulsions between carboxylate anions and inability to interrelate with negatively charged proteinaceous moieties, leading to a decline in interfacial layers thickness (70 nm) and thus less effective steric stabilization (Alba and Kontogiorgos 2017; Siew et al. 2008). Increased solution ionic strength upon divalent cations (Ca^{2+}) or monovalent cations' (Na^{+}) addition reduces the sugar beet pectin emulsifying activity and emulsion stabilizing effect (Nakauma et al. 2008; Ngouémazong et al. 2015). The sugar beet pectin emulsifying activity affected by salts addition is primarily reflected in increased initial emulsion droplet size rather than deterioration in its adsorption affinity towards the oil phase (Nakauma et al. 2008). At low pH values, added cations screen the negative charges of the sugar beet pectin carboxylate moiety and thus reduce the electrostatic repulsion favouring the cation-induced crosslinking of pectin chains in the adsorbed layer (Nakauma et al. 2008; Ngouémazong et al. 2015; Siew et al. 2008). Consequently, an increase of the layer thickness and, thus, enlargement in the emulsion droplets diameter occurs, promoting further bridging flocculation between droplets leading to the creation of coarser unstable emulsions upon storage (Leroux et al. 2003; Nakauma et al. 2008; Siew et al. 2008).

Considering the presented discussion, it is evident that sugar beet pectin action as an emulsifier can be elucidated only by observing the common influences of all structural elements as well as environmental conditions. The suggested mechanism explaining sugar beet pectin interfacial activity and thus emulsifying properties are presented in Fig. 48.1. First proposed hypothesis for sugar beet pectin adsorption at

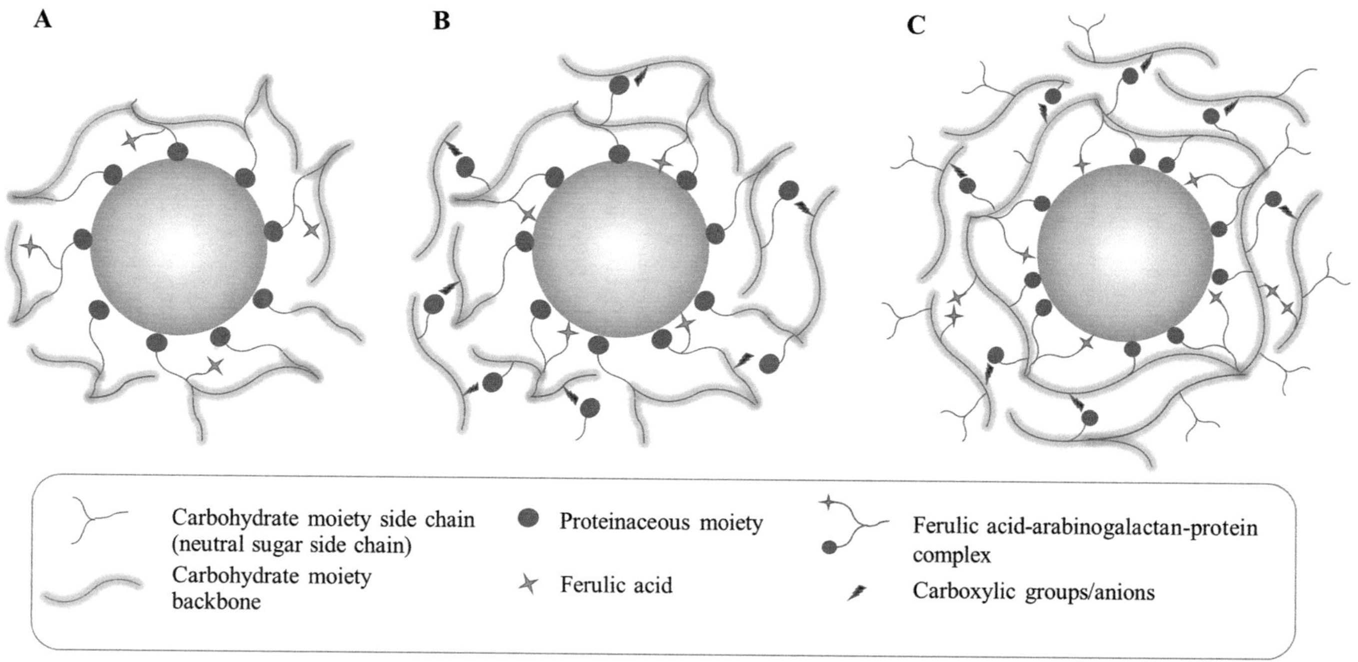

Fig. 48.1 Proposed mechanisms of sugar beet pectin adsorption on oil–water interface and stabilization of oil–water emulsions. (**a**) Loop and tail adsorption model, (**b**) multilayer adsorption model, (**c**) ferulic acid–arabinogalactan–protein complex adsorption model

the oil–water interface was similar to gum Arabic (Leroux et al. 2003). Considering the conformational difference between sugar beet pectin and gum Arabic, Leroux et al. (2003) proposed "loop and tail" adsorption model (Fig. 48.1a) where sugar beet pectin proteinaceous moiety behaves as an anchor by adsorbing onto oil droplets' surface, reducing the interfacial tension, while carbohydrate moiety represents a tail conferring physical stability to the produced simple emulsions. The authors also indicate the possibility of Ca^{2+} cross-linking among sugar beet pectin chains causing bridging flocculation and deterioration of emulsion stability, which can be hindered by a greater number of acetyl groups in pectin carbohydrate moiety (Leroux et al. 2003). Further findings reveal that, apart from the paramount role of proteinaceous moiety, the role of an auxiliary anchor can be ascribed to both ferulic acid and acetyl groups along the sugar beet pectin carbohydrate moiety (Funami et al. 2007; Nakauma et al. 2008). The proposed loops and tails adsorption model was confirmed by Siew et al. (2008) and additionally complemented given that the studied pectin adsorbed layer thickness implied multilayer adsorption of sugar beet pectin at the oil–water interface. The suggested multilayer adsorption model (Fig. 48.1b) was elucidated by electrostatic complexation among the positively charged proteinaceous moiety (present in a higher extent) and the negatively charged carboxylic groups from Gal A residues of the carbohydrate moiety allowing effective electrostatic stabilization of emulsion (Alba and Kontogiorgos 2020; Ngouémazong et al. 2015; Siew et al. 2008).

Nevertheless, according to the latest mechanism explaining emulsifying properties of sugar beet pectin, proteinaceous moiety and ferulic acid from ferulic acid–arabinogalactan–protein complex represent main anchors for adsorption on oil droplet surface (Fig. 48.1c) (Chen et al. 2016b). However, sugar beet pectin emulsifying activity is additionally affected by proteinaceous moiety and ferulic acid content and availability (Williams et al. 2005). Based on the suggested mechanism, besides GalA, methyl and acetyl groups in the backbone (Chen et al. 2016b), emulsion stabilizing effect of sugar beet pectin is additionally ascribed to the neutral sugar side chains existence in RG-I (especially arabinose and galactose) and their interactions with polygalacturonic acid (main chain) (Chen et al. 2016b), proteins, and/or ferulic acid between adsorbed pectin chains (Ngouémazong et al. 2015), resulting in the formation of an adsorbed hydrated layer which enables effective steric stabilization of emulsion (Fig. 48.1c).

Further enhancement in sugar beet pectin emulsifying properties was achieved by conducting diverse treatments and modifications such as isopropanol fractionation (Karnik and Wicker 2018), ultrasonic treatment (Yang et al. 2020), controlled dry-heating (Wang et al. 2021), enzymatic polymerization using laccase (Jung and Wicker 2012) and horseradish peroxidase (Zhang et al. 2015), genipin cross-linking (Lin et al. 2020), ferulic acid (Liu et al. 2020) and β-cyclodextrin addition (Liu et al. 2021), esterification using octenyl succinic anhydride (Chen et al. 2015c), formation of conjugates with whey protein isolate (Guo et al. 2019) and sodium caseinate (Zhang and Wolf 2019, 2021), formation of multilayer emulsifier systems with β-lactoglobulin (Katsuda et al. 2008).

Sugar beet pectin emulsifying potential is widely applied for the encapsulation of numerous oils, further applicable as carriers for bioactive compounds delivery in foods systems as disclosed in Sect. 48.5.1.

48.4.2 Gelling Properties

The capability of sugar beet pectin towards gel formation is highly dependent on structural elements comprising its macromolecular structure. Besides the initially believed influence of methyl groups arrangement over the HG domain and their amount expressed as DM on sugar beet pectin gelling ability (Harel et al. 1998), further research also revealed significant contribution of neutral sugar side chains and acetyl groups presence (DA) to the corresponding influence. Additionally, as intrinsic factors, GalA content and molecular weight effect on gelling were also observed (Buchholt et al. 2004). As regards external factors, gelation is affected by pH alongside temperature, nature, and concentration of cosolute or cation (Reichembach and de Oliveira Petkowicz 2021).

Based on the DM, pectin is classified as low-methyl-esterified pectin, LMP, (DM $< 50\%$) or high-methyl-esterified pectin, HMP (DM $> 50\%$) (BeMiller 2019), and consequently is susceptible to different gelation mechanisms. HMP usually form acidic gels in acidic environment (pH 2.8–3.6) and sucrose presence ($>55\%$) where hydrophobic interactions among methyl-ester groups and hydrogen bonds among carboxyl and hydroxyl groups form junction zones (Reichembach and de Oliveira Petkowicz 2021). LMP gelation occurs according to the egg-box model in a wider pH range and requires presence of bivalent cations (usually Ca^{2+}). In the mentioned model, junction zones represent electrostatic interactions between the ionized carboxyl groups in adjacent chains and Ca^{2+} (Reichembach and de Oliveira Petkowicz 2021). Consensus on the sugar beet pectin classification according to DM was not established since a wide range of DM values was reported depending on the applied extraction method (Tables 48.1 and 48.2). However, a definite conclusion is that sugar beet pectin in native state is not susceptible to either of mentioned gelling mechanisms, which is assigned to the high content of methyl esters (Harel et al. 1998), acetyl groups, and neutral sugar side chains (Funami et al. 2007) introducing steric hindrance, likewise low molecular weight (Williams et al. 2005; Yapo et al. 2007a). Consequently, modifications of the macromolecular structure of sugar beet pectin are required to enable gel formation according to egg-box model. Different approaches for performing demethylation and deacetylation of sugar beet pectin were studied. Application of enzymes (fungal and plant pectin esterase) was effective for sugar beet pectin demethylation and deacetylation, but required long incubation time (Buchholt et al. 2004). Acid (H_2SO_4, HCl) usage proved to be inefficient regarding significant DM and DA reduction and gel formation (Buchholt et al. 2004; Zhang et al. 2021). Alkali (NH_3, NaOH) application at low temperatures 3–5 °C and pH values 11–12 is the most efficient method for reducing DM and DA values of sugar beet pectin concomitantly, preserving high molecular weight and GalA content (Harel et al. 1998; Mata et al. 2009b; Zhang et al. 2021). The sugar beet pectin modified by alkali under stated conditions could be cross-linked by Ca^{2+} to create stable elastic gels for further application.

48.4.3 Viscosity

The intrinsic viscosity and apparent viscosity of sugar beet pectin are frequently determined for acquiring insight into pectin conformation in the solution further responsible for limited oil droplets movement bringing enhanced emulsion stability (Ngouémazong et al. 2015). Sugar beet pectin viscosity is primarily ascribed to molecular weight, followed by DM establishing increasing tendency with higher values of the mentioned intrinsic factors. Nevertheless, presence of proteins in the structure was also mentioned as a factor of influence regarding viscosity increase, regardless of reduced molecular weight of sugar beet pectin obtained by SWE or high hydrostatic pressure treatment (Peighambardoust et al. 2021; Peng et al. 2016). Moreover, the exact role of branched neutral side chains on sugar beet pectin viscosity was not fully explored, but relates random coil conformation of high molecular weight RG-I domain with a low intrinsic viscosity (Morris et al. 2010). Opposite interpretations of the corresponding structural component influence are present in literature for other pectins (Reichembach and de Oliveira Petkowicz 2021). In addition, external factors influence, primarily pectin concentration followed by pH (Peng et al. 2015; Peng et al. 2016), temperature (Chen et al. 2015a; Peighambardoust et al. 2021), and presence of sugars and salt (Reichembach and de Oliveira Petkowicz 2021) cannot be disregarded. Increase in sugar beet pectin concentration in solution generally leads to viscosity increase. Many studies described sugar beet pectin solution as non-Newtonian fluid with pseudoplastic behaviour at low (1–2%) (Huang et al. 2018; Peighambardoust et al. 2021; Peng et al. 2015; Peng et al. 2016; Lv et al. 2013) as well as higher concentrations (10–15%) (Chen et al. 2015a), while fewer described it acting as Newtonian fluid (Kaya et al. 2021) at a constant temperature. Considering pH, strong acidic environment (pH 3) contributes to viscosity increase (Peng et al. 2015) due to enhanced intermolecular hydrogen bonding (Peng et al. 2015). Conversely, temperature rise reflects as decrease in apparent viscosity resulting from reduced intermolecular interactions (Chen et al. 2015a, Peighambardoust et al. 2021).

48.4.4 Prebiotic Properties

Sugar beet pectin represents a source of pectic oligosaccharides (POS) considered as an emerging group of prebiotics. The POS prebiotic effect primarily reflects in improved growth and activity of targeted intestinal microflora (*Bifidobacterium* and *Lactobacillus* strains) deriving short-chain fatty acids (SCFA) with numerous health-promoting effects (Gullón et al. 2013; Prandi et al. 2018; Larsen et al. 2019). Diverse poly- and oligosaccharides comprise the POS group obtained from sugar beet pectin: glucooligosaccharides (GOS), galactooligosaccharides (GalOS), arabinooligosaccharides (AraOS), and oligogalacturonides (OGalA) (Prandi et al. 2018). Their structural variations (DM, molecular weight, distribution of free and methylated carboxyl groups, neutral sugars) are related to manifested physiological properties and consequently induced health-promoting effects (Holck et al. 2011; Larsen et al. 2019). Various approaches were applied for POS obtention from sugar beet pectin and pulp, among which enzymatic hydrolysis with different enzymes (Holck et al. 2011; Martínez et al. 2009) and multienzyme complexes (Prandi et al. 2018) is the most common followed by POS purification and fractionation by membrane separation.

48.5 Sugar Beet Pectin Uses

48.5.1 Sugar Beet Pectin as Encapsulation Agent

Encapsulation represents the envelopment of a bioactive compound (core material) inside resistant coating (wall material), and considering the obtained particle size (capsule) can be referred as macro-encapsulation (>500 μm), microencapsulation (0.2–500 μm) and nano-encapsulation (<0.2 μm) (Ruiz Canizales et al. 2019). Biopolymers available as wall materials for encapsulation are proteins, lipids, and polysaccharides (maltodextrin, starch, pectin, chitosan, alginate, and gums). Great diversity in polysaccharides sources, their biodegradability, biocompatibility, nontoxicity, and thermal resistance allows their frequent use in different food products. Additionally, coupling of two wall materials (protein/polysaccharide, lipid/polysaccharide, polysaccharide/polysaccharide) can also enhance the physico-chemical characteristics of encapsulated particles (Ruiz Canizales et al. 2019).

Pectin GRAS status (FDA 2013), ability to stabilize emulsions, form gels and bind cations, coupled with availability and low cost indicates its potential as encapsulation material in food systems. Moreover, an increase in pectin hydrophobicity, linked to the increase in DM, allows pectin interaction with highly hydrophobic compounds, their targeted and controlled release (Rehman et al. 2019). Pectins from different sources, among them sugar beet pectin, were mostly applied as wall materials in both micro- and nano-encapsulation of bioactive compounds through spray-drying, emulsions formation, liposomes preparation, hydrogel formation, nanocomplex formation, and coacervation (Rehman et al. 2019). Applications of sugar beet pectin for bioactive compounds encapsulation in food systems are summarized in Fig. 48.2., while the main findings classified according to sugar beet pectin role in encapsulation are highlighted in the subsequent section.

First studies exploring the potential of sugar beet pectin in encapsulation were based on its emulsifying properties, thus involving the emulsion-forming and spray-drying techniques. Drusch (2007) used sugar beet pectin as emulsifying wall constituent, coupled with glucose syrup, for the microencapsulation of fish oil by spray-drying. Median oil droplet size below 2 μm and a maximum viscosity of 179 mPas was achieved in emulsions with up to 50% oil and 2.2% sugar beet pectin.

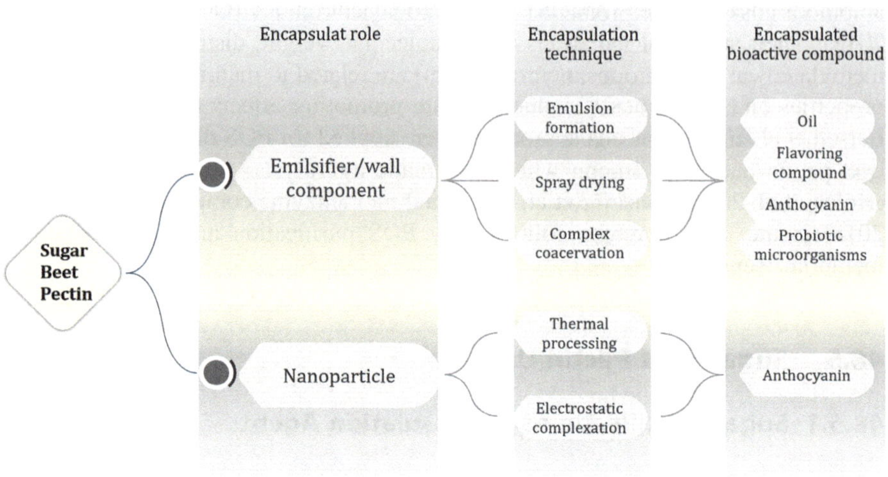

Fig. 48.2 Applications of sugar beet pectin for bioactive compounds encapsulation in food systems

Investigated physico-chemical characteristics of microencapsulated fish oil reveal good microencapsulation efficiency, and thus oxidative stability of microcapsules. However, the author concluded that the sugar beet pectin application may be limited by the maximum oil load of the capsules (Drusch 2007). In a similar extended study, Drusch et al. (2007) used different types of *n-octenylsuccinate*-derivatized starch, gum Arabic, sugar beet pectin, sodium caseinate, and/or glucose syrup as emulsifying wall constituents for fish oil microencapsulation by spray-drying, aiming to define factors affecting microencapsulated oil oxidative stability. Sugar beet pectin-based microcapsules exhibited high quantity of surface oil and consequently increased oil droplet size after reconstruction with a decline in pore volume compared to other wall materials. Still, sugar beet pectin-based microcapsules showed higher oxidative stability than gum Arabic-based microcapsules. These researchers concluded that lower molecular weight emulsifying wall constituents have better stabilizing performance on the oil–water interface in the parent emulsions which, together with the surface composition of the dried microcapsule, have a great impact on microencapsulated oil oxidative stability (Drusch et al. 2007).

Moreover, in another study, Polavarapu et al. (2011) investigated the microencapsulation of fish oil and fish oil–extra virgin olive oil (1:1) using sugar beet pectin-stabilized emulsions (pH 3) and spray-drying. Although microencapsulation efficiencies were higher than 90%, impairment in microcapsule wall integrity, enlargement of oil droplet size and lipid oxidation are detected during storage (25 °C, up to 3 months), expressing poor sugar beet pectin performance as a wall material because of the residual divalent cations' presence (Cu^{2+}, Fe^{2+}) which catalyse the autoxidation of unsaturated fatty acids (Polavarapu et al. 2011). Hence, another investigation was conducted by the mentioned researchers with the introduction of EDTA as a chelating agent, aiming to improve oil oxidative stability.

A significantly higher protective effect was achieved in emulsions and spray-dried microcapsules formulated with EDTA compared to microcapsules without EDTA under ambient and accelerated storage conditions. The greatest oxidative stability was attained in fish oil–extra virgin olive oil emulsions with EDTA proving its great role in maximizing the oxidative stability of fish oil in sugar beet pectin-stabilized emulsions and spray-dried microcapsules (Polavarapu et al. 2012).

d-Limonene, a food flavouring component, was microencapsulated by spray-drying using sucrose ester, polyglycerol ester, and sugar beet pectin as emulsifiers along with gum Arabic and maltodextrin as wall materials (Paramita et al. 2010). Sugar beet pectin-stabilized emulsions, irrespective of the wall material used, exhibited good stability with optimal particle size and high viscosity reflecting in increased *d*-Limonene retention (86% and 94%), compared to other applied emulsifiers. Despite the higher surface oil content of the spray dried microcapsules, the authors concluded that the microencapsulation of *d*-Limonene in maltodextrin/sugar beet pectin system is effective considering the low amount of sugar beet pectin used (Paramita et al. 2010).

Berg et al. (2012) encapsulated blueberry extract rich in anthocyanins by spray-drying using pectins from several sources (apple, citrus, sugar beet, and amidated pectin) and caffeine, with an aim to enhance anthocyanins retention from the shellac-coated maltodextrin granulates in simulated gastric fluid. The structure of spray-dried microcapsules and coating layer of the granulates were not affected by pectins' addition (Berg et al. 2012). However, the researchers found the reverse connection between water-binding capacity of spray-dried microcapsules and initial anthocyanins release. Lower anthocyanin release was achieved for microcapsules with sugar beet pectin and citrus pectin with high degree of esterification, which showed higher water binding capacities as compared to other used pectins (Berg et al. 2012).

Zhang et al. (2016) developed solid/oil/water emulsions using soybean oil and sugar beet pectin, with and without calcium addition, as a delivery system for *Lactobacillus salivarius* NRRL B-30514. Encapsulation efficiency of *L. salivarius* reached 86% with emulsion droplets size below 17 μm, enabling their good viability during storage (2 weeks, 4 °C), pasteurization (63 °C, 30 min), and the in vitro gastric and intestinal digestions. Furthermore, the authors indicate that Ca^{2+}addition induced a cross-linking between sugar beet pectin on emulsion droplets leading to higher survival of viable *L. salivarius* after simulated gastrointestinal digestions, confirming the great potential of these solid/oil/water emulsions as a probiotic delivery system in food products (Zhang et al. 2016).

Lan et al. (2021) studied the microencapsulation of hemp seed oil using complex coacervation technique with pea protein isolate and sugar beet pectin as wall materials. The authors investigated the impact of wall/core ratios (1:1, 2:1, and 4:1) and coacervation formation pH (3.5 and 2.5) on physico-chemical properties of spray-dried hemp seed oil microcapsules (Lan et al. 2021). Results reveal that lower microcapsule wall/core ratio reflected in lower encapsulation efficiency and oxidative stability of hemp seed oil. Microcapsules of pea protein isolate–sugar beet pectin complex coacervates formed at pH 2.5 had greater encapsulation efficiency, but lower oxidative stability than that formed at pH 3.5. The authors highlighted that determination of wall/core ratio and coacervation formation pH is crucial for achieving the equilibrium between technical performance and the oxidative stability of the core material (Lan et al. 2021).

An interesting study regarding garlic essential oil encapsulation by spray-drying using β-cyclodextrin (β-CD) and sugar beet pectin as wall materials was conducted by Emadzadeh et al. (2021). Different core/wall ratios (1:6, 2:6, 3:6) and sugar beet pectin/β-CD ratios (1:1, 1:2, 2:1) were tested and their effect concerning the optimized microcapsules' physico-chemical characteristics, their release in a gastrointestinal tract model system, followed by practical application in acidic food beverage were assessed. In samples with core/wall and sugar beet pectin/β-CD ratios of 3:6–1:2 and 1:6–1:2, the highest encapsulation efficiency was attained. Encapsulation enhanced the thermal stability of garlic essential oil, while sugar beet pectin presence reinforced the microcapsule wall structure enhancing the retention of garlic essential oil in the gastrointestinal tract and release behaviour in acidic food beverage during storage (Emadzadeh et al. 2021).

Apart from emulsion formation and spray-drying as the most frequently used encapsulation techniques, Arroyo-Maya and McClements (2015) used protein isolate/sugar beet pectin nanoparticles for anthocyanin encapsulation by thermal processing and electrostatic complexation. Higher encapsulation efficiency was achieved when anthocyanin was added before heating, rather than after thermal processing (Arroyo-Maya and McClements 2015). Although heat stability was improved, the resultant encapsulated anthocyanin exhibited lower antioxidant activity and colour stability compared to non-encapsulated anthocyanin, showing low effectiveness of applied nano-encapsulation technique (Arroyo-Maya and McClements 2015).

Potential of sugar beet pectin application in production of interpenetrating polymer network hydrogels as a delivery system for probiotics was explored by Yan et al. (2021). Interpenetrating polymer network hydrogels were assembled from soy protein isolate and sugar beet pectin by enzymatic method and used for encapsulation of *Lactobacillus paracasei LS14*. Storage modulus (G′) and hardness of soy protein isolate/sugar beet pectin hydrogels decreased and slight disruption in hydrogel structure was observed upon *Lb. paracasei LS14* encapsulation. The highest viability of encapsulated *Lb. paracasei LS14* was attained in hydrogels with 10% soy protein isolate and 3.5% sugar beet pectin induced by 10 U/g laccase (Yan et al. 2021). Additionally, after 21 days of storage, lyophilized gels displayed greater storage stability but lower probiotic viability than hydrogels. The authors demonstrated that corresponding gels can be applied as probiotics carriers in probiotic-containing foods (Yan et al. 2021).

The application of sugar beet pectin as encapsulation agent is also studied in more complex systems such as robust W/O/W emulsion stabilized by nanoparticles for co-encapsulation of betanin and curcumin (Tang et al. 2021). Low sugar beet pectin quantities applied (1.2–4%) and promising results achieved in encapsulation studies conducted so far reveal that sugar beet pectin tends to be one of the most applicable biopolymers in the encapsulation of bioactive compounds for food systems.

48.5.2 Sugar Beet Pectin in Food Packaging

Pectin as a biopolymer with versatile properties and relatively low cost represents a convenient polymeric matrix applicable in food packaging. Furthermore, GRAS status of pectin (FDA 2013) and the possibility to be metabolized in the human body together with food enables the creation of pectin-based edible films and coatings (Espitia et al. 2014). Common methods for pectin-based edible films production are casting, extrusion, spraying, and knife-coating (Espitia et al. 2014), while the manufacturing of pectin-based coatings is most commonly accomplished by dipping the food in the coating-forming solution and subsequent drying (Reichembach and de Oliveira Petkowicz 2021). Pectin-based edible films and coatings provide a good barrier to oil, O_2, CO_2 and volatile (metabolites) losses, owing to the creation of well-arranged hydrogen bonded network (Hassan et al. 2018), act as product quality improvers and hinder the microbial growth of spoilage flora and pathogens (Campos et al. 2011). However, considering pectin's hydrophilic character, pectin-based films and coatings are susceptible to moisture and display weak water barrier properties (Chaichi et al. 2017). To overcome this drawback, implementation of hydrophobic additives (lipids) as an inherent part of pectin-based film structure is proposed. Additionally, other biopolymers such as proteins and polysaccharides (carboxymethyl cellulose, alginate, and chitosan) were included in pectin-based film matrix to strengthen the structure and to resemble the physical properties of conventional packaging films (Lazaridou and Biliaderis 2020). Divergency in pectin structures and adjustable chemical and physical properties make pectin-based edible film or coating preparations convenient vehicles for delivering active compounds, and thus usage in active food packaging (Kumar et al. 2020).

Extensive research on pectin potential in edible film and coating preparations has been conducted in recent years (Espitia et al. 2014; Kumar et al. 2020; Lazaridou and Biliaderis 2020; Reichembach and de Oliveira Petkowicz 2021). Many studies involved the use of commercial pectins without indicating the pectin sources, making it difficult to distinguish the impact of pectin source on their performance in edible film and coating preparations. Studies concerning sugar beet pectin usage in food packaging are reviewed in the subsequent section.

Composite preparation using different levels of sugar beet pectin and pectin-extracted sugar beet pulp was investigated by Liu et al. (2012). Authors evaluated the required amount of residual pectin in pectin-extracted sugar beet pulp for the preparation of composites, likewise the effect of pectin content (5, 15, 25, 35 g/g d. m.) on composites properties. Composites were prepared in a twin-screw extruder using water and glycerol as plasticizers. During compounding, pectin was plasticized and acted as binder at low concentrations and matrix polymer at high concentrations in the resulting composites. Authors revealed that pectin-extracted sugar beet pulp, which still contains a certain amount of residual pectin, could be used as resource in composite preparations (Liu et al. 2012). The combination of sugar beet pectin, yellow tea, rice, and fenugreek proteins in edible film formation was analysed by Ostrman et al. (2018), assessing the physico-chemical, mechanical, bioactive, and sensory properties of the resulting edible films. Sugar beet pectin and protein solutions (1:1), previously prepared in yellow tea extract, were used for edible films preparation with glycerol as plasticizer. The edible film containing only sugar beet pectin presented better mechanical and sensory properties compared to films with both sugar beet pectin and protein. Nevertheless, rice protein inclusion yielded edible films with the highest total polyphenolic content and antioxidant capacity. Results implied that sugar beet pectin and rice protein are applicable in the preparation of edible films with optimal functional and mechanical properties (Ostrman et al. 2018). Still, further research regarding the inclusion of sugar beet pectin as structuring component in biopolymer-based films and coatings is required to fully exploit its potential as an inexpensive and readily available material, enable large scale reproduction, and broaden the applicability of pectin-based films and coatings intended for traditional and active food packaging.

48.5.3 Sugar Beet Pectin Gels in Heavy Metals Removal

Sugar beet pectin was established as biosorbent alternative to apple and citrus pectins and algal biomass with the advantage of being abundant, obtained from dried raw material, and thus season-independent (Mata et al. 2009a). One of the earliest investigations on the usage of sugar beet pectin in heavy metals removal from water was conducted by Dronnet et al. (1996). In the corresponding study, batch binding efficiency of Cu^{2+}, Pb^{2+}, Zn^{2+}, Cd^{2+}, Ni^{2+}, and Ca^{2+} with sugar beet and citrus pectins having similar DM was estimated. Both pectins used followed the same order of selectivity $Cu^{2+}{\sim}Pb^{2+}{>>}Zn^{2+}{>}Cd^{2+}{\sim}Ni^{2+}{\geq}Ca^{2+}$. However, a decline in the affinity of Ca^{2+}, Ni^{2+} and Zn^{2+} towards sugar beet pectin occurred in the presence of 0.1 M $NaNO_3$ as the supporting salt, compared to citrus pectin. This discrepancy was assigned to the presence of acetyl groups close to ionic sites on sugar beet pectin (Dronnet et al. 1996).

Further studies involved changes in sugar beet pectin structure to address difficulties in gel formation and make it more competitive with citrus and apple pectins and algal biomass. Harel et al. (1998) subjected sugar beet pectin to demethylation, using ammonia enabling gel formation at concentrations of 2% w/v or higher in the presence of Ca^{2+}. The capability of sugar beet pectin calcium gels to bind Cd^{2+}in an aqueous solution was examined and compared with commercial citrus pectin and calcium alginate. Sugar beet pectin gel beads showed similar cadmium-binding capacity with algal alginate, but much lower compared to citrus pectin. However, desorption of Cd^{2+} from sugar beet pectin gel beads by 0.3 M $CaCl_2$ as desorbent reached 81% (Harel et al. 1998). A step further in the effort to enhance sugar beet pectin gels biosorption ability was the immobilization of *P. putida* cells into sugar beet pectin gel matrix. The suitability of sugar beet pectin gels as an immobilization matrix for viable microorganisms was confirmed, but utilization of these immobilized-cell systems in heavy metals removal remains unexplored (Harel et al. 1998). However, sugar beet pectin gels, with immobilized biomass of the brown algae *Fucus vesiculosus*, were convenient for gold and lead biosorption, with additional reduction of Au^{3+} to Au (Mata et al. 2007).

Sugar beet pectin hydro- and xerogels batch biosorption and desorption potential towards Cd^{2+}, Pb^{2+}, Cu^{2+} was studied by Mata et al. (2009b) and compared with alginate. Rates of metal biosorption recorded for sugar beet pectin gels had the following order: $Cu^{2+}>Pb^{2+}>Cd^{2+}$, showing greater efficiency compared to alginate. As suggested by the authors, ion exchange with calcium present in gel structure and chelation or complexation with carboxyl groups were the principal mechanisms involved in heavy metals binding. As regards desorption, 0.1 M HNO_3 was established as the most effective desorbent in corresponding metals recovery and reutilization of sugar beet pectin gels, unlike HCl and H_2SO_4. Additionally, sugar beet pectin xerogels with their compact structure, better conservation, and reutilization possibility are established as more suitable for application in heavy metals removal than hydrogels (Mata et al. 2009b).

The potential of sugar beet pectin xerogels in the continuous biosorption of heavy metals was further explored by Mata et al. (2009a, 2010). A fixed-bed column and serial columns with sugar beet pectin xerogels as biosorbents were used in copper removal from aqueous solution (Mata et al. 2009a). Influence of treatment conditions such as flow rate, bed height, inlet metal concentration and feeding system (drop and reverse) on biosorption parameters (saturation time, amount of adsorbed and treated metal, column performance, and metal uptake) were assessed (Mata et al. 2009a). Determined optimal treatment conditions for copper biosorption with sugar beet pectin xerogels in fixed-bed column were: 3 g of biomass, 25 mg/L metal, 2 mL/min feed flowrate and a reverse feeding system (Mata et al. 2009a). A decline in saturation time was observed with feed flow rate and inlet metal concentration increment, while an increase in bed height increased saturation time. By applying a reverse feeding system, formation of preferential flow channels in the columns was prevented, thus diminishing their great influence on biosorption parameters. Overall metal uptake in serial columns was comparable to a single column with equal biomass quantity. Total desorption of copper from the column was achieved using 0.1 M HNO_3 (Mata et al. 2009a). Endurance and effectiveness of sugar beet pectin xerogels towards Cd^{2+}, Pb^{2+} and Cu^{2+} removal after multiple batch sorption-desorption cycles involving gels regeneration step with 1 M $CaCl_2$ was further examined (Mata et al. 2010). Results revealed great stability of sugar beet pectin xerogels as biosorbents in successive sorption-desorption cycles using 0.1 M HNO_3 as desorbent avoiding significant losses in biosorbent mass and biosorption capacity. Introduction of regeneration step with 1 M $CaCl_2$ after each desorption reinforced the structure of sugar beet pectin xerogels, enabling an increase in Cd^{2+} uptake and aligning it for Pb^{2+} and Cu^{2+} (Mata et al. 2010). Established excellent reusability of sugar beet pectin xerogels in continuous sorption-desorption-regeneration cycles could provide guidance towards their industrial application.

In the most recent research by Ma et al. (2016), the potential of sugar beet pectin in Hg^{2+} removal was estimated, taking into account variations in treatment conditions such as pH, time, temperature, and initial Hg^{2+} concentration. An increase in Hg^{2+} adsorption capacity was accomplished by applying higher initial pH and higher initial Hg^{2+}concentration reaching maximal adsorption capacity of 23.6 mg/g. According to the authors, Hg^{2+} removal was primarily associated with monolayer adsorption of Hg^{2+} onto O–H band of hydroxyl and carboxyl groups in sugar beet pectin, as confirmed by FTIR spectrometry, showing a great potential of sugar beet pectin in mercury removal from wastewater.

48.6 Future Prospects

A room for improvement in edible films and coatings preparation using sugar beet pectin, alone or in combination with other biopolymers, still exists and can diminish large amounts of waste generation induced by the growing food market. Additionally, the potential of sugar beet pectin pectic oligosaccharides as emerging prebiotics should be further explored and commercialized throughout the development of health beneficial food products as a response to arising consumers' demands.

48.7 Conclusion

Sugar beet pectin's role as a natural ingredient was recognized and established as a driving force for extensive research in an attempt to further expand its usage in diverse industries. In this regard, the arising problem is the achievement of optimal sugar beet pectin functionality to meet different industry demands. Sugar beet pectin structure comprises of highly branched rhamnogalacturonan I with reduced homogalacturonan share and length, having a high degree of acetylation. Although straightforward sugar beet pectin structure and function relationship cannot be established, a paramount link between protein and ferulic acid content and emulsifying properties as well as the gelling ability and acetyl groups, methyl groups, and neutral sugar content is observed. Conventional sugar beet pectin isolation method includes acidic extraction, while promising results were obtained for emerging green extraction methods from which microwave-assisted extraction is closest to industrial application. By tailoring sugar beet pectin structure with the application of different extraction methods and conditions desired properties can be reached. However, the existence of the cooperative impact of multiply structural elements and characteristics on sugar beet pectin functionality cannot be disregarded when producing tailor-made pectin. Considering that properties dictate its usage, sugar beet pectin emulsifying properties enabled its frequent usage in preparation of oil–water emulsions as wall material in encapsulation of bioactive compounds, preparation of edible films, and coatings. Furthermore, sugar beet pectin gelling properties, improved after modification, facilitate its application in heavy metals binding and removal.

References

Abou-Elseoud WS, Hassan EA, Hassan ML (2021) Extraction of pectin from sugar beet pulp by enzymatic and ultrasound-assisted treatments. Carbohydr Polym Technol Appl 2:100042

Adetunji LR, Adekunle A, Orsat V, Raghavan V (2017) Advances in the pectin production process using novel extraction techniques: a review. Food Hydrocoll 62:239–250

Adiletta G, Brachi P, Riianova E, Crescitelli A, Miccio M, Kostryukova N (2020) A simplified biorefinery concept for the valorization of sugar beet pulp: ecofriendly isolation of pectin as a step preceding torrefaction. Waste Biomass Valori 11:2721–2733

Alba K, Kontogiorgos V (2017) Pectin at the oil-water interface: relationship of molecular composition and structure to functionality. Food Hydrocoll 68:211–218

Alba K, Kontogiorgos V (2020) Emulsification properties of pectin. In: Kontogiorgos V (ed) Pectin: technological and physiological properties. Springer Nature Switzerland AG, Cham, pp 83–97

Almohammed F, Koubaa M, Khelfa A, Nakaya M, Mhemdi H, Vorobiev E (2017) Pectin recovery from sugar beet pulp enhanced by high-voltage electrical discharges. Food Bioprod Process 103:95–103

Arroyo-Maya IJ, McClements DJ (2015) Biopolymer nanoparticles as potential delivery systems for anthocyanins: fabrication and properties. Food Res Int 69:1–8

Babbar N, Dejonghe W, Gatti M, Sforza S, Kathy E (2016) Pectic oligosaccharides from agricultural by-products: production, characterization and health benefits. Crit Rev Biotechnol 36(4): 594–606

Bai L, Huan S, Li Z, McClements DJ (2017) Comparison of emulsifying properties of food-grade polysaccharides in oil-in-water emulsions: gum arabic, beet pectin, and corn fiber gum. Food Hydrocoll 66:144–153

BeMiller JN (2019) Carbohydrate chemistry for food scientists, 3rd edn. Elsevier Inc., AACC International, UK, pp 303–312

Berg S, Bretz M, Hubbermann EM, Schwarz K (2012) Influence of different pectins on powder characteristics of microencapsulated anthocyanins and their impact on drug retention of shellac coated granulate. J Food Eng 108:158–165

Bohn K (1998) Composition of sugarbeet and sugarcane and chemical behavior of constituents in processing. In: van der Poel PW, Schiweck H, Schwartz T (eds) Sugar technology beet and cane sugar manufacture. Verlag Dr. Bartens KG, Berlin, pp 115–151

Buchholt HC, Christensen TMIE, Fallesen B, Ralet M-C, Thibault J-F (2004) Preparation and properties of enzymatically and chemically modified sugar beet pectins. Carbohydr Polym 58: 149–161

Campos CA, Gerschenson LN, Flores SK (2011) Development of edible films and coatings with antimicrobial activity. Food Bioprocess Technol 4(6):849–875

Chaichi M, Hashemi M, Badii F, Mohammadi A (2017) Preparation and characterization of a novel bionanocomposite edible film based on pectin and crystalline nanocellulose. Carbohydr Polym 157:167–175

Chen H-M, Fu X, Luo Z-G (2015a) Properties and extraction of pectin-enriched materials from sugar beet pulp by ultrasonic-assisted treatment combined with subcritical water. Food Chem 168:302–310

Chen H-M, Fu X, Abbasi AM, Luo Z-G (2015b) Preparation of environment-friendly pectin from sugar beet pulp and assessment of its emulsifying capacity. Int J Food Sci Technol 2015:1324–1330

Chen H-M, Fu X, Luo Z-G (2015c) Esterification of sugar beet pectin using octenyl succinic anhydride and its effect as an emulsion stabilizer. Food Hydrocoll 49:53–60

Chen H, Qiu S, Gan J, Liu Y, Zhu Q, Yin L (2016a) New insights into the functionality of protein to the emulsifying properties of sugar beet pectin. Food Hydrocoll 57:262–270

Chen H-M, Fu X, Luo Z-G (2016b) Effect of molecular structure on emulsifying properties of sugar beet pulp pectin. Food Hydrocoll 54:99–106

Chen H, Qiu S, Liu Y, Zhu Q, Yin L (2018) Emulsifying properties and functional compositions of sugar beet pectins extracted under different conditions. J Dispers Sci Technol 39(4):484–490

Concha Olmos J, Zúñiga Hansen ME (2012) Enzymatic depolymerization of sugar beet pulp: production and characterization of pectin and pectic-oligosaccharides as a potential source for functional carbohydrates. Chem Eng J 192:29–36

Dronnet VM, Renard CMCG, Axelos MAV, Thibault JF (1996) Characterisation and selectivity of divalent metal ions binding by citrus and sugar-beet pectins. Carbohydr Polym 30:253–263

Drusch S (2007) Sugar beet pectin: a novel emulsifying wall component for microencapsulation of lipophilic food ingredients by spray-drying. Food Hydrocoll 21:1223–1228

Drusch S, Serfert Y, Scampicchio M, Schmidt-Hansberg B, Schwarz K (2007) Impact of physicochemical characteristics on the oxidative stability of fish oil microencapsulated by spray-drying. J Agric Food Chem 55:11044–11051

Emadzadeh B, Ghorani B, Naji-Tabasi S, Charpashlo E, Molaveisi M (2021) Fate of β-cyclodextrin-sugar beet pectin microcapsules containing garlic essential oil in an acidic food beverage. Food Biosci 42:101029

Espitia PJP, Du W-X, Avena-Bustillos RJ, Soares NFF, McHugh TH (2014) Edible films from pectin: physical-mechanical and antimicrobial properties—a review. Food Hydrocoll 35:287–296

FDA (2013) Direct food substances affirmed as generally recognized as safe. Subpart B–listing of specific substances affirmed as GRAS. Section 184, p 1588

Fishman ML, Chau HK, Cooke PH, Hotchkiss AT Jr (2008) Global structure of microwave-assisted flash-extracted sugar beet pectin. J Agric Food Chem 56:1471–1478

Funami T, Zhang G, Hiroe M, Noda S, Nakauma M, Asai I, Cowman MK, Al-Assaf S, Phillips GO (2007) Effects of the proteinaceous moiety on the emulsifying properties of sugar beet pectin. Food Hydrocoll 21(8):1319–1329

Funami T, Nakauma M, Ishihara S, Tanaka R, Inoue T, Phillips GO (2011) Structural modifications of sugar beet pectin and the relationship of structure to functionality. Food Hydrocoll 25(2): 221–229

Gullón B, Gómez B, Martínez-Sabajanes M, Yáñez R, Parajó JC, Alonso JL (2013) Pectic oligosaccharides: manufacture and functional properties. Trends Food Sci Technol 30:153–161

Guo X, Guo X, Meng H, Zhang B, Yu S (2017) Using the high temperature resistant pH electrode to auxiliarily study the sugar beet pectin extraction under different extraction conditions. Food Hydrocoll 70:105–113

Guo X, Guo X, Meng H, Chen X, Zeng Q, Yu S (2019) Influences of different pectins on the emulsifying performance of conjugates formed between pectin and whey protein isolate. Int J Biol Macromol 123:246–254

Harel P, Mignot L, Sauvage JP, Junter GA (1998) Cadmium removal from dilute aqueous solution by gel beads of sugar beet pectin. Ind Crop Prod 7(2–3):239–247

Hassan B, Chatha SAS, Hussain AI, Zia KM, Akhtar N (2018) Recent advances on polysaccharides, lipids and protein based edible films and coatings: a review. Int J Biol Macromol 109:1095–1107

Holck J, Hjernø K, Lorentzen A, Vigsnæs LK, Hemmingsen L, Licht TR, Mikkelsen JD, Meye AS (2011) Tailored enzymatic production of oligosaccharides from sugar beet pectin and evidence of differential effects of a single DP chain length difference on human faecal microbiota composition after *in vitro* fermentation. Process Biochem 46:1039–1049

Huang X, Li D, Wang L-J (2017) Characterization of pectin extracted from sugar beet pulp under different drying conditions. J Food Eng 211:1–6

Huang X, Li D, Wang L-j (2018) Effect of particle size of sugar beet pulp on the extraction and property of pectin. J Food Eng 218:44–49

Jafarzadeh-Moghaddam M, Shaddel R, Peighambardoust SH (2021) Sugar beet pectin extracted by ultrasound or conventional heating: a comparison. J Food Sci Technol 58:2567–2578

Jung J, Wicker L (2012) Laccase mediated conjugation of sugar beet pectin and the effect on emulsion stability. Food Hydrocoll 28:168–173

Karnik D, Wicker L (2018) Emulsion stability of sugar beet pectin fractions obtained by isopropanol fractionation. Food Hydrocoll 74:249–254
Katsuda MS, Mcclements DJ, Miglioranza LHS, Decker EA (2008) Physical and oxidative stability of fish oil-in-water emulsions stabilized with β-lactoglobulin and pectin. J Agric Food Chem 56: 5926–5931
Kaya B, Okur I, Alpas H, Oztop MH (2021) High hydrostatic pressure assisted extraction of pectin from sugar beet pulp. Int J Food Sci Technol. https://doi.org/10.1111/ijfs.15079
Kirby AR, MacDougall AJ, Morris VJ (2006) Sugar beet pectin–protein complexes. Food Biophys 1:51–56
Kirby AR, MacDougall AJ, Morris VJ (2008) Atomic force microscopy of tomato and sugar beet pectin molecules. Carbohydr Polym 71:640–647
Kumar M, Tomar M, Saurabh V, Mahajan T, Punia S, Contreras MDM, Rudra SG, Kaur C, Kennedy JF (2020) Emerging trends in pectin extraction and its anti-microbial functionalization using natural bioactives for application in food packaging. Trends Food Sci Technol 105:223–237
Lan Y, Ohm J-B, Chen B, Rao J (2021) Microencapsulation of hemp seed oil by pea protein isolate-sugar beet pectin complex coacervation: influence of coacervation pH and wall/core ratio. Food Hydrocoll 113:106423
Lara-Espinoza C, Carvajal-Millán E, Balandrán-Quintana R, López-Franco Y, Rascón-Chu A (2018) Pectin and pectin-based composite materials: beyond food texture. Molecules 23(4):942
Lara-Espinoza C, Sanchez-Villegas JA, Lopez-Franco Y, Carvajal-Millan E, Troncoso-Rojas R, Carvallo-Ruiz T, Rascon-Chu A (2021) Composition, physicochemical features, and covalent gelling properties of ferulated pectin extracted from three sugar beet (*Beta vulgaris* L.) cultivars grown under desertic conditions. Agronomy 11(1):40
Larsen N, Bussolo de Souza C, Krych L, Barbosa Cahú T, Wiese M, Kot W, Hansen KM, Blennow A, Venema K, Jespersen L (2019) Potential of pectins to beneficially modulate the gut microbiota depends on their structural properties. Front Microbiol 10:223
Lazaridou A, Biliaderis CG (2020) Edible films and coatings with pectin. In: Kontogiorgos V (ed) Pectin: technological and physiological properties. Springer Nature Switzerland AG, Cham, pp 99–123
Leroux J, Langendorff V, Schick G, Vaishnav V, Mazoyer J (2003) Emulsion stabilizing properties of pectin. Food Hydrocoll 17:455–462
Levigne S, Ralet M-C, Thibault J-F (2002) Characterisation of pectins extracted from fresh sugar beet under different conditions using an experimental design. Carbohydr Polym 49:145–153
Li D-Q, Jia X, Wei Z, Liu Z-Y (2012) Box–Behnken experimental design for investigation of microwave-assisted extracted sugar beet pulp pectin. Carbohydr Polym 88:342–346
Li D-Q, Du G-M, Jing W-W, Li J-F, Yan J-Y, Liu Z-Y (2015) Combined effects of independent variables on yield and protein content of pectin extracted from sugar beet pulp by citric acid. Carbohydr Polym 129:108–114
Li J, Zhang L, Li J-F, Li D-Q (2019) Comparative studies of combined influence of variables on the esterification degree of pectin extracted by sulfuric acid and citric acid. Adv Polym Technol 2019:6313241
Lin J, Yu S, Ai C, Zhang T, Guo X (2020) Emulsion stability of sugar beet pectin increased by genipin crosslinking. Food Hydrocoll 101:105459
Liu B, Zhang J, Liu L, Hotchkiss AT (2012) Utilization of pectin extracted sugar beet pulp for composite application. J Biobased Mater Bioenergy 6:185–192
Liu Z, Pi F, Guo X, Guo X, Yu S (2019) Characterization of the structural and emulsifying properties of sugar beet pectins obtained by sequential extraction. Food Hydrocoll 88:31–42
Liu Z, Guo X, Meng H (2020) Added ferulic acid enhances the emulsifying properties of pectins from different sources. Food Hydrocoll 100:105439
Liu Z, Li Y, Geng S, Mo H, Liu B (2021) Fabrication of food-grade pickering high internal phase emulsions stabilized by the mixture of β-cyclodextrin and sugar beet pectin. Int J Biol Macromol 182:252–263

Lv C, Wang Y, Wang L-J, Li D, Adhikari B (2013) Optimization of production yield and functional properties of pectin extracted from sugar beet pulp. Carbohydr Polym 95:233–240

Ma S, Yu S-J, Zheng X-L, Wang X-X, Bao Q-D, Guo X-M (2013) Extraction, characterization and spontaneous emulsifying properties of pectin from sugar beet pulp. Carbohydr Polym 98:750–753

Ma X, Li D, Wu Z, Zhang H, Chen X, Liu Z (2016) Mercury removal by adsorption on pectin extracted from sugar beet pulp: optimization by response surface methodology. Chem Eng Technol 39(2):371–377

Mao Y, Lei R, Ryan J, Arrutia Rodriguez F, Rastall B, Chatzifragkou A, Winkworth-Smith C, Harding SE, Ibbett R, Binner E (2019) Understanding the influence of processing conditions on the extraction of rhamnogalacturonan-I "hairy" pectin from sugar beet pulp. Food Chem X 2:100026

Maravić N, Šereš Z, Nikolić I, Dokić P, Kertész S, Dokić L (2019) Emulsion stabilizing capacity of sugar beet fibers compared to sugar beet pectin and octenyl succinate modified maltodextrin in the production of O/W emulsions: individual and combined impact. LWT Food Sci Technol 108:392–399

Martínez M, Gullón B, Yáñez R, Alonso JL, Parajó JC (2009) Direct enzymatic production of oligosaccharide mixtures from sugar beet pulp: experimental evaluation and mathematical modeling. J Agric Food Chem 57:5510–5517

Mata YN, Torres E, Blázquez ML, Ballester A, Gonzalez F, Muñoz JA (2007) Lead and gold removal using sugar-beet pectin gels with and without immobilized *Fucus vesiculosus*. Adv Mater Res 20:599–602

Mata YN, Blázquez ML, Ballester A, González F, Muñoz JA (2009a) Optimization of the continuous biosorption of copper with sugar-beet pectin gels. J Environ Manag 90:1737–1743

Mata YN, Blázquez ML, Ballester A, González F, Muñoz JA (2009b) Sugar-beet pulp pectin gels as biosorbent for heavy metals: preparation and determination of biosorption and desorption characteristics. Chem Eng J 150:289–301

Mata YN, Blázquez ML, Ballester A, González F, Muñoz JA (2010) Studies on sorption, desorption, regeneration and reuse of sugar-beet pectin gels for heavy metal removal. J Hazard Mater 178:243–248

Matora AV, Korshunova VE, Shkodina OG, Zhemerichkin DA, Ptitchkin NM, Morris ER (1995) The application of bacterial enzymes for extraction of pectin from pumpkin and sugar beet. Food Hydrocoll 9(1):43–46

Mesbahi G, Jamalian J, Farahnaky A (2005) A comparative study on functional properties of beet and citrus pectins in food systems. Food Hydrocoll 19:731–738

Michel F, Thibault J-F, Mercier C, Heitz F, Pouillaude F (1985) Extraction and charcterization of pectins from sugar beet pulp. J Food Sci 50:1499–1500

Morris GA, Ralet M-C, Bonnin E, Thibault J-F, Harding SE (2010) Physical characterisation of the rhamnogalacturonan and homogalacturonan fractions of sugar beet (*Beta vulgaris*) pectin. Carbohydr Polym 82:1161–1167

Nakauma M, Funami T, Noda S, Ishihara S, Al-Assaf S, Nishinari K, Phillips GO (2008) Comparison of sugar beet pectin, soybean soluble polysaccharide, and gum arabic as food emulsifiers. 1. Effect of concentration, pH, and salts on the emulsifying properties. Food Hydrocoll 22:1254–1267

Ngouémazong ED, Christiaens S, Shpigelman A, Van Loey A, Hendrickx M (2015) The emulsifying and emulsion-stabilizing properties of pectin: a review. Compr Rev Food Sci Food Saf 14(6):705–718

Nuñez A, Fishman ML, Fortis LL, Cooke PH, Hotchkiss AT Jr (2009) Identification of extensin protein associated with sugar beet pectin. J Agric Food Chem 57:10951–10958

Ochoa-Villarreal M, Aispuro-Hernández E, Vargas-Arispuro I, Martínez-Téllez MÁ (2012) Plant cell wall polymers: function, structure and biological activity of their derivatives. In: De Souza Gomes A (ed) Polymerization. IntechOpen, London, pp 63–86. https://doi.org/10.5772/46094

Oosterveld A, Beldman G, Schols HA, Voragen AGJ (2000) Characterization of arabinose and ferulic acid rich pectic polysaccharides and hemicelluloses from sugar beet pulp. Carbohydr Res 328:185–197
Ostrman M, Vojvodić Cebin A, Bušić A, Vrtar S, Komes D (2018) Development and characterization of edible films based on pectin from sugar beet pulp. In: Proceedings 9th international congress of food technologists, biotechnologists and nutritionists. Croatian Society of Food Technologists, Biotechnologists and Nutritionists, Zagreb, Croatia, pp 37–42
Pacheco MT, Villamiel M, Moreno R, Moreno FJ (2019) Structural and rheological properties of pectins extracted from industrial sugar beet by-products. Molecules 24(3):392
Paramita V, Furuta T, YoShii H (2010) Microencapsulation efficacy of *d*-limonene by spray drying using various combinations of wall materials and emulsifiers. Food Sci Technol Res 16(5): 365–372
Peighambardoust SH, Jafarzadeh-Moghaddam M, Pateiro M, Lorenzo JM, Domínguez R (2021) Physicochemical, thermal and rheological properties of pectin extracted from sugar beet pulp using subcritical water extraction process. Molecules 26(5):1413
Peng X-Y, Mu T-H, Zhang M, Sun H-N, Chen J-W, Yu M (2015) Optimisation of production yield by ultrasound-/microwave-assisted acid method and functional property of pectin from sugar beet pulp. Int J Food Sci Technol 50:758–765
Peng X-Y, Mu T-H, Zhang M, Sun H-N, Chen J-W, Yu M (2016) Effects of pH and high hydrostatic pressure on the structural and rheological properties of sugar beet pectin. Food Hydrocoll 60:161–169
Pi F, Liu Z, Guo X, Guo X, Meng H (2019) Chicory root pulp pectin as an emulsifier as compared to sugar beet pectin. Part 1: influence of structure, concentration, counterion concentration. Food Hydrocoll 89:792–801
Pińkowska H, Wolak P, Krzywonos M, Złocińska A (2021) Comprehensive study of hydrothermal extraction of pectin from sugar beet pulp. Waste Biomass Valori 12:4587–4598
Polavarapu S, Oliver CM, Ajlouni S, Augustin MA (2011) Physicochemical characterisation and oxidative stability of fish oil and fish oil–extra virgin olive oil microencapsulated by sugar beet pectin. Food Chem 127(4):1694–1705
Polavarapu S, Oliver CM, Ajlouni S, Augustin MA (2012) Impact of extra virgin olive oil and ethylenediaminetetraacetic acid (EDTA) on the oxidative stability of fish oil emulsions and spray-dried microcapsules stabilized by sugar beet pectin. J Agric Food Chem 60:444–450
Prandi B, Baldassarre S, Babbar N, Bancalari E, Vandezande P, Hermans D, Bruggeman G, Gatti M, Elst K, Sforza S (2018) Pectin oligosaccharides from sugar beet pulp: molecular characterization and potential prebiotic activity. Food Funct 9:1557–1569
Ralet M-C, Thibault J-F (2009) Chapter 1. Hydrodynamic properties of isolated pectin domains: a way to figure out pectin macromolecular structure? In: Schols HA, Vissers RGF, Voragen AGJ (eds) Pectins and pectinases. Wageningen Academic Publishers, Wageningen, pp 19–34
Ralet MC, Cabrera JC, Bonnin E, Quéméner B, Hellìn P, Thibault J-F (2005) Mapping sugar beet pectin acetylation pattern. Phytochemistry 66(15):1832–1843
Rehman A, Ahmad T, Aadil RM, Spotti MJ, Bakry AM, Khan IM, Zhao L, Riaz T, Tong Q (2019) Pectin polymers as wall materials for the nano-encapsulation of bioactive compounds. Trends Food Sci Technol 90:35–46
Reichembach LH, de Oliveira Petkowicz CL (2021) Pectins from alternative sources and uses beyond sweets and jellies: an overview. Food Hydrocoll 118:106824
Remoroza C, Buchholt HC, Gruppen H, Schols HA (2014) Descriptive parameters for revealing substitution patterns of sugarbeet pectins using pectolytic enzymes. Carbohydr Polym 101: 1205–1215
Ropartz D, Ralet M-C (2020) Pectin structure. In: Kontogiorgos V (ed) Pectin: technological and physiological properties. Springer Nature Switzerland AG, Cham, pp 17–36
Ruiz Canizales J, Velderrain Rodríguez GR, Domínguez Avila JA, Preciado Saldaña AM, Villegas Ochoa MA, González Aguilar GA (2019) Encapsulation to protect different bioactives to be

used as nutraceuticals and food ingredients. In: Mérillon JM, Ramawat KG (eds) Bioactive molecules in food. Springer Nature Switzerland AG, Cham, pp 2163–2181

Schmidt US, Schmidt K, Kurz T, Endreß HU, Schuchmann HP (2015) Pectins of different origin and their performance in forming and stabilizing oil-in-water-emulsions. Food Hydrocoll 46: 59–66

Schols HA, Voragen AGJ (1996) Complex pectins: structure elucidation using enzymes. Progr Biotechnol 14:47–55

Schols HA, Coenen G-J, Voragen AGJ (2009) Chapter 1. Revealing pectin's structure. In: Schols HA, Vissers RGF, Voragen AGJ (eds) Pectins and pectinases. Wageningen Academic Publishers, Wageningen, pp 19–34

Siew CK, Williams PA (2008) Role of protein and ferulic acid in the emulsification properties of sugar beet pectin. J Agric Food Chem 56:4164–4171

Siew CK, Williams PA, Cui SW, Wang Q (2008) Characterization of the surface-active components of sugar beet pectin and the hydrodynamic thickness of the adsorbed pectin layer. J Agric Food Chem 56:8111–8120

Sun R, Hughes S (1998) Extraction and physico-chemical characterization of pectins from sugar beet pulp. Polym J 30(8):671–677

Tang X-Y, Wang Z-M, Meng H-C, Lin J-W, Guo X-M, Zhang T, Chen H-L, Lei C-Y, Yu S-J (2021) Robust W/O/W emulsion stabilized by genipin-cross-linked sugar beet pectin-bovine serum albumin nanoparticles: co-encapsulation of betanin and curcumin. J Agric Food Chem 69:1318–1328

Vincken J-P, Schols HA, Oomen RJFJ, McCann MC, Ulvskov P, Voragen AGJ, Visser RGF (2003) If homogalacturonan were a side chain of rhamnogalacturonan I. Implications for cell wall architecture. Plant Physiol 132(4):1781–1789

Voragen AGJ, Coenen G-J, Verhoef RP, Schols HA (2009) Pectin, a versatile polysaccharide present in plant cell walls. Struct Chem 20:263–275

Wang Z, Meng H, Zhang T, Guo X (2021) Investigation into the polymerization and changes of physicochemical properties of sugar beet pectin through controlled dry-heating. Food Hydrocoll 110:106212

Westphal Y, Kühnel S, de Waard P, Hinz SWA, Schols HA, Voragen AGJ, Gruppen H (2010) Branched arabino-oligosaccharides isolated from sugar beet arabinan. Carbohydr Polym 345: 1180–1189

Williams PA, Sayers C, Viebke C, Senan C (2005) Elucidation of the emulsification properties of sugar beet pectin. J Agric Food Chem 53:3592–3597

Yan W, Jia X, Zhang Q, Chen H, Zhu Q, Yin L (2021) Interpenetrating polymer network hydrogels of soy protein isolate and sugar beet pectin as a potential carrier for probiotics. Food Hydrocoll 113:106453

Yang Y, Chen D, Yu Y, Huang X (2020) Effect of ultrasonic treatment on rheological and emulsifying properties of sugar beet pectin. Food Sci Nutr 8:4266–4275

Yapo BM (2011) Rhamnogalacturonan-I: a structurally puzzling and functionally versatile polysaccharide from plant cell walls and mucilages. Polym Rev 51:391–413

Yapo BM, Robert C, Etienne I, Wathelet B, Paquot M (2007a) Effect of extraction conditions on the yield, purity and surface properties of sugar beet pulp pectin extracts. Food Chem 100:1356–1364

Yapo BM, Wathelet B, Paquot M (2007b) Comparison of alcohol precipitation and membrane filtration effects on sugar beet pulp pectin chemical features and surface properties. Food Hydrocoll 21:245–255

Zhang J, Wolf B (2019) Physico-chemical properties of sugar beet pectin-sodium caseinate conjugates via different interaction mechanisms. Foods 8:192

Zhang J, Wolf B (2021) Impact of type of sugar beet pectin–sodium caseinate interaction on emulsion properties at pH 4.5 and pH 7. Foods 10:631

Zhang L, Shi Z, Shangguan W, Fang Y, Nishinari K, Phillips GO, Jiang F (2015) Emulsification properties of sugar beet pectin after modification with horseradish peroxidase. Food Hydrocoll 43:107–113

Zhang Y, Lin J, Zhong Q (2016) S/O/W emulsions prepared with sugar beet pectin to enhance the viability of probiotic *Lactobacillus salivarius* NRRL B-30514. Food Hydrocoll 52:804–810

Zhang T, Wang Z, Yu S, Guo X, Ai C, Tang X, Chen H, Lin J, Zhang X, Meng H (2021) Effects of pH and temperature on the structure, rheological and gel-forming properties of sugar beet pectins. Food Hydrocoll 116:106646

GPSR Compliance

The European Union's (EU) General Product Safety Regulation (GPSR) is a set of rules that requires consumer products to be safe and our obligations to ensure this.

If you have any concerns about our products, you can contact us on ProductSafety@springernature.com

In case Publisher is established outside the EU, the EU authorized representative is:

Springer Nature Customer Service Center GmbH
Europaplatz 3
69115 Heidelberg, Germany

Batch number: 10755324

Printed by Printforce, the Netherlands